Flora of North America

North of Mexico

Edited by FLORA OF NORTH AMERICA EDITORIAL COMMITTEE

VOLUME 19

Magnoliophyta: Asteridae, part 6: Asteraceae, part 1

ASTERALES, part 1 (Aster order)

NEW YORK OXFORD · OXFORD UNIVERSITY PRESS · 2006

Oxford University Press, Inc., publishes works that further
Oxford University's objective of excellence
in research, scholarship, and education.

Oxford New York
Auckland Cape Town Dar es Salaam Hong Kong Karachi
Kuala Lumpur Madrid Melbourne Mexico City Nairobi
New Delhi Shanghai Taipei Toronto

With offices in
Argentina Austria Brazil Chile Czech Republic France Greece
Guatemala Hungary Italy Japan Poland Portugal Singapore
South Korea Switzerland Thailand Turkey Ukraine Vietnam

Library of Congress Cataloging-in-Publication Data
(Revised for Volumes 19–21)
Flora of North America north of Mexico
edited by Flora of North America Editorial Committee.
Includes bibliographical references and indexes.
Contents: v. 1. Introduction—v. 2. Pteridophytes and gymnosperms—
v. 3. Magnoliophyta: Magnoliidae and Hamamelidae—
v. 22. Magnoliophyta: Alismatidae, Arecidae, Commelinidae (in part), and Zingiberidae—
v. 26. Magnoliophyta: Liliidae: Liliales and Orchidales—
v. 23. Magnoliophyta: Commelinidae (in part): Cyperaceae—
v. 25. Magnoliophyta: Commelinidae (in part): Poaceae, part 2—
v. 4. Magnoliophyta: Caryophyllidae (in part): part 1—
v. 5. Magnoliophyta: Caryophyllidae (in part): part 2—
v. 19. Magnoliophyta: Asteridae (in part): Asteraceae, part 1
v. 20. Magnoliophyta: Asteridae (in part): Asteraceae, part 2
v. 21. Magnoliophyta: Asteridae (in part): Asteraceae, part 3

ISBN-13: 978-0-19-530563-0 (v. 19)
ISBN-10: 0-19-530563-9 (v. 19)
1. Botany—North America.
2. Botany—United States.
3. Botany—Canada.
I. Flora of North America Editorial Committee.
QK110.F55 2002   581.97 92-30459

Printing number: 9 8 7 6 5 4 3 2 1
Printed in the United States of America
on acid-free paper

Flora of North America

Contributors to Volumes 19–21

Geraldine A. Allen	Arthur Haines	Thomas F. Patterson
Justin W. Allison	Neil A. Harriman	Donald J. Pinkava
Loran C. Anderson	Elizabeth M. Harris	A. Michael Powell
Susan J. Bainbridge	Ronald L. Hartman	Robert E. Preston
Gary I. Baird	Walter C. Holmes	Roland P. Roberts
Bruce G. Baldwin	Phillip E. Hyatt	Edward E. Schilling
Theodore M. Barkley†	John P. Janovec	Randall W. Scott
Randall J. Bayer	Dale E. Johnson	John C. Semple
Mark W. Bierner	Vesna Karaman-Castro	Leila M. Shultz
A. Linn Bogle	Brian R. Keener	Beryl B. Simpson
David J. Bogler	David J. Keil	Kunsiri Chaw Siripun
Kristin R. Brodeur	Robert W. Kiger	Alan R. Smith
Luc Brouillet	John C. La Duke	Pamela S. Soltis
Gregory K. Brown	Eric E. Lamont	David M. Spooner
Judith M. Canne-Hilliker	Michelle R. Leonard	Phyllis Spurr
Gerald D. Carr	Timothy K. Lowrey	John L. Strother
Robert L. Carr	Alison McKenzie Mahoney	Scott D. Sundberg†
Kenton L. Chambers	Staci Markos	Debra K. Trock
Raymund Chan	John S. Mooring	Billie L. Turner
Donna M. Cherniawsky	James D. Morefield	Matt W. Turner
Jerry G. Chmielewski	David R. Morgan	Lowell E. Urbatsch
Curtis Clark	Caleb A. Morse	Linda E. Watson
W. Dennis Clark	David F. Murray	William A. Weber
Jennifer A. Clevinger	Amy Trauth Nare	Ronald R. Weedon
Rachel E. Cook	Guy L. Nesom	Mark A. Wetter
Patricia B. Cox	Kurt M. Neubig	Molly A. Whalen
W. S. Davis	Jörg Ochsmann	R. David Whetstone
Robert D. Dorn	Robert J. O'Kennon	Dieter H. Wilken
A. Michele Funston	Robert Orndufft†	Steven J. Wolf
L. D. Gottlieb	James C. Parks†	Sharon C. Yarborough

Editors for Volumes 19–21

Theodore M. Barkley†, Lead and Taxon Editor	Luc Brouillet, Taxon Editor	Helen Jeude, Technical Editor	John L. Strother, Taxon Editor
Kanchi Gandhi, Nomenclatural Editor	Robert W. Kiger, Bibliographic Editor	Kay Yatskievych, Managing Editor	James L. Zarucchi, Editorial Director

Volume Composition

Pat Harris	Martha J. Hill	Ruth T. King	Kristin Pierce

Pluchea odorata
var. odorata

Contents

Dedication   vii

Founding Member Institutions   viii

Donors   ix

Editorial Committee   x

Project Staff   xi

Contributors   xii

Reviewers   xv

Acknowledgments   xvi

Preface for Volume 19   xvii

Introduction   xix

ASTERIDAE, in part

ASTERALES, in part

  Asteraceae   3
    Mutisieae   70
    Cynareae   82
    Arctotideae   195
    Vernonieae   200
    Cichorieae   214
    Calenduleae   379
    Gnaphalieae   384
    Inuleae   471
    Plucheeae   475
    Anthemideae   485

Literature Cited   560

Index   561

Theodore M. "Ted" Barkley
1934–2004

Courtesy of the Rotary Club of Fort Worth

The Flora of North America Board and the Compositae Editorial Team are honored to dedicate the three volumes of the Asteraceae to the late Dr. Theodore "Ted" M. Barkley. An influential member of the FNA editorial board since its founding, Dr. Barkley passed away in the summer of 2004 while still actively involved in preparing the current volumes for publication. A dedicated and renowned seneciologist, he worked on his groundsel treatments to the last minute. After nearly four decades at Kansas State University, where he trained many taxonomists and co-edited, among others, the *Flora of the Great Plains*, Ted retired in 1998 to the Botanical Research Institute of Texas to serve as lead editor of the three Asteraceae volumes. In the years that followed, thanks to his leadership and managerial abilities as well as to his sense of humor, he obtained the financial resources and developed a robust program to ensure publication of the largest family in the North American flora. Ted will be remembered by all of us for his delight at playing with language, bad puns intended, and for his culture, wisdom, pragmatism, loyalty to friends, and dedication to the FNA project and to the Compositae volumes in particular. But mostly, he will be remembered as a scholar and a gentleman.

vii

FOUNDING MEMBER INSTITUTIONS

Flora of North America Association

Arnold Arboretum
Jamaica Plain, Massachusetts

Agriculture and Agri-Food Canada
Ottawa, Ontario

Canadian Museum of Nature
Ottawa, Ontario

Carnegie Museum of Natural
  History
Pittsburgh, Pennsylvania

Field Museum of Natural History
Chicago, Illinois

Fish and Wildlife Service
United States Department of the
  Interior
Washington, D.C.

Harvard University Herbaria
Cambridge, Massachusetts

Hunt Institute for Botanical
  Documentation
Carnegie Mellon University
Pittsburgh, Pennsylvania

Jacksonville State University
Jacksonville, Alabama

Jardin Botanique de Montréal
Montréal, Québec

Kansas State University
Manhattan, Kansas

Missouri Botanical Garden
St. Louis, Missouri

New Mexico State University
Las Cruces, New Mexico

New York State Museum
Albany, New York

Northern Kentucky University
Highland Heights, Kentucky

The New York Botanical Garden
Bronx, New York

The University of British Columbia
Vancouver, British Columbia

The University of Texas
Austin, Texas

Université de Montréal
Montréal, Québec

University of Alaska Fairbanks
Fairbanks, Alaska

University of Alberta
Edmonton, Alberta

University of California
Berkeley, California

University of California
Davis, California

University of Idaho
Moscow, Idaho

University of Illinois
Urbana-Champaign, Illinois

University of Iowa
Iowa City, Iowa

The University of Kansas
Lawrence, Kansas

University of Michigan
Ann Arbor, Michigan

University of Oklahoma
Norman, Oklahoma

University of Ottawa
Ottawa, Ontario

University of Southwestern
  Louisiana
Lafayette, Louisiana

University of Western Ontario
London, Ontario

University of Wyoming
Laramie, Wyoming

Utah State University
Logan, Utah

For their support of the preparation of this volume, we gratefully acknowledge and thank:

Chanticleer Foundation

National Science Foundation
(Award DEB-0206645)

The Fairweather Foundation

ChevronTexaco

ESRI

*Project Staff — past and present
involved with the preparation of Volumes 19–21*

Justin W. Allison, *Administrative Assistant*
Barbara Alongi, *Illustrator*
Michael Blomberg, *Scanning Specialist*
Trisha K. Consiglio, *GIS Analyst*
Shellie Davis Eldredge, *Editorial Assistant*
Sheila Flinchpaugh, *Illustrator*
Bee F. Gunn, *Illustrator*
Pat Harris, *Editorial Assistant and Compositor*
Linny Heagy, *Illustrator*
Claire A. Hemingway, *Technical Editor*
Martha J. Hill, *Technical Editor and Compositor*
Helen Jeude, *Senior Technical Editor*
Fred Keusenkothen, *Scanning Supervisor*
Ruth T. King, *Editorial Assistant and Compositor*
Marjorie C. Leggitt, *Illustrator*
Asha McElfish, *Editorial Volunteer*
John Myers, *Illustrator and Illustration Compositor*
Amy Trauth Nare, *NSF Summer Intern*
Guy L. Nesom, *Staff Botanist*
Kristin Pierce, *Editorial Assistant and Compositor*
Heidi H. Schmidt, *Editorial Assistant*
Hong Song, *Programmer*
Linda Ann Vorobik, *Illustrator*
Yevonn Wilson-Ramsey, *Illustration Coordinator and Illustrator*
George Yatskievych, *Technical Adviser*

Contributors to Volumes 19–21

Geraldine A. Allen
University of Victoria
Victoria, British Columbia

Justin W. Allison
Botanical Research Institute of
*   Texas*
Fort Worth, Texas

Loran C. Anderson
Florida State University
Tallahassee, Florida

Susan J. Bainbridge
University of California
Berkeley, California

Gary I. Baird
Brigham Young University-Idaho
Rexburg, Idaho

Bruce G. Baldwin
University of California
Berkeley, California

Theodore M. Barkley†
Botanical Research Institute of
*   Texas*
Fort Worth, Texas

Randall J. Bayer
Australian National Herbarium
Centre for Plant Biodiversity
*   Research*
Canberra, Australia

Mark W. Bierner
Boyce Thompson Arboretum
Superior, Arizona

A. Linn Bogle
University of New Hampshire
Durham, New Hampshire

David J. Bogler
Missouri Botanical Garden
St. Louis, Missouri

Kristin R. Brodeur
Oxford, Alabama

Luc Brouillet
Université de Montréal
Montréal, Québec

Gregory K. Brown
University of Wyoming
Laramie, Wyoming

Judith M. Canne-Hilliker
University of Guelph
Guelph, Ontario

Gerald D. Carr
Oregon State University
Corvallis, Oregon

Robert L. Carr
Eastern Washington University
Cheney, Washington

Kenton L. Chambers
Oregon State University
Corvallis, Oregon

Raymund Chan
University of Hawaii at Manoa
Honolulu, Hawaii

Donna M. Cherniawsky
The Provincial Museum of Alberta
Edmonton, Alberta

Jerry G. Chmielewski
Slippery Rock University
Slippery Rock, Pennsylvania

Curtis Clark
California State Polytechnic
*   University*
Pomona, California

W. Dennis Clark
Arizona State University
Tempe, Arizona

Jennifer A. Clevinger
Walsh University
North Canton, Ohio

Rachel E. Cook
Chicago Botanic Garden
Glencoe, Illinois

Patricia B. Cox
TVA Heritage Program
Knoxville, Tennessee

W. S. Davis
Louisville, Kentucky

Robert D. Dorn
Lingle, Wyoming

A. Michele Funston
Missouri Botanical Garden
St. Louis, Missouri

L. D. Gottlieb
Ashland, Oregon

Arthur Haines
New England Wild Flower Society
Garden in the Woods
Framingham, Massachusetts

Neil A. Harriman
University of Wisconsin-Oshkosh
Oshkosh, Wisconsin

Elizabeth M. Harris
University of Central Florida
Orlando, Florida

Ronald L. Hartman
University of Wyoming
Laramie, Wyoming

Walter C. Holmes
Baylor University
Waco, Texas

Philip E. Hyatt
USDA Forest Service
Atlanta, Georgia

John P. Janovec
Botanical Research Institute of
  Texas
Fort Worth, Texas

Dale E. Johnson
Timber Press
Portland, Oregon

Vesna Karaman-Castro
Louisiana State University
Baton Rouge, Louisiana

Brian R. Keener
The University of Alabama
Tuscaloosa, Alabama
The University of West Alabama
Livingston, Alabama

David J. Keil
California Polytechnic State
  University
San Luis Obispo, California

Robert W. Kiger
Carnegie Mellon University
Pittsburgh, Pennsylvania

John C. La Duke
University of North Dakota
Grand Forks, North Dakota

Eric E. Lamont
The New York Botanical Garden
Bronx, New York

Michelle R. Leonard
Sudbury, Ontario

Timothy K. Lowrey
University of New Mexico
Albuquerque, New Mexico

Alison McKenzie Mahoney
Minnesota State University-
  Mankato
Mankato, Minnesota

Staci Markos
University of California
Berkeley, California

John S. Mooring
Santa Clara University
Santa Clara, California

James D. Morefield
Department of Conservation
  & Natural Resources
Nevada Natural Heritage Program
Carson City, Nevada

David R. Morgan
University of West Georgia
Carrollton, Georgia

Caleb A. Morse
The University of Kansas
Lawrence, Kansas

David F. Murray
University of Alaska,
  Museum of the North
Fairbanks, Alaska

Amy Trauth Nare
Fort Worth, Texas

Guy L. Nesom
Botanical Research Institute of
  Texas
Fort Worth, Texas

Kurt M. Neubig
Florida Museum of Natural
  History
Gainesville, Florida

Jörg Ochsmann
Göttingen, Germany

Robert J. O'Kennon
Botanical Research Institute of
  Texas
Fort Worth, Texas

Robert Ornduff†
University of California
Berkeley, California

James C. Parks†
Millersville University
Millersville, Pennsylvania

Thomas F. Patterson
South Texas Community College
McAllen, Texas

Donald J. Pinkava
Arizona State University
Tempe, Arizona

A. Michael Powell
Sul Ross State University
Alpine, Texas

Robert E. Preston
Jones & Stokes
Sacramento, California

Roland P. Roberts
Towson University
Towson, Maryland

Edward E. Schilling
University of Tennessee
Knoxville, Tennessee

Randall W. Scott
Northern Arizona University
Flagstaff, Arizona

John C. Semple
University of Waterloo
Waterloo, Ontario

Leila M. Shultz
Utah State University
Logan, Utah

Beryl B. Simpson
University of Texas
Austin, Texas

Kunsiri Chaw Siripun
Kasetsart University
Kamphaengsaen Campus, Nakhon
Pathom, Thailand

Alan R. Smith
University of California
Berkeley, California

Pamela S. Soltis
University of Florida
Gainesville, Florida

David M. Spooner
University of Wisconsin-Madison
Madison, Wisconsin

Phyllis L. Spurr
University of Northern Kentucky
Highland Heights, Kentucky

John L. Strother
University of California, Berkeley
Berkeley, California

Scott D. Sundberg†
Oregon State University
Corvallis, Oregon

Debra K. Trock
California Academy of Sciences
San Francisco, California

Billie L. Turner
The University of Texas at Austin
Austin, Texas

Matt W. Turner
Austin, Texas

Lowell E. Urbatsch
Louisiana State University
Baton Rouge, Louisiana

Linda E. Watson
Miami University
Oxford, Ohio

William A. Weber
University of Colorado Museum
Boulder, Colorado

Ronald R. Weedon
Chadron State College
Chadron, Nebraska

Mark A. Wetter
University of Wisconsin-Madison
Madison, Wisconsin

Molly A. Whalen
Flinders University
Adelaide, Australia

R. David Whetstone
Jacksonville State University
Jacksonville, Alabama

Dieter H. Wilken
Santa Barbara Botanic Garden
Santa Barbara, California

Steven J. Wolf
California State University
Stanislaus
Turlock, California

Sharon C. Yarborough
Yerington, Nevada

Reviewers

Ray Angelo
New England Botanical Club
Cambridge, Massachusetts

Susan J. Bainbridge
University of California
Berkeley, California

Bruce Bennett
NatureServe Yukon
Whitehorse, Yukon

David E. Boufford
Harvard University Herbaria
Cambridge, Massachusetts

Larry E. Brown
Spring Branch Science Center
  Herbarium
Houston, Texas

Kenton L. Chambers
Oregon State University
Corvallis, Oregon

Tom S. Cooperrider
Kent State University
Kent, Ohio

Craig C. Freeman
The University of Kansas
Lawrence, Kansas

Arthur Haines
New England Wild Flower Society
Garden in the Woods
Framingham, Massachusetts

Robert Kral
Cairo, Georgia

David F. Murray
University of Alaska,
  Museum of the North
Fairbanks, Alaska

Richard W. Spellenberg
New Mexico State University
Las Cruces, New Mexico

Edward G. Voss
University of Michigan Herbarium
Ann Arbor, Michigan

Dieter H. Wilken
Santa Barbara Botanic Garden
Santa Barbara, California

Richard P. Wunderlin
University of South Florida
Tampa, Florida

Acknowledgments

Members of the Flora of North America Association (FNAA), especially those involved in the preparation and production of the three Asteraceae volumes, extend special and heartfelt gratitude to:

Sy Sohmer, Director of the Botanical Research Institute of Texas (BRIT), for inviting Dr. Barkley and the Asteraceae project to locate there;

The Botanical Institute of Texas and staff for warmly and graciously welcoming and housing the Flora of North America (FNA) Editorial Center at BRIT;

Barney Lipscomb, editor of *Sida*, who went to incredible lengths to make sure all new species and combinations were published promptly, ensuring that these taxa could be included in the Asteraceae volumes;

Guy Nesom, BRIT staff botanist and part-time botanist for FNA, who provided many insights concerning numerous treatments, helped with many questions, and provided additional botanical support for the technical staff at BRIT and elsewhere;

Claire A. Hemingway, Mary Ann Schmidt, and Kay Yatskievych for spirited and rewarding discussions regarding finer points of editing and style; and

Justin W. Allison, BRIT Administrative Assistant, for his steadfast competence in communications, statistics, and tracking.

In addition, we appreciate the efforts of many individuals who provided advice, corrections, and support for the Asteraceae team. Some are members of FNAA; they went well beyond their normal jobs in providing extra aid and encouragement. They include: Bruce G. Baldwin, Richard G. Beidleman, Frédéric Coursol, Kanchi Gandhi, Werner Greuter, Vernon L. Harms, Larry Hufford, Philip Jenkins, Robert W. Kiger, Leslie R. Landrum, John McNeill, Sue Meades, David F. Murray, Dan H. Nicolson, Mike Oldham, Jackie M. Poole, John F. Pruski, Alan R. Smith, Richard W. Spellenberg, Robert Vogt, Dieter H. Wilken, George Yatskievych, and James L. Zarucchi.

To the authors who contributed treatments to the Asteraceae volumes, we extend our appreciation and thanks for their hard work, their scholarly contributions, and their enduring patience with relentless rounds of editings and questions.

Many botanists are fortunate to have a significant other who is supportive beyond the call of duty. We wish to acknowledge the crucial support provided the late Theodore M. Barkley and Scott D. Sundberg by their respective spouses, Mary Barkley and Linda K. Hardison, as both were putting finishing touches to their contributions during their last months.

Taxon editors Brouillet and Strother extend special recognition and thanks to Helen Jeude. Throughout the project, she was not only our stalwart "tech" editor, she was also our advisor, comforter, and friend.

Preface for Volume 19

Since the publication of *Flora of North America* volume 5, the ninth volume to have been published in the *Flora* series, in early 2005, Guy Baillargeon, Barney L. Lipscomb, Jay A. Raveill (Taxon Editor), Michael A. Vincent (Taxon Editor), and Kay Yatskievych (Managing Editor and Production Coordinator) have become members of the Flora of North America Association Board of Directors. The Board succeeded the former Editorial Committee as the result of a reorganization finalized in 2003, but for the sake of continuity of citation, authorship of *Flora* volumes is still to be cited as "Flora of North America Editorial Committee, eds."

Sadly, in late 2005, two long-standing members of the FNAA Board, John W. Thieret and Grady L. Webster, passed away. They will be greatly missed.

The vast majority of editorial processing for the three, simultaneously-published Asteraceae volumes of the *Flora* was undertaken at the Botanical Research Institute of Texas in Fort Worth. Considerable editorial processing of treatments and writing of suprageneric descriptions and keys were done at the Université de Montréal and at the University of California, Berkeley. The maps were prepared at the editorial center at the Missouri Botanical Garden in St. Louis based on taxon distribution statements found in the treatments along with additional data for the indicators showing occurrence in Alaska, Greenland, and the larger Canadian provinces and territories. Pre-press production for the Asteraceae volumes, including typesetting and layout, plus coordination for all aspects of planning, executing, scanning, and labeling the illustrations, was done at the the the St. Louis center.

In addition to her duties as the project's Illustration Coordinator, Yevonn Wilson-Ramsey prepared the illustrations for all genera in Arctotideae, Calenduleae, Inuleae, and Mutisieae, most taxa of Anthemideae (except those credited here to others), taxa of Cichorieae (*Agoseris*, *Krigia*, *Launaea*, *Malacothrix*, *Microseris*, *Nothocalaïs*, *Phalacroseris*, *Prenanthella*, *Sonchus*, *Stebbinsoseris*, *Tragopogon*, *Uropappus*, and *Youngia*), and *Antennaria* (Gnaphalieae). Other illustrations were prepared by Barbara Alongi, for Plucheeae (including the color frontispiece depicting *Pluchea odorata* var. *odorata*) and taxa of Cichorieae (*Arnoseris*, *Cichorium*, *Hedypnois*, *Helminthotheca*, and *Hieracium*) and Gnaphalieae (*Anaphalis*, *Euchiton*, *Facelis*, *Gamochaeta*, *Gnaphalium*, *Helichrysum*, *Pseudognaphalium*, *Omalotheca*, and *Xerochrysum*); by Sheila Flinchpaugh, for *Pleiacanthus* (Cichorieae); by Bee F. Gunn, for Vernonieae and taxa of Cichorieae (*Calycoseris*, *Chaetadelpha*, *Chondrilla*, *Crepis*, *Hypochaeris*, *Ixeris*, *Lapsana*, *Lapsanastrum*, *Leontodon*, *Lygodesmia*, *Munzothamnus*, *Pinaropappus*, *Prenanthes*, *Rafinesquia*, *Shinnersoseris*, and *Stephanomeria*); by Linny Heagy, for *Taraxacum* (Cichorieae) and taxa of Anthemideae (*Achillea* and *Tripleurospermum*) and Gnaphalieae (*Ancistrocarphus*, *Diaperia*, *Filago*, *Herpervax*, *Logfia*, *Micropsis*, *Micropus*, *Psilocarphus*, and *Stylocline*); by

Marjorie C. Leggitt, for taxa of Cichorieae (*Lactuca*, *Mulgedium*, *Mycelis*, *Picris*, *Pyrrhopappus*, *Rhagadiolus*, *Scolymus*, *Scorzonera*, *Tolpis*, and *Urospermum*); by John Myers, for Cynareae and taxa of Anthemideae (*Oncosiphon*, *Pentzia*, and *Picrothamnus*) and Cichorieae (*Anisocoma*, *Atrichoseris*, and *Glyptopleura*); and by Linda Ann Vorobik, for *Artemisia* (Anthemideae). Illustrations were scanned by Michael Blomberg and Fred Keusenkothen. Composition and labeling of all artwork was completed by John Myers assisted by Heidi H. Schmidt.

The Flora of North America Association remains deeply grateful to the many people who continue to help create and sustain the *Flora*.

Introduction

Scope of the Work

Flora of North America North of Mexico is a synoptic account of the plants of North America north of Mexico: the continental United States of America (including the Florida Keys and Aleutian Islands), Canada, Greenland (Kalâtdlit-Nunât), and St. Pierre and Miquelon. The *Flora* is intended to serve both as a means of identifying plants within the region and as a systematic conspectus of the North American flora.

The *Flora* will be published in 30 volumes. Volume 1 contains background information that is useful for understanding patterns in the flora. Volume 2 contains treatments of ferns and gymnosperms. Families in volumes 3–26, the angiosperms, are arranged according to the classification system of A. Cronquist (1981). Bryophytes will be covered in volumes 27–29. Volume 30 will contain the cumulative bibliography and index.

The first two volumes were published in 1993, Volume 3 in 1997, and Volumes 22, 23, and 26, the first three of five covering the monocotyledons, appeared in 2000, 2002, and 2002, respectively. Volume 4, the first part of the Caryophyllales, was published in late 2003. Volume 25, the second part of the Poaceae, was published in mid-2003. Volume 5, completing the Caryophyllales plus Polygonales and Plumbaginales, was published in early 2005. The correct bibliographic citation for the *Flora* is: Flora of North America Editorial Committee, eds. 1993+. *Flora of North America North of Mexico*. 12+ vols. New York and Oxford.

Volumes 19–21 treat 2413 species in 418 genera contained in 14 tribes of Asteraceae. For additional statistics, please refer to Table 1.

Contents · General

The *Flora* includes accepted names, selected synonyms, literature citations, identification keys, descriptions, chromosome numbers, phenological information, summaries of habitats and geographic ranges, and other biological observations. Economic uses, weed status, and conservation status are provided from specified sources. Each volume contains a bibliography and an index to the taxa included in that volume. The treatments, written and reviewed by experts from throughout the systematic botanical community, are based on original observations of herbarium specimens and, whenever possible, on living plants. These observations are supplemented by critical reviews of the literature.

Table 1. *Statistics for Volumes 19–21 of Flora of North America.*

Tribe	Total Genera	Total Species	Endemic Genera	Endemic Species	Introduced Genera	Introduced Species	Conservation Taxa
Volume 19							
Mutisieae	7	14	1	4	0	0	0
Cynareae	17	116	0	48	14	50	30
Arctotideae	3	4	0	0	3	4	0
Vernonieae	6	25	1	18	2	2	0
Cichorieae	49	229	7	112	21	64	22
Calenduleae	4	7	0	0	4	7	0
Gnaphalieae	19	111	1	50	5	17	5
Inuleae	3	5	0	0	3	5	0
Plucheeae	3	12	0	3	0	3	0
Anthemideae	26	99	1	37	17	38	4
Volume 20							
Astereae	77	719	33	561	1	5	175
Senecioneae	29	167	8	117	6	20	43
Volume 21							
Heliantheae	148	746	40	470	6	22	131
Eupatorieae	27	159	5	98	0	3	16
Total for Asteraceae	418	2413	97	1518	82	240	426

Italic = introduced

Basic Concepts

Our goal is to make the *Flora* as clear, concise, and informative as practicable so that it can be an important resource for both botanists and nonbotanists. To this end, we are attempting to be consistent in style and content from the first volume to the last. Readers may assume that a term has the same meaning each time it appears and that, within groups, descriptions may be compared directly with one another. Any departures from consistent usage will be explicitly noted in the treatments (see also References).

Treatments are intended to reflect current knowledge of taxa throughout their ranges worldwide, and classifications are therefore based on all available evidence. Where notable differences of opinion about the classification of a group occur, appropriate references are mentioned in the discussion of the group.

Documentation and arguments supporting significantly revised classifications are published separately in botanical journals before publication of the pertinent volume of the *Flora*. Similarly, all new names and new combinations are published elsewhere prior to their use in the *Flora*. No nomenclatural innovations will be published intentionally in the *Flora*.

Taxa treated in full include extant and recently extinct native species, hybrids that are well established (or frequent), and waifs or cultivated plants that are found frequently outside cultivation and give the appearance of being naturalized. Taxa mentioned only in discussions include waifs or naturalized plants now known only from isolated old records and some nonnative, economically important or extensively cultivated plants, particularly when they are relatives of native species. Excluded names and taxa are listed at the ends of appropriate sections, e.g., species at the end of genus, genera at the end of family.

Treatments are intended to be succinct and diagnostic but adequately descriptive. Characters and character states used in the keys are repeated in the descriptions. Descriptions of related taxa at the same rank are directly comparable.

With few exceptions, taxa are presented in taxonomic sequence. If an author is unable to produce a classification, the taxa are arranged alphabetically, and the reasons are given in the discussion.

Treatments of hybrids follow that of one of the putative parents. Hybrid complexes are treated at the ends of their genera, after the descriptions of species.

We have attempted to keep terminology as simple as accuracy permits. Common English equivalents usually have been used in place of Latin or Latinized terms or other specialized terminology, whenever the correct meaning could be conveyed in approximately the same space, e.g., "pitted" rather than "foveolate," but "striate" rather than "with fine longitudinal lines." See *Categorical Glossary for the Flora of North America Project* (R. W. Kiger and D. M. Porter 2001; also available online at http://huntbot.andrew.cmu.edu) for standard definitions of generally used terms. Very specialized terms are defined, and sometimes illustrated, in the relevant family or generic treatments.

References

Authoritative general reference works used for style are *The Chicago Manual of Style*, ed. 14 (University of Chicago Press 1993); *Webster's New Geographical Dictionary* (Merriam-Webster 1988); and *The Random House Dictionary of the English Language*, ed. 2, unabridged (S. B. Flexner and L. C. Hauck 1987). *B-P-H/S. Botanico-Periodicum-Huntianum/Supplementum* (G. D. R. Bridson and E. R. Smith 1991) has been used for abbreviations of serial titles, and *Taxonomic Literature*, ed. 2 (F. A. Stafleu and R. S. Cowan 1976–1988) and its supplements by F. A. Stafleu and E. A. Mennega (1992+) have been used for abbreviations of book titles.

Graphic Elements

All genera and approximately 30 percent of the species in this volume are illustrated. Illustration panels have been enlarged for this and subsequent volumes in the series. The illustrations may show diagnostic traits or complex structures. Most illustrations have been drawn from herbarium specimens selected by the authors. In some cases living material or photographs have been used. Data on specimens that were used and parts that were illustrated have been recorded. This information, together with the archivally preserved original drawings, is deposited in the Missouri Botanical Garden Library and is available for scholarly study.

Specific Information in Treatments

Keys

Dichotomous keys are included for all ranks below family if two or more taxa are treated. For dioecious species, keys are designed for use with either staminate or pistillate plants. Keys are designed also to facilitate identification of taxa that flower before leaves appear. More than one key may be given, and for some groups tabular comparisons may be presented in addition to keys.

Nomenclatural Information

Basionyms of accepted names, with author and bibliographic citations, are listed first in synonymy, followed by any other synonyms in common recent use, listed in alphabetical order, without bibliographic citations.

Vernacular names in common use are given in the appropriate language. In general, such names have not been coined for use in the *Flora*. Those preferred by governmental or conservation agencies are listed if known.

The last names of authors of taxonomic names have been spelled out. The conventions of *Authors of Plant Names* (R. K. Brummitt and C. E. Powell 1992) have been used as a guide for including first initials to discriminate individuals who share surnames. Exceptions include "Alph. Wood" instead of "A. W. Wood" and "K. F. Parker" instead of "K. L. Parker" (Brummitt, pers. comm.)

If only one infraspecific taxon within a species occurs in the flora area, nomenclatural information (literature citation, basionym with literature citation, relevant other synonyms) is given for the species, as is information on the number of infraspecific taxa in the species and their distribution worldwide, if known. A description and detailed distributional information are given only for the infraspecific taxon.

Descriptions

Character states common to all taxa are noted in the description of the taxon at the next higher rank. For example, if flowers are unisexual for all species treated within a genus, that character state is given in the generic description. Characters used in keys are repeated in the descriptions. Characteristics are given as they occur in plants from the flora area. Characteristics that occur only in plants from outside the flora area may be given within square brackets, or instead may be noted in the discussion following the description. In families with one genus and one or more species, the family description is given as usual, the genus description is condensed, and the species are described as usual. Any special terms that may be used when describing members of a genus are presented and explained in the genus description.

In reading descriptions, the reader may assume, unless otherwise noted, that: the plants are green, photosynthetic, and reproductively mature; woody plants are perennial; stems are erect; roots are fibrous; leaves are simple and petiolate; flowers are bisexual, radially symmetric, and pediceled; perianth parts are hypogynous, distinct, and free; and ovaries are superior. Because measurements and elevations are almost always approximate, modifiers such as "about," "circa," or "±" are usually omitted.

Unless otherwise noted, dimensions are length × width. If only one dimension is given, it is length or height. All measurements are given in metric units. Measurements usually are based on dried specimens but these should not differ significantly from the measurements found in fresh or living material.

Chromosome numbers generally are given only if published, documented counts are available from North American material or from an adjacent region. No new counts are published intentionally in the *Flora*. Chromosome counts from nonsporophyte tissue have been converted to the *2n* form. The base number ($x = $) is given for each genus. This represents the lowest known haploid count for the genus unless evidence is available that the base number differs.

Flowering time and often fruiting time are given by season, sometimes qualified by early, mid, or late, or by months. Elevations over 50 m generally are rounded to the nearest 100 m; those 50 m and under are rounded to the nearest 10 m. Mean sea level is shown as 0 m, with the understanding that this is approximate. Elevation often is omitted from herbarium specimen labels, particularly for collections made where the topography is not remarkable, and therefore precise elevation is sometimes not known for a given taxon.

The term "introduced" is defined broadly to refer to plants that were released deliberately or accidentally into the flora and that now exist as wild plants in areas in which they were not recorded as native in the past. The distribution of non-native plants is often poorly documented and presence of the plants in the flora may be ephemeral.

If a taxon is globally rare or if its continued existence is threatened in some way, the words "of conservation concern" appear before the statements of elevation and geographic range.

Criteria for taxa of conservation concern are based on NatureServe's (formerly The Nature Conservancy)—see http://www.natureserve.org—designations of global rank (G-rank) G1 and G2:

G1 Critically imperiled globally because of extreme rarity (5 or fewer occurrences or fewer than 1000 individuals or acres) or because of some factor(s) making it especially vulnerable to extinction.

G2 Imperiled globally because of rarity (5–20 occurrences or fewer than 3000 individuals or acres) or because of some factor(s) making it very vulnerable to extinction throughout its range.

The occurrence of species and infraspecific taxa within political subunits of the *Flora* area is depicted by dots placed on the outline map to indicate occurrence in a state or province. For the 48 contiguous states of the United States and the smaller Canadian provinces, a single dot is used in those units where a taxon is known to occur. In the case of Greenland, the larger Canadian provinces and territories, and the main area of Alaska, a dot's position can vary to indicate more northern, southern, or central/scattered distributions (also western or eastern only for Alaska). In the case of Alaska, the occurrence of a taxon in the Aleutian Islands and/or the panhandle area adjacent to British Columbia may also be indicated. However, the dots for these areas may not be readily seen due to the small map size. The Nunavut boundary on the maps has been provided by the GeoAccess Division, Canada Centre for Remote Sensing, Earth Science. Authors are expected to have seen at least one specimen documenting each geographic unit record and have been urged to examine as many specimens as possible from throughout the range of each taxon. Additional information about taxon distribution may be presented in the discussion.

Distributions are stated in the following order: Greenland; St. Pierre and Miquelon; Canada (provinces and territories in alphabetic order); United States (states in alphabetic order); Mexico (11 northern states may be listed specifically, in alphabetic order); West Indies; Bermuda; Central America (Belize, Costa Rica, El Salvador, Guatemala, Honduras, Nicaragua, Panama); South America; Europe, or Eurasia; Asia (including Indonesia); Africa; Pacific Islands; Australia; Antarctica.

Discussion

The discussion section may include information on taxonomic problems, distributional and ecological details, interesting biological phenomena, economic uses, and toxicity (see "Caution," below).

Selected References

Major references used in preparation of a treatment or containing critical information about a taxon are cited following the discussion. These, and other works that are referred to in discussion or elsewhere, are included in Literature Cited at the end of Volume 21.

CAUTION

The Flora of North America Editorial Committee **does not encourage, recommend, promote, or endorse** any of the folk remedies, culinary practices, or various utilizations of any plant described within these volumes. Information about medicinal practices and/or ingestion of plants, or of any part or preparation thereof, has been included only for historical background and as a matter of interest. Under no circumstances should the information contained in these volumes be used in connection with medical treatment. Readers are strongly cautioned to remember that many plants in the flora are toxic or can cause unpleasant or adverse reactions if used or encountered carelessly.

Key to boxed codes following accepted names:

- C̄ of conservation concern
- Ē endemic to the flora area
- F̄ illustrated
- Ī introduced to the flora area
- W̄ weedy, based mostly on R. H. Callihan et al. (1995) and/or D. T. Patterson et al. (1989)

Flora of North America

187. ASTERACEAE Martinov

· Composite Family

Compositae Adanson

Theodore M. Barkley†

Luc Brouillet

John L. Strother

Annuals, biennials, perennials, subshrubs, shrubs, vines, or trees. Roots usually taproots, sometimes fibrous. **Stems** usually erect, sometimes prostrate to ascending (underground stems sometimes woody caudices or rhizomes, sometimes fleshy). **Leaves** usually alternate or opposite, sometimes in basal rosettes, rarely in whorls; rarely stipulate, usually petiolate, sometimes sessile, sometimes with bases decurrent onto stems; blades usually simple (margins sometimes 1–2+ times pinnatifid or palmatifid), rarely compound. **Inflorescences** indeterminate *heads* (also called capitula); each head usually comprising a surrounding *involucre* of *phyllaries* (involucral bracts), a *receptacle*, and (1–)5–300+ *florets*; individual heads sessile or each borne on a *peduncle*; heads borne singly or in usually determinate, rarely indeterminate, arrays (cymiform, corymbiform, racemiform, spiciform, etc.); involucres sometimes subtended by *calyculi* (sing. calyculus); phyllaries borne in 1–5(–15+) series proximal to (i.e., outside of or abaxial to) the florets; receptacles usually flat to convex, sometimes conic or columnar, either paleate (bearing *paleae* or receptacular bracts that individually subtend some or all of the florets) or epaleate (lacking paleae); epaleate receptacles sometimes bristly or hairy or bearing subulate enations among the florets. **Florets** bisexual, pistillate, functionally staminate, or neuter (also called neutral); sepals highly modifed (instead of ordinary sepals, each ovary usually bears a *pappus* of bristles, awns, and/or scales, sometimes in combination within a single pappus); petals connate, corollas (3–)5-merous, ± actinomorphic or zygomorphic (one or both kinds in a single head, see descriptions of *radiate, discoid, liguliflorous, disciform,* and *radiant* following); stamens (4–)5, alternate with corolla lobes, filaments inserted on corollas, usually distinct, anthers introrse, usually connate and forming tubes around styles (rarely filaments connate and anthers distinct; e.g., Heliantheae, Ambrosiinae); ovaries inferior, 2-carpellate, and 1-locular with 1 basally attached, anatropous ovule; styles 1 in each bisexual, functionally staminate, or pistillate floret; each style usually ringed at base by a nectary, distally 2-branched with stigmatic papillae borne on adaxial face of each branch in 2 separate or contiguous lines or in 1 continuous band (styles usually not branched in functionally staminate florets), style branches apically truncate or appendaged beyond the stigmatic bands or lines, appendages usually papillate to hirsute distally on abaxial (or abaxial and adaxial) faces. **Fruits** (technically *cypselae*, historically called achenes) usually dry with relatively thick, tough pericarps, sometimes beaked (rostrate) and/or winged (alate), often dispersed with aid from *pappi*. **Seeds** 1 per fruit, exalbuminous; embryos straight.

Genera ca. 1500, species ca. 23,000 (418 genera, 2413 species in the flora): nearly worldwide, especially rich in numbers of species and/or in numbers of plants in arid and semiarid regions of subtropical and lower to middle temperate latitudes.

Asteraceae (Compositae, "composites," or "comps") have long been recognized as a natural group, and circumscription of the group has never been controversial (although some authors have divided the traditional family into three or more families). A. Cronquist (1981) placed Asteraceae as the only family in the order Asterales within subclass Asteridae, associated with the Gentianales, Rubiales, Dipsacales, and Calycerales and relatively distant from Campanulales. On recent molecular phylogenetic data, the Angiosperm Phylogeny Group (2003; see references there for details; classification abbreviated APGII hereafter) has suggested that Asteraceae are better treated as part of a more widely defined Asterales within the asterids II informal clade (or campanulid clade; see W. S. Judd and R. G. Olmstead 2004). Judd and Olmstead summarized the higher-order relationships of Asteraceae as follows (in order of decreasing inclusiveness; synapomorphies in parentheses): asterids (ovules unitegmic and tenuinucellate, iridoid chemistry); core asterids (sympetaly, stamen number equal to petal number, stamen epipetaly, mostly 2–3-carpellate gynoecia); campanulids (early sympetaly), comprising eight unassigned families plus Aquifoliales, which is sister to Dipsacales, Apiales, and Asterales (last three sharing frequently inferior ovaries, polyacetylenes); and Asterales, which appears to be sister to Dipsacales-Apiales (K. Bremer et al. 2004). The order Asterales (valvate petals, lack of apotracheal parenchyma, storage of inulin, ellagic acid present, and, possibly, the presence of a plunger or brush pollen presentation mechanism) now includes the following families (fide APGII): Alseuosmiaceae, Argophyllaceae, Calyceraceae, Campanulaceae (optionally including Lobeliaceae), Goodeniaceae, Menyanthaceae, Pentaphragmaceae, Phellinaceae, Rousseauaceae, and Stylidiaceae. Within Asterales, Asteraceae is part of a clade (corollas with more or less fused lateral veins joining midvein near lobe apices, thick integuments, no endosperm haustorium) with the Menyanthaceae (cosmopolitan with Southern Hemisphere genera) basal to a more nested clade (inferior ovaries, possibly connate anthers, pollen exine with bifurcating columellae) comprising Asteraceae, Goodeniaceae (mainly Australia), and Calyceraceae (South America), the last being the immediate sister to Asteraceae (highly modified, persistent calyces, corolla venation patterns, unilocular and uniovulate gynoecia, pollen with intercolpar depressions, specialized fruits). Aggregation of flowers into heads with involucres appears to have been a parallel phenomenon in Calyceraceae and Asteraceae, given the determinate nature of the former and indeterminate (racemose) organization of the latter. Some traits typical of Asteraceae predate evolution of the family as a distinct clade. Relationships of Asteraceae and Calyceraceae have been discussed by M. H. G. Gustafsson and Bremer (1995). Synapomorphies of the Asteraceae clade include: calyces modified to structures called pappi, anthers connate (forming tubes) and styles modified to function as brushes in a specialized pollen presentation mechanism, ovaries each containing a single basal ovule, and production of sesquiterpene lactones.

K. Bremer et al. (2004) gave an Early Cretaceous origin for the Asteridae and the basal campanulids, and a Late Cretaceous origin for the Asterales. Bremer and M. H. G. Gustafsson (1997) also hypothesized a Late Cretaceous ancestry of Asterales in East Gondwanaland (Australasia), with later expansion into West Gondwanaland (South America-Antarctica), where the Asteraceae originated before the final separation of South America and Antarctica. Similarly, M. L. DeVore and T. F. Stuessy (1995) argued that the close relationships of Asteraceae to Goodeniaceae and Calyceraceae, plus the basal position of Barnadesioideae K. Bremer & R. K. Jansen (Asteraceae), indicated a South America-Antarctica-Australia origin for the complex. After reviewing previous hypotheses, they proposed a late Eocene origin for the complex and suggested a South American origin for the Asteraceae based on the basal position of the South American Barnadesioideae (see also Stuessy et al. 1996, on Barnadesioideae origin in southern South America in the Oligocene) and their sister relationship to Calyceraceae. Fossil pollen data (both Mutisieae and Asteroideae types—notably Heliantheae in the broad sense— among earliest reports) reviewed by A. Graham (1996) appear to indicate an Eocene origin for Asteraceae in South America, with migration to North America at least by the Oligocene, possibly as early as the late Eocene. More recently, M. S. Zavada and S. E. de Villiers (2000; and references therein) reported Asteraceae pollen (assignable to Mutisieae in the broad sense) from the Paleocene-Eocene of South Africa, suggesting an earlier, West Gondwana (southern Africa or Australia) origin for the family. Such data indicate that some tribes of Asteraceae may have arrived in North America via long-distance dispersal or island hopping well before closure of the isthmus of Panama. They also have a bearing on the possible times of radiation of some tribes in North America, particularly Heliantheae in the broad sense and Eupatorieae, which originated in the continent (including Mexico and parts of Central America), and those that came to North America from or through South America such as Mutisieae, Vernonieae, some Plucheeae, and Astereae. Other tribes, such as Cynareae, Cichorieae, some Gnaphalieae, and Anthemideae, may have reached North America from Eurasia, possibly via Beringia (or as Amphi-Atlantic disjuncts), at a later time.

The bases of a tribal classification within Asteraceae were established in the nineteenth century, primarily through the work of H. Cassini (especially in articles scattered through the 61 volumes of F. Cuvier 1816–1845; Cassini included synopses of his tribes as part of his entry for *Zoegea*, i.e., zyégée in French; the articles have been collected in three volumes by R. M. King and H. W. Dawson 1975), C. F. Lessing (1832), A. P. de Candolle (1828–1838, 1836–1838), and, particularly, G. Bentham (1873). In the twentieth century, the tribal system of Cassini, as elaborated by Bentham, was widely followed with only slight modifications (see S. Carlquist 1976; A. Cronquist 1955, 1977; C. Jeffrey 1978; G. Wagenitz 1976b; see also J. Small 1919 and, for alternate views on Heliantheae-Eupatorieae, H. Robinson 1996).

A molecular phylogenetic study by R. K. Jansen and J. D. Palmer (1987) established that a South American clade (later named Barnadesioideae) is basal within Asteraceae. Both cladistic morphologic analyses (e.g., K. Bremer 1994, 1996) and mostly chloroplast-DNA molecular phylogenies (e.g., Jansen et al. 1991, 1992; K. J. Kim et al. 1992; Kim and Jansen 1995; R. J. Bayer and J. R. Starr 1998; P. K. Eldenäs et al. 1999; B. G. Baldwin et al. 2002) have deepened our knowledge of tribal interrelationships within Asteraceae and led to the recent proposal of a phylogenetic classification for the family with 10 subfamilies and 35 tribes (J. L. Panero and V. A. Funk 2002).

Treatment of Asteraceae here differs from some of the recently proposed classifications in that some groups continue to be traditionally circumscribed (e.g., Mutisieae in the broad sense, Heliantheae in the broad sense, including Helenieae and excluding Eupatorieae). Where appropriate and so far as practicable, new taxonomies are acknowledged in our discussions of individual tribes (which see). In North America, the following subfamilies and tribes, as defined by J. L. Panero and V. A. Funk (2002), are represented (tribes with no native representatives are marked by asterisks): Mutisioideae-Mutisieae in the strict sense, Gochnatioideae-Gochnatieae, and Hecastocleioideae-Hecastocleideae (all included in Mutisieae here, which see), Carduoideae (Cardueae = Cynareae), Cichorioideae (*Arctoteae, Cichorieae, Vernonieae), and Asteroideae [Senecioneae, *Calenduleae, Gnaphalieae, Anthemideae, Astereae, Plucheeae, *Inuleae, Eupatorieae, and the following segregates of Heliantheae in the broad sense (all treated here within or as subtribes of a fairly traditionally circumscribed Heliantheae): Bahieae, Chaenactideae, Coreopsideae, Helenieae, Heliantheae in the strict sense, Madieae, *Millereae, Perityleae, Polymnieae, and Tageteae)].

Asa Gray produced the first broadly influential floristic synthesis of North American Asteraceae. Other authors who made important contributions to floristics of North American Asteraceae in the nineteenth and first half of the twentieth centuries were S. F. Blake, N. L. Britton, R. S. Ferris, M. L. Fernald, E. L. Greene, H. M. Hall, M. E. Jones, D. D. Keck, P. A. Rydberg, J. K. Small, and S. Watson. Some of those authors had narrower concepts of genera and species than had their predecessors and they freely recognized new taxa in Asteraceae (mostly genera and species). Floristics of North American Asteraceae in the second half of the twentieth century was especially influenced by A. Cronquist (e.g., 1955, 1980, 1994; H. A. Gleason and Cronquist 1991), who usually favored traditional generic circumscriptions.

In the last 20 years or so, developments in molecular systematics have led to revisions of generic limits in some tribes of Asteraceae and, sometimes, to a return to generic concepts that had been suggested earlier but largely ignored. More or less worldwide, taxonomies in some tribes or parts of tribes have included segregate genera that have been revived or newly published. Most of the innovations will be summarized in the forthcoming Asterales volume of K. Kubitzki et al. (1990+). The generic circumscriptions adopted here incorporate recent taxonomic findings relevant to North America, insofar as our contributors have accepted them. As a result, many of the genera treated herein have never been presented in a major flora before, and some species are included within genera with which they were not associated traditionally. Thus, the Flora brings together much new knowledge and many new names. In most instances, circumscriptions of species have turned out to be conventional. So far as practicable, recently named species from North America have been accounted for within relevant treatments herein.

With 418 genera and 2413 species (Table 1), Asteraceae is, numerically, the largest family in the flora of North America north of Mexico. Members of the family are found in diverse habitats, from the High Arctic tundra and polar deserts to the Sonoran warm-desert scrub, and from alpine habitats to salt marshes. Asteraceae are particularly conspicuous elements of warm-desert and intermountain grasslands, as well as of desert scrubs, notably the intermountain desert scrub where *Artemisia* dominates (M. G. Barbour and N. L. Christensen 1993). Among other conspicuous species, members of *Solidago* and *Symphyotrichum* form a very showy part of the fall flowering in eastern North America, and members of Heliantheae sometimes produce striking displays in the American West (e.g., *Gaillardia* spp., *Lasthenia* spp., members of Madiinae).

Much has been published, not only on systematics (at various levels), but on biology, chemistry, and economic and medical uses of Asteraceae worldwide, particularly in proceedings (from conferences and symposia) edited by V. H. Heywood et al. (1977), T. J. Mabry and G. Wagenitz (1990), and D. J. N. Hind et al. (1995, 1996).

Relatively few North American species of Asteraceae are economically important or widely used ethnobotanically. The only major Asteraceae crop of North American origin is the sunflower, *Helianthus annuus*, which is valued for its seed oil and is appreciated in the horticultural trade. Other crop plants from native species worth mention are *Helianthus tuberosus*, the Jerusalem artichoke, and *Parthenium argentatum*, the guayule, a source of rubber. *Echinacea* spp. are touted as health plants. Members of several genera of Asteraceae native to the flora are grown for their ornamental value, notably species of *Coreopsis* (tickseeds), *Echinacea* (coneflowers), *Helianthus* (sunflowers), *Liatris* (blazingstars and gayfeathers), *Rudbeckia* (black-eyed Susans), *Solidago* (goldenrods), and *Symphyotrichum* ("asters" of the trade).

Many species of Asteraceae have been introduced into North America, mainly from Europe and Asia, some deliberately for medicines, foods, or horticulture, others accidentally (often with seeds or other agricultural products or by other means). Few, if any, of the introduced taxa are thought to be noxious at the continental level, but some (e.g., *Acroptilon*) are considered noxious in large parts of their ranges within the flora. *Taraxacum officinale* is a common lawn weed that (in terms of dollars spent and herbicides applied in weed control) has an economic and ecologic impact disproportionate to the actual harm it causes; other weedy introduced Asteraceae are of little economic consequence. Some native Asteraceae are toxic to cattle and other livestock and are therefore considered weeds. And some native species of open habitats (e.g., *Symphyotrichum pilosum*) are often considered weeds because they invade fields left fallow. Ragweeds (especially *Ambrosia artemisiifolia* and *A. trifida*) range over nearly the whole continent and their wind-blown pollens cause late-summer allergic reactions (hayfever) for a large number of people. Because ragweeds have a large impact on human health, they have a significant, negative economic impact.

In contrast to Orchidaceae, for which a wealth of excellent, well-illustrated popular books are available, few popular field guides on Asteraceae of North America have been published. The guide by T. M. Antonio and S. Masi (2001) deserves notice for its maps, color photographs, and useful information.

Composites (members of Asteraceae) share some unusual morphologic traits and some morphologic terms are used in particular ways as applied here to them.

For treatments of composites here, "perennials" are herbaceous and differ from annuals and biennials in living longer than two years and differ from subshrubs, shrubs, and trees in not developing woody aerial stems.

In most composites, leaf venation comprises a midrib plus more or less equal lateral nerves or veins; such leaves are described as pinnately nerved. Venation in leaf blades of some composites often consists of a midrib plus relatively strong lateral veins that diverge at or just distal to bases of blades. Such leaves are described as 3-nerved, 3(–5)-nerved, 5-nerved, etc., and, as appropriate, the phrases "from bases" or "distal to bases" may be added for clarification.

Composites often have subsessile to sessile or sunken glandular hairs that consist of multicellular bases supporting globular elements that usually contain resinous or sticky substances. Such structures have been called glands, glandular hairs, glandular trichomes, punctae, resin dots, and so on. Sometimes, the glands are embedded in epidermal depressions or pits. Epidermes with glands more or less sunk into or embedded within the surface have been called glandular-punctate and/or punctate-glandular. The glands may be colorless (translucent) or yellowish to dark brown or orange and are sometimes more prominent on dried specimens than in living plants. In keys and descriptions here, gland-dotted refers to the presence of such glandular hairs, whether sessile or in depressions or pits (as appropriate, "in pits" or "sessile" may be added for clarification).

Inflorescences of composites are called *heads* (or capitula, sing. capitulum). Heads may be borne singly (i.e., not clearly associated with other heads on the same plant) or associated in arrays. The arrays of heads on composites correspond to arrays of individual flowers (inflorescences) on plants of other families; arrays of heads are sometimes called capitulescences. Terms for architectural structures of arrays of heads are parallel to terms for kinds of inflorescences: cymiform, corymbiform, paniculiform, racemiform, spiciform, thyrsiform, etc.

In *radiate* heads, peripheral florets (*ray florets*) in one or more series have corollas with zygomorphic limbs and may be pistillate, or styliferous and sterile, or neuter; the central florets (*disc florets*) in radiate heads have ± actinomorphic corollas and may be bisexual or functionally staminate. In *liguliflorous* heads, all florets are bisexual and (usually) fertile and have zygomorphic corollas (*ligulate florets*); liguliflorous heads are characteristic of Cichorieae and are found in no other composites. In *discoid* heads, all florets have ± actinomorphic corollas and all are either bisexual and fertile or all are either functionally staminate or pistillate (in monoecious or dioecious taxa, e.g., *Baccharis* spp.). In *disciform* heads, all florets have ± actinomorphic corollas, and peripheral florets (in one or more series) are usually pistillate and usually have relatively slender (often filiform) corollas. Such peripheral pistillate florets are generally thought to be derived by reduction from ray florets, and plants with disciform heads are generally thought to be derived from ancestors with radiate heads. The central florets of disciform heads are usually bisexual, sometimes functionally staminate. By tradition and for simplicity, both the peripheral, pistillate florets and the inner, bisexual or functionally staminate florets in disciform heads may be referred to as "disc" florets. In *radiant* heads, all florets have ± actinomorphic corollas and the peripheral florets usually have much enlarged corollas and may be bisexual, pistillate, or neuter; the central florets of radiant heads are usually bisexual. Some composites have peripheral, bisexual florets with slightly to strongly zygomorphic corollas (e.g., some members of *Chaenactis, Lessingia, Thymophylla,* et al.); heads of such plants do not quite conform to any of the five types just described and such heads may be referred to as "quasi-radiate" or "quasi-radiant." Some florets in heads of some Mutisieae have 2-lipped corollas and those heads may be called "quasi-radiate" or "quasi-liguliflorous." The term *eradiate* is used to refer collectively to discoid, disciform, and radiant heads.

Heads with all florets of one sexual form (bisexual, pistillate, or functionally staminate) are called *homogamous* (discoid and liguliflorous heads are homogamous, some radiant heads may be homogamous) and heads with florets of two or more sexual forms are called *heterogamous* (radiate and disciform heads are heterogamous, some radiant heads may be heterogamous).

Phyllaries collectively constitute an *involucre*, usually number 5–21(–50+), usually are unequal (outermost usually shorter than the inner), and usually are arranged ± imbricately (overlapping like shingles) in 3–5(–15+), usually ± spiral series. Sometimes, the phyllaries are ± equal in 1–2 series; they are rarely wanting (e.g., *Psilocarphus* spp.). Phyllaries may be herbaceous or chartaceous to scarious and are often medially herbaceous with chartaceous to scarious borders and/or apices. The phyllaries "proper" are sometimes immediately subtended by a *calyculus* (pl. calyculi) of (1–)3–15+ distinct, usually shorter bractlets in 1(–3+) series (e.g., *Coreopsis* spp., *Taraxacum* spp.).

Receptacles may bear *paleae* (i.e., some or all florets are individually subtended by a bractlet called a palea or receptacular bract). Collectively paleae have been called "chaff" and paleate receptacles have been described as "chaffy." Receptacles that bear paleae are referred to as *paleate* and receptacles that never bear paleae are referred to as *epaleate*. Epaleate receptacles sometimes bear subulate enations (e.g., some *Gaillardia* spp.) or bristles or subulate to linear scales (e.g., some Cynareae), or fine hairs (e.g., some Anthemideae). Epaleate receptacles (and paleate receptacles that have shed their paleae) may be smooth or pitted (alveolate, foveolate, etc.).

The terms *tube*, *throat*, and *limb* have been variously used in descriptions of corollas of composites. Here, in ± actinomorphic corollas of bisexual and functionally staminate disc florets, the *tube* is the part of the corolla proximal to the insertion of the staminal filaments, and the *limb* is the part that is distal to insertion of the filaments. The limb comprises, proximally, the *throat* and, distally, the *lobes*. The distinction between tube and throat hinges on insertion of filaments, not on external morphology.

The relatively flat portion of a corolla of a ligulate floret from a liguliflorous head (i.e., members of Cichorieae) is called a *ligule*; it terminates in 5 teeth or lobes. The relatively flat portion of a corolla of a ray floret is called a *lamina*; it terminates in 0–3(–4) teeth or lobes. More or less bilabiate corollas are characteristic of some members of Mutisieae and are seldom found in members of other tribes.

Fruits of composites have been called "achenes" because they resemble true achenes. Achenes are dry, hard, single-seeded fruits derived from unicarpellate, superior ovaries. Ovaries of composites are bicarpellate and inferior. Fruits derived from ovaries of composites are called *cypselae* (sing. *cypsela*, a term coined by C. de Mirbel in 1815). Morphology of an ovary of a composite at flowering is often markedly different from the morphology of the mature fruit (cypsela) derived from that ovary. References to cypselae in keys and descriptions here almost always refer to mature fruits, not to ovaries at flowering.

Shapes of cypselae have been used in distinguishing among species, genera, and even subtribes of composites. In most composites, cypselae are ± isodiametric in cross section. In some composites, cypselae are characteristically ± lenticular to elliptic in cross section. Such cypselae are said to be *compressed* (or laterally flattened) if the longer axis of the cross section is ± parallel to a radius of the head (e.g., *Verbesina* spp.). Cypselae are said to be *obcompressed* (or radially flattened) if the shorter axis of the cross section is ± parallel to a radius of the head (e.g., *Coreopsis* spp.).

In composites, *pappi* (sing. pappus) are found where calyces are usually found on inferior ovaries; pappi have been shown to be greatly modified calyces. They show a great range of diversity and are often diagnostic for recognition of taxa, especially at rank of genus and below. The forms of individual pappus elements intergrade. For keys and descriptions here, the following distinctions are made: cross sections of bristles and *awns* are ± circular or polygonal and have the longer diameter of the cross section no more than 3 times the shorter diameter. Pappus elements with "flatter" cross sections (i.e., longer diameter more than 3 times the shorter diameter) are called *scales*, regardless of relative overall lengths and widths of the elements. As used here, "subulate scale" and "setiform scale" mean much the same as "flattened bristle" of some authors. Pliable to stiff pappus bristles with diameters less than ca. 50 mm are called *fine bristles*; pliable to stiff bristles with diameters 50–100 mm are called *coarse bristles*. Rigid pappus elements with ± circular or polygonal cross sections greater than 100 mm in diameter are called *awns*. Bristles, awns, and scales may be smooth or finely to coarsely barbed or plumose. A scale of a pappus may terminate in one or more bristlelike or awnlike appendages; such scales are said to be *aristate*.

In keys and descriptions, "pappus" and "pappi" usually refer to structures found on cypselae (mature fruits), not to "immature pappi" of ovaries at flowering. Sometimes pappi of ovaries that do not form fruits (e.g., in functionally staminate florets of some tarweeds) may be taxonomically useful and may be referred to in descriptions and keys.

SELECTED REFERENCES   Antonio, T. M. and S. Masi. 2001. The Sunflower Family in the Upper Midwest.... Indianapolis. Bayer, R. J. and J. R. Starr. 1998. Tribal phylogeny of the Asteraceae based on two non-coding chloroplast sequences, the *trn*L intron and the *trn*L/F intergenic spacer. Ann. Missouri Bot. Gard. 85: 242–256.   Bentham, G. 1873. Notes on the classification, history, and geographical distribution of Compositae. J. Linn. Soc., Bot 13: 335–577.   Bremer, K. 1987. Tribal interrelationships of the Asteraceae. Cladistics 3: 210–253.   Bremer, K. 1994. Asteraceae: Cladistics and Classification. Portland.   Bremer, K. 1996. Major clades and grades of the Asteraceae. In: D. J. N. Hind et al., eds. 1996. Proceedings of the International Compositae Conference, Kew, 1994. 2 vols. Kew. Vol. 1, pp. 1–7.   Carlquist, S. 1966. Wood anatomy of Anthemidae, Ambrosieae, Calenduleae, and Arctotideae (Compositae). Aliso 6(2): 1–23.   Carlquist, S. 1976. Tribal interrelationships and phylogeny of the Asteraceae. Aliso 8: 465–492.   Cronquist, A. 1955. Compositae. In: C. L. Hitchcock et al. 1955–1969. Vascular Plants of the Pacific Northwest. Seattle. Vol. 5.   Cronquist, A. 1955b. Phylogeny and taxonomy of the Compositae. Amer. Midl. Naturalist 53: 478–511.   Cronquist, A. 1977. The Compositae revisited. Brittonia 29: 137–153.   Cronquist, A. 1980. Asteraceae. In: A. E. Radford et al., eds. 1980+. Vascular Flora of the Southeastern United States. 2+ vols. Chapel Hill. Vol. 1.   Cronquist, A. 1994. Asteraceae. In: A. Cronquist et al., eds. 1972+. Intermountain Flora. Vascular Plants of the Intermountain West, U.S.A. 5+ vols. in 6+. New York and London. Vol. 5, pp. 5–471.   Eldenäs, P. K., M. Källersjö, and A. A. Anderberg. 1999. Phylogenetic placement and circumscription of tribes Inuleae s. str. and Plucheeae (Asteraceae): Evidence from sequences of chloroplast gene *ndh*F. Molec. Phylogen. Evol. 13: 50–58.   Heywood, V. H., J. B. Harbourne, and B. L. Turner, eds. 1977. The Biology and Chemistry of the Compositae. 2 vols. London, New York, and San Francisco.   Hind, D. J. N., H. J. Beentje, P. D. S. Caligari, and S. A. L. Smith, eds. 1996. Proceedings of the International Compositae Conference, Kew, 1994. 2 vols. Kew.   Hind, D. J. N., C. Jeffrey, and G. V. Pope, eds. 1995. Advances in Compositae Systematics. Kew.   Jansen, R. K. et al. 1991. Phylogeny and character evolution in the Asteraceae based on chloroplast DNA restriction site mapping. Syst. Bot. 16: 98–115.   Jansen, R. K. et al. 1992. Chloroplast DNA variation in the Asteraceae: Phyologenetic and evolutionary implications. In: D. E. Soltis et al., eds. 1992. Molecular Systematics of Plants. New York. Pp. 252–294.   Jeffrey, C. 1978. Compositae. In: V. H. Heywood, ed. 1978. Flowering Plants of the World. Oxford. Pp. 263–268.   Jeffrey, C. 1995. Compositae systematics 1975–1993. Developments and desiderata. In: D. J. N. Hind et al., eds. 1995. Advances in Compositae Systematics. Kew. Pp. 3–22.   Kim, K. J. et al. 1992. Phylogenetic implications of *rbc*L sequence variation in the Asteraceae. Ann. Missouri Bot. Gard. 79: 428–445.   Kim, K. J. and R. K. Jansen. 1995. *ndh*F sequence evolution and the major clades in the sunflower family. Proc. Natl. Acad. Sci. U.S.A. 92: 10379–10383.   King, R. M. and H. W. Dawson, eds. 1975. Cassini on Compositae.... 3 vols. New York.   Mabry, T. J. and G. Wagenitz, eds. 1990. Research advances in the Compositae. Pl. Syst. Evol., Suppl. 4.   Panero, J. L. and V. A. Funk. 2002. Toward a phylogenetic subfamilial classification for the Compositae (Asteraceae). Proc. Biol. Soc. Wash. 115: 909–922.   Robinson, H. 1996. Recent studies in the Heliantheae and Eupatorieae. In: D. J. N. Hind et al., eds. 1996. Proceedings of the International Compositae Conference, Kew, 1994. 2 vols. Kew. Vol. 1, pp. 627–653.   Small, J. 1919. The Origin and Development of the Compositae. London. [New Phytol. Repr. 11.]   Stuessy, T. F., T. Sang, and M. L. DeVore. 1996. Phylogeny and biogeography of subfamily Barnadesioideae with implications for early evolution of the Compositae. In: D. J. N. Hind et al., eds. 1996. Proceedings of the International Compositae Conference, Kew, 1994. 2 vols. Kew. Vol. 1, pp. 463–490.   Turner, B. L. 1996+. The Comps of Mexico: A Systematic Account of the Family Asteraceae. 2+ vols. Huntsville, Tex. [Phytologia Mem. 10, 11.]   Wagenitz, G. 1976b. Systematics and phylogeny of the Compositae (Asteraceae). Pl. Syst. Evol. 125: 29–46.

Key to Tribes of Asteraceae

Following is a synoptic key to tribes into which genera of composites of the flora area are placed. Keys to genera within each tribe will be found in the accounts of the individual tribes. Because some traits in the key to tribes and in keys to genera within tribes may be difficult to assess, we have also provided a key to artificial groups of composites and keys to genera within those artificial groups. Those keys will be found following the key to tribes.

1. Heads liguliflorous (sap usually milky) . . . . 187e. Cichorieae, v. 19, p. 214
1. Heads disciform, discoid, radiate, radiant, or quasi-radiate, -radiant, or -liguliflorous (not truly liguliflorous; sap rarely milky).
    2. Anther bases usually rounded or obtuse to acute (sometimes sagittate, not tailed).
        3. Styles (in bisexual, fertile florets) usually distally dilated and ± cylindric, style branches linear (sometimes adhering almost to minutely parted tips or essentially lacking), style-branch appendages essentially 0 . . . . 187c. Arctotideae, v. 19, p. 195
        3. Styles usually ± filiform (not distally dilated), style branches mostly lanceolate or linear (not adhering almost to tips), style-branch appendages mostly clavate, deltate, lanceolate, penicillate, or terete, sometimes filiform to linear, or essentially 0.

[4. Shifted to left margin.—Ed.]

4. Style-branch appendages usually terete to clavate (lengths usually 2–5+ times lengths of stigmatic lines; leaves usually opposite, sometimes whorled or alternate; heads discoid; corollas white, ochroleucous, or pink to purplish (not yellow)............ 187n. Eupatorieae, v. 21, p. 459

4. Style-branch appendages essentially 0 or ± penicillate or deltate to lanceolate, linear, or filiform (lengths mostly 0–3+ times lengths of stigmatic areas or lines).

    5. Margins of leaf blades usually 1–3-palmately or -pinnately divided (ultimate lobes usually linear to filiform), sometimes dentate or entire; phyllaries usually in 3–5+ series, unequal, and scarious or margins and/or apices notably scarious, sometimes in 1–2 series, subequal, and herbaceous, margins and apices little, if at all, scarious; pappi usually 0 (cypselar wall tissues sometimes produced as pappus-like, entire or toothed wings, coronas, or tubes), if pappi present, usually of scales, rarely of bristles...... .... 187j. Anthemideae, v. 19, p. 485

    5. Margins of leaf blades entire, dentate, lobed, palmatifid, pinnatifid, or dissected (ultimate lobes sometimes linear to filiform); phyllaries in 1–2 series and ± equal to subequal or in 3–8+ series and unequal, usually herbaceous to chartaceous, sometimes scarious or with margins and/or apices notably scarious; pappi usually of scales and/or awns and/or bristles, sometimes 0.

        6. Heads usually discoid (quasi-radiant or quasi-radiate in *Stokesia*); corollas white, ochroleucous, or pink to cyanic (not yellow); style branches lance-linear to ± lanceolate, abaxially hirsutulous, adaxially stigmatic (continuously) from bases nearly to apices, appendages essentially 0 .... 187d. Vernonieae, v. 19, p. 200

        6. Heads usually radiate or discoid, sometimes disciform, rarely ± radiant; corollas (some or all) usually yellow to orange, sometimes white, ochroleucous, or pink to cyanic, red, or purplish to brown; style branches ± linear to lanceolate, abaxially usually glabrous and smooth to papillate, sometimes hirsutulous, adaxially stigmatic (continuously or in 2 lines) from bases to appendages or apices, appendages usually penicillate or deltate to lanceolate, sometimes essentially 0.

           7. Leaves alternate; phyllaries in 1–2 series and equal to subequal in most spp. (sometimes coherent, often subtended by calyculi), rarely in 3+ series and unequal (e.g., *Lepidospartum*); receptacles epaleate; style-branch appendages usually penicillate or essentially 0; cypselae usually ± columnar to prismatic, seldom compressed, obcompressed, or flattened; pappi usually of (30–100+) fine (never plumose) bristles, rarely of subulate scales (e.g., *Tetradymia* spp.) or 0 .... 187l. Senecioneae, v. 20, p. 540

           7. Leaves whorled, opposite, or alternate; phyllaries subequal in 1–2 series or unequal in 3–5+ series; receptacles paleate or epaleate; style-branch appendages usually deltate to lanceolate, seldom penicillate; cypselae sometimes compressed, obcompressed, or flattened; pappi 0 or of scales and/or bristles and/or awns.

               8. Leaves alternate (sometimes mostly basal); calyculi 0; phyllaries in 3–5+ series and unequal in most spp., in 1–2 series and subequal in some spp. (mostly linear to oblanceolate); receptacles usually epaleate (except *Eastwoodia*, *Rigiopappus*, and some *Baccharis* spp.); style-branch appendages glabrous adaxially; pappi (rarely 0) usually of bristles, sometimes of scales, rarely of awns 0 .... 187k. Astereae, v. 20, p. 3

               8. Leaves usually wholly or partially opposite (at least proximally), sometimes mostly whorled or alternate (sometimes mostly basal); calyculi 0 or of 1–15+ bractlets; phyllaries usually in 3–5+ series and unequal (then usually lanceolate to ovate or broader), sometimes in 1–2 series and subequal (then usually linear to lanceolate); receptacles paleate or epaleate; style-branch appendages usually abaxially and adaxially papillate to hispidulous; pappi (sometimes 0) usually of scales (scales sometimes aristate, sometimes plumose), sometimes of awns or smooth, barbellulate to barbellate, or plumose bristles (e.g., *Bebbia*) .... 187m. Heliantheae, v. 21, p. 3

[2. Shifted to left margin.—Ed.]

2. Anther bases ± tailed.
   9. Corollas all 2-lipped . . . . 187a. Mutisieae (in part; *Acourtia, Trixis*), **v. 19, p. 70**
   9. Corollas actinomorphic or zygomorphic (not all 2-lipped, usually none 2-lipped).
     10. Cypselae clavate (distally stipitate-glandular) . . . 187a. Mutisieae (in part; Adenocaulon), **v. 19, p. 70**
     10. Cypselae mostly columnar, ellipsoid, obpyramidal, or prismatic, seldom clavate (if clavate, not distally stipitate-glandular).
       11. Heads aggregated in second-order heads (florets usually 1–3 per individual head).
         12. Shrubs (40–70 cm); cypselae glabrous or glabrate . . . .
           . . . . 187a. Mutisieae (in part; *Hecastocleis*), **v. 19, p. 70**
         12. Perennials (100–200 cm); cypselae villous . . . .
           . . . . 187b. Cynareae (in part; *Echinops*), **v. 19, p. 82**
       11. Heads borne singly or in corymbiform, paniculiform, or racemiform arrays (not aggregated in second-order heads; florets usually 3–300+ per individual head).
         13. Heads homogamous (sometimes heterogamous and disciform or radiant, plants often prickly-spiny and thistlelike and/or phyllary apices spinose or ± expanded into distinct, often fimbriate-fringed, pectinate, and/or prickly appendages) . . . . 187b. Cynareae (in part), **v. 19, p. 82**
         13. Heads usually heterogamous (plants not prickly-spiny and thistlelike; sometimes homogamous and unisexual in Gnaphalieae; homogamous in *Gochnatia* of Mutisieae: shrubs with leaves abaxially white, tomentose, and adaxially green, glabrate or glabrous).
           14. Cypselae usually polymorphic within heads (straight, arcuate, contorted, or ± coiled), bodies usually tuberculate, ridged, and/or winged (blue-black and drupelike in *Chrysanthemoides*); pappi 0 . . . 187f. Calenduleae, **v. 19, p. 379**
           14. Cypselae usually monomorphic, bodies usually smooth or ribbed; pappi usually of bristles, sometimes of scales or 0.
             15. Heads radiate (ray corollas usually yellow) or quasi-radiate ("quasi-ray" corollas 2-lipped in Mutisieae).
               16. Leaves basal and/or cauline (basal often withering before flowering); heads usually borne in corymbiform or paniculiform arrays, sometimes borne singly (stems ± leafy) . . . . 187h. Inuleae, **v. 19, p. 471**
               16. Leaves all or mostly basal; heads mostly borne singly (stems scapiform) . . . . . . 187a. Mutisieae (in part; *Chaptalia, Leibnitzia*), **v. 19, p. 70**
             15. Heads usually disciform or discoid, rarely quasi-radiate (then "quasi-ray" corollas usually pink to purplish, sometimes whitish or ochroleucous, rarely yellowish; whitish corollas of peripheral pistillate florets in *Sachsia* of Plucheeae, sometimes with minute, 3-toothed laminae).
               17. Phyllaries usually (12–30+) in 3–10+ series (then scarious or margins and/or apices usually notably scarious), sometimes (3–10) in 1–2 series or 0 (plants often 1–10 cm and ± densely woolly) . . . . 187g. Gnaphalieae, **v. 19, p. 384**
               17. Phyllaries (12–30+) in 3–6+ series (usually ± herbaceous to chartaceous, sometimes indurate, margins and/or apices seldom notably scarious) . . . . 187i. Plucheeae, **v. 19, p. 475**

Key to Artificial Groups of Genera of Asteraceae

In the following key, "radiate heads" have ray florets; "eradiate heads" lack ray florets and may be disciform, discoid, or radiant. Ray florets have zygomorphic corollas with laminae; the laminae may be showy (as in some species of *Helianthus*) or inconspicuous (as in some species of *Erigeron*). Usually, we have included plants with inconspicuous ray laminae in keys to genera of both radiate and eradiate groups.

Some plants have questionably paleate or epaleate receptacles. Epaleate receptacles of some plants are notably pitted and have fimbriate to deeply lacerate pit borders; such receptacles have sometimes been interpreted as paleate. Plants with notably lacerate pit borders are usually keyed here as both paleate and epaleate.

Some plants with pappi of conspicuous bristles often have the bristles subtended by minute, inconspicuous scales. Although such plants technically belong to groups with pappi "wholly, or partially, of awns or scales," they are usually also keyed here in groups characterized as having pappi "wholly of bristles," because the scales are easily overlooked. As well, some pappus elements are borderline between being called subulate or setiform scales or being called "flattened bristles." Consequently, some plants that technically belong to groups with pappi of scales are keyed both in groups with pappi "wholly of bristles" and in groups with pappi "wholly, or partially, of awns or scales."

1. Plants usually with milky sap; heads liguliflorous (corollas all ligulate, i.e., ± strap-shaped, 5-lobed) . . . . Group 1 (Cichorieae), **v. 19, p. 214**
1. Plants usually with clear sap, rarely with milky sap; heads discoid, disciform, radiant, or radiate (corollas of bisexual and/or functionally staminate florets mostly actinomorphic or nearly so).
   2. "Ray" corollas (or all corollas) ± 2-lipped (each with a 3-toothed, abaxial lamina opposite 2 curled, filiform to linear, adaxial lobes 1–3+ mm long) . . . . Group 2, **v. 19, p. 14**
   2. Rays none or ray corollas not distinctly 2-lipped (ray corollas rarely with 1–2 teeth to 1+ mm opposite their laminae).
      3. Receptacles paleate (i.e., some or all inner florets each subtended by a palea, a receptacular bract).
         4. Heads radiate (1 or more peripheral florets pistillate, neuter, or styliferous and sterile, their corollas zygomorphic; other florets bisexual or functionally staminate, their corollas actinomorphic or nearly so).
            5. Pappi none or nearly so . . . . Group 3, **v. 19, p. 14**
            5. Pappi of awns, bristles, and/or scales (on some or all ovaries or cypselae).
               6. Pappi wholly of bristles . . . . 292. *Acmella* (in part), **v. 21, p. 132**
               6. Pappi wholly or partially of awns (the awns often barbed) and/or scales (the scales relatively broad to narrow, sometimes subulate to setiform, sometimes barbellate to plumose and/or distally aristate) . . . . Group 4, **v. 19, p. 19**
         4. Heads eradiate (corollas of all florets ± actinomorphic, sometimes corollas of outermost, peripheral florets with reduced or enlarged limbs, the heads discoid, disciform, or radiant).
            7. Pappi none or nearly so . . . . Group 5, **v. 19, p. 24**
            7. Pappi of awns, bristles, and/or scales (on some or all ovaries or cypselae).
               8. Pappi wholly of bristles (the bristles smooth or barbellulate to plumose) . . . . Group 6, **v. 19, p. 27**
               8. Pappi wholly or partially of awns (the awns often barbed) and/or scales (the scales relatively broad to narrow, sometimes subulate to setiform, sometimes barbellate to plumose and/or distally aristate) . . . . Group 7, **v. 19, p. 28**
      3. Receptacles epaleate (lacking paleae, sometimes bristly, setose, or hairy, or pitted and pit borders ± laciniate).
         9. Heads eradiate (corollas of all florets ± actinomorphic, sometimes corollas of outermost, peripheral florets with reduced or enlarged limbs, the heads discoid, disciform, or radiant).
            10. Pappi none or nearly so . . . . Group 12, **v. 19, p. 51**
            10. Pappi of awns, bristles, and/or scales (on some or all ovaries or cypselae).

11. Pappi wholly of bristles (the bristles smooth or barbellulate to
    plumose) . . . . Group 13, **v. 19, p. 53**
11. Pappi wholly or partially of awns (the awns often barbed) and/or scales
    (the scales relatively broad to narrow, sometimes subulate to setiform,
    sometimes barbellate to plumose and/or distally aristate) . . . . Group 14, **v. 19, p. 64**

[9. Shifted to left margin.—Ed.]

9. Heads radiate (1 or more peripheral florets pistillate, neuter, or styliferous and sterile, their
   corollas zygomorphic; other florets bisexual or functionally staminate, their corollas actinomorphic or nearly so).
   12. Pappi none or nearly so. . . . . Group 8, **v. 19, p. 30**
   12. Pappi well-developed on some or all ovaries or cypselae.
       13. Pappi wholly or partially of awns (the awns often barbed) and/or scales (the scales
           relatively broad to narrow, sometimes subulate to setiform, sometimes barbellate
           to plumose and/or distally aristate). . . . . Group 11, **v. 19, p. 45**
       13. Pappi wholly of bristles (the bristles smooth or barbellulate to plumose).
           14. Ray corollas white, or pink to purple (with little, if any, yellow) . . . . Group 9, **v. 19, p. 35**
           14. Ray corollas brown, orange, red, or yellow . . . . Group 10, **v. 19, p. 40**

Key to Genera of Group 2

"Ray" corollas or all corollas distinctly ± 2 lipped.

1. Leaves: abaxial faces glabrous or with scattered, straight, sometimes glandular, hairs and/or
   double hairs.
   2. Perennials (caudices brown-woolly); corollas pink, lavender, or white . . . . 2. *Acourtia* (in part), **v. 19, p. 72**
   2. Shrubs; corollas yellow . . . . 3. *Trixis* (in part), **v. 19, p. 75**
1. Leaves: abaxial faces with dense, white or yellowish wool.
   3. Heads chasmogamous, (produced well after rosette leaves); cypselae glabrous
      or with inflated, blunt hairs. . . . . 6. *Chaptalia* (in part), **v. 19, p. 78**
   3. Heads chasmogamous (early spring, before and concurrently with first rosette leaves)
      or cleistogamous (summer and fall); cypselae with slender, sharp-pointed
      hairs . . . . 7. *Leibnitzia* (in part), **v. 19, p. 80**

Key to Genera of Group 3

Heads radiate; receptacles paleate; pappi none or nearly so.

1. Leaves alternate, margins of leaf blades usually 1–3-palmately or -pinnately lobed (ultimate
   lobes usually linear to filiform), sometimes dentate; phyllaries usually in 3–5+ series,
   unequal, scarious or margins and/or apices notably scarious.
   2. Heads in compact to open (± flat-topped), corymbiform or compound-corymbiform
      arrays; rays 3–5(–12); disc florets (5–)15–30+ (cypselae obcompressed) . . . . 113. *Achillea*, **v. 19, p. 492**
   2. Heads borne singly or in lax, corymbiform arrays; rays 5–30+; disc florets 40–300+
      (cypselae mostly not, sometimes weakly, obcompressed).
      3. Disc corolla tubes ± cylindric (bases ± saccate or spurred, ± clasping apices of
         ovaries and/or cypselae); cypsela ribs or nerves (weak) 2 lateral and 1 adaxial.
         4. Phyllaries 16–24 in 2–3+ series; rays orange, yellow, or white with yellow bases
            . . . . 114. *Cladanthus*, **v. 19, p. 495**
         4. Phyllaries 22–45+ in 3–4+ series; rays white. . . . . . . 115. *Chamaemelum* (in part), **v. 19, p. 496**
      3. Disc corolla tubes ± compressed or cylindric (bases sometimes proximally dilated,
         not saccate or spurred); cypsela ribs 9–10, or 0, or 2 lateral (sometimes ± winged)
         plus 3–10 finer ribs on each face.
         5. Annuals (biennials); rays usually white, rarely yellow or pink . . . . 122. *Anthemis*, **v. 19, p. 537**
         5. Perennials; rays yellow . . . . 128. *Cota*, **v. 19, p. 547**
1. Leaves mostly opposite, and/or sometimes alternate or whorled, margins of leaf blades
   usually entire, dentate, or lobed, sometimes 1–2-palmatifid or -pinnatifid (ultimate lobes
   sometimes linear to filiform); phyllaries usually in 3–5+ series and unequal, sometimes in 1–
   2 series and ± equal to subequal, usually herbaceous to chartaceous, sometimes with margins and/or apices notably scarious.

[6. Shifted to left margin.—Ed.]

6. Annuals; heads obscurely radiate (ray laminae minute; cypselae shed with 2 adjacent, ±
   fleshy paleae; Arizona) . . . . 247. *Parthenice*, v. 21, p. 23
6. Annuals, biennials, perennials, subshrubs, or shrubs; heads usually conspicuously radiate
   (laminae showy; cypselae seldom shed with accessory structures).
   7. Plants often with tack-glands or pit-glands on stems, leaves, and/or phyllaries; phyllar-
      ies (or paleae functioning as phyllaries) usually in 1+ series (each often wholly or partly
      investing ovary of a subtended floret); paleae often in 1 series between ray and disc
      florets, often connate in a ring, sometimes each disc floret subtended by a palea; ray
      laminae often flabellate (lobe lengths often ¹/₂+ laminae).
      8. Ray cypselae obcompressed (each mostly or completely enveloped by a phyllary).
         9. Annuals, 1–20 cm; disc florets 1(–2) . . . . 345. *Hemizonella*, v. 21, p. 296
         9. Annuals or perennials, 2–150 cm; disc florets 3–120+.
            10. Perennials (rhizomatous); disc corollas white . . . . 343. *Holozonia* (in part), v. 21, p. 294
            10. Annuals; disc corollas yellow (sometimes reddish with age).
               11. Disc pappi of 10, apically obtuse scales . . . . . 332. *Achyrachaena* (in part), v. 21, p. 258
               11. Disc pappi 0.
                  12. Calyculi 0 or of 2–5 bractlets; ray florets 5; disc florets 6, function-
                     ally staminate . . . . 334. *Lagophylla*, v. 21, p. 260
                  12. Calyculi 0; ray florets 3–27; disc florets 4–120+, bisexual . . . . .
                     . . . . 335. *Layia* (in part), v. 21, p. 262
      8. Ray cypselae compressed, ± terete, ± 3-angled in cross section, or ± obcompressed
         (each ± ¹/₂ enveloped by a phyllary).
         13. Annuals; styles of disc florets hairy proximal to minute branches (receptacles
            paleate throughout, ray corollas white with abaxial purple lines) . . . .
            . . . . 333. *Blepharipappus* (in part), v. 21, p. 259
         13. Annuals, perennials, subshrubs, or shrubs; styles of disc florets glabrous
            proximal to branches.
            14. Perennials (± scapiform; disc pappi of subulate, ciliate-plumose scales)
               . . . . 331. *Raillardella* (in part), v. 21, p. 256
            14. Annuals, perennials (leafy-stemmed), subshrubs, or shrubs; disc pappi 0 or
               of scales, scales seldom both subulate and ciliate-plumose.
               15. Annuals or perennials; peduncle bracts without terminal pit-glands,
                  tack-glands, or spines; rays yellow; ray cypselae usually compressed,
                  rarely terete (cross sections usually ± 3-angled, then abaxial sides rela-
                  tively broad, ± rounded, adaxial sides ± 2-faced, angles between those
                  faces 15–70°; each ray cypsela usually completely or mostly enveloped
                  by a phyllary).
                  16. Disc pappi 0 . . . . 350. *Madia* (in part), v. 21, p. 303
                  16. Disc pappi of 5–21 scales (scales sometimes subulate to setiform,
                     bristlelike) . . . . 344. *Kyhosia* (in part), v. 21, p. 295
               15. Annuals, subshrubs, or shrubs; peduncle bracts sometimes each with
                  terminal pit-gland, tack-gland, or spine (or apiculus); rays yellow, whit-
                  ish, or rose; ray cypselae terete to subterete or ± obcompressed (cross
                  sections nearly circular with adaxial sides ± flattened to slightly bulg-
                  ing, or ± 3-angled, then abaxial sides usually ± broadly 2-faced, angles
                  between those faces usually 90+° and adaxial sides ± flattened to slightly
                  bulging; in *Centromadia* spp., distal leaves spine-tipped, each cypsela
                  ± enveloped by a phyllary, cypselae sometimes compressed).
                  17. Annuals; leaves filiform to narrowly linear, margins often strongly
                     revolute; peduncle bracts usually with tack-glands; ray corolla lobes
                     (at least the lateral) often spreading (lengths often ¹/₂–⁵/₆ of total
                     laminae).
                     18. Ray cypselae beaked; tack-glands absent . . . .
                        . . . . 336. *Osmadenia* (in part), v. 21, p. 269
                     18. Ray cypselae not beaked; tack-glands present . . . .
                        . . . . 337. *Calycadenia* (in part), v. 21, p. 270

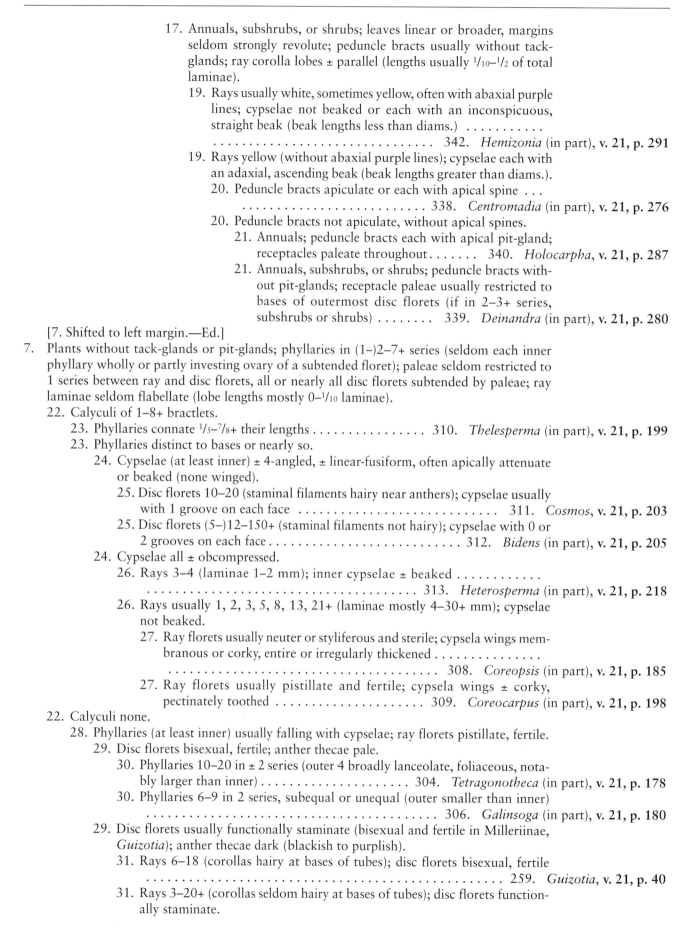

17. Annuals, subshrubs, or shrubs; leaves linear or broader, margins
    seldom strongly revolute; peduncle bracts usually without tack-
    glands; ray corolla lobes ± parallel (lengths usually $^{1}/_{10}$–$^{1}/_{2}$ of total
    laminae).
    19. Rays usually white, sometimes yellow, often with abaxial purple
        lines; cypselae not beaked or each with an inconspicuous,
        straight beak (beak lengths less than diams.) . . . .
        . . . . 342. *Hemizonia* (in part), **v. 21, p. 291**
    19. Rays yellow (without abaxial purple lines); cypselae each with
        an adaxial, ascending beak (beak lengths greater than diams.).
        20. Peduncle bracts apiculate or each with apical spine . . .
            . . . . 338. *Centromadia* (in part), **v. 21, p. 276**
        20. Peduncle bracts not apiculate, without apical spines.
            21. Annuals; peduncle bracts each with apical pit-gland;
                receptacles paleate throughout . . . . 340. *Holocarpha*, **v. 21, p. 287**
            21. Annuals, subshrubs, or shrubs; peduncle bracts with-
                out pit-glands; receptacle paleae usually restricted to
                bases of outermost disc florets (if in 2–3+ series,
                subshrubs or shrubs) . . . . 339. *Deinandra* (in part), **v. 21, p. 280**
[7. Shifted to left margin.—Ed.]
7. Plants without tack-glands or pit-glands; phyllaries in (1–)2–7+ series (seldom each inner
   phyllary wholly or partly investing ovary of a subtended floret); paleae seldom restricted to
   1 series between ray and disc florets, all or nearly all disc florets subtended by paleae; ray
   laminae seldom flabellate (lobe lengths mostly 0–$^{1}/_{10}$ laminae).
   22. Calyculi of 1–8+ bractlets.
       23. Phyllaries connate $^{1}/_{5}$–$^{7}/_{8}$+ their lengths . . . . 310. *Thelesperma* (in part), **v. 21, p. 199**
       23. Phyllaries distinct to bases or nearly so.
           24. Cypselae (at least inner) ± 4-angled, ± linear-fusiform, often apically attenuate
               or beaked (none winged).
               25. Disc florets 10–20 (staminal filaments hairy near anthers); cypselae usually
                   with 1 groove on each face . . . . 311. *Cosmos*, **v. 21, p. 203**
               25. Disc florets (5–)12–150+ (staminal filaments not hairy); cypselae with 0 or
                   2 grooves on each face . . . . 312. *Bidens* (in part), **v. 21, p. 205**
           24. Cypselae all ± obcompressed.
               26. Rays 3–4 (laminae 1–2 mm); inner cypselae ± beaked . . . .
                   . . . . 313. *Heterosperma* (in part), **v. 21, p. 218**
               26. Rays usually 1, 2, 3, 5, 8, 13, 21+ (laminae mostly 4–30+ mm); cypselae
                   not beaked.
                   27. Ray florets usually neuter or styliferous and sterile; cypsela wings mem-
                       branous or corky, entire or irregularly thickened . . . .
                       . . . . 308. *Coreopsis* (in part), **v. 21, p. 185**
                   27. Ray florets usually pistillate and fertile; cypsela wings ± corky,
                       pectinately toothed . . . . 309. *Coreocarpus* (in part), **v. 21, p. 198**
   22. Calyculi none.
       28. Phyllaries (at least inner) usually falling with cypselae; ray florets pistillate, fertile.
           29. Disc florets bisexual, fertile; anther thecae pale.
               30. Phyllaries 10–20 in ± 2 series (outer 4 broadly lanceolate, foliaceous, nota-
                   bly larger than inner) . . . . 304. *Tetragonotheca* (in part), **v. 21, p. 178**
               30. Phyllaries 6–9 in 2 series, subequal or unequal (outer smaller than inner)
                   . . . . 306. *Galinsoga* (in part), **v. 21, p. 180**
           29. Disc florets usually functionally staminate (bisexual and fertile in Milleriinae,
               Guizotia); anther thecae dark (blackish to purplish).
               31. Rays 6–18 (corollas hairy at bases of tubes); disc florets bisexual, fertile
                   . . . . 259. *Guizotia*, **v. 21, p. 40**
               31. Rays 3–20+ (corollas seldom hairy at bases of tubes); disc florets function-
                   ally staminate.

32. Heads borne singly, usually pedunculate; fruits (perigynia) smooth or bullate to tuberculate (1–4 mm) . . . . 256. *Melampodium* (in part), **v. 21, p. 34**

32. Heads often in clusters of 2–3, mostly sessile; fruits ± prickly (4–8 mm) . . . . 257. *Acanthospermum* (in part), **v. 21, p. 36**

28. Phyllaries persistent (in fruit); ray florets pistillate and fertile, or styliferous and sterile, or neuter.

[33. Shifted to left margin.—Ed.]

33. Receptacles spheric to high-conic or columnar (mostly 8–20+ mm).

34. Phyllaries subequal or unequal (outer longer than inner); ray florets 3–21+, neuter; disc florets 100–200+, bisexual, fertile; stigmatic papillae usually in 2 lines.

35. Involucres (early flowering) hemispheric to rotate, 15–30+ mm diam.; phyllaries 15–30+ in 2–3 series, subequal; cypselae ± 4-angled, not strongly compressed, margins not pectinate or ciliate . . . . 261. *Rudbeckia* (in part), **v. 21, p. 44**

35. Involucres rotate, 8–12+ mm diam.; phyllaries 14–28+ in 2 series, unequal (outer notably longer than inner); cypselae strongly compressed, abaxial margin of each usually pectinate or ciliate . . . . 262. *Ratibida* (in part), **v. 21, p. 60**

34. Phyllaries subequal or unequal (outer usually shorter, rarely longer, than inner); ray florets 3–40+, usually pistillate and fertile, sometimes styliferous and sterile, or neuter; disc florets 4–200+, usually bisexual and fertile, sometimes functionally staminate; stigmatic papillae usually continuous, rarely in 2 lines.

36. Ray florets usually 5–21 (more in "double" cultivars), corollas usually yellow to orange, sometimes purple, red, or whitish (usually persistent, sessile, becoming papery) . . . . 266. *Zinnia* (in part), **v. 21, p. 71**

36. Ray florets usually (2–)5–35, corollas usually yellow to orange, sometimes white (seldom sessile, laminae usually borne on tubes, never persistent and becoming papery).

37. Disc florets functionally staminate (only ray florets produce cypselae).

38. Phyllaries 8–10 in 2 series . . . . 268. *Lindheimera* (in part), **v. 21, p. 75**

38. Phyllaries 12–45+ in (2–)3–4 series . . . . 270. *Berlandiera* (in part), **v. 21, p. 83**

37. Disc florets bisexual, fertile.

39. Leaves basal and cauline, alternate; involucres 12–40 mm diam.; rays dark purple to pale pink, white, or yellow . . . . 272. *Echinacea*, **v. 21, p. 88**

39. Leaves cauline, opposite; involucres 10–15 mm diam.; rays yellow . . . . 291. *Pascalia*, **v. 21, p. 131**

33. Receptacles mostly flat to convex or conic (mostly 0–5 mm).

40. Leaves mostly cauline and alternate (proximal sometimes opposite), or mostly opposite (distal sometimes alternate); ray florets usually neuter or styliferous and sterile.

41. Disc corollas yellow (bases often dilated, clasping tops of ovaries) . . . . . 263. *Zaluzania*, **v. 21, p. 63**

41. Disc corollas yellow to orange or brown-purple (bases not clasping tops of ovaries).

42. Receptacle paleae each completely investing and falling with a cypsela (each forming a hardened perigynium) . . . . 296. *Sclerocarpus* (in part), **v. 21, p. 137**

42. Receptacle paleae sometimes conduplicate, ± enfolding cypselae (not forming perigynia).

43. Heads borne singly (peduncles usually distally dilated, fistulose) . . . . 297. *Tithonia* (in part), **v. 21, p. 138**

43. Heads borne singly or in corymbiform, paniculiform, racemiform, or thyrsiform arrays (peduncles rarely, if ever, notably dilated or fistulose).

44. Cypselae flattened, thin-margined . . . . 298. *Simsia* (in part), **v. 21, p. 140**

44. Cypselae ± compressed, biconvex, or 3- or 4-angled, often obpyramidal.

45. Shrubs (leaves often lobed, lobes usually 3–9, ± linear) . . . . 301. *Viguiera* (in part), **v. 21, p. 172**

45. Annuals or perennials (leaves not lobed).

46. Annuals; leaf blades lanceolate to linear; involucres 5–6 mm diam.; phyllaries 11–17 . . . . 299. *Helianthus* (in part), **v. 21, p. 141**

46. Annuals or perennials; leaf blades lance-linear, lanceolate, ovate, rhombic, or rhombic-ovate; involucres 6–14 mm diam.; phyllaries 14–25 . . . . 300. *Heliomeris*, **v. 21, p. 169**

[40. Shifted to left margin.—Ed.]

40. Leaves usually cauline and opposite, sometimes mostly basal and/or mostly alternate; rays florets pistillate and fertile (if neuter, leaves mostly basal or alternate).

47. Disc florets functionally staminate.

48. Anther thecae green, staminal filaments hairy; Arizona. . . . . 260. *Guardiola*, **v. 21, p. 42**

48. Anther thecae dark or pale (not green), staminal filaments not hairy; e United States.

49. Ray florets 2–6, corollas pale yellow to whitish; disc florets 12–30+; cypselae (patently inserted on receptacles) 3–6-ribbed or -nerved (finely striate between ribs, apices often minutely beaked). . . . . 258. *Polymnia* (in part), **v. 21, p. 39**

49. Ray florets 7–13, corollas yellow; disc florets 40–80; cypselae (obliquely inserted on receptacles) 30–40-ribbed or -nerved (not beaked).

50. Cypselae each shed separate from its subtending phyllary. . . . . . 255. *Smallanthus*, **v. 21, p.33**

50. Cypselae each enclosed within and shed within a perigynium (formed from an inner, subtending phyllary).

51. Heads borne singly, usually pedunculate; fruits (perigynia) smooth or bullate to tuberculate (1–4 mm) . . . . 256. *Melampodium* (in part), **v. 21, p. 34**

51. Heads often in clusters of 2–3, mostly sessile; fruits ± prickly (4–8 mm) . . . . 257. *Acanthospermum* (in part), **v. 21, p. 36**

47. Disc florets bisexual and fertile.

52. Stigmatic papillae in 2 lines.

53. Leaf blades simple; heads borne singly; phyllaries ± connate, ± carinate . . . . 359. *Eriophyllum* (in part), **v. 21, p. 353**

53. Leaf blades 1–2-pinnately or -pedately lobed; heads in corymbiform to paniculiform arrays; phyllaries distinct, not carinate.

54. Leaves cauline, opposite; rays 5–8, pale yellow (fading white) . . . . 309. *Coreocarpus* (in part), **v. 21, p. 198**

54. Leaves mostly basal, alternate; rays 8, white . . . 351. *Hymenopappus* (in part), **v. 21, p. 309**

52. Stigmatic papillae usually continuous, sometimes none (functionally staminate florets), rarely in 2 lines.

55. Ray florets usually 5–21, corollas usually yellow to orange (usually persistent, sessile, becoming papery) . . . . 264. *Heliopsis* (in part), **v. 21, p. 67**

55. Ray florets (2–)5–35, corollas usually yellow to orange, sometimes white (seldom sessile, laminae usually borne on tubes, never persistent and becoming papery).

56. Disc florets functionally staminate (only ray florets produce cypselae).

57. Phyllaries 8–10 in 2 series . . . . 267. *Chrysogonum* (in part), **v. 21, p. 74**

57. Phyllaries 12–45+ in (2–)3–4 series . . . . 269. *Silphium* (in part), **v. 21, p. 77**

56. Disc florets bisexual, fertile.

58. Leaves mostly cauline, mostly opposite.

59. Perennials (prostrate; cypselae often rostrate, each with apical boss or neck) . . . . 286. *Sphagneticola* (in part), **v. 21, p. 126**

59. Annuals or perennials (not prostrate; cypselae not rostrate).

60. Cypselae 3–4-angled (weakly or not at all compressed or obcompressed, epidermes usually thick, corky).

61. Rays 20–40, white or whitish (paleae linear-filiform, not conduplicate) . . . . 289. *Eclipta* (in part), **v. 21, p. 128**

61. Rays 13–21, yellow to orange (paleae lanceolate to ovate, conduplicate) . . . . 291. *Pascalia* (in part), **v. 21, p. 131**

60. Cypselae (all or at least disc) strongly compressed or obcompressed or flattened (epidermes seldom thick and corky).

62. Perennials (coarse, 10–150 cm; larger leaves mostly 10–50 cm); involucres 10–50 mm diam.; phyllaries 22–32+ in ± 3 series . . . . 280. *Helianthella* (in part), **v. 21, p. 114**

62. Annuals or perennials (mostly 10–30+ cm; larger leaves mostly 2–10 cm); involucres 3–6+ mm diam.; phyllaries 8–15+ in 1–3 series . . . . 292. *Acmella* (in part), **v. 21, p. 132**

[58. Shifted to left margin.—Ed.]
58. Leaves mostly basal, or basal and cauline, or cauline, mostly alternate.
　　63. Cypselae prismatic, or nearly so, 3–4-angled.
　　　　64. Leaves mostly basal (cauline usually notably smaller than basal) .... 273. *Balsamorhiza*, **v. 21, p. 93**
　　　　64. Leaves basal and cauline, or mostly cauline.
　　　　　　65. Leaves mostly cauline (blades narrowly oblong to linear, 5–25 mm wide)
　　　　　　. . . . . 274. *Scabrethia* (in part), **v. 21, p. 99**
　　　　　　65. Leaves basal and cauline, or mostly cauline (blades mostly 30–120 mm wide).
　　　　　　　　66. Leaves mostly elliptic, lanceolate, or oblong (basal and cauline, basal
　　　　　　　　usually notably larger than cauline, cauline mostly sessile). . . . .
　　　　　　　　. . . . . 275. *Wyethia* (in part), **v. 21, p. 100**
　　　　　　　　66. Leaves mostly orbiculate, ovate, or rounded-deltate (mostly cauline, mostly
　　　　　　　　petiolate, proximal and distal usually ±similar) . . . . . 276. *Agnorhiza* (in part), **v. 21, p. 104**
　　63. Cypselae compressed to flattened.
　　　　67. Cypselae winged . . . . 277. *Verbesina* (in part), **v. 21, p. 106**
　　　　67. Cypselae sometimes thin-edged (margins sometimes ciliate or corky-thickened, never
　　　　truly winged).
　　　　　　68. Perennials (scapiform); leaves all or mostly basal; involucres 20–30+ mm diam.
　　　　　　. . . . . 278. *Enceliopsis* (in part), **v. 21, p. 112**
　　　　　　68. Perennials (rarely scapiform), subshrubs, or shrubs; leaves usually cauline, some-
　　　　　　times basal and cauline; involucres 4–30 mm diam.
　　　　　　　　69. Perennials (rhizomatous); leaves linear to filiform . . . .
　　　　　　　　. . . . . 279. *Phoebanthus* (in part), **v. 21, p. 113**
　　　　　　　　69. Perennials or shrubs (not rhizomatous); leaves mostly deltate, elliptic, lan-
　　　　　　　　ceolate, or ovate (and most intermediate shapes, not linear to filiform).
　　　　　　　　　　70. Ray florets 8–21, pistillate and fertile . . . . 280. *Helianthella* (in part), **v. 21, p. 114**
　　　　　　　　　　70. Ray florets 8–25(–40), neuter . . . . 282. *Encelia*, **v. 21, p. 118**

Key to Genera of Group 4

Heads radiate; receptacles paleate; pappi wholly, or partially, of awns or scales.

1. Plants often with tack-glands or pit-glands on stems, leaves, and/or phyllaries; phyllaries (or
　paleae functioning as phyllaries) usually in 1+ series (each often wholly or partly investing
　ovary of a subtended floret); paleae often in 1 series between ray and disc florets, often
　connate in a ring, sometimes each disc floret subtended by a palea; ray corolla laminae often
　flabellate (lobes often ¹⁄₂+ lengths of laminae; pappus scales often plumose and/or woolly).
　2. Ray cypselae obcompressed (each mostly or completely enveloped by a phyllary).
　　3. Perennials (rhizomatous); disc corollas white . . . . 343. *Holozonia* (in part), **v. 21, p. 294**
　　3. Annuals; disc corollas yellow (sometimes reddish with age).
　　　4. Pappi of apically obtuse scales . . . . 332. *Achyrachaena*(in part), **v. 21, p. 258**
　　　4. Pappi of apically acute scales . . . . 335. *Layia* (in part), **v. 21, p. 262**
　2. Ray cypselae compressed, ± terete, ± 3-angled in cross section, or ± obcompressed (each
　± ¹⁄₂ enveloped by a phyllary).
　　5. Annuals (styles of discs hairy proximal to minute branches; receptacles paleate
　　　throughout, ray corollas white with abaxial purple lines, pappi of subulate,
　　　plumose scales) . . . . 333. *Blepharipappus* (in part), **v. 21, p. 259**
　　5. Annuals, perennials, subshrubs, or shrubs (styles of discs glabrous proximal to
　　　branches).
　　　6. Perennials (± scapiform); disc pappi of subulate, ciliate-plumose scales . . . .
　　　. . . . . 331. *Raillardella* (in part), **v. 21, p. 256**
　　　6. Annuals, perennials (leafy-stemmed), subshrubs, or shrubs; disc pappi of scales
　　　(scales seldom both subulate and ciliate-plumose).
　　　　7. Annuals or perennials; peduncle bracts without terminal pit-glands, tack-
　　　　glands, or spines; ray corollas yellow; ray cypselae usually compressed, rarely
　　　　terete (cross sections usually ± 3-angled, then abaxial sides relatively broad,
　　　　± rounded, adaxial sides ± 2-faced, angles between those faces 15–70°;
　　　　each ray cypsela usually completely or mostly enveloped by a phyllary).

8. Perennials.
   9. Involucres campanulate to hemispheric; anthers ± dark purple; ray
       cypselae not beaked . . . . 344. *Kyhosia* (in part), **v. 21, p. 295**
   9. Involucres campanulate, ellipsoid, or globose; anthers yellow to
       brownish; ray cypselae beaked . . . . 347. *Anisocarpus* (in part), **v. 21, p. 299**
8. Annuals.
   10. Leaves all alternate, margins entire; heads borne singly; pappi of
       3–5 scales . . . . 146. *Rigiopappus*, **v. 20, p. 48**
   10. Leaves: proximal opposite, distal alternate, margins entire or
       toothed; heads borne singly or in ± umbelliform or corymbiform
       arrays; pappi of 5–12 scales.
       11. Anthers yellow to brownish . . . . 346. *Harmonia*, **v. 21, p. 297**
       11. Anthers ± dark purple . . . . 348. *Jensia*, **v. 21, p. 301**
7. Annuals, subshrubs, or shrubs; peduncle bracts sometimes each with ter-
   minal pit-gland, tack-gland, or spine (or apiculus); ray corollas yellow,
   whitish, or rose; ray cypselae terete to subterete or ± obcompressed (cross
   sections nearly circular with adaxial sides ± flattened to slightly bulging, or
   ± 3-angled, then abaxial sides usually ± broadly 2-faced, angles between
   those faces usually 90+° and adaxial sides ± flattened to slightly bulging; in
   Centromadia spp., distal leaves spine-tipped, each cypsela ± enveloped by
   a phyllary, cypselae sometimes compressed).
   12. Annuals; leaves filiform to narrowly linear, margins often strongly revo-
       lute; peduncle bracts often with tack-glands; ray corolla lobes (at least
       the lateral) often spreading (lengths often $1/2$–$5/6$ of total laminae).
       13. Ray cypselae beaked; tack-glands absent. . . . 336. *Osmadenia* (in part), **v. 21, p. 269**
       13. Ray cypselae not beaked; tack-glands present . . . .
       . . . . 337. *Calycadenia* (in part), **v. 21, p. 270**
   12. Annuals, subshrubs, or shrubs; leaves linear or broader, margins sel-
       dom strongly revolute; peduncle bracts usually without tack-glands;
       ray corolla lobes ± parallel (lengths usually $1/10$–$1/2$ of total laminae).
       14. Rays usually white, sometimes yellow, often with abaxial purple
          lines; cypselae not beaked or each with an inconspicuous, straight
          beak (beak lengths less than diams.) . . . . 341. *Blepharizonia* (in part), **v. 21, p. 289**
       14. Rays yellow (without abaxial purple lines); cypselae each with an
          adaxial, ascending beak (beak lengths greater than diams.)
          15. Peduncle bracts apiculate or each with an apical spine . . . .
          . . . . 338. *Centromadia* (in part), **v. 21, p. 276**
          15. Peduncle bracts not apiculate, without apical spines . . . . . .
          . . . . 339. *Deinandra* (in part), **v. 21, p. 280**
1. Plants without tack-glands or pit glands; phyllaries in (1–)2–7+ series (seldom each inner
  phyllary wholly or partly investing ovary of a subtended floret); paleae seldom restricted to
  1 series between ray and disc florets, all or nearly all disc florets subtended by paleae;
  laminae of ray corollas seldom flabellate (lobes mostly 0–$1/10$ lengths of laminae); pappi
  usually of awns and/or scales (seldom plumose).
  16. Calyculi of 1–8+ bractlets.
    17. Phyllaries (excluding calyculi) 3–4(–6) in 1(–2) series; disc florets 3–4+ (function-
       ally staminate) . . . . 314. *Dicranocarpus*, **v. 21, p. 219**
    17. Phyllaries (5–)8–34+ in ± 2 series; disc florets (5–)12–150+.
      18. Phyllaries connate $1/5$–$7/8$+ their lengths. . . . . 310. *Thelesperma* (in part), **v. 21, p. 199**
      18. Phyllaries usually distinct, rarely connate ± $1/10$ their lengths.
        19. Cypselae (at least inner) ± 4-angled, ± linear-fusiform, often apically
          attenuate or beaked (none winged).
          20. Disc florets 10–20 (staminal filaments hairy proximal to anther
            collars); cypselae usually with 1 groove on each face . . . .
            . . . . 311. *Cosmos* (in part), **v. 21, p. 203**
          20. Disc florets (5–)12–150+ (staminal filaments not hairy); cypselae with
            0 or 2 grooves on each face . . . . 312. *Bidens* (in part), **v. 21, p. 205**

19. Cypselae all ± obcompressed.
    21. Ray florets 3–4 (laminae 1–2 mm); inner cypselae ± beaked . . . . . .
        . . . . 313. *Heterosperma* (in part), **v. 21, p. 218**
    21. Ray florets usually 1, 2, 3, 5, 8, 13, 21+ (laminae mostly 4–30+ mm);
        cypselae not beaked.
        22. Cypselae rarely winged (margins sometimes thickened, winged in
            B. aristosa and *B. polylepis*); pappi of barbellate awns . . . .
            . . . . 312. *Bidens* (in part), **v. 21, p. 205**
        22. Cypselae (some or all) usually ± winged; pappi of 2, bristly cusps
            or scales (in *Coreopsis*) or of 1–2 barbellate awns (in *Coreocarpus*).
            23. Ray florets usually neuter, or styliferous and sterile; wings of
                cypselae membranous or corky, entire or irregularly thickened
                . . . . 308. *Coreopsis* (in part), **v. 21, p. 185**
            23. Ray florets usually pistillate and fertile; wings of cypselae
                ± corky, pectinately toothed . . . . 309. *Coreocarpus* (in part), **v. 21, p. 198**
[16. Shifted to left margin.—Ed.]
16. Calyculi none.
    24. Phyllaries (at least inner) usually falling with cypselae; ray florets pistillate, fertile.
        25. Disc florets usually functionally staminate; anther thecae dark (blackish to
            purplish; shoulders of cypselae may bear 1–3 pappus-like, triangular to ovate, or
            ± subulate enations; cypselae shed with subtending phyllary plus 2 contiguous
            discflorets and their investing paleae) . . . . 246. *Parthenium* (in part), **v. 21, p. 20**
        25. Disc florets bisexual, fertile; anther thecae pale (cypselae sometimes shed with
            accessory structures).
            26. Phyllaries 10–20 in ± 2 series (outer 4 broadly lanceolate, foliaceous, notably
                larger than inner) . . . . 304. *Tetragonotheca* (in part), **v. 21, p. 178**
            26. Phyllaries 6–30+ in 2–5 series (subequal or unequal, outer smaller than inner).
                27. Annuals (ray cypselae often each shed together with subtending phyllary
                    and 2 adjacent paleae) . . . . 306. *Galinsoga* (in part), **v. 21, p. 180**
                27. Perennials (cypselae shed separate from paleae) . . . . 305. *Tridax*, **v. 21, p. 179**
    24. Phyllaries persistent (in fruit); ray florets pistillate and fertile, or styliferous and sterile,
        or neuter.
        28. Receptacles spheric to high-conic or columnar (mostly 8–20+ mm).
            29. Phyllaries subequal or unequal (outer longer than inner); ray florets 3–21+,
                neuter; disc florets 100–200+, bisexual, fertile; stigmatic papillae usually in 2
                lines.
                30. Involucres (early flowering) hemispheric to rotate, 15–30+ mm diam.; phyl-
                    laries 15–30+ in 2–3 series, subequal; cypselae ± 4-angled, not strongly
                    compressed, margins not pectinate or ciliate . . . . 261. *Rudbeckia* (in part), **v. 21, p. 44**
                30. Involucres rotate, 8–12+ mm diam.; phyllaries 14–28+ in 2 series, unequal
                    (outer notably longer than inner); cypselae strongly compressed, abaxial
                    margin of each usually pectinate or ciliate . . . . 262. *Ratibida* (in part), **v. 21, p. 60**
            29. Phyllaries subequal or unequal (outer usually shorter, rarely longer, than in-
                ner); ray florets 3–40+, usually pistillate and fertile, sometimes styliferous and
                sterile, or neuter; disc florets 4–200+, usually bisexual and fertile, sometimes
                functionally staminate; stigmatic papillae usually continuous, rarely in 2 lines.
                31. Anther thecae pale; stigmatic papillae in 2 lines . . . .
                    . . . . 383. *Amblyolepis* (in part), **v. 21, p. 420**
                31. Anther thecae usually dark; stigmatic papillae usually continuous, rarely in
                    2 lines.
                    32. Ray florets usually 5–21 (more in "double" cultivars), corollas usually
                        yellow to orange, sometimes purple, red, or whitish (usually persistent,
                        sessile, becoming papery) . . . . 266. *Zinnia* (in part), **v. 21, p. 71**
                    32. Ray florets usually (2–)5–35, corollas usually yellow to orange, some-
                        times white (seldom sessile, laminae usually borne on tubes, never
                        persistent and becoming papery).

33. Disc florets functionally staminate (only ray florets produce
cypselae).
34. Phyllaries 8–10 in 2 series. . . . . 268.　*Lindheimera* (in part), **v. 21, p. 75**
34. Phyllaries 12–45+ in (2–)3–4 series. . . . . . 270.　*Berlandiera* (in part), **v. 21, p. 83**
33. Disc florets bisexual, fertile.
35. Leaves basal and cauline, alternate; involucres 12–40 mm
diam.; rays dark purple to pale pink, white, or yellow . . . .
. . . . 272.　*Echinacea* (in part), **v. 21, p. 88**
35. Leaves cauline, opposite; involucres 10–15 mm diam.; rays
yellow. . . . . 291.　*Pascalia* (in part), **v. 21, p. 131**
[28. Shifted to left margin.—Ed.]
28. Receptacles mostly flat to convex or conic (mostly 0–5 mm).
36. Leaves mostly cauline and alternate (proximal sometimes opposite), or mostly opposite
(distal sometimes alternate); ray florets usually neuter or styliferous and sterile; pappi
usually fragile or readily falling, sometimes persistent, of scales or awns.
37. Receptacle paleae each completely investing and falling with cypsela (each forming
a hardened perigynium) . . . . 296.　*Sclerocarpus* (in part), **v. 21, p. 137**
37. Receptacle paleae sometimes conduplicate, ± enfolding cypselae (not forming
perigynia).
38. Heads borne singly (peduncles usually distally dilated, fistulose) . . . . . .
. . . . 297.　*Tithonia* (in part), **v. 21, p. 138**
38. Heads borne singly or in corymbiform, paniculiform, racemiform, or thyrsiform
arrays (peduncles rarely, if ever, notably dilated or fistulose).
39. Cypselae flattened, thin-margined . . . . 298.　*Simsia* (in part), **v. 21, p. 140**
39. Cypselae ± compressed, biconvex, or 3- or 4-angled, often obpyramidal.
40. Cypselae glabrous or glabrate.
41. Shrubs (leaves often lobed, lobes usually 3–9, ± linear) . . . .
. . . . 301.　*Viguiera* (in part), **v. 21, p. 172**
41. Annuals or perennials (leaves not lobed) . . . . 299.　*Helianthus* (in part), **v. 21, p. 141**
40. Cypselae usually ± strigose, sometimes glabrous or glabrate (pappi of
2–6+ persistent, readily falling, or tardily falling, scales).
42. Shrubs; involucres 5–9 mm diam. . . . . 302.　*Bahiopsis*, **v. 21, p. 174**
42. Annuals or perennials; involucres (5–)7–40+ mm diam.
43. Pappi readily falling . . . . 299.　*Helianthus* (in part), **v. 21, p. 141**
43. Pappi persistent or tardily falling . . . . 301.　*Viguiera* (in part), **v. 21, p. 172**
36. Leaves usually cauline and opposite, sometimes mostly basal and/or mostly alternate;
ray florets pistillate and fertile (if neuter, leaves mostly basal or alternate); pappi usually
persistent, of awns, bristles, and/or scales.
44. Stigmatic papillae in 2 lines.
45. Pappi of 1–2 retrorsely barbellate awns . . . . 309.　*Coreocarpus* (in part), **v. 21, p. 198**
45. Pappi of 2–50 smooth to barbellate scales.
46. Disc cypselae 4–5-angled.
47. Leaves mostly basal or basal and cauline, alternate, margins usually 1–
2-pinnately or -palmati-pinnately lobed; phyllaries distinct . . . .
. . . . 351.　*Hymenopappus* (in part), **v. 21, p. 309**
47. Leaves mostly cauline, mostly alternate (proximal sometimes oppo-
site), margins entire or toothed; phyllaries ± connate. . . . .
. . . . 359.　*Eriophyllum* (in part), **v. 21, p. 353**
46. Disc cypselae not angled.
48. Rays 2–4; pappi of ca. 50, setiform scales . . . . 329.　*Pseudoclappia* (in part), **v. 21, p. 252**
48. Rays 5–15+; pappi of 6–10+ scales . . . . 384.　*Gaillardia* (in part), **v. 21, p. 421**
44. Stimatic papillae usually continuous, sometimes none, rarely in 2 lines.
49. Ray florets usually 5–21, corollas usually yellow to orange, sometimes purple,
red, or whitish (usually persistent, sessile, becoming papery).
50. Leaf margins serrate to coarsely toothed . . . . 264.　*Heliopsis* (in part), **v. 21, p. 67**
50. Leaf margins entire . . . . 265.　*Sanvitalia*, **v. 21, p. 70**

[49. Shifted to left margin.—Ed.]

49. Ray florets usually (2–)5–35, corollas usually yellow to orange, sometimes white (seldom sessile, laminae usually borne on tubes, never persistent and becoming papery).

51. Disc florets functionally staminate (only ray florets produce cypselae).
  52. Phyllaries 8–10 in 2 series . . . . 267. *Chrysogonum* (in part), **v. 21, p. 74**
  52. Phyllaries 12–45+ in (2–)3–4 series.
    53. Ray florets 8–35 (in 1–3 series); cypselae shed alone without accessory structures . . . . 269. *Silphium* (in part), **v. 21, p. 77**
    53. Ray florets usually (2–)8(–13); cypselae each shed together with subtending phyllary and 2–4 adjacent paleae and disc florets . . . . 271. *Engelmannia*, **v. 21, p. 87**
51. Disc florets bisexual, fertile.
  54. Leaves mostly basal, or basal and cauline, or cauline, mostly alternate.
    55. Cypselae prismatic, or nearly so, 3–4-angled.
      56. Leaves mostly cauline (blades narrowly oblong to linear, 5–25 mm wide) . . . . 274. *Scabrethia* (in part), **v. 21, p. 99**
      56. Leaves basal and cauline, or mostly cauline (blades mostly 30–120 mm wide).
        57. Leaves mostly elliptic, lanceolate, or oblong (basal and cauline, basal usually notably larger than cauline, cauline mostly sessile) . . . . 275. *Wyethia* (in part), **v. 21, p. 100**
        57. Leaves mostly orbiculate, ovate, or rounded-deltate (mostly cauline, mostly petiolate, proximal and distal usually ± similar) . . . . 276. *Agnorhiza* (in part), **v. 21, p. 104**
    55. Cypselae compressed to flattened.
      58. Cypselae winged; pappi persistent, of 2(–3) scales (scales often aristate or subulate, without additional scales) . . . . 277. *Verbesina* (in part), **v. 21, p. 106**
      58. Cypselae sometimes thin-edged (margins sometimes ciliate or corky-thickened, never truly winged); pappi usually of (1–)2 subulate scales or bristlelike awns plus 2–4+ shorter scales (rarely of 2–3 aristate scales without additional scales).
        59. Perennials (scapiform); leaves all or mostly basal; involucres 20–30+ mm diam . . . . 278. *Enceliopsis* (in part), **v. 21, p. 112**
        59. Perennials (rarely scapiform), subshrubs, or shrubs; leaves usually cauline, sometimes basal and cauline; involucres 4–30 mm diam.
          60. Perennials (rhizomatous); leaves linear to filiform . . . . 279. *Phoebanthus* (in part), **v. 21, p. 113**
          60. Perennials or shrubs (not rhizomatous); leaves mostly deltate, elliptic, lanceolate, or ovate (and most intermediate shapes, not linear to filiform).
            61. Ray florets 8–21, pistillate and fertile . . . . 280. *Helianthella* (in part), **v. 21, p. 114**
            61. Ray florets 8–40, neuter, or styliferous and sterile.
              62. Subshrubs or shrubs (glabrous or ± scabrellous, usually vernicose); phyllaries 12–40 in 2–4+ series (subequal or unequal, outer longer) . . . . 281. *Flourensia* (in part), **v. 21, p. 117**
              62. Annuals, perennials, or shrubs (glabrous or canescent, hirtellous, scabrellous, strigose, or tomentose, often gland-dotted or glandular-puberulent to stipitate-glandular, seldom vernicose); phyllaries 18–30(–50+) in 2–3+ series (subequal or unequal, outer shorter).
                63. Perennials (*E. nutans*), subshrubs, or shrubs; pappi sometimes fragile, of 2 weak, villous scales . . . . 282. *Encelia* (in part), **v. 21, p. 118**
                63. Annuals or perennials; pappi usually persistent, of 2 subulate scales . . . . 283. *Geraea* (in part), **v. 21, p. 122**

[54. Shifted to left margin.—Ed.]
54. Leaves mostly cauline, mostly opposite.
    64. Pappi usually coroniform or cyathiform (cypselae often rostrate, each with apical boss
        or neck).
        65. Subshrubs or shrubs (erect); cypselae (some or all) strongly compressed, notably
            winged . . . . 285. *Wedelia* (in part), **v. 21, p. 125**
        65. Perennials (prostrate); cypselae strongly biconvex to plumply 3–4-angled (not
            compressed, not winged, epidermes usually corky, often tuberculate) . . . .
            . . . . 286. *Sphagneticola* (in part), **v. 21, p. 126**
    64. Pappi usually of 2–4+ awns, bristles, and/or scales (not cyathiform, cypselae not
        rostrate).
        66. Some or all cypselae winged (each bordered by wing of membranous or corky
            tissue different from that of body of cypsela).
            67. Heads in glomerules or borne singly (sessile or subsessile in axils); cypselae
                winged (rays, not discs, wings lacerate) . . . . 287. *Synedrella*, **v. 21, p. 127**
            67. Heads borne singly or in corymbiform, dichasiiform, or paniculiform arrays
                (not sessile); cypselae winged (rays and discs, wings not lacerate).
                68. Phyllaries 9–30 in 1–4 series (outer 2–5 similar to others, unlike foliage);
                    pappi of 2(–3) persistent (often aristate or subulate) scales without addi-
                    tional scales . . . . 277. *Verbesina* (in part), **v. 21, p. 106**
                68. Phyllaries 26–38+ in 3–4+ series (outer 2–6+ similar to foliage in shape,
                    texture, and indument); pappi of 2–3 fragile or persistent awns or subulate
                    scales plus 2–8+ distinct or basally connate, erose or lacerate scales
                    (often each cypsela with additional seta on inner shoulder) . . . . 288. *Jefea*, **v. 21, p. 127**
        66. Cypselae sometimes sharp-edged (not winged).
            69. Cypselae 3–4-angled (weakly or not at all compressed or obcompressed,
                epidermes usually thick, corky).
                70. Rays 20–40, white or whitish (paleae linear-filiform, not conduplicate)
                  . . . . 289. *Eclipta* (in part), **v. 21, p. 128**
                70. Rays 7–30, yellow to orange (paleae lanceolate to ovate, conduplicate).
                  71. Leaves elliptic, linear, oblanceolate, obovate, or ovate, glabrous or
                    puberulent to villous and/or sericeous (outer phyllaries elliptic, oblan-
                    ceolate, or ovate) . . . . 290. *Borrichia*, **v. 21, p. 129**
                71. Leaves lanceolate to lance-linear, sparsely scabrous (outer phyllaries
                    lance-linear to linear) . . . . 291. *Pascalia* (in part), **v. 21, p. 131**
            69. Cypselae (all or at least disc) strongly compressed or obcompressed or flat-
                tened (epidermes seldom thick and corky).
                72. Annuals or perennials (mostly 5–30+ cm; larger leaves mostly 1–5+ cm);
                  involucres 3–8 mm diam.; phyllaries 5 in 1(–2) series . . . 293. *Calyptocarpus*, **v. 21, p. 133**
                72. Perennials (coarse, 10–150 cm; larger leaves mostly 5–50 cm); involucres
                  10–50 mm diam.; phyllaries 12–35 in 2–5 series.
                  73. Leaf blades oblanceolate to lanceolate or lance-linear (longer usually
                    8–50 cm), margins entire . . . . 280. *Helianthella* (in part), **v. 21, p. 114**
                73. Leaf blades rounded-deltate to ovate or lance-ovate (longer usually 5–
                    8 cm), margins coarsely serrate . . . . 294. *Lasianthaea*, **v. 21, p. 133**

Key to Genera of Group 5
Heads eradiate; receptacles paleate; pappi none or nearly so.

1. Leaves mostly opposite (distal sometimes alternate; sometimes whorled).
    2. Phyllaries absent ("involucre" sometimes consisting of 1 series of paleae).
        3. Receptacles ± obovoid; disc florets functionally staminate . . . . 101. *Psilocarpus*, **v. 19, p. 456**
        3. Receptacles flat or convex; disc florets bisexual {corollas yellow to red-orange, some
            times purplish) . . . . 350. *Madia* (in part), **v. 21, p. 303**

2. Phyllaries present.
    4. Heads bisexual or unisexual; anthers distinct.
        5. Pistillate and functionally staminate florets in separate heads (cypselae shed within hardened, often prickly, spiny, tuberculate, or winged, perigynia, forming "burs" or nutlike structures).
            6. Staminate heads: phyllaries partially or wholly connate; receptacles ± flat or convex; pistillate heads: phyllaries 12–30(–80+) in 1–8+ series, outer (1–)5–8 distinct or ± connate, the rest ± connate (becoming indurate, their distinct tips forming straight or hooked spines, tubercles, or wings) . . . . 244. *Ambrosia* (in part), **v. 21, p. 10**
            6. Staminate heads: phyllaries distinct to bases; receptacles conic to columnar; pistillate heads: phyllaries 30–75+ in 6–12+ series, outer 5–8 distinct, the rest proximally connate (becoming indurate, their distinct tips usually forming hooked spines) . . . . 245. *Xanthium* (in part), **v. 21, p. 19**
        5. Pistillate and functionally staminate florets usually together in same heads (sometimes some heads staminate; cypselae not enclosed within perigynia).
            7. Cypselae strongly obcompressed to -obflattened (subtended by accrescent phyllaries, margins corky-winged and ± irregularly toothed) . . . . 248. *Dicoria* (in part), **v. 21, p. 24**
            7. Cypselae sometimes ± obcompressed (not subtended by accrescent phyllaries, margins not corky-winged and toothed).
                8. Heads in (bracteate) racemiform or spiciform arrays . . . . . 249. *Iva* (in part), **v. 21, p. 25**
                8. Heads in (± ebracteate) paniculiform arrays. . . . . 250. *Cyclachaena* (in part), **v. 21, p. 28**
    4. Heads bisexual; anthers connate.
        9. Corollas white or pinkish (phyllaries 10–15+, prominently 2–3-nerved). . . . . 402. *Isocarpha*, **v. 21, p. 490**
        9. Corollas usually yellow to orange or ochroleucous, sometimes whitish.
          10. Calyculi present.
             11. Calyculi of (3–)5–13(–21+), ± herbaceous (sometimes foliaceous) bractlets or bracts (sometimes surpassing phyllaries); cypselae rarely winged . . . . 312. *Bidens* (in part), **v. 21, p. 205**
             11. Calyculi usually of 1–8+ bractlets; capselae often winged.
                12. Calyculi of 1–3+ bractlets; phyllaries usually distinct, rarely connate ± $^1/_{10}$ their lengths . . . . 309. *Coreocarpus* (in part), **v. 21, p. 198**
                12. Calyculi of 3–8+ bractlets; phyllaries connate $^1/_5$–$^7/_8$+ their lengths . . . . 310. *Thelesperma* (in part), **v. 21, p. 199**
           10. Calyculi absent.
             13. Disc florets functionally staminate (e United States) . . . . 258. *Polymnia* (in part), **v. 21, p. 39**
             13. Disc florets bisexual, fertile.
                14. Shrubs; phyllaries in ± 1 series, proximally connate (heads with 1–2 flowers, borne in headlike glomerules) . . . . 295. *Lagascea*, **v. 21, p. 136**
                14. Annuals, perennials, or subshrubs; phyllaries in (1–)2–5 series, distinct or inner ± connate.
                    15. Phyllaries in ± 2 series, ± equal . . . . 309. *Coreocarpus*, **v. 21, p. 198**
                    15. Phyllaries in 1–5 series, subequal or unequal.
                      16. Receptacles flat to convex or ± conic; cypselae usually winged . . . . 277. *Verbesina* (in part), **v. 21, p. 106**
                      16. Receptacles conic; cypselae not winged . . . . 292. *Acmella* (in part), **v. 21, p. 132**
1. Leaves mostly alternate (proximal sometimes opposite).
    17. Annuals; involucres absent, vestigial, or inconspicuous, often simulated by leaves or paleae, sometimes glandular.
        18. Stems and leaves villous to hispid, glandular-pubescent distally . . . . 350. *Madia* (in part), **v. 21, p. 303**
        18. Stems and leaves usually arachnoid-sericeous to lanuginose, sometimes glabrescent, usually eglandular.

19. Bisexual florets (1–)2–10(–11), pappi of (11–)13–28+ bristles visible in heads; functionally staminate florets 0.

    20. Receptacles fungiform to obovoid (heights 0.4–1.6 times diams.); most pistillate paleae ± saccate, each ± enclosing a floret, apices blunt; innermost paleae spreading in fruit; cypselae dimorphic (outer longer than inner) . . . . 97. *Logfia* (in part), **v. 19, p. 443**

    20. Receptacles cylindric to clavate (heights 5–15 times diams.); most pistillate paleae open to ± folded (at most, each enfolding, not enclosing, a floret; apices acuminate to aristate); innermost paleae erect to ascending in fruit; cypselae monomorphic (outer ± equaling inner) . . . . 98. *Filago* (in part), **v. 19, p. 447**

19. Bisexual florets 0 or 2–7, pappi 0; functionally staminate florets 0 or 2–12, pappi 0, or of 1–10(–13) bristles hidden in heads.

    21. Pistillate paleae saccate most of lengths (each enclosing a floret, outermost rarely open); pappi 0, or of 1–10(–13) bristles.

        22. Staminate paleae 5(–7), ± spreading proximally, enlarged in fruit (apices incurved to uncinate); pistillate paleae with 3, ± parallel (prominent) nerves; cypselae: corolla scars apical . . . . 104. *Ancistrocarphus*, **v. 19, p. 465**

        22. Staminate paleae 0, or 1–4, erect, not enlarged in fruit (apices erect); pistillate paleae with 5+, reticulate (and prominent) or ± parallel (and obscure) nerves; cypselae: corolla scars subapical to ± lateral.

            23. Pistillate paleae (obcompressed to terete, not galeate): wings ± erect (and apical); receptacles cylindric to clavate (heights 2.8–8 times diams.); phyllaries 0 or 1–4 (similar to paleae); cypselae: corolla scars subapical . . . . 99. *Stylocline* (in part), **v. 19, p. 450**

            23. Pistillate paleae (compressed, galeate): wings ± erect (and lateral) or inflexed (and subapical); receptacles depressed-spheric to obovoid (heights 0.5–1.8 times diams.); phyllaries 4–6 (unlike paleae); cypselae: corolla scars ± lateral . . . . 100. *Micropus* (in part), **v. 19, p. 454**

    21. Pistillate paleae open most of lengths, flat or concave to loosely folded (not enclosing florets); pappi 0.

        24. Bisexual paleae saccate, each enclosing a floret, apices 2-fid or 3-fid; cypselae (at least outer) strigose; coastal Texas . . . . 103. *Micropsis*, **v. 19, p. 463**

        24. Bisexual or staminate paleae flat to concave, not enclosing florets, apices entire; cypselae glabrous; c, w North America.

            25. Receptacles glabrous; pistillate paleae falling (all or the inner together); staminate (or bisexual) paleae: bodies ± spatulate (apices scarcely enlarged); c North America . . . . 102. *Diaperia*, **v. 19, p. 460**

            25. Receptacles bristly; pistillate paleae persistent; staminate paleae: bodies obovate (apices enlarged); California, Oregon . . . . 105. *Hesperevax*, **v. 19, p. 467**

[17. Shifted to left margin.—Ed.]

17. Annuals, biennials, perennials, subshrubs, or shrubs; involucres present.

  26. Plants usually aromatic; phyllary margins and apices scarious.

    27. Heads in broad, paniculiform arrays, or in narrow, racemiform or spiciform arrays; disc florets 2–20(–30+) . . . . 119. *Artemisia* (in part), **v. 19, p. 503**

    27. Heads borne singly or in lax, corymbiform arrays; disc florets 60–250+.

      28. Perennials; stems villous to strigoso-sericeous; leaf blades 2–3-pinnately lobed . . . . 115. *Chamaemelum* (in part), **v. 19, p. 496**

      28. Subshrubs; stems often tomentose to lanate, sometimes glabrate or glabrous and gland-dotted; leaf blades 1-pinnately lobed (± vermiform) . . . . 116. *Santolina*, **v. 19, p. 497**

  26. Plants usually not aromatic; phyllary margins and apices herbaceous.

    29. Disc florets bisexual, fertile.

      30. Subshrubs; leaf blades linear to filiform; involucres obconic . . . . 315. *Varilla*, **v. 21, p. 221**

      30. Annuals, perennials, subshrubs, or shrubs; leaf blades suborbiculate, deltate, ovate, cordate-ovate, elliptic, lanceolate, linear, oblanceolate, rhombic, or spatulate; involucres turbinate, campanulate, hemispheric, or broader.

31. Cypselae prismatic, or nearly so, 3–4-angled . . . . 276. *Agnorhiza* (in part), **v. 21, p. 104**
31. Cypselae compressed to flattened.
    32. Annuals or perennials; cypselae winged . . . . 277. *Verbesina* (in part), **v. 21, p. 106**
    32. Perennials, subshrubs, or shrubs; cypselae sometimes thin-edged
        (margins ciliate, never winged) . . . . 282. *Encelia* (in part), **v. 21, p. 118**
[29. Shifted to left margin.—Ed.]
29. Disc florets pistillate and functionally staminate (sometimes in separate, unisexual heads).
    33. Pistillate and functionally staminate florets in separate heads (cypselae shed within hardened, often prickly, spiny, tuberculate, or winged, perigynia, forming "burs" or nutlike structures).
        34. Staminate heads: phyllaries partially or wholly connate; receptacles ± flat or convex; pistillate heads: phyllaries 12–30(–80+) in 1–8+ series, outer (1–)5–8 distinct or ± connate, the rest ± connate (becoming indurate, their distinct tips forming straight or hooked spines, tubercles, or wings) . . . . 244. *Ambrosia* (in part), **v. 21, p. 10**
        34. Staminate heads: phyllaries distinct to bases; receptacles conic to columnar; pistillate heads: phyllaries 30–75+ in 6–12+ series, outer 5–8 distinct, the rest proximally connate (becoming indurate, their distinct tips usually forming hooked spines) . . . . 245. *Xanthium* (in part), **v. 21, p. 19**
    33. Pistillate and functionally staminate florets usually together in same heads (sometimes some heads staminate; cypselae not enclosed within perigynia).
        35. Cypselae shed with accessory structures (each with at least 2 paleae, sometimes florets and/or a phyllary as well); anthers ± connate . . . . 246. *Parthenium* (in part), **v. 21, p. 20**
        35. Cypselae usually shed free of accessory structures; anthers weakly coherent or distinct.
            36. Subshrubs or shrubs (phyllaries, paleae, and cypselae ± villous) . . . . . 251. *Oxytenia*, **v. 21, p. 29**
            36. Annuals or perennials (rarely woody at bases; phyllaries, paleae, and cypselae glabrous or strigillose and/or hispidulous).
                37. Heads usually in loose, (± bracteate or ebracteate) paniculiform arrays (sometimes 3–6+ distal to axil of each bract); herbaceous phyllaries usually 5; lobes of functionally staminate corollas soon reflexed . . . . . . 253. *Hedosyne*, **v. 21, p. 30**
                37. Heads mostly borne singly (in leaf axils or remote from axils, ± scattered); herbaceous phyllaries usually 3+; lobes of functionally staminate corollas usually erect at flowering . . . . 254. *Chorisiva*, **v. 21, p. 31**

Key to Genera of Group 6

Heads eradiate; receptacles paleate; pappi wholly of bristles.

1. Leaves all or mostly opposite.
    2. Involucres cylindric; corollas white or purple to blue, lavender, or reddish . . . . 416. *Chromolaena* (in part), **v. 21, p. 544**
    2. Involucres ± hemispheric or broader to ovoid; corollas yellow to orange (lobes sometimes reddish).
        3. Involucres ± hemispheric or broader; receptacles flat to convex; corolla throats narrowly funnelform to cylindric; cypselae obpyramidal; pappi of 2–12 barbellulate bristles . . . . 284. *Melanthera* (in part), **v. 21, p. 123**
        3. Involucres ± hemispheric to ovoid; receptacles conic; corolla throats campanulate; cypselae ellipsoid or obovoid; pappi of 1–3 awnlike bristles . . . . 292. *Acmella* (in part), **v. 21, p. 132**
1. Leaves alternate.
    4. Annuals; involucres 0 or inconspicuous.
        5. Bisexual florets 0; functionally staminate florets 2–6, pappi (0–)1–10(–13) bristles (hidden in heads).
            6. Pistillate paleae (obcompressed to terete, not galeate): wings ± erect (and apical); receptacles cylindric to clavate (heights 2.8–8 times diams.); phyllaries 0 or 1–4 (similar to paleae); cypselae: corolla scars subapical . . . . 99. *Stylocline* (in part), **v. 19, p. 450**

6. Pistillate paleae (compressed, galeate): wings ± erect (and lateral) or inflexed (and subapical); receptacles depressed-spheric to obovoid (heights 0.5–1.8 times diams.); phyllaries 4–6 (unlike paleae); cypselae: corolla scars ± lateral . . . . 100. *Micropus* (in part), **v. 19, p. 454**
5. Bisexual florets (1–)2–10(–11), pappi of (11–)13–28+ bristles visible in heads; functionally staminate florets 0.
    7. Receptacles fungiform to obovoid (heights 0.4–1.6 times diams.); most pistillate paleae ± saccate, each ± enclosing a floret, apices blunt; innermost paleae spreading in fruit; cypselae dimorphic (outer longer than inner) . . . . 97. *Logfia* (in part), **v. 19, p. 443**
    7. Receptacles cylindric to clavate (heights 5–15 times diams.); most pistillate paleae open to ± folded (at most, each enfolding, not enclosing a floret; apices acuminate to aristate); innermost paleae erect to ascending in fruit; cypselae monomorphic (outer ± equaling inner) . . . . 98. *Filago* (in part), **v. 19, p. 447**
4. Perennials, subshrubs, or shrubs; involucres conspicuous.
    8. Subshrubs or shrubs . . . . 171. *Lorandersonia* (in part), **v. 20, p. 177**
    8. Perennials.
        9. Phyllary tips spiny or hooked-spiny.
            10. Leaf margins spiny; phyllary tips spiny . . . . 8. *Carlina*, **v. 19, p. 84**
            10. Leaf margins not spiny; phyllary tips hooked-spiny . . . . 16. *Arctium* (in part), **v. 19, p. 168**
        9. Phyllary tips not spiny or hooked-spiny.
            11. Corollas yellow . . . . 276. *Agnorhiza* (in part), **v. 21, p. 104**
            11. Corollas white, pinkish, lavender to dark magenta, pinkish purple, or ± blue.
                12. Cypselae 5-angled or -grooved; pappi of 1(–5+), ± glandular setae . . . . 413. *Hartwrightia* (in part), **v. 21, p. 540**
                12. Cypselae ca. 10-ribbed; pappi of 35–40 barbellulate to barbellate bristles . . . . 411. *Carphephorus* (in part), **v. 21, p. 535**

Key to Genera of Group 7
Heads eradiate; receptacles paleate; pappi wholly, or partially, of awns or scales.

1. Florets 1(–2) per head (heads in globose, second-order heads or headlike glomerules).
    2. Leaf margins spiny (plants thistlelike) . . . . 9. *Echinops* (in part), **v. 19, p. 85**
    2. Leaf margins ± serrate (not spiny, plants not thistlelike) . . . . 295. *Lagascea* (in part), **v. 21, p. 136**
1. Florets (5–)20–800+ per head (heads borne singly or in arrays, not in globose, second-order heads or headlike glomerules).
    3. Phyllaries 0 (outer receptacle paleae forming an "involucre"; inner paleae 0 or each subtending a floret).
        4. Annuals (inner receptacle paleae falling, distinct, each subtending a floret) . . . . 335. *Layia* (in part), **v. 21, p. 262**
        4. Perennials (inner receptacle paleae 0).
            5. Leaves mostly basal (in rosettes), opposite, faces sericeous . . . . 331. *Raillardella* (in part), **v. 21, p. 256**
            5. Leaves basal and cauline, proximal opposite, distal alternate, faces hirsute to strigose or pubescent . . . . 347. *Anisocarpus* (in part), **v. 21, p. 299**
    3. Phyllaries in (1–)2–7+ series (seldom each phyllary wholly or partly investing ovary of a subtended floret; receptacle paleae usually each subtending a floret).
        6. Calyculi of (0–)1–21+ bractlets or bracts (sometimes surpassing phyllaries); phyllaries usually in ± 2 series, usually ± equal; disc cypselae ± obcompressed to flat (often winged), or ± unequally 3–4-angled and linear-fusiform (pappi of 1–2 usually retrorsely, sometimes antrorsely, barbellate or ciliate, rarely smooth, subulate scales or awns).
            7. Phyllaries connate $^1/_5$–$^7/_8$+ their lengths . . . . 310. *Thelesperma* (in part), **v. 21, p. 199**
            7. Phyllaries usually distinct, rarely connate ± $^1/_{10}$ their lengths.

8. Cypselae (some or all) usually ± winged; pappi of 1–2 barbellate awns
. . . . 309. *Coreocarpus* (in part), **v. 21, p. 198**

8. Cypselae rarely winged; pappi usually of (1–)2–4(–8) usually barbellate or
ciliate, rarely smooth, awns . . . . 312. *Bidens* (in part), **v. 21, p. 205**

[6. Shifted to left margin.—Ed.]

6. Calyculi 0; phyllaries in 1–7+ series; disc cypselae seldom obcompressed or 4-angled and
linear-fusiform (usually not winged).

9. Phyllaries falling together with cypselae (plus 2 contiguous disc florets and their invest-
ing paleae). . . . . 246. *Parthenium* (in part), **v. 21, p. 20**

9. Phyllaries persistent (in fruit).

10. Receptacles columnar . . . . 261. *Rudbeckia* (in part), **v. 21, p. 44**

10. Receptacles mostly flat to convex or conic.

11. Leaves basal and cauline, mostly alternate; pappi usually fragile or readily
falling. . . . . 413. *Hartwrightia* (in part), **v. 21, p. 540**

11. Leaves usually cauline and opposite, sometimes mostly basal and/or mostly
alternate; pappi usually persistent (readily falling in *Melanthera*).

12. Disc corollas lavender, pink, purple, or white; anther thecae cream or purple
. . . . 391. *Marshallia*, **v. 21, p. 456**

12. Disc corollas usually orange to yellow, sometimes brown, pink, purple,
red, or white; anther thecae pale or dark (not violet).

13. Stigmatic papillae in 2 lines.

14. Paleae conduplicate (each ± clasping a cypsela)

15. Pappi of 5–8 linear-lanceolate, erose-margined scales
. . . . 166. *Eastwoodia*, **v. 20, p. 169**

15. Pappi of 15–25+ plumose, setiform scales. . . . . 303. *Bebbia*, **v. 21, p. 177**

14. Paleae subulate or setiform (not conduplicate and each clasping a
cypsela).

16. Corollas white, pinkish, cream, or yellow; cypselae clavate to
± cylindric or compressed, obscurely 8–20-angled; pappi of
(1–)4–20, ± erose scales in 1–4 series . . . 378. *Chaenactis* (in part), **v. 21, p. 400**

16. Corollas yellow or orange to red, purplish, or brown; cypselae
obpyramidal to clavate, ± 4-angled; pappi of 6–10+ medially
thickened, laterally scarious scales in 1–2 series (all, some, or
none aristate) . . . . 384. *Gaillardia* (in part), **v. 21, p. 421**

13. Stigmatic papillae usually continuous, sometimes none (functionally
staminate florets), rarely in 2 lines.

17. Leaves mostly cauline, mostly opposite.

18. Corollas white or whitish; pappi readily falling, of 2–12
barbellate awns (or bristles) . . . . 284. *Melanthera* (in part), **v. 21, p. 123**

18. Corollas yellow to orange; pappi usually persistent and
coroniform or cyathiform (each an erose, fimbriate, or
lacerate cup, with or without additional awns or bristles, borne
on rostra), or of 2–4+ awns and/or scales.

19. Pappi usually of 2–3 (often aristate or subulate) scales (not
cyathiform; cypselae not rostrate) . . . 277. *Verbesina* (in part), **v. 21, p. 106**

19. Pappi usually cyathiform (fimbriate cups plus 0–3 coarse
bristles or awns, borne on rostra; cypselae often rostrate)
. . . . 285. *Wedelia* (in part), **v. 21, p. 125**

17. Leaves mostly basal, or basal and cauline, or cauline, mostly
alternate.

20. Cypselae winged; pappi of 2(–3) scales (scales often aristate or
subulate, without additional scales) . . . . 277. *Verbesina* (in part), **v. 21, p. 106**

20. Cypselae sometimes thin-edged (margins sometimes ciliate or
corky-thickened, never truly winged); pappi usually of
(1–)2, subulate scales or bristlelike awns plus 2–4+ shorter
scales (rarely of 2–3, aristate scales without additional scales).

[21. Shifted to left margin.—Ed.]

21. Subshrubs or shrubs (glabrous or ± scabrellous, usually vernicose); phyllaries 12–40 in 2–
4+ series (subequal or unequal, outer longer). 281. *Flourensia* (in part), **v. 21, p. 117**
21. Annuals, perennials, or shrubs (glabrous or canescent, hirtellous, scabrellous, strigose, or
tomentose, often gland-dotted or glandular-puberulent to stipitate-glandular, seldom
vernicose); phyllaries 18–30(–50+) in 2–3+ series (subequal or unequal, outer shorter).
  22. Perennials (*E. nutans*), subshrubs, or shrubs; pappi sometimes fragile, of 2 weak,
    bristlelike (villous) scales. . . . . 282. *Encelia* (in part), **v. 21, p. 118**
  22. Annuals or perennials; pappi usually persistent, of 2 awns or subulate scales . . .
    . . . . 283. *Geraea* (in part), **v. 21, p. 122**

Key to Genera of Group 8
Heads radiate; receptacles epaleate; pappi none or nearly so.

1. Cypselae usually polymorphic within heads (straight, arcuate, contorted, or ± coiled), sometimes ± beaked (bodies usually tuberculate, ridged, and/or winged, sometimes drupelike).
  2. Shrubs or trees; cypselae fleshy (drupelike) . . . . 83. *Chrysanthemoides*, **v. 19, p. 379**
  2. Annuals, perennials, or shrubs; cypselae not fleshy (often tuberculate or ridged and/or
    winged).
    3. Disc florets bisexual (some or all fertile) . . . . 84. *Dimorphotheca*, **v. 19, p. 380**
    3. Disc florets all functionally staminate.
      4. Cypselae arcuate to ± coiled, abaxially tuberculate, sometimes winged . . . .
      . . . . 85. *Calendula*, **v. 19, p. 381**
      4. Cypselae triquetrous-prismatic to clavate, ± tuberculate and/or winged . . . .
      . . . . 86. *Osteospermum*, **v. 19, p. 382**
1. Cypselae usually monomorphic, sometimes dimorphic, within heads, rarely, if ever, beaked
(bodies smooth, papillate, muricate, ribbed, nerved, or rugulose, sometimes winged).
  5. Margins of leaf blades usually 1–3-palmately or -pinnately lobed (ultimate lobes
    usually linear to filiform), sometimes dentate or entire; phyllaries in (2–)3–5+ series,
    unequal, and scarious or margins and/or apices notably scarious.
    6. Cypselae dimorphic: outer (ray) usually 3-angled and ± winged (except
      Mauranthemum); inner (disc) ± compressed-prismatic or ± flattened (angles winged),
      or ± quadrate (1 or 2 angles sometimes winged), or columnar (not winged).
      7. Subshrubs or shrubs (plants often persisting after cultivation) . . . .
      . . . . 131. *Argyranthemum*, **v. 19, p. 552**
      7. Annuals.
        8. Stems and leaves hirtellous to pilosulous (some hairs gland-tipped, plants
        sticky, viscid) . . . . 130. *Heteranthemis*, **v. 19, p. 551**
        8. Stems and leaves glabrous or hairy (hairs not gland-tipped, plants not
        viscid).
          9. Rays proximally white or red to purple, distally yellow or white; disc
          corollas proximally ochroleucous, distally red to purple (phyllaries
          ± carinate) . . . . 132. *Ismelia*, **v. 19, p. 552**
          9. Rays mostly white (usually yellowish at bases) or mostly yellow (some-
          times paler distally); disc corollas ± yellow (phyllaries not carinate).
          10. Involucres 8–12(–15+) mm diam.; rays white (usually yellowish at
          bases, drying pinkish), laminae oblong to ovate (6–12+ mm); disc
          corolla lobes (2–)5 (without resin sacs) . . . .
          . . . . 133. *Mauranthemum* (in part), **v. 19, p. 554**
          10. Involucres 15–25+ mm diam.; rays mostly yellow, sometimes paler
          distally, laminae linear, oblong, or ovate (8–25 mm); disc corolla
          lobes 5 (each with a resin sac) . . . . 134. *Glebionis*, **v. 19, p. 554**
    6. Cypselae ± monomorphic, outer and inner similar (none notably winged).

11. Annuals.
    12. Leaves usually irregularly 1-pinnately lobed or toothed; involucres ± hemi-
        spheric, 8–12(–15+) mm diam.; disc florets 60–100+; cypsela ribs 7–10
        . . . . 133. *Mauranthemum* (in part), **v. 19, p. 554**
    12. Leaves (1–)2–3-pinnately lobed; involucres 4–14 mm diam.; disc florets
        120–750+; cypsela ribs 3–5.
        13. Cypselae obconic, slightly compressed (usually asymmetric, apices
            oblique), ribs 5, faces smooth between ribs (pericarps without resin
            sacs, or resin sacs within lateral ribs) . . . . 124. *Matricaria* (in part), **v. 19, p. 540**
        13. Cypselae trigonous, ± compressed, ribs 3–5 (0–2 abaxial, 2 lateral, 1
            adaxial, usually whitish, relatively thick, smooth), faces smooth or rug-
            ose to tuberculate between ribs (pericarps usually with 2–3, sometimes
            1–5 mostly abaxial-apical, resin sacs) . . 129. *Tripleurospermum* (in part), **v. 19, p. 548**
11. Biennials or perennials.
    14. Heads usually in lax to dense, corymbiform arrays, rarely borne singly;
        pappi usually coroniform, rarely 0 . . . . 112. *Tanacetum* (in part), **v. 19, p. 489**
    14. Heads usually borne singly or in 2s or 3s, sometimes in corymbiform
        arrays; pappi usually 0, sometimes crowns of 6–12 irregular teeth.
        15. Ray florets styliferous and sterile . . . . 136. *Leucanthemella*, **v. 19, p. 557**
        15. Ray florets pistillate and fertile.
            16. Cypselae trigonous, ± compressed, ribs 3–5 (0–2 abaxial, 2 lateral,
                1 adaxial, usually whitish, relatively thick, smooth), faces smooth
                or rugose to tuberculate between ribs (pericarps usually with 2–3,
                sometimes 1–5 mostly abaxial-apical, resin sacs) . . . .
                . . . . 129. *Tripleurospermum* (in part), **v. 19, p. 548**
            16. Cypselae ± columnar, cylindro-obconic, or obovoid, ribs 5–10 (peri-
                carps without apical resin sacs).
                17. Plants (10–)50–130(–200+) cm; disc corolla tubes ± cylindric
                    (proximally swollen, becoming spongy in fruit); cypselae
                    ± columnar to obovoid, ribs ± 10 (pericarps with myxogenic
                    cells) . . . . 137. *Leucanthemum*, **v. 19, p. 557**
                17. Plants mostly (2.5–)5–40 cm; disc corolla tubes ± cylindric
                    (not swollen, not becoming spongy in fruit); cypselae cylindro-
                    obconic, ribs 5–8(–10) (pericarps without myxogenic cells).
                    18. Leaf margins entire; involucres 4–6.5 mm diam.;
                        phyllaries 20–26(+) in 2(–3+) series (receptacles ± villous)
                        . . . . 120. *Hulteniella* (in part), **v. 19, p. 534**
                    18. Leaf margins usually pinnati-palmately lobed (lobes 3–7),
                        ultimate margins coarsely crenate, dentate, or entire;
                        involucres 13–29 mm diam.; phyllaries 22–44 in 3(–4)
                        series (receptacles glabrous) . . . . 121. *Arctanthemum*, **v. 19, p. 535**

[5. Shifted to left margin.—Ed.]

5. Margins of leaf blades entire, dentate, lobed, palmatifid, pinnatifid, or dissected (ultimate
   lobes sometimes linear to filiform); phyllaries in 1–2 series and ± equal to subequal, or in 3–
   8+ series and unequal, usually herbaceous to chartaceous, sometimes with margins and/or
   apices notably scarious.
   19. Leaves alternate; phyllaries in 1–2 series and equal to subequal (sometimes coherent,
       often subtended by calyculi); style-branch appendages usually penicillate or essentially
       none; cypselae usually ± columnar to prismatic, seldom compressed, obcompressed, or
       flattened.
       20. Perennials.
           21. Rays usually whitish or bluish, pinkish, purplish, or reddish (often proximally
               pale and distally darker); disc floret style-branches: stigmatic areas in 2 lines
               . . . . 219. *Pericallis* (in part), **v. 20, p. 607**
           21. Rays yellow; disc floret style-branches: stigmatic areas continuous . . . .
               . . . . 224. *Doronicum* (in part), **v. 20, p. 611**

20. Annuals.
    22. Disc florets functionally staminate (not producing cypselae, styles not divided); leaf blades linear or pinnately lobed, ultimate margins entire, faces glabrous or sparsely floccose-tomentose . . . . 242. *Blennosperma*, **v. 20, p. 640**
    22. Disc florets bisexual (fertile, styles 2-branched); leaf blades spatulate or oblanceolate to lanceolate or linear, margins toothed or entire, faces glabrous or glabrate . . . . 243. *Crocidium* (in part), **v. 20, p. 641**
[19. Shifted to left margin.—Ed.]
19. Leaves whorled, opposite, or alternate; phyllaries in 1–2 series and subequal, or in 3–6+ series and unequal; style-branch appendages usually deltate to lanceolate, seldom penicillate or none; cypselae sometimes compressed, obcompressed, or flattened.
  23. Leaves opposite (at least proximally).
    24. Leaves and/or phyllaries dotted or streaked with pellucid (schizogenous) glands containing strong-scented (lemon or spicy) oils; ray florets borne on bases of subtending phyllaries . . . . 316. *Pectis* (in part), **v. 21, p. 222**
    24. Leaves and/or phyllaries rarely dotted or streaked (never with pellucid, schizogenous glands containing strong-scented oils, plants sometimes with sessile or stipitate, surface glands and may be otherwise strong-scented); ray florets borne on receptacles.
      25. Leaves (often ± succulent) sessile or nearly so, blades usually oblong to linear or filiform (not lobed); cypselae clavate to cylindric and 8–15-ribbed.
        26. Heads in tight, corymbiform aggregations or arrays; phyllaries 2–6(–9) in ± 1 series, subequal; rays 1(–2) . . . . 327. *Flaveria* (in part), **v. 21, p. 247**
        26. Heads borne singly; phyllaries 12–15 in 3+ series, unequal; rays 3–10 . . . . 330. *Jaumea* (in part), **v. 21, p. 253**
      25. Leaves (seldom succulent) petiolate or sessile, blades often lobed; cypselae usually obpyramidal to obconic, sometimes columnar or flattened, seldom clavate or cylindric, often 4–5-angled (not both clavate to cylindric and 8–15-ribbed; sometimes cylindric and 5–10-nerved).
        27. Phyllaries often ± conduplicate and navicular; disc corollas 4-lobed; cypselae strongly flattened to subcylindric, ± callous-margined, often ciliate . . . . 352. *Perityle* (in part), **v. 21, p. 317**
        27. Phyllaries usually flat to weakly navicular; disc corollas (4–)5-lobed; cypselae narrowly clavate or columnar to obconic or obpyramidal (not strongly compressed or flattened, callous-margined, and ciliate).
          28. Leaves usually sessile, sometimes obscurely petiolate, blades (or lobes) linear, margins entire or 1–2-pinnately lobed . . . . 354. *Lasthenia* (in part), **v. 21, p. 336**
          28. Leaves usually petiolate, sometimes sessile, blades rounded-deltate or cordate to ovate, narrowly trullate, lanceolate, linear, obovate, oblanceolate, or cuneate to spatulate, margins entire, toothed, or distally 3-lobed (not 1–2-pinnately lobed).
            29. Phyllaries 20–40+ in 3–4+ series; rays 12–34, yellow . . . . 370. *Venegasia* (in part), **v. 21, p. 385**
            29. Phyllaries 5–18 in 1–3 series; rays 5–12, yellow, or white or pinkish with reddish veins.
              30. Leaf blades either linear and margins entire, or narrowly cuneate to spatulate and margins usually distally 3-lobed; phyllaries 5–8 in 1 series; rays either yellow, or white or pinkish with reddish veins . . . . 365. *Syntrichopappus* (in part), **v. 21, p. 379**
              30. Blades lanceolate, narrowly trullate, oblanceolate, obovate, margins entire or denticulate; phyllaries 5–18 in 1–3 series; rays yellow.
                31. Phyllaries 5–12 in 1 series . . . . 361. *Arnica* (in part), **v. 21, p. 366**
                31. Phyllaries 14–18 in ± 3 series . . . . 362. *Jamesianthus* (in part), **v. 21, p. 377**
  23. Leaves alternate or mostly alternate (sometimes proximal opposite).
    32. Rays white, pink, purple, or blue.

33. Receptacles conic.
    34. Leaves usually pinnatifid to dentate, sometimes entire; rays without midstripe abaxially; cypselae not or slightly compressed.
        35. Plants gland-dotted; cypselae oblong to narrowly obovoid, slightly compressed, 2-nerved, sometimes gland-dotted . . . . 141. *Egletes*, **v. 20, p. 35**
        35. Plants eglandular; cypselae columnar to prismatic, usually 4-angled, 4–12-ribbed (ribs relatively thick), eglandular . . . .
           . . . . 188. *Aphanostephus* (in part), **v. 20, p. 351**
    34. Leaves usually entire, sometimes ± toothed; rays usually with pink or purplish midstripe abaxially; cypselae strongly compressed or flattened.
        36. Perennials 5–20 cm (short-rhizomatous); leaves mostly basal, margins entire or crenate-serrate; rays 35–90 in 3–4 series (closing at night); cypselae obconic, glabrous to short-strigose or ciliate-margined
           . . . . 139. *Bellis*, **v. 20, p. 22**
        36. Annuals, biennials, or perennials, 5–50 cm (tap- or fibrous-rooted); leaves basal and cauline, margins usually entire, sometimes toothed; rays 10–65(–85) in 1 series; cypselae obovoid to oblanceoloid-obovoid, glabrous or glochidiate-hairy . . . . 177. *Astranthium*, **v. 20, p. 203**
33. Receptacles flat to convex.
    37. Leaves mostly basal or basal and cauline, margins 1–2-pinnately lobed (lobes usually filiform); rays 8, white; cypselae obpyramidal (4- or 5-angled, faces 1–4-ribbed) . . . . 351. *Hymenopappus* (in part), **v. 21, p. 309**
    37. Leaves basal (usually withering by flowering) or mostly cauline, margins entire or toothed, sometimes 1-pinnately lobed (lobes not usually filiform); rays (1–)3–70, white or pink to blue or purple; cypselae linear, fusiform, clavate, oblanceoloid, obovoid, or obconic, ± flattened or ± compressed to ± cylindric, sometimes 2–18-nerved.
        38. Phyllaries 5–8 in 1 series, subequal; rays white or pinkish with reddish veins . . . . 365. *Syntrichopappus* (in part), **v. 21, p. 379**
        38. Phyllaries 5–50 in 2–6 series, equal to unequal; rays white or blue to purple (not with reddish veins).
            39. Disc corolla lobes 4 . . . . 352. *Perityle* (in part), **v. 21, p. 317**
            39. Disc corolla lobes 5.
                40. Heads borne singly (terminal); phyllaries in 2–3 series, equal to subequal, faces eglandular . . . . . . 179. *Chaetopappa* (in part), **v. 20, p. 206**
                40. Heads in loose, corymbiform arrays; phyllaries in 2–6 series, unequal, (herbaceous) faces stipitate-glandular . . . .
                . . . . 213. *Psilactis* (in part), **v. 20, p. 462**
[32. Shifted to left margin.—Ed.]
32. Rays usually yellow to orange, sometimes red.
    41. Disc corollas usually 4-lobed; cypselae usually callous-margined and ciliate . . . .
    . . . . 352. *Perityle* (in part), **v. 21, p. 317**
41. Disc corollas 5-lobed; cypselae not callous-margined or notably ciliate.
    42. Subshrubs or shrubs.
        43. Heads (± sessile) in compact glomerules (glomerules densely clustered in terminal, corymbiform arrays); involucres 1.5–2 mm diam.; rays 4–6; disc florets 4–6 . . . . 156. *Gymnosperma*, **v. 20, p. 94**
        43. Heads (± pedunculate) borne singly or in corymbiform to paniculiform arrays (not in glomerules); involucres 5–25 mm diam.; rays 5–60+; disc florets 20–300.
            44. Phyllaries in 5–8+ series, apices (terete to filiform or subulate) recurved or straight, or looped to hooked or patent . . . . 202. *Grindelia* (in part), **v. 20, p. 424**
            44. Phyllaries in 1–4 series, apices usually straight, not recurved, looped, hooked, or patent.

45. Leaf blades usually 1–2(–3)-pinnately lobed, ultimate margins entire or toothed, faces ± woolly; involucres campanulate to hemispheric, 3–12+ mm diam.; phyllaries in 1+ series (sometimes ± connate) . . . . 359. *Eriophyllum* (in part), **v. 21, p. 353**

45. Leaf blades rounded-deltate or cordate to ovate, entire or toothed, abaxial faces puberulent or gland-dotted; involucres hemispheric to globose, 12–25+ mm diam.; phyllaries in 3–4 series (outer rotund to oblong, herbaceous-foliaceous) . . . . 370. *Venegasia* (in part), **v. 21, p. 385**

[42. Shifted to left margin.—Ed.]

42. Annuals, biennials, or perennials.

46. Ray florets usually neuter, rarely pistillate and fertile or styliferous and sterile (disc corolla lobes often shaggily hairy, hairs ± moniliform) . . . . 384. *Gaillardia* (in part), **v. 21, p. 421**

46. Ray florets pistillate, fertile.

47. Ray florets in 2 series (staminodes often present), adaxial faces of inner laminae marked basally with purple-brown fans $^1/_4$–$^1/_8$ their lengths . . . . . . 27. *Arctotis* (in part), **v. 19, p. 198**

47. Ray florets in 1 series (without staminodes), adaxial faces of laminae not marked basally with purple-brown.

48. Phyllaries 25–100+ in (3–)4–9+ series, apices (terete to filiform or subulate) recurved or straight, or looped to hooked or patent . . . . . 202. *Grindelia* (in part), **v. 21, p. 424**

48. Phyllaries 5–40 in 1–4 series, apices usually straight, not recurved, looped, hooked, or patent.

49. Leaf blades simple, margins entire, spinulose-toothed, or sinuate-dentate (not lobed).

50. Phyllaries 30–40 in 2–4 series (annuals; leaf blades oblanceolate, margins entire or spinulose-toothed, faces glabrous, gland-dotted; rays 14–63) . . . . 203. *Xanthocephalum* (in part), **v. 20, p. 436**

50. Phyllaries 2–34 in 1–2(–3) series.

51. Leaf blades linear to lanceolate, margins entire; disc florets 10–21, functionally staminate. . . . . 154. *Amphiachyris* (in part), **v. 20, p. 87**

51. Leaf blades narrowly oblong, oblanceolate, or lance-linear, margins sinuate-dentate or entire; disc florets 10–100+, bisexual, fertile.

52. Leaf blades narrowly oblong, oblanceolate, or lance-linear, margins sinuate-dentate or entire; heads usually borne singly (sometimes in corymbiform arrays); phyllaries in 1 series, distinct; cypselae obcompressed or ± prismatic, 2–4-angled . . . . 357. *Monolopia* (in part), **v. 21, p. 349**

52. Leaf blades lance-linear to broadly ovate, margins entire; heads borne singly or in cymiform arrays; phyllaries in 2 series, distinct or connate; cypselae narrowly obpyramidal, weakly ribbed or striate . . . . 388. *Baileya* (in part), **v. 21, p. 444**

49. Leaf blades 1–2(–3)-pinnately or -ternately lobed, pinnatifid, or distally 3-lobed.

53. Phyllaries in 1 series.

54. Leaf blades distally 3-lobed. . . . . 365. *Syntrichopappus* (in part), **v. 21, p. 379**

54. Leaf blades usually 1–2(–3)-pinnately lobed.

55. Ray cypselae obcompressed (disc corollas with rings of hairs at bases of limbs) . . . . 358. *Pseudobahia*, **v. 21, p. 351**

55. Ray cypselae usually prismatic, 4–5-angled (disc corollas without rings of hairs) . . . . 359. *Eriophyllum* (in part), **v. 21, p. 353**

53. Phyllaries in 2–3+ series.

56. Cypselae obpyramidal (lengths usually 3+ times diams.) . . . . 374. *Amauriopsis*, **v. 21, p. 392**

56. Cypselae stoutly obconic to obpyramidal (lengths usually 1–2, rarely to 3.5 times diams.).

57. Phyllary margins usually notably membranous to scarious; disc corollas usually whitish, sometimes purplish or yellowish; cypselae usually 4-angled and 12–16-ribbed . . . . 351. *Hymenopappus* (in part), **v. 21, p. 309**

57. Phyllary margins seldom scarious; disc corollas yellow or partly yellow-brown proximally; cypselae sometimes weakly ribbed or striate (not both 4-angled and 12–16-ribbed).

    58. Receptacles usually hemispheric, globoid, ovoid, conic (seldom flat); ray florets 3–16 (corollas withering, falling early or tardily) . . . . 386. *Hymenoxys* (in part), **v. 21, p. 435**

    58. Receptacles flat to convex; ray florets usually 5–7 or 20–55 (corollas usually marcescent) . . . . . . 388. *Baileya* (in part), **v. 21, p. 444**

Key to Genera of Group 9

Heads radiate; receptacles epaleate; ray corollas white, pink, or purple (with little, if any, yellow); pappi wholly of bristles (without awns or scales).

1. Cauline leaves opposite (at least proximal).
    2. Leaves sessile; leaves and phyllaries dotted with pellucid (schizogenous) glands containing strong-scented oils . . . . 316. *Pectis* (in part), **v. 21, p. 222**
    2. Leaves petiolate or sessile; leaves and phyllaries sparsely tomentose to woolly or glabrescent (without schizogenous glands) . . . . 365. *Syntrichopappus* (in part), **v. 21, p. 379**
1. Cauline leaves all alternate.
    3. Phyllaries in 1–2 series, equal.
        4. Pappi persistent, either 1 apically plumose bristle (plus short-toothed cups) or 1–12 bristles (often alternating with easily overlooked, laciniate scales) . . . . 180. *Monoptilon* (in part), **v. 20, p. 210**
        4. Pappi falling or persistent (fragile), usually 20–100+ barbellulate or smooth (barbellate in *Pericallis*) bristles.
            5. Disc florets (all or mostly) functionally staminate (not producing cypselae) . . . . 241. *Petasites* (in part), **v. 20, p. 635**
           5 Disc florets (all or mostly) bisexual, fertile.
               6. Phyllary margins not interlocking; ray corollas usually whitish or bluish, pinkish, purplish, or reddish (often proximally pale and distally darker); pappi readily falling, usually of 20–40+ bristles (discs; rays sometimes of 2 subulate to setiform scales or 0) . . . . 219. *Pericallis* (in part), **v. 20, p. 607**
               6. Phyllary margins interlocking; ray corollas usually white (with little, if any, yellow), rarely purplish to reddish; pappi persistent or readily falling, usually of 30–80+ bristles.
                   7. Calyculi of (0–)1–8+ bractlets; disc corolla tubes shorter than or equaling throats; style branches stigmatic in 2 lines (stamen filaments usually distally expanded into swollen collars); cypselae 5-ribbed or -angled . . . . 215. *Senecio* (in part), **v. 20, p. 544**
                   7. Calyculi none; disc corolla tubes equaling or longer than throats; style branches with continuous stigmatic areas (stamen filaments cylindric, not distally expanded); cypselae 10-ribbed or -nerved . . . . 227. *Tephroseris* (in part), **v. 20, p. 615**
    3. Phyllaries in 2–7+ series, unequal to subequal (pappi persistent, bristles usually barbellate, sometimes barbellulate).
        8. Subshrubs or shrubs (clambering, sprawling, or vinelike).
            9. Shrubs (clambering, sprawling, or vinelike); rays pale rose-purple to pale pink (Atlantic coastal plain) . . . . 211. *Ampelaster*, **v. 20, p. 460**
            9. Subshrubs or shrubs (not clambering, sprawling, or vinelike); rays white, pink, red-purple, purple, or violet.

10. Plants often thorny (thorns green); rays white; disc corolla veins orange, resinous . . . . 190. *Chloracantha* (in part), **v. 20, p. 358**

10. Plants not thorny; rays white, pink, red-purple, purple, or violet; disc corolla veins not orange-resinous.

    11. Stems, leaves, and phyllaries resinous; leaf margins entire; phyllaries 18–24, midnerves expanded, faces glabrous . . . . 152. *Neonesomia* (in part), **v. 20, p. 85**

    11. Stems and phyllaries not resinous (leaves sometimes resinous in *Hazardia*); leaf margins entire or toothed or serrate (spinulose or bristly); phyllaries (15–)20–90+, midnerves not expanded, faces glabrous or hairy, sometimes stipitate-glandular.

        12. Ray florets neuter; cypselae cuneiform to linear, not compressed . . . . 208. *Corethrogyne* (in part), **v. 20, p. 450**

        12. Ray florets bisexual, fertile; cypselae obscurely cordate, deltoid, ovoid, ellipsoid, oblong, obovoid, cylindro-obconic, fusiform, or linear, usually ± compressed.

            13. Cypselae dimorphic (ray 3-sided, disc compressed), obscurely cordate, ellipsoid, oblong, or obovoid . . 195. *Xanthisma* (in part), **v. 20, p. 383**

            13. Cypselae monomorphic (usually ± compressed), deltoid, ovoid, obovoid, cylindro-obconic, fusiform, or linear.

                14. Cypselae 7–10-nerved . . . . 192. *Herrickia* (in part), **v. 20, p. 361**

                14. Cypselae 2–5-ribbed or -nerved.

                    15. Phyllaries in 5–9 series, not keeled; pappi reddish brown . . . . 206. *Hazardia* (in part), **v. 20, p. 445**

                    15. Phyllaries in 2–6 series, keeled; pappi stramineous to tawny.

                        16. Cypselae flattened, 2–3(–4)-nerved; pappi of outer, shorter bristles or scales plus inner, longer bristles . . . . 150. *Ionactis* (in part), **v. 20, p. 82**

                        16. Cypselae compressed, 4–5-ribbed; pappi of ± unequal bristles . . . . 199. *Xylorhiza* (in part), **v. 20, p. 406**

[8. Shifted to left margin.—Ed.]

8. Annuals, biennials, or perennials.

    17. Annuals or biennials.

        18. Disc corolla veins orange-resinous.

             19. Heads borne singly . . . . 179. *Chaetopappa* (in part), **v. 20, p. 206**

             19. Heads usually in corymbiform or paniculiform arrays, sometimes borne singly.

                20. Disc corolla throats slightly indurate and inflated . . . . 186. *Erigeron* (in part), **v. 20, p. 256**

                20. Disc corolla throats narrowly funnelform (not indurate and inflated).

                    21. Leaf faces often stipitate-glandular or gland-dotted; phyllaries lacking orange to brown midnerves; cypselae densely sericeous, ± strigillose, or glabrous, often stipitate-glandular and/or gland-dotted . . . . 142. *Laënecia* (in part), **v. 20, p. 36**

                    21. Leaf faces eglandular; phyllaries with orange to brownish midnerves; cypselae glabrous or strigillose, eglandular . . . . 187. *Conyza* (in part), **v. 20, p. 348**

        18. Disc corolla veins not orange-resinous.

             22. Pappus bristles usually 5 . . . . 145. *Pentachaeta* (in part), **v. 20, p. 46**

             22. Pappus bristles usually 20–80+ (ray sometimes 0).

                23. Stems glabrous (or distally hairy in lines only), eglandular; ray florets 90–110 in 4–5 series (laminae filiform) . . . . 214. *Symphyotrichum* (in part), **v. 20, p. 465**

                23. Stems glabrous or hairy, often stipitate-glandular as well; ray florets (4–)7–80 in 1 series (laminae usually strap-shaped).

                    24. Heads in corymbiform arrays; phyllaries in 2–3 series; ray pappi none . . . . 213. *Psilactis* (in part), **v. 20, p. 462**

                    24. Heads usually borne singly (at ends of branches, leafy stems, or scapiform peduncles), sometimes in cymiform or corymbiform arrays; phyllaries in 3–12+ series; ray pappi usually present (usually none in *Arida*).

25. Cypselae 2(–3)-nerved . . . . 176. *Townsendia* (in part), **v. 20, p. 193**
25. Cypselae smooth (at most obscurely nerved) or 4–13-ribbed or -nerved per face.
    26. Leaves deeply 1–2-pinnatifid (some or all lobes or teeth acute and bristle-tipped) . . . . 196. *Machaeranthera*, **v. 20, p. 394**
    26. Leaves usually entire, toothed, or lobed (if 1–2-pinnatifid, teeth or lobes often rounded, sometimes apiculate, mostly not bristle-tipped).
        27. Cypselae ± obconic, not compressed, smooth (at most obscurely nerved); pappus bristles distinct or basally connate, tan to reddish . . . . 209. *Lessingia*, **v. 20, p. 452**
        27. Cypselae linear to narrowly obovoid or narrowly oblong, 4–13-ribbed or -nerved; pappus bristles distinct, white to tawny.
            28. Plants hairy, sometimes stipitate-glandular; leaves entire or toothed; ray pappi of 40–50 bristles . . . . 197. *Dieteria* (in part), **v. 20, p. 395**
            28. Plants glabrous and leaves entire or toothed (ciliate or teeth bristle-tipped or apiculate), or plants hairy, sometimes stipitate-glandular, and leaves 1–2-pinnatifid; ray pappi usually 0 (if 20–30 bristles, leaves 1–2-pinnatifid) . . . . 198. *Arida* (in part), **v. 20, p. 401**

[17. Shifted to left margin.—Ed.]
17. Perennials.
  29. Plants colonial; stems branched, lateral branches strongly ascending, commonly modified to green thorns; leaves early withering; phyllaries (1–)3(–5)-nerved (usually wet sites in arid, sw United States) . . . . 190. *Chloracantha* (in part), **v. 20, p. 358**
  29. Plants sometimes colonial; stems single or clustered, simple or branched (not becoming thorny); at least distal leaves persistent through flowering; phyllaries usually 1-nerved, seldom 3-nerved.
    30. Cypselae usually obconic or obovoid, sometimes lanceoloid, flattened or compressed, margins ribbed (sometimes also 1–2 nerves on faces).
      31. Leaves cauline; phyllaries keeled.
        32. Plants 10–160 cm, minutely stipitate-glandular distally; proximal leaves withering by flowering, proximalmost scalelike, cauline distally increasing in size to mid stems, mid and distal blades lanceolate or lance-ovate to elliptic; heads in racemiform or corymbiform arrays (pappi of outer, shorter and inner, longer bristles in 3 series) . . . . 143. *Eucephalus* (in part), **v. 20, p. 39**
        32. Plants 4–30(–70) cm, sometimes stipitate-glandular; leaves persistent to flowering, mostly equal in size and shape, blades spatulate (proximal), linear, narrowly oblong, or elliptic-lanceolate; heads borne singly or in 2s or 3s, or in corymbiform arrays . . . . 150. *Ionactis* (in part), **v. 20, p. 82**
      31. Leaves basal and/or cauline; phyllaries not keeled.
        33. Leaf faces eglandular; corolla lobes lanceolate . . . . 138. *Aster*, **v. 20, p. 20**
        33. Leaf faces often gland-dotted; corolla lobes deltate or lance-deltate.
            34. Plants taprooted or with branched caudices; heads borne singly; leaves usually entire, rarely toothed or lobed; phyllaries unequal, 1-nerved (nerves not golden-resinous); disc corolla throats funnelform; cypselae glabrous or hairy (hairs glochidiform) . . . . 176. *Townsendia* (in part), **v. 20, p. 193**
            34. Plants rhizomatous, sometimes taprooted; heads borne singly or in corymbiform arrays; leaves entire, ± dentate, or pinnatifid; phyllaries equal or unequal, 1–3-nerved (nerves golden-resinous); disc corolla throats usually tubular, sometimes strongly inflated-indurate; cypselae glabrous, strigose, or sericeous . . . . 186. *Erigeron* (in part), **v. 20, p. 256**
    30. Cypseale ± narrowly obconic, obovoid, oblanceoloid, ellipsoid, lanceloid, or fusiform, to cylindric or linear, ± compressed or terete, usually 3–12+-nerved or -ribbed on faces (margins not ribbed).

[35. Shifted to left margin.—Ed.]

35. Cypselae ± dimorphic (ray 3-sided and rounded abaxially, disc ± compressed); pappi of relatively coarse (± flattened) bristles.

    36. Stems simple; leaves mostly basal, margins serrate or serrulate; involucres depressed-hemispheric; cypselae 3–9-ribbed per face; pappus bristles coarsely barbellate . . . . 195. *Xanthisma* (in part), **v. 20, p. 383**

    36. Stems usually branched; leaves basal (persistent or withering by flowering) and cauline (distally ± reduced) or mostly cauline, margins pinnately lobed or pinnatifid, toothed, or entire; involucres turbinate, campanulate, or hemispheric; cypselae 8–13-nerved per face; pappus bristles barbellulate. . . . . 198. *Arida* (in part), **v. 20, p. 401**

35. Cypselae monomorphic; pappi of relatively fine bristles.

    37. Anther bases tailed . . . . 110. *Sachsia* (in part), **v. 19, p. 477**

    37. Anther bases rounded, obtuse, or acute (not tailed).

        38. Plants rhizomatous, sometimes with caudices; heads usually in paniculiform or racemiform arrays, rarely borne singly.

            39. Stems spreading-hirsute, eglandular; heads in narrow wand-shaped, paniculiform arrays; phyllary midribs translucent and swollen; rays 7–9 (corollas white to pale cream). . . . . 163. *Solidago* (in part), **v. 20, p. 107**

            39. Stems usually glabrous, often distally hairy in lines, sometimes ± densely hairy, sometimes distally stipitate-glandular; heads in ± open or dense (not wand-shaped), paniculiform arrays; phyllary midnerves not translucent and swollen; rays 8–75 (corollas white, pink, blue, or purple).

                40. Phyllaries usually unequal, sometimes subequal, proximally indurate, distally with defined green zones, sometimes distally herbaceous or outer wholly foliaceous, sometimes stipitate-glandular . . . . . 214. *Symphyotrichum* (in part), **v. 20, p. 465**

                40. Phyllaries subequal, herbaceous (without definite distal green zones, not foliaceous), stipitate-glandular.

                    41. Stems ± densely villous; leaves cauline, blades 1-nerved (venation reticulate), lanceolate to elliptic, abaxial faces glabrate to ± strigose, adaxial sparsely villous (distal stipitate-glandular); phyllaries often purplish, apices of outer acuminate; cold, wet soils, montane and boreal North America . . . . 210. *Canadanthus*, **v. 20, p. 458**

                    41. Stems glabrous; leaves basal and cauline, blades 3-nerved, linear, faces glabrous (distal stipitate-glandular); phyllaries green, apices of outer acute; damp, alkaline areas, deserts and dry prairies, w North America . . . . 212. *Almutaster*, **v. 20, p. 461**

        38. Plants taprooted, with caudices, or rhizomatous; heads usually in corymbiform (or flat-topped, racemiform) arrays or borne singly, sometimes grouped in loose, corymbiform arrays.

            42. Plants usually tapooted, sometimes with caudices (plus also rhizomatous from fibrous roots in *Chaetopappa*); stems usually 1 (sometimes 2–5+ in clumps), branched or simple; heads borne singly or in loose, corymbiform arrays.

                43. Stems and leaves usually densely white-tomentose, sometimes glabrate; ray florets neuter; cypselae cuneiform or linear; pappi reddish to brownish (bristles relatively coarse, California) . . . . 208. *Corethrogyne* (in part), **v. 20, p. 450**

                43. Stems and leaves glabrous, glabrate, canescent, villous, or tomentose; ray florets pistillate, fertile; cypselae fusiform, cylindric, obovoid, or linear; pappi hyaline or white to tawny.

                    44. Stems simple; leaf margins entire; phyllaries mostly foliaceous (margins sometimes proximally indurate); rays white (maturing or drying bluish or purplish).

                      45. Leaves basal and cauline (crowded), blades 1-nerved, linear-oblanceolate to lanceolate; phyllaries not keeled; cypselae 5-nerved . . . . 179. *Chaetopappa* (in part), **v. 20, p. 206**

                      45. Leaves mostly basal (rosettes), blades 3-nerved, linear to oblanceolate; phyllaries often ± keeled; cypselae 5–10-nerved (nerves raised). . . . . 191. *Oreostemma*, **v. 20, p. 359**

44. Stems branched or simple; leaf margins entire or toothed (teeth spine-tipped); phyllaries proximally white-indurate, distally green or herbaceous (usually not foliaceous); rays white, blue, violet, or purple.

46. Stems mostly simple (scapiform); leaves mostly basal (rosettes, often marcescent), margins entire or irregularly serrate (teeth apiculate or ± spinulose); phyllaries squarrose; cypselae 8–10-ribbed (canyons, rock faces, Utah) . . . . 192. *Herrickia* (in part), **v. 20, p. 361**

46. Stems mostly branched; leaves basal and cauline, margins entire or toothed (teeth spinose-tipped); phyllaries appressed, spreading, or reflexed; cypselae 4-nerved, 4–6-ribbed, or smooth.

47. Stems mostly single; cauline leaf blades lanceolate to oblanceolate; phyllaries not keeled . . . . 197. *Dieteria* (in part), **v. 20, p. 395**

47. Stems clustered; cauline leaf blades usually spatulate to obovate or oblong, rarely elliptic; phyllaries keeled . . . . 199. *Xylorhiza* (in part), **v. 20, p. 406**

[42. Shifted to left margin.—Ed.]

42. Plants rhizomatous or with caudices; stems 1–5+, usually simple; heads usually in corymbiform arrays, sometimes borne singly (then plants long-rhizomatous, rays pink).

48. Phyllaries flat, not keeled, midveins orange-resinous or swollen and translucent.

49. Plants 40–200 cm; leaves basal (withering, reduced) and cauline, blades 1-nerved, lanceolate to elliptic; phyllary midveins orange-resinous; rays 2–10(–16) . . . . 144. *Doellingeria*, **v. 20, p. 43**

49. Plants 10–40 cm; leaves basal (persistent, well developed) and cauline (reduced), blades usually 1-nerved, sometimes ± 3-nerved, linear to linear-lanceolate, phyllary midveins swollen, translucent; ray florets 10–20 . . . . 163. *Solidago* (in part), **v. 20, p. 107**

48. Phyllaries ± rounded, sometimes ± keeled, midveins not swollen.

50. Rhizomes with swollen apical buds; heads in ± loose, corymbiform arrays or borne singly (nodding in bud); phyllaries lance-ovate to linear (membranous, proximally not indurate, green along midnerves); cypselae 5–8-nerved (lateral 2 thicker), ± densely gland-dotted (e North America) . . . . 149. *Oclemena*, **v. 20, p. 78**

50. Rhizomes none or not apically swollen; heads in corymbiform arrays (erect in bud); phyllaries ovate, oblong, lanceolate, or linear-lanceolate (proximally indurate, distally with sharply delimited green apical zones); cypselae 7–12(–18)-nerved, eglandular.

51. Involucres cylindric; rays 1–6, white; disc corollas white or cream; cypselae ± densely strigose . . . . 160. *Sericocarpus*, **v. 20, p. 101**

51. Involucres cylindro-campanulate or campanulate; rays 5–60, white to purple; disc corollas yellow; cypselae glabrous or ± densely strigillose.

52. Stems and leaves usually stipitate-glandular, sometimes eglandular and glaucous; leaves mostly cauline, entire or spinulose-serrate, glabrous or scabrellous; phyllaries usually keeled, sometimes rounded, apices acute to long-acuminate; rays 8–27; disc corolla tubes shorter than throats (w Cordilleras) . . . . 192. *Herrickia* (in part), **v. 20, p. 361**

52. Stems and leaves usually eglandular, sometimes stipitate-glandular (e North America only), not glaucous; leaves basal and/or cauline, serrate (teeth sometimes spinose, then blades linear, grasslike, se North America) or entire, hairy or glabrous; phyllaries usually rounded, sometimes keeled, apices obtuse to acute; rays 5–60; disc corolla tubes shorter or longer than throats . . . . 193. *Eurybia*, **v. 20, p. 365**

Key to Genera of Group 10.

Heads radiate; receptacles epaleate; ray corollas yellow, orange, red, or brown; pappi wholly of bristles.

1. Leaves opposite (at least proximally, or if mostly basal and subopposite or alternate, cypselae obcompressed, each shed with a subtending, linear, membranous scale).
    2. Leaves and/or phyllaries dotted or streaked with pellucid (schizogenous) glands containing strong-scented oils.
        3. Annuals or perennials; ray florets borne on bases of subtending phyllaries; style branches of bisexual disc florets knob-like . . . . 316. *Pectis* (in part), **v. 21, p. 222**
        3. Subshrubs or shrubs; ray florets borne on receptacles; style branches of bisexual disc florets linear . . . . 319. *Chrysactinia*, **v. 21, p. 232**
    2. Leaves and/or phyllaries not dotted or streaked (never with pellucid, schizogenous glands containing strong-scented oils, plants sometimes with sessile or stipitate, surface glands and may be otherwise strong-scented).
        4. Leaves (somewhat succulent) sessile or nearly so, filiform to linear, margins entire; phyllaries 2–5 in 1 series . . . . 325. *Haploësthes*, **v. 21, p. 245**
        4. Leaves (seldom succulent) usually petiolate, sometimes ± sessile, blades cordate, deltate, elliptic, lanceolate, narrowly trullate, linear, oblanceolate, oblong, obovate, ovate, cuneate, or spatulate, margins entire, toothed, or distally 3-lobed; phyllaries 5–23 in 1–3 series.
            5. Leaves all or nearly all opposite (distalmost cauline sometimes alternate and usually smaller).
                6. Phyllaries 8–23 in (1–)2 series (subequal); pappi persistent, of 10–50 bristles . . . . 361. *Arnica* (in part), **v. 21, p. 366**
                6. Phyllaries 14–18+ in ± 3 series (unequal); pappi fragile, of 6–8+ bristles . . . . 362. *Jamesianthus* (in part), **v. 21, p. 377**
            5. Leaves alternate and either proximal opposite or leaves mostly basal (sometimes subopposite).
                7. Phyllaries 8–22 in 2–3 series; cypselae obcompressed (each shed together with a subtending, linear, membranous scale, margins ciliate); pappi of distinct bristles . . . . 364. *Bartlettia*, **v. 21, p. 378**
                7. Phyllaries 5–8 in 1 series; cypselae narrowly obconic, clavate, or fusiform (margins not ciliate); pappi of basally connate or coherent bristles . . . . 365. *Syntrichopappus* (in part), **v. 21, p. 379**
1. Leaves all alternate.
    8. Style-branch appendages usually penicillate or essentially 0.
        9. Phyllaries in (2–)3–7+ series, unequal to subequal (distinct, calyculi 0); pappi of (± coarse) barbellate bristles.
            10. Annuals (pilosulous to hispid and stipitate-glandular, viscid); involucres 3–8 mm diam.; ray laminae 2–5(–7) mm . . . . 107. *Dittrichia*, **v. 19, p. 473**
            10. Perennials; involucres 10–40 mm diam.; ray laminae 10–30+ mm . . . . 108. *Inula* (in part), **v. 19, p. 473**
        9. Phyllaries in 1–2 series, equal to subequal (sometimes coherent, often subtended by calyculi); pappi of (fine) smooth or barbellulate bristles.
            11. Shrubs or vines.
                12. Vines (usually ± twining and climbing; corollas orange to ± brick-red; Florida) . . . . 220. *Pseudogynoxys*, **v. 20, p. 608**
                12. Shrubs.
                    13. Leaves (or lobes) linear (± evenly distributed) . . . . . 215. *Senecio* (in part), **v. 20, p. 544**
                    13. Leaves lance-elliptic or lanceolate to lance-linear (clustered distally on branches; Arizona and New Mexico) . . . . 226. *Barkleyanthus*, **v. 20, p. 614**
            11. Annuals, biennials, perennials, or subshrubs.
                14. Annuals.
                    15. Receptacles dome-shaped to conic (heights equaling or greater than diameters) . . . . 243. *Crocidium* (in part), **v. 20, p. 641**

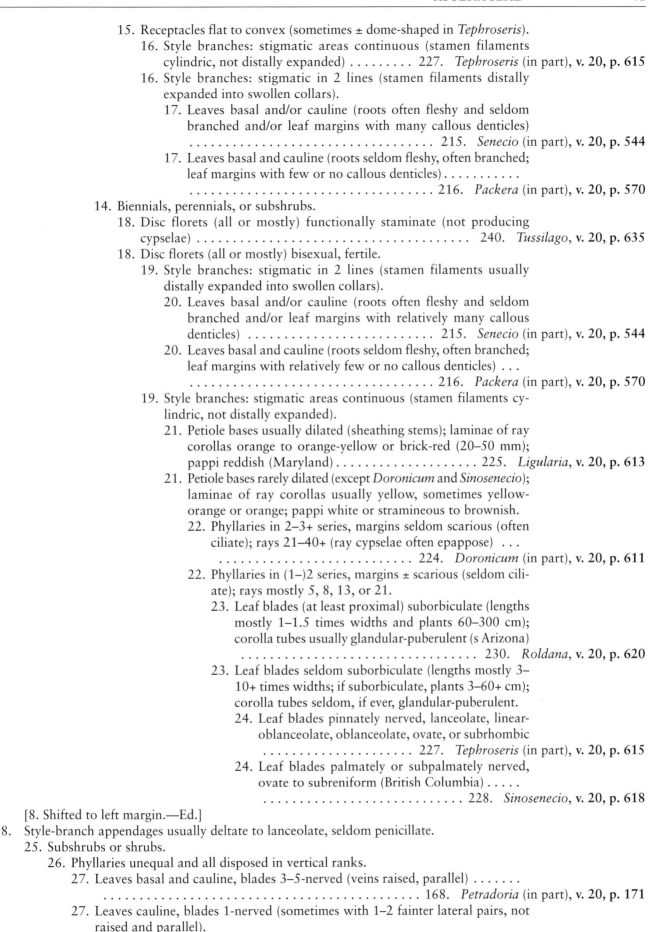

15. Receptacles flat to convex (sometimes ± dome-shaped in *Tephroseris*).
    16. Style branches: stigmatic areas continuous (stamen filaments cylindric, not distally expanded) . . . . 227. *Tephroseris* (in part), **v. 20, p. 615**
    16. Style branches: stigmatic in 2 lines (stamen filaments distally expanded into swollen collars).
        17. Leaves basal and/or cauline (roots often fleshy and seldom branched and/or leaf margins with many callous denticles) . . . . 215. *Senecio* (in part), **v. 20, p. 544**
        17. Leaves basal and cauline (roots seldom fleshy, often branched; leaf margins with few or no callous denticles) . . . . 216. *Packera* (in part), **v. 20, p. 570**
14. Biennials, perennials, or subshrubs.
    18. Disc florets (all or mostly) functionally staminate (not producing cypselae) . . . . 240. *Tussilago*, **v. 20, p. 635**
    18. Disc florets (all or mostly) bisexual, fertile.
        19. Style branches: stigmatic in 2 lines (stamen filaments usually distally expanded into swollen collars).
            20. Leaves basal and/or cauline (roots often fleshy and seldom branched and/or leaf margins with relatively many callous denticles) . . . . 215. *Senecio* (in part), **v. 20, p. 544**
            20. Leaves basal and cauline (roots seldom fleshy, often branched; leaf margins with relatively few or no callous denticles) . . . . 216. *Packera* (in part), **v. 20, p. 570**
        19. Style branches: stigmatic areas continuous (stamen filaments cylindric, not distally expanded).
            21. Petiole bases usually dilated (sheathing stems); laminae of ray corollas orange to orange-yellow or brick-red (20–50 mm); pappi reddish (Maryland) . . . . 225. *Ligularia*, **v. 20, p. 613**
            21. Petiole bases rarely dilated (except *Doronicum* and *Sinosenecio*); laminae of ray corollas usually yellow, sometimes yellow-orange or orange; pappi white or stramineous to brownish.
                22. Phyllaries in 2–3+ series, margins seldom scarious (often ciliate); rays 21–40+ (ray cypselae often epappose) . . . . 224. *Doronicum* (in part), **v. 20, p. 611**
                22. Phyllaries in (1–)2 series, margins ± scarious (seldom ciliate); rays mostly 5, 8, 13, or 21.
                    23. Leaf blades (at least proximal) suborbiculate (lengths mostly 1–1.5 times widths and plants 60–300 cm); corolla tubes usually glandular-puberulent (s Arizona) . . . . 230. *Roldana*, **v. 20, p. 620**
                    23. Leaf blades seldom suborbiculate (lengths mostly 3–10+ times widths; if suborbiculate, plants 3–60+ cm); corolla tubes seldom, if ever, glandular-puberulent.
                        24. Leaf blades pinnately nerved, lanceolate, linear-oblanceolate, oblanceolate, ovate, or subrhombic . . . . 227. *Tephroseris* (in part), **v. 20, p. 615**
                        24. Leaf blades palmately or subpalmately nerved, ovate to subreniform (British Columbia) . . . . 228. *Sinosenecio*, **v. 20, p. 618**
[8. Shifted to left margin.—Ed.]
8. Style-branch appendages usually deltate to lanceolate, seldom penicillate.
  25. Subshrubs or shrubs.
    26. Phyllaries unequal and all disposed in vertical ranks.
      27. Leaves basal and cauline, blades 3–5-nerved (veins raised, parallel) . . . . 168. *Petradoria* (in part), **v. 20, p. 171**
      27. Leaves cauline, blades 1-nerved (sometimes with 1–2 fainter lateral pairs, not raised and parallel).

28. Involucres cylindric, 5–6 mm; phyllaries 8–12, stramineous, flat, midnerves orange-resinous; rays 1–3; disc florets 2–5; cypselae densely strigoso-sericeous . . . . 161. *Chrysoma* (in part), **v. 20, p. 105**
28. Involucres cylindric, obconic, or hemispheric, 4–15 mm; phyllaries 13–30, green to tan, flat to keeled, midnerves sometimes evident, sometimes enlarged subapically and glandular (not orange-resinous); rays (1–)6–8; disc florets 4–15; cypselae glabrous or densely hairy . . . .
. . . . 171. *Lorandersonia* (in part), **v. 20, p. 177**
[26. Shifted to left margin.—Ed.]
26. Phyllaries equal or unequal and usually disposed in spirals.
29. Plants spinescent (at least with age) . . . . 174. *Amphipappus* (in part), **v. 20, p. 185**
29. Plants not spinescent.
30. Basal leaves pinnatifid (lobes bristle-tipped); pappus bristles basally flattened
. . . . 195. *Xanthisma* (in part), **v. 20, p. 383**
30. Basal leaves not pinnatifid; pappus bristles not basally flattened.
31. Leaves entire or toothed (bases clasping or subclasping); heads usually in spiciform, racemiform, paniculiform, subumbelliform, or corymbiform arrays, rarely borne singly.
32. Leaves (teeth sometimes bristly) glabrous or densely tomentose (hairs not flagelliform); heads in spiciform, racemiform, narrowly paniculiform, or corymbiform arrays; phyllaries in 5–9 series; pappi of 20–30 reddish brown bristles in 1–2 series . . . . 206. *Hazardia* (in part), **v. 20, p. 445**
32. Leaves (teeth not bristly) densely short-woolly (hairs flagelliform); heads in subumbelliform to paniculiform arrays; phyllaries in 3–5 series; pappi of outer, triangular scales plus 20–40 inner bristles in (2–)3 series; e North America . . . . 182. *Chrysopsis* (in part), **v. 20, p. 213**
31. Leaves entire (bases not clasping); heads borne singly or in cymiform, spiciform, corymbiform, or racemiform (sometimes paniculiform or thyrsiform) arrays.
33. Shrubs.
34. Involucres hemispheric, obconic, or cylindric; disc corolla lobes equal; cypselae prismatic . . . . 148. *Ericameria* (in part), **v. 20, p. 50**
34. Involucres turbinate; disc corolla lobes unequal; cypselae ± turbinate
. . . . 152. *Neonesomia* (in part), **v. 20, p. 85**
33. Subshrubs.
35. Stems prostrate to erect, mat-forming, branched; leaves cauline (crowded); heads borne singly . . . . 167. *Nestotus*, **v. 20, p. 169**
35. Stems erect, not mat-forming, branched or simple; leaves basal and cauline or mostly cauline (then not crowded); heads in spiciform, racemiform, or corymbiform arrays, or glomerate and/or pedunculate-solitary in flat-topped or multi-storied, corymbiform arrays.
36. Plants rhizomatous; leaves 1–3(–5)-nerved, linear to lanceolate, gland-dotted; heads usually glomerate, sometimes pedunculate-solitary, in flat-topped or multi-storied, corymbiform arrays
. . . . 158. *Euthamia* (in part), **v. 20, p. 97**
36. Plants stoutly taprooted; leaves 1-nerved, oblanceolate, eglandular or obscurely gland-dotted; heads (not glomerate) in racemiform or loose, spiciform or corymbiform arrays . . . . . 165. *Columbiadoria*, **v. 20, p. 167**

[25. Shifted to left margin.—Ed.]
25. Annuals, biennials, or perennials.
    37. Pappi of easily overlooked setae or scales plus inner, longer bristles in 2+ series.
        38. Plants taprooted or with simple caudices (and fibrous-rooted); heads borne singly or in 2s or 3s; phyllaries in 2–3 series, equal or subequal, flat, usually 1–3-nerved (nerves golden-resinous); cypselae 2-nerved, not resinous . . . . . 186. *Erigeron* (in part), **v. 20, p. 256**
        38. Plants taprooted, rhizomatous, or with branched caudices; heads usually in corymbiform, ± paniculiform, or subumbelliform arrays, sometimes borne singly; phyllaries in 3–5 series, unequal, usually thickened or keeled (not in *Bradburia*), 1-nerved (nerves not golden-resinous); cypselae smooth or 1–14-nerved or -ribbed, nerves or ribs sometimes resinous.
            39. Stems and leaves silky-sericeous, rarely glabrate; leaves sessile, blades 3–11-parallel-nerved (nerves sunken), linear to lanceolate or ovate (often grasslike), margins entire; involucres turbinate . . . . 183. *Pityopsis* (in part), **v. 20, p. 222**
            39. Stems and leaves whitish-strigose, pilose, or hispid, or arachnose to woolly (hairs flagelliform, soft), or glabrous; leaves sessile or basal petiolate, blades usually 1-nerved (veins reticulate, raised abaxially), spatulate, ovate-oblanceolate, ovate, elliptic, elliptic-oblong, oblanceolate, linear-lanceolate, or (usually distal) linear, margins entire, serrate, or dentate (sometimes coarsely ciliate); involucres campanulate or turbinate.
                40. Plants proximally woolly, distally arachnose or pilose (hairs flagelliform); basal leaves sessile . . . . 182. *Chrysopsis* (in part), **v. 20, p. 213**
                40. Plants sparsely to ± densely hispid, strigose, or pilose (hairs not flagelliform); basal leaves petiolate.
                    41. Perennials (taproots relatively short and/or caudices woody); basal petioles ciliate; cauline leaves much reduced distally, not clasping; heads borne singly or in lax, paniculiform arrays; cypselae monomorphic, ray pappi of outer, shorter, setiform scales or bristles plus inner, longer bristles . . . . 181. *Bradburia,* **v. 20, p. 211**
                    41. Perennials (caudices woody); basal petioles long-strigoso-ciliate; cauline leaves ± reduced distally, sometimes clasping or subclasping; heads borne singly or in corymbiform, sometimes paniculiform arrays; cypselae dimorphic or monomorphic, ray pappi 0, or of outer, linear-lanceolate scales plus inner, longer bristles . . . . 185. *Heterotheca* (in part), **v. 20, p. 230**
    37. Pappi of bristles in 1–4 series (seldom with notably shorter bristles, scales, or setae).
        42. Cypselae dimorphic (rays often ± 3-angled, discs ± compressed).
            43. Heads borne singly or in corymbiform arrays (peduncles not cobwebby); involucres 4–10 mm; disc corolla throats gradually ampliate, ± funnelform; style-branch appendages lanceolate . . . . 195. *Xanthisma* (in part), **v. 20, p. 383**
            43. Heads borne singly or (2–3) in paniculiform or subcorymbiform-cymiform arrays (peduncles often cobwebby); involucres 7–16 mm; disc corolla throats abruptly ampliate, funnelform; style-branch appendages deltate . . . . 204. *Rayjacksonia,* **v. 20, p. 437**
        42. Cypselae monomorphic (all ± compressed or all ± 3-angled).
            44. Leaves 3-nerved (nerves ± parallel), faces thin-arachnose (in minute, abaxial lacunae) . . . . 184. *Croptilon* (in part), **v. 20, p. 228**
            44. Leaves usually 1-nerved (sometimes 1–5-nerved), faces glabrous or ± hairy (not thin-arachnose in abaxial lacunae).
                45. Cypselae compressed, each edge 1-nerved . . . . 142. *Laënnecia* (in part), **v. 20, p. 36**
                45. Cypselae sometimes ± compressed, each edge not 1-nerved.
                    46. Annuals.
                        47. Heads borne singly (at tips of branches); involucres eglandular; cypselae oblanceoloid or fusiform and (3–)5-ribbed or -nerved.

48. Involucres campanulate to turbinate; phyllaries equal or subequal; cypselae oblanceoloid, 3–5-nerved, not beaked; pappus bristles (3–)5–20 (usually in multiples of 5) in 1 series . . . . 145. *Pentachaeta* (in part), **v. 20, p. 46**

48. Involucres ± cylindric to turbinate or obconic; phyllaries unequal; cypselae ± fusiform, 5-nerved, beaked; pappus bristles (12–)30–40 in (1–)2 series (outer shorter) . . . . 147. *Tracyina*, **v. 20, p. 50**

47. Heads borne singly or in corymbiform arrays; involucres glandular; cypselae clavate and 3-nerved or obconic and obscurely nerved.

49. Disc florets functionally staminate . . . . 207. *Benitoa* (in part), **v. 20, p. 450**

49. Disc florets bisexual, fertile (corollas of peripheral florets sometimes palmately expanded, resembling rays) . . . . 209. *Lessingia* (in part), **v. 20, p. 452**

[46. Shifted to left margin.—Ed.]

46. Perennials.

50. Heads usually glomerate and/or sometimes pedunculate-solitary in flat-topped or multistoried, corymbiform arrays . . . . 158. *Euthamia* (in part), **v. 20, p. 97**

50. Heads borne singly or in spiciform, racemiform, paniculiform, or corymbiform arrays (if glomerate, not in flat-topped, corymbiform arrays).

51. Plants rhizomatous (often colonial); heads in dense corymbiform or paniculiform arrays.

52. Stems and leaves sometimes stipitate-glandular or gland-dotted; heads in rounded, club-shaped, wand-shaped, or pyramid-shaped paniculiform (often secund) arrays, or in flat-topped, corymbiform arrays; involucres cylindric to campanulate, 3–12 mm, sometimes stipitate-glandular; phyllary midveins usually swollen, translucent, apices often with green zone, sometimes reflexed; rays 3–15(–21); disc corolla lobes lanceolate, erect to reflexed, style-branch appendages triangular; cypselae obconic, sometimes ± compressed, 5–8-nerved . . . . 163. *Solidago* (in part), **v. 20, p. 107**

52. Stems and leaves stipitate-glandular; heads in dense, flat-topped, corymbiform arrays; involucres campanulate to hemispheric, 10–11 mm, stipitate-glandular; phyllary midveins not swollen, apices green-tipped and spreading-reflexed; rays 12–20; disc corolla lobes triangular, spreading, style-branch appendages linear; cypselae fusiform, distinctly compressed, 12–16-nerved (nerves whitish, raised; w United States) . . . . 164. *Oreochrysum*, **v. 20, p. 166**

51. Plants usually taprooted or with caudices, sometimes also from spreading roots (stems single or clustered); heads usually borne singly, sometimes (2–5) in loose, corymbiform, cymiform, or paniculiform arrays, or in spiciform, racemiform, paniculiform, or corymbiform arrays, or glomerate (some *Oönopsis*).

53. Heads usually in rounded, club-shaped, wand-shaped, or pyramid-shaped, paniculiform (often secund) arrays, or flat-topped, corymbiform arrays, or in spiciform, racemiform, or cymiform arrays, sometimes borne singly.

54. Plants with caudices (stems clustered); leaves basal and cauline; heads in rounded, club-shaped, wand-shaped, or pyramid-shaped paniculiform (often secund) arrays, or flat-topped, corymbiform arrays; involucres campanulate to cylindric, 3–12 mm; phyllary midveins usually swollen and translucent; cypselae 5–8-nerved . . . . 163. *Solidago* (in part), **v. 20, p. 107**

54. Plants taprooted; leaves mostly cauline; heads usually in spiciform, racemiform, or cymiform arrays, sometimes borne singly; involucres campanulate, 11–13 mm; phyllary midveins not swollen; cypselae 4–5-nerved; California, Oregon . . . . 206. *Hazardia* (in part), **v. 20, p. 445**

53. Heads borne singly or (2–15) in spiciform, racemiform, paniculiform, or corymbiform arrays.

55. Pappi brownish.

56. Stems and leaves glabrous or sparsely tomentose, eglandular; leaves basal (usually withering by flowering) and cauline, blades narrowly oblanceolate to lanceolate or linear, margins entire; heads borne singly, or (2–12) in glomerules or in loose, corymbiform arrays (subtended by little-reduced distal leaves); rays 6–25; cypselae prismatic or narrowly turbinate . . . . 200. *Oönopsis* (in part), **v. 20, p. 410**

56. Stems and leaves loosely tomentose to woolly, sometimes gland-dotted or stipitate-glandular; leaves basal (persistent) and cauline, basal blades oblanceolate to elliptic or linear, cauline lanceolate, margins entire, spinulose-dentate or -serrate, or shallowly laciniate; heads borne singly or (2–15, ± sessile) in racemiform, spiciform, or loose, corymbiform arrays (at ends of scapiform stems or peduncles); rays 10–80; cypselae subcylindric-fusiform . . . . 201. *Pyrrocoma* (in part), **v. 20, p. 413**

55. Pappi whitish or stramineous.

57. Plants 1–2 cm (± pulvinate), not mat-forming; leaves 1-nerved; heads ± sessile; phyllary margins ± scarious . . . . 176. *Townsendia* (in part), **v. 20, p. 193**

57. Plants 1–30(–60) cm, sometimes mat-forming; leaves 1–5-nerved; margins of outer phyllaries herbaceous (proximally indurate).

58. Stems eglandular or stipitate-glandular, sometimes resinous; leaf margins entire, faces glabrous, scabrous, villous, or lanate, usually stipitate-glandular, sometimes eglandular; phyllaries unequal; rays 5–17; cypselae usually sericeous, sometimes glabrous . . . 170. *Stenotus*, **v. 20, p. 174**

58. Stems densely stipitate-glandular (viscid); leaf margins entire, toothed, or lobed, faces glabrous or scabrous, sometimes stipitate-glandular; phyllaries subequal (outer foliaceous); rays 11–23(–35); cypselae glabrous or villous . . . . 172. *Tonestus* (in part), **v. 20, p. 181**

Key to Genera of Group 11

Heads radiate; receptacles epaleate; pappi wholly, or partially, of awns or scales.

1. Leaves and/or phyllaries dotted or streaked with pellucid (schizogenous) glands containing strong-scented oils.
    2. Leaves opposite (blade margins proximally bristly-ciliate; ray florets borne on bases of subtending phyllaries; style branches of bisexual florets knob-like) . . . 316. *Pectis* (in part), **v. 21, p. 222**
    2. Leaves opposite or mostly alternate (blades often lobed, margins sometimes bristly-ciliate; ray florets borne on receptacles, not on bases of subtending phyllaries; style branches of bisexual florets linear).
        3. Phyllaries distinct to bases or nearly so.
            4. Rays yellow to orange (pappus scales multiaristate) . . . . 317. *Dyssodia*, **v. 21, p. 230**
            4. Rays whitish with pinkish or purplish stripes (pappi of scales plus bristles) . . . . 318. *Nicolletia*, **v. 21, p. 231**
        3. Phyllaries ± connate $^1/_3$–$^7/_8$+ their lengths (margins of outer sometimes distinct to bases).
            5. Calyculi 0; pappi of 2–5(–10) scales in ± 1 series (usually 0–5+ oblong to lanceolate, erose-truncate to laciniate, plus 0–2+, longer, subulate or aristate, some or all sometimes connate) . . . . 321. *Tagetes* (in part), **v. 21, p. 235**
            5. Calyculi usually of (1–)5–8(–22) bractlets, rarely 0; pappi usually of 8–20 scales in 2 series (rarely coroniform).
                6. Bractlets of calyculi subulate or pectinate . . . . 322. *Dysodiopsis*, **v. 21, p. 237**
                6. Bractlets of calyculi deltate to subulate, not pectinate.
                    7. Plants (20–)30–70+ cm; involucres (7–18 ×) 5–12 mm; phyllaries weakly connate $^1/_3$–$^2/_3$ their lengths . . . . 323. *Adenophyllum*, **v. 21, p. 237**
                    7. Plants mostly (1–)5–30 cm; involucres (4–6 ×) 2–7 mm; phyllaries strongly connate $^2/_3$–$^7/_8$ their lengths (margins of outer sometimes distinct to bases) . . . . 324. *Thymophylla*, **v. 21, p. 239**

1. Leaves and phyllaries rarely dotted or streaked (never with pellucid, schizogenous glands containing strong-scented oils, plants sometimes with sessile or stipitate, surface glands, sometimes strong-scented).
    8. Leaves opposite, or proximal opposite (sometimes 2–3+ pairs) and distal alternate.
        9. Cypselae compressed or flattened, usually callous-margined and ± ciliate.
            10. Phyllaries connate ²/₃+ their lengths . . . . 354. *Lasthenia* (in part), **v. 21, p. 336**
            10. Phyllaries distinct to bases.
                11. Annuals, perennials, subshrubs, or shrubs, 2–45(–75) cm (usually glabrous or hairy other than woolly); leaves often lobed; involucres campanulate, cylindric, funnelform, or hemispheric, 3–15 mm diam.; disc florets 5–200+
                . . . . 352. *Perityle* (in part), **v. 21, p. 317**
                11. Annuals 1–3(–5) cm (woolly); involucres campanulate, 3–5 mm diam.; disc florets 7–12+ . . . . 356. *Eatonella*, **v. 21, p. 348**
        9. Cypselae (at least disc) mostly columnar or obpyramidal and 4–5-angled, or clavate, columnar, cylindric, or obconic and 10–15-ribbed (not both compressed and with ± ciliate margins).
            12. Disc cypselae mostly clavate, columnar, cylindric, or obconic and 10–15-ribbed.
                13. Heads in tight, corymbiform aggregations or arrays; rays 0, 1, or 3–5.
                    14. Rays 3–5; pappi of 5 erose scales alternating with 5 setiform scales or bristles (sometimes all 10 elements basally connate) . . . . . 326. *Sartwellia*, **v. 21, p. 246**
                    14. Rays 0 or 1; pappi of 2–4 hyaline scales . . . . 327. *Flaveria* (in part), **v. 21, p. 247**
                13. Heads borne singly; rays 2–4 or 10–15.
                    15. Perennials; pappi of 1–5 scales . . . . 330. *Jaumea* (in part), **v. 21, p. 253**
                    15. Subshrubs or shrubs; pappi of 12–25, or ca. 50 scales.
                        16. Rays 10–15; pappi of 12–25 subulate scales in 1 series . . . .
                        . . . . 328. *Clappia*, **v. 21, p. 251**
                        16. Rays 2–4; pappi of ca. 50 setiform scales in 3–4 series . . . .
                        . . . . 329. *Pseudoclappia* (in part), **v. 21, p. 252**
            12. Disc cypselae mostly columnar or obpyramidal and 4–5-angled (sometimes obovoid, smooth in *Lasthenia*; 2- or 4-angled in *Monolopia*).
                17. Ray laminae ca. 0.5 mm (inconspicuous); disc cypselae 2-angled (pappi of 2–7 scales) . . . . 357. *Monolopia* (in part), **v. 21, p. 349**
                17. Ray laminae mostly 1–25+ mm (usually conspicuous).
                    18. Disc corolla lobes lance-linear, lance-oblong, or linear (equal or unequal).
                        19. Disc corollas creamy to bright yellow . . . 371. *Hymenothrix* (in part), **v. 21, p. 387**
                        19. Disc corollas pinkish, purplish, or whitish. . . 372. *Palafoxia* (in part), **v. 21, p. 388**
                    18. Disc corolla lobes mostly deltate to lance-deltate or lance-ovate.
                        20. Leaves usually sessile, sometimes petiolate; phyllaries distinct or partially connate (pappus scales not notably medially thickened).
                            21. Leaves opposite ± throughout, faces usually glabrous or glabrate (sparsely woolly to villous in *L. minor* and *L. platycarpha*) . . . . 354. *Lasthenia* (in part), **v. 21, p. 336**
                            21. Leaves opposite proximally, distal usually alternate, faces usually densely to sparsely woolly . . . . . . 359. *Eriophyllum* (in part), **v. 21, p. 353**
                        20. Leaves usually petiolate; phyllaries distinct (pappus scales usually notably medially thickened).
                            22. Phyllaries 4–9(–12), margins often purplish or yellowish)
                            . . . . 367. *Schkuhria* (in part), **v. 21, p. 381**
                            22. Phyllaries 6–18+ (margins rarely purplish, not yellowish).
                                23. Annuals, biennials, or perennials, 10–80+ cm; leaves all or mostly opposite (if perennials, blades or lobes lanceolate to oblong, 2–20+ mm wide), or all or mostly alternate (proximal opposite) . . . . 368. *Bahia* (in part), **v. 21, p. 383**
                                23. Perennials, 3–20+ cm; leaves all or mostly opposite (blades or lobes lanceolate to lance-linear, mostly 1–8 mm wide)
                                . . . . 369. *Picradeniopsis*, **v. 21, p. 384**

[8. Shifted to left margin.—Ed.]

8. Leaves alternate.
   24. Pappi partially of scales (scales subtended by, subtending, or alternating with bristles).
      25. Rays mostly white, sometimes wholly or partially blue, lilac, maroon, pink, purple, or violet.
         26. Phyllary mid-nerves usually orange.
            27. Plants not colonial (stems sometimes ± clustered); stems often stipitate-glandular; basal leaves persistent or withering by flowering; heads borne singly or (2–10) in corymbiform arrays; pappi usually of outer setae or scales plus 5–40(–50) inner bristles . . . . 186. *Erigeron* (in part), **v. 20, p. 256**
            27. Plants colonial; stems eglandular; basal leaves withering by flowering; heads in corymbiform or diffuse, paniculiform arrays; pappi of 2–3 awns plus shorter bristles or scales, or wholly of minute scales . . 189. *Boltonia* (in part), **v. 20, p. 353**
         26. Phyllary mid-nerves not orange.
            28. Pappi of scales alternating with bristles.
               29. Phyllaries 10–50 in 2–6 series; disc florets 5–25 . . . . 179. *Chaetopappa* (in part), **v. 20, p. 206**
               29. Phyllaries 10–14 in (1–)2 series; disc florets 28–40 . . . . 180. *Monoptilon* (in part), **v. 20, p. 210**
            28. Pappi of scales subtending bristles.
               30. Leaf blades lance-elliptic, linear, oblong, or spatulate . . . . 150. *Ionactis* (in part), **v. 20, p. 82**
               30. Leaf blades cordate-deltate to orbiculate or polygonally lobed (plants often persisting after cultivation) . . . . 219. *Pericallis* (in part), **v. 20, p. 607**
      25. Rays mostly yellow to orange, sometimes partially (rarely wholly) blue, brown, lilac, maroon, pink, purple, red, or violet.
         31. Pappi of (25–40) bristles subtending (8–15) subulate scales . . . . 202. *Grindelia* (in part), **v. 20, p. 424**
          31. Pappi of scales (often setiform, inconspicuous) subtending bristles.
            32. Heads borne singly or in 2s or 3s; phyllaries in 2–3(–4) series, usually 1- or 3-nerved (nerves golden-resinous), equal or subequal; cypselae 2-ribbed (each margin thickened, not resinous) . . . . 186. *Erigeron* (in part), **v. 20, p. 256**
            32. Heads usually in corymbiform, ± paniculiform, or subumbelliform arrays, sometimes borne singly; phyllaries in 3–5 series, 1-nerved (nerves not golden-resinous), unequal to subequal; cypselae smooth or 1–14-nerved or -ribbed (nerves or ribs sometimes resinous).
               33. Leaves sessile, blades 3–11-parallel-nerved (nerves sunken), mostly lanceolate, linear, or ovate (often grasslike), margins entire, faces usually silky-sericeous, rarely glabrate; involucres turbinate . . . . 183. *Pityopsis* (in part), **v. 20, p. 222**
               33. Leaves sessile or petiolate (basal), blades usually 1-nerved, mostly elliptic, elliptic-oblong, lance-linear, linear, oblanceolate, ovate, ovate-oblanceolate, or spatulate, margins entire, dentate, or serrate (sometimes ciliate), faces usually arachnose, hispid, pilose, strigose, or woolly, sometimes glabrate or glabrous; involucres campanulate, hemispheric, or turbinate.
                  34. Anthers tailed (leaf blades oblong to narrowly oblanceolate, 1–3 cm × 2–7 mm, bases clasping, margins entire, ± revolute; phyllaries lance-linear to linear, 3–5 mm, pilosulous; rays 10–30, laminae 1.5–2+ mm) . . . . 106. *Pulicaria*, **v. 19, p. 471**
                  34. Anthers not tailed.
                     35. Plants proximally woolly, distally arachnose or pilose; basal leaves sessile . . . . 182. *Chrysopsis* (in part), **v. 20, p. 213**
                     35. Plants sparsely to ± densely hispid, strigose, or pilose; basal leaves petiolate.

36. Basal leaves: petioles ciliate; cauline leaves much reduced distally (bases not clasping); heads borne singly or in lax, paniculiform arrays; ray pappi of outer, setiform scales (0.3–1 mm) plus 20–35 inner bristles (3–6 mm) . . . . 181. *Bradburia* (in part), **v. 20, p. 211**

36. Basal leaves: petioles long-strigoso-ciliate; cauline leaves ± reduced distally (bases sometimes clasping or subclasping); heads usually in corymbiform, sometimes paniculiform, arrays, sometimes borne singly; ray pappi 0, or of outer, linear to linear-lanceolate or triangular scales (0.2–1 mm) plus 30–45 inner bristles (3–11 mm) . . . . 185. *Heterotheca* (in part), **v. 20, p. 230**

[24. Shifted to left margin.—Ed.]

24. Pappi wholly of scales (scales sometimes aristate, not associated with distinct bristles).

37. Styles (in bisexual, fertile florets) usually distally dilated and ± cylindric, style branches linear (sometimes adhering almost to minutely parted tips or essentially lacking, style-branch appendages essentially none; plants usually persisting after cultivation).

38. Phyllaries connate $^1/_3$–$^3/_4$ their lengths; ray laminae 5-nerved, 4-lobed or -toothed . . . . 25. *Gazania*, **v. 19, p. 196**

38. Phyllaries distinct; ray laminae 4-nerved, 3-lobed or -toothed.

39. Ray florets neuter, corollas adaxially yellow (sometimes drying bluish) or ± bluish; pappi of 7–8+ scales ca. 0.5 mm (usually hidden by cypsela hairs) . . . . 26. *Arctotheca*, **v. 19, p. 197**

39. Ray florets pistillate, corollas adaxially whitish to purplish, or yellow to orange (then purple at bases); pappi (0 or) of 5–8 scales 0.5–4 mm (usually not hidden by cypsela hairs) . . . . 27. *Arctotis* (in part), **v. 19, p. 198**

37. Styles usually ± filiform (not distally dilated), style branches mostly lanceolate or linear (not adhering almost to tips, style-branch appendages either deltate, lanceolate, penicillate, or terete, or essentially none).

40. Rays mostly white, sometimes wholly or partially blue, lilac, maroon, pink, purple, or violet.

41. Leaves (1–)2–3-pinnately lobed.

42. Perennials; heads usually in lax to dense, corymbiform arrays, rarely borne singly . . . . 112. *Tanacetum* (in part), **v. 19, p. 489**

42. Annuals; heads borne singly or in open, corymbiform arrays . . . . 124. *Matricaria* (in part), **v. 19, p. 540**

41. Leaves usually not lobed, sometimes 1-pinnately lobed.

43. Phyllaries mostly 5–13 in 1–2+ series; rays 5–8.

44. Biennials and perennials; pappus scales 12–18 . . . . 351. *Hymenopappus* (in part), **v. 21, p. 309**

44. Annuals; pappus scales 20–40 (setiform, basally coherent or connate, falling together or in groups) . . . . 365. *Syntrichopappus* (in part), **v. 21, p. 379**

43. Phyllaries (4–)13–80 in 2–7+ series; rays (1–)10–67.

45. Leaves all or mostly basal (blades linear, entire; heads borne singly; receptacles villous) . . . . 120. *Hulteniella* (in part), **v. 19, p. 534**

45. Leaves basal and cauline or mostly cauline.

46. Phyllaries usually 13 or 21 in (1–)2 series, subequal (leaf blades palmately nerved, cordate-deltate to orbiculate or polygonally lobed, margins dentate to denticulate, faces sparsely hairy; plants often persisting after cultivation) . . . . 219. *Pericallis* (in part), **v. 20, p. 607**

46. Phyllaries 4–80 in 2–7+ series, subequal to unequal.

47. Heads in corymbiform to paniculiform arrays (cypselae usually strongly compressed, often winged) . . . 189. *Boltonia* (in part), **v. 20, p. 353**

47. Heads usually borne singly (sometimes clusters of 3–6 in *Gutierrezia*).

48. Cypselae ellipsoid, oblanceolate, obovate, or obovoid, strongly compressed to ± flattened (usually hairy, hair tips often glochidiform).
    49. Cypsela margins sometimes thickened (not winglike); pappi of 12–35 lanceolate or subulate to setiform scales . . . . 176. *Townsendia* (in part), **v. 20, p. 193**
    49. Cypsela margins sometimes ± winglike (wings glabrous); pappi of 2 barbellate awns . . . . 178. *Dichaetophora*, **v. 20, p. 205**
48. Cypselae angular-columnar, or ± columnar, clavate, cylindric, or oblanceoloid, usually subterete, sometimes slightly compressed (sometimes with hair tips glochidiform).
    50. Leaves oblanceolate to lance-linear, usually pinnatifid, sometimes entire . . . . . 188. *Aphanostephus* (in part), **v. 20, p. 351**
    50. Leaves usually filiform, lanceolate, linear, oblanceolate, oblong, or spatulate, toothed or entire (margins sometimes ciliate).
        51. Shrubs (leaves mostly clustered distally on stems) . . . . 135. *Nipponanthemum*, **v. 19, p. 555**
        51. Annuals.
            52. Plants 2–17 cm; leaves glabrous or sparsely pilose . . . . 145. *Pentachaeta* (in part), **v. 20, p. 46**
            52. Plants 12–30 cm; leaves glabrous or minutely hairy, gland-dotted (glutinous) . . . . 155. *Gutierrezia* (in part), **v. 20, p. 88**

[40. Shifted to left margin.—Ed.]

40. Rays mostly yellow to orange, sometimes partially (rarely wholly) blue, brown, lilac, maroon, pink, purple, red, or violet.
53. Ray corollas marcescent (usually becoming reflexed, dry, persisting past flowering).
    54. Heads usually in close corymbiform or glomerulate clusters; involucres mostly campanulate, cylindric, or obconic; rays 1–8; disc florets 5–25+ . . . . 390. *Psilostrophe*, **v. 21, p. 453**
    54. Heads borne singly; involucres campanulate, hemispheric, or rotate; rays 2–55; disc florets 10–250+.
        55. Leaves basal and cauline; blades lance-linear to broadly ovate, sometimes pinnately lobed or pinnatifid, faces usually floccose-woolly; phyllaries 8–13, or 21–34, in 2 series (abaxial faces floccose-tomentose); rays 5–7 or 20–55; disc florets 10–20 or 40–100+ . . . . 388. *Baileya* (in part), **v. 21, p. 444**
        55. Leaves all basal, or basal-proximal, or basal and cauline; blades mostly oblanceolate to linear or filiform, sometimes lobed, faces glabrous or ± hairy, eglandular or ± gland-dotted; phyllaries 11–60+ in 3 series (abaxial faces ± hairy); rays 7–27; disc florets 20–250+ . . . . 389. *Tetraneuris* (in part), **v. 21, p. 447**
53. Ray corollas not marcescent (usually withering and falling after flowering).
56. Leaves mostly 2–3-pinnately lobed.
    57. Perennials; phyllaries (20–)30–60 in (2–)3–5+ series; rays 10–21+; disc florets 60–300+ . . . . 112. *Tanacetum* (in part), **v. 19, p. 489**
    57. Subshrubs; phyllaries 8–16 in 2 series; rays 4–9; disc florets 10–25 . . . . 360. *Constancea*, **v. 21, p. 362**
56. Leaves usually not lobed, sometimes 3-lobed or 1-pinnately lobed (1–2-ternately lobed in some species of *Bahia*).
58. Disc florets functionally staminate (not producing fruits).
    59. Annuals; rays 5–15.
        60. Phyllaries 12–15 in 1–2(–3) series . . . . 154. *Amphiachyris* (in part), **v. 20, p. 87**
        60. Phyllaries 22–35 in 5–6 series . . . . 207. *Benitoa* (in part), **v. 20, p. 450**
    59. Perennials or shrubs; rays 1–5.
        61. Shrubs; disc florets 3–7; pappus scales 15–20 . . . . 174. *Amphipappus* (in part), **v. 20, p. 185**
        61. Perennials; disc florets 6–15; pappus scales 2–6 . . . . 386. *Hymenoxys* (in part), **v. 21, p. 435**

58. Disc florets bisexual, fertile (producing fruits).
[62. Shifted to left margin.—Ed.]

62. Disc corollas often brown-purple to red-brown or tipped with brown-purple to red-brown
(tubes much shorter than abruptly much-dilated, urceolate to campanulate throats, lobes
often shaggily hairy, hairs ± moniliform).

    63. Stems not winged (receptacles usually with setiform enations; style-branch apices
± attenuate). . . . . 384. *Gaillardia* (in part), **v. 21, p. 421**

    63. Stems often winged (by decurrent leaf bases; receptacles rarely with setiform enations;
style-branch apices penicillate or truncate) . . . . 385. *Helenium* (in part), **v. 21, p. 426**

62. Disc corollas usually uniformly yellow to cream, sometimes purplish to reddish (tubes much
shorter than to about equaling slightly dilated, funnelform to cylindric throats, lobes not
shaggily hairy with moniliform hairs).

    64. Receptacles deeply pitted (each cypsela nested within a 5–6-sided cell) . . . . . 382. *Balduina*, **v. 21, p. 419**

    64. Receptacles smooth or ± pitted (cypselae not nested within cells; outer disc florets rarely
subtended by paleae in *Amblyolepis*).

        65. Phyllaries (7–)20–60 in 2–4+ series, mostly unequal (laminae of ray corollas
usually coiling in age).

            66. Pappus scales 18–22.

                67. Shrubs (20–40 cm); rays 5–14 . . . . 173. *Acamptopappus* (in part), **v. 20, p. 184**

                67. Perennials (pulvinate, 1–2 cm); rays 13–21 . . . . . . 176. *Townsendia* (in part), **v. 20, p. 193**

            66. Pappus scales 4–12.

                68. Cypselae clavate to linear (compressed, lenticular in cross section), silky
hairy; pappi of 4 quadrate to spatulate scales (leaves usually thinly lanate
to densely woolly and/or gland-dotted, glandular-puberulent, glandular-
villous, or stipitate-glandular) . . . . 377. *Hulsea*, **v. 21, p. 396**

                68. Cypselae clavate, cylindric, oblanceoloid, oblong, or ovoid, terete or com-
pressed, usually (3–)5-ribbed, 5–8-nerved, or 4–6 sided, mostly glabrous
or sparsely strigillose or strigose; pappi of 5–20 irregular or awn-like scales.

                    69. Leaf margins entire or spinulose toothed; cypselae terete or slightly
compressed with rounded edges or 4–6-sided, without prominent
nerves, glabrous or slightly strigose . . . . . 203. *Xanthocephalum* (in part), **v. 20, p. 436**

                    69. Leaf margins entire; cypselae clavate, cylindric, oblanceoloid, oblong,
or ovoid, sometimes compressed, usually (3–)5-ribbed or 5–8-nerved,
mostly glabrous or sparsely strigillose or strigose.

                        70. Annuals; receptacles flat to convex (without hooked hairs); disc
corollas 3- or 5-lobed; cypselae (3–)5-ribbed. . . . .
. . . . 145. *Pentachaeta* (in part), **v. 20, p. 46**

                        70. Annuals, perennials, or subshrubs; receptacles flat to conic (with
hooked hairs); disc corollas 5-lobed; cypselae 5–8-nerved. . . . .
. . . . 155. *Gutierrezia* (in part), **v. 20, p. 88**

        65. Phyllaries 5–15(–40) in 1–2 series, mostly subequal to equal (laminae of ray corol-
las rarely coiling in age).

            71. Pappus scales 30–40 (setiform, basally coherent or connate, falling together or
in groups; annuals; leaves often 3-lobed near tips; ray laminae 3–5 mm) . . .
. . . . 365. *Syntrichopappus* (in part), **v. 21, p. 379**

            71. Pappus scales 2–16.

                72. Cypselae stoutly obconic or obpyramidal (lengths usually 1–2, rarely to
3.5 times diams.).

                  73. Annuals, biennials, or perennials; leaves all or mostly cauline (often 1–
2-terately lobed) . . . . 368. *Bahia* (in part), **v. 21, p. 383**

                  73. Perennials; leaves basal or basal and cauline (not lobed) . . . .
. . . . 375. *Platyschkuhria*, **v. 21, p. 394**

                72. Cypselae narrowly clavate or columnar to obconic or obpyramidal (lengths
usually 3+ times diams., if stouter, usually ± compressed).

74. Phyllaries 17–21 in 2 series (inner hyaline, scalelike; herbage notably sweet scented) . . . . 383. *Amblyolepis* (in part), **v. 21, p. 420**
74. Phyllaries 8–12 in 2 series, or 6–40 in 2 series (inner herbaceous to scarious or scarious-margined; herbage not notably sweet-scented).
    75. Leaf blades sometimes pinnately lobed (lobes mostly filiform, linear, or oblong); phyllaries: outer connate or distinct, inner distinct . . . . 386. *Hymenoxys* (in part), **v. 21, p. 435**
    75. Leaf blades (some or all) pinnately lobed (lobes mostly deltate to obovate); phyllaries: all basally connate . . . . 387. *Plateilema*, **v. 21, p. 444**

Key to Genera of Group 12
Heads eradiate; receptacles epaleate; pappi none or nearly so.

1. Leaves all or mostly opposite (distal sometimes alternate).
   2. Corollas yellow.
      3. Heads disciform (peripheral florets pistillate) . . . . 354. *Lasthenia* (in part), **v. 21, p. 336**
      3. Heads discoid (florets all bisexual).
         4. Florets (5–)20–100+ per head (heads borne singly or in open arrays, not crowded in headlike or tight glomerules).
            5. Leaf blades usually lobed (mostly 5–30 mm); phyllaries 8–16 in 2–3 series, distinct. . . . . 352. *Perityle* (in part), **v. 21, p. 317**
            5. Leaf blades usually triangular-hastate to narrowly deltate, seldom notably lobed (30–120 mm); phyllaries 15–21 in 1(–2) series, wholly or partially connate . . . . 353. *Pericome* (in part), **v. 21, p. 334**
         4. Florets 1–5(–15) per head (heads often crowded in headlike or tight glomerules).
            6. Leaves usually glabrous (often succulent). . . . . 327. *Flaveria* (in part), **v. 21, p. 247**
            6. Leaves usually hispid, sericeous, strigose, villous, or woolly.
               7. Shrubs. . . . . 331. *Lagascea* (in part), **v. 21, p. 136**
               7. Annuals. . . . . 350. *Madia* (in part), **v. 21, p. 303**
   2. Corollas mostly white, sometimes blue, lavender, pink, or purple.
      8. Cypselae obcompressed or -obflattened (subtended by accrescent phyllaries, margins corky-winged, ± toothed) . . . . 248. *Dicoria* (in part), **v. 21, p. 24**
      8. Cypselae mostly prismatic or columnar (if obcompressed, not subtended by accrescent phyllaries, margins not corky-winged and toothed).
         9. Style-branch appendages usually terete to clavate (lengths usually 2–5+ stigmatic lines).
            10. Florets 5(–6) . . . . 397. *Stevia* (in part), **v. 21, p. 483**
            10. Florets 20–125.
               11. Plants 20–120 cm (terrestrial); involucres campanulate, 3–6 mm diam.; phyllaries usually 2-nerved . . . . 396. *Ageratum* (in part), **v. 21, p. 481**
               11. Plants 10–30 cm (aquatic); involucres hemispheric or broader, 6–9 mm diam.; phyllaries not notably nerved . . . . 400. *Shinnersia*, **v. 21, p. 488**
         9. Style-branch appendages essentially none or deltate, lanceolate, linear, filiform, or penicillate (lengths mostly 0–3 times stigmatic lines).
            12. Anthers connate. . . . . 352. *Perityle* (in part), **v. 21, p. 317**
            12. Anthers distinct.
               13. Heads in (bracteate) racemiform or spiciform arrays (1–2 per bract); pistillate florets usually 1–8+, rarely 0; functionally staminate florets 3–20+, corolla lobes soon reflexed . . . . 249. *Iva* (in part), **v. 21, p. 25**
               13. Heads in (usually ebracteate) paniculiform arrays; pistillate florets 5; functionally staminate florets 5–10(–20+), corolla lobes erect . . . . 250. *Cyclachaena* (in part), **v. 21, p. 28**

1. Leaves alternate.
    14. Corollas mostly white, sometimes blue, lavender, pink, or purple.
        15. Annuals.
            16. Heads disciform (usually sessile); cypselae ± obovate to oblanceolate or oblong-cuneate, obcompressed or -obflattened (winged, usually producing apical spines) . . . . 127. *Soliva* (in part), **v. 19, p. 545**
            16. Heads discoid (usually pedunculate); cypselae clavate to ± cylindric or compressed (obscurely 8–20-angled, not winged or with spines) . . . .
            . . . . 378. *Chaenactis* (in part), **v. 21, p. 400**
        15. Perennials, subshrubs, or shrubs.
            17. Phyllaries in 6+ series (apices often with ± dentate or fringed, linear to ovate appendages, sometimes spine-tipped) . . . . 24. *Centaurea* (in part), **v. 19, p. 181**
            17. Phyllaries usually in 1–3 series (apices without appendages or spine tips).
                18. Style-branch appendages usually terete to clavate (lengths usually 2–5+ stigmatic lines).
                    19. Phyllaries 5(–6) in ± 1 series; florets 5(–6) . . . . 397. *Stevia* (in part), **v. 21, p. 483**
                    19. Phyllaries 12–15 in 2–3 series; florets 7–10 . . . . 413. *Hartwrightia* (in part), **v. 21, p. 540**
                18. Style-branch appendages essentially none or deltate, lanceolate, linear, filiform, or penicillate (lengths mostly 0–3 times stigmatic lines).
                    20. Leaves basal and cauline, blades ovate to ± triangular (bases mostly truncate to cordate or hastate), margins coarsely dentate or lobulate to denticulate or entire (distalmost leaves often linear or scalelike); cypselae clavate to obovoid, faces distally stipitate-glandular . . . .
                    . . . . 5. *Adenocaulon* (in part), **v. 19, p. 77**
                    20. Leaves mostly cauline, blades lanceolate to oblanceolate, margins laciniately pinnately lobed; cypselae pyriform, ± obcompressed, densely gland-dotted . . . . 252. *Leuciva* (in part), **v. 21, p. 29**
    14. Corollas mostly yellow (sometimes pale or reddish, or red-brown to purplish).
        21. Leaf margins ± spiny (plants ± thistlelike; phyllaries: apical appendages prominently veiny, spiny-dentate or -lobed, spine-tipped) . . . . 23. *Carthamus* (in part), **v. 19, p. 178**
         21. Leaf margins not spiny (plants not thistlelike; phyllary apices sometimes apiculate, not spine-tipped).
            22. Receptacles usually ± globose (leaves seldom lobed, stems usually winged by decurrent leaf bases) . . . . 385. *Helenium* (in part), **v. 21, p. 426**
            22. Receptacles usually flat to convex (if ± conic, leaves usually 1–2-pinnately lobed, leaf bases not decurrent on stems).
                23. Heads disciform (peripheral florets pistillate).
                    24. Pistillate corollas lacking . . . . 126. *Cotula* (in part), **v. 19, p. 543**
                    24. Pistillate corollas ± tubular (distally 3–5-lobed or truncate).
                        25. Heads usually in lax to dense, corymbiform arrays, rarely borne singly; involucres mostly hemispheric or broader, (3–)5–22+ mm diam.; phyllaries (20–)30–60+ in (2–)3–5+ series . . . .
                        . . . . 112. *Tanacetum* (in part), **v. 19, p. 489**
                      25. Heads usually in paniculiform, racemiform, or spiciform arrays, sometimes in subcapitate clusters, sometimes borne singly; involucres hemispheric to campanulate, obconic, ovoid, or turbinate, 1.5–5(–12) mm diam.; phyllaries 5–20+ in 1–4(–7) series.
                          26. Heads borne singly or (2–20+) in usually corymbiform, rarely paniculiform, arrays or in subcapitate clusters; disc floret corollas bright yellow or ochroleucous (leaves usually gland-dotted) . . . . 118. *Sphaeromeria* (in part), **v. 19, p. 499**
                        26. Heads (usually 20–200+, 2–12+ in *Picrothamnus*) in paniculiform, racemiform, or spiciform arrays, rarely borne singly; disc floret corollas usually pale yellow, rarely red.

27. Subshrubs or shrubs (thorny); disc florets functionally
staminate (corollas ± villous) . . . . 117. *Picrothamnus*, **v. 19, p. 498**
27. Annuals, perennials, subshrubs, or shrubs (not thorny);
disc florets bisexual and fertile, or functionally staminate
(corollas usually glabrous, sometimes hairy, not villous)
. . . . 119. *Artemisia* (in part), **v. 19, p. 503**

[23. Shifted to left margin.—Ed.]

23. Heads discoid (all florets bisexual, fertile).
  28. Subshrubs or shrubs.
    29. Heads in paniculiform, racemiform, or spiciform arrays . . . . . 119. *Artemisia* (in part), **v. 19, p. 503**
    29. Heads borne singly or in loose, corymbiform arrays.
      30. Phyllaries 5–6 in 1–2 series; florets 5–9 . . . . 237. *Tetradymia* (in part), **v. 20, p. 629**
      30. Phyllaries 22–40 in 3–4+ series; florets (20–)40–100(–300+).
        31. Phyllaries 22–40 in 3–4+ series, deltate-ovate to elliptic or obovate; cypselae
          narrowly obovoid to oblong or elliptic, ± terete or flattened, ribs 5 (faces
          gland-dotted between ribs) . . . . 125. *Pentzia* (in part), **v. 19, p. 543**
        31. Phyllaries 25–100+ in (3–)4–9+ series, mostly filiform, linear, or lanceolate;
          cypselae ellipsoid to obovoid, ± compressed, sometimes ± 3–4-angled (faces
          smooth, striate, ribbed, furrowed, or rugose, glabrous) . . . .
          . . . . 202. *Grindelia* (in part), **v. 20, p. 424**
  28. Annuals or perennials.
    32. Leaves 0–1-pinnately lobed.
      33. Cypselae ± columnar to obovoid, ribs ± 10 (style-branch appendages rings of
        papillae or essentially 0) . . . . 137. *Leucanthemum* (in part), **v. 19, p. 557**
      33. Cypselae ellipsoid to obovoid, ± compressed, sometimes ± 3–4-angled (style-
        branch appendages linear or lanceolate to ± deltate) . . . . 202. *Grindelia* (in part), **v. 20, p. 424**
    32. Leaves 1–2-pinnately lobed.
      34. Perennials . . . . 112. *Tanacetum* (in part), **v. 21, p. 480**
      34. Annuals.
        35. Phyllaries lance-triangular to ± ovate or elliptic (carinate); cypselae colum-
          nar to prismatic, ribs 4 . . . . 123. *Oncosiphon*, **v. 19, p. 539**
        35. Phyllaries oblong or ovate to spatulate or linear-spatulate (not carinate);
          cypselae obconic, slightly compressed, ribs 5 . . . . 124. *Matricaria* (in part), **v. 19, p. 540**

Key to Genera of Group 13
Heads eradiate; receptacles epaleate; pappi wholly of bristles.

1. Leaves opposite (at least proximally) or whorled . . . . Group 13a, **v. 19, p. 54**
1. Leaves alternate.
  2. Phyllaries usually in 1–2 series and equal to subequal (sometimes coherent, often sub-
    tended by calyculi; in 2–4+ series and unequal in *Lepidospartum*, shrubs with leaves of
    flowering stems filiform to acerose or scalelike and pappi of ca. 150 bristles in 3–4
    series) . . . . Group 13b, **v. 19, p. 55**
  2. Phyllaries usually in 3–10+ series and unequal, sometimes 1–2 series and subequal
    (distinct, calyculi 0).
    3. Corollas white, ochroleucous, or pink to purplish, sometimes red (then plants prickly-
      spiny) . . . . Group 13c, **v. 19, p. 56**
    3. Corollas cream or yellowish to yellow, orange, or brick-red (sometimes whitish,
      greenish, reddish, or purplish, or red-tipped) . . . . Group 13d, **v. 19, p. 59**

Key to Genera of Group 13a

1. Corollas yellow to orange.
  2. Leaves and phyllaries dotted or streaked with pellucid (schizogenous) glands containing strong-scented oils . . . . 320. *Porophyllum* (in part), **v. 21, p. 233**
  2. Leaves and/or phyllaries rarely dotted or streaked (never with pellucid, schizogenous glands, sometimes stipitate-glandular) . . . . 361. *Arnica* (in part), **v. 21, p. 366**
1. Corollas white, ochroleucous, or pink to purplish.
  3. Involucres narrowly cylindric, (1–)2–3 mm diam.; phyllaries 4, or 5(–6) in ± 1–2 series; florets 4, or 5(–6).
    4. Subshrubs or shrubs; phyllaries 5(–6); florets 5(–6) . . . . 397. *Stevia* (in part), **v. 21, p. 483**
    4. Vines; phyllaries 4; florets 4 . . . . 417. *Mikania*, **v. 21, p. 545**
  3. Involucres campanulate, cylindric, ellipsoid, hemispheric, or obconic, (2–)3–7(–25) mm diam.; phyllaries (5–)8–45(–65+) in (1–)2–8+ series; florets (3–)10–125(–200+).
    5. Cypselae 8–10-ribbed.
      6. Leaf blades deltate, lance-elliptic, lance-linear, lanceolate, lance-ovate, lance-rhombic, linear, oblong, obovate, ovate, rhombic-ovate, spatulate, suborbiculate, margins crenate, dentate, entire, laciniate-dentate, lobed, or serrate; style bases enlarged, hairy . . . . 403. *Brickellia* (in part), **v. 21, p. 491**
      6. Leaf blades linear (distal sometimes scalelike), margins entire; style bases not enlarged, glabrous . . . . 406. *Asanthus*, **v. 21, p. 509**
    5. Cypselae (3–)4–5(–8)-ribbed.
      7. Pappi of 2–6+, coarsely barbellate bristles . . . . 399. *Trichocoronis*, **v. 21, p. 487**
      7. Pappi of (5–)10–80+, barbellulate, barbellate, or plumose bristles (or setiform scales).
        8. Involucres cylindric (3–4+ mm diam.); pappus bristles plumose (basally coherent or connate, falling together or in groups) . . . . 409. *Carminatia*, **v. 21, p. 511**
        8. Involucres usually obconic to hemispheric, sometimes campanulate, cylindric, or ellipsoid (2–7 mm diam.); pappus bristles smooth or barbellulate to barbellate (not plumose).
          9. Phyllaries ± equal.
            10. Receptacles conic . . . . 394. *Conoclinium*, **v. 21, p. 478**
            10. Receptacles flat or convex.
              11. Phyllaries 2- or 3-nerved, or not notably nerved, or pinnately nerved; style bases usually puberulent (glabrous in *Eupatorium capillifolium*); cypselae usually gland-dotted . . . . 392. *Eupatorium* (in part), **v. 21, p. 462**
              11. Phyllaries 3-nerved, or 0- or 2-nerved; style bases glabrous; cypselae sometimes gland-dotted.
                12. Involucres 2–3 mm diam.; phyllaries 7–16 in 1–2 series; florets 3–13 . . . . 415. *Koanophyllon* (in part), **v. 21, p. 542**
                12. Involucres 3–6 mm diam.; phyllaries ca. 30 in 2–3 series; florets 10–60 . . . . 418. *Ageratina*, **v. 21, p. 547**
          9. Phyllaries unequal (outer shorter).
            13. Style bases usually puberulent (glabrous in *Eupatorium capillifolium*); cypselae usually glabrous and gland-dotted, sometimes scabrellous on ribs.
              14. Leaves mostly opposite (sometimes whorled, distal sometimes alternate) . . . . 392. *Eupatorium* (in part), **v. 21, p. 462**
              14. Leaves mostly whorled (3–7 per node), rarely opposite . . . . 393. *Eutrochium*, **v. 21, p. 474**
            13. Style bases usually glabrous (hirsute in *Flyriella*); cypselae glabrous or hirsute, hirtellous, hispidulous, hispidulo-strigose, puberulent, or scabrellous (sometimes gland-dotted).

[15. Shifted to left margin.—Ed.]

15. Annuals or perennials; involucres 2–5+ mm diam.; florets 10–30.
    16. Perennials, 20–60 cm (viscid); corollas white to ochroleucous, throats ± cylindric (± contracted distally, lengths 4–6 times diams.) . . . . 405. *Flyriella*, **v. 21, p. 507**
    16. Annuals or perennials, 30–120+ cm (not viscid, stems usually puberulent, hairs curled); corollas bluish, pinkish, purplish, or white, throats funnelform (not contracted distally, lengths 2.5–4 times diams.) . . . . 414. *Fleischmannia*, **v. 21, p. 540**
15. Perennials, subshrubs, or shrubs; involucres (2–)4–7 mm diam.; florets (3–)25–50.
    17. Phyllaries usually readily falling, 18–65+ in 4–6+ series, 3–5-nerved; cypselae (3–)5-ribbed, scabrellous, usually gland-dotted. . . . . 416. *Chromolaena* (in part), **v. 21, p. 544**
    17. Phyllaries usually persistent, 7–35 in (1–)2–4 series, 2- or 4-nerved, 3-nerved, or obscurely nerved; cypselae 5(–7)-ribbed, hispidulous, hispidulo-strigose, puberulent, or sparsely scabrellous (sometimes gland-dotted).
        18. Phyllaries 2- or 4-nerved; corollas white to yellowish white; pappi readily falling or fragile . . . . 404. *Brickelliastrum*, **v. 21, p. 507**
        18. Phyllaries 3-nerved or obscurely nerved; corollas usually blue, lavender, or pinkish, sometimes white; pappi persistent.
            19. Involucres 5–7 mm diam.; phyllaries 30–35; florets 30–50 . . . . . 395. *Tamaulipa*, **v. 21, p. 480**
            19. Involucres 2–3 mm diam.; phyllaries 7–16; florets 3–13 . . . . 415. *Koanophyllon* (in part), **v. 21, p. 542**

Key to Genera of Group 13b

1. Heads disciform.
    2. Leaf blades palmately or palmati-pinnately nerved, mostly deltate to ovate or orbiculate (plants polygamodioecious, stems of "staminate" plants wither soon after flowering, stems of "pistillate" plants elongate after flowering) . . . . 241. *Petasites* (in part), **v. 20, p. 635**
    2. Leaf blades pinnately nerved, mostly ovate to lanceolate (sometimes pinnately lobed or dissected).
        3. Pistillate florets sometimes 1–3 in 1 series (*Senecio mohavensis*) . . . . 215. *Senecio* (in part), **v. 20, p. 544**
        3. Pistillate florets 10–100+ in 1–3 series . . . . 217. *Erechtites* (in part), **v. 20, p. 602**
1. Heads discoid.
    4. Subshrubs, shrubs (treelets), or vines.
        5. Vines (California, Oregon) . . . . 221. *Delairea* (in part), **v. 20, p. 608**
        5. Subshrubs or shrubs (treelets).
            6. Phyllaries 4–6 in 1–2 series and equal or subequal; florets 4–9 . . . . 237. *Tetradymia* (in part), **v. 20, p. 629**
            6. Phyllaries mostly 5, 8, 13, or 21 in (1–)2 series and equal or subequal, or 8–13 or 12–23+ in 2–4+ series and unequal (outer shorter); florets 3–80+.
                7. Phyllaries mostly 5, 8, 13, or 21 in (1–)2 series and equal or subequal . . . . 215. *Senecio* (in part), **v. 20, p. 544**
                7. Phyllaries 8–23+ in 2–4+ series and unequal (leaves of flowering stems filiform to acerose or scalelike) . . . . 238. *Lepidospartum* (in part), **v. 20, p. 632**
    4. Annuals, biennials, or perennials.
        8. Plants ± velutinous or villous (hairs purplish); style-branch appendages ± filiform (hispidulous, 1–2 mm; perennials, s Florida) . . . . 222. *Gynura*, **v. 20, p. 610**
        8. Plants rarely velutinous or villous (hairs rarely purplish); style-branch appendages essentially 0 (or deltoid to conic, mostly 0.1–1 mm; style-branch apices usually truncate-penicillate or truncate to rounded-truncate).
            9. Phyllaries (4–)5(–6; yellow; Wyoming) . . . . 239. *Yermo*, **v. 20, p. 634**
            9. Phyllaries (4–)5–30+ (usually green).

[10. Shifted to left margin.—Ed.]

10. Corollas usually yellow, sometimes orange-yellow or orange, rarely orange-red, purplish, or reddish.
  11. Heads in racemiform or subthyrsiform arrays; phyllaries (4–)5(–8); florets (4–)5(–8) . . . . 236. *Rainiera*, **v. 20, p. 628**
  11. Heads usually in corymbiform, sometimes cymiform, racemiform, or subumbelliform, arrays or borne singly; phyllaries (5–)8–30+; florets (5–)20–80+.
    12. Calyculi usually of 1–8+ bractlets, sometimes 0; style-branch apices usually truncate-penicillate.
      13. Leaves basal and/or cauline (roots often fleshy and seldom branched and/or leaf margins with relatively many callous denticles) . . . . 215. *Senecio* (in part), **v. 20, p. 544**
      13. Leaves basal and cauline (roots seldom fleshy, often branched; leaf margins with relatively few or no callous denticles) . . . . 216. *Packera* (in part), **v. 20, p. 570**
    12. Calyculi 0; style-branch apices usually rounded-truncate or with deltoid to conic appendages.
      14. Involucres 3–8 mm diam.; florets 11–26 . . . . 235. *Luina*, **v. 20, p. 627**
      14. Involucres 7–15+ mm diam.; florets 20–80+.
        15. Leaf blades pinnately nerved, lanceolate, linear-oblanceolate, oblanceolate, ovate, or subrhombic . . . . 227. *Tephroseris* (in part), **v. 20, p. 615**
        15. Leaf blades palmately nerved, ± reniform to orbiculate . . . . . 234. *Cacaliopsis*, **v. 20, p. 627**
10. Corollas usually white or ochroleucous to greenish or whitish, sometimes lavender, pinkish, or purplish.
  16. Leaf blades palmately or palmati-pinnately nerved, mostly deltate to ovate or orbiculate (plants polygamodioecious, stems of "staminate" plants wither soon after flowering, stems of "pistillate" plants elongate after flowering) . . . . 241. *Petasites* (in part), **v. 20, p. 635**
  16. Leaf blades palmately, palmati-pinnately, or pinnately nerved, mostly ovate, obovate, oblanceolate, or elliptic to lanceolate, lance-linear, or linear, sometimes cordate, deltate, hastate, reniform, or orbiculate (plants not polygamodioecious).
    17. Annuals . . . . 218. *Emilia*, **v. 20, p. 605**
    17. Perennials.
      18. Phyllaries 5–8; florets 4–8.
        19. Heads in racemiform to subpaniculiform arrays (Alaska) . . . 229. *Parasenecio*, **v. 20, p. 619**
        19. Heads usually in corymbiform, sometimes paniculiform, arrays.
          20. Basal and proximal cauline leaf blades ovate to elliptic (deeply 3–4-pinnatisect); calyculi of 1–3 bractlets (s Arizona, New Mexico) . . . . 231. *Psacalium* (in part), **v. 20, p. 621**
          20. Basal and proximal cauline leaf blades mostly cordate, deltate, elliptic, hastate, ovate, or reniform, sometimes lanceolate or lance-linear (not 3–4-pinnatisect); calyculi 0 (e North America) . . . . 232. *Arnoglossum*, **v. 20, p. 622**
      18. Phyllaries 7–21; florets 10–80+.
        21. Plants (50–)60–240 cm; calyculi of 4–9+ bractlets . . . . 223. *Hasteola*, **v. 20, p. 610**
        21. Plants 10–70 cm; calyculi 0 or of 1–5+ bractlets.
          22. Leaf blades ovate to subrhombic (1–3 × 1–2 cm) . . . . 227. *Tephroseris* (in part), **v. 20, p. 615**
          22. Leaf blades ovate to nearly cordate (5–15+ × 3–10+ cm) . . . . 233. *Rugelia*, **v. 20, p. 625**

Key to Genera of Group 13c

1. Heads aggregated in second-order heads (florets 1 per individual head, each cluster subtended by ovate to orbiculate, prickly-margined bracts; subshrubs or shrubs) . . . . 1. *Hecastocleis* (in part), **v. 19, p.71**
1. Heads borne singly or in corymbiform, paniculiform, or racemiform arrays (not aggregated in second-order heads; florets usually 3–300+ per individual head).
  2. Corollas usually 2-lipped (at least outer; inner sometimes nearly actinomorphic).

3. Abaxial leaf faces glabrous or with scattered, straight, sometimes glandular, hairs and/or double hairs; cypselae not beaked . . . . 2. *Acourtia* (in part), **v. 19, p. 72**

3. Abaxial leaf faces usually tomentose, tomentulose, or covered with dense wool (thinly gray-tomentose in *Leibnitzia*); cypselae distally ± tapered or ± constricted into necks or beaks.

    4. Heads chasmogamous (produced well after rosette leaves); cypselae glabrous or pubescent (hairs relatively short, apices rounded or apiculate) . . . .
. . . . 6. *Chaptalia* (in part), **v. 19, p. 78**

    4. Heads chasmogamous (early spring, produced before and concurrently with first rosette leaves) or cleistogamous (produced after rosette leaves); cypselae strigose to hispid (hairs relatively long, apices sharp-pointed) . . . .
. . . . 7. *Leibnitzia* (in part), **v. 19, p. 80**

[2. Shifted to left margin.—Ed.]

2. Corollas actinomorphic (not two-lipped).

  5. Plants often prickly-spiny and thistlelike and/or phyllary apices spinose or ± expanded into distinct, often fimbriate-fringed, pectinate, and/or prickly appendages.

    6. Leaf margins spiny.

      7. Stems winged.

        8. Receptacles not bristly (deeply pitted); cypsela attachments basal; pappus bristles basally connate . . . . 10. *Onopordum*, **v. 19, p. 87**

        8. Receptacles bristly (bearing setiform scales or "flattened bristles," not pitted); cypsela attachments slightly lateral; pappus bristles usually distinct, sometimes basally connate . . . . 12. *Carduus*, **v. 19, p. 91**

      7. Stems not or rarely winged (some *Cirsium* spp.).

        9. Corollas ± purple; cypsela attachments lateral; pappi of distinct, minutely barbed (not plumose) flattened bristles ("setiform scales") . . . .
. . . . 24. *Centaurea* (in part), **v. 19, p. 181**

        9. Corollas white or purplish to red; cypsela attachments basal or oblique-basal; pappi of basally connate, plumose flattened bristles ("setiform scales").

          10. Involucres 35–100+ mm diam. (largest leaves 60–150 cm; receptacles becoming fleshy) . . . . 11. *Cynara* (in part), **v. 19, p. 89**

          10. Involucres 10–50 mm diam. (largest leaves usually 20–50, sometimes to 110, cm; receptacles usually not notably fleshy) . . . . 13. *Cirsium*, **v. 19, p. 95**

    6. Leaf margins not spiny (tips sometimes ± spinose-apiculate).

      11. Heads disciform or radiant (peripheral florets usually neuter).

        12. Heads disciform (phyllary appendages dentate or fringed, spiny or not; receptacles bearing setiform scales) . . . . 24. *Centaurea* (in part), **v. 19, p. 181**

        12. Heads radiant.

          13. Phyllary appendages 0 (biennials or perennials; spines on phyllary apices caducous) . . . . 19. *Mantisalca* (in part), **v. 19, p. 173**

          13. Phyllary appendages present.

            14. Annuals; leaf margins mostly entire or denticulate to serrulate; cypsela attachments oblique-basal; involucres 20–40 mm diam.; phyllary bodies linear, margins entire, appendages fimbriate; corollas of peripheral florets 30–70 mm . . . . 21. *Plectocephalus*, **v. 19, p. 175**

            14. Annuals, biennials, or perennials; leaf margins entire or toothed to pinnately lobed; involucres 10–25(–40) mm diam.; cypsela attachments lateral; phyllary bodies oblong to ovate or obovate, margins fimbriate, appendages fimbriate; corollas of peripheral florets 15–30(–45) mm . . . . 24. *Centaurea* (in part), **v. 19, p. 181**

      11. Heads discoid (all florets bisexual and fertile).

        15. Cypsela attachments ± lateral.

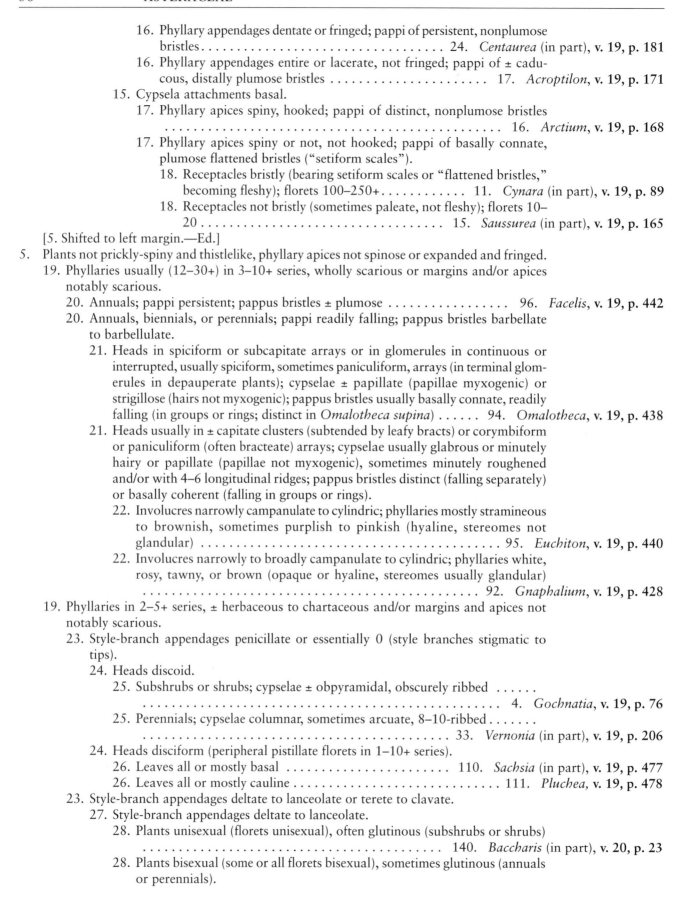

  16. Phyllary appendages dentate or fringed; pappi of persistent, nonplumose
    bristles . . . . 24. *Centaurea* (in part), **v. 19, p. 181**
  16. Phyllary appendages entire or lacerate, not fringed; pappi of ± cadu-
    cous, distally plumose bristles . . . . 17. *Acroptilon*, **v. 19, p. 171**
 15. Cypsela attachments basal.
  17. Phyllary apices spiny, hooked; pappi of distinct, nonplumose bristles
    . . . . 16. *Arctium*, **v. 19, p. 168**
  17. Phyllary apices spiny or not, not hooked; pappi of basally connate,
    plumose flattened bristles ("setiform scales").
    18. Receptacles bristly (bearing setiform scales or "flattened bristles,"
      becoming fleshy); florets 100–250+ . . . . 11. *Cynara* (in part), **v. 19, p. 89**
    18. Receptacles not bristly (sometimes paleate, not fleshy); florets 10–
      20 . . . . 15. *Saussurea* (in part), **v. 19, p. 165**
[5. Shifted to left margin.—Ed.]
5. Plants not prickly-spiny and thistlelike, phyllary apices not spinose or expanded and fringed.
 19. Phyllaries usually (12–30+) in 3–10+ series, wholly scarious or margins and/or apices
  notably scarious.
  20. Annuals; pappi persistent; pappus bristles ± plumose . . . . 96. *Facelis*, **v. 19, p. 442**
  20. Annuals, biennials, or perennials; pappi readily falling; pappus bristles barbellate
    to barbellulate.
    21. Heads in spiciform or subcapitate arrays or in glomerules in continuous or
      interrupted, usually spiciform, sometimes paniculiform, arrays (in terminal glom-
      erules in depauperate plants); cypselae ± papillate (papillae myxogenic) or
      strigillose (hairs not myxogenic); pappus bristles usually basally connate, readily
      falling (in groups or rings; distinct in *Omalotheca supina*) . . . . . . 94. *Omalotheca*, **v. 19, p. 438**
    21. Heads usually in ± capitate clusters (subtended by leafy bracts) or corymbiform
      or paniculiform (often bracteate) arrays; cypselae usually glabrous or minutely
      hairy or papillate (papillae not myxogenic), sometimes minutely roughened
      and/or with 4–6 longitudinal ridges; pappus bristles distinct (falling separately)
      or basally coherent (falling in groups or rings).
      22. Involucres narrowly campanulate to cylindric; phyllaries mostly stramineous
        to brownish, sometimes purplish to pinkish (hyaline, stereomes not
        glandular) . . . . 95. *Euchiton*, **v. 19, p. 440**
      22. Involucres narrowly to broadly campanulate to cylindric; phyllaries white,
        rosy, tawny, or brown (opaque or hyaline, stereomes usually glandular)
        . . . . 92. *Gnaphalium*, **v. 19, p. 428**
 19. Phyllaries in 2–5+ series, ± herbaceous to chartaceous and/or margins and apices not
  notably scarious.
  23. Style-branch appendages penicillate or essentially 0 (style branches stigmatic to
    tips).
    24. Heads discoid.
      25. Subshrubs or shrubs; cypselae ± obpyramidal, obscurely ribbed . . . . . .
        . . . . 4. *Gochnatia*, **v. 19, p. 76**
      25. Perennials; cypselae columnar, sometimes arcuate, 8–10-ribbed . . . .
        . . . . 33. *Vernonia* (in part), **v. 19, p. 206**
    24. Heads disciform (peripheral pistillate florets in 1–10+ series).
      26. Leaves all or mostly basal . . . . 110. *Sachsia* (in part), **v. 19, p. 477**
      26. Leaves all or mostly cauline . . . . 111. *Pluchea*, **v. 19, p. 478**
  23. Style-branch appendages deltate to lanceolate or terete to clavate.
    27. Style-branch appendages deltate to lanceolate.
      28. Plants unisexual (florets unisexual), often glutinous (subshrubs or shrubs)
        . . . . 140. *Baccharis* (in part), **v. 20, p. 23**
      28. Plants bisexual (some or all florets bisexual), sometimes glutinous (annuals
        or perennials).

29. Perennials . . . . 162. *Brintonia*, **v. 20, p. 106**
29. Annuals.
   30. Plants (gracile) 2–14 cm; leaves linear to filiform, margins entire, abaxial faces glabrous or sparsely pilose; pappi of (3–)5 bristles . . . . 145. *Pentachaeta* (in part), **v. 20, p. 46**
   30. Plants (1–)2–100 cm; leaves ovate or obovate to lanceolate, linear, or subulate, margins entire, toothed, or pinnately lobed, abaxial faces glabrous or sparsely tomentose or woolly; pappi of 3–55 (sometimes basally connate) bristles . . . . 209. *Lessingia* (in part), **v. 20, p. 452**
[27. Shifted to left margin.—Ed.]
27. Style-branch appendages terete to clavate (lengths usually 2–5+ stigmatic lines).
  31. Cypselae (3–)4–5(–8)-ribbed.
    32. Pappi of (5–)10–80+ barbellulate, barbellate, or plumose bristles (or setiform scales) . . . . 395. *Tamaulipa* (in part), **v. 21, p. 480**
    32. Pappi of 1–5 ± glandular setae . . . . 413. *Hartwrightia* (in part), **v. 21, p. 540**
  31. Cypselae 8–10-ribbed.
    33. Leaves mostly cauline (at flowering; mostly petiolate, sometimes sessile) . . . . 412. *Garberia*, **v. 21, p. 538**
    33. Leaves basal or basal and cauline (cauline mostly sessile).
      34. Heads usually in spiciform or racemiform, rarely corymbiform or thyrsiform, arrays; pappi of 12–40 coarsely barbellate to plumose bristles. . . . . . . 410. *Liatris*, **v. 21, p. 512**
      34. Heads in corymbiform to paniculiform arrays; pappi of 35–40 barbellulate to barbellate (subequal) bristles . . . . 411. *Carphephorus* (in part), **v. 21, p. 535**

Key to Genera of Group 13d

1. Corollas all 2-lipped (shrubs) . . . . 3. *Trixis* (in part), **v. 19, p. 75**
1. Corollas actinomorphic.
  2. Stems winged (heads in spiciform arrays; phyllaries 12–30+ in 3–6+ series, usually ± herbaceous to chartaceous, sometimes indurate, margins and/or apices seldom notably scarious; corollas yellowish) . . . . 109. *Pterocaulon*, **v. 19, p. 476**
  2. Stems not winged.
    3. Phyllaries usually (12–30+) in 3–10+ series, scarious or margins and/or apices usually notably scarious, sometimes (3–10) in 1–2 series.
      4. Heads usually discoid (unisexual or nearly so, staminate or pistillate; plants unisexual or nearly so; predominantly pistillate heads rarely with 1–9 central, functionally staminate florets; predominantly staminate heads rarely with 1–4+ peripheral, pistillate florets; involucres mostly 6–10 mm).
        5. Plants (0.2–)4–25(–70) cm; basal leaves usually present at flowering (withering before in *A. geyeri*); pappus bristles (at least pistillate) usually basally connate or coherent . . . . 87. *Antennaria*, **v. 19, p. 388**
        5. Plants mostly 20–80(–120+) cm; basal leaves usually withering before flowering; pappus bristles distinct or basally connate . . . . 90. *Anaphalis*, **v. 19, p. 426**
      4. Heads usually disciform (plants not unisexual; heads mostly alike, each with 4–200+ pistillate and 1–200+ bisexual or functionally staminate florets; heads rarely discoid in *Xerochrysum*, which has involucres 10–30 mm and brightly colored phyllaries in 3–8+ series).
        6. Pistillate florets fewer than bisexual.
          7. Subshrubs; heads in glomerules in corymbiform arrays; involucres campanulate, 4–8 mm . . . . 89. *Helichrysum*, **v. 19, p. 425**
          7. Annuals, biennials, or perennials; heads borne singly or (2–3) in loose, corymbiform arrays; involucres ± hemispheric, 10–30 mm . . . . 91. *Xerochrysum*, **v. 19, p. 427**
        6. Pistillate florets more numerous than bisexual.

8. Heads usually in glomerules in corymbiform or paniculiform arrays, sometimes in terminal clusters; cypselae usually glabrous, sometimes with papilliform hairs (hairs not myxogenic), sometimes minutely roughened and/or with 4–6 longitudinal ridges; pappus bristles usually distinct (falling separately), sometimes basally coherent (falling in groups or rings) . . . . 88. *Pseudognaphalium*, **v. 19, p. 415**

8. Heads usually in glomerules in continuous or interrupted, usually spiciform, sometimes paniculiform, arrays (in terminal glomerules in depauperate plants); cypselae ± papillate (papilliform hairs myxogenic); pappus bristles basally connate, readily falling (in rings) . . . . 93. *Gamochaeta*, **v. 19, p. 431**

[3. Shifted to left margin.—Ed.]

3. Phyllaries in 3–5+ series, usually ± herbaceous to chartaceous, sometimes indurate, margins and/or apices seldom notably scarious.

9. Leaves and/or phyllaries dotted or streaked with pellucid (schizogenous) glands containing strong-scented oils . . . . 320. *Porophyllum*, **v. 21, p. 233**

9. Leaves and/or phyllaries rarely dotted or streaked (never with pellucid, schizogenous glands containing strong-scented oils, plants sometimes with sessile or stipitate, surface glands and strong-scented).

10. Subshrubs, shrubs, or treelets.

11. Plants unisexual (florets unisexual), often glutinous . . . . . . 140. *Baccharis* (in part), **v. 20, p. 23**

11. Plants bisexual (some or all florets bisexual), sometimes glutinous, gland-dotted, or stipitate-glandular.

12. Phyllaries disposed in vertical ranks and unequal.

13. Leaves basal and cauline, blades 3-nerved; heads in glomerate clusters grouped in flat-topped, corymbiform arrays; phyllaries yellowish, sometimes distally green . . . . 157. *Bigelowia*, **v. 20, p. 95**

13. Leaves cauline, blades 1-nerved (sometimes with 1–2 fainter lateral pairs); heads (sometimes clustered) in paniculiform, corymbiform, or cymiform arrays, or borne singly; phyllaries stramineous, tan, or green, distally green or purplish.

14. Leaf faces gland-dotted (in pits); heads in dense, cymiform arrays . . . . 161. *Chrysoma* (in part), **v. 20, p. 105**

14. Leaf faces gland-dotted (sessile) or stipitate glandular; heads borne singly or in condensed, cymiform clusters, grouped in paniculiform or corymbiform arrays, or in congested, cymiform to corymbiform arrays.

15. Heads in congested, cymiform to corymbiform arrays; disc florets 4–15; cypselae oblong to obconic . . . . 171. *Lorandersonia* (in part), **v. 20, p. 177**

15. Heads borne singly or in condensed, cymiform clusters grouped in paniculiform or corymbiform arrays; disc florets (2–)5–6 (–40); cypselae subcylindric . . . . . 175. *Chrysothamnus* (in part), **v. 20, p. 187**

12. Phyllaries disposed in spirals and equal or unequal.

16. Shrubs or treelets.

17. Heads borne singly; pappi of 30–60 bristles subtending 15–20 subulate-aristate scales, or of ca. 120 bristles . . . . 363. *Peucephyllum* (in part), **v. 21, p. 378**

17. Heads sometimes borne singly, sometimes in clusters or glomerate, usually in spiciform, racemiform, paniculiform, cymiform, or corymbiform arrays; pappi of 15–60 bristles (ca. 150 bristles in 3–4 series in 238. *Lepidospartum*, in part, shrubs with leaves of flowering stems filiform or acerose to scalelike).

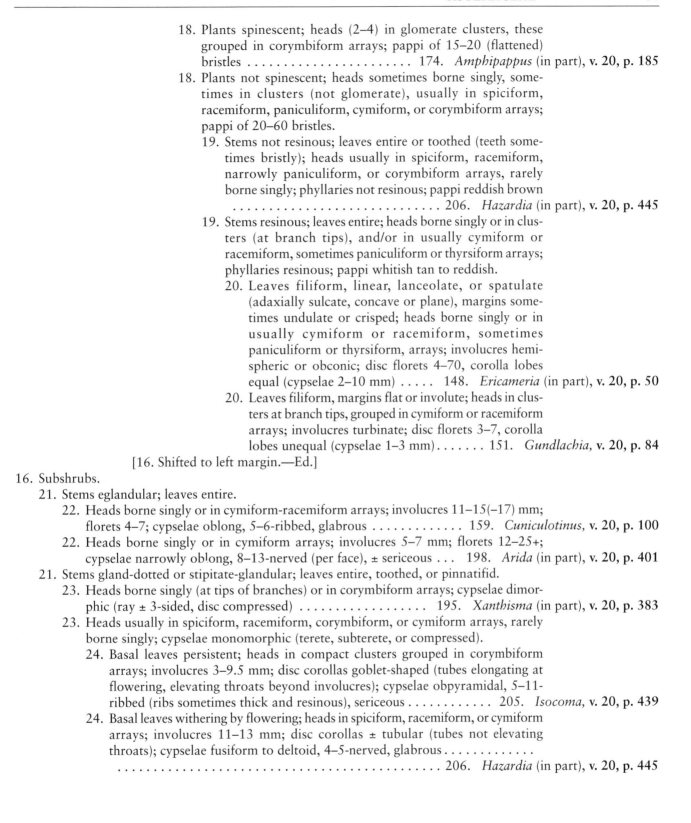

18. Plants spinescent; heads (2–4) in glomerate clusters, these grouped in corymbiform arrays; pappi of 15–20 (flattened) bristles . . . . 174. *Amphipappus* (in part), **v. 20, p. 185**

18. Plants not spinescent; heads sometimes borne singly, sometimes in clusters (not glomerate), usually in spiciform, racemiform, paniculiform, cymiform, or corymbiform arrays; pappi of 20–60 bristles.

    19. Stems not resinous; leaves entire or toothed (teeth sometimes bristly); heads usually in spiciform, racemiform, narrowly paniculiform, or corymbiform arrays, rarely borne singly; phyllaries not resinous; pappi reddish brown . . . . 206. *Hazardia* (in part), **v. 20, p. 445**

    19. Stems resinous; leaves entire; heads borne singly or in clusters (at branch tips), and/or in usually cymiform or racemiform, sometimes paniculiform or thyrsiform arrays; phyllaries resinous; pappi whitish tan to reddish.

        20. Leaves filiform, linear, lanceolate, or spatulate (adaxially sulcate, concave or plane), margins sometimes undulate or crisped; heads borne singly or in usually cymiform or racemiform, sometimes paniculiform or thyrsiform, arrays; involucres hemispheric or obconic; disc florets 4–70, corolla lobes equal (cypselae 2–10 mm) . . . . . 148. *Ericameria* (in part), **v. 20, p. 50**

        20. Leaves filiform, margins flat or involute; heads in clusters at branch tips, grouped in cymiform or racemiform arrays; involucres turbinate; disc florets 3–7, corolla lobes unequal (cypselae 1–3 mm) . . . . 151. *Gundlachia*, **v. 20, p. 84**

[16. Shifted to left margin.—Ed.]

16. Subshrubs.

  21. Stems eglandular; leaves entire.

    22. Heads borne singly or in cymiform-racemiform arrays; involucres 11–15(–17) mm; florets 4–7; cypselae oblong, 5–6-ribbed, glabrous . . . . 159. *Cuniculotinus*, **v. 20, p. 100**

    22. Heads borne singly or in cymiform arrays; involucres 5–7 mm; florets 12–25+; cypselae narrowly oblong, 8–13-nerved (per face), ± sericeous . . . 198. *Arida* (in part), **v. 20, p. 401**

  21. Stems gland-dotted or stipitate-glandular; leaves entire, toothed, or pinnatifid.

    23. Heads borne singly (at tips of branches) or in corymbiform arrays; cypselae dimorphic (ray ± 3-sided, disc compressed) . . . . 195. *Xanthisma* (in part), **v. 20, p. 383**

    23. Heads usually in spiciform, racemiform, corymbiform, or cymiform arrays, rarely borne singly; cypselae monomorphic (terete, subterete, or compressed).

        24. Basal leaves persistent; heads in compact clusters grouped in corymbiform arrays; involucres 3–9.5 mm; disc corollas goblet-shaped (tubes elongating at flowering, elevating throats beyond involucres); cypselae obpyramidal, 5–11-ribbed (ribs sometimes thick and resinous), sericeous . . . . 205. *Isocoma*, **v. 20, p. 439**

        24. Basal leaves withering by flowering; heads in spiciform, racemiform, or cymiform arrays; involucres 11–13 mm; disc corollas ± tubular (tubes not elevating throats); cypselae fusiform to deltoid, 4–5-nerved, glabrous . . . . 206. *Hazardia* (in part), **v. 20, p. 445**

[10. Shifted to left margin.—Ed.]
10. Annuals, biennials, or perennials.
 25. Annuals or biennials.
  26. Heads usually discoid (sometimes radiant in *Lessingia*).
   27. Leaf blades rounded-deltate, reniform, rounded-rhombic, or suborbiculate, margins toothed or entire, faces usually lanate, pilose, or tomentose, sometimes glandular-pubescent and/or furfuraceous as well; pappi of 35–150 bristles in 1–4 series . . . . 380. *Psathyrotes* (in part), **v. 21, p. 416**
   27. Leaf blades ovate, obovate, oblong, lanceolate, oblanceolate, or linear, faces glabrous, canescent, or puberulent, abaxial glabrous or sparsely tomentose to woolly, sometimes stipitate-glandular as well; pappi of 3–55 bristles in 1–3 series.
    28. Leaf adaxial faces glabrous, puberulent, or canescent; involucres turbinate; phyllaries usually spreading to reflexed, rarely appressed; corollas yellow (limbs of peripheral florets not expanded); pappus bristles white to tawny, distinct . . . . 197. *Dieteria* (in part), **v. 20, p. 395**
    28. Leaf adaxial faces sparsely tomentose to woolly; involucres hemispheric, obconic, campanulate, or narrowly cylindric; phyllaries erect or recurved; corollas (white, pink, lavender, or) yellow (limbs of peripheral florets frequently palmately expanded, heads ± radiant); pappus bristles tan to reddish, distinct or connate . . . . 209. *Lessingia*, **v. 20, p. 452**
  26. Heads usually disciform.
   29. Annuals (eglandular); disc corollas without prominent orange veins, style appendages lanceolate or linear.
    30. Heads borne singly; peripheral pistillate (or reduced ray) florets in 1 series; pappi (0 or) of (3–)5 bristles (or subulate scales, not surpassing corollas at flowering) . . . . 145. *Pentachaeta* (in part), **v. 20, p. 46**
    30. Heads in paniculiform arrays (borne singly in small plants); peripheral pistillate florets in 2–5+ series; pappi of (20–)25–40(–55) bristles (surpassing corollas at flowering) . . . . 214. *Symphyotrichum* (in part), **v. 20, p. 465**
   29. Annuals or biennials (usually gland-dotted or stipitate-glandular, sometimes eglandular, in *Conyza*); disc corollas sometimes with prominent orange veins (*Erigeron*), style appendages deltate.
    31. Biennials; stems ascending; heads borne singly (at ends of branches); disc corolla throats somewhat inflated (white-indurate); pappi of outer, shorter setae plus 15–20 inner, barbellate bristles . . . . 186. *Erigeron* (in part), **v. 20, p. 256**
    31. Annuals; stems erect; heads usually in corymbiform, paniculiform or racemiform arrays, rarely borne singly; disc corollas narrowly funnelform (throats neither inflated nor indurate); pappi of 10–30 barbellate bristles.
     32. Leaf faces often stipitate-glandular or gland-dotted; phyllaries lacking orange to brown midnerves; cypselae densely sericeous, ± strigillose, or glabrous, often stipitate-glandular and/or gland-dotted . . . . 142. *Laënnecia* (in part), **v. 20, p. 36**
     32. Leaf faces eglandular; phyllaries with orange to brownish midnerves; cypselae glabrous or strigillose, eglandular . . . . 187. *Conyza* (in part), **v. 20, p. 348**
 25. Perennials.
  33. Heads disciform.
   34. Cauline leaf margins entire or spinulose-serrate; heads borne singly or 2–3 in ± ± corymbiform arrays; cypselae subcylindric-fusiform, 3–4-angled, ± compressed, with 10–12 faint nerves; pappi of 15–60, rigid, unequal, smooth bristles in 1 series . . . . 201. *Pyrrocoma* (in part), **v. 20, p. 413**
   34. Cauline leaf margins entire, dentate, or pinnatifid (lobed); heads in corymbiform arrays; cypselae oblong or oblong-obovoid to elliptic or obovoid, ± compressed or flattened, 2(–4)-nerved or ± nerved on edges; pappi of outer setae or scales plus 5–40 bristles, or of 30–40+ bristles in 2 series (outer usually shorter).

35. Plants ± densely white-tomentose (at least some surfaces); phyllaries 1-nerved (nerves not golden-resinous); disc corollas yellowish, throats narrowly funnelform, not indurate (nerves pale, not resinous); pappi of 30–40+ bristles in 2 series (outer bristles usually shorter) . . . . 142. *Laënnecia* (in part), **v. 20, p. 36**

35. Plants ± hirsute or pilose; phyllaries usually 1–3-nerved (nerves golden-resinous); disc corollas yellow, strongly constricted basally, throats sometimes strongly inflated-indurate (nerves often orange-resinous); pappi of outer setae plus 7–25 inner bristles or of 7–25 bristles . . . .
. . . . 186. *Erigeron* (in part), **v. 20, p. 256**

[33. Shifted to left margin.—Ed.]

33. Heads discoid.
  36. Style-branch appendages usually abaxially and adaxially papillate to hispidulous.
    37. Stems erect; heads borne singly or in corymbiform arrays; cypselae narrowly obconic; pappi of 70–90, distinct bristles . . . . 366. *Psathyrotopsis*, **v. 21, p. 380**
    37. Stems erect or spreading; heads borne singly; cypselae cylindro-fusiform to obpyramidal; pappi of 35–150, distinct or basally connate bristles . . . .
. . . . 380. *Psathyrotes* (in part), **v. 21, p. 416**
  36. Style-branch appendages glabrous adaxially.
    38. Phyllary midnerves translucent and swollen (at least basally, not resinous) . . . . .
. . . . 163. *Solidago* (in part), **v. 20, p. 107**
    38. Phyllary midnerves usually not notably swollen (orange-resinous in *Erigeron*).
      39. Phyllary and corolla nerves orange-resinous; pappi of outer, shorter setae or scales plus inner, longer bristles . . . . 186. *Erigeron* (in part), **v. 20, p. 256**
      39. Phyllary and corolla nerves not orange-resinous, or only corolla nerves orange-resinous; pappi of equal or unequal bristles.
        40. Phyllaries subequal, foliaceous . . . . 169. *Toiyabea*, **v. 20, p. 172**
        40. Phyllaries usually unequal (sometimes subequal, then outer not foliaceous), outer ± herbaceous (sometimes foliaceous), or chartaceous, or proximally indurate.
          41. Phyllaries keeled, distally with relatively small green zones or green along midveins; pappus bristles in 2–3 series (outer notably shorter or relatively few).
            42. Leaves cauline (proximal withering by flowering), margins without coarse spreading cilia near bases; phyllary margins often reddish, sometimes hyaline, abaxial faces glabrous or glabrate to woolly, sometimes stipitate-glandular . . . . . . 143. *Eucephalus* (in part), **v. 20, p. 39**
            42. Leaves basal and cauline (basal and proximal withering by flowering), margins with coarse spreading cilia near bases or on petioles; phyllary margins not reddish, scarious, abaxial faces ± hispid or stipitate-glandular . . . . 185. *Heterotheca* (in part), **v. 20, p. 230**
          41. Phyllaries usually flat, sometimes keeled, distally herbaceous or green; pappi of equal or unequal bristles in 1–3 series (outer not notably shorter).
            43. Leaf faces densely scabrous and short-stipitate-glandular; phyllaries ± keeled proximally (Esmeralda County, Nevada) . . . .
. . . . 172. *Tonestus* (in part), **v. 20, p. 181**
            43. Leaf faces glabrous or canescent, ± puberulent, hispidulous, tomentose, or villous, and/or sometimes ± stipitate-glandular or gland-dotted; phyllaries sometimes keeled.

[44. Shifted to left margin.—Ed.]

44. Leaves basal and cauline, basal and proximal petiolate, distal sessile; style-branch appendages lanceolate.
    45. Plants densely stipitate-glandular, with caudices; stems densely clustered, simple; leaf blades obovate or oblong to broadly oblanceolate; phyllaries keeled; disc corolla throats cylindric; cypselae white strigoso-hirsute . . . . 194. *Triniteurybia*, **v. 20, p. 382**
    45. Plants sparsely, if at all, stipitate-glandular, taprooted; stems single, usually branched; leaf blades lanceolate to oblanceolate; phyllaries flat; disc corolla throats funnelform; cypselae sparsely appressed-hairy . . . . 197. *Dieteria* (in part), **v. 20, p. 395**
44. Leaves cauline, sessile; style-branch appendages triangular.
    46. Phyllary margins scarious; disc corolla lobes unequal; cypselae obpyramidal, sericeous; pappi of 40–50 tawny bristles . . . . 205. *Isocoma* (in part), **v. 20, p. 439**
    46. Phyllary margins not scarious; disc corolla lobes equal; cypselae prismatic, narrowly turbinate, fusiform, or deltoid, glabrous or sparsely scabrous; pappi of 15–30 reddish brown to brownish bristles.
        47. Stems glabrous or scabrous, eglandular; leaf bases not subclasping, margins entire; cypselae prismatic or narrowly turbinate; pappus bristles barbellate . . . .
           . . . . 200. *Oönopsis* (in part), **v. 20, p. 410**
        47. Stems scabrous to sparsely tomentulose, distally stipitate-glandular; leaf bases subclasping, margins serrate; cypselae fusiform to deltoid; pappus bristles smooth
           . . . . 206. *Hazardia* (in part), **v. 20, p. 445**

Key to Genera of Group 14
Heads eradiate; receptacles epaleate; pappi wholly, or partially, of awns or scales.

1. Florets 1(–2) or (1–)2–4(–5+) per head and heads in second-order heads or in headlike clusters.
   2. Subshrubs or shrubs.
      3. Leaves alternate, blades narrowly lance-linear, margins serrate (± thickened, spiny near bases); corollas purplish . . . . 1. *Hecastocleis* (in part), **v. 19, p. 71**
      3. Leaves opposite, blades lance-ovate to ovate, margins ± serrate (not notably thickened or spiny); corollas yellow . . . . 295. *Lagascea* (in part), **v. 21, p. 136**
   2. Annuals or perennials.
      4. Annuals, 1–5 cm; leaves opposite (basal and cauline, crowded, seemingly whorled)
         . . . . 307. *Dimeresia*, **v. 21, p. 182**
      4. Perennials, mostly 10–200 cm; leaves alternate (basal or basal and cauline, basal usually crowded).
         5. Plants mostly 100–200 cm (thistlelike); leaf blades 1–3-pinnately lobed, ultimate margins toothed (prickly or spiny) . . . . 9. *Echinops* (in part), **v. 19, p. 85**
         5. Plants mostly 10–120 cm (not thistlelike); leaf blades not lobed, margins usually toothed, rarely entire (not prickly or spiny).
            6. Heads (1–)10–40 per cluster, each cluster subtended by (2–)3, ± deltate bracts; pappi of 5(–6), 1-aristate scales (look closely for squamiform, gradually to abruptly tapering base of each arista), no scales tipped with plicate aristae . . . . 29. *Elephantopus*, **v. 19, p. 202**
            6. Heads 1–5+ per cluster, each cluster subtended by 1–2, lanceolate to spatulate or linear bracts; pappi of 6–10, ± laciniate to aristate scales, 2(–3+) of the aristate scales each with its awnlike arista plicate (2-folded) distally
               . . . . 30. *Pseudelephantopus*, **v. 19, p. 204**

1. Florets (2–)10–100(–1000) per head and heads borne singly or in usually corymbiform, paniculiform, racemiform, or spiciform arrays (heads ± crowded in *Orochaenactis*, annuals with pappi of 11–17 basally connate, oblanceolate scales, and in *Eriophyllum mohavense*, annuals with pappi of 12–14 distinct, linear to spatulate scales).

  7. Leaves opposite (at least proximal, cauline sometimes mostly alternate) or whorled.

     8. Corollas mostly yellow (sometimes with purple, red, or red-brown).

       9. Leaves and/or phyllaries dotted or streaked with pellucid (schizogenous) glands containing strong-scented oils . . . . 321. *Tagetes* (in part), **v. 21, p. 235**

       9. Leaves and/or phyllaries rarely dotted or streaked (never with pellucid, schizogenous glands containing strong-scented oils, plants sometimes with sessile or stipitate, surface glands and sometimes otherwise strong-scented).

         10. Phyllaries often ± conduplicate and navicular; disc corollas usually 4-lobed; cypselae strongly flattened or weakly 3–4-angled, usually callous-margined, often ciliate.

           11. Leaf blades usually lobed (mostly 5–30 mm); phyllaries 8–16 in 2–3 series, distinct . . . . 352. *Perityle*, **v. 21, p. 317**

           11. Leaf blades usually triangular-hastate to narrowly deltate, seldom notably lobed (30–120 mm); phyllaries 15–21 in 1(–2) series, wholly or partially connate . . . . 353. *Pericome*, **v. 21, p. 334**

         10. Phyllaries usually flat to weakly navicular (none in *Carlquistia* in which receptacular paleae function as an involucre); disc corollas (4–)5-lobed; cypselae usually obpyramidal to obconic, sometimes columnar, seldom clavate or cylindric, often 4–5-angled (seldom strongly compressed or flattened, callous-margined, and ciliate).

           12. Pappus scales (9–17) subulate, plumose . . . . 349. *Carlquistia*, **v. 21, p. 302**

           12. Pappus scales not both subulate and plumose.

             13. Leaves usually sessile, sometimes obscurely petiolate (rarely truly petiolate); pappus scales not notably medially thickened.

               14. Leaves glabrous (often granular-glandular, not woolly) . . . . 355. *Amblyopappus*, **v. 21, p. 348**

               14. Leaves ± woolly or tomentose (usually stems and/or phyllaries as well) . . . . 359. *Eriophyllum* (in part), **v. 21, p. 353**

             13. Leaves usually petiolate; pappus scales usually notably medially thickened.

               15. Annuals; leaves all or mostly opposite; pappus scales 8+ . . . . 367. *Schkuhria* (in part), **v. 21, p. 381**

               15. Annuals or biennials; leaves proximally opposite, mostly alternate; pappus scales 11–18.

                 16. Leaves lobed (lobes linear, 1–2 mm wide); disc corolla lobes lance-linear, lance-oblong, or linear (lengths mostly 2+ times widths); pappus scales distinct, narrowly lanceolate to subulate, some or all ± aristate . . . . 371. *Hymenothrix* (in part), **v. 21, p. 387**

                 16. Leaves lance-oblong to linear (not lobed); disc corolla lobes mostly deltate, lance-deltate, lanceolate, or ovate (lengths mostly 1–2 times widths); pappus scales basally connate (falling together), oblanceolate, obtuse . . . 379. *Orochaenactis*, **v. 21, p. 414**

     8. Corollas mostly white to cream or blue, lavender, pink, or purple.

       17. Style-branch appendages essentially 0 or ± penicillate or deltate to lanceolate, linear, or filiform (lengths mostly 0–3+ times lengths of stigmatic areas or lines).

         18. Phyllaries often ± conduplicate and navicular; disc corollas usually 4-lobed; cypselae strongly flattened or weakly 3–4-angled, usually callous-margined, often ciliate . . . . 352. *Perityle* (in part), **v. 21, p. 317**

         18. Phyllaries usually flat to weakly navicular; disc corollas 5-lobed; cypselae obpyramidal, 4-angled (not callous-margined and ciliate).

19. Leaf blades broadly lanceolate to linear (not lobed); corollas usually
pinkish or purplish, sometimes whitish; cypselae densely to sparsely
hairy (hairs straight) . . . . 372. *Palafoxia* (in part), **v. 21, p. 388**

19. Leaf blades (at least mid-cauline) 3- or 5-lobed or -foliolate (blades or
leaflets broadly to narrowly oblong to ovate); corollas whitish; cypselae
sparsely hairy (hairs curled) . . . . 373. *Florestina*, **v. 21, p. 392**

17. Style-branch appendages usually terete to clavate (lengths usually 2–5+ times
lengths of stigmatic lines).

20. Involucres narrowly cylindric, (1–)2–3 mm diam.; phyllaries 5(–6) in ± 1–
2 series; florets 5(–6) . . . . 397. *Stevia* (in part), **v. 21, p. 483**

20. Involucres campanulate, cylindric, ellipsoid, hemispheric, or obconic, (2–)
3–7(–25) mm diam.; phyllaries (5–)8–45(–65+) in (1–)2–8+ series; florets
(3–)10–125(–200+).

21. Cypselae 8–10-ribbed (pappi of 1–5+ muticous, erose, lacerate, or
lanceolate to subulate scales plus 9–12+ aristate scales) . . . .
. . . . 398. *Carphochaete*, **v. 21, p. 486**

21. Cypselae (3–)4–5(–8)-ribbed.

22. Pappi of 8–13 plumose, setiform scales (coherent or ± connate,
falling together or in groups) . . . . 409. *Carminatia* (in part), **v. 21, p. 511**

22. Pappi usually of 2–6(–12) muticous or aristate to subulate scales
plus 0–6(–12) setiform scales, rarely coroniform (*Ageratum*).

23. Phyllaries unequal; receptacles flat to convex (not warty).

24. Leaves mostly sessile (or nearly so), blades linear; cypselae
± fusiform . . . . 407. *Malperia* (in part), **v. 21, p. 509**

24. Leaves petiolate, blades ovate, deltate, or rhombic to
lanceolate; cypselae prismatic . . . . 408. *Pleurocoronis*, **v. 21, p. 510**

23. Phyllaries ± equal; receptacles convex to conic or hemispheric
(sometimes warty).

25. Leaves whorled (4 or 6 per node), blades linear; heads
borne singly (plants ± aquatic) . . . . 401. *Sclerolepis*, **v. 21, p. 488**

25. Leaves mostly opposite (distal sometimes alternate), blades
elliptic, lanceolate, or oblong; heads usually in cymiform
to corymbiform arrays, sometimes borne singly.

26. Leaves petiolate; involucres 3–6 mm diam.; phyllaries
usually 2-nerved; pappi usually of 5–6 aristate scales,
rarely coroniform . . . . 396. *Ageratum* (in part), **v. 21, p. 481**

26. Leaves sessile; involucres 3–4(–5) mm diam.; phyllaries obscurely 3–4-nerved; pappi of 2–6 setiform scales
. . . . 399. *Trichocoronis*, **v. 21, p. 487**

[7. Shifted to left margin.—Ed.]

7. Leaves alternate.

27. Leaf margins prickly to spiny (plants ± thistlelike).

28. Stems winged.

29. Cypsela attachments basal; pappus scales basally connate . . . .
. . . . 10. *Onopordum* (in part), **v. 19, p. 87**

29. Cypsela attachments slightly lateral; pappus scales usually distinct, sometimes
basally connate . . . . 12. *Carduus* (in part), **v. 19, p. 91**

28. Stems not or rarely winged (some *Cirsium* spp.).

30. Leaves variegated (with white veins or mottlings; stamen filaments connate)
. . . . 14. *Silybum*, **v. 19, p. 164**

30. Leaves not variegated (stamen filaments distinct).

31. Corollas white or purplish to red; cypsela attachments basal; pappi of
basally connate, plumose, setiform scales ("flattened bristles").

32. Involucres 35–100+ mm diam. (larger leaves 60–150 cm; receptacles
becoming fleshy) . . . . 11. *Cynara* (in part), **v. 19, p. 89**

32. Involucres 10–50 mm diam. (larger leaves usually 20–50, sometimes to
110 cm; receptacles usually not notably fleshy) . . . . . . 13. *Cirsium* (in part), **v. 19, p. 95**

31. Corollas yellow to orange, red, or ± purple; cypsela attachments lateral; pappi 0 or of distinct, barbellulate, setiform scales ("flattened bristles") or subulate scales.
    33. Heads discoid (all florets bisexual, fertile); cypselae 4-angled . . . . 23. *Carthamus* (in part), **v. 19, p. 178**
    33. Heads disciform (peripheral florets neuter); cypselae terete, 20-ribbed . . . . 24. *Centaurea* (in part), **v. 19, p. 181**

[27. Shifted to left margin.—Ed.]
27. Leaf margins not prickly or spiny (tips sometimes spinose or apiculate; plants not thistlelike).
  34. Corollas mostly white or blue, lavender, pink, or purple.
    35. Anthers tailed (styles swollen or with rings of hairs proximal to branches).
      36. Heads discoid (all florets bisexual and fertile).
        37. Cypsela attachments basal (receptacles becoming fleshy; florets 100–250+; pappus scales basally connate) . . . . 11. *Cynara* (in part), **v. 19, p. 89**
        37. Cypsela attachments ± lateral.
          38. Phyllary appendages dentate or fringed (often spiny); pappi of persistent, nonplumose scales . . . . 24. *Centaurea* (in part), **v. 19, p. 181**
          38. Phyllary appendages entire or lacerate, not fringed; pappi of readily falling, distally plumose bristles . . . . 17. *Acroptilon*, **v. 19, p. 171**
      36. Heads disciform or radiant (peripheral florets usually neuter).
        39. Heads disciform.
          40. Phyllary appendages dentate or fringed, often spiny . . . . 24. *Centaurea* (in part), **v. 19, p. 181**
          40. Phyllary appendages none (apices acute, entire) . . . . 22. *Crupina*, **v. 19, p. 177**
        39. Heads radiant.
          41. Phyllary appendages none (apices spiny).
            42. Biennials or perennials; spines on phyllary apices soon falling; cypsela apices not coronate . . . . 19. *Mantisalca* (in part), **v. 19, p. 173**
            42. Annuals; spines on phyllary apices persistent; cypsela apices coronate . . . . 20. *Volutaria*, **v. 19, p. 174**
          41. Phyllary appendages present.
            43. Cypselae compressed (oblong; attachment scars rimmed, rims whitish, swollen), apices denticulate . . . . 18. *Amberboa*, **v. 19, p. 172**
            43. Cypselae ± terete (barrel-shaped; attachment scars not rimmed), apices entire . . . . 24. *Centaurea* (in part), **v. 19, p. 181**
    35. Anthers not tailed (sometimes sagittate; styles not swollen or with rings of hairs proximal to branches).
      44. Phyllaries 18–70+ in 3–8+ series.
        45. Heads pseudo-radiant or -radiate (corollas of peripheral, bisexual florets enlarged, zygomorphic); phyllary margins (at least of the outer), pectinately spinose-toothed . . . . 28. *Stokesia*, **v. 19, p. 201**
        45. Heads discoid; phyllary margins not pectinately spinose-toothed.
          46. Annuals (perhaps persisting); cypselae not ribbed . . . . . 31. *Cyanthillium*, **v. 19, p. 204**
          46. Perennials (sometimes functionally annuals); cypselae 8–10-ribbed.
            47. Heads each subtended by 3–8+, ± foliaceous bracts; pappi readily falling . . . . 32. *Centratherum*, **v. 19, p. 206**
            47. Heads not each subtended by foliaceous bracts; pappi persistent . . . . 33. *Vernonia*, **v. 19, p. 206**
      44. Phyllaries 5–15(–21) in 1–2(–3) series.
        48. Style-branch appendages usually terete to clavate (lengths usually 2–5+ times lengths of stigmatic lines).
          49. Phyllaries 5(–6) in 1 series; florets 5(–6) . . . . 397. *Stevia* (in part), **v. 21, p. 483**
          49. Phyllaries 12–15 in 2–3 series; florets 7–10 . . . . 413. *Hartwrightia* (in part), **v. 21, p. 540**
        48. Style-branch appendages mostly deltate to lanceolate, linear, or filiform, sometimes essentially none (lengths mostly 0–3+ times lengths of stigmatic areas or lines).

50. Heads in (usually secund) spiciform arrays; florets 3 . . . . . . 153. *Thurovia*, **v. 20, p. 86**
50. Heads borne singly or in corymbiform or cymiform arrays; florets 8–70.
    51. Cypselae stoutly obconic to obpyramidal, usually 4-angled and 12–16-ribbed (lengths usually 1–2, rarely to 3.5 times diams.; phyllary margins usually notably membranous to scarious) . . . . 351. *Hymenopappus* (in part), **v. 21, p. 309**
    51. Cypselae narrowly clavate to cylindric or compressed, obscurely 8–20-angled, or ± quadrangular with 8–12 obscure nerves (lengths usually 3+ times diams., if stouter, usually ± compressed; phyllary margins not notably membranous or scarious).
        52. Leaves mostly basal; heads borne singly; florets 10–30+ (perennials 2–9 cm, 10–20+ cm across; leaf blades 1- or 3-nerved, cordate, elliptic, ovate, or rounded, margins entire or distally ± crenate, revolute to ± plane, faces ± strigose and gland-dotted, adaxial sometimes glabrescent) . . . . 376. *Chamaechaenactis*, **v. 21, p. 395**
        52. Leaves basal and cauline or mostly cauline; heads borne singly or in ± cymiform arrays; florets 8–70+ . . . . 378. *Chaenactis* (in part), **v. 21, p. 400**
[34. Shifted to left margin.—Ed.]
34. Corollas mostly yellow to orange (sometimes with purple, red, or red-brown).
    53. Shrubs; pappi of 12–30 scales (sometimes with bristles as well).
        54. Phyllaries 4–6, or 8–18, in 2 series, ± equal.
            55. Florets 4–8 (cypselae copiously pilose, hairs white, 4–14 mm, sometimes obscuring pappi) . . . . 237. *Tetradymia* (in part), **v. 20, p. 629**
            55. Florets 12–21 (cypselae hirsute, hairs tawny to reddish, 0.5–1 mm; pappi often of both scales and bristles) . . . . 363. *Peucephyllum* (in part), **v. 21, p. 378**
        54. Phyllaries 7–60+ in 3–5 series, mostly unequal.
            56. Phyllaries 40–60+ in 3–5 series (often in vertical ranks) . . . . 175. *Chrysothamnus* (in part), **v. 20, p. 187**
            56. Phyllaries 7–25 in 3 series.
                57. Involucres campanulate, globose, or hemispheric, 4–13 × 7–16 mm; florets 13–45 . . . . 173. *Acamptopappus* (in part), **v. 20, p. 184**
                57. Involucres turbino-cylindric, 4–5.5 × 2–3 mm; florets 3–7 (functionally staminate) . . . . 174. *Amphipappus* (in part), **v. 20, p. 185**
    53. Annuals, biennials, perennials, subshrubs, or shrubs; pappi mostly of 1–12(–20+) scales (scales sometimes subtending 20–100+ bristles; if shrubs, pappus scales 1 or 3–5).
        58. Shrubs (stems ± tomentose; leaf blades obscurely pinnate; heads borne singly or in loose, corymbiform arrays; pappi of 1 or 3–5 scales) . . . . 125. *Pentzia* (in part), **v. 19, p. 543**
        58. Annuals, biennials, perennials, or subshrubs.
            59. Leaves usually 1–3-pinnate (usually aromatic when crushed); phyllary margins usually notably scarious; style-branch appendages essentially none; pappus scales usually rudimentary (3–5, subulate in *Sphaeromeria*).
                60. Annuals; heads discoid. . . . . 124. *Matricaria* (in part), **v. 19, p. 540**
                60. Perennials; heads disciform (peripheral florets pistillate).
                    61. Heads usually in lax to dense, corymbiform arrays, rarely borne singly; involucres (3–)5–10 mm diam.; florets 60–300+ . . . . 112. *Tanacetum* (in part), **v. 19, p. 489**
                    61. Heads usually (8–20+) in tight, capitate arrays; involucres (2–)3–4 mm diam.; florets 30–50+ . . . . 118. *Sphaeromeria* (in part), **v. 19, p. 499**
            59. Leaves usually not lobed, sometimes 1–4-pinnate (seldom notably aromatic when crushed); phyllary margins seldom notably scarious; style-branch appendages mostly deltate to lanceolate or linear; pappus scales usually conspicuous (sometimes associated with bristles).

[62. Shifted to left margin.—Ed.]

62. Phyllaries 25–125 in 3–9 series; pappi of scales only or of 5–20 scales plus 20–100+ bristles.

   63. Phyllary apices often looped, hooked, patent, recurved, straight, or incurved; pappi fragile or readily falling, of 8–15 subulate to setiform scales . . . . . 202. *Grindelia* (in part), **v. 20, p. 424**

   63. Phyllary apices usually erect and straight, sometimes spreading or reflexed; pappi persistent or readily falling, usually of scales plus bristles.

      64. Heads usually in corymbiform, ± paniculiform, or subumbelliform arrays, sometimes borne singly; phyllaries in 3–5 series, unequal, usually thickened or keeled, 1-nerved (nerves not golden-resinous); cypselae 3-angled or 4–12-ribbed . . . . 185. *Heterothecaa* (in part), **v. 20, p. 230**

      64. Heads borne singly or in 2s or 3s; phyllaries in 2–3 series, equal or subequal, flat, usually 1–3-nerved (nerves golden-resinous); cypselae 2-nerved . . . . 186. *Erigeron* (in part), **v. 20, p. 256**

62. Phyllaries 6–25(–60) in 2–3 series; pappi mostly wholly of scales (usually 5, 10, 15, or 20, subulate-aristate scales in *Pentachaeta*).

   65. Cypselae narrowly clavate to oblanceoloid, or columnar to obconic or obpyramidal (lengths usually 3+ times diams., if stouter, usually ± compressed).

      66. Leaves usually sessile, sometimes obscurely petiolate (rarely truly petiolate); pappus scales often aristate, not medially thickened . . . . 145. *Pentachaeta* (in part), **v. 20, p. 46**

      66. Leaves usually petiolate (± sessile in some spp. of *Chaenactis*); pappus scales sometimes aristate or lacerate, often notably medially thickened.

         67. Pappus scales 4–20, not lacerate (usually erose, sometimes uniaristate) . . . . 378. *Chaenactis* (in part), **v. 21, p. 400**

         67. Pappus scales 5, lacerate (divisions bristlelike) . . . . 381. *Trichoptilium*, **v. 21, p. 418**

   65. Cypselae stoutly obconic to obpyramidal (lengths usually 1–2, rarely to 3.5, times diams.).

      68. Phyllaries usually strongly reflexed in fruit; receptacles mostly globose (sometimes with setiform enations); disc corollas often brown-purple to red-brown or tipped with brown-purple to red-brown (tubes much shorter than abruptly much-dilated, urceolate to campanulate throats, lobes often shaggily hairy, hairs ± moniliform).

         69. Stems not winged (receptacles usually with setiform enations; style-branch apices ± attenuate) . . . . 384. *Gaillardia* (in part), **v. 21, p. 421**

         69. Stems often winged (by decurrent leaf bases; receptacles rarely with setiform enations; style-branch apices penicillate or truncate) . . . . . 385. *Helenium* (in part), **v. 21, p. 426**

      68. Phyllaries mostly erect to spreading in fruit; receptacles flat, conic, domed, hemispheric, or ovoid (smooth or pitted, without setiform enations); disc corollas usually uniformly yellow to cream, sometimes purplish to reddish (tubes much shorter than to about equaling slightly dilated, funnelform to cylindric throats, lobes not shaggily hairy with moniliform hairs).

         70. Leaf blades simple or 1–2-pinnately lobed (lobes mostly filiform, linear, or oblong), ultimate margins entire or toothed, faces glabrous or hairy, usually ± gland-dotted (often in pits); phyllaries persistent (or inner falling), usually (6–)6–30(–40) in 2 series and unequal (outer usually connate), sometimes 28–50 in 2–3 series and subequal (usually spreading to erect in fruit) . . . . 386. *Hymenoxys* (in part), **v. 21, p. 435**

         70. Leaf blades mostly oblanceolate to linear or filiform, sometimes lobed, ultimate margins usually entire, sometimes toothed, faces glabrous or ± hairy, eglandular or ± gland-dotted; phyllaries usually persistent, 11–60+ in 2–3 series (mostly spreading to erect in fruit, distinct, herbaceous; outer with or without scarious margins, abaxial faces ± hairy; mid usually same number as, alternating with, and similar to outer, almost always with ± scarious margins; inner narrower than others, margins scarious) . . . . 389. *Tetraneuris* (in part), **v. 21, p. 447**

187a. ASTERACEAE Martinov tribe MUTISIEAE Cassini, J. Phys. Chim. Hist. Nat. Arts 88: 199. 1819

Annuals, perennials, subshrubs, or shrubs (sometimes clambering) [trees, vines]. **Leaves** basal and/or cauline; alternate [opposite]; petiolate or sessile; margins entire, dentate, pinnately lobed, or pinnatifid [pinnately compound, spiny, tipped with tendrils]. **Heads** homogamous (discoid in *Gochnatia, Hecastocleis*; quasi-discoid, -radiate, or -liguliflorous in *Acourtia, Trixis*) or heterogamous (disciform in *Adenocaulon*, disciform or quasi-radiate or -liguliflorous in *Chaptalia* and *Leibnitzia*; sometimes cleistogamous in *Chaptalia* and *Leibnitzia*); borne singly (sometimes on scapiform stems) or in corymbiform, paniculiform, or racemiform arrays (aggregated in second-order heads, florets 1–3 per individual head in *Hecastocleis*). **Calyculi** usually 0 (of 3–7 bractlets in *Trixis*; second-order heads subtended by leaflike bracts in *Hecastocleis*). **Phyllaries** persistent or tardily falling, in 1–5+ series, distinct or connate, unequal to subequal, usually ± herbaceous, margins and apices seldom notably scarious. **Receptacles** flat to convex, epaleate [paleate] (usually ± foveolate to alveolate, margins of sockets sometimes ± membranous). **Florets** of 1, 2, or 3+ kinds in a head (some not readily assignable to usual ray- and disc-floret categories, ± 3 combinations in the flora): (1) all florets bisexual and fertile, corollas actinomorphic (*Gochnatia, Hecastocleis*) or zygomorphic (often 2-lipped; *Acourtia, Trixis*); (2) outer florets pistillate and inner bisexual, corollas actinomorphic (*Adenocaulon*); and (3) outer florets pistillate, corollas usually zygomorphic (raylike or liguliform to 2-lipped), sometimes lacking limbs (reduced to tubes), inner florets bisexual or functionally staminate, corollas zygomorphic (2-lipped) or actinomorphic (*Chaptalia, Leibnitzia*); anther bases ± tailed, apical appendages ovate or lanceolate to linear [none]; styles (bisexual, fertile florets) abaxially glabrous or papillate to hairy, branches ± linear, adaxially stigmatic in 2 lines from bases nearly to apices (± continuous around apices in *Adenocaulon*), apices rounded or truncate, appendages essentially none. **Cypselae** ± monomorphic within heads, usually cylindric, clavate, fusiform, or turbinate, sometimes ± beaked, bodies smooth or ribbed (glabrous, glabrate, sericeous, or velutinous; sometimes distally stipitate-glandular, *Adenocaulon*); **pappi** 0 or persistent, of smooth or barbellulate to plumose bristles or subulate to setiform and plumose scales.

Genera ca. 76, species ca. 1000 (7 genera, 14 species in the flora): North America, mostly South America, Asia, Africa, Pacific Islands (Hawaii), Australia.

Mutisieae as traditionally circumscribed is now considered to be polyphyletic. Molecular work has shown that *Barnadesia* and its allies, a South American group formerly placed in Mutisieae, constitute the basal clade of the family; they are now treated as a distinct subfamily, Barnadesioideae (K. Bremer and R. K. Jansen 1992). Similarly, J. L. Panero and V. A. Funk (2002) segregated two additional subfamilies from Mutisieae (based on *Gochnatia* and on *Hecastocleis*). Mutisieae in the broad sense are centered in the New World tropics and subtropics and are relatively poorly represented elsewhere.

Gerbera jamesonii Bolus ex Hooker f. (Transvaal Daisy; perennials, leaves in basal rosettes, petioles 10–20+ cm, blades usually pinnately lobed, 12–25+ cm, abaxially lanate, peduncles scapiform, to 50+ cm, ray corollas red to orange, pink, or yellow, laminae 2–3+ cm) has been noted as established in Florida (http://www.plantatlas.usf.edu).

SELECTED REFERENCES Hansen, H. V. 1990. Phylogenetic studies in the *Gerbera*-complex (Compositae, tribe Mutisieae, subtribe Mutisiinae). Nordic J. Bot. 9: 469–485. Kim, H. G., D. J. Lookerman, and R. K. Jansen. 2002. Systematic implications of *ndb*F sequence variation in the Mutisieae (Asteraceae). Syst. Bot. 27: 598–609. Simpson, B. B. and C. E. Anderson. 1978. Mutisieae. In: N. L. Britton et al., eds. 1905+. North American Flora.... 47+ vols. New York. Ser. 2, part 10, pp. 1–13.

1. Leaves: abaxial faces glabrous or with scattered, straight, sometimes glandular, hairs and/
   or double hairs.
   2. Leaf blades narrowly linear-lanceolate (spiny near bases, ± thickened at margins);
      corollas actinomorphic; style-branch apices rounded .... 1. *Hecastocleis*, p. 71
   2. Leaves broadly lanceolate or ovate (sometimes linear-lanceolate in *Trixis*; not spiny at
      bases or thickened at margins); corollas 2-lipped; style-branch apices truncate.
      3. Perennials (caudices brown-woolly); corollas pink, lavender, or white .... 2. *Acourtia*, p. 72
      3. Shrubs; corollas yellow .... 3. *Trixis*, p. 75
1. Leaves: abaxial faces usually tomentose, tomentulose, or covered with dense wool (thinly
   gray-tomentose in *Leibnitzia*).
   4. Shrubs or subshrubs; leaves (stiff): adaxial faces dark green; corollas actinomorphic
      (heads discoid) .... 4. *Gochnatia*, p. 76
   4. Perennials; leaves (soft): adaxial faces light green; corollas actinomorphic (heads discoid
      or disciform) or actinomorphic and zygomorphic (heads disciform or quasi-radiate).
      5. Leaf blades triangular to broadly triangular, bases cordate to hastate; heads in
         paniculiform arrays; corollas of inner florets actinomorphic .... 5. *Adenocaulon*, p. 77
      5. Leaf blades elliptic to obovate, bases cuneate; heads borne singly; corollas of inner
         florets zygomorphic (2-lipped) or nearly actinomorphic.
         6. Heads chasmogamous (produced well after rosette leaves); cypselae glabrous
            or pubescent (hairs relatively short, apices rounded or apiculate) .... 6. *Chaptalia*, p. 78
         6. Heads chasmogamous (early spring, before and concurrently with first rosette
            leaves) or cleistogamous (produced after rosette leaves); cypselae strigose to
            hispid (hairs relatively long, apices sharp-pointed) .... 7. *Leibnitzia*, p. 80

1. HECASTOCLEIS   A. Gray, Proc. Amer. Acad. Arts 17: 220. 1882 • [Greek *hecastos*,
each, and *cleios*, to shut up, alluding to one floret enclosed in each involucre] $\boxed{\text{E}}$

Beryl B. Simpson

Subshrubs or shrubs, 40–80(–150+) cm (usually with tufts of hairs in axils of leaves). **Leaves** cauline; sessile; blades linear, bases cuneate, margins entire or proximally spiny, faces tomentulose, glabrescent. **Heads** discoid, clustered in second-order heads (each cluster 15–25 mm diam. and subtended by ovate to orbiculate, prickly-margined bracts). **Involucres** (each enclosing 1 floret) cylindric to fusiform, 10 mm. **Phyllaries** in 2–3 series, ovate to lanceolate or linear, unequal, apices acute to cuspidate. **Receptacles** flat, smooth, glabrous, epaleate. **Florets** 1, bisexual, fertile; corollas reddish purple to greenish white, actinomorphic (lobes 5, lance-linear, glabrous); anther basal appendages slightly fimbriate, apical appendages lanceolate, acute; style branches relatively stout (0.1–0.5 mm), apices rounded (abaxial faces papillose to pilose). **Cypselae** ± terete (± 4 mm), not beaked, obscurely 4–5-nerved, faces glabrescent, not glandular-hairy; **pappi** of 6 unequal, lanceolate or multitoothed scales (sometimes ± coalescent and forming lacerate crowns). *x* = 8.

Species 1: sw United States.

SELECTED REFERENCE Williams, M. J. 1977. *Hecastocleis shockleyi* A. Gray. Mentzelia 3: 18.

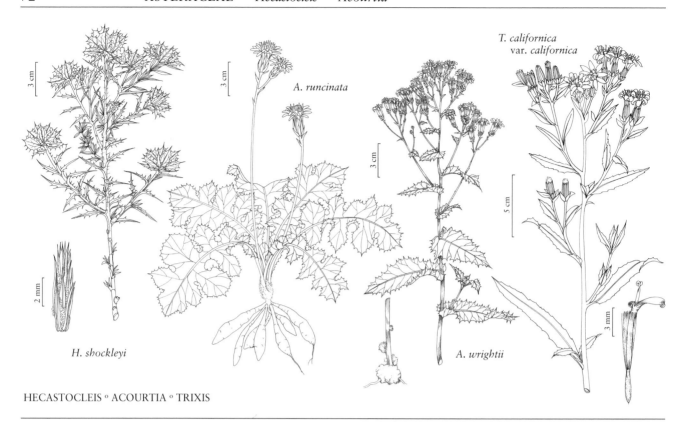

T. californica
var. *californica*

A. runcinata

H. shockleyi

A. wrightii

HECASTOCLEIS ○ ACOURTIA ○ TRIXIS

1. Hecastocleis shockleyi A. Gray, Proc. Amer. Acad.
Arts 17: 221. 1882 • Prickle-leaf E F

Leaves: blades 1–4 cm, bases ± clasping, margins thickened, usually with 3–6 spines 1–3 mm near bases, apices mucronate or acute. **Phyllaries** usually lanate on margins. **Corollas** pinkish purple in bud, white at flowering, ca. 10 mm. **Cypselae** brown, 5 mm; **pappi** 1–2 mm. $2n = 16$.

Flowering Apr–Jun. Granite, basalt, and caliche soils, low desert scrub; 1700–2000 m; Calif., Nev.

Hecastocleis shockleyi is known from southwestern Nevada and adjacent California.

2. ACOURTIA D. Don, Trans. Linn. Soc. London 16: 203. 1830 • Desertpeony [For Mrs. A'Court, a British amateur botanist]

Beryl B. Simpson

Perennials, (2.5–)5–50(–150+) cm (caudices brown-woolly, aerial stems glabrate or resinous-punctate). **Leaves** basal, cauline, or both; shortly petiolate or sessile; blades elliptic-oblong, lanceolate, oblong, oblong-lanceolate, oblong-oblanceolate, orbiculate, ovate, ovate-elliptic, or rhombic-orbiculate (thin and chartaceous to thick and coriaceous), bases cuneate to cordate or clasping, margins entire or lobed or pinnately parted, dentate, or serrate, faces usually minutely stipitate-glandular and/or hirtellous. **Heads** quasi-radiate [discoid] (see florets), borne singly or in paniculiform or corymbiform arrays. **Involucres** turbinate or obconic to campanulate, 6–17+ mm. **Phyllaries** in 1–7 series, lanceolate to oblanceolate or linear, unequal (rigid, margins scarious), apices obtuse to acute, acuminate, or mucronate. **Receptacles** concave, flat, or convex, usually foveolate, alveolate, or reticulate, pubescent, sometimes paleate (paleae apically pubescent). **Florets** 3–25(–80), bisexual, fertile; corollas pink to lavender or white [yellow], zygomorphic (2-lipped; outer lip liguliform, 3-toothed, inner usually smaller, 2-lobed, lobes often curled); anther basal appendages entire, elongate, rounded, apical appendages lanceolate; style branches relatively short, apices blunt-penicillate (abaxial faces usually glabrous, i.e., without collecting hairs). **Cypselae** ± fusiform or terete to cylindric, 4–10 mm, not beaked, usually ± ribbed, faces glabrous or stipitate-glandular; **pappi** of 40–60(–80+) tan or white, ± barbellate to nearly smooth bristles in 1–3(–9) series. *x* = 27.

Species ca. 41 (5 in the flora): warm regions of North America, Mexico, Central America.

Acourtia consists of two clades, one with species that have scapiform stems and the other with species that have leafy flowering stems. From about 1873 to 1973, *Acourtia* species were treated as members of *Perezia*, usually as *Perezia* sect. *Acourtia* (D. Don) A. Gray. J. L. Reveal and R. M. King (1973) reestablished *Acourtia* for the leafy-stemmed North American species, and B. L. Turner (1978) added the scapiform species. Molecular evidence (H. G. Kim et al. 2002) indicated *Acourtia* is most closely related to *Proustia* and *Trixis* and not to *Perezia*.

SELECTED REFERENCES Bacigalupi, R. 1931. A monograph of the genus *Perezia*, section *Acourtia*, with a provisional key to the section *Euperezia*. Contr. Gray Herb. 97: 1–81. Cabrera R., L. 1992. Systematics of *Rzedowskiela* Gen. Nov. (Asteraceae: Mutisieae: Nassauviinae). Ph.D. dissertation. University of Texas. Cabrera R., L. 2001. Six new species of *Acourtia* (Asteraceae) and a historical account of *Acourtia mexicana*. Brittonia 53: 416–429. Reveal, J. L. and R. M. King. 1973. Re-establishment of *Acourtia* D. Don (Asteraceae). Phytologia 27: 228–232. Turner, B. L. 1978. Taxonomic study of the scapiform species of *Acourtia* (Asteraceae–Mexico). Phytologia 38: 456–468. Turner, B. L. 1993c. New taxa, new combinations, and nomenclatural comments on the genus *Acourtia* (Asteraceae: Mutisieae). Phytologia 74: 385–412.

1. Leaves basal (blades pinnately lobed); heads borne singly (or 2–3, on scapiform peduncles) . . . . 1. *Acourtia runcinata*
1. Leaves cauline (blades not pinnately lobed); heads borne singly (at tips of branches) or in corymbiform or paniculiform arrays.
   2. Plants 2.5–30 cm; leaf blades rhombic-orbiculate to suborbiculate (hollylike, lengths about equaling widths) . . . . 2. *Acourtia nana*
   2. Plants 30–150 cm; leaf blades elliptic, elliptic-oblong, lance-oblong, oblong-lanceolate, oblong-ovate, oval, ovate, or ovate-elliptic (not hollylike, lengths mostly greater than widths).
      3. Leaf blades oval to ovate-elliptic; phyllaries acuminate; florets 3–6 . . . . 3. *Acourtia thurberi*
      3. Leaf blades elliptic, elliptic-oblong, oblong-lanceolate, or oblong-ovate; phyllaries obtuse to acute or mucronate; florets 8–20.
         4. Leaf margins dentate to denticulate; phyllary apices obtuse to shortly acute; florets 8–12 . . . . 4. *Acourtia wrightii*
         4. Leaf margins spinulose-denticulate; phyllary apices acute to acuminate or mucronate; florets 10–20 . . . . 5. *Acourtia microcephala*

1. **Acourtia runcinata** (D. Don) B. L. Turner, Phytologia 38: 460. 1978 • Featherleaf desertpeony, desert paeonia ⊡

Clarionia runcinata Lagasca ex D. Don, Trans. Linn. Soc. London 16: 207. 1830

Plants 5–35 cm. **Leaves** basal; petioles 0.5–9 cm; blades oblong-oblanceolate, 2.5–23 cm, bases cuneate, margins pinnately lobed (lobes ovate), spinulose-dentate, apices acute, faces hirtellous and/or finely stipitate-glandular. **Heads** borne singly or 2–3 together (on scapiform peduncles). **Involucres** turbinate, 14–17 mm. **Phyllaries** in 2–4 series, lanceolate to subulate (3–15 mm), margins glandular-hairy, apices acuminate, abaxial faces glandular. **Receptacles** reticulate, hispidulous. **Florets** 25–53; corollas pink or lavender-pink, 12–22 mm. **Cypselae** fusiform to linear-fusiform, 4–8 mm, stipitate-glandular and/or hispidulous; **pappi** tan or white, 1–17 mm. *2n* = ca. 54.

Flowering year round (mostly Mar–Aug). Juniper forests, oak woodlands, dry matorral, desert scrub on calcareous, sandy clay, and gypsiferous soils; 0–1600 m;

Tex.; Mexico (Chihuahua, Coahuila, Hidalgo, Nuevo León, San Luis Potosí, Tamaulipas).

Acourtia runcinata is the most widespread species of the genus, within which it is unique in possessing fasciculate, tuberous-fusiform roots. In the flora, it grows in central and southwestern Texas.

2. **Acourtia nana** (A. Gray) Reveal & R. M. King, Phytologia 27: 230. 1973 • Desert holly, dwarf desertpeony

Perezia nana A. Gray, Mem. Amer. Acad. Arts, n. s. 4: 111. 1849

Plants 2.5–30 cm, (divaricately branching). **Leaves** cauline; sessile; blades rhombic-orbiculate to suborbiculate, 10–50 mm, bases cuneate, margins coarsely and irregularly prickly-dentate, faces glabrous (reticulate). **Heads** borne singly (at branch tips). **Involucres** campanulate, 14–17 mm. **Phyllaries** in 4 series, broadly ovate, margins glandular, apices acute to mucronate, abaxial faces glabrous. **Receptacles** reticulate (sockets separated by squarish, apically pubescent paleae 1 mm). **Florets** 15–24; corollas lavender-pink or white, 10–17 mm. **Cypselae** subcylindric, 3–7.5 mm, densely stipitate-glandular; **pappi** white or tawny, 10–15 mm. $2n = 54$.

Flowering (Mar–)Apr(–Jun). Gravel, sandstone, silty, or caliche soils in desert scrub; 0–1800 m; Ariz., N.Mex., Tex.; Mexico (Chihuahua, Coahuila, Nuevo León, San Luis Potosí, Sonora, Zacatecas).

Acourtia nana grows primarily in the trans-Pecos and western Edwards Plateau.

3. **Acourtia thurberi** (A. Gray) Reveal & R. M. King, Phytologia 27: 231. 1973 • Thurber's desertpeony

Perezia thurberi A. Gray, Pl. Nov. Thurber., 324. 1854

Plants 40–150 cm (stems sulcate to striate distally, densely glandular). **Leaves** cauline and/or basal; sessile; blades ovate to ovate-elliptic, 1.5 (cauline)–18 (basal) cm, bases shortly sagittate or clasping, margins acerose-denticulate, faces densely glandular-puberulent. **Heads** in subcongested corymbiform arrays. **Involucres** obconic to campanulate, 7–9 mm. **Phyllaries** in 2–3 series, oblong-oblanceolate, apices acuminate, abaxial faces densely glandular-hairy. **Receptacles** alveolate, glandular. **Florets** 3–6; corollas lavender-pink (purple), 7–12 mm. **Cypselae** subcylindric to subfusiform, 3–7 mm, glandular; **pappi** bright white, 8–9 mm (rigid). $2n = 54$.

Flowering Oct–Nov. Gravel and caliche soils in warm Sonoran desert scrub; 100–200 m; Ariz.; N.Mex.; Mexico (Chihuahua, Durango, Sonora).

4. **Acourtia wrightii** (A. Gray) Reveal & R. M. King, Phytologia 27: 232. 1973 • Brownfoot [F]

Perezia wrightii A. Gray, Smithsonian Contr. Knowl. 3(5): 127. 1852

Plants 30–120 cm. **Leaves** cauline; sessile; blades oblong-lanceolate to elliptic-oblong, 2.5–13 cm, bases sagittate or clasping, margins dentate to denticulate, faces minutely stipitate-glandular and/or hirtellous (reticulate). **Heads** in corymbiform arrays. **Involucres** turbinate, 5–8 mm. **Phyllaries** in 2–3 series, linear to lanceolate, margins fimbrillate-glandular, apices obtuse to acute, abaxial faces glabrous or glandular-hairy. **Receptacles** reticulate (edges of sockets glandular). **Florets** 8–12; corollas pink or purple, 9–20 mm. **Cypselae** linear-fusiform, 2–6 mm, glandular-puberulent; **pappi** bright white, 9–12 mm. $2n = 54$.

Flowering Jun–Nov. Gravel, caliche, or sandy loamy soils in open desert (Lower Sonoran Desert); 400–1400 m; Ariz., N.Mex., Tex., Utah; Mexico (Chihuahua, Coahuila, Durango, Nuevo León, San Luis Potosí, Sonora, Zacatecas).

5. **Acourtia microcephala** de Candolle in A. P. de Candolle and A. L. P. P. de Candolle, Prodr. 7: 66. 1838

Perezia microcephala (de Candolle) A. Gray

Plants 60–120 cm. **Leaves** cauline; sessile; blades oblong-ovate to elliptic, 5–15 cm, bases cordate, clasping, auriculate, margins spinulose-denticulate, faces densely glandular-scabrid. **Heads** in paniculiform arrays. **Involucres** turbinate to campanulate, 6–8(–10+) mm. **Phyllaries** in 3–4 series, linear to oblanceolate, apices acute to acuminate or mucronate, margins and abaxial faces glandular-hairy. **Receptacles** alveolate, glandular-hairy. **Florets** 10–20; corollas lavender-pink, 6–12 mm. **Cypselae** cylindric, 1.5–6 mm, bristly glandular-puberulent; **pappi** tawny, 6–10 mm. $2n = 54$.

Flowering May–Jul. Gravel and caliche soils in scrub oak, oak woodlands, and chaparral; 60–1300 m; Calif.; Mexico (Baja California).

In the flora, *Acourtia microcephala* is restricted to the dry areas of California from Monterey to San Diego counties and on the Channel Islands.

3. TRIXIS P. Browne, Civ. Nat. Hist. Jamaica, 312, plate 33, fig. 1. 1756, • Threefold

[Greek *trixos*, 3-fold, describing the 3-cleft corolla]

David J. Keil

Shrubs [herbs], 20–300+ cm (often rhizomatous, stems hairy, hairs gland-tipped and glandless). **Leaves** cauline; sessile or petiolate; blades elliptic, lanceolate, linear, or linear-lanceolate, bases cuneate, margins dentate or entire, faces (at least abaxial) usually glandular and ± strigose. **Heads** quasi-radiate (see florets), borne singly or in corymbiform or paniculiform arrays. **Involucres** ± cylindric, 12–15 mm (subtended by calyculi of 3–7 spreading or ascending bractlets). **Phyllaries** (reflexed in fruit) in 1–2 series, linear (± keeled), apices acute. **Receptacles** flat, smooth, hairy, epaleate. **Florets** 4–25[–60], bisexual, fertile; corollas yellow (aging white), zygomorphic (2-lipped; outer lip liguliform, lobes 3, lance-deltate, inner lip smaller, lobes ± filiform, recurved); anther basal appendages entire, apical appendages oblong, acute; style branches relatively short, apices truncate-penicillate. **Cypselae** subcylindric to fusiform, often ± beaked, ribs 5, faces glandular-hairy; **pappi** of 60–80+ dull white [tawny], finely barbed bristles. *x* = 9.

Species ca. 65 (2 in the flora): North America, Mexico, West Indies, Central America, South America.

SELECTED REFERENCE Anderson, C. E. 1972. A monograph of the Mexican and Central American species of *Trixis* (Compositae). Mem. New York Bot. Gard. 22(3): 1–68.

1. Leaves usually ascending, almost parallel with stems, stomates on both faces, induments: abaxial faces glandular and sparsely strigose (sometimes only on veins), adaxial faces glandular and strigose to pilose; papilla-like double hairs of cypselae producing mucilage when wetted . . . . 1. *Trixis californica*
1. Leaves usually spreading, stomates on abaxial faces, induments: margins and, sometimes, abaxial midveins with glandless and glandular hairs, otherwise glabrous; papilla-like double hairs of cypselae not producing mucilage when wetted . . . . 2. *Trixis inula*

1. Trixis californica Kellogg, Proc. Calif. Acad. Sci. 2: 182, fig. 53. 1862 • California or American threefold

F

Varieties 2 (1 in the flora): sw United States, n Mexico.

1a. Trixis californica Kellogg var. **californica** F

Plants 20–200 cm. **Leaves** usually ascending, almost parallel with stems; petioles winged, 1.5–5 mm; blades linear-lanceolate to lanceolate, 2–11 cm, bases acute, margins entire or denticulate (often revolute), apices acute, induments: abaxial faces glandular-puberulent and sparsely strigose (sometimes only on veins), adaxial faces glandular and strigose to pilose; stomates on both faces. **Heads** usually in corymbiform or paniculiform arrays, rarely borne singly. **Peduncles** 5–45 mm, often ± inflated distally. **Calyculi** of 5–7 linear to lanceolate or ovate bractlets 5–16 mm. **Phyllaries** usually 8, linear, 8–14 mm, apices acute. **Florets** 11–25; corolla tubes 6–9 mm, outer lips 5–8 mm, inner 4–5.5 mm. **Cypselae** 6–10.5 mm, papilla-like double hairs producing mucilage when wetted; **pappi** 7.5–12 mm. *2n* = 54.

Flowering Jan–Nov. Dry, rocky slopes, washes, desert flats, thorn scrub; 200–1600(–2200) m; Ariz., Calif., N.Mex., Tex.; Mexico (Baja California, Baja California Sur, Chihuahua, Coahuila, Durango, Nuevo León, San Luis Potosí, Sinaloa, Sonora, Tamaulipas, Zacatecas).

2. Trixis inula Crantz, Inst. Rei Herb. 1: 329. 1766 • Mexican trixis, hierba del aire, tropical threefold

Plants 30–300+ cm (much branched). **Leaves** usually spreading; petioles 1–3 mm; blades linear-lanceolate to lanceolate or elliptic, 2.5–16.5 cm, bases attenuate to ± truncate, margins entire or denticulate (± flat), induments: margins and sometimes abaxial midveins with glandless and glandular hairs, otherwise glabrous; stomates on abaxial faces. **Heads** in corymbiform or paniculiform arrays. **Peduncles** 1–20(–25) mm, little, if at all, inflated distally. **Calyculi** of 3–5 linear to lanceolate (rarely ovate) bractlets 2.5–17 mm. **Phyllaries** usually 8, linear to oblong, 8–13(–15) mm, apices acute. **Florets** (8–)10–15; corolla tubes 5–9 mm, outer lips 4–7.5 mm, inner 3.5–5.5 mm. **Cypselae** 4.5–9 mm, papilla-like double hairs not producing mucilage when wetted; **pappi** 7–11 mm. **2n = 54.**

Flowering Mar–Nov (following rains). Open, sandy sites, thorn scrub, palm groves, thickets, roadsides; 0–100 m; Tex.; Mexico; West Indies; Central America; n South America.

The earliest botanical name for the plants here called *Trixis inula* is *Inula trixis* Linnaeus, 1760. In 1763, Linnaeus changed the name from *Inula trixis* to *Perdicium radiale*, which was based on the same type and is illegitimate under today's rules of botanical nomenclature because the epithet *trixis* should have been used in 1763. The name *Trixis radialis* (Linnaeus) Kuntze, 1891, was based on *Perdicium radiale* of Linnaeus and has been used by some botanists; it is superfluous and illegitimate under today's rules.

4. GOCHNATIA   Kunth in A. von Humboldt et al., Nov. Gen. Sp. 4(fol.): 15. 1818; 4(qto.): 19. 1820 • [For Frédéric Karl Gochnat, d. 1816, a botanist who worked with Cichorieae]

Beryl B. Simpson

Subshrubs or shrubs [herbs, trees], 100–400[–1000+] cm (sometimes dioecious). **Leaves** cauline; petiolate [sessile]; blades lance-elliptic to lanceolate [elliptic, linear, ovate] (stiff), bases ± cuneate [rounded], margins entire [denticulate to dentate] (often revolute), abaxial faces usually tomentose [minutely stipitate-glandular or gland-dotted and/or hirtellous], adaxial faces usually glabrous or glabrescent. **Heads** discoid [quasi-discoid or disciform], in ± corymbiform or glomerulate arrays [borne singly]. **Involucres** obconic or campanulate to cylindric, 5–8 [–15] mm. **Phyllaries** in 3–10(–15) series (outer often intergrading with peduncular bractlets), ovate to lanceolate, unequal, apices obtuse to acute. **Receptacles** flat, smooth or alveolate, usually glabrous, rarely glandular-hairy, epaleate. **Florets** 4–25[–150], bisexual or unisexual, fertile [pistillate or functionally staminate]; corollas whitish [cream-colored to yellow], actinomorphic (lobes 5, lanceolate to lance-linear, glabrous); anther basal appendages laciniate [entire], apical appendages lanceolate, rounded to apiculate; style branches relatively short, apices rounded. **Cypselae** ± obpyramidal [cylindric or turbinate], not beaked, obscurely ribbed, faces usually ± sericeous to velutinous and/or glandular-hairy; **pappi** of 20–30+ whitish to stramineous, smooth or ± barbellate bristles and/or setiform-subulate scales in 1–2 series.

Species ca. 68 (1 in the flora): mostly tropical and warm North America and South America, West Indies; two species in tropical mountains of southeast Asia.

In an analysis of DNA sequences from a sample of species in Gochnatiinae, H. G. Kim et al. (2002) found that *Gochnatia* is monophyletic and the subtribe is not. K. Bremer (1994) concluded that *Gochnatia* is paraphyletic to other genera in Mutisieae. N. Roque and D. J. N. Hind (2001) segregated one species from *Gochnatia* as a monotypic genus, *Ianthopappus*.

SELECTED REFERENCES Cabrera, A. L. 1971. Revisión del género *Gochnatia* (Compositae). Revista Mus. La Plata, Secc. Bot. 12: 1–160. Roque, N. and M. S. F. Silvestre Capelato. 2001. Generic delimitation of *Gochnatia*, *Richterago*, and *Ianthopappus* (Compositae–Mutisieae) based on pollen morphology. Grana 40: 197–204.

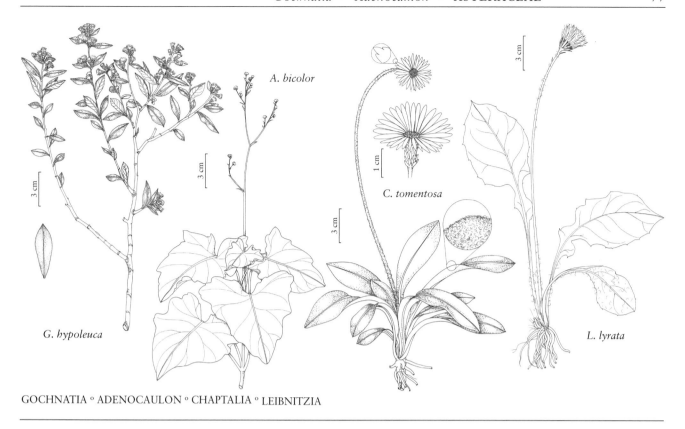

A. bicolor

C. tomentosa

G. hypoleuca

L. lyrata

GOCHNATIA ° ADENOCAULON ° CHAPTALIA ° LEIBNITZIA

1. Gochnatia hypoleuca (de Candolle) A. Gray, Proc. Amer. Acad. Arts 19: 57. 1883 • Shrubby bullseye, chomonque F

Moquinia hypoleuca de Candolle in A. P. de Candolle and A. L. P. P. de Candolle, Prodr. 7: 23. 1838

Stems densely leafy, rigid, young branches densely lanose. **Leaves:** blades 2–5 cm. **Corollas** 6–8 mm. **Cypselae** 2.5–4 mm; **pappi:** bristles unequal, longest 5–6 mm. Flowering Dec–Mar. Gravel and caliche soils in dry scrub vegetation; 30–70 m; Tex.;

Mexico (Coahuila, Durango, Hidalgo, Michoacán, Nuevo León, San Luis Potosí, Tamaulipas, Zacatecas).

Gochnatia hypoleuca occurs in southern Texas and in Mexico (to 4000 m). It is closely related to *Gochnatia obtusata* [treated by Cabrera as *Gochnatia hypoleuca* subsp. *obtusata* (S. F. Blake) Cabrera], a species of southern Mexico.

5. ADENOCAULON Hooker, Bot. Misc. 1: 19, plate 15. 1829 • [Greek *aden*, gland, and *kaulos*, stem]

David J. Keil

Perennials (perhaps flowering first year, usually rhizomatous), (10–)30–60(–100+) cm (aerial stems usually proximally tomentose and distally stipitate-glandular). **Leaves** basal and cauline; petiolate (proximal) or sessile (distal); blades ovate to ± triangular [suborbiculate], bases mostly truncate to cordate or hastate, margins coarsely dentate or lobulate to denticulate or entire, abaxial faces tomentose, adaxial faces glabrate (distalmost leaves often linear or scalelike). **Heads** disciform, borne in open paniculiform arrays. **Involucres** cuplike to saucer-shaped (inconspicuous), 4–8+ mm diam. **Phyllaries** in 1–2 series, ovate to oblong, ± equal (reflexed in fruit), apices obtuse. **Receptacles** convex, smooth, glabrous, epaleate. **Florets:** outer 3–7 florets pistillate, corollas white or ochroleucous, obscurely zygomorphic (± 2-lipped; staminodes 4–5, minute; style branches relatively short); inner 2–10 florets functionally staminate, corollas white or ochroleucous, actinomorphic (lobes 5); anther basal appendages (relatively short) entire, apical appendages narrowly triangular; style branches none (apices of styles obtuse to acute). **Cypselae** clavate to obovoid, not beaked, ± ribbed, faces distally stipitate-glandular; **pappi** 0. $x = 23$.

Species 5 (1 in the flora): North America, Central America, s South America, e Asia.

Adenocaulon is well adapted for animal dispersal. The prominent glandular hairs on the cypselae are very sticky and cling readily to fabrics, fur, and feathers. The intercontinental disjunctions within the range of *Adenocaulon* are probably a result of dispersal of cypselae on the feathers of birds. Bird dispersal may also account for the intracontinental disjunctions of *A. bicolor* (in the Pacific Northwest, the Black Hills, and the northern Great Lakes region). An alternate hypothesis for the disjunctions of *A. bicolor* is a former transcontinental periglacial distribution that was broken apart by Holocene climatic changes.

SELECTED REFERENCES Bittmann, M. 1990. Die Gattung *Adenocaulon* (Compositae): I. Morphologie. Candollea 45: 389–420. Bittmann, M. 1990b. Die Gattung *Adenocaulon* (Compositae): II. Ökologie, Verbreitung und Systematik. Candollea 45: 493–518.

1. **Adenocaulon bicolor** Hooker, Bot. Misc. 1: 19, plate 15. 1829 • Trail plant E F

Aerial stems usually leafy only near bases, openly branched. **Leaves:** petioles winged; blades 3–nerved, 3–25 cm. **Phyllaries** 5–6 (–10), 1–2 mm. **Peripheral florets:** corollas soon falling, 0.5–1.2 mm. **Inner florets:** corollas tardily falling, 1–2.3 mm. **Cypselae** 5–9 mm. $2n = 46$.

Flowering Jun–Oct. Woods, forests, usually in shade; 0–2000 m; Alta., B.C., Ont.; Calif., Idaho, Mich., Minn., Mont., Oreg., S.Dak., Wash., Wis., Wyo.

Adenocaulon bicolor is a common forest herb from southwestern Canada to central California. It is disjunct in the Black Hills (eastern Wyoming, western South Dakota) and the Great Lakes region (southern Ontario, northern Michigan). Reports of the species from Minnesota and Wisconsin are unverified.

6. CHAPTALIA Ventenat, Descr. Pl. Nouv., plate 61. 1802, name conserved • [For J. A. C. Chaptal, 1756–1831, who invented the wine-making process called chaptalization]

Guy L. Nesom

Perennials, 3–40(–100) cm (fibrous-rooted, sometimes rhizomatous; stems ± scapiform). **Leaves** basal; sessile or ± petiolate; blades elliptic or elliptic-obovate to obovate, obovate-elliptic, ovate, or sublyrate, bases cuneate, margins entire or denticulate, serrulate, or dentate to lobed, abaxial faces usually covered with dense wool, adaxial faces glabrous or glabrescent. **Heads** quasi-radiate or ± disciform (see florets; chasmogamous, produced well after rosette leaves, erect in flowering, nodding or erect in bud and again in fruit), borne singly. **Involucres** cylindric to campanulate, (7–)9–15+ mm (larger in fruit). **Phyllaries** in 2–5+ series, lanceolate to lance-linear, unequal, apices acute. **Receptacles** flat to convex, foveolate or smooth, glabrous, epaleate. **Florets** (dimorphic or trimorphic in 1–2 outer, pistillate zones plus 1 inner, bisexual or functionally staminate zone): outer pistillate-zone florets 9–38(–90+) in 1–2(–3) series, fertile, corollas creamy white to purple (sometimes with adaxial midstripe), zygomorphic (liguliform or bilabiate, inner lip often bifurcate, limbs sometimes reduced, style branches terete and linear to flattened and oblong); inner pistillate-zone florets 0 or 3–50 in 1–2 series, fertile, corollas usually ± zygomorphic (with reduced laminae and inner lips), sometimes reduced to tubes;

innermost florets usually bisexual and fertile, sometimes functionally staminate, 15–40, corollas whitish to pinkish, usually zygomorphic (2-lipped, lobes recurved or coiling), sometimes nearly actinomorphic (lobes ± equal or lobes ± 0); anther basal appendages entire, apical appendages lanceolate; style branches relatively short (abaxially pilose). **Cypselae** fusiform, often slightly flattened, distally ± tapered into necks or beaks (0.5–1.6[–3] times bodies), ribs mostly 4–12, faces glabrous or hairy (hairs duplex, relatively short, apices rounded to apiculate), not glandular; **pappi** of 50+ stramineous to pinkish, barbellulate bristles. $x = 24$.

Species ca. 60 (3 in the flora): North America, Central America, South America.

SELECTED REFERENCES Nesom, G. L. 1984. Taxonomy and distribution of *Chaptalia dentata* and *C. albicans* (Asteraceae: Mutisieae). Brittonia 36: 396–401. Nesom, G. L. 1995. Revision of *Chaptalia* (Asteraceae: Mutisieae) from North America and continental Central America. Phytologia 78: 153–188.

1. Heads erect in bud, flowering, and fruit; peduncles dilated distally; laminae of pistillate corollas 0.2–0.3 mm wide . . . . 1. *Chaptalia albicans*
1. Heads nodding in bud and fruit, erect in flowering; peduncles not dilated distally; laminae of pistillate corollas 0.2–1.5 mm wide.
    2. Leaves petiolate, blades mostly obovate to sublyrate; laminae of pistillate corollas evenly cream-colored (turning to crimson); central florets bisexual, fertile; uplands, Texas, New Mexico . . . . 2. *Chaptalia texana*
    2. Leaves sessile, blades elliptic to elliptic-obovate; laminae of pistillate corollas creamy white with purple, abaxial midstripe; central florets functionally staminate; coastal plain, North Carolina to Texas . . . . 3. *Chaptalia tomentosa*

1. Chaptalia albicans (Swartz) Ventenat ex B. D. Jackson, Index Kew 1: 506. 1893 • White sunbonnet

Tussilago albicans Swartz, Prodr., 113. 1788; *Chaptalia leiocarpa* (de Candolle) Urban

Leaves sessile or nearly so; blades obovate to obovate-elliptic, 2–14 cm, margins retrorsely serrulate to denticulate-apiculate, abaxial faces white-tomentose, adaxial faces green, glabrous or glabrate. **Heads** erect in bud, flowering, and fruit. **Peduncles** ebracteate, 6–15 cm in flowering, 12–37 cm in fruit, dilated distally. **Florets:** outer pistillate, corollas creamy white, rarely purple tinged, laminae 0.2–0.3 mm wide; inner florets bisexual, fertile. **Cypselae** 8.4–11.2 mm, beaks filiform, lengths 0.5–0.6+ times bodies, faces glabrous or sparsely glandular (usually only along the nerves). $2n = 24$, ca. 29.

Flowering (Mar–)Apr–Jul(–Nov). Grassy areas or open savannas, sometimes near evergreen oaks; 0–50 m; Fla.; Mexico; West Indies; Central America.

Chaptalia albicans was treated by A. Cronquist (1980) as *C. dentata* (Linnaeus) Cassini. The latter is known only from the West Indies (G. Nesom 1984) and contrasts with *C. albicans* in having outer pistillate florets with corollas white to greenish (versus white) and laminae (0.2–)0.4–0.7 mm (versus 0.2–0.3 mm) wide, style branches of pistillate florets (0.5–)0.7–0.9 mm (versus 0.8–1.3 mm), cypselae 5.5–7.5(–8.5) mm (versus 8.4–11.2 mm), orange (versus white) carpodia, and pappus bristles 5.8–8.5 mm (versus 7.8–10.5 mm).

2. Chaptalia texana Greene, Leafl. Bot. Observ. Crit. 1: 191. 1906 • Silverpuff

Chaptalia nutans (Linnaeus) Polakowski var. *texana* (Greene) Burkart

Leaves petiolate (petioles $^1/_8$–$^1/_3$ lengths of blades); blades obovate to ovate or elliptic or sublyrate, 3–21 cm, margins lobed to denticulate, abaxial faces thinly gray-tomentose, adaxial faces green-glabrate. **Heads** nodding in bud and fruit, erect in flowering. **Peduncles** ebracteate or bracts 1–2, 13–34 cm at flowering, 16–46 cm in fruit, not dilated distally. **Florets:** outer pistillate, corollas evenly cream colored, turning crimson, laminae 0.2–0.8 mm wide; inner florets bisexual, fertile. **Cypselae** 11.5–13 mm, beaks filiform, lengths 1–1.6 times bodies, faces sparsely to moderately papillate (bodies and beaks). $2n = 24$.

Flowering Mar–Jun. Slopes in thin, rocky (limestone) soils, usually in woods with abundant oaks; 200–1500 m; N.Mex., Tex.; Mexico (Chihuahua, Coahuila, Nuevo León, San Luis Potosí, Tamaulipas, and others).

3. Chaptalia tomentosa Ventenat, Descr. Pl. Nouv., plate 61. 1802 • Woolly sunbonnet [E] [F]

Leaves sessile; blades elliptic to elliptic-obovate, 5–18(–24) cm, margins denticulate, abaxial faces densely white-tomentose, adaxial faces green, glabrous or glabrate. **Heads** nodding in bud and fruit, erect in flowering. **Peduncles** ebracteate, 5–20 cm at flowering, 25–40 cm in fruit, not dilated distally. **Florets:** outer pistillate, corollas creamy white (with purple, abaxial midstripe), laminae 0.9–1.5 mm wide; inner florets functionally staminate. **Cypselae** 3.8–5.1 mm, beaks stout, lengths 0.2–0.25 times bodies, faces (bodies only) glabrous, beaks hairy (hairs spreading-ascending, swollen-apiculate). $2n = 48$.

Flowering Dec–Apr(–May in North Carolina). Coastal plain pinelands, sandy soils in grass-sedge bogs, along ditches, open areas, thin woods; 0–50 m; Ala., Fla., Ga., La., Miss., N.C., S.C., Tex.

---

7. LEIBNITZIA Cassini in F. Cuvier, Dict. Sci. Nat. ed. 2, 25: 420. 1822 • [For G. W. Leibnitz, 1646–1716, philosopher, political advisor, mathematician, and scientist]

Guy L. Nesom

Perennials, 5–60+ cm (fibrous-rooted; stems 1–11, scapiform, sometimes bracteate). **Leaves** basal; petiolate; blades elliptic to obovate, oblanceolate, or lyrate, bases cuneate, margins usually sinuately lobed to dentate, sometimes entire or pinnatifid, abaxial faces thinly gray-tomentose [covered with dense wool], adaxial faces glabrous or glabrescent (vernal leaves appearing after or concurrently with first heads). **Heads** quasi-radiate (see florets), borne singly (erect in bud, flowering, and fruit). **Involucres** cylindric to campanulate, 9–20+ mm. **Phyllaries** in 3–4+ series, lanceolate to lance-linear, unequal, apices acute to acuminate. **Receptacles** flat to convex, foveolate to alveolate, glabrous, epaleate. **Florets:** outer 6–15 pistillate, fertile, corollas usually pinkish to purplish (vernal with relatively broad laminae ± equaling tubes and bifurcate inner lips, autumnal with relatively narrow, greatly shortened laminae); inner 6–20+ florets bisexual, fertile, corollas usually whitish (vernal funnelform, 2-lipped, lobes 5, recurved or coiled, autumnal narrowly tubular, barely 2-lipped or nearly actinomorphic, lobes 5, erect); anther basal appendages entire, apical appendages lanceolate; style branches relatively short, apices rounded to truncate (abaxial faces pilose). **Cypselae** ± fusiform (somewhat flattened), distally ± constricted into relatively broad necks or narrow beaks $^1/_5$–$^1/_3$+ lengths of bodies, ribs [5–]8+, faces strigose to hispid (hairs duplex, relatively long, apices sharp-pointed), not glandular; **pappi** of 50–80+ stramineous, barbellulate bristles. $x = 23$.

Species 6 (1 in the flora). United States, Mexico, Asia.

Two species of *Leibnitzia* occur in North America (one is restricted to western Mexico); the remaining four are Asian. The species are characterized by dimorphic heads: those produced in the spring appear before or with the leaves and are chasmogamous and quasi-radiate; the fall forms appear after the leaves and are cleistogamous (not exposing any florets but setting full fruit). The florets also differ morphologically between the chasmogamous and cleistogamous heads. The widespread American species (*L. lyrata*) has been placed within *Chaptalia*; species of *Leibnitzia* are distinguished from all *Chaptalia* species by the slender, sharp-pointed hairs on their cypselae.

SELECTED REFERENCE Nesom, G. L. 1983. Biology and taxonomy of American *Leibnitzia* (Asteraceae). Brittonia 35: 126–139.

1. **Leibnitzia lyrata** (Schultz-Bipontinus) G. L. Nesom, Phytologia 78: 169. 1995 • Seemann's sunbonnet F

Gerbera lyrata Schultz-Bipontinus in B. Seemann, Bot. Voy. Herald, 313. 1856, based on *Chaptalia lyrata* D. Don, Trans. Linn. Soc. London 16: 243. 1830, not (Willdenow) Sprengel 1826; *C. alsophila* Greene; *C. leucocephala* Greene; *Leibnitzia seemannii* (Schultz-Bipontinus) G. L. Nesom

Leaves: petioles ⅛–⅝ lengths of blades; blades 3–19 cm. **Peduncles** 8–64 cm in fruit, often dilated distally. **Heads:** vernal chasmogamous, autumnal cleistogamous. **Cypselae** tan to purplish, 6–10 mm. $2n = 46$.

Flowering (chasmogamous) Mar–early Jun, (cleistogamous) Aug–Nov. Grassy, partially open areas in pine, pine-oak, or oak woods, often in disturbed sites; 2000–2500 m (to 3400 m in Mexico); Ariz., N.Mex.; Mexico (Chihuahua, Coahuila, Durango, Nuevo León, San Luis Potosí, Zacatecas, and southward).

Plants of *Leibnitzia lyrata* with cleistogamous heads are encountered and collected much more often than those with chasmogamous heads.

The combination *Leibnitzia lyrata* proposed by G. L. Nesom (1995) appeared to be based on *Chaptalia lyrata* D. Don (an illegitimate later homonym) but was actually based on *Gerbera lyrata* Schultz-Bipontinus, which was cited in synonymy by Nesom.

187b. ASTERACEAE Martinov tribe CYNAREAE Lamarck & de Candolle, Syn. Pl. Fl. Gall., 267. 1806

Cardueae Cassini

Annuals or perennials (sometimes coarse and/or robust, often prickly-spiny and thistlelike [sub-shrubs, shrubs, or trees]; rarely dioecious, e.g., some *Cirsium* spp.). **Leaves** basal and/or cauline; alternate; ± petiolate or sessile; (leaf bases often decurrent on stems) margins usually lobed to dissected, sometimes dentate or entire (usually spiny). **Heads** mostly homogamous (usually discoid, sometimes disciform or radiant, then peripheral florets usually pistillate or neuter, sometimes bisexual or with staminodes), borne singly or in corymbiform, paniculiform, or racemiform arrays (heads with 1 floret each aggregated into second-order heads in *Echinops*). **Calyculi** 0 (involucres sometimes closely subtended by leaflike peduncle bracts). **Phyllaries** usually persistent [readily falling], in (1–)3–5+ series, usually distinct, usually unequal, usually herbaceous (sometimes fleshy), margins (entire or denticulate to pectinate, sometimes spiny) and apices seldom notably scarious (apices often spinose or ± expanded into distinct, often fimbriate-fringed, pectinate, and/or spiny appendages). **Receptacles** flat to convex, usually epaleate (often pitted and often bristly-setose or densely hairy). **Ray florets** 0 (corollas of peripheral florets in radiant heads often notably enlarged, usually 5-lobed, sometimes zygomorphic and raylike or ± 2-lipped). **Peripheral (pistillate) florets** 0 or (in disciform heads) in 1–3+ series; corollas (usually present) usually yellow, sometimes ochroleucous or cyanic. **Disc florets** bisexual and fertile (rarely functionally staminate); corollas yellow, cyanic, or white, usually actinomorphic, lobes 5, usually narrowly triangular to ± linear, seldom deltate (sometimes unequal, corollas then ± zygomorphic); anther bases ± tailed, apical appendages usually oblong (filaments sometimes papillate to pilose; connate in *Silybum*); styles (bisexual, fertile florets) distally enlarged or swollen, usually dilated and/or with rings of hairs at or near point of bifurcation, abaxially smooth or papillate to hairy (at least distally, sometimes ± throughout), "branches" often connate, adaxially continuously stigmatic ± to tips, apices rounded to acute, appendages essentially none. **Cypselae** usually monomorphic within heads (often thick-walled, hard, nutlike, receptacular attachments basal or lateral, bases sometimes each with an elaiosome), usually ellipsoid, obovoid, or ovoid, sometimes rounded-prismatic, terete, 4–5-angled, or ± compressed, rarely beaked, bodies usually smooth, sometimes rugose or 10- or 20-nerved (glabrous or puberulent to villous; often with apical umbo and/or crown in addition to pappus); **pappi** (rarely 0) readily falling or persistent, usually of fine to coarse, barbellate to plumose bristles, sometimes of scales, sometimes both bristles and scales.

Genera 83, species 2500 (17 genera, 116 species in the flora): mostly Old World, especially Mediterranean; some species widely introduced.

The circumscription for Cynareae adopted here is the traditional one and includes the three elements (Cynareae in the narrow sense, Carlineae, and Echinopeae) recognized as tribally distinct by M. Dittrich (1977[1978]). Work by K. Bremer (1987) supported the Dittrich scheme. A traditional circumscription of Cynareae was maintained by J. L. Panero and V. A. Funk (2002).

SELECTED REFERENCES Dittrich, M. 1977[1978]. Cynareae—systematic review. In: V. H. Heywood et al., eds. 1977[1978]. The Biology and Chemistry of the Compositae. 2 vols. London. Vol. 2, pp. 999–1015. Garcia-Jacas, N., A. Susanna, T. Garnatje, and R. Vilatersana. 2001. Generic delimitations and phylogeny of the subtribe Centaureinae (Asteraceae): A combined nuclear and chloroplast DNA analysis. Ann. Bot. (Oxford) 87: 503–515. Howell, J. T. 1959. Distributional data on weedy thistles in western North America. Leafl. W. Bot. 9: 17–32. Moore, R. J. and C. Frankton. 1974. The Thistles of Canada. Ottawa. [Canada Dept. Agric., Res. Branch, Monogr. 10.] Susanna, A., N. Garcia-Jacas, D. E. Soltis, and P. S. Soltis. 1995. Phylogenetic relationships in tribe Cardueae (Asteraceae) based on ITS sequences. Amer. J. Bot. 82: 1056–1068. Wagenitz, G. and F. H. Hellwig. 1996. Evolution of characters and phylogeny of the Centaureae. In: D. J. N. Hind et al., eds. 1996. Proceedings of the International Compositae Conference, Kew, 1994. 2 vols. Kew. Vol. 1, pp. 491–510.

1. Leaf margins spiny.
    2. Florets 1 per head (heads in globose, second-order heads) . . . . 9. *Echinops*, p. 85
    2. Florets 3–250+ per head (heads borne singly or in ± open arrays, not in globose second-order heads).
        3. Stems winged.
            4. Receptacles bearing setiform scales ("flattened bristles"), (not pitted); cypselar attachments slightly lateral; pappus bristles usually distinct, sometimes basally connate . . . . 12. *Carduus*, p. 91
            4. Receptacles not bristly (deeply pitted); cypselar attachments basal; pappus bristles basally connate . . . . 10. *Onopordum*, p. 87
        3. Stems not or rarely winged (some *Cirsium* spp.).
            5. Leaves variegated (stamen filaments connate) . . . . 14. *Silybum*, p. 164
            5. Leaves not variegated (stamen filaments distinct).
                6. Corollas yellow to orange, red, or ± purple; cypsela attachments lateral; pappi 0 or of distinct, minutely barbed (not plumose) setiform scales ("flattened bristles") or subulate scales.
                    7. Heads discoid (all florets fertile); receptacles bearing subulate scales; cypselae 4-angled . . . . 23. *Carthamus*, p. 178
                    7. Heads disciform (peripheral florets sterile); receptacles bristly ("flattened bristles"), cypselae terete, 20-ribbed . . . . 24. *Centaurea* (in part), p. 181
                6. Corollas white or purplish to red; cypsela attachments basal or oblique-basal; pappi of basally connate, plumose setiform scales ("flattened bristles").
                    8. Receptacles scaly (sometimes bristly); cypsela attachments oblique-basal . . . . 8. *Carlina*, p. 84
                    8. Receptacles densely bristly-setose; cypselar attachments basal.
                        9. Involucres 35–100+ mm diam. (largest leaves 60–150 cm; receptacles becoming fleshy) . . . . 11. *Cynara* (in part), p. 89
                        9. Involucres 10–50 mm diam. (largest leaves 20–50(–110) cm; receptacles usually not notably fleshy) . . . . 13. *Cirsium*, p. 95
1. Leaf margins not spiny (tips sometimes ± spinose-apiculate).
    10. Heads discoid (all florets bisexual and fertile).
        11. Cypselar attachments ± lateral.
            12. Phyllary appendages entire or lacerate, not fringed; pappi of ± caducous, distally plumose bristles . . . . 17. *Acroptilon*, p. 171
            12. Phyllary appendages dentate or fringed; pappi 0 or if persistent, then non-plumose bristles or scales . . . . 24. *Centaurea* (in part), p. 181
        11. Cypselar attachments basal.
            13. Phyllary apices spiny, hooked; bristles of pappus distinct, not plumose . . . . 16. *Arctium*, p. 168
            13. Phyllary apices spiny or not, not hooked; setiform scales ("flattened bristles") of pappus basally connate, plumose.
                14. Receptacles densely long-bristly ("flattened bristles") (becoming fleshy); florets 100–250+ . . . . 11. *Cynara* (in part), p. 89
                14. Receptacles usually subulate-scaly, sometimes bristly or naked (not fleshy); florets 10–20 . . . . 15. *Saussurea*, p. 165
    10. Heads radiant or disciform (peripheral florets usually neuter).
        15. Heads disciform.
            16. Phyllary appendages dentate or fringed, spiny or not; receptacles bristly ("flattened bristles") . . . . 24. *Centaurea* (in part), p. 181
            16. Phyllary appendages 0 (apices acute, entire); receptacles bearing subulate scales . . . . 22. *Crupina*, p. 177
        15. Heads radiant.
            17. Phyllary appendages 0.

18. Biennials or perennials; spines on phyllary apices caducous; cypsela apices not coronate .... 19. *Mantisalca*, p. 173
18. Annuals; spines on phyllary apices persistent; cypsela apices coronate .... 20. *Volutaria*, p. 174

[17. Shifted to left margin.—Ed.]
17. Phyllary appendages present.
  19. Cypselae compressed (oblong; attachment scars rimmed, rims whitish, swollen), apices denticulate .... 18. *Amberboa*, p. 172
  19. Cypselae ± terete (barrel-shaped; attachment scars not rimmed), apices entire.
    20. Annuals; leaf margins mostly entire or denticulate to serrulate; cypsela attachment oblique-basal; involucres 20–40 mm diam.; phyllary bodies linear, margins entire, appendages fimbriate; corollas of peripheral florets 30–70 mm .... 21. *Plectocephalus*, p. 175
    20. Annuals, biennials, or perennials; leaf margins entire or toothed to pinnately lobed; involucres 10–25(–40) mm diam.; cypsela attachment lateral; phyllary bodies oblong to ovate or obovate, margins fimbriate, appendages fimbriate; corollas of peripheral florets 15–30(–45) mm .... 24. *Centaurea* (in part), p. 181

8. CARLINA Linnaeus, Sp. Pl. 2: 828. 1753; Gen. Pl. ed. 5, 360. 1754 • Carline-thistle [For Charles V, 1500–1558, Holy Roman Emperor] ⊡

David J. Keil

Biennials [annuals or perennials, shrubs, or dwarf trees], 10–80 cm, herbage spiny. **Stems** erect or spreading, simple to branched distally or throughout, branches ascending to spreading. **Leaves** basal and cauline [all cauline]; petiolate or sessile; blade margins dentate to pinnately lobed, ± spiny, faces ± tomentose or glabrate. **Heads** discoid, borne singly or in corymbiform arrays. **Involucres** hemispheric to campanulate. **Phyllaries** many in several series, outer ovate to lanceolate, bases appressed; (at least outer ± leaflike); middle smaller, scarious, margins spiny-dentate or -lobed, apical appendages spine-tipped, inner apices spreading, membranous, white to stramineous, raylike, entire. **Receptacles** flat to conic, epaleaete scaly or bristly (each floret surrounded by ± connate, membranous scales dissected into linear lobes). **Florets** many; corollas yellow to maroon, tubular-funnelform, lobes triangular; anther bases tailed, linear-oblong, apical appendages acute; style branches: fused portions ca. 0.5 mm, with basal nodes glabrous or minutely hairy, distally puberulent, distinct portions tapered, not or scarcely separating. **Cypselae** cylindric to fusiform, not ribbed, without apical rim, hairy with forked, 2-celled hairs, attachment scars basal-oblique; **pappi** readily falling, of many persistent bristles in 1 series, basally connate in groups of 3–10, plumose. $x = 9, 10$.

Species 28 (1 in the flora): introduced; Eurasia, Mediterranean region, Macaronesia.

SELECTED REFERENCES Meusel, H. and A. Kästner. 1990. Lebensgeschichte der Gold- und Silberdisteln. Monographie der Mediterran–Mitteleuropaischen Compositen-Gattung *Carlina* Band I. Merkmalsspektren und Lebensraume der Gattung. Denkschr. Oesterr. Akad. Wiss., Math.-Naturwiss. Kl. 127: 1–294. Meusel, H. and A. Kästner. 1994. Lebensgeschichte der Gold- und Silberdisteln. Monographie der Mediterran–Mitteleuropaischen Compositen-Gattung *Carlina* Band II. Artenvielfalt und Stammesgeschichte der Gattung. Denkschr. Oesterr. Akad. Wiss., Math.-Naturwiss. Kl. 128: 1–657. Meusel, H., A. Kästner, and E. Vitek. 1996. The evolution of *Carlina*—A hypothesis based on ecogeography. In: D. J. N. Hind et al., eds. 1996. Proceedings of the International Compositae Conference, Kew, 1994. 2 vols. Kew. Vol. 1, pp. 723–737.

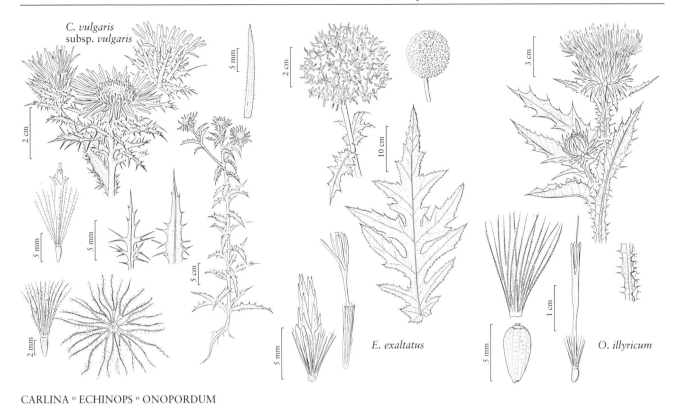

CARLINA ° ECHINOPS ° ONOPORDUM

1. Carlina vulgaris Linnaeus, Sp. Pl. 2: 828. 1753
[F] [I]

Subspecies 2 (1 in the flora): introduced; Eurasia.

1a. Carlina vulgaris Linnaeus subsp. **vulgaris** [F] [I]

Carlina vulgaris Linnaeus subsp. *longifolia* (C. Reichenbach) Nyman

Plants 10–80 cm. **Stems** simple to distally much branched, arachnoid-tomentose. **Leaves:** basal tapering to winged petioles, blades oblong-oblanceolate, 7–15 cm, margins with spine-tipped teeth and lobes, abaxially ± arachnoid-tomentose, adaxially ± glabrate; cauline leaves sessile, ± clasping, progressively smaller, margins with spine-tipped lobes, faces ± glabrate. **Heads** usually 2–5 in corymbiform clusters. **Involucres** 15–30 mm diam. **Phyllaries:** outer leaflike, middle green or purplish, spiny-fringed, arachnoid, inner with appressed bases, tips yellow or straw-colored, acuminate, spreading, slender, these resembling laminae of ray corollas. **Corollas** 7–9 mm, yellow to maroon-purple. **Cypselae** brownish, 2–4 mm, sericeous; **pappi** of ca. 30 bristles, basally connate in 10 bundles, 8–10 mm. $2n = 20$ (Denmark).

Flowering Jul–Sep. Roadsides, fields, waste places; 0–500 m; introduced; N.J., N.Y.; Eurasia.

Carlina vulgaris is widely distributed across Europe and parts of Asia, often as a weed. Although it has a limited distribution at present in North America, it has the potential to become a serious weed problem as have several other Eurasian thistles.

9. ECHINOPS Linnaeus, Sp. Pl. 2: 814. 1753; Gen. Pl. ed. 5, 356. 1754 • Globe thistle [Greek *echinos*, hedgehog, spiny, and *ops*, face, appearance, alluding to spiny heads] [I]

David J. Keil

Perennials, 100–200 cm, herbage usually ± tomentose, spiny. **Stems** usually erect, simple or branched. **Leaves** basal and cauline; sessile or petiolate; blade margins dentate to pinnately 1–3-pinnately lobed or divided, lobes and teeth spiny, faces ± tomentose, sometimes glandular.

Heads discoid, many, each with 1 floret, sessile, in pedunculate, spheric secondary heads. **Secondary involucres** of reflexed, laciniate-pinnatifid bracts. **Primary involucres** ellipsoid, subtended by bristles. **Phyllaries** many in several series, unequal, lanceolate (outer) to linear (inner), entire, apices sometimes expanded and fringed, not spine-tipped. **Receptacles** turbinate, bearing elongate subulate scales. **Florets** 1 per primary head; corollas white to greenish, blue-gray, blue, or purple, tubes elongate, throats very short, lobes linear; anther bases sharply tailed, apical appendages narrowly triangular, acute; style branches: fused portions with minutely hairy rings, distinct portions divergent, linear-oblong. **Cypselae** ± cylindric, 4-angled, apices ± truncate, without crowns, densely villous with long, stiff, appressed or ascending, multicellular hairs, attachment scars basal; **pappi** of many, short, ± connate [or distinct] scales. x = 13, 14, 15, 16.

Species ca. 120 (3 in the flora): introduced; Eurasia, Africa.

1. Lobes of leaf blades linear or narrowly oblong . . . . 3. *Echinops ritro*
1. Lobes of leaf blades lanceolate to triangular.
   2. Adaxial leaf faces glandular . . . . 1. *Echinops sphaerocephalus*
   2. Adaxial leaf faces glabrous or sparsely strigose . . . . 2. *Echinops exaltatus*

1. Echinops sphaerocephalus Linnaeus, Sp. Pl. 2: 814. 1753 • Great globe-thistle, boulette commune [I]

Plants 100–200 cm. **Stems** simple to much branched, ± glandular, ± tomentose. **Leaves:** basal and proximal cauline leaves shortly winged-petiolate, distal cauline sessile, clasping; blades oblong-elliptic to narrowly obovate, margins ± subentire or 1–2-pinnately lobed, lobes lanceolate to triangular, margins revolute, spiny-dentate, spine-tipped, spines slender, 2–4 mm; abaxial faces densely gray- or white-tomentose, adaxial faces green, glandular-scabrous. **Secondary heads** 3–6 cm diam. **Involucres** 15–25 mm. **Outer phyllaries** ± glandular, inner phyllary apices attenuate, expanded, fringed. **Corollas** white to pale blue, 12–14 mm, tube ca. 5.5 mm, lobes ca. 7 mm. **Cypselae** 7–10 mm; **pappi** of ± connate, ciliate scales 1–1.5 mm. $2n$ = 30, 32.

Flowering summer (Jun–Aug). Disturbed sites; 0–1700 m; introduced; Man., Ont., Que., Sask.; Calif., Colo., Conn., Ill., Ind., Iowa, Ky., Maine, Md., Mass., Mich., N.H., N.Y., Pa., Vt., Va., Wash., W.Va., Wis.; Eurasia.

Echinops sphaerocephalus is sometimes cultivated, and sometimes it escapes from cultivation. The species has been reported from Saskatchewan and may be introduced there; that appears questionable.

2. Echinops exaltatus Schrader, Hort. Gott. 2: 15, plate 9. 1811 • Tall globe-thistle, boulette de Hongrie [F] [I]

Echinops commutatus Juratzka

Plants 40–150 cm. **Stems** simple or more commonly much branched, white tomentose, especially distally. **Leaves:** basal and proximal cauline leaves winged-petiolate, mid and distal cauline clasping; blades ovate to elliptic or obovate, margins 1–2-pinnately lobed, lobes lanceolate to triangular, spiny-dentate, spine-tipped, spines slender, 2–3 mm; abaxial faces white-tomentose, adaxial green, glabrous to sparsely strigose. **Secondary heads** 3.5–6 cm diam. **Involucres** 20–30 mm. **Phyllaries** eglandular, ciliate, apices slender, slightly recurved. **Corollas** white to pale blue, 8–15 mm, tubes 2–6 mm, lobes 5–6 mm. **Cypselae** 7–8 mm; **pappi** of ± connate, ciliate scales ± 1 mm. $2n$ = 30.

Flowering summer (Jun–Sep). Disturbed sites; 0–500 m; introduced; N.B., Ont., Que.; Wash.; Europe.

Echinops exaltatus sometimes escapes from cultivation.

3. Echinops ritro Linnaeus, Sp. Pl. 2: 815. 1753 [I]

Subspecies 5 (1 in the flora): introduced; Europe, w Asia.

3a. Echinops ritro Linnaeus subsp. **ruthenicus** (M. Bieberstein) Nyman, Consp. Fl. Eur. 2: 399. 1879 • Southern globe-thistle [I]

Echinops ruthenicus M. Bieberstein, Fl. Taur.-Caucas. 3: 597. 1819

Plants 100–150 cm. **Stems** simple or distally ± branched, ± white-tomentose, proximally sometimes ± glabrate. **Leaves:** basal and proximal cauline winged-petiolate, mid cauline ± clasping; blades elliptic to obovate, margins 2-pinnately divided nearly to midvein, lobes linear to oblong, spiny-serrate, spine-tipped, abaxial faces white-tomentose, adaxial faces green, sparsely viscid-glandular; distal smaller, less divided. **Secondary heads** 3.5–4.5 cm diam. **Involucres** 15–20 mm. **Inner phyllaries** lavender, apices attenuate, serrulate, scabrous, not glandular. **Corollas** blue to purple (white). **Cypselae** ca. 6 mm; **pappi** of ± connate, minutely barbed scales ca. 1 mm. *2n* = 32 (Greece).

Flowering summer (Jun–Aug). Disturbed sites; 0–500 m; introduced; N.Y., Wash.; Europe; Asia.

Subspecies *ruthenicus* sometimes escapes from cultivation.

10. ONOPORDUM Linnaeus, Sp. Pl. 2: 827. 1753; Gen. Pl. ed. 5, 359. 1754 • Cotton thistle, onoporde [Greek *onopordon*, name for cotton thistle] [I]

David J. Keil

Biennials, 50–400+ cm, coarse, prickly. **Stems** usually erect, ± branched, spiny-winged. **Leaves** basal and cauline; winged-petiolate (basal) or sessile (cauline); blade bases narrowing, margins pinnately lobed or divided and dentate, teeth and lobes tipped with stout spines. **Heads** discoid, borne singly or in corymbiform arrays; (peduncles 0 or spiny winged). **Involucres** hemispheric to ovoid or spheric. **Phyllaries** many in 8–10+ series, linear to ovate, entire, tapered to stiff spines, middle and outer often spreading or reflexed. **Receptacles** flat to convex, epaleate, not bristly, alveolate with apically fringed pits. **Florets** many; corollas white or purple, actinomorphic or weakly zygmorphic, tubes slender, throats cylindric or narrowly goblet-shaped, lobes linear; anther bases acute-tailed, apical appendages subulate; style branches: fused portions with minutely hairy nodes, long, cylindric, minutely papillate, distinct portions minute. **Cypselae** ± cylindric, 4–5-angled, usually ± transversely roughened, glabrous, attachment scars basal; **pappi** falling in ring, of many barbed or plumose bristles, basally connate. *x* = 17.

Species 25–60 (3 in the flora): introduced; Eurasia.

SELECTED REFERENCE Dress, W. J. 1966. Notes on the cultivated Compositae 9. *Onopordum.* Baileya 14: 74–86.

1. Herbage green, ± sticky-glandular . . . . 3. *Onopordum tauricum*
1. Herbage ± canescent-tomentose.
   2. Leaves dentate to shallowly pinnatifid; phyllaries linear, bases 2–2.5 mm wide . . . . 1. *Onopordum acanthium*
   2. Leaves shallowly to ± deeply 1–2 pinnatifid; phyllaries lanceolate to ovate, bases 3–8 mm wide . . . . 2. *Onopordom illyricum*

1. **Onopordum acanthium** Linnaeus, Sp. Pl. 2: 827.
1753 • Scotch thistle, Onoporde acanthe [I] [w]

Subspecies 3 (1 in the flora): introduced; Eurasia.

1a. **Onopordum acanthium** Linnaeus subsp.
acanthium [I] [w]

Plants 50–400 cm, herbage canescent-tomentose throughout or ± glabrescent. **Stems** appressed-hairy; wings to 15 mm wide. **Leaves** 10–60 cm, margins dentate to shallowly pinnatifid, lobes 8–10 pairs, broadly triangular, densely tomentose, especially on abaxial faces. **Heads** mostly in clusters of 2–3, at tips of branches. **Involucres** ± spheric, 20 mm diam. (excluding spines), bases truncate to concave. **Phyllaries** linear, bases 2–2.5 mm wide, puberulent, ± cobwebby-tomentose, spines to 6 mm, adaxially glabrous. **Corollas** purple or white, 22–25 mm, lobes glabrous. **Cypselae** 4–5 mm, transversely roughened; **pappi** of many pink to reddish, minutely barbed bristles 7–9 mm. $2n = 34$.

Flowering summer (Jun–Sep). Grasslands, woodlands, riparian areas, deserts, disturbed ground, roadsides; 0–2200 m; introduced; B.C., N.B., N.S., Ont., Que.; Ala., Ariz., Calif., Colo., Conn., Del., Fla., Idaho, Ill., Ind., Iowa, Kans., Ky., Md., Mass., Mich., Minn., Mo., Mont., Nebr., Nev., N.J., N.Mex., N.Y., Ohio, Okla., Oreg., Pa., R.I., Tex., Utah, Vt., Va., Wash., W.Va., Wis., Wyo.; Eurasia, introduced in Australia

Scotch thistle is the national emblem of Scotland. Although it is sometimes cultivated as an ornamental, Scotch thistle is considered to be a noxious weed both in Canada and the United States. Infestations severely degrade rangelands, and dense stands are practically impenetrable because of the spiny nature and large size of the plant. This species has also invaded rangelands in Australia.

2. **Onopordum illyricum** Linnaeus, Sp. Pl. 2: 827.
1753 • Illyrian thistle [F] [I] [w]

Plants 50–250 cm, herbage canescent-tomentose throughout. **Stems:** wings 0.5–2 cm wide. **Leaves** 10–50 cm, margins shallowly to ± deeply 1–2-pinnatifid with 8–10 pairs of triangular lobes. **Heads** mostly borne singly at branch tips. **Involucres** 30–60 mm diam (excluding spines) ± spheric, bases truncate to concave. **Phyllaries** lanceolate to ovate, bases 3–8 mm wide, glabrous or ± cob-

webby-tomentose, spines to 5 mm. **Corollas** purple, 25–35 mm. **Cypselae** 4–5 mm; **pappi** of many whitish, plumose bristles 10–12 mm. $2n = 34$ (France).

Flowering summer (Jun–Aug). Grasslands, fields, roadsides, oak woodlands; 200–500 m; introduced; Calif.; s Europe (Mediterranean region); introduced in Australia.

Illyrian thistle is considered to be a noxious weed in California where efforts to eradicate it from the state's flora have been implemented. This species has also invaded rangelands in Australia, where it is introduced.

3. **Onopordum tauricum** Willdenow, Sp. Pl. 3: 1687.
1803 • Taurian thistle, bull cottonthistle [I] [w]

Plants 50–200 cm, herbage glandular-puberulent, sticky throughout. **Stems:** wings 0.5–2 cm wide. **Leaves** 10–30 cm, margins shallowly to deeply 1–2-pinnatifid, with 6–8 pairs of acutely triangular lobes, thinly arachnoid tomentose when young. **Heads** mostly borne singly at branch tips. **Involucres** ± spheric, 20–50 mm diam (excluding spines), base truncate to concave. **Phyllaries** lanceolate, bases 3–4 mm wide, glabrous or glandular-puberulent, sometimes ± cobwebby-tomentose, spines to 4 mm. **Corollas** purplish pink, 25–30 mm. **Cypselae** 5–6 mm; **pappi** of many whitish to tan, scabrous or minutely barbed bristles 8–10 mm. $2n = 34$ (Russia).

Flowering summer (Jun–Sep). Grasslands, arid woodlands, riparian areas, roadsides, agricultural lands; 600–2200 m; introduced; Calif., Colo.; s Europe; sw Asia.

Taurian thistle is a noxious weed in California and Colorado. In southeastern Colorado it sometimes grows with *Onopordum acanthium*. Putative hybrids have been observed in this area.

11. CYNARA Linnaeus, Sp. Pl. 2: 827. 1753; Gen. Pl. ed. 5, 359. 1754 • [Greek *kynara*, artichoke] [I]

David J. Keil

Annuals or perennials, 50–250 cm, herbage ± arachnoid-tomentose. **Stems** ± erect, simple or branched, (leafy), stout. **Leaves** basal and cauline; petiolate (basal and proximal cauline) or sessile (distal cauline); blade margins 1–3-pinnately lobed or divided, sometimes essentially compound, spineless or with slender to very stout marginal spines, cauline progressively smaller and less divided distally, distalmost bractlike, abaxial faces pilose to densely gray-tomentose, adaxial glabrous or thinly tomentose, sometimes glandular. **Heads** discoid, borne singly or in few-headed, terminal, cymiform arrays. **Involucres** hemispheric or ovoid, sometimes constricted distally, 5–15 cm diam. **Phyllaries** many in 5–8+ series, unequal; outer lanceolate to broadly ovate, leathery, margins entire, with appressed bases and spreading apical appendages, acute to broadly obtuse or truncate, spine-tipped or spineless; inner scarious. **Receptacles** concave to flat or convex, epaleate, densely long-bristly. **Florets** many; corollas white, blue, or purple, tubes very slender, throats abruptly expanded, cylindric, lobes linear; anther bases long-sagittate, fringed, apical appendages oblong; style branches: fused portions long, cylindric, minutely papillate, distinct portions minute. **Cypselae** ± cylindric to obpyramidal, ± 4-angled, finely ribbed, sometimes ± flattened, glabrous, apices truncate, smooth, attachment scars basal; **pappi** falling in rings of many (white or brownish), stiff bristles in 3–7 series, connate at bases, plumose proximally, often merely barbed distally. $x = 34$.

Species 8 (1 in the flora): introduced; Mediterranean region, Macaronesia, w Asia.

SELECTED REFERENCE Wiklund, A. 1992. The genus *Cynara* L. (Asteraceae–Cardueae). Bot. J. Linn. Soc. 109: 75–123.

1. Cynara cardunculus Linnaeus, Sp. Pl. 2: 827. 1753

• Cardoon, artichoke, artichoke thistle [F] [I] [W]

Taproots fleshy. **Stems** glabrous to densely arachnoid-tomentose. **Leaves:** basal blades 30–200 cm, margins deeply 1–2-pinnately lobed or divided to nearly compound, lobes oblong to lanceolate, entire to coarsely toothed, teeth and lobes innocuous to prominently spine-tipped, spines 1–30 mm, often clustered along petiole and at base of lobes, abaxial faces densely gray- or white-tomentose, adaxial faces thinly cobwebby-tomentose; cauline leaves often short-decurrent as spiny wings. **Involucres** often purplish tinged, 30–150 × 40–150 mm excluding spreading phyllary tips, constricted distally or not. **Phyllaries** lanceolate to broadly ovate, bases appressed, spreading apices obtuse to acute or acuminate, spineless or tipped with spines 1–9 mm or truncate, abruptly mucronate, and spineless or minutely spine-tipped. **Corollas** blue or purple (rarely white), 3–5 cm; styles long-exserted. **Cypselae** 4–8 mm; **pappus** bristles 2–4 cm.

Subspecies 2 (2 in the flora): introduced; California; Mediterranean region, Macaronesia.

Cynara cardunculus is a species of considerable economic importance. The globe artichoke, formerly treated as *C. scolymus*, was included as a horticulturally derived form of *C. cardunculus* (A. Wiklund 1992). The artichoke and the cardoon, another horticultural race of *C. cardunculus*, have been cultivated for centuries–the former for edible phyllary bases and receptacles, and the latter for edible stems and leaf rachises. That species has a darker side, however. Wild type races (artichoke thistles) are invasive and tenacious weeds that have infested Mediterranean climate areas of California, South America, South Africa, and Australia. Wiklund recognized two subspecies of *C. cardunculus*: subsp. *cardunculus* includes the artichoke, cardoon, and various wild types; subsp. *flavescens* includes some of the most invasive weedy members of the species. It is not certain that all of the weedy artichoke thistles in California are members of the latter subspecies.

1. Middle involucral bracts broadly obtuse, truncate or emarginate to long-acuminate, apices with or without very narrow yellowish margins, spineless or tipped with slender spines to 9 mm . . . .
. . . . 1a. *Cynara cardunculus* subsp. *cardunculus*
1. Middle involucral bracts acute to short-acuminate, apices with yellowish margins 0.5–1 mm wide, tipped with stout spines to 2–5 mm . . . .
. . . . 1b. *Cynara cardunculus* subsp. *flavescens*

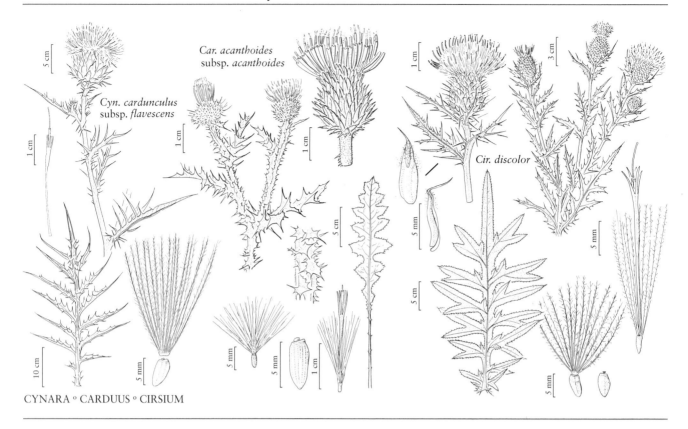

Cyn. cardunculus
subsp. *flavescens*

Car. acanthoides
subsp. *acanthoides*

Cir. discolor

CYNARA ∘ CARDUUS ∘ CIRSIUM

1a. Cynara cardunculus Linnaeus subsp. **cardunculus**
[I] [W]

Cynara cardunculus Linnaeus var.
scolymus (Linnaeus) Fiori; *C.
scolymus* Linnaeus

Leaves armed with spines 1–3 cm
or unarmed or nearly so in some
cultivated forms. **Middle phyllaries** acuminate at apex with point
22–38 mm and spine tip 6–9 mm,
or in some cultivated forms broadly
obtuse to truncate and mucronate with or without spine
tip 1–2 mm, distal margins with or without indistinct
yellowish margins. $2n = 34$.

Flowering spring–summer (Apr–Jul). Disturbed areas in sea bluffs, grasslands, coastal scrub, open woodlands, roadsides; 0–500 m; introduced; Calif.; e Mediterranean region.

The artichoke, selected for reduced spines on both
leaves and phyllaries, and for enlarged heads, sometimes
escapes in disturbed habitats. These plants are usually
propagated vegetatively for uniform crop characteristics. When grown from seed, however, reversions to the
much spinier wild types occur spontaneously and may
have given rise to some of the forms of artichoke thistles.
Additionally, intermediates may arise through hybridization of cultivated and wild races.

1b. Cynara cardunculus Linnaeus subsp. **flavescens**
Wiklund, Bot. J. Linn. Soc. 109: 120, fig. 15A–D, F–J.
1992   [F] [I] [W]

Leaves armed with spines 1–3 cm.
Middle phyllaries acute to short-
acuminate at apex with point 10–
21 mm and spine tip 2–5(–6) mm,
distal margins with prominent
yellowish margins 0.5–1+ mm.

Flowering spring–summer
(Apr–Jul). Disturbed areas in
grasslands, coastal scrub, chapar-
ral, riparian habitats, fallow fields, roadsides; 0–
500 m; introduced; Calif.; w Mediterranean region,
Macaronesia.

Artichoke thistle is a noxious weed that tends to
spread aggressively in overgrazed range lands and may
invade undisturbed natural vegetation as well. Dense
infestations form pure stands, sometimes several hec-
tares in extent. The long, needlelike spines deter herbi-
vores and the large rosettes suppress the growth of other
plants. Vigorous root sprouts can develop from fragments of the deep taproots left in the soil by cultivation
or mechanical clearing.

12. CARDUUS Linnaeus, Sp. Pl. 2: 820. 1753; Gen. Pl. ed. 5, 358. 1754 • Plumeless thistle, chardon [From ancient name of thistlelike plant] 1

David J. Keil

Annuals or biennials [perennials], 30–200(–400) cm, spiny, ± tomentose, sometimes glabrate. **Stems** erect, simple to much branched, (spiny-winged). **Leaves** basal and cauline; petiolate or sessile; blade margins spiny dentate, often 1–2-pinnately lobed, faces glabrous or hairy, eglandular. **Heads** discoid, borne singly or 2–20 in dense clusters or corymbiform arrays. (**Peduncles** naked or leafy-bracteate, spiny-winged or not winged.) **Involucres** cylindric to spheric. **Phyllaries** many in 7–10+ series, linear to broadly ovate, bases appressed, margins entire, apices ascending to spreading or reflexed, acute, spine-tipped. **Receptacles** flat, epaleate, bearing setiform scales ("flattened bristles"). **Florets** several–many; corollas white to pink or purple, ± bilateral, tubes long, slender, throats short, campanulate, abruptly expanded from tubes, lobes linear; anther bases sharply short-tailed, apical appendages oblong; style branches: fused portions with slightly, minutely puberulent, swollen basal nodes, distally papillate or glabrous, distinct portions very short. **Cypselae** ovoid, slightly compressed, faces smooth, glabrous, attachment scars slightly lateral; **pappi** persistent or falling in rings, of many minutely barbed, basally connate bristles or setiform, minutely barbed scales ("minutely flattened bristles"). *x* = 8, 9, 10, 11, 13.

Species ca. 90 (5 in the flora): introduced; Eurasia, Africa.

SELECTED REFERENCES Batra, S. W. T., J. R. Coulson, P. H. Dunn, and P. E. Boldt. 1981. Insects and Fungi Associated with *Carduus* Thistles (Compositae). Washington. [U.S.D.A. Techn. Bull. 1616.] Desrochers, A. M., J. F. Bain, and S. I. Warwick. 1988. A biosystematic study of the *Carduus nutans* complex in Canada. Canad. J. Bot. 66: 1621–1631. Desrochers, A. M., J. F. Bain, and S. I. Warwick. 1988b. The biology of Canadian weeds. 89. *Carduus nutans* L. and *Carduus acanthoides* L. Canad. J. Plant Sci. 68: 1053–1068. Franco, J. do A. 1976. *Carduus.* In: T. G. Tutin et al., eds. 1964–1980. Flora Europaea. 5 vols. Cambridge. Vol. 4, pp. 220–232. Kazmi, S. M. A. 1964. Revision der Gattung *Carduus* (Compositae). Teil II. Mitt. Bot. Staatssaml. München 5: 279–550.

1. Phyllary appendages 2–7 mm wide, usually wider than appressed bases; peduncles often elongate, distally wingless; heads often nodding, usually borne singly or in leafy corymbiform arrays; involucres 20–70 mm diam. . . . . 3. *Carduus nutans*
1. Phyllary appendages 0.5–1.5 mm wide, usually narrower than appressed bases; peduncles short, if present, usually winged throughout or wingless only near tip; heads erect, 1–many, often clustered at branch tips; involucres 7–30 mm diam.
   2. Involucres spheric or hemispheric.
      3. Corollas 13–20 mm; heads 18–25 mm; involucres 14–20 mm; abaxial leaf faces glabrate except for long, curled, septate hairs along veins . . . . 1. *Carduus acanthoides*
      3. Corollas 11–16 mm; heads 15–18 mm; involucres 12–17 mm; abaxial leaf faces sparsely to densely tomentose with fine, nonseptate hairs and often with curled, septate hairs along veins as well . . . . 2. *Carduus crispus*
   2. Involucres cylindric or narrowly ellipsoid.
      4. Heads 1–5 at ends of branches; phyllaries not scarious-margined, ± persistently tomentose, distally scabrous on margins and faces . . . . 4. *Carduus pycnocephalus*
      4. Heads 5–20 at ends of branches; phyllaries scarious-margined, glabrous or sparingly tomentose, distally ciliolate or glabrous . . . . 5. *Carduus tenuiflorus*

1. Carduus acanthoides Linnaeus, Sp. Pl. 2: 821. 1753

• Plumeless or welted thistle, chardon épineux

F I W

Subspecies 2 (1 in the flora): introduced; Eurasia.

1a. Carduus acanthoides Linnaeus subsp. **acanthoides**

F I W

Annuals or biennials, 30–200 (–400) cm. **Stems** openly branching, villous with curled, septate hairs to nearly glabrous, teeth of wings to 1.5 cm, wing spines to 8 mm. **Leaves:** basal 10–30 cm, tapering to winged petioles, margins spiny toothed to deeply 1–2×-pinnately divided; cauline sessile, gradually smaller; marginal spines to 5 mm; abaxial leaf faces glabrate except for long, curled, septate hairs along veins; adaxial faces sparsely hairy or glabrate. **Heads** borne singly or in corymbiform arrays of 2–5, sometimes also in upper axils, 18–25 mm. **Peduncles** spiny-winged throughout, to 4(–10) cm. **Involucres** hemispheric, 14–20 × 10–25 mm. **Phyllaries** linear to narrowly lanceolate, outer and middle with appressed bases 1–1.5 mm wide and appressed to spreading appendages 0.5–1 mm wide, distally glabrous or minutely ciliolate, spine tips 1–2 mm, the inner phyllaries with unarmed, straight or twisted tips. **Corollas** purple or ± white, 13–20 mm. **Cypselae** golden to brown, 2.5–3 mm; **pappus** bristles 11–13 mm. $2n = 22$.

Flowering summer–early fall (Jun–Oct). Aggressive weed of waste ground, pastures, roadsides, fields; 0–3000 m; introduced; B.C., N.B., N.S., Ont., Que.; Calif., Colo., Idaho, Ill., Iowa, Kans., Md., Mich., Minn., Mo., Mont., Nebr., N.J., N.Y., N.C., N.Dak., Pa., R.I., S.Dak., Vt., Va., W.Va., Wis., Wyo.; Eurasia.

Carduus acanthoides has been reported as growing in Connecticut, Delaware, Kentucky, Massachusetts, Maine, Ohio, Oklahoma, Texas, and Washington; I have not seen specimens from those states. It was collected only once in New Brunswick in 1992.

Carduus acanthoides is a serious weed in pasturelands, where it displaces and suppresses growth of other plants and limits access of grazing animals to more palatable plants (A. M. Desrochers et al. 1988). It sometimes grows in dense stands.

2. Carduus crispus Linnaeus, Sp. Pl. 2: 821. 1753

• Welted or curled thistle, chardon crépu   I

Annuals or biennials, 30–150 cm. **Stems** openly branching, villous with curled, septate hairs to nearly glabrous, spiny wings to 1.5 cm wide, wing spines to 3 mm. **Leaves:** basal tapering to winged petioles, blades 10–20 cm, margins spiny-toothed to ± shallowly pinnately lobed; cauline sessile, gradually smaller, margins often more deeply divided, marginal spines to 3 mm; abaxial leaf faces ± tomentose with long, one-celled hairs and/or long, curled, septate hairs along veins or glabrate; adaxial faces sparsely hairy or glabrate. **Heads** borne singly or in groups of 2–5, 15–18 mm. **Peduncles** spiny-winged to near apex or throughout, to 4 cm. **Involucres** ± spheric, 12–17 × 12–17 mm. **Phyllaries** narrowly lanceolate, outer and middle with appressed bases ca. 1 mm wide and appressed to spreading appendages 0.5–1 mm wide, spine tips 1–1.5 mm, inner with unarmed, straight tips. **Corollas** purple or ± white, 11–16 mm, lobes ca. 3.5 times length of throat. **Cypselae** light brown to gray-brown, 2.5–3.8 mm; **pappus** bristles 11–13 mm. $2n = 16$ (Sweden).

Flowering summer–fall (Jul–Sep). Weed of waste ground, pastures, roadsides, fields; 0–500 m; introduced; B.C., N.B., N.S., Ont., Que.; N.J., Pa.; Eurasia.

Canadian distributions above follow R. J. Moore and C. Frankton (1974); I have not seen those specimens. *Carduus crispus* has been reported also from Arkansas, Connecticut, Iowa, Illinois, Maine, Maryland, Massachusetts, Missouri, New York, North Dakota, Ohio, Rhode Island, Vermont, Virginia, and West Virginia; I have not seen specimens from those states.

Two subspecies of *Carduus crispus* have been recognized (S. M. A. Kazmi 1964); those are not differentiated here.

Carduus crispus closely resembles the much more common *C. acanthoides*. Some published records of *C. crispus* are probably *C. acanthoides*. Although the degree of spininess and tough versus brittle stems were used as key characters (A. Cronquist 1980; H. A. Gleason and A. Cronquist 1991) to differentiate the two taxa, both characters are subjective, and the second is impractical with dry material.

3. Carduus nutans Linnaeus, Sp. Pl. 2: 821. 1753
• Musk or nodding thistle, chardon penché ☐I☐ ☐W☐

Carduus macrocephalus Desfontaines; *C. macrolepis* Petermann; *C. nutans* subsp. *leiophyllus* (Petrovič) Stojanov & Stefanoff; *C. nutans* subsp. *macrocephalus* (Desfontaines) Nyman; *C. nutans* var. *macrocephalus* (Desfontaines) B. Boivin; *C. nutans* subsp. *macrolepis* (Petermann) Kazmi; *C. nutans* var. *vestitus* (Hallier) B. Boivin; *C. thoermeri* Weinmann

Annuals or biennials, 40–200+ cm. **Stems** glabrous to tomentose; teeth of wings to 10 mm, wing spines 2–10 mm. **Leaves:** basal tapering to winged petioles, blades 10–40 cm, margins 1–2×-pinnately lobed; cauline sessile, shorter, margins less divided, glabrous or ± hairy. **Heads** borne singly or in corymbiform arrays, sometimes a few axillary, at least terminal head usually conspicuously pedunculate, often nodding, 20–40 mm. **Peduncles** 2–30 cm, unwinged distally or throughout, finely tomentose. **Involucres** hemispheric, 20–60 mm × 20–70 mm. **Phyllaries** lanceolate to ovate, outer and middle with appressed bases 2–4 mm wide and spreading to reflexed, appendages 2–7 mm wide, proximally glabrous or ± tomentose, distally glabrous to minutely scabridulous, spine tips 1–4 mm, inner phyllaries with unarmed, straight or twisted tips. **Corollas** purple, 15–28 mm, lobes 2.5–3 times longer than throat. **Cypselae** golden to brown, 4–5 mm; **pappus** bristles 13–25 mm. *2n* = 16.

Flowering late spring–summer (May–Sep). Aggressive weed of waste ground, pastures, roadsides, fields; 0–3000 m; introduced; Alta., B.C., Man., N.B., Nfld. and Labr. (Nfld.), N.S., Ont., Que., Sask.; Ala., Ariz., Ark., Calif., Colo., D.C., Ga., Idaho, Ill., Ind., Iowa, Kans., Ky., La., Md., Miss., Mo., Mont., Nebr., Nev., N.J., N.Mex., N.Y., N.C., N.Dak., Ohio, Okla., Oreg., Pa., S.C., S.Dak., Tenn., Tex., Utah, Va., Wash., W.Va., Wis., Wyo.; Eurasia.

Although reported from Connecticut, Delaware, Indiana, Massachusetts, Michigan, New Hampshire, Ohio, and Rhode Island, I have seen no specimens of *Carduus nutans* from those places.

Carduus nutans is part of a variable complex that has been treated as one to several species or as a single species with several subspecies or varieties. The New World plants apparently represent multiple introductions, probably representing more than one of these taxa. Various intermediates are evident, and many specimens cannot be reliably assigned. Insufficient evidence exists to reliably apply the names of the various segregate entities to North American material. In a biosystematic study, two subspecies of *C. nutans* were differentiated in Canada (A. M. Desrochers et al. 1988). Subspecies *nutans* was characterized as having arachnoid phyllaries with the terminal appendage only slightly wider than the appressed phyllary base, moderately to densely pubescent leaf bases, and a head diameter of 1.5–3.5 cm. Subspecies *leiocephalus* in contrast has glabrous phyllaries with the terminal appendage definitely wider than the base, glabrous or slightly pubescent bases, and heads 1.8–7 cm in diameter. Subspecies *nutans* was distributed in eastern Canada from Newfoundland to southern Ontario and subsp. *leiocephalus* from Ontario to British Columbia. Whether the results of the study (Desrochers et al.) are applicable to all the populations of musk thistles occurring in the United States has not been determined.

Hybrids between *Carduus acanthoides* and *C. nutans* (*C.* ×*orthocephalus* Wallroth) have been documented from Ontario and Wisconsin and probably occur at other sites where the parental taxa co-occur.

Nodding thistle is one of the most serious weeds in North America. It is unpalatable to wildlife and livestock and often forms dense, impenetrable stands in pastures and rangelands. It readily colonizes disturbed sites in many different habitats. A single large terminal head can produce as many as 1200 cypselae. Efforts to control musk thistle infestations with *Rhinocyllus conicus*, a European seed head weevil, have met with some success, but concerns have been raised because this parasite also attacks native *Cirsium* species.

4. Carduus pycnocephalus Linnaeus, Sp. Pl. ed. 2, 2: 1151. 1763 • Italian thistle ☐I☐

Subspecies 2 (1 in the flora): introduced; Eurasia (Mediterranean region).

4a. Carduus pycnocephalus Linnaeus subsp. pycnocephalus ☐I☐

Annuals, 20–200 cm. **Stems** simple to openly branched, loosely tomentose with fine single-celled hairs and villous with curled, septate hairs; teeth of wings to 10 mm, wing spines to 20 mm. **Leaves:** basal tapering to winged petioles, blades 10–25 cm, margins pinnately 2–5-lobed, abaxial faces ± tomentose, adaxial faces tomentose and pilose, ± glabrate; cauline sessile, shorter, margins less divided, distally reduced to bracts. **Heads** borne singly or clustered in ± tight groups of 2–5 at ends of branches and sometimes in upper axils, sessile or short-pedunculate, 20–25 mm. **Peduncles** winged throughout or distally unwinged, 0–2 cm, tomentose. **Involucres** cylindric to ellipsoid (appearing campanulate when pressed), 17–22 mm × 7–15 mm (diam.). **Phyllaries** linear-lanceolate, with appressed, loosely tomentose bases 2–3 mm wide and ascending, linear appendages 0.5–1.5 mm wide, not scarious-margined, distally scabrous on midribs and margins, spine tips 1–3 mm, the inner straight, erect, with unarmed or minutely armed tips. **Corollas** ± purple, 14–16 mm; lobes ca. 3 times longer than throat. **Cypselae** golden to brown, 4–6 mm, finely 20-nerved; **pappus** bristles 15–20 mm. $2n$ = ca. 54 (Chile), 60, 62.

Flowering spring–early summer (Mar–Jul). Aggressive weed of waste ground, rangelands, pastures, roadsides, fields; 0–1000 m; introduced; Ala., Ark., Calif., Idaho, Miss., N.Y., Oreg., Pa., S.C., Tex.; Europe (Mediterranean region).

Carduus pycnocephalus var. *pycnocephalus* has been reported from New York; I have not seen a specimen from there.

Italian thistle is a serious rangeland pest in much of California, especially near the coast where it sometimes forms pure stands, both in full sun and in partial shade. Populations increase under grazing pressure as more palatable plants are preferentially consumed.

Old herbarium records indicate that *Carduus pycnocephalus* was introduced on ballast into several east-coast ports (e.g., Philadelphia, Mobile) in the 1800s; the lack of subsequent collections suggests that conditions were unsuitable for the species to become permanently established.

The only published chromosome counts for *Carduus pycnocephalus* from North American material are from California specimens (A. M. Powell et. al. 1974). Published chromosome counts ($2n$ = 18, 31, 32, 54, 60, 64, 80) for *C. pycnocephalus* from a variety of Old World localities indicate that this is a complex species in need of further investigation.

5. **Carduus tenuiflorus** Curtis, Fl. Londin. 2(6,61): plate 55. 1789 · Slender-flowered thistle [I]

Carduus pycnocephalus Linnaeus var. *tenuiflorus* (Curtis) Fiori

Annuals, 20–200 cm. **Stems** simple or openly branched, loosely tomentose with fine single-celled hairs and villous with curled, septate hairs; teeth of wings to 25 mm, wing spines to 15 mm. **Leaves:** basal tapered to winged petioles, blades 10–25 cm, margins pinnately 6–10-lobed, abaxial faces tomentose, adaxial faces loosely tomentose and villous or ± glabrate; cauline sessile, shorter, less divided. **Heads** clustered in ± tight arrays of 5–20+ at ends of stems, usually sessile, 15–22 mm × 7–12 mm. **Involucres** cylindric to ellipsoid (appearing campanulate when pressed), 15–20 × 7–12 mm. **Phyllaries** linear-lanceolate, bases appressed, 2–2.5 mm wide, ± glabrate, and ascending, appendages 0.5–1.5 mm wide, narrowly scarious-margined, distally glabrous or minutely ciliolate, spine tips 1–2 mm, inner phyllaries with erect, straight, unarmed tips. **Corollas** pinkish, 10–14 mm; lobes 1.5–2.5 times longer than throat. **Cypselae** brown, 4–5 mm, finely 10–13-nerved; **pappus** bristles 10–15 mm. $2n$ = 54.

Flowering spring–early summer (Apr–Jul). Aggressive weed of waste ground, pastures, roadsides, fields; 0–1000 m; introduced; Calif., Oreg., Pa.; s Europe (Mediterranean region).

Carduus tenuiflorus has been reported from New Jersey, Texas, and Washington; I have not seen specimens from those states.

Carduus pycnocephalus and *C. tenuiflorus* are similar annuals with small, usually tightly clustered heads. The number of heads per capitulescence is usually ultimately greater in *C. tenuiflorus*, but early season plants of this species often have only a few heads. At the end of the growing season the fruiting heads of *C. tenuiflorus* are aggregated in dense, subspheric clusters. Stem wings tend to be more pronounced in *C. tenuiflorus*. Fresh corollas of *C. pycnocephalus* are rose-purple whereas those of *C. tenuiflorus* have a more pinkish tinge, but this difference is subtle and not reliable on herbarium material. The phyllaries of *C. tenuiflorus* are membranous-margined, more or less glabrate, and lack the short, stiff, upwardly appressed trichomes of *C. pycnocephalus*. All published chromosome counts for *Carduus tenuiflorus* from both Old and New World material are the same.

The two species sometimes grow in mixed populations and at times appear to intergrade. Hybridization has been reported in Europe (S. W. T. Batra et al. 1981) and is suspected to occur in California. Hybrids between *C. pycnocephalus* and *C. tenuiflorus* have been designated *Carduus* ×*theriotii* Rouy.

13. CIRSIUM Miller, Gard. Dict. abr. ed. 4, vol. 1. 1754 • Thistle, chardon [Greek *kirsion*, thistle]

David J. Keil

Annuals, biennials, or perennials, 5–400 cm, spiny. **Stems** (1–several) erect, branched or simple, sometimes narrowly spiny-winged. **Leaves** basal and cauline; finely bristly-dentate to coarsely dentate or 1–3 times pinnately lobed, teeth and lobes bristly-tipped, faces green and glabrous or densely gray-canescent, usually eglandular. **Heads** discoid, borne singly, terminal and in distal axils, or in racemiform, spiciform, subcapitate, paniculiform, or corymbiform arrays. (**Peduncles** with ± reduced leaflike bracts.) **Involucres** cylindric to ovoid or spheric, (1–6 ×) 1–8 cm. **Phyllaries** many in 5–20 series, subequal or weakly to strongly, outer and middle with bases appressed and apices spreading to erect, usually spine-tipped, innermost usually with erect, flat, often twisted, entire or dentate, usually spineless apices (distal portion of phyllary midveins in many species with elongate, glutinous resin gland, usually milky in fresh material but dark brown to black when dry). **Receptacles** flat to convex, epaleate, covered with tawny to white bristles or setiform scales. **Florets** 25–200+; corollas white to pink, red, yellow or purple, ± bilateral, tubes long, slender, distally bent, throats short, abruptly expanded, cylindric, lobes linear; (filaments distinct) anther bases sharply short-tailed, apical appendages linear-oblong; style tips elongate (as measured in descriptions including the slightly swollen nodes, long cylindric fused portions of style branches and very short distinct portions). **Cypselae** ovoid, ± compressed, with apical rims, smooth, not ribbed, glabrous, basal attachment scars slightly angled; **pappi** persistent or falling in rings, in 3–5 series of many flattened, plumose bristles or plumose, setiform scales (longer bristles shorter than corollas except in *C. foliosum* and *C. arvense*). $x = 17$.

Species ca. 200 (62 in the flora): North America, Eurasia, n Africa.

Only three genera in Cynareae are represented by native species in the New World, and of these *Cirsium* is by far the most widely distributed and diverse. Native species of *Cirsium* range from sea level to alpine and from boreal regions of Canada to the tropics of Central America. Members of the genus occur in a myriad of habitats including swamps, meadows, forests, prairies, sand dunes, and deserts.

Preliminary molecular phylogenetic studies by D. G. Kelch and B. G. Baldwin (2003) indicated that this diversity is the product of a rapid evolutionary diversification based upon a single initial introduction from Eurasia. Relationships among the North American species are apparently complex, and molecular studies have only begun to provide an outline of phylogeny for these plants. Although there has been a remarkable evolutionary and morphologic diversification in North American *Cirsium*, it has not been accompanied by very much divergence in the base sequences of genes commonly used to elucidate phylogenetic relationships. This suggests either that the diversification has been very rapid or that genetic markers in North American *Cirsium* mutate more slowly than in most other lineages.

Chromosomal diversification has accompanied the morphologic radiation of North American *Cirsium*. Many New World *Cirsium* species share the chromosomal base number of $x = 17$ that also predominates in most Eurasian species. Among the North American thistles, however, is a mostly descending dysploid series with chromosome numbers ranging from $n = 18$ to $n = 10$. Very few instances of polyploidy are known among New World *Cirsium*.

Cirsium species of remarkably different morphologies often are able to hybridize. Although in some hybrid combinations fertility is reduced, in others the formation of complex hybrid swarms indicates a lack of breeding barriers and the potential for emergence of novel character combinations. In the absence of adequate sampling and field observations, hybrids may go unrecognized—treated as distinct taxa or as variants of non-hybrid taxa, or left occupying the indeterminate folders of herbaria. In other cases hybridization has been invoked without much evidence as an explanation for *Cirsium* variants encountered in herbaria or in the field. Hybrid combinations are listed herein when evidence is convincing. Additional hybrids are likely to be found where the ranges of *Cirsium* species overlap. I have seen no documentation of hybridization between native American *Cirsium* species and introduced Eurasian taxa.

Much of the geographic range currently occupied by New World *Cirsium* species was greatly affected by the events of the Quaternary. Large areas were glaciated and other areas were vastly different during glacial episodes. The ancestors of thistles that currently occupy the high mountains of western North America were undoubtedly displaced elevationally and/or latitudinally during the recurrent glacial and interglacial episodes of the Pleistocene. Taxa that are currently isolated may have been in contact during glacial episodes with the opportunity for hybridization and genetic interchange. Episodes of prehistoric hybridization may have led to some of the character combinations found in modern American thistles, particularly in the western half of the continent. Current isolation and localized selection or genetic drift apparently have promoted differentiation of populations separated on mountaintop islands.

One of the most challenging aspects for a taxonomist studying New World *Cirsium* is the presence of species complexes that are apparently evolutionary works in progress. Some of the thistles, especially in the mountainous western part of North America, are frustratingly polymorphic with much overlapping variability and intergradation of characters. Early taxonomists, basing their work on a limited sampling of the morphologic diversity, named many of the forms as species, and the literature is rife with species names. The infilling that results from more collectors visiting more localities within the ranges of these complexes has blurred the boundaries between many of the proposed species and often added forms that do not "fit" the characteristics of named species. As I faced the challenges of preparing this treatment, I recognized that maintaining some of the named entities as species would, for consistency, require a further proliferation of species names. I have chosen to go the other way. Instead of proposing yet more ill-defined microspecies, I have chosen to recognize that the groups in question are rapidly evolving, only partially differentiated assemblages of races that have not reached the level of stability that is usually associated with the concept of species. Certainly much of such variation within the genus deserves a level of taxonomic recognition, or at least should be mentioned, but for those assemblages I think it much more prudent to recognize varieties—entities that may be expected to freely intergrade—rather than species.

Many problems remain to be worked out in North American *Cirsium*. Further investigation will undoubtedly reveal the need for refinement or major revision within some of the species groups. Studies that focus on variation within and among populations and on the biological basis for the variations are much needed. The field is open and the challenges are many.

Preparation of a workable key to *Cirsium* species has been frustratingly difficult. Extensive and overlapping ranges of variation in morphologic characteristics often require that a species be keyed two or more times. The resulting key is longer and more complex than I would prefer, and I have no doubt ignored, overlooked, or been completely unaware of variants that will not key out. Caveat clavitor!

The reputation of *Cirsium* has suffered greatly as a result of the introduction to North America of a few invasive weedy species from Eurasia. *Cirsium vulgare* (bull thistle) and *C. arvense* (Canada thistle—a misnomer) have long been despised as noxious weeds. In recent years *C. palustre* (European swamp thistle) has joined their ranks. Additionally, weedy

Eurasian species of *Carduus*, *Onopordum*, *Centaurea*, etc., add to the public perception that all thistles are bad. Most North American native *Cirsium* are not at all weedy, and many are strikingly attractive plants. All are spiny plants that command respect, but they deserve a better reputation as one of North America's evolutionary success stories.

Native *Cirsium* species have come under threat from biocontrol programs instituted to suppress populations of weedy introduced thistles. Beginning in 1968 the seedhead weevil *Rhinocyllus conicus* has been widely introduced in various areas of the United States and Canada, primarily to control weedy species of *Carduus*. S. M. Louda et al. (1997) reported that *R. conicus* has crossed over to several native species of *Cirsium*. They observed that the number of viable cypselae in infested heads was greatly reduced; e.g., heads of *C. canescens* infested by *R. conicus* produced 14.1 percent of the number of viable cypselae as in uninfested heads. Not all taxa are impacted as much as *C. canescens*, particularly those with later flowering phenology (Louda 1998). R. W. Pemberton (2000) reported that 22 *Cirsium* taxa in North America are known hosts of *R. conicus*. I suspect that the number is higher. During my field work I have observed that the heads of many *Cirsium* species are heavily parasitized, although I have not determined which of these are infested by *R. conicus* and which by native seedhead parasites. The long-term impacts of *R. conicus* and other biocontrol agents on native thistles, particularly rare taxa, remain to be determined.

SELECTED REFERENCES Hsi, Y.-T. 1960. Taxonomy, Distribution and Relationships of the Species of *Cirsium* Belonging to the Series *Undulata*. Ph.D. dissertation. University of Minnesota. Kelch, D. G. and B. G. Baldwin. 2003. Phylogeny and ecological radiation of New World thistles (*Cirsium*, Cardueae–Compositae) based on ITS and ETS rDNA sequence data. Molec. Ecol. 12: 141–151. Moore, R. J. and C. Frankton. 1969. Cytotaxonomy of some *Cirsium* species of the eastern United States, with a key to eastern species. Canad. J. Bot. 47: 1257–1275. Petrak, F. 1917. Die nordamerikanischen Arten der Gattung *Cirsium*. Beih. Bot. Centralbl. 35(2): 223–567.

Key to Groups of *Cirsium* Species

1. Plants of Great Plains, e North America, and Greenland . . . . Group 1, p. 97
1. Plants of nw Canada, Pacific Coast, Intermountain Region, sw Deserts, and Rocky Mountains.
   2. Involucres 3–5 cm . . . . Group 2, p. 100
   2. Involucres 1–3 cm . . . . Group 3, p. 102

Group 1
Cirsium species of Great Plains, eastern North America, and Greenland

1. Involucres usually 1–2.5 cm.
   2. Plants dioecious or nearly so; common invasive weed . . . . 2. *Cirsium arvense*
   2. Plants hermaphroditic, with bisexual florets.
      3. Bases of mid cauline leaves long-decurrent as spiny wings.
         4. Heads crowded at stem tips; peduncles 0–1 cm; invasive weed in northern forests, wetlands . . . . 3. *Cirsium palustre*
         4. Heads in open paniculiform arrays, borne singly at tips of slender peduncles 1–15 cm; New Mexico . . . . 33. *Cirsium wrightii*
      3. Bases of mid cauline leaves not decurrent or sometimes short-decurrent, forming spiny wings to 3 cm.
         5. Longer pappus bristles exceeding corollas by 1–8 mm; corollas very slender, throat scarcely wider than tube; abaxial faces of outer and middle phyllaries without glutinous ridge; heads sessile in dense mass at tip of thick fleshy stem, overtopped by crowded distal cauline leaves; w Canada . . . . 55. *Cirsium foliosum*
         5. Longer pappus bristles shorter than corollas; corolla throat noticeably wider than tube; abaxial faces of outer and middle phyllaries with elongate glutinous ridge (milky when fresh, dark when dry, sometimes very narrow); heads evidently pedunculate, not overtopped by crowded distal cauline leaves.

[6. Shifted to left margin.—Ed.]

6. Apices of outer and middle phyllaries erect, ± appressed, tipped by erect or ascending spines or cusps.
    7. Abaxial leaf face densely white tomentose with non-septate trichomes.
        8. Leaves 4–8 cm wide, main spines 1–2 mm; Greenland . . . . 4. *Cirsium helenioides*
        8. Leaves 0.5–4 cm wide, main spines usually 3–5 mm; se United States . . . .
            . . . . 15. *Cirsium virginianum* (in part)
    7. Abaxial leaf face thinly tomentose or glabrate and villous with septate trichomes.
        9. Leaves deeply pinnatifid; widespread in e North America . . . . 8. *Cirsium muticum* (in part)
        9. Leaves unlobed to shallowly pinnatifid; Virginia to Georgia . . . . 9. *Cirsium repandum* (in part)
6. Apices of at least outer phyllaries widely spreading, tipped by spreading spines or short cusps.
    10. Phyllary spines 0–1(–3) mm.
        11. Leaves thick, ± rigid, abaxially white-tomentose; New Jersey to Florida . . . .
            . . . . 15. *Cirsium virginianum* (in part)
        11. Leaves thin, flexible, abaxially thinly tomentose, ± glabrate.
            12. Peduncles leafy-bracted; widespread, e North America . . . . 8. *Cirsium muticum* (in part)
            12. Peduncles essentially naked; Virginia to Florida, Louisiana . . . . 16. *Cirsium nuttallii*
    10. Phyllary spines (1–)2–9 mm.
        13. Leaf bases decurrent.
            14. Adaxial leaf faces green, glabrous or thinly arachnoid; phyllary spines 2–7 mm, stout, broad-based; w Great Plains . . . . 29. *Cirsium pulcherrimum*
            14. Adaxial leaf faces gray-tomentose; phyllary spines 1–3 mm, slender; sandy shores and dunes around Great Lakes... . . . . 23. *Cirsium pitcheri* (in part)
        13. Leaf bases not decurrent (except sometimes in *C. pitcheri* or *C. texanum*); phyllary spines slender.
            15. Involucres 1.2–2 cm.
                16. Mid and distal cauline leaves linear to narrowly elliptic, bases tapered, cuneate, not decurrent; widespread, se United States . . . . 14. *Cirsium carolinianum*
                16. Mid and distal cauline leaves ovate, broadly sessile, sometimes auriculate-clasping or short-decurrent; Louisiana, Oklahoma, Texas, adventive in Missouri . . . . 17. *Cirsium texanum*
            15. Involucres 2–3.5(–4) cm.
                17. Leaves gray- or white-tomentose on both faces, divided nearly to midvein into linear lobes; corollas usually white; beaches and dunes around Great Lakes . . . . 23. *Cirsium pitcheri* (in part)
                17. Leaves adaxially green, ± glabrate, abaxially white-tomentose, undivided or shallowly to deeply pinnatifid; corollas lavender to purple (rarely white); widespread.
                    18. Stems villous with septate trichomes or distally white-tomentose; widespread, e North America . . . . 7. *Cirsium discolor* (in part)
                    18. Stems gray- or white-tomentose throughout; n Great Plains, occasionally eastward . . . . 18. *Cirsium flodmanii* (in part)

1. Involucres 2.5–5 cm.
    19. Adaxial leaf faces covered with short, ± appressed, bristlelike spines; common weed
        . . . . 1. *Cirsium vulgare*
    19. Adaxial leaf faces without bristlelike spine.
        20. Corollas red; w Texas . . . . 44. *Cirsium turneri*
        20. Corollas white to yellow, lavender, or purple.
            21. Each head closely subtended by involucre-like ring of spiny-margined bracts
                about as long as involucre; Maine to Florida, w to Texas . . . . 11. *Cirsium horridulum*
            21. Heads not individually subtended by involucre-like ring of spiny-margined
                bracts, sometimes cluster of several heads collectively surrounded by crowded
                distal cauline leaves that overtop them.
                22. Abaxial face of outer and middle phyllaries without glutinous ridge; heads
                    surrounded and overtopped by distal cauline leaves; widespread, w North
                    America, disjunct on Mingan Archipelago, Quebec . . . . 54. *Cirsium scariosum*
                22. Abaxial faces of outer and middle phyllaries with elongate glutinous ridge
                    (milky when fresh, dark when dry); heads in most species elevated above
                    cauline leaves.
                    23. Spines (or sharp, cusplike tips) of outer and middle phyllaries erect,
                        ± appressed.
                        24. Phyllary spines 0–0.5 mm . . . . 8. *Cirsium muticum* (in part)
                        24. Phyllary spines 0.5–0.6 mm.
                            25. Apices of innermost phyllaries usually scarious, often
                                expanded, ± erose-dentate.
                                26. Heads evidently pedunculate above distal cauline leaves;
                                    leaves usually less than 5 times longer than wide; midwest,
                                    e United States, s Canada . . . . 12. *Cirsium pumilum*
                                26. Heads ± crowded, surrounded and overtopped by distal
                                    cauline leaves; leaves usually 5+ times longer than wide;
                                    Canada, Black Hills of South Dakota, Wyoming . . . . .
                                    . . . . 53. *Cirsium drummondii*
                            25. Apices of innermost phyllaries linear-attenuate.
                                27. Stems usually branched, distal ¹/₂ usually leafy; adaxial
                                    leaf faces shaggy villous with septate trichomes, abaxial
                                    faces villous with septate trichomes, loosely arachnoid
                                    when young; Virginia to Georgia . . . . 9. *Cirsium repandum* (in part)
                                27. Stems usually simple, distal ¹/₂ nearly naked with only few
                                    bractlike leaves; adaxial leaf faces glabrous or sparingly
                                    villous with coarse, multicellular trichomes, abaxial faces
                                    loosely arachnoid when young, often ± glabrate; coastal,
                                    North Carolina to Louisiana . . . . 10. *Cirsium lecontei*
                    23. Spines (or short, cusplike tips) of outer and middle phyllaries some-
                        times erect, sometimes spreading.
                        28. Phyllary spines 0–0.5 mm; widespread in e North America
                            . . . . 8. *Cirsium muticum* (in part)
                        28. Phyllary spines 2–12 mm.
                            29. Stems thinly tomentose when young, ± glabrate or tomentum
                                persisting distally; adaxial leaf faces green, glabrate.
                                30. Peduncles with much reduced bracts; usually some roots
                                    with tuberlike enlargements; Louisiana, Oklahoma, Texas
                                    . . . . 13. *Cirsium engelmannii*
                                30. Peduncles leafy-bracted; roots without tuberlike enlarge-
                                    ments.
                                    31. Cauline leaves usually unlobed or shallowly pinnati-
                                        fid, or when more deeply lobed, lobes relatively wide;
                                        margins flat; apices of innermost phyllaries usually
                                        dilated and ± erose or serrulate; widespread, ne United
                                        States . . . . 6. *Cirsium altissimum*

31. Cauline leaves deeply pinnatifid into narrow, linear-lanceolate lobes; margins ± revolute; apices of innermost phyllaries attenuate, not dilated; widespread in e North America . . . . 7. *Cirsium discolor* (in part)

29. Stems uniformly and persistently gray- or white-tomentose; adaxial leaf faces thinly to densely tomentose when young, sometimes green and glabrate in age.

32. Cauline leaves not decurrent or with decurrent wing 0–1 cm.

33. Cauline leaves elliptic to oblanceolate, shallowly lobed to pinnatifid; cypselae 3–5 mm, apical collar stramineous; root sprouts arising from horizontal runner roots; n Great Plains, occasionally eastward . . . . 18. *Cirsium flodmanii* (in part)

33. Cauline leaves ovate to lanceolate, subentire to coarsely toothed or shallowly lobed; cypselae 6–7 mm, apical collar colored like body; root sprouts arising from deep taproots; Great Plains, occasionally eastward . . . . 19. *Cirsium undulatum*

32. Cauline leaves decurrent 1–5+ cm.

34. Spines of phyllaries 5–12 mm; corollas pale lavender or rarely white; Great Plains . . . . 24. *Cirsium ochrocentrum*

34. Spines of phyllaries usually 2–4 mm; corollas usually dull white, rarely lavender or purple.

35. Leaves less deeply divided, lobes linear or oblong; phyllary spines 2–4 mm; Great Plains to c Rocky Mountains . . . . 22. *Cirsium canescens*

35. Leaves divided nearly to midvein into linear lobes; phyllary spines 1–2(–3) mm; beaches and dunes around Great Lakes. . . . . 23. *Cirsium pitcheri* (in part)

Group 2
Large-headed *Cirsium* species of Pacific Coast, Intermountain Region, southwestern Deserts, and Rocky Mountains

1. Adaxial leaf faces with slender ± appressed bristlelike spines; common weed. . . . . 1. *Cirsium vulgare*
1. Adaxial leaf faces without bristlelike spines.

2. Abaxial faces of outer and middle phyllaries with elongate glutinous ridge (milky when fresh, dark when dry, sometimes very narrow).

3. Corolla lobes 1.5+ times as long as tube; style 1.5–4 mm; se California to s Colorado, Arizona, New Mexico . . . . 43. *Cirsium arizonicum* (in part)

3. Corolla lobes shorter than to equaling tube; style 2–8 mm.

4. Herbage glabrous or ± villous or tomentose with septate trichomes; fine, nonseptate trichomes usually 0.

5. Phyllaries strongly imbricate, outer and middle ovate or lanceolate, appressed, ascending to erect, spines 2–3 mm; Canada, Black Hills of South Dakota, Wyoming, Colorado. . . . . 53. *Cirsium drummondii*

5. Phyllaries subequal or imbricate, outer and middle with short, appressed bases and long, linear, stiffly spreading to ascending apices; spines 3–35 mm.

6. Outer phyllaries entire; spines 3–5 mm; n Rocky Mountains. . . . . 48. *Cirsium hookerianum* (in part)

6. Outer phyllaries usually pinnately spiny; spines 7–35 mm; Rocky Mountains and high peaks of intermountain region . . . . . 51. *Cirsium eatonii* (in part)

4. Herbage ± densely tomentose with fine, non-septate trichomes; septate trichomes 0 (except sometimes in *C. hookerianum*).
  7. Glutinous ridges on phyllaries narrow, inconspicuous, sometimes absent; n Rocky Mountains . . . . 48. *Cirsium hookerianum* (in part)
  7. Glutinous ridges on phyllaries well developed.
    8. Mid and distal cauline leaves evidently decurrent, spiny wings to 5 cm.
      9. Phyllary spines usually 2–4 mm; corollas ochroleucous (rarely lavender-tinged) . . . . 22. *Cirsium canescens*
      9. Phyllary spines usually 5–12 mm; corollas pale lavender . . . . 24. *Cirsium ochrocentrum* (in part)
    8. Mid and distal cauline leaves not or scarcely decurrent.
      10. Corollas red, pink, or reddish purple; Arizona, New Mexico . . . . 24. *Cirsium ochrocentrum* (in part)
      10. Corollas white to lavender or purple.
        11. Corollas creamy white, rarely lavender-tinged; e Oregon, Washington, w Idaho . . . . 5. *Cirsium brevifolium*
        11. Corollas lavender to purple; widespread.
          12. Cauline leaves elliptic to oblanceolate, shallowly lobed to pinnatifid; cypselae 3–5 mm, apical collar stramineous; root sprouts arising from horizontal runner roots; se British Columbia to n Colorado, Great Plains . . . . 18. *Cirsium flodmanii*
          12. Cauline leaves ovate to lanceolate, subentire to coarsely toothed or shallowly lobed; cypselae 6–7 mm, bodies and apical collars concolorous; root sprouts arising from deep taproots; widespread . . . . 19. *Cirsium undulatum*

[2. Shifted to left margin.—Ed.]
2. Abaxial faces of outer and middle phyllaries without elongate glutinous ridge (sometimes present in *C. arizonicum*).
13. Corollas pink, red, or reddish purple.
  14. Corolla lobes shorter than throat; n California, sw Idaho, nw Nevada . . . . . . 45. *Cirsium andersonii*
  14. Corolla lobes longer than throat.
    15. Corolla tube 3.5–5 mm; w Texas . . . . 44. *Cirsium turneri*
    15. Corolla tube 7–18 mm.
      16. Biennials from taproots; stems usually solitary; style tips 4–5 mm; s Oregon, California, Nevada . . . . 40. *Cirsium occidentale* (in part)
      16. Perennials from taprooted caudices or runner roots; stems 1–several; style tips 1.5–4.5 mm; se California to s Colorado, Arizona, New Mexico . . . . 43. *Cirsium arizonicum* (in part)
13. Corollas white to lavender or purple.
  17. Outer and middle phyllaries lanceolate to ovate, appressed, spines 1–12 mm.
    18. Biennials or monocarpic perennials from taproots; usually from moist sites; widespread . . . . 54. *Cirsium scariosum*
    18. Perennials from runner roots; usually from dry sites; coastal c, n California . . . . 56. *Cirsium quercetorum*
  17. Outer and middle phyllaries with short appressed bases and stiffly spreading to ascending, lanceolate to linear-acicular apices; spines 3–35 mm.
    19. Outer phyllaries subequal, rigidly spreading, with spines 5–10 mm; middle and inner phyllaries imbricate, with bodies appressed and apices spreading, spineless; Santa Clara County, California . . . . 57. *Cirsium praeteriens*
    19. Outer and middle phyllaries subequal or imbricate, with spreading to incurved-ascending, lanceolate to linear-acicular apices; middle and inner phyllaries with bodies appressed or not, and at least the middle with apices deflexed to spreading or ascending, spine-tipped.

[20. Shifted to left margin.—Ed.]

20. Adaxial leaf faces thinly arachnoid-tomentose to densely felty with fine, non-septate trichomes, sometimes glabrate.
    21. Phyllaries spiny-ciliate; coastal dunes, s, c California . . . . 58. *Cirsium rhothophilum*
    21. Phyllaries usually entire.
        22. Heads borne singly or in corymbiform arrays; trichomes usually all fine, non-septate; California . . . . 40. *Cirsium occidentale* (in part)
        22. Heads usually crowded in racemiform arrays; some trichomes on stems and leaves often septate; nw United States, w Canada . . . . 48. *Cirsium hookerianum* (in part)
20. Adaxial leaf faces glabrous or villous along midveins with septate trichomes.
    23. Corolla lobes filiform with knoblike tips; style included or exserted only 1–2 mm beyond corolla lobes; British Columbia to coastal s California, Montana . . . . . . 47. *Cirsium brevistylum*
    23. Corolla lobes linear but not filiform, not knobbed at tip; style exserted well beyond corolla lobes.
        24. Leaves and stems glabrous.
            25. Basal and proximal cauline leaves 2–5 cm wide, strongly undulate; Rocky Mountains and high peaks of intermountain region . . . . 51. *Cirsium eatonii* (in part)
            25. Basal and proximal cauline leaves 6–12 cm wide, not strongly undulate; plants of hanging gardens in sw Utah . . . . 61. *Cirsium joannae*
        24. Leaves and/or stems villous and/or tomentose.
            26. Basal and proximal cauline leaves plane to moderately undulate and shallowly to ± deeply divided into 5–10 pairs of usually well separated, linear to broadly triangular lobes; British Columbia to w Oregon . . . . 46. *Cirsium edule*
            26. Basal and proximal cauline leaves strongly undulate and deeply divided into 10–20 pairs of closely spaced, usually narrow lobes; Rocky Mountains and high peaks of intermountain region . . . . 51. *Cirsium eatonii* (in part)

Group 3

Small-headed *Cirsium* species of Pacific Coast, Intermountain Region, southwestern deserts, and Rocky Mountains

1. Plants dioecious or nearly so; common invasive weed . . . . 2. *Cirsium arvense*
1. Plants with bisexual florets.
    2. Bases of mid cauline leaves long-decurrent as spiny wings.
        3. Heads crowded at stem tips, peduncles 0–1 cm; invasive weed in wetlands, northern forests, British Columbia . . . . 3. *Cirsium palustre*
        3. Heads in open paniculifrom arrays, borne singly on slender peduncles 1–15 cm; se Arizona, New Mexico . . . . 33. *Cirsium wrightii*
    2. Bases of mid cauline leaves not decurrent or decurrent as spiny wings to 5 cm.
        4. Heads nodding; adaxial leaf faces glandular or not.
            5. Adaxial leaf faces densely puberulent with mixture of short, multicellular, ± glandular trichomes and finer, arachnoid, non-septate trichomes; serpentine wetlands, coast ranges, c California . . . . 59. *Cirsium fontinale*
            5. Adaxial leaf faces glabrous.
                6. Involucres densely tomentose; Rocky Mountains and high peaks of intermountain region . . . . 51. *Cirsium eatonii* (in part)
                6. Involucres glabrous.
                    7. Phyllaries maroon, drying dark brown or blackish; corollas rich rose-purple; s New Mexico . . . . 62. *Cirsium vinaceum*
                    7. Phyllaries green, drying green or light brown; corollas dull white to pink or purple; n Arizona, s Utah.

8. Involucres 1.4–2 cm (excluding recurved outer phyllary apices that extend below involucre base); phyllary apices lance-ovate, rather abruptly contracted into recurved spines 3–25 mm, margins sparingly tomentose or glabrate; ne Arizona, se Utah . . . 60. *Cirsium rydbergii* (in part)
8. Involucres 2.5–4 cm (including spreading to curved-ascending phyllary apices); phyllary spices linear, spines 5–12 mm, margins scabridulous-ciliolate; ne Arizona, se Utah . . . . 61. *Cirsium joannae* (in part)

[4. Shifted to left margin.—Ed.]

4. Heads usually erect; adaxial leaf faces usually not glandular.
  9. Margins of outer phyllaries hispidulous-ciliolate, spiny-fringed, pinnately spiny, or with expanded, scarious appendages.
    10. Heads usually not closely subtended by clustered leafy bracts, often each subtended by 1 leaf or borne on peduncles with much reduced bracts.
      11. Midribs of outer and middle phyllaries forming an elongate glutinous ridge (milky when fresh, dark when dry); margins of outer phyllaries minutely spiny-ciliate.
        12. Corollas deep purple; Arizona, New Mexico . . . . 27. *Cirsium grahamii*
        12. Corollas pale rose-purple; coastal c California . . . . 35. *Cirsium hydrophilum* (in part)
      11. Midribs of phyllaries not glandular or with narrow, inconspicuous glutinous ridge; margins of outer phyllaries usually ± conspicuously fringed or spiny-ciliate.
        13. Phyllaries long-acicular; involucres usually ± densely arachnoid with fine, non-septate trichomes; heads often ± sessile in tight clusters at tips of main stem and branches; British Columbia to w Oregon . . . . 46. *Cirsium edule* (in part)
        13. Phyllaries not long-acicular; involucres glabrous or loosely floccose, or arachnoid with coarse, septate trichomes; heads borne singly or in various arrays.
          14. Arrays racemiform, spiciform, or subcapitate, or plants forming low, rounded mounds; plants ± fleshy; usually ± wet habitats, widespread . . . . 54. *Cirsium scariosum* (in part)
          14. Arrays ± openly branched; plants usually not fleshy.
            15. Plants stout, usually 100–300 cm; stems 2–10 cm diam. near base, hollow; middle and inner phyllaries entire; San Joaquin Valley, California . . . . 34. *Cirsium crassicaule* (in part)
            15. Plants slender, usually 10–110 cm; stems usually less than 2 cm diam. at base, not hollow; middle and inner phyllaries often with fringed appendages.
              16. Biennials; barren stony habitats, w Colorado . . . . 31. *Cirsium perplexans*
              16. Perennials, monocarpic or polycarpic; various habitats.
                17. Corollas 16–20 mm; Colorado, Utah, Wyoming . . . . 30. *Cirsium clavatum* (in part)
                17. Corollas 18–28 mm; Washington to n California . . . . 32. *Cirsium remotifolium* (in part)
    10. Heads usually closely subtended by clustered ± leafy bracts.
      18. Leaf faces closely gray-white felty-tomentose, margins very prominently undulate; dunes, headlands, s, c California . . . . 58. *Cirsium rhothophilum*
      18. Leaf faces glabrous to arachnoid-tomentose adaxially, sometimes white- or gray-tomentose abaxially, margins flat to undulate.
        19. Corollas pale creamy yellow to bright yellow.
          20. Biennials; basal and proximal cauline leaves absent at flowering; leaves unlobed or pinnatifid with lobes well separated, not overlapping; heads loosely to densely clustered at tips of main stem and branches, often also in distal leaf axils; involucres ± arachnoid, but not obscured by dense, woolly pubescence; corollas ± pale yellow; montane meadows, Colorado, Arizona, New Mexico . . . . 50. *Cirsium parryi*

20. Perennials; basal and proximal cauline leaves present at flowering; leaves regularly pinnatifid with closely spaced, ± overlapping lobes; heads borne in massive, heavy, commonly nodding terminal clusters or spikes; involucres ± concealed by dense, woolly pubescence; corollas bright yellow; subalpine and alpine, Colorado . . . . 51. *Cirsium eatonii* (in part)

19. Corollas white to purple.

    21. Involucres ± glabrous or very thinly arachnoid.

        22. Capitulescence open, many-headed, corymbiform or paniculiform; inner phyllaries erect or twisted, not expanded; San Joaquin Valley, California . . . . 34. *Cirsium crassicaule* (in part)

        22. Capitulescence racemiform, spiciform, or subcapitate, or plants forming low, rounded mounds; inner phyllaries often with expanded, erose, scarious tips.

            23. Outer phyllaries erose to lacerate or spiny-fringed; mountains of w, c Montana . . . . 49. *Cirsium longistylum*

            23. Outer phyllaries usually not erose to lacerate; usually ± wet habitats, widespread . . . . 54. *Cirsium scariosum* (in part)

    21. Involucres conspicuously arachnoid-tomentose.

        24. Heads sessile or subsessile, crowded in dense subcapitate to spiciform arrays; Rocky Mountains and highpeaks of intermountain region . . . . 51. *Cirsium eatonii* (in part)

        24. Heads sessile to evidently pedunculate, not crowded in dense subcapitate to spiciform arrays.

            25. Distal leaves ± stiff, wickedly spiny, spines stout, often 10–15 mm; most outer phyllaries spiny-margined; coastal n, c California . . . . 42. *Cirsium andrewsii*

            25. Distal leaves thin, ± weakly spiny, spines slender, usually less than 8 mm; few outer phyllaries spiny-margined.

                26. Style tips conspicuously exserted beyond corolla lobes; British Columbia to w Oregon . . . . 46. *Cirsium edule* (in part)

                26. Style tips included or exserted only 1–2 mm beyond corolla lobes; British Columbia to coastal s California, Montana . . . . 47. *Cirsium brevistylum* (in part)

[9. Shifted to left margin.—Ed.]

9. Margins of outer phyllaries usually entire (bracts subtending head usually spiny).

    27. Corolla lobes twice as long as corolla throat or longer; se California to s Colorado, Arizona, New Mexico . . . . 43. *Cirsium arizonicum*

    27. Corolla lobes shorter than to slightly exceeding corolla throat.

        28. Corollas bright pink, red, or rich purple.

             29. Heads usually crowded at branch tips; peduncles usually 0–4 cm; corollas abruptly expanded from tube to throat; usually wetland sites; California, Nevada, Oregon. . . . . 36. *Cirsium douglasii*

             29. Heads usually borne singly; peduncles 1–30 cm; corollas gradually expanded from tube to throat; usually dry sites; California, Nevada, Oregon . . . . 40. *Cirsium occidentale* (in part)

        28. Corollas white to pale pink, lavender, or purple.

             30. Middle and outer phyllaries usually spreading to reflexed.

                31. Plants lush, lax; herbage essentially glabrous or glabrescent throughout; heads in paniculiform arrays; hanging gardens and canyon bottoms; n Arizona, s Utah.

32. Involucres 1.4–2 cm (excluding recurved outer phyllary apices that extend below involucre base); phyllary apices lance-ovate, rather abruptly contracted into recurved spines 3–25 mm, margins sparingly tomentose or glabrate; ne Arizona, se Utah . . . . 60. *Cirsium rydbergii* (in part)

32. Involucres 2.5–4 cm (including spreading to curved-ascending phyllary apices); phyllary apices linear, spines 5–12 mm, margins scabridulous-ciliolate; ne Arizona, se Utah . . . . 61. *Cirsium joannae* (in part)

[31. Shifted to left margin.—Ed.]

31. Plants usually ± erect; herbage usually ± pubescent; heads in various arrays; widespread.

33. Leaves deeply 2–3–pinnately divided, lobes linear to linear-lanceolate; abaxial leaf faces glabrous to thinly tomentose and villous along major veins; e Utah, w Colorado . . . . 52. *Cirsium ownbeyi*

33. Leaves shallowly to deeply pinnatifid; abaxial leaf faces usually thinly to densely tomentose.

34. Spines of principal phyllaries usually 7–20+ mm.

35. Perennials.

36. Phyllaries thinly arachnoid or glabrate; e Idaho and ne Utah to Wyoming and nw Nebraska . . . . 29. *Cirsium pulcherrimum* (in part)

36. Phyllaries densely and persistently arachnoid with slender trichomes connecting adjacent phyllaries; coastal bluffs, dunes, c California . . . . 40. *Cirsium occidentale* (in part)

35. Biennials or short-lived monocarpic perennials.

37. Main spines of cauline leaves usually less than 8 mm; California . . . . 40. *Cirsium occidentale* (in part)

37. Main spines of cauline leaves usually 8–15 mm; deserts of sw United States . . . . 41. *Cirsium neomexicanum*

34. Spines of principal phyllaries usually 1–6 mm.

38. At least some inner phyllaries usually fringed or spiny-toothed.

39. Corollas 16–20 mm; Colorado, Utah, Wyoming . . . . 30. *Cirsium clavatum* (in part)

39. Corollas 18–28 mm; Washington to n California . . . . 32. *Cirsium remotifolium* (in part)

38. Phyllaries all entire.

40. Biennials; heads usually in open corymbiform arrays.

41. Abaxial face of outer and middle phyllaries with elongate glutinous ridge (milky when fresh, dark when dry); cauline leaves with decurrent wings 1–3 cm; Oregon and e California to w Wyoming . . . . 38. *Cirsium inamoenum* (in part)

41. Abaxial face of outer and middle phyllaries without elongate glutinous ridge; cauline leaves not decurrent or spiny decurrent wings to 2 cm; California . . . . 40. *Cirsium occidentale* (in part)

40. Perennials; heads usually in ± congested flat-topped or racemiform arrays.

42. Principal cauline leaves decurrent 1.5–3.5 cm; e Idaho and ne Utah to Wyoming and nw Nebraska . . . . 29. *Cirsium pulcherrimum* (in part)

42. Principal cauline leaves clasping or decurrent 1–10 mm.

43. Abaxial leaf faces densely gray-white tomentose; stems and leaves tomentose with fine, non-septate trichomes, septate trichomes absent; s Oregon, n California . . . . 20. *Cirsium ciliolatum* (in part)

43. Abaxial leaf faces ± green, thinly tomentose; stems and leaves with mixture of fine, non-septate trichomes and coarser, septate trichomes, especially along stems and on midveins on abaxial leaf faces; widespread, n California to e Wyoming . . . . . 39. *Cirsium cymosum* (in part)

[30. Shifted to left margin.—Ed.]

30. Middle and usually outer phyllaries appressed or stiffly ascending (sometimes spines abruptly spreading).

　　44. Leaves all basal or crowded on very short, densely leafy stems; heads ± sessile, closely subtended by rosette leaves.

　　　　45. Biennials or short-lived monocarpic perennials, taprooted; usually ± wet habitats, widespread . . . . 54. *Cirsium scariosum* (in part)

　　　　45. Perennials from creeping rootstocks; usually ± dry habitats; coastal c, n California . . . . 56. *Cirsium quercetorum* (in part)

　　44. Leaves basal and evidently cauline or all cauline; heads sessile or pedunculate, evidently raised above rosette leaves.

　　　　46. Phyllaries spines 10–20+ mm.

　　　　　　47. Plants persistently tomentose; heads usually not closely subtended by well-developed leaves . . . . 24. *Cirsium ochrocentrum* (in part)

　　　　　　47. Plants sparsely arachnoid-tomentose, soon glabrescent; heads closely subtended by well-developed leaves; usually ± wet habitats, widespread . . . 54. *Cirsium scariosum* (in part)

　　　　46. Phyllary spines usually less than 10 mm.

　　　　　　48. Corollas usually 25–50 mm.

　　　　　　　　49. Biennials or short-lived monocarpic perennials, taprooted.

　　　　　　　　　　50. Cauline leaves evidently decurrent . . . . 38. *Cirsium inamoenum* (in part)

　　　　　　　　　　50. Cauline leaves not or only slightly decurrent. sometimes auriculate-clasping.

　　　　　　　　　　　　51. Heads sessile or short-pedunculate, often crowded in spiciform or racemiform arrays; usually ± wet habitats, widespread . . . . 54. *Cirsium scariosum* (in part)

　　　　　　　　　　　　51. Heads usually evidently pedunculate, usually in ± open, corymbiform arrays; usually dry sites, Oregon and California to w Wyoming . . . . 39. *Cirsium cymosum* (in part)

　　　　　　　　49. Perennials, taprooted or from creeping runner roots.

　　　　　　　　　　52. Corollas purple.

　　　　　　　　　　　　53. Phyllaries usually ovate; abaxial face of outer and middle phyllaries without elongate glutinous ridge; coastal n, c California . . . . 56. *Cirsium quercetorum* (in part)

　　　　　　　　　　　　53. Phyllaries usually lanceolate; abaxial face of outer and middle phyllaries with elongate glutinous ridge (milky when fresh, dark when dry).

　　　　　　　　　　　　　　54. Cauline leaves elliptic to oblanceolate, shallowly lobed to pinnatifid; cypselae 3–5 mm, apical collar straw-colored; root sprouts arising from horizontal runner roots; se British Columbia to n Colorado, Great Plains . . . . 18. *Cirsium flodmanii* (in part)

　　　　　　　　　　　　　　54. Cauline leaves ovate to elliptic, oblong, or lanceolate, subentire to coarsely toothed or shallowly lobed; cypselae 6–7 mm, apical collar colored like body; root sprouts, if any, arising from deep taproots.

　　　　　　　　　　　　　　　　55. Involucres of larger heads often exceeding 3 cm; corolla lobes 6.5–13 mm; widespread . . . . 19. *Cirsium undulatum* (in part)

　　　　　　　　　　　　　　　　55. Involucres of larger heads commonly less than 3 cm; corolla lobes 5.5–9.5 mm; w Colorado, ne New Mexico, se Utah . . . . 21. *Cirsium tracyi* (in part)

　　　　　　　　　　52. Corollas white to pink or lavender.

　　　　　　　　　　　　56. Larger leaves regularly pinnatifid with 8–15 pairs of lobes 0.5–2 cm; cauline leaves evidently decurrent.

　　　　　　　　　　　　　　57. Principal leaf spines 5–20 mm; Wyoming to New Mexico, adventive in California . . . . 24. *Cirsium ochrocentrum* (in part)

　　　　　　　　　　　　　　57. Principal leaf spines 3–5 mm; Colo., Utah, Wyoming . . . . 26. *Cirsium barnebyi* (in part)

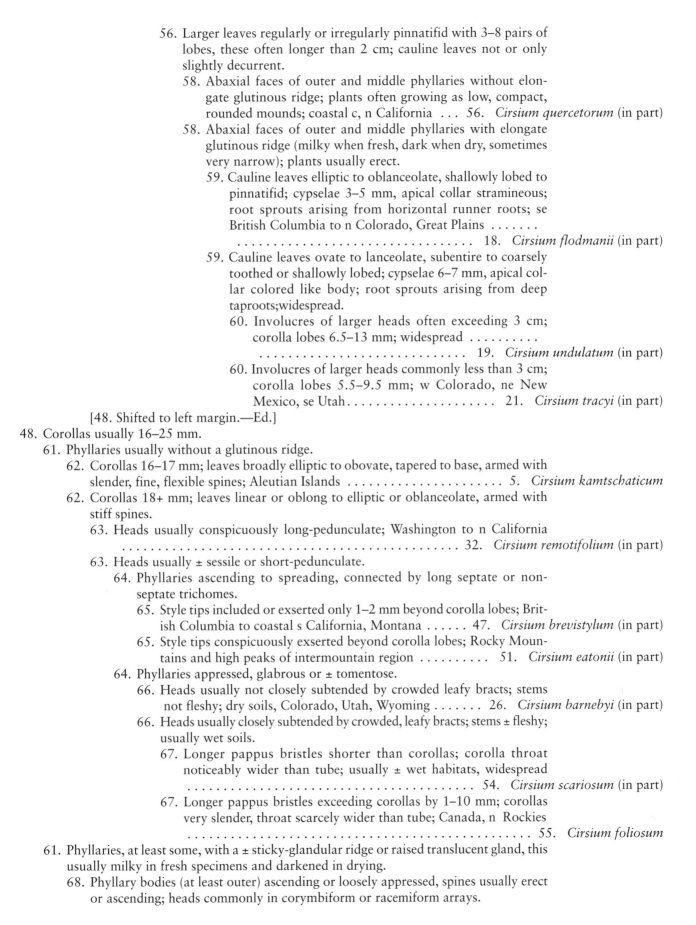

56. Larger leaves regularly or irregularly pinnatifid with 3–8 pairs of lobes, these often longer than 2 cm; cauline leaves not or only slightly decurrent.

  58. Abaxial faces of outer and middle phyllaries without elongate glutinous ridge; plants often growing as low, compact, rounded mounds; coastal c, n California ... 56. *Cirsium quercetorum* (in part)

  58. Abaxial faces of outer and middle phyllaries with elongate glutinous ridge (milky when fresh, dark when dry, sometimes very narrow); plants usually erect.

    59. Cauline leaves elliptic to oblanceolate, shallowly lobed to pinnatifid; cypselae 3–5 mm, apical collar stramineous; root sprouts arising from horizontal runner roots; se British Columbia to n Colorado, Great Plains . . . . 18. *Cirsium flodmanii* (in part)

    59. Cauline leaves ovate to lanceolate, subentire to coarsely toothed or shallowly lobed; cypselae 6–7 mm, apical collar colored like body; root sprouts arising from deep taproots;widespread.

      60. Involucres of larger heads often exceeding 3 cm; corolla lobes 6.5–13 mm; widespread . . . . 19. *Cirsium undulatum* (in part)

      60. Involucres of larger heads commonly less than 3 cm; corolla lobes 5.5–9.5 mm; w Colorado, ne New Mexico, se Utah. 21. *Cirsium tracyi* (in part)

[48. Shifted to left margin.—Ed.]

48. Corollas usually 16–25 mm.

  61. Phyllaries usually without a glutinous ridge.

    62. Corollas 16–17 mm; leaves broadly elliptic to obovate, tapered to base, armed with slender, fine, flexible spines; Aleutian Islands . . . . 5. *Cirsium kamtschaticum*

    62. Corollas 18+ mm; leaves linear or oblong to elliptic or oblanceolate, armed with stiff spines.

      63. Heads usually conspicuously long-pedunculate; Washington to n California . . . . 32. *Cirsium remotifolium* (in part)

      63. Heads usually ± sessile or short-pedunculate.

        64. Phyllaries ascending to spreading, connected by long septate or non-septate trichomes.

          65. Style tips included or exserted only 1–2 mm beyond corolla lobes; British Columbia to coastal s California, Montana . . . . . . 47. *Cirsium brevistylum* (in part)

          65. Style tips conspicuously exserted beyond corolla lobes; Rocky Mountains and high peaks of intermountain region . . . . 51. *Cirsium eatonii* (in part)

        64. Phyllaries appressed, glabrous or ± tomentose.

          66. Heads usually not closely subtended by crowded leafy bracts; stems not fleshy; dry soils, Colorado, Utah, Wyoming . . . . 26. *Cirsium barnebyi* (in part)

          66. Heads usually closely subtended by crowded, leafy bracts; stems ± fleshy; usually wet soils.

            67. Longer pappus bristles shorter than corollas; corolla throat noticeably wider than tube; usually ± wet habitats, widespread . . . . 54. *Cirsium scariosum* (in part)

            67. Longer pappus bristles exceeding corollas by 1–10 mm; corollas very slender, throat scarcely wider than tube; Canada, n Rockies . . . . 55. *Cirsium foliosum*

  61. Phyllaries, at least some, with a ± sticky-glandular ridge or raised translucent gland, this usually milky in fresh specimens and darkened in drying.

    68. Phyllary bodies (at least outer) ascending or loosely appressed, spines usually erect or ascending; heads commonly in corymbiform or racemiform arrays.

69. Cauline leaves conspicuously decurrent; Oregon and e California
    to w Wyoming . . . . 38. *Cirsium inamoenum* (in part)
69. Cauline leaves auriculate-clasping, sometimes also short-decurrent.
    70. Pubescence all of fine, non-septate arachnoid trichomes,
        often dense and persistent on leaves; s Oregon, n California
        . . . . 20. *Cirsium ciliolatum* (in part)
    70. Pubescence usually mixture of fine, non-septate arachnoid
        trichomes and coarser, septate trichomes, usually ± loose and
        irregularly deciduous from leaves in age; Oregon, n Cali-
        fornia to w Wyoming . . . . 39. *Cirsium cymosum* (in part)
    [68. Shifted to left margin.—Ed.]
68. Phyllary bodies tightly appressed, spines ascending to abruptly spreading; heads borne
    singly or in paniculiform or corymbiform arrays, or in few-headed arrays at stem tips.
    71. Leaves glabrous to thinly tomentose on both faces, ± glabrate adaxially.
        72. Pubescence usually mixture of fine, non-septate arachnoid trichomes and coarser,
            septate trichomes, usually ± loose and irregularly deciduous from leaves in age;
            Oregon and n California to w Wyoming. . . . . 39. *Cirsium cymosum* (in part)
        72. Pubescence, when present, of fine, non-septate trichomes; septate trichomes usually
            absent.
            73. Polycarpic perennials 20–100 cm; mountains of e Utah, w Colorado
                . . . . 30. *Cirsium clavatum* (in part)
            73. Biennials or monocarpic perennials.
                74. Plants 100–220 cm; phyllary spines 1–2 mm; wet soils, coastal c California
                    . . . . 35. *Cirsium hydrophilum* (in part)
                74. Plants 20–100 cm; phyllary spines 2–7 mm; dry slopes, Oregon, e California
                    to w Wyoming . . . . 38. *Cirsium inamoenum* (in part)
    71. Leaves densely tomentose on abaxial faces, often on both faces.
        75. Inner phyllaries usually dark purple to blackish near tips.
            76. Heads 1–6, borne singly or few at branch tips; Arizona, Colorado, New Mexico,
                Utah . . . . 28. *Cirsium wheeleri* (in part)
            76. Heads 10–many, ± crowded at branch tips; moist soils, California, Nevada,
                Oregon. . . . . 36. *Cirsium douglasii*
        75. Inner phyllaries greenish to brown or stramineous.
            77. Basal leaves 8–20+ cm wide; e California to sw Utah, n Arizona . . . . . . 37. *Cirsium mohavense*
            77. Basal leaves 0.7–7 cm wide.
                78. Cauline leaves not decurrent; Arizona, Colorado, New Mexico, Utah
                    . . . . 28. *Cirsium wheeleri* (in part)
                78. Cauline leaves evidently decurrent.
                    79. Larger leaves regularly pinnatifid with 8–15 pairs of closely spaced
                        lobes; both leaf faces ± densely tomentose; Colorado, Utah, Wyoming
                        . . . . 26. *Cirsium barnebyi* (in part)
                    79. Larger leaves unlobed or with 5–8(–many) pairs of lobes, these usually
                        separated by ± broad, U-shaped sinuses; adaxial leaf faces often
                        glabrous or glabrate.
                        80. Corollas dull white or faintly lavender-tinged (sometimes bright
                            pink-purple in var. *davisii*); phyllary spines usually fine; Oregon
                            and e California to w Wyoming . . . . 38. *Cirsium inamoenum* (in part)
                        80. Corollas pink to purple; phyllary spines usually stout, ± flattened;
                            e Idaho and ne Utah to Wyoming and nw Nebraska
                            . . . . 29. *Cirsium pulcherrimum* (in part)

1. **Cirsium vulgare** (Savi) Tenore, Fl. Napol. 5: 209. 1835 • Bull or common or spear thistle, gros chardon, chardon vulgaire ou lancéolé, piqueux I W

Carduus vulgaris Savi, Fl. Pis. 2: 241. 1798; *C. lanceolatus* Linnaeus 1753, not *Cirsium lanceolatum* Hill 1769

Biennials, 30–200(–300) cm; taproots. **Stems** 1–many, erect or ascending, branches few–many, ascending, villous with septate trichomes. **Leaves:** blades oblong-lanceolate to obovate, 15–40 × 6–15 cm, margins plane or revolute, coarsely 1–2-pinnatifid with rigidly divergent lobes, sometimes merely spinose-dentate, lobes triangular to lanceolate, entire to spiny-dentate, main spines 2–10 mm, abaxial faces gray-tomentose, villous with septate trichomes along veins, adaxial green, covered with short appressed bristlelike spines, sometimes tomentose when young; basal present or absent at flowering, petioles winged, bases tapered; principal cauline winged-petiolate, mid and distal becoming sessile, well distributed or not, progressively reduced distally, at least distal decurrent as long spiny wings; distal cauline often more deeply lobed than proximal, main lobes rigidly spiny, margins spinulose, otherwise entire. **Heads** few–many in corymbiform or paniculiform arrays. **Peduncles** 1–6 cm. **Involucres** hemispheric to campanulate, 3–4 × 2–4 cm, loosely arachnoid-tomentose. **Phyllaries** in 10–12 series, strongly imbricate, linear-lanceolate (outer) to linear (inner), outer and middle appressed, (bases stramineous), margins entire, abaxial faces without glutinous ridge, apices radiating, greenish, spines 2–5 mm; apices of inner phyllaries flat, serrulate to minutely erose. **Corollas** purple (rarely white), 25–35 mm, tubes 18–25 mm, throats 5–6 mm, lobes 5–7 mm; style tips 3.5–6 mm. **Cypselae** light brown with darker streaks, 3–4.5 mm, apical collar not differentiated; **pappi** 20–30 mm. $2n = 68$.

Flowering mostly summer (Jun–Sep), year round in areas with mild climates. Invasive weed of disturbed sites, pastures, meadows, forest openings, roadsides; 0–2200 m; introduced; St. Pierre and Miquelon; Alta., B.C., Man., N.B., Nfld. and Labr. (Nfld.), N.S., Ont., P.E.I., Que., Sask.; Ala., Alaska, Ariz., Ark., Calif., Colo., Conn., Del., D.C., Fla., Ga., Idaho, Ill., Ind., Iowa, Kans., Ky., La., Maine, Md., Mass., Mich., Minn., Miss., Mo., Mont., Nebr., Nev., N.H., N.J., N.Mex., N.Y., N.C., N.Dak., Ohio, Okla., Oreg., Pa., R.I., S.C., S.Dak., Tenn., Tex., Utah, Vt., Va., Wash., W.Va., Wis., Wyo.; Eurasia.

Native to Eurasia, *Cirsium vulgare* is the only thistle in North America with bristlelike spines borne on the adaxial leaf faces. These structures are variously described in the literature as trichomes ("spreading hirsute," "scabrous-hispid," "coarsely hispid," "rigid, rather pungent setae," "prickly-hairy"), prickles, or spines ("setose-spinulose," "appressed and dense spines"). My examination of cleared leaves of *C. vulgare* indicated that these structures are not epidermal outgrowths (trichomes or prickles) but emerge from fine veinlets within the tissues of the leaf. As such, they are properly treated as spines.

Bull thistle is a noxious weed that has invaded disturbed habitats across the continent. Distasteful to livestock, it can increase in heavily grazed pastures. It occurs in a wide variety of habitats.

2. **Cirsium arvense** (Linnaeus) Scopoli, Fl. Carniol. ed. 2, 2: 126. 1772 • Canada or creeping or field thistle, chardon du Canada ou des champs, cirse des champs I W

Serratula arvensis Linnaeus, Sp. Pl. 2: 820. 1753; *Breea arvensis* (Linnaeus) Lessing; *Carduus arvensis* (Linnaeus) Robson; *Cirsium arvense* var. *argenteum* (Peyer ex Vest) Fiori; *C. arvense* var. *horridum* Wimmer & Grabowski; *C. arvense* var. *integrifolium* Wimmer & Grabowski; *C. arvense* var. *mite* Wimmer & Grabowski; *C. arvense* var. *vestitum* Wimmer & Grabowski; *C. incanum* (S. G. Gmelin) Fischer ex M. Bieberstein; *C. setosum* (Willdenow) Besser ex M. Bieberstein

Perennials, dioecious or nearly so, 30–120(–200) cm; colonial from deep-seated creeping roots producing adventitious buds. **Stems** 1–many, erect, glabrous to appressed gray-tomentose; branches 0–many, ascending. **Leaves:** blades oblong to elliptic, 3–30 × 1–6 cm, margins plane to revolute, entire and spinulose, dentate, or shallowly to deeply pinnatifid, lobes well separated, lance-oblong to triangular-ovate, spinulose to few-toothed or few-lobed near base, main spines 1–7 mm, abaxial faces glabrous to densely gray-tomentose, adaxial green, glabrous to thinly tomentose; basal absent at flowering, petioles narrowly winged, bases tapered; principal larger cauline proximally winged-petiolate, distally sessile, well distributed, gradually reduced, not decurrent; distal cauline becoming bractlike, entire, toothed, or lobed, spinulose or not. **Heads** 1–many, borne singly or in corymbiform or paniculiform arrays at tips of main stem and branches. **Peduncles** 0.2–7 cm. **Involucres** ovoid in flower, ± campanulate in fruit, 1–2 × 1–2 cm, arachnoid tomentose, ± glabrate. **Phyllaries** in 6–8 series, strongly imbricate, (usually purple-tinged), ovate (outer) to linear (inner), abaxial faces with narrow glutinous ridge, outer and middle appressed, entire, apices ascending to spreading, spines 0–1 mm (fine); apices of inner phyllaries flat,

± flexuous, margins entire to minutely erose or ciliolate. **Corollas** purple (white or pink); staminate 12–18 mm, (remaining longer than pappus when head is fully mature), tubes 8–11 mm, throats 1–1.5 mm, lobes 3–5 mm; pistillate 14–20 mm, (overtopped by pappi in fruit), tubes 10–15 mm, throats ca. 1 mm, lobes 2–3 mm; style tips 1–2 mm. **Cypselae** brown, 2–4 mm, apical collar not differentiated; **pappi** 13–32 mm, exceeding corollas. $2n = 34$.

Flowering summer (Jun–Oct). Disturbed sites, fields, pastures, roadsides, forest openings; 0–2600 m; introduced; Greenland; St. Pierre and Miquelon; Alta., B.C., Man., N.B., Nfld. and Labr. (Nfld.), N.W.T., N.S., Ont., P.E.I., Que., Sask., Yukon; Ala., Alaska, Ariz., Ark., Calif., Colo., Conn., Del., D.C., Idaho, Ill., Ind., Iowa, Kans., Ky., Maine, Md., Mass., Mich., Minn., Mo., Mont., Nebr., Nev., N.H., N.J., N.Mex., N.Y., N.C., N.Dak., Ohio, Oreg., Pa., R.I., S.Dak., Tenn., Tex., Utah, Vt., Va., Wash., W.Va., Wis., Wyo.; native, Eurasia.

Cirsium arvense is one of the most economically important agricultural weeds in the world. It was introduced to North America in the 1600s and soon was recognized as a problem weed. Weed control legislation against the species was passed by the Vermont legislature in 1795 (R. J. Moore 1975). Canada thistle is now listed as a noxious weed in most areas where it occurs. It has very high seed production, and the runner roots readily survive the fragmentation that accompanies cultivation.

Numerous variants of *Cirsium arvense* have been named based upon such features as pubescence, extent of leaf division, and spininess. Although extreme variants can be strikingly different, they are connected by such a web of intermediates that there seems to be little value in according any of them formal taxonomic recognition.

SELECTED REFERENCE Moore, R. J. 1975. The biology of Canadian weeds. 13. *Cirsium arvense* (L.) Scop. Canad. J. Bot. 55: 1033–1048.

3. **Cirsium palustre** (Linnaeus) Scopoli, Fl. Carniol. ed. 2, 2: 128. 1772 • European swamp or marsh thistle, cirse ou chardon des marais [1]

Carduus palustris Linnaeus, Sp. Pl. 2: 822. 1753

Biennials or monocarpic perennials, 30–200(–300) cm; clusters of fibrous roots. **Stems** single, erect, villous to tomentose with jointed trichomes, distally tomentose with fine, unbranched trichomes; branches 0–few, ascending, (short). **Leaves:** blades narrowly elliptic to oblanceolate, 15–30+ × 3–10 cm, margins shallowly to very deeply pinnatifid, narrow lobes separated by broad sinuses, spiny-dentate to lobed, main spines 2–6 mm, abaxial villous to tomentose with jointed trichomes, sometimes also thinly tomentose with fine unbranched trichomes, adaxial faces villous with septate trichomes or glabrate; basal often present at flowering, petioles spiny-winged, bases tapered; cauline many, sessile, gradually reduced and becoming widely spaced above, bases long-decurrent with prominently spiny wings; distal cauline deeply pinnatifid with few-toothed spine-tipped lobes. **Heads** few–many in dense clusters at branch tips. **Peduncles** 0–1 cm. **Involucres** ovoid to campanulate, 1–1.5 × 0.8–1.3 cm, thinly cobwebby tomentose with fine unbranched trichomes. **Phyllaries** in 5–7 series, strongly imbricate, greenish, or with purplish tinge, lanceolate to ovate (outer) or linear-lanceolate (inner), margins thinly arachnoid-ciliate, abaxial faces with narrow glutinous ridge, outer and middle appressed, entire, apices acute, mucronate or spines erect or spreading, weak, 0.3–1 mm; apices of inner phyllaries purplish, linear-attenuate, scarious, flat. **Corollas** lavender to purple (white), 11–13 mm, tubes 5–7 mm, throats 2–3 mm, lobes 3–4.5 mm; style tips 1.5–2 mm. **Cypselae** tan to stramineous, 2.5–3.5 mm, apical collars 0.1–0.2 mm, shiny; **pappi** 9–11 mm. $2n = 34$.

Flowering summer (Jul–Aug). Marshes, wet forests; 10–800 m; introduced; St. Pierre and Miquelon; B.C., Nfld. and Labr. (Nfld.), N.S., Ont., Que.; Mass., Mich., N.H., N.Y., Wis.; Europe.

Cirsium palustre is a noxious weed, native to Europe, that invasively spreads through wetland communities, forming impenetrable spiny stands as it displaces native species. The range of this pernicious weed in North America is rapidly expanding. It has the potential to spread into boreal forest areas across the continent; in Europe it grows nearly to the Arctic Circle. The rapid spread of *C. palustre* in Michigan (E. G. Voss 1972–1996, vol. 3) is indicative of its invasiveness. Spontaneous hybrids between *C. palustre* and *C. arvense* have been reported from England and other European countries (W. A. Sledge 1975) and can be expected wherever these species grow together in North America.

4. **Cirsium helenioides** (Linnaeus) Hill, Hort. Kew., 64. 1768 (as helenoides) • Melancholy thistle [C]

Carduus helenioides Linnaeus, Sp. Pl. 2: 825. 1753; *Cirsium heterophyllum* (Linnaeus) Hill

Perennials, 40–120 cm; runner roots. **Stems** single, erect, ± arachnoid-tomentose; branches 0 or few, ascending. **Leaves** oblong to broadly lanceolate, 20–40 × 4–8 cm, finely spinulose-dentate or proximal cauline pinnatifid, lobes undivided, finely spinulose-dentate, main spines 1–2 mm, abaxial faces white-tomentose (with non-septate trichomes), adaxial glabrous; basal present at flowering, petiolate, bases tapered; cauline sessile, reduced distally, bases clasping, not decurrent; distal (few, well separated), oblong or linear, the uppermost reduced to linear bracts. **Heads** borne singly or less commonly 2–5 in terminal clusters. **Peduncles** 2–10(–30) cm (elevated above distal leaves). **Involucres** broadly ovoid, 2–3 × 2–3.5 cm, glabrous or loosely arachnoid. **Phyllaries** in 8–10 series, imbricate, green, ovate or lanceolate (outer) to linear-lanceolate (inner), abaxial faces with a prominent elongate glutinous ridge, outer and middle tightly appressed, margins entire, apices with ascending, weak spines 0–1 mm; apices of inner phyllaries attenuate, flat. **Corollas** red-purple, 25–30 mm, tubes 10–23 mm, throats 8–14 mm (noticeably wider than tubes), lobes 7–10 mm; style tips 4–5 mm. **Cypselae** light brown, 3–5 mm, bodies and apical collars concolorous; **pappi** 20–30 mm. $2n$ = 34.

Flowering summer (Jul–Aug). Fjordlands; 0–50 m; of conservation concern; Greenland; Iceland; Europe; Asia.

Cirsium helenioides is one of only two species of the genus that have native populations in the Old World and the flora area. Neither reaches the North American mainland.

The conservation status of *Cirsium helenioides* is not known; it is known in the flora area only from a single fjord and possibly should be considered of conservation concern.

5. **Cirsium kamtschaticum** Ledebour ex de Candolle in A. P. de Candolle and A. L. P. P. de Candolle, Prodr. 6: 644. 1838 • Kamchatka thistle

Perennials, 25–200 cm; rhizomes stout. **Stems** single, erect, ± glabrous to variably tomentose with coarse, jointed, multicellular trichomes and/or fine smooth trichomes; branches 0–few, ascending. **Leaves:** blades broadly elliptic to obovate, 15–40 × 7–15 cm, subentire to coarsely pinnatifid ¹/₂–²/₃ length to midveins, lobes few, lanceolate to triangular-ovate, shallowly lobed or dentate, main spines bristlelike, fine, innocuous, 3–6 mm, abaxial glabrous to villous with septate trichomes or thinly tomentose with jointed trichomes, adaxial faces glabrous or loosely tomentose along midveins; basal usually absent at flowering, winged-petiolate, ciliate with fine, flexible spines to 8 mm; principal cauline well distributed, little reduced, bases broadly tapered to clasping, short-decurrent; distalmost moderately reduced. **Heads** 1–few, in spiciform or subcapitate arrays. **Peduncles** 0–1 cm. **Involucres** hemispheric to broadly campanulate, 1.5–2 × 2–3.5 cm, ± densely arachnoid. **Phyllaries** in 5–7 series, subequal, green or tinged purple, linear or linear-lanceolate, abaxial faces without glutinous ridge, outer and middle erect or outer spreading, entire, apices long-acuminate, spines 0–2 mm; apices of inner phyllaries straight or flexuous, flat. **Corollas** pink to purple, 16–17 mm, tubes 8–9 mm, throats 3–4 mm, lobes 4–5 mm. **Style tips** 3–4 mm. **Cypselae** brown, 4 mm, apical collars not well differentiated; **pappi** 12–15 mm. $2n$ = 68 (Japan).

Flowering summer (Jul–Sep). Meadows and tundra; 0–100 m; Alaska; Asia (Japan, Siberia).

Cirsium kamtschaticum grows in the western Aleutian Islands, eastern Siberia, Sahkalin, the Kurile Islands and northern Japan (Hokkaido). It is one of only two species of the genus that have native populations in the Old World and the flora area. Neither reaches the North American mainland.

6. **Cirsium altissimum** (Linnaeus) Sprengel, Syst. Veg. 3: 373. 1826 • Tall or roadside thistle [E]

Carduus altissimus Linnaeus, Sp. Pl. 2: 824. 1753; *Cirsium altissimum* var. *biltmoreanum* Petrak; *C. iowense* (Pammell) Fernald

Biennials or short-lived monocarpic perennials, (50–)100–300(–400) cm; taproots and often a cluster of coarse fibrous roots, roots without tuberlike enlargements. **Stems** single, erect, villous with septate trichomes, sometimes ± glabrate, sometimes distally thinly tomentose; branches few–many, ascending. **Leaves:** blades oblanceolate to elliptic or ovate, 10–40 × 1–13 cm, margins flat, finely spiny-toothed and otherwise undivided to coarsely toothed or shallowly pinnatifid, lobes broadly triangular, main spines 1–5 mm, abaxial faces white-tomentose, adaxial faces green, glabrate to villous with septate trichomes; basal usually absent at flowering, winged-petiolate, bases tapered; principal cauline well distributed, gradually reduced, bases narrowed, sometimes weakly clasping; distal cauline well developed. **Heads** 1–many, in corymbiform or paniculiform arrays, (± elevated above principal cauline leaves), not subtended by ring of spiny bracts. **Peduncles** 0–5 cm (leafy-bracted). **Involucres** ovoid to broadly cylindric or campanulate, (2–)2.5–3.5(–4) × (1.5–)2–3(–4) cm, thinly arachnoid. **Phyllaries** in 10–20 series, strongly imbricate, greenish with subapical darker central zone, ovate (outer) to lanceolate (inner), abaxial faces with a narrow glutinous ridge (milky when fresh, dark when dry), outer and middle entire, bodies appressed, spines slender, abruptly spreading, 3–4 mm; apices of inner phyllaries spreading, narrow, flattened, entire, spines spreading, slender, 3–4 mm; apices of inner phyllaries spreading, narrow, flattened, ± dilated, ± erose or finely serrulate. **Corollas** pink to purple (white), 20–35 mm, tubes 10–16 mm, throats 5–12 mm, lobes 5–9 mm. **Style tips** 4–6 mm. **Cypselae** tan to dark brown, 4–5.5 mm, apical collars stramineous, 0.5–1 mm; **pappi** 12–24 mm. $2n = 18$.

Flowering summer–fall (Jun–Oct). Prairies, woodlands, disturbed sites, often in damp soil; 50–700 m; Ala., Ark., Del., D.C., Fla., Ga., Ill., Ind., Iowa, Kans., Ky., La., Md., Mass., Mich., Minn., Miss., Mo., Nebr., N.Y., N.C., N.Dak., Ohio, Okla., Pa., S.C., S.Dak., Tenn., Tex., W.Va., Wis.

Plants of *Cirsium altissimum* ranging from southern Minnesota to Texas often have more deeply divided leaves than do populations in other portions of the species' range. Some botanists (e.g., R. J. Moore and C. Frankton 1969; D. S. Correll and M. C. Johnston 1970) have treated those plants as *C. iowense*. Others (e.g., R. E. Brooks 1986; H. A. Gleason and A. Cronquist 1991; G. B. Ownbey and T. Morley 1991) have treated them as *C. altissimum*. Still others considered them to be derivatives of hybridization between *C. altissimum* and *C. discolor* (J. T. Kartesz and C. A. Meacham 1999) and treated them as *C. ×iowense*. Indeed the existence of these plants blurs the distinction between *C. altissimum* and *C. discolor*, and herbarium specimens are often difficult to assign.

Natural hybrids between *Cirsium altissimum* and *C. discolor* are well documented (R. A. Davidson 1963; G. B. Ownbey and Hsi Y.-T. 1963; Ownbey 1964; S. Dabydeen 1997). Ownbey and Dabydeen both reported that apparent F₁ hybrids between the two species have low seed set in comparison with the parental taxa. W. L. Bloom (1977) reported that the chromosomes of the two species differ by several rearrareaments. Dabydeen reported a count of $2n = 19$ with multiple meiotic irregularities for an apparent F₁ hybrid. However, the presence of numerous individuals and populations seemingly intermediate between *C. altissimum* and *C. discolor* indicates that although F₁ hybrids have low fertility, long-term processes may have stabilized hybrid derivatives of higher fertility. Ownbey and Hsi reported mitotic counts of $2n = 18$ and 20 from a population that they treated as *C. altissimum*. In their discussion they noted that their plants represented "the segregate called *C. iowense*" and had been collected a short distance from that taxon's type locality. R. J. Moore and C. Frankton (1969) reported a chromosome number of $2n = 18$ for a plant from Texas that they considered to be *C. iowense*. Further investigation of morphologic variation, chromosome number, meiotic behavior, and fertility is needed of populations named as *C. iowense* to determine how those plants should be treated.

7. **Cirsium discolor** (Muhlenberg ex Willdenow) Sprengel, Syst. Veg. 3: 373. 1826 • Field thistle, chardon discolore E F

Cnicus discolor Muhlenberg ex Willdenow, Sp. Pl. 3: 1670. 1803; *Carduus discolor* (Muhlenberg ex Willdenow) Nuttall

Biennials or sometimes perennials, 80–200 cm; taproots and often cluster of coarse fibrous roots, roots without tuberlike enlargements. **Stems** single, erect, villous with septate trichomes, sometimes ± glabrate, distally ± tomentose; branches few–many, ascending. **Leaves:** blades oblanceolate to elliptic or ovate, 10–25(–50) × 1–13 (–25) cm, usually deeply divided more than halfway to midveins, proximal sometimes undivided, lobes linear-lanceolate, margins revolute, ascending, entire or spinulose to remotely few toothed or sharply lobed, main spines 1–5 mm, abaxial faces white-tomentose, adaxial faces green, villous with septate trichomes or glabrate; basal usually absent at flowering, winged-petiolate, bases tapered; principal cauline well distributed, gradually reduced, bases narrowed, sometimes weakly clasping, not decurrent; distal cauline well developed. **Heads** 1–many in corymbiform or paniculiform arrays. **Peduncles** 0–5 cm (not overtopped by crowded distal cauline leaves). **Involucres** ovoid to broadly cylindric or campanulate, 2–3.5(–4) × 1.5–3 cm, thinly arachnoid. **Phyllaries** in 10–12 series, strongly imbricate, greenish with subapical darker central zone, ovate (outer) to lanceolate (inner), abaxial faces with narrow glutinous ridge, outer and middle bodies appressed, margins entire, spines abruptly spreading to deflexed, slender, 3–9 mm; spines slender, 3–9 mm; apices of inner phyllaries spreading, narrow, flattened, finely serrulate. **Corollas** pink to purple (white), 25–32 mm, tubes 12–16 mm, throats 7–10 mm, (noticeably wider than tubes), lobes 6–9 mm; style tips 4–6 mm. **Cypselae** tan to brownish, 4–5 mm, apical collars straw-colored, 0.5–75 mm; **pappi** 18–25 mm. $2n = 20, 21, 22$.

Flowering summer–fall (Jun–Oct). Tallgrass prairies, deciduous woodlands, forest openings, disturbed sites, often in damp soil; 5–800 m; Man., Ont., Que., Sask.; Ala., Ark., Conn., Del., D.C., Ga., Ill., Ind., Iowa, Kans., Ky., La., Maine, Md., Mass., Mich., Minn., Mo., Nebr., N.H., N.Y., N.C., Ohio, Pa., R.I., S.C., S.Dak., Tenn., Vt., Va., W.Va., Wis.

Cirsium discolor is widespread in eastern North America from the prairies of southeastern Saskatchewan, western Minnesota, and Iowa south to northern Louisiana and east across southern Canada to the New England states and the southern Appalachians. It hybridizes with both *C. altissimum* (discussed thereunder) and *C. muticum* (G. B. Ownbey 1951b, 1964; W. L. Bloom

1977). Meiosis in first-generation hybrids between *C. discolor* and *C. muticum* is usually irregular (Bloom) and most pollen grains are infertile (Ownbey 1951b; Bloom). The presence of a small number of viable cypselae in heads of putative F$_1$ hybrids (Ownbey 1951b) indicates that some F$_2$ hybrids or backcrosses are formed.

8. **Cirsium muticum** Michaux, Fl. Bor.-Amer. 2: 89. 1803 • Swamp thistle, dunce-nettle, horsetops, chardon mutique [E]

Carduus muticus (Michaux) Persoon; *Cirsium muticum* var. *monticola* (Fernald) Fernald

Biennials, 30–230 cm; taproots fleshy. **Stems** single, erect, villous with septate trichomes or glabrate, distally sometimes thinly tomentose; branches few–many, ascending. **Leaves** ovate to broadly elliptic or obovate, 15–55 × 4–20 cm, deeply pinnatifid, to ⁷/₈ to midribs, lobes linear to lanceolate, acute to acuminate, irregularly few toothed or lobed, main spines 2–3 mm, abaxial faces thinly tomentose or glabrate, villous with septate trichomes on the veins, adaxial faces thinly pilose; basal usually absent at flowering, petioles spiny-winged, bases tapered; principal cauline petiolate or sessile, gradually reduced distally, bases sometimes ± clasping, not decurrent; distal cauline bractlike with narrowly linear lobes, often spinier than the proximal. **Heads** 1–many in ± open corymbiform or paniculiform arrays. **Peduncles** 0–15 cm (sometimes overtopped by distal cauline leaves, not subtended by involucre-like ring of bracts). **Involucres** ovoid to broadly cylindric or campanulate, 1.7–3 × 1–3 cm, arachnoid. **Phyllaries** in 8–12 series, strongly imbricate, dull green with darker subapical patch, ovate (outer) to linear-lanceolate (inner), abaxial faces with narrow glutinous ridge, outer and middle appressed, bodies minutely spinulose, apices obtuse to acute, spines erect (sometimes appearing as spreading in dry specimens), 0–0.5 mm; apices of inner phyllaries straight or ± flexuous, flattened. **Corollas** lavender or purple (white), 16–32 mm, tubes 7–15 mm, throats 4.5–10 mm (noticeably wider than tubes), lobes 4–8 mm; style tips 3.5–5 mm. **Cypselae** dark brown, 4.5–5.5 mm, apical collars yellow, 0.3 mm; **pappi** 12–20 mm. **2*n*** = 20, 21, 22, 23, 30.

Flowering summer (Jul–Sep). Wet soil in meadows, prairies, marshes, swamps, bogs, open woods; 0–1500+ m; St. Pierre and Miquelon; Man., N.B., Nfld. and Labr., N.S., Ont., P.E.I., Que., Sask.; Ala., Ark., Conn., Del., Fla., Ga., Ill., Ind., Iowa, Ky., La., Maine, Md., Mass., Mich., Minn., Mo., N.H., N.J., N.Y., N.C., N.Dak., Ohio., Okla., Pa., R.I., S.C., Tenn., Tex., Vt., Va., W.Va., Wis.

Cirsium muticum is very widely distributed across the eastern half of North America from the prairies of southeastern Saskatchewan across southern Canada to Newfoundland and south in the United States from North Dakota and Maine to southeastern Texas and northern Florida. It is more common in the northern half of this range and extends from the coastal plain to the Appalachian highlands. The widely scattered populations in coastal lowlands in the southern United States may be relics of the glacial distribution of the species.

Cirsium muticum is known to hybridize with *C. discolor* (discussed thereunder) and *C. flodmanii*. Draining and modification of wetlands have affected populations of *C. muticum* in some areas.

9. **Cirsium repandum** Michaux, Fl. Bor.-Amer. 2: 89. 1803 • Sand-hill or coastal-plain thistle [E]

Carduus repandus (Michaux) Persoon

Perennials, 20–80 cm; creeping roots deep-seated, sometimes appearing as taprooted biennials. **Stems** 1–several, spreading to erect, (usually very leafy in distal ¹/₂), loosely arachnoid, and villous with jointed, multicellular trichomes; branches 0–few from above middle, ascending. **Leaves** linear to oblong or oblanceolate, 6–16 × 1–3.5 cm, unlobed to sinuate-dentate or shallowly pinnatifid, main spines 1–4 mm, fine, faces ± green, shaggy-villous with septate trichomes, abaxial loosely arachnoid when young; basal and proximal cauline usually absent at time of flowering; mid and distal nearly uniform in size or gradually reduced, bases clasping; distalmost cauline ± bractlike. **Heads** 1–5, in corymbiform arrays. **Peduncles** 0–2 cm. **Involucres** ovoid or cylindric to campanulate, 2–4 × 1.5–4 cm, loosely arachnoid, ± glabrate. **Phyllaries** in 6–9 series, imbricate, lanceolate (outer) to linear (inner), abaxial faces with glutinous ridge, outer and middle tightly appressed, bodies scabrous or spinulose, spines erect or weakly ascending, 1–4 mm; apices of inner phyllaries long-acuminate, spineless. **Corollas** light purple, 33–40 mm, tubes 14–15 mm, throats 12–15 mm, lobes 7–9 mm; style tips 4.5–6 mm. **Cypselae** light brown, 3.5–4 mm, apical collars yellowish, ca. 0.8 mm; **pappi** 15–30 mm. **2*n*** = 30.

Flowering spring–summer (May–Jul). Sandhills, pine barrens, roadsides; 0–150 m; Ga., N.C., S.C., Va.

Cirsium repandum occurs on the Atlantic coastal plain. R. J. Moore and C. Frankton (1969) suggested that *Cirsium repandum* originated through ancient hybridization between *C. pumilum* var. *pumilum* and *C. horridulum*. They noted that an artificial hybrid (2*n* = 32) between *C. repandum* (2*n* = 30) and *C. horridulum* (2*n* = 34) had a mosaic of features of the parental taxa.

10. Cirsium lecontei Torrey & A. Gray, Fl. N. Amer. 2: 458. 1843 • Black or Le Conte's thistle [C][E]

Carduus lecontei (Torrey & A. Gray) Pollard

Perennials but sometimes appearing biennial, 35–110 cm; taproots, sometimes with root sprouts. **Stems** 1–few, erect, distal ½ nearly naked, loosely arachnoid; branches 0–5(–10), stiffly ascending. **Leaves:** blades linear to oblong or narrowly elliptic, 15–25 × 1–4 cm, coarsely toothed to shallowly pinnatifid, lobes undivided or coarsely few-toothed, main spines 3–6 mm, abaxial faces often ± glabrate, loosely arachnoid when young, adaxial glabrous or sparingly villous with coarse, multicellular trichomes; basal sometimes absent at flowering, petiolate; principal cauline sessile, progressively reduced distally, bases clasping or ± decurrent; distal cauline few, widely separated, bractlike. **Heads** borne singly or less commonly 2–5(–10) in open, corymbiform arrays. **Peduncles** 5–30 cm (elevated above cauline leaves, not subtended by ring of involucre-like bracts). **Involucres** broadly cylindric to campanulate, 2.5–4 × 1.5–4 cm, loosely arachnoid, ± glabrate. **Phyllaries** in 6–10 series, imbricate, ovate or lanceolate (outer) to linear-lanceolate (inner), abaxial faces with prominent glutinous ridge, outer and middle tightly appressed, margins spinulose-serrulate, spines ascending, 0.5–2 mm; apices of inner flat, linear-acuminate. **Corollas** pink-purple, 22–45 mm, tubes 10–23 mm, throats 8–14 mm, lobes 7–10 mm; style tips 4–5 mm. **Cypselae** light brown, 5–5.75 mm, apical collars paler than body, ca. 0.75 mm; **pappi** 20–40 mm. $2n = 28, 32$.

Flowering spring–summer (May–Aug). Sandy pinelands of coastal plain, often in damp soil; of conservation concern; 0–150 m; Ala., Fla., La., Miss., N.C., S.C.

Cirsium lecontei occurs on the southern coastal plain. R. J. Moore and C. Frankton (1969) suggested that it originated as a derivative of ancient hybridization between the ancestors of *C. horridulum* and *C. nuttallii*. They further suggested a relationship between *C. lecontei* and *C. grahamii* of Arizona and hypothesized an ancient dispersal from the southeastern coastal plain to the western cordillera. Although such relationships are possible, I have seen little support for them in my examination of these taxa. I think it is more likely that *C. lecontei*, *C. horridulum*, and *C. nuttallii* originated from a common stock, and that the resemblances between *C. lecontei* and *C. grahamii* are a result of convergence.

11. Cirsium horridulum Michaux, Fl. Bor.-Amer. 2: 90. 1803 • Bristly or horrid or yellow or bull thistle [F]

Biennials or **perennials**, (± fleshy), 15–250 cm; stout taproots and a fascicle of fleshy lateral roots, often perennating by root sprouts. **Stems** 1–several, usually erect, often stout, glabrous to densely tomentose; branches 0–many, spreading to ascending, short, stout. **Leaves:** blades linear to oblanceolate or oblong-elliptic, 10–40 × 2–10 cm, unlobed and spiny-dentate to deeply pinnatifid, lobes spiny-dentate or coarsely lobed, main spines stout, 5–30 mm, abaxial faces subglabrous to loosely tomentose, adaxial glabrous to ± densely villous with septate trichomes; basal present at flowering, spiny winged-petiolate, bases often tapered; principal cauline sessile, well distributed, often not much reduced distally, bases often ± auriculate-clasping; distal cauline often spinier than the proximal. **Heads** 1–20 in subcapitate to corymbiform arrays (each closely subtended by an involucre-like ring of spiny-margined bracts). **Peduncles** 0–5 cm. **Involucres** hemispheric to campanulate, 3–5 × 3–8 cm. **Phyllaries** in 5–9 series, subequal to imbricate, light green to stramineous, lanceolate to linear, distally often with reddish margins, abaxial faces without glutinous ridge, often ± thinly tomentose, often scabridulous in submarginal bands; outer and middle appressed-ascending, bodies usually reddish-tinged, margins setulose-ciliolate, apices acuminate, spines 1–2 mm, weak; apices of inner straight, flat. **Corollas** white to yellow, pink, purple, or red, 30–47 mm, tubes 11–30 mm, throats 6–10 mm, lobes 7–10 mm; style tips 3–5 mm. **Cypselae** straw-colored to tan, 4–6 mm, apical collars weakly differentiated; **pappi** 25–35 mm. $2n = 32, 33, 34, 35$.

Varieties 4 (3 in the flora): e United States, Mexico.

Although several variants have been given taxonomic recognition as species, these seem at most races. Flower color varies greatly, sometimes within populations and sometimes on a populational or regional basis. Herbarium specimens are sometimes difficult to assign to variety.

1. Stems densely tomentose; involucres ± densely tomentose... 11a. *Cirsium horridulum* var. *horridulum*
1. Stems glabrous or sparsely tomentose; involucres glabrous.
  2. Leaves shallowly to deeply pinnatifid, main spines 10–30 mm .... 11b. *Cirsium horridulum* var. *megacanthum*
  2. Leaves spinose-dentate to shallowly pinnatifid, main spines mostly 5–10 mm .... 11c. *Cirsium horridulum* var. *vittatum*

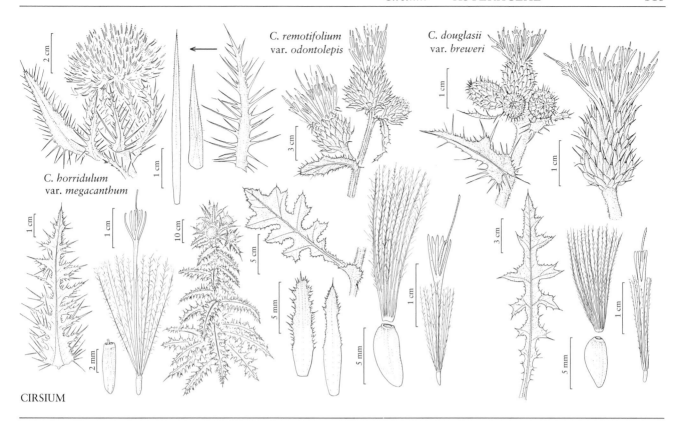

C. remotifolium
var. *odontolepis*

C. douglasii
var. *breweri*

C. horridulum
var. *megacanthum*

CIRSIUM

11a. Cirsium horridulum Michaux var. **horridulum**
• Horrid thistle E

Carduus spinosissimus Walter 1788, not *Cirsium spinosissimum* (Linnaeus) Scopoli 1772; *C. horridulum* var. *elliottii* Torrey & A. Gray

Plants 15–180 cm. **Stems** densely tomentose. **Leaves:** blades oblanceolate or oblong-elliptic, 10–30 cm, shallowly to deeply pinnatifid, main spines mostly 5–10 mm, abaxial faces loosely tomentose. **Heads** 1–10. **Involucres** 4–5 × 5–7 cm, ± densely tomentose. **Phyllaries:** outer and middle bodies scabridulous or minutely spinulose, marginal spinules usually 1 mm or shorter. **Corollas** yellow or crimson to red-purple. $2n$ = 32, 34.

Flowering winter–spring (Apr–Jun). Fields, meadows, pinelands, roadsides, often weedy, often in damp soil; 0–300 m; Ala., Conn., Del., Fla., Ga., La., Maine, Md., Mass., Miss., N.H, N.J., N.Y., N.C., Pa., R.I., S.C., Tenn., Tex., Va.

Variety *horridulum* occurs on the Atlantic and Gulf coastal plains fom southern Maine to Florida and west to eastern Texas and into the piedmont in much of the southeastern United States. Artificial hybrids have been produced between var. *horridulum* and *Cirsium repandum* (R. J. Moore and C. Frankton 1969), and natural hybrids with *C. pumilum* var. *pumilum* have been reported.

11b. Cirsium horridulum Michaux var. **megacanthum** (Nuttall) D. J. Keil, Sida 21: 214. 2004 • Bigspine thistle E F

Cirsium megacanthum Nuttall, Trans. Amer. Philos. Soc., n. s. 7: 421. 1841 (as Circium)

Plants 100–250 cm. **Stems** glabrous to sparsely tomentose. **Leaves:** blades oblanceolate or oblong-elliptic, 20–40 cm, shallowly to deeply pinnatifid, main spines 10–30 mm, abaxial faces glabrous or nearly so. **Heads** 10–20. **Involucres** 4–5 × 5–8 cm, glabrous. **Phyllaries:** outer and middle bodies scabridulous or minutely spinulose, marginal spinules usually 1 mm or shorter. **Corollas** pink to purple. $2n$ = 32, 34.

Flowering spring (Mar–Jun). Meadows, pastures, roadsides, forest openings, low ground, often in damp soil; 0–100 m; Ala., Ark., Fla., La., Miss., Okla., Tex.

Variety *megacanthum* occurs on the coastal plain and lower piedmont from northern Florida to eastern Texas and southeastern Oklahoma. Nuttall described this thistle as "one of the most terribly armed plants in the genus."

11c. Cirsium horridulum Michaux var. **vittatum** (Small) R. W. Long, Rhodora 72: 45. 1970

• Florida thistle [E]

Carduus vittatus Small, Bull. New York Bot. Gard. 3: 439. 1905; *C. smallii* (Britton) H. E. Ahles; *Cirsium smallii* Britton; *C. vittatum* (Small) Small

Plants 30–90 cm. **Stems** glabrous to sparsely tomentose. **Leaves:** blades linear to oblong, 10–30 cm, spinose-dentate to shallowly pinnatifid, main spines mostly 5–10 mm, abaxial faces glabrous or nearly so. **Heads** 1–10. **Involucres** 3–4 × 3–6 cm, glabrous. **Phyllaries:** outer and middle bodies densely scabridulous, marginal spinules 0.1–0.5 mm. **Corollas** purple. $2n$ = 32, 33, 34, 35.

Flowering winter–summer (Feb–Aug). Meadows, pastures, old fields, pinelands, usually in damp soil; 0–100 m; Ala., Fla., Ga., La., Miss., N.C., S.C.

Variety *vittatum* grows on the coastal plain.

12. Cirsium pumilum (Nuttall) Sprengel, Syst. Veg. 3: 375. 1826 • Pasture thistle [E]

Carduus pumilus Nuttall, Gen. N. Amer. Pl. 2: 130. 1818

Biennials or monocarpic perennials, 25–100 cm; sometimes perennating by root sprouts. **Stems** erect, villous with septate trichomes and sometimes thinly arachnoid tomentose; branches 0–few, distal, ascending. **Leaves:** blades oblong-elliptic, 5–30+ × 2–10 cm, ± undulate, shallowly to deeply pinnatifid, lobes ovate to broadly triangular, usually separated by broad sinuses, spinose-dentate or lobed, main spines 1.5–7 mm, slender to stout, abaxial faces villous with septate trichomes, at least along veins, sometimes thinly arachnoid, adaxial faces villous with septate trichomes and thinly arachnoid tomentose or subglabrous; basal often present at flowering, bases tapered; principal cauline sessile, moderately reduced distally, bases often auriculate-clasping; distal reduced, similar to proximal. **Heads** 1–few, borne singly at tips of main stem and branches, often closely subtended by 1–several bracts. **Peduncles** 0–15+ cm (above distal leaves), leafy-bracted. **Involucres** broadly cylindric to ovoid, 3.5–5 cm, 2.5–3 cm diam. (appearing much wider and hemispheric to campanulate in pressed specimens), loosely arachnoid on phyllary margins or glabrate. **Phyllaries** in 8–10 series, imbricate, ovate or lanceolate (outer) to lance-linear (inner), abaxial faces with ± prominent glutinous ridge, outer and middle appressed, apices ascending to spreading, spines 1.5–6 mm; apices of

middle and inner narrowed and scabrid-denticulate, innermost spineless, tapered and entire or with expanded, erose-denticulate, flexuous tips. **Corollas** pink to purple (white), 40–60 mm, tubes 20–35 mm, throats 11–15 mm, lobes 7–10 mm; style tips 3–7 mm. **Cypselae** stramineous to brown, 3.5–5 mm, apical collars yellow or colored like body; **pappi** 35–45 mm. $2n$ = 30.

Varieties 2 (2 in the flora): Canada, United States.

Flowers of *Cirsium pumilum* are reportedly sweetly scented.

1. Spines of outer phyllaries stout, 3–6 mm; plants usually with several long branches; leaves usually lobed nearly to the midvein; roots rarely and only slightly tuberous-thickened; cypselae usually 3.5–4 mm . . . . 12a. *Cirsium pumilum* var. *pumilum*
1. Spines of outer phyllaries slender, 1.5–3 mm; plants simple or distally with a few short branches; leaves usually shallowly lobed; roots often tuberous-thickened; cypselae usually 4.5–5 mm . . . . 12b. *Cirsium pumilum* var. *hillii*

12a. Cirsium pumilum (Nuttall) Sprengel var. **pumilum** [E]

Cirsium odoratum (Muhlenberg ex W. P. C. Barton) Petrak

Plants mostly 30–100 cm; branches 3–5+, long. **Roots** usually slender, not or rarely slightly tuberous-thickened. **Leaves** usually lobed nearly to midveins, sinuses deep, U-shaped, main spines ± stout, 3–7 mm. **Heads** 5.5–6 cm. **Phyllaries:** outer 3–5 mm wide at base, spines stout, 3–6 mm. **Corollas** usually 40–45 mm. **Cypselae** usually 3.5–4 mm. $2n$ = 30.

Flowering summer (Jun–Sep). Fields, pastures, open woods, roadsides; 0–600 m; Conn., Del., Maine, Md., Mass., N.H., N.J., N.Y., N.C., Ohio, Pa., R.I., S.C., Vt., Va., W.Va.

Variety *pumilum* grows on the Atlantic coastal plain from Maine to South Carolina and ranges inland to Pennsylvania and Ohio. It is known to hybridize with *Cirsium horridulum* var. *horridulum* (R. J. Moore and C. Frankton 1966).

12b. Cirsium pumilum (Nuttall) Sprengel var. **hillii** (Canby) B. Boivin, Naturaliste Canad. 94: 646. 1972 • Hill's thistle ⬚E⬚

Cnicus hillii Canby, Gard. & Forest 4: 101. 1891; *Cirsium hillii* (Canby) Fernald; *C. pumilum* subsp. *hillii* (Canby) R. J. Moore & Frankton

Plants mostly 25–60 cm, branches 0–4, short. **Roots** often tuberous-thickened. **Leaves** usually shallowly lobed, main spines ± fine, 3–6 mm. **Heads** 5–7 cm. **Phyllaries:** outer 3–6 mm wide at base, spines slender, 1.5–3 mm. **Corollas** usually 45–55 mm. **Cypselae** usually 4.5–5 mm. $2n = 30$.

Flowering summer (Jun–Aug). Sandy or gravelly soils, prairies, limestone barrens, pastures, pine barrens, open woods, and oak savannas; 200–400 m; Ont.; Ill., Ind., Iowa, Mich., Minn., Wis.

Variety *hillii* occurs mainly in prairie areas from Minnesota and Iowa east to southern Ontario, Michigan, and northern Indiana. It has often been treated as a species distinct from *Cirsium pumilum*. As R. J. Moore and C. Frankton (1966) pointed out, the differences separating these taxa are, for the most part, metric characters that show considerable overlap. Some specimens, especially those from Ohio and western Pennsylvania, are difficult to place, and scattered individuals within the area of var. *pumilum* would readily be assigned to var. *hillii* were they growing in the area of that taxon. Moore and Frankton chose to recognize the infraspecific taxa within *C. pumilum* as subspecies, a rank that is seldom employed in North American *Cirsium* taxonomy.

Hill's thistle is a taxon that has attracted the attention of conservationists because of declining populations. Although the U.S. Fish and Wildlife Service determined in 1999 that this plant does not warrant elevation to Candidate Species status, it is recognized as a species of concern in several states. Throughout much of its range the populations have declined due to habitat destruction and fragmentation that have accompanied agricultural development and urbanization.

13. Cirsium engelmannii Rydberg, Fl. Rocky Mts., 1069. 1917 • Blackland or Engelmann's thistle ⬚E⬚

Cirsium virginianum (Linnaeus) Michaux var. *filipendulum* A. Gray, Manual ed. 2, 233. 1856, not *C. filipendulum* Lange 1861; *C. terrae-nigrae* Shinners

Biennials or monocarpic perennials, 40–200 cm; taproots and clusters of coarse fibrous roots that often have tuberlike thickenings. **Stems** single, erect, often branched above middle, thinly arachnoid-tomentose, ± glabrate; branches few, ascending. **Leaves:** blades elliptic or ovate, 5–20 × 1–10 cm, usually deeply pinnatifid, lobes narrowly to broadly triangular, sinuses broad, rounded (basal and distal cauline sometimes less divided, lobes linear-lanceolate), margins revolute, spreading, entire or spinulose to remotely few-toothed or sharply lobed, main spines slender, 1–5 mm, abaxial faces white-tomentose, adaxial green, villous with septate trichomes or glabrate; basal usually absent at flowering, winged-petiolate, bases tapered; principal cauline well distributed, gradually reduced, bases narrowed, sometimes weakly clasping; distal reduced, widely separated, distalmost bractlike. **Heads** 1–10+, borne at tips of main stem and branches. **Peduncles** 2–20+ cm, essentially naked with much reduced bracts. **Involucres** ovoid to broadly cylindric or campanulate, 2.5–3.5 × 2–3 cm, thinly arachnoid. **Phyllaries** in 10–12 series, strongly imbricate, greenish with subapical darker central zone, ovate (outer) to lanceolate (inner), abaxial faces with narrow glutinous ridge; outer and middle entire, bodies appressed, spines abruptly spreading to deflexed, slender, 2–4 mm; apices of inner phyllaries narrow, flexuous, flattened, entire or finely erose. **Corollas** pink to purple (white), 32–38 mm, tubes 15–20 mm, throats 6–9 mm, lobes 8–11 mm; style tips 5–6 mm. **Cypselae** brown, 5–6 mm, apical collars yellow, ca. 1 mm; **pappi** 25–30 mm. $2n = 18$ (as *C. terrae-nigrae*), 20 + 1B.

Flowering spring–summer (May–Jul). Tallgrass prairies, old fields, roadsides, oak savannas, forest edges, in calcareous clay or rarely sandy soils; 50–200 m; La., Okla., Tex.

Cirsium engelmannii occurs mostly in the blackland prairies of eastern Texas. It ranges northward into southeastern Oklahoma and eastward to northwestern Arkansas.

14. **Cirsium carolinianum** (Walter) Fernald & B. G. Schubert, Rhodora 50: 229. 1948 • Carolina or purple or soft or smallhead thistle [E]

Carduus carolinianus Walter, Fl. Carol., 195. 1788; *Cirsium flaccidum* (Small) Petrak

Biennials, 50–180 cm; taproots short with many slender, fibrous lateral roots. **Stems** usually single, erect, glabrous to ± tomentose, sometimes sparsely villous with septate trichomes; branches few, usually distal, ascending. **Leaves:** blades linear to oblanceolate or elliptic, 10–30 × 1–5 cm, unlobed and spinulose to irregularly dentate or pinnatifid, lobes narrowly to broadly triangular, sometimes coarsely toothed or lobed toward base, acuminate, main spines slender, 1–5 mm, abaxial faces gray-tomentose, adaxial glabrous or sparsely villous with septate trichomes; basal often present at flowering, petioles slender, winged, bases long-tapered; principal cauline relatively few (10–25), petiolate or distal sessile, mostly restricted to proximal ¹/₂ of stems, progressively reduced distally, bases tapered, not decurrent; distal cauline widely separated, linear to narrowly elliptic, reduced, becoming ± bractlike, merely spinulose to irregularly dentate or shallowly lobed. **Heads** (1–)2–9(–many), in paniculiform arrays. **Peduncles** slender, 1–15 cm (not overtopped by distal leaves). **Involucres** narrowly ovoid to campanulate, 1.2–2 × 1.2–2 cm, thinly arachnoid-ciliate. **Phyllaries** in 7–10 series, imbricate, green, linear to lanceolate (outer) or linear to linear-lanceolate (inner), abaxial faces with narrow, glutinous ridge; outer and middle ascending to appressed, bodies entire, apices widely spreading (at least the outer), spines ascending to spreading (at least the outer), slender, 1–4 mm; apices of inner phyllaries flat, often twisted, acuminate. **Corollas** pink-purple (white), 15–20 mm, tubes 5–9 mm, throats 5–7 mm (noticeably wider than tubes), lobes 4–5 mm; style tips 4 mm. **Cypselae** light brown, 3–4 mm, apical collars yellowish, 0.5–1 mm; **pappi** 12–14 mm. $2n$ = 20, 22.

Flowering spring–summer (Apr–Jul). Wooded areas, openings, fields, roadsides; 50–300 m; Ala., Ark., Ga., Ill., Ind., Ky., La., Miss., Mo., N.C., Ohio, Okla., S.C., Tenn., Tex.

Cirsium carolinianum is widely distributed in the southeastern United States: on the Gulf coastal plain from Texas to Alabama north through the Ouachita and Ozark highlands to southeastern Missouri; in the Ohio River Valley from southernmost Illinois to southern Ohio and northern Kentucky; and in the southern Appalachians and Piedmont from Alabama and Tennessee to southern Virginia. *Cirsium carolinianum*, though widespread, is a taxon of conservation concern over part of its range. The replacement of open woods by dense forests brought about by fire suppression has greatly reduced available habitat.

15. **Cirsium virginianum** (Linnaeus) Michaux, Fl. Bor.-Amer. 2: 90. 1803 • Virginia thistle [E]

Carduus virginianus Linnaeus, Sp. Pl. 2: 824. 1753; *C. revolutus* Small; *Cirsium revolutum* (Small) Petrak

Biennials or perennials, 60–150 (–200) cm; crown sprouts from cluster of fibrous roots, these often tuberous-thickened. **Stems** usually single, erect, thinly appressed-tomentose, ± glabrate in age; branches 0–few in distal ¹/₃, ascending. **Leaves** very numerous, firm-textured, blades 3–15 cm, thick, ± rigid, linear or linear-elliptic, 0.5–2 cm wide and spinulose, or narrowly ovate, 2–4 cm wide, deeply lobed, lobes remote, spreading, separated by broad sinuses, few toothed or lobed, margins often revolute, main spines slender, 3–5(–9) mm, abaxial faces white-tomentose, adaxial green, glabrous or thinly tomentose; basal usually absent at flowering, winged-petiolate; proximal cauline usually absent at flowering, well separated, winged-petiolate; middle and distal numerous (30–70+), sessile, well distributed, gradually reduced distally, bases tapered, not decurrent; distal linear, entire or few lobed, ca. 1 cm. **Heads** 1–10+ in open, corymbiform or paniculiform arrays. **Peduncles** 10–15 cm (not overtopped by distal leaves). **Involucres** ovoid to cylindric or narrowly campanulate, 1.7–2.4 × 1–2 cm, glabrous or outer phyllaries very thinly tomentose. **Phyllaries** in 8–13 series, strongly imbricate, light green to brownish with dark apices, ovate (outer) to narrowly linear-elliptic (inner), abaxial faces with evident, narrow glutinous ridge; outer and middle appressed, bodies entire, apices erect or spreading, muticous to short-spinose, spines ascending to spreading, weak, 1–2 mm; apices of inner all straight and entire or innermost ± flexuous, erose. **Corollas** purple, 21–26 mm, tubes 8.5–11 mm, throats 6–8 mm (noticeably wider than tubes), lobes 4–8 mm; style tips 3.5–5 mm. **Cypselae** dark brown, 4–5 mm, apical collars yellowish, 0.5–1; **pappi** 17–20 mm. $2n$ = 28.

Flowering summer–fall (Aug–Oct). Moist savannas, pine barrens, coastal plain bogs; 0–150 m; Del., Fla., Ga., N.J., N.C., S.C., Va.

Cirsium virginianum occurs on the Atlantic coastal plain from Delaware to Florida.

16. **Cirsium nuttallii** de Candolle in A. P. de Candolle and A. L. P. P. de Candolle, Prodr. 6: 651. 1838 (as nuttalii) • Nuttall's thistle E

Carduus glaber Nuttall, Gen. N. Amer. Pl. 2: 129. 1818, not *Cirsium glabrum* de Candolle 1805; *C. nuttallii* (de Candolle) Pollard

Biennials, 20–350 cm; taprooted. **Stems** usually single, erect, glabrous or villous with septate trichomes; branches few–many, ascending. **Leaves:** blades narrowly to broadly elliptic, (10–)15–60 × (2–)5–15 cm, thin, ± flexible, deeply pinnatifid, lobes narrow, spreading, coarsely dentate or lobed, main spines 2–5 mm, abaxial faces thinly tomentose but often wholly glabrate in age, adaxial glabrous or sparsely villous with septate trichomes; basal often absent at flowering, petioles slender, winged, bases tapered; principal cauline becoming sessile and gradually reduced distally, bases spiny-lobed, sometimes decurrent; distal reduced to linear bracts. **Heads** few–many, in open corymbiform or paniculiform arrays. **Peduncles** 1–15 cm, essentially naked (not overtopped by crowded distal leaves). **Involucres** hemispheric to campanulate, 1.5–2.5 × 1–2.5 cm, thinly arachnoid or glabrate. **Phyllaries** in 6–10 series, strongly imbricate, green or brownish, ovate or elliptic (outer) to linear-lanceolate (inner), abaxial faces with narrow glutinous ridge; outer and middle appressed, bodies entire, spines abruptly spreading, slender, 1–2(–3) mm; apices of inner often flexuous, flat, attenuate. **Corollas** white to pink, lavender, or purple, 17–25 mm, tubes 5–11 mm, throats 4–7 mm (noticeably wider than tubes), lobes 5–7 mm; style tips 3–4.5 mm. **Cypselae** dark brown, 3–4 mm, apical collars stramineous, 0.5 mm; **pappi** 17–21 mm (longer bristles shorter than corollas). *2n* = 24, 26, 28.

Flowering summer (Jun–Aug). Roadsides, ditches, woodlands, usually in damp soil; 0–100 m; Ala., Fla., Ga., La., Miss., N.C., S.C., Tex., Va.

Cirsium nuttallii occurs on the southern coastal plain from southeastern Virginia to southern Florida and west to eastern Louisiana.

17. **Cirsium texanum** Buckley, Proc. Acad. Nat. Sci. Philadelphia 13: 460. 1862 • Texas or Texas purple or southern thistle

Cirsium austrinum (Small) E. D. Schulz; *C. helleri* (Small) Cory; *C. texanum* var. *stenolepis* Shinners

Annuals or biennials, 20–200 cm; taprooted. **Stems** usually single, erect, tomentose to ± glabrate; branches 0–many, usually restricted to distal part, ascending. **Leaves:** blades oblong to elliptic, 7–30 × 2–12 cm, unlobed and merely spinulose to irregularly dentate or shallowly to deeply pinnatifid, lobes ± triangular, separated by narrow to wide sinuses, sometimes coarsely dentate or lobed proximally, obtuse to acute, main spines slender to stout, 1–5 mm, abaxial faces arachnoid tomentose, adaxial glabrous or thinly arachnoid; basal often absent at flowering, petioles slender, ± winged; cauline progressively reduced, proximal petiolate, mid and distal broadly sessile, bases ± auriculate-clasping or decurrent 1–3 cm; distalmost linear to lanceolate, bractlike, irregularly dentate or shallowly lobed. **Heads** 1–many, in openly paniculiform arrays. **Peduncles** slender, 3–30 cm (not overtopped by crowded distal leaves). **Involucres** ovoid to hemispheric, 1.5–2 × 1.5–2 cm, thinly arachnoid, glabrate. **Phyllaries** in 8–10 series, imbricate, green, lanceolate (outer) to linear (inner), abaxial faces with prominent glutinous ridge; outer and middle appressed, bodies entire, acute, spines spreading, slender, 1–5 mm; apices of inner often flexuous, flat, scabrid-ciliolate, acuminate. **Corollas** white to pink-purple, 20–25 mm, tubes 7–10 mm, throats 6–8 mm (noticeably wider than tubes), lobes 4–7 mm; style tips 3–4 mm. **Cypselae** brown, 3–5 mm, apical collars not differentiated; **pappi** 15–16 mm. *2n* = 22, 23, 24.

Flowering spring–summer (Apr–Jul). Roadsides, pastures, fields, shrub-tree savannas; 0–1000 m; Ark., La., Mo., N.Mex., Okla., Tex.; Mexico (Coahuila, Durango, Nuevo León, San Luis Potosí, Tamaulipas).

Cirsium texanum ranges from the Chihuahuan Desert regions of trans-Pecos Texas and adjacent southeastern New Mexico across the plains of Texas and southern Oklahoma to southwestern Arkansas and southwestern Louisiana and south into north-central Mexico. D. S. Correll and M. C. Johnston (1970) suggested hybridization between *Cirsium texanum* and *C. undulatum* to explain anomalous specimens in the Edwards Plateau and trans-Pecos regions of western Texas.

18. Cirsium flodmanii (Rydberg) Arthur, Torreya 12: 34. 1912 • Prairie or Flodman's thistle, chardon de Flodman E

Carduus flodmanii Rydberg, Mem. N. Y. Bot. Gard. 1: 451. 1900; *Cirsium oblanceolatum* (Rydberg) K. Schumann

Perennials 30–140 cm; horizontal runner roots that produce root sprouts. **Stems** 1–several, erect, gray- or white-tomentose; branches 0–few, ascending. **Leaves:** blades oblong-oblanceolate to narrowly elliptic, 4–40 × 1–10 cm, bases usually not decurrent, finely spiny-toothed and undivided to coarsely toothed or deeply pinnatifid, lobes broadly triangular to linear-lanceolate, often revolute-margined, main spines 1–7 mm, abaxial faces white-tomentose, adaxial faces green, thinly tomentose, ± glabrate; basal usually absent or withered at flowering, winged petiolate; principal cauline proximally winged-petiolate, distally sessile, well distributed, gradually reduced, bases usually not decurrent; distal cauline well developed. **Heads** erect, borne singly and terminal on main stem and branches, or few in corymbiform arrays from distal axils (not subtended by ring of spiny-margined bracts). **Peduncles** 0–5 cm (elevated above distal leaves). **Involucres** ovoid to broadly campanulate, 2–3.5 × 2.5–3.5 cm, thinly arachnoid. **Phyllaries** in 7–12 series, strongly imbricate, greenish with subapical darker central zone, ovate or lanceolate (outer) to linear (inner), abaxial faces with prominent glutinous ridge; outer and middle entire, bodies appressed, entire, acute, spines abruptly spreading, slender, 2–4 mm; apices of inner spreading, flexuous, narrow, flattened, finely serrulate, ± scabrous. **Corollas** purple (white), 23–36 mm, tubes 12–15 mm, throats 6–8.5 mm, lobes 5–9 mm; style tips 4–7 mm. **Cypselae** light brown, 3–5 mm, apical collars stramineous, 0.5–1 mm; **pappi** (white or tawny) 20–30 mm. $2n = 22, 24$.

Flowering summer (Jun–Sep). Tallgrass, mixedgrass, shortgrass prairies, meadows, pastures, often in damp soil; 100–2400 m; Alta., Man., Ont., Que., Sask.; Colo., Ill., Iowa, Kans., Mich., Minn., Mont., Nebr., N.Dak., S.Dak., Wis., Wyo.

Cirsium flodmanii ranges from Saskatchewan and Alberta south through the northern Great Plains and intermountain valleys of Montana and Wyoming to northeastern Colorado and east through the prairies to Minnesota and Iowa, and in widely scattered locations eastward to northern Illinois, southern Wisconsin, southern Ontario, and southern Quebec. It is known to hybridize with *C. muticum* and *C. undulatum*. Hybrids between *C. flodmanii* and *C. undulatum* are highly sterile with numerous meiotic irregularities (S. Dabydeen 1987).

19. Cirsium undulatum (Nuttall) Sprengel, Syst. Veg. 3: 374. 1826 • Wavyleaf or gray or pasture thistle

Carduus undulatus Nuttall, Gen. N. Amer. Pl. 2: 130. 1818; *Cirsium megacephalum* (A. Gray) Cockerell ex Daniels; *C. undulatum* var. *megacephalum* (A. Gray) Fernald

Perennials, 20–230 cm; deeply seated runner roots that produce adventitious buds. **Stems** 1–several, erect or ascending, densely gray-tomentose; branches 0–few, usually above middle, ascending. **Leaves:** blades elliptic to oblong or ovate, 10–40 × 1–10 cm, margins strongly undulate, coarsely dentate or shallowly to deeply lobed, lobes ascending to spreading, ± triangular, well separated to closely spaced, spinulose and coarsely dentate or usually cleft into 2–3 lanceolate to triangular, often entire-margined, spine-tipped divisions, main spines (yellowish), 2–12+ mm, abaxial densely gray-tomentose, adaxial faces thinly tomentose; basal sometimes present at flowering, winged-petiolate; principal cauline becoming sessile and progressively reduced distally, widest at base, bases ± auriculate-clasping to short-decurrent; distal reduced, spinier. **Heads** 1–10+, terminal on branches, in leafy, ± corymbiform arrays. **Peduncles** 0–25+ cm. **Involucres** ovoid to hemispheric or broadly campanulate, 2.5–4.5 × 1.5–4.5 cm, loosely arachnoid on phyllary margins or glabrate. **Phyllaries** in 8–12 series, imbricate, ovate to lanceolate (outer) to linear-lanceolate (inner), abaxial faces with prominent glutinous ridge; outer and middle appressed, spines spreading, 1.5–5 mm; apices of inner narrow, often flexuous, flat, ± entire, spineless or weakly spiny. **Corollas** lavender to pink, purple, or white, 24–50 mm, tubes 12–28 mm, throats 6–14 mm, lobes 6.5–13 mm; style tips 5–7.5 mm. **Cypselae** light to dark brown, 6–7 mm, bodies and apical collars concolorous, narrow; **pappi** 20–38 mm (usually scabridulous). $2n = 26$.

Flowering spring–autumn (May–Oct). Mixedgrass prairie, shortgrass prairie, Palouse prairie, sagebrush deserts, pinyon-juniper woodlands, openings in montane coniferous forests, often in disturbed areas; 100–2800 m; Alta., B.C., Man., Sask.; Ariz., Calif., Colo., Ga., Idaho, Ill., Ind., Iowa, Kans. Mich., Minn., Mo., Mont., Nebr., N.Mex., N.Dak., Okla., Oreg., Pa., S.Dak., Tex., Utah, Wash., Wis., Wyo.; Mexico (Chihuahua, Coahuila, Durango, Sonora).

Cirsium undulatum is widely distributed in the western half of North America from the dry plains and plateaus of the Pacific Northwest eastward across the Great Plains to Manitoba and the Dakotas and south to Texas, New Mexico, and northwestern Mexico. It occurs in scattered localities in the Rocky Mountains and northeastern Great Basin region. At least some of the few widely scattered records from the eastern United States are probably introductions. *Cirsium undulatum* is both widespread and variable. Plants of the Great Plains region tend to be low-growing with a few large heads and elongate corollas. Plants of the Pacific Northwest are usually taller and produce smaller, more numerous heads with shorter corollas. A detailed study of this species might reveal races worthy of recognition as infraspecific taxa.

Wavyleaf thistle is listed by California as a noxious weed. However, most reports of *Cirsium undulatum* in California are based upon misidentifications of *C. canescens*. *Cirsium undulatum* is known to hybridize with *C. flodmanii*, *C. hookerianum*, and *C. scariosum* var. *coloradense*. J. T. Howell (1960b) reported that *C. undulatum* was suspected to hybridize with *C. brevifolium* in the Pacific Northwest.

20. **Cirsium ciliolatum** (L. F. Henderson) J. T. Howell, Leafl. W. Bot. 9: 9. 1959 • Ashland thistle C E

Cirsium undulatum (Nuttall) Sprengel var. *ciliolatum* L. F. Henderson, Bull. Torrey Bot. Club 27: 348. 1900 (as Circium); *C. howellii* Petrak

Perennials 60–200 cm, arachnoid, tomentose; runner roots producing adventitious buds. **Stems** 1–several, erect, thinly arachnoid to densely white-tomentose; branches 0–few, ascending. **Leaves:** blades oblong-elliptic, 10–30 × 3–12 cm, margins finely spiny-toothed and otherwise undivided to coarsely dentate, shallowly lobed, or deeply laciniate-pinnatifid, lobes broadly triangular to linear-lanceolate, main spines 1–6 mm, abaxial white-tomentose, adaxial faces ± green, thinly to densely arachnoid-tomentose; basal often present at flowering, winged petiolate; principal cauline well distributed, proximally winged-petiolate, distally sessile, gradually reduced, bases auriculate-clasping or short-decurrent as spiny wings; distal cauline few, reduced to linear bracts. **Heads** borne singly and terminal on main stem and branches, or few in corymbiform arrays. **Peduncles** 0–15 cm. **Involucres** ovoid to hemispheric, 1.5–2.3 × 1.5–3 cm, thinly arachnoid, ± glabrate. **Phyllaries** in 5–7 series, strongly imbricate, greenish with subapical darker central zone, lanceolate (outer) to linear (inner), abaxial faces with prominent glutinous ridge; outer and middle bodies appressed, entire, apices entire or finely serrulate, spines ascending, slender, 1–3 mm; apices of inner spreading to erect, narrow, ± flattened, finely serrulate, ± scabrous. **Corollas** dull white to lavender, 15–25 mm, tubes 7–11 mm, throats 5–7 mm, lobes 5–7 mm; style tips 5–7 mm. **Cypselae** brown, 3.5–7 mm, apical collars tan, 0.1–0.2 mm; **pappi** white or tawny, 15–20 mm.

Flowering spring–summer (May–Aug). Grassy areas, open woodlands; of conservation concern; 500–1400 m; Calif., Oreg.

Cirsium ciliolatum is restricted to a few sites in and around the Klamath Range of Jackson County, Oregon, and Siskiyou County, California. It is listed by the state of California as endangered. It is closely related to *C. undulatum* and perhaps should be treated as a variety of that species. Pending a comprehensive study of the variation within *C. undulatum*, I have maintained *C. ciliolatum* as a distinct species.

21. **Cirsium tracyi** (Rydberg) Petrak, Beih. Bot. Centralbl. 35(2): 424. 1917 • Tracy's thistle E

Carduus tracyi Rydberg, Bull. Torrey Bot. Club 32: 133. 1905; *Cirsium acuatum* (Osterhout) Cockerell; *C. floccosum* (Rydberg) Petrak; *C. undulatus* Nuttall var. *tracyi* (Rydberg) S. L. Welsh

Perennials, 50–200+ cm; taprooted. **Stems** 1–several, erect or ascending, thinly gray-tomentose or ± glabrate; branches few to many, ascending. **Leaves:** blades elliptic to oblong, 8–40 × 1–12 cm, margins weakly to strongly undulate, spinose-dentate or shallowly to deeply lobed, lobes ascending to spreading, ± triangular, mostly well separated, spinulose and coarsely dentate or cleft into 2–3 lanceolate to triangular, often entire-margined, spine-tipped divisions, main spines 2.5–7+ mm, abaxial faces densely gray-tomentose, adaxial thinly tomentose; basal sometimes present at flowering, winged-petiolate; principal cauline becoming sessile and progressively reduced distally, widest at bases, bases ± auriculate-clasping to short-decurrent; distal cauline reduced, often spinier. **Heads** 1–many, terminal on branches and often in leaf axils, in leafy, ± corymbiform arrays. **Peduncles** 0–10+ cm. **Involucres** ovoid to hemispheric or broadly campanulate, 2–3 × 1.7–3.5 cm, loosely arachnoid on phyllary margins or glabrate. **Phyllaries** in 6–10 series, imbricate, ovate to lanceolate (outer) to linear-lanceolate (inner), margins entire, abaxial faces with prominent glutinous ridge; outer and middle appressed, spines spreading, slender to stout, 2–6 mm; apices of inner often flexuous, narrow, flat, ± entire, spineless or tipped with weak spines. **Corollas** white to lavender or pink-purple, 23–30 mm, tubes 9–14 mm, throats 5.5–10.5 mm, lobes 5.5–9.5 mm; style tips 4–7 mm. **Cypselae** light to dark brown, 6–7 mm, apical collars colored like body or rarely yellowish, narrow; **pappi** 20–23 mm, usually noticeably shorter than corolla. $2n = 24$.

Flowering late spring–summer (Jun–Aug). Dry slopes, sagebrush deserts, pinyon-juniper woodlands, openings in montane coniferous forests, often in disturbed areas; 1400–2900 m; Colo., N.Mex., Utah.

Cirsium tracyi occurs from eastern Utah and western Colorado south in the Colorado Plateau and southern Rocky Mountains to northwestern New Mexico. Large-headed plants of *Cirsium tracyi* and small-headed individuals of *C. undulatum* are sometimes difficult to distinguish. P. L. Barlow-Irick (unpubl.) found that although there is much overlap in floral measurements of *C. tracyi* and *C. undulatum*, the means for some of these characters are statistically significant. Corolla lobes of *C. tracyi*, for instance, average about 7 mm and those of *C. undulatum* about 10 mm. The species differ in chromosome number as well.

22. **Cirsium canescens** Nuttall, Trans. Amer. Philos. Soc., n. s. 7: 420. 1841 (as Circium) • Platte or prairie thistle E

Cirsium nebraskense (Britton) Lunell; *C. plattense* (Rydberg) Cockerell ex Daniels; *C. nelsonii* (Pammel) Petrak

Biennials or monocarpic perennials, 20–100 cm; taproots long. **Stems** usually 1, erect, ± densely gray-tomentose with fine, non-septate trichomes; branches 0 or few, usually above middle in distal 1/2, ascending. **Leaves:** blades oblong to elliptic or obovate, 10–25(–40) × 2–6(–12) cm, coarsely dentate or shallowly lobed to deeply pinnatifid, lobes well separated, triangular to linear or oblong, often revolute-margined, ascending to spreading, spinulose to spinose-dentate, main spines 2–3(–10) mm, faces gray-tomentose, more densely abaxially, sometimes glabrate adaxially; basal usually present at flowering, winged-petiolate; principal cauline progressively reduced distally, bases decurrent as spiny wings 1–5 cm, sometimes with expanded auricles; distal cauline usually much reduced, less lobed. **Heads** 1–10+, terminal on branches or in distal axils, in openly corymbiform to racemiform arrays. **Peduncles** 0–10 cm. **Involucres** hemispheric to broadly campanulate, usually truncate or indented at base, 3–4 × 2.5–4 cm in first-formed heads, often smaller (1.5–2 cm) in later ones, loosely arachnoid on phyllary margins or glabrate. **Phyllaries** in 6–9 series, imbricate, ovate-lanceolate (outer) to linear-lanceolate (inner), abaxial faces with prominent glutinous ridge; bodies of outer and middle appressed, acute, spines ascending to spreading, 2–4(–8) mm; apices of inner expanded and flat, often twisted, scabrid-margined, and erose, spineless. **Corollas** dull white or lavender-tinged, 20–35 mm, tubes 10–17 mm, throats 6–11 mm, lobes 4–9 mm; style tips 5–8 mm. **Cypselae** light brown, 5–7 mm, sometimes with darker streaks, apical collar very narrow, lighter colored; **pappi** 18–30 mm, usually noticeably shorter than corolla. $2n = 34, 36$.

Flowering spring–summer (May–Aug). Sandy or gravelly soils in short-grass prairie, often in disturbed areas, mountain meadows, grassy slopes in montane coniferous forests; 1100–3800 m; Calif., Colo., Mo., Mont., Nebr., Nev., S.Dak., Wyo.

Cirsium canescens grows in the northern Great Plains from eastern Montana and Wyoming to eastern Colorado and Nebraska; an upland race occurs in the Rocky Mountains of eastern Colorado. It has been reported from Iowa, North Dakota, and Ohio; I have not seen specimens from those states. It is adventive in northeastern California.

Cirsium canescens hybridizes locally with *C. scariosum* and *C. parryi*. Further investigations may reveal that high-elevation forms of *C. canescens* from the mountains of Colorado are worthy of taxonomic recognition. These plants flower later than the low elevation forms of the Great Plains and occur in rather different ecologic conditions, but I have found no features that readily distinguish them. Populations of *C. canescens* have been particularly affected by the seedhead weevil *Rhinocyllus conicus*, introduced to North America to control weedy species of *Carduus* (S. M. Louda et al. 1997; Louda 1998).

23. **Cirsium pitcheri** (Torrey ex Eaton) Torrey & A. Gray, Fl. N. Amer. 2: 456. 1843 • Dune or sand-dune thistle E

Cnicus pitcheri Torrey ex Eaton, Man. Bot. ed. 5, 180. 1829

Biennials or short-lived monocarpic perennials, 20–100 cm; taproots long. **Stems** 1 or few, erect, densely gray-tomentose; branches 0 to several, ascending to spreading. **Leaves:** blades elliptic to obovate, 10–30 × 8–14 cm, deeply divided nearly to midveins, lobes ascending to spreading, linear, remote, margins revolute, entire or minutely spinulose, main spines 1–2 mm, faces gray-tomentose, more densely so abaxially; basal present or withered at flowering, petiolate; principal cauline well distributed, bases decurrent as linear-lobed to spiny wings 1–3 cm; distal cauline well developed. **Heads** 1–20+ in corymbiform arrays. **Peduncles** 0–5 cm. **Involucres** ovoid to campanulate, 2–3 × 2–3 cm, loosely arachnoid on phyllary margins or glabrate. **Phyllaries** in 6–8 series, imbricate, ovate-lanceolate (outer) to linear-lanceolate (inner), abaxial faces with narrow glutinous ridge; outer and middle appressed, acute, spines ascending to spreading, slender, 1–2(–3) mm; apices of inner often flexuous, flattened, spineless, scabrid. **Corollas** dull white or pinkish-tinged (rarely rich purple), 20–30 mm, tubes 8.5–15 mm, throats 4.5–10 mm, lobes 3–8 mm; style tips 3.5–5.5 mm. **Cypselae** light brown, sometimes with darker streaks, 6–7.5 mm, apical collars lighter colored, very narrow; **pappi** 15–30 mm, usually noticeably shorter than corolla. $2n = 34$.

Flowering spring–summer (May–Sep). Sand dunes and beaches; 180–200 m; Ont.; Ill., Ind., Mich., Wis.

Cirsium pitcheri is endemic to beach and dune habitats around lakes Huron, Michigan, and Superior. It has been extirpated from portions of its former range at the southern end of Lake Michigan. It is threatened by foot traffic, off-road vehicular activity, and clearing and development of beachside habitats. It is in the Center for Plant Conservation's National Collection of Endangered Plants.

24. Cirsium ochrocentrum A. Gray, Mem. Amer. Acad. Arts, n. s. 4: 110. 1849

Perennials, 30–90 cm; crown sprouts or runner roots producing adventitious buds. **Stems** 1–20+, erect or ascending, densely gray-tomentose with non-septate trichomes; branches 0 or few, usually in distal $^1/_2$, ascending. **Leaves:** blades oblong to narrowly elliptic, 10–30 × 2–8 cm, strongly undulate, margins coarsely dentate or shallowly to deeply pinnatifid with 8–15 pairs of lobes 0.5–2 cm, often revolute, lobes ± triangular, closely spaced, spreading, spinose-dentate and cleft into 2–5 spine-tipped divisions, main spines 5–20 mm, yellowish, abaxial faces densely white-tomentose, adaxial thinly gray-tomentose; basal usually present at flowering, winged-petiolate; principal cauline sessile, progressively reduced distally, bases ± auriculate to long-decurrent as spiny wings; distal cauline usually much reduced, less lobed. **Heads** 1–few, in leafy, ± corymbiform arrays. **Peduncles** 0–4 cm. **Involucres** ovoid to hemispheric or broadly campanulate, 2.5–4.5 × 2.5–4.5 cm in first-formed heads, often smaller in later ones, loosely arachnoid on phyllary margins or glabrate. **Phyllaries** in 5–10 series, imbricate, ovate (outer) to linear-lanceolate (inner), margins entire, abaxial faces with narrow glutinous ridge; outer and middle appressed, spines spreading, 3–12 mm; apices of inner often flexuous, expanded and flat, scabrid-margined, sometimes erose, spineless. **Corollas** white or pale lavender to purple, pink, or red, 25–45 mm, tubes 8–25 mm, throats 6–17 mm, lobes 6–15 mm; style tips 2–8 mm. **Cypselae** light brown, sometimes with lighter or darker streaks, 6–9 mm, apical collars colored like the body, narrow; **pappi** (white or tawny), 20–40 mm, usually noticeably shorter than corolla. $2n = 30, 31, 32, 34$.

Varieties 2 (2 in the flora): w United States, n Mexico.

1. Stems densely leafy, nodes crowded; leaves often long-decurrent; corollas white or pale lavender to purple . . . . 24a. *Cirsium ochrocentrum* var. *ochrocentrum*
1. Stems leafy, nodes usually well separated; distal cauline leaves clasping, or if decurrent spiny wing usually less than 1 cm; corollas red, pink, or reddish purple (rarely white) . . . . 24b. *Cirsium ochrocentrum* var. *martinii*

24a. Cirsium ochrocentrum A. Gray var. **ochrocentrum** • Yellowspine thistle E

Stems densely leafy, nodes crowded. **Leaves:** distal cauline bases often long-decurrent. **Corollas** white or pale lavender to purple. $2n = 32, 34$.

Flowering spring–summer (May–Sep). Short-grass prairies, desert grasslands, sagebrush steppes, pinyon-juniper, mesquite woodlands, often in disturbed areas; 400–2200 m; Ariz., Calif., Colo., Kans., Nebr., N.Mex., Okla., S.Dak., Tex., Utah, Wyo.

Variety *ochrocentrum* grows in the Great Plains from South Dakota to New Mexico and Texas. It has been introduced in California; perhaps the Utah and Arizona records are introductions as well.

24b. Cirsium ochrocentrum A. Gray var. **martinii** (Barlow-Irick) D. J. Keil, Sida 21: 215. 2004 • Martin's thistle

Cirsium ochrocentrum A. Gray subsp. *martinii* Barlow-Irick, Novon 9: 320, fig. 2. 1999

Stems leafy, nodes usually well separated. **Leaves:** distal cauline bases clasping, or if decurrent, spiny wings usually less than 1 cm. **Corollas** red, pink, or reddish purple (rarely white). $2n = 30, 31, 32, 34$.

Flowering spring–autumn (May–Sep). Desert grasslands, arid shrublands, pine-, oak-, juniper-, or mesquite-dominated woodlands, often in disturbed areas, grassy slopes in montane pine forests; 1300–2200 m; Ariz., N.Mex.; Mexico (Chihuahua, Coahuila, Sonora, Durango).

Variety *martinii* occurs from east-central and southeastern Arizona to western New Mexico and is scattered southward into north-central Mexico.

25. Cirsium brevifolium Nuttall, Trans. Amer. Philos. Soc., n. s. 7: 421. 1841 (as Circium) • Palouse thistle [E]

Cirsium palousense (Piper) Piper

Perennials, 25–120 cm; taproots with horizontal root sprouts. **Stems** 1–several, erect, thinly gray-tomentose with fine, non-septate trichomes; branches 0–many, ascending. **Leaves** oblanceolate or elliptic, 15–45 × 2–10 cm, unlobed and merely spinulose to dentate or deeply pinnatifid, lobes well separated, linear to triangular-ovate, merely spinulose to few toothed or lobed near base, margins often revolute, main spines 2–3(–6) mm, abaxial faces densely gray-tomentose, adaxial green, thinly tomentose or ± glabrate; basal often present at flowering, narrowly winged-petiolate; principal cauline well distributed, gradually reduced distally, bases of proximal cauline winged-petiolate or sessile, bases of distal cauline expanded and ± clasping, margins sometimes spinier than those of proximal; distalmost cauline becoming bractlike, often unlobed or less deeply divided than proximal. **Heads** borne singly and terminal on main stems and branches or few from distal axils in corymbiform or paniculiform arrays. **Peduncles** 1–8 cm. **Involucres** hemispheric to campanulate, 2.5–3.5 × 2–4 cm, glabrous or loosely floccose. **Phyllaries** in 6–10 series, strongly imbricate, greenish to brown, ovate to lanceolate (outer) to linear (inner), abaxial faces with prominent glutinous ridge; outer and middle appressed, bodies entire, spines abruptly spreading, fine, 2–3(–5) mm; apices of inner commonly flexuous or reflexed, flat, scarious. **Corollas** creamy white, rarely lavender-tinged, 22–28 mm, tubes 8–13 mm, throats 7–11 mm, lobes 4–6 mm; style tips 5–6 mm. **Cypselae** brown, 5–6 mm, apical collars yellowish, 0.5–1 mm; **pappi** 18–22 mm. $2n = 22, 26$.

Flowering summer (Jun–Oct). Palouse prairie; 600–1300 m; Idaho, Oreg., Wash.

Cirsium brevifolium occurs in the Palouse prairie region of eastern Washington, eastern Oregon, and western Idaho.

26. Cirsium barnebyi S. L. Welsh & Neese, Brittonia 33: 296, fig. 3. 1981 • Barneby's thistle [C][E]

Perennials, 30–50 cm; caudices and woody taproots. **Stems** 1–few, erect, gray-tomentose or glabrate; branches few, above middle, ascending. **Leaves:** blades oblong-elliptic, 10–35 × 2–7 cm, strongly undulate, margins shallowly to deeply lobed, lobes 8–15 pairs, linear-lanceolate to broadly triangular, closely spaced, spreading, coarsely spinose-dentate or cleft into 2–5 spine-tipped divisions, main spines 3–5 mm, faces densely gray-white-tomentose; basal usually present at flowering, winged-petiolate; principal cauline becoming sessile and progressively reduced distally, bases decurrent as spiny wings to 5 cm; distal cauline usually much reduced, less lobed. **Heads** 1–20+, borne singly or clustered at branch tips, in leafy, ± corymbiform arrays. **Peduncles** 0–4 cm. **Involucres** ovoid to hemispheric or campanulate, 1.7–2 × 1.5–2 cm, loosely arachnoid on phyllary margins or glabrate. **Phyllaries** in 6–9 series, imbricate, ovate (outer) to linear-lanceolate (inner), entire, abaxial faces with narrow glutinous ridge; outer and middle appressed, spines ascending to spreading, stramineous, 2–7 mm; apices of inner often flexuous, narrow, flat, entire, spineless, glabrous. **Corollas** lavender to pink-purple, 18–28 mm, tubes 7–9 mm, throats 4–8 mm, lobes 5–11 mm; style tips 3.5–5 mm. **Cypselae** tan to brown, 5–5.5 mm, apical collars colored like body, narrow; **pappi** 15–23 mm.

Flowering summer (Jun–Sep). Dry juniper woodlands, sagebrush scrub, on shale, limestone, sandstone; of conservation concern; 1600–2600 m; Colo., Utah, Wyo.

Cirsium barnebyi occurs from the southern Rocky Mountains of southwestern Wyoming, northeastern Utah, and northwestern Colorado.

27. Cirsium grahamii A. Gray, Smithsonian Contr. Knowl. 5(6): 102. 1853 (as grahami) • Graham's thistle

Biennials, 50–100 cm; taproots slender and fascicles of thick fibrous roots. **Stems** 1, erect, thinly arachnoid and/or puberulent to short-pilose, sometimes ± glabrate; branches 0–4, ascending. **Leaves:** blades oblanceolate to oblong-elliptic, 20–30 × 3–8 cm, spinulose and otherwise entire or coarsely dentate to deeply pinnatifid, lobes entire or coarsely few toothed or lobed, main spines slender, 3–6 mm, abaxial ± persistently gray-tomentose, sometimes pilose along veins, adaxial faces thinly arachnoid and

± glabrate; basal often present at flowering, sessile or narrowly winged-petiolate; principal cauline gradually winged-petiolate or sessile, reduced distally, bases sometimes clasping or short-decurrent; distal cauline ascending, becoming bractlike, narrow, lobed or not. **Heads** 1–5. **Peduncles** 10–30 cm. **Involucres** hemispheric, 2–3 × 2–4 cm, thinly arachnoid or glabrous. **Phyllaries** in ca. 8 series, imbricate, proximally brownish, distally dark purplish, lanceolate to linear, margins of outer hispidulous-ciliolate, spiny fringed, pinnately spiny or with scarious appendages, abaxial faces with prominent, glutinous ridge; outer and middle appressed or only apices spreading, bodies minutely spinulose-denticulate, spines erect to ascending, 1.5–2.5 mm; apices of inner phyllaries often flexuous, flat, scabridulous. **Corollas** deep purple, 22–30 mm, tubes 13–18 mm, throats 4–5 mm, lobes 5–8 mm; style tips 4–4.5 mm. **Cypselae** tan with dark speckles to dark purplish brown, 4–5.5 mm, apical collars not differently colored; **pappi** 13–18 mm. $2n$ = 32 (Mexico).

Flowering Jul–Sep. Oak woodlands, coniferous forests, meadows, often in damp soil; 1400–2600 m; Ariz., N.Mex.; Mexico (Chihuahua, Durango, Sonora).

Cirsium grahamii occurs in the mountains of southeastern Arizona and southwestern New Mexico. It forms hybrid swarms with *C. parryi* and *C. scariosum* var. *coloradense* in the White Mountains of Arizona.

28. Cirsium wheeleri (A. Gray) Petrak, Bot. Tidsskr. 31: 67. 1911 • Wheeler's thistle

Cnicus wheeleri A. Gray, Proc. Amer. Acad. Arts 19: 56. 1883; *Cirsium blumeri* Petrak; *C. olivescens* (Rydberg) Petrak; *C. perennans* (Greene) Wooton & Standley; *C. wheeleri* var. *salinense* S. L. Welsh

Perennials, slender, 15–60 cm; taprooted with deep-seated root sprouts. **Stems** 1–few, erect, closely gray-tomentose; branches 0–few, ascending. **Leaves:** blades lanceolate to narrowly elliptic, 10–25 × 1–4 cm, unlobed and merely spinulose or pinnately lobed about halfway to midveins, often terminal lobes long-tapered, lobes short, lanceolate to triangular, entire to few toothed or lobed, well separated by wide, U-shaped sinuses, main spines slender, 2–5 mm, abaxial faces gray-tomentose, adaxial green, glabrous to thinly tomentose; basal usually present at flowering, winged-petiolate; principal cauline winged-petiolate proximally, mid and distal sessile, progressively reduced distally, bases not or scarcely decurrent, sometimes distal weakly clasping; distalmost often reduced to lanceolate or linear, long-acuminate bracts. **Heads** 1–6, borne singly or few at branch tips in corymbiform arrays. **Peduncles** 0–10 cm. **Involucres** hemispheric to subcylindric, 1.5–2.2 × 1.5–2.5 cm, thinly floccose-tomentose or glabrate. **Phyllaries** in 6–9 series, imbricate, pale green with darker apices, brownish when dry, lanceolate (outer) to linear-lanceolate (inner), margins of outer entire, abaxial faces with narrow glutinous ridge; outer and middle appressed proximally, apices spreading to ascending, bodies entire or rarely spinulose, spines slender, 3–7 mm; apices of inner often dark purple or blackish, flexuous, scarious, entire to pectinate-fringed, tapered or expanded. **Corollas** white or pink to pale purple, 20–28 mm, tubes 9–14 mm, throats 5–7.5 mm, lobes 5–10 mm; style tips 2.5–6 mm. **Cypselae** stramineous with brownish streaks, 6.5–7 mm, apical collars colored like body, narrow; **pappi** 15–20 mm. $2n$ = 28.

Flowering summer–fall (Jul–Oct). Coniferous forests, pine-oak, juniper-dominated woodlands, meadows; 1200–2900 m; Ariz., Colo., N.Mex., Tex., Utah; Mexico (Chihuahua, Sonora).

Cirsium wheeleri occurs from the mountains of the Colorado Plateau of central Utah and southwestern Colorado south through the highlands of Arizona and New Mexico to southwestern Texas and northwestern Mexico. The recently described *C. wheeleri* var. *salinense* is a minor variant with subentire leaves that is scattered through much of the range of the species.

29. Cirsium pulcherrimum (Rydberg) K. Schumann, Just's Bot. Jahresber. 29(1): 566. 1903 (as *pulcherrimus*) • Wyoming thistle E

Carduus pulcherrimus Rydberg, Bull. Torrey Bot. Club 28: 510. 1901

Perennials polycarpic, 15–60(–90) cm; deep-seated woody taproots and caudices. **Stems** 1–few, erect or ascending, arachnoid-tomentose or ± glabrate; branches 0–5+, usually in distal 1/2, ascending. **Leaves:** blades linear to oblong, oblanceolate, or elliptic, 5–25 × 0.6–7 cm, unlobed and merely spinulose or spiny-dentate to regularly pinnatifid, lobes 5–8 (–many) pairs, well separated, usually with broad, U-shaped sinuses to crowded, linear to triangular-ovate, ascending-spreading to retrorse, merely spinulose to coarsely dentate or few lobed, main spines 2–7 mm, ± slender, abaxial faces gray to white, usually densely arachnoid-tomentose, sometimes ± glabrate, sometimes villous with septate trichomes along veins, adaxial green, glabrous or less commonly thinly to densely gray-tomentose; basal often present at flowering, spiny winged-petiolate; principal cauline well distributed, gradually reduced distally, proximal usually winged-petiolate, mid and distal sessile, bases decurrent as spiny wings 1.5–3.5 cm; distalmost reduced, ± bractlike. **Heads**

1–few, borne singly or in 2–3-headed clusters in ± congested flat-topped or racemiform arrays at tips of main stem and branches, sometimes also in distal axils. **Peduncles** 0–15 cm. **Involucres** ovoid to campanulate, 1.8–2.7 × 1–2 cm, thinly arachnoid-tomentose or glabrate. **Phyllaries** in 6–7 series, ± imbricate, green or with dark subapical patch or appendage, linear to linear-lanceolate, margins entire, abaxial faces with narrow glutinous ridge; outer and middle bases appressed, apical appendages spreading to stiffly ascending, linear-lanceolate to acicular, entire, spines spreading or ascending, stout, 2–7 mm, often flattened; apices of inner stiffly erect or sometimes flexuous, narrow, flat. **Corollas** pink to purple (creamy white), 18–25 mm, tubes 7–9 mm, throats 5.5–7.5 mm, lobes 4–8 mm; style tips 3–5.5 mm. **Cypselae** tan to dark brown, 5–6 mm, apical collars yellow, narrow; **pappi** 14–16 mm. $2n = 34$.

Varieties 2 (2 in the flora): w, c United States.

Cirsium pulcherrimum is closely related to *C. clavatum*. In southeastern Wyoming and northern Colorado some plants combine foliage and involucral characters of *C. pulcherrimum* var. *pulcherrimum* and *C. clavatum* var. *americanum*. The inheritance of these characters needs to be examined at the population level to determine whether the intermediates are hybrids or the products of past introgression or incomplete differentiation.

1. Distal leaf face gray- to white-tomentose; cypselae without stramineous apical collar . . . .
. . . . 29a. *Cirsium pulcherrimum* var. *aridum*
1. Distal leaf face usually green, glabrous or ± glabrate, but sometimes persistently tomentose; cypselae often with narrow stramineous apical collar . . . . 29b. *Cirsium pulcherrimum* var. *pulcherrimum*

29a. Cirsium pulcherrimum (Rydberg) K. Schumann var. **aridum** (Dorn) D. J. Keil, Sida 21: 215. 2004 • Cedar Rim thistle [C] [E]

Cirsium aridum Dorn, Vasc. Pl. Wyoming ed. 2, 304, plate [p. 305]. 1992

Stems often clustered from branched rootstock, 15–40 cm, bases not thickened. **Adaxial leaf faces** gray- to white-tomentose. **Heads** 1–4, in compact arrays. **Cypselae** without stramineous apical collars.

Flowering summer (Jun–Aug). Barren slopes in shallow, stony soil in very open, arid grasslands; of conservation concern; 2000–2200 m; Wyo.

In describing variety *aridum*, Dorn suggested that it "seems to belong to the *C. hookerianum* group," but I fail to see this relationship. Its affinities clearly lie with *Cirsium pulcherrimum*. It is usually differentiable from var. *pulcherrimum* by the characters above, but it grows

and intergrades with var. *pulcherrimum* on the Sweetwater Plateau in Fremont County. W. Fertig (unpubl.) has studied the ecology and distribution of var. *aridum* [as *C. aridum*] and the very similar pubescent forms of var. *pulcherriumum* [as *C. pulcherrimum*].

Variety *aridum* occurs in scattered localities on barren hills in Carbon, Fremont, and Sweetwater counties. It is a rare taxon, but it has not received official recognition by governmental agencies. However it is in the Center for Plant Conservation's National Collection of Endangered Plants.

29b. Cirsium pulcherrimum (Rydberg) K. Schumann var. **pulcherrimum** • Wyoming thistle [E]

Stems from branched or unbranched rootstocks, 15–60(–90) cm, bases often thickened. **Adaxial leaf faces** glabrous to thinly gray-tomentose. **Heads** often 4+, in ± elongate arrays. **Cypselae** often with stramineous apical collars. $2n = 34$.

Flowering summer (Jun–Aug). Grasslands, sagebrush scrub, coniferous forest openings, roadsides, often in stony soil; 1100–2400 m; Colo., Idaho, Mont., Nebr., Utah, Wyo.

Variety *pulcherrimum* occurs from eastern Montana and northeastern Utah across Wyoming in the Wyoming Basin, Rocky Mountains, and western Great Plains to northern Colorado and northwestern Nebraska. It is likely that it occurs in southeastern Idaho as well. In the online Atlas of the Vascular Flora of Wyoming, several localities are plotted for var. *pulcherrimum* in Lincoln County, close to the Idaho state line. R. E. Brooks (1986) included Idaho in the range of *C. pulcherrimum*; I have not seen documentation for such occurrence. *Cirsium pulcherrimum* (i.e., var. *pulcherrimum*) is included in the list of Nebraska Plants of Special Concern.

Variety *pulcherrimum* is known to hybridize with *Cirsium eatonii* var. *murdockii* in Wyoming.

30. Cirsium clavatum (M. E. Jones) Petrak, Beih. Bot. Centralbl. 35(2): 310. 1917 [E]

Cnicus clavatus M. E. Jones, Proc. Calif. Acad. Sci., ser. 2, 5: 704. 1895

Biennials or monocarpic or polycarpic perennials, 20–100 cm; taproots sometimes with branched caudices. **Stems** 1–several, erect or ascending, glabrous or thinly arachnoid-tomentose; branches 0–10+, slender, usually arising in distal ¹/₂, ascending. **Leaves:** blades oblong to oblanceolate or elliptic, 5–40 × 3–11 cm, unlobed and merely spinulose-dentate or more commonly regularly deeply pinnatifid, lobes well separated to crowded, linear to triangular-ovate, ascending-spreading to retrorse, merely spinulose to coarsely dentate or proximally few-lobed, main spines 2–5(–7) mm, slender, abaxial faces green to gray, glabrous or thinly to densely arachnoid-tomentose, sometimes glabrate, often villous with septate trichomes along veins, adaxial green, glabrous; basal usually present at flowering, sessile or petiolate; principal cauline well distributed, proximal usually winged-petiolate, mid sessile, decurrent as spiny wings 1–3 cm; distal cauline ± reduced. **Heads** few–many, borne singly or clustered in corymbiform, paniculiform, or racemiform arrays at tips of main stem and branches, sometimes also in distal axils not closely subtended by clustered leafy bracts. **Peduncles** 0–30 cm. **Involucres** ovoid to campanulate, 1.5–3 × 1–3 cm, glabrous to thinly arachnoid-tomentose and/or villous-ciliate, with long septate trichomes connecting adjacent phyllaries. **Phyllaries** in 5–6 series, imbricate or subequal, outer green or with maroon to dark brown subapical patch or appendage, linear to ovate, abaxial faces with narrow glutinous ridge that may be concealed by trichomes; outer and middle with bases appressed, apical appendages erect or ascending, ovate to linear-lanceolate or acicular, entire or spinulose to broadly expanded, scarious, and erose-dentate, apical appendages, spines erect or ascending, 1–5 mm, ± flattened; apices of inner sometimes flexuous or reflexed, narrow, flat, entire or ± expanded, scarious and lacerate-dentate. **Corollas** creamy white to pale pinkish, 16–20 mm, tubes 6.5–9 mm, throats 4–7.5 mm, lobes 4–6 mm; style tips 3.5–5 mm. **Cypselae** tan to dark brown, 5–6 mm, apical collars not or scarcely differentiated; **pappi** 14–16 mm.

Varieties 3 (3 in the flora): central Rocky Mountains.

Cirsium clavatum is a polymorphic and variable species.

1. Involucres densely villous or tomentose with long, septate trichomes connecting adjacent phyllaries .... 30c. *Cirsium clavatum* var. *osterhoutii*
1. Involucres glabrous or thinly arachnoid-tomentose with fine, non-septate trichomes.
   2. Some or all of the phyllaries usually with dilated, scarious, erose to fringed appendages; mostly Colorado and Wyoming .... 30b. *Cirsium clavatum* var. *americanum*
   2. Phyllaries usually entire; w Colorado and Utah .... 30a. *Cirsium clavatum* var. *clavatum*

30a. Cirsium clavatum (M. E. Jones) Petrak var. **clavatum** • Fish Lake thistle [E]

Cirsium clavatum var. *markaguntense* S. L. Welsh

Abaxial leaf faces glabrous to ± tomentose. **Involucres** ovoid to campanulate, 1.5–2 × 1–2 cm, glabrous or thinly arachnoid-tomentose with fine, non-septate trichomes. **Phyllaries** in 5–6 series, strongly imbricate, green or with dark subapical patch, lanceolate to ovate, abaxial faces with narrow glutinous ridge; outer and middle bases appressed, apical appendages erect or ascending, linear-lanceolate, entire, usually not scarious or fringed, acicular-acuminate, spines erect or ascending, 3–6 mm; apices of inner sometimes flexuous or reflexed, narrow, flat, and entire or minutely toothed to slightly expanded and erose.

Flowering summer (Jun–Sep). Sagebrush scrub, aspen groves, meadows, openings in montane coniferous forests; 2100–3400 m; Colo., Utah.

Cirsium clavatum grows from the Colorado Plateau of central Utah eastward into the Rocky Mountains of western Colorado. R. J. Moore and C. Frankton (1965) suggested that *C. clavatum* (i.e., var. *clavatum* here) is a derivative of hybridization between *C. eatonii* and *C. centaureae* (i.e., *C. clavatum* var. *americanum* here). S. L. Welsh (1983) noted that the distribution of var. *clavatum* is distinct from those of var. *americanum* and *C. eatonii*. It is certainly possible that ancient hybridization may have contributed to the origin of var. *clavatum*, but there is no indication that the modern plants over most of its range are hybrids. The morphology of certain plants from southeastern Utah indicates a possibility of hybridization between *C. clavatum* and *C. eatonii* var. *eatonii*. The close relationship between var. *clavatum* and var. *americanum* is evident in the many overlapping vegetative features. The recently described *C. clavatum* var. *markaguntense* S. L. Welsh is a minor variant with subentire glabrous leaves.

30b. Cirsium clavatum (M. E. Jones) Petrak var. **americanum** (A. Gray) D. J. Keil, Sida 21: 211. 2004 • Rocky Mountain fringed thistle [E]

Cnicus carlinoides Schrank var. *americanus* A. Gray, Proc. Amer. Acad. Arts 10: 48. 1874; *Cirsium centaureae* (Rydberg) K. Schumann; *C. griseum* (Rydberg) K. Schumann; *C. laterifolium* (Osterhout) Petrak; *C. modestum* (Osterhout) Cockerell; *C. scapanolepis* Petrak; *C. spathulifolium* Rydberg

Abaxial leaf faces ± tomentose. **Involucres** ovoid to campanulate, 1.5–3 × 1.5–3 cm, glabrous or thinly arachnoid-tomentose with fine, non-septate trichomes. **Phyllaries** in 5–6 series, strongly imbricate to subequal, green or with maroon to dark brown subapical patch or appendage, linear to ovate, abaxial faces with narrow glutinous ridge; outer and middle bases appressed, apical appendages erect or ascending, linear-lanceolate to obovate, entire or more commonly dilated, scarious, and erose-toothed to fringed, acicular-acuminate to broadly obtuse, spines erect or ascending, 1–6 mm, often flattened; apices of inner sometimes flexuous or reflexed, narrow, flat, and entire or more commonly ± expanded, scarious, and lacerate-dentate. $2n = 34, 36$ (as *C. centaureae*).

Flowering summer (Jun–Sep). Oak scrub, sagebrush scrub, grasslands, juniper-pine woodlands, aspen groves, openings in montane coniferous forests; 2100–3100 m; Colo., Utah, Wyo.

Variety *americanum* is scattered in the Rocky Mountains of south-central Wyoming and Colorado. The varietal epithet *americanum* was based upon syntypes from both Colorado and California. The latter are referable to *Cirsium remotifolium*.

Variations in leaf characteristics, head size, and phyllary features combined with a narrow species concept led Osterhout and Rydberg to propose several species and hybrids for what I am treating as one taxon. Variety *americanum* certainly is polymorphic and in need of further study focused on local and regional variation. It is possible that some of the local variants in Colorado may deserve taxonomic recognition, but I have been unable to detect consistent patterns in the variation. Some specimens from Route and Jackson counties there closely approach var. *clavatum*. Others from southeastern Wyoming approach *Cirsium pulcherrimum* var. *pulcherrimum* and may be hybrids or derivatives of past hybridization.

30c. Cirsium clavatum (M. E. Jones) Petrak var. **osterhoutii** (Rydberg) D. J. Keil, Sida 21: 212. 2004 • Osterhout's thistle [E]

Carduus osterhoutii Rydberg, Bull. Torrey Bot. Club 32: 131. 1905; *Cirsium osterhoutii* (Rydberg) Petrak; *C. araneans* Rydberg

Abaxial leaf faces ± tomentose. **Involucres** ovoid to campanulate, 2–2.8 × 1.5–3 cm, villous-ciliate with long septate trichomes connecting adjacent phyllaries and/or thinly arachnoid-tomentose. **Phyllaries** in 5–6 series, imbricate or subequal, green or with maroon to dark brown subapical patch or appendage, linear to linear-lanceolate, abaxial faces with narrow glutinous ridge that is often concealed by trichomes; outer and middle bases appressed, apical appendages erect or stiffly ascending, linear-lanceolate to acicular, entire or spinulose-dentate, spines erect or ascending, 2–5 mm; apices of inner sometimes flexuous or reflexed, narrow, flat, and entire or expanded, scarious, and lacerate-dentate.

Flowering summer (Jul–Aug). Openings in montane coniferous forests, subalpine, alpine; 3000–3600 m; Colo.

The densely pubescent involucre of var. *osterhoutii* is a feature that links those plants with *Cirsium eatonii* var. *eriocephalum*. Some specimens tentatively assigned to *C. clavatum* var. *osterhoutiii* may be derivatives of past introgression with *C. eatonii* var. *eriocephalum*. Congested heads and strongly undulate leaves with numerous closely spaced lobes are features that suggest such a relationship. Further studies at both population and regional levels are needed to establish the nature of the variation patterns. Where I have observed var. *osterhoutii* in the field, populations appeared to be morpologically stable, and there was no evidence of current introgression with *C. eatonii*.

31. Cirsium perplexans (Rydberg) Petrak, Beih. Bot. Centralbl. 35(2): 441. 1917 • Adobe Hills thistle [C][E]

Carduus perplexans Rydberg, Bull. Torrey Bot. Club 32: 132. 1905; *Cirsium vernale* (Osterhout) Cockerell

Biennials, slender, 20–100 cm; taprooted. **Stems** usually 1, erect, thinly arachnoid-tomentose, sparsely pilose distally with short, jointed trichomes; branches few to many, often arising from proximal nodes, ascending. **Leaves:** blades oblong to elliptic, 15–30 × 2–6 cm, often unlobed and merely spinulose or spiny-dentate, sometimes pinnatifid ca. halfway to midveins, lobes separated by broad sinuses, undivided to coarsely few-dentate, main spines slender, 2–5(–10) mm, abaxial faces ± persistently thinly gray-tomentose, adaxial green, glabrous to thinly tomentose, sometimes sparsely pilose on midveins; basal sometimes present at flowering, sessile or short winged-petiolate; principal cauline sessile, progressively reduced, becoming bractlike distally, mid and distal bases broadly clasping; distal reduced to linear or lanceolate bracts. **Heads** few–many, in ± openly branched corymbiform or paniculiform arrays; not closely subtended by clustered leafy bracts. **Peduncles** (0–)3–20 cm. **Involucres** hemispheric to subspheric, 1.3–2.5 × (1–)1.5–2.5 cm, glabrous to loosely floccose. **Phyllaries** in 5–8(–10) series, strongly imbricate, green with darker green to brown subapical patch, broadly ovate or oblong (outer) to lanceolate (inner), abaxial faces with prominent to obscure glutinous ridge; outer and middle appressed, spines or terminal appendages spreading to reflexed, bodies entire or with expanded, ± scarious, ± pectinately fringed terminal appendages, tips merely mucronate or with weak spines spreading to reflexed, 1–3 mm; apices of inner often flexuous, flat, scarious, serrulate to expanded and pectinately fringed. **Corollas** lavender to reddish purple, (16–)19–22 mm, tubes 6–9 mm, throats 5–8 mm, lobes 5–7 mm, style tips 5–6 mm. **Cypselae** dark brown, 4–5 mm, apical collars stramineous or not differently colored, very narrow; **pappi** 15–17 mm.

Flowering spring–summer (May–Aug). Barren shale hillsides, gypsiferous clay soils, open, nearly unvegetated sites in areas of pinyon-juniper woodlands, sagebrush scrub, saltbush scrub, or Gambel oak brush, roadsides; 1800–2100 m; Colo.

Cirsium perplexans occurs in a few scattered sites at relatively low elevations in the Rocky Mountains of west-central Colorado. In view of this restricted distribution, the common name used by governmental agencies, Rocky Mountain thistle, is misleading; one would expect a species so named to be widely distributed in the Rocky Mountains. The name Adobe Hills thistle is descriptive of the habitat.

32. **Cirsium remotifolium** (Hooker) de Candolle in A. P. de Candolle and A. L. P. P. de Candolle, Prodr. 6: 655. 1838 • Remote-leaved or fewleaf thistle

E F

Carduus remotifolius Hooker, Fl. Bor.-Amer. 1: 302. 1833

Perennials, 20–150 cm, monocarpic; taprooted or polycarpic, perennating by runner roots. **Stems** usually 1, erect, finely arachnoid-tomentose, sometimes villous with septate trichomes below nodes; branches 0–10+, slender, usually arising in distal ¹/₂, ascending. **Leaves:** blades linear-oblong to oblanceolate or elliptic, 7–30 × 1–15 cm, unlobed and spinulose to dentate or shallowly to deeply pinnatifid, lobes well separated, linear to triangular-ovate, dentate to deeply lobed, main spines 2–5 mm, slender, abaxial faces green to gray, thinly to densely arachnoid-tomentose, sometimes glabrate, sometimes villous with septate trichomes along veins, adaxial green, glabrous; basal sometimes present at flowering, sessile or winged-petiolate; principal cauline mostly in proximal ¹/₂, winged-petiolate or sessile, bases narrowed, sometimes auriculate; distal well separated, progressively reduced, becoming bractlike, often unlobed or less deeply divided than the proximal, sometimes spinier than proximal, bases often distally expanded and auriculate-clasping. **Heads** few–many, borne singly or in openly branched in corymbiform, racemiform, or paniculiform arrays on main stem and branches, sometimes also in distal axils, not closely subtended by clustered leaf bracts. **Peduncles** (0–)2–15 cm. **Involucres** ovoid to hemispheric or campanulate, 1.5–2.5 × 1.5–3.5 cm, glabrous to arachnoid-floccose. **Phyllaries** in 6–8 series, subequal to strongly imbricate, green, linear to obovate (outer) to linear (inner), abaxial faces with inconspicuous glutinous ridge; outer and middle bases appressed, margins entire to spinulose-dentate or broad, scarious, lacerate-dentate, spines absent or ascending to spreading, 1–2 mm; apices of inner sometimes flexuous or reflexed, narrow, flat, entire or expanded, scarious, and lacerate-dentate. **Corollas** creamy white to purple, 18–28 mm, tubes 7–12 mm, throats 5–12 mm, lobes 3.5–7 mm, style tips 4–6 mm. **Cypselae** tan to dark brown, 4.5–5.5 mm, apical collars differentiated or not; **pappi** 13–23 mm. $2n = 32$.

Varieties 3 (3 in the flora): w United States

Cirsium remotifolium occurs from the Coast Ranges and valleys of the Pacific Northwest to the western slopes of the Cascade and Klamath ranges, south in the California North Coast Ranges to the San Francisco Bay region. It is closely related to the *C. clavatum* complex of the Rocky Mountains region. Both have a similar growth habit and some forms variably express the character of broadly scarious, lacerate-toothed phyllary margins. Gray, in naming *Cnicus carlinoides* var. *americanus,* included as syntypes both California and Colorado specimens. F. Petrak (1917) treated both the West Coast plants and those of the Rocky Mountains as *Cirsium* subsect. *Americana,* recognizing *C. remotifolium* with several infraspecific taxa plus two other species, *C. callilepis* and *C. amblylepis* from the West Coast, and four additional species from the Rocky Mountains. A. Cronquist (1955) rejected Petrak's subspecies, treating *C. remotifolium* in a restricted sense, limited to plants of Washington and Oregon without dilated phyllary tips, and circumscribed *C. centaureae* broadly to include the Rocky Mountains and West Coast plants with dilated phyllary tips. Because of the frequent presence of dilated phyllary tips in *C. remotifolium* in the restricted sense, Cronquist acknowledged the likelihood of past introgression with *C. centaureae* in the broad sense.

J. T. Howell (1960b) recognized three species: *Cirsium remotifolium, C. acanthodontum,* and *C. callilepis,* the latter with four varieties collectively corresponding to the West Coast representatives of *C. centaureae* (in the sense of Cronquist). Because of the great similarity of the various West Coast plants and their intergradation, I see no value in recognizing two or more species.

The West Coast and Rocky Mountains plants are clearly related, but are separated by the Great Basin region and there is little chance of current genetic interchange. As is often the case with American *Cirsium,* genetic enrichment from past hybridization with other nearby species within their respective areas has likely been fertile ground for evolutionary diversification. Different species have contributed genes in the Pacific states and in the Rockies. I have chosen to recognize two geographically-based species complexes, each with intergrading races here treated as varieties. I treat the West Coast plants as *C. remotifolium* and the Rocky Mountains plants as *C. clavatum.*

1. Phyllary margins ciliate with tiny spreading to recurved spines. . . . .
 . . . . 32c. *Cirsium remotifolium* var. *rivulare*
1. Phyllary margins unappendaged or dilated, scarious, and ± lacerate-toothed.
 2. Phyllaries narrowly oblong or linear, often ± subequal, all or most without scarious-dilated margins . . . .
 . . . . . . 32a. *Cirsium remotifolium* var. *remotifolium*
 2. Phyllaries oblong to obovate, often strongly graduated, most or all with dilated, scarious, erose to lacerate-dentate margins . . . .
 . . . . 32b. *Cirsium remotifolium* var. *odontolepis*

32a. Cirsium remotifolium (Hooker) de Candolle var. **remotifolium** • Remote-leaved thistle [E]

Phyllaries imbricate or often subequal, narrowly oblong or linear, entire or sometimes weakly expanded and scarious. **Corollas** cream-colored, 20–28 mm, tubes 6–12 mm, throats 8–10 mm, lobes 4.5–6.5 mm.

Flowering late spring–summer (May–Aug). Fields, meadows, forest openings, open woods, brushy slopes; 40–1400 m; Calif., Oreg., Wash.

Variety *remotifolium* occurs primarily west of the Cascade Range in Washington and Oregon and on coastal-facing slopes in northwestern California. Intermediates with var. *odontolepis* are known through much of that range.

32b. Cirsium remotifolium (Hooker) de Candolle var. **odontolepis** Petrak, Beih. Bot. Centralbl. 35(2): 298. 1917 • Pacific fringed or fringe-scaled thistle [E] [F]

Cirsium amblylepis Petrak; *C. americanum* (A. Gray) K. Schuman var. *callilepis* (Greene) Jepson; *C. callilepis* (Greene) Jepson; *C. callilepis* var. *oregonense* (Petrak) J. T. Howell; *C. callilepis* var. *pseudocarlinoides* (Petrak) J. T. Howell; *C. remotifolium* subsp. *oregonense* Petrak; *C. remotifolium* subsp. *pseudocarlinoides* Petrak

Phyllaries usually strongly linear to oblong or obovate, margins and apices expanded and scarious, erose to ± lacerate-toothed. **Corollas** cream-colored or purple, 18–25 mm, tubes 7.5–9 mm, throats 6.5–10.5 mm, lobes 3.5–6.5 mm.

Flowering summer (Jun–Sep). Grasslands, meadows, stream banks, brushy slopes, open coniferous or mixed conifer-hardwood forests; 25–2000 m; Calif., Oreg.

Variety *odontolepis* occurs from the Coast Ranges and valleys of Oregon and the western Cascade Range south through the Siskiyou Area and the California North Coast Ranges to the San Francisco Bay region. California populations of var. *odontolepis* usually have cream-colored corollas. Some Oregon populations have cream-colored corollas, and others, particularly in the Cascade Range, have purple corollas. The latter have been called subsp. *oregonense.* In other features these plants are very similar and have overlapping patterns of variation. Further study is needed to determine whether the purple-flowered plants are worthy of taxonomic recognition.

There is much variation in the features of the phyllaries of var. *odontolepis*. Four names, *Cirsium callilepis*, *C. remotifolium* var. *odontolepis*, *C. remotifolium* subsp. *pseudocarlinoides*, and *C. amblylepis* all have as their type location Mt. Tamalpais in Marin County, California, near the southern limit of the distribution of *C. remotifolium*. The type of var. *odontolepis* has phyllary tips recurved or spreading. Other populations have narrow to broad phyllaries with tips erect to spreading, and variously erose to laciniate-toothed. Phyllary pubescence varies as well. Plants with phyllaries intermediate between those of var. *odontolepis* and var. *remotifolium*, and between those of var. *odontolepis* and var. *rivulare* are known.

Interspecific hybridization is suspected to have contributed to the diversity of forms of var. *odontolepis*. J. T. Howell (1949) speculated that the form named by Petrak as *Cirsium amblylepis* from Mt. Tamalpais may have originated through hybridization with *C. quercetorum*. Some specimens from central Oregon may be derivatives from hybridization with *C. edule*.

32c. Cirsium remotifolium (Hooker) de Candolle var. **rivulare** Jepson, Man. Fl. Pl. Calif., 1164. 1925 • Klamath thistle E

Cirsium acanthodontum S. F. Blake; *C. oreganum* Piper

Phyllaries imbricate or subequal, linear, margins and/or apices regularly spinulose-serrulate, sometimes weakly expanded and scarious. **Corollas** usually purple (infrequently cream-colored), 20–25 mm, tubes 8–12 mm, throats 5–6 mm, lobes 6–8 mm. $2n = 32$.

Flowering late spring–summer (May–Aug). Sea bluffs, river valleys, meadows, grasslands, open coniferous or mixed coniferous-hardwood forests; 10–1200 m; Calif., Oreg.

Plants of var. *rivulare* occur in the western Klamath Mountains of southwestern Oregon and northwestern California and the adjacent coastal plain. Although these plants are usually slender and erect with well separated leaves, they assume a compact growth form in maritime habitats.

33. Cirsium wrightii A. Gray, Smithsonian Contr. Knowl. 5(6): 101. 1853 • Wright's marsh thistle C

Biennials or monocarpic perennials, 100–300 cm; taproots short with many slender, fibrous lateral roots. **Stems** usually 1, erect, glabrous to ± tomentose; branches many, usually restricted to distal part of stem, ascending. **Leaves:** blades oblong to elliptic, 10–60 × 5–20 cm, unlobed and merely spinulose to irregularly dentate or shallowly to deeply pinnatifid, lobes ± broadly triangular, separated by wide sinuses, obtuse to acute, sometimes coarsely toothed or lobed, main spines slender, 1–3 mm, faces thinly arachnoid, soon glabrescent; basal often present at flowering, petioles slender, ± winged; cauline progressively reduced, proximal petiolate, mid and distal sessile, long-decurrent; distalmost linear to narrowly elliptic, bractlike, spinulose to irregularly dentate or shallowly lobed. **Heads** many, in openly paniculiform arrays, borne singly at tips of peduncles. **Peduncles** slender, 1–15 cm. **Involucres** ovoid to hemispheric, 1–2 × 1–2 cm, thinly arachnoid, glabrate. **Phyllaries** in 8–9 series, strongly imbricate, green, lanceolate (outer) to linear (inner), abaxial faces with prominent glutinous ridge; outer and middle appressed, bodies entire, apices acute, spines spreading, slender, ca. 1 mm; apices of inner often flexuous, acuminate, flat, scabrid-ciliolate. **Corollas** white to pink-purple, 19–21 mm, tubes 9–10 mm, throats 4–4.5 mm, lobes 5–7 mm; style tips 2–3.5 mm. **Cypselae** brown, ca. 4.5 mm, apical collars stramineous, 0.2 mm; **pappi** 15–16 mm.

Flowering summer–fall (Aug–Oct). Springs, seeps, marshes, stream banks, often in alkaline soil; of conservation concern; 1100–2600 m; Ariz., N.Mex., Tex.; Mexico (Chihuahua, Sonora).

Wright's thistle occurs from the mountains of south-central New Mexico eastward to the cienegas of the adjacent southwestern Great Plains. *Cirsium wrightii* is listed by the state of New Mexico as a species of concern. The one known site in Cochise County, Arizona, is apparently historic.

Hybrids are known between *Cirsium wrightii* and *C. vinaceum* in the Sacramento Mountains of New Mexico. I have observed hummingbird visits to the heads of both species, though *C. wrightii* shows none of the apparent adaptations to hummingbirds (P. L. Barlow-Irick 2002) that are seen in such taxa as *C. occidentale* var. *candidissimum*, *C. andersonii*, and *C. arizonicum*.

34. Cirsium crassicaule (Greene) Jepson, Fl. W. Calif., 506. 1901 • Slough thistle C E

Carduus crassicaulis Greene, Proc. Acad. Nat. Sci. Philadelphia 44: 357. 1893

Annuals or biennials, (60–)100–300 cm; taprooted. **Stems** usually 1, erect, stout, (hollow, 2–10 cm diam. at base), openly branched distally, thinly arachnoid, villous with jointed trichomes, at least proximally. **Leaves:** blades elliptic to broadly oblanceolate, 15–70 × 30–150+ cm, flat, pinnatifid ¹/₂–²/₃ distance to midvein, larger usually with broad sinuses, lobes broad, few lobed or dentate, main spines 3–8 mm, abaxial faces gray-tomentose, adaxial thinly arachnoid-tomentose, sometimes midveins with jointed trichomes; basal present or withered at flowering, winged-petiolate; principal cauline sessile, progressively reduced distally, bases clasping or short-decurrent 1–2 cm; distal cauline reduced, becoming bractlike, sometimes spinier than proximal. **Heads** 1–several at branch tips, closely subtended by clustered leafy bracts or not, collectively forming open, corymbiform or paniculiform arrays. **Peduncles** 0–15 cm. **Involucres** ovoid to campanulate, 1.5–3 × 1.5–3 cm, ± glabrous. **Phyllaries** in 5–7 series, weakly unequal, dark green to brownish, lanceolate (outer) to linear (inner), abaxial faces without (or with very obscure) glutinous ridge; outer and middle appressed or apices spreading, at least outer irregularly spiny-fringed, finely serrulate, spines slender, 3–5 mm; apices of inner erect, abaxial faces gray-tomentose, ± twisted. **Corollas** pale rose-purple (white), 19–26 mm, tubes 9–12 mm, throats 4–6 mm, lobes 5–9 mm; style tips 3.5–4.5 mm. **Cypselae** dark brown, 5–5.5 mm, collars narrow, ± stramineous; **pappi** 15–20 mm. $2n = 32$.

Flowering spring (Apr–Jun). Freshwater marshes, canal banks; 5–100 m; Calif.

Cirsium crassicaule is known only from a few sites in the San Joaquin Valley. Some populations are threatened by habitat modification and development pressures.

35. Cirsium hydrophilum (Greene) Jepson, Fl. W. Calif., 507. 1901 C E

Carduus hydrophilus Greene, Proc. Acad. Nat. Sci. Philadelphia 44: 358. 1893; *Cirsium vaseyi* (A. Gray) Jepson var. *hydrophilum* (Greene) Petrak

Biennials or monocarpic perennials, 100–220 cm; taprooted. **Stems** 1–several, erect, (hollow), openly branched distally or throughout, thinly arachnoid with fine, non-septate trichomes, glabrate. **Leaves:** blades elliptic to broadly oblanceolate, 10–40+ cm, pinnatifid ¹/₂–²/₃ distance to midveins, larger usually with broad sinuses, lobes broad, few lobed or dentate, main spines 2–9 mm, abaxial faces ± gray-tomentose, sometimes ± glabrate, adaxial thinly arachnoid-tomentose, soon glabrescent; basal present or withered at flowering, winged-petiolate; principal cauline sessile, progressively reduced distally, bases auriculate-clasping or shortly decurrent; distal cauline reduced, bractlike, often spinier than proximal. **Heads** borne singly or few at branch tips, sometimes subtended by clustered, ± leafy bracts, collectively forming ± open, many-headed paniculiform arrays. **Peduncles** 0–10+ cm. **Involucres** ovoid to campanulate, 1.5–2.5 × 1.5–3 cm, thinly arachnoid, glabrate. **Phyllaries** in 6–9 series, imbricate, dark green to brownish, lanceolate (outer) to linear (inner), abaxial faces with narrow glutinous ridge; outer and middle appressed, apices spreading, finely serrulate, spines slender, 1–2 mm; apices of inner erect, ± flexuous. **Corollas** pale rose-purple, 18–23 mm, tubes 8–10 mm, throats 5–6 mm, lobes 5–7 mm; style tips 3.5–4.5 mm. **Cypselae** dark brown to black, 4–5 mm, collars very narrow, stramineous; **pappi** ca. 15 mm. $2n = 32$.

Varieties 2 (2 in the flora): California.

1. Heads usually 2.5–3 cm; cypselae ca. 5 mm, oblong; tidal marshes . . . .
. . . . 35a. *Cirsium hydrophilum* var. *hydrophilum*
1. Heads usually 3–3.5 cm; cypselae 4–5 mm, oblong or elliptic; serpentine springs . . . .
. . . . 35b. *Cirsium hydrophilum* var. *vaseyi*

35a. Cirsium hydrophilum (Greene) Jepson var. **hydrophilum** • Suisun thistle C E

Heads usually 2.5–3 cm. **Cypselae** oblong, ca. 5 mm. $2n = 32$.

Flowering spring–summer (May–Sep). Tidal marshes; of conservation concern; 0–5 m; Calif.

Suisun thistle is known only from the Suisun Marsh (Solano County) in the Sacramento River delta.

Variety *hydrophilum* is listed by the U.S. Fish and Wildlife Service as endangered.

35b. Cirsium hydrophilum (Greene) Jepson var. **vaseyi**
(A. Gray) J. T. Howell, Leafl. W. Bot. 9: 11. 1959
  • Mount Tamalpais or Vasey's thistle [C] [E]

Cnicus breweri A. Gray var. *vaseyi*
A. Gray in A. Gray et al., Syn. Fl.
N. Amer. 1(2): 404. 1884; *Cirsium
montigenum* Petrak; *C. vaseyi* (A.
Gray) Jepson

Heads usually 3–3.5 cm. **Cypselae**
oblong or elliptic, 4–5 mm. **2n** =
32.

Flowering spring–summer
(May–Jul). Spring-fed marshy meadows on serpentine
parent material in areas of chaparral and mixed
evergreen forest; of conservation concern; 300–500 m;
Calif.

Variety *vaseyi* is endemic to the slopes of Mt.
Tamalpais in the southern North Coast Range of Marin
County, California.

36. Cirsium douglasii de Candolle in A. P. de Candolle
and A. L. P. P. de Candolle, Prodr. 6: 643. 1838
  • California swamp or Douglas's thistle [E] [F]

**Biennials or short-lived mono-
carpic perennials, 60–250 cm;
taprooted. Stems** 1–several,
erect or ascending, densely gray-
tomentose; branches few–many,
ascending to spreading. **Leaves:**
blades oblong-elliptic to obovate,
10–60 × 2–15 cm, unlobed or
shallowly to deeply pinnatifid,
lobes lanceolate to ovate-triangular, ascending to spreading, entire to coarsely dentate or lobed, main spines
slender to stout, 2–30 mm, faces densely gray-
tomentose, rarely glabrate; basal present at flowering,
petiolate; principal cauline well distributed, proximal
winged-petiolate, distal sessile, bases auriculate-
clasping or decurrent as a spiny wing 1–3 cm; distalmost
well separated, bractlike. **Heads** 10–many, often
crowded at branch tips, collectively forming paniculiform
arrays. **Peduncles** 0–4(–8) cm. **Involucres** ovoid to hemispheric, 1.5–3 cm, 2–4.5 cm diam, loosely arachnoid on
phyllary margins or glabrate. **Phyllaries** in 6–8 series,
imbricate, often with dark purple patch near tip, ovate-
lanceolate (outer) to linear-lanceolate (inner), abaxial
faces with linear to elliptic glutinous ridge; outer and
middle appressed, entire, spines spreading, 1–9 mm;
apices of inner often purple-tinged, often flexuous, flattened, spineless, scabrid. **Corollas** rose-purple (white
or pinkish-tinged), 18–21 mm, tubes 8–9 mm, throats
5–6 mm (abruptly expanded), lobes 5–6 mm; style tips
3–4.5 mm. **Cypselae** dark brown to black, 5–6 mm,
apical collars not differentiated; **pappi** 15–20 mm.

Varieties 2 (2 in the flora): w United States.

1. Distal cauline leaves with larger spines mostly less
   than 7 mm . . . . 36b. *Cirsium douglasii* var. *breweri*
1. Distal cauline leaves with larger spines usually
   7–20 mm . . . . 36a. *Cirsium douglasii* var. *douglasii*

36a. Cirsium douglasii de Candolle var. **douglasii** [E]

Plants to 150 cm. **Cauline leaves**
usually deeply lobed, spines usually 7–20 mm. **Heads** mostly 2.5–
3.5 cm. **2n** = 34.

Flowering summer (Jun–Aug).
Springs, seeps, streamsides,
coastal bluffs, coniferous and
hardwood forests, often on serpentine; 10–1200 m; Calif.

Variety *douglasii* grows in the Coast Ranges from
Humboldt to Monterey and San Benito counties.

36b. Cirsium douglasii de Candolle var. **breweri**
(A. Gray) D. J. Keil & C. E. Turner, Phytologia 73:
313. 1992 • Brewer's thistle [E] [F]

Cnicus breweri A. Gray, Proc. Amer.
Acad. Arts 10: 43. 1874; *Cirsium
breweri* (A. Gray) Jepson; *C.
douglasii* var. *canescens* (Petrak) J.
T. Howell

Plants to 250 cm. **Cauline leaves**
usually unlobed or shallowly
lobed, spines usually less than
2–7 mm. **Heads** mostly 2–3 cm.
2n = 30.

Flowering summer (Jun–Aug). Wet soil, streams, fens,
marshes, springs in montane coniferous forest areas,
often on serpentine; 1200–1900 m; Calif., Nev., Oreg.

Variety *breweri* grows from the eastern North Coast
Ranges of California and the Klamath and Siskiyou
ranges east to the northen Sierra Nevada.

37. Cirsium mohavense (Greene) Petrak, Bot. Tidsskr. 31: 68. 1911 • Mojave thistle C E

Carduus mohavensis Greene, Proc. Acad. Nat. Sci. Philadelphia 44: 361. 1893; *Cirsium rusbyi* (Greene) Petrak; *C. virginense* S. L. Welsh

Biennials or perennials, 30–250 cm; taprooted. **Stems** 1–several, erect, proximally simple, distally branched, ± densely gray-tomentose; branches 0–many, ascending to spreading. **Leaves**: blades oblong-elliptic to oblanceolate, 10–60 × 2–15 cm, unlobed and merely spinulose or spiny-dentate or shallowly to deeply pinnatifid, lobes linear-lanceolate to ovate-triangular, spreading, entire to coarsely dentate, main spines slender to stout, 3–30 mm, faces ± gray-tomentose, sometimes ± glabrate; basal often present at flowering, winged-petiolate; principal cauline decreasing distally, proximal winged-petiolate, distal sessile, bases decurrent as spiny wings 1–5 cm; distalmost well separated, bractlike. **Heads** 1–many, in corymbiform or paniculiform arrays. **Peduncles** 0–15 cm. **Involucres** ovoid to hemispheric, 1.5–2.5 × 1.5–2 cm, loosely arachnoid on phyllary margins or glabrate. **Phyllaries** in 5–8 series, imbricate, (inner greenish to brown or stramineous), lanceolate or ovate (outer) to linear-lanceolate (inner), entire, abaxial faces with narrow glutinous ridge; outer and middle appressed, spines spreading, 3–7 mm; apices of inner often flexuous, flattened, spineless, scabrid. **Corollas** white to pink or lavender, 16–25 mm, tubes 7–12 mm, throats 4–7 mm, lobes 4–8 mm, style tips 3–4 mm. **Cypselae** stramineous to dark brown, 3–6 mm, apical collars 0.2–0.3 mm, yellowish; **pappi** 14–16 mm. $2n$ = 30, 32.

Flowering summer–fall (Jun–Oct). Wet soil, streams, springs, meadows in desert and desert woodland areas; of conservation concern; -50–2200 m; Ariz., Calif., Nev., Utah.

Cirsium mohavense ranges from scattered sites in eastern California east in the Basin and Range Province of southern Nevada to southwestern Utah and nortwestern Arizona, mostly in Mojave Desert region. When Welsh proposed *Cirsium virginense* for a geographically limited group of plants from southwestern Utah and northwestern Arizona (and subsequently discovered in extreme southeastern Nevada), he indicated that its relationship to other western thistles was unknown. Subsequently, he indicated (S. L. Welsh 1983; Welsh et al. 1993) that the affinities of the taxon apparently lie with *C. mohavense*, but he did not attempt to distinguish *C. virginense* from *C. mohavense* (in the strict sense) because the latter was not known to occur in Utah. A. Cronquist (1994) attempted the distinction. The only character he used in his key was life span of the plants: biennial (*C. mohavense*) versus perennial, spreading by

creeping roots (*C. virginense*). In the descriptions of the two taxa he elaborated on this character, indicating that *C. mohavense* is single-stemmed and *C. virginense* often multistemmed. In the remaining features the plants are very similar or overlap extensively.

Distinction of two taxa on the basis of duration is impractical and probably inaccurate. Specimens commonly lack roots, and in those specimens in which bases are present, I have seldom been able to make any distinction between biennial taproots and perennial taproots. In particular I have seen no evidence of creeping roots. I am not aware of any study of either taxon that documents the life history of the plants. Some specimens of *C. mohavense* (in the strict sense) appear to have perennial bases like those attributed to *C. virginense* by Cronquist. For instance, a specimen of *C. mohavense* from Death Valley (*Thorne & Ratcliff 2287*, BRY) is indistinguishable from specimens of *C. virginense* (e.g., *Atwood 13374*, BRY) from Nevada and Utah. Both have a branched root crown with multiple rosettes and nearly identical leaves and heads.

38. Cirsium inamoenum (Greene) D. J. Keil, Sida 21: 214. 2004 E

Carduus inamoenus Greene, Fl. Francisc., 479. 1897, based on *C. undulatus* Nuttall var. *nevadensis* Greene, Proc. Acad. Nat. Sci. Philadelphia 44: 361. 1893, not *C. nevadensis* Greene 1896, not *Cirsium nevadense* (Greene) Petrak 1917

Biennials or monocarpic **perennials**, 20–100 cm; deeply taprooted. **Stems** 1–several, erect, thinly to densely gray-tomentose with fine, non-septate trichomes; branches 0–many, ascending. **Leaves**: blades oblanceolate or elliptic, 10–35 × 1–7 cm, unlobed and spinulose to dentate or deeply pinnatifid, usually 5–8 pairs of lobes, well separated, linear to lance-triangular, spinulose to few toothed or lobed, main spines 2–7 mm, abaxial faces densely gray-tomentose or sometimes ± glabrate, adaxial gray to ± green, thinly tomentose or ± glabrate; basal sometimes present at flowering, narrowly winged-petiolate; principal cauline well distributed, gradually reduced, sometimes spinier than basal; proximal winged-petiolate, mid sessile, bases spiny-winged, decurrent 1–3 cm; distal becoming bractlike, often unlobed or less deeply divided than proximal. **Heads** 1–many, in open corymbiform arrays or crowded near stem tips. **Peduncles** 0–25 cm. **Involucres** ovoid or hemispheric to campanulate, 2–3 × 1.5–5 cm, glabrous or loosely floccose to densely arachnoid. **Phyllaries** in 6–10 series, strongly imbricate or sometimes subequal, greenish to brown, ovate to linear-lanceolate (outer) to linear (inner), entire, abaxial faces with narrow or scarcely developed glutinous ridge; outer and mid appressed or apices ascending to spreading, linear, bodies entire, spines ascending to abruptly spreading, usually fine, 2–6 mm; apices of inner narrow, spine-tipped or spineless. **Corollas** dull white or faintly lavender-tinged to bright pink-purple, 19–31 mm, tubes 7–13 mm, throats 6.5–9.5 mm, lobes 4–8 mm; style tips 3.5–7 mm. **Cypselae** brown, 5–8 mm, apical collars not differentiated; **pappi** 12–25 mm. $2n = 32, 34, 36$.

Varieties 2 (2 in the flora): w United States.

Cirsium inamoenum is a variable complex across the northern Great Basin and adjacent mountains. A. Cronquist (1994) treated this complex as a single species under the name *C. subniveum* without infraspecific taxa and including taxa that formerly had been assigned to *C. utahense* (e.g., J. T. Howell 1960b). Some populations consist of small-headed, white-flowered plants with strong involucres and short, appressed phyllaries. Others have larger heads, white or lavender to pink-purple corollas, and phyllaries with longer, ascending to spreading tips. My treatment of this complex as *C. inamoenum* is similarly inclusive as was Cronquist's treatment of *C. subniveum*, except that I believe *C. humboldtense*, which Cronquist included, is probably a derivative of hybridization between *C. subniveum* and *C. eatonii* var. *peckii*. I have observed such hybrids on the slopes of Steens Mountain in Harney County, Oregon, and the type of *C. humboldtense* closely resembles some of the introgressants. I have examined several other specimens that are likely the products of hybridization of *C. inamoenum* with other varieties of *C. eatonii*.

I have chosen to recognize racial differentiation within *Cirsium inamoenum* at the rank of variety. The main difference between *Cirsium inamoenum* var. *inamoenum* and var. *davisii* is corolla color. Unfortunately this feature is sometimes difficult to determine on herbarium specimens, and many collectors fail to include corolla color on specimen labels. Some geographic overlap occurs between var. *davisii*, which has a distribution centered in northeastern Utah, southeastern Idaho, and adjacent southwestern Wyoming, and the more widespread var. *inamoenum*.

Plants of northeastern Oregon, southeastern Washington, and adjacent western Idaho often have large heads and densely tomentose foliage. These were named *Cirsium wallowense* by Peck. Similar plants occur sporadically in other portions of the range of *Cirsium inamoenum* var. *inamoenum* and I chose not to recognize these northwestern populations as a third variety. Additional study might clarify the relationships of these plants.

Some specimens of *Cirsium inamoenum* in central Nevada and Utah approach *C. neomexicanum*. It seems likely that these species have interacted in the past.

1. Corollas white or pale lavender . . . .
. . . . 38a. *Cirsium inamoenum* var. *inamoenum*
1. Corollas lavender to rich pink-purple . . . .
. . . . 38b. *Cirsium inamoenum* var. *davisii*

38a. Cirsium inamoenum (Greene) D. J. Keil var. **inamoenum** • Greene's thistle [E]

Involucres 1.5–5 cm diam. **Corollas** white or pale lavender. $2n = 32, 34, 36$ (as *C. subniveum*).

Flowering summer (Jun–Aug). Arid slopes, roadsides, grasslands, sagebrush scrub, pinyon-juniper woodlands, montane coniferous forests; 750–2800 m; Calif., Idaho, Nev., Oreg., Utah, Wash., Wyo.

Variety *inamoenum* grows from the rainshadow slopes of the Sierra Nevada and Cascades eastward across the northern Basin and Range province to the mountains of southern Idaho, western Wyoming, and northeastern Utah. On the slopes of Steens Mountain in eastern Oregon it forms hybrids with *Cirsium eatonii* var. *peckii*. In the Snake Range it apparently hybridizes with *C. eatonii* var. *viperinum*.

38b. Cirsium inamoenum (Greene) D. J. Keil var. **davisii** (Cronquist) D. J. Keil, Sida 21: 214. 2004 • Davis's thistle [E]

Cirsium davisii Cronquist, Leafl. W. Bot. 6: 46. 1950

Involucres 1.5–3.5 cm diam. **Corollas** lavender to rich pink-purple.

Flowering summer (Jun–Aug). Arid slopes, roadsides, grasslands, sagebrush scrub, dry woodlands, montane forests; 1400–3100 m; Idaho, Nev., Utah, Wyo.

Variety *davisii* occurs in the northern Great Basin region and adjacent mountains from northern Nevada and southern Idaho to eastern Wyoming and northern Utah.

39. Cirsium cymosum (Greene) J. T. Howell, Amer. Midl. Naturalist 30: 37. 1943 [E]

Carduus cymosus Greene, Fl. Francisc., 480. 1897; *Cirsium botrys* Petrak; *C. triacanthum* Petrak

Biennials or perennials, 25–120 cm, pubescence a mixture of fine, non-septate arachnoid trichomes and coarser, septate trichomes, especially along stems and on midveins on abaxial leaf faces, usually ± loose and irregularly deciduous from leaves in age; taprooted. **Stems** usually 1, erect, ± gray-tomentose, sometimes villous with septate trichomes; branches 0–10+, usually arising in distal ¹/₂, ascending, usually reaching a ± common height. **Leaves:** blades linear-oblong to oblanceolate or elliptic, 10–30 × 3–7 cm, shallowly to deeply pinnatifid with 3–8 pairs of lobes, longer than 2 cm, lobes well separated, linear to triangular-ovate, dentate to lobed proximally, main spines slender, 2–7 mm, faces green to gray, thinly to densely arachnoid-tomentose with fine, non-septate trichomes, sometimes villous with septate trichomes along veins, usually ± loose and irregularly deciduous from leaves in age; basal often present at flowering, sessile or winged-petiolate; principal cauline mostly in proximal ¹/₂, winged-petiolate or sessile, bases narrowed, auriculate, veins often prominently raised on abaxial faces; distal sessile, auriculate-clasping or short-decurrent 1–10 mm, progressively reduced becoming bractlike, often unlobed or less deeply divided and sometimes spinier than proximal. **Heads** borne singly, terminal on main stem and branches, sometimes also in distal axils, erect, not subtended by well-developed leaves, collectively forming corymbiform or racemiform arrays. **Peduncles** (0–)2–15 cm. **Involucres** ovoid to hemispheric or campanulate, 2–3 × 1.5–3.5 cm, ± arachnoid-floccose, often glabrate. **Phyllaries** in 8–10 series, subequal to strongly imbricate, green, linear to lanceolate (outer) to linear (inner), entire, abaxial faces with inconspicuous to prominent glutinous ridge; outer and mid bodies loosely spreading to ascending or appressed, apices subappressed to ascending or spreading, flat, spines ascending to spreading, fine, 2–4 mm; apices of inner commonly flexuous or reflexed, narrow, flat, scarious. **Corollas** creamy white to purplish, 20–31 mm, tubes 8–14 mm, throats 5.5–10 mm, lobes 6–7 mm; style tips 4–6 mm. **Cypselae** tan to dark brown, 5–7.5 mm, apical collars not differentiated; **pappi** 16–25 mm.

Varieties 2 (2 in the flora): w United States.

Past floras have treated *Cirsium cymosum* and *C. canovirens* as separate species. In my examination of these plants across their combined ranges I realized that they are connected by numerous intermediates and that I could find no characters that consistently distinguish them.

1. Larger heads 20–35 mm diam.; outer phyllaries elongate, often nearly as long as inner; glutinous ridge narrow, weakly developed . . . . 39a. *Cirsium cymosum* var. *cymosum*
1. Larger heads 15–25 mm diam.; outer phyllaries usually much shorter than inner phyllaries; glutinous ridge prominent, well developed, appearing dark brown on dry specimens . . . . 39b. *Cirsium cymosum* var. *canovirens*

39a. Cirsium cymosum (Greene) J. T. Howell var. **cymosum** • Peregrine thistle [E]

Larger heads 20–35 mm diam. **Outer phyllaries** elongate, often nearly as long as inner; glutinous ridge narrow, weakly developed. **2***n* = 30, 34.

Flowering spring–summer (Apr–Aug). Grassy areas, sagebrush steppe, California woodlands, open coniferous or conifer-hardwood forests, roadsides; 200–2000 m; Calif., Nev., Oreg.

Variety *cymosum* ranges from the Siskiyou Range south through the California North Coast Range to Mount Diablo in the northern South Coast Range, and east into the southern Cascades and the northwestern Great Basin.

39b. Cirsium cymosum (Greene) J. T. Howell var. **canovirens** (Rydberg) D. J. Keil, Sida 21: 212. 2004 • Graygreen thistle [E]

Carduus canovirens Rydberg, Mem. New York Bot. Gard. 1: 450. 1900; *Cirsium canovirens* (Rydberg) Petrak

Larger heads 15–25 mm diam. **Outer phyllaries** usually much shorter than inner; glutinous ridge prominent, well developed, appearing dark brown on dry specimens. **2***n* = 34 (as *C. canovirens*).

Flowering spring–summer (Apr–Aug). Grasslands, sagebrush steppe, pinyon-juniper woodlands, dry coniferous forests, roadsides; 600–2600 m; Calif., Idaho, Mont., Nev., Oreg., Wyo.

Variety *canovirens* occurs from the dry mountains and valleys of eastern Oregon and the rain shadow slopes of the northern Sierra Nevada eastward across the northern Great Basin to Idaho, southern Montana, and western Wyoming. D. J. Keil and C. E. Turner (1993) recognized a polymorphic *Cirsium canovirens* that included *C. subniveum* (here treated as *C. inamoenum*). My subsequent investigations indicate that the merger of those taxa was erroneous, based in part on misidentified specimens.

40. Cirsium occidentale (Nuttall) Jepson, Fl. W. Calif.,
509. 1901 • Western thistle E F

Carduus occidentalis Nuttall, Trans.
Amer. Philos. Soc., n. s. 7: 418.
1841

Biennials, 5–400 cm; taproots.
Stems usually 1, thinly to densely
gray- or white-tomentose, sometimes ± glabrate; branches few–
many, usually from above mid or
near base in compact, moundlike
dwarf plants, ascending to spreading. **Leaves:** blades
oblong–elliptic to oblanceolate, 6–40 × 1.5–10+ cm,
shallowly to deeply pinnatifid, lobes usually rigidly
spreading, undivided or with 1–2 pairs of coarse teeth
or lobes, main spines 5–15 mm, both faces gray- to white-
tomentose, sometimes ± glabrate or adaxial faces green,
thinly arachnoid-tomentose; basal sometimes present at
flowering, petiolate or sessile and bases tapered, spiny-
winged; principal cauline much reduced distally, sessile,
bases decurrent or not, as spiny wings; distal much
reduced, linear, ± bractlike. **Heads** 1–many in loose to
tight clusters (barely raised above rosette in dwarf
plants). **Peduncles** 1–30 cm. **Involucres** ovoid to spheric,
1.5–5 × 1.5–8 cm, arachnoid to ± loosely tomentose,
often adjacent phyllaries connected by conspicuous
arachnoid trichomes, sometimes glabrous or glabrate.
Phyllaries in 7–10 series, subequal to strongly imbricate,
green or stramineous to purple-tinged, linear to narrowly
lanceolate, abaxial faces without glutinous ridge; outer
and mid bodies appressed, entire, apices deflexed to
spreading or ascending, short-triangular to elongate,
linear-acicular, spines spreading to reflexed, 1–10+ mm;
apices of inner erect, often flexuous, flat. **Corollas** white
to lavender, pink, rose-purple, or red, 18–40 mm, tubes
8–18 mm, throats 5–7 mm, lobes 5–10 mm; style tips
4–5 mm. **Cypselae** ± brown, 5–6 mm, apical collars not
differentiated; **pappi** 15–30 mm.

Varieties 7 (7 in the flora): w United States.

1. Plants compact, rounded, moundlike; heads usually not much elevated above leaves. . . . .
   . . . . 40c. *Cirsium occidentale* var. *compactum*
1. Plants usually erect; principal heads usually conspicuously pedunculate.
   2. Corollas white to light purple or rose. . . . .
      . . . . . . 40d. *Cirsium occidentale* var. *californicum*
   2. Corollas deep purple to bright pink or red.

[3. Shifted to left margin.—Ed.]
3. Plants densely white-tomentose; phyllaries persistently white-tomentose (except spines); outer phyllaries usually very long, spreading to reflexed
   . . . . 40g. *Cirsium occidentale* var. *candidissimum*
3. Plants variably tomentose, sometimes ± glabrate;
   phyllaries ± arachnoid to floccose-tomentose,
   sometimes green and glabrate; outer phyllaries
   short to long, ascending to spreading or reflexed.
   4. Involucres usually about as long as wide or
      wider than long; phyllaries densely and persis-
      tently arachnoid with fine trichomes connect-
      ing tips of adjacent phyllaries.
      5. Phyllary apices ± imbricate, the proximal
         usually shorter than medial and distal, lan-
         ceolate to linear-acicular, 0.5–15 mm; co-
         rollas bright purple . . . 40a. *Cirsium occidentale*
         var. *occidentale*
      5. Phyllary apices subequal, all long- acicular,
         1.5–3 cm; corollas light to deep reddish
         purple . . . 40b. *Cirsium occidentale* var. *coulteri*
   4. Involucres usually longer than wide; phyllaries
      tomentose or glabrate, sparingly or not arach-
      noid with fine trichomes connecting tips of
      adjacent phyllaries.
      6. Corollas 20–24 mm, deep reddish purple;
         s Santa Lucia Mountains of San Luis Obispo
         County, California . . . . 40e. *Cirsium occidentale*
         var. *lucianum*
      6. Corollas 23–35 mm, bright pink to red;
         widespread . . . . 40f. *Cirsium occidentale*
         var. *venustum*

40a. Cirsium occidentale (Nuttall) Jepson var.
occidentale • Cobwebby thistle E

Plants erect, usually 30–150 cm
or taller. **Leaf faces** usually
densely tomentose abaxially, less
so and sometimes glabrate
adaxially. **Heads** usually long-
pedunculate, sometimes in tight
clusters at ends of peduncles,
elevated well above proximal
leaves. **Involucres** usually wider
than long, 4–5 cm diam., ± densely and persistently
arachnoid with fine trichomes connecting tips of adjacent phyllaries. **Phyllaries** usually ± imbricate, outer
ascending or spreading or reflexed, mid phyllary apices
ascending to spreading, straight or distally curved, usually 1–2 cm × 1–2 mm. **Corollas** ± bright purple,
usually 25–35 mm. *2n* = 28, 29, 30.

Flowering spring–summer (Mar–Jul). Coastal scrub,
chaparral, oak woodlands, stabilized dunes, roadsides;
0–200 m; Calif.

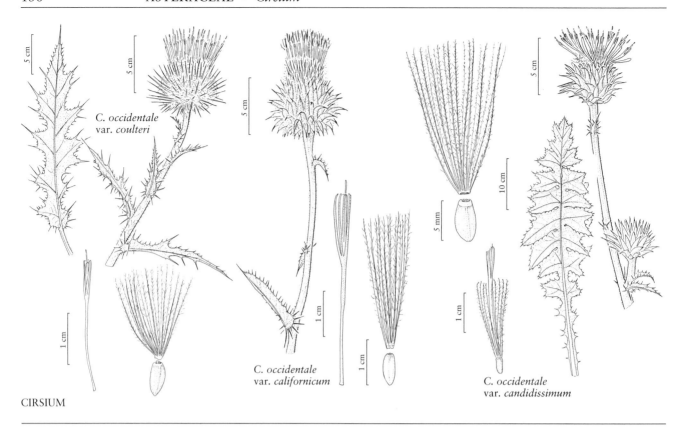

C. *occidentale*
var. *coulteri*

C. *occidentale*
var. *californicum*

C. *occidentale*
var. *candidissimum*

CIRSIUM

Variety *occidentale* occupies a variety of habitats in the coastal zone of southern and central California. Considerable variation occurs from population to population in head size, flower color, and pubescence. It sometimes occurs together with and appears to intergrade with var. *coulteri*. Where there has been no hybridization, the two may be strikingly dissimilar, but individuals of some populations cannot be assigned with confidence to either variety.

40b. Cirsium occidentale (Nuttall) Jepson var. **coulteri** (Harvey & A. Gray) Jepson, Fl. W. Calif., 509. 1901
· Coulter's thistle C E F

Cirsium coulteri Harvey & A. Gray, Mem. Amer. Acad. Arts, n. s. 4: 110. 1849

Plants erect or bushy, usually 30–150+ cm, variably tomentose, sometimes ± glabrate. **Leaf faces** gray-tomentose or adaxial ± glabrate. **Heads** sometimes in tight clusters at ends of peduncles, usually long-pedunculate, usually elevated well above proximal leaves. **Involucres** usually about as wide as tall, 4–5 cm diam., densely arachnoid with fine trichomes connecting tips of adjacent phyllaries. **Phyllaries** subequal, outer ascending to spreading or reflexed, mid apices ascending to stiffly spreading, straight, acicular,

usually 1.5–3 × 1–2 mm. **Corollas** light purple to rich reddish purple, usually 25–33 mm.

Flowering spring–summer (Mar–Jun). Coastal slopes and ridges, dunes, coastal scrub, grassland, oak woodlands; of conservation concern; 0–700 m; Calif.

Variety *coulteri* grows in the coastal zone of central and southern California. The epithet *coulteri* was for many years misapplied by California botanists. W. L. Jepson ([1923–1925]) used it at the varietal level for a range of plants that included parts of what I am calling vars. *venustum* and *californicum*. P. A. Munz (1959) and various other authors recognized *Cirsium coulteri* as a species, primarily in the context that I recognize as var. *venustum*. J. T. Howell (1959b) examined the type of *C. coulteri* and concluded that it and *C. occidentale* (in the strict sense) are synonymous. I have examined photos of the types of both *C. occidentale* and *C. coulteri* and have reached a different conclusion. The type of *C. occidentale* (BM) has heads with ± imbricate, comparatively short phyllary appendages. The type of *C. coulteri* (TCD) has long, acicular, subequal phyllaries. D. J. Keil and C. E. Turner (1993) treated the plants with the two head types all as var. *occidentale*, but I believe it is preferable to recognize separate varieties. Some intermediates are known, but var. *coulteri* is usually readily separable from var. *occidentale*.

40c. Cirsium occidentale (Nuttall) Jepson var. **compactum** Hoover, Vasc. Pl. San Luis Obispo Co., 310. 1970 (as compacta) • Compact cobwebby thistle [C] [E]

Plants compactly branched, forming low rounded mounds, 5–100 cm, densely gray-tomentose. **Leaf faces** gray-tomentose or adaxial ± glabrate. **Heads** borne singly or in ± tight clusters, short-pedunculate, closely subtended and often overtopped by basal and large cauline leaves. **Involucres** wider than long, 5–8 cm diam., densely arachnoid with fine trichomes connecting tips of adjacent phyllaries. **Phyllaries:** imbricate to subequal, outer and mid apices ± spreading, straight, usually 1–2 cm × usually 1–2 mm. **Corollas** dark rose-purple, 25–30 mm. $2n = 30$.

Flowering winter–summer (Feb–Jul). Coastal sea bluffs, dunes in grassland and coastal scrub; 0–50 m; of conservation concern; Calif.

Variety *compactum* is locally common along the immediate coast of northern San Luis Obispo County and formerly ranged as far north as San Francisco. Populations referable to this race occur on the California Channel Islands and on the mainland in Monterey County. Because of the very narrow ecologic zone occupied by these plants they are particularly vulnerable to habitat loss. Variety *compactum* is closely related to vars. *occidentale* and *coulteri*.

40d. Cirsium occidentale (Nuttall) Jepson var. **californicum** (A. Gray) D. J. Keil & C. E. Turner, Phytologia 73: 315. 1992 • California thistle [E] [F]

Cirsium californicum A. Gray in War Department [U.S.], Pacif. Railr. Rep. 4(5): 112. 1857; *C. bernardinum* (Greene) Petrak; *C. californicum* var. *bernardinum* (Greene) Petrak; *C. californicum* subsp. *pseudoreglense* Petrak

Plants erect, usually 50–200 cm, thinly to densely gray-tomentose, sometimes glabrate. **Leaf faces** abaxially green to gray, adaxially gray. **Heads** in ± open clusters, short- to long-pedunculate, elevated well above proximal leaves. **Involucres** usually about as wide as long, 1.5–5 cm, subglabrous to densely arachnoid. **Phyllaries** usually imbricate, mid apices appressed to loosely spreading or ascending, sometimes twisted, usually less than 1 cm (but sometimes much longer, 1–3 mm. **Corollas** white to light purple or rose, 18–35 mm. $2n = 28, 29, 30$ (as *C. californicum*).

Flowering spring–summer (Apr–Aug). Pine-oak woodlands, riparian woodlands, chaparral, openings in mixed evergreen forests, roadsides; 100–2200 m; Calif.

Variety *californicum* occurs in both coastal and interior mountains of California from the northern South Coast Range and the northern Sierra Nevada to the mountains of southwestern California. Considerable variation exists in head size, corolla color, and in length and display of phyllary appendages. In several areas of its range, the predominantly white- to light purple-flowered var. *californicum* occurs with red-flowered var. *venustum*. These plants are highly interfertile (H. Wells 1983; D. J. Keil and C. E. Turner 1992). Introgressive hybridization among them has resulted in a variety of emergent phenotypes and may have contributed to the variation within var. *californicum*.

40e. Cirsium occidentale (Nuttall) Jepson var. **lucianum** D. J. Keil, Sida 21: 214. 2004 • Cuesta Ridge thistle [C] [E]

Plants erect, 30–200 cm, densely gray-tomentose. **Heads** in openly branched arrays, long-pedunculate, elevated well above proximal leaves. **Involucres** about as wide as long, 2–4 cm, floccose-arachnoid, without fine trichomes connecting tips of adjacent phyllaries. **Phyllaries** imbricate, outer ascending to spreading or reflexed, mid apices ascending to spreading, straight or distally curved, usually 5–8 × 1–3 mm. **Corollas** dark reddish purple, 20–24 mm.

Flowering spring–summer (Apr–Jul). Chaparral, openings in closed cypress conifer forests, mixed evergreen forests, oak woodlands; 500–1500 m; of conservation concern; Calif.

Variety *lucianum* occupies a narrow corridor along and adjacent to the main ridge of the southern Santa Lucia Mountains of San Luis Obispo County. D. J. Keil and C. E. Turner (1993) treated these plants as an atypical race of var. *californicum*. They resemble small-headed plants of the latter but differ in their dark, reddish purple corollas. They approach the ranges of var. *californicum* and var. *venustum* but are not known to grow with either. They may represent a stabilized emergent form derived by prehistoric hybridization between var. *californicum* and var. *venustum*.

40f. Cirsium occidentale (Nuttall) Jepson var. **venustum** (Greene) Jepson, Man. Fl. Pl. Calif., 1167. 1925 • Venus thistle [E]

Carduus venustus Greene, Proc. Acad. Nat. Sci. Philadelphia 44: 359. 1893; *Cirsium occidentale* subsp. *venustum* (Greene) Petrak; *C. proteanum* J. T. Howell

Plants usually erect, usually 50–300 cm, variably tomentose, sometimes ± glabrate. **Heads** sometimes in tight clusters at ends of peduncles, usually long-pedunculate, elevated well above proximal leaves. **Involucres** usually longer than wide, 2–6 cm, subglabrous to densely arachnoid, usually without fine trichomes connecting tips of adjacent phyllaries. **Phyllaries** imbricate, outer and mid apices ascending to rigidly spreading or reflexed, straight, 5–20+ × usually 2–3 mm. **Corollas** usually ± red (white, pink, rarely purple), 23–35 mm. $2n = 30$.

Flowering spring–summer (Apr–Jul). Foothill oak-pine woodlands, grasslands, chaparral, pinyon-juniper woodlands, Joshua tree woodlands, roadsides; 200–2300 m; Calif., Nev.

Variety *venustum* has the widest ecological range of the races of *C. occidentale*. Populations occur within a few miles of the California coast in the North and South Coast Ranges and western Transverse Range and range eastward across the state into scattered sites in the Sierra Nevada to the higher elevations of the arid mountains of the western Mojave Desert and adjacent areas of the southwestern Great Basin Desert. Most populations of these plants can be recognized by their striking red to reddish pink corollas. The heads are sometimes visited by hummingbirds as well as by a variety of insects. Intermediates have been documented between var. *venustum* and vars. *californicum* and *candidissimum*.

40g. Cirsium occidentale (Nuttall) Jepson var. **candidissimum** (Greene) J. F. Macbride, Contr. Gray Herb. 53: 22. 1918 • Snowy thistle [E] [F]

Carduus candidissimus Greene, Proc. Acad. Nat. Sci. Philadelphia 44: 359. 1893; *Cirsium occidentale* subsp. *candidissimum* (Greene) Petrak; *C. pastoris* J. T. Howell

Plants erect or ± bushy, usually 40–200 cm, (branches often stiffly spreading), densely white-tomentose. **Leaf faces** densely white-tomentose. **Heads** sometimes in tight clusters at ends of peduncles, short- to long-pedunculate, elevated well above proximal leaves. **Involucres** as wide as to longer than wide, usually 2–6 cm, persistently and densely white-tomentose. **Phyllaries** imbricate or subequal, outer usually very long, spreading to reflexed, mid apices usually 1.5–3 cm × 2–3 mm, rigidly spreading or reflexed. **Corollas** red (rarely white or pink), 26–40 mm. $2n = 30, 31, 32, 33, 34, 60$.

Flowering summer (Apr–Sep). Coastal scrub, grassy openings in montane coniferous forests, arid woodlands, sagebrush scrub, roadsides; 150–1500 m; Calif., Nev., Oreg.

Variety *candidissimum* is a very attractive taxon with snowy, white-tomentose foliage and involucres and bright red corollas. It comes into contact with var. *venustum* in the North Coast Ranges of California, and individuals of some populations are difficult to place with confidence into either variety. Some plants of northern Nevada are also intermediate in their features between var. *candidissimum* and var. *venustum*.

Most populations of var. *candidissimum* occur in the mountains of northern California. Upland populations range from the Klamath Range of southwestern Oregon to the northern North Coast Ranges of northern California and eastward through the Siskiyou and Cascade ranges and the northern Sierra Nevada to mountains of the Basin and Range Province in northeastern California and adjacent northwestern Nevada. An isolated occurrence in the Blue Mountains of northeastern Oregon may represent an introduction.

A remarkable lowland population from the Carmel Highlands range of central coastal California occurs on shrub-covered hillsides adjacent to coastal redwood groves. Its habitat is markedly different from that of the upland populations, but I have found no morpologic basis for recognizing it as a distinct taxon. One plant in the population was observed to have purple instead of red corollas–perhaps a result of past hybridization with var. *occidentale*.

Heads of var. *candidissimum* are actively visited by hummingbirds as well as a variety of insects (P. L. Barlow-Irick 2002).

41. Cirsium neomexicanum A. Gray, Smithsonian Contr. Knowl. 5(6): 101. 1853 (as neo-mexicanum) • Desert or New Mexico thistle

Cirsium arcuum A. Nelson; *C. humboldtense* Rydberg; *C. neomexicanum* var. *utahense* (Petrak) S. L. Welsh; *C. undulatum* (Nuttall) Sprengel var. *albescens* D. C. Eaton; *C. utahense* Petrak

Biennials, 40–290 cm; taprooted. **Stems** usually 1, erect, thinly gray-tomentose, sometimes ± glabrate; branches few–many, usually from above middle, ascending. **Leaves:** blades oblong–elliptic to oblanceolate, 6–35 × 1.5–7 cm, shallowly to deeply pinnatifid, lobes usually rigidly spreading, undivided or with 1–2 pairs of coarse teeth or lobes, main spines 5–15 mm, faces gray-tomentose, sometimes glabrate; basal often present at flowering, winged-petiolate or sessile, bases tapered, spiny-winged; principal cauline sessile, much reduced distally, bases decurrent as spiny wings less than 5 cm; distal much reduced, ± bractlike, sometimes scarcely more than a cluster of long spines. **Heads** 1–6 (many on large individuals), borne singly or in corymbiform arrays. **Peduncles** (2.5–)5–30 cm, bracted. **Involucres** shallowly hemispheric or campanulate, 2–3 × 2.5–5 cm, arachnoid to ± loosely tomentose, sometimes glabrous. **Phyllaries** in 7–10 series, imbricate to subequal, linear to narrowly lanceolate, abaxial faces with narrow or no glutinous ridge; outer and mid bodies appressed, entire or minutely spinulose, apices deflexed to spreading or ascending, long, flat, spines spreading to reflexed, 4–15 mm; apices of inner erect, often flexuous, flat. **Corollas** white to pale lavender or pink, 18–27 mm, tubes 8–14 mm, throats 4–7 mm, lobes 5–9 mm; style tips 4–5 mm. **Cypselae** dark brown, 5–6 mm, apical collars not differentiated; pappi 15–20 mm. *2n* = 30 (as *C. utahense*), 32; 30 + 1 I.

Flowering spring–summer (Mar–Jul). Canyons, slopes, roadsides in deserts, dry grasslands, and arid woodlands dominated by pinyon pines, junipers, oaks, Joshua trees; 300–2100 m; Ariz., Calif., Colo., Nev., N.Mex., Tex., Utah; Mexico (Sonora).

Desert thistle is widespread in the Mojave and Sonoran deserts and ranges into the southern Great Basin desert, western Chihuahuan desert, and into adjacent mountains of Utah, southwestern Colorado, Arizona, and New Mexico.

The name *Cirsium utahense* has been widely applied in the past to plants that are here recognized as *C. inamoenum.* S. L. Welsh (1983) treated it as a variety of *C. neomexicanum.* I have examined the type of *C. utahense* and can find no basis for distinguishing it from *C. neomexicanum* at any rank. The desert thistle is closely related to *C. occidentale.*

42. **Cirsium andrewsii** (A. Gray) Jepson, Fl. W. Calif., 506. 1901 • Franciscan thistle [C][E]

Cnicus andrewsii A. Gray, Proc. Amer. Acad. Arts 10: 45. 1874

Biennials (or short-lived monocarpic perennials), 60–200 cm; taprooted. **Stems** several, erect to spreading, thinly arachnoid, soon glabrous; branches ± fleshy, usually much branched proximally, spreading to ascending. **Leaves:** blades ± elliptic, 30–75 × 10–20 cm, shallowly to deeply pinnatifid, lobes oblong to ovate, unlobed or with several prominent secondary lobes or large teeth,

obtuse to acute, main spines 2–7 mm, abaxial gray arachnoid-tomentose, adaxial faces thinly arachnoid, glabrate; basal often present at flowering, spiny winged-petiolate; principal cauline sessile, bases clasping with broad, spiny-margined auricles, reduced distally, spinier than proximal; distal much reduced, spines 7–20 mm. **Heads** several–many, in congested corymbiform arrays. **Peduncles** 0–7 cm. **Involucres** ovoid to hemispheric or campanulate, 1.5–3 × 1.5–5 cm, sparsely to densely arachnoid, finely short-ciliate. **Phyllaries** in ca. 6 series, dark green or brown or with stramineous margins and a darker central zone, imbricate, linear-lanceolate (outer) to linear (inner), abaxial faces without glutinous ridge; outer and mid bodies appressed, spiny-ciliate, apices long-spreading to ascending long-acuminate, spines straight, stout, 5–15 mm; apices of inner straight or twisted, long, entire, flat or spine-tipped. **Corollas** dark reddish purple, 17–24 mm, tubes 8–11 mm, throats 3.5–6 mm, lobes 5–7 mm; style tips 3–4 mm. **Cypselae** dark brown, 4–5 mm, apical collars narrow; **pappi** 15 mm. *2n* = 32.

Flowering spring–summer (May–Sep). Headlands, ravines, seeps near coast, sometimes on serpentine; of conservation concern; 0–100 m; Calif.

Cirsium andrewsii occurs along the coast of north-central California from San Mateo to Marin counties. It reportedly hybridizes with *C. quercetorum* (F. Petrak 1917; J. T. Howell 1960b).

43. **Cirsium arizonicum** (A. Gray) Petrak, Bot. Tidsskr. 31: 68. 1911 • Arizona thistle [F]

Cnicus arizonicus A. Gray, Proc. Amer. Acad. Arts 10: 44. 1874

Perennials, 30–150 cm; taprooted caudices or runner roots. **Stems** 1–several, erect or ascending, glabrous to thinly arachnoid-tomentose with fine non-septate trichomes and/or villous with septate trichomes, sometimes ± glabrate; branches 0–many, ascending. **Leaves:** blades oblong-elliptic, 3–40 × 1–13 cm, unlobed and spinulose to shallowly lobed or divided nearly to midvein, lobes few–many, ovate to linear-acuminate, often again lobed or divided, main spines 2–30 mm, abaxial faces green, glabrous to densely gray tomentose, sometimes midveins villous with septate trichomes, adaxial green, glabrous to gray-tomentose, sometimes glabrate; basal sometimes present at flowering, unlobed to deeply spiny-lobed, winged-petiolate or sessile; principal cauline sessile, well distributed, gradually diminished distally, bases sometimes decurrent as spiny wings to 2.5 cm or clasping; distalmost sometimes ± bractlike. **Heads** 1–100+, erect, in corymbiform or paniculiform arrays. **Peduncles** 0–15 cm. **Involucres** cylindric or ovoid to campanulate, 1.5–4 × 1–2.5 cm (body), loosely arachnoid or ± glabrous. **Phyllaries** in 7–9 series, imbricate, green or the inner reddish to rich reddish purple, ovate or lanceolate (outer) to linear (inner), margins of outer entire, abaxial faces often with narrow, inconspicuous glutinous ridge; outer and mid bodies appressed, short, entire, apices spreading to ascending, inconspicuous to long, narrow, entire or minutely ciliolate, spines erect to reflexed (outer) to ascending (inner), slender to stout, cylindric or basally flattened, 1–30 mm; apices of inner unarmed or with straight or flexuous spines, short, flat. **Corollas** pink to red, lavender, or purple (white), 25–31 mm, tubes 7–12.5 mm, throats 1.5–8.5 mm, lobes 10–17 mm; style tips 1–4 mm. **Cypselae** brown, 3.5–7 mm, apical collars stramineous, 0.2–0.3 mm; **pappi** 17–28 mm. $2n$ = 30, 32, 34.

Varieties 5 (5 in the flora): sw United States, nw Mexico.

The *Cirsium arizonicum* complex is widely distributed from the Sierra Nevada, White Mountains, and New York Mountains of eastern California across the mountains of the southern Great Basin and Colorado Plateau to the mountains of eastern Colorado, Arizona, and New Mexico. This group of plants comprises a series of intergrading races with intricately overlapping patterns of variation. For plants that I am treating as *C. arizonicum* (in the broad sense), F. Petrak (1917) recognized three species, one with a variety and two subspecies plus his unstated type subspecies and variety. R. J. Moore and C. Frankton (1974b) revised the complex, recognizing six species, three of them newly described, for the plants I treat as *C. arizonicum* plus *C. turneri*, which I do not include in *C. arizonicum*. P. L. Barlow-Irick (2002), in a work focused on statistical analyses of variation patterns, recognized six species also, but circumscribed very differently from those of Moore and Frankton. Two of the species proposed by Barlow-Irick have not been formally described.

I have wrestled with how to treat these plants since beginning my research for this treatment. After careful consideration of the complex patterns of variation among members of the *C. arizonicum* complex, I acknowledged the futility of trying to distinguish more than one species. Any character combinations that I or others have attempted to use to distinguish species break down hopelessly when enough specimens are examined. Instead I have chosen to recognize that in this complex, as in several others, the plants in question are a work of evolution in progress. *Cirsium arizonicum* is a rapidly evolving, only partially differentiated assemblage of races that have not reached the level of stability that is usually associated with the concept of species. Certainly there is much variation within the group that deserves a level of taxonomic recognition, or at least should be mentioned, but I think it much more prudent to recognize

varieties–entities that may be expected to freely intergrade–rather than species. The geographic area where these plants occur, the highlands of the American Southwest, has had a turbulent history in the Quaternary with major shifts in climate, vegetation, and elevational zonation accompanying the vicissitudes of glacial and interglacial episodes. The complicated patterns of variation in *C. arizonicum* reflect both that history and the geographic and topographic complexity of the region.

Heads of *Cirsium arizonicum* are visited by hummingbirds as well as a variety of insects (P. L. Barlow-Irick 2002). Hummingbirds are the most common visitors, but hummingbirds and bees are both apparently effective pollinators in *C. arizonicum*.

SELECTED REFERENCES Barlow-Irick, P. L. 2002. Biosystematic Analysis of the *Cirsium arizonicum* Complex of the Southwestern United States. Ph.D. dissertation. University of New Mexico. Moore, R. J. and C. Frankton. 1974b. The *Cirsium arizonicum* complex of the southwestern United States. Canad. J. Bot. 52: 543–551.

1. Corollas bright red or reddish pink.
　2. Leaves glabrous on both faces . . . .
　　. . . . . 43b. *Cirsium arizonicum* var. *rothrockii*
　2. Leaves ± tomentose, at least on the abaxial faces . . . . . 43a. *Cirsium arizonicum* var. *arizonicum*
1. Corollas lavender to reddish purple.
　3. Stems and abaxial leaf midveins villous to tomentose with septate trichomes; leaves conspicuously decurrent; leaves deeply divided, lobes many, narrow, closely spaced, each tipped by a very slender spine 5–12 mm; northeastern Arizona and northwestern New Mexico . . .
　　. . . . . 43d. *Cirsium arizonicum* var. *chellyense*
　3. Stems and abaxial leaf midveins glabrous to tomentose with fine, non-septate trichomes; septate trichomes usually absent; leaf divisions various.
　　4. Principal marginal leaf spines 3–10 mm; New Mexico, northeastern Arizona, southeastern Utah, and southwestern Colorado . . . . . 43c. *Cirsium arizonicum* var. *bipinnatum*
　　4. Principal marginal leaf spines 5–30 mm; southeastern California and southwestern Nevada . . . .
　　　. . . . . 43e. *Cirsium arizonicum* var. *tenuisectum*

43a. **Cirsium arizonicum** (A. Gray) Petrak var. **arizonicum** [E]

Cirsium arizonicum var. *nidulum* (M. E. Jones) S. L. Welsh; *C. nidulum* (M. E. Jones) Petrak

Stems glabrous to tomentose, usually without septate trichomes. **Leaf blades** unlobed to deeply divided, faces usually ± tomentose, at least abaxially, usually without septate trichomes; main marginal spines 2–15 mm, often ± stout; cauline bases narrowed, truncate or clasping, but not or only slightly decurrent. **Involucres** cylindric to campanulate. **Phyllary spines** 1–20 mm, slender to very stout. **Corollas** usually bright red or pink, 24–34 mm, tubes 7–12, throats 2–5 mm, lobes 13–18 mm; style tips–2.5 mm. $2n$ = 30, 32, 34 (as *C. arizonicum* × *C. nidulum*).

Flowering spring–fall (May–Oct). Pine-oak-juniper woodlands, montane coniferous forests, subalpine; 1200–3600 m; Ariz., Calif., Nev., N.Mex., Utah.

Variety *arizonicum* is widely distributed across the uplands of the American Southwest from eastern California across the Basin and Range Province to the Colorado Plateau and the mountains of Arizona and New Mexico. It is the most widely distributed and most variable variety of *Cirsium arizonicum*. Viciously spiny plants with deeply divided leaves from the northern part of the range of var. *arizonicum* often have been treated as *C. nidulum*. These plants form a continuum with the less spiny plants of more southerly distribution, however, and separation of two varieties (or species) is arbitrary.

Leaves of some plants of var. *arizonicum* from Coconino County, Arizona, resemble those of the sympatric *C. wheeleri*. I have not seen hybrids between these taxa, but it is likely that some genetic interchange between the two has taken place in the past.

43b. **Cirsium arizonicum** (A. Gray) Petrak var. **rothrockii** (A. Gray) D. J. Keil, Sida 21: 210. 2004 • Rothrock's thistle [E]

Cnicus rothrockii A. Gray, Proc. Amer. Acad. Arts 17: 220. 1882; *Cirsium rothrockii* (A. Gray) Petrak

Stems glabrous. **Leaf blades** ± deeply divided or the distal unlobed and merely spiny-dentate, faces glabrous or the basal abaxially sometimes thinly tomentose, rarely with septate trichomes abaxially along the midveins; main marginal spines often stout, 5–15 mm; cauline bases usually narrowed, not decurrent. **Involucres** cylindric to narrowly ovoid. **Phyllary spines** slender to very stout, 5–30 mm. **Corollas** bright red, 28–32 mm, tubes 10–12, throats 1.5–3 mm, lobes 15–17 mm; style tips 1–2.5 mm. $2n$ = 30 (as *C. rothrockii*).

Flowering summer–fall (Jun–Oct). Rocky slopes, embankments, pine-oak-juniper-cypress woodlands, montane coniferous forests; 1500–2800 m; Ariz., N.Mex.

Variety *rothrockii* is endemic to Chiricahua, Dos Cabseos, Huachuca, Peloncillo, Pinal, Pinaleno, Sierra Ancha, and White mountains of southeastern Arizona and adjacent New Mexico, where it largely replaces var. *arizonicum*. The name *Cirsium rothrockii* has been misapplied to similarly glabrous plants of the Four Corners region (e.g., S. L. Welsh 1983), but those plants are assigned here to other varieties.

43c. **Cirsium arizonicum** (A. Gray) Petrak var. **bipinnatum** (Eastwood) D. J. Keil, Sida 21: 209. 2004 • Four Corners thistle [E]

Cnicus drummondii Torrey & A. Gray var. *bipinnatus* Eastwood, Zoë 4: 8. 1893; *Cirsium bipinnatum* (Eastwood) Rydberg; *C. calcareum* (M. E. Jones) Wooton & Standley; *C. calcareum* var. *bipinnatum* (Eastwood) S. L. Welsh; *C. calcareum* var. *pulchellum* (Greene ex Rydberg) S. L. Welsh; *C. diffusum* (Eastwood) Rydberg; *C. pulchellum* (Greene ex Rydberg) Wooton & Standley; *C. pulchellum* subsp. *bipinnatum* (Eastwood) Petrak; *C. pulchellum* subsp. *diffusum* (Eastwood) Petrak; *C. pulchellum* var. *glabrescens* Petrak

Stems glabrous to tomentose, usually without septate trichomes. **Leaf blades** unlobed to deeply divided, faces glabrous to tomentose on one or both, usually without septate trichomes; main marginal spines slender to stout, 3–10 mm; cauline bases narrowed to truncate or clasping or decurrent as spiny wings to 30 mm. **Involucres** cylindric or ovoid to campanulate; spines of mid phyllaries 1–6 mm, slender to very stout. **Corollas** usually lavender to purple or reddish purple, 22–34 mm, tubes 7–12, throats 4–9 mm, lobes 8–18 mm; style tips 1.2–4.5 mm.

Flowering spring–summer (May–Oct). Canyons, rocky slopes, desert scrub, pine-oak-juniper woodlands, openings in montane coniferous forests; 900–3100 m; Ariz., Colo., N.Mex., Utah.

Variety *bipinnatum* grows in the Colorado Plateau of New Mexico, northeastern Arizona, southeastern Utah, and southwestern Colorado south to the Sacramento Mountains of southern New Mexico. It is a polymorphic assemblage of races and local populations with overlapping patterns of variation. Extremes are diverse and strikingly different, and various authors have attempted to segregate variants as species, subspecies, or varieties. Apparently independent segregation of character states of multiple variable characters has resulted in a complex array of local morphotypes. Leaves range from wholly glabrous to densely tomentose on one or both faces. The bases of cauline leaves vary from sessile and not at all decurrent to prominently decurrent with spiny wings to 3 cm. Leaf lobing and spininess are similarly variable. Heads may be solitary at the branch tips or densely clustered. I have been unable to detect patterns that can consistently separate the many forms in any meaningful way.

43d. Cirsium arizonicum (A. Gray) Petrak var. **chellyense** (R. J. Moore & Frankton) D. J. Keil, Sida 21: 209. 2004 • Navajo thistle [C][E]

Cirsium chellyense R. J. Moore & Frankton, Canad. J. Bot. 52: 547, plate 2. 1974; *C. chuskaense* R. J. Moore & Frankton; *C. navajoense* R. J. Moore & Frankton

Stems glabrous or villous to tomentose with septate trichomes. **Leaf blades** deeply 1–2× divided, (lobes many, narrow, closely spaced), faces glabrous or abaxial villous along midveins with septate trichomes; main marginal spines usually very slender, 5–12 mm; cauline bases usually decurrent as spiny wings to 30 mm. **Involucres** cylindric to narrowly ovoid. **Phyllary spines** slender to ± stout, 1–20 mm. **Corollas** usually reddish purple, 25–32 mm, tubes 7–11, throats 4–7.5 mm, lobes 11–16 mm; style tips 2.5–3.5 mm. $2n$ = 34 (as *C. chuskaense*).

Flowering summer (Jun–Sep). Desert scrub, grasslands, pine-oak-juniper woodlands, ponderosa pine forests; of conservation concern; 1600–2800 m; Ariz., N.Mex.

R. J. Moore and C. Frankton (1974b) recognized three narrowly defined new species in the *Cirsium arizonicum* complex, all from a small area in the Defiance Plateau and Chuska Mountains in northeastern Arizona and northwestern New Mexico. The characters used to differentiate these taxa break down when additional specimens are examined, and they are all combined here as a single variety of *C. arizonicum*.

43e. Cirsium arizonicum (A. Gray) Petrak var. **tenuisectum** D. J. Keil, Sida 21: 210. 2004 • Desert mountains thistle [C][E][F]

Stems thinly arachnoid tomentose, ± glabrate, without septate trichomes. **Leaf blades** deeply divided, often nearly to the midveins, abaxial faces arachnoid tomentose or sometimes glabrate, without septate trichomes, adaxial thinly arachnoid or glabrate; main marginal spines often stout, 5–30 mm; cauline bases narrowed to truncate or ± clasping, but not or only slightly decurrent. **Involucres** cylindric to campanulate. **Phyllary spines** stout, 5–25 mm. **Corollas** reddish purple, 25–35 mm, tubes 10–13, throats 5–8 mm, lobes 10–13.5 mm; style tips 1–2 mm. $2n$ = 34 (as *C. nidulum*).

Flowering summer–fall (Jun–Nov). Rocky slopes, drainages, roadsides, pine-oak-juniper woodlands, montane coniferous forests; of conservation concern; 1500–2800 m; Calif., Nev.

Variety *tenuisectum* occurs in the New York Mountains of southeastern California and the Spring Mountains of southwestern Nevada. The name *Cirsium nidulum* has long been misapplied to the plants here treated as var. *tenuisectum* (e.g., P. A. Munz 1959; J. T. Howell 1960b; R. J. Moore and C. Frankton 1974b; D. J. Keil and C. E. Turner 1993). My examination of the type of *C. nidulum* indicates that the taxon is properly treated as a synonym of *C. arizonicum* var. *arizonicum*.

Cirsium eatonii var. *clokeyi* occurs in close proximity to *C. arizonicum* var. *tenuisectum* in the ski area of upper Lee Canyon in the Spring Mountains. I did not see any evidence of hybridization during my brief field survey in that area.

44. Cirsium turneri Warnock, SouthW. Naturalist 5: 101. 1960 • Cliff thistle

Perennials 15–45 cm; stout, branched caudices. **Stems** 5–30+, horizontal or hanging from cliff sides, thinly appressed gray-tomentose and villous with septate trichomes; branches 0–few, distal, ascending. **Leaves:** blades oblong-elliptic to oblanceolate, 5–30 × 1–5 cm, shallowly to deeply pinnatifid, lobes spreading, triangular, coarsely dentate or lobed, obtuse to acute, main spines slender, 4–10 mm, abaxial faces green or gray-tomentose, villous with septate trichomes along midveins, ± glabrate, adaxial green and glabrous or thinly tomentose, ± glabrate; basal often present at flowering, spiny winged-petiolate; principal cauline sessile, gradually reduced distally;

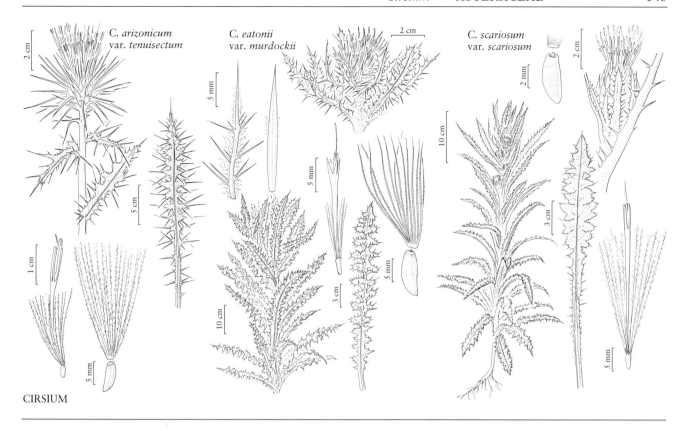

C. arizonicum var. tenuisectum   C. eatonii var. murdockii   C. scariosum var. scariosum

CIRSIUM

distal oblong, bases ± clasping, usually less deeply lobed and often spinier than proximal. **Heads** 1–6+, borne singly or in condensed corymbiform arrays. **Peduncles** 0–1 cm. **Involucres** cylindric to narrowly campanulate, 3.5–4.5 × 1.5–2 cm, loosely arachnoid, glabrate, finely short-ciliate. **Phyllaries** in 5–6 series, imbricate, linear-lanceolate (outer) to linear (inner), entire, abaxial faces without glutinous ridge, apices red to reddish purple, stiffly ascending, long-acuminate, spines straight, 1–10 mm, ± flattened; apices of inner stramineous to red, straight or flexuous. **Corollas** red to reddish purple, 26–27 mm, tubes 3.5–5 mm, throats 7.5–9.5 mm, lobes 12–14 mm; style tips ca. 3 mm. **Cypselae** stramineous, 5–6 mm, apical collars not differentiated; **pappi** 20–25 mm.

Flowering summer (Jun–Sep). Crevices in limestone or basaltic cliffs; 900–1500 m; Tex.; Mexico (Chihuahua, Coahuila).

Cirsium turneri is known from the mountains of the Big Bend area of trans-Pecos Texas and adjacent areas of northern Mexico.

45. Cirsium andersonii (A. Gray) Petrak, Bot. Tiddsskr. 31: 68. 1911 • Anderson's or rose thistle E

Cnicus andersonii A. Gray, Proc. Amer. Acad. Arts 10: 44. 1874

Perennials (but often appearing biennial), (15–)40–70(–100) cm; rootstocks producing erect, taprooted caudices and rosettes. **Stems** usually 1, erect, subglabrous to puberulent and/or tomentose; branches 0–several, stiffly ascending. **Leaves**: blades ± elliptic, 8–35 × 4–8 cm, divided about halfway to midveins, lobes spreading, triangular, coarsely dentate or with a few broad lobes, obtuse to acute, main spines 1–5 mm, abaxial faces green or gray, thinly tomentose, adaxial green and glabrous to sparingly pilose; basal often present at flowering, spiny winged-petiolate; main cauline reduced distally, bases clasping; distal much reduced, linear-oblong, usually less deeply lobed and often spinier than proximal. **Heads** 1–6, borne singly or in corymbiform, racemiform, or spiciform arrays. **Peduncles** 0–20 cm. **Involucres** broadly cylindric to narrowly campanulate, 3–5 × 2–4 cm, loosely arachnoid or ± glabrous, finely short-ciliate. **Phyllaries** in 6–8 series, imbricate, outer green, inner purple to red, linear-lanceolate (outer) to linear (inner), abaxial faces without glutinous ridge; outer and mid bodies short, appressed, entire or spinulose-ciliate, apices long-spreading to ascending, entire or spinulose-ciliate or rarely with expanded, fringed appendages, spines straight, weak, 1–3 mm; apices of inner red to purple, straight or rarely twisted, long, flat, entire. **Corollas** red to reddish purple, 30–45 mm, tubes 10–20 mm, throats 10–16 mm, lobes 9–11 mm; style tips 3.5–5 mm. **Cypselae** brown, 6–7 mm, apical collars narrow; **pappi** 25–40 mm. $2n$ = 32, 64.

Flowering summer (Jul–Sep). Moist to dry soils, openings in montane woodlands, montaine coniferous forests, aspen groves; 1100–2900 m; Calif., Idaho, Nev.

Cirsium andersonii grows in the Cascade Range of northern California south through the Sierra Nevada of eastern California and western Nevada. It has been reported from the mountains of southwestern Idaho, but I have not seen specimens from there.

Heads of *Cirsium andersonii* are actively visited by hummingbirds as well as a variety of insects (P. L. Barlow-Irick 2002).

46. **Cirsium edule** Nuttall, Trans. Amer. Philos. Soc., n. s. 7: 420. 1841 (as Circium) • Edible thistle, cardon E

Biennials or monocarpic perennials, 20–350 cm; taprooted. **Stems** usually 1, erect, simple to openly branched in distal $^1/_2$, ± villous with jointed trichomes, sometimes finely arachnoid, sometimes ± glabrate; branches 0–many, ascending. **Leaves:** blades oblong to elliptic or oblanceolate, 5–50 × 1–10 cm, plane to moderately undulate, coarsely dentate to deeply pinnatifid, lobes 5–10 well separated, linear, narrowly to broadly triangular, spinulose to spiny-dentate or shallowly lobed, main spines 3–10 mm, abaxial faces thinly to densely villous along major veins with septate trichomes, sometimes thinly arachnoid-tomentose, sometimes glabrescent, adaxial glabrous to sparsely villous or shaggy-tomentose along midveins with septate trichomes; basal often absent at flowering, spiny winged-petiolate or sessile; principal cauline well distributed, only gradually reduced, bases auriculate-clasping; distal moderately to strongly reduced, thin, often spinier than the proximal. **Heads** 1–many, erect, often crowded and ± sessile in tight clusters at stem tips, closely subtended by clusters of leafy bracts or not, collectively forming corymbiform or paniculiform arrays. **Peduncles** 0–5(–30) cm. **Involucres** narrowly ovoid to hemispheric or campanulate, 1.5–3.5 × 1.5–4 cm (including spines), loosely to densely arachnoid with fine, non-septate trichomes. **Phyllaries** in 4–8 series, subequal, green or often purplish, bodies short, appressed, abaxial faces without glutinous ridge, apices stiffly radiating to ascending, straight or flexuous, narrowly linear, plane to acicular, spines straight, slender, 1–10+ mm; outermost spiny-fringed or entire, mid entire or minutely serrulate; apices of inner straight, sometimes expanded and erose, flat. **Corollas** purple (pink or white), (15–)18–22(–33) mm, tubes 7–11 mm, throats (4–)5–8.5(–13) mm, lobes linear but not filiform, not knobbed at tips, (2–)4.5–7(–10) mm; style tips 3–4(–5) mm, conspicuously exserted beyond corolla lobes. **Cypselae** dark brown, 3.5–6.5 mm, apical collars not differentiated; **pappi** 9–19(–25) mm.

Varieties 3 (3 in the flora): w North America.

The edible thistle has had a convoluted nomenclatural history. The labels on the type collection (BM) bear the following information: "*Circium* [sic] * *edule*" and "R. Mountains & plains of Columbia." When Nuttall published the name, he listed the range as "The plains of Oregon and the Blue Mountains." The type specimen closely resembles plants from western Oregon, but plants of *Cirsium edule* are not known to occur in the Blue Mountains. J. T. Howell (1943) noted that the name *C. edule* had long been in use for two distinctly different species, one with a long, slender corolla tube, short lobes, and a barely exserted style and the other with a stouter corolla with a shorter tube, longer lobes, and a long-exserted style. He applied the name *C. edule* to the species with the slender corolla, and took up the name *C. macounii* for the second species. After examining the type of *C. edule*, A. Cronquist (1953) pointed out that Howell had erred in applying that name to the short-styled species, and described the latter as *C. brevistylum*. Cronquist expressed doubt as to the collection locality of the type of *C. edule*, focusing on the Blue Mountains and not noting the duality of the location data on the specimen and in Nuttall's publication. It is likely that the plants Nuttall observed in the Blue Mountains were *C. brevistylum*.

J. T. Howell (1960b) resurrected the name *Cirsium hallii* to apply to a group of Oregon thistles growing west of the Cascade Range, attributing to it a type locality of Salem, Oregon. Howell (1943) had noted that he had borrowed "the type" of *C. hallii* from the Gray Herbarium. This species had been described (as *Cnicus hallii*) by Gray based upon three syntypes (one each cited from California, Utah, and Oregon). The specimen examined by Howell was apparently the Oregon collection by [Elihu] Hall that Gray cited. The Utah and California specimens are different taxa. After examining photographs of the holotype of *C. edule* and the Oregon specimen of *C. hallii*, and various specimens collected in the area that Howell described as the range of *C. hallii*, I have concluded that *C. hallii* and *C. edule* are clearly conspecific.

Cirsium edule is a polymorphic species much in need of an in-depth field-based investigation. R. J. Moore and C. Frankton (1962) noted that in the northern part of its range, *C. edule* occurs mostly at elevations from 300 to over 2100 m. However, along the Oregon coast the species occurs on sea bluffs a few meters above the surf. Populations from montane sites are often rather different in appearance from those of lowland areas, and coastal plants differ from those of nearby more interior areas. Both montane and strictly coastal plants tend to be compact with heads tightly crowded and usually with very densely arachnoid involucres. Plants of non-montane interior sites tend to be taller and more openly branched. Plants of interior sites in southern Washington and Oregon have smaller heads with less densely arachnoid involucres than those farther to the north or along the seashore. The spiny tips of the phyllaries may be ascending or may radiate from the head forming a dense, spiny ball.

Hybridization may have played a role in the diversification of *Cirsium edule*. Hybrids between *C. edule* var. *macounii* and *C. brevistylum* in southern Canada have been named as *C. ×vancouveriense* R. J. Moore & C. Frankton. *Cirsium edule* and *C. brevistylum* overlap extensively in parts of their ranges and hybrids may occur throughout their area of sympatry. Some of the variation in the southern part of the range of *C. edule* may be a result of past introgression with various forms of *C. remotifolium*.

1. Heads mostly borne singly; peduncles 10–30 cm
   . . . . 46c. *Cirsium edule* var. *wenatchense*
1. Heads crowded at stem tips; peduncles usually 0–1 cm.
   2. Phyllary apices plane to acicular, ascending or ± spreading, spines 1–5 mm . . . .
      . . . . 46a. *Cirsium edule* var. *edule*
   2. Phyllary apices long-acicular, widely spreading, spines 5–15 mm . . . .
      . . . . 46b. *Cirsium edule* var. *macounii*

46a. Cirsium edule Nuttall var. **edule** • Hall's thistle

E

Cirsium hallii (A. Gray) M. E. Jones

Heads usually crowded at stem tips. **Peduncles** 0–1(–4) cm. **Involucres** 1.5–2.5 × 1.5–3 cm, thinly to densely arachnoid. **Phyllary apices** ascending or ± spreading, plane to acicular, spines 1–5 mm. **Corollas** 14–21 mm, tubes 5–11 mm, throats 4–8 mm, lobes 4–6 mm; style tips 3–4 mm. **Cypselae** light to dark brown, 3.5–4.5 mm; **pappi** 9–13 mm.

Flowering summer (Jun–Aug). Sea bluffs, roadsides, damp soil at edge of woods, openings in conifer or conifer-hardwood forests; 0–900 m; Oreg., Wash.

Variety *edule* is known from the coast and coast ranges of Oregon and Washington. Some coastal populations of var. *edule* approach montane forms of var. *macounii* in stature, head size, and involucre pubescence. Some of the variation in var. *edule* may be a result of past introgression with *Cirsium remotifolium*. Populations of var. *edule* from within the ranges of *C. remotifolium* vars. *odontolepis*, *remotifolium*, and *rivulare* often show features of those taxa.

46b. Cirsium edule Nuttall var. **macounii** (Greene) D. J. Keil, Sida 21: 213. 2004 • Macoun's thistle

E

Carduus macounii Greene, Ottawa Naturalist 16: 38. 1902; *Cirsium macounii* (Greene) Petrak

Heads often densely crowded at stem tips. **Peduncles** 0–5(–10) cm. **Involucres** 2–4 × 2.5–4 cm, usually densely arachnoid. **Phyllary apices** widely spreading, long-acicular, spines 5–15 mm. **Corollas** (15–)18–22 mm, tubes 7–11 mm, throats (4–)5–8.5 mm, lobes (2–)4.5–7 mm; style tips 4–5 mm. **Cypselae** dark brown, 5–6.5 mm; **pappi** 20–25 mm. $2n = 34$ (as *C. edule*).

Flowering Jun–Oct. Damp soil, mostly montane meadows, forests, alpine; 0–1900 m; B.C.; Alaska, Oreg., Wash.

Variety *macounii* ranges from the coastal mountains of southeastern Alaska and British Columbia to Vancouver Island, the Olympic Peninsula, and northern Cascades of Washington, and the mountains of north-western Oregon.

46c. Cirsium edule Nuttall var. **wenatchense** D. J. Keil, Sida 21: 213. 2004 • Wenatchee thistle

C E

Heads mostly borne singly. **Peduncles** 10–30 cm. **Involucres** 3–4 × 4–5 cm, moderately arachnoid. **Phyllary apices** widely spreading, long-acicular, spines 5–15 mm. **Corollas** 29–33 mm, tubes 9–11 mm, throats 8–12 mm, lobes 9–10 mm; style tips 3–4 mm. **Cypselae** dark brown, 4.5–6 mm; **pappi** 20–25 mm.

Flowering summer (Jul–Aug). Stream banks, rocky slopes; of conservation concern; 600–1200 m; Wash.

Variety *wenatchense* is known only from the Wenatchee Mountains of central Washington. Little is known of its ecology.

47. Cirsium brevistylum Cronquist, Leafl. W. Bot. 7: 26. 1953 • Indian or clustered or short-style thistle E

Annuals or biennials, 20–350 cm; taprooted. **Stems** usually 1, erect, simple or branched in distal ¹/₂, loosely to densely villous or viscid-pilose with jointed trichomes, often arachnoid below heads; branches 0–many, ascending. **Leaves:** blades oblong to elliptic or oblanceolate, 15–35 × 2–10 cm, flat to ± undulate, coarsely dentate to shallowly pinnatifid, lobes broadly triangular, spinulose to spiny-dentate or shallowly lobed, main spines slender, 3–7 mm, abaxial faces thinly gray-tomentose, villous along major veins, sometimes glabrescent, adaxial sparsely villous or viscid-pilose along midveins with jointed trichomes; basal often absent at flowering, spiny winged-petiolate; principal cauline well distributed, gradually reduced, proximal winged-petiolate, mid and distal sessile, bases clasping or short-decurrent; distal moderately to strongly reduced, often spinier than the proximal. **Heads** 1–many, ± erect, usually crowded in subcapitate to tight corymbiform arrays, closely subtended by clustered ± leafy bracts. **Peduncles** 0–1(–30) cm. **Involucres** hemispheric to campanulate, 2.5–3.5 cm, 2.5–4 cm diam., loosely to densely arachnoid, phyllaries connected by long septate or non-septate trichomes. **Phyllaries** radiating in 5–10 series, subequal, green, linear-acicular, outermost margins sometimes spiny-fringed, otherwise all entire or minutely serrulate, abaxial faces without glutinous ridge; outer and mid bases short-appressed, apices stiffly radiating to ascending, long, very narrow, spines straight, slender, 3–5 m; apices of inner straight, flat. **Corollas** white to pink or purple, very slender, 20–25 mm, tubes 10–17 mm, throats 4–5 mm, lobes filiform with knoblike tips, 3–5 mm; style tips 2–4 mm, included or exserted (only 1–2 mm beyond corolla lobes). **Cypselae** brown, 3–4.5 mm, apical collars stramineous, 0.2 mm; **pappi** 10–22 mm. $2n = 34$.

Flowering spring–summer (Apr–Sep). Coastal meadows, marshes, swamps, riparian woodlands, moist sites in coastal scrub, chaparral, coastal woodlands, mixed conifer-hardwood forests, or coniferous forests; 0–1000 m; B.C.; Calif., Idaho, Mont., Oreg., Wash.

Cirsium brevistylum occurs in the coast ranges and adjacent coastal slope from southwestern British Columbia to southern California. In the Pacific Northwest its range extends inland to the northern Rocky Mountains of southern British Columbia, Idaho, and northwestern Montana, and the Blue and Wallowa ranges of eastern Oregon. It is absent from the central and southern Cascade Range.

In older literature the name *Cirsium edule* was widely misapplied to this species. A. Cronquist (1953) pointed out that the type of *C. edule* has corolla and style features quite different from those of the plants that had been called by that name and established the name *C. brevistylum*, based upon the notably short styles of this species. Hybrids of *C. brevistylum* with *C. edule* have been named *C. ×vancouveriense* R. J. Moore & C. Frankton.

48. Cirsium hookerianum Nuttall, Trans. Amer. Philos. Soc., n. s. 7: 418. 1841 (as Circium) • Hooker's or white thistle E

Cirsium kelseyi (Rydberg) Petrak

Biennials or monocarpic (sometimes polycarpic?) perennials, 20–150 cm; taprooted. **Stems** usually 1 and erect, less commonly several and ascending, simple to sparingly short-branched in distal ¹/₂, variably villous with jointed trichomes, and/or finely arachnoid, or ± glabrate; branches on distal stems 0–many, short, ascending. **Leaves:** blades linear-oblong to elliptic, 5–25 × 1–8 cm, subentire to coarsely dentate or deeply pinnatifid, lobes lance-oblong to broadly triangular, spinulose to spiny-dentate or shallowly lobed, main spines 2–10 mm, abaxial faces usually ± densely gray- or white-tomentose with felted arachnoid trichomes, ± villous to tomentose along major veins with septate trichomes, sometimes glabrous or glabrate, adaxial ± green, glabrous to thinly arachnoid, often ± villous or tomentose with septate trichomes; basal often present at flowering, spiny winged-petiolate or sessile; principal cauline well distributed, proximally winged-petiolate, distally sessile, gradually reduced, bases sometimes short-decurrent; distal ± reduced, often narrower than proximal, sometimes with non-pigmented bases, sometimes pectinately spiny. **Heads** 1–many, borne singly or crowded in spiciform, racemiform, subcapitate, or sometimes more openly branched corymbiform arrays. **Peduncles** 0–8+ cm. **Involucres** (green or often purplish), broadly ovoid, 2–3.3 × 1.5–4 cm, loosely to densely villous with septate trichomes to tomentose and/or arachnoid. **Phyllaries** in 4–8 series, imbricate to subequal, bases short-appressed, entire, abaxial faces with or without narrow glutinous ridge, apices stiffly spreading to ascending, linear, long, plane, spines straight, slender, 3–5 mm; apices of inner flexuous, sometimes expanded and erose. **Corollas** white, ochroleucous, or occasionally pink, 20–28 mm, tubes 10–13 mm, throats 6.5–9 mm, lobes 5–7 mm; style tips 3–5.5 mm. **Cypselae** dark brown, 5–6.5 mm, apical collars not differentiated; **pappi** 18–22 mm. $2n = 34$.

Flowering summer (Jun–Sep). Moist soil, grasslands, aspen parkland, forest edges and openings, subalpine, alpine meadows; 600–2900 m; Alta., B.C.; Idaho, Mont., Wash., Wyo.

Cirsium hookerianum occurs from the Canadian Coast Ranges of British Columbia east to the northern Cascade Range and the northern Rocky Mountains. The relationship between *C. hookerianum, C. kelseyi*, which I have tentatively included in *C. hookerianum*, and *C. longistylum* needs further investigation. A case could be made for including all three in an expanded concept of *C. hookerianum*, but more investigation of the variation patterns is needed before this is done. Certainly *C. kelseyi* is better treated within or as a close ally of *C. hookerianum* than in *C. scariosum* (var. *scariosum*), where R. J. Moore and C. Frankton (1974) synonymized it. *Cirsium hookerianum* is known to hybridize with *C. undulatum.*

49. **Cirsium longistylum** R. J. Moore & Frankton, Canad. J. Bot. 41: 1562, plate 1. 1963 • Long-style thistle C E

Perennials monocarpic, 40–150 cm; taprooted. **Stems** usually 1, erect, less commonly several, ascending, simple to sparingly short-branched in distal 1/2, less commonly openly branched, villous with jointed trichomes; branches on distal stems 0–many, short, ascending. **Leaves:** blades linear to oblong or elliptic, 10–30+ × 1–10 cm, margins flat to undulate, subentire to coarsely dentate or shallowly to deeply pinnatifid, lobes lance-oblong to broadly triangular, spinulose to spiny-dentate or shallowly lobed, main spines 3–12 mm, abaxial faces green and subglabrous to gray- or white-tomentose with felted arachnoid trichomes, ± villous to tomentose along major veins with septate trichomes, rarely glabrous or glabrate, adaxial ± green, glabrous or villous with septate trichomes; basal often present at flowering, spiny winged-petiolate or sessile; principal cauline well distributed, proximally winged-petiolate, distally sessile, gradually reduced, bases sometimes short-decurrent (0–2 cm); distal ± reduced, often narrower than the proximal, sometimes with non-pigmented bases. **Heads** several–many, erect, usually in racemiform or spiciform arrays, usually closely subtended by clustered ± leafy bracts. **Peduncles** 0–15+ cm. **Involucres** (green), broadly ovoid, 1.5–2.5 × 1.5–2.5 cm, loosely villous with septate trichomes, sparingly if at all arachnoid. **Phyllaries** in 4–8 series, subequal, ± lanceolate, bases appressed, apices ascending, linear to broadly expanded, erose to lacerate or spiny-fringed, spines straight, slender, 2–3 mm, abaxial faces with or without narrow glutinous ridge; apices of inner

flexuous, sometimes expanded and erose. **Corollas** white, ochroleucous, 19–23 mm, tubes 6.5–8.5 mm, throats 7.5–11 mm, lobes 4–5 mm; style tips 4–5.5 mm, conspicuously exserted. **Cypselae** brown, 5.5–6.5 mm, apical collars not differentiated; **pappi** 17–20 mm. **2n** = 34.

Flowering summer (Jun–Aug). Moist soil, roadsides, meadows, forest edges and openings; of conservation concern; 1500–2400 m; Mont.

Cirsium longistylum is endemic to the Big Belt, Castle, Elkhorn, and Little Belt ranges of west-central Montana. It is highly variable, and several authors have suggested that it has introgressed with one or more other species (R. J. Moore and C. Frankton 1963; J. M. Poole and B. L. Heidel 1993; S. J. Brunsfeld and C. T. Baldwin, unpubl.). It is closely related to *C. hookerianum*, and the two probably share a common ancestry or a history of hybrid interactions dating back to the Pleistocene. *Cirsium longistylum* is perhaps also affected by modern or historic introgression involving *C. scariosum* var. *scariosum.*

50. **Cirsium parryi** (A. Gray) Petrak, Bot. Tidsskr. 31: 68. 1911 • Parry thistle E

Cnicus parryi A. Gray, Proc. Amer. Acad. Arts 10: 47. 1874; *Cirsium gilense* (Wooton & Standley) Wooton & Standley; *C. inornatum* (Wooton & Standley) Wooton & Standley; *C. pallidum* (Wooton & Standley) Wooton & Standley; *C. parryi* subsp. *mogollonicum* Schaack & G. A. Goodwin

Biennials, 50–200+ cm; taprooted. **Stems** 1, erect, puberulent to pilose with jointed trichomes, sometimes also thinly arachnoid; branches 0–many, ascending, often nodding at tips. **Leaves:** blades oblong to lanceolate or oblanceolate, 10–30 × 2–5 cm, margins flat to undulate, spinulose and otherwise entire to coarsely dentate or shallowly to deeply pinnatifid, lobes well separated, spinulose to coarsely few-dentate, main spines slender to stout, 1–15 mm, one or both faces thinly pilose, sometimes thinly arachnoid, green and ± glabrescent at maturity; basal usually absent at flowering, sessile or winged-petiolate; principal cauline well distributed, proximal absent at flowering, moderately reduced distally, winged-petiolate or sessile (proximal), sessile and auriculate-clasping to slightly decurrent 0–2 cm; distal well developed, spreading, lobed or unlobed. **Heads** 1–many, ± erect, loosely to densely clustered at tip of main stem and branches in subcapitate to racemiform arrays, often also in distal leaf axils, closely subtended by clusters of unlobed to deeply dissected, often very spiny bracts. **Peduncles** 0–4 cm. **Involucres** hemispheric to subspheric, 1.5–2.5 × 1.5–3 cm, glabrous to finely arachnoid and/or pilose, often long pilose-ciliate with arachnoid trichomes connecting adjacent phyllaries. **Phyllaries** in 5–8 series, imbricate to subequal, proximally greenish, distally darker, becoming brownish, linear to narrowly lanceolate, outer often nearly as long as inner, abaxial faces with poorly developed glutinous ridge; outer and mid bases appressed, apices loosely ascending to spreading, bodies entire to spiny-ciliate or terminal appendages expanded, ± scarious, pectinately fringed, spines straight, 2–6 mm; apices of inner flat or spine-tipped, sometimes expanded and fimbriate. **Corollas** ochroleucous to ± yellow (rarely white or purple), 11–17 mm, tubes 5.5–11 mm, throats 2–4 mm, lobes 3–5 mm; style tips 2–4 mm. **Cypselae** tan to dark brown, 4–6 mm, apical collars narrow, not differently colored; **pappi** 9–15 mm. $2n = 34$.

Flowering summer–fall (Jul–Oct). Stream banks, montane meadows, damp soil in montane coniferous forests; 2100–3700 m; Ariz., Colo., N.Mex.

Cirsium parryi ranges from the Rocky Mountains of central and southern Colorado south to the San Francisco Peaks, Pinaleno Mountains, and White Mountains of Arizona, and the Mogollon and Sacramento ranges of southern New Mexico. Within this broad range several minor variants have been recognized at the species level. The features that supposedly distinguish *C. gilense*, *C. inornatum*, and *C. pallidum* vary widely and inconsistently through the range of the species. In like manner the characters used by Schaack and Goodwin to distinguish subsp. *mogollonicum* fall well within the variation of the species as a whole and do not seem adequate to separate subsp. *mogollonicum* from the rest of *C. parryi* at any taxonomic rank. *Cirsium parryi* hybridizes with *C. grahamii* in Arizona and *C. canescens* in Colorado.

51. **Cirsium eatonii** (A. Gray) B. L. Robinson, Rhodora 13: 240. 1911 (as eatoni) • Mountaintop or Eaton's thistle E F

Cnicus eatonii A. Gray, Proc. Amer. Acad. Arts 19: 56. 1883 (as eatoni), based on *Cirsium eriocephalum* A. Gray [not Wallroth] var. *leiocephalum* D. C. Eaton in S. Watson, Botany (Fortieth Parallel), 196. 1871

Perennials, 10–150 cm; taprooted caudices. **Stems** 1–several, (fleshy), erect or ascending, simple to sparingly branched in distal ¹/₂, sometimes openly branched, glabrous to villous or tomentose with septate trichomes, sometimes ± glabrate; branches on distal stems 0–many, ascending. **Leaves:** blades oblong, 10–30 × 1–5 cm, margins usually strongly undulate, unlobed and spiny-dentate or shallowly to deeply pinnatifid with 10–20 pairs of lobes,

teeth or lobes closely spaced, often overlapping, lance-oblong to broadly triangular, deeply 3-lobed, ± spiny-dentate, main spines 2–12 mm, abaxial faces glabrous or villous with septate trichomes along midveins to densely arachnoid-tomentose, adaxial glabrous or villous with septate trichomes along midveins; basal often present at flowering, spiny winged-petiolate or sessile; principal cauline many, well distributed, proximally ± winged-petiolate, distally sessile, gradually reduced; distal not much reduced, often closely subtending heads. **Heads** 1–many, erect or nodding, closely subtended by spiny-fringed bracts, usually sessile or short-pedunculate and crowded in subcapitate, spiciform, or racemiform (less commonly in openly branched) arrays. **Peduncles** 0–14+ cm. **Involucres** green or suffused with dark purple, broadly ovoid to campanulate, 2–5 × 1.5–5 cm (appearing wider when pressed), loosely to densely villous or tomentose with septate trichomes and/or arachnoid-tomentose with finer, non-septate trichomes. **Phyllaries** in 4–5 series, subequal, bases short-appressed, abaxial faces without or with very narrow glutinous ridge, apices usually stiffly ascending to spreading, linear-acicular, tapering to spines 7–35 mm; outer usually pinnately spiny, sometimes entire; apices of inner straight, plane or spine-tipped. **Corollas** ochroleucous or yellow to lavender, pink, or purple, 15–35 mm, tubes 3.5–10 mm, throats 5–14 mm, lobes (linear), 4–12.5 mm; style tips 3–6 mm, conspicuously exserted beyond corolla lobes. **Cypselae** dark brown, 5.5–7 mm, apical collars stramineous or not differentiated; **pappi** 12–25 mm.

Varieties 7 (7 in the flora): w United States; Rocky Mountains and high peaks of Great Basin desert region.

Cirsium eatonii is a polymorphic species widely distributed in a high elevation archipelago across the central Rocky Mountains and the Intermountain Region. During Pleistocene glacial episodes, the progenitors of this species complex undoubtedly occupied lower elevation sites and likely had more contiguous populations. Post-glacial isolation of these populations in allopatric high elevation sites has allowed them to differentiate to a greater or lesser extent. Prehistoric or recent introgressive hybridization with other thistle species probably has contributed to the diversification of the complex (R. J. Moore and C. Frankton 1965). Several of the races recognized here as varieties have been treated in the past as species (e.g., *C. clokeyi*, *C. peckii*). Their current geographic isolation and more or less distinctive features might support such recognition, but application of this approach across the complex would result in a proliferation of microspecies.

1. Involucres densely tomentose, individual phyllaries ± obscured by pubescence.
    2. Arrays nodding; corollas yellow or pink to pale purple . . . . . 51d. *Cirsium eatonii* var. *eriocephalum*
    2. Arrays erect; corollas white or pink to deep purple.
        3. Corolla throats 8–11.5 mm; se Oregon, nw Nevada . . . . 51g. *Cirsium eatonii* var. *peckii*
        3. Corolla throats 3.5–8 mm.
            4. Plants 50–150 cm, strictly erect; corollas pink or pale to deep purple; s Colorado . . . .
                . . . . 51e. *Cirsium eatonii* var. *hesperium*
            4. Plants 10–65 cm, ascending to erect; corollas white to pink or lavender; Idaho and Montana to Utah and n Colorado
                . . . . 51f. *Cirsium eatonii* var. *murdockii*
1. Involucres glabrous or thinly tomentose, individual phyllaries evident.
    5. Corollas usually 25–35 mm.
        6. Longest pappus bristles 16–18 mm . . . .
            . . . . 51b. *Cirsium eatonii* var. *clokeyi*
        6. Longest pappus bristles 20–25 mm . . . .
            . . . . 51c. *Cirsium eatonii* var. *viperinum*
    5. Corollas usually 17–25 mm.
        7. Most outer phyllaries with lateral spines; Nevada and Utah . . . .
            . . . . 51a. *Cirsium eatonii* var. *eatonii*
        7. Most outer phyllaries without lateral spines; nw Nevada and se Oregon . . . . .
            . . . . 51g. *Cirsium eatonii* var. *peckii*

51a. Cirsium eatonii (A. Gray) B. L. Robinson var. **eatonii** E

Cirsium eatonii var. *harrisonii* S. L. Welsh

Stems ascending to erect, slender, 10–50 cm. **Leaf faces** glabrous or nearly so. **Heads** usually short-pedunculate, in erect, short, racemiform or spiciform arrays, rarely openly branched. **Involucres** 2–2.5 cm, thinly arachnoid-tomentose or glabrate, individual phyllaries evident. **Phyllaries** sometimes suffused with dark purple; outer with few–many lateral spines; apical spines slender to stout. **Corollas** purple, 17–26 mm, tubes 5–10 mm, throats 4.5–10.5 mm, lobes 5.5–7.5 mm. **Pappi** 15–19 mm.

Flowering summer (Jul–Sep). Rocky slopes, canyons, pinyon-juniper woodlands to alpine, montane coniferous forests, subalpine forests, alpine slopes; 2100–3500 m; Nev., Utah.

Variety *eatonii* is distributed on various of the sky islands in the Basin and Range province of Nevada and Utah. Habitats vary from shaded forest understory sites to forest openings or open exposed sites.

51b. Cirsium eatonii (A. Gray) B. L. Robinson var. **clokeyi** (S. F. Blake) D. J. Keil, Sida 21: 212. 2004
• Clokey or Spring Mountains or white-spine thistle
C E

Cirsium clokeyi S. F. Blake, Proc. Biol. Soc. Wash. 51: 8. 1938

Stems erect or ascending, stout, 40–150 cm. **Leaf faces** glabrous or nearly so. **Heads** usually short-pedunculate in erect, racemiform arrays or sometimes long-pedunculate in openly corymbiform arrays. **Involucres** 3–5 cm, glabrous or thinly arachnoid tomentose, individual phyllaries evident. **Phyllaries** green or purplish-tinged; outer pectinately spiny 1/2 their length with many lateral spines; apical spines stout. **Corollas** purple, 24–33 mm, tubes 3.5–7 mm, throats 11–14 mm, lobes 8–12.5 mm. **Pappi** 16–18 mm. $2n = 34$ (as *C. clokeyi*).

Flowering summer (Jul–Sep). Gravelly slopes, ravines, montane coniferous forests, subalpine forests, alpine scree; of conservation concern; 2300–3500 m; Nev.

Variety *clokeyi* is endemic to the Spring Range of Clark County. Its range overlaps that of *Cirsium arizonicum* var. *tenuisectum* but no hybrids between the two are known.

51c. Cirsium eatonii (A. Gray) B. L. Robinson var. **viperinum** D. J. Keil, Sida 21: 212. 2004
• Snake Range thistle E

Stems erect or ascending, 25–40 cm. **Leaf faces** glabrous or nearly so. **Heads** 1–5, subsessile or short-pedunculate, in erect, racemiform or corymbiform arrays. **Involucres** 3–5 cm, thinly arachnoid with non-septate trichomes, individual phyllaries evident. **Phyllaries** green or purplish-tinged; outer with numerous lateral spines; apical spines stout. **Corollas** lavender to purple, 29–35 mm, tubes 9–12.5 mm, throats 9–12 mm, lobes 9–11 mm. **Pappi** 20–25 mm.

Flowering summer (Jul–Sep). Rocky subalpine slopes, open bristlecone pine forests; 3300–3500 m; Nev.

Variety *viperinum* is apparently endemic to upper elevations of the Snake Range of White Pine County, Nevada. Heads of var. *viperinum* are similar in size to those of var. *clokeyi*. These taxa can be distinguished readily by the features in the key. Ranges of the two varieties are separated by about 340 km. Hybridization with *Cirsium inamoenum* is suspected based upon apparently intermediate specimens.

51d. Cirsium eatonii (A. Gray) B. L. Robinson var. **eriocephalum** (A. Gray) D. J. Keil, Sida 21: 1645. 2005 • Mountain or alpine thistle E

Cnicus eriocephalus A. Gray, Proc. Amer. Acad. Arts 10: 46. 1874, based on *Cirsium eriocephalum* A. Gray, Proc. Acad. Nat. Sci. Philadelphia 15: 69. 1864, not Wallroth 1841; *Carduus hookerianus* Nuttall var. *eriocephalus* (A. Gray) A. Nelson; *Cirsium scopulorum* (Greene) Cockerell ex Daniels

Stems erect or ascending, 30–150 cm. **Leaf faces** thinly to densely arachnoid-tomentose, especially abaxially, and villous or tomentose with septate trichomes, especially along midveins, adaxially ± glabrate. **Heads** sessile or short-pedunculate in usually nodding, densely woolly-tomentose, spiciform or racemiform arrays. **Involucres** 2.5–3.5 cm, densely tomentose with septate trichomes (individual phyllaries obscured by tomentum). **Phyllaries:** outer with stiff lateral spines; apical spines slender. **Corollas** yellow (northern populations) or pink to pale purple (southern populations), 13–18 mm, tubes 6–9 mm, throats 3–5.5 mm, lobes 3–7 mm. **Pappi** 10–15 mm. $2n$ = 34 (as *C. scopulorum*).

Flowering summer (Jul–Sep). Forest openings, alpine and subalpine meadows, windswept alpine ridges; 2200–3800 m; Colo., N.Mex., Utah.

51e. Cirsium eatonii (A. Gray) B. L. Robinson var. **hesperium** (Eastwood) D. J. Keil, Sida 21: 212. 2004 • Tall mountain thistle E

Cnicus hesperius Eastwood, Bull. Calif. Acad. Sci., ser. 3, 1: 122. 1898; *Cirsium hesperium* (Eastwood) Petrak

Stems strictly erect, 50–150 cm. **Leaf faces** glabrous or nearly so or abaxially finely arachnoid-tomentose and/or villous to tomentose with septate trichomes on one or both faces. **Heads** usually sessile in stiffly erect, tight spiciform arrays, sometimes also sessile in distal leaf axils. **Involucres** 2–2.5 cm, tomentose with septate trichomes; outer phyllaries with a few stiff lateral spines; spines of phyllaries slender. **Corollas** pink or pale to deep purple, 14–21 mm, tubes 4.5–10 mm, throats 3.5–5 mm, lobes 3.5–7 mm. **Pappi** 8–17 mm.

Flowering summer (Jul–Sep). Rocky slopes, subalpine meadows, forest openings; 2700–3400 m; Colo.

Variety *hesperium* is distributed in the San Juan Mountains and Spanish Peaks area of southern Colorado. It differs from var. *eriocephalum* in its stiffly erect arrays. Plants from the Spanish Peaks area approach var. *eriocephalum*.

51f. Cirsium eatonii (A. Gray) B. L. Robinson var. **murdockii** S. L. Welsh, Great Basin Naturalist 42: 200. 1982 • Northern mountain thistle C E F

Cirsium murdockii (S. L. Welsh) Cronquist; *C. polyphyllum* (Rydberg) Petrak; *C. tweedyi* (Rydberg) Petrak

Stems erect or ascending, stout, 10–75 cm. **Leaf faces** glabrous or nearly so or abaxial ± villous and/or tomentose with septate or non-septate trichomes. **Heads** sessile or short-pedunculate in erect, few-headed, subcapitate or spiciform arrays. **Involucres** 2–3 cm, thinly to densely tomentose with septate or non-septate trichomes (phyllaries sometimes obscured by tomentum). **Phyllaries** green or purplish-tinged; outer with few lateral spines; apical spines slender. **Corollas** white to pink or lavender, 17–23 mm, tubes 6–10 mm, throats 5–8 mm, lobes 3–6 mm. **Pappi** 11–18 mm. $2n$ = 34 (as *C. tweedyi*).

Flowering summer (Jul–Sep). Talus slopes, rocky subalpine and alpine ridges, openings in subalpine forests, subalpine meadows; of conservation concern; 2300–3900 m; Colo., Idaho, Mont., Nev., Utah, Wyo.

Variety *murdockii* grows in the central Rocky Mountains from central Idaho and southern Montana to western Wyoming and northern Colorado. It also grows in the Ruby Mountains of northern Nevada and the Uintah Mountains of northeastern Utah. Considerable variation occurs in quantity and quality of pubescence of the involucres of var. *murdockii*. In some individuals the pubescence is so dense that it obscures the phyllary bodies whereas in others the individual phyllaries are clearly visible. Most commonly the pubescence consists primarily of septate trichomes, but sometimes finer non-septate trichomes predominate. Both types of trichomes may be present.

Variety *murdockii* is known to hybridize with *Cirsium pulcherrimum* var. *pulcherrimum* and *C. scariosum* var. *scariosum* in Wyoming.

51g. Cirsium eatonii (A. Gray) B. L. Robinson var. **peckii** (L. F. Henderson) D. J. Keil, Sida 21: 212. 2004 • Steens Mountain or ghost thistle [E]

Cirsium peckii L. F. Henderson, Madroño 5: 97. 1939

Stems erect or ascending, stout, 10–150 cm. **Leaf faces** glabrous or nearly so or abaxial ± villous with septate trichomes. **Heads** several–many, usually subsessile or short-pedunculate, crowded in erect, spiciform or racemiform arrays, less commonly openly branched. **Involucres** 2–4 cm, loosely to densely villous with septate trichomes and thinly arachnoid with non-septate trichomes. **Phyllaries** green; outer with few or no lateral spines; apical spines stout. **Corollas** pink to purple, 21–25 mm, tubes 6.5–8 mm, throats 8–11.5 mm, lobes 5.5–7 mm. **Pappi** 17–22 mm. $2n = 34$ (as *C. peckii*).

Flowering summer (Jun–Aug). Grasslands, juniper woodlands, grass-sagebrush steppes, subalpine slopes, roadsides; 1300–2900 m; Nev., Oreg.

Variety *peckii* occurs from Steens Mountain and the Pueblo Mountains of Harney County, Oregon, south to the Black Rock Range and Jackson Mountains of Humboldt County, Nevada. On the lower portion of its distribution along Steens Mountain Loop Road at an elevation of about 1900 m, var. *peckii* and *Cirsium inamoenum* grow together and freely hybridize, forming a complex hybrid swarm. Intermediates variably combine the features of the parent taxa, with the habit more openly branched than typical for *C. eatonii* var. *peckii*, variably arachnoid tomentose leaves, ± glandular phyllaries, and lavender to pale pink flowers. Variety *peckii* may range as far south as the West Humboldt Mountains. The type of *Cirsium humboldtense* Rydberg (*Carduus nevadensis* Greene) closely resembles some of the hybrids between *C. inamoenum* and *C. eatonii* var. *peckii*.

52. Cirsium ownbeyi S. L. Welsh, Great Basin Naturalist 42: 200. 1982 • Ownbey's thistle [E]

Perennials, 30–70 cm; taproots and branched caudices with persistent dark-brown leaf bases. **Stems** 1–several, erect, simple or sparingly branched in distal 1/2, glabrous or thinly arachnoid and sparingly villous with jointed trichomes. **Leaves:** blades oblong to elliptic or oblanceolate, 15–30+ × 2–7 cm, deeply 2–3-pinnately divided, lobes linear to linear-lanceolate, spinulose to spiny-dentate or shallowly lobed, main spines slender, 2–8 mm, abaxial faces glabrous to thinly tomentose and villous along major

veins, soon glabrescent, adaxial glabrous; basal present at flowering, narrowly spiny winged-petiolate; principal cauline well distributed, proximal winged-petiolate, mid and distal sessile, gradually reduced, bases decurrent as spiny wings 1–3 cm; distalmost reduced to spiny-pectinate bracts. **Heads** 1–few, erect, ± crowded in corymbiform arrays. **Peduncles** 0–4 cm. **Involucres** ovoid, 1.8–2.5 cm, 1.5–2.5 cm diam., loosely arachnoid, glabrate. **Phyllaries** in 5–6 series, imbricate, green, linear-lanceolate, abaxial faces without or with poorly developed glutinous ridge; outer and mid bases appressed, apices stiffly radiating to ascending, long, very narrow, entire, spines slender, 3–10 mm; apices of inner straight, flexuous. **Corollas** white to pink or pink-purple, 16–20 mm, tubes 6–8 mm, throats 5–6 mm, lobes 5–7 mm; style tips 3.5–4.5 mm. **Cypselae** brown, ca. 4 mm, apical collars not differentiated; **pappi** 13–17 mm.

Flowering summer (Jun–Aug). Stony soils in sparsely vegetated areas of pinyon-juniper woodlands, sagebrush scrub, arid grasslands, and riparian scrub, in dry sites or sometimes on seeps; 1500–2400 m; Colo., Utah, Wyo.

Cirsium ownbeyi is endemic to the eastern side of the Uintah Mountains in northeastern Utah, southwestern Wyoming, and northwestern Colorado. It is in the Center for Plant Conservation's National Collection of Endangered Plants.

53. Cirsium drummondii Torrey & A. Gray, Fl. N. Amer. 2: 459. 1843 • Drummond's or dwarf or short-stemmed thistle [E]

Cirsium coccinatum Osterhout

Biennials or monocarpic perennials, acaulescent or caulescent, 5–110 cm; taproots stout. **Stems** erect, stout, fleshy, leafy, simple or distally branched, villous or tomentose with long, septate trichomes; branches usually short, stout, ascending. **Leaves:** blades oblong-elliptic to oblanceolate, 15–30+ × 3–7 cm, usually shallowly to deeply pinnatifid, lobes ovate to broadly triangular, spreading, usually separated by broad U-shaped sinuses, spinose-dentate or coarsely lobed, main spines 2–5(–8) mm, slender, abaxial faces villous with septate trichomes, at least along veins, sometimes thinly arachnoid, adaxial villous with septate trichomes; basal often present at flowering, spiny winged-petiolate; principal cauline winged-petiolate or sessile, not much reduced distally; distal reduced, similar to proximal, crowded around heads. **Heads** 1–5(–9), borne singly or crowded in corymbiform arrays at tips of main stems, often closely subtended and overtopped by 1–several distal leaves. **Peduncles** 0–5(–10) cm, leafy-bracted. **Involucres** broadly ovoid to hemispheric, 3.5–5 × 3.5–5 cm (appearing much wider and ± campanulate in pressed specimens), loosely arachnoid on phyllary margins or glabrate. **Phyllaries** in 4–6 series, strongly imbricate, ovate or broadly lanceolate (outer) to lance-linear (inner), abaxial faces with ± narrow glutinous ridge; outer and mid appressed, spines erect to ascending, 2–3 mm; apices of mid and inner narrowed and scabrid-denticulate, innermost spineless, with expanded, flexuous, erose-denticulate tips. **Corollas** purple (white), 30–48 mm, tubes 17–30 mm, throats 6.5–11 mm, lobes 5–7 mm; style tips 5–7 mm. **Cypselae** stramineous to light brown, 3.5–5.5 mm, apical collar yellow, narrow; **pappi** 30–42 mm.

Flowering summer (Jun–Aug). Dry to moist soil, prairies, pastures, meadows, forest edges, woodland openings, roadsides; 300–2300 m; Alta., B.C., Man., N.W.T., Ont., Sask.; Colo., S.Dak., Wyo.

Cirsium drummondii is widely distributed across Canada from the Northwest Territories to British Columbia and Ontario. The name *C. drummondii* has been misapplied to a wide range of plants across the western United States that are now treated as one or another variety of the polymorphic *C. scariosum*. The only documented modern occurrences of *C. drummondii* in the United States are in the Black Hills of South Dakota and adjacent Wyoming. Specimens collected by Hall and Harbour (*342*) are the only ones of *C. drummondii* known from Colorado. Somewhat similar plants from northern Nevada are treated here as *C. scariosum* var. *toiyabense*. During Pleistocene glaciations the ancestors of *C. drummondii* undoubtedly occupied a more southerly distribution and very likely came into direct contact with populations of *C. scariosum*. The observed similarities between *C. drummondii* and *C. scariosum* var. *toiyabense* are probably a relict of that ancient contact.

54. **Cirsium scariosum** Nuttall, Trans. Amer. Philos. Soc., n. s. 7: 420. 1841 (as Circium) • Meadow or elk thistle, chardon écailleux [F]

Cirsium hookerianum Nuttall var. *scariosum* (Nuttall) B. Boivin

Biennials or monocarpic perennials, acaulescent, short caulescent and forming low rounded mounds, or caulescent and erect, 0–200 cm; taprooted. **Stems** absent, or with crowded branches from near base, or simple and erect, often fleshy and thickened, glabrous to thinly gray-tomentose, often villous with septate trichomes. **Leaves:** blades linear to elliptic, 5–20 × 3–7 cm, plane to strongly undulate, unlobed or shallowly to deeply pinnatifid, lobes linear-lanceolate to broadly triangular, closely spaced, spreading, spinose-dentate or lobed, main spines slender to stout, 2–15+ mm, abaxial faces glabrous or thinly to densely tomentose, ± villous with septate

trichomes along the veins, glabrate or trichomes persistent, adaxial thinly arachnoid tomentose and soon glabrescent; basal often present at flowering, sessile or winged-petiolate; cauline many in caulescent forms, reduced distally or not, winged-petiolate or distal sessile; distal often well developed, similar to proximal, sometimes much narrower and bractlike. **Heads** 1–many, erect, borne singly or often densely crowded in spiciform, racemiform, or subcapitate arrays, especially in acaulescent or short-caulescent plants, often closely subtended by distalmost leaves. **Peduncles** 0–10 cm, leafy-bracted. **Involucres** ovoid to hemispheric, 2–4 × 1.5–6 cm, loosely arachnoid on phyllary margins or glabrate. **Phyllaries** in 5–10 series, imbricate, ovate or lanceolate (outer) to linear or linear-lanceolate (inner), margins (outer) entire or scarious-fringed, abaxial faces without glutinous ridge; outer and mid appressed, spines erect to spreading 0.5–13 mm; apices of mid and inner narrowed and scabro-denticulate or with expanded, erose-dentate tips, spineless or tipped with flattened spines. **Corollas** white or pale lavender to purple, 20–40 mm, tubes 7–24 mm, throats 4–12 mm (noticeably larger than tubes), lobes 4–10 mm; style tips 3.5–8 mm. **Cypselae** light to dark brown, 4–6.5 mm, apical collars usually colored like body; **pappi** 17–35 mm, white to tan. $2n = 34, 36$.

Varieties 8 (8 in the flora): w North America, disjunct to e Quebec (Mingan Archipelago).

Cirsium scariosum is a widely distributed complex of intergrading races distributed from southwestern Canada to northwestern Mexico. These plants range from acaulescent rosettes with a tight cluster of sessile heads to tall, erect, unbranched plants, or moundlike, more or less openly branched herbs. Acaulescent and caulescent plants sometimes occur in the same population.

Members of this complex have been variously treated in the past. F. Petrak (1917) recognized ten species plus several subspecies for the taxa I am treating here as *C. scariosum* (in the broad sense). In floras, the names *C. drummondii* and *C. foliosum* have been widely misapplied to these plants (R. J. Moore and C. Frankton 1964). The latter two species, while clearly related to *C. scariosum*, have a range restricted mostly to Canada. Moore and Frankton (1967) attempted to bring order to the complex and recognized four species for plants that I include here in *C. scariosum*: *C. acaulescens*, *C. congdonii*, *C. coloradense*, and *C. scariosum* in the restricted sense. Moore and Frankton substituted the prior name *C. tioganum* for *C. acaulescens*. Unfortunately they did not extend their study widely enough and did not include some members of the complex in their investigations. S. L. Welsh (1982) proposed *C. scariosum* var. *thorneae* from Utah and lumped the various species recognized by Moore and Frankton within a highly polymorphic var. *scariosum*. After consulting with A. Cronquist and studying his manuscript treatment of

Cirsium for the *Intermountain Flora*, D. J. Keil and C. E. Turner (1993) also accepted a broadly construed *C. scariosum*. Cronquist (1994) treated *C. scariosum* as an extremely variable species that included the four species recognized by Moore and Frankton plus the variety proposed by Welsh. Cronquist chose to not recognize infraspecific taxa.

In the present treatment I have examined these plants from a biogeographic perspective with the goal of discerning regional patterns of variation. The large number of specimens available has allowed me to examine distributional patterns in relation to the topography and biogeographic history of the regions where this species occurs. My field studies also have provided me with observations that help to explain some of the anomalous specimens represented in herbaria. Although the variation within and between populations is sometimes amazing, more-or-less differentiated geographic races can be discerned. Because of the extraordinary and overlapping patterns of variation across the range of *Cirsium scariosum*, the following key to varieties should be regarded as at best an approximation.

1. Plants acaulescent (occasional short-caulescent individuals sometimes present in a population).
  2. Corollas pink to purple; Sierra Nevada of w Nevada and e California to San Bernardino Mountains of s California . . . .
    . . . . 54e. *Cirsium scariosum* var. *congdonii*
  2. Corollas white to faintly pink-or lilac-tinged; widespread.
    3. Abaxial leaf faces usually gray-tomentose; widespread, Colorado to s Oregon, n California . . . .
      . . . . . 54c. *Cirsium scariosum* var. *americanum*
    3. Abaxial leaf faces usually green, glabrous or glabrate; s. California . . . .
      . . . . 54g. *Cirsium scariosum* var. *citrinum*
1. Plants caulescent (occasional acaulescent individuals sometimes present in a population).
  4. Larger leaf spines 1–3 cm . . . .
    . . . . 54d. *Cirsium scariosum* var. *thorneae*
  4. Larger leaf spines usually shorter.
    5. Corollas purple.
      6. Corolla lobes 5.5–8 mm; sw. Idaho, n Nevada, se Oregon . . . .
        . . . . . 54f. *Cirsium scariosum* var. *toiyabense*
      6. Corolla lobes 3.5–6 mm; e Oregon to sw Montana . . . .
        . . . . . 54b. *Cirsium scariosum* var. *scariosum*
    5. Corollas white to faintly pink- or lilactinged.
      7. Stems usually proximally branched, plants often forming low, rounded mound; heads usually borne on short to ± elongate lateral branches; corollas 26–36 mm.

8. Apices of inner phyllaries acuminate and entire or rarely toothed; s California . . . . . . 54g. *Cirsium scariosum* var. *citrinum*
8. Apices of inner phyllaries usually expanded as a scarious, erose-toothed appendage; ne California, se Oregon . . . . 54h. *Cirsium scariosum* var. *robustum*

[7. Shifted to left margin.—Ed.]

7. Stems usually erect, proximally unbranched; heads usually sessile or short-pedunculate in subcapitate to spiciform or racemiform arrays, usually closely subtended by numerous distal leaves; corollas 20–29 mm.
  9. Heads usually ± tightly clustered at stem tips, closely subtended and often overtopped by crowded distal leaves; distal leaves ± thin, usually fringed with numerous weak spines, often ± unpigmented proximally or tinged pink to purplish . . . . 54a. *Cirsium scariosum* var. *scariosum*
  9. Heads usually in ± leafy, racemiform arrays, usually subtended by ± reduced, bractlike distal leaves; distal leaves firm, strongly spiny, usually green throughout . . . .
    . . . . 54b. *Cirsium scariosum* var. *coloradense*

54a. Cirsium scariosum Nuttall var. **scariosum** E F

Cirsium butleri (Rydberg) Petrak; *C. lacerum* (Rydberg) Petrak; *C. magnificum* (A. Nelson) Petrak; *C. minganense* Victorin

Plants erect, caulescent or occasionally acaulescent, 15–200 cm. **Stems** usually simple, very leafy, glabrous or villous to tomentose with septate trichomes. **Leaves:** blades linear to oblong, oblanceolate, or narrowly elliptic, pinnately lobed or often unlobed, larger leaf spines usually 1 cm or less, abaxial faces glabrous to gray-tomentose, adaxial glabrous or villous with septate trichomes; distal narrow, ± thin, often unpigmented proximally or tinged pink or purplish, spines numerous, weak. **Heads** 1–10+, sessile or short-pedunculate, tightly clustered at stem tips, usually subtended and ± overtopped by crowded, distal leaves. **Involucres** 2–3.5 cm. **Phyllaries:** outer and mid lanceolate to ovate, spines slender to stout, 1–8 mm; apices of inner acuminate and entire or abruptly expanded into scarious, erose-toothed appendages. **Corollas** white to purple, 20–28 mm, tubes 9–14.5 mm, throats 6–10 mm, lobes 3.5–6 mm; style tips 3–6.5 mm. **Cypselae** 4.5–6.5 mm; **pappi** 17–25 mm. $2n$ = 34 (as *C. foliosum*), 36.

Flowering summer (Jun–Sep). Moist, sometimes saline soils, meadows, ditches, stream banks, forest openings, sagebrush zone to subalpine forests; 0 (Quebec) or

600–2800 m; Alta., B.C., Que.; Calif., Colo., Idaho, Mont., Oreg., Utah, Wash., Wyo.

The presence of *Cirsium scariosum* on the islands of the Mingan Archipelago in Quebec, some 3200 km east of the Rocky Mountains populations, has led to alternative hypotheses regarding the disjunction. Frère Marie-Victorin (1925) hypothesized that the disjunct distribution of *C. minganense* from what he called *C. foliosum* (Hooker) Candolle was a result of migration during deglaciation (18,000 to ca. 8000 BP) from a glacial refugium in western North America to eastern Canada in the barren habitats along the receding ice front. Later (1938) he presented a second hypothesis that Pleistocene glacial events had divided a preglacial range into vicariant populations that survived in separate refugia in western and eastern regions. R. J. Moore and C. Frankton (1967) argued that the disjunction is modern, resulting from a chance introduction of *C. scariosum* from western North America to Quebec in the early twentieth century. They reached this conclusion because early collectors that had visited the Mingan Archipelago had failed to collect this conspicuous thistle.

Hybrids are known between *Cirsium scariosum* var. *scariosum* and *C. eatonii* var. *murdockii* in northern Wyoming.

54b. Cirsium scariosum Nuttall var. **coloradense** (Rydberg) D. J. Keil, Sida 21: 215. 2004 • Colorado thistle [E]

Carduus coloradensis Rydberg, Bull. Torrey Bot. Club 32: 132. 1905; *Cirsium coloradense* (Rydberg) Cockerell ex Daniels; *C. erosum* (Rydberg) K. Schumann; *C. olivescens* (Rydberg) Petrak; *C. tioganum* (Congdon) Petrak var. *coloradense* (Rydberg) Dorn

Plants usually erect, caulescent (rarely acaulescent), 20–150 cm. **Stems** usually simple, proximally unbranched, very leafy, ± villous with septate trichomes and/or thinly arachnoid tomentose, often glabrate. **Leaves:** blades oblanceolate or narrowly elliptic, pinnately lobed, longer spines ± stout, usually 1 cm or shorter, abaxial faces glabrous to thinly gray-tomentose, adaxial glabrous; distal narrow, firm, green throughout or unpigmented proximally. **Heads** 3–10+, ± sessile or pedunculate, usually in spiciform or racemiform arrays, subtended by ± reduced bractlike distal leaves that often do not overtop the heads. **Peduncles** 0–18 cm. **Involucres** 2–3 cm. **Phyllaries:** outer and mid lanceolate to narrowly ovate, spines slender to stout, 1–5 mm; apices of inner acuminate and entire or serrate, or abruptly expanded into scarious, erose-toothed appendages. **Corollas** white (rarely purple), 22–29 mm, tubes 11–16 mm, throats 4–5 mm, lobes

6–9.5 mm; style tips 4–6 mm. **Cypselae** 4–6 mm; **pappi** 18–25 mm. **2***n* = 34, 36? (as *C. foliosum*).

Flowering summer (Jun–Sep). Wet soil, forests, meadows, roadsides; 1900–2500 m; Ariz., Colo., N.Mex., Utah, Wyo.

Variety *coloradense* is common in the mountains of southern and central Colorado with outlying populations in northern New Mexico, northeastern Arizona, central Utah, and southeastern Wyoming. In Colorado it is largely allopatric with the usually acaulescent var. *americanum*. Some plants from the White Mountains and San Francisco Peaks of Arizona and from Gunnison County, Colorado, approach var. *thorneae* in having deeply divided, extremely spiny distal leaves that overtop the heads. Putative hybrids between var. *coloradense* and *Cirsium grahamii* have been documented in Apache County, Arizona, and between var. *coloradense* and *C. undulatum* in Las Animas County, Colorado.

54c. Cirsium scariosum Nuttall var. **americanum** (A. Gray) D. J. Keil, Sida 21: 215. 2004 • Dinnerplate or sessile or stemless thistle [F]

Cirsium acaule (Linnaeus) Scopoli var. *americanum* A. Gray, Proc. Acad. Nat. Sci. Philadelphia 15: 68. 1864; *C. acaulescens* (A. Gray) K. Schumann; *C. americanum* (A. Gray) K. Schumann; *C. coloradense* (Rydberg) Cockerell ex Daniels subsp. *acaulescens* (A. Gray) Petrak; *C. coloradense* subsp. *longissimum* (A. Heller) Petrak; *C. drummondii* Torrey & A. Gray var. *acaulescens* (A. Gray) J. F. Macbride; *C. drummondii* subsp. *latisquamum* Petrak; *C. drummondii* var. *oregonense* Petrak; *C. drummondii* subsp. *vexans* Petrak; *C. tioganum* (J. W. Congdon) Petrak

Plants usually acaulescent or nearly so (with dense rosettes of leaves and cluster of sessile or subsessile heads), less commonly caulescent and to 45 cm. **Stems** absent or short, stout, fleshy, usually unbranched, very leafy, villous or tomentose with septate trichomes. **Leaves:** blades linear to oblong, oblanceolate, or narrowly elliptic, pinnately lobed or often unlobed, longer spines slender or stout, usually 1 cm or less, sometimes unpigmented proximally or tinged pink or purplish, abaxial faces gray-tomentose with fine, non-septate trichomes and/or villous with septate trichomes, adaxial glabrous or thinly villous with septate trichomes. **Heads** 1–many, sessile or subsessile, crowded. **Involucres** 1.5–3 cm. **Phyllaries:** outer and mid lanceolate to ovate, spines slender to stout, 1–12 mm; apices of inner acuminate and entire or abruptly expanded into scarious, erose-toothed appendage. **Corollas** white or pink-tinged (rarely purple), 22–30 mm, tubes 11.5–16.5 mm, throats 5–10.5 mm, lobes 5–6.5 mm; style tips 3.5–6.5 mm.

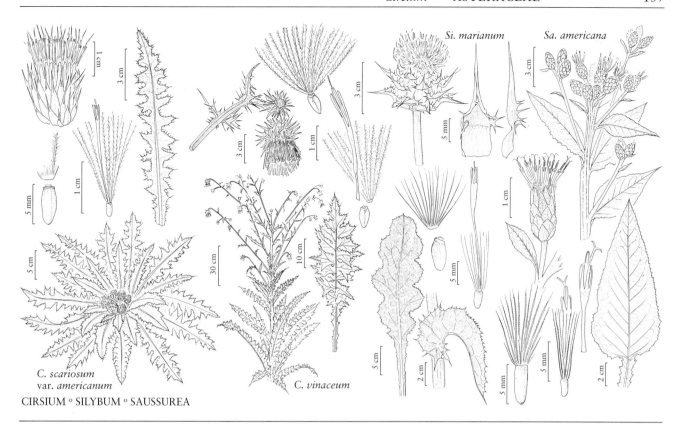

C. *scariosum*
var. *americanum*

C. *vinaceum*

Si. marianum

Sa. americana

CIRSIUM ∘ SILYBUM ∘ SAUSSUREA

Cypselae 4.5–6 mm; pappi 20–25 mm. **2n** = 34 (as *C. foliosum*).

Flowering summer (Jun–Aug). Seasonally damp, sometimes saline soil in meadows, grasslands, open forests, sagebrush scrub; 1100–3800 m; Calif., Colo., Idaho, Nev., Oreg., Utah, Wyo.; Mexico (Baja California).

Variety *americanum* is a variable taxon widely distributed in mountains and valleys of the western United States from the eastern slope of the Cascades and Sierra Nevada across the northern Great Basin to the central Rocky Mountains. Within some populations dominated by acaulescent forms, occasional short, caulescent individuals or even taller plants may occur. These approach var. *coloradense* in the Colorado Rockies and resemble var. *scariosum* in the valleys of Oregon. Short caulescent individuals within populations of var. *coloradense* and var. *scariosum* approach var. *americanum*. Although the extremes are strikingly different, the genetic differences are apparently slight. Acaulescent individuals from the San Francisco Peaks and White Mountains of Arizona are treated here as var. *coloradense*. In the mountains of Colorado putative hybrids have been documented between *Cirsium scariosum* var. *americanum* and *C. canescens* and between *C. scariosum* var. *americanum* and *C. undulatum*.

54d. Cirsium scariosum Nuttall var. **thorneae** S. L. Welsh, Great Basin Naturalist 42: 201. 1982

• Thorne's thistle [E]

Plants erect, caulescent (rarely nearly acaulescent), 20–130 cm. **Stems** distally short-branched to openly much-branched throughout, leafy, glabrous or villous with septate trichomes. **Leaves:** blades oblong, oblanceolate, or narrowly elliptic, usually pinnately lobed more than halfway to midveins, abaxial faces glabrous or gray-tomentose, adaxial glabrous; distal usually deeply divided, fiercely armed with stout spines, the longer 1–3 cm. **Heads** 1–10+, sessile or pedunculate, solitary or crowded near tip of main stem or branches, usually subtended and ± overtopped by distal leaves. **Involucres** (broadly ovoid to hemispheric) 2.5–3.5 cm. **Phyllaries:** outer and mid lanceolate to narrowly ovate, spines slender to stout, 2–8 mm; apices of inner usually abruptly expanded into scarious, erose-toothed appendages. **Corollas** white to dull purple, 29–34 mm, tubes 14–18 mm, throats 6.5–9 mm, lobes 7–8.5 mm; style tips 5–6.5 mm. **Cypselae** 4.5–5 mm; **pappi** 20–27 mm.

Flowering summer (Jun–Sep). Meadows, streamsides, valley bottoms, often in saline soils; 1500–2200 m; Colo., Idaho, Nev., Utah.

Variety *thorneae* grows mostly in the Basin and Range province of Utah with populations in eastern Nevada, southern Idaho, and western Colorado.

54e. Cirsium scariosum Nuttall var. **congdonii** (R. J. Moore & Frankton) D. J. Keil, Sida 21: 215. 2004

• Rosette thistle [E]

Cirsium congdonii R. J. Moore & Frankton, Canad. J. Bot. 45: 1738. 1967

Plants acaulescent or nearly so (with dense rosettes of leaves and cluster of sessile or subsessile heads). **Stems** absent or very short, simple, stout, fleshy, very leafy. **Leaves:** blades oblong to oblanceolate or elliptic, shallowly pinnately lobed or sometimes unlobed, longer spines slender, less than 1 cm, abaxial faces green and ± glabrous to white-tomentose, adaxial faces green, glabrous or villous with septate trichomes. **Heads** 1–many, sessile or subsessile, crowded. **Involucres** 2–3 cm. **Phyllaries:** outer and mid lanceolate to ovate, spines slender, 1–4 mm; apices of inner linear-acuminate, often twisted, entire or minutely toothed. **Corollas** pink to reddish purple, 22–30 mm, tubes 10–15 mm, throats 5–7.5 mm, lobes 4–7 mm; style tips 3–5 mm. **Cypselae** 4–4.5 mm; pappi 17–28 mm. **2*n*** = 34 (as *C. congdonii*).

Flowering summer (Jun–Aug). Meadows, springs, stream banks; 1500–3100 m; Calif., Nev.

Variety *congdonii* grows from the Sierra Nevada of western Nevada and eastern California to the San Bernardino Mountains of southern California.

54f. Cirsium scariosum Nuttall var. **toiyabense** D. J. Keil, Sida 21: 216. 2004 • Toiyabe thistle [E]

Plants erect, subacaulescent to caulescent, 5–55 cm. **Stems** usually simple, leafy, glabrous to villous or tomentose with septate trichomes, sometimes thinly arachnoid tomentose. **Leaves:** blades oblong to oblanceolate or elliptic, pinnately lobed or often unlobed, longer spines less than 1 cm, abaxial faces green and ± glabrous to white-tomentose, adaxial faces green, glabrous or villous with septate trichomes. **Heads** 1–10+, sessile or short-pedunculate, crowded at stem tips, usually subtended and ± overtopped by distal leaves. **Involucres** 2–3 cm. **Phyllaries:** outer and mid lanceolate to ovate, spines slender, 2–4 mm; apices of inner acuminate and entire or abruptly expanded into scarious, erose-toothed appendages. **Corollas** rose-purple, 23–31 mm, tubes 11–16 mm, throats 4.5–8.5 mm, lobes 5.5–8.5 mm; style tips 3.5–4.5 mm. **Cypselae** 4–6 mm; pappi 22–25 mm.

Flowering spring–summer (May–Aug). Meadows, pastures, springs; 1250–2300 m; Idaho, Nev., Oreg.

Variety *toiyabense* occurs in the western Basin and Range province from northern Nevada to southeastern Oregon and southwestern Idaho.

54g. Cirsium scariosum Nuttall var. **citrinum** (Petrak) D. J. Keil, Sida 21: 215. 2004

• La Graciosa thistle [C]

Cirsium quercetorum (A. Gray) Jepson var. *citrinum* Petrak, Beih. Bot. Centralbl. 35(2): 363. 1917; *Carduus validus* Greene; *Cirsium drummondii* Torrey & A. Gray subsp. *latisquamum* Petrak; *C. loncholepis* Petrak

Plants acaulescent to bushy and mounding or erect, 5–100 cm. **Stems** usually 1, branched distally or openly so throughout, leafy, glabrous or thinly arachnoid-tomentose. **Leaves:** blades oblanceolate to elliptic, deeply pinnately lobed, longer spines slender to stout, usually 1 cm or shorter, abaxial faces usually green, glabrous or glabrate or thinly to densely tomentose, adaxial faces glabrous to sparingly villous with septate trichomes or thinly arachnoid-tomentose. **Heads** 2–many, subsessile or short-pedunculate, in ± congested corymbiform to subcapitate arrays at stem tips (in age clustered axillary heads often developing), subtended and ± overtopped by distal leaves or these ± reduced. **Involucres** 2.5–4 cm. **Phyllaries:** outer and mid lanceolate to ovate, spines 1–6 mm, slender and weak or broad and flat; apices of inner stiffly erect or thin and often contorted, acuminate and entire or rarely toothed. **Corollas** white or lightly purple-tinged (purple), (21–)26–35 mm, tubes (11–)14–22 mm, throats 6–9 mm, lobes (4–)5.5–8 mm; style tips 4–6 mm. **Cypselae** 4.5–5 mm; pappi 23–31 mm. **2*n*** = 34 (as *Cirsium loncholepis*).

Flowering mostly spring–summer (Apr–Jul, occasionally as late as Nov in coastal sites). Wet ground, meadows, pastures, springs, marshes, both coastal and interior; of conservation concern; 0–2500 m; Calif.; Mexico (Baja California?).

Variety *citrinum* is the southern California race of *Cirsium scariosum*. As is the case in other varieties of this complex, taxonomic recognition has been given to local populations, named on the basis of limited samples. After examining specimens gathered from various populations in southern California, I realized that although exemplars of local populations may have distinctive features, there are so many intermediates that recognizing any of the local races becomes arbitrary. Individuals treated as *Cirsium loncholepis* (from coastal sites bordering the mouth of the Santa Maria River and adjacent regions in San Luis Obispo and Santa Barbara counties) are often indistinguishable from plants of upland populations of the San Emigdio Mountains (Kern and Ventura counties) in the vicinity of the headwaters of the Cuyama River, a tributary of the Santa Maria River. Other individuals from the San Emigdio Mountains cannot be distinguished from plants of other upland and lowland southern California sites ranging south to San Diego County.

Cirsium loncholepis was recognized by the California Department of Fish and Game in 1990 as a threatened species, and in 2000 by the U.S. Fish and Wildlife Service as endangered. Its conservation status is being reevaluated in light of my conclusions about its taxonomy. I believe that the combined taxon, *C. scariosum* var. *citrinum*, should still be of conservation concern. Many of the collections from southern California are from highly developed areas, and the number and size of extant populations need to be evaluated.

Interspecific hybridization involving var. *citrinum* is known or suspected in several instances. A putative hybrid between it and *Cirsium occidentale* var. *occidentale* from dunes in San Luis Obispo County combines the leaf pubescence and arachnoid involucre of *C. occidentale* with features of *C. scariosum*. An apparent hybrid between var. *citrinum* and the strikingly different *C. rhothophilum* was collected from coastal dunes in northern Santa Barbara County. A population of unusual, purple-flowered individuals from Price Canyon northeast of Los Alamos in Santa Barbara County may be the product of past hybridization with another species. These plants are tall (1–1.6 m according to label data) and their stems and distal leaf faces are villous with septate trichomes.

54h. Cirsium scariosum Nuttall var. **robustum** D. J. Keil, Sida 21: 215. 2004 • Shasta Valley thistle

E

Plants caulescent, 25–70 cm. **Stems** (1) often very stout, branched distally or throughout, leafy, glabrous, villous with septate trichomes, or arachnoid-tomentose. **Leaves** oblanceolate to elliptic, deeply pinnately lobed, longer spines slender or stout, usually 1 cm or shorter, abaxial thinly to densely arachnoid-tomentose, villous with septate trichomes along midveins, adaxial villous with septate trichomes or thinly arachnoid-tomentose. **Heads** 3–many, evidently pedunculate, in corymbiform or subcapitate arrays at stem tips (in age clustered axillary heads sometimes developing), subtended and ± overtopped by distal leaves or these ± reduced. **Involucres**

2.5–4 cm. **Phyllaries:** outer and mid lanceolate to ovate, spines slender to ± broad and flat, 1–6 mm; apices of inner linear-acuminate or more commonly expanded as scarious, erose-toothed appendages, often contorted. **Corollas** white, 30–36 mm, tubes 14–22 mm, throats 7–12 mm, lobes 5–10 mm; style tips 6–8 mm. **Cypselae** 4–6.5 mm; **pappi** 22–32 mm.

Flowering summer (Jun–Jul). Wet ground, meadows, pastures, marshes; 900–1900 m; Calif., Oreg.

Variety *robustum* is known only from northern California (Siskiyou County) and south central Oregon (Klamath and Lake counties). It differs from var. *scariosum* in its larger, evidently pedunculate heads.

55. Cirsium foliosum (Hooker) de Candolle in A. P. de Candolle and A. L. P. P. de Candolle, Prodr. 6: 654. 1838 • Leafy or foliose or elk thistle E

Carduus foliosus Hooker, Fl. Bor.-Amer. 1: 303. 1833

Biennials or monocarpic perennials, 25–70+ cm; taprooted. **Stems** usually 1, erect, stout, ± fleshy, simple, very leafy, densely villous or tomentose with septate trichomes. **Leaves:** blades linear-oblong to oblanceolate (elliptic), 5–20(–25) × 1–4(–7) cm, subentire to dentate or pinnatifid, lobes lance-oblong to triangular, spinulose to spiny-dentate or shallowly lobed, main spines slender, 2–5(–10) mm, abaxial faces often thinly gray- or white-tomentose with felted arachnoid trichomes, ± villous along major veins with septate trichomes, adaxial green, glabrous to thinly arachnoid, often ± villous with septate trichomes; basal usually present at flowering, spiny winged-petiolate or sessile; principal cauline well distributed, proximally winged-petiolate, distally sessile, not or only slightly reduced; distal often narrower than proximal. **Heads** few–many, erect, sessile or subsessile, crowded in dense, woolly, leafy-bracted, subcapitate arrays, closely subtended and overtopped by crowded leafy bracts. **Peduncles** 0–1 cm. **Involucres** broadly ovoid, 2–2.5 × 1.5–2 cm, green, glabrous to densely villous with septate trichomes on margins. **Phyllaries** in 4–6 series, imbricate, lanceolate or ovate (outer) to linear-lanceolate (inner), bases appressed, margins of outer entire, abaxial faces without glutinous ridge, apices appressed to ascending, spines straight, slender, 2–3 mm; apices of inner erect, straight. **Corollas** white to pale pink, 21–25 mm, tubes 12–14 mm, throats (very slender, scarcely larger than tubes) 6–7 mm, lobes 3–4 mm; style tips 2.5–3 mm, short exserted. **Cypselae** light brown, 4–5.5 mm, apical collars yellow, narrow; **pappi** 23–29 mm, exceeding corollas. $2n = 34$.

Flowering summer (Jul–Aug). Moist soil, grasslands, meadows, edges and openings in boreal forest, subalpine forests and alpine slopes; 150–2600 m; Alta., B.C., N.W.T., Yukon; Wyo.

Cirsium foliosum occurs in the northern Rockies from Wyoming to the Yukon and eastward to the Slave River area in the Northwest Territories and northeastern Alberta. Reports for Alaska are unconfirmed (R. Lipkin, Alaska Natural Heritage Program, pers. comm.). The name *Cirsium foliosum* has been misapplied to a wide range of plants across the western United States that now are treated as one or another variety of the polymorphic *C. scariosum*. The only documented occurrences of *C. foliosum* in the lower 48 states are in the mountains of northern Wyoming. Somewhat similar plants from other mountain areas of the western United States are treated as *C. scariosum* var. *scariosum*. During Pleistocene glaciations the ancestors of *C. foliosum* undoubtedly occupied a more southerly distribution and very likely came into direct contact with ancestral populations of *C. scariosum*. The observed similarities between *C. foliosum* and *C. scariosum* var. *scariosum* may be a relic of hybridization in that ancient contact zone. On the other hand, the corolla features of *C. foliosum* suggest that this is a self-pollinating species, perhaps derived from an ancestral population similar to the modern *C. scariosum* var. *scariosum*.

56. **Cirsium quercetorum** (A. Gray) Jepson, Fl. W. Calif., 507. 1901 • Brownie or Alameda County thistle E

Cnicus quercetorum A. Gray, Proc. Amer. Acad. Arts 10: 40. 1874; *Cirsium quercetorum* var. *walkerianum* (Petrak) Jepson; *C. quercetorum* var. *xerolepis* Petrak; *C. walkerianum* Petrak

Perennials, subacaulescent and forming compact, rounded mounds, 5–20 cm, or ± erect and to 70(–90) cm; runner roots producing adventitious buds. **Stems** 1–10+, erect or ascending, glabrous to thinly gray-tomentose, sometimes villous with septate trichomes; branches 0 or few, ascending. **Leaves:** blades elliptic to obovate, 5–20 × 3–7 cm, strongly undulate, shallowly to deeply pinnatifid with 3–8 pairs of lobes, lobes linear-lanceolate to broadly triangular, (often longer than 2 cm), closely spaced, spreading, spinose-dentate or lobed, main spines slender to stout, 2–15 mm, abaxial faces thinly to densely tomentose, ± villous with septate trichomes along veins, glabrescent or trichomes persistent, adaxial thinly arachnoid-tomentose and soon glabrescent; basal usually present at flowering, petiolate; principal cauline petiolate, progressively reduced distally, bases sometimes decurrent as spiny wings to 1 cm; distal reduced, similar to proximal. **Heads** 1–few, erect, ± crowded, often closely subtended by distalmost leaves.

Peduncles 0–10 cm, leafy-bracted. **Involucres** ovoid to hemispheric or broadly campanulate, 2.5–5 (in first-formed heads, often smaller in later heads) × 2.5–6 cm, loosely arachnoid on phyllary margins or glabrate. **Phyllaries** in 5–10 series, imbricate, ovate or lanceolate (outer) to linear-lanceolate (inner), margins of outer entire, abaxial faces without glutinous ridge; outer and mid appressed, spines erect or ascending, (0–)1–2(–10) mm; apices of mid and inner narrowed and scabrido-denticulate or with expanded, spinuloso-serrate or -dentate tips, spineless or spine-tipped. **Corollas** white or pale lavender to purple, 25–35 mm, tubes 10–20 mm, throats 7–10 mm, lobes 5–8 mm; style tips 2.5–4.5 mm. **Cypselae** brown, 5–6.5 mm, apical collars colored like body; **pappi** 20–40 mm. $2n = 32$.

Flowering spring–summer (Apr–Aug). Usually dry sites, coastal bluffs, grasslands, oak woodlands, coastal scrub; 0–400 m; Calif.

Cirsium quercetorum occurs in the north and south Coast ranges of California from Mendocino to San Luis Obispo counties. It overlaps in range and habitat with several other thistle species and has been reported to hybridize with *C. andrewsii*, *C. douglasii*, *C. occidentale*, *C. remotifolium* var. *odontolepis*, and *C. fontinale* var. *fontinale* (F. Petrak 1917; J. T. Howell 1960b). Considerable variation occurs within the range of *C. quercetorum*, and two of the variants have been given taxonomic recognition as vars. *walkerianum* and *xerolepis*. Additional study over the range of the species is needed to determine whether these or other variants should be recognized formally.

Cirsium quercetorum appears to be related to the polymorphic *C. scariosum* complex. The perennial habit with runner roots of *C. quercetorum* consistently distinguishes it from the monocarpic *C. scariosum*.

57. **Cirsium praeteriens** J. F. Macbride, Contr. Gray Herb. 53: 19. 1918 • Lost or Palo Alto thistle C E

Biennials or perennials, probably more than 100 cm; rootstock unknown. **Stems** stout, erect, loosely arachnoid with fine trichomes and villous with jointed trichomes; branching unknown. **Leaves:** blades elliptic to oblanceolate, 15–30+ × 6–8+ cm, divided halfway or more to midveins, lobes linear-lanceolate, rigidly spreading, entire or trifid, acuminate, main spines stout, 5–15 mm, abaxial faces tomentose with fine, non-septate trichomes, villous along major veins with septate trichomes, adaxial glabrescent or sparsely tomentose, villous along veins; basal not observed; cauline well distributed, distally not much reduced, sessile, bases clasping, not decurrent.

Heads 1–5, terminal and in distal axils in spiciform arrays. **Peduncles** 0–1 cm. **Involucres** hemispheric to broadly campanulate, 3–4 × 4–5+ cm, arachnoid. **Phyllaries** in 6–8 series, narrowly lanceolate to linear, outer subequal, rigidly spreading, spines 5–10 mm, inner ± imbricate, bodies appressed, glutinous ridge absent, apices spreading, margins spinulose or scabrid, apices of mid and inner flattened, spineless, scabrid. **Corollas** white, 30–33 mm, tubes 16 mm, throats 9–12 mm, lobes 5.5–9 mm; style tips 6 mm. **Cypselae** light brown, 6 mm, collars also light brown, ca. 0.75 mm; **pappi** 25–33 mm.

Flowering summer (Jun–Jul). Habitat unknown; of conservation concern; 0–100 m; Calif.

Cirsium praeteriens is known only from Santa Clara County, where J. W. Congdon collected it in Palo Alto in 1897 and 1901. It is presumed extinct.

58. **Cirsium rhothophilum** S. F. Blake, J. Wash. Acad. Sci. 21: 336. 1931 • Surf thistle [C] [E]

Carduus maritima Elmer, Bot. Gaz. 39: 45. 1905, not *Cirsium maritimum* Makino 1910

Biennials or short-lived, usually monocarpic perennials, 10–100 cm; taprooted with simple or branched caudices. **Stems** 1–several, spreading to erect, bushlike or forming low rounded mounds, gray-tomentose with appressed feltlike trichomes; branches 0–several, inclined to ascending, stiff. **Leaves:** blades elliptic to ovate, 10–25 cm, strongly undulate, usually broadly pinnatifid, lobes entire or coarsely few-toothed or -lobed, main spines abrupt, 1–4 mm, faces gray-white-tomentose with appressed feltlike, non-septate trichomes; basal present or withered at flowering, winged-petiolate; principal cauline well distributed, winged-petiolate to sessile, gradually reduced, bases clasping with expanded auricles; distal reduced, spines to 8 mm. **Heads** 1–many, erect, terminal on branches in subcapitate to congested, corymbiform arrays, closely subtended by clustered, ± leafy bracts. **Peduncles** 0–7 cm. **Involucres** hemispheric or campanulate, 3–4 × 4–6 cm, densely arachnoid. **Phyllaries** in 8–10 series, imbricate, linear, abaxial faces without glutinous ridge; outer and mid bases short-appressed, margins spiny-ciliate, apices long, spreading to erect, spines straight, 2–5 mm; apices of inner flattened or spine-tipped, serrate to scabrid, sometimes pectinately fringed. **Corollas** white to pale yellow, 20–34 mm, tubes 11–15 mm, throats 5–8 mm, lobes 5–8 mm; style tips 3–4 mm. **Cypselae** light-brown to black, 5–7 mm, apical collars whitish, 0.2–0.3 mm; **pappi** 15–25 mm. $2n = 34$.

Flowering mostly spring–summer (Apr–Aug), occasionally year round. Coastal dunes and bluffs; of conservation concern; 0–20 m; Calif.

Cirsium rhothophilum is endemic to the dunes of southern San Luis Obispo and northern Santa Barbara counties. It rarely forms hybrids with *C. occidentale* var. *occidentale* and *C. scariosum* var. *citrinum.*

59. **Cirsium fontinale** (Greene) Jepson, Fl. W. Calif., 505. 1901 • Fountain thistle [C] [E]

Cnicus fontinalis Greene, Bull. Calif. Acad. Sci. 2: 151. 1886

Monocarpic perennials, 50–220 cm; woody tap-rooted caudices with numerous coarse, fibrous lateral roots, sometimes forming new rosettes from root sprouts. **Stems** 1–several, erect, loosely arachnoid-tomentose, glandular-pilose; branches several to many, spreading. **Leaves:** blades oblanceolate to oblong or elliptic, 20–70 × 5–16 cm, strongly undulate, shallowly to deeply pinnatifid, lobes coarsely toothed or with triangular secondary lobes, main spines 3–15 mm, abaxial faces more densely tomentose, adaxial densely glandular with short, multicellular trichomes and ± tomentose with fine, arachnoid, non-septate trichomes; basal often present at flowering, petiolate; principal cauline well distributed, becoming sessile, gradually reduced distally, bases auriculate-decurrent with broad, spiny-margined wings 3 cm or less; distal cauline leaves progressively reduced, ± bractlike. **Heads** few–many, ± nodding, in paniculiform or corymbiform arrays, bracts leafy or much reduced. **Peduncles** 0–7 cm. **Involucres** (green to purple), hemispheric or campanulate, 1.5–3 × 2–5 cm, glabrous or puberulent. **Phyllaries** in 6–10 series, imbricate, ovate to lanceolate, abaxial faces without glutinous ridge; outer and mid bases appressed, short, bodies entire to erose or ciliolate (rarely outer spine-margined), apices spreading to recurved or reflexed, long, dilated, adaxially puberulent, spines 1–6 mm; apices of inner erect or reflexed, flattened or tipped with short spines. **Corollas** white to pinkish or lavender, 14.5–22 mm, tubes 5–8 mm, throats 4.5–9 mm, lobes 3–8 mm; style tips 2.5–4.5 mm. **Cypselae** brownish, 3.4–5 mm, apical collars tan; **pappi** 12–15 mm.

Varieties 3 (3 in the flora): c California.

1. Cypselae 4.1–5 mm; phyllary spines 1–2 mm
. . . . 59b. *Cirsium fontinale* var. *fontinale*
1. Cypselae 3.4–4.1 mm; phyllary spines 2–6 mm.
  2. Longer spines of cauline leaves 10–18 mm; most phyllaries without marginal spines. . .
. . . . 59a. *Cirsium fontinale* var. *campylon*
  2. Longer spines of cauline leaves 4–7 mm; many outer phyllaries with marginal spines . . . . .
. . . . 59c. *Cirsium fontinale* var. *obispoense*

59a. Cirsium fontinale (Greene) Jepson var. **campylon** (H. Sharsmith) Pilz ex D. J. Keil & C. E. Turner, Phytologia 73: 313. 1992 • Mt. Hamilton thistle C E

Cirsium campylon H. Sharsmith, Madroño 5: 85, fig. 1. 1939

Basal leaves: longer spines 4–15 mm, distal faces equally glandular and tomentose. **Cauline leaves:** longer spines 10–18 mm. **Distalmost stem bracts** typically separated from heads. **Phyllaries** 65–85, 25–35 of them reflexed; apical spines 3–6 mm; marginal spines usually absent. **Cypselae** 3.4–4.1 mm. $2n = 34$.

Flowering spring-summer (Apr–Aug). Serpentine seeps in areas of chaparral, valley grasslands, foothill woodlands; of conservation concern; 300–800 m; Calif.

Variety *campylon* is endemic to the Mt. Hamilton Range in Alameda, Santa Clara, and Stanislaus counties.

59b. Cirsium fontinale (Greene) Jepson var. **fontinale** C E

Basal leaves: longer spines 4–6 mm; distal faces more densely glandular than tomentose. **Cauline leaves:** longer spines 8–12 mm. **Distalmost stem bracts** usually tightly subtending heads. **Phyllaries** 100–120, 35–45 of them reflexed; apical spines 1–2 mm; marginal spines often present. **Cypselae** 4.1–5 mm. $2n = 34, 34 + 1B$.

Flowering spring-summer (Apr–Aug). Serpentine seeps; of conservation concern; ca. 120 m; Calif.

Variety *fontinale* is known only from the vicinity of Crystal Springs Reservoir, San Mateo County.

59c. Cirsium fontinale (Greene) Jepson var. **obispoense** J. T. Howell, Leafl. W. Bot. 2: 71. 1938 • Chorro Creek bog thistle C E

Basal leaves: longer spines 5–9 mm; distal faces more densely glandular than tomentose. **Cauline leaves:** longer spines 4–7 mm. **Distalmost stem bracts** tightly subtending heads or separated from them. **Phyllaries** 70–120, 28–38 of them reflexed; apical spines 2–3 mm; marginal spines usually absent. **Cypselae** 3.8–4 mm. $2n = 34$.

Flowering winter–summer (Jan–Sep). Serpentine seeps, coastal live oak woodlands, grasslands, riparian areas; of conservation concern; 40–300 m; Calif.

Variety *obispoense* is known only from the southern Santa Lucia and San Luis ranges of San Luis Obispo County.

60. Cirsium rydbergii Petrak, Beih. Bot. Centralbl. 35(2): 315. 1917 • Rydberg's or alcove thistle E

Cirsium lactucinum Rydberg

Perennials, 100–300 cm; caudices and taproots, spreading by creeping roots. **Stems** 1–several, erect or ascending to lax and hanging, glabrous or thinly tomentose; branches 0–few, ascending. **Leaves:** blades elliptic, 30–90+ × 10–40 cm, 1–2 times pinnately lobed, lobes linear to ovate, strongly undulate, main spines slender, 5–15 mm, faces often glaucous, glabrous or thinly tomentose and soon glabrescent; basal present at flowering, petiolate or winged-petiolate; proximal cauline winged-petiolate; mid sessile, much reduced, less deeply lobed, bases clasping, short-decurrent 0–2 cm; distal linear or lanceolate, bractlike, very spiny. **Heads** few–many, erect or nodding in clusters at tips of distal branches in paniculiform arrays, not closely subtended by clustered leafy bracts. **Peduncles** 0.5–6 cm. **Involucres** hemispheric, 1.4–2 × 1–2 cm, phyllary margins thinly tomentose or glabrate. **Phyllaries** in 5–8 series, strongly imbricate, (green, drying green or light brown), ovate to lance-oblong, abaxial faces with or without poorly developed glutinous ridge; outer and mid bases appressed, margins entire, not scabridulous-ciliolate, apices spreading or reflexed, green to brownish, lance-ovate, elongate, flattened, spines slender, 3–25 mm; apices of inner straight, entire. **Corollas** dull white to pink or purple, 16–20 mm, tubes 7–8.5 mm, throats 4–6.5 mm, lobes 4.5–6 mm; style tips 2.5 mm. **Cypselae** gray or brown, 3.7–4.5 mm, apical collars not differentiated; **pappi** 10–15 mm. $2n = 34$.

Flowering spring–summer (May–Sep). Hanging gardens, seeps, stream banks; 1000–1500 m; Ariz., Utah.

Cirsium rydbergii is endemic to the Colorado Plateau of northern Arizona and southeastern Utah.

61. Cirsium joannae S. L. Welsh, N. D. Atwood & L. C. Higgins in S. L. Welsh et al., Utah Fl. ed. 3, 168. 2003 • Joanna's thistle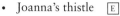

Perennials, 100–150 cm; caudices taprooted. **Stems** 1, fleshy, erect, openly branched in distal ¹/₂, glabrous; branches on distal stems several, ascending. **Leaves:** blades oblong, 10–50 × 6–12 cm, margins usually not strongly undulate, coarsely dentate or shallowly to deeply pinnatifid with 10–15 pairs of lobes, teeth or lobes ± closely spaced, not much overlapping, narrowly to broadly triangular, spiny-dentate or larger shallowly 3–5-lobed, main spines 2–12 mm, abaxial faces glabrous, adaxial glabrous; basal present at flowering, petiolate or spiny winged-petiolate; principal cauline many, well distributed, proximally ± winged-petiolate, distally sessile, gradually reduced, less divided, bases auriculate-claping and/or decurrent as spiny wings to 2 cm; distal much reduced. **Heads** several–many, erect or nodding, usually sessile or short-pedunculate, crowded in subcaptitate or short, spiciform clusters at branch tips, collectively forming open, paniculiform arrays. **Peduncles** 0–2 cm **Involucres** green, ovoid to campanulate (not including spreading phyllary apices), 2.5–4 × 2.5–3.5 cm, appearing glabrous. **Phyllaries** in 5–7 series, unequal, outer longer than inner, bases short-appressed, abaxial faces without glutinous ridge, minutely scabridulous, apices green, thick, spreading to curved-ascending, proximally flattened, linear, spines stout, 5–12 mm; outer entire or pinnately spiny, inner entire, scabridulous-ciliolate; apices of inner flexuous, sometimes slightly expanded and minutely erose. **Corollas** lavender-pink, 20–26 mm, tubes 5–8.5 mm, throats 5.2–8 mm, lobes linear, 9–10 mm; style tips 3.5–4.2 mm, conspicuously exserted beyond corolla lobes. **Cypselae** dark brown, 4.5–5 mm, apical collars not differeniated; **pappi** ca. 20 mm.

Flowering summer (Aug–Sep). Hanging gardens with *Lobelia*, *Abies*, and *Adiantum*; 1700 m; Utah.

Cirsium joannae is endemic to Zion National Park in southwestern Utah. It apears to be most closely related to *C. rydbergii*, which occurs in similar habitats in southeastern Utah and northeastern Arizona.

62. Cirsium vinaceum (Wooton & Standley) Wooton & Standley, Contr. U.S. Natl. Herb. 19: 751. 1915 • Sacramento Mountains thistle C E F

Carduus vinaceus Wooton & Standley, Contr. U.S. Natl. Herb. 16: 196. 1913

Perennials, 100–200 cm; creeping roots. **Stems** 1, erect, (dark maroon), glabrous; branches several to many, stiffly ascending, distally nodding. **Leaves:** blades elliptic, 10–50 × 5–20 cm, 1–2 times pinnately lobed or divided, lobes lanceolate to ovate, main spines slender to stout, 3–10 mm, faces glossy green, glabrous; basal present at flowering, petiolate or winged-petiolate, divided nearly to midveins; proximal cauline winged-petiolate; mid and distal sessile, progressively reduced, less deeply lobed, bases auriculate-clasping; distalmost linear or lanceolate, bractlike, very spiny. **Heads** many, nodding, borne 1–few at branch tips, collectively forming open, paniculiform arrays. **Peduncles** 0.5–15+ cm. **Involucres** (excluding spreading tips) broadly ovoid to hemispheric or campanulate, 2–3 × 2–3 cm, glabrous. **Phyllaries** in 8–10 series, strongly imbricate, (dark maroon, drying dark brown or blackish), ovate or lanceolate (outer) to linear-lanceolate (inner), abaxial faces with poorly developed glutinous ridge; outer and mid bases appressed, margins entire, apices spreading to reflexed, elongate, ovate to lanceolate, 5–20 mm, flattened, ciliolate, adaxially minutely villous with septate trichomes, spines slender, 1–3 mm; apices of inner flexuous, entire. **Corollas** rich rose-purple, 20–26 mm, tubes 4–5 mm, throats 7–10 mm, lobes 10–11 mm; style tips 2–2.5 mm. **Cypselae** brown, ca. 5 mm, apical collars not differentiated; **pappi** (brown), 18–20 mm.

Flowering spring–summer (May–Sep). Wet soil around calcareous springs and seeps, stream banks, montane meadows, coniferous forest margins; of conservation concern; 2300–2900 m; N.Mex.

Cirsium vinaceum is perhaps most closely related to *C. rydbergii*. Heads of *C. vinaceum* are actively visited by hummingbirds and by several kinds of bees (K. A. Burks 1994). Hybrids are known between *C. vinaceum* and *C. wrightii*.

Cirsium vinaceum is endemic to the Sacramento Mountains, Otero County. It is in the Center for Plant Conservation's National Collection of Endangered Plants.

Excluded species:

Cirsium canum (Linnaeus) Allioni was reported in 1924 from Massachusetts by C. H. Knowlton and W. Deane (Rhodora 26: 84) based on an 1899 collection from Weston. It has not been included in subsequent local or regional floras and apparently is not established in the flora area.

Cirsium scabrum (Poiret) Bonnet & Barratte was reported [as *Cnicus giganteus* (Desfontaines) Willdenow] in 1900 by A. Eastwood (Zoë 5: 59) based on a 1900 collection from Santa Cruz County, California. J. T. Howell (1959, 1960b) noted that *Cirsium scabrum* had not become naturalized. It apparently is not established in the flora area.

14. SILYBUM Adanson, Fam. Pl. 2: 116, 605. 1763, name conserved • Milk thistle [Greek *silybon*, a kind of thistle] ☐

David J. Keil

Annuals or biennials, taprooted, 15–300 cm, herbage glabrous, puberulent, or slightly tomentose, spiny. **Stems** erect, usually simple. **Leaves** basal and cauline; petiolate (basal and proximal cauline) or sessile (distal cauline); blades adaxially variegated, margins dentate and often coarsely pinnately lobed, teeth and lobes spine-tipped, glabrous or puberulent. **Heads** discoid, borne singly, terminal and in distal axils. (**Peduncles** with reduced leaflike bracts.) **Involucres** ovoid to spheric, 15–60 mm diam. **Phyllaries** many in 4–6 series, unequal, outer and mid with appressed bases and spreading, lanceolate to ovate, spiny-fringed, terminal appendages, at least mid spine-tipped, innermost with erect, flat, entire, spineless apices. **Receptacles** flat, epaleate, covered with whitish bristles. **Florets** 25–100+; corollas pink to purple, tubes slender, distally bent, abruptly expanded into short throats, lobes linear; stamen filaments connate, anther bases sharply short-tailed, anther appendages oblong; style branches: fused portions with slightly swollen subterminal nodes, distally cylindric, distinct portions minute. **Cypselae** ovoid, slightly compressed, not ribbed, apices with smooth, entire rims, glabrous, basal attachment scars slightly angled; **pappi** falling in rings, outer of many minutely barbed, basally connate, subulate scales, inner of minute smooth bristles. $x = 17$.

Species 2 (1 in the flora): introduced; Mediterranean region.

1. Silybum marianum (Linnaeus) Gaertner, Fruct. Sem. Pl. 2: 378. 1791 • Blessed milkthistle, chardon Marie ☐ ☐

Carduus marianus Linnaeus, Sp. Pl. 2: 823. 1753

Stems glabrous or slightly tomentose. **Leaves:** basal wing-petioled, blades 15–60+ cm, margins coarsely lobed; cauline leaves clasping, progressively smaller and less divided, bases spiny, coiled, auriculate. **Phyllary** appendages spreading, ovate, 1–4 cm including long-tapered spine tips. **Corollas** 26–35 mm; tubes 13–25 mm, throats campanulate, 2–3 mm, lobes 5–9 mm. **Cypselae** brown and black spotted, 6–8 mm; **pappus** scales 15–20 mm. $2n = 34$.

Flowering Feb–Jun (west), Jul–Sep (north). Roadsides, pastures, waste areas; sometimes cultivated; 0–800 m; introduced; Alta., B.C., N.B., N.S., Ont., Que., Sask.; Ala., Ariz., Ark., Calif., Conn., Ind., La., Mich., Miss., Nev., N.H., N.J., N.Mex., N.Y., N.C., Ohio, Oreg., Pa., Tenn., Tex., Vt., Va., Wash., W.Va.; s Europe (Mediterranean region).

Silybum marianum is sometimes cultivated as an ornamental, a minor vegetable, or as a medicinal herb. Young shoots can be boiled and eaten like cabbage and young leaves can be added to salads. The seeds can be used as a coffee substitute. Extracts of *S. marianum* are used as an herbal treatment for liver ailments.

15. SAUSSUREA de Candolle, Ann. Mus. Natl. Hist. Nat. 16: 156, 196, plates 10–13. 1810, name conserved • Saw-wort [For Nicolas Théodore (1767–1845) and Horace Bénédict (1740–1799) de Saussure, Swiss naturalists]

David J. Keil

Perennials, 5–120+ cm; herbage tomentose or glabrescent, not spiny. **Stems** erect or ascending, simple or branched. **Leaves** basal or cauline (sometimes cauline only), sessile or petiolate; blade margins entire or dentate to pinnately lobed, faces glabrous to densely tomentose, glandular or eglandular. **Heads** discoid, borne singly or in corymbiform arrays. **Involucres** ovoid to campanulate or ± turbinate. **Phyllaries** many in 3–5(–10+) series, subequal to strongly unequal, appressed or not, ovate to lanceolate, margins entire, toothed, or lobed, apices obtuse or acute, appendaged or not, not spine-tipped. **Receptacles** flat or convex, epaleate, smooth, usually subulate-scaly, sometimes bristly or naked. **Florets** 10–20; corollas white to blue or purple, tubes slender, abruptly expanded to throats, lobes linear; anther bases short-tailed, apical appendages linear, acute; style branches: fused portions with minutely hairy subterminal nodes, distinct portions oblong to linear, short-papillate. **Cypselae** oblong, ± angled, cylindric or 4–5-angled, ribs (when present) smooth or roughened, apices entire, glabrous or minutely glandular, attachment scars basal; **pappi** usually of 2 series, outer of readily falling, short bristles, inner persistent or falling as unit, of basally connate, usually longer, plumose bristles. $x = 13, 14, 16, 17, 18, 19$?.

Species 300–400 (6 in the flora): North America, Eurasia, 1 in Australia.

Saussurea is a notoriously difficult, largely Asiatic genus with species boundaries often indistinct.

SELECTED REFERENCE Lipschitz, S. J. 1979. Rod *Saussurea* DC. (Asteraceae). Leningrad.

1. Outer and mid phyllaries with toothed or lobed appendages . . . . 6. *Saussurea amara*
1. Outer and mid phyllaries entire, without appendages.
   2. Proximal leaves ovate to lanceolate, usually more than 30 mm wide, bases broadly obtuse to truncate or cordate; plants 30–120 cm.
      3. Cauline leaves usually more than 20, finely to ± coarsely serrate or dentate; s Alaska and Yukon to California, Idaho, and Montana . . . . 1. *Saussurea americana*
      3. Cauline leaves usually 15 or fewer, coarsely laciniate-dentate; nw Alaska . . . . 2. *Sausserea triangulata*
   2. Proximal leaves linear to elliptic, 2–25 mm wide, bases acute to acuminate; plants 3–40 cm.
      4. Phyllaries subequal, linear to lanceolate; receptacles naked . . . . 4. *Saussurea nuda*
      4. Phyllaries strongly unequal, the outer ovate to lanceolate, conspicuously shorter than inner; receptacles scaly.
         5. Tips of outer and mid phyllaries acute; Alaska and nw Canada . . . . . . 3. *Saussurea angustifolia*
         5. Tips of outer and mid phyllaries ± rounded; Rocky Mountains . . . . 5. *Saussurea weberi*

1. Saussurea americana D. C. Eaton, Bot. Gaz. 6: 283. 1881 • American saw-wort [E] [F]

Plants 30–120+ cm; rootstocks short, stout; herbage loosely tomentose when young, glabrescent, sometimes ± glandular. **Stems** 1–many, leafy, simple or with ascending branches. **Leaves** cauline, usually more than 20, well distributed, proximal and mid with winged petioles to 6 cm, wings sometimes decurrent 1–2 cm on stems, blades lanceolate to triangular-ovate, 5–15 cm, bases cordate to truncate or tapering, margins sharply dentate, apices acute; mid and distal usually sessile, smaller, narrower, bases tapering. **Heads** 5–30+ in tight to open corymbiform arrays; (peduncles 0–5 cm). **Involucres** 10–15 mm. **Phyllaries** in ca. 5 series, strongly unequal, outer ± ovate, inner lanceolate, abaxial faces pale green, distally dark purplish to nearly black, loosely tomentose. **Receptacles** naked. **Florets** 8–21; corollas usually pale lavender-blue to dark purple (rarely white), 11–13 mm; tubes 5–6.5 mm, throats 1.5–2 mm, lobes 3.5–4 mm. **Cypselae** 4–6 mm; **pappus** bristles brownish, outer 3–7 mm, inner 9–10 mm.

Flowering Jul–Aug. Moist canyons, meadows, streamsides in montane forests; 1000–2600 m; Alta., B.C., Yukon; Alaska, Calif., Idaho, Mont., Oreg., Wash.

Saussurea americana is closely related to an Asian species, *S. foliosa* Ledebour.

2. Saussurea triangulata Trautvetter & C. A. Meyer in A. T. von Middendorff, Reise Siber. 1(2,3): 58. 1856

Plants 30–70 cm; from rhizomes. **Stems** leafy, simple or with ascending branches. **Leaves** basal and cauline, well distributed or distally much smaller, proximal and mid with winged petioles to 15 cm, not decurrent on stems, proximal petioles ciliate with multicellular trichomes, distal petioles loosely tomentose when young, glabrescent; blades lanceolate to triangular-ovate, 6–13 cm, bases truncate or subcordate to obtuse or subacute, margins coarsely laciniate-dentate, ciliate with multicellular trichomes, apices acute or acuminate, faces glabrous; mid and distal progressively smaller, becoming sessile, narrower, bases obtuse to acuminate, cauline usually 15 or fewer. **Heads** 4–11, in congested corymbiform to capitate terminal arrays (peduncles 0–4 cm). **Involucres** 10–12 mm. **Phyllaries** in 4–5 series, unequal, appressed throughout, the outer ovate, inner narrowly lanceolate, abaxially dark purplish to nearly black, with or without central green or tan patch, margins loosely arachnoid-tomentose,

apices acute to obtuse or rounded, faces glabrate. **Receptacles** scaly. **Florets** 10–17; corollas light purple, goblet-shaped, 9–11 mm; tubes 4–5.5 mm, throat 2 mm, lobes 3–3.5 mm; anthers dark purple, 4–5 mm; style branches recurved, 1–1.5 mm. **Cypselae** ca. 3 mm; pappi brownish, outer bristles very slender, 0.5–4 mm, inner 14–17 bristles stouter, 8–9 mm, all plumose. $2n$ = 26 (Russia).

Flowering Jul–Aug. Slopes, ridges, sometimes in shade; 300–400 m; Alaska; Russian Far East (Chukotka).

In North America *Saussurea triangulata* is known only from the Waring Mountains of northwestern Alaska, where it was discovered in the late 1990s. I provisionally recognize the Waring Mountains plants as a disjunct population of *S. triangulata* pending examination of additional material from both Alaskan and Siberian populations.

3. Saussurea angustifolia (Linnaeus) de Candolle, Ann. Mus. Natl. Hist. Nat. 16: 200. 1810 • Narrow-leaved saw-wort, common saussurea

Serratula alpina Linnaeus var. *angustifolia* Linnaeus, Sp. Pl. 2: 817. 1753; *S. angustifolia* (Linnaeus) Willdenow

Plants 3–50 cm; herbage loosely tomentose when young, ± glabrescent. **Stems** arising from slender rhizomes, usually simple or few, ascending. **Leaves** basal and cauline, smaller distally, sessile, blades linear to narrowly elliptic, 3–12(–25) cm, bases acute, margins entire or remotely dentate, apices acute. **Heads** 2–10+ in open or crowded corymbiform arrays; (peduncles 1–5 cm). **Involucres** 9–14 mm. **Phyllaries** strongly unequal, the outer ± ovate, inner lanceolate, abaxial faces dark green, usually also tinged dark purplish, pilose or loosely tomentose; tips of outer and mid phyllaries acute. **Receptacles** scaly. **Florets** 8–22; corollas purple, 11–15 m, tubes 6–7.5 mm, throats 2–2.5 mm, lobes 3–5. mm; anthers darker purple. **Cypselae** 3–4 mm; **pappus** bristles brownish, outer 1–2 mm, the inner 9–10 mm.

Varieties 3 (3 in the flora): nw Canada, Alaska, Russian Far East (Chukotka).

Extreme forms of the varieties of *Saussurea angustifolia* are distinctive, ranging from slender, erect, subglabrous specimens of var. *angustifolia* to dwarf, densely pubescent forms of vars. *viscida* and *yukonensis*. The extremes are connected by intermediates. As is indicated by the synonymy, little unanimity exists in the interpretation of these taxa. I have chosen to follow S. L. Welsh (1974) in treating *S. densa* as a variety of *S. angustifolia* rather than as a distinct species. The extent to which the differences among these taxa are genetic or environmentally induced has not been investigated.

1. Outer phyllaries usually 1.5–2 times longer than wide; stems 10–40(–50) cm; leaves mostly 2–8 mm wide ...... 3a. *Saussurea angustifolia* var. *angustifolia*
1. Outer phyllaries usually 2–4 times longer than wide; stems usually 3–15 cm; leaves often wider than 8 mm.
    2. Leaves tomentose throughout or villous and viscid-glandular on margins and abaxially on midvein; plants mostly 5–15 cm; mostly coastal and insular Alaska .... 3b. *Saussurea angustifolia* var. *viscida*
    2. Leaves ± arachnoid-tomentose, not or scarcely glandular-viscid, bases usually white tomentose; plants mostly 3–14 cm; interior Alaska and Yukon .... 3c. *Saussurea angustifolia* var. *yukonensis*

3a. Saussurea angustifolia (Linnaeus) de Candolle var. **angustifolia**

Plants 10–40(–50) cm. **Stems** usually simple, ± thinly arachnoid. **Leaves** mostly 2–8 mm wide; basal blades linear or narrowly elliptic, margins entire or denticulate, glabrous or abaxially thinly arachnoid; cauline linear or oblong, crowded or well separated, usually not surpassing heads. **Heads** crowded or in ± open arrays. **Phyllaries** unequal in 3–5 series; outer usually 1.5–2 times longer than wide. $2n$ = 26, 52.

Flowering Jul–Aug. Moist to dry sites, arctic and alpine tundras, bogs, riverbanks, grassy slopes, willow thickets, spruce forests; 10–1300 m; B.C., N.W.T., Nunavut, Yukon; Alaska; Russian Far East (Chukotka).

3b. Saussurea angustifolia (Linnaeus) de Candolle var. **viscida** (Hultén) S. L. Welsh, Great Basin Naturalist 28: 152. 1968 • Sticky saw-wort

Saussurea viscida Hultén, Fl. Alaska Yukon 10: 1627, fig. 6c. 1950

Plants 5–15 cm. **Stems** simple, lightly arachnoid and villous with viscid hairs. **Leaves** often wider than 8 mm; basal lanceolate to elliptic, margins denticulate, arachnoid throughout or viscid-villous on margins and abaxially on midvein; cauline leaves narrower, crowded, usually surpassing heads. **Heads** crowded. **Phyllaries** unequal in 2–3 series; outer phyllaries usually 2–4 times longer than wide. $2n$ = unknown.

Flowering Jul–Aug. Tundra, insular and coastal regions; 10–100 m; Alaska; Russian Far East (Chukotka).

3c. Saussurea angustifolia (Linnaeus) de Candolle var. **yukonensis** A. E. Porsild, Bull. Natl. Mus. Canada 101: 28. 1945 [E]

Saussurea angustifolia (Willdenow) de Candolle subsp. *yukonensis* (A. E. Porsild) W. J. Cody; *S. viscida* Hultén var. *yukonensis* (A. E. Porsild) Hultén

Plants 3–14 cm. **Stems** simple, lightly arachnoid. **Leaves** often wider than 8 mm; basal lanceolate to elliptic, margins denticulate, arachnoid throughout; cauline narrower, crowded, usually surpassing heads. **Heads** crowded. **Phyllaries** unequal in 3–5 series; outer usually 2–4 times longer than wide.

Flowering Jul–Aug. Usually dry sites, arctic and alpine tundras; 10–1300 m; N.W.T., Yukon; Alaska.

4. Saussurea nuda Ledebour, Icon. Pl. 1: 15, plate 61. 1829 • Chaffless or dwarf saw-wort

Saussurea alpina (Linnaeus) de Candolle; *S. alpina* var. *ledebourii* A. Gray; *S. densa* (Hooker) Rydberg; *S. nuda* subsp. *densa* (Hooker) G. W. Douglas; *S. nuda* var. *densa* (Hooker) Hultén

Plants 5–60 cm; branched caudices; herbage subglabrous to loosely tomentose when young, ± glabrescent, at least proximally. **Stems** usually simple. **Leaves** basal and cauline ± smaller distally, tapered to winged petioles to 7 cm, blades elliptic or lanceolate to ovate, 5–15 cm, bases obtuse to acute, margins subentire to sinuate, dentate or denticulate, apices acute to acuminate; distal cauline sessile, ± decurrent. **Heads** 3–20+ in corymbiform to subcapitate arrays; (peduncles 0–5 cm). **Involucres** 10–15 mm. **Phyllaries** in 3–4 series subequal or weakly imbricate, linear to lanceolate, abaxial faces dark green, often tinged dark purplish, loosely villous or ± tomentose. **Receptacles** naked. **Florets** 15–20; corollas purple, 8–11 mm, tubes 4–6 mm, throats 1.5–2 mm, lobes 3–3.5 mm; anthers dark purple. **Cypselae** stramineous, 6–7 mm; pappi of white to brownish, outer bristles 1–3 mm, inner 9–10 mm. $2n$ = 26 (as *S. densa*).

Flowering Jun–Aug. Coastal dunes, estuaries, alpine tundra; 0–100, 2000–2800 m; Alta.; B.C.; Alaska, Mont.; Russian Far East.

North American populations of *Saussurea nuda* occur in two distinctly different sets of habitats. Plants from coastal Alaska have been recognized as var. *nuda* and dwarfed alpine plants from the northern Rockies as var. *densa*. According to E. Hultén (1941–1950, vol. 10), the only differences between var. *densa* and var. *nuda* are the "more densely denticulated leaves and very congested inflorescences" of the former. Notwithstanding the very different habitats, some coastal Alaskan s pecimens [e.g., *Ward 53* (ALA)] are indistinguishable from the alpine forms. In Siberia, *S. nuda* is a polymorphic species occurring from coastal sites to alpine areas (S. J. Lipschitz 1979).

Saussurea ×*tschuktschorum* Lipschitz is apparently a hybrid between *S. angustifolia* and *S. nuda*. It resembles *S. angustifolia*; it has naked receptacles like *S. nuda*.

5. Saussurea weberi Hultén, Svensk Bot. Tidskr. 53: 202, fig. 1. 1959 [E]

Plants 5–22 cm; branched caudices. **Stems** usually simple. **Leaves** basal and cauline not much smaller distally, tapered to winged petioles 1–7 cm; blades elliptic or lanceolate to ovate, 2–8 cm, bases acute, margins entire or denticulate, apices acute, thinly tomentose when young, glabrescent, abaxially minutely resin-gland-dotted; distal cauline sessile. **Heads** 1–15 in condensed subcapitate arrays; (peduncles 0–1 cm). **Involucres** 10–12 mm. **Phyllaries** in 3–4 series, unequal, ovate to lanceolate, abaxial faces dark green and tinged dark purplish, long-pilose; tips of outer and mid phyllaries ± rounded. **Receptacles** scaly. **Florets** 9–10; corollas purple, 10–12 mm, tubes 4.5–6 mm, throats 1.5–2 mm, lobes 3.5–4 mm; anthers dark purple. **Cypselae** 3–5 mm; **pappi** of white or tawny, outer bristles 1–4 mm, inner 8–11 mm. $2n = 26$.

Flowering Jul–Aug. Gravelly alpine tundras, scree slopes; 2800–5200 m; Colo., Mont., Wyo.

Saussurea weberi appears to be closely related to the *S. angustifolia* complex.

6. Saussurea amara (Linnaeus) de Candolle, Ann. Mus. Natl. Hist. Nat. 16: 200. 1810 • Tall saw-wort [I]

Serratula amara Linnaeus, Sp. Pl. 2: 819. 1753

Varieties 3 (1 in the flora): introduced; c Asia.

6a. Saussurea amara (Linnaeus) de Candolle var. **glomerata** (Poiret) Trautvetter, Bull. Soc. Imp. Naturalistes Moscou 39(1): 369. 1866 [I]

Saussurea glomerata Poiret in J. Lamarck et al., Encycl., suppl. 5: 71. 1817

Plants 15–60 cm; stout taproots; herbage ± scabrous, ± glaucous. **Stems** erect, simple or branched, unwinged. **Leaves:** basal and proximal cauline long-petiolate, smaller distally, blades elliptic, 5–20 cm, bases acute, margins entire to sinuate dentate or shallowly lobed, apices acuminate; mid and distal cauline sessile or short-petiolate, ± bractlike, entire. **Heads** 2–8 in corymbiform cymes at tips of main stems and branches; (peduncles to 1–5 cm). **Involucres** ca. 15 mm. **Phyllaries** in 3–4 series, strongly unequal, outer ± lanceolate with dark green toothed or lobed appendages, mid ± linear-oblong with expanded, rounded, pink, membranous, toothed appendages, innermost linear, with or without reduced appendages. **Receptacles** scaly. **Florets** 8–18; corollas pink or pale purple, ca. 17 mm; anthers dark purple. **Cypselae** ca. 3 mm; pappi off-brownish, outer bristles 1–2 mm, inner ca. 10 mm. $2n = 26$ (Russia, as *S. glomerata*).

Flowering Jul–Sep. Barnyard and garden weed; 900–1000 m; introduced; Alta.; Asia.

Saussurea amara var. *glomerata* was reported (as *S. glomerata*) from the vicinity of Debolt, Alberta (H. Groh 1943) and was included in *The Thistles of Canada* (R. J. Moore and C. Frankton 1974). I am not aware of any additional stations, and the persistence of these plants in the Debolt area is unknown.

16. ARCTIUM   Linnaeus, Sp. Pl. 2: 816. 1753; Gen. Pl. ed. 5, 357. 1754 • Burdock, clotbur, bardane [Greek *arktion*, from *arktos*, bear, perhaps alluding to rough involucre] [I]

David J. Keil

Biennials or (monocarpic) perennials, 50–300 cm; herbage not spiny. **Stems** erect, openly branched, branches ascending. **Leaves** basal and cauline; long-petiolate; gradually smaller distally; blade margins entire or dentate (pinnately lobed or dissected), faces abaxially resingland-dotted, adaxially often tomentose. **Heads** discoid, in leafy-bracted racemiform to paniculiform or corymbiform arrays. (**Peduncles** 0 or 1–9 cm.) **Involucres** spheric to ovoid. **Phyllaries** many in 9–17 series, outer and mid narrowly linear, bases appressed, margins entire, apices stiffly radiating, hooked-spiny tipped, inner linear, ascending or erect, straight tipped.

Receptacles ± flat, epaleate, bearing subulate scales. **Florets** (5–)20–40+; corollas pink to ± purple, glabrous or glandular-puberulent, tubes elongate, throats campanulate, lobes narrowly triangular, ± equal; anther bases tailed, apical appendages ovate, obtuse to acute; style branches: fused portions distally hairy-ringed, distinct portions oblong, acute or obtuse. **Cypselae** obovoid, ± compressed, rough or ribbed, glabrous, attachment scars basal; **pappi** falling, of many bristles in 2–4 series. ***x*** = 18.

Species 10 (3 in the flora): introduced; Eurasia, n Africa, widely introduced worldwide.

At maturity the dry heads of *Arctium* species are readily caducous with the enclosed cypselae, and the hooked phyllary tips cling easily to fur or fabrics. Animal dispersal is a major factor in the spread of burdock species across North America. The burs are a major problem when they become entangled in the wool of sheep and fur of dogs and other animals.

Published chromosome reports for *Arctium* other than *n* = 18 are probably in error because of difficulty in interpretation of somatic chromosomes (R. J. Moore and C. Frankton 1974).

SELECTED REFERENCES Arènes, J. 1950. Monographie du genre *Arctium* L. Bull. Jard. Bot. État Bruxelles 29: 67–156. Duistermaat, H. 1996. Monograph of *Arctium* L. (Asteraceae). Gorteria, suppl. 3: 1–143.

1. Heads usually sessile to short-pedunculate in racemiform or paniculiform clusters
   . . . . 3. *Arctium minus*
1. Heads usually long-pedunculate in corymbiform clusters.
   2. Involucre 2.5–4 cm diam.; phyllary apices glabrous or loosely cobwebby; corollas
      glabrous . . . . 1. *Arctium lappa*
   2. Involucre 1.5–2.5 mm diam.; phyllary apices densely cobwebby; corollas minutely
      glandular-puberulent . . . . 2. *Arctium tomentosum*

1. Arctium lappa Linnaeus, Sp. Pl. 2: 816. 1753

• Great burdock, grande bardane [1]

Plants to 100–300 cm. **Basal leaves:** petioles solid, 15–36 cm, glabrous or thinly cobwebby; blades 25–80 × 20–70 cm, coarsely dentate to subentire, abaxially thinly gray-tomentose, adaxially green, sparsely short-hairy to nearly glabrous. **Heads** usually in corymbiform clusters, long-pedunculate. **Peduncles** 2.5–6 cm. **Involucres** 25–45 mm diam. **Phyllaries** linear to linear-lanceolate, glabrous to loosely cobwebby, inner usually stramineous (sometimes purplish), margins with minute spreading or reflexed hairs. **Florets** 40+; corollas purple (occasionally white), 9–14 mm, glabrous. **Cypselae** light brown, often with darker spots, 6–7.5 mm; **pappus** bristles 2–5 mm. **2*n*** = 32 (Japan), 34 (China), 36 (Japan); (Sweden).

Flowering summer–early fall (Jul–Oct). Waste places, roadsides, fields, forest clearings; 0–2200 m; introduced; Alta., B.C., Man., N.B., Ont., Que., Sask.; Ala., Ariz., Calif., Colo., Conn., Ill., Maine, Mass., Mich., Minn., Nev., N.H., N.Y., N.Dak., Pa., R.I., Utah, Vt., Wash., Wis.; Eurasia.

BONAP lists Georgia, Idaho, Indiana, Iowa, Maryland, Montana, New Jersey, North Carolina, Ohio, Oregon, and Wyoming; I have not seen specimens.

Roots and young leaves of *Arctium lappa* are edible and can be used in a variety of food preparations. Extracts of *Arctium* species purportedly have health benefits and are sold as food supplements. This species is sometimes cultivated as a minor crop.

2. Arctium tomentosum Miller, Gard. Dict. ed. 8, *Arctium* no. 3. 1768 (as tomentosis) • Woolly burdock, bardane tomenteuse [1]

Plants to 250 cm. **Basal leaves:** petioles hollow or solid, 10–15 cm, glandular-hairy; blades 30–40 × 16–28 cm, coarsely dentate to subentire, abaxially white-tomentose, adaxially green, sparsely short-hairy. **Heads** usually in corymbiform clusters, long-pedunculate. **Peduncles** 1.5–12 cm. **Involucres** 15–25 mm diam., densely cobwebby (rarely glabrate). **Phyllaries** linear to linear-lanceolate, inner usually purplish, margins with minute spreading or reflexed glandular hairs. **Florets** 30+; corollas rose-purple, (occasionally white), 9–13 mm, limb minutely glandular. **Cypselae** light brown, 5–8 mm; **pappus** bristles 1–3 mm. **2*n*** = 36.

Flowering summer–early fall (Jul–Oct). Waste places, roadsides, fields, forest clearings; 0–1600 m; introduced; Alta., Man., N.B., Nfld. and Labr. (Nfld.), N.S., Ont., P.E.I., Que., Sask.; Colo., Conn., Maine, Mass., Minn., Mo., N.H., N.Dak., Ohio, S.Dak., Vt.; Eurasia.

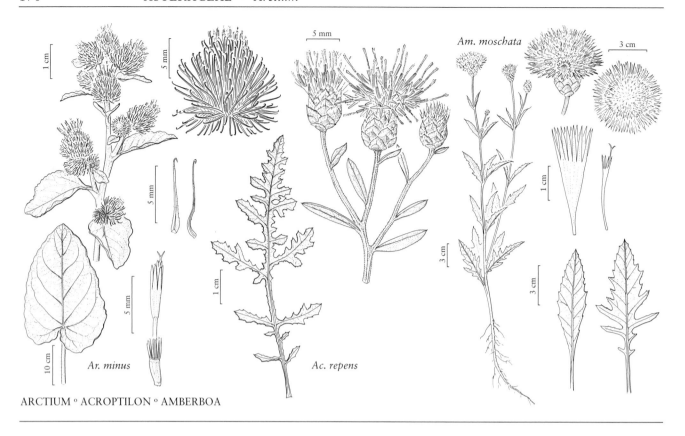

Ar. minus

Ac. repens

Am. moschata

ARCTIUM ° ACROPTILON ° AMBERBOA

Arctium tomentosum has been reported from Illinois, Iowa, Maryland, New Jersey, Oregon, Pennsylvania, South Carolina, and Wisconsin; I have not seen specimens. The involucres of *Arctium tomentosum* are usually very densely cobwebby.

Exceptional forms of *A. tomentosum* have nearly glabrous involucres. Forms of *A. minus* with especially cobwebby involucres have been misidentified as *A. tomentosum*; they lack the corymbiform capitulescence and glandular corollas of the latter.

3. **Arctium minus** (Hill) Bernhardi, Syst. Verz., 154. 1800 • Common or lesser burdock, petite bardane, cibourroche, chou bourache, bourrier [F] [I]

Lappa minor Hill, Veg. Syst. 4: 28. 1762

Plants to 50–300 cm. **Basal leaves:** petioles hollow (sometimes only at base), 15–50 cm, thinly to densely cobwebby; blades 30–60 × 15–35 cm, coarsely dentate to subentire (rarely deeply dissected), abaxially ± thinly gray-tomentose, adaxially green, sparsely short-hairy. **Heads** in racemiform or paniculiform clusters, sessile to pedunculate. **Peduncles** 0–9.5 cm. **Involucres** 15–40 mm diam. **Phyllaries** linear to linear-lanceolate, glabrous to densely cobwebby, inner often purplish tinged, margins often minutely serrate with fine teeth, puberulent with glandular and or eglandular hairs. **Florets** 30+; corollas purple, pink, or white, 7.5–12 mm, glabrous or limb glandular-puberulent. **Cypselae** dark brown or with darker spots, 5–8 mm; **pappus** bristles 1–3.5 mm. **2***n* = 32 (Germany), 36 (as *A. nemorosum*).

Flowering summer–early fall (Jul–Sep). Waste places, roadsides, fields, forest clearings; 0–2200 m; introduced; St. Pierre and Miquelon; Alta., B.C., Man., N.B., Nfld. and Labr. (Nfld.), N.S, Ont., P.E.I, Que., Sask.; Ala., Ariz., Ark., Calif., Colo., Conn., Del., D.C., Ga., Idaho, Ill., Ind., Iowa, Kans., Ky., Maine, Md., Mass., Mich., Minn., Miss., Mo., Mont., Nebr., Nev., N.H., N.J., N.Mex., N.Y., N.C., N.Dak., Ohio, Okla., Oreg., Pa., R.I., S.C., S.Dak., Tenn., Tex., Utah, Vt., Va., Wash., W.Va., Wis., Wyo.; Eurasia.

Arctium minus has been reported from Delaware and Texas; I have not seen specimens.

Arctium minus is a complex species with many variants that have been recognized at ranks ranging from forma to species (J. Arènes 1950). Some North American workers (e.g., R. J. Moore and C. Frankton 1974) have often distinguished plants with involucres more than 3 cm diameter that equal or overtop the corollas as *A. nemorosum*. Arènes treated those plants as a s ubspecies of *A. minus*. *Arctium nemorosum* was recognized as a species distinct from *A. minus* (H. Duistermaat 1996), with a different and more restricted circumscription than that used by North American workers.

Although most of the characters that Duistermaat used to separate those *A. nemorosum* from *A. minus* overlap extensively, the consistently wider mid phyllaries of *A. nemorosum* (1.7–2.5 mm wide versus 0.6–1.6 mm in *A. minus*) supposedly distinguish the species. None of the North American specimens examined in preparation of this treatment had the wide phyllaries of *A. nemorosum* in the sense of Duistermaat, who stated that she had seen no material of this taxon from the American continent. Some American authors have taken up the name *Arctium vulgare* in place of *A. nemorosum* and applied *A. vulgare* (dubbed woodland burdock) to the larger-headed North American plants. Duistermaat considers *A. vulgare* to be a synonym of *A. lappa*.

17. ACROPTILON Cassini in F. Cuvier, Dict. Sci. Nat. ed. 2, 50: 464. 1827

• Russian knapweed, creeping knapweed, Russian centaurea, mountain-bluet, Turkestan thistle, centaurée de Russie [Greek *akron*, tip, and *ptilon*, feather, describing the pappus bristles] [I]

David J. Keil

Perennials, 30–100 cm, not spiny. **Stems** erect, branched distally or throughout, branches ascending, not winged. **Leaves:** basal and proximal cauline; sessile or petiolate; blade margins coarsely dentate or 1–2-pinnately lobed, margins of mid and distal cauline dentate or entire, faces loosely tomentose or glabrate, puberulent and resin-gland-dotted. **Heads** discoid, in (leafy-bracted) corymbiform or paniculiform arrays. **Involucres** ovoid to subspheric, constricted, 6–10 mm diam. **Phyllaries** many in 6–8 series, outer round to ovate, bases tightly appressed, margins entire, apices widely scarious, obtuse or acute, inner lanceolate, margins entire, apices acute to acuminate, those of innermost bristly-ciliate or -plumose. **Receptacles** flat, epaleate, bearing setiform scales ("flattened bristles"). **Florets** 15–36; corollas blue, pink, or white, tubes very slender, usually bent distally, throats abruptly expanded, lobes linear; anther bases short-tailed, apical appendages oblong; style branches: fused portions with slightly swollen basal nodes minutely hairy, distally papillate, distinct portions very short, apices triangular. **Cypselae** obovoid, slightly compressed, smooth or with indistinct ribs, glabrous, attachment scars sub-basal; **pappi** ± falling, of many unequal setiform scales ("flattened bristles"), proximally barbed, distally plumose (at least the longer). $x = 13$.

Species 1: introduced; Eurasia.

In most American floristic literature *Acroptilon* has been included within *Centaurea*, from which it differs by the subbasal rather than lateral attachment scars on the cypselae and the absence of sterile outer florets. The chromosome base number $x = 13$ is higher than that in most species of *Centaurea* in the strict sense. Molecular phylogenetic studies of the relationships of Cynareae genera (A. Susanna et al. 1995) support the segregation of *Acroptilon* from *Centaurea*.

SELECTED REFERENCE Watson, A. K. 1980. The biology of Canadian weeds. 43. *Acroptilon* (*Centaurea*) *repens* (L.) DC. Canad. J. Pl. Sci. 60: 993–1004.

1. Acroptilon repens (Linnaeus) de Candolle in A. P. de Candolle and A. L. P. P. de Candolle, Prodr. 6: 663. 1838 [F] [I] [W]

Centaurea repens Linnaeus, Sp. Pl. ed. 2, 2: 1293. 1763; *Acroptilon picris* (Pallas ex Willdenow) C. A. Meyer; *C. picris* Pallas ex Willdenow

Creeping roots usually dark brown or black, with scaly adventitious buds. **Stems** ± cobwebby-tomentose. **Leaves:** basal and proximal cauline often deciduous by flowering, blades oblong, 4–15 cm; mid and distal linear to linear-lanceolate or oblong, 1–7 cm. **Involucres** 9–17 mm, loosely cobwebby. **Phyllaries:** apices of inner acute or acuminate, densely short-pilose. **Corollas** 11–14 mm, tubes 6.5–7.5 mm, throats 2–3.5 mm, lobes 3–3.5 mm. **Cypselae** ivory to grayish or brown, 2–4 mm; **pappus** bristles white, 6–11 mm. *2n* = 26.

Flowering late spring–summer (May–Sep). Fields, roadsides, riverbanks, ditch banks, clearcuts, cultivated ground; 0–2300 m; introduced; Alta., B.C., Man., Ont., Sask.; Ariz., Calif., Colo., Idaho, Iowa, Kans., Minn., Mont., Nebr., Nev., N.Mex., N.Dak., Okla., Oreg., S.Dak., Tex., Utah, Wash., Wyo.; Mexico (Baja California); c Asia.

Acroptilon repens has been reported also from Arkansas, Illinois, Indiana, Kentucky, Louisiana, Michigan, Missouri, New York, Ohio, Virginia, and Wisconsin; I have not seen specimens from those states.

Acroptilon repens is a serious weed pest, especially in the western United States. It is a strong competitor in infested areas, often forming dense colonies, and has allelopathic effects on other plants growing nearby. It is very difficult to control or eradicate once it becomes established. It reproduces vigorously from seed and spreads from adventitious buds borne on deep-seated runner roots. Root fragments readily regenerate as new individuals after cultivation. In addition, Russian knapweed is very poisonous to horses, causing neurological symptoms. Because of its bitter taste, it is usually avoided by grazing animals, and consequently it tends to spread when more palatable plants are consumed.

18. AMBERBOA (Persoon) Lessing, Syn. Gen. Compos., 8. 1832, name conserved • [Pre-Linnaean genus name *Amberboi* Vaillant, cited by Linnaeus in his original publication of *Centaurea*] [I]

David J. Keil

Centaurea Linnaeus subg. *Amberboa* Persoon, Syn. Pl. 2: 481. 1807

Annuals or **biennials**, 20–70 cm; herbage not prickly, glabrate. **Stems** erect, usually branched from near bases. **Leaves** basal and cauline (distal smaller); petiolate (basal and proximal cauline) or sessile (distal cauline); blade margins dentate or ± lobed (basal) or entire to lobed (cauline). **Heads** ± radiant, borne singly. (**Peduncles** slender.) **Involucres** ovoid, 12–16 mm diam. **Phyllaries** many in several series, bases appressed, margins scarious, apices obtuse, inner with oblong, scarious appendages, these entire or spiny. **Receptacles** flat, epaleate, bearing setiform scales ("flattened bristles"). **Florets** many; corollas white to pink, purple, or yellow; outer sterile, corollas expanded, raylike, bilateral, 5–many-lobed; inner fertile, corollas actinomorphic; anther bases tailed, apical appendages oblong; styles branches: fused portions with basal nodes minutely hairy, distinct portions minute. **Cypselae** oblong, compressed, (apices denticulate), faces ribbed, wrinkled, with long, ascending hairs, basal attachment scars lateral, surrounded by whitish, swollen rims); **pappi** persistent, of many scales in several series, distinct, narrow [rarely 0]. *x* = 16.

Species 6 (1 in the flora): introduced; Mediterranean region to c Asia.

Amberboa has often been included within *Centaurea*, from which it differs by cypselae each with a denticulate apex and a conspicuous rim around the basal scar. The chromosome base number *x* = 16 is higher than that in most species of *Centaurea* in a strict sense. Molecular phylogenetic studies of the relationships of Cynareae genera (A. Susanna et al. 1995) place *Amberboa* as sister to the remaining genera of the Centaureinae.

SELECTED REFERENCES Gabrielian, E. and C. E. Jarvis. 1996. *Amberboa moschata, A. glauca,* and *A. amberboi* (Asteraceae: Cardueae). A note on their taxonomy and typification of their names. Taxon 45: 213–215. 1996. Iljin, M. M. 1932. A critical survey of the genus *Amberboa* Less. Izv. Bot. Sada Akad. Nauk S.S.S.R. 30: 101–116.

1. Amberboa moschata (Linnaeus) de Candolle in A. P. de Candolle and A. L. P. P. de Candolle, Prodr. 6: 560. 1838 • Sweet-sultan [F] [I]

Centaurea moschata Linnaeus, Sp. Pl. 2: 909. 1753

Stems with slender branches. **Leaf blades** oblanceolate, 10–25 cm, margins dentate to lyrate-pinnatifid; cauline blades linear to lanceolate, margins dentate to pinnately dissected into linear-lanceolate segments. **Heads** long-pedunculate, ca. 5 cm diam. **Involucres** thinly hairy. **Phyllaries** green, rounded, outer scarious margined, inner with oblong, scarious appendages. **Florets** fragrant; outer conspicuously expanded, many lobed. **Cypselae** dark brown, 3.5–4 mm, glabrous; **pappus** scales 3.5–4 mm. $2n$ = 32 (from cultivated material).

Flowering winter–early summer (Jan–Jul). Escaped or persistent from cultivation in disturbed sites; 0–1200 m; introduced; B.C.; c Asia.

British Columbia is cited on the basis of a 1936 collection from ballast in Vancouver. *Amberboa moschata* has been reported from California, Illinois, Indiana, Iowa, Maine, and Utah; I have not seen specimens from those states.

Amberboa moschata has been cultivated as a garden ornamental since the 1600s. It occasionally grows as a casual garden escape in the United States and southern Canada. Wild forms of *Amberboa moschata* from Turkey and Armenia are described (E. Gabrielian and C. E. Jarvis 1996) as having pink or lilac marginal flowers, and a closely related species *A. amberboi* from central Asia as yellow-flowered. Corollas in garden forms are white, pink, purple, or yellow.

19. MANTISALCA Cassini, Bull. Sci. Soc. Philom. Paris 1818: 142. 1818 • [Anagram of specific epithet *salmantica*] [I]

David J. Keil

Biennials or perennials, 50–100 cm, herbage not spiny. **Stems** erect, branched. **Leaves** basal and cauline; petiolate (basal) or sessile (cauline); blade margins ± lobed (proximal) or dentate or lobed (distal), sparsely hirsute (basal and proximal cauline) or glabrous (distal cauline). **Heads** radiant, borne singly. **Involucres** ovoid to spheric, 10–15 mm diam. **Phyllaries** many in 6–8 series, unequal, appressed, ovate, margins entire, apices obtuse to acute, narrowly membranous fringed, each with a short deciduous spine. **Receptacles** flat, epaleate, long-bristly. **Florets** many; outer neuter, corollas expanded and ± raylike, ± bilateral, staminodes present; inner fertile, corollas purple (rarely white), radial, tubes very slender, throats narrowly funnelform, lobes linear; anther bases tailed, apical appendages oblong; style branches: fused portions with minutely hairy nodes, distinct portions minute. **Cypselae** ± barrel-shaped, ± compressed, with elaiosomes ribbed, transversely roughened, apices without prominent collars, smooth, faces glabrous, basal attachment scars oblique or lateral; **pappi** of several series, outer of stiff, scabrous bristles, inner a single abaxial scale. x = 9, 10, 11.

Species 4 (1 in the flora): introduced; Mediterranean region.

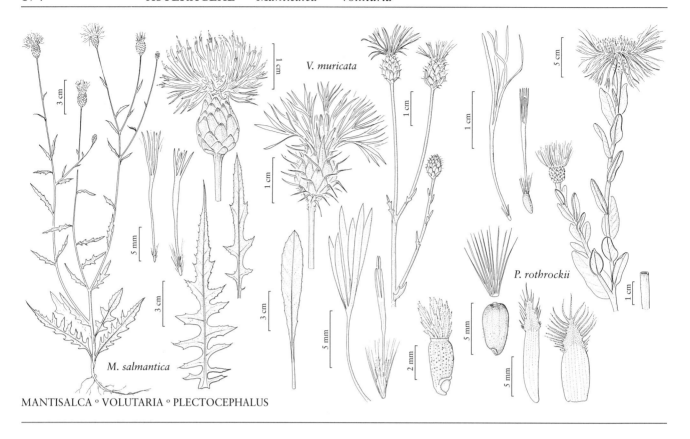

MANTISALCA ° VOLUTARIA ° PLECTOCEPHALUS

1. Mantisalca salmantica (Linnaeus) Briquet &
Cavillier, Biblioth. Universelle Rev. Suisse, pér. 5, 5: 111.
1930 • Dagger-flower F I

Centaurea salmantica Linnaeus, Sp.
Pl. 2: 918. 1753

Herbage cobwebby-tomentose
proximally, glabrous distally.
Stems usually 1, leafless distally.
Leaves: basal blades oblong, 10–
25 cm, margins pinnately lobed;
cauline linear to lanceolate,
smaller distally, dentate to pin-
nately dissected. **Heads** long-pedunculate. **Involucres**
ovoid, distally narrowed. **Phyllaries** greenish or

stramineous, apically blackish, spine tips deciduous,
spreading or reflexed, 1–3 mm. **Corollas** ± purple (rarely
white). **Cypselae** dark brown, 3–4 mm; **pappus** bristles
brownish white or reddish, 2–3 mm. $2n$ = 18 (Italy), 20
(North Africa), 22 (Europe).

Flowering spring–early summer (May–Jun). Disturbed
sites; 0–1700 m; introduced; Ariz., Calif.; Europe; n
Africa.

Mantisalca salmantica is native to the Mediterranean
region. It is considered an uncommon introduction into
disturbed sites.

20. VOLUTARIA Cassini, Bull. Sci. Soc. Philom. Paris 1816: 200. 1816 • [Latin *voluta*,
twisted, spiral, and *-aria*, possession, alluding to spirally coiled corolla lobes of original
species] I

David J. Keil

Cyanopsis Cassini; *Volutarella* Cassini

Annuals, to 50 cm; taprooted, not spiny. **Stems** erect, branched, leafy; branches few–many,
ascending. **Leaves** basal and cauline; winged-petiolate (basal and proximal cauline) or sessile

(mid and distal cauline); blade margins entire to dentate or pinnately divided, faces villous (hairs septate), minutely glandular. **Heads** radiant, borne singly or in few-headed corymbiform arrays. **Involucres** ovoid, 10–15 mm diam. **Phyllaries** many in several series, unequal, appressed, ovate to lanceolate, margins entire, apices acute, tipped by ascending, spreading or reflexed, flattened spines. **Receptacles** flat, epaleate, bristly. **Florets:** peripheral neuter; corollas pink to purple, [blue, or yellow], spreading, lobes (5–6), linear; inner fertile, corollas pink to purple, [blue, yellow, or oarea, colored like the outer or not], tubes slender, throats narrowly cylindric, lobes linear-oblong; anther bases tailed, apical appendages lanceolate; style branches: fused portions with minutely hairy nodes, distinct portions short, linear. **Cypselae** ± barrel-shaped, weakly compressed, ribbed, apices with prominent collars, faces pitted, attachment scars lateral (excavated, surrounded by prominent rims, with eliasomes); **pappi** persistent, of several series of many distinct, white to tawny, narrow scales. $x = 8, 12, 13, 14$.

Species 16 (1 in the flora): introduced; Europe (Mediterranean region), w Asia, Macaronesia, ne Africa.

1. **Volutaria muricata** (Linnaeus) Maire in É. Jahandiez et al., Cat. Pl. Maroc, 3. 817. 1934 • Morocco knapweed F I

Centaurea muricata Linnaeus, Sp. Pl. 2: 918. 1753; *Amberboa muricata* (Linnaeus) de Candolle; *Cyanopsis muricata* (Linnaeus) Dostál; *Volutarella muricata* (Linnaeus) Bentham & Hooker f.

Leaves: proximal oblanceolate, margins denticulate to deeply pinnatifid, distal smaller, linear, ± entire or lobed, faces villous with septate hairs, minutely resin-gland-dotted. **Heads** 3–5 cm diam. **Peduncles** 5–15 cm, distally leafless, finely arachnoid-tomentose, especially near tips. **Involucres** 15–18 mm.

Phyllaries in 5–7 series, minutely spinulose-serrulate, finely pilose, outer and mid ovate, appressed, dark-margined, abruptly tipped by spreading or reflexed, flattened spine tips 3–5 mm, inner lanceolate, with erect or flaring, flattened, ± spineless tips. **Florets:** corolla lobes of sterile 10–15 mm, spreading; corollas of fertile 13–14 mm, tubes 5–6 mm, throats 2–4 mm, lobes 5–6 mm. **Cypselae** pale gray-brown, scars with cartilaginous rims; **pappi** of irregularly toothed scales 1–2.5 mm, outer shorter and narrower than inner. $2n = 24$ (n Africa, as *Cyanopsis muricata*).

Flowering Mar–Jun. Disturbed sites, sparingly established, in coastal regions, probably escaped from cultivation; 0–100 m; introduced; Calif.; sw Europe; n Africa.

21. PLECTOCEPHALUS D. Don in R. Sweet, Brit. Fl. Gard., ser. 2, 1: plate 51. 1830

• Basketflower [Greek *plektos*, woven, and *kephale*, head, alluding to interwoven fringes of phyllaries]

David J. Keil

Centaurea Linnaeus sect. *Plectocephalus* (D. Don) de Candolle

Annuals, 30–200 cm, not spiny. **Stems** erect, branched. **Leaves** basal and cauline; petiolate or sessile; blade margins entire or dentate, faces puberulent, minutely glandular-punctate. **Heads** radiant, borne singly or in open cymiform arrays. **Peduncles** fistulose. **Involucres** ovoid to hemispheric or campanulate, 30–60 mm diam. **Phyllaries** many in 8–10+ series, unequal, narrow, bodies linear, appressed, entire, apices expanded into erect to spreading, narrowly triangular, fringed appendages. **Receptacles** flat, epaleate, bristly. **Florets** many, peripheral neuter; corollas pink to purple, ± zygomorphic, elongate and expanded; inner fertile, corollas pink, purple, cream, or pale yellow, zygomorphic or actinomorphic, ± bent at junction of tubes and throats, tubes elongate, very slender, throats cylindric, lobes linear; anther bases tailed, apical appendages oblong; style branches: fused portions with minutely hairy nodes, distinct portions minute. **Cypselae:** basal attachment scars oblique (with small elaiosome on one side), obovoid or ± barrel-shaped, ± compressed, weakly ribbed, glabrous or puberulent with 2-celled hairs; **pappi** readily falling, of 1–3 series of stiff, minutely barbed bristles. *x* = 13.

Species 4 (2 in the flora): North America, Mexico, South America, Africa.

Molecular phylogenetic studies have been informative regarding the relationships of members of Cynareae, Centaureinae. *Centaurea* as traditionally recognized is polyphyletic. *Plectocephalus* is part of the basal grade in Centaureinae and is not closely related to *Centaurea* in a narrow sense (A. Susanna et al. 1995). Morphologically *Plectocephalus* is difficult to separate from the more derived and highly diverse *Centaurea*. *Plectocephalus* cypselae have obliquely basal attachment scars and the hilum of the seeds is basal. *Centaurea* cypselae have lateral attachment scars and hilum of the seed is lateral. The ovaries of *Plectocephalus* initially have 2-celled hairs; these may be absent from mature cypselae. *Centaurea* produces only 1-celled hairs on ovaries.

SELECTED REFERENCE   Roalson, E. H. and K. W. Allred. 1998. A clarification of *Centaurea americana* and *Centaurea rothrockii* (Compositae: Cardueae). New Mexico Botanist 7: 3–5.

1. Mid phyllaries with (4–)5–7(–8) pairs of lobes, distally light to dark stramineous
    . . . . 1. *Plectocephalus americanus*
1. Mid phyllaries with (9–)10–13(–15) pairs of lobes, distally medium brown to dark brown
    . . . . 2. *Plectocephalus rothrockii*

1. **Plectocephalus americanus** (Nuttall) D. Don in R. Sweet, Brit. Fl. Gard., ser. 2, 1: plate 51. 1830
    • American basketflower, powderpuff or thornless thistle, American star-thistle, cardo del valle

Centaurea americana Nuttall, J. Acad. Nat. Sci. Philadelphia 2: 117. 1821

Plants 50–200 cm. **Stems** usually 1, erect, sparingly branched, glabrous, minutely scabrous and glandular. **Leaves** scabrous; basal sessile or winged-petiolate, usually absent at anthesis, blades oblanceolate to narrowly obovate, 10–20 cm, margins entire or sparingly denticulate; cauline sessile, usually not much smaller except among heads, blades ovate to lanceolate, mostly 5–10 cm, entire or serrulate. **Involucres** broadly hemispheric, 30–50 mm. **Phyllaries:** bodies pale green, broadly elliptic (outer) to linear (inner), apices with appendages erect to spreading, whitish to stramineous (less commonly pale brown to purple), fringed with 9–15 slender spinelike teeth 2–3 mm, teeth not conspicuously ciliate; mid with (4–)5–7(–8) pairs of lobes; faces glabrous or loosely cobwebby-tomentose. **Corollas** of neutral florets pink-purple (rarely white), 35–50 mm, enlarged, raylike; of bisexual florets pinkish, 20–25 mm. **Cypselae** grayish brown to black, 4–5 mm, glabrous or with white hairs near bases; **pappus** bristles unequal, stiff, 6–14 mm. *2n* = 26.

Flowering Feb–Aug. Prairies, fields, open woods, grasslands, roadsides, other disturbed sites; 0–2100 m; Ariz., Ark., Kans., La., Mo., Okla., Tex.; Mexico (Chihuahua, Coahuila, Nuevo León, San Luis Potosí, Tamaulipas).

Plectocephalus americanus is an attractive and showy plant and has been in cultivation for many years. It occasionally escapes from cultivation outside its native range.

2. **Plectocephalus rothrockii** (Greenman) D. J. N. Hind, Bot. Mag. 13: 5. 1996 • Mexican or Rothrock's basketflower, Rothrock's knapweed   [F]

Centaurea rothrockii Greenman, Bot. Gaz. 37: 221. 1904

Plants 30–150 cm. **Stems** usually 1, erect, sparingly branched, loosely tomentose, sometimes glabrate. **Leaves** loosely tomentose, soon glabrate; basal petiolate, usually absent at anthesis, smaller than cauline, blades oblanceolate, tapered to base, margins entire or denticulate; cauline sessile, often clasping, usually not much smaller except among heads, blades ovate to lanceolate or oblong, margins entire or serrulate to denticulate; distally reduced to bracts, sometimes appendaged like phyllaries. **Involucres** broadly hemispheric, 20–35 mm. **Phyllaries:** bodies stramineous to greenish, ovate (outer) to oblong (inner), appendages erect to spreading, ± brown, fringed with 15–25 slender, ciliate, spinelike teeth 2–4 mm; mid with (9–)10–13(–15) pairs of lobes; faces tomentose or glabrescent. **Corollas** of neutral florets pink to purple (rarely white), 30–70 mm, enlarged, raylike; of bisexual yellowish, 20–30 mm. **Cypselae** dark brown, ca. 4 mm, glabrous; **pappus** bristles unequal, stiff, 5–6 mm.

Flowering Jul–Nov. Damp soil near streams, roadsides, open pine-oak woodlands and forests; 1300–2900 m; Ariz., N.Mex.; Mexico (Chihuahua, Durango, Sinaloa, Sonora).

Plectocephalus rothrockii closely resembles *P. americanus*. Although the ranges of the two taxa approach each other in southeastern Arizona and adjacent Chihuahua, they are locally allopatric, separated by habitat differences, and are not known to grow together. *Plectocephalus americanus* is a species of grassland habitats of the southern Great Plains; *P. rothrockii* is largely restricted to moister canyon sites in the Sierra Madre Occidental of Mexico and associated ranges of the American southwest. Mexican basketflower is sometimes cultivated, and can be expected to occur outside its native range.

22. CRUPINA (Persoon) de Candolle, Ann. Mus. Natl. Hist. Nat. 16: 157. 1810

• [Pre-Linnaean generic name of unknown derivation] ⊡

David J. Keil

Centaurea Linnaeus subg. *Crupina* Persoon, Syn. Pl. 2: 488. 1807

Annuals, 20–100 cm, not spiny. **Stems** erect, openly branched distally. **Leaves** basal and cauline (distally reduced to bracts); sessile (basal) or petiolate (cauline); blade margins entire or toothed to pinnately divided (basal) or 1–2-pinnately divided (cauline). **Heads** radiant, borne singly or clustered at branch tips. **Involucres** cylindric to ovoid. **Phyllaries** overlapping in 4–6 series, unequal, oblong-lanceolate, acute, unappendaged. **Receptacles** flat, epaleate, bearing subulate scales. **Florets** 3–15, outer 2–14 neuter, inner 1–2 fertile; corollas purple, ± bilateral, tubes slender, gradually expanded into narrowly funnelform throats, lobes linear; anther bases short-tailed, apical appendages narrowly triangular; style branches: fused portions with minutely puberulent nodes, distinct portions very short, triangular. **Cypselae** cylindric [or ± compressed], bases puberulent, faces smooth, not ribbed, distally softly pubescent, attachment scars basal [or lateral]; **pappi** persistent, present only on fertile florets, outer 1–2 series of numerous stiff, minutely barbed bristles, inner of 5–10 short lacerate-dentate scales. $x = 28, 30$.

Species 3–4 (1 in the flora): introduced; Europe, Asia, n Africa.

Crupina is a member of subtribe Centaureinae. Molecular phylogenetic studies (A. Susanna et al. 1995; N. Garcia-Jacas et al. 2001) place *Crupina* in a basal grade of genera related to *Centaurea*. *Crupina* differs from *Centaurea* by having scaly (rather than bristly) receptacles, and in its cypselar structure (J. Briquet 1930), trichome types (J. Briquet 1930b), unique pollen morphology (G. Wagenitz 1955), and high chromosome base number.

SELECTED REFERENCES Briquet, J. 1930. Carpologie du genre *Crupina*. Candollea 4: 241–278. Briquet, J. 1930b. Les émergences et trichomes de *Crupina*. Candollea 4: 191–201.

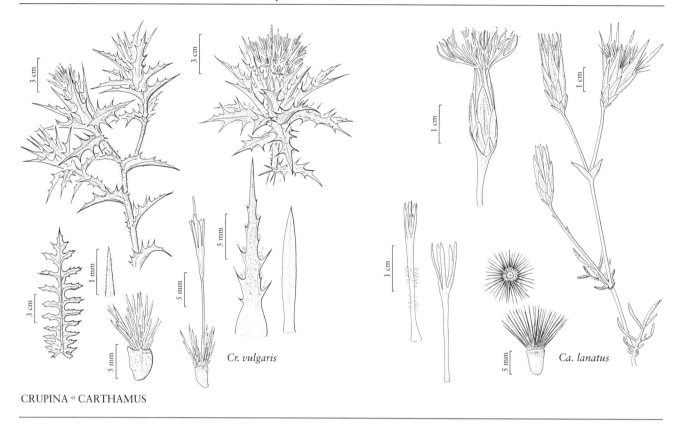

Cr. vulgaris

Ca. lanatus

CRUPINA ∘ CARTHAMUS

1. Crupina vulgaris Persoon ex Cassini in F. Cuvier, Dict. Sci. Nat. ed 2, 12: 68. 1818 • Bearded creeper, common crupina F I W

Centaurea crupina Linnaeus, Sp. Pl. 2: 909. 1753

Plants to 8 cm. **Stems** leafy to branches, openly branched, ridged. **Leaves:** basal often absent by anthesis, blades oblong to obovate, margins entire or dentate to deeply pinnately divided, scabrous, ciliate with glochidiate-tipped bristles; cauline 1–3.5 cm, 1–2-pinnately divided into linear lobes, ciliate with glochidiate-tipped bristles. **Peduncles** slender, 0.5–8 cm, bractless or with 1–several scalelike to linear bracts. **Involucres** sometimes falling at maturity with enclosed cypselae, 8–20 mm. **Sterile florets** 2–4. **Fertile florets** usually 1; corollas ca. 14 mm, tube with branched hairs. **Cypselae** usually 1 per head, barrel-shaped, 3–6 × 1.5–3.5 mm, attachment scar basal; **pappus bristles** in 2 unequal series, widely spreading, blackish brown, 5–10 mm; **pappus scales** erect, 0.5–1 mm. $2n$ = 28 (Greece), 30 (Yugoslavia).

Flowering spring–summer (Jun–Aug). Grassy areas, roadsides, open woodlands, forests; 200–1000 m; introduced; Calif., Idaho, Mass., Oreg., Wash.; Europe; w Asia.

I have seen material only from California; occurrences in the other states listed are documented.

Crupina vulgaris is listed as a noxious weed by governmental agencies both within and outside of its documented range. It is a highly invasive species that infests pastures, rangelands, hayfields, and disturbed sites, sometimes forming dense pure stands. It is unpalatable to livestock and tends to increase as more desirable forage is consumed. Since it was first reported in west-central Idaho in 1969, it has spread rapidly in western rangelands, forming infestations in several states.

23. CARTHAMUS Linnaeus, Sp. Pl. 2: 830. 1753; Gen. Pl. ed. 5, 361. 1754 • Distaff thistle [Arabic *qartam*, safflower] I

David J. Keil

Annuals or perennials, 30–180 cm, herbage glabrous to ± glandular and/or ± tomentose. **Stems** usually erect, branched distally or throughout, (leafy). **Leaves** basal and cauline or all cauline;

basal and proximal cauline winged-petiolate, distal cauline sessile, sometimes clasping; blade margins dentate to pinnately lobed, ± spiny, faces glabrous or ± glandlar and/or ± tomentose. **Heads** discoid, borne singly or in few-headed cymiform arrays. **Involucres** ovoid, constricted distally. **Phyllaries** many in 4–5 series, linear to ovate (at least outer ± leaflike), bases appressed, apical appendages more herbaceous, prominently veiny, spiny-dentate or -lobed, spine-tipped. **Receptacles** convex to conic, epaleate, bearing subulate scales. **Florets** 15–60+; corollas yellow to red or ± purple, tubes very slender, throats gradually or abruptly expanded, ± cylindric or short-campanulate, lobes linear; anther bases short-tailed, apical appendages oblong; style branches: fused portions with slightly swollen basal nodes minutely hairy, distally minutely papillate, distinct portions very short. **Cypselae** oblong to obpyramidal, ± 4-angled, apices with smooth or dentate rims, faces usually ± roughened (outer) or smooth (inner), glabrous, attachment scars lateral; **pappi** 0 or (usually only inner cypselae) ± persistent, of many, usually unequal, narrow scales overlapping in several series. *x* = 10, 12.

Species 14 (4 in the flora): introduced; United States; Mediterranean region.

SELECTED REFERENCES Ashri, A. and P. F. Knowles. 1959. Further notes on *Carthamus* in California. Leafl. W. Bot. 9: 5–8. Ashri, A. and P. F. Knowles. 1960. Cytogenetics of safflower (*Carthamus* L.) species and their hybrids. Agron. J. 52: 11–17. Hanelt, P. 1963. Monographisch Übersicht der Gattung *Carthamus* L. (Compositae). Feddes Repert. Spec. Nov. Regni Veg. 67: 41–180. Hanelt, P. 1976. *Carthamus*. In: T. G. Tutin et al., eds. 1964–1980. Flora Europaea. 5 vols. Cambridge. Vol. 4, pp. 302–303. Khidir, M. O. and P. F. Knowles. 1970. Cytogenetic studies of *Carthamus* species (Compositae) with 32 pairs of chromosomes. I. Intrasectional hybridization. Amer. J. Bot. 57: 123–129. Khidir, M. O. and P. F. Knowles. 1970b. Cytogenetic studies of *Carthamus* species (Compositae) with 32 pairs of chromosomes. II. Intersectional hybridization. Canad. J. Genet. Cytol. 12: 90–99.

1. Corollas pink or pale purple; cauline leaves ± deeply lobed . . . . 1. *Carthamus leucocaulos*
1. Corollas yellow to red; cauline leaves toothed to deeply lobed.
　　2. Cauline leaves toothed, otherwise undivided, weakly spiny; corollas yellow to red; cypselae white . . . . 4. *Carthamus tinctorius*
　　2. Cauline leaves ± divided, very spiny; corollas yellow; cypselae brown.
　　　　3. Anthers yellow, pollen yellow; corollas bright yellow; outer phyllaries usually not more than 1.5 times as long as inner; herbage ± cobwebby; stems stramineous . . . . 2. *Carthamus lanatus*
　　　　3. Anthers white with purple lines, pollen white; corollas pale yellow; outer phyllaries usually 2 times as long as inner; herbage ± sparsely hairy; stems white or stramineous . . . . 3. *Carthamus creticus*

1. **Carthamus leucocaulos** Sibthorp & Smith, Fl. Graec. Prodr. 2(1): 160. 1813 • White-stem distaff thistle I W

Plants 30–100 cm, herbage glabrescent. **Stems** rigidly erect, shiny white, often much-branched distally. **Leaves** basal and cauline; basal often absent at anthesis, petioles winged, blades pinnately divided into 6–8 pairs of narrow lobes; cauline spreading or recurved, lanceolate to ovate, rigid, clasping, 3–7-veined from base, margins pinnately divided into 2–3 pairs of short, spine-tipped lobes, apices spine-tipped. **Involucres** ovoid, 10–13 mm, usu-ally ± glabrous. **Outer phyllaries** ascending or ± spreading, very shiny, 40–50 mm, 2.5–3.5 times as long as inner, terminal appendages spreading to ascending, spiny-lobed, prominently spine-tipped. **Corollas** pink or pale purple, 13–17 mm, throats abruptly expanded; anthers white or pink with purple stripes; pollen white. **Cypselae** brown, 3–5 mm, outer roughened; **pappus** scales 5–7 mm. *2n* = 20 (Greece).

Flowering summer (Jun–Aug). Disturbed sites; 0–200 m; introduced; Calif.; Europe (Greece, islands of Aegean Sea).

Between 1969 and 1990 an infestation of this noxious weed was documented in Sonoma County. Efforts to eradicate it were apparently successful.

2. Carthamus lanatus Linnaeus, Sp. Pl. 2: 830. 1753

• Woolly distaff thistle [F] [I] [W]

Plants 40–180 cm, herbage ± densely glandular, loosely cobwebby to ± woolly. **Stems** rigidly erect, openly branched distally, stramineous. **Leaves** basal and cauline; basal often absent at anthesis, petioles winged, blades 10–15 cm, margins pinnately 1–2-divided into linear or lanceolate spine-tipped lobes; cauline spreading or recurved, lanceolate to ovate, rigid, clasping, 3–7-veined from base, margins with narrow spine-tipped lobes, spinose-acuminate. **Involucres** ovoid, body 25–35 mm, usually ± tomentose. **Outer phyllaries** ascending or ± spreading, 35–50 mm, usually not more than 1.5 times as long as inner, terminal appendages spreading to ascending, linear-lanceolate, spiny lobed, prominently spine-tipped. **Corollas** yellow, sometimes red- or black-veined, 25–35 mm, throats gradually expanded; anthers yellow; pollen yellow. **Cypselae** brown, 4–6 mm, the outer roughened; **pappus** scales 1–13 mm. $2n = 44$.

Flowering Apr–Sep(–Nov). Roadsides, grain fields, pastures; 0–1100 m; introduced; Ariz., Calif., Mass., N.J., Okla., Oreg., Tex.; Europe.

Native to the Mediterranean region, *Carthamus lanatus* is a viciously spiny noxious weed, sometimes forming nearly impenetrable stands. In rangelands it is known to injure the eyes and mouths of livestock, and it tends to spread when more palatable plants are consumed. Because of the close relationship between the cultivated safflower (*Carthamus tinctorius*) and its weedy relatives, biocontrol has not been an option for controlling weedy species such as *C. lanatus*.

3. Carthamus creticus Linnaeus, Sp. Pl. ed. 2, 2: 1163. 1763 • Smooth distaff thistle [I] [W]

Carthamus baeticus Boissier & Reuter; *C. lanatus* Linnaeus subsp. *baeticus* (Boissier & Reuter) Nyman; *C. lanatus* subsp. *creticus* (Linnaeus) Holmboe

Plants 40–100 cm, herbage ± sparsely hairy. **Stems** rigidly erect, openly branched above, stramineous. **Leaves** basal and cauline; basal often absent at anthesis, petioles winged, blades pinnately 1–2-divided into linear or lanceolate spine-tipped lobes, cauline spreading or recurved, lanceolate to ovate, rigid, clasping, margins spiny-lobed, spine-tipped. **Involucres** ovoid, 20–25 mm, very thinly cobwebby or becoming glabrous. **Outer phyllaries** ascending or ± spreading, 35–55 mm, usually 2 times as

long as inner, terminal appendages spreading to ascending, spiny-lobed, prominently spine-tipped. **Corollas** pale yellow, 25–35 mm, throats abruptly expanded; anthers white with purple stripes; pollen white. **Cypselae** brown, 4–6 mm, outer roughened; **pappus** scales 1–10 mm. $2n = 64$.

Flowering summer (Jun–Aug). Fields, roadsides; 0–500 m; introduced; B.C.; Calif., Nev., Oreg., S.C.; Europe.

Carthamus creticus has been reported for British Columbia in all recent floras, as *C. lanatus* subsp. *baeticus*. It is native to the Mediterranean region.

Apparently an allohexaploid derived by hybridization between *Carthamus leucocaulos* ($2n = 20$) and *C. lanatus* ($2n = 44$) (M. O. Khidir and P. F. Knowles 1970b), *C. creticus* is similar to *C. lanatus* and was treated as a subspecies of the latter (P. Hanelt 1963, 1976).

Most American botanists have recognized this taxon at species rank, using the name *Carthamus baeticus* ascribed to (Boissier & Reuter) Nyman, based on the assumption that Nyman (Consp. Fl. Eur., 419. 1879) had proposed a new combination at the species level based on *Kentrophyllum baeticum* Boissier & Reuter. P. Hanelt (1963) used the name *C. lanatus* subsp. *creticus* for this taxon and treated both *C. creticus* and *C. baeticus* as synonyms. However, Hanelt (1976) substituted *C. lanatus* subsp. *baeticus* as the name for the taxon, ascribing the combination to the same Nyman publication. Hanelt (pers. comm.) has indicated that the contradictory nomenclatural citations were a result of Nyman's peculiar way of presenting taxa that he considered to be subspecies: "in the work of Nyman the small-printed taxa subsumed under a 'true' species name and characterized by an asterisk had to be accepted as subspecies." As a subspecific epithet, *baeticus* (1879) has nomenclatural priority over *creticus* (1914), hence Hanelt's 1976 use of the former.

Like *Carthamus lanatus*, *C. creticus* is a noxious weed that can severely degrade infested rangelands.

4. Carthamus tinctorius Linnaeus, Sp. Pl. 2: 830. 1753 • Safflower [I]

Plants 30–100+ cm, herbage ± glabrous. **Stems** ± stramineous, glabrous. **Leaves** usually all cauline, dark green; blades lanceolate to elliptic or broadly ovate, 2–8.5 cm, margins dentate with minutely spine-tipped teeth, veiny, shiny. **Involucres** ovoid, 20–40 mm diam., ± glabrous. **Outer phyllaries** spreading to reflexed, 1.5–2 times longer than inner, terminal appendages minutely spiny-toothed, minutely spine-tipped. **Corollas** yellow to red, 20–30

mm, throats abruptly expanded; anthers yellow to red; pollen yellow to red. **Cypselae** white, 7–9 mm, slightly roughened; **pappus** scales absent or if present, 1–4 mm. **2***n* = 24.

Flowering late spring–summer (May–Aug). Escaped from cultivation in disturbed sites; 0–900 m; introduced; Alta., B.C.; Ariz., Calif., Colo., Idaho, Ill., Iowa, Kans., Mass., Mont., Nebr., N.Mex., N.Dak. Ohio, Oreg., Utah, Wash.; Europe.

Carthamus tinctorius is apparently native originally to the eastern Mediterranean; it is known only in cultivation and as escapes today. Safflower has been reported from Texas; I have not seen the specimen.

Safflower is cultivated as an oil seed, a source of vegetable dye, as birdseed, and as an ornamental. It is one of the earliest known crop plants, with cultivation dating back to prehistoric times. In the United States safflower is grown principally in California and Arizona; it has been a successful crop in every state west of the 100th meridian.

Carthamus oxyacantha M. Bieberstein (wild safflower) was collected in 1978 in Monterey County, California. It is considered by the United States Department of Food and Agriculture to be a noxious weed subject to eradication if found. In central and southern Asia it is a pernicious weed of agricultural lands and other disturbed ground. *Carthamus oxyacantha* most closely resembles cultivated safflower; it has smaller heads and much spinier leaves. Its cypselae are usually darkly pigmented, smaller (4–5 mm versus 5.5–9 mm), and almost always lack pappi.

24. CENTAUREA Linnaeus, Sp. Pl. 2: 909. 1753; Gen. Pl. ed. 5, 389. 1754 • Knapweed, star thistle, cornflower, centaurée [Greek *kentaurieon*, ancient plant name associated with Chiron, a centaur famous for knowledge of medicinal plants] [1]

David J. Keil

Jörg Ochsmann

Acosta Adanson; *Cnicus* Linnaeus; *Grossheimia* Sosnowsky & Takhtajan; *Jacea* Miller; *Leucacantha* Nieuwland & Lunell

Annuals, biennials, or perennials, 20–300 cm, glabrous or tomentose. **Stems** erect, ascending, or spreading, simple or branched. **Leaves** basal and cauline; petiolate or sessile; proximal blade margins often ± deeply lobed, (spiny in *C. benedicta*), distal ± smaller, often entire, faces glabrous or ± tomentose, sometimes also villous, strigose, or puberulent, often glandular-punctate. **Heads** discoid, disciform, or radiant, borne singly or in corymbiform arrays. **Involucres** cylindric or ovoid to hemispheric. **Phyllaries** many in 6–many series, unequal, proximal part appressed, body margins entire, distal parts expanded into erect to spreading, usually ± dentate or fringed, linear to ovate appendages, spine-tipped or spineless. **Receptacles** flat, epaleate, bristly. **Florets** 10–many; outer usually sterile, corollas slender and inconspicuous to much expanded, ± bilateral; inner fertile, corollas white to blue, pink, purple, or yellow, bilateral or radial, often bent at junction of tubes and throats, lobes linear-oblong, acute; anther bases tailed, apical appendages oblong; style branches: fused portions with minutely hairy nodes, distinct portions minute. **Cypselae** ± barrel-shaped, ± compressed, smooth or ribbed, apices entire (denticulate in *C. benedicta*), glabrous or with fine, 1-celled hairs, attachment scars lateral (with or without elaiosomes); **pappi** 0 or ± persistent, of 1–3 series of smooth or minutely barbed, stiff bristles or narrow scales. *x* = 8, 9, 10, 11, 12, 13, 15.

Species ca. 500 (20 in the flora): introduced; Eurasia, n Africa, widely introduced worldwide.

Taxonomic limits of *Centaurea* have been controversial. The genus has great morphologic diversity, and studies have revealed much cytologic (e.g., N. Garcia-Jacas et al. 1996) and palynologic (e.g., G. Wagenitz 1955) variation as well. During the nineteenth and twentieth centuries, various taxonomists attempted, with limited success, to divide *Centaurea* into smaller genera or workable infrageneric taxa. The relations of several satellite genera have been controversial as well.

Recent molecular phylogenetic studies (A. Susanna et al. 1995; N. Garcia-Jacas et al. 2000, 2001) have begun to clarify relationships within *Centaurea* and between *Centaurea* and other genera. These studies make it clear that *Centaurea* as traditionally defined is polyphyletic, and that generic boundaries should be realigned if monophyletic taxa are to be recognized. Some taxa traditionally included within *Centaurea* (e.g., the two native North American species, *Centaurea americana* and *C. rothrockii*) fall outside the redefined generic boundaries and are here treated in *Plectocephalus*. Others usually placed into segregate genera (e.g., *Cnicus benedictus*) are firmly nested within *Centaurea*. Because the type species of *Centaurea* (*C. centaurium* Linnaeus, an African species) falls outside the main lineage of the genus, a proposal has been made to conserve *Centaurea* with a different type species (W. Greuter et al. 2001), thereby maintaining the nomenclatural stability of most of the numerous species that do fall within the principal *Centaurea* clade.

Although several *Centaurea* species are widely established as members of the North American flora, and some of these are widely distributed invasive weeds, some of the taxa listed by J. T. Kartesz and C. A. Meacham (1999) are apparently waifs and not permanent members of the flora. These taxa are discussed informally immediately below.

Centaurea aspera Linnaeus (rough star thistle) is known from nineteenth-century collections from ballast piles in New York; it does not appear to be established as a member of the North American flora. It can readily be distinguished from the similar *C. diluta*: the phyllary appendages are divided into palmately radiating clusters of short spines.

Centaurea babylonica Linnaeus has been reported from California as a waif (F. Hrusa et al. 2002); apparently it is not established as a permanent member of the flora. It is a tall (to 3 m), yellow-flowered *Centaurea* with numerous heads clustered in spiciform arrays. The phyllaries are leathery, and appendages are absent or reduced to spines 1 mm or shorter.

Centaurea cineraria Linnaeus (dusty miller), an Italian species known from casual garden escapes (California, Maryland, New York), probably is not permanently established as a member of the North American flora. It is a perennial with pinnately or bipinnately divided, densely gray-tomentose leaves, usually solitary, radiant heads somewhat larger than those of *C. stoebe*, and purple corollas.

Centaurea eriophora Linnaeus, reported from California and Colorado (J. T. Kartesz and C. A. Meacham 1999), would key below to *C. sulphurea*. It differs from *C. sulphurea* in having densely arachnoid-tomentose involucres. The California report is based on an early twentieth-century collection from Los Angeles County (*A. Davidson 2334*, UC). It is unlikely that this species is permanently established as a member of the California flora. The Colorado report is apparently erroneous, referenced to a report of *Jacea pratensis* Lamarck (= *Centaurea jacea* Miller), a wholly different plant, in W. A. Weber and R. C. Wittmann (1992).

A population of *Centaurea trichocephala* M. Bieberstein ex Willdenow (featherhead or hairy-head knapweed) was found in the late 1970s in a degraded pasture in eastern Washington (B. F. Roché and C. T. Roché 1991). A weed-control program was instituted, and the plants were successfully eradicated. Although it is apparently not established anywhere in North America, *C. trichocephala* is listed as a noxious weed in Oregon. These plants resemble *C. phrygia* in having elongate, pectinate-fringed phyllary appendages. In *C. trichocephala* the linear-filiform, featherlike appendages are much narrower than the phyllary bodies. Plants of the species spread by horizontal roots. According to Roché and Roché, *C. trichocephala* is apparently self-sterile; the Oregon plants spread clonally and formed no seeds.

Although *Cnicus* has usually been recognized as a distinctive monotypic genus, it has been merged into *Centaurea* by various authors (e.g., K. Bremer 1994; G. Wagenitz and F. H. Hellwig 1996). Recent molecular systematic studies (N. Garcia-Jacas et al. 2000) provide additional evidence that it is nested within *Centaurea*.

SELECTED REFERENCES Garcia-Jacas, N., A. Susanna, V. Mozaffarian, and R. Ilarslan. 2000. The natural delimitation of *Centaurea* (Asteraceae: Cardueae): ITS sequence analysis of the *Centaurea jacea* group. Pl. Syst. Evol. 223: 185–199. Moore, R. J. 1972. Distribution of native and introduced knapweeds (*Centaurea*) in Canada and the United States. Rhodora 74: 331–346. Roché, B. F. and C. T. Roché. 1991. Identification, introduction, distribution, ecology, and economics of *Centaurea* species. In: L. F. James et al., eds. 1991. Noxious Range Weeds. Boulder, San Francisco, and Oxford. Pp. 274–291. Wagenitz, G. 1955. Pollenmorphologie und Systematik in der Gattung *Centaurea* L. s.l. Flora 142: 213–279.

1. Corollas yellow.
    2. Phyllary appendages broadly scarious, ± covering phyllary bodies, lacerate fringed, spineless or tipped by weak spines 1–2 mm . . . . 4. *Centaurea macrocephala*
    2. Phyllary appendages spiny fringed, not covering phyllary bodies, tipped with spines 5–25 mm.
        3. Heads sessile, each closely subtended and ± concealed by involucrelike cluster of expanded, foliar bracts . . . . 16. *Centaurea benedicta*
        3. Heads evidently pedunculate, or if sessile, each not concealed by involucrelike cluster of expanded, foliar bracts.
            4. Corollas usually longer than 25 mm . . . . 20. *Centaurea sulphurea*
            4. Corollas usually 10–20 mm.
                5. Central spines of principal phyllaries 5–10 mm . . . . 18. *Centaurea melitensis*
                5. Central spines of principal phyllaries 10–25 mm . . . . 19. *Centaurea solstitialis*
1. Corollas white to pink, blue, or purple.
    6. Principal phyllaries tipped with spines.
        7. Central spines of principal phyllaries 10–25 mm.
            8. Young leaves ± gray tomentose when young; involucre bodies 6–8 mm diam.; florets 25–40 . . . . 14. *Centaurea calcitrapa*
            8. Young leaves minutely bristly, green; involucre bodies 8–14 mm diam.; florets many . . . . 15. *Centaurea iberica*
        7. Central spines of principal phyllaries usually 1–5 mm.
            9. Involucres 8–15 mm diam. . . . . 17. *Centaurea diluta*
            9. Involucres 3–5 mm diam.
                10. Corollas 12–13 mm; involucres 10–13 mm . . . . 12. *Centaurea diffusa*
                10. Corollas 7–9 mm; involucres 7–8 mm . . . . 13. *Centaurea virgata*
    6. Principal phyllaries not spine-tipped.
        11. Annuals.
            12. Proximal leaves lanceolate, acute; phyllary appendages white to dark brown or black, teeth ± 1 mm; pappi 2–4 mm . . . . 1. *Centaurea cyanus*
            12. Proximal leaves oblong, obtuse; phyllary appendages silvery white to brown, teeth 1.5–2 mm; pappi 6–8 mm . . . . 2. *Centaurea depressa*
        11. Perennials.
            13. Phyllary appendages tapering to long, often recurved, pectinately dissected, filiform tips . . . . 10. *Centaurea phrygia*
            13. Phyllary appendages obtuse to acute, erect or ascending.
                14. Involucres 10–13 mm . . . . 11. *Centaurea stoebe*
                14. Involucres 15–25 mm.
                    15. Phyllary appendages evidently decurrent along phyllary margins.
                        16. Leaves decurrent, entire or remotely dentate, or proximal sometimes pinnately lobed . . . . 3. *Centaurea montana*
                        16. Leaves not decurrent, 1–2-pinnately divided or distal entire . . . . 5. *Centaurea scabiosa*
                    15. Phyllary appendages not or only slightly decurrent along phyllary margins.

[17. Shifted to left margin.—Ed.]

17. Phyllary appendages roundish, seldom triangular, scarious, light to dark brown, ± undivided to irregularly lacerate.
    18. Pappus absent . . . . 6. *Centaurea jacea*
    18. Pappus present, 3–5 mm, of many white bristles . . . . 17. *Centaurea diluta*
17. Phyllary appendages ± triangular, brown to black, ± wholly pectinately fimbriate; pappus present or absent.
    19. Heads discoid; peripheral florets not expanded, showy; pappus blackish, shorter than 1 mm; green parts of phyllaries ± totally covered by black appendages, involucres appearing totally black . . . . 7. *Centaurea nigra*
    19 Heads radiant; peripheral florets expanded, showy, raylike; pappus absent or rudimentary, when present usually not black; green part of phyllaries sometimes evident or appendages light to dark brown.
        20. Heads relatively broad, the pressed involucres usually as wide as or wider than long; green parts of phyllaries usually fully covered by ± brown, variably pectinately fimbriate appendages, involucres light to dark brown . . . . 8. *Centaurea ×moncktonii*
        20. Heads relatively narrow, the pressed involucres usually longer than wide; green parts of phyllaries not fully covered by black appendages, involucres black and green . . . . 9. *Centaurea nigrescens*

1. **Centaurea cyanus** Linnaeus, Sp. Pl. 2: 911. 1753
    • Bachelor's-button, garden cornflower, cornflower, bluebottle, bluebonnets, blaver, blue-poppy, thimbles, brushes, corn pinks, witch's bells, hurtsickle, bleuet, barbeau, casse lunette [1]

Leucacantha cyanus (Linnaeus) Nieuwland & Lunell

Annuals, 20–100 cm. **Stems** usually 1, erect, ± openly branched distally, loosely tomentose. **Leaves** ± loosely gray-tomentose; basal leaf blades linear-lanceolate, 3–10 cm, margins entire or with remote linear lobes, apices acute; cauline linear, usually not much smaller except among heads, usually entire. **Heads** radiant, in open, rounded or ± flat-topped cymiform arrays, pedunculate. **Involucres** campanulate, 12–16 mm. **Phyllaries:** bodies green, ovate (outer) to oblong (inner), tomentose or becoming glabrous, margins and erect appendages white to dark brown or black, scarious, fringed with slender teeth ± 1 mm. **Florets** 25–35; corollas blue (white to purple), those of sterile florets raylike, enlarged, 20–25 mm, those of fertile florets 10–15 mm. **Cypselae** stramineous or pale blue, 4–5 mm, finely hairy; **pappi** of many unequal stiff bristles, 2–4 mm. $2n = 24$ (Russia).

Flowering spring–summer (May–Sep). Grasslands, woodlands, forests, roadsides, other disturbed sites; 50–2400 m; introduced; Greenland; Alta., B.C., Man., N.B., Nfld. and Labr. (Nfld.), N.S., Ont., P.E.I., Que., Yukon; Ala., Alaska, Ariz., Ark., Calif., Colo., Conn., Del., D.C., Fla., Ga., Idaho, Ill., Ind., Iowa, Kans., Ky., La., Maine, Md., Mass., Mich., Minn., Miss., Mo., Mont., Nebr., Nev., N.H., N.J., N.Mex., N.Y., N.C., N.Dak., Ohio, Okla., Oreg., Pa., R.I., S.C., S.Dak., Tenn., Tex., Utah, Vt., Va., Wash., W.Va., Wis., Wyo.; s Europe.

Centaurea cyanus is a commonly cultivated garden ornamental. Its cypselae are often included in wildflower seed mixes and it naturalizes readily in many areas.

2. **Centaurea depressa** M. Bieberstein, Fl. Taur.-Caucas. 2: 346. 1808 • Low cornflower [1]

Annuals, 20–60 cm. **Stems** usually several–many from base, spreading, ± openly branched distally, loosely gray-tomentose. **Leaves** ± loosely gray-tomentose; basal and proximal cauline petiolate, blades oblong, 5–10 cm, margins entire or pinnatifid with terminal segment largest, apices obtuse; mid and distal cauline sessile, linear-lanceolate to oblong, blades usually not much smaller, entire, mucronate. **Heads** radiant, borne singly, pedunculate. **Involucres** ovoid to campanulate, 15–20 mm. **Phyllaries:** bodies green, ovate (outer) to oblong (inner), glabrous, margins and erect appendages silvery white to brown, scarious, fringed with slender teeth 1.5–2 mm. **Florets** 25–35; corollas of sterile florets spreading, dark blue, 25–30 mm, enlarged, those of fertile florets purple, ca. 15 mm. **Cypselae** brown, 4.5–6 mm, puberulent near attachment scar, otherwise glabrous; **pappi** of outer series of unequal stiff bristles 2–8 mm, inner series of slender scales ca. 1.5 mm. $2n = 16$ (Armenia).

Flowering spring–summer (May–Jul). Disturbed ground; 50–1400 m; introduced; Md., Nev.; sw, c Asia.

3. **Centaurea montana** Linnaeus, Sp. Pl. 2: 911. 1753
   • Mountain cornflower or bluet, centaurée des
   montagnes Ⓘ

Perennials, 25–80 cm, from rhizomes or stolons. **Stems** 1–several, erect, simple or sparingly branched, villous with septate hairs and thinly arachnoid-tomentose with long, simple hairs. **Leaves** thinly villous and ± tomentose, glabrate; proximal leaves winged-petiolate, blades 10–30 cm, margins entire or remotely dentate to pinnately lobed; mid and distal leaves sessile, blades decurrent, ovate to oblong or lanceolate, entire or remotely denticulate. **Heads** radiant, borne singly or in few-headed corymbiform arrays; (peduncles to 7 cm). **Involucres** ovoid to ± campanulate, 20–25 mm. **Principal phyllaries:** bodies greenish, ovate to lanceolate, scarious-margined, appendages appressed, brown to black, unarmed, decurrent on phyllary margins, pectinate-fringed, puberulent; innermost phyllaries sometimes unappendaged. **Florets** 35–60+; sterile florets 10–20, corollas blue (white, purple, or pink), 2.5–4.5 cm, corolla tube elongate. **Disc florets** 25–40+; corollas purple, ca. 20 mm; anthers dark blue-purple. **Cypselae** ± brown, 5–6 mm, sericeous; **pappi** of bristles 0.5–1.5 mm. $2n = 24$ (Germany), 40 (Russia), 44 (France).

Flowering summer (Jun–Aug). Escaped from cultivation, roadsides, woodlands, sagebrush scrub; 0–1400 m; introduced; St. Pierre and Miquelon; B.C., N.B., Nfld. and Labr. (Nfld.), Ont., Que.; Alaska, Idaho, Maine, Mich., Minn., Mont., N.H., N.Y., Oreg., Pa., Utah, Wash., Wis.; Europe.

Centaurea montana is a very handsome plant, native to the mountains of Europe, now widely cultivated as an ornamental.

4. **Centaurea macrocephala** Puschkarew ex Willdenow, Sp. Pl. 3: 2298. 1803 • Globe centaurea, big-head knapweed, yellow bachelor's button or cornflower, centaurée à gros capitules Ⓘ

Grossheimia macrocephala (Puschkarew ex Willdenow) Sosnowsky & Takhtajan

Perennials, 50–170 cm. **Stems** usually several, erect, unbranched or sparingly branched distally, villous with septate hairs, thinly arachnoid-tomentose, fistulose proximal to heads. **Leaves** short-villous and thinly arachnoid, ± glabrate, resin-gland-dotted; basal and proximal cauline petiolate, blades oblanceolate to narrowly ovate, 10–30 cm, margins entire or shallowly dentate; cauline sessile, shortly decurrent, not much smaller except those crowded proximal to heads, blades lanceolate to ovate, 5–10 cm, entire, often ± undulate, apices acute. **Heads** disciform or weakly radiant, borne singly, sessile, closely subtended by clusters of reduced leaves. **Involucres** ovoid to hemispheric, 25–35 mm. **Phyllaries:** bodies pale green or stramineous, ovate or broadly lanceolate, glabrous, appendages erect to spreading, brown, scarious, abruptly expanded, 1–2 cm wide, ± covering phyllary bodies, lacerate fringed, sometimes tipped by weak spines 1–2 mm, glabrous. **Florets** many; corollas yellow; corollas of sterile florets slightly expanded, ca. 4 mm; corollas of disc florets ca. 3.5 mm. **Cypselae** 7–8 mm; **pappi** of many setiform scales ("flattened bristles"), 5–8 mm. $2n = 18$ (Russia).

Flowering summer (Jun–Sep). Garden escape in meadows, grassy clearings; 400–2000 m; introduced; Ont., Que.; Colo., Mich., Wash., Wis.; e Europe; w Asia.

Although *Centaurea macrocephala* is cultivated as an ornamental and for cut flowers in many areas, it has been declared a noxious weed by the state of Washington because of its potential status as an invader.

5. **Centaurea scabiosa** Linnaeus, Sp. Pl. 2: 913. 1753
   • Greater knapweed, hardheads, centaurée scabieuse
   Ⓘ

Perennials, 30–150 cm. **Stems** 1–several, branches ascending, glabrous to ± hirsute. **Leaves** minutely hispid, resin-gland-dotted; basal and proximal cauline petiolate, blades 10–25 cm, margins usually 1–2-pinnately divided into linear or oblong segments; mid and distal cauline smaller, entire or once dissected. **Heads** borne singly or few in open cymiform arrays, pedunculate. **Involucres** ovoid to hemispheric, becoming campanulate, 15–25 mm. **Phyllaries:** bodies dark green, ovate (outer) to oblong-lanceolate (inner), glabrous or finely arachnoid, margins and erect appendages black, ± fringed distally with slender teeth, inner phyllaries with brownish scarious, expanded, erose dissected appendages. **Florets** many; corollas reddish purple (white), those of sterile florets 35–40 mm, often conspicuously enlarged, those of fertile florets 20–25 mm. **Cypselae** brown, 4.5–5 mm, puberulent; **pappi** of many unequal stiff bristles, white, 4–5 mm. $2n = 20$ (Russia), 40.

Flowering summer (Jun–Sep). Disturbed sites, pastures, sparingly escaped from cultivation, probably not persisting in all areas where reported; 50–1800 m; introduced; B.C., N.B., Ont., Que. ; Conn., Idaho, Iowa, Ky., Maine, Mont., N.H., N.J., N.Y., N.Dak., Ohio, Pa., Utah, Wyo.; Europe.

6. Centaurea jacea Linnaeus, Sp. Pl. 2: 914. 1753
   • Brown or brown-ray knapweed, centaurée jacée,
   jacée des prés   [I]

Jacea pratensis Lamarck

Perennials, 30–150 cm. **Stems** 1–few, erect or ascending, openly branched distally, villous to scabrous with septate hairs, loosely tomentose, ± glabrate. **Leaves:** basal and proximal cauline petiolate, blades oblanceolate or elliptic, 5–25 cm, margins entire or shallowly dentate to irregularly pinnately lobed; distal cauline sessile, not decurrent, gradually smaller, blades linear to lanceolate, entire or dentate. **Heads** radiant, in few-headed corymbiform arrays, leafy-bracted pedunculate. **Involucres** ovoid to campanulate or hemispheric, 15–18 mm, usually about as wide as high. **Principal phyllaries:** bodies lanceolate to ovate, loosely tomentose or glabrous, usually concealed by expanded appendages, appendages usually light brown, erect, overlapping, ± concave, usually roundish, margins pale, broad, entire to coarsely dentate, membranous. **Inner phyllaries:** tips truncate, irregularly dentate or lobed. **Florets** 40–100+; corollas purple (rarely white), those of sterile florets ± expanded, exceeding corollas of fertile florets, those of fertile florets 15–18 mm. **Cypselae** tan, 2.5–3 mm, finely hairy; **pappi** absent. $2n = 22, 44$.

Flowering summer–fall (Jun–Oct). Roadsides, fields, pastures, waste ground; 50–1300 m; introduced; Greenland; B.C., N.B., Ont., Que.; Calif., Conn., Del., Idaho, Ill., Ind., Iowa, Ky., Maine, Md., Mass., Mich., Mont., N.H., N.J., N.Y., Ohio, Oreg., Pa., R.I., Utah, Vt., Va., Wash., W.Va., Wis.; Eurasia.

Brown knapweed is listed as a noxious weed in Washington.

The *Centaurea jacea* complex has been the subject of much controversy. The plants are widely distributed in Europe and variable in readily noticeable characters of the heads, florets, and cypselae. Several entities are commonly recognized, usually at the species level. The various named taxa are apparently all more or less interfertile, and natural hybridization has resulted in a plethora of intermediates that variously combine the features of the parental types. The numerous intermediates have been considered to be interspecific hybrids in some treatments or alternatively have been named as species or as infraspecific taxa within one or another of the parental species. The nomenclatural tangles are daunting, complicated by misapplication of names and the inadequate indexing of infraspecific names.

In an elegant biosystematic study of the representatives of this complex in England, E. M. Marsden-Jones and W. B. Turrill (1954) demonstrated the hazards of attempting to apply different names to all of the numerous intermediates. Despite the clear evidence that the entities are part of one biological species, Marsden-Jones and Turrill chose, for nomenclatural convenience, to treat the English plants as three species (*Centaurea jacea, C. nigra,* and *C. nemoralis*) with numerous interspecific hybrids rather than as a single variable species, thereby leaving a large number of sexually reproducing forms unassignable to species.

Further biosystematic studies of the *Centaurea jacea* complex involving additional races were carried out by C. Gardou (1972). She demonstrated that there are at least 18 diploid cytotypes within this complex plus a number of tetraploids. Most of the diploids have discrete geographic ranges in Europe. Some diploid members of the complex are apomictic, others autogamous, and still others outcrossers. Hybrids among the diploids are variably interfertile. Various tetraploids have arisen; some resemble one or another of the diploid races; many apparently allotetraploid races variously combine features of the diploids. The tetraploids of various origins are fully interfertile, have much wider ecologic tolerances than the diploids, and have spread widely. Introgression is common. Some of the diploids easily hybridize with tetraploids. The many intermediates are difficult to classify. Gardou's conclusion was that if one were strictly to apply the biological species concept, one would have to consider the complex, which she treated as sect. *Jacea* Cassini, to be a single species. Gardou did not, however, offer a taxonomic treatment of the complex that reflected her conclusion.

J. Dostál's (1976) treatment of *Centaurea* for *Flora Europaea* represents an opposing approach, with numerous species recognized in the *C. jacea* complex and distributed into three sections within what he treated as subg. *Jacea* (Miller) Hayek. He further divided some of these species into subspecies. This typological approach may be useful in sorting the variable entities into convenient pigeonholes, but it artificially applied a semblance of order to an unruly complex of interbreeding races.

North American botanists have usually recognized three or four species within the *Centaurea jacea* complex: *C. jacea, C. nigra, C. nigrescens* and/or *C. dubia* (under several different names), and *C. ×pratensis,* the last a collective for the various intermediates between *C. jacea* and *C. nigra.* According to G. Wagenitz (1987), the illegitimate name *C. pratensis* Thuillier refers to a dubious, ill-defined taxon of presumably hybrid origin that is also treated as *C. thuillieri, C. debeauxii* subsp. *thuillieri,* and as a subspecies of *C. nigra.* For spontaneous interspecific hybrids between *C. jacea* and *C. nigra* the name *C. ×moncktonii* C. E. Britton is accepted by C. A. Stace (1991) and in other recent European floras. The European sources of the North American plants have not been determined, and it is likely that multiple introductions have occurred. The taxon recognized as *C. nemoralis* Jordan by E. M. Marsden-Jones and

W. B. Turrill (1954) has usually been included by American botanists in *C. nigra* (e.g., R. J. Moore and C. Frankton 1974). Both diploid and tetraploid counts have been published from North American material.

Neither E. M. Marsden-Jones and W. B. Turrill (1954) nor C. Gardou (1972) discussed the relationship of *Centaurea nigrescens* to *C. jacea* and *C. nigra*, though Gardou by implication included *C. nigrescens* in her conclusion that sect. *Jacea* is a single biological species. *Centaurea nigrescens* has been variously merged with *C. jacea* and *C. nigra* in past treatments or maintained as a distinct species, sometimes with multiple subspecies (e.g., J. Dostál 1976). H. A. Gleason and A. Cronquist (1991) and several other recent authors treated *C. nigrescens* as a synonym of *C. dubia* Suter.

We have chosen here to follow the traditional approach for North American material of recognizing three species, *Centaurea jacea*, *C. nigra*, and *C. nigrescens* plus a nothospecies, *C. ×moncktonii*, though, as indicated above, these could as well be treated as a single species, *C. jacea*, comprising broadly inclusive subspecies. For those who prefer the latter approach, the respective names are *Centaurea jacea* subsp. *jacea*, *C. jacea* subsp. *nigra* (Linnaeus) Bonnier & Layens, and *C. jacea* subsp. *nigrescens* (Willdenow) Čelakovský. The hybrids between *C. jacea* and *C. nigra* may be treated as the nothosubspecies *C. jacea* subsp. ×*pratensis* (W. D. J. Koch) Čelakovský (as subsp.); the epithet *pratensis* is legitimate at the subspecific level.

We do not attempt here to differentiate for North American material the various subspecies of *Centaurea jacea* that have been recognized in European floras (e.g., J. Dostál 1976).

7. Centaurea nigra Linnaeus, Sp. Pl. 2: 911. 1753

• Black knapweed, lesser knapweed, centaurée noire ⊡

Centaurea jacea Linnaeus subsp. *nigra* (Linnaeus) Bonnier & Layens; *C. nemoralis* Jordan

Perennials, 30–150 cm. Stems 1–few, erect or ascending, openly branched distally, villous to scabrous with septate hairs and loosely tomentose, ± glabrate. Leaves: basal and proximal cauline petiolate, blades oblanceolate or elliptic, 5–25 cm, margins entire or shallowly dentate to irregularly pinnately lobed; distal cauline sessile, not decurrent, gradually smaller, blades linear to lanceolate, entire or dentate. Heads discoid, in few-headed corymbiform arrays, borne on leafy-bracted peduncles. Involucres ovoid to campanulate or hemispheric, 15–18 mm, usually ± as wide as high. Principal phyllaries: bodies lanceolate to ovate, loosely tomentose or glabrous, bases usually ± concealed by expanded appendages,

appendages erect, overlapping, dark brown to black, flat, margins pectinately dissected into numerous wiry lobes. Inner phyllaries: tips truncate, irregularly dentate or lobed. Florets 40–100+, all fertile; corollas purple (rarely white), 15–18 mm. Cypselae tan, 2.5–3 mm, finely hairy; pappi of many blackish, unequal, sometimes deciduous bristles 0.5–1 mm. $2n = 22, 44$.

Flowering summer–fall (Jun–Oct). Roadsides, fields, clearings, waste areas; 0–300 m; introduced; St. Pierre and Miquelon; B.C., N.B., Nfld. and Labr. (Nfld.), N.S., Ont., P.E.I., Que.; Calif., Conn., Del., Idaho, Ill., Iowa, Ky., Maine, Md., Mass., Mich., Mo., Mont., N.H., N.J., N.Y., Ohio, Oreg., Pa., R.I., Vt., Va., Wash., W.Va., Wis.; Europe.

Black knapweed is listed as a noxious weed in Colorado and Washington.

SELECTED REFERENCE Ockendon, D. L., S. M. Walters, and T. P. Whiffen. 1969. Variation within *Centaurea nigra*. Proc. Bot. Soc. Brit. Isles 7: 549–552.

8. Centaurea ×moncktonii C. E. Britton, Bot. Soc. Exch. Club Brit. Isles 6: 172. 1921 (as monktonii)

• Meadow or protean knapweed ⊡

Centaurea debeauxii Godron & Grenier subsp. *thuillieri* Dostál; *C. jacea* Linnaeus var. *pratensis* W. D. J. Koch; *C. jacea* subsp. ×*pratensis* (W. D. J. Koch) Čelakovský; *C. nigra* Linnaeus var. *radiata* de Candolle; *C. thuillieri* (Dostál) J. Duvigneaud & Lambinon

Perennials, 30–150 cm. Stems 1–few, erect or ascending, openly branched distally, villous to scabrous with septate hairs and loosely tomentose, ± glabrate. Leaves: basal and proximal cauline petiolate, blades oblanceolate or elliptic, 5–25 cm, margins entire or shallowly dentate to irregularly pinnately lobed; distal cauline sessile, not decurrent, blades linear to lanceolate, gradually smaller, entire or dentate. Heads usually radiant (rarely discoid), in few-headed corymbiform arrays, borne on leafy-bracted peduncles. Involucres ovoid to campanulate or hemispheric, 15–18 mm, usually ± as wide as high. Principal phyllaries: bodies lanceolate to ovate, loosely tomentose or glabrous, usually concealed by expanded appendages, appendages erect, overlapping, light to dark brown, flat or ± concave, margins varying from coarsely dentate to pectinately dissected into ± wiry lobes. Inner phyllaries: tips truncate, irregularly dentate or lobed. Florets 40–100+, all fertile or the peripheral sterile; corollas purple (rarely white), those of sterile florets ± expanded and exceeding corollas of fertile florets, those of fertile florets 15–18 mm. Cypselae tan, 2.5–3 mm, finely hairy; pappi 0 or of many unequal, sometimes caducous bristles 0.5–1 mm. $2n = 22$ (England), 44.

Flowering spring–fall (May–Nov). Roadsides, riverbanks, pastures, meadows, forest openings, waste areas; 0–1000 m; introduced; B.C., Nfld. and Labr., N.S., Ont., Que.; Calif., Conn., Idaho, Ill., Maine, Mass., Mich., Minn., Mo., Mont., N.H., N.J., N.Y., Ohio, Oreg., Pa., R.I., Vt., Va., Wash., Wis.; Europe.

Centaurea ×*moncktonii* is native to Europe or originated in North America from European ancestry.

Meadow knapweeds represent an array of mutually interfertile intermediates derived by hybridization and backcrossing among the various cytotypes of the *Centaurea jacea* complex. The plants variously combine features of *C. jacea* and *C. nigra*, and perhaps *C. nigrescens* as well. The hybrid complex includes both diploids and tetraploids. Extremes approach the parental types. Meadow knapweeds are often present without either parent in the immediate vicinity. They are considered to be noxious weeds in British Columbia, Idaho, Oregon, and Washington.

Centaurea pratensis J. L. Thuillier, sometimes applied to plants that belong here, is not a legitimate name.

9. **Centaurea nigrescens** Willdenow, Sp. Pl. 3: 2288. 1803 • Tyrol or short-fringed or Vochin knapweed, centaurée noirâtre [I]

Centaurea dubia Suter subsp. *nigrescens* (Willdenow) Hayek; *C. dubia* subsp. *vochinensis* (Bernhardi ex Reichenbach) Hayek; *C. jacea* Linnaeus subsp. *nigrescens* (Willdenow) Čelakovsky; *C. transalpina* Schleicher ex de Candolle; *C. vochinensis* Bernhardi ex Reichenbach

Perennials, 30–150 cm. **Stems** 1–few, erect or ascending, openly branched distally, villous to scabrous with septate hairs and loosely tomentose, ± glabrate. **Leaves:** basal and proximal cauline, petiolate, blades oblanceolate or elliptic, 5–25 cm, margins entire or shallowly dentate to irregularly pinnately lobed; distal cauline sessile, not decurrent, blades linear to lanceolate, gradually smaller, entire or dentate. **Heads** radiant or discoid, in few-headed corymbiform arrays, borne on leafy-bracted peduncles. **Involucres** 15–18 mm, subcylindric to ovoid or campanulate, usually longer than wide, even when pressed. **Principal phyllaries:** bodies lanceolate to ovate, loosely tomentose or glabrous, usually not fully covered by narrow appendages, these erect, overlapping, dark brown to black, flat, margins pectinately dissected into 6–8 pairs of wiry lobes. **Inner phyllaries:** tips truncate, irregularly dentate or lobed. **Florets** 40–100+, all fertile or peripheral sterile; corollas purple (rarely white), those of sterile florets ± expanded and exceeding corollas of fertile florets, those of fertile florets 15–18 mm. **Cypselae** tan, 2.5–3 mm, finely hairy; **pappi** 0 or of many

unequal, sometimes caducous bristles 0.5–1 mm. **2n** = 22 (Hungary), 44 (Hungary; Italy).

Flowering summer–fall (Jun–Oct). Roadsides, fields, waste areas; 0–1000 m; introduced; B.C., Ont., Que.; Calif., Conn., Del., Fla., Ill., Ind., Mass., Mo., Mont., N.H., N.J., N.Y., Ohio, Oreg., Pa., R.I., Vt., Va., Wash., W.Va., Wis., Wyo.; Europe.

Tyrol knapweed is considered to be a noxious weed in Washington and Oregon.

In recent years there has been much controversy regarding the name(s) to be applied to the North American Tyrol knapweeds. The names *Centaurea vochinensis*, *C. nigrescens*, and *C. dubia* have all been used in twentieth-century North American floras, and J. T. Kartesz and C. A. Meacham (1999) have accepted *C. transalpina* as well. R. J. Moore (1972) tentatively accepted two species, *C. nigrescens* and *C. dubia*, placing *C. transalpina* and *C. vochinensis* as synonyms through application beneath both species. Moore discussed the considerable similarities and practical difficulties of differentiating the taxa. H. A. Gleason and A. Cronquist (1991) recognized *C. dubia* as including *C. nigrescens* and *C. vochinensis*. E. G. Voss (1972–1996, vol. 3) recognized *C. nigrescens* as including *C. dubia* and *C. vochinensis*. Kartesz and Meacham accept *C. nigrescens* as a species, including *C. vochinensis*; they also accept *C. transalpina* with *C. dubia* as a synonym. In our investigation of the North American Tyrol knapweeds we have not been able to distinguish more than one (admittedly variable) entity. At the species level the correct name for this taxon is *Centaurea nigrescens*.

Centaurea dubia Suter, sometimes applied to plants that belong here, is not a valid name.

10. **Centaurea phrygia** Linnaeus, Sp. Pl. 2: 910. 1753 • Wig knapweed [I]

Centaurea austriaca Willdenow

Perennials, 15–80 cm. **Stems** few-many, erect, simple or branched. **Leaves** ± arachnoid-tomentose; basal and proximal cauline winged-petiolate, blades lanceolate to ovate, 3–15 cm, margins entire or dentate; distal cauline sessile, sometimes clasping, not decurrent, well developed. **Heads** usually radiant, usually borne singly. **Involucres** ovoid to ± spheric, 15–20 mm. **Principal phyllaries:** bodies lanceolate to ovate, loosely tomentose or glabrous, appendages brown or blackish, lanceolate to ovate, ± covering bodies of adjacent phyllaries, tips often recurved, elongate, featherlike, pectinately dissected into long, filiform lobes. **Inner phyllaries:** tips erect, ovate or orbiculate, irregularly dentate or lobed. **Florets** many, the peripheral sterile; corollas pink or purple, those of sterile much expanded and exceeding corollas of fertile florets, those of fertile 20–25 mm. **Cypselae** tan, 3–4 mm, finely hairy; **pappi** 0 or of many unequal bristles 0.5–2 mm. **2n** = 22 (Russia), 44 (Slovenia).

Flowering summer (Jul–Sep). Disturbed sites; 100–300 m; introduced; Fla., Ill., Mo., N.J., N.Y., Ohio, Pa., Vt., Va., W.Va.; Europe.

According to R. J. Moore (1972), reports of *Centaurea nervosa* Willdenow [*C. uniflora* Turra subsp. *nervosa* (Willdenow) Bonnier & Layens] from New York were based on a specimen referable to *C. phrygia* subsp. *phrygia*. Moore called these plants *C. austriaca* Willdenow, which J. Dostál (1976) treated as a synonym of *C. phrygia* subsp. *phrygia*.

Specimens of *Centaurea phrygia* are sometimes misidentified as *C. nigrescens* (or one or another of its synonyms) or as *C. nigra*. The elongate, often recurved, setose-ciliate tips of the phyllary appendages are a readily recognizable characteristic of this species. Considerable morphologic variation occurs in vegetative features and head dimensions in American material of the species, and it is possible that one or more of the specimens we have identified as *C. phrygia* represent an extreme variant of one of the members of the *C. jacea* complex. J. Dostál (1976) recognized 10 subspecies of *C. phrygia* in Europe. We have chosen not to assign the sparse North American material to subspecies.

11. **Centaurea stoebe** Linnaeus, Sp. Pl. 2: 914. 1753 [F] [I] [W]

Subspecies 3 (1 in the flora): introduced; Europe.

Native to southeastern Europe, *Centaurea stoebe* has been introduced to the whole of Europe, as far north as southern Sweden.

The nomenclature of *Centaurea stoebe* in the broad sense has been a source of confusion in European literature for about 200 years. The names used in that group (*C. stoebe*, *C. rhenana*, *C. maculosa*, *C. biebersteinii*) have been applied to different taxa by different authors with varying circumscriptions. Different species concepts were used in western and eastern Europe. Unfortunately this fact was not taken into account properly in the treatment by J. Dostál (1976).

Recent studies have shown that the American plants are identical with plants introduced to the whole of Europe (J. Ochsmann 2001). Subsp. *micranthos*, a tetraploid perennial, is clearly distinct from the diploid, biennial plants native to central Europe known as *C. stoebe* Linnaeus subsp. *stoebe*, *C. rhenana* Boreau, or *C. maculosa* Lamarck. In most American literature the name *Centaurea maculosa* Lamarck has been misapplied to *C. stoebe* subsp. *micranthos*. W. A. Weber (1987, 1990) treated this taxon as *Acosta maculosa* (Lamarck) Holub. The treatment of about 100 species of *Centaurea* sect. *Acrolophus* Cassini as the genus *Acosta*

by J. Holub (1972) and others is supported by neither morphologic nor molecular characters and is not widely accepted in Europe.

SELECTED REFERENCE Ochsmann, J. 2001. On the taxonomy of spotted knapweed (*Centaurea stoebe* L.). In: L. Smith, ed. 2001. Proceedings of the First International Knapweed Symposium of the Twenty-first Century, March 15–16, 2001, Coeur d'Alene, Idaho. Albany, Calif. Pp. 33–41.

11a. **Centaurea stoebe** Linnaeus subsp. **micranthos** (S. G. Gmelin ex Gugler) Hayek, Repert. Spec. Nov. Regni Veg. Beih. 30(2): 766. 1931 • Spotted knapweed, centaurée maculée ou tachetée [F] [I] [W]

Centaurea maculosa Lamarck subsp. *micranthos* S. G. Gmelin ex Gugler, Ann. Hist.-Nat. Mus. Natl. Hung. 6: 167. 1908; *C. biebersteinii* de Candolle

Perennials, 30–150 cm. **Stems** usually many, loosely gray-tomentose, branches ascending. **Leaves** loosely gray-tomentose or becoming glabrous, resin-gland-dotted; basal and proximal cauline petiolate, blades 10–15 cm, margins 1–2 times divided into linear or oblong segments; mid and distal cauline smaller, entire or dissected. **Heads** usually many in open cymiform arrays, pedunculate. **Involucres** ovoid, 10–13 mm. **Phyllaries:** bodies pale green or pink tinged, ovate (outer) to oblong (inner), with several prominent parallel veins, glabrous or finely tomentose, erect appendages decurrent on distal phyllary margins, dark brown or black, scarious, fringed with slender teeth. **Florets** 30–40; corollas pink to purple (white), those of sterile florets 15–25 mm, slender or somewhat enlarged, those of fertile florets 12–15 mm. **Cypselae** whitish or pale brown, 3–3.5 mm, finely hairy; **pappi** in 1–2 series, of many white, unequal, stiff bristles to 5 mm. **2n** = 36.

Flowering summer (Jun–Sep). Roadsides, fields, open forests; 50–2800 m; introduced; Alta., B.C., N.B., N.S., Ont., Que.; Ala., Ariz., Ark., Calif., Colo., Conn., Del., Fla., Ga., Idaho, Ill., Ind., Iowa, Kans., Ky., La., Maine, Md., Mass., Mich., Minn., Mo., Mont., Nebr., Nev., N.H., N.J., N.Mex., N.Y., N.C., N.Dak., Ohio, Oreg., Pa., R.I., S.C., S.Dak., Tenn., Utah, Vt., Va., Wash., W.Va., Wis., Wyo.; Europe.

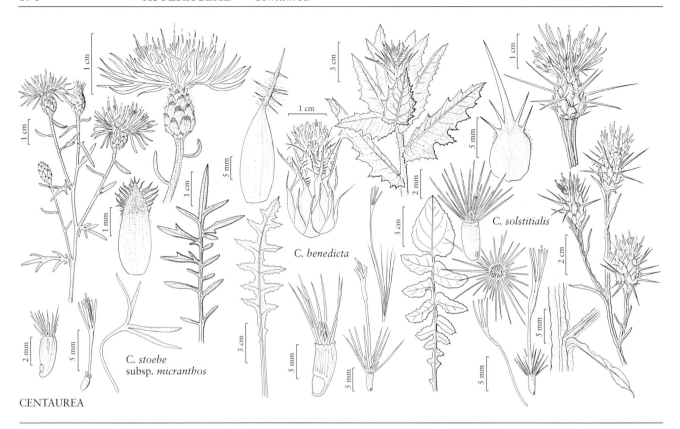

C. benedicta

C. solstitialis

C. stoebe
subsp. *micranthos*

CENTAUREA

12. Centaurea diffusa Lamarck in J. Lamarck et al., Encycl. 1: 675. 1785 • Diffuse or tumble or white knapweed, centaurée diffuse [1]

Acosta diffusa (Lamarck) Soják

Annuals or perennials, 20–80 cm. **Stems** 1–several, much-branched throughout, puberulent and ± gray tomentose. **Leaves** hispidulous and ± short-tomentose; basal and proximal cauline petiolate, often absent at anthesis, blades 10–20 cm, margins bipinnately dissected into narrow lobes; mid cauline sessile, bipinnately dissected; distal much smaller, entire or pinnately lobed. **Heads** disciform, in open paniculiform arrays. **Involucres** narrowly ovoid or cylindric, 10–13 × 3–5 mm. **Principal phyllaries:** bodies pale green, ovate to lanceolate, glabrous or finely tomentose, with a few prominent parallel veins, margins and erect appendages fringed with slender stramineous spines, each phyllary tipped by spine 1–3 mm. **Inner phyllaries** lanceolate, ± acute, appendage lacerate or spine-tipped. **Florets** 25–35; corollas cream white (rarely pink or pale purple), those of sterile florets 12–13 mm, slender, inconspicuous, those of fertile florets 12–13 mm. **Cypselae** dark brown, ca. 2–3 mm; **pappi** 0 or less than 0.5 mm, only rudimentary. *2n* = 18, 36.

Flowering summer (Jun–Aug). Disturbed sites in grasslands, woodlands, open coniferous forests; 100–2200 m; introduced; Alta., B.C., Ont., Que., Sask., Yukon; Ariz., Calif., Colo., Conn., Idaho, Ill., Ind., Iowa, Ky., Mass., Mich., Mo., Mont., Nebr., Nev., N.H., N.J., N.Mex., Oreg., R.I., Tenn., Utah, Wash., Wyo.; Europe.

Centaurea diffusa is native to southeastern Europe and casually adventive in central and western Europe.

Centaurea diffusa readily hybridizes with *C. stoebe* subsp. *micranthos* and is often confused with their fertile hybrid (*C.* ×*psammogena* G. Gáyer); the latter can be recognized by its cypselae bearing pappi and having conspicuously radiant heads. Morphologically the hybrids are extremely variable; they may be intermediate or may closely resemble one or the other of the parents. Conspicuously radiant heads and pappi are always present; appendages of the phyllaries are brown to black, or rarely stramineous; spines are absent or short and *2n* = 18. *Centaurea* ×*psammogena* is known from waste places, roadsides, railway tracks; 50–2500 m; B.C., Ont., Que.; Colo., Mass., Mich., Mo., N.C., Oreg., Tenn., Wash. It may occur spontaneously where the ranges of the parent species overlap; they may also be distributed separately. In mixed stands it replaces *C. diffusa* by introgression. Hybrids are often misidentified as *C. diffusa*.

13. Centaurea virgata Lamarck in J. Lamarck et al., Encycl. 1: 670. 1785 [I] [W]

Subspecies 2 (1 in the flora): introduced; w Asia.

13a. Centaurea virgata Lamarck subsp. **squarrosa** (Boissier) Gugler, Ann. Hist.-Nat. Mus. Natl. Hung. 6: 248. 1908 • Squarrose knapweed [I] [W]

Centaurea virgata var. *squarrosa* Boissier, Fl. Orient. 3: 651. 1875, based on *C. squarrosa* Willdenow, Sp. Pl. 3: 2319. 1803, not Roth 1800

Perennials, 20–50 cm, scabrous. **Stems** several, diffusely much-branched distally, finely tomentose, minutely resin-gland-dotted. **Leaves** finely tomentose, minutely resin-gland-dotted; basal and proximal cauline petiolate, often absent at anthesis, blades 10–15 cm, margins 1–2 times pinnately divided into linear segments; mid and distal cauline sessile, smaller, pinnately divided or simple. **Heads** disciform, in paniculiform arrays, sessile or short-pedunculate. **Involucres** narrowly ovoid or cylindric, 7–8 × 3–5 mm, falling at maturity with enclosed cypselae. **Principal phyllaries:** bodies pale green or stramineous, ovate to lanceolate, sometimes purple-tinged, glabrous or finely tomentose, sparsely resin-gland-dotted, margins scarious, appendages spreading, fringed with slender stramineous spines, each tipped by spreading to reflexed spine 1–3 mm. **Inner phyllaries** lanceolate, ± acute. **Florets** 10–14(–16); corollas pink or pale purple, those of sterile florets very slender, 3-lobed, linear, not exceeding disc corollas, those of fertile florets 7–9 mm. **Cypselae** light brown or stramineous, 2.5–3.5 mm, glabrous; **pappi** of white bristles 1–2.5 mm or sometimes very reduced. $2n = 36$.

Flowering summer (Jun–Sep). Rangelands, pastures, open forests, roadsides; 1000–2000 m; introduced; Calif., Colo., Mich., Nev., Oreg., Utah, Wyo.; w Asia.

Squarrose knapweed is an invasive pest in rangelands of western North America. At maturity the fruiting heads readily fall from the plant and can become lodged in the fur of animals. *Centaurea virgata* subsp. *squarrosa* has been declared a noxious weed in several western states.

In the *Synthesis of the North American Flora* (J. T. Kartesz and C. A. Meacham 1999) *Centaurea virgata* is incorrectly listed as a synonym of *C. triumfettii* Allioni, a very different plant. This was apparently based on a misreading of *C. variegata* Lamarck (not a legitimate name), a synonym of *C. triumfettii*, as *C. virgata*.

14. Centaurea calcitrapa Linnaeus, Sp. Pl. 2: 917. 1753 • Purple star-thistle, caltrops, chausse-trappe, centaurée chausse-trappe [I]

Annuals, biennials, or short-lived perennials, 20–100 cm. **Stems** 1–several, often forming rounded mounds, puberulent to loosely tomentose. **Leaves** puberulent to loosely gray-tomentose, becoming ± glabrous, minutely resin-gland-dotted; proximal leaves petiolate, blades 10–20 cm, 1–3 times pinnately dissected, rosette with central cluster of spines; mid sessile, not decurrent, blades ovate, usually less than 10 cm, narrowly lobed; distal blades linear to oblong, entire to shallowly lobed. **Heads** disciform, borne singly or in leafy cymiform arrays, sessile or short-pedunculate. **Involucres** ovoid, 15–20 × 6–8 mm. **Principal phyllaries:** bodies greenish or stramineous, ovate, scarious-margined, appendages stramineous, spiny fringed at base, each tipped by a stout spreading spine 10–25 mm. **Inner phyllaries:** appendages truncate, spineless. **Florets** 25–40; corollas purple, all ± equal, 15–24 mm; sterile corollas slender. **Cypselae** white or brown-streaked, 2.5–3.4 mm, glabrous; **pappi** 0. $2n = 20$.

Flowering summer–autumn (Jun–Nov). Pastures, fields, roadsides; 0–1700 m; Ont.; Ala., Ariz., Calif., D.C., Fla., Ga., Ill., Iowa, Md., Mass., N.J., N.Mex., N.Y., Oreg., Pa., Utah, Va., Wash.; Europe; Africa.

Centaurea calcitrapa is native to southern Europe and northern Africa. It is listed as a noxious weed in Arizona, California, Nevada, New Mexico, Oregon, and Washington. These plants are unpalatable and increase on rangelands as more desirable forage plants are consumed. Dense stands are impenetrable because of the vicious spines on the mature involucres.

Centaurea ×*pouzinii* de Candolle, an apparently stabilized hybrid between *Centaurea aspera* ($2n = 22$) and *C. calcitrapa* ($2n = 20$), has been reported from California. A chromosome count of $2n = 42$ has been reported from California material of this nothospecies (A. M. Powell et al. 1974). *Centaurea* ×*pouzinii* can be distinguished from *C. calcitrapa* by its shorter spines and by cypselae with a short pappus. Reports of *C. calcitrapoides* Linnaeus from North America are apparently based on this hybrid.

15. Centaurea iberica Treviranus ex Sprengel, Syst. Veg.
3: 406. 1826 • Iberian star thistle or knapweed [I]

Annuals, biennials, or short-lived perennials, 20–200 cm. **Stems** 1–several, divaricately much branched, often forming rounded mound, puberulent to loosely tomentose. **Leaves** hispidulous to loosely tomentose, ± glabrate, minutely resin-gland-dotted; proximal leaves petiolate, blades 10–20 cm, margins 1–2 times pinnately lobed or dissected, rosette with central cluster of spines; mid sessile, not decurrent, blades ± lanceolate, shorter; distal blades linear to oblong, entire to coarsely dentate or shallowly lobed. **Heads** disciform, borne singly or in leafy cymiform arrays, sessile or short-pedunculate. **Involucres** ovoid to hemispheric, (10–)13–18 mm. **Principal phyllaries:** bodies greenish or stramineous, ovate, scarious–margined, appendages stramineous, spiny–fringed at base, each tipped by stout spreading spine (0.5–)1–3 cm. **Inner phyllaries:** appendages truncate, spineless. **Florets** many; corollas white, pink, or pale purple, those of sterile florets slender, 15–20 mm, those of fertile florets 15–20 mm. **Cypselae** white- or brown-streaked, 3–4 mm, glabrous; **pappi** of white bristles 1–2.5(–3) mm. $2n$ = 16, 20.

Flowering summer (Jun–Sep). Roadsides, pastures, fields; 0–1500 m; introduced; Calif., Kans., Oreg., Wash., Wyo.; Europe; Asia.

Centaurea iberica is native to southeastern Europe through central Asia.

Iberian star thistle is considered to be a noxious weed in several states of the western United States. Weed control measures in Oregon and Washington have apparently eradicated the species in those states. *Centaurea iberica* is very similar to *C. calcitrapa,* from which it differs by its pappose cypselae and often more robust habit. The Kansas and Wyoming plants were originally reported as *C. calcitrapa* (R. L. McGregor 1986).

16. Centaurea benedicta (Linnaeus) Linnaeus, Sp. Pl.
ed. 2, 2: 1296. 1763 • Blessed thistle, chardon bénit
[F] [I]

Cnicus benedictus Linnaeus, Sp. Pl.
2: 826. 1753

Annuals, to 60 cm. **Stems** often spreading or prostrate, usually branched throughout, usually reddish, ± loosely tomentose. **Leaves** mostly cauline, sessile and often short-decurrent or proximal tapering to winged petioles, blades lanceolate to oblanceolate, 6–25 cm, margins coarsely dentate or pinnately lobed, lobes and teeth armed with short, weak spines, faces sparsely to densely hairy with jointed multicellular hairs and slender cobwebby hairs, resin-gland-dotted. **Heads** disciform, borne singly, sessile, each subtended by involucre-like cluster of leaf-like bracts. **Involucres** ± spheric, 20–40 mm. **Phyllaries** in several series, tightly overlapping, outer ovate with tightly appressed bases and spreading spine tips, inner lanceolate, tipped by pinnately divided spines more than 5 mm. **Florets** many; corollas yellow, those of sterile florets linear, 3-lobed, not exceeding disc corollas, very slender, those of disc florets 19–24 mm. **Cypselae** cylindric, slightly curved, 8–11 mm, with 20 prominent ribs, tipped by a 10-dentate rim, glabrous, attachment scars lateral; **pappi** of 2 series of awns, outer 9–10 mm, smooth or ± roughened, inner 2–5 mm, roughened with short spreading hairs. $2n$ = 22.

Flowering spring–summer (Apr–Aug). Roadsides, fields, waste places, sometimes cultivated; 0–1300 introduced; N.B., N.S., Ont.; Ala., Ariz., Ark. Calif., Conn., Fla., Ga., Ill., Md., N.J., N.Y., N.C., Oreg., S.C., Tenn., Tex., Utah, Va., Wash., Wis.; Europe; Asia; widely introduced worldwide.

Centaurea benedicta is native to the Mediterranean region and Asia Minor. F. K. Kupicha (1975) recognized two varieties of *Cnicus benedictus*: var. *benedictus* and var. *kotschyi* Boissier. A combination apparently has not been made for var. *kotschyi* in *Centaurea*. I have not determined whether one or both races are represented in North American plants of *Centaurea benedicta.*

Blessed thistle is cultivated in many areas of the world as a medicinal herb. The leaves, stems, and flowers are all used in herbal preparations for digestive and liver ailments.

17. Centaurea diluta Aiton, Hort. Kew. 3: 261. 1789
• North African knapweed [I]

Annuals or perennials, to 200 cm. **Stems** simple proximally, openly branched distally, glabrous or thinly hairy. **Leaves** thinly pubescent, basal and proximal cauline petiolate, blades 10–15 cm, margins coarsely pinnately lobed; mid cauline sessile or short-petiolate, short-decurrent, blades obovate or narrowly oblong, 2–8 cm, entire to pinnately lobed; distal cauline oblong, entire to irregularly lobed. **Heads** radiant, in open cymiform arrays, pedunculate. **Involucres** ovoid, 8–15 mm diam. **Principal phyllaries:** bodies greenish, ovate, scarious-margined, appendages stramineous to brown, scarious, fringed with slender teeth, tipped by slender spines 1–5 mm. **Innermost phyllaries** unarmed with brown, expanded, lacerate appendages. **Florets** many; corollas pink-purple, those of sterile florets 25–30 mm, enlarged, raylike, those of fertile florets ± 20 mm. **Cypselae** tan, 3–3.5 mm; **pappi** many, white, unequal bristles 3–5 mm. $2n = 20$.

Flowering spring (Apr–Jun). Escaped from cultivation in disturbed sites; 0–100 m; introduced; Calif., Mo., N.Y.; sw Europe; n Africa.

18. **Centaurea melitensis** Linnaeus, Sp. Pl. 2: 917. 1753 • Tocalote, Maltese star thistle or centaury, Napa thistle, croix de Malte [I]

Annuals, 10–100 cm, herbage loosely gray-tomentose and villous with jointed multicellular hairs, sometimes minutely scabrous, minutely resin-gland-dotted. **Stems** 1–few, few–many branched distally. **Leaves:** basal and proximal cauline petiolate or tapering to base, usually absent at anthesis, blades oblong to oblanceolate, 2–15 cm, margins entire to dentate or pinnately lobed; cauline long-decurrent, blades linear to oblong or oblanceolate, 1–5 cm, entire or dentate. **Heads** disciform, 1–few at branch tips, borne singly or in open leafy corymbiform arrays, sometimes clustered in distal axils, sessile or pedunculate. **Involucres** ovoid, 10–15 mm, loosely cobwebby-tomentose or becoming glabrous. **Principal phyllaries:** bodies ± stramineous, ovate, appendages purplish, spiny-fringed at base, each tipped by slender spine 5–10 mm. **Inner phyllaries:** appendages entire, acute or spine-tipped. **Florets** many; corollas yellow, those of sterile florets 10–12 mm, slender, inconspicuous, those of fertile florets 10–12 mm. **Cypselae** dull white or light brown, ca. 2.5 mm, finely hairy; **pappi** of many white, unequal, stiff bristles 2.5–3 mm. $2n = 24$.

Flowering mostly spring–summer (Apr–Jul). Roadsides, fields, pine-oak woodlands, chaparral, agricultural areas; 0–1500 m; widely introduced; B.C.; Ala., Ariz., Calif., Ga., Idaho, Ill., Mass., Miss., Mo., Nev., N.J., N.Mex., Oreg., Pa., Tex., Utah, Wash., Wis.; Mexico (Baja California); Europe; Asia; Africa.

Centaurea melitensis is native to the Mediterranean region. It is listed as a noxious weed in New Mexico.

19. **Centaurea solstitialis** Linnaeus, Sp. Pl. 2: 917. 1753 • Yellow or Barnaby star-thistle, St. Barnaby's thistle, centauré du solstice [F] [I] [W]

Annuals, 10–100 cm. **Stems** simple or often branched from base, forming rounded bushy plants, gray-tomentose. **Leaves** gray-tomentose and scabrous to short-bristly; basal and proximal cauline petiolate or tapered to base, usually absent at anthesis, blades 5–15 cm, margins pinnately lobed or dissected; cauline long-decurrent, blades linear to oblong, 1–10 cm, entire. **Heads** disciform, borne singly or in open leafy arrays, long-pedunculate. **Involucres** ovoid, 13–17 mm, loosely cobwebby-tomentose or becoming glabrous. **Principal phyllaries:** bodies pale green, ovate, appendages stramineous to brown, each with palmately radiating cluster of spines, and stout central spine 10–25 mm. **Inner phyllaries:** appendages scarious, obtuse or abruptly spine tipped. **Florets** many; corollas yellow, all ± equal, 13–20 mm; sterile florets slender, inconspicuous. **Cypselae** dimorphic, 2–3 mm, glabrous, outer dark brown, without pappi, inner white or light brown, mottled; **pappi** of many white, unequal bristles 2–4 mm, fine. $2n = 16$.

Flowering mostly summer–autumn (Jun–Oct), sometimes year-round in frostfree coastal habitats. Roadsides, fields, pastures, woodlands; 0–2000 m; widely introduced; Alta., Man., Ont., Sask.; Ariz., Calif., Colo., Conn., Del., Fla., Idaho, Ill., Ind., Iowa, Kans., Ky., Md., Mass., Mich., Minn., Mo., Mont., Nebr., Nev., N.H., N.J., N.Mex., N.Y., N.C., N.Dak., Ohio, Okla., Oreg., Pa., R.I., S.C., S.Dak., Tenn., Tex., Utah, Va., Wash., W.Va., Wis., Wyo.; s Europe.

Centaurea solstitialis is a serious weed pest, especially in the western United States, where it has invaded millions of acres of rangelands, and it is listed as a noxious weed in eleven western states and two Canadian provinces. It is a strong competitor in infested areas, often forming dense colonies. It is very difficult to control or eradicate once it becomes established. In addition, yellow star-thistle is poisonous to horses; when ingested over a prolonged period it causes a neurological disorder called equine nigropallidal encephalomalacia, or "chewing disease." Although its bitter taste and spiny heads usually deter grazing animals, horses sometimes will seek it out. Yellow star-thistle tends to spread in rangelands when more palatable plants are consumed.

20. Centaurea sulphurea Willdenow, Enum. Pl., 930. 1809 • Sicilian star-thistle, sulphur-colored Sicilian thistle, sulphur knapweed [I]

Annuals, 10–100 cm. **Stems** simple to openly branched, branches ascending, villous to hispid with septate hairs and loosely tomentose. **Leaves** ± villous to hispid with septate hairs, minutely resin-gland- dotted; basal winged-petiolate, blades oblong to oblanceolate, 10–15 cm, margins pinnately lobed, lobes acute, finely dentate; cauline sessile, long-decurrent with narrow wings, linear-oblong to oblanceolate, 1–6 cm, entire or distally serrate with short, spine-tipped teeth. **Heads** disciform, borne singly or in open, few-headed corymbiform arrays, long-pedunculate. **Involucres** ovoid, 12–30 mm, distally constricted. **Principal phyllaries:** bodies greenish or stramineous, ovate to elliptic, glabrous, appendages spreading to reflexed, brown to blackish purple, each with palmately radiating cluster of spines, central spine stout, 1–2.5 cm, base dark brown to black, distally stramineous. **Inner phyllaries:** appendages acute or spine-tipped. **Florets** many; corollas yellow, all ± equal, 25–35 mm; corollas of sterile florets slender, inconspicuous. **Cypselae** dark brown, 5–8 mm, glabrous; **pappi** of many, brown to blackish, unequal bristles 6–7 mm. $2n$ = 24.

Flowering spring–summer (May–Jul). Disturbed sites, grasslands, woodlands, pastures, roadsides; 0–300 m; Calif.; sw Europe.

Centaurea sulphurea is considered to be a noxious weed by the state of California.

Excluded species:

Centaurea bovina. J. T. Kartesz and C. A. Meacham (1999) listed *C. bovina* Velenovský from Massachusetts, referencing *Rhodora* 1924, perhaps based on collections reported as *C. diffusa* from Norfolk County (C. H. Knowlton and W. Deane 1924). Knowlton and Deane made no mention of *C. bovina.* According to J. Dostál (1976), *C. bovina* differs from *C. diffusa* in having smaller involucres (6–7 mm versus 7–10 mm; 3.5 mm versus 4–5 mm diameter), and purple versus pink flowers). Examination of one of the Norfolk County collections (*Churchill s.n.*, MIN) revealed no differences from *C. diffusa.*

According to R. Angelo (pers. comm.), "The Kartesz citation is puzzling. This taxon [*C. bovina*] is not cited in the 1924 volume of *Rhodora* or anywhere in the first 50 years of *Rhodora* according to the index for that volume and our 50 year index (which is quite comprehensive).

"Also, there are no specimens identified as this taxon in the New England Botanical Club (NEBC) herbarium or from Massachusetts in the Harvard University Herbaria collections. We also looked for re-identifications in the other taxa in *Centaurea* in the NEBC and found no specimens that were earlier identified as *C. bovina.*

"*The Vascular Plants of Massachusetts*[...] (1999) of [B. A.] Sorrie and [P.] Somers does not list this taxon, nor does the *Flora of the Northeast*[...] (1999) by [D. W.] Magee and [H. E.] Ahles."

Centaurea paniculata. According to our herbarium studies, reports of *C. paniculata* Linnaeus (Jersey knapweed) from North America are apparently referable to *C. stoebe* subsp. *micranthos. Centaurea paniculata* is quite similar to *C. stoebe* in habit; it differs clearly by its narrowly ovoid or cylindric heads.

187c. ASTERACEAE Martinov tribe ARCTOTIDEAE Cassini, J. Phys. Chim. Hist. Nat. Arts 88: 159. 1819 ☐

Annuals, perennials [shrubs] (sometimes prickly, sap sometimes milky, e.g., *Gazania*). **Leaves** basal and/or cauline; alternate; petiolate or sessile; margins usually pinnately lobed to dissected, sometimes dentate or entire (sometimes revolute and/or prickly). **Heads** usually heterogamous (usually radiate) [homogamous, discoid], borne singly (on scapiform peduncles) [in corymbiform, racemiform, or umbelliform arrays, sometimes aggregated in second-order heads]. **Calyculi** 0. **Phyllaries** usually persistent, in (2–)3–6+ series, distinct or ± connate, usually unequal, usually herbaceous [sometimes fleshy], margins and/or apices sometimes notably scarious (at least inner; sometimes prickly-ciliate). **Receptacles** flat to conic, epaleate [paleate] (often foveolate). **Ray florets** in 1–2 series, usually pistillate and fertile or styliferous and sterile, sometimes neuter (sometimes with 4–5 staminodes); corollas mostly white, yellow, orange, blue, red, or purple (rarely with 2 teeth opposite the 3- or 4-toothed laminae). **Disc florets** usually bisexual and fertile, sometimes functionally staminate; corollas yellow or dark brown to purple, actinomorphic, lobes [4–]5, narrowly triangular or lanceolate to deltate; anther bases obtuse to acute (sometimes ± sagittate, not tailed), apical appendages ± ovate to lanceolate; styles (in bisexual, fertile florets) proximally glabrous, usually distally dilated and ± cylindric (usually hispidulous near bases of cylinders), branches linear (sometimes adhering almost to minutely parted tips or essentially lacking), adaxially continuously stigmatic from bases to apices, apices rounded to acute, appendages essentially none. **Cypselae** usually monomorphic within heads, mostly ellipsoid, obovoid, or ovoid, ± terete, angled, or ± flattened, not beaked, bodies usually ribbed, sometimes winged (often sericeous, tomentulose, or woolly, sometimes glabrous); **pappi** (sometimes 0) usually persistent, usually of 4–8[–16+] scales in ± 2+ series, sometimes coroniform.

Genera 16, species ca. 200 (3 genera, 4 species in the flora): introduced; mostly Old World (especially s Africa); some species widely introduced as horticultural escapes.

The circumscription for Arctotideae adopted here follows that of K. Bremer (1994) and is narrower than that of N. T. Norlindh (1977[1978]). Arctotideae have been thought to be closely allied to members of Cynareae, chiefly on similarities in styles of some taxa and prickly, thistly habits in some taxa; such a relationship has been rejected by recent workers. J. L. Panero and V. A. Funk (2002) associated Arctotideae with their subfamily Cichorioideae, along with tribes Vernonieae, Liabeae, Cichorieae, and Gundelieae.

Haplocarpha lyrata Harvey (perennials, leaves in basal rosettes, 8–15 cm, blades usually lyrate, abaxially lanate, peduncles scapiform, 10–25+ cm, ray corollas mostly yellow, abaxially green or red) has been noted as established in Florida (www.plantatlas.usf.edu).

SELECTED REFERENCES Funk, V. A., R. Chan, and S. C. Keeley. 2004. Insights into the evolution of the tribe Arctoteae (Compositae: subfamily Cichorioideae s.s.) using *trn*L-F, *ndh*F, and ITS. Taxon 53: 637–655. Norlindh, N. T. 1977[1978]. Arctoteae—systematic review. In: V. H. Heywood et al., eds. 1977[1978]. The Biology and Chemistry of the Compositae. 2 vols. London. Vol. 2, pp. 943–959.

1. Phyllaries connate ¹/₃–³/₄ their lengths; laminae of ray corollas 5-nerved, 4-lobed or -toothed . . . . 25. *Gazania*, p. 196
1. Phyllaries distinct; laminae of ray corollas 4-nerved, 3-lobed or -toothed.
   2. Ray florets neuter, corollas adaxially yellow (sometimes drying bluish) or ± bluish; pappi of 7–8+ scales ca. 0.5 mm (usually hidden by hairs on cypselae) . . . . 26. *Arctotheca*, p. 197
   2. Ray florets pistillate, corollas adaxially whitish to purplish, or yellow to orange (then purple at bases); pappi 0 or of 5–8 scales 0.5–4 mm (usually not hidden by hairs on cypselae) . . . . 27. *Arctotis*, p. 198

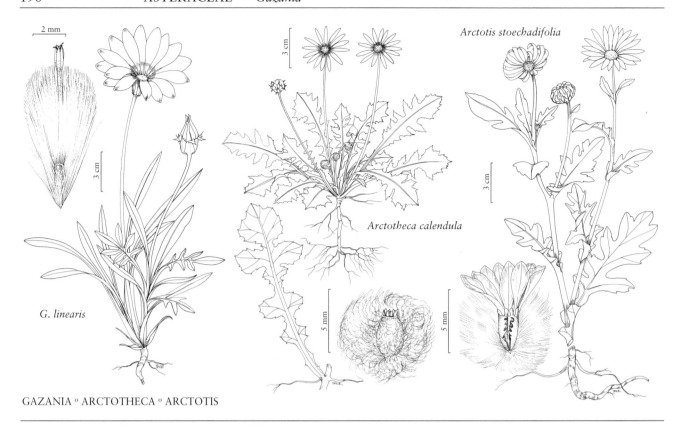

Arctotis stoechadifolia

Arctotheca calendula

G. linearis

GAZANIA ∘ ARCTOTHECA ∘ ARCTOTIS

25. GAZANIA Gaertner, Fruct. Sem. Pl. 2: 451, plate 173, fig. 2. 1791, name conserved

• Treasure-flower [Greek *gaza,* riches or royal treasure, alluding to splendor of flowers; or for Theodorus of Gaza (1398–1478), who translated the works of Theophrastus] [I]

Alison McKenzie Mahoney

Perennials [annuals, shrubs], [5–]10–35 cm (often cespitose, not prickly, sometimes with milky sap). **Stems** usually erect (often ± congested; rootstocks often woody). **Leaves** usually mostly basal, sometimes cauline as well; petiolate or sessile; blades linear to lanceolate, spatulate or oblanceolate, margins entire or pinnately lobed, abaxial faces white-woolly, adaxial usually glabrate or glabrous, sometimes arachnose. **Involucres** ± campanulate, turbinate, or cylindric, [5–]10–15+ mm diam. **Phyllaries** in 2–4 series, connate $^1/_2$–$^3/_4$ their lengths, margins ± scarious, apices acute, abaxial faces glabrous [arachnose to tomentose]. **Receptacles** conic or convex, deeply alveolate (pits enclosing cypselae, their margins often ciliate). **Ray florets** neuter; corollas yellow, orange, or red to maroon (usually each with darker abaxial stripe and a darker adaxial spot or blotch near base), laminae 5-veined, 4-toothed. **Disc florets** bisexual, fertile; corollas yellow to orange. **Cypselae** obovoid, ribs 0, faces villous; **pappi** persistent, of 7–8[–12+], lanceolate to subulate-aristate scales in 2 series (± hidden by hairs on cypselae). $x = 9$.

Species ca. 20 (1 in the flora): introduced; South Africa, Namibia, tropical East Africa; cultivated and/or introduced elsewhere.

1. Gazania linearis (Thunberg) Druce, Rep. Bot. Soc. Exch. Club Brit. Isles 4: 624. 1917 [F] [I]

Gorteria linearis Thunberg, Prodr. Pl. Cap., 162. 1800; *Gazania longiscapa* de Candolle

Stems sometimes proximally woody. **Leaves** mostly basal; blades either linear to lanceolate and not lobed, 10–20(–38) cm × 6–10 mm, or oblanceolate to oblong and pinnately lobed, 10–20 cm × 25–50 mm, or both; bases usually attenuate, margins usually entire, sometimes ± prickly, revolute, midveins prominent, abaxial faces white-villous, adaxial faces glabrate to arachnose. **Heads** 3.5–8 cm diam. (across rays). **Peduncles** scapiform, (6–)10–30(–35) cm. **Phyllaries:** outer lanceolate, margins prickly-ciliate; the inner with margins undulate, ciliate, with ± submarginal dark stripe, cuspidate. **Ray florets** 13–18; corolla laminae yellow or orange, usually each with dark abaxial stripe and adaxial basal blotch or spot, (20–)35–42 × 10 mm. **Cypselae** 1–2 mm; **pappi** of 7–8 scales 3–4 mm.

Flowering Oct–Jul. Roadsides, waste places, especially in urban coastal areas; 0–400(–900) m; introduced; Calif., N.Mex.; Africa.

---

26. ARCTOTHECA   J. C. Wendland, Bot. Beob., 41. 1798; Hort. Herrenhus. 1: 8, plate 6. 1798 • Capeweed [Greek *arktos*, brown bear, and *theke*, case, capsule, container, alluding to dense, woolly tomentum of cypselae of some species]   [I]

Alison McKenzie Mahoney

Perennials (sometimes flowering first year), (4–)8–20(–30+) cm (usually stoloniferous and/or rhizomatous). **Stems** decumbent or creeping to ± erect. **Leaves** basal and cauline; petiolate; blades mostly obovate, margins usually pinnatifid to pinnatisect (often lyrate to runcinate) [entire], abaxial faces ± white-woolly, adaxial sparsely puberulent to arachnose (glabrate). **Involucres** hemispheric to campanulate, 10–15[–20+] mm diam. **Phyllaries** in 3–6+ series, distinct, ovate or oblong to linear, unequal, margins ± scarious, apices ± acute to acuminate, abaxial faces glabrous or arachnose. **Receptacles** flat, alveolate or smooth, epaleate. **Ray florets** neuter (staminodes sometimes present); corollas adaxially yellow (sometimes drying bluish) or ± bluish, laminae 4-veined, 3-toothed. **Disc florets** bisexual, fertile; corollas yellow or purplish to brownish. **Cypselae** ± obovoid or ± prismatic (4-angled), ribs 3–5, faces usually villous or woolly [glabrate]; **pappi** usually present, persistent, of 7–9+ scales (borne within callous borders crowning cypselae) [sometimes none]. *x* = 9.

Species 4 (1 in the flora): introduced; South Africa, Mozambique.

1. Arctotheca calendula (Linnaeus) Levyns, J. S. African Bot. 8: 284. 1942 [F] [I] [W]

Arctotis calendula Linnaeus, Sp. Pl. 2: 922. 1753; *Cryptostemma calendula* (Linnaeus) Druce

Plants usually stoloniferous. **Leaves** obovate, (2–)5–20(–30+) × (1–)2–5(–7) cm, margins pinnatisect (lyrate to runcinate), remotely prickly, abaxial faces white-pannose, adaxial faces sparsely puberulent to arachnose, usually glandular as well. **Heads** 4–7 cm diam. (across the rays). **Phyllaries:** outer reflexed, apices mucronate, white-woolly; inner appressed, margins hyaline, apices rounded, glabrous. **Ray florets** 11–17(–25); corolla laminae abaxially greenish to purplish, adaxially yellow (drying to basally ochroleucous, apically blue, forming bull's eye around disc), 10–25 × 2–4 mm, sparsely puberulent, glandular. **Disc florets:** corollas yellow proximally, bluish distally. **Cypselae** dark brown, 3 mm, densely woolly; **pappi** ca. 0.5 mm (usually hidden by hairs on cypselae).

Flowering Oct–Aug. Roadsides, old fields, other disturbed habitats; 0–300 m; introduced; Calif.; Africa.

Most populations of *Arctotheca calendula* are sterile and spread aggressively by stolons; at least three populations in the flora are fertile and highly invasive. The species is listed by the California Exotic Pest Plant Council (CalEPPC) as a weed with the potential to spread explosively (Red Alert, CDFA A).

27. ARCTOTIS Linnaeus, Sp. Pl. 2: 922. 1753; Gen. Pl. ed. 5, 394. 1754 • African-daisy

[Greek *arktos*, brown bear, and *ous*, *otos*, ear, perhaps alluding to shape of pappus scales]   ⊡

Alison McKenzie Mahoney

Annuals or perennials [subshrubs], [5–]10–100 cm (hirsute to arachnose or woolly, sometimes glandular as well). **Stems** erect [decumbent]. **Leaves** basal and/or cauline; petiolate or sessile; blades mostly spatulate to oblanceolate, margins entire, sinuate-dentate, or pinnately lobed, abaxial faces tomentose to sparsely arachnose. **Involucres** hemispheric to campanulate, 10–25 mm diam. **Phyllaries** in 3–6+ series, distinct, outer linear to oblong, inner ovate to elliptic, margins ± scarious. **Receptacles** flat, alveolate (pit margins often ciliate), epaleate. **Ray florets** pistillate, fertile (staminodes often present); corollas white, cream, yellow, orange, pink, purple, violet, or blue (sometimes dark abaxially), laminae 4-veined, 3-toothed. **Disc florets** usually bisexual and fertile, sometimes functionally staminate; corollas yellow or brown to violet or purple. **Cypselae** ovoid to obovoid, abaxial faces ± rounded, 3–5-ribbed or -winged (wings sometimes toothed and inflexed into 2 longitudinal grooves), adaxial faces flattened, apices sometimes forming callous borders, faces (smooth or transversely rugose) glabrous or pilose to sericeous; **pappi** usually present, persistent, usually of 5–8 ovate to oblong scales in 2 series, sometimes coroniform, sometimes none. $x = 9$.

Species ca. 60 (2 in the flora): introduced; s Africa, cultivated and/or introduced elsewhere.

1. Cypselae ± ovoid, 1.3–1.5 mm, glabrous; pappi 0 or coroniform.................... 1. *Arctotis fastuosa*
1. Cypselae ± obovoid, 2–3 mm, sericeous (hairs from bases) and tomentulose (on faces);
   pappi of 5–8 hyaline scales 0.5–4 mm .... 2. *Arctotis stoechadifolia*

1. Arctotis fastuosa Jacquin, Pl. Hort. Schoenbr. 2: 20, plate 166. 1797 • Monarch-of-the-veld ⊡

Venidium fastuosum (Jacquin) Stapf

Annuals, 10–45(–80) cm, hirsute to woolly. **Leaves:** basal lanceolate to oblanceolate, 4–9 cm × 15–25 mm, margins pinnatifid to pinnatisect; distal cauline smaller (sessile, bases clasping, margins entire). **Peduncles** (10–)18–22 cm, with 3 or more leaves. **Phyllaries:** outer spreading linear (bases ± broad), abaxial faces arachnose to hirsute; inner appressed ± lanceolate (apices rounded, hyaline, ciliate). **Ray florets** 35–50 in 2 series; corolla laminae 30–55, 4–7 mm, adaxial faces orange to yellow (outer unmarked or with smaller or less distinct marks, inner marked basally with purple-brown fans ¹/₈–¹/₄ their lengths). **Cypselae** ovoid, 1.3–1.5 mm, glabrous; **pappi** 0 or coroniform. $2n = 18$.

Flowering spring–summer. Roadsides, urban waste places; 0–500 m; introduced; Calif.; Africa.

2. Arctotis stoechadifolia P. J. Bergius, Descr. Pl. Cap., 324. 1767 • Blue-eyed African-daisy ⨍⊡

Arctotis grandis Thunberg; *A. stoechadifolia* var. *grandis* (Thunberg) Lessing; *A. venusta* Norlindh

Annuals or short-lived perennials, (20–)40–70(–100) cm, arachnose, pannose, or woolly. **Leaves:** basal and cauline, obovate, 5–20 cm × 10–45 mm, margins usually entire to undulate, sometimes dentate, lyrate, or pinnately lobed (basal and proximal cauline usually withering before flowering, their bases petioliform); distal cauline smaller (sessile, bases clasping). **Peduncles** (6–)10–20(–30) cm, sometimes with 1 or 2 leaves (1–3 cm). **Phyllaries:** outer appressed, linear to linear-lanceolate, abaxial faces arachnose (appendages 1–3 mm, blunt, woolly); inner appressed, ± lanceolate (appendages 4–6 mm, rounded to acute, ciliate). **Ray florets** 22–30 in 1 series; corolla laminae (17–)20–30 × 2–4 mm, abaxial faces violet, adaxial white (sometimes yellowish proximally). **Cypselae** obovoid, 2–3 mm, sericeous (hairs from bases) and tomentulose (on faces); **pappi** of 5–8, ovate to oblong, hyaline scales 0.5–4 mm (outer shorter). $2n = 18$.

Flowering (Jan–)Apr–Nov. Roadsides, waste places, especially in sandy soils near coasts; 0–300 m; introduced; Calif.; Africa.

N. T. Norlindh (1964, 1965) treated *Arctotis venusta* as separate from *A. stoechadifolia*. In the sense of Norlindh, the latter is a rare endemic of sand dunes of the Southwestern Cape of Africa and *A. venusta* is common, more widespread, sometimes weedy in its native range, and cultivated elsewhere as an ornamental. Specimens from California examined for this treatment most closely resemble *A. venusta* in having tap roots with erect stems not rooting at nodes and phyllaries with blunt, woolly appendages less than 3 mm.

187d. Asteraceae Martinov tribe **Vernonieae** Cassini, J. Phys. Chim. Hist. Nat. Arts 88: 203. 1819

Annuals, biennials, perennials, or shrubs [trees or lianas] (sap rarely milky). **Leaves** usually cauline, sometimes basal or basal and cauline; alternate (rarely subopposite distally) [opposite]; usually petiolate, sometimes sessile (or petioles winged); margins usually ± dentate, sometimes entire [lobed or dissected]. **Heads** homogamous (discoid, pseudo-radiant or -liguliflorous in *Stokesia*), usually in corymbiform, paniculiform, or scorpioid arrays, sometimes borne singly or in glomerules [aggregated in second-order heads]. **Calyculi** 0. **Phyllaries** usually persistent [readily falling], in 2–8+ series, distinct, unequal, herbaceous to chartaceous, margins and/or apices sometimes scarious. **Receptacles** flat to convex, usually epaleate (often foveolate, sometimes setose). **Ray florets** 0 (corollas of peripheral florets enlarged, zygomorphic, ± raylike in *Stokesia*). **Disc florets** bisexual, fertile; corollas white, ochroleucous, or pink to cyanic [yellow]; anther bases ± sagittate [tailed], apical appendages ovate to lanceolate; styles abaxially hirsutulous (at least distally), branches lance-linear to ± lanceolate, adaxially continuously stigmatic from bases nearly to apices, apices acute, appendages essentially none. **Cypselae** ± monomorphic within heads, columnar to clavate, fusiform, or prismatic, sometimes compressed, not beaked, bodies smooth, nerved, or ribbed (glabrous or hirsutulous to strigillose, sometimes resin-gland-dotted as well); **pappi** usually persistent, usually in 2 series (outer series of shorter, stouter bristles or narrow scales, inner of longer, usually barbellate bristles), sometimes in 1 series (bristles or scales, scales often aristate).

Genera 100–140, species ca. 1300 (6 genera, 25 species in the flora): mostly tropics and warm-temperate regions of New World and Old World.

Most members of Vernonieae are herbs, subshrubs, or shrubs (*Vernonia arborea* Buchanan-Hamilton of tropical Asia may form trees to 33 m). They are characterized by discoid heads of bisexual florets with purple to pink or white corollas, calcarate anthers, attenuate, abaxially hirsutulous style branches stigmatic ± uniformly (rather than in two lines or bands) nearly to their tips, and pollen grains with regular, polygonal, patterns of ± spiny to smooth ridges. Centers of species concentration for the tribe are found in Africa, Madagascar, South America, and Antilles. In the flora, most species are found in the eastern and southern states of the United States. The plants are often associated with open, prairie or savanna-like areas.

Treating clades recognized by J. L. Panero and V. A. Funk (2002) as corresponding to tribes, Vernonieae is sister to Liabeae (none in the flora) and is included with Arctotideae (introduced), Cichorieae, and Gundelieae (none in the flora) within Cichorioideae.

Historically, 80% or so of the species in the tribe were included in *Vernonia*. H. Robinson (1999) has argued for resurrections and recircumscriptions of some old genera and recognition of some "new" genera, resulting in a *Vernonia* of ca. 20 species.

Stokesia laevis and some *Vernonia* species are grown as ornamentals. Some *Vernonia* species have been used medicinally in folk remedies and some may be locally troublesome as weeds (e.g., *V. baldwinii*).

SELECTED REFERENCE Robinson, H. 1999. Generic and subtribal classification of American Vernonieae. Smithsonian Contr. Bot. 89.

1. Heads pseudo-radiant (corollas of peripheral, bisexual florets enlarged, zygomorphic); margins of phyllaries (at least the outer), pectinately spinose-toothed . . . . 28. *Stokesia*, p. 201
1. Heads ± discoid; margins of phyllaries not pectinately spinose-toothed.

[2. Shifted to left margin.—Ed.]

2. Heads sessile, borne in congested clusters; florets (1–)4(–5) in each head.
    3. Heads (1–)10–40 per cluster, each cluster subtended by (2–)3 ± deltate bracts; pappi of 5(–6) 1-aristate scales (look closely for squamiform, gradually to abruptly tapering base of each arista), no scales tipped with plicate aristae . . . . 29. *Elephantopus*, p. 202
    3. Heads 1–5+ per cluster, each cluster subtended by 1–2 lanceolate to spatulate or linear bracts; pappi of 6–10 ± laciniate to aristate scales, 2(–3+) of aristate scales each with awnlike arista plicate (2-folded) distally . . . . 30. *Pseudelephantopus*, p. 204
2. Heads mostly pedunculate, not borne in congested clusters; florets 9–100+ in each head.
    4. Annuals (perhaps persisting); cypselae not ribbed . . . . 31. *Cyanthillium*, p. 204
    4. Perennials or functionally annuals; cypselae 8–10-ribbed.
        5. Heads each subtended by 3–8+, ± foliaceous bracts; pappi caducous . . . . 32. *Centratherum*, p. 206
        5. Heads not each subtended by foliaceous bracts; pappi persistent . . . . 33. *Vernonia*, p. 206

28. STOKESIA L'Héritier, Sert. Angl., 27. 1789 • [For Jonathan Stokes, 1755–1831, English physician and botanist] E

John L. Strother

Perennials, 2–5+ dm; perhaps rhizomatous. **Leaves** basal and cauline; proximal petiolate, blades ovate to lanceolate or lance-linear; distal ± sessile, blades ovate or elliptic to lanceolate or lance-linear, bases ± clasping, margins entire or spinose-toothed; all with apices rounded to acute, faces glabrous or glabrate, resin-gland-dotted. **Heads** pseudo-radiant (see here at corollas), ± pedunculate, not individually bracteate; borne singly or in loose, ± corymbiform arrays 6–12 cm diam. **Involucres** ± hemispheric, 25–45 mm diam. **Phyllaries** 25–35+ in 5–7 series, the outer with appressed, ± chartaceous bases, distally ± foliaceous, margins pectinately spiny-toothed (at least at base), inner ± chartaceous throughout, mostly entire, faces ± tomentulose and resin-gland-dotted. **Florets** 12–35(–70+); corollas usually blue to purplish blue (rarely white or lilac), tubes longer than funnelform throats, lobes 5, lance-linear (in peripheral florets adaxial sinus much deeper than others and corollas zygomorphic, ± raylike or ligulelike, in central florets corollas ± actinomorphic). **Cypselae** ± columnar, 3–4-angled, glabrous; **pappi** caducous, of 4–5 scales. *x* = 7.

Species 1: se United States.

1. Stokesia laevis (Hill) Greene, Erythea 1: 3. 1893 E F

Carthamus laevis Hill, Hort. Kew., 57, plate 5. 1768

Stems tomentulose, glabrescent. **Leaves:** basal with petioles 3–12 cm, narrowly winged, blades 8–15 × 1–5 cm; cauline sessile, ± clasping, blades 5–12 × 1–3 cm. **Involucres** 25–45 × 25–45 mm. **Phyllaries:** outer 15–35+ mm, foliaceous portions elliptic to spatulate or linear, margins ± spiny, inner oblong to linear, 10–15+ mm, margins mostly entire, tips spiny. **Corollas** 15–25+ mm (outer) or 12–15+ mm (inner). **Cypselae** 5–8 mm; **pappi** 8–12 mm. *2n* = 14.

Flowering Jun–Sep. Openings in woodlands, bogs; 10–100 m; Ala., Fla., Ga., La., Miss., S.C.

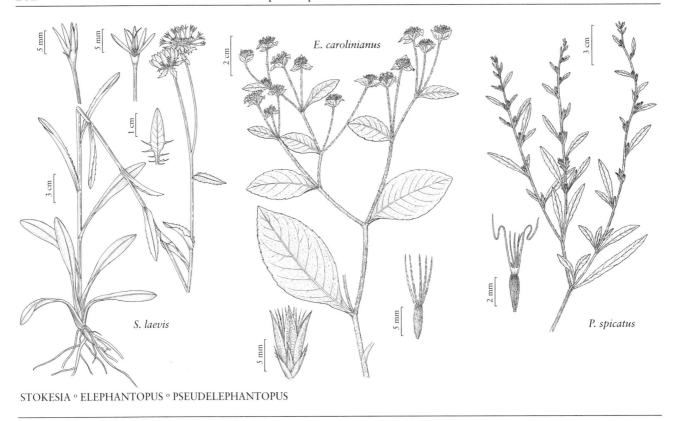

STOKESIA ∘ ELEPHANTOPUS ∘ PSEUDELEPHANTOPUS

29. ELEPHANTOPUS Linnaeus, Sp. Pl. 2: 814. 1753; Gen. Pl. ed. 5, 355. 1754

• Elephant's foot [Greek *elephantos*, elephant, and *pous*, foot; probably alluding to rosettes of basal leaves in original species]

John L. Strother

Perennials, (1–)2–8(–12+) dm; often rhizomatous or stoloniferous. **Leaves** mostly basal or mostly cauline at flowering; sessile or petiolate, petioles ± winged (often clasping at bases); blades mostly elliptic, ovate, or obovate to lanceolate, oblanceolate, or spatulate (rarely orbiculate), bases ± cuneate, margins usually toothed (rarely entire), apices obtuse to acute, abaxial or both faces usually resin-gland-dotted. **Heads** ± discoid, sessile, not individually bracteate, in clusters of (1–)10–40+ in corymbiform-paniculiform arrays 6–15(–25) cm diam. (each cluster subtended by 2–3 ± deltate bracts). **Involucres** ± cylindric, 1–3+ mm diam. **Phyllaries** 8 in 4 decussate pairs, the outer 4 ovate, inner 4 lanceolate, all ± chartaceous, margins entire, tips ± spinose to apiculate, abaxial faces of inner 4 usually dotted distally with resin glands. **Florets** (1–)4(–5+); corollas white or pink to purple, tubes longer than abruptly funnelform throats, lobes 5, lance-linear, unequal (abaxial sinus deepest). **Cypselae** ± clavate, sometimes ± flattened, 10-nerved or -ribbed, strigillose to hirsutulous; **pappi** persistent, of 5(–6), 1-aristate scales (look closely for squamiform, gradually to abruptly tapering base of each arista). *x* = 11.

Species 12–15+ (4 in the flora): mostly warm-temperate, subtropical, and tropical regions worldwide, sometimes as naturalized ruderals.

Pseudelephantopus spicatus is sometimes treated as a member of *Elephantopus*.

SELECTED REFERENCE Clonts, J. A. 1972. A Revision of the Genus *Elephantopus* Including *Orthopappus* and *Pseudelephantopus* (Compositae). Ph.D. thesis. Mississippi State University.

1. Leaves at flowering all or mostly cauline . . . . 1. *Elephantopus carolinianus*
1. Leaves at flowering mostly basal (cauline leaves much smaller or wanting).
    2. Inner phyllaries 9–11+ mm; pappi 6–8 mm . . . . 2. *Elephantopus tomentosus*
    2. Inner phyllaries 6–8 mm; pappi 3–4.5 mm.
        3. Inner phyllaries ± densely strigose-villous with hairs (0.3–)0.5–1 mm; cypselae 3–
            3.5 mm . . . . 3. *Elephantopus elatus*
        3. Inner phyllaries sparsely strigose or hispidulous with hairs 0.05–0.3(–0.5) mm;
            cypselae 2.5–3 mm . . . . 4. *Elephantopus nudatus*

1. Elephantopus carolinianus Raeuschel, Nomencl. Bot. ed. 3, 256. 1797 [E] [F]

Plants (1–)3–8(–12+) dm. **Leaves** mostly cauline at flowering; blades broadly elliptic or ovate to lanceolate, 6–12(–18+) cm × 30–80(–120+) mm (including petioles), both faces sparsely pilose to hirsute. **Bracts** rounded-deltate to lance-deltate, (5–)10–15(–25+) × (4–)6–12+ mm. **Inner phyllaries** 8–10 mm, sparsely hispidulous to pilosulous, hairs 0.1–0.3 mm. **Cypselae** 2.5–4 mm; **pappi** 4–5 mm. *2n* = 22.

Flowering Aug–Sep(–Oct). Open or shaded, damp to wet places in pine forests and mixed forests, often on sandy soils; 10–700 m; Ala., Ark., Del., D.C., Fla., Ga., Ill., Ind., Kans., Ky., La., Md., Miss., Mo., N.J., N.C., Ohio, Okla., Pa., S.C., Tenn., Tex., Va., W.Va.

2. Elephantopus tomentosus Linnaeus, Sp. Pl. 2: 814. 1753 [E]

Plants (1–)3–6+ dm. **Leaves** mostly basal at flowering; blades usually obovate to oblanceolate or spatulate (rarely ± orbiculate), 3–8(–10+) cm × 4–12(–25+) mm (including petioles), abaxial faces ± densely pilose, adaxial sparsely pilose to hirsute. **Bracts** rounded-cordate, 8–15(–20+) × 7–15(–18+) mm. **Inner phyllaries** 9–11+ mm, pilosulous, hairs 0.3–0.6(–0.9) mm. **Cypselae** (3–)4–5 mm; **pappi** 6–8 mm. *2n* = 22.

Flowering Aug–Sep. Open or shaded, dry to wet places in pine forests and mixed forests, often on sandy soils; 10–600 m; Ala., Ark., Fla., Ga., Ky., La., Md., Miss., N.C., Okla., S.C., Tenn., Tex., Va.; Mexico (probably introduced).

3. Elephantopus elatus Bertoloni, Mem. Reale Accad. Sci. Ist. Bologna 2: 607. 1850 [E]

Plants (1–)6–7+ dm. **Leaves** mostly basal at flowering; blades oblanceolate, 9–14(–20+) cm × 20–35(–45+) mm (including petioles), abaxial faces pilose to hirsute, adaxial sparsely pilose to hirsute. **Bracts** rounded-deltate to lance-deltate, 8–12+ × 6–8+ mm. **Inner phyllaries** 6–8 mm, ± densely strigose to villous, hairs (0.3–)0.5–1 mm. **Cypselae** 3–3.5 mm; **pappi** 3–4 mm. *2n* = 22.

Flowering Aug–Sep. Open or shaded, dry to wet places in pine forests and mixed forests, usually on sandy soils; 0–50 m; Ala., Fla., Ga., La., Miss., S.C.

4. Elephantopus nudatus A. Gray, Proc. Amer. Acad. Arts 15: 47. 1880 [E]

Plants (1–)3–11+ dm. **Leaves** mostly basal at flowering; blades mostly oblanceolate to spatulate, sometimes elliptic, 7–15(–20+) cm × 20–35(–45+) mm (including petioles), both faces strigose or pilose to hirsute. **Bracts** rounded-deltate to lance-deltate, 6–15+ × 4–9+ mm. **Inner phyllaries** 6–7.5 mm, sparsely strigose or hispidulous with hairs 0.05–0.3(–0.5) mm. **Cypselae** 2.5–3 mm; **pappi** 3–4.5 mm. *2n* = 22.

Flowering Aug–Oct. Open or shaded, dry to wet places in pine forests and mixed forests, often on sandy soils; 0–100 m; Ala., Ark., Del., Fla., Ga., La., Md., Miss., N.C., S.C., Tex., Va.

30. PSEUDELEPHANTOPUS Rohr, Skr. Naturhist.-Selsk. 2: 214. 1792 (as Pseudo-Elephantopus), orthography conserved · False elephant's foot [Greek *pseudo-*, false or resembling, and generic name *Elephantopus*] ⊡

John L. Strother

Perennials (sometimes suffrutescent), 2–6(–10+) dm; usually ± rhizomatous. **Leaves** mostly cauline at flowering; sessile or petiolate, petioles ± winged (often clasping at bases); blades mostly obovate to oblanceolate, spatulate, or linear, bases cuneate, margins usually toothed (rarely entire), apices obtuse to acute, abaxial faces usually resin-gland-dotted. **Heads** ± discoid, sessile, not individually bracteate, in clusters of 1–5+ in spiciform or paniculo-spiciform arrays 5–10 mm diam., each cluster subtended by 1–2 lanceolate to spatulate or linear bracts. **Involucres** ± cylindric to fusiform, 2–3+ mm diam. **Phyllaries** 8 in 4 decussate pairs, the outer ovate, inner lanceolate, all ± chartaceous, margins entire, tips ± spinose or cuspidate (abaxial faces of inner 4 usually resin-gland-dotted distally). **Florets** (2–)4(–5+); corollas white or pink to purple, tubes longer than abruptly funnelform throats, lobes 5, lance-linear, unequal (abaxial sinus deepest). **Cypselae** ± clavate, sometimes ± flattened, 8–10-nerved or -ribbed, closely strigillose to hirsutulous; **pappi** persistent, of 6–10 ± laciniate to aristate scales, 2(–3+) of aristate scales each with awnlike arista plicate (2-folded) distally. *x* = 14?.

Species 2–3 (1 in the flora): introduced; Florida, Neotropics; introduced in Paleotropics.

1. Pseudelephantopus spicatus (Jussieu ex Aublet) C. F. Baker, Trans. Acad. Sci. St. Louis 12: 55. 1902 [F] [I]

Elephantopus spicatus Jussieu ex Aublet, Hist. Pl. Guian. 2: 808. 1775

Leaf blades mostly 3–15(–20+) cm × 10–30(–45+) mm (including petioles), both faces sparsely pilose or hirsute, often glabrescent, abaxial resin-gland-dotted. **Bracts** 15–45+ × 2–5+ mm. **Inner phyllaries** 9–12 mm, sparsely hispidulous to pilosulous (hairs 0.1–0.3 mm), often glabrescent. **Cypselae** 7–8 mm; pappi 1–9+ mm (including aristae). *2n* = 26, 28 [3 reports, all from outside flora].

Flowering Jan (probably year-round). Open places on sandy soils; 0–10+ m; introduced; Fla.; Mexico; West Indies; Central America; South America; Asia; Pacific Islands.

31. CYANTHILLIUM Blume, Bijdr. Fl. Ned. Ind., 889. 1826 · [Origin uncertain; probably Greek *cyanos*, blue, and *anthyllion*, little flower, alluding to corollas] ⊡

John L. Strother

Annuals (perhaps persisting), 2–6(–12+) dm. **Leaves** mostly cauline (at flowering); petioles ± winged; blades ovate to trullate, deltate, oblanceolate, or spatulate, bases ± cuneate, margins serrate, apices rounded to acute, abaxial faces ± hirtellous to densely piloso-strigillose, resin-gland-dotted, adaxial faces ± scabrellous or glabrate. **Heads** discoid, ± pedunculate, not subtended by foliaceous bracts, (12–)40–100+ in ± corymbiform arrays (6–)10–15+ cm diam. **Involucres** ± campanulate to turbinate or hemispheric, 4–5 mm diam. **Phyllaries** 24–32+ in 3–4+ series, the outer subulate to lanceolate, inner ± lanceolate, all ± chartaceous, margins entire, tips apiculate to spinose, abaxial faces ± strigillose, ± resin-gland-dotted. **Florets** 13–20(–24+);

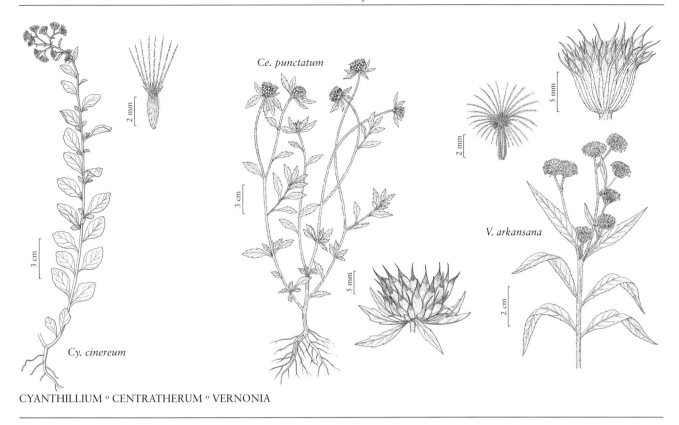

Ce. punctatum

V. arkansana

Cy. cinereum

CYANTHILLIUM ∘ CENTRATHERUM ∘ VERNONIA

corollas lavender to pink or purplish, tubes longer than funnelform throats, lobes 5, lance-linear, ± equal. **Cypselae** ± columnar, not ribbed, ± strigillose; **pappi** of ± 20 ± persistent outer scales, plus ± 20 caducous inner bristles. $x = 9$ (18?).

Species 1–2 (1 in the flora): introduced; Florida; probably paleotropical in origin, now widely established in tropical and warm-temperate regions as naturalized ruderals.

1. Cyanthillium cinereum (Linnaeus) H. Robinson, Proc. Biol. Soc. Wash. 103: 252. 1990 [F] [I]

Conyza cinerea Linnaeus, Sp. Pl. 2: 862. 1753; *Vernonia cinerea* (Linnaeus) Lessing

Leaf blades 20–35(–50+) × 12–25(–30+) mm (including petioles). **Peduncles** 3–10(–20+) mm. **Bracts** ± linear at proximal nodes, wanting distally. **Phyllaries** 3–4 mm. **Cypselae** 1.5–2 mm; **pappi** white, outer scales 0.1–0.3 mm, inner bristles 3–4 mm. $2n = 36$.

Flowering year-round. Disturbed places; 0–10 m; introduced; Fla.; Mexico; West Indies; Central America; South America; Asia; Africa; Indian Ocean Islands; Pacific Islands.

32. CENTRATHERUM Cassini, Bull. Sci. Soc. Philom. Paris 1817: 31. 1817; in F. Cuvier, Dict. Sci. Nat. ed. 2, 7: 383. 1817 • Larkdaisy [Latin *centrum*, center, and *atherum*, prickle or awn, perhaps alluding to spine-tipped middle phyllaries of original species]

John L. Strother

Perennials (or functionally annuals), 1–3(–8+) dm (stems sometimes rooting at proximal nodes). **Leaves** cauline; sessile or petiolate, petioles ± winged; blades ovate to obovate, lanceolate, or linear, bases ± cuneate, margins toothed, apices acute, abaxial faces usually ± hirtellous to strigillose or tomentose, sometimes nearly glabrous, adaxial faces sparsely scabrellous or glabrate, both usually resin-gland-dotted. **Heads** discoid (pedunculate, each subtended by 3–8+, ± foliaceous bracts), borne singly. **Involucres** ± hemispheric, 6–12(–18+) mm diam. **Phyllaries** 24–50+ in 4–8+ series, each proximally firm, distally ± scarious, the outer ovate to deltate or lanceolate, inner oblong to lanceolate, margins entire, tips rounded to acute, usually apiculate to seta-tipped or attenuate-spinose, abaxial faces glabrous or sparsely strigillose to tomentose and usually ± resin-gland-dotted distally. **Florets** 30–50(–100+); corollas usually lavender to purplish (rarely white), tubes longer than funnelform throats, lobes 5, lance-linear, ± equal. **Cypselae** ± columnar to plumply clavate, 8–10-ribbed, glabrous, often resin-gland-dotted; **pappi** caducous, of 20–40 lance-linear to subulate scales. $x = 16$.

Species 2–4 (1 in the flora): Florida, Neotropics, Pacific Islands (Philippines), Australia.

SELECTED REFERENCES Gleason, H. A. 1922. *Centratherum*. In: N. L. Britton et al., eds. 1905+. North American Flora.... 47+ vols. New York. Vol. 33, pp. 49–50. Kirkman, L. K. 1981. Taxonomic revision of *Centratherum* and *Phyllocephalum* (Compositae: Vernonieae). Rhodora 83: 1–24.

1. Centratherum punctatum Cassini in F. Cuvier, Dict. Sci. Nat. ed. 2, 7: 384. 1817 F

Plants 1–5+ dm. **Stems** usually erect. **Leaf blades** obovate to lanceolate, 2–6(–8+) cm × 8–25(–40+) mm. **Involucres** 8–15+ × 8–15+ mm. **Phyllaries** 30–50+ in 4–8+ series, glabrous or sparsely tomentulose. **Florets** 30–60+. **Cypselae** 1–2.5+ mm; **pappi** stramineous, 1–3+ mm. $2n = 32, 64$.

Flowering year-round. Disturbed places; 0–10+ m; Fla.; Mexico; West Indies; Central America; South America.

33. VERNONIA Schreber, Gen. Pl. 2: 541. 1791, name conserved • Ironweed [For William Vernon, d. 1711, English botanist]

John L. Strother

Perennials, 2–20(–30+) dm (rhizomatous or not). **Leaves** usually mostly cauline (rarely mostly basal or basal and cauline); sessile or petiolate; blades ovate, elliptic, lanceolate, oblanceolate, spatulate, linear, or filiform, bases usually ± cuneate (rounded-truncate in *V. pulchella*), margins usually toothed (rarely entire), apices acute to attenuate, abaxial faces usually ± scabrellous to strigillose or tomentose to pannose, sometimes glabrate or glabrous, usually resin-gland-dotted (sometimes ± pitted), adaxial faces ± scabrellous or glabrate, sometimes resin-gland-dotted (rarely pitted). **Heads** discoid, ± pedunculate, not subtended by foliaceous bracts, (6–)40–100+ in ± corymbiform to paniculiform arrays (6–)10–25+ cm diam. **Involucres** ± campanulate to obconic or hemispheric, 3–8(–11+) mm diam. **Phyllaries** 18–70+ in 4–7+ series, the outer ovate to lanceolate or subulate, inner ± lanceolate to oblong, all ± chartaceous, margins entire, often ciliolate, tips rounded (then sometimes apiculate), or acuminate, subulate, or filiform, faces glabrous or sparsely strigillose to tomentose, sometimes ± gland-dotted. **Florets** 9–30(–65+); corollas usually purplish or pink (rarely white), tubes longer than funnelform throats, lobes 5, lance-linear, ± equal. **Cypselae** ± columnar, sometimes arcuate, 8–10-ribbed, glabrous or ± strigillose to hirtellous, often resin-gland-dotted; **pappi** persistent, of 20–30+ outer, erose to subulate scales or bristles plus 20–40+ inner, longer, subulate to setiform scales or bristles. *x* = 17.

Species 20 or so (17 in the flora): mainly c, e North America, n Mexico, 2–3 species in South America.

The circumscription of *Vernonia* adopted here follows that of H. Robinson (1999).

Vernonias hybridize; almost every one of the species recognized here has been noted as sometimes hybridizing with one or more others. Putative hybrid plants are usually intermediate between parentals in some traits; such plants may not "key" satisfactorily to any of the species treated here. Some putative hybrids have been named. *Vernonia guadalupensis* is "without much doubt a hybrid of *V. baldwinii* Torrey and *V. lindheimeri* Engelmann & Gray" (L. H. Shinners 1950); *V. vulturina* Shinners (known only from the type collection) may be a product of *V. baldwinii* × *V. marginata*; *V. ×georgiana* Bartlett may refer to *V. acaulis* × *V. angustifolia*. Additional putative hybrids (S. B. Jones 1964) are *V. ×concinna* Gleason (*V. ovalifolia* × *V. angustifolia*), *V. ×dissimilis* Gleason (*V. altissima* × *V. angustifolia*), and *V. ×recurva* Gleason (*V. pulchella* × *V. angustifolia*).

In the key and descriptions, "l/w = " refers to lengths divided by widths for blades of leaves; lengths of phyllaries include subulate to filiform tips (if any).

SELECTED REFERENCES Gleason, H. A. 1922b. *Vernonia*. In: N. L. Britton et al., eds. 1905+. North American Flora.... 47+ vols. New York. Vol. 33, pp. 52–95. Jones, S. B. 1964. Taxonomy of the narrow-leaved *Vernonia* of the southeastern United States. Rhodora 66: 382–401. Jones, S. B. and W. Z. Faust. 1978. *Vernonia*. In: N. L. Britton et al., eds. 1905+. North American Flora.... 47+ volumes. New York. Ser. 2, part 10, pp. 180–195. Shinners, L. H. 1950. Notes on Texas Compositae. IV. Field & Lab. 18: 25–32.

1. Phyllary tips usually subulate to filiform, sometimes acuminate.
  2. Leaves mostly basal (cauline leaves much smaller with narrower blades) . . . . 1. *Vernonia acaulis*
  2. Leaves mostly cauline (basal leaves wanting at flowering or ± like cauline).
    3. Involucres 11–15 mm diam.; phyllaries (50–)60–70+; florets 50–100+ . . . . . . 2. *Vernonia arkansana*
    3. Involucres 4–8(–10) mm diam.; phyllaries 22–46(–60+); florets 12–45(–65).
      4. Involucres ± campanulate to obconic; florets 12–24(–30+).
        5. Leaf blades (mid stem) lance-linear to filiform, 5–12 cm × 2–4(–8+) mm
        . . . . 3. *Vernonia angustifolia*
        5. Leaf blades (mid stem) oblanceolate to lance-linear, 3–7 cm × (5–)10–20+
        mm . . . . 4. *Vernonia pulchella*
      4. Involucres ± hemispheric; florets 30–45(–65).
        6. Leaf blade l/w = 2.5–3.5(–4); pappi stramineous to whitish, outer bristles
        or subulate scales intergrading with inner ones . . . . 5. *Vernonia glauca*
        6. Leaf blade l/w = (3.3–)4–6+; pappi fuscous to purplish, outer scales
        contrasting with inner bristles . . . . 6. *Vernonia noveboracensis*
1. Phyllary tips usually acute to rounded or rounded-apiculate, seldom acuminate.
  7. Phyllaries densely sericeo-tomentose to pannose.
    8. Leaf blades adaxially densely sericeo-tomentose; florets 40–50+ . . . . 7. *Vernonia larsenii*
    8. Leaf blades adaxially usually glabrate, sometimes sparsely arachno-tomentose;
    florets 12–24+ . . . . 8. *Vernonia lindheimeri*
  7. Phyllaries usually glabrous or puberulent to scabrellous (rarely sparsely arachno-
  tomentose).

[9. Shifted to left margin.—Ed.]

9. Leaf blades abaxially glabrate (and pitted; best seen at 10× or greater, pits containing awl-shaped hairs or glands 0.1–0.5+ mm).
  10. Involucres obconic to hemispheric; cypselae 2–3 mm . . . . 9. *Vernonia texana*
  10. Involucres campanulate; cypselae 3–5 mm.
    11. Leaf blades 1–3+ mm wide; florets 10–12+ . . . . 10. *Vernonia lettermannii*
    11. Leaf blades (2.5–)5–20(–40) mm wide; florets (10–)15–25+.
      12. Peduncles (3–)10–35 mm; involucres (7–)9–11 mm; phyllary tips ± acuminate; inner pappi 8–9+ mm . . . . 11. *Vernonia marginata*
      12. Peduncles 1–8(–12+) mm; involucres 5–7(–8+) mm; phyllary tips acute or rounded-apiculate; inner pappi 5–7+ mm . . . . 12. *Vernonia fasciculata*
9. Leaf blades abaxially scabrellous, scaberulous, or puberulous to tomentose or pannose (not pitted).
  13. Leaf blades 2–8+ mm wide . . . . 13. *Vernonia blodgettii*
  13. Leaf blades 15–40(–75+) mm wide.
    14. Stems glabrous . . . . 14. *Vernonia flaccidifolia*
    14. Stems puberulent to tomentose (sometimes glabrescent).
      15. Leaf blades abaxially usually scabrellous (with appressed, awl-shaped hairs), sometimes glabrescent, not or sparsely resin-gland-dotted . . . . 15. *Vernonia gigantea*
      15. Leaf blades abaxially usually puberulous to tomentose or pannose (with ± erect, ± curled hairs), seldom glabrescent, conspicuously resin-gland-dotted.
        16. Involucres broadly campanulate to urceolate, (6–)7–10+ × 5–9+ mm; (phyllaries seldom resin-gland-dotted); florets 30–55 . . . . 16. *Vernonia missurica*
        16. Involucres broadly campanulate to hemispheric, 4–6(–8+) × 4–7+ mm; (phyllaries often resin-gland-dotted); florets (15–)20–25(–35+) . . . . 17. *Vernonia baldwinii*

1. **Vernonia acaulis** (Walter) Gleason, Bull. New York Bot. Gard. 4: 222. 1906 [E]

Chrysocoma acaulis Walter, Fl. Carol., 196. 1788

Plants 4–6(–10) dm. **Stems** ± puberulent, glabrescent. **Leaves** mostly basal; blades oblong-elliptic to spatulate (basal, the distal narrower), 11–25(–30) cm × (21–)60–75+ mm, l/w = 2–4(–5+), abaxially usually glabrous or glabrate, sometimes hirsutulous on veins, usually resin-gland-dotted, adaxially ± scabrellous or glabrate. **Heads** in loose, corymbiform arrays. **Peduncles** 5–50 mm. **Involucres** campanulate to hemispheric, 5–10 × 6–7 mm. **Phyllaries** 40–60 in 5–6 series, glabrate, margins ciliolate, the outer lance-linear, 2–4 mm, inner lance-ovate to lance-subulate, 5–9 mm, tips subulate to filiform. **Florets** 30–40(–55+). **Cypselae** 2.5–3+ mm; **pappi** whitish to stramineous, outer scales ca. 20, 0.6–1 mm, contrasting with 25–40, 5–7+ mm inner bristles. *2n* = 34.

Flowering Jul–Aug. Open woods, pine savannas, bogs, disturbed places; 10–300 m; Fla., Ga., N.C., S.C.

2. **Vernonia arkansana** de Candolle in A. P. de Candolle and A. L. P. P. de Candolle, Prodr. 7: 264. 1838 [E] [F]

Vernonia crinita Rafinesque

Plants 8–12(–20) dm. **Stems** scabrellous to pilosulous, glabrescent. **Leaves** mostly cauline; blades lance-linear, 6–14(–20) cm × 7–15(–25) mm, l/w = 7–10(–18), abaxially scabrellous (hairs awl-shaped), resin-gland-dotted, adaxially scaberulous, glabrescent, resin-gland-dotted. **Heads** in ± corymbiform arrays. **Peduncles** 2–5 cm. **Involucres** ± hemispheric, 11–15 × 11–15 mm. **Phyllaries** (50–)60–70+ in 5–6 series, scabrellous (and resin-gland-dotted), margins ciliolate, the outer lance-ovate, 3–8+ mm, inner lanceolate, 12–15 mm, tips subulate to filiform. **Florets** 50–100+. **Cypselae** 3–5 mm; **pappi** fuscous to purplish, outer scales 25–30+, 0.5–1 mm, contrasting with 25–30+, 5–7 mm inner bristles. *2n* = 34.

Flowering Aug–Oct. Leas, roadsides, stream banks, in sand or on limestone; 200–300 m; Ark., Kans., Mo., Okla.

Vernonia arkansana was published in April, *V. crinita* in October 1838.

3. **Vernonia angustifolia** Michaux, Fl. Bor.-Amer. 2: 94. 1803 [E]

Vernonia angustifolia subsp. *mohrii* (S. B. Jones) S. B. Jones & W. Z. Faust; *V. angustifolia* subsp. *scaberrima* (Nuttall) S. B. Jones & W. Z. Faust; *V. angustifolia* var. *scaberrima* (Nuttall) A. Gray; *V. scaberrima* Nuttall

Plants 5–10+ dm. **Stems** sparsely appressed-puberulent, glabrescent. **Leaves** mostly cauline; blades (mid stem) lance-linear to filiform, 5–12 cm × 2–4(–8+) mm, l/w = (8–)12–30(–60+), abaxially glabrous but for scattered hairs on midribs or scabrellous (hairs awl-shaped), resin-gland-dotted or not, adaxially scabrellous. **Heads** in ± corymbiform to paniculiform arrays. **Peduncles** 8–25 mm. **Involucres** ± campanulate to obconic, 5–7(–10) × 4–6(–9) mm. **Phyllaries** 22–45+ in 5–6 series, sparsely scabrellous, glabrescent, margins ciliolate, the outer lanceolate to subulate, 1.5–3 mm, inner lance-ovate to lanceolate, 5–9+ mm, tips acuminate to subulate. **Florets** 12–20(–30). **Cypselae** 2.5–3 mm; **pappi** stramineous to purplish, outer scales 25–30, 0.5–1.2+ mm, contrasting with 35–40+, 5.5–6+ mm inner bristles. $2n = 34$.

Flowering Jun–Aug. Pine barrens, oak woodlands; 10–50 m; Ala., Fla., Ga., La., Miss., N.C., S.C.

4. **Vernonia pulchella** Small, Bull. Torrey Bot. Club 25: 145. 1898 [E]

Plants 4–7+ dm. **Stems** sparsely appressed-puberulent. **Leaves** mostly cauline; blades (mid stem) mostly oblanceolate to lance-linear, rarely ovate, 3–7 cm × (5–)10–20+ mm, l/w = (2.5–)3.5–6+ (bases rounded-truncate), abaxially scabrellous (hairs awl-shaped), resin-gland-dotted, adaxially scabrellous. **Heads** in ± corymbiform to paniculiform arrays. **Peduncles** 5–35 mm. **Involucres** ± campanulate, 6–11 × 5–9 mm. **Phyllaries** 40–50+ in 5–6 series, sparsely puberulent, margins arachno-ciliate, the outer subulate to filiform, 3–7 mm, inner lance-attenuate to filiform, 9–10+ mm, tips subulate to filiform. **Florets** 14–25(–30+). **Cypselae** 3–3.5 mm; **pappi** stramineous, outer scales 25–30, 0.6–1.1 mm, contrasting with 30+, 6–7+ mm inner bristles. $2n = 34$.

Flowering Jul–Sep. Low, wet places in pine woods; 20–50 m; Ga., S.C.

5. **Vernonia glauca** (Linnaeus) Willdenow, Sp. Pl. 3: 1633. 1803 [E]

Serratula glauca Linnaeus, Sp. Pl. 2: 818. 1753

Plants 6–10+ dm. **Stems** sparsely appressed-puberulent, glabrescent. **Leaves** mostly cauline; blades lanceolate to oblanceolate, 10–15(–18+) cm × 30–45(–70+) mm, l/w = 2.5–3.5(–4), abaxially scabrellous (hairs awl-shaped), resin-gland-dotted or not, adaxially glabrous or sparsely puberulent. **Heads** in corymbiform to paniculiform arrays. **Peduncles** 2–35 mm. **Involucres** ± hemispheric, 5–8 × 7–8 mm. **Phyllaries** 40–60 in 5–6 series, sparsely puberulent to tomentulose, glabrescent, margins ciliolate, the outer lance-deltate to subulate, 1–3 mm, inner oblong, 4–7+ mm, tips acuminate to subulate or filiform. **Florets** 30–45+. **Cypselae** 3–4 mm; **pappi** stramineous to whitish, outer subulate scales or bristles 30, 0.5–1.5+ mm, intergrading with 30+, 6–8 mm inner subulate scales or bristles. $2n = 34$.

Flowering Jul–Sep. Dry fields, marshes; 10–400 m; Ala., Del., D.C., Ga., Md., Mass., N.J., N.C., Pa., S.C., Va.

6. **Vernonia noveboracensis** (Linnaeus) Michaux, Fl. Bor.-Amer. 2: 95. 1803 [E] [F]

Serratula noveboracensis Linnaeus, Sp. Pl. 2: 818. 1753; *Vernonia harperi* Gleason; *V. noveboracensis* var. *tomentosa* Britton

Plants 8–12(–20) dm. **Stems** puberulent, glabrescent. **Leaves** mostly cauline; blades ± lanceolate, 9–15(–25+) cm × 15–45(–60+) mm, l/w = (3.3–)4–6+, abaxially scabrellous, tomentose, or pannose, resin-gland-dotted, adaxially scabrellous, often resin-gland-dotted. **Heads** in corymbiform to paniculiform arrays. **Peduncles** 2–35 mm. **Involucres** ± hemispheric, 6–10 × 7–10 mm. **Phyllaries** 35–60+ in 4–6+ series, sparsely tomentulose, glabrescent, margins ciliolate, the outer lanceolate to subulate, 1–3 mm, inner oblong, 7–9+ mm, tips subulate to filiform. **Florets** 30–45(–65). **Cypselae** 3.5–4+ mm; **pappi** fuscous to purplish, outer scales 20, 0.2–0.6+ mm, contrasting with 30–40+, 5–7+ mm inner bristles. $2n = 34$.

Flowering Aug–Oct. Abandoned fields, marshes, roadsides; 10–600 m; Ala., Conn., Del., D.C., Fla., Ga., Md., Mass., N.J., N.Y., N.C., Pa., R.I., S.C., Tenn., Va., W.Va.

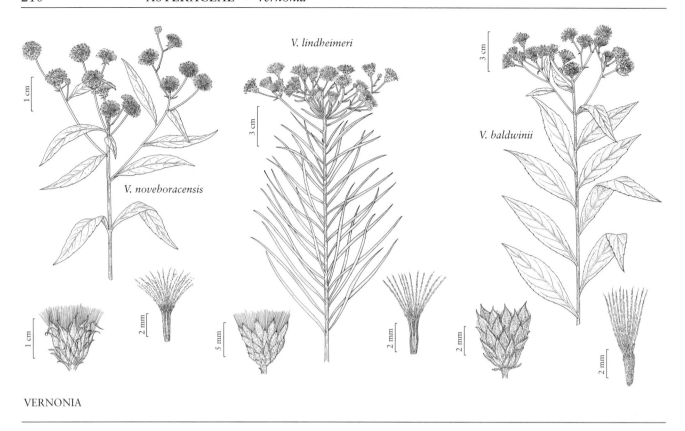

V. lindheimeri

V. noveboracensis

V. baldwinii

VERNONIA

7. **Vernonia larseniae** B. L. King & S. B. Jones, Brittonia 27: 84. 1975 (as larsenii)

Vernonia lindheimeri A. Gray & Engelmann var. *leucophylla* Larsen, Ann. Missouri Bot. Gard. 15: 333, plate 45. 1928

Plants 7–10+ dm. **Stems** tomentose to pannose. **Leaves** mostly cauline; blades linear, 5–9(–12+) cm × 3–5(–8+) mm, l/w = 16–32, abaxially tomentose, adaxially densely sericeo-tomentose. **Heads** in corymbiform arrays. **Peduncles** 10–60 mm. **Involucres** ± hemispheric, 8–12 × 7–12 mm. **Phyllaries** 50–60+ in 5–6+ series, sericeo-tomentose to pannose, including margins, the outer lance-deltate, 1–3 mm, the inner lanceolate, 7–11+ mm, tips acute. **Florets** 40–50+. **Cypselae** 3.5–4+ mm; **pappi** stramineous, outer subulate scales or bristles 20+, 0.6–1.5+ mm, intergrading with 30+, 6–7+ mm inner subulate scales or bristles. **2n** = 34.

Flowering Jun. Arroyos, on limestone; 800–900 m; Tex.; Mexico (Coahuila).

8. **Vernonia lindheimeri** A. Gray & Engelmann, Proc. Amer. Acad. Arts 1: 46. 1847 [F]

Plants 2–5(–8+) dm. **Stems** ± pannose. **Leaves** mostly cauline; blades linear, 5–8 cm × 2–4 mm, l/w = 17–25(–40), abaxially densely sericeo-tomentose, adaxially usually glabrate, sometimes sparsely arachno-tomentose. **Heads** in corymbiform arrays. **Peduncles** 5–35 mm. **Involucres** ± obconic to campanulate, 6–10 × 5–8 mm. **Phyllaries** 40–50+ in 5–6+ series, sericeo-tomentose, including margins, the outer lance-ovate, 1–2 mm, the inner lanceolate to lance-linear, 6–7(–9+) mm, tips acute. **Florets** 12–24+. **Cypselae** 3–4 mm; **pappi** stramineous to purplish, outer scales 20+, 0.6–1.1 mm, contrasting with 20+, 6–7+ mm inner bristles. **2n** = 34.

Flowering Jun–Aug. Calcareous soils, rocky banks; 300 m; Ark., Tex.; Mexico (Coahuila).

9. Vernonia texana (A. Gray) Small, Fl. S.E. U.S., 1338. 1903 E

Vernonia angustifolia Michaux var. *texana* A. Gray in A. Gray et al., Syn. Fl. N. Amer. 1(2): 91. 1884

Plants 4–8(–10+) dm. **Stems** puberulent, glabrescent. **Leaves** basal and cauline; blades ovate-lanceolate (basal) to narrowly lanceolate (distal), 5–12+ cm × (5–)12–25+ mm, l/w = 3–5 (basal) or 8–17 (distal), abaxially glabrate (pitted, awl-shaped hairs in pits), adaxially puberulent and/or scabrellous. **Heads** in open, paniculiform-scorpioid arrays. **Peduncles** 1–35 mm. **Involucres** ± obconic to hemispheric, 4.5–6 × 5–7 mm. **Phyllaries** 35–45+ in 5–6+ series, glabrescent, margins arachno-ciliolate, the outer lance-deltate, 1–32 mm, inner oblong to linear, 4–5(–6) mm, tips acute or rounded-apiculate. **Florets** 12–24+. **Cypselae** 2–3 mm; **pappi** usually whitish to stramineous (rarely purplish), outer scales or bristles 20+, 0.3–1+ mm, contrasting or intergrading with 20+, 6–7+ mm inner subulate scales or bristles. **2n** = 34.

Flowering Jun–Aug. Pinelands, scrub oak woodlands, sandy or sandy-clay soils; 60–200 m; Ark., La., Miss., Okla., Tex.

10. Vernonia lettermannii Engelmann ex A. Gray, Notes Compositae, 78. 1880 (as lettermani) E

Plants (2–)5–6+ dm. **Stems** glabrous or glabrate. **Leaves** mostly cauline; blades filiform, 5–9+ cm × 1–3+ mm, l/w = 25–50+, abaxially glabrate (pitted, awl-shaped hairs in pits), adaxially puberulent, glabrescent (sometimes pitted). **Heads** in corymbiform arrays. **Peduncles** 6–20+ mm. **Involucres** narrowly campanulate, 7–9(–10+) × 4–5+ mm. **Phyllaries** 30–40+ in 5–6+ series, sparsely tomentose, glabrescent, margins arachno-ciliolate, the outer lanceolate to subulate, 1–3 mm, inner oblong to lanceolate, 5–7+ mm, tips acute. **Florets** 10–12+. **Cypselae** 3–4 mm; **pappi** fuscous to purplish, outer scales 25+, 0.1–0.5 mm, contrasting with 35+, 6–7+ mm inner bristles. **2n** = 34.

Flowering Jul–Aug. Flood plains, terraces; 100–200 m; Ark., Okla.

11. Vernonia marginata (Torrey) Rafinesque, Atlantic J. 1: 146. 1832

Vernonia altissima Nuttall var. *marginata* Torrey, Ann. Lyceum Nat. Hist. New York 2: 210. 1827

Plants 3–5(–8+) dm. **Stems** puberulent, glabrescent. **Leaves** mostly cauline; blades narrowly lanceolate to lance-linear, 5–15+ cm × (2.5–)8–12+ mm, l/w = 8–20(–30+), abaxially glabrate (pitted, awl-shaped hairs in pits), adaxially puberulent, glabrescent (sometimes pitted). **Heads** in corymbiform arrays. **Peduncles** (3–)10–35 mm. **Involucres** narrowly campanulate, (7–)9–11 × 5–6 mm. **Phyllaries** 35–40+ in 5–6+ series, glabrescent, margins arachno-ciliolate, the outer lance-ovate, 1–2 mm, inner oblong to linear-oblong, 6–9+ mm, tips ± acuminate. **Florets** 10–25+. **Cypselae** 4–5 mm; **pappi** stramineous to purplish, outer subulate scales or bristles 30+, 0.6–2+ mm, intergrading with 40+, 8–9+ mm inner subulate scales or bristles. **2n** = 34.

Flowering Jul–Aug. Ditches, sandy flats, stream banks; 600–2000 m; Colo., Kans., N.Mex., Okla., Tex.; Mexico (Coahuila).

12. Vernonia fasciculata Michaux, Fl. Bor.-Amer. 2: 94. 1803 E

Vernonia fasciculata subsp. *corymbosa* (Schweinitz) S. B. Jones; *V. fasciculata* var. *corymbosa* (Schweinitz) B. G. Schubert

Plants 3–12+ dm. **Stems** puberulent, glabrescent. **Leaves** mostly cauline; blades ± lanceolate, 5–12(–20+) cm × 5–18(–40+) mm, l/w = 5–9(–14+), abaxially glabrate (pitted, with awl-shaped hairs in pits), adaxially scabrellous, resin-gland-dotted (sometimes pitted). **Heads** in congested, corymbiform arrays. **Peduncles** 1–8(–12+) mm. **Involucres** ± campanulate, 5–7(–8+) × 4–6 mm. **Phyllaries** 25–35+ in 4–5+ series, glabrescent, margins arachno-ciliolate, the outer lance-ovate, 1–3 mm, inner oblong to linear-oblong, 5–7+ mm, tips acute or rounded-apiculate. **Florets** 12–25+. **Cypselae** 3.5–4 mm; **pappi** fuscous to purplish, outer subulate scales or bristles 20–30, 0.5–3+ mm, intergrading with 35–45+, 5–7+ mm inner subulate scales or bristles. **2n** = 34.

Flowering Jul–Sep. Bottomlands, ditches, low prairies; 100–1200 m; Man.; Colo., Ill., Ind., Iowa, Kans., Ky., Mass., Minn., Mo., Nebr., N.Dak., Ohio, Okla., S.Dak., Wis.

13. Vernonia blodgettii Small, Fl. S.E. U.S., 1160, 1338. 1903

Plants 2–3(–5+) dm. **Stems** glabrous or glabrate. **Leaves** mostly cauline; blades lance-linear to linear, 3–7+ cm × 2–8+ mm, l/w = 7–15+, abaxially scabrellous (hairs appressed, awl-shaped), resin-gland-dotted, adaxially scabrellous, glabrescent, resin-gland-dotted. **Heads** in open, corymbiform to paniculiform arrays. **Peduncles** 12–35 mm. **Involucres** campanulate to obconic, 6–8 × 6–10 mm. **Phyllaries** 30–40+ in 4–5 series, puberulent, margins arachno-ciliate, the outer lance-ovate, 2–3 mm, inner lance-oblong to linear-oblong, 6–7 mm, tips acute or rounded-apiculate. **Florets** 18–25+. **Cypselae** 2.5–3+ mm; **pappi** stramineous to whitish, outer scales 25–30, 0.5–1.1 mm, contrasting with 30–40+, 5–7+ mm inner bristles. $2n = 34$.

Flowering year-round. Damp, peaty or sandy soils; 0–10 m; Fla.; West Indies (Bahamas).

14. Vernonia flaccidifolia Small, Bull. Torrey Bot. Club 25: 144. 1898 E

Plants 10–20+ dm. **Stems** glabrous. **Leaves** mostly cauline; blades ± lanceolate, 10–25(–30+) cm × 15–35(–60+) mm, l/w = 5–7+, abaxially puberulous, not resin-gland-dotted, adaxially scabrellous, glabrescent, not resin-gland-dotted. **Heads** in paniculi-form-scorpioid arrays. **Peduncles** 1–8(–15) mm. **Involucres** broadly campanulate to hemispheric, 4–5+ × 4–5 mm. **Phyllaries** 30–45+ in 5–6 series, glabrescent, margins ciliolate, the outer lance-ovate, 1–3 mm, inner lance-oblong to oblong, 3.5–5 mm, tips rounded. **Florets** 15–25+. **Cypselae** 3–3.5 mm; **pappi** stramineous to whitish, outer scales 25–35, 0.8–1.1 mm, contrasting with 35–45+, 5–6+ mm inner bristles. $2n = 34$.

Flowering Aug. Disturbed places in woodlands; 90–400 m; Ala., Ga., Tenn.

15. Vernonia gigantea (Walter) Branner & Coville, Rep. (Annual) Arkansas Geol. Surv. 4: 189. 1891 E

Chrysocoma gigantea Walter, Fl. Carol., 196. 1788; *Vernonia altissima* Nuttall; *V. altissima* var. *taeniotricha* S. F. Blake; *V. gigantea* subsp. *ovalifolia* (Torrey & A. Gray) Urbatsch; *V. ovalifolia* Torrey & A. Gray

Plants 8–20(–30+) dm. **Stems** puberulent, glabrescent. **Leaves** mostly cauline; blades ± lanceolate, 12–25+ cm × 20–60+ mm, l/w = (3.5–)4–7+, abaxially scabrellous (hairs awl-shaped), sometimes glabrescent, not or sparsely resin-gland-dotted, adaxially strigillose, glabrescent, not resin-gland-dotted. **Heads** in corymbiform-scorpioid arrays. **Peduncles** 1–12(–20+) mm. **Involucres** broadly campanulate to hemispheric, 4–5+ × 4–5 mm. **Phyllaries** 30–40+ in 4–5 series, glabrate, margins ciliolate, the outer lance-ovate, 1–2 mm, inner oblong, 3.5–5 mm, tips acute or rounded-apiculate. **Florets** (9–)18–24(–30). **Cypselae** 2.5–3.5 mm; **pappi** usually purplish, sometimes stramineous, outer scales 20–25, 0.5–1 mm, contrasting with 35–40+, 4.5–6+ mm inner bristles. $2n = 34$.

Flowering Jun–Sep. Flood plains; 10–300 m; Ont.; Ala., Ark., Fla., Ga., Ill., Ind., Kans., Ky., La., Md., Mich., Miss., Mo., Nebr., N.C., Ohio, Okla., Pa., S.C., Tenn., Tex., Va., W.Va.

16. Vernonia missurica Rafinesque, Herb. Raf., 28. 1833 E

Plants 6–12(–20+) dm. **Stems** puberulent. **Leaves** mostly cauline; blades elliptic to lance-ovate or lanceolate, 6–16(–20+) cm × 18–48+ mm, l/w = 2.5–4(–6+), abaxially usually puberulent to tomentose or pannose (hairs ± erect, ± curled), seldom glabrate, resin-gland-dotted, adaxially scabrellous, glabrescent, not resin-gland-dotted. **Heads** in corymbiform-scorpioid arrays. **Peduncles** 3–35 mm. **Involucres** broadly campanulate to urceolate, (6–)7–10+ × 5–9+ mm. **Phyllaries** 50–70+ in 6–7 series, sparsely scabrellous, glabrescent (seldom resin-gland-dotted), margins ciliolate, the outer lanceolate, 1–2 mm, inner linear-oblong to oblong, 6–7(–9+) mm, tips acute or rounded-apiculate. **Florets** 30–55+. **Cypselae** 3.5–4; **pappi** stramineous to whitish, outer scales 25–30, 0.5–1.1 mm, contrasting with 35–40+, 6–8+ mm inner bristles. $2n = 34$.

Flowering Jul–Sep. Prairies, loamy to sandy soils; 30–200 m; Ala., Ark., Ga., Ill., Ind., Iowa, Kans., Ky., La., Mich., Miss., Mo., Okla., Tenn., Tex.

17. **Vernonia baldwinii** Torrey, Ann. Lyceum Nat. Hist.
New York 2: 211. 1827 (as baldwini) E F

Vernonia baldwinii subsp. *interior* (Small) W. Z. Faust; *V. baldwinii* var. *interior* (Small) B. G. Schubert; *V. interior* Small

Plants 6–10(–15) dm. **Stems** puberulent to ± tomentose. **Leaves** mostly cauline; blades elliptic to lance-ovate or lanceolate, 8–15 (–18+) cm × 20–45(–75+) mm, l/w = 2–5, abaxially usually puberulent to tomentose or pannose (hairs ± erect, ± curled), seldom glabrate, resin-gland-dotted, adaxially scabrellous, glabrescent, not resin-gland-dotted. **Heads** in corymbiform-scorpioid arrays. **Peduncles** 1–25 mm. **Involucres** broadly campanulate to hemispheric, 4–6(–8+) × 4–7+ mm. **Phyllaries** 45–65+ in 5–6 series, usually puberulent (often resin-gland-dotted distally), sometimes glabrescent, margins ciliolate, the outer lance-ovate, 1–2 mm, inner oblong to lanceolate, 5–8+ mm, tips rounded-apiculate to acute (sometimes recurved). **Florets** (15–)20–25(–35+). **Cypselae** 2.5–3 mm; **pappi** fuscous to purplish, outer scales 25–30, 0.2–1 mm, contrasting with 35–40+, 5–7+ mm inner bristles. $2n = 34$.

Flowering Jun–Nov. Disturbed places, grasslands, flood plains, forest margins, prairies; 10–1100 m; Ark., Colo., Ill., Iowa, Kans., Ky., La., Mich., Mo., Nebr., Okla., Tex.

Regarding *Vernona baldwinii* and *V. interior*, L. H. Shinners (1950) wrote, "The tips of the phyllaries vary from loosely appressed to squarrose, and from puberulent to almost completely glabrous on the inner face. The geographic distribution of the two extremes is nearly identical. I consider the two to be merely forms of one species." I concur.

187e. Asteraceae Martinov tribe Cichorieae Lamarck & de Candolle, Syn. Pl. Fl. Gall., 255. 1806 (as Cichoraceae)

Lactuceae Cassini

Annuals, biennials, perennials, subshrubs, or shrubs [trees, vines] (sap usually milky). **Leaves** basal and/or cauline; alternate (proximal opposite in *Shinnersoseris*) [opposite]; petiolate or sessile; margins usually dentate or pinnately lobed (then frequently runcinate), sometimes prickly, sometimes entire, rarely much divided (bases often clasping). **Heads** homogamous (liguliflorous), usually in corymbiform or paniculiform arrays, sometimes borne singly (on scapiform peduncles), sometimes subsessile in axillary clusters on stems or among leaves of basal rosettes [aggregated in second-order heads]. **Calyculi** 0 or of 1–15+ bractlets in 1–3+ series. **Phyllaries** usually persistent, usually in 3–5+ series, distinct, and unequal, sometimes in 1–2 series, distinct or connate, and subequal to equal, margins (seldom prickly) and/or apices sometimes notably scarious (phyllaries sometimes enfold and fall with subtended florets or cypselae, e.g., in *Rhagadiolus*). **Receptacles** flat to convex, epaleate (sometimes bristly-setose) or paleate (paleae enfold and fall with subtended cypselae in *Scolymus*). **Florets** ligulate (bisexual, fertile); corollas usually yellow to orange, sometimes cyanic or white (zygomorphic, 5-toothed); anther bases usually tailed, apical appendages ovate to lanceolate or hardly developed (pollen sometimes brightly colored); styles abaxially usually papillate to hirsute (mostly distally), branches filiform to stout, adaxially continuously stigmatic from bases almost to apices, apices rounded to acute, appendages essentially 0. **Cypselae** usually monomorphic within heads, ± clavate, columnar, ellipsoid, fusiform, or prismatic, often compressed, obcompressed, or flattened, often beaked or apically tapered, bodies smooth, muricate, rugose, or tuberculate, often ribbed, sometimes winged (glabrous or hairy); **pappi** (rarely 0) persistent or readily falling, usually of fine to coarse, often barbellate, sometimes plumose bristles, sometimes of awns or scales, sometimes combinations of bristles, awns, and/or scales (scales often aristate).

Genera ca. 100, species ca. 1600 (49 genera, 229 species in the flora): nearly worldwide, mostly in Old World, mostly at temperate latitudes; some species widely introduced.

SELECTED REFERENCES Jansen, R. K. et al. 1991b. Systematic implications of chloroplast DNA variation in the subtribe Microseridinae (Asteraceae: Lactuceae). Amer. J. Bot. 78: 1015–1027. Kim, S. C., D. J. Crawford, and R. K. Jansen. 1996. Phylogenetic relationships among the genera of the subtribe Sonchinae (Asteraceae): Evidence from ITS sequences. Syst. Bot. 21: 417–432. Lee, J. and B. G. Baldwin. 2004. Subtribes of principally North American genera of Cichorieae (Compositae). Novon 14: 309–313. Lee, J., B. G. Baldwin, and L. D. Gottlieb. 2002. Phylogeny of *Stephanomeria* and related genera (Compositae–Lactuceae) based on 18S-26S nuclear rDNA ITS and ETS sequences. Amer. J. Bot. 89: 160–168. Lee, J., B. G. Baldwin, and L. D. Gottlieb. 2003. Phylogenetic relationships among the primarily North American genera of Cichorieae (Compositae) based on analysis of 18S–26S rDNA ITS and ETS sequences. Syst. Bot. 28: 616–626. Price, H. J. and K. Bachmann. 1975. DNA content and evolution in the Microseridinae. Amer. J. Bot. 62: 262–267. Stebbins, G. L. 1953. A new classification of the tribe Cichorieae, family Compositae. Madroño 12: 33–64. Tomb, A. S. 1974. Chromosome numbers and generic relationships in subtribe Stephanomeriinae (Compositae: Cichorieae). Brittonia 26: 203–216. Whitton, J., R. S. Wallace, and R. K. Jansen. 1995. Phylogenetic relationships and patterns of character change in the tribe Lactuceae (Asteraceae) based on chloroplast DNA restriction site variation. Canad. J. Bot. 73: 1058–1073.

1. Cypselae beaked (outer sometimes beakless in *Hypochaeris*).
   2. Stems scapiform.
      3. Heads usually in spiciform, paniculiform, cymiform, or corymbiform arrays (sometimes borne singly).
         4. Leaf margins spiny; calyculi 0; phyllaries in 3–5+ series, unequal; pappi of crisped (frizzy), outer and straight, coarse, inner bristles . . . . 47. *Launaea* (in part), p. 272
         4. Leaf margins not spiny; calyculi of 5–12 bractlets; phyllaries in 1(–2) series, equal; pappi of uniform bristles . . . . 36. *Crepis* (in part), p. 222

3. Heads borne singly.
 5. Pappi of outer scales and inner bristles, or of aristate scales.
  6. Cypsela beaks relatively long or 0; pappi of outer scales and inner, plumose bristles; annuals or perennials; leaves oblanceolate to oblong . . . . 52. *Leontodon* (in part), p. 294
  6. Cypsela beaks relatively short; pappi of lanceolate, apically notched, aristate scales, aristae smooth; annuals; leaves linear to narrowly lanceolate, grasslike. . . . . 65. *Uropappus* (in part), p. 322
 5. Pappi of bristles.
  7. Receptacles paleate; corollas pink, purple, lavender, or nearly white; perennials . . . . 81. *Pinaropappus*, p. 374
  7. Receptacles usually epaleate (if paleate, corollas yellow; some *Agoseris*); corollas usually yellow to orange, rarely ochroleucous, pink, pinkish, purplish, or white; annuals, biennials, or perennials.
   8. Calyculi of (6–)8–18(–20) bractlets in 1–3 series; phyllaries equal; perennials . . . . 37. *Taraxacum*, p. 239
   8. Calyculi 0; phyllaries equal or unequal; annuals, biennials, or perennials.
    9. Leaves entire, toothed, or pinnately lobed (not spiny); heads borne singly; corollas yellow, orange, pinkish, purplish, or white; pappi of ± barbellate bristles; annuals or perennials . . . . 66. *Agoseris* (in part), p. 323
    9. Leaves often pinnately lobed, ultimate margins dentate (spiny); heads borne singly or in spiciform or paniculiform arrays; corollas yellow to ochroleucous; pappi of crisped (frizzy), outer and straight, coarse, inner bristles; annuals or biennials . . . . 47. *Launaea* (in part), p. 272

[2. Shifted to left margin.—Ed.]
2. Stems leafy.
10. Pappi of plumose bristles or subulate scales; annuals or biennials.
 11. Corollas white, sometimes abaxially rose- or purple-veined; calyculi of spreading to reflexed, unequal bractlets; annuals . . . . 70. *Rafinesquia*, p. 348
 11. Corollas yellow, orange, purple, or pinkish to purplish; calyculi 0; annuals or biennials.
  12. Leaves linear to lance-linear or lance-attenuate, entire; peduncles inflated distally; involucres campanulate; pappi of subulate scales; biennials (winter annuals); heads borne singly; corollas yellow or purple . . . . 59. *Tragopogon*, p. 303
  12. Leaves mostly obovate to oblong-obovate, the distal ovate to linear, usually pinnately lobed, sometimes dentate or entire; peduncles little, if at all, inflated distally; involucres urceolate; pappi of bristles; annuals; heads borne singly or in corymbiform arrays; corollas yellow, sometimes abaxially striped with red . . . . 53. *Urospermum*, p. 296
10. Pappi of smooth or barbellate bristles, or of subulate scales or awns, or of aristate scales; annuals, biennials, or perennials.
 13. Perennials.
  14. Involucres narrowly cylindric, 1–2+ mm diam.; phyllaries (4–)5; florets 5; cypselae obovoid to lanceoloid. . . . . 43. *Mycelis* (in part), p. 257
  14. Involucres cylindric to campanulate, 2–5 mm diam.; phyllaries 5–21; florets 6–150+; cypselae cylindric to fusiform or lanceoloid.
   15. Corollas usually bluish; cypselae lanceoloid . . . . 44. *Mulgedium* (in part), p. 258
   15. Corollas yellow to white; cypselae cylindric to fusiform.
    16. Phyllaries 8–21+; pappi: outer crowns of spreading, white hairs plus 2–3+ inner series of rufous to stramineous bristles. . . . . 82. *Pyrrhopappus*, p. 376
    16. Phyllaries 5–10; pappi of white bristles.

17. Calyculi of 3–4, minute bractlets; involucres cylindric; cypselae cylindric, 5+-ribbed (without rings of scales); pappi of 40–50+ bristles . . . . 38.　*Chondrilla*, p. 252

17. Calyculi of 3–10+, deltate to lanceolate bractlets; involucres campanulate to cylindric; cypselae fusiform, 10-ribbed or -winged (with rings of scales at bases of beaks); pappi of 20–30 bristles . . . . 39.　*Ixeris*, p. 254

[13. Shifted to left margin.—Ed.]

13. Annuals or biennials.
　18. Peduncles inflated distally (fistulose); pappi of aristate scales . . . . 65.　*Uropappus* (in part), p. 322
　18. Peduncles not inflated distally; pappi of bristles or subulate scales (not aristate).
　　19. Receptacles paleate; outer cypselae beakless; all or inner pappus bristles plumose
　　. . . . . 54.　*Hypochaeris*, p. 297
　　19. Receptacles epaleate; all cypselae beaked; pappus bristles usually barbellulate or smooth (sometimes plumose in *Helminthotheca*).
　　　20. Calyculi 0; leaves spiny . . . . 47.　*Launaea* (in part), p. 272
　　　20. Calyculi of 2–16 bractlets; leaves not spiny.
　　　　21. Cypselae dimorphic, outer gibbous, inner ellipsoid to fusiform, not ribbed (rugulose); pappi of subulate to setiform scales . . . . 55.　*Helminthotheca*, p. 300
　　　　21. Cypselae monomorphic, ellipsoid, oblong, obovoid, or lanceoloid, or cylindric to fusiform, ribbed; pappi of bristles (sometimes with minute outer crowns as well).
　　　　　22. Cypselae ellipsoid to oblong or obovoid to lanceoloid; pappi of minute outer crowns plus inner bristles (sometimes 2–3+ series of bristles in *Lactuca*).
　　　　　　23. Involucres narrowly cylindric, 1–2+ mm diam.; calyculi of 2–4 bractlets in 1 series . . . . 43.　*Mycelis* (in part), p. 257
　　　　　　23. Involucres campanulate to cylindric, 2–5 mm diam.; calyculi of 3–10+ bractlets in 2–3 series . . . . 45.　*Lactuca*, p. 259
　　　　　22. Cypselae fusiform, not or little compressed; pappi of bristles.
　　　　　　24. Stems dotted with tack-glands; pappus bristles basally connate (falling in rings), white . . . . 61.　*Calycoseris*, p. 307
　　　　　　24. Stems eglandular or glandular (without tack-glands); pappus bristles distinct or inner basally connate, white to tawny.
　　　　　　　25. Heads usually in cymiform, corymbiform, or paniculiform arrays, sometimes borne singly; corollas usually yellow or orange, sometimes white, pink, or reddish; cypselae 10–20-ribbed . . . . 36.　*Crepis* (in part), p. 222
　　　　　　　25. Heads borne singly; corollas white to pale yellow; cypselae 4–5-angled or -ribbed (angles roughened) . . . . 75.　*Glyptopleura*, p. 361
1. Cypselae beakless.
　26. Stems scapiform.
　　27. Pappi 0 or coroniform.
　　　28. Perennials; leaves (± fleshy) linear to oblanceolate, margins entire; calyculi 0
　　　. . . . . 80.　*Phalacroseris*, p. 374
　　　28. Annuals or biennials; leaves obovate or oblanceolate to spatulate, margins ± dentate or lobed; calyculi of 1–10+ bractlets or 0.
　　　　29. Leaf margins pinnately lobed; cypselae oblong, 10–13-ribbed . . . 40.　*Lapsanastrum*, p. 254
　　　　29. Leaf margins ± dentate to toothed; cypselae obovoid or subcylindric to weakly clavate, 4–5- or 8–10-ribbed.
　　　　　30. Leaves oblanceolate to spatulate; heads borne singly or 2–3; peduncles naked, inflated distally; involucres broadly campanulate to urceolate; phyllaries basally connate; cypselae obovoid, 8–10-ribbed . . . . . . 49.　*Arnoseris*, p. 276
　　　　　30. Leaves obovate; heads in corymbiform arrays; peduncles sometimes bracteate, not inflated distally; involucres cylindro-campanulate; phyllaries distinct; cypselae subcylindric or weakly clavate, 4–5-ribbed (ribs corky) . . . . 62.　*Atrichoseris*, p. 309

27. Pappi of bristles and/or scales, or of aristate scales.
  31. Heads usually in corymbiform, paniculiform, or cymiform arrays; calyculi of 3–12 bractlets.
    32. Leaves entire or dentate to pinnatifid (often lyrate or runcinate); cypselae 10–20-ribbed; pappi of usually distinct, white to tawny bristles . . . . 36. *Crepis* (in part), p. 222
    32. Leaves pinnately lobed (lyrate); cypselae 11–13-ribbed (ribs spiculate); pappi of basally connate, white bristles . . . . 41. *Youngia*, p. 255
  31. Heads usually borne singly (sometimes in spiciform or paniculiform arrays in *Launaea*); calyculi 0.
    33. Pappi of subulate or aristate scales.
      34. Heads erect; cypselae narrowed distally; pappi of lustrous, white subulate or aristate scales (bodies narrowly lanceolate to subulate) . . . .
. . . . 67. *Nothocalaïs* (in part), p. 335
      34. Heads nodding (at least in bud); cypsela apices truncate; pappus scales silvery to yellowish, brownish, or blackish (rarely white), plumose, barbellate, or barbellulate.
        35. Involucres fusiform to ovoid or globose; pappi of deltate, lanceolate, linear, oblong, orbiculate, or ovate aristate scales, aristae plumose, barbellate, or barbellulate . . . . 68. *Microseris* (in part), p. 338
        35. Involucres campanulate; pappi of narrowly lanceolate aristate scales, aristae barbellulate . . . . 69. *Stebbinsoseris* (in part), p. 346
    33. Pappi wholly or partly of bristles (in *Leontodon*, pappi of outer cypselae of scales, of inner cypselae of bristles; sometimes scales + bristles in *Krigia*).
      36. Pappi of outer cypselae of scales, of inner cypselae of plumose bristles
. . . . 52. *Leontodon* (in part), p. 294
      36. Pappi of bristles or of outer scales plus inner bristles.
        37. Phyllary margins papery (wider than midribs); receptacles paleate; pappus bristles plumose. . . . . 63. *Anisocoma*, p. 309
        37. Phyllary margins not papery; receptacles usually epaleate (sometimes paleate in *Agoseris*); pappus bristles mostly smooth, barbellulate, or ± barbellate.
          38. Leaves spiny; corollas yellow to ochroleucous; pappi of crisped (frizzy), outer and coarse, straight, inner bristles . . . . . 47. *Launaea* (in part), p. 272
          38. Leaves not spiny; corollas usually yellow to orange, sometimes pink, purple, or red; pappi of ± barbellate bristles.
            39. Phyllaries in 1–2 series; cypselae 10–20-nerved or -ribbed; pappi of 5 outer scales plus 5–45 inner bristles. . . . . 76. *Krigia* (in part), p. 362
            39. Phyllaries in 2–5(–7) series; cypselae 8–10(–15)-ribbed; pappi of bristles in 1–6 series.
              40. Phyllaries hairy or glabrous; cypselae usually beaked (6–31 mm) . . . . 66. *Agoseris* (in part), p. 323
              40. Phyllaries usually glabrous; cypselae not beaked (5–10 mm)
                41. Plants taprooted; pappi of 30–80, whitish bristles
. . . . 67. *Nothocalaïs* (in part), p. 335
                41. Plants rhizomatous (usually in bogs); pappi of 24–48, brownish bristles . . . . 68. *Microseris* (in part), p. 338
[26. Shifted to left margin.—Ed.]
26. Stems leafy.
  42. Pappi 0, or of scales or aristate scales, or of awns plus bristles, or coroniform.
    43. Pappi 0.
      44. Stems winged, wings spiny; receptacles paleate. . . . . 34. *Scolymus* (in part), p. 220
      44. Stems not winged; receptacles epaleate.
        45. Heads borne singly; florets 5–35; cypselae ± monomorphic (1.3–1.7 mm)
. . . . 76. *Krigia* (in part), p. 362
        45. Heads in corymbiform or thyrsiform arrays; florets 5–15; cypselae heteromorphic (3–25 mm, outer longer than inner).

46. Leaves ovate to suborbiculate, coarsely dentate; heads in corymbiform to thyrsiform arrays; cypselae subcylindric (curved), terete to slightly compressed, ± 20-ribbed .... 42. *Lapsana*, p. 257
46. Leaves ovate-lanceolate to linear, entire or dentate to pinnately lobed; heads in ± corymbiform arrays; cypselae ± terete (outer straight or arcuate, inner straight to coiled), not ribbed .... 56. *Rhagadiolus*, p. 300

43. Pappi of scales or aristate scales, or of awns or subulate scales plus bristles, or coroniform.
47. Corollas usually blue, sometimes pink or white; cypselae prismatic, 3–5-angled; pappi coroniform (erose scales) .... 35. *Cichorium*, p. 221
47. Corollas orange or yellow to white or pink to lavender; cypselae columnar or cylindric to fusiform or obconic, 6–15-ribbed; pappi usually of scales (scales sometimes aristate), sometimes of awns or subulate scales plus bristles, or coroniform (*Krigia wrightii*, annuals, cypselae 1.3–1.6 mm).
48. Phyllaries enfolding outer cypselae.
49. Peduncles inflated distally; calyculi of 3–10+, deltate to lanceolate bractlets; phyllaries 5–13+ in 1 series; florets 8–30+; cypselae cylindric to fusiform, 12–15-ribbed; pappi of outer cypselae coroniform (short scales), of inner cypselae of lance-aristate to subulate-aristate scales (0–5+ outer and 5+ inner) .... 57. *Hedypnois*, p. 302
49. Peduncles not inflated distally; calyculi of 8–13, linear to filiform bractlets; phyllaries 20–25+ in 2+ series; florets 30–100+; cypselae columnar, 6–8(–10)-ribbed; pappi of setiform scales plus smooth or barbellate bristles .... 50. *Tolpis*, p. 277
48. Phyllaries not enfolding outer cypselae.
50. Heads ± nodding (at least in bud); pappi of aristate scales.
51. Perennials; involucres campanulate, fusiform, ovoid, or globose; pappus scales deltate, lanceolate, linear, oblong, orbiculate, or ovate, aristae plumose, barbellate, or barbellulate .... 68. *Microseris* (in part), p. 338
51. Annuals; involucres campanulate; pappus scales narrowly lanceolate, aristae barbellulate .... 69. *Stebbinsoseris* (in part), p. 346
50. Heads erect; pappi of setiform, subulate, or aristate scales or of awns or subulate scales plus bristles.
52. Perennials; leaves linear to linear-lanceolate, entire; corollas pale lavender to white; cypselae 5-angled or -ridged; pappi of awns or subulate scales plus bristles .... 77. *Chaetadelpha*, p. 368
52. Annuals, biennials, or perennials; leaves ovate-lanceolate or oblong to lanceolate or linear, entire or pinnately lobed; corollas white, yellow, or purplish; cypselae sometimes 10-nerved; pappi of setiform, subulate, or aristate scales.
53. Leaves oblong or lanceolate to oblanceolate or linear, spiny; involucres campanulate to urceolate; receptacles paleate; corollas yellow; cypselae obovoid, ribs 0; pappi (0 or) of 2–4 setiform to aristate scales plus 0–4 muticous to lanceolate scales .... 34. *Scolymus* (in part), p. 220
53. Leaves ovate-lanceolate to lanceolate or linear, not spiny; involucres ovoid to cylindric; receptacles epaleate; corollas whitish to yellow or purplish; cypselae columnar to obclavate or fusiform, not compressed, usually 10-nerved; pappi of 28–50+ plumose to barbellate, subulate to setiform scales .... 60. *Scorzonera*, p. 306

[42. Shifted to left margin.—Ed.]
42. Pappi wholly or mostly of bristles (sometimes with minute outer scales).
54. Shrubs; California .... 71. *Munzothamnus*, p. 349
54. Annuals, biennials, or perennials.
55. Corollas yellow, orange, ochroleucous, pinkish to reddish, or white (sometimes abaxially pinkish, reddish, or purplish).

56. Leaves spiny (on margins, bases often auriculate); phyllaries in 3–5+ series.
    57. Involucres 3–5 mm diam.; phyllaries 18–25; cypselae weakly compressed, cylindric to prismatic or fusiform, 4–5-ribbed (or grooved), ribs muricate; pappi of crisped (frizzy, often basally coherent or connate), outer plus straight, coarse, inner bristles . . . . 47. *Launaea* (in part), p. 272
    57. Involucres 5–15+ mm diam.; phyllaries 27–50; cypselae compressed, oblong or oblanceoloid to elliptic, 2–4(–5)-ribbed, smooth, rugose, or tuberculate; pappi of distinct outer bristles plus basally coherent or connate inner bristles . . . . 48. *Sonchus*, p. 273
56. Leaves not spiny (bases not auriculate); phyllaries in 1–2+, or 2–3, or 4–6 series.
    58. Calyculi 0.
        59. Heads borne singly; corollas yellow to orange; pappi (usually 2-seriate): outer of 5+ scales, inner of 5–45 bristles . . . . 76. *Krigia* (in part), p. 362
        59. Heads in corymbiform to paniculiform arrays; corollas yellow to white; pappi (1 or 2-seriate): outer pappi 0, or of minute teeth plus 0–6 bristles, inner (or only) pappi of 15–35, basally coherent bristles . . . . . . 64. *Malacothrix*, p. 310
    58. Calyculi usually of 3–16+ bractlets (sometimes intergrading with phyllaries).
        60. Perennials; phyllaries in 2+ series; cypselae columnar or prismatic to ± urceolate . . . . 51. *Hieracium*, p. 278
        60. Annuals, biennials, or perennials; phyllaries in 1–2 series; cypselae subcylindric to fusiform.
            61. Corollas yellow or orange (white, pink, reddish), not abaxially reddish; cypselae 10–20-ribbed, sometimes spiculate; pappi persistent, of distinct bristles . . . . 36. *Crepis* (in part), p. 222
            61. Corollas yellow, often abaxially reddish; cypselae 5–10-ribbed, transversely rugulose; pappi falling, of basally connate bristles . . . 58. *Picris*, p. 302
    [55. Shifted to left margin.—Ed.]
55. Corollas usually purple, lavender, pink, or blue, sometimes white, rarely yellow.
  62. Annuals.
    63. Pappus bristles distinct, plumose (at least distally) . . . . 72. *Stephanomeria* (in part), p. 350
    63. Pappus bristles basally connate, smooth or barbellulate.
      64. Stems glandular-puberulent to glabrescent; leaves basal and cauline, blades spatulate to oblanceolate or minute, margins irregularly dentate to runcinate, often spinulose; peduncles minutely bracteate; cypselae columnar (tapering slightly proximally), 5-ribbed . . . . 73. *Prenanthella*, p. 359
      64. Stems glabrous; leaves cauline (opposite proximally), blades linear, margins entire; peduncles ebracteate; cypselae subcylindric (apices abruptly constricted), 10-ribbed . . . . 78. *Shinnersoseris*, p. 368
  62. Perennials.
    65. Heads (usually nodding) in racemiform, thyrsiform, corymbiform, or paniculiform arrays; corollas usually white to pink or lavender, rarely yellow . . . . 46. *Prenanthes*, p. 264
    65. Heads (erect) usually borne singly (sometimes in ± corymbiform to paniculiform arrays); corollas usually bluish, pink to lavender, or rose, rarely white.
      66. Leaves oblong, elliptic, or ovate to lanceolate or linear; phyllaries 8–13+ in 1–2 series; florets (1–)15–50+; cypselae compressed, lanceoloid . . . . 44. *Mulgedium* (in part), p. 258
      66. Leaves linear to oblong (distal sometimes bractlike or subulate distally); phyllaries 3–10 in ± 1 series; florets 3–16; cypselae not compressed, columnar.
        67. Stems rigid, spine-tipped (thorny); leaves cauline, entire; phyllaries 3–5; florets 3–5 . . . . 74. *Pleiacanthus*, p. 361
        67. Stems not spine-tipped; leaves basal and cauline, entire, dentate, or pinnately lobed; phyllaries 4–10; florets (4–)5–16.
            68. Cypselae subcylindric, subterete, angled, or sulcate; pappus bristles barbellate . . . . 79. *Lygodesmia*, p. 369
            68. Cypselae columnar, 5-angled; pappus bristles (wholly or partly) plumose . . . . 72. *Stephanomeria* (in part), p. 350

34. SCOLYMUS Linnaeus, Sp. Pl. 2: 813. 1753; Gen. Pl. ed. 5, 355. 1754 • [Greek *skolymus*, a kind of thistle or artichoke; allusion unclear, perhaps for perceived similarity to *Cynara*, globe artichoke] ☐

John L. Strother

Annuals, biennials, or perennials, 10–80(–200+) cm; taprooted. **Stems** usually 1, erect, branched distally, glabrous or hairy (internodes winged, margins spiny, faces ± scabrellous and/or sparsely arachnose). **Leaves** basal and cauline (mostly cauline at flowering); basal ± petiolate, distal sessile; blades oblong or lanceolate to oblanceolate or linear, margins pinnately lobed to dentate, usually strongly spiny (faces ± scabrellous and/or sparsely arachnose). **Heads** borne singly (axillary and terminal) or in ± spiciform arrays. **Peduncles** usually 0 (heads in axils of ± foliaceous bracts; if peduncles produced, not inflated, strongly bracteate). **Calyculi** 0. **Involucres** campanulate to urceolate, 8–12[–16] mm diam. (larger in fruit). **Phyllaries** 24–30+ in 3+ series, lanceolate to lance-linear (± flat proximally), unequal to subequal, margins little, if at all, scarious, apices spine-tipped. **Receptacles** conic to hemispheric, smooth, glabrous, paleate; paleae winged (each palea enfolding, adnate to, and shed with cypsela). **Florets** 30–60+; corollas yellow. **Cypselae** (each shed with its enfolding palea) brownish, obcompressed, ± obovoid, not beaked, ribs 0, glabrous; **pappi** 0 or persistent (fragile), of 2–4[–5] whitish to stramineous, subequal, setiform-aristate scales plus 0–4 muticous to lanceolate scales in ± 1 series. *x* = 10.

Species 3 (2 in the flora): introduced; Europe, Middle East, n Africa, Atlantic Islands.

1. Annuals; margins of stem wings and leaf blades usually white and thickened; pappi 0 . . . . 1. *Scolymus maculatus*
1. Biennials or perennials; margins of stem wings and leaf blades little, if at all, white or thickened; pappi of 2–4 setiform-aristate scales plus 0–4 muticous to lanceolate scales . . . . 2. *Scolymus hispanicus*

1. Scolymus maculatus Linnaeus, Sp. Pl. 2: 813. 1753
☐

Annuals. Stem wings ± continuous, margins spinose, usually white and thickened. **Leaf blades** 40–200 × 20–80 mm, margins usually white and thickened. **Involucres** 12–18 × 8–12 mm, larger in fruit. **Phyllaries** ovate-lanceolate to lanceolate, glabrous. **Cypsela/palea units** ± obovate, 2.5–4 mm; **pappi** 0. *2n* = 20.

Flowering Jul. Disturbed sites; 0–100 m; introduced; N.C.; Europe.

2. Scolymus hispanicus Linnaeus, Sp. Pl. 2: 813. 1753
• Golden thistle, Spanish salsify or oyster [F][I][W]

Biennials or perennials. Stem wings not continuous, margins spinose, little, if at all, white or thickened. **Leaf blades** 40–200 × 15–70 mm, margins little, if at all, white or thickened. **Involucres** 15–20 × 8–10 mm, larger in fruit. **Phyllaries** lanceolate, glabrous. **Cypsela/palea units** ± orbiculate to ovate, 3–5 mm; **pappi** of 2–4 setiform-aristate scales plus 0–4 muticous to lanceolate scales. *2n* = 20.

Flowering Jul. Disturbed sites; 0–100 m; introduced; Ala., Calif., N.Y., Pa.; Europe.

In the Mediterranean region, *Scolymus hispanicus* is used (or has been used) as a medicinal herb and root vegetable. It is considered to be a noxious weed in the United States.

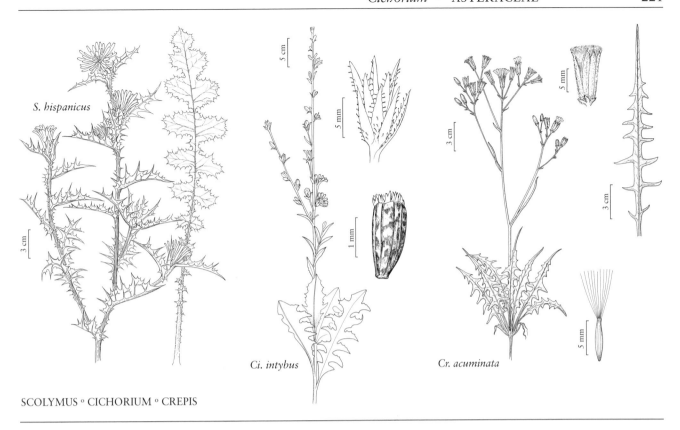

S. hispanicus

Ci. intybus

Cr. acuminata

SCOLYMUS ∘ CICHORIUM ∘ CREPIS

35. CICHORIUM Linnaeus, Sp. Pl. 2: 813. 1753; Gen. Pl. ed. 5, 354. 1754 • [Ancient Arabic name] [I]

John L. Strother

Perennials [annuals, biennials], [2–]10–120+ cm; taprooted. **Stems** usually 1, erect, branched distally or throughout, setose or hispid to pilose, or glabrous. **Leaves** basal and cauline; usually sessile; basal blades oblanceolate to lance-linear, margins usually runcinate-pinnate to dentate, rarely entire; cauline similar, smaller, margins dentate or entire. **Heads** mostly in glomerules (axillary and nearly sessile), some borne singly (on ± elongate peduncles). **Peduncles** (dimorphic: most 0–2 mm, some 12–85+ mm): the longer often slightly inflated distally, not bracteate. **Calyculi** 0 (or interpreted as outer phyllaries). **Involucres** ± cylindric, 3–5+ mm diam. **Phyllaries** 10–15+ in 2+ series, lance-ovate to lanceolate or linear, unequal, margins little, if at all, scarious, apices obtuse to acute. **Receptacles** flat, pitted, ± hispid, usually epaleate. **Florets** 8–25+; corollas usually blue [purple], sometimes pink or white. **Cypselae** brownish, ± prismatic (3–5-angled), not beaked, faces smooth, glabrous; **pappi** persistent, coroniform (of 40–60+, whitish, subequal, erose scales in 1–2 series). $x = 9$.

Species 6 (1 in the flora): introduced; Europe, n Africa, Asia; introduced also in South America, Africa, Pacific Old World.

The salad green known as endive (*Cichorium endivia* Linnaeus) may sometimes be found as an escape from gardens or agricultural plantings. It differs from *C. intybus* in having purple corollas, cypselae 1.5–2.5 mm, and pappi 0.6–1+ mm.

1. Cichorium intybus Linnaeus, Sp. Pl. 2: 813. 1753

• Chicory, chicorée F I

Perennials (sometimes flowering first year). **Leaves:** blades of basal 5–35+ × 1–8(–12+) cm; cauline similar, smaller, narrower, distal mostly linear. **Peduncles** mostly 0–2 mm, some narrowly clavate, 12–45(–85+) mm. **Phyllaries:** outer 5–6 lance-ovate to lanceolate, 4–7 mm, basally cartilaginous, distally herbaceous, inner 8+ lance-linear to linear, 6–12 mm, herbaceous, all usually with some gland-tipped hairs 0.5–0.8 mm on margins near bases or on abaxial faces toward tips. **Cypselae** 2–3 mm; **pappi** 0.01–0.2 mm. $2n = 18$.

Flowering Apr–Jul. Disturbed sites; 0–1500 m; introduced; St. Pierre and Miquelon; Alta., B.C., Man., N.B., Nfld. and Labr., N.S., Ont., P.E.I., Que., Sask.; Ark., Calif., Conn., Ill., Ind., Iowa, Kans., Maine, Mass., Mich., Mo., Nev., N.H., N.Y., N.C., Pa., R.I., Tex., Utah, Vt.; Europe; Asia; introduced also in Africa, South America.

Leaves of *Cichorium intybus* are sometimes used as salad greens; the roasted roots are sometimes ground and used as an addition to (or adulterant of) coffee.

36. CREPIS Linnaeus, Sp. Pl. 2: 805. 1753; Gen. Pl. ed. 5, 350. 1754 • Hawksbeard, crépis [Greek *krepis*, slipper or sandal, possibly alluding to shape of cypselae, a name of a plant in writings by Theophrastus]

David J. Bogler

Annuals, biennials, or perennials, 3–120 cm; usually taprooted, sometimes rhizomatous (roots deep or shallow, woody or fibrous, caudices often woody). **Stems** 1–20+, erect to decumbent, simple (sometimes scapiform) or branched, usually striate, glabrous or hairy, often densely hispid or setose (hairs often stipitate-glandular). **Leaves** basal (often in rosettes) and cauline; petiolate (at least basal, petioles ± winged); basal blades mostly elliptic, ovate, or lanceolate to linear, or spatulate to oblanceolate, often lyrate or runcinate, margins entire, dentate, serrate, toothed, or pinnately lobed, lobes sometimes toothed; cauline usually present, lobed or entire, usually reduced in size and lobing distally. **Heads** (erect) usually in cymiform, corymbiform, or paniculiform arrays, sometimes borne singly. **Peduncles** not inflated distally, not bracteate. **Calyculi** of 5–12, reduced, subulate to lanceolate or deltate bractlets in ± 1 series, mostly unequal, glabrous, tomentulose, or setose. **Involucres** cylindric to campanulate (sometimes becoming turbinate in fruit), 4–15 mm diam. **Phyllaries** 5–18 in 1–2 series, lanceolate, equal or subequal, (bases becoming thickened and keeled, keels sometimes pronounced in fruit) margins green to yellowish, often scarious, apices acute to acuminate, abaxial faces glabrous, tomentose, or setose, sometimes stipitate-glandular, adaxial glabrous or with appressed hairs. **Receptacles** flat or convex, usually pitted, glabrous or hairy, epaleate [paleate, paleae narrow, thin]. **Florets** 5–100+; corollas usually yellow or orange, sometimes white, pink, or reddish. **Cypselae** monomorphic or dimorphic, yellow, brown, green, red, and/or black, subcylindric or fusiform, terete or subterete, usually curved, apices tapered or beaked, ribs 10–20, sometimes spiculate-roughened, faces glabrous or hispidulous; **pappi** persistent or falling, of 80–150, usually distinct, sometimes basally connate, white to tawny, coarse to fine, ± equal (or outer shorter), barbellulate bristles in 1–2 series. $x = 3, 4, 5, 6, 11$.

Species ca. 200 (24 in the flora): North America, Eurasia, Africa; introduced nearly worldwide.

Crepis is generally recognized by the rosettes of coarse, often pinnately lobed leaves, erect heads, epaleate receptacles, calyculate involucres, yellow corollas, subcylindric or fusiform, ribbed cypselae, and pappi of barbellulate bristles. The taxonomy and evolutionary relationships of *Crepis* were studied by E. B. Babcock (1947) and his associates. Their work was thorough and important because of the effort to incorporate cytogenetic information in the evolutionary analysis. Extensive survey of chromosome number and karyotype indicated two major ploidy groups in *Crepis*, corresponding to New World and Old World species complexes. Of the 12 species of *Crepis* native to North America, 10 are polyploids with $x = 11$. The core diploid populations commonly occupy discrete ecologic zones and are thought to be entirely distinct from one another, yet they are interconnected by a continuous complex series of intergrading polyploid forms that are partly or completely apomictic (Babcock). The polyploids are of two forms, autopolyploids that are similar to the diploids, and allopolyploids that combine the characteristics of two or more diploid species. The allopolyploid forms of hybrid origin may exhibit the characteristics of multiple parental species and therefore are difficult to classify. Some of the heterogeneous apomictic populations, or groups of populations, have been grouped together and recognized as subspecies; those taxa are often difficult to identify and further study is clearly needed. Despite these difficulties, the subspecific taxa of Babcock were tentatively included in the present study. The Old World species are mostly diploid ($n = 3, 4, 5,$ or 6). Babcock concluded that there was a progressive decrease in the chromosome numbers, from $n = 6$ to $n = 3$. Along with the decrease is a corresponding increase in chromosome asymmetry and reduction in chromosome length.

SELECTED REFERENCE Babcock, E. B. 1947. The genus *Crepis*. Pt. 1: The taxonomy, phylogeny, distribution, and evolution of *Crepis*. Pt. II: Systematic treatment. Univ. Calif. Publ. Bot. 21, 22.

1. Annuals or biennials (perennials; taproots usually shallow).
    2. Stems branched (dichotomously, heads sessile in axils); phyllaries lanceolate (inner becoming indurate, often enclosing and partially fused to cypselae) . . . . 24. *Crepis zacintha*
    2. Stems branched (not dichotomously; phyllaries lanceolate to lance-linear (free from cypselae).
        3. Cypselae not beaked (apices sometimes ± narrowed).
            4. Stems (at least proximally) hispid and stipitate-glandular (viscid) . . . . 18. *Crepis pulchra*
            4. Stems glabrate, glabrescent, hispid, ± setose, tomentose, or tomentulose (not viscid).
                5. Adaxial faces of phyllaries ± appressed-hairy (hairs white, shiny, 0.1–0.2+ mm).
                    6. Annuals; abaxial faces of phyllaries tomentose to hispidulous; cypselae reddish or purplish brown, 3–4 mm; pappi 4–5 mm . . . . 22. *Crepis tectorum*
                    6. Biennials; abaxial faces of phyllaries ± canescent-tomentose; cypselae yellowish or reddish brown, 4–7 mm; pappi 5–7 mm . . . . 5. *Crepis biennis*
                5. Adaxial faces of phyllaries glabrous.
                    7. Involucres 5–8 mm; phyllaries glandular-setose (setae black, in 2 rows); cypselae 1.5–2.5 mm . . . . 7. *Crepis capillaris*
                    7. Involucres 8–10 mm; phyllaries glabrous or glabrate; cypselae 2.5–4 mm . . . . 14. *Crepis nicaeënsis*
        3. Cypselae (at least inner) beaked.
            8. Cypselae dimorphic.
                9. Stems scapiform; heads 1–2 (borne singly); corollas pink or white . . . . 19. *Crepis rubra*
                9. Stems branched; heads 3–10+; corollas mostly yellow, usually reddish purple abaxially . . . . 9. *Crepis foetida*
            8. Cypselae usually monomorphic.
                10. Stems coarsely setose or hispid (setae yellowish); calyculi of 10–14 bractlets (not reflexed); cypselae reddish brown, beaks 1–2 mm . . . . 21. *Crepis setosa*
                10. Stems glabrate or hispid and/or tomentose, sometimes sparsely setose (setae black); calyculi of 5–12 bractlets (reflexed); cypselae pale brown or yellowish, beaks 2–5 mm . . . . 23. *Crepis vesicaria*

1. Perennials (taproots and caudices becoming woody).
    11. Plants glabrous.
        12. Stems arcuate or decumbent, scapiform; heads 2–3; cypselae beaked (beak lengths nearly 2 times bodies) . . . . 6. *Crepis bursifolia*
        12. Stems ± erect or ascending, usually branched; heads 5–10(–100); cypselae seldom beaked (beaks relatively short).
            13. Stems in dense clumps (plants often rhizomatous), simple or branched proximally; leaves 2–9 × 0.5–2.5 cm; involucres 8–13 mm; cypselae subcylindric to fusiform, apices sometimes tapered or narrowed, not beaked, ribs 10–13, broad, smooth; alpine habitats . . . . 13. *Crepis nana*
            13. Stems in loose clumps (plants taprooted, roots vertical), branched dichotomously distally; leaves 1–4 × 0.5–1.5 cm; involucres 8–10 mm; cypselae fusiform, apices beaked (beaks 1–2 mm), ribs 10, narrow, minutely spiculate-roughened; stream banks, gravel bars . . . . 8. *Crepis elegans*
    11. Plants usually ± hairy, sometimes glabrous.
        14. Leaves usually entire or weakly dentate, sometimes closely dentate, serrate, or pinnately lobed.
            15. Stems scapiform; leaves mostly basal (rosettes), cauline leaves reduced; involucres turbinate-campanulate, 10–12 × 8–12 mm . . . . 20. *Crepis runcinata*
            15. Stems branched distally; leaves mostly cauline (blades broadly oblanceolate to elliptic); involucres cylindro-campanulate, 10–15 × 6–12 mm . . . . 16. *Crepis pannonica*
        14. Leaves usually pinnately lobed or sharply serrate.
            16. Stems usually densely setose, stipitate-glandular (setae 1–3 mm) . . . . 12. *Crepis monticola*
            16. Stems usually tomentose or tomentulose, sometimes glabrate or bristly-setose (setae or hairs to 1 mm).
                17. Phyllaries tomentose to tomentulose and/or setose (setae blackish, green, or whitish); cypselae dark to olive, greenish, or reddish brown, yellowish, or blackish, weakly ribbed or striate.
                    18. Plants 5–35 cm; heads 1–9; involucres 11–21 × 5–10 mm; phyllaries densely tomentose or setose . . . . 11. *Crepis modocensis*
                    18. Plants 20–80 cm; heads 15–20+; involucres 9–17 × 4–7 mm; phyllaries tomentulose and coarsely green-setose . . . . 4. *Crepis barbigera*
                17. Phyllaries usually glabrous, tomentose or tomentulose, sometimes stipitate-glandular or sparsely setose (setae black); cypselae yellowish or reddish brown or dark to blackish green, distinctly ribbed.
                    19. Phyllaries 5–8; florets 5–10(–15).
                        20. Heads 7–10(–30) in corymbiform arrays; phyllaries densely tomentulose near margins (strongly keeled, medians usually glabrous); cypselae reddish brown . . . . 17. *Crepis pleurocarpa*
                        20. Heads 30–70(–100+) in compound, corymbiform arrays; phyllaries usually glabrous, sometimes evenly tomentose (not strongly keeled); cypselae yellowish or brown . . . . 1. *Crepis acuminata*
                    19. Phyllaries 7–18; florets 6–40.
                        21. Leaf lobes narrowly lanceolate or linear; cypselae dark or blackish green, apices tapered, not beaked . . . . 2. *Crepis atribarba*
                        21. Leaf lobes deltate or broadly lanceolate; cypselae yellowish or brownish, apices narrowed to strongly tapered.
                            22. Plants 25–60 cm; heads (10–)20–60, in ± flat-topped, compound, corymbiform arrays; involucres narrowly cylindric, 3–5 mm diam.; florets 7–12 . . . . 10. *Crepis intermedia*
                            22. Plants 8–40 cm; heads (1–)2–22, in corymbiform, cymiform, or paniculiform arrays; involucres cylindric, 5–15 mm diam.; florets 9–40.

[23. Shifted to left margin.—Ed.]

23. Stems hispid, sometimes stipitate-glandular distally; leaves: faces gray-tomentose; phyllaries sometimes stipitate-glandular .... 15. *Crepis occidentalis*
23. Stems sparsely to densely tomentose, often stipitate-glandular proximally; leaves: faces sparsely to densely tomentose, stipitate-glandular (midribs red in fresh specimens): phyllaries conspicuously stipitate-glandular .... 3. *Crepis bakeri*

1. Crepis acuminata Nuttall, Trans. Amer. Philos. Soc., n. s. 7: 437. 1841 • Longleaf or tapertip hawksbeard E F

Crepis acuminata subsp. *pluriflora* Babcock & Stebbins; *C. angustata* Rydberg; *C. seselifolia* Rydberg

Perennials, 20–65 cm (taproots deep, woody, caudices swollen, branched, often covered by old leaf bases). **Stems** 1–5, erect, stout, branched near or beyond middles, tomentulose (at least proximally). **Leaves** basal and cauline; petiolate; blades elliptic to lanceolate, 8–40 × 0.5–6(–11) cm, margins deeply pinnately lobed, lobes 5–10 pairs, usually lobed (± halfway to midveins, lobes entire), apices long-acuminate, faces ± tomentulose. **Heads** 30–70(–100+), in compound, corymbiform arrays. **Calyculi** of 5–7, triangular, tomentulose bractlets 1–2 mm. **Involucres** cylindro-campanulate, 8–16 × 2–3 mm. **Phyllaries** 5–8, (medially green) lanceolate, 8–12 mm, (margins yellowish, often scarious), apices acute (ciliate), abaxial faces usually glabrous, sometimes sparsely tomentulose, adaxial glabrous. **Florets** 5–10(–15); corollas yellow, 10–18 mm. **Cypselae** pale yellowish brown, subcylindric, 6–9 mm, apices ± narrowed (not beaked), ribs 12; **pappi** white, 6–9 mm. $2n = 22, 33, 44, 55, 88$.

Flowering May–Aug. Dry rocky hillsides, ridges, grassy flats, open pine woods; 1000–3300 m; Ariz., Calif., Colo., Idaho, Iowa, Mont., Nebr., Nev., N.Mex., Oreg., Utah, Wash., Wyo.

Crepis acuminata is identified by the narrow, pinnately lobed leaves cleft about half way to the midrib and with long-acuminate apices, heads with relatively few florets, relatively small involucres, and glabrous phyllaries. The fertile diploid form of this species is most widespread (E. B. Babcock 1947). In addition, there are apomictic, polyploid populations. The latter often are more variable in leaf size, shape, and indument, and can be difficult to distinguish from *C. pleurocarpa* and *C. intermedia*.

2. Crepis atribarba A. Heller, Bull. Torrey Bot. Club 26: 314. 1899 (as atrabarba) • Slender or dark hawksbeard E

Crepis exilis Osterhout; *C. exilis* subsp. *originalis* Babcock & Stebbins; *C. occidentalis* Nuttall var. *gracilis* D. C. Eaton

Perennials, 15–70 cm (taproots slender, caudices swollen, often covered by old leaf bases). **Stems** 1–2, erect, slender, usually branched distal to middles, glabrous or tomentulose. **Leaves** basal and cauline; petiolate; blades lanceolate to linear, 10–35 × 0.5–6 cm, margins deeply pinnately lobed (lobes narrowly lanceolate or linear, usually entire or toothed), apices acuminate, faces tomentulose to glabrate. **Heads** 3–30, in corymbiform arrays. **Calyculi** of 5–10, narrowly triangular to lanceolate, tomentose bractlets 1–3 mm. **Involucres** cylindro-campanulate, 10–12 × 4–7 mm. **Phyllaries** 8–13, lanceolate, 10–12 mm (margins yellow, scarious, eciliate), apices acute, abaxial faces usually tomentulose, sometimes glabrous, often with coarse, green or blackish setae, adaxial glabrous or with fine, appressed hairs. **Florets** 6–35; corollas yellow, 10–18 mm. **Cypselae** dark or blackish green, subcylindric, 3–10 mm, apices tapered, not beaked, ribs 12–15 (distinct); **pappi** whitish, 5–9 mm. $2n = 22, 33, 44, 55, 88$.

Flowering May–Jul. Dry, open, grassy places, sagebrush slopes, pine forests, gravelly stream banks; 200–3000 m; Alta., B.C., Sask.; Colo., Idaho, Mont., Nebr., Nev., Oreg., Utah, Wash., Wyo.

Crepis atribarba is generally recognized by the deeply pinnately lobed leaves with linear lobes, fine tomentulose indument on stems and leaves, setose phyllaries, and dark green, strongly ribbed cypselae. It is a variable mixture that includes polyploid, apomictic forms and hybrids with *C. acuminata* and other species. The typical form is recognized by its short stature, narrow pinnately lobed, tomentulose leaves, stems with 3–10 heads, and phyllaries with scattered, black, eglandular setae. Larger, more robust forms with stems 30–70 cm, 10–30+ heads, narrower involucres, and few or no black setae have been recognized as subsp. *originalis*. The latter was considered by E. B. Babcock (1947) to represent the original diploid form of the species; it is difficult to distinguish in practice.

3. Crepis bakeri Greene, Erythea 3: 73. 1895 [E]

Perennials, 10–30 cm (taproots thick, caudices swollen, often covered by old leaf bases). **Stems** 1–3, erect (often reddish), stout, mostly simple, sparsely to densely tomentose, often stipitate-gland-ular proximally. **Leaves** basal and cauline; petiolate (at least basal); blades elliptic, runcinate, 8–20 × 2–5 cm, margins pinnately lobed (lobes broadly lanceolate, coarsely dentate, midribs often reddish), apices acute, faces sparsely to densely tomentose, stipitate-glandular. **Heads** 2–22 (1–3 per branch), in cymiform arrays. **Calyculi** of 8–10, deltate or lanceolate, tomentose bractlets 3–8 mm. **Involucres** cylindric, 11–21 × 5–15 mm. **Phyllaries** 10–14, lanceolate, 10–14 mm (margins yellowish), apices acute, abaxial faces glabrous or ± tomentose, sometimes setose and stipitate-glandular, adaxial glabrous or with fine hairs. **Florets** 11–40; corollas yellow, 16–20 mm. **Cypselae** dark or pale brown to yellowish, fusiform, 6–11 mm, apices ± tapered, ribs 10–13; **pappi** whitish, 6–13 mm. $2n = 22, 33, 44, 55$.

Subspecies 3 (3 in the flora): w United States.

Crepis bakeri is generally recognized by the low stature, dense rosettes of pinnately lobed leaves with coarsely dentate lobes, tomentose stems and leaves, stipitate-glandular hairs distally on stems, relatively large involucres, and densely flowered heads. It is considered closely related to *C. occidentalis*. Three somewhat weakly defined subspecies were recognized by E. B. Babcock (1947).

1. Involucres narrowly cylindric or turbinate, 18–21 mm in fruit; calyculus bractlets deltate (longest much shorter than phyllaries); pappi longer than cypselae . . . . 3c. *Crepis bakeri* subsp. *idahoensis*
1. Involucres broadly cylindric, 13–20 mm in fruit; calyculus bractlets lanceolate (longest ± ¹/₂ lengths of phyllaries); pappi ± equal to or shorter than cypselae.
  2. Involucres 16–20 mm in fruit; cypselae 8–10.5 mm, apices somewhat narrow, not strongly tapered; pappi 9–10.5 mm . . . . 3a. *Crepis bakeri* subsp. *bakeri*
  2. Involucres 13–17 mm in fruit; cypselae 6–9 mm, apices strongly tapered; pappi 6–9 mm . . . . 3b. *Crepis bakeri* subsp. *cusickii*

3a. Crepis bakeri Greene subsp. **bakeri** • Baker's hawksbeard [E]

Plants 10–30 cm. **Leaves** 8–12 × 3–4 cm, deeply lobed, lobes lanceolate to elliptic, faces canescent-tomentulose. **Heads** 2–13. **Calyculi**: bractlets lanceolate (longest ± ¹/₂ phyllaries). **Involucres** broadly cylindric, 16–20 mm in fruit. **Cypselae** dark brown, 8–10.5 mm, apices slightly tapered; **pappi** 9–10.5 mm. $2n = 44$.

Flowering May–Jul. Dry open slopes; 500–1900 m; Calif., Nev., Oreg., Wash.

3b. Crepis bakeri Greene subsp. **cusickii** (Eastwood) Babcock & Stebbins, Publ. Carnegie Inst. Wash. 504: 140. 1938 • Cusick's hawksbeard [E]

Crepis cusickii Eastwood, Bull. Torrey Bot. Club 30: 502. 1903

Plants 8–16 cm. **Leaves** 8–12 × 2–2.5 cm, deeply lobed, lobes triangular, dentate, faces tomentose. **Heads** 1–4(–10). **Calyculi**: bractlets lanceolate (longest ± ¹/₂ phyllaries). **Involucres** broadly cylindric, 13–17 mm in fruit. **Cypselae** dark brown, 6–9(–10) mm, narrowed and strongly tapered at apices; **pappi** 6–9 mm. $2n = 22, 33$.

Flowering Jun–Jul. Dry open places, sagebrush scrub; 1200–2200 m; Calif., Oreg., Utah.

Subspecies *cusickii* is usually smaller than subsp. *bakeri*.

3c. Crepis bakeri Greene subsp. **idahoensis** Babcock & Stebbins, Publ. Carnegie Inst. Wash. 504: 141, fig. 22o–q. 1938 • Idaho hawksbeard [C][E]

Plants 25–30 cm. **Leaves** 15–18 × 5–5.5 cm, shallowly lobed, lobes deltate, sharply dentate, faces glabrate. **Heads** 7–22. **Calyculi**: bractlets deltate (longest much shorter than phyllaries). **Involucres** narrowly cylindric or turbinate, 18–21 mm in fruit. **Cypselae** reddish brown, 8 mm, apices narrow, not strongly tapered; **pappi** 12–13 mm. $2n = 55$.

Flowering May–Jul. Dry open places; of conservation concern; 400–2200 m; Calif., Idaho.

Plants of subsp. *idahoensis* are generally larger and more robust than the other subspecies, with more heads per stem. They are possibly allopolyploids, with *Crepis occidentalis* or *C. monticola* in their lineage (E. B. Babcock 1947).

4. Crepis barbigera Leiberg ex Coville, Contr. U.S. Natl. Herb. 3: 565, plate 26. 1896 [E]

Perennials, 20–80 cm (taproots slender, caudices swollen). Stems 1–3(–5), erect, branched (branches strict with relatively few secondary branches), sparsely to densely tomentulose. Leaves basal and cauline; petiolate (petiole bases broadened, clasping); blades elliptic-lanceolate, 10–40 × 2–7 cm, margins deeply pinnately lobed (lobes mostly lanceolate or falcate, usually entire, rarely with 1–2 teeth), apices attentuate, faces tomentose or glabrate, sometimes setose. Heads 15–20+, in congested, corymbiform arrays. Calyculi of 5–7, lanceolate, greensetose bractlets 2–5 mm. Involucres cylindric, 9–17 × 4–7 mm. Phyllaries 6–10, lanceolate, 12–15 mm (margins yellowish, scarious), apices acute, abaxial faces tomentulose and coarsely green-setose, adaxial glabrous or with fine hairs. Florets 8–25; corollas yellow, 18–20 mm. Cypselae dark brown to olive, subcylindric, 8–9 mm, apices tapered, ribs 10–12 (strong); pappi whitish or yellowish white, 6–9 mm. $2n = 44, 55, 88$.

Flowering May–Jul. Open rocky places, sandy slopes, dry pine-oak woods, sagebrush slopes, foothills and plains; 100–2000 m; Idaho, Oreg., Wash.

Crepis barbigera is recognized by its relatively tall stature, deeply pinnately lobed leaves, tomentulose stems, and phyllaries with coarse, green, eglandular setae. It is a complex of polyploid, apomictic forms, combining characteristics of *C. atribarba*, *C. acuminata*, and *C. modocensis*, from which the species is presumed to have been derived by intercrossing (E. B. Babcock 1947).

5. Crepis biennis Linnaeus, Sp. Pl. 2: 807. 1753
· Rough hawksbeard [I]

Biennials, 20–120 cm (taproots branched). Stems 1, erect, slender to robust, branched proximally or near middles, ± setulose or glabrescent. Leaves basal and cauline; petiolate (at least basal); blades oblanceolate to runcinate, 5–25 × 1.5–7.5 cm, margins pinnately lobed or dentate (terminal lobes triangular), apices ± acute, faces slightly scabrous (hairs yellow, fine). Heads 12–14, in simple or compound, corymbiform arrays. Calyculi of 7–9, lance-linear, glabrous or tomentulose bractlets 3–6 mm.

Involucres campanulate, 8–13 × 5–9 mm. Phyllaries 10–17 (pale to dark green or nearly black) linear-lanceolate, 10–11 mm, (margins scarious) apices acute to obtuse (ciliate), abaxial faces ± canescent-tomentose, adaxial often with yellowish or black, appressed hairs. Florets 30–100; corollas yellow, 12–18 mm. Cypselae yellowish or reddish brown, fusiform, 4–7 mm, apices narrowed (not beaked), ribs 13–20; pappi white, 5–7 mm (somewhat unequal). $2n = 40$.

Flowering Jun–Aug. Meadows and fields; 500–1200 m; introduced; Nfld. and Labr. (Nfld.); Mich., N.Y., Ohio, Pa., Va., Vt.; Europe.

Crepis biennis is recognized by its biennial habit, pinnately lobed leaves with triangular lobes, and relatively short hairs on the adaxial faces of the phyllaries. It has been reported from Newfoundland but apparently does not persist there.

6. Crepis bursifolia Linnaeus, Sp. Pl. 2: 805. 1753
· Italian hawksbit [I]

Perennials, 5–35 cm (taproots stout, caudices covered by old leaf bases). Stems 2–9+, arcuate or decumbent, slender, scapiform, cymosely branched distally, glabrous. Leaves mostly basal; petiolate; blades oblanceolate, lyrate, 2.5–25 × 0.5–6 cm, margins pinnately lobed (lateral lobes lanceolate, dentate, acute, terminal lobes usually larger), apices obtuse or acute, faces glabrous. Heads 2–3 (peduncles slender), in cymiform arrays. Calyculi of 10–14, (lax) linear, tomentose or glandular-pubescent bractlets 2–5 mm. Involucres cylindric, 8–11 × 3–4 mm. Phyllaries 10–12 (reflexed, medially yellowish) lanceolate (bases keeled, margins dark greenish, sometimes scarious), apices acute (ciliate), abaxial faces tomentulose, adaxial with fine hairs. Florets 30–60; corollas light yellow, greenish abaxially, 10–11 mm. Cypselae pale brown, fusiform, 6–7 mm, beaked, beaks pale (lengths nearly 2 times bodies), ribs 10; pappi white, 3–4 mm. $2n = 8$.

Flowering Apr–Sep. Waste places, lawns; 0–100 m; introduced; Calif.; Europe.

Crepis bursifolia is identified by the dense basal rosettes of glabrous, lyrate leaves with dentate lateral lobes and relatively large terminal lobes, relatively few heads on slender peduncles, and cypselae with relatively thin beaks two times lengths of the bodies. It is an aggressive weed.

7. Crepis capillaris (Linnaeus) Wallroth, Linnaea 14: 657. 1840 • Smooth hawksbeard, crépis capillaire I

Lapsana capillaris Linnaeus, Sp. Pl. 2: 812. 1753; *Crepis cooperi* A. Gray; *C. virens* Linnaeus

Annuals or biennials, 10–90 cm (taproots shallow). **Stems** 1(–6+), erect to ± procumbent, usually simple (usually with single stout leader, sometimes multiple with slender laterals), hispid proximally or throughout. **Leaves:** basal and cauline; petiolate (petiole bases clasping); blades lanceolate or oblanceolate, runcinate or lyrate, 5–30 × 1–4.5 cm, margins pinnately divided to sharply dentate (lobes remote, unequal), apices obtuse or acute, mucronate, faces glabrous or sparsely hispid (hairs yellow; proximal cauline auriculate and clasping). **Heads** 10–15(–30+), in corymbiform arrays. **Calyculi** of 8, linear, tomentulose or stipitate-glandular bractlets 2–4 mm. **Involucres** cylindric to turbinate, 5–8 × 3–6 mm. **Phyllaries** 8–16, lanceolate, 6–7 mm (margins scarious), apices acute, abaxial faces stipitate-glandular and glandular setose (setae black, usually in 2 rows), adaxial glabrous. **Florets** 20–60; corollas deep yellow (reddish abaxially), 8–12 mm (hairy). **Cypselae** brownish yellow, fusiform, 1.5–2.5 mm, apices narrowed (not beaked), ribs 10 (glabrous or scabrous); **pappi** white (fluffy), 3–4 mm (scarcely surpassing phyllaries). $2n = 6$.

Flowering May–Nov. Meadows, pastures, lawns, roadsides, fields, waste places; 0–1300 m; introduced; Alta., B.C., N.B., N.S., Ont., Que.; Alaska, Ark., Calif., Colo., Conn., Del., D.C., Idaho, Ill., Ind., Iowa, Ky., Maine, Md., Mass., Mich., Miss., Mo., Mont., Nev., N.H., N.J., N.Y., N.C., N.Dak., Ohio, Oreg., Pa., R.I., Tenn., Tex., Utah, Vt., Va., Wash., W.Va., Wis.; Europe.

Crepis capillaris is recognized by its shallow root system, dense rosettes of coarsely dentate or pinnately lobed leaves, erect slender stems, auriculate-based cauline leaves, relatively small heads, phyllaries with double rows of black setae, and fluffy white pappi. It is weedy and can become a serious lawn pest. It is one of only three species of *Crepis* with $2n = 6$; E. B. Babcock (1947) considered it to be advanced in the genus.

8. Crepis elegans Hooker, Fl. Bor.-Amer. 1: 297. 1833 • Elegant hawksbeard E

Perennials, 6–30 cm (taproots deep, caudices stout). **Stems** 5–20, erect or ascending (often reddish brown, in loose clumps), ± dichotomously branched distally, glabrous. **Leaves** basal and cauline; petiolate (petiole bases clasping); blades spatulate or elliptic to ovate, 1–4 × 0.5–1.5 cm, margins coarsely dentate or entire, apices acute, faces glabrous. **Heads** 10–100+, in dense paniculiform arrays. **Calyculi** of 7–8 (blackish green), ovate, glabrous bractlets 1–2 mm. **Involucres** cylindric, 8–10 × 2–3 mm. **Phyllaries** 8–10, (blackish green) oblong, 8–10 mm, (margins scarious) apices acute or obtuse, faces glabrous. **Florets** 6–10; corollas yellow, 6–8 mm. **Cypselae** golden brown, fusiform (subterete or flattened), 4–5 mm, beaked (beaks delicate), ribs 10 (narrow, spiculate-roughened); **pappi** white, 4–5 mm. $2n = 14$.

Flowering Jun–Sep. Stream banks, gravelly flats, sandbars, roadsides; 1300–2000 m; Alta., B.C., N.W.T., Ont., Yukon; Alaska, Mont., Wyo.

Crepis elegans is recognized by its loose, cespitose habit, relatively small spatulate leaves, blackish green, glabrous phyllaries, and beaked cypselae. It is thought to be closely related to *C. nana*, and, possibly, derived from it (E. B. Babcock 1947).

9. Crepis foetida Linnaeus, Sp. Pl. 2: 807. 1753 • Stinking or roadside hawksbeard I

Annual, biennials, or perennials, 10–50 cm (roots fibrous, shallow). **Stems** 1(–3+), erect to decumbent or prostrate, branched proximally or distally, hispid and/or setose. **Leaves** basal and cauline; petiolate; blades oblanceolate, runcinate, 3–13 × 1–3 cm, margins denticulate to pinnately lobed (lobes deltate to lanceolate, often sharply serrate, terminal relatively large), apices acute, faces hispid to villous (cauline sessile, blades ovate to lanceolate or linear, runcinate, bases auriculate, margins deeply pinnately lobed, lobes linear). **Heads** 3–10+, in cymiform arrays. **Calyculi** of 8–10, linear to lanceolate, densely hispid bractlets 2–5 mm (becoming lax). **Involucres** cylindric to turbinate, 7–16 × 4–13 mm. **Phyllaries** 8–12, lanceolate (bases strongly keeled, enclosing marginal cypselae, margins green), apices acute to attenuate, abaxial faces hispid or setose, adaxial with fine hairs. **Florets** 80–100+; corollas yellow (usually reddish purple adaxially), 9–16 mm. **Cypselae** (dimorphic) subcylindric, outer stout, 7–9 mm, nearly beakless, inner 12–17 mm, beaks 2–5 mm; **pappi** dull white, 3–7 mm. $2n = 10$.

Flowering Apr–Sep. Seashores, plains, hills, and mountains; 80–1200 m; introduced; Fla., Ga., Mass., N.C., Wis.; Eurasia.

Crepis foetida is polymorphic; it is recognized by its annual or biennial habit, usually erect and hispid or setose stems, sharply runcinate leaves, hispid or setose involucres, and dimorphic cypselae.

10. Crepis intermedia A. Gray in A. Gray et al., Syn. Fl. N. Amer. 1(2): 432. 1884 • Limestone or small-flower hawksbeard E

Crepis acuminata Nuttall var. *intermedia* (A. Gray) Jepson

Perennials, 25–60 cm (taproots stout or slender, caudices swollen, simple or branched, covered with brown leaf bases). **Stems** 1–2, erect, branched (proximal branches elongate, branched distally), ± tomentose-canescent. **Leaves** basal and cauline; petiolate (petiole bases clasping); blades elliptic-lanceolate, 10–40 × 2–9 cm, margins pinnately lobed (lobes remote or close, entire or dentate), apices acute or acuminate, faces densely or sparsely gray-tomentose. **Heads** (10–)20–60, in ± flat-topped, compound, corymbiform or paniculiform arrays. **Calyculi** of 6–8, narrowly triangular, tomentulose bractlets 2–4 mm. **Involucres** narrowly cylindric, 10–16 × 3–5 mm. **Phyllaries** 7–10, (medially green) lanceolate, 10–13 mm (margins scarious), apices acute, abaxial faces ± tomentulose, sometimes with greenish eglandular setae, adaxial with fine hairs. **Florets** 7–12; corollas yellow, 14–30 mm. **Cypselae** yellow or golden brown, subcylindric, 6–9 mm, tapered distally, ribs 10–12 (smooth); **pappi** dusky white, 7–10 mm. $2n = 33, 44, 55, 88$.

Flowering May–Jul. Open rocky ridges, dry slopes, open forests; 800–3900 m; Alta., B.C., Sask.; Ariz., Calif., Colo., Idaho, Mont., Nev., N.Mex., Oreg., Utah, Wash., Wyo.

Crepis intermedia is a somewhat unnatural group of polyploid apomicts that combines the features of multiple species, including *C. acuminata*, *C. pleurocarpa*, *C. modocensis*, and *C. atribarba* (E. B. Babcock 1947). The plants are usually over 25 cm, with leaves deeply pinnately lobed (gray-tomentose, cleft about halfway to midribs), with acuminate apices, and more or less flat-topped arrays of heads. The leaves vary greatly in size and lobing and are always gray-tomentose. The number of heads per plant is usually more than 20.

11. Crepis modocensis Greene, Erythea 3: 48. 1895 • Modoc or Siskiyou hawksbeard E

Perennials, 5–35 cm (taproots slender, caudices branched). **Stems** 1–4, erect, slender to stout, simple or sparsely branched, glabrate to tomentose and bristly-setose. **Leaves** basal and cauline; petiolate; basal blades lanceolate, 7–25 × 2–4 cm, margins deeply pinnately lobed (lobes lanceolate, dentate, teeth mucronate), apices acuminate, faces tomentulose (at least when young). **Heads** 1–9, borne singly or 2–9 in cymiform arrays. **Calyculi** of 8–10, lanceolate, tomentose and often setose bractlets 2–4 mm. **Involucres** cylindric, 11–21 × 5–10 mm. **Phyllaries** 8–18, (medially green) lanceolate, 10–16 mm, (bases keeled, margins yellowish, often scarious), apices acute, abaxial faces often densely, blackish or whitish tomentose or setose, sometimes glabrous, adaxial with fine (shiny) hairs. **Florets** 10–60; corollas yellow, 13–22 mm. **Cypselae** blackish or greenish, reddish, reddish brown, or yellowish, subcylindric to fusiform, 7–12 mm, apices tapered or beaked (beaks 1–3 mm), ribs 10 (strong to weak); **pappi** dusky white, 5–13 mm. $2n = 22, 33, 44, 55, 66, 88$.

Subspecies 4 (4 in the flora): w North America.

Crepis modocensis is recognized by its tomentose or coarsely bristly stems and petioles, rosettes of deeply pinnately lobed leaves, rather large heads with relatively many phyllaries, and blackish cypselae.

1. Setae of stems and petioles yellowish, ± straight, (those of phyllaries blackish or 0); cypselae tapered, not distinctly beaked.
   2. Stems branching near middle; involucres 11–16 mm; pappi 5–10 mm . . . . 11a. *Crepis modocensis* subsp. *modocensis*
   2. Stems low, branching proximally; involucres 13–21 mm; pappi 9–13 mm . . . . 11d. *Crepis modocensis* subsp. *subacaulis*
1. Setae of stems and petioles (and phyllaries) whitish, conspicuously curled; cypselae beaked (beaks 1–3 mm).
   3. Plants 15–30 cm; involucres 12–17 mm; pappi 7–10 mm . . . 11c. *Crepis modocensis* subsp. *rostrata*
   3. Plants 6–20 cm; involucres 11–13 mm; pappi 5–7 mm . . . . . . 11b. *Crepis modocensis* subsp. *glareosa*

11a. Crepis modocensis Greene subsp. **modocensis**
  • Modoc or Siskiyou hawksbeard E

Crepis scopulorum Coville

Plants 10–35 cm; setae yellowish, ± straight. **Stems** (slender, always with primary axes, branched near or beyond middles) sparsely setose (setae yellowish). **Leaves:** petioles setose; blades (narrowly elliptic), 10–20 × 2–4 cm. **Involucres** 11–16 mm. **Phyllaries** sparsely setose (setae blackish). **Cypselae** deep greenish to blackish or deep reddish brown, 7–12 mm, tapered distally, not distinctly beaked; **pappi** 5–10 mm. **2n** = 22

Flowering May–Jul. Dry open ridges, rocky slopes; 900–2500 m; B.C.; Calif., Colo., Idaho, Mont., Nev., Oreg., Utah, Wash., Wyo.

Subspecies *modocensis* includes a diploid form and apomictic polyploid forms (E. B. Babcock 1947).

11b. Crepis modocensis Greene subsp. **glareosa**
  (Piper) Babcock & Stebbins, Publ. Carnegie Inst.
  Wash. 504: 154. 1938 C E

Crepis glareosa Piper, Bull. Torrey Bot. Club 28: 42. 1901

Plants 6–20 cm; setae whitish, conspicuously curled (1–2 mm). **Stems** (stout, branched at bases or proximal to middles), canescent-tomentulose, sparsely setose. **Leaves:** petioles setose; blades 4–12 × 1.5–4 cm (tomentulose to glabrate, midribs setose). **Involucres** 11–13 mm. **Phyllaries** densely setose or tomentose at bases. **Cypselae** greenish or yellowish, 6–7 mm, beaked (beaks 2–3 mm); **pappi** 5–7 mm.

Flowering May–Jul. Dry open places, alpine slopes; of conservation concern; 1500–2500 m; Wash.

Subspecies *glareosa* is known only from Kittitas County.

11c. Crepis modocensis Greene subsp. **rostrata**
  (Coville) Babcock & Stebbins, Publ. Carnegie Inst.
  Wash. 504: 152. 1938 E

Crepis rostrata Coville, Contr. U.S. Natl. Herb. 3: 564, plate 25. 1896

Plants 15–30 cm; setae whitish, conspicuously curled or crisped (1–2 mm). **Stems** densely setose or tomentose. **Leaves:** petioles setose; blades 10–15 × 3–4 cm (tomentulose or glabrate, midribs setose). **Involucres** 12–17 mm. **Phyllaries** densely setose or tomentose. **Cypselae** greenish black, 7–10 mm, beaked (beaks 1–2 mm, ribs alternating strong and weak); **pappi** 7–10 mm. **2n** = 22, 33, 44.

Flowering May–Jul. Dry open places, rocky ridges; 1000–1100 m; B.C.; Wash.

Subspecies *rostrata* is distinguished mainly by the dense curly setae of the involucres and cypselae with relatively short, coarse beaks.

11d. Crepis modocensis Greene subsp. **subacaulis**
  (Kellogg) Babcock & Stebbins, Publ. Carnegie Inst.
  Wash. 504: 148. 1938 E

Crepis occidentalis Nuttall var. *subacaulis* Kellogg, Proc. Calif. Acad. Sci. 5: 50. 1873

Plants 6–30 cm; setae yellowish, usually ± straight, sometimes bent (ca. 1 mm). **Stems** (stout, branched proximally) setose (setae yellowish). **Leaves:** petioles setose; blades 10–15 × 3–4 cm (margins pinnately lobed, lobes dentate). **Heads** 1–6. **Involucres** 13–21 mm. **Phyllaries** sparsely setose (setae blackish) or glabrous. **Cypselae** blackish to brownish or dark reddish, slightly narrowed, not beaked (ribs strong); **pappi** 9–13 mm.

Flowering May–Jul. Wet meadows, steep slopes; 1800–2100 m; Calif., Mont., Nev., Oreg.

12. Crepis monticola Coville, Contr. U.S. Natl. Herb. 3: 562, plate 22. 1896 • Mountain hawksbeard E

Crepis occidentalis Nuttall var. *crinita* A. Gray

Perennials, 10–35 cm (taproots vertical, caudices simple or branched). **Stems** 1–3, erect (reddish brown), stout, branched proximally, densely setose and stipitate-glandular (setae 1–3 mm). **Leaves** basal and cauline; petiolate; blades elliptic or oblanceolate, 10–25 × 2–4 cm, margins pinnately lobed or sharply serrate (lobes lanceolate, acuminate), apices acute, faces villous or coarsely setose, stipitate-glandular (cauline 2–3, bases clasping, margins dentate or serrate). **Heads** 2–20, in loose cymiform arrays. **Calyculi** of 3–10, narrowly lanceolate to linear, densely setose and stipitate-glandular bractlets 3–5 mm. **Involucres** cylindro-campanulate, 14–24 × 5–15 mm. **Phyllaries** 7–12, lanceolate, 14–20 mm (margins yellowish, not scarious), apices long-acuminate, abaxial faces densely and coarsely setose, adaxial glabrous or with fine, appressed, yellowish hairs. **Florets** 16–20; corollas yellow, 16–21 mm. **Cypselae** reddish brown, fusiform, 5.5–9 mm, apices narrowed (not beaked), ribs 13; **pappi** creamy white, 9–13 mm (outer bristles shorter and finer). **2n** = 22, 33, 44, 55, 77, 88.

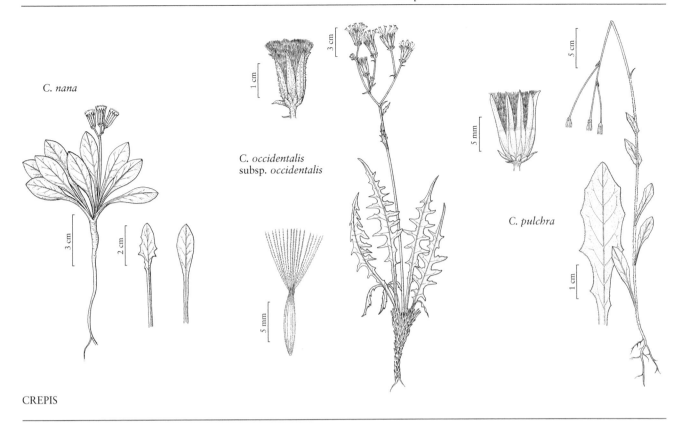

C. nana

C. occidentalis
subsp. occidentalis

C. pulchra

CREPIS

Flowering May–Jul. Coniferous forests, thickets, open woods, valleys and foothills, dry gravelly open areas; 700–2400 m; Calif., Oreg.

Crepis monticola is recognized by the densely stipitate-glandular stems and leaves, and long-acuminate phyllaries.

13. Crepis nana Richardson in J. Franklin et al., Narr. Journey Polar Sea, 746. 1823 • Dwarf alpine hawksbeard [F]

Crepis nana var. *lyratifolia* (Turczaninow) Hultén; *C. nana* subsp. *ramosa* Babcock

Perennials, 10–20 cm (taproots often with creeping rhizomes, caudices relatively short). **Stems** 1–10+, erect or ascending (in dense clumps), simple or proximally branched, glabrous. **Leaves** basal and cauline; petiolate (at least basal); blades (often purplish), orbiculate to spatulate, less often lyrate or runcinate, 2–9 × 0.5–2.5 cm, (bases abruptly 0) margins entire or pinnately lobed, apices obtuse to acute, faces glabrous (glaucous). **Heads** 5–80+ (among or beyond leaves), in cymiform arrays. **Calyculi** of 5–10 (dark green or blackish), lanceolate, glabrous bractlets 2–3 mm. **Involucres** cylindric, 8–13 × 3–4 mm. **Phyllaries** 8–10, (dark green or purple medially) oblong, 10–11 mm, (margins scarious, eciliate) apices acute, faces glabrous. **Florets** 9–12; corollas yellow, purple-tinged abaxially, 9–12 mm. **Cypselae** golden brown, subcylindric to fusiform, 4–7 mm, apices sometimes tapered (not beaked), ribs 10–13 (broad, smooth); **pappi** (falling) bright white, 4–6 mm. $2n = 14$.

Flowering May–Sep. Talus slopes, rocky alpine places, sandy stream banks, gravel bars, exposed sites in shrub communities; 300–4000 m; Alta., B.C., Nfld. and Labr., N.W.T., Nunavut, Yukon; Alaska, Calif., Colo., Idaho, Mont., Nev., Oreg., Utah, Wash., Wyo.; Asia (Russia).

Crepis nana occurs in North America and northern Asia. It is recognized by the tufted, cespitose habit, elongate roots and rhizomes, and occurrence in alpine habitats. In the typical form, the plants are tufted, the stems are not leafy, and the heads are borne among the leaves. Taller specimens with elongated, leafy branches and heads borne well beyond the basal leaves are sometimes recognized as subsp. *ramosa*; these characteristics appear to be part of the normal range of variation for the species.

Crepis nana is closely related to *C. elegans*, differing mainly in the shape of the cypselae. The cypselae of *C. nana* are almost always more columnar, wider at bases, and with broader ribs, than those of *C. elegans*.

The name *Crepis nana* subsp. *clivicola* Legge is invalid.

14. Crepis nicaeënsis Balbis ex Persoon, Syn. Pl. 2: 376. 1807 • Turkish hawksbeard [I]

Annuals or biennials, 25–110 cm (taproots shallow). **Stems** usually 1, erect, branched from middles or distally, densely hispidulous proximally. **Leaves** basal and cauline; petiolate (petioles winged, bases clasping); blades oblanceolate, runcinate, 16–19 × 2–4 cm, margins pinnately lobed, dentate, or finely denticulate, apices obtuse, faces finely hispidulous (hairs yellowish. **Heads** 2–15, in corymbiform arrays. **Calyculi** of 7–9, linear, glabrous or glabrate bractlets 3–6 mm. **Involucres** campanulate, 8–10 × 4–6 mm. **Phyllaries** 10–15, lanceolate (strongly tapered), 7–8 mm (bases strongly keeled, partly enclosing marginal cypselae, margins scarious), apices obtuse (ciliate), abaxial faces glabrous or glabrate, adaxial glabrous. **Florets** 50–60; corollas yellow (sometimes reddish distally), 10–11 mm. **Cypselae** golden brown, fusiform, 2.5–4 mm, ribs 10 (broad); **pappi** bright white, 4–5 mm. $2n = 8$.

Flowering May–Aug. Waste places, meadows; 1400–1600 m; introduced; B.C.; Mass., Mich., Ohio, Vt., Wash.; Europe.

Crepis nicaeënsis is distinguished by the annual or biennial habit, shallow root system, hispid stems, and glabrate phyllaries enclosing outer cypselae. It is similar in habit to *C. biennis*, which differs in its larger heads and 13–20-ribbed cypselae; it is considered closely related to *C. capillaris* (E. B. Babcock 1947).

15. Crepis occidentalis Nuttall, J. Acad. Nat. Sci. Philadelphia 7: 29. 1834 • Gray or western hawksbeard [E] [F]

Psilochenia occidentalis (Nuttall) Nuttall

Perennials, 8–40 cm; taproots deep, caudices swollen, (often covered with old leaf bases). **Stems** 1–3, erect, stout, branched from bases or beyond, hispid, tomentose, or tomentulose, sometimes stipitate-glandular distally. **Leaves** basal and cauline; petiolate; blades elliptic, runcinate, (5–)8–20 × 2–5 cm, margins pinnately-lobed to sinuously dentate (lobes broadly lanceolate, often dentate), apices acute or acuminate, faces gray-tomentose, sometimes stipitate-glandular. **Heads** 2–30, in loose corymbiform arrays. **Calyculi** of 6–8, lanceolate or linear, glabrate to tomentose bractlets 2–6 mm. **Involucres** cylindric, 11–19 × 5–10 mm. **Phyllaries** 7–13, lanceolate, 12–15 mm, (bases thickened, keeled, margins green, often scarious) apices acute or acuminate, abaxial faces gray-tomentose, sometimes setose (setae black or greenish) or stipitate-glandular, adaxial glabrous or with fine hairs. **Florets** 10–40; corollas yellow, 18–22 mm. **Cypselae** golden or dark brown, subcylindric, 6–10 mm, apices tapered (not beaked), ribs 10–18, strong and rounded; **pappi** yellowish white, 10–12 mm (bristles unequal). $2n = 22, 33, 44, 55, 66, 77, 88$.

Subspecies 4 (4 in the flora): w North America.

Crepis occidentalis is recognized by the old, brown leaf bases persisting on caudices, by stems, leaves, and phyllaries gray-tomentose, and by loose, corymbiform arrays with relatively few, relatively large heads. It is widespread and polymorphic. Some specimens have coarse setae or black, stipitate glands on the phyllaries in addition to the tomentose indument, the stipitate glands sometimes extending proximally on stems. Four intergrading subspecies were recognized by E. B. Babcock (1947). The sexual diploid forms are found in subsp. *occidentalis* and occur in northern California and adjacent Nevada. The other subspecies are polyploid and apomictic (Babcock).

1. Phyllaries sparsely to densely stipitate-glandular.
    2. Phyllaries (peduncles and distal cauline leaves) stipitate-glandular (lacking large dark or black glandular setae); phyllaries 7–8 or 10–13; florets 18–30 .... 15a. *Crepis occidentalis* subsp. *occidentalis*
    2. Phyllaries (peduncles and distal cauline leaves) stipitate-glandular (and with dark or black, glandular setae); phyllaries 8, florets 10–14 .... 15c. *Crepis occidentalis* subsp. *costata*
1. Phyllaries 8–12, usually eglandular, if glandular, phyllaries 8.
    3. Plants 10–40 cm (stems with definite primary axes, branched distally; phyllaries mostly 8; leaves coarsely dentate or pinnately lobed (lobes closely spaced) .... 15d. *Crepis occidentalis* subsp. *pumila*
    3. Plants 5–20 cm (stems branched proximally; phyllaries 8–12; leaves deeply pinnately lobed (lobes remotely spaced, lanceolate, or linear, entire or dentate .... 15b. *Crepis occidentalis* subsp. *conjuncta*

15a. Crepis occidentalis Nuttall subsp. **occidentalis** • Largeflower hawksbeard [E] [F]

Plants 10–40 cm. **Stems** stipitate-glandular. **Leaves** 10–20 × 2–4 cm, sharply dentate to pinnately lobed (lobes dentate; distal cauline leaves stipitate-glandular). **Heads** 10–30. **Peduncles** tomentulose, stipitate-glandular. **Phyllaries** 10–13, sparsely to densely stipitate-glandular (lacking dark or long black setae). **Florets** 18–30. **Cypselae** golden brown. $2n = 22, 33$.

Flowering May–Jun. Arid rocky hillsides, sagebrush scrub; 1000–2200 m; Alta., B.C., Sask.; Ariz., Calif., Colo., Idaho, Mont., Nev., N.Mex., Oreg., S.Dak., Utah, Wash., Wyo.

15b. Crepis occidentalis Nuttall subsp. **conjuncta** Babcock & Stebbins, Publ. Carnegie Inst. Wash. 504: 134, fig. 22a–e. 1938 [E]

Plants 5–20 cm. Stems (branched proximally) tomentose (not stipitate-glandular). Leaves 10–18 × 3–4 cm, deeply pinnately lobed (lobed remote, lanceolate or linear, entire or dentate). Heads 2–9. Peduncles tomentulose, eglandular. Phyllaries 8–12, tomentulose, eglandular, often glabrous distally. Florets 12–40. Cypselae dark brown.

Flowering Jun–Jul. Ridgetops, black shale hills, volcanic aggregate, gravelly soils; 1400–2100 m; Calif., Colo., Mont., Oreg., Wash., Wyo.

15c. Crepis occidentalis Nuttall subsp. **costata** (A. Gray) Babcock & Stebbins, Publ. Carnegie Inst. Wash. 504: 124. 1938 [E]

Crepis occidentalis var. *costata* A. Gray in W. H. Brewer et al., Bot. California 1: 435. 1876

Plants 8–40 cm. Stems tomentose, sometimes stipitate-glandular distally. Leaves 5–20 × 2–3.5 cm, pinnately lobed (lobes dentate; distal cauline stipitate-glandular and with large dark or black setae). Heads 15–30. Peduncles stipitate-glandular. Phyllaries 7–8, stipitate-glandular and with dark or black, glandular setae. Florets 10–14. Cypselae golden brown. $2n = 44$.

Flowering Jun–Jul. Grassy banks, dry rocky hillsides, black shale or sandstone, juniper-oak woods; 1200–2500 m; B.C., Sask.; Calif., Colo., Idaho, Mont., Nev., Oreg., Utah., Wash., Wyo.

The major distinguishing characteristic of subsp. *costata* appears to be the relatively large, dark or black setae on the phyllaries. As defined by E. B. Babcock (1947), subsp. *costata* is a series of polyploid apomictic forms that are difficult to distinguish from subsp. *occidentalis*, as well as other species such as *C. intermedia* and *C. bakeri*. Reports for Arizona and South Dakota were not confirmed for this treatment.

15d. Crepis occidentalis Nuttall subsp. **pumila** (Rydberg) Babcock & Stebbins, Publ. Carnegie Inst. Wash. 504: 128. 1938 [E]

Crepis pumila Rydberg, Mem. New York Bot. Gard. 1: 462. 1900

Plants 10–40 cm. Stems (with definite primary axes, branched distally) tomentulose, eglandular. Leaves 10–20 × 2–5 cm, coarsely dentate or pinnately lobed (lobes closely spaced. Heads 5–15. Peduncles tomentulose, stipitate-glandular. Phyllaries mostly 8, usually ± tomentose (at least proximally), rarely eglandular. Florets 12–20. Cypselae brown.

Flowering Jun–Jul. Serpentine slopes, open pine woods; 800–1800 m; B.C.; Calif., Idaho, Mont., Nev., Oreg., Utah, Wash.

16. Crepis pannonica (Jacquin) K. Koch, Linnaea 23: 689. 1851 [I]

Hieracium pannonicum Jacquin, Collectanea 5: 148. 1796

Perennials, 30–130 cm (taproots stout, caudices branched). Stems 1, erect, branched distally, setulose and sometimes glandular proximally. Leaves mostly cauline; petiolate or sessile; blades broadly oblanceolate to elliptic, 15–30 × 4–6 cm, margins closely toothed (teeth corneous apically and sometimes along margins), apices acute, faces scabrous, ± setulose and stipitate-glandular; (cauline sessile, ovate or elliptic, bases auriculate). Heads 1–8, borne singly or (on racemiform branches) in compound, paniculiform or corymbiform arrays. Calyculi of 10–12, lanceolate to linear, sparsely tomentulous bractlets 3–5 mm. Involucres cylindro-campanulate, 10–15 × 6–12 mm. Phyllaries 12–15, lanceolate, 9–10 mm, (bases strongly thickened, margins green), apices (dark) acute, abaxial faces tomentulose, sometimes with setae, adaxial glabrous. Florets 50–90; corollas yellow, 15–18 mm (sparsely tomentulose). Cypselae brown, fusiform, 5–6 mm, apices narrowed (not beaked), ribs 15–20; pappi white (soft), 5–8 mm. $2n = 8$.

Flowering Jun–Aug. Dry, open, grassy areas, pastures; 1000–2000 m; introduced; Conn.; Europe.

Crepis pannonica is recognized by its leafy, erect stems, broadly elliptic or obovate leaves with closely dentate margins, auriculate distal leaves, and tomentulose phyllaries.

17. Crepis pleurocarpa A. Gray, Proc. Amer. Acad. Arts 17: 221. 1882 • Naked hawksbeard [E]

Crepis acuminata Nuttall var. *pleurocarpa* (A. Gray) Jepson; *C. intermedia* A. Gray var. *pleurocarpa* (A. Gray) A. Gray

Perennials, 15–60 cm (taproots slender, caudices swollen). **Stems** 1–3, slender or stout, branched proximally (sparingly) or distally, glabrate to tomentulose. **Leaves** basal and cauline; petiolate (petioles relatively broadly winged); blades elliptic or oblanceolate, often runcinate, 7–28 × 0.5–7 cm, margins pinnately lobed to dentate (lobes remote, lanceolate to narrowly triangular, often recurved), apices attenuate, faces usually tomentulose to glabrate, sometimes glandular. **Heads** 7–10(–30), in corymbiform arrays. **Calyculi** of 5–6, deltate to lanceolate, tomentulose bractlets 1.5–4 mm. **Involucres** cylindro-campanulate, 8–16 × 3–5 mm. **Phyllaries** 5 (–10), (deep green or black) lanceolate, 10–16 mm, (bases becoming strongly keeled and swollen, often glabrous, margins yellowish, conspicuously and densely tomentulose), apices acute to strongly acuminate, abaxial faces densely tomentulose adaxial with fine hairs. **Florets** 5–10(–12); corollas yellow, 15–20 mm. **Cypselae** deep reddish brown, subcylindric, 5–8 mm, apices constricted, ribs 10 (prominent); **pappi** yellowish white, 6–12 mm (bristles unequal). $2n = 22, 33, 44, 55, 77, 88$.

Flowering Jun–Aug. Streams in mixed conifer forests, road cuts, steep rocky serpentine slopes; 400–2200 m; Calif., Idaho, Nev., Oreg., Wash.

Crepis pleurocarpa is distinguished by its narrow, acuminate, silvery leaves, 5(–10), strongly keeled phyllaries with conspicuous white, tomentose margins, strongly ribbed cypselae, and relatively few florets per head. Otherwise, it is very similar to *C. acuminata* and *C. intermedia*.

18. Crepis pulchra Linnaeus, Sp. Pl. 2: 806. 1753 • Smallflower hawksbeard [F] [I]

Annuals, 5–100 cm (taproots slender). **Stems** 1, erect, simple, proximally hispid and stipitate-glandular (viscid), distally glabrous. **Leaves** basal and cauline; petiolate; blades oblanceolate or runcinate, 1–24 × 1–5 cm, (bases attenuate) margins deeply pinnately lobed to denticu-late (lobes triangular, terminal lobes largest), apices obtuse to acute, faces densely stipitate-glandular (viscid). **Heads** 10–40, in loose, corymbiform arrays. **Calyculi** of 5–7, ovate or lanceolate, glabrous bractlets 1–2 mm.

Involucres cylindric (turbinate in fruit), 8–12 × 3–5 mm. **Phyllaries** 10–14, (green medially) lanceolate, 8–10 mm, (bases strongly keeled and thickened, margins scarious), apices acute, faces glabrous. **Florets** 15–30; corollas light yellow, 5–12 mm. **Cypselae** (monomorphic or dimorphic) green to yellowish brown, subcylindric, outer 5–6 mm, inner 4–5 mm, apices attenuate (not beaked), ribs 10–12; **pappi** dusky white (very fine, fluffy), 4–5 mm. $2n = 8$.

Flowering Apr–Aug. Dry open habitats, rolling grasslands, pastures, abandoned fields, waste areas, railroads, roadsides; 0–3000 m; introduced; Ont.; Ala., Ark., D.C., Ga., Ill., Ind., Ky., La., Md., Miss., Mo., N.C., Ohio, Okla., Oreg., S.C., Tenn., Tex., Va., W.Va.; Eurasia.

Crepis pulchra is identified by its annual habit; solitary, erect, glandular, and viscid stems; narrowly oblanceolate, runcinate, hispid leaves with relatively large terminal segments; glabrous and strongly keeled phyllaries; sometimes dimorphic cypselae; and fluffy, dusky white pappi.

19. Crepis rubra Linnaeus, Sp. Pl. 2: 806. 1753 • Red hawksbeard [I]

Annuals, 4–40 cm (taproots shallow). **Stems** 1–8, decumbent to ascending, scapiform, branched proximally, glabrate to tomentulose. **Leaves** basal and cauline; petiolate; blades (at least basal) oblanceolate or runcinate, 2–15 × 0.5–3 cm, (bases attenuate) margins pinnately lobed to dentate, apices acute, faces hirsute. **Heads** 1(–2), usually borne singly (peduncles scapiform). **Calyculi** of 8–10, ovate-lanceolate, glabrous bractlets 4–8 mm. **Involucres** cylindro-campanulate, 11–15 × 4–7 mm. **Phyllaries** 8–14, (dark medially), lanceolate, 10–12 mm, (margins yellowish) apices acute, abaxial faces sparsely to densely stipitate-glandular, adaxial with fine, appressed hairs. **Florets** 40–100; corollas pink or white, 16–17 mm. **Cypselae** dimorphic, dark brown, fusiform: outer curved, 8–9 mm, coarsely beaked, inner straight, 12–21 mm, finely beaked, ribs 10 (sharply spiculate); **pappi** yellowish white to dusky white (fine), 5–8 mm. $2n = 10$.

Flowering Apr–Jun. Rocky fields, waste places; 200–300 m; introduced; Calif.; Eurasia; also introduced widely.

Native to the Mediterranean region and Asia Minor, *Crepis rubra* is widely cultivated throughout the world and occasionally escapes. It can be easily recognized by its annual habit, scapiform stems, relatively large, often single heads, and pink or white corollas. Wild plants are shorter than cultivated ones.

20. Crepis runcinata (E. James) Torrey & A. Gray, Fl. N. Amer. 2: 487. 1843 [F]

Hieracium runcinatum E. James, Account Exped. Pittsburgh 1: 453. 1823

Perennials, 15–65 cm (taproots relatively long, caudices swollen). **Stems** 1–3, erect or ascending, scapiform, branched near middles, glabrous or hispid, sometimes stipitate-glandular distally. **Leaves** mostly basal (rosettes); petiolate; blades elliptic, lanceolate, linear, oblanceolate, obovate, or spatulate, 3–30 × 0.5–8 cm (bases attenuate) margins usually entire or weakly dentate, sometimes serrate, dentate, or pinnately lobed, apices rounded, faces glabrous or hispid to hispidulous (sometimes glaucous). **Heads** (1–)3–15(–30), borne singly or in ± corymbiform arrays. **Calyculi** of 5–12, narrowly triangular, glabrous or tomentulose bractlets 1–3 mm. **Involucres** turbinate-campanulate, 7–21 × 8–12 mm. **Phyllaries** 10–16, lanceolate or oblong, 8–10 mm, (bases keeled and thickened, margins scarious) apices usually acute, sometimes attenuate or obtuse (often ciliate-tufted), abaxial faces glabrous or tomentulose, sometimes stipitate-glandular, adaxial glabrous. **Florets** 20–50; corollas golden yellow, 9–18 mm. **Cypselae** dark to golden reddish or yellowish brown, fusiform, 3.5–8 mm, tapered distally or beaked, ribs 10–13 (strong); **pappi** white, 4–9 mm.

Subspecies 7 (7 in the flora): North America, n Mexico.

Crepis runcinata is recognized by its basal rosettes of weakly dentate or almost entire leaves, scapiform stems, branching near middles, and reduced cauline leaves. The stems and leaves are usually glabrous. Multiple subspecies were described by E. B. Babcock (1947); the variation is continuous. Babcock suggested that this is the only American species that shows a relationship to Asian species.

1. Phyllaries eglandular.
  2. Leaves obovate, oblanceolate, or spatulate, 1.5–4 cm wide . . . . 20d. *Crepis runcinata* subsp. *glauca*
  2. Leaves narrowly oblanceolate or linear, 0.5–2 cm wide . . . . 20c. *Crepis runcinata* subsp. *barberi*
1. Phyllaries usually stipitate-glandular or finely glandular-hispid.
  3. Teeth of leaves prominently white-tipped; phyllaries broadly lanceolate or oblong (California, Nevada, Oregon).

4. Involucres 19–21 mm; phyllaries broadly lanceolate, apices long-acuminate; cypselae ± distinctly beaked . . . . 20b. *Crepis runcinata* subsp. *andersonii*
4. Involucres 10–13 mm; phyllaries oblong, apices obtuse or acute; cypselae not beaked . . . . 20g. *Crepis runcinata* subsp. *imbricata*
[3. Shifted to left margin.—Ed.]
3. Teeth of leaves not prominently white-tipped (or only minutely so); phyllaries lanceolate.
  5. Leaves coarsely dentate or pinnately lobed (glaucous) . . . . . . 20e. *Crepis runcinata* subsp. *hallii*
  5. Leaves remotely toothed or serrate, or pinnately lobed, or entire.
    6. Leaves 0.5–3.5 cm wide . . . . 20a. *Crepis runcinata* subsp. *runcinata*
    6. Leaves 2.5–8 cm wide . . . . 20f. *Crepis runcinata* subsp. *hispidulosa*

20a. Crepis runcinata (E. James) Torrey & A. Gray subsp. runcinata • Fiddleleaf or naked-stem or scapose or dandelion hawksbeard [E] [F]

Plants 20–60 cm. **Leaves:** petioles narrowly winged; blades narrowly obovate, elliptic, lanceolate, or spatulate, sometimes runcinate, 0.5–3.5 cm wide, margins entire or toothed to pinnately lobed (lobes or teeth remote, not prominently white-tipped), faces glabrous or hispidulous. **Heads** 1–12. **Involucres** 8–10 mm. **Phyllaries** lanceolate, strongly stipitate-glandular, apices acute. **Cypselae** dark brown, 3.5–7.5 mm, tapered, not beaked; **pappi** 4–6 mm. $2n = 22$.

Flowering Jun–Jul. Moist meadows, low wet areas, swales, bogs; 400–2700 m; Alta., B.C., Man., Sask.; Colo., Idaho, Minn., Mont., Nebr., Nev., N.Mex., N.Dak., Oreg., S.Dak., Utah, Wash., Wyo.

20b. Crepis runcinata (E. James) Torrey & A. Gray subsp. andersonii (A. Gray) Babcock & Stebbins, Publ. Carnegie Inst. Wash. 504: 104. 1938 • Anderson's hawksbeard [E]

Crepis andersonii A. Gray, Proc. Amer. Acad. Arts 6: 553. 1865; *Crepis runcinata* var. *andersonii* (A. Gray) Cronquist

Plants 25–50 cm. **Leaves:** petioles broadly winged; blades oblanceolate, 2–3 cm wide, margins strongly and coarsely serrate or toothed (teeth prominently white-tipped), faces glabrous or hispidulous. **Heads** 6–20. **Involucres** 19–21 mm. **Phyllaries** broadly lanceolate, apices long-acuminate, faces usually stipitate-glandular.

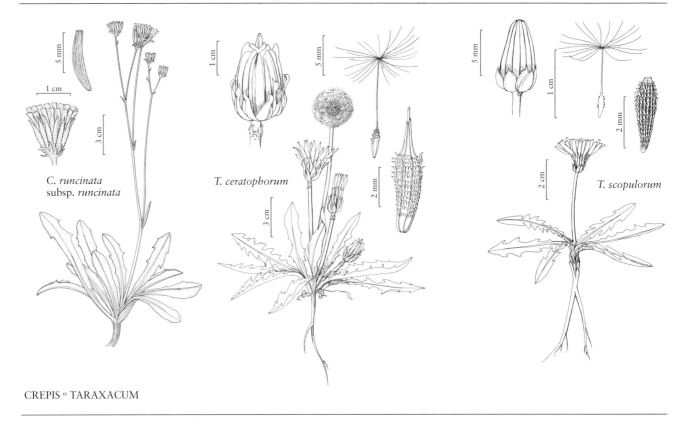

C. runcinata
subsp. *runcinata*

T. ceratophorum

T. scopulorum

CREPIS ° TARAXACUM

Cypselae pale yellow to reddish brown, 6–8 mm, ± distinctly beaked; **pappi** 6–9 mm. **2*n*** = 22.

Flowering May–Jul. Alkaline seeps, grasslands, moist alkaline valley bottoms; 1200–1500 m; Calif., Nev.

Subspecies *andersonii* is identified mainly by the relatively large involucres with densely stipitate-glandular phyllaries and the leaf margins with sharp, prominently white-tipped teeth.

20c. Crepis runcinata (E. James) Torrey & A. Gray subsp. **barberi** (Greenman) Babcock & Stebbins, Publ. Carnegie Inst. Wash. 504: 100. 1938
   • Barber's hawksbeard

Crepis barberi Greenman, Proc. Amer. Acad. Arts 40: 52. 1904

Plants 35–65 cm. **Leaves:** petioles narrowly winged; blades narrowly oblanceolate or linear, 0.5–2 cm wide, margins entire or retrorsely toothed or pinnately lobed (lobes lanceolate), faces glabrous (glaucous). **Heads** 3–7. **Involucres** 11–17 mm. **Phyllaries** lanceolate, apices acute, faces glabrous or sparsely tomentose proximally, eglandular. **Cypselae** dark brown, 5–7 mm, not beaked; **pappi** 7–8 mm.

Flowering Jun–Sep. Alkaline flats; 1200–1300 m; Ariz., Nev., N.Mex.; Mexico (Chihuahua).

Subspecies *barberi* is usually distinguished by relatively narrow leaves that are often entire or weakly dentate, scapiform stems, and glabrous phyllaries.

20d. Crepis runcinata (E. James) Torrey & A. Gray subsp. **glauca** (Nuttall) Babcock & Stebbins, Publ. Carnegie Inst. Wash. 504: 98. 1938 • Smooth hawksbeard [E]

Crepidium glaucum Nuttall, Trans. Amer. Philos. Soc., n. s. 7: 436. 1841; *Crepis runcinata* var. *glauca* (Nuttall) B. Boivin

Plants 20–50 cm. **Leaves:** petioles narrowly winged; blades narrowly obovate, oblanceolate, or spatulate, 1.5–4 cm wide, margins dentate, faces glabrous (glaucous). **Heads** 3–15. **Involucres** 7–12 mm. **Phyllaries** lanceolate, apices attenuate, faces glabrous or slightly tomentose, eglandular. **Cypselae** golden brown, 3.5–5.5 mm, not beaked; **pappi** 5–6 mm. **2*n*** = 22

Flowering May–Aug. Alkaline, wet or dry meadows; 1600–2300 m; Alta., Man., Sask.; Ariz., Colo., Idaho, Mont., Nebr., Nev., N.Mex., N.Dak., S.Dak., Tex., Utah, Wyo.

20e. Crepis runcinata (E. James) Torrey & A. Gray subsp. **hallii** Babcock & Stebbins, Publ. Carnegie Inst. Wash. 504: 104, fig. 13. 1938 • Hall's or meadow hawksbeard E

Plants 20–60 cm. **Leaves:** petioles narrowly winged; blades oblanceolate or narrowly obovate, 1.5–3 cm wide, margins coarsely dentate or pinnately lobed (teeth not prominently white-tipped), faces glabrous (glaucous). **Heads** 5–14. **Involucres** 9–13 mm. **Phyllaries** lanceolate, apices acute, faces strongly stipitate-glandular. **Cypselae** chestnut brown, 4.5–6.5 mm, narrowed, not beaked; **pappi** 6–7 mm. **2***n* = 22

Flowering Jul–Aug. Moist alkaline meadows, seeps, valley bottoms; 1200–2500 m; Calif., Nev.

Subspecies *hallii* is similar to subsp. *runcinata*, differing mainly in the dentate leaf margins. It differs from subsp. *glauca* mainly in having stipitate-glandular phyllaries.

20f. Crepis runcinata (E. James) Torrey & A. Gray subsp. **hispidulosa** (Howell ex Rydberg) Babcock & Stebbins, Publ. Carnegie Inst. Wash. 504: 96. 1938 E

Crepis runcinata var. *hispidulosa* Howell ex Rydberg, Mem. New York Bot. Gard. 1: 461. 1900

Plants 25–50 cm. **Leaves:** petioles narrowly winged; blades obovate, 2.5–8 cm wide, margins remotely toothed, pinnately lobed, or entire (teeth not prominently white-tipped, apices obtuse), faces glabrous or glandular-hispid. **Heads** 10–30. **Involucres** 8–12 mm. **Phyllaries** lanceolate, apices acute, faces strongly or finely glandular-hispid. **Cypselae** brown, 3.5–5 mm, not beaked; **pappi** 4–8 mm. **2***n* = 22.

Flowering Jun–Jul. Dry or moist alkaline meadows; 1300–2500 m; Alta., Sask.; Colo., Idaho, Mont., N.Dak., Oreg., Utah, Wash., Wyo.

Subspecies *hispidulosa* is distinguished by its relatively large, obtuse leaves and relatively numerous heads with glandular-hispid phyllaries.

20g. Crepis runcinata (E. James) Torrey & A. Gray subsp. **imbricata** Babcock & Stebbins, Publ. Carnegie Inst. Wash. 504: 102, fig. 11. 1938 E

Plants 15–30 cm. **Leaves:** petioles narrowly winged; blades oblanceolate or elliptic, 1.5–3.5 cm wide, margins strongly and closely dentate (teeth prominently white-tipped), faces glabrous. **Heads** 3–7. **Involucres** 10–13 mm. **Phyllaries** oblong, apices obtuse or acute (ciliate), faces strongly stipitate-glandular. **Cypselae** reddish brown, 4.5–5 mm, narrowed, not beaked; **pappi** 5–7 mm.

Flowering May–Jun. Wet meadows, alkaline seeps; 1200–1900 m; Nev., Oreg.

21. Crepis setosa Haller f., Arch. Bot. (Leipzig): 1(2): 1. 1797 • Bristly hawksbeard I

Annuals, 8–80 cm (taproots shallow). **Stems** 1, erect (often reddish), stout (fistulose), simple or branched proximally, coarsely setose or hispid (at least distally, setae yellowish). **Leaves** basal and cauline; petiolate; blades oblanceolate, often runcinate or lyrate, 5–30 × 1–8 cm, margins dentate to pinnately lobed (terminal lobes often relatively large), apices acute to obtuse, faces finely hispid (coarsely setose along midribs; cauline leaves lanceolate, bases sagittate with acuminate lobes, margins dentate to deeply laciniate proximally). **Heads** 10–20, in paniculiform or cymiform arrays. **Calyculi** of 10–14, linear, coarsely setose bractlets 2–4 mm. **Involucres** cylindro-campanulate, 6–10 × 4–10 mm. **Phyllaries** 12–16, lanceolate, 6–7 mm, (bases strongly keeled and thickened, margins green to yellowish), apices acuminate, abaxial faces coarsely setose or hispid, adaxial with fine hairs. **Florets** 10–20; corollas yellow, sometimes reddish abaxially, 8–10 mm. **Cypselae** reddish brown, fusiform, 3–5 mm, beaked (beaks 1–2 mm), ribs 10 (rounded, spiculate near bases of beaks); **pappi** white (fine, soft), 4 mm. **2***n* = 8.

Flowering May–Nov. Openings in mixed conifer forest, disturbed areas, lawns; 50–500 m; introduced; Calif., Conn., Mo., Mont., N.Y., Ohio, Oreg., Pa., Tenn., Tex., Vt., Wis.; Europe.

Crepis setosa is recognized by its annual habit, shallow roots, coarsely setose stems, leaves, and involucres, the relatively large runcinate leaves, sagittate-laciniate cauline leaves, finely beaked cypselae, and white, fine pappus bristles.

22. Crepis tectorum Linnaeus, Sp. Pl. 2: 807. 1753
• Narrowleaf hawksbeard, crépis des troits 1 w

Annuals, 10–100 cm (taproots shallow). **Stems** 1, erect (fistulose), branched distally or from bases, tomentulose and/or hispid. **Leaves** basal and cauline; petiolate; blades lanceolate to oblanceolate, often coarsely runcinate, 5–15 × 1–4 cm, margins entire, denticulate, or dentate to pinnately lobed (lobes remote, coarse, unequal, apices acute to acuminate, abaxial faces glabrous or tomentose, adaxial glabrous (proximal cauline sessile, bases auriculate, distal usually linear, entire). **Heads** 5–20(–100+), in paniculiform or corymbiform arrays. **Calyculi** of ± 12, subulate, tomentose and hispidulous bractlets 2–5 mm (often becoming scarious). **Involucres** cylindro-campanulate, 6–9 × 7–8 mm. **Phyllaries** 12–15, lanceolate, 5–9 mm, (bases becoming keeled and thickened, margins scarious), apices acute to attenuate (white-ciliate, tomentulose), abaxial faces tomentose to hispidulous, adaxial with fine, appressed hairs. **Florets** 30–70; corollas yellow (without red on ligules), 10–13 mm. **Cypselae** dark reddish or purplish brown, fusiform, 3–4 mm, apices constricted (not beaked), ribs 10 (rounded, minutely spiculate); **pappi** white (fine, soft), 4–5 mm. $2n = 8$.

Flowering May–Sep. Dry, sandy, pine woods, disturbed places, abandoned fields, forest clearings, wooded slopes, dry streambeds; 100–300 m; introduced; Greenland; Alta., B.C., Man., N.B., Nfld. and Labr. (Labr.), N.W.T., N.S., Ont., Que., Sask., Yukon; Alaska, Calif., Conn., D.C., Ill., Ind., Iowa, Maine, Md., Mass., Mich., Minn., Mo., Mont., Nebr., N.J., N.Y., N.C., N.Dak., Ohio, Oreg., Pa., R.I., Wash., Wis., Wyo.; Europe; introduced, Asia.

Crepis tectorum is recognized by its annual habit, keeled phyllaries with minute hairs on adaxial faces, and dark reddish or purplish brown cypselae. It is widespread, often abundant, occurs in a great variety of habitats, and is considered a noxious weed in some states.

23. Crepis vesicaria Linnaeus, Sp. Pl. 2: 805. 1753
• Beaked or weedy hawksbeard 1

Annuals, biennials, or perennials, 3–120 cm (taproots slender to thick, caudices swollen). **Stems** 1, erect to arcuate or decumbent (green or purple proximally), usually much branched, glabrate to hispid and/or tomentose, sometimes sparsely setose (setae black). **Leaves** basal and cauline; petiolate (bases clasping); blades oblanceolate to ovate, often runcinate, 10–35 × 2–8 cm, margins pinnately lobed to toothed (terminal lobes relatively large), apices obtuse or acute, faces usually hirsute (hairs sometimes only on veins) or glabrous (cauline sessile, bases auriculate, clasping, margins ± toothed). **Heads** 10–20, in lax, corymbiform arrays. **Calyculi** of 5–12, ovate to linear-lanceolate, glabrous bractlets 3–4 mm (reflexed in fruit, scarious). **Involucres** cylindro-campanulate (becoming turbinate or urceolate in fruit), 5–14 × 5–6 mm. **Phyllaries** 7–16, (reflexed at maturity) lanceolate, 10–12 mm, (margins green to yellowish), apices obtuse or acute (ciliate), abaxial faces tomentose and often stipitate-glandular, adaxial with fine, appressed hairs. **Florets** 50–70; corollas yellow (reddish abaxially), 6–15 mm. **Cypselae** (monomorphic or dimorphic) pale brown or yellowish, fusiform, 4–9 mm, outer wider with apices attenuate (not beaked), inner gradually tapered, beaked (beaks 2–5 mm, ± equal to bodies), ribs 10 (narrow); **pappi** white (fine, soft), 3–6 mm. $2n = 8, 16$.

Flowering Feb–Oct. Sandy clearings, hillsides; 0–300 m; introduced; B.C.; Calif., Conn., N.Y., N.C., Oreg., Pa.; Europe; introduced, South America.

Native to the Mediterranean region of western Europe, *Crepis vesicaria* is recognized by its annual or biennial habit, pinnately lobed leaves, reflexed calyculi, tomentose and glandular phyllaries, and slender, long-beaked inner cypselae. It is polymorphic; subspecies are recognized in Europe. E. B. Babcock (1947) identified the North American plants as subsp. *taraxaciflora* (Thuiller) Thellung, which some Europeans (T. G. Tutin et al. 1964–1980, vol. 4) have listed as a synonym of subsp. *haenseleri* (Boissier ex de Candolle) P. D. Sell.

24. Crepis zacintha (Linnaeus) Babcock, Univ. Calif. Publ. Bot. 19: 404. 1941 • Striped hawksbeard 1

Lapsana zacintha Linnaeus, Sp. Pl. 2: 811. 1753; *Rhagadiolus zacintha* (Linnaeus) Allioni; *Zacintha verrucosa* Gaertner

Annuals, 20–30 cm (taproots shallow). **Stems** 1–3, erect, simple or dichotomously branched proximally or distally, hispid proximally, glabrous distally. **Leaves** basal and cauline; petiolate; blades lyrate, 10–20 × 3–4 cm, pinnately lobed (lateral lobes remote, triangular, terminal lobes relatively large, ovate or truncate), apices obtuse, faces hispid (hairs pale, relatively short). **Heads** borne singly (sessile in axils of branches). **Calyculi** of 5, lanceolate, glabrous or proximally tomentulose bractlets 3–6 mm. **Involucres** cylindric, 5–7 × 3–7 mm. **Phyllaries** 10, lanceolate (proximal ½ of each becoming indurate, swollen and angular in fruit, enclosing and partially fused to cypsela, margins green), apices bent at right angles (ciliate), abaxial faces basally tomentulose, adaxial glabrous. **Florets** ca. 30; corollas yellow, deep purplish red abaxially, 7 mm. **Cypselae** (dimorphic) yellowish, ribs 10 (smooth), outer strongly compressed and triangular, obconic, 2–2.5 mm, strongly constricted proximally, apices truncate, inner obconic, ca. 2.5 mm, tapered proximally, apices constricted; **pappi** white (fine, soft), 1.5 mm. $2n = 6$.

Flowering Jun–Jul. Grassy slopes, fields, gravelly waste areas, roadsides; 200–300 m; introduced; Tex.; Europe.

Crepis zacintha is recognized by its annual habit, lyrate-pinnatifid leaves mostly in rosettes, cymiform-dichotomous branching, single heads in axils of branches, indurate and bent phyllaries, and dimorphic, angular cypselae. Because of its unique morphology, it sometimes has been assigned to a separate, monotypic genus (*Zacintha* Miller).

37. TARAXACUM F. H. Wiggers, Prim. Fl. Holsat., 56. 1780, name conserved

• Dandelion, pissenlit [Arabic to Persian *talkh chakok*, a bitter herb]

Luc Brouillet

Perennials, (10–)30–400(–600+ in fruit) cm (sexual or apomictic); taprooted or with branched caudices. **Stems** (1–10+) erect or ascending, scapiform (terete), simple (hollow), glabrous or villous proximal to heads. **Leaves** basal (in rosettes, erect or patent to nearly horizontal); petiolate or sessile; blades oblong to obovate or oblanceolate to linear-oblanceolate, runcinate or lyrate (bases cuneate to ± attenuate), margins subentire to dentate or pinnately lobed (apices rounded or obtuse to acute or acuminate, faces glabrous or glabrate to sparsely villous, pilose, or villosulous). **Heads** borne singly. **Calyculi** persistent, of (6–)8–18(–20) broadly ovate to lanceolate bractlets in (1–)2–3 series, distinct (appressed before flowering, recurved to spreading or reflexed in fruit), unequal (shorter than phyllaries, margins scarious, ciliate or not, apices corniculate, callous, or neither). **Involucres** campanulate to cylindro-campanulate or urceolate to cylindric, 8–40 mm diam. **Phyllaries** 7–25 in 2(–3) series, weakly coherent proximally in buds (interlocking folded margins), distinct later, erect (sometimes slightly spreading) in flower, closing at fruit maturation, reflexed at dispersal (exposing globes of cypselae with fully spread pappi), ± equal, herbaceous, glabrous; inner lanceolate to linear-lanceolate, margins scarious, ciliate or not, apices acuminate, sometimes corniculate, callous, or flat. **Receptacles** ± flat, epaleate. **Florets** (15–)20–150; corollas yellow, sometimes greenish, rarely cream or pale pink [white], often purplish- or gray-striped abaxially (anthers yellow or yellow-cream, sometimes darker; styles yellow or greenish, sometimes grayish to blackish). **Cypselae** straw-colored to olive, brown, or red to pale or dark gray, bodies oblanceoloid to obovoid, ± flattened (distally ± swollen, forming discrete, conic, or terete "cones" supporting beaks [without cones]), beaked [beakless], ribs 4–12(–15), faces muricate (at least distally) [nearly smooth], glabrous; **pappi** persistent, of 50–105+ distinct, white to cream-colored or yellowish to sordid, equal, barbellulate bristles in 1 series. $x = 8$.

Species 60(–2000) (15 in the flora): North America, South America, Eurasia; worldwide weeds (e.g., *Taraxacum officinale*, *T. erythrospermum*).

The type of the genus, *Taraxacum officinale*, is conserved. This name is linked to the (very general) description of *Leontodon taraxacum* Linnaeus. A. J. Richards (1985) typified *T. officinale*, via *L. taraxacum*, on a specimen that is apparently referable to *T. campylodes* Haglund, a microspecies of sect. *Crocea* restricted to Lapland, which thus became the basis of sect. *Taraxacum*. J. Kirschner and J. Štěpánek (1987) underlined that this typification of *T. officinale* does not reflect usage of the name, which raises considerable ambiguity as to its application, because Richards essentially defined a new content for it. The species usually referred to as *T. officinale* must now be referred to sect. *Ruderalia* (Kirschner and Štěpánek);

no name was proposed that would correspond closely with the species currently called *T. officinale*. A proposal to conserve the name *T. officinale* with a neotype that would preserve its common usage for this widespread entity has been suggested; this has yet to be discussed fully.

Taraxacum Zinn (1757) (= *Leontodon* Linnaeus) is a rejected name.

The genus has been monographed by H. Handel-Mazzetti (1907) and by R. Doll (1974). Infrageneric nomenclature has recently been reviewed by A. J. Richards (1985) and by J. Kirschner and J. Štěpánek (1987, 1997). The European species were treated by Richards and P. D. Sell (1973) and much work has been done since; there is no overall treatment for Asia; Russian authors have covered Siberia. The number of species in the genus depends on the disposition of agamic microspecies within species complexes, which varies greatly among authors, particularly in Europe [e.g., A. A. Dudman and Richards (1997) recognized 105 species for Great Britain and Ireland]. North American *Taraxacum*, particularly in the boreal and arctic zones, has been investigated by numerous researchers, many of whom incorporated new taxa described by H. Dahlstedt (1906); only works touching North America north of Mexico are mentioned here. Obviously, Scandinavian and Russian works also were significant (e.g., Dahlstedt; Doll 1977; M. L. Fernald 1933; E. L. Greene 1901b; G. Haglund 1943, 1946, 1948, 1949; M. P. Porsild 1930; P. A. Rydberg 1901), but often in a manner limited geographically or taxonomically, and no complete review exists. Most often, the taxonomy of the genus has been presented within the context of floras (e.g., S. G. Aiken et al., http://www.mun.ca/biology/delta/arcticf/_ca/www/asta.htm, with excellent photographs of Arctic species; T. W. Böcher et al. 1978; A. Cronquist 1955, 1994; Fernald 1950; H. A. Gleason and A. Cronquist 1991; E. Hultén 1955, 1968; A. E. Porsild 1950b, 1957, 1964; A. E. Porsild and W. J. Cody 1980; H. J. Scoggan 1978–1979, part 4; Rydberg 1900c). The result of all these efforts has not been a clarification of the North American situation, but rather a taxonomy and nomenclature in utter confusion (Cronquist 1994). The current treatment does not solve all nomenclatural and taxonomic problems, many of which will depend for their ultimate solution on work done in Europe.

I have adopted a broad definition of *Taraxacum* species for North America, broader at least than what is usually seen in European treatments. For instance, the species most familiar to North Americans were introduced from Europe (*T. officinale* and *T. erythrospermum*; see below for a justification of the use of these names), possibly several times, and represent variable agamic complexes, but this variation appears continuous and multidimensional. There seems to be no utility for the users in describing a multitude of narrowly defined microspecies. For the native arctic and western alpine species, the impact of the Pleistocene glaciations, which covered much of the territory now occupied by those species except for ice-free parts of Alaska and Yukon, must be considered. It is likely that most populations spread recently from southern or Beringian refugia after the ice withdrew and that the number of species that migrated is restricted. Isolation in the Rocky Mountains and adjacent areas may explain some of the phenotypic diversity, but not enough to warrant a large number of narrowly defined, endemic entities. The situation in eastern North America (Greenland, Labrador, Newfoundland, and adjacent areas) may have been influenced by the amphi-Atlantic dispersal of some taxa. Again, given the small number of such species in the North American flora, all concentrated in that region, it is unlikely that the number of species actually present would reach the number that has been described for the area. Therefore, at the present time, delimitation of readily distinguishable taxa appears more useful than trying to dissect finely the variation present into microspecies that would have little experimental validation.

Another reason for using broad species limits is provided by population genetics. For instance, in Europe, S. B. J. Menken et al. (1995) showed that diploid and triploid members of *Taraxacum* sect. *Ruderalia* are less genetically isolated than formerly supposed and form a cohesive unit, because of the exchange of genetic material between ploidy levels despite the fact that the latter are usually agamic. The molecular study of genetic variation by L. M. King (1993) in introduced asexual *Taraxacum* taxa in North America also shows the importance of hybridization to explain variation, in addition to mutations, another important factor (King and B. A. Schaal 1990). M. T. Brock (2004) also documented gene exchange between the introduced agamic *T. officinale* and native diploid populations of *T. ceratophorum* in Colorado. This is cause for conservation concern in areas where introduced dandelions, notably the common dandelion, invade populations of native species, such as in the Gulf of Saint Lawrence area or the western Cordilleras. It is also possible that the prolific common dandelions not only genetically assimilate but also competitively displace native populations, which might be the case for some populations of *T. laurentianum* in western Newfoundland.

A. A. Dudman and A. J. Richards (1997) described some of the sources of phenotypic plasticity (or drying artifacts) in *Taraxacum* that may affect the identification (or delimitation) of species: juvenile and shaded leaves usually are less divided than older, sun-exposed or stressed ones, and the terminal lobes usually are smaller; some traits described as characteristic of a species may occur on only some leaves of a rosette; ligule color may change in dried material; cypsela size, though mostly consistent within species, may vary considerably within a head, the outer often being shorter; finally, cypsela color changes with maturity and insolation, and fades on specimens, and in some groups, the variation in color is such that this trait may lose its significance in delimiting entities. R. J. Taylor (1987) also emphasized the importance of phenotypic plasticity in weedy dandelion morphologic variation.

There is a spontaneous mutant form of *Taraxacum erythrospermum* (called *T. laevigatum* forma *scapifolium* F. C. Gates & S. F. Prince) in which one or more lobed and dentate leaves (or bracts), progressively reduced distally, are present on the scape or peduncle. Also, calyculus bracts are more or less modified to enlarged, lobed and dentate bracts, instead of the usual bractlets. The phyllaries appear unaffected. The form is genetically determined, as it bred true. This shows that scapes of dandelions are modified stems where leaf expression is repressed, and that calyculi are indeed distinct in origin from the involucres and should be considered as a separate structure and not as an external series of the involucre, as is often done in descriptions.

Evolution and population biology in *Taraxacum*, notably with respect to breeding systems, apomixis, and variation, has been the object of numerous studies (e.g., J. C. M. den Nijs and S. B. J. Menken 1994; J. Hughes and A. J. Richards 1988, 1989; L. M. King 1993; King and B. A. Schaal 1990; J. C. Lyman and N. C. Ellstrand 1998; M. Mogie and H. Ford 1988; Mogie and Richards 1983; Richards 1970, 1970b, 1973, 1989, 1996; O. T. Solbrig 1971; R. J. Taylor 1987). Molecular phylogenetic studies have not been effective so far in solving problems of relationships within *Taraxacum* (e.g., J. Kirschner et al. 2003).

Chromosome counts of North American *Taraxacum* species are few and mainly come from A. W. Johnson and J. G. Packer (1968), T. Mosquin and D. E. Hayley (1966), G. A. Mulligan (1984), and Packer and G. D. McPherson (1974). I have not been able to examine all vouchers, and it has been difficult sometimes to attribute reports to species. The same problem exists with Russian chromosome number reports and I prefer not to include them here (see the website of S. G. Aiken et al. for such references).

The synonymy provided below must be taken with caution. Few types were seen (though many photographs were).

Taraxacum species have been used medicinally (mostly as a diuretic) and in alimentation (as greens and to make wine); they are particularly rich sources of vitamin C (E. Small and P. M. Catling 1999).

North American species of *Taraxacum* fall within the following sections: *Ruderalia* Kirschner, H. Øllgaard & Štěpánek (*T. officinale*, *T. latilobum*); *Erythrosperma* Dahlstedt (*T. erythrospermum*); *Palustria* Dahlstedt (*T. palustre*); *Spectabilia* Dahlstedt (*T. lapponicum*, *T. spectabile*); *Borealia* Handel-Mazzetti (*T. californicum*, *T. ceratophorum*, *T. laurentianum*, *T. trigonolobum*); and *Arctica* Jurtzev (*T. alaskanum*, *T. carneocoloratum*, *T. holmenianum*, *T. hyparcticum*, *T. phymatocarpum*, *T. scopulorum*).

D. F. Brunton (1989) reported the presence of small populations of an undetermined sect. *Spectabilia* species in wet ditches while collecting *Taraxacum palustre* in Ontario and New York state; he did not collect vouchers (Brunton, pers. comm.). In part on this basis, H. A. Gleason and A. Cronquist (1991, 2004) included *T. spectabile* Dahlstedt (Bot. Not. 1905: 159) in the northeastern North American flora, from southern Ontario and New York state. Apart from an old specimen from New York state, I have been unable to locate vouchers of the species for New York or Ontario. Establishments of the species in North America needs to be more rigorously documented. The taxon can be recognized by its hairy, abaxially purple-spotted and veined leaves, by its non-corniculate phyllaries and calyculus bractlets, and by its ovate to lanceolate bractlets that are reflexed at flowering.

I am indebted to S. G. Aiken, with whom I was able to discuss *Taraxacum* while she was preparing her flora of the Canadian Arctic Archipelago, and to her, D. F. Murray, and R. Elven for having shared unpublished information assembled for the Panarctic Flora project. They cannot be blamed for the treatment presented here, because the interpretation of data is wholly mine.

SELECTED REFERENCES Handel-Mazzetti, H. 1907. Monographie der Gattung *Taraxacum*. Leipzig and Vienna. Richards, A. J. and P. D. Sell. 1973. *Taraxacum*. In: T. G. Tutin et al., eds. 1964–1980. Flora Europaea. 5 vols. Cambridge. Vol. 4, pp. 332–343.

1. All or some phyllary apices notably horned; calyculus bractlets notably horned (sect. *Borealia* in part).
   2. Involucres 18–26 mm; calyculus bractlets 15–21 × 3.5–5.5 mm; phyllaries 18–25; florets ca. 150+; cypsela beaks 10–17 mm; leaves usually erect, sometimes patent; coastal meadows, Gulf of St. Lawrence, Quebec, Newfoundland . . . . 7. *Taraxacum laurentianum*
   2. Involucres 8–20 mm; calyculus bractlets 5–12 × 1.5–5 mm; phyllaries 12–16(–20); florets 40–150; cypsela beaks 4–10(–14) mm; leaves usually horizontal to patent, sometimes ± erect; boreal and low arctic North America, Rocky Mountains.
      3. Leaves horizontal or patent to ± erect, oblanceolate-obovate (largest in distal ¹/₅) to oblanceolate (often runcinate), usually ± strongly toothed proximally, less so on distal ¹/₅–¹/₄, sometimes lobed proximally or ± regularly and deeply, apices rounded or truncate to obtuse (rarely acute), apiculate; calyculus bractlets 14–20; involucres 15–20 mm; florets 120–150; Aleutian Islands and islands of the Bering Sea, Alaska . . . . 9. *Taraxacum trigonolobum*
      3. Leaves horizontal to patent (sometimes erect), narrowly oblanceolate to ± linear-oblanceolate or linear-oblong (often ± runcinate), margins irregularly to regularly, ± deeply lobed to lacerate, often toothed or subentire, teeth ± well developed, apices obtuse to sometimes acute, sometimes mucronate; involucres (5–)8–19 (–21) mm; calyculus bractlets 12–16(–20); florets 40–85+; widespread, boreal and Low Arctic, w Mountains . . . . 6. *Taraxacum ceratophorum*
1. Phyllary apices usually hornless (sometimes callous or horns relatively small); calyculus bractlets usually hornless or horns relatively small.

[4. Shifted to left margin.—Ed.]

4. Plants often 5–75 cm; leaves 10+, horizontal to erect, lobes (when present) mostly retrorse, triangular to lanceolate; calyculus bractlets soon reflexed to recurved (appressed to spreading in *T. palustre*); phyllaries 13–19; florets (40–)50–100+; ± widespread introduced taxa (temperate and boreal zones), or boreal to low Arctic native species (often moist habitats, seaside meadows or streamsides in eastern boreal zone).

  5. Plants 5–6.5(–10 in fruit) cm; leaves ± 10, 2.5–9 × 0.2–1.1 cm, toothed to sometimes shallowly lobed; calyculus bractlets appressed to spreading, margins widely scarious; florets ca. 50; flowering early spring (mainly wet ditches) . . . . 4. *Taraxacum palustre*

  5. Plants (1–)5–40(–75 in fruit), leaves 10–20+, 4–47.5 × (0.7–)1–10 cm; calyculus bractlets soon reflexed to recurved, or spreading and eventually reflexed, margins narrowly or not scarious; florets 40–100+; flowering spring to early summer (north) or nearly year-round (south).

   6. Leaves lacerate (lobes often lanceolate), petioles ± slightly winged distally; cypselae brick red to reddish brown or purple . . . . 3. *Taraxacum erythrospermum*

   6. Leaves usually shallowly to deeply lobed or toothed (occasionally lacerate in *T. officinale*, then often some early leaves less deeply so), petioles narrowly to broadly winged; cypselae olive or olive-brown to tan, reddish brown, or grayish.

    7. Leaves sessile to ± broadly winged-petiolate, bases usually cuneate, sometimes attenuate, lobed or dentate; calyculus bractlets 10–12, margins usually not scarious; phyllary margins not or narrowly scarious; cypselae tan to reddish brown, cones conic; e boreal to low Arctic . . . . 5. *Taraxacum lapponicum*

    7. Leaves petiolate, petioles ± narrowly winged, bases usually attenuate, margins usually lobed and ± dentate; calyculus bractlets 12–18, margins narrowly scarious; phyllary margins scarious; cypselae olive to tan or grayish, cones terete; temperate to boreal regions.

     8. Terminal leaf lobes ± as large as distal laterals, teeth few, sometimes 0; calyculus bractlets lanceolate; florets 40–100+; cypsela bodies (2–)2.5–2.8 (–4) mm, beaks 7–9 mm, faces proximally smooth to ± tuberculate; widespread temperate to boreal weed . . . . 1. *Taraxacum officinale*

     8. Terminal leaf lobes broader than laterals, teeth 5–7 on lobes and sinuses; calyculus bractlets ovate to broadly lanceolate; florets ca. 150; cypsela bodies 2.8–3.8 mm, beaks 8–12 mm, tuberculate proximally; seaside grassy slopes and taluses or clifftops, Gulf of St. Lawrence, and adjacent areas . . . . .
     . . . . 2. *Taraxacum latilobum*

4. Plants 1–12(–30 in fruit) cm; leaves usually fewer than 10 (10–20 in *T. californicum*), horizontal to patent (rarely erect), lobes often straight, deltate to triangular; calyculus bractlets appressed to spreading (becoming reflexed in fruit); phyllaries 7–14(–16); florets 20–50+; w montane and high alpine, or Arctic regions.

  9. Leaves not lobed or lobes slightly retrorse or straight, lanceolate to triangular-acuminate, teeth narrowly to broadly triangular; calyculus bractlets ovate-lanceolate, margins scarious; involucres green, tips purplish gray, broadly campanulate; San Bernardino Mountains, California. . . . . 8. *Taraxacum californicum*

  9. Leaves usually regularly lobed (dentate to denticulate or entire in *T. phymatocarpum*), usually not dentate; outer calyculus bractlets ovate to broadly ovate, margins narrowly or not scarious; involucres dark green, mostly narrowly campanulate; Rocky Mountains and Arctic (sect. *Arctica*).

   10. Corollas pink (sometimes ± bronze when fresh) or cream-colored to white or pink distally, abaxially pinkish-striped.

11. Leaves deeply and regularly lobed to subentire; calyculus bractlets horned; involucres narrowly campanulate, 15–30 mm; corollas cream-colored to white or pink distally, outer abaxially pinkish-striped, 15–20 × 1.2–3 mm; cypselae tan or straw-colored to brown, sometimes grayish, muricate in distal $^1/_2$; pappi white or sordid . . . . 10. *Taraxacum hyparcticum*

11. Leaves sometimes shallowly lobed; calyculus bractlets hornless; involucres broadly campanulate, (10–)12–16 mm; corollas pink (± bronze when fresh), outer 13–14 × 2.4–2.6 mm; cypselae grayish, greenish, or yellowish, muricate in distal $^1/_4$; pappi yellowish; Yukon, Alaska . . . . 15. *Taraxacum carneocoloratum*

[10. Shifted to left margin.—Ed.]

10. Corollas pale to dark yellow.

    12. Leaves oblanceolate to linear-oblanceolate, margins entire or toothed to denticulate (sometimes some nearly runcinate); corollas pale yellow, sometimes lemon-colored; cypselae dark brown, grayish or blackish, muricate $^1/_2$–$^3/_4$+ . . . . 11. *Taraxacum phymatocarpum*

    12. Leaves runcinate, regularly lobed, lobes straight, deltate to triangular; corollas yellow to dark yellow; cypselae usually yellowish to brown or reddish brown, sometimes grayish, muricate in distal $^1/_2$ or less.

        13. Plants 1–5 cm; leaves linear-oblanceolate to narrowly oblanceolate, (1–)1.5–4 × 0.5–1 cm; outer corollas 7.5–8.8 mm; cypsela cones 0.5–0.6 mm . . . . . . 12. *Taraxacum scopulorum*

        13. Plants (1.5–)3–15(–20) cm; leaves oblanceolate to oblong-oblanceolate or narrowly oblong, (1.5–)2–11.6 × 0.3–2.2; outer corollas 11–20 mm; cypsela cones 0.8–1.1 mm.

            14. Plants 5–15(–20) cm; leaves dark green; calyculus bractlets 10–14; phyllaries ca. 14, apices sometimes callous and/or horned; cypselae yellowish or straw-colored, beaks slender; Arctic . . . . 13. *Taraxacum holmenianum*

            14. Plants (1.5–)3–9(–16 in fruit) cm; leaves green; calyculus bractlets 7–9; phyllaries 7–8, apices not callous, hornless; cypselae maroon to brown or reddish brown, sometimes grayish, beaks stout . . . . 14. *Taraxacum alaskanum*

1. Taraxacum officinale F. H. Wiggers, Prim. Fl. Holsat., 56. 1780 • Common dandelion, pissenlit officinal [I]

Leontodon taraxacum Linnaeus, Sp. Pl. 2: 798. 1753; *Taraxacum officinale* var. *palustre* Blytt; *T. sylvanicum* R. Doll

Plants (1–)5–40(–60) cm; taproots seldom branched. **Stems** 1–10+, erect or ascending, sometimes ± purplish (usually equaling or surpassing leaves), glabrous or sparsely villous, slightly more so distally. **Leaves** 20+, horizontal to erect; petioles ± narrowly winged; blades oblanceolate, oblong, or obovate (often runcinate), (4–)5–45 × (0.7–)1–10 cm, bases attenuate to narrowly cuneate, margins usually shallowly to deeply lobed to lacerate or toothed, lobes retrorse, broadly to narrowly triangular to nearly lanceolate, acute to long-acuminate, terminals ± as large as distal laterals, ultimate margins toothed or entire (secondary lobules irregular, perpendicular to retrorse), teeth minute to pronounced apices acute to acuminate or obtuse, faces glabrous or sparsely villous (commonly on midveins). **Calyculi** of 12–18, reflexed, sometimes ± glaucous, lanceolate bractlets in 2 series, 6–12 × 2.8–3.5 mm, margins very narrowly white-scarious, sometimes villous-ciliate distally, apices acuminate, hornless. **Involucres** green to dark green or brownish green, tips dark gray or purplish, campanulate, 14–25 mm. **Phyllaries** 13–18 in 2 series, lanceolate, 2–2.8 mm wide, margins scarious (proximal $^2/_3$) to narrowly scarious, apices acuminate, erose-scarious, usually hornless (seldom appendaged), callous. **Florets** 40–100+; corollas yellow (orange-yellow), 15–22 × 1.7–2 mm (outer). **Cypselae** olivaceous or olive-brown, or straw-colored to grayish, bodies oblanceoloid, (2–)2.5–2.8(–4) mm, cones shortly terete, 0.5–0.9 mm, beaks slender, 7–9 mm, ribs 4–12, sharp, faces proximally smooth to ± tuberculate, muricate in distal $^1/_3$; pappi white to sordid, 5–6(–8) mm. $2n = 24, 40, [16, 32]$.

Flowering nearly year-round (fall–spring, south; spring or summer, north). Often damp low places, lawns, roadsides, waste grounds, disturbed banks and shores; 0–2000+ m; introduced; Greenland; St. Pierre and Miquelon; Alta., B.C., Man., N.B., Nfld. and Labr., N.W.T., N.S., Nunavut, Ont., P.E.I., Que., Sask., Yukon; Ala., Alaska, Ariz., Ark., Calif., Colo., Conn., Del., D.C., Fla., Ga., Idaho, Ill., Ind., Iowa, Kans., Ky., La., Maine, Md., Mass., Mich., Minn., Miss., Mo., Mont., Nebr., Nev., N.H., N.J., N.Mex., N.Y., N.C., N.Dak., Ohio, Okla., Oreg., Pa., R.I., S.C., S.Dak., Tenn., Tex., Utah,

Vt., Va., Wash., W.Va., Wis., Wyo.; Europe; also introduced in Mexico; introduced nearly worldwide.

Taraxacum officinale is the most widespread dandelion in temperate North America, though its abundance decreases in the arid south. It is a familiar weed of lawns and roadsides. It is also the species most commonly used for medicinal and culinary purposes (e.g., E. Small and P. M. Catling 1999).

Phenotypic and genotypic variation of this species have been studied in North America (L. M. King 1993; King and B. A. Schaal 1990; J. C. Lyman and N. C. Ellstrand 1998; O. T. Solbrig 1971; R. J. Taylor 1987), but results of those studies did not lead to the recognition of microspecies.

Specimens of *Taraxacum officinale* with deeply lobed leaves are sometimes difficult to distinguish from those of *T. erythrospermum* when fruits are missing (see also R. J. Taylor 1987). Usually, however, early leaves of the former are much less deeply lobed than those of the latter, which are more consistently lacerate throughout development, though broadly winged initially. The two taxa are easily distinguished in fruit, the red cypselae of *T. erythrospermum* standing out from the dull olive ones of *T. officinale*.

In northeastern North America, *Taraxacum officinale* and *T. lapponicum* often are confused, which has led to reports of the common dandelion farther north than I have been able to verify (it has yet to be collected from the Nunavik region of Quebec, for instance). The characters in the key above help separate the two taxa.

The typification by A. J. Richards (1985) would leave the common dandelion of both Europe and North America without a valid name (J. Kirschner and J. Štěpánek 1987). For the time being, with the nomenclatural situation still not resolved, I am following traditional usage of the name *Taraxacum officinale*.

2. **Taraxacum latilobum** de Candolle in A. P. de Candolle and A. L. P. P. de Candolle, Prodr. 7: 146. 1838 • Large-lobed dandelion, pissenlit à lobes larges 〔E〕

Plants 9–75 cm; taproots seldom branched. **Stems** 1–7+, erect to ascending, ± purplish, glabrate, sometimes ± sparsely villous distally. **Leaves** 10+, erect to patent; petioles ± narrowly winged; blades broadly oblanceolate to narrowly obovate (often runcinate), 15–47.5 × 2.5–9 cm, bases attenuate, margins usually shallowly lobed to sometimes lacerate (mostly proximally), lobes retrorse or straight, broadly deltate to triangular, sometimes antrorsely curved apically, acute to acuminate, teeth 5–7 on lobes and in sinuses, irregular, triangular or sometimes lanceolate, terminals broader than laterals,

apices obtuse to bluntly short-caudate or -acuminate (rarely acute), faces glabrous or glabrate to sparsely pilose or villous. **Calyculi** of ca. 18, reflexed to recurved, ovate to broadly lanceolate bractlets in 3 series, 8–11 × 2–5 mm, margins narrowly scarious, sometimes proximally more widely so, apices acuminate, hornless, ± scarious-erose, tips often purplish and blackish. **Involucres** dark green, often purplish-tinged, campanulate, (13–)15–23 mm. **Phyllaries** 14–18 in 2 series, lanceolate to linear (outer) or ovate to lance-ovate (inner), 1.2–4 mm wide, scarious, narrowly (outer, distal part of inner) or widely (proximal part of inner), apices long-acuminate, hornless, scarious, erose, hyaline, purplish-grayish. **Florets** ca. 150; corollas yellow (outer dark gray striped abaxially, also purplish), 13–17 × 1–1.3 mm. **Cypselae** olive-tan to tan, bodies oblanceoloid, 2.8–3.8 mm, cones terete, 0.9–1 mm, beaks slender, 8–12 mm, ribs 5, wide (with 2–3 rows of tubercles or spines), faces proximally tuberculate, muricate in distal ⅓–½, distalmost spines sometimes very sharp, fused in pairs and flattened; **pappi** white to creamy, 6.5–7 mm.

Flowering summer. Seaside calcareous slopes and grassy taluses or clifftops; 0–30 m; Nfld. and Labr. (Nfld.), Que.; Maine.

Taraxacum latilobum is known only from the Gulf of St. Lawrence and adjacent areas.

3. **Taraxacum erythrospermum** Andrzejowski ex Besser, Enum. Pl., 75. 1822 • Red-seeded dandelion, pissenlit à graines rouges 〔I〕

Taraxacum laevigatum (Willdenow) de Candolle var. *erythrospermum* (Andrzejowski ex Besser) J. Weiss; *T. officinale* F. H. Wiggers var. *erythrospermum* (Andrzejowski ex Besser) Babington; *T. scanicum* Dahlstedt

Plants (1–)5–30(–60) cm; taproots seldom branched. **Stems** 1–15+, ascending to erect, pinkish to reddish, (± equaling foliage), glabrous or sparsely villous, usually more densely so distally. **Leaves** 20+, horizontal to erect; petioles ± slightly winged distally; blades obovate to oblanceolate (runcinate), 5–25 × 1–4 cm, bases attenuate, margins lacerate, lobes retrorse, triangular to nearly lanceolate, acute to long-acuminate, terminals about as large as distal laterals, teeth usually few, rarely 0, irregular, straight to retrorse, minute to pronounced or secondary lobules, apices usually acute or acuminate, sometimes obtuse, rarely rounded, faces glabrous or glabrate to sparsely villosulous (mainly midveins). **Calyculi** of 16–18, reflexed to recurved, sometimes glaucous or purplish, lanceolate bractlets in 2 series, 3.8–10 × 1–2 mm, margins white to purplish, narrowly scarious, apices acute or long-acuminate, erose, hornless.

Involucres green, tips reddish gray, urceolate (closed) to cylindro-campanulate (open), 10–25 mm. **Phyllaries** 18–19 in 2 series, lanceolate-linear, 1.2–2.1 mm wide, margins scarious, slightly revolute at green edge, apices long-acuminate, erose-scarious, usually at least some horned. **Florets** ca. 70–75+; corollas sulphur yellow, outer with abaxial purplish or grayish stripe, 12–16 × 1–1.5 mm. **Cypselae** brick red to reddish brown or reddish purple, bodies oblanceoloid to obovoid, (2.2–) 2.5–3(–4) mm, cones terete, 0.8–1.3 mm, beaks slender, (5–)7–8.5 mm, ribs ca. 15, sharp, faces proximally ± tuberculate (sometimes barely so) to conspicuously muricate in distal ¹/₂; **pappi** white to sordid, 4–7 mm. $2n$ = 16, 24, 32 (Europe).

Flowering nearly year-round (except summer, south, to spring, north). Waste grounds, roadsides, lawns; 0–1500+ m; introduced; Alta., B.C., Man., N.B., N.W.T., N.S., Ont., Que., Sask., Yukon; Ala., Alaska, Ariz., Ark., Ga., Iowa, Maine, Mass., Mich., Miss., Nebr., N.J., N.Mex., N.C., N.Dak., Ohio, Okla., Oreg., Pa., Tex., Wash., W.Va., Wis., Wyo.; Europe; also introduced in Mexico.

Early leaves of *Taraxacum erythrospermum* sometimes may be broadly winged along the midvein, making distinction from *T. officinale* difficult; usually, its later leaves become more deeply lobed with time.

The name *Taraxacum laevigatum* has been used for *L. erythrospermum* in North America, following H. Handel-Mazzetti (1907). L. H. Shinners (1949) questioned that usage. The name is listed in the index of *Flora Europaea* (A. J. Richards and P. D. Sell 1973) as an unassigned synonym; it could be related to three different entities of sect. *Spectabilia*. And, it is not mentioned by other modern students of the group. Therefore, (1) given that the North American entity has not been identified with a particular Eurasian taxon; (2) to avoid using a microspecies name such as *T. scanicum*; and (3) despite the lack of typification of the name, I am using *T. erythrospermum* as a place holder until nomenclatural issues are resolved. This clearly associates the taxon with the section to which it belongs.

4. **Taraxacum palustre** (Lyons) Symons, Syn. Pl. Ins. Brit., 172. 1798 • Marsh dandelion, pissenlit palustre [I]

Leontodon palustris Lyons, Fasc. Pl. Cantabr., 48. 1763 (as palustre); *Taraxacum turfosum* (Schultz-Bipontinus) van Soest

Plants 5–6.5(–10 in fruit) cm; taproots seldom branched. **Stems** 1–5+, decumbent to ascending, purple, (rarely exceeding foliage before fruiting), sparsely villous or glabrate to ± densely villous distally. **Leaves** 10+, horizontal to ± erect; petioles often purplish (midveins also), ± narrowly winged; blades oblanceolate to linear-oblanceolate, 2.5–9 × 0.2–1.1 cm, bases attenuate to long-cuneate, margins usually toothed, sometimes pinnately, shallowly lobed, lobes fewer than 10 per side, remote, (and teeth) straight to retrorse, narrow, deltate to narrowly triangular, often acuminate, apices obtuse to acute, faces glabrous or sparsely villous (particularly along midveins). **Calyculi** 10–15, appressed to spreading, pale to dark purplish green, ovate to elliptic bracklets in 2 series, 6–8.5 × 2–5 mm, margins ± purplish, widely scarious, apices acuminate to caudate, hornless. **Involucres** green to grayish green, campanulate, 12–16 mm. **Phyllaries** 14–16 in 2 series, lanceolate to lance-linear, 1.5–2.5 mm wide, margins scarious to narrowly scarious in proximal ¹/₂, apices long-acuminate, erose-scarious, hornless. **Florets** ca. 50; corollas yellow, outer abaxially gray-striped, 13–14 × 1.4–2.2 mm. **Cypselae** straw-colored to olivaceous, bodies narrowly obovoid, 2.8–4 mm, cones terete, 0.8–1 mm, beaks slender, 7–9 mm, ribs ca. 6, sharp, faces slightly muricate in distal ¹/₃; **pappi** white to cream, 4.5–6.5 mm. $2n$ = 24, 32, 40 (reported for complex in Europe).

Flowering early spring. Wet ditches, roadsides, and waste grounds of temperate climates; 10–100 m; introduced; Ont., Que.; Mich., N.Y.; Europe.

This small dandelion has only recently been reported from North America (D. F. Brunton 1989). It has now spread east into Quebec, where it is known beyond the Montreal area, west into Michigan, and south into northern New York state. It is mostly spreading in wet ditches along highways and flowers in early spring.

It clearly belongs to sect. *Palustria*. Until it is firmly associated with a European species, I am using the name *Taraxacum palustre*. The name *T. cognatum* Štěpánek & Kirschner, which designates a microspecies from central Europe, has been applied to a North American specimen sent by D. F. Brunton to J. Kirschner.

5. **Taraxacum lapponicum** Kihlman ex Handel-Mazzetti, Monogr. Taraxacum, 73. 1907 • Lapland dandelion, pissenlit de Laponie

Taraxacum alukense R. Doll; *T. ambigens* Fernald; *T. atroglaucum* M. P. Christensen; *T. campylodes* G. E. Haglund; *T. croceum* Dahlstedt; *T. curvidens* M. P. Christensen; *T. cyclocentrum* M. P. Christensen; *T. davidssonii* M. P. Christensen; *T. dilutisquameum* M. P. Christensen; *T. firmum* Dahlstedt; *T. latispinulosum* M. P. Christensen; *T. naevosum* Dahlstedt; *T. obtusatum* Dahlstedt; *T. pleniflorum* M. P. Christensen; *T. spectabile* Dahlstedt; *T. torngatense* Fernald

Plants (3–)5–37(–47 in fruit) cm; taproots sometimes branched. **Stems** 1–5, erect to ascending, purplish, (usually exceeding leaves), glabrate to sparsely villous proximally, ± densely villous distally. **Leaves** 5–20, erect to patent, sometimes horizontal; sessile (bases sometimes as wide as blade) to ± broadly winged petiolate (occasionally slender on young or deeply shaded specimens, usually at least some ± winged); blades oblanceolate or oblong-oblanceolate (often runcinate), (3–)3.5–27 × 0.6–5 cm, bases usually cuneate, sometimes attenuate, margins regularly and shallowly lobed, sometimes ± deeply (not lacerate) to dentate, lobes mostly retrorse, usually triangular, sometimes deltate or lanceolate, ± acuminate, often replaced by teeth, teeth 0–5 on lobes or irregular, often coarse, triangular to lanceolate, ± acuminate, apices obtuse or acute to short-acuminate, faces glabrous or glabrate, midveins often sparsely villous. **Calyculi** of 10–12, spreading to eventually reflexed, often pale, sometimes purplish-tinged, lance-ovate to broadly lanceolate (thin) bractlets in 3 series, 5.5–9.5 × 1.8–2.7(–4.2) mm, margins not or narrowly scarious, apices long-acuminate, hornless. **Involucres** green to dark green, campanulate to cylindro-campanulate, 12–22 mm. **Phyllaries** 14–18 in 2 series, lanceolate, 1.4–3.2 mm wide, margins not or very narrowly scarious (outer) to narrowly scarious in proximal 1/2 (inner), apices long-acuminate, hornless, tips ± hyaline, blackish and/or purplish, scarious. **Florets** 60–110+; corollas yellow (outer abaxially gray and/or purplish striped), 15–22 × 1.1–1.7 mm. **Cypselae** tan to reddish brown, bodies oblanceoloid, 2.8–3.5 mm, cones conic, 0.5–0.8 mm, beaks slender, 6.5–12 mm, ribs 4–5 prominent (to 13–15 fine), faces proximally smooth to occasionally slightly tuberculate, muricate in distal 1/4–1/3 (spines usually sparse); **pappi** white to cream, 5–7.5 mm. $2n = 32$.

Flowering early summer. Arctic marshes, snow patches, moist areas or seepage slopes, with high organic contents, shores of rivers and brooks (south); 0–1100 m; Greenland; Nfld. and Labr., Nunavut, Que.; Eurasia.

Taraxacum lapponicum is an amphi-Atlantic taxon whose North American distribution lies in the eastern Arctic and boreal zones.

This species has the most controversial taxonomy. Barely distinguishable morphotypes have received species names, and the phenotypic plasticity of the taxon has not been taken into account in that process. In particular, it is often found in more shaded habitats than other dandelions in the south of its range, particularly when growing along streams in eastern Canada. The correct name to be applied to this species is controversial, and many European names have been applied to North American plants in the complex, sometimes without regard to the type. Here, I am recognizing a more widely defined species, using the name *Taraxacum lapponicum*, the one most often used in North America (correctly or not), pending more rigorous experimental study of variation in the complex.

The name *Taraxacum islandiciforme* Dahlstedt ex M. P. Chistiansen is invalid.

6. **Taraxacum ceratophorum** (Ledebour) de Candolle in A. P. de Candolle and A. L. P. P. de Candolle, Prodr. 7: 146. 1838 • Horned dandelion, pissenlit tuberculé F

Leontodon ceratophorus Ledebour, Icon. Pl. 1: 9, plate 34. 1829; *Taraxacum ambigens* Fernald var. *fultius* Fernald; *T. angulatum* G. E. Haglund; *T. arctogenum* Dahlstedt; *T. brachyceras* Dahlstedt; *T. carthamopsis* A. E. Porsild; *T. coverum* R. Doll; *T. dumetorum* Greene; *T. eriophorum* Rydberg; *T. eurylepium* Dahlstedt; *T. groenlandicum* Dahlstedt; *T. hyperboreum* Dahlstedt; *T. integratiforme* R. Doll; *T. integratum* G. E. Haglund; *T. lacerum* Greene; *T. lateritium* Dahlstedt; *T. longii* Fernald; *T. mackenziense* A. E. Porsild; *T. malteanum* Dahlstedt ex G. E. Haglund; *T. maurolepium* G. E. Haglund; *T. microcerum* R. Doll; *T. ovinum* Greene; *T. pellianum* A. E. Porsild; *T. pseudonorvegicum* Dahlstedt ex G. E. Haglund; *T. umbriniforme* R. Doll; *T. umbrinum* Dahlstedt ex G. E. Haglund

Plants (1–)6–50 cm; taproots branched. **Stems** 1–10+, ascending to erect, ± purplish (at least proximally), densely villous (young) becoming glabrescent, sparsely villous to glabrate or glabrous proximally, ± densely villous distally. **Leaves** ± 10, horizontal to patent, sometimes erect; sessile or petioles ± broadly winged (bases barely narrowed compared to blades); blades narrowly oblanceolate to linear-oblanceolate or linear-oblong (often ± runcinate), 4–30 × (0.4–)0.5–5 cm, bases cuneate to attenuate, margins lobed ± deeply to lacerate, irregularly to regularly, often toothed, merely denticulate, or subentire, lobes retrorse, straight or antrorse, deltate to triangular, acute to acuminate, teeth 0–1 on lobes, often more or mostly in sinuses, apices obtuse to sometimes acute, sometimes mucronate, faces glabrous or glabrate to very sparsely villous. **Calyculi** of 12–16 (–20), appressed to spreading, pale, ovate to elliptic or lance-ovate to lanceolate (sometimes thin) bractlets in 2–3 series, 5–12 × (0.9–)1.5–5 mm, margins hyaline, white or purplish, scarious, apices caudate to acuminate, ± strongly horned, callous, or occasionally some (rarely all) hornless, tips obtuse to rounded, scarious, erose. **Involucres** dark green, sometimes ± glaucous, campanulate to ± hemispheric, (5–)8–19(–21) mm. **Phyllaries** (10–)12–14(–17) in 2 series, lanceolate to ovate-lanceolate (inner), 1.5–4.5 mm wide, margins scarious or not (outer), inner broadly scarious in proximal 1/2, apices usually horned, occasionally hornless, horns sometimes exceeding apices, tips white to purplish, scarious, erose. **Florets** 40–85+; corollas yellow, drying cream to whitish (outer abaxially gray or purple-striped on drying), 10–22 × 1–2.8 mm. **Cypselae** olivaceous to olive brown, tan to olivaceous tan, brown to reddish brown, grayish brown or straw-colored, bodies oblanceoloid to obovoid, 2.5–4(–5) mm wide, cones conic or narrowly conic to broadly terete, 0.5–0.9 mm, beaks slender, 4.5–14 mm, ribs 5, large (bearing 10–15 narrower ones), faces proximally tuberculate or sometimes nearly smooth (usually with at least some tubercules) to muricate in distal ¹/₃–¹/₂, sometimes wholly muricate; **pappi** white to cream, 5–7.5(–8) mm. **2n** = 16, [24], 32, 40, 48.

Flowering spring–summer. Wet to moist areas, calcareous or igneous rocks, gravel, sand, or clay, wet meadows, shores of streams, sandy or gravelly seashores, seepage slopes, early-melting snowbeds (south); 0–3000 m; Greenland; Alta., B.C., Man, Nfld. and Labr., N.W.T., Nunavut, Ont., Que., Sask., Yukon; Alaska, Calif., Colo., Mont., Nev., N.Mex., Oreg., Utah, Wash., Wyo.; Eurasia.

Taraxacum ceratophorum is the most widespread native dandelion in North America, ranging from the low Arctic and boreal zone to the western Cordilleras, in the montane and alpine zones.

This complex has been subdivided into many microspecies in North America, most of which appear unworthy of recognition. In the Quebec-Labrador Peninsula, *Taraxacum ceratophorum* grades continuously into what has been called *T. hyperboreum*. Inclusion of *T. hyperboreum* bridges the gap between typical *T. ceratophorum* and *T. lacerum*. *Taraxacum lacerum* stands out by its very lacerate leaves, but intermediates exist and it is impossible to draw a firm boundary. The lacerate Newfoundland form, *T. longii*, may be a spontaneous mutation within the range of *T. ceratophorum*. If *T. lacerum* were recognized, we would have to place *T. longii* within the former based on leaf morphology, though leaf orientation would be odd there (*T. lacerum* tends to have ascending leaves, and *T. longii* leaves that are flatter on the substrate). A more thorough morphometric and biosystematic study of this complex is warranted. Nonetheless, I have recognized two segregates (*T. laurentianum*, *T. trigonolobum*) that stand out from the continuum otherwise observed in the complex.

7. **Taraxacum laurentianum** Fernald, Rhodora 35: 375, plate 272, figs. 5–9. 1933 • Gulf of St. Lawrence dandelion, pissenlit du golfe du Saint-Laurent [C][E]

Plants 10–42 cm; taproots branched. **Stems** 1–5+, erect to ascending, purplish, (usually exceeding leaves), glabrate to sparsely villous, sometimes more densely so distally. **Leaves** 5–10+; usually erect, sometimes patent; sessile to ± broadly winged-petiolate; blades oblanceolate (some younger leaves ± runcinate at least proximally), 10–30 × 2–4 cm, bases cuneate to attenuate, margins

sometimes ± deeply lobed (younger leaves), usually toothed, larger teeth well developed, lobelike, retrorse to antrorse, lanceolate to triangular or sometimes deltate, sometimes double, acuminate, apices obtuse, faces adaxially sparsely pilose or glabrate (denser on midveins), abaxially glabrous or glabrate. **Calyculi** 15–18, ascending, later spreading and recurving, greenish or purplish, broadly lanceolate to narrowly ovate (herbaceous) bractlets in 2–3 series, 15–21 × 3.5–5.5 mm, margins hyaline, often purplish, narrowly scarious, apices long- to short-acuminate, strongly horned, tips purplish, scarious, erose. **Involucres** green, broadly campanulate, 18–26 mm. **Phyllaries** 18–25 in 2 series, lanceolate (outer) to ovate-lanceolate (inner), 2.5–5.2 mm wide, margins scarious narrowly or not (outer) to broadly so proximally and narrowly distally (inner), apices long-acuminate (sometimes tapered), strongly horned, tips scarious, purplish black, erose. **Florets** ca. 150+; corollas yellow (abaxially gray-striped, becoming purplish in drying), 14–25 × 0.8–1.5+ mm. **Cypselae** grayish-olivaceous to tan or olivaceous straw, bodies oblanceoloid to obovoid, 3.2–4 mm, cones narrowly conic, 0.9–1.2 mm, beaks slender, 10–17 mm, ribs 5 (wide)–14 (narrow), faces proximally ± tuberculate (sometimes ribs smooth), muricate in distal ¹/₃–¹/₂; **pappi** creamy, 8–9.5 mm. **2n** = 40 [unpublished].

Flowering summer. Calcareous, seashore meadows and gravelly seashores (coastal), turfy areas of seaside taluses, sandy riverside meadows in boreal forest zone; of conservation concern; 0–5 m; Nfld. and Labr. (Nfld.), Que.

Though very similar to *Taraxacum ceratophorum*, this Gulf of St. Lawrence endemic stands out by its statistically significantly larger heads and other parts (L. Brouillet, unpubl. data). It also occupies a habitat atypical of normal *T. ceratophorum* in the area. It is possible that the species is under threat from its introduced congener *T. officinale*, and from human activities in its fragile, easily disturbed coastal habitat.

8. **Taraxacum californicum** Munz & I. M. Johnston, Bull. Torrey Bot. Club 52: 227. 1925 • California dandelion [C][E]

Taraxacum ceratophorum (Ledebour) de Candolle var. *bernardinum* Jepson

Plants 3.5–10(–20) cm (to 30 cm in fruit); taproots branched. **Stems** 1–10+, ascending to decumbent (at and after flowering) (occasionally erect), usually purplish, (mostly at or below foliage before fruiting) glabrous, often cobwebby-villous basally. **Leaves** 10–20, horizontal to patent; petioles usually broadly, sometimes narrowly winged;

blades oblanceolate to narrowly oblong, sometimes linear-oblanceolate, 5–12 × 1–2(–3) cm, bases cuneate, margins toothed or denticulate, occasionally with basal lobes, rarely some entire, lobes straight or slightly retrorse, lanceolate to triangular-long-acuminate, teeth narrowly to broadly triangular, straight to retrorse, apices obtuse to rounded or ± acute, faces glabrate or glabrous. **Calyculi** 9–12, appressed, green, sometimes ± hyaline, often purplish, ovate-lanceolate to broadly ovate bractlets in 2 series, 5–7 × 1.7–4.3 mm, margins hyaline, purplish (at least distally) or white, scarious, apices acuminate to long-acuminate or widely caudate, tips ± rounded, erose, hornless. **Involucres** green, tips purplish gray, broadly campanulate, 11–16 mm. **Phyllaries** 12–16 in 2 series, lanceolate to lanceolate-linear, 2–3.5 mm wide, margins not or narrowly scarious, sometimes broadly so basally, hyaline, apices rounded, erose, scarious. **Florets:** corollas pale yellow, outer abaxially striped pale purplish, 9–10 × 1.5–2. **Cypselae** pale brown, bodies obovoid, flattened, 2.2–2.7 mm, cones narrowly conic, 0.5–0.7 mm, beaks slender, 6–9.5, ribs 15–16, narrow, faces muricate distal $^2/_3$–$^3/_4$; **pappi** creamy, 4.5–5.5 mm. $2n = 16$.

Flowering late spring–summer. Moist alpine meadows in yellow pine forest zone; 1900–2400 m; Calif.

Taraxacum californicum is known only from the San Bernardino Range. It is easily distinguished from *T. ceratophorum* by the lack of horns on the phyllaries and bractlets of calyculi.

9. Taraxacum trigonolobum Dahlstedt, Ark. Bot. 20A(1): 8, fig. 5. 1926

Plants 8–41(–56.5 in fruit) cm; taproots seldom branched. **Stems** 1–5, ascending to erect, purplish, glabrous or glabrescent proximally, ± densely villous distally. **Leaves** fewer than 10, horizontal or patent to ± erect; petioles widely winged; blades oblanceolate-obovate (largest in distal $^1/_5$) to oblanceolate (often runcinate, particularly early leaves), (3.5–)6–29.5 × (1.3–)2.1–5.5 cm, bases cuneate to attenuate, margins sometimes lobed proximally or ± regularly and deeply, usually ± strongly toothed proximally (less so on distal $^1/_2$–$^1/_5$), lobes retrorse to straight or antrorse, triangular to deltate, ± acuminate, teeth mostly proximal, rarely on distal $^1/_4$–$^1/_5$ or terminal lobes, small to coarse, irregular (1–5 on lobes), apices rounded or truncate to obtuse, rarely acute, apiculate, faces glabrous or glabrate (very sparsely pilose, mainly on midveins). **Calyculi** 14–20, appressed (to spreading), often pale green, ovate to elliptic-ovate bractlets in 2–3 series, 5–8 × 2.8–4.5 mm, margins narrowly white, scarious (occasionally outer ± ciliate), apices acuminate

to caudate, some strongly horned, tips dark-scarious. **Involucres** dark green, ± glaucous, broadly campanulate, 15–20 mm. **Phyllaries** 14–20 in 2 series, lanceolate, 2.1–2.8 mm wide, margins ± narrowly scarious (wider proximally, or not in outer series), apices strongly horned, tips purplish with dark center, scarious, ± erose. **Florets** 120–150; corollas yellow, outer grayish-purplish striped abaxially, 19–25 × 1.3–2.3 mm. **Cypselae** brown to reddish brown, bodies oblanceoloid, 3–3.5 mm, cones conic, 0.8–1 mm, beaks slender, 7.5–10 mm, ribs 4–5 prominent (to 15 fine), faces proximally smooth (faintly tuberculate), muricate in distal $^1/_4$; **pappi** yellowish or cream to white, 7–8 mm.

Flowering summer. Herbaceous vegetation, lakeshores, moist banks, alluvial shores, moist meadows, moist heaths, grassy cliff tops; 0–600 m; Alaska; Russian Far East (Kamchatka).

Taraxacum trigonolobum is known only from the Aleutian Islands and islands of the Bering Sea, on the American side. It is probably the taxonomically weakest entity that I am recognizing in this complex, but its distribution and leaf morphology, among other attributes, are distinctive enough for recognition at the present time.

10. Taraxacum hyparcticum Dahlstedt, Ark. Bot. 4(8): 17, fig. 3. 1905 • High-Arctic dandelion

Plants (3–)5–12(–30) cm; taproots occasionally branched. **Stems** 1–5, ascending to erect, pinkish to reddish or purplish, (barely exceeding foliage), glabrous or glabrate (rarely sparsely villous and villous distally). **Leaves** fewer than 10, usually patent, rarely erect; petioles sometimes narrowly winged; blades oblanceolate (often ± runcinate), 2–12+ × (0.3–)0.5–1.2 cm, bases attenuate, margins lobed deeply and regularly to denticulate or subentire, lobes 5–6 pairs, straight to retrorse, triangular to deltate, teeth triangular 0–1 on lobes or 1–4 shallow pairs if subentire, acute to ± obtuse, apices obtuse to ± acute, faces glabrous. **Calyculi** of 10–14, spreading to appressed (thinner than phyllaries), dark green, broadly ovate or ovate to oblong bractlets in 2–3 series, 6–10 × 3.4–5 mm, margins narrowly hyaline, scarious, apices abruptly acuminate to caudate, strongly horned, tip ± scarious, erose. **Involucres** dark green to bluish black or dark purplish green, narrowly campanulate, 15–30 mm. **Phyllaries** 8–14 in 2 series, lanceolate to lance-ovate, 1.5–3.5 mm wide, margins narrowly (outer) to widely scarious in proximal $^1/_3$ (inner), narrowly so distally, apices long-acuminate, sometimes callous, hornless (rarely very small horns), tips white-scarious, erose, rounded. **Florets** 25–50; corollas cream-colored to white or pink-tinged distally, outer pinkish-striped abaxially,

15–20 × 1.2–3 mm. **Cypselae** tan or straw-colored to brown (or reddish brown), sometimes grayish, bodies obovoid to oblanceoloid, 3–3.7(–4) mm, mostly broad, cones conic, 0.4–0.7 mm, beaks slender, 3.5–4.5 mm, ribs 10–13 (5–6 prominent), mostly broad, faces proximally tuberculate, muricate in distal ^{1}/$_2$; **pappi** whitish or sordid, 5–5.5(–7) mm. **2*n*** = [24, 32, 40? some erroneous reports in literature under this name from Eurasia].

Flowering summer. Dry, moderately drained areas in tundra, raised sand terraces, low center polygons on old surfaces, sandy, eroded knolls, marine/lacustrine deposits, rocky streambeds, dry slopes; 20–1010 m; Greenland; N.W.T., Nunavut; Alaska; Eurasia.

Taraxacum hyparcticum is sporadic in Arctic Eurasia (coastal Russia from Russian Far East west to Novaya Zemlya); it is mainly high-arctic. It is characterized by its small stature and large heads with white to yellowish cream ligules.

11. **Taraxacum phymatocarpum**   J. Vahl in G. C. Oeder et al., Fl. Dan. 13(39): 6, plate 2298. 1840
   • Northern dandelion

Plants 2–12(–30) cm (longer in fruit); taproots branched. **Stems** 1–(2–5), erect to ascending, reddish, glabrous. **Leaves** fewer than 10, usually horizontal, sometimes patent, rarely erect; petioles slightly winged; blades oblanceolate to linear-oblanceolate (sometimes some nearly runcinate), 1.5–8 × 0.2–0.8(–1.3) cm, bases attenuate, margins usually toothed to denticulate in 1–5 pairs or entire, sometimes shallowly lobed, lobes (and teeth) retrorse or straight, triangular, apices acute to obtuse, faces usually glabrous, sometimes very sparsely villous (mostly along midveins). **Calyculi** of 8–14, appressed to spreading, later reflexed to revolute, sometimes purplish-tinged, broadly ovate or ovate to lance-ovate bractlets in 2–3 series, 3–5 × 1.8–2.5 mm, margins not or barely scarious, apices usually acuminate, sometimes acute, hornless, tips slightly scarious-erose. **Involucres** dark to blackish green, narrowly campanulate (urceolate when closed), 9–14 mm. **Phyllaries** 8–12 in 2 series, oblong-lanceolate or broadly lanceolate to lance-ovate (inner), 1.8–3 mm wide, margins not or narrowly scarious (outer) to widely so in proximal ^{1}/$_4$–^{1}/$_2$, occasionally purplish, apices sometimes callous, hornless, tips scarious, erose. **Florets** 35–50; corollas pale yellow (sometimes lemon), sometimes drying pinkish (not striped adaxially), outer abaxially gray to purple striped, 10–12 × 2–3.1 mm. **Cypselae** dark brown, grayish to blackish, bodies oblanceoloid, flattened, (3–)4–4.5 mm, cones conic, 0.5–0.9 mm, beaks stout, (2–)3–5 (^{3}/$_4$+

length of body), ribs 15 (5 prominent), faces proximally at least tuberculate, muricate ^{1}/$_2$–^{3}/$_4$+; **pappi** creamy or white, 4–5.5(–7.5) mm. **2*n*** = 24, 32, 40.

Flowering summer. Slopes to stream banks in tundra, usually dry to drained areas, usually calcareous, rocky taluses, gravel, sand, clay, exposed gravelly-turfy limestone barrens (south); 0–700 m; Greenland; Nfld. and Labr. (Nfld.), N.W.T., Nunavut; Alaska; Eurasia.

A single disjunct population of *Taraxacum phymatocarpum* was found by M. L. Fernald and colleagues on Burnt Cape, Northern Peninsula, Newfoundland; otherwise the species is arctic. This diminutive species is characterized by its mostly unlobed leaves, pale yellow ligules, and usually dark cypselae.

12. **Taraxacum scopulorum**   (A. Gray) Rydberg, Mem. New York Bot. Gard. 1: 455. 1900   • Alpine dandelion   E F

Taraxacum officinale F. H. Wiggers var. *scopulorum* A. Gray in A. Gray et al., Syn. Fl. N. Amer. 1(2): 440. 1884; *Leontodon scopulorum* (A. Gray) Rydberg

Plants 1–5 cm; taproots branched. **Stems** 1–3(–5), ascending to erect, reddish or purplish, glabrous. **Leaves** fewer than 10, horizontal; petioles ± narrowly winged; blades linear-oblanceolate to narrowly oblanceolate (usually runcinate), (1–)1.5–4 × 0.5–1 cm, bases attenuate, margins lobed regularly, deeply, in 3–5(–7) pairs (occasionally dentate, rarely entire or subentire), lobes triangular, straight or often retrorse (also teeth), acute to acuminate or obtuse, teeth 0 or occasionally 1–3 on lobes (or replacing lobes), apices acuminate, faces glabrous or sparsely arachnoid-villous (particularly young). **Calyculi** (6–)8–10, appressed to spreading, soon reflexed or revolute (thinner than phyllaries), pruinose, sometimes purplish (particularly adaxially), narrowly to broadly ovate bractlets in (1–)2 series, 2.5–3.8 × 1.8–2.6 mm, margins not to ± scarious proximally, apices acuminate to broadly, shortly caudate, hornless, not callous, tips slightly scarious (purple), entire or slightly erose. **Involucres** dark to blackish green, pruinose, sometimes purplish, narrowly campanulate to cylindro-campanulate or cylindric, (6–)10–12(–14) mm. **Phyllaries** 8–12 in 2 series, lanceolate (outer) to ovate (inner), 1.5–3.1 mm wide, margins not scarious (outer) or broadly so, narrowing in distal ^{1}/$_2$ (green zone often darker marginally), apices acuminate, hornless, not callous, entire or tips slightly scarious-erose. **Florets** (ca. 15–)35–50; corollas yellow (sometimes gray-striped abaxially), 7.5–8.8 × 1.3–2 mm. **Cypselae** pale reddish brown or reddish-tinged or straw-colored to dark gray, bodies oblanceoloid, 2.8–3.5 mm, cones conic, 0.5–0.6 mm, beaks stout, 2.8–4.5, ribs 7 (wide) to 14 narrow

(3–5 prominent), faces proximally smooth to tuberculate, muricate in distal ¹/₂; **pappi** white, 4.8–5.5 mm. **2*n*** = 16? (as *pumilum*).

Flowering summer. Dry to moderately drained areas, gravel, sand, silt, clay, with low organic content, ridges, arctic seashores, barrens, flood plains, high alpine tundra (south); 40–3400 m; Greenland; Alta., B.C., N.W.T., Nunavut; Colo., Idaho, Utah, Wyo.

Taraxacum scopulorum is known from the western Canadian Arctic Archipelago and from high-alpine summits in the western Cordilleras. This small species is characterized by lobed leaves, yellow ligules, and cypselae that are paler and less muricate than those of *T. phymatocarpum*.

The name *Taraxacum lyratum* Ledebour has been used for this species since the work of H. Handel-Mazzetti (1907), who annotated the type of *T. scopulorum* (GH). That type originated from the mountains of central Asia, and in his monograph Handel-Mazzetti included Kamchatka and northern Alaska in its range; I have not seen specimens of the species from Alaska. Also, his concept of *T. lyratum* seems to include heterogeneous elements, notably what is called here *T. alaskanum*. At the present time, I prefer to treat this species as restricted to North America, where the name *T. scopulorum* clearly applies to this very distinct, diminutive dandelion. It appears to be closely related to *T. phymatocarpum*, but the latter has mostly entire or slightly dentate leaves (some leaves sometimes slightly deltate-lobed), as opposed to the wholly deltate-lobed ones of *T. scopulorum*. A. E. Porsild (1957) illustrated this species under the name *T. pumilum*.

13. Taraxacum holmenianum Sahlin, Folia Geobot. Phytotax. 18: 445. 1983 • Holmen's dandelion Ⓔ

Taraxacum pumilum Dahlstedt, Ark. Bot. 4(8): 27, fig. 5. 1905, not Gaudichaud 1825

Plants 5–15(–20) cm; taproots branched. **Stems** 1–7+, ascending, purple, glabrous or glabrate. **Leaves** ca. 10+, horizontal to widely patent, rarely ± erect, (dark green); petioles winged mostly narrowly; blades oblanceolate to oblong-oblanceolate (runcinate), (1.5–)2–6(–9) × 0.3–1.3 cm, bases attenuate, margins lobed regularly, ± alternately, shallowly to deeply, lobes usually straight to slightly retrorse, sometimes slightly antrorse, deltate to triangular, acute to acuminate, apices acute to acuminate, faces glabrous. **Calyculi** of 10–14, appressed to slightly spreading, purplish, ovate to lance-ovate bracklets in 2 series, 3–6 (–7.5) × 1.8–2.5(–3) mm, not or narrowly scarious, hyaline purplish, apices long-acuminate to short-caudate, tips scarious, erose. **Involucres** dark olive green,

sometimes purplish, tips purplish, campanulate, 11–15 (–20) mm. **Phyllaries** ca. 14 in 2 series, lanceolate to lance-ovate, 1–3 mm wide, margins not scarious or narrowly to widely scarious in proximal ¹/₂, sometimes distally, apices sometimes callous, sometimes also some small-horned, tips scarious, erose. **Florets** 30–50(–60); corollas dark yellow (outer with distinct, central greenish gray or pinkish stripe, tips dark), 15–20 × 1–3 mm. **Cypselae** yellowish or straw-colored, bodies obovoid to oblanceoloid, 3.5–4.2 mm wide, cones narrowly conic, 0.9–1.1 mm, beaks slender, 4.5–5 mm, ribs 12–16 (3–4 prominent), sharp, faces proximally tuberculate, muricate in distal ¹/₂; **pappi** white to yellowish, 5.5–6.5 mm. **2*n*** = 16.

Flowering summer. Tundra hummocks and polygons, open slopes, on clay or gravel; 0–300+ m; Greenland; N.W.T., Nunavut.

Taraxacum holmenianum is a strictly Arctic species. It is distinguished by its deep yellow ligules.

14. Taraxacum alaskanum Rydberg, Bull. Torrey Bot. Club 28: 512. 1901 • Alaska dandelion

Taraxacum kamtschaticum Dahlstedt; *T. pseudokamtschaticum* Jurtzev

Plants (1.5–)3–9(–16, mostly in fruit) cm; taproots sometimes branched. **Stems** 1–3+, ascending, proximally purplish, glabrous or glabrate. **Leaves** fewer than 10, horizontal to patent (green); petioles slender or decurrent lines from bases; blades oblanceolate or narrowly oblong (usually runcinate), (1.5–)2.2–11.6 × 0.4–2.2 cm, bases attenuate, margins usually lobed regularly, ± deeply, in 3–5(–6) pairs, occasionally (younger) only toothed or denticulate, lobes usually retrorse, occasionally straight or antrorse, triangular to narrowly triangular, terminals often largest, acuminate to acute-rounded, sometimes toothed, teeth 0(–1) on lobes or sinuses, triangular, apices obtuse to acute, faces glabrous. **Calyculi** of 7–9, spreading, becoming reflexed to revolute, often purplish, particularly adaxially, ovate to lance-ovate or elliptic bracklets in 2 series, 2.5–4.5 × 1.7–2.7 mm, margins not or narrowly scarious, hyaline, apices acuminate to caudate, tips sometimes flaring, scarious, erose, hornless. **Involucres** dark green, often glaucous, often purplish, particularly adaxially, cylindro- to narrowly campanulate, 9–14 mm. **Phyllaries** 7–8 in 3 series, lanceolate, 1.4–2.6(–4) mm wide, margins not scarious (some outer) to narrowly scarious, apices long-acuminate, tips purplish or grayish, often flared, scarious, hornless. **Florets** 30–55+; corollas yellow (gray-striped abaxially, sometimes becoming orange purplish with age on drying), outer 11–14 × 1.5–2.4 mm. **Cypselae** maroon to brown or reddish brown, sometimes grayish, bodies oblanceoloid (sometimes narrowly), 3–3.8 mm, cones conic, 0.8–1 mm, beaks stout, 3–6 mm, ribs 15 narrow (6 prominent), faces proximally ± tuberculate, ± muricate in distal $^1/_3$–$^1/_2$ (or less); **pappi** white to yellowish, 4–6.5 mm. $2n = 24, 32$ (as *T. kamtschaticum*).

Flowering summer. Alpine slopes and tundra, arctic tundra, rich arctic seaside bluffs; 0–2200 m; Yukon; Alaska; Russian Far East.

In all the specimens examined from Alaska, it is not possible to find consistent differences between *Taraxacum alaskanum* and *T. kamtschaticum* (as applied in Alaska; or *T. pseudokamtschaticum*). Assignment to one or the other species appears random. The leaf character used in keys to separate the entities does not work, or at least is not matched by specimens as determined; cypsela color varies within species, and the slight difference noted is not sufficient to warrant separation. Neither is there a significant size difference between coastal and inland material. The name *T. sibiricum* Dahlstedt has been applied mistakenly by American authors to this entity.

15. **Taraxacum carneocoloratum** A. Nelson, Amer. J. Bot. 32: 290. 1945 • Pink dandelion [E]

Plants 3.5–9 cm; taproots branched. **Stems** 1–3, ascending to erect, purplish proximally to completely (barely exceeding foliage), glabrous. **Leaves** fewer than 10(–15), horizontal to patent or ± erect; petioles narrowly winged (mostly distally); blades oblanceolate (sometimes runcinate), 2.8–8 × 0.5–1.4 cm, bases attenuate, margins lobed shallowly (about $^1/_2$ width of blades or less) or toothed, lobes straight or retrorse, sometimes antrorse, triangular or deltate to lanceolate, obtuse to acute or acuminate, teeth 0 on lobes, apices obtuse, faces glabrous. **Calyculi** of 10–12, appressed (to spreading), very widely ovate (outer) to ovate, dark green, often purple-tipped bractlets in 2(–3) series, 4–5.5 × 2.5–4.5 mm, white-scarious to not scarious, apices acuminate to caudate, hornless, scarious-erose. **Involucres** dark green, broadly campanulate, (10–)12–16 mm. **Phyllaries** 12–14 in 2 series, ovate to lance-ovate, 2.8–5.5 mm wide, margins not scarious (some outer) to broadly so (at least proximally, inner), apices acuminate, inner scarious and erose, hornless. **Florets** 20–30+; corollas pink purplish to pinkish cream (± bronze when fresh), outer 13–14 × 2.4–2.6 mm. **Cypselae** grayish, greenish, or yellowish, bodies oblanceoloid, ca. 3–4+ mm, cones [mature not seen], beaks stout, ribs [mature not seen], faces proximally smooth, muricate in distal $^1/_4$; **pappi** yellowish, ca. 7 mm.

Flowering summer. High alpine, gravelly areas and scree slopes, ridge crests, dry substrates; 500–2500 m; Yukon; Alaska.

Taraxacum carneocoloratum may be associated with unglaciated areas of Alaska and Yukon, where it is infrequent. It is easily distinguished in bloom by its pink ligules (± bronze when fresh).

38. CHONDRILLA Linnaeus, Sp. Pl. 2: 796. 1753; Gen. Pl. ed. 5, 348. 1754 • Gum or Spanish succory [Name used by Dioscorides for plant that exudes milky juice or gum] [I]

L. D. Gottlieb

Perennials, 40–150 cm; taprooted. **Stems** 1–6, erect or ascending, much branched, basally setose, distally glabrous. **Leaves** basal and cauline; petiolate (basal and proximal cauline, petioles winged); blades (at least basal) oblanceolate, pinnatifid (often runcinate) or coarsely and irregularly toothed (apices acute; distal cauline reduced, entire). **Heads** (terminal and axillary) usually borne singly (sometimes clustered along stems and branches). **Peduncles** (nearly 0) not inflated, not bracteate. **Calyculi** of 3–4, minute bractlets. **Involucres** cylindric, 2.5–5 mm diam. **Phyllaries** 5–9 in ± 1 series, linear-lanceolate, equal. **Receptacles** pitted, glabrous, epaleate. **Florets** 7–15; corollas (soon withering) yellow. **Cypselae** tan to black, nearly cylindric, beaks slender or stout, ribs 5+ (with alternating grooves), faces glabrous; **pappi** persistent (on expanded discs at tips of beaks), of 40–50+, distinct, white, smooth bristles in 1 series. $x = 5$.

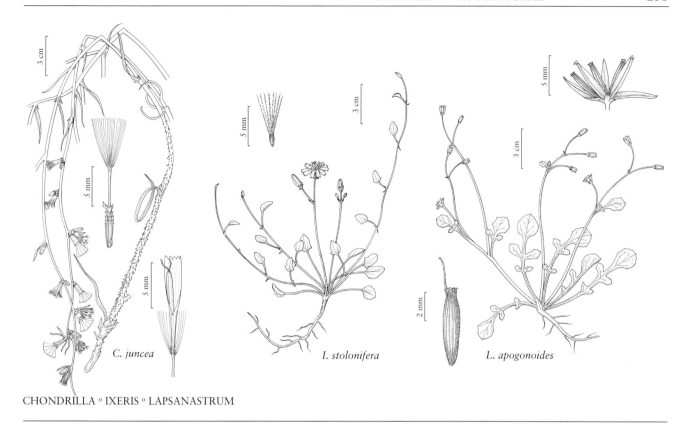

CHONDRILLA ○ IXERIS ○ LAPSANASTRUM

C. juncea   I. stolonifera   L. apogonoides

Species ca. 25 (1 in the flora): introduced; Eurasia.

SELECTED REFERENCE Heap, J. W. 1993. Control of rush skeletonweed (*Chondrilla juncea*) with herbicides. Weed Technol. 7: 954–959.

1. Chondrilla juncea Linnaeus, Sp. Pl. 2: 796. 1753

• Rush skeletonweed F I W

Stems with retrorse, coarse, bristly hairs on basal 10–15 cm, distally glabrous. **Leaves:** basal withered before flowering, blades 5–13 × 1.5–3 cm. **Involucres** 9–12 mm. **Phyllaries** tomentose. **Cypselae:** bodies 3–4 mm, beaks 5–6 mm, ribs with antrorse tubercles distally; **pappi** 5–6 mm. $2n = 15$.

Flowering Jul–Oct. Roadsides, rangelands, grain fields, waste places and other disturbed ground; 0–600 m; introduced; B.C., Ont.; Calif., Del., D.C., Ga., Idaho, Ill., Md., Mich., Mont., N.J., Oreg., Pa., Va., Wash., W.Va.; Eurasia; Africa; Australia.

Chondrilla juncea is native to the Mediterranean region of Europe, North Africa, and Asia Minor. It is a weed in North America (not listed as noxious at the federal level). Its deep and extensive root system competes strongly for soil moisture and nutrients and makes control difficult because it helps the plants survive drought, cultivation, grazing, and most selective herbicides. The large, stiff branches and stems interfere with harvesting. The species is said to be "the most serious weed of Australian wheat-growing regions" (F. D. Panetta and J. Dodd 1987). It also infests millions of acres in California, Idaho, Oregon, and Washington.

Chondrilla juncea is an obligate apomict; its seeds are formed by a parthenogenetic process (E. Battaglia 1949). Nevertheless, the species is highly variable in morphology and biochemical traits.

39. IXERIS (Cassini) Cassini in F. Cuvier, Dict. Sci. Nat. ed. 2, 25: 62. 1822

• [No etymology in protologue]   [I]

John L. Strother

Taraxacum F. H. Wiggers subg. *Ixeris* Cassini, Bull. Sci. Soc. Philom. Paris 1821: 173. 1821

Perennials [annuals or biennials], (5–)10–30+ cm; taprooted (sometimes rhizomatous; stolons produced in some taxa). **Stems** (excluding stolons) 1–5+, erect (often scapiform), usually branched distally, glabrous. **Leaves** basal and cauline; petiolate; blades elliptic, oblong, orbiculate, or ovate [lanceolate to oblanceolate or linear], margins entire or denticulate to dentate [pinnately lobed] (faces glabrous [± scabrous]). **Heads** borne singly or in loose, corymbiform [paniculiform] arrays. **Peduncles** not inflated distally, usually bracteate. **Calyculi** of 3–10+, deltate to lanceolate bractlets. **Involucres** campanulate to cylindric, 2–5[–8+] mm diam. **Phyllaries** 5–10[–13+] in ± 1 series, linear [lanceolate], equal to subequal, margins sometimes scarious, apices obtuse to acute. **Receptacles** convex, weakly pitted, glabrous, epaleate. **Florets** 6–12+; corollas yellow [white or bluish]. **Cypselae** reddish brown [black], bodies ± fusiform, sometimes ± compressed, ribs (or wings) 10, beaks equal to or shorter than bodies, faces glabrous; **pappi** persistent, of 20–30, white [brown], ± equal, smooth or barbellulate bristles in 1 series. $x = 8$.

Species ca. 20 (1 in the flora): introduced; Southeast Asia.

1. Ixeris stolonifera A. Gray, Mem. Amer. Acad. Arts, n. s. 6: 396. 1858 [F] [I]

Lactuca stolonifera (A. Gray) Bentham ex Maximowicz

Plants stoloniferous. **Aerial stems** ± scapiform. **Leaf blades** 5–30+ × 5–30+ mm, margins often with 1(–2), ± retrorse teeth near bases. **Peduncles** (1–)5–15+ cm. **Involucres** 8–10 mm. **Cypselae** 6 mm; **pappi** 4–7 mm. $2n = 16, 24$.

Flowering Jun–Aug. Disturbed sites; 0–100 m; introduced; Del., N.J., N.Y., Pa.; Asia (Japan).

The combination *Lactuca stolonifera* has been mistakenly attributed to "(A. Gray) Bentham & Hooker f."

40. LAPSANASTRUM J. H. Pak & K. Bremer, Taxon 44: 19. 1995 • [*Lapsana*, generic name, and Latin *-astrum*, indicating inferiority or an incomplete resemblance]   [I]

David J. Bogler

Annuals [biennials], 10–20 cm; taprooted [occasionally stoloniferous]. **Stems** 1–20+, prostrate, simple or sparingly branched distally, slender, glabrous. **Leaves** basal and cauline; petiolate or sessile; basal blades oblanceolate, lyrate-pinnately lobed (lateral lobes relatively wide, terminals relatively large); cauline relatively few, reduced. **Heads** in loose, corymbiform arrays or borne singly. **Peduncles** not inflated, minutely bracteate or naked. **Calyculi** of 2–4 scalelike bractlets (to 1 mm, equal), glabrous. **Involucres** cylindro-campanulate, 4–8 mm diam. **Phyllaries** 5[–7] in 1 series, broadly to narrowly lanceolate, equal, margins not scarious, apices acute to acuminate, faces glabrous. **Receptacles** ± flat, smooth, glabrous, epaleate. **Florets** 5–8; corollas yellow (glabrous [pubescent]). **Cypselae** golden brown, oblong, adaxially flattened, (often 2-

winged with 1–2 apical hooklike projections [hooks 0 in some species]), not beaked, ribs 10–13, ciliate; **pappi** 0. $x = 8$.

Species 4 (1 in the flora): introduced; Asia.

Members of *Lapsanastrum* were formerly included in *Lapsana*. Cladistic analysis by J. H. Pak and K. Bremer (1995) indicated that they are distinct from *Lapsana communis* and are more closely related to *Youngia*, supporting their recognition as a distinct genus.

SELECTED REFERENCE Pak, J. H. and K. Bremer. 1995. Phylogeny and reclassification of the genus *Lapsana* (Asteraceae: Lactuceae). Taxon 44: 13–21.

1. Lapsanastrum apogonoides (Maximowicz) J. H. Pak & K. Bremer, Taxon 44: 19. 1995 • Japanese nipplewort F I

Lapsana apogonoides Maximowicz, Bull. Acad. Imp. Sci. Saint-Pétersbourg 18: 288. 1873

Stems slender, weak, flattened. **Leaves:** basal blades 40–100 × 10–25 mm, dentate, apices obtuse. **Phyllaries** broadly lanceolate, thin, scarcely keeled, 4–5 mm, stellately spreading in fruit. **Corollas** 6–10 mm. **Cypselae** straight, 4–5 mm. $2n = 16$.

Flowering May–Jun. Cultivated fields, disturbed areas, low boggy areas; 10–100 m; introduced; Oreg.; Asia (China, Japan, Korea).

Lapsanastrum apogonoides is distinguished by its prostrate habit, weak stems, lyrate leaves, and 5 phyllaries spreading in fruit.

41. YOUNGIA Cassini, Ann. Sci. Nat. (Paris) 23: 88. 1831 • [For "deux Anglais célèbres, l'un comme poète, l'autre comme physicien," both named Young; the poet may have been Edward Young (also dramatist), 1683–1765; the physician may have been Thomas Young (also physicist and Egyptologist), 1773–1829] I

Phyllis L. Spurr

Annuals, biennials [perennials], (10–)20–90+ cm; taprooted. **Stems** 1–5+, erect (often scapiform), usually branched distally, sometimes throughout, proximally glabrous, puberulent, or tomentose. **Leaves** all or mostly basal; petiolate (petiole bases often dilated, ± clasping); blades oblong or ovate to oblanceolate, margins usually pinnately lobed (± lyrate), ultimate margins denticulate. **Heads** (4–150) in corymbiform to paniculiform arrays. **Peduncles** (filiform) not distally inflated, seldom bracteate. **Calyculi** of 3–5+, deltate to ovate (membranous) bractlets. **Involucres** cylindric to campanulate, 2–3+ mm diam. **Phyllaries** usually 8 in 1–2 series, lanceolate to linear, ± equal (reflexed in fruit), margins ± scarious, apices obtuse to acute. **Receptacles** flat to convex, ± pitted, glabrous, epaleate. **Florets** 8–25+; corollas yellow, sometimes abaxially purplish (anther bases with linear, acute auricles). **Cypselae** ± reddish brown, ± fusiform and compressed [± terete], weakly or not beaked, ribs 11–13, ± spiculate to scabrellous on ribs; **pappi** (borne on discs at tips of cypselae) persistent (fragile) [falling], of 40–60+, basally coherent [distinct], white [yellowish or grayish], subequal, smooth to barbellulate bristles in ± 1 series. $x = 5$ or 8.

Species ca. 30 (1 in the flora): introduced; Asia; introduced also in South America, Europe, Africa, Pacific Islands, Australia.

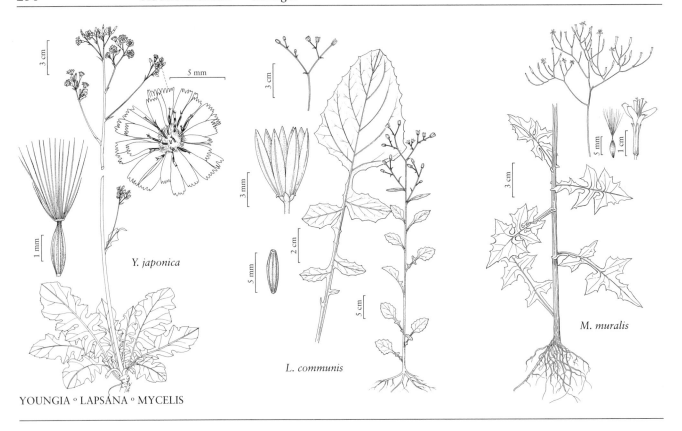

Y. japonica

L. communis

M. muralis

YOUNGIA ∘ LAPSANA ∘ MYCELIS

Youngia americana Babcock (based on a specimen from Alaska) has not been used as an accepted name for plants in the flora area; it was treated as a synonym of *Crepis nana* var. *lyratifolia* (Turczaninow) Hultén by E. Hultén (1968).

SELECTED REFERENCE Babcock, E. B. and G. L. Stebbins. 1937. The genus *Youngia*. Publ. Carnegie Inst. Wash. 484.

1. Youngia japonica (Linneaus) de Candolle in A. P. de Candolle and A. L. P. P. de Candolle, Prodr. 7: 194. 1838 • Oriental false hawksbeard [F] [I]

Prenanthes japonica Linnaeus, Mant. Pl., 107. 1767; *Crepis japonica* (Linnaeus) Bentham; *Youngia japonica* subsp. *elstonii* (Hochreutiner) Babcock & Stebbins

Stems terete, fistulose. **Leaves:** petioles 1–10 cm, glabrous, puberulent, or densely hairy (hairs often brownish, crinkled); blades 3–12(–25) × 2–4(–6) cm, lateral lobes 0–20, mostly gradually reduced proximally, terminal lobes elliptic, ovate, obovate, or oblong-truncate, larger than laterals, apices obtuse or acute. **Peduncles** 1–5(–15) mm. **Phyllaries** 3.5–6 mm, bases and midribs becoming ± spongy, abaxial faces glabrous, glabrate, or hairy (hairs appressed, shining). **Florets:** corollas mostly 4.5–6.5 mm; anthers dark green (drying purplish); styles and style-branches yellow. **Cypselae** 1.5–2.5 mm, bases hollow, lightly calloused; **pappi** 2.5–3.5 mm, slightly surpassing phyllaries. $2n = 16$.

Flowering spring–summer (year-round south). Waste places, lawns, etc.; 0–2400 m; introduced; Ala., Ark., D.C., Fla., Ga., Ky., La., Md., Miss., N.Y., N.C., Pa., S.C., Tenn., Tex., Va.; se Asia; introduced also in Mexico, Central America, South America, Europe, Africa, Pacific Islands, Australia.

Youngia japonica is now considered a pantropical weed. Relatively few specimens in the flora match what Babcock and Stebbins called subsp. *elstonii*, with cauline leaves almost as large as the basal and with conspicuous, lobed bracts at the bases of the proximalmost branches of the capitulescence. In subsp. *japonica*, to which most of our specimens are referred, the cauline leaves are much reduced or lacking, as are the bracts of the capitulescence.

42. LAPSANA Linnaeus, Sp. Pl. 2: 811. 1753; Gen. Pl. ed. 5, 353. 1754 • Nipplewort

[Greek *lapsanae*, a vegetable mentioned by Dioscorides, perhaps actually *Raphanus*, with lyrate leaves resembling those of *Lapsana*] ⊡

David J. Bogler

Annuals [biennials], 15–150 cm; fibrous-rooted. **Stems** 1, erect, simple or branched, glabrate to sparsely or densely pilose, hairs often stipitate-glandular. **Leaves** basal and cauline (not in rosettes); narrowly winged-petiolate; blades ovate to suborbiculate (thin), margins entire, dentate, or lyrate-pinnatifid proximally (terminal lobes larger than laterals, faces glabrate to sparsely hirsute; distal sessile, lanceolate, reduced). **Heads** in open, corymbiform to thyrsiform arrays. **Peduncles** (slender) slightly inflated distally, ebracteate. **Calyculi** of 4–5 subulate or scalelike, glabrous bractlets. **Involucres** cylindric to campanulate, 2–5 mm diam. **Phyllaries** 8–10 in 1 series, linear-oblong, subequal, (strongly keeled) margins green, not scarious, apices acute, faces glabrous. **Receptacles** flat, smooth, glabrous, epaleate. **Florets** 8–15; corollas yellow. **Cypselae** dimorphic (outer much longer than inner), tan to golden brown, subcylindric, curved, terete to slightly compressed, not beaked, ± 20-ribbed, glabrous; **pappi** 0. $x = 7$.

Species 1: introduced; Europe, Asia.

Lapsana formerly included about 9 species, some from eastern Asia. Based on cladistic analysis of morphologic characters, the eastern Asian species have been removed to *Lapasanastrum*, a strongly supported monophyletic group characterized by spreading phyllaries and distinctive fruit anatomy (J. H. Pak and K. Bremer 1995).

SELECTED REFERENCE Pak, J. H. and K. Bremer. 1995. Phylogeny and reclassification of the genus *Lapsana* (Asteraceae: Lactuceae). Taxon 44: 13–21.

1. Lapsana communis Linnaeus, Sp. Pl. 2: 811. 1753 F ⊡

Leaves: blades 1–15(–30) × 1–7 (–10) cm. **Heads** 5–25(–100+). **Calyculi:** bractlets keeled in fruit, 0.5–1 mm. **Involucres** 5–10 × 3–4 mm. **Phyllaries** 3–9 mm. **Corollas** 7–10 mm. **Cypselae** 3–5 mm. $2n = 12, 14, 16$.

Flowering Apr–Sep. Mesic woods, sheltered waste areas, roadsides, stream banks; 50–1900 m; introduced; Greenland, B.C., Ont., Que., Sask.; Alaska, Ariz., Ark., Calif., Colo., Conn., Idaho, Ill., Ind., Ky., Maine, Md., Mass., Mich., Mo., N.J., N.Y., N.C., N.Dak., Ohio, Okla., Oreg., Pa., R.I., Tenn., Tex., Utah, Vt., Va., Wash., W.Va., Wis.; Eurasia.

Lapsana communis is widely distributed in North America. It is easily recognized by the abruptly constricted lyrate leaves with relatively large terminal lobes, heads of relatively small flowers with yellow corollas, keeled phyllaries, and epappose cypselae. It is aggressively weedy and often found in shady disturbed sites. The milky juice of *L. communis* is said to be soothing to sensitive skin, particularly on the nipples of nursing mothers.

43. MYCELIS Cassini in F. Cuvier, Dict. Sci. Nat. ed. 2, 33: 483. 1824 • [No etymology in protologue; no readily discernible meaning from Greek or Latin roots] ⊡

John L. Strother

Annuals or perennials, (10–)40–90+ cm; taprooted. **Stems** usually 1, usually erect, branched distally, glabrous. **Leaves** basal and cauline (mostly cauline at flowering); proximal ± petiolate, distal ± sessile; blades oblanceolate to spatulate (bases often clasping), margins pinnately lobed (lyrate to runcinate, terminal lobes ± deltate) and ± sharply dentate (faces glabrous). **Heads** in paniculiform to thyrsiform arrays. **Peduncles** not inflated distally, sometimes bracteolate. **Calyculi** of 2–4 (often spreading to patent), ± deltate to lanceolate bractlets in 1 series. **Involucres** narrowly cylindric, 1–2+ mm diam. **Phyllaries** (4–)5 in 1(–2) series (reflexed in fruit), linear, equal, margins little, if at all, scarious, apices rounded. **Receptacles** flat to convex, weakly pitted, glabrous, epaleate. **Florets** 5; corollas yellow. **Cypselae:** bodies blackish to reddish, compressed, obovoid to lanceoloid, beaks whitish, stout, ribs 5–7 on each face, faces scabrellous; **pappi** persistent (borne on discs at tips of beaks), white; outer of 12–20+, minute setae, inner of 60–80+, white, subequal, barbellulate bristles in 1–2+ series. *x* = 9.

Species 1: introduced; Europe.

1. Mycelis muralis (Linnaeus) Dumortier, Fl. Belg., 60. 1827 [F] [I]

Prenanthes muralis Linnaeus, Sp. Pl. 2: 797. 1753; *Lactuca muralis* (Linnaeus) Gaertner

Leaf blades 35–120(–180+) × 10–50(–80+) mm. **Peduncles** filiform, 5–25+ mm. **Involucres** 7–11+ mm. **Phyllaries:** abaxial faces glabrous. **Cypselae:** bodies 2.5–3 mm, beaks 0.5–1 mm; **pappi:** outer 0.1–0.2 mm, inner 5–6 mm. *2n* = 18.

Flowering Jun–Oct. Disturbed sites, openings in mixed forests, calcareous or sandy soils; 10–400 m; introduced; B.C., Ont., Que.; Maine, Mass., Mich., Minn., N.H., N.Y., Oreg., Vt., Wash.; Europe.

44. MULGEDIUM Cassini in F. Cuvier, Dict. Sci. Nat. ed. 2, 33: 296. 1824 • [Latin *mulgere*, to milk, alluding to milky sap]

John L. Strother

Perennials [annuals or biennials], 15–100+ cm; ± rhizomatous. **Stems** usually 1, usually erect, branched distally, glabrous or glabrate. **Leaves** basal and cauline or mostly cauline; petiolate (basal) or sessile; blades oblong, elliptic, or ovate to lanceolate or linear, margins entire or dentate to pinnately lobed (faces glabrous, often glaucous). **Heads** borne singly or in corymbiform to paniculiform arrays. **Peduncles** not inflated distally, usually bracteate. **Calyculi** of 3–13+, deltate to lanceolate bractlets (sometimes intergrading with phyllaries). **Involucres** cylindric, 2–5 [–8+] mm diam. **Phyllaries** 8–13+ in 1–2 series, lanceolate to linear, subequal to equal, margins little, if at all, scarious, apices acute. **Receptacles** flat, pitted, glabrous, epaleate. **Florets** (10–)15–50+; corollas usually bluish [yellow]. **Cypselae** reddish brown to brown-mottled or slatey [blackish], bodies ± compressed, lanceoloid, beaks 0 (or gradually set off from and ± concolorous with bodies), ribs 4–6 on each face, faces glabrous [scabrid]; **pappi** persistent (borne on discs at tips of cypselae or beaks), of 80–120+, whitish, ± equal, barbellulate to nearly smooth bristles in 2–3+ series. *x* = 9.

Species 15 or so (1 in the flora): North America, Europe, Asia.

In referring a species long included in *Lactuca* to *Mulgedium*, I was influenced by K. Bremer (1994).

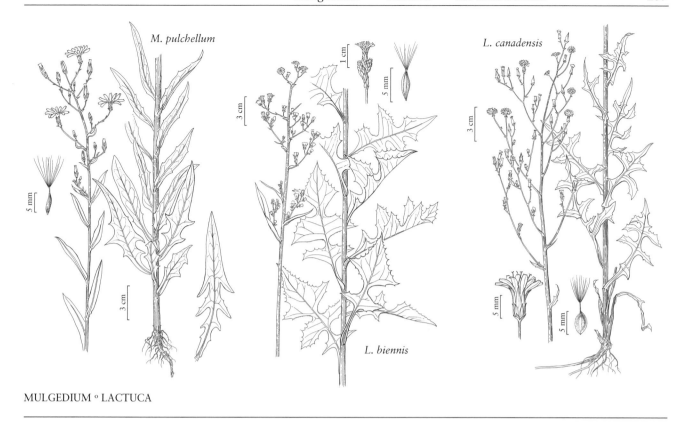

M. pulchellum

L. canadensis

L. biennis

MULGEDIUM ° LACTUCA

1. Mulgedium pulchellum (Pursh) G. Don in R. Sweet, Hort. Brit. ed. 3, 418. 1839 [E] [F]

Sonchus pulchellus Pursh, Fl. Amer. Sept. 2: 502. 1813; *Lactuca pulchella* (Pursh) de Candolle; *L. tatarica* (Linnaeus) C. A. Meyer subsp. *pulchella* (Pursh) Stebbins

Leaf blades 3–12(–18) cm × 5–25(–35+) mm. **Involucres** 12–15+ mm. **Cypselae:** bodies 4–5+ mm, beaks 0–1 mm; **pappi** 7–9(–12+) mm. **2n** = 18.

Flowering Jun–Sep. Calcareous sites, clearings in forests or shrublands, meadows, roadsides, stream banks, other wet sites; 800–3200 m; Alta., B.C., Man., N.W.T., Ont., Que., Sask.; Alaska, Ariz., Calif., Colo., Idaho, Ill., Ind., Iowa, Kans., La., Maine, Mich., Minn., Mo., Mont., Nebr., Nev., N.Mex., N.Y., N.Dak., Ohio, Okla., Oreg., Pa., S.Dak., Tex., Utah, Wash., Wis., Wyo.

The type of *Mulgedium pulchellum* may be conspecific with that of *M. tataricum* (Linnaeus) de Candolle, a Eurasian species. Or, if "perennial" plus "Fl. blue" constitutes sufficient description for valid publication of the name *Lactuca oblongifolia* Nuttall (1813), then a new combination in *Mulgedium* based on that name may be appropriate for what is here called *M. pulchellum*. Presence of *Mulgedium pulchellum* in Texas is based on a single, early collection.

45. LACTUCA Linnaeus, Sp. Pl. 2: 795. 1753; Gen. Pl. ed. 5, 348. 1754 • Lettuce, laitue

[No etymology in protologue; traceable to Latin *lac*, milk, alluding to the milky sap]

John L. Strother

Annuals or biennials, 15–450+ cm; taprooted. **Stems** usually 1, usually erect, branched distally or throughout, glabrous or hairy (sometimes hispid to setose). **Leaves** basal and cauline or mostly cauline (at flowering); sessile or petiolate; blades orbiculate, ovate, oblong, or lanceolate to oblanceolate, linear, or filiform, margins entire or denticulate to pinnately lobed (faces glabrous or hairy, often ± setose). **Heads** borne singly or in corymbiform to paniculiform arrays.

Peduncles not inflated distally, sometimes bracteate. **Calyculi** of 3–10+, deltate to lanceolate bractlets in 2–3 series (sometimes intergrading with phyllaries). **Involucres** campanulate to cylindric, 2–5[–8+] mm diam. **Phyllaries** 5–13+ in ± 2 series (erect or reflexed in fruit), lanceolate to linear, usually subequal to equal, margins sometimes scarious, apices obtuse to acute. **Receptacles** flat to convex, pitted, glabrous, epaleate. **Florets** 6–50+; corollas yellow, bluish, or whitish. **Cypselae** reddish brown, tan, whitish, or purplish to blackish, bodies compressed to flattened, elliptic to oblong, beaks stout (0.1–1 mm, gradually or weakly set off from bodies) or filiform (2–6 mm, sharply set off from bodies), ribs 1–9 on each face, faces often transversely rugulose, usually glabrous; **pappi** persistent (borne on discs at tips of cypselae or beaks), obscurely double (spp. 1–2), each a minute, erose corona 0.05–0.2 mm subtending 40–80+, white or fuscous, ± equal, barbellate to barbellulate bristles in 1–2 series, or simple (spp. 3–10) of 80–120+, white, ± equal, barbellulate to nearly smooth bristles in 2–3+ series. $x = 9$.

Species ca. 75 (10 in the flora): North America, Mexico, Central America, Eurasia, Africa.

The common head and leaf lettuces of home gardens and commerce are derived from *Lactuca sativa*.

1. Corollas usually bluish, sometimes white, rarely yellow, not or seldom deliquescent; cypselae ± compressed-lanceoloid to compressed-fusiform, beaks 0 or ± stout (gradually set off from bodies of cypselae), 0.1–0.5(–1) mm.
    2. Florets (15–)20–30(–50+); pappi ± fuscous . . . . 1. *Lactuca biennis*
    2. Florets 10–15(–25+); pappi white . . . . 2. *Lactuca floridana*
1. Corollas usually yellow, sometimes bluish (or drying bluish) or white, usually deliquescent; cypselae ± flattened, ± elliptic to obovate or oblanceolate, beaks ± filiform (sharply set off from bodies of cypselae), 1–4+ mm.
    3. Faces of cypselae 1(–3)-nerved.
        4. Involucres 10–12 mm; cypsela bodies 2.5–3.5 mm . . . . 3. *Lactuca canadensis*
        4. Involucres 12–20 mm; cypsela bodies 4.5–6 mm.
            5. Leaves on proximal $^1/_3$–$^1/_2$ of each stem, blades spatulate to lance-linear . . .
            . . . . 4. *Lactuca graminifolia*
            5. Leaves on proximal ($^1/_3$–)$^1/_2$–$^3/_4$ of each stem, blades ovate or oblanceolate to spatulate.
                6. Margins of leaf blades entire or denticulate, seldom prickly; florets 12–24+ . . . . 5. *Lactuca hirsuta*
                6. Margins of leaf blades ± toothed, ± prickly; florets 20–50+ . . . . 6. *Lactuca ludoviciana*
    3. Faces of cypselae (3–)5–9-nerved.
        7. Cypselae purplish to blackish, ± elliptic . . . . 7. *Lactuca virosa*
        7. Cypselae pale brown to grayish or whitish, ± obovate or ± oblanceolate.
            8. Blades of undivided leaves lanceolate to linear or filiform; heads in racemiform or spiciform arrays . . . . 8. *Lactuca saligna*
            8. Blades of undivided leaves mostly ± oblong or mostly obovate to orbiculate; heads in corymbiform or paniculiform arrays.
                9. Blades of cauline leaves usually ± oblong, sometimes narrowly obovate to lanceolate, midribs usually prickly setose, rarely smooth; phyllaries usually reflexed in fruit . . . . 9. *Lactuca serriola*
                9. Blades of cauline leaves ovate to orbiculate, midribs usually smooth, rarely prickly setose; phyllaries usually erect in fruit . . . . 10. *Lactuca sativa*

1. **Lactuca biennis** (Moench) Fernald, Rhodora 42: 300. 1940 [E] [F]

Sonchus biennis Moench, Methodus, 545. 1794

Annuals or biennials, (15–)75–200 (–300+) cm. **Leaves** on proximal $^2/_3$–$^3/_4$ of each stem; blades of undivided cauline leaves ovate to lanceolate, margins entire or denticulate, midribs sometimes sparsely piloso-setose. **Heads** in paniculiform arrays. **Involucres** 7–12+ mm. **Phyllaries** usually reflexed in fruit. **Florets** (15–)20–30(–50+); corollas bluish or whitish, sometimes yellowish, seldom deliquescent. **Cypselae:** bodies brown (often mottled), ± compressed-ellipsoid, 4–5+ mm, beaks ± stout, 0.1–0.5+ mm, faces (4–)5–6-nerved; **pappi** ± fuscous, 4–6+ mm. **2n = 34.**

Flowering Jul–Oct. Swamps, stream banks, woods; 900–1500 m; St. Pierre and Miquelon; Alta., B.C., Man., N.B., Nfld. and Labr., N.S., Ont., P.E.I., Que., Sask., Yukon; Alaska, Calif., Colo., Conn., Del., D.C., Idaho, Ill., Ind., Iowa, Ky., Maine, Md., Mass., Mich., Minn., Mont., N.H., N.J., N.Mex., N.Y., N.C., N.Dak., Ohio, Oreg., Pa., R.I., S.Dak., Tenn., Utah, Vt., Va., Wash., W.Va., Wis., Wyo.

The type of *Lactuca terrae-novae* Fernald is probably conspecific with that of *L. biennis*. The type of *L. biennis* may be conspecific with that of *L. floridana*.

2. **Lactuca floridana** (Linnaeus) Gaertner, Fruct. Sem. Pl. 2: 362. 1791 [E]

Sonchus floridanus Linnaeus, Sp. Pl. 2: 794. 1753; *Lactuca floridana* var. *villosa* (Jacquin) Cronquist; *Mulgedium floridanum* (Linnaeus) de Candolle

Annuals or biennials, 25–150 (–200+) cm. **Leaves** on proximal $^2/_3$–$^3/_4$ of each stem; blades of undivided cauline leaves oblong, ovate, or elliptic, margins entire or denticulate, midribs sometimes sparsely pilose. **Heads** in (± pyramidal) paniculiform arrays. **Involucres** (8–)10–12+ mm. **Phyllaries** usually reflexed in fruit. **Florets** 10–15(–25+); corollas bluish or whitish, seldom deliquescent. **Cypselae:** bodies brown (often mottled), ± compressed-lanceoloid to -fusiform, 4–5 mm, beaks ± stout, 0.1–0.5(–1) mm, faces 5–6-nerved; **pappi** white, 4–5 mm. **2n = 34.**

Flowering (Jun–)Aug–Sep(–Oct). Moist to wet places, margins of thickets and woods; 10–200 m; Man., Ont.; Ala., Ark., Del., D.C., Fla., Ga., Ill., Ind., Iowa, Kans., Ky., La., Md., Mass., Mich., Miss., Mo., Nebr., N.J.,

N.Y., N.C., Ohio, Okla., Pa., S.C., S.Dak., Tenn., Tex., Va., W.Va., Wis.

The "double" pappi of *Lactuca floridana* (and *L. biennis*) are very similar to pappi found in species assigned to *Cicerbita* Wallroth (ca. 35 spp., Europe, Asia, Africa). Return of the species to *Cicerbita* as *C. floridana* (Linnaeus) Wallroth may have merit.

3. **Lactuca canadensis** Linnaeus, Sp. Pl. 2: 796. 1753 [F]

Lactuca canadensis var. *latifolia* Kuntze; *L. canadensis* var. *longifolia* (Michaux) Farwell; *L. canadensis* var. *obovata* Wiegand; *L. sagittifolia* Elliott

Biennials, (15–)40–200(–450+) cm. **Leaves** on proximal $^1/_2$–$^3/_4$ of each stem; blades of undivided cauline leaves oblong, obovate, or lanceolate to spatulate or lance-linear, margins entire or denticulate, midribs sometimes sparsely pilose. **Heads** in ± corymbiform to paniculiform arrays. **Involucres** 10–12+ mm. **Phyllaries** usually reflexed in fruit. **Florets** 15–20+; corollas bluish or yellowish, usually deliquescent. **Cypselae:** bodies brown (often mottled), ± flattened, elliptic, 5–6 mm, beaks ± filiform, 1–3 mm, faces 1(–3)-nerved; **pappi** white, 5–6 mm. **2n = 34.**

Flowering Jun–Oct. Roadsides, swamps, salt marshes, thickets; 0–2200 m; B.C., Man., N.B., N.S., Ont., P.E.I., Que., Yukon; Ala., Ariz., Ark., Calif., Colo., Conn., Del., D.C., Fla., Ga., Idaho, Ill., Ind., Iowa, Kans., Ky., La., Maine, Md., Mass., Mich., Minn., Miss., Mo., Mont., Nebr., Nev., N.H., N.J., N.Mex., N.Y., N.C., N.Dak., Ohio, Okla., Oreg., Pa., R.I., S.C., S.Dak., Tenn., Tex., Utah, Vt., Va., Wash., W.Va., Wis., Wyo.; Mexico; Central America; Eurasia.

4. **Lactuca graminifolia** Michaux, Fl. Bor.-Amer. 2: 85. 1803

Biennials, 25–90(–150+) cm. **Leaves** on proximal $^1/_3$–$^1/_2$ of each stem; blades of undivided cauline leaves spatulate to lance-linear, margins entire or denticulate, midribs sometimes setose. **Heads** in ± paniculiform arrays. **Involucres** 12–20+ mm. **Phyllaries** usually reflexed in fruit. **Florets** 15–20+; corollas bluish to purplish, usually deliquescent. **Cypselae:** bodies brown (often mottled), ± flattened, elliptic, 5–6 mm, beaks ± filiform, 2–4 mm, faces 1(–3)-nerved; **pappi** white, 5–9 mm. **2n = 34.**

Flowering (Feb–)Apr–Jun(–Sep). Sandy ridges, pine forests, canyons; 10–1700 m; Ala., Ariz., Colo., Fla.,

Ga., La., Miss., N.J., N.Mex., N.C., S.C., Tex.; Mexico (Chihuahua); Central America.

Plants in the western populations of *Lactuca graminifolia* have been called *L. graminifolia* var. *arizonica* McVaugh; they may belong to a Mexican species.

5. Lactuca hirsuta Muhlenberg ex Nuttall, Gen. N. Amer. Pl. 2: 124. 1818 [E]

Lactuca hirsuta var. *albiflora* (Torrey & A. Gray) Shinners; *L. hirsuta* var. *sanguinea* (Bigelow) Fernald

Biennials, 15–80(–120) cm. **Leaves** on proximal ¹⁄₃–²⁄₃ of each stem; blades of undivided cauline leaves ± ovate, margins denticulate (sometimes ± ciliate), midribs usually piloso-setose. **Heads** in corymbiform to paniculiform arrays. **Involucres** 12–18+ mm. **Phyllaries** usually reflexed in fruit. **Florets** 12–24+; corollas usually yellow, sometimes drying bluish, usually deliquescent. **Cypselae:** bodies brown, ± flattened, elliptic, 4.5–5+ mm, beaks ± filiform, 2.5–3.5 mm, faces 1(–3)-nerved; **pappi** white, 6.5–8(–10+) mm. **2*n*** = 34.

Flowering Jul–Aug(–Sep). Openings in woods; 10–100 m; N.S., Ont., P.E.I., Que.; Ala., Ark., Conn., D.C., Ga., Ill., Ind., Ky., La., Maine, Md., Mass., Mich., Miss., Mo., N.H., N.J., N.Y., N.C., Ohio, Pa., R.I., S.C., Tenn., Tex., Vt., Va., W.Va.

The type of *Lactuca hirsuta* may be conspecific with that of *L. graminifolia*.

6. Lactuca ludoviciana (Nuttall) Riddell, W. J. Med. Phys. Sci. 8: 491. 1835 [E]

Sonchus ludovicianus Nuttall, Gen. N. Amer. Pl. 2: 125. 1818; *Lactuca campestris* Greene

Biennials, 15–150 cm. **Leaves** on proximal ¹⁄₂–³⁄₄ of each stem; blades of undivided cauline leaves obovate or oblanceolate to spatulate, margins denticulate (piloso-ciliate), midribs usually piloso-setose. **Heads** in paniculiform arrays. **Involucres** 12–15+ mm. **Phyllaries** usually reflexed in fruit. **Florets** 20–50+; corollas usually yellow, sometimes bluish, usually deliquescent. **Cypselae:** bodies brown to blackish (usually mottled), ± flattened, elliptic, 4.5–5+ mm, beaks ± filiform, 2.5–4.5 mm, faces 1(–3)-nerved; **pappi** white, 5–7(–11) mm. **2*n*** = 34.

Flowering Jun–Sep. Openings in woods, stream banks, prairies; 100–1400 m; B.C., Man., Sask.; Ariz., Ark., Calif., Colo., Idaho, Ill., Ind., Iowa, Kans., Ky., La., Minn., Mont., Nebr., N.Mex., N.Dak., Okla., Oreg., S.Dak., Tex., Utah, Wash., Wis., Wyo.

7. Lactuca virosa Linnaeus, Sp. Pl. 2: 795. 1753 [I]

Biennials, 20–120(–200+) cm. **Leaves** on proximal ¹⁄₂–²⁄₃ of each stem; blades of undivided cauline leaves obovate to spatulate, margins denticulate, midribs usually prickly-setose. **Heads** in paniculiform arrays. **Involucres** 12–15 mm. **Phyllaries** usually reflexed in fruit. **Florets** 10–15; corollas yellow, usually deliquescent. **Cypselae:** bodies purplish to blackish, ± flattened, ± elliptic, 3.5–4 mm, beaks ± filiform, 2.5–3.5 mm, faces 5–7-nerved; **pappi** white, 5–6 mm. **2*n*** = 18.

Flowering May, Oct. Disturbed sites; 10–400 m; introduced; Ala., Calif., D.C.; Europe.

8. Lactuca saligna Linnaeus, Sp. Pl. 2: 796. 1753 [I]

Annuals (perhaps persisting), 15–70 (–100+) cm. **Leaves** on proximal ¹⁄₂–³⁄₄+ of each stem; blades of undivided cauline leaves ± linear to filiform, margins entire or denticulate, midribs usually prickly-setose. **Heads** in racemiform to spiciform arrays. **Involucres** 6–9(–13+) mm. **Phyllaries** usually erect in fruit. **Florets** 6–12(–20+); corollas yellow (sometimes abaxially bluish), usually deliquescent. **Cypselae:** bodies pale brown, ± flattened, elliptic to oblanceolate, 2.5–3.5 mm, beaks ± filiform, (2–)5–6 mm, faces 5–7-nerved; **pappi** white, 5–6 mm. **2*n*** = 18.

Flowering Aug–Oct. Disturbed sites; 10–1500 m; introduced; Ont., Que.; Ala., Ariz., Ark., Calif., Del., D.C., Ga., Ill., Ind., Iowa, Kans., Ky., La., Maine, Md., Mass., Mich., Mo., Nebr., Nev., N.J., N.Mex., N.Y., N.C., Ohio, Okla., Oreg., Pa., S.C., Tenn., Tex., Va., Wash., W.Va., Wis.; Europe; introduced also in Mexico.

9. Lactuca serriola Linnaeus, Cent. Pl. II, 29. 1756 [F] [I]

Annuals (perhaps persisting), (15–)30–70(–100+) cm. **Leaves** on proximal ¹⁄₂–³⁄₄ of each stem; blades of undivided cauline leaves usually ± oblong, sometimes obovate to lanceolate, margins denticulate, usually prickly, midribs usually prickly-setose, rarely smooth. **Heads** in paniculiform arrays. **Involucres** 9–10(–12) mm. **Phyllaries** usually reflexed in fruit. **Florets** 12–20; corollas yellow, usually deliquescent. **Cypselae:** bodies pale grayish to

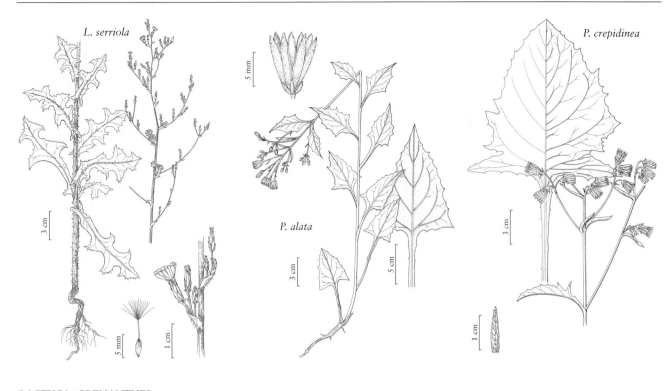

L. serriola

P. crepidinea

P. alata

LACTUCA ∘ PRENANTHES

tan, ± flattened, oblanceolate, 2.5–3.5 mm, beaks ± filiform, 2.5–4 mm, faces (3–)5–9-nerved; **pappi** white, (3–)4–5 mm. $2n = 18$.

Flowering (May–)Jul–Sep(–Oct). Roadsides, disturbed sites; 10–2300 m; introduced; Alta., B.C., Man., N.B., N.S., Ont., P.E.I., Que., Sask.; Ala., Ariz., Ark., Calif., Colo., Conn., Del., D.C., Fla., Ga., Idaho, Ill., Ind., Iowa, Kans., Ky., La., Maine, Md., Mass., Mich., Minn., Miss., Mo., Mont., Nebr., Nev., N.H., N.J., N.Mex., N.Y., N.C., N.Dak., Ohio, Okla., Oreg., Pa., R.I., S.C., S.Dak., Tenn., Tex., Utah, Vt., Va., Wash., W.Va., Wis., Wyo.; Europe; also introduced nearly worldwide.

The name *Lactuca scariola* Linnaeus is evidently illegitimate; it is a superfluous name based on the same type as *L. serriola*. Plants included here in *L. serriola* with most leaf blades obovate to lanceolate (not lobed) and lacking prickles on midribs have been called *L. scariola* subsp. or var. *integrata* or *L. serriola* forma *integrifolia* or *L. serriola* var. *integrata*.

10. Lactuca sativa Linnaeus, Sp. Pl. 2: 795. 1753 ⊞

Annuals or biennials, (15–)30–70(–100+) cm. **Leaves** on proximal $^1/_2$–$^3/_4$ of each stem; blades of undivided cauline leaves ovate to orbiculate, margins entire or denticulate, seldom prickly, midribs rarely prickly-setose. **Heads** usually in corymbiform, sometimes in paniculiform, arrays. **Involucres** 8–13+ mm. **Phyllaries** usually erect in fruit. **Florets** 7–15(–30+); corollas yellow (sometimes streaked with violet), usually deliquescent. **Cypselae:** bodies pale grayish to whitish or tan, ± flattened, obovate, 3–4 mm, beaks ± filiform, 3–5 mm, faces 5–9-nerved; **pappi** white, 3.5–4+ mm. $2n = 18$.

Flowering mostly Jul–Sep. Disturbed sites, abandoned plantings; 10–1000 m; introduced; Ont.; Ala., Calif., Conn., Del., D.C., Idaho, Ill., Ind., Maine, Mass., Mich., Mo., N.H., N.J., N.Mex., N.Y., N.C., N.Dak., Ohio, Okla., Oreg., Pa., R.I., Vt., Wash., W.Va.; Eurasia; introduced also in Mexico; introduced or ephemeral nearly worldwide.

46. PRENANTHES Linnaeus, Sp. Pl. 2: 797. 1753; Gen. Pl. ed. 5, 349. 1754 (as Prenantes) • Rattlesnakeroot, cankerweed, gall-of-the-earth, prenanthe [Greek *prenes*, drooping, and *anthos*, flower, alluding to drooping heads]

David J. Bogler

Perennials, 5–250 cm; taprooted, often producing offshoots connected by slender rhizomes. **Stems** 1–5, usually erect, sometimes decumbent (*P. bootii*), usually simple (leafy), usually glabrous proximally, tomentulose distally. **Leaves** basal and cauline; petiolate or sessile; blades deltate to triangular, or ovate to oblanceolate, or oblong to linear, or spatulate, margins often pinnately or palmately lobed (sometimes deeply cleft and appearing compound), ultimate margins entire or coarsely serrate or dentate (apices acute, obtuse, or rounded; distal leaves reduced in size and lobing. **Heads** (usually nodding at flowering) in racemiform, paniculiform, thyrsiform, or corymbiform arrays. **Peduncles** not inflated distally, bracteate. **Calyculi** of 2–12, triangular to linear-lanceolate or subulate, unequal bractlets. **Involucres** narrowly cylindric to campanulate (bases often attenuate), 2–14 mm diam. **Phyllaries** 3–15 in 1 series, (yellow green or green to purple or blackish) subulate or linear to lanceolate or elliptic, equal, margins scarious, apices acute, faces glabrous, tomentulose, hispid, or coarsely setose. **Receptacles** slightly convex, smooth, glabrous, epaleate. **Florets** 4–38; corollas usually creamy white, pink, or lavender, rarely yellow or red (glabrous). **Cypselae** golden brown to light tan, narrowly subcylindric, or fusiform to oblanceoloid, or oblong to linear, subterete or angled, apices truncate, not beaked, faces finely 5–12-ribbed, usually glabrous; **pappi** persistent, of 30–50, dull white to yellow or tan, rarely reddish brown, ± equal, barbellulate bristles in 1 series. $x = 8$.

Species ca. 26–30 (14 in the flora): North America, n Asia, sc Africa (1 species).

Prenanthes is recognized by the erect and simple habit, deeply divided proximal leaves, whitish, yellow, or pinkish corollas in nodding heads, calyculate involucres, and pappi of barbellulate bristles. Leaf shape, size, and degree of lobing are often used for distinguishing species but are sometimes exceptionally variable. The proximal leaves are usually different in size, shape, and lobing from the distal leaves. Other taxonomic characters include size and habit, corolla color, number of florets per head, and phyllary number, color, and indument. The cypselae and pappi tend to be uniform.

Molecular ITS studies by S. C. Kim et al. (1996) suggested that *Prenanthes,* as here circumscribed, may be polyphyletic; additional sampling including North American taxa is needed to confirm the relationships of *Prenanthes* and recognition of *Nabalus* Cassini at the genus level.

SELECTED REFERENCES Johnson, M. F. 1979. The genus *Prenanthes* L. (Cichorieae–Asteraceae) in Virginia. Castanea 45: 24–30. Milstead, W. L. 1964. A Revision of the North American Species of *Prenanthes*. Ph.D. dissertation. Purdue University. Singhurst, J. R., R. J. O'Kennon, and W. C. Holmes. 2004. The genus *Prenanthes* (Asteraceae: Lactuceae) in Texas. Sida 21: 181–191.

1. Phyllaries glabrous.
    2. Florets and phyllaries (4–)5(–6) . . . . 3. *Prenanthes altissima*
    2. Florets and phyllaries 6–20.
        3. Heads in elongate, slender, spikelike, narrowly racemiform arrays; proximal leaves oblong to linear, pinnately lobed, lobes narrow and at right angles; corollas pinkish
            . . . . 5. *Prenanthes autumnalis*
        3. Heads in racemiform, paniculiform, or narrowly thyrsiform arrays; proximal leaves ovate to deltate, palmately lobed, irregularly dentate or entire; corollas whitish, pinkish, or pale yellow.

4. Phyllaries ± purplish or maroon (dark on old specimens); corollas whitish to
   pinkish; pappi usually reddish brown (rarely yellowish) . . . . 2. *Prenanthes alba*
4. Phyllaries green to tan, dark green, or blackish; corollas white or pale yellow;
   pappi pale yellow.
   5. Proximal leaves palmately 3(–5)-lobed (often deeply cleft, compound, lobes
      lanceolate, sinuses angular); involucres campanulate (attenuate at bases);
      calyculus bractlets triangular; corollas pale yellow . . . . 14. *Prenanthes trifoliolata*
   5. Proximal leaves entire or shallowly lobed (bases hastate or sagittate to
      truncate or rounded); involucres cylindro-campanulate or narrowly
      campanulate (rounded, not attentuate at bases); calyculus bractlets subulate
      to narrowly lanceolate; corollas white.
      6. Plants 5–25 cm; stems decumbent; alpine regions, n Appalachians
         . . . . 7. *Prenanthes bootii*
      6. Plants 8–75 cm; stems erect; n Rocky Mountains . . . . 12. *Prenanthes sagittata*
1. Phyllaries tomentulose, hispid, or coarsely setose.
   7. Proximal leaves petiolate, petioles ± winged, blades spatulate, usually unlobed, apices
      usually rounded or obtuse; mid cauline leaves sessile, often clasping; heads ± ascending.
      8. Midstems roughly hispid or coarsely setose; proximalmost leaves withered or
         deciduous by flowering; leaves ± hispid abaxially; corollas pale yellow to creamy
         white . . . . 4. *Prenanthes aspera*
      8. Midstems glabrous; proximalmost leaves persistent to flowering; leaves glabrous;
         corollas usually pinkish (sometimes white or lavender) . . . . 10. *Prenanthes racemosa*
   7. Proximal leaves petiolate, petioles winged, blades deltate or triangular to elliptic or
      ovate (not spatulate), sometimes lobed, apices acute or obtuse; mid cauline leaves sessile
      or petiolate; heads nodding.
      9. Proximal leaves sagittate, lobed; phyllaries hispid or tomentulose.
         10. Plants 80–150 cm; proximal leaves sagittate, bases sagittate; phyllaries hispid
             along midveins; Edwards Plateau, Texas . . . . 8. *Prenanthes carrii*
         10. Plants 15–80 cm; proximal leaves triangular or irregularly elliptic, bases truncate
             or hastate; phyllaries finely tomentulose; Pacific Northwest to Alaska . . . .
             . . . . 1. *Prenanthes alata*
      9. Proximal leaves deltate, ovate, or elliptic, lobed or unlobed; phyllaries coarsely
         setose.
         11. Proximal leaves deltate, ovate, or hastate, margins entire or coarsely serrate;
             phyllaries (9–)12–15, dark green to blackish; florets 15–38 . . . . 9. *Prenanthes crepidinea*
         11. Proximal leaves deltate to ovate or elliptic, margins palmately or pinnately
             lobed or unlobed, dentate or entire; phyllaries 5–10, green or purple to lavender;
             florets 5–15.
             12. Proximal leaves coarsely dentate or serrate (unlobed); phyllaries purple or
                 lavender . . . . 6. *Prenanthes barbata*
             12. Proximal leaves 3–5-lobed; phyllaries usually green (sometimes purple in
                 P. serpentaria).
                 13. Leaves about twice as long as wide, pinnately lobed, lobes and sinuses
                     rounded; heads in broad paniculiform or corymbiform arrays, at least
                     some branches elongate; phyllaries green or often purple, with
                     appressed, green or tan, coarse setae, sometimes reduced to a single
                     seta per phyllary; florets 8–14; widespread in e, s United States
                     . . . . 13. *Prenanthes serpentaria*
                 13. Leaves about as long as wide, palmately lobed, lobes angular; heads
                     and sinuses angular; heads in narrow paniculiform or thyrsiform arrays,
                     branches short; phyllaries dark green, with erect, coarse setae; florets
                     5–7(–13); mid to high elevations in Blue Ridge Mountains of North
                     Carolina, Tennessee, Virginia . . . . 11. *Prenanthes roanensis*

1. Prenanthes alata (Hooker) D. Dietrich, Syn. Pl. 4: 1309. 1847 • Western or wing-leaved rattlesnakeroot, white or western white lettuce [E] [F]

Nabalus alatus Hooker, Fl. Bor.-Amer. 1: 294, plate 102. 1833; *Sonchus hastatus* Lessing, Linnaea 6: 99. 1831, not *Prenanthes hastata* Thunberg 1784; *P. lessingii* Hultén

Plants 15–80 cm; taproots slender to thickened, tuberous. **Stems** 1(–10), erect or ascending, green to purple, usually simple, glabrous or glabrate proximally, tomentulose distally. **Leaves:** proximal present at flowering; petioles winged (2–6 cm); blades triangular to irregularly elliptic, 3–25 × 1–7 cm, thin or slightly coriaceous, bases abruptly constricted, truncate to slightly hastate, margins irregularly dentate or coarsely serrate, apices acute to acuminate, faces glabrate; distal sessile, ovate to lanceolate, reduced. **Heads** (10–17) in broad, corymbiform arrays, lateral branches often elongate and overtopping main stems. **Involucres** narrowly campanulate, 10–13 × 5–6 mm. **Calyculi** of 2–3, dark green, lanceolate or subulate bractlets 1–3 mm, glabrous or finely tomentulose. **Phyllaries** 8, green to dark green, lanceolate, 8–11 mm, margins scarious, apices acute, finely tomentulose. **Florets** 7–16; corollas white to purplish, 9–16 mm. **Cypselae** brown to light tan, subcylindric, 4–7 mm, weakly 7–10-ribbed; **pappi** pale yellow to dull white, 8–10 mm. *2n* = 16.

Flowering Jun–Sep. Stream banks, mountain springs, seeps, cliffs near shore, moist shady places; 0–1500 m; Alta., B.C.; Alaska, Idaho, Mont., Oreg., Wash.

Prenanthes alata is recognized by its relatively small size, elongate and winged petioles, triangular-hastate leaf blades, heads in broad corymbiform arrays, and dark green, finely tomentulose phyllaries.

2. Prenanthes alba Linnaeus, Sp. Pl. 2: 798. 1753 • White rattlesnakeroot, prenanthe blanche [E]

Nabalus albus (Linnaeus) Hooker

Plants 20–175 cm; taproots short and thickened, fibrous. **Stems** erect, often mottled purple or nearly all purple (often with stout bases, 8–12 mm diam.), proximally glabrous, tomentulose distally. **Leaves:** proximal present at flowering; petioles narrowly winged (to 18 cm); blades usually ovate to triangular or cordate, 4–30 × 3–18 cm, thin to coriaceous, bases often hastate, margins usually coarsely dentate or serrate, sometimes deeply 3-lobed or parted, lobes acute, faces glabrous adaxially, pale to whitish and sometimes hirsute abaxially; distal reduced. **Heads** (3–8, in clusters) in paniculiform arrays (densest near apices). **Involucres** cylindric to campanulate, 13–15 × 3–5 mm. **Calyculi** of 5–7, triangular to lanceolate bractlets 1–3 mm, glabrous. **Phyllaries** (6–)8(–9), ± purplish or maroon, lanceolate, 10–13 mm, margins scarious, minutely ciliate, apices acute, faces glabrous. **Florets** 7–9(–13); corollas whitish to pale pink, lavender or red, 9–15 mm. **Cypselae** brown or tan, elliptical to linear, 3.5–6 mm; **pappi** usually reddish brown, sometimes rusty, rarely yellowish, 6–7 mm. *2n* = 32.

Flowering Aug–Oct. Sandy oak-scrub, open oak-hickory woods, deciduous forests, dunes, creek banks, road cuts; 100–200 m; Man., Ont., Que., Sask.; Ark., Conn., Del., Ga., Ill., Ind., Iowa, Ky., Maine, Md., Mass., Mich., Minn., Mo., N.H., N.J., N.Y., N.C., N.Dak., Ohio, Pa., R.I., S.Dak., Tenn., Vt., Va., W.Va., Wis.

Prenanthes alba is recognized by the purplish stems, relatively large, coarse, ovate or triangular leaves, relatively long, winged petioles, glabrous and often purple phyllaries, and usually reddish brown pappi. The leaves are variable, occasionally deeply 3-lobed. W. L. Milstead (1964) recognized specimens with a pale yellow pappi as "subsp. *pallida*," distributed on the east coast in New Jersey, Virginia, and North Carolina. Because pappus color tends to fade somewhat on herbarium specimens, that character is difficult to assess on older specimens. The name was not validly published.

3. Prenanthes altissima Linnaeus, Sp. Pl. 2: 797. 1753 • Tall rattlesnakeroot, prenanthe élevée [E]

Nabalus altissimus (Linnaeus) Hooker; *Prenanthes altissima* var. *cinnamomea* Fernald; *P. altissima* var. *hispidula* Fernald

Plants 40–250 cm; taproots thickened, knotty, tuberous. **Stems** erect, greenish to purplish, glabrous proximally, often tomentulose distally. **Leaves:** proximal present at flowering; petioles winged; blades usually ovate or triangular, 4–15 × 2–16 cm, thin, bases truncate to hastate or cordate, margins entire or shallowly dentate, often deeply 3-lobed, faces glabrous or with scattered hairs on veins; distal reduced in size and lobing. **Heads** in narrow or spreading, paniculiform arrays. **Involucres** cylindric, 9–14 × 2–3 mm. **Calyculi** of 4–6, blackish, triangular bractlets 1–4 mm, glabrous. **Phyllaries** (4–)5(–6), pale green, often blackish at bases and apices, linear to lanceolate, 10–12 mm, glabrous or sparsely hairy. **Florets** (4–)5(–6); corollas pale yellow to greenish yellow, 7–15 mm. **Cypselae** brown to tan, subcylindric, subterete, 4–5 mm, indistinctly 5–10-ribbed; **pappi** usually whitish or pale yellow, sometimes reddish brown, 5–6 mm. *2n* = 16.

Flowering Aug–Nov. Open deciduous hardwood or mixed woods, shaded slopes, bluffs, disturbed areas, roadsides; 50–800 m; N.B., N.S., Ont., P.E.I., Que.; Ala., Ark., Conn., Del., Ga., Ill., Ind., Ky., La., Maine, Md., Mass., Mich., Mo., N.H., N.J., N.Y., N.C., Ohio, Okla., Pa., R.I., S.C., Tenn., Tex., Vt., Va., W.Va.

Prenanthes altissima is recognized by its narrow involucres with 5 pale green, glabrous phyllaries, (4–)5 (–6) florets, and pale yellow to greenish yellow corollas. Pappi in this species are most commonly whitish or pale yellow. Specimens with reddish brown to orange pappi have been recognized as var. *cinnamomea*, found in Arkansas, Louisiana, and Missouri. Specimens with densely hairy stems and pale yellow pappi have been recognized as var. *hispidula*, found mostly in New York, New England, and adjacent Canada.

4. Prenanthes aspera Michaux, Fl. Bor.-Amer. 2: 83. 1803 • Rough rattlesnakeroot E

Nabalus asper (Michaux) Torrey & A. Gray

Plants 35–170 cm; taproots fusiform, thickened, tuberous, with smaller lateral roots. **Stems** erect, green and mottled purple, glabrous proximally, midstems hispid or setose, scabrous to coarsely setose distally. **Leaves:** proximalmost usually withered by flowering; petiolate (midcauline sessile, clasping); blades spatulate, 4–11 × 2–5 cm, firm, coriaceous, brittle, bases attenuate, margins entire or weakly dentate, apices acute to obtuse, faces glabrate to roughly hispid abaxially; distal reduced. **Heads** in erect, racemiform or narrowly paniculiform arrays. **Involucres** cylindric to campanulate, 12–17 mm. **Calyculi** of 6–12, green, triangular to lanceolate bractlets 2–5 mm, coarsely setose. **Phyllaries** 6–10, yellow green to tan, linear-lanceolate, 8–15 mm, coarsely setose. **Florets** 8–19; corollas pale yellow to creamy white, 9–17 mm. **Cypselae** tan, subcylindric, subterete, 5–6 mm, irregularly 10–12-ribbed; **pappi** pale yellow or brown, 7–8 mm. $2n = 16$.

Flowering Aug–Oct. Tall-grass prairies, dry prairies, dry rocky woods; 50–600 m; Ark., Ga., Ill., Ind., Iowa, Kans., Ky., La., Minn., Miss., Mo., Nebr., Ohio, Okla., Pa., S.Dak., Tenn., Wis.

Prenanthes aspera is easily recognized by its narrow, erect habit, unlobed, spatulate, weakly dentate leaves, basal leaves withered by flowering, heads in narrow, spiciform arrays, and densely setose phyllaries. The distribution closely matches that of the tall grass prairie, and like undisturbed prairies, this species is now rare or endangered in some states.

5. Prenanthes autumnalis Walter, Fl. Carol., 193. 1788 • Slender rattlesnakeroot E

Nabalus virgatus de Candolle; *Prenanthes virgata* Michaux

Plants 40–140 cm; taproots thickened, tuberous. **Stems** 1–5, erect, mostly green with mottled purple, slender, simple, glabrous. **Leaves:** proximal present at flowering; petiolate (petioles winged); blades oblong to linear, 7–18 × 0.5–6 cm, coriaceous, bases attenuate, margins deeply pinnately lobed or divided, lobes narrow and at right angles, apices acute to acuminate, faces glabrous or slightly glabrate along veins; distally sessile and much reduced. **Heads** (1–2 in pedunculate lateral clusters) in spiciform or racemiform arrays. **Involucres** cylindro-campanulate, 10–13 × 3–4 mm. **Calyculi** of 6–8, green to purple, triangular to subulate bractlets 1–4 mm. **Phyllaries** 7–8, pink or purple, linear to elliptic, 8–13 mm, glabrous. **Florets** 8–10; corollas usually pinkish, sometimes lavender or white, 11–17 mm. **Cypselae** dark green, elliptic to linear, subterete to angular, 3.5–5.5 mm, indistinctly 8–10-ribbed; **pappi** pale yellow to tan, 5 mm.

Flowering Aug–Oct. Coastal plain, low savannas, sandy pinelands, moist places, swales; 0–200 m; Del., Fla., Ga., Md., Miss., N.J., N.C., S.C., Va.

Prenanthes autumnalis is recognized by its erect, slender habit, pinnately lobed proximal leaves, narrowly racemiform or spiciform arrays of heads, glabrous phyllaries, and usually pinkish corollas.

6. Prenanthes barbata (Torrey & A. Gray) Milstead ex Cronquist, Brittonia 29: 223. 1977 • Barbed rattlesnakeroot E

Nabalus fraseri de Candolle var. *barbatus* Torrey & A. Gray, Fl. N. Amer. 2: 481. 1843; *Prenanthes serpentaria* Pursh var. *barbata* (Torrey & A. Gray) A. Gray

Plants 50–150 cm; taproots thickened, tuberous, with lateral side roots. **Stems** erect, purplish, glabrous proximally, finely tomentulose or setose distally. **Leaves:** proximal withered by flowering; petiolate (petioles 2–3 cm, winged; blades ± elliptic, 4–10 × 1–4 cm, coriaceous, bases narrowed (not hastate), margins coarsely and irregularly dentate or serrate (unlobed), apices acute, faces glabrous or papillose-hispid; mid cauline cuneate or sessile and clasping; distal sessile and reduced. **Heads** (15–50) in (much-branched) paniculiform or thyrsiform arrays (nodding). **Calyculi** of 7–11, dark green to purple, triangular to lance-ovate bractlets 2–5 mm, hispid or setose along midveins. **Involucres** cylindric to campanulate, 12–16 × 4–6 mm. **Phyllaries** 6–8, usually purple or lavender, subulate to lanceolate, 11–14 mm, margins scarious, coarsely hispid or setose along midveins. **Florets** 10–15; corollas white to yellowish white, 13–16 mm. **Cypselae** tan, subcylindric, subterete to 5-angled, 8–10 mm, indistinctly 8–10-ribbed; **pappi** yellowish white, 7–8 mm.

Flowering Aug–Oct. Sandy oak-hickory-pine woods, savannas, prairies, pine barrens; 10–100 m; Ala., Ark., Ga., Ky., La., Okla., Tenn., Tex.

Prenanthes barbata is identified by its erect habit, irregularly dentate leaves, heads with 6–8 purple phyllaries, coarsely setose calyculi and phyllaries, and white to yellowish white corollas.

7. Prenanthes bootii (de Candolle) D. Dietrich, Syn. Pl. 4: 1309. 1847 • Boot's rattlesnakeroot [C] [E]

Nabalus bootii de Candolle in A. P. de Candolle and A. L. P. P. de Candolle, Prodr. 7: 241. 1838

Plants 5–25 cm; taproots short, thick, tuberous. **Stems** decumbent to erect, mottled purple, simple, glabrous proximally, tomentulose distally. **Leaves:** proximal present at flowering; petiolate (petioles 2–8 cm, not winged); blades ovate to deltate, 2–8 × 0.5–3 cm, bases hastate or sagittate, margins entire or weakly dentate, faces glabrous; distal reduced, elliptic to lanceolate. **Heads** (10–20) in narrow, racemiform to thyrsiform arrays (nodding). **Calyculi** of 4–5, dark green to blackish, subulate bractlets 2–5 mm, glabrous. **Involucres** cylindro-campanulate, 10–11 × 5–6 mm. **Phyllaries** 8–11, dark green to almost black, lanceolate to subulate, 8–12 mm, faces glabrous. **Florets** 9–20; corollas white, 7–13 mm. **Cypselae** light tan to yellow, subcylindric, subterete to angled, 5–6 mm, indistinctly 7–10-ribbed; **pappi** pale yellow, 6–8 mm. $2n = 32$.

Flowering Jul–Aug. Alpine areas above treeline, mountains; of conservation concern; 1500–2000 m; Maine, N.H., N.Y., Vt.

Prenanthes bootii is recognized by its relatively short, decumbent habit, deltate to hastate proximal leaves, entire or weakly dentate margins, glabrous and blackish green phyllaries, white corollas, and alpine habitat.

8. Prenanthes carrii Singhurst, O'Kennon & W. C. Holmes, Sida 21: 187, fig. 2. 2004 • Carr's rattlesnakeroot [C] [E]

Plants 80–150 cm; taproots tuberous, with lateral roots. **Stems** erect, simple, glabrous or strigose proximally, tomentose distally. **Leaves:** proximal usually present at flowering; petiolate (petioles 2.2–3 cm); blades (light green) sagittate to ovate, 13–25 × 7–12 cm, thin, bases attenuate, margins coarsely and irregularly dentate, apices acute to rounded, faces glabrous or lightly setose along veins; distal reduced to bracts. **Heads** in paniculiform arrays. **Involucres** cylindric to campanulate, 4–9 × 2–3 mm. **Calyculi** of 5–12, linear-subulate to narrowly lanceolate bractlets 2–4 mm, hispid. **Phyllaries** 8, green to rose, linear-subulate to lanceolate, 9–11 mm, (apices minutely ciliate) faces glabrate (midribs sparingly hispid). **Florets** 9–11; corollas white to creamy, 11.5–13.5 mm. **Cypselae** golden yellow to tan, subcylindric, angled to terete, 6–7 mm, prominently 12–15-ribbed; **pappi** white to tan or yellow, 7–8 mm.

Flowering Aug–Nov. Rich woodlands, canyons; of conservation concern; 300–900 m; Tex.

Prenanthes carrii is recognized primarily by its relatively tall size and long-petiolate, sagittate proximal and mid-cauline leaves, which are similar to those of *P. alata* and *P. sagittata*, species found far to the north. It is known only from the southwestern Edwards Plateau. It is thought to be closely related to *P. barbata* (J. R. Singhurst et al. 2004).

9. Prenanthes crepidinea Michaux, Fl. Bor.-Amer. 2: 84. 1803 • Nodding rattlesnakeroot [E] [F]

Nabalus crepidineus (Michaux) de Candolle

Plants 100–300 cm; taproots fusiform, thick and tuberous. **Stems** erect, green to tan, glabrous proximally, tomentose distally. **Leaves:** proximal present at flowering; petiolate (petioles winged, 1–13 cm, margins sometimes serrate); blades deltate to broadly ovate, 8–23 × 3–14 cm, coriaceous, bases hastate or sagittate, margins entire or coarsely dentate to serrate (teeth often large and remote, faces glabrous or scabrous, hirsute along veins; distal petiolate or sessile, ovate or elliptic, entire. **Heads** in broad, open, (leafy) paniculiform arrays (branches often elongate, nodding). **Calyculi** of 18–20 dark green, triangular to lanceolate bractlets 2–5 mm, coarsely setose. **Involucres** campanulate, 12–16 × 7–14

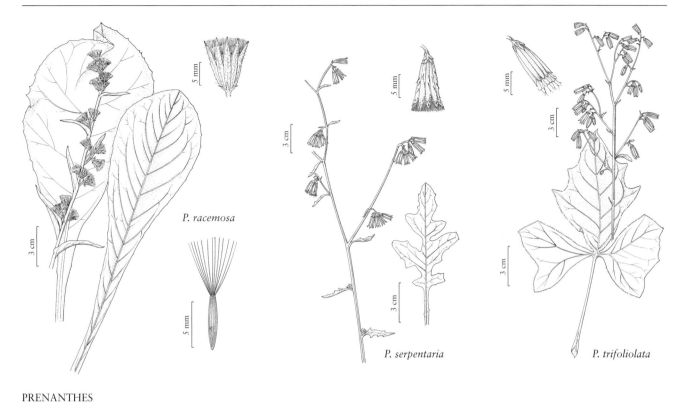

P. racemosa

P. serpentaria

P. trifoliolata

PRENANTHES

mm. **Phyllaries** 9–15, dark green to blackish, lanceolate to elliptic, 10–16 mm, margins scarious, moderately to densely, ± coarsely setose on midribs. **Florets** 15–38; corollas usually white, sometimes yellow, 9–15 mm. **Cypselae** golden brown, oblong to linear, subterete to angled, 5–6 mm, unequally 10–12-ribbed; **pappi** tan, 6–8 mm. $2n = 32$

Flowering Aug–Sep. Moist, rich, deciduous woods, lowland or upland woods, thickets, low prairies, wet areas in rich soil; 300–1200 m; Ark., Del., Ill., Ind., Iowa, Ky., Mich., Minn., Mo., N.Y., Ohio, Pa., Tenn., W.Va., Wis.

Prenanthes crepidinea is generally recognized by its tall and robust habit, large deltate or ovate proximal leaves, coarsely dentate margins with relatively large, remote teeth, heads borne in open, paniculiform arrays, dark green or blackish and moderately to densely setose phyllaries, and white or yellow corollas. In some areas *P. crepidinea* and *P. serpentaria* overlap in leaf morphology and corolla color, perhaps indicative of past hybridization, and can be difficult to distinguish. *Prenanthes serpentaria* is generally smaller in stature, has more elongate, pinnately lobed leaves with rounded sinuses, more sparsely setose phyllaries, and yellow or greenish yellow corollas. *Prenanthes crepidinea* appears to favor moist, bottomland sites, in contrast to the drier and acidic upland habitats preferred by *P. serpentaria*. *Prenanthes crepidinea* appears to hybridize with *P. aspera* in southeastern Missouri (J. A. Steyermark 1963).

10. **Prenanthes racemosa** Michaux, Fl. Bor.-Amer. 2: 84. 1803 • Glaucous or purple rattlesnakeroot, glaucous white lettuce, prenanthe à grappe [E] [F]

Nabalus racemosus (Michaux) Hooker; *Prenanthes racemosa* subsp. *multiflora* Cronquist; *P. racemosa* var. *pinnatifida* A. Gray

Plants 30–175 cm; taproots fusiform, tuberous. **Stems** erect, green or light purple, simple, (stout), glabrous and glaucous proximally, setose or hispid distally. **Leaves:** proximal usually present at flowering; petioles broadly winged, (1–15 cm); blades broadly oblanceolate to spatulate, 4–25 × 1–8 cm, coriaceous, bases attenuate, clasping, margins entire or weakly denticulate, apices obtuse or rounded, faces glabrous; mid cauline sessile, clasping; distal sessile and reduced. **Heads** (ascending) in (elongate) narrowly racemiform or paniculiform arrays. **Involucres** campanulate, 11–12 × 4–7 mm. **Calyculi** of 8, dark green to purple, narrowly triangular-subulate bractlets 2–4 mm, coarsely setose. **Phyllaries** 7–14, green to purple, lanceolate to linear, 10–12 mm, margins scarious, sparsely to densely setose. **Florets** 9–29; corollas usually pinkish, sometimes white or lavender, 7–13 mm. **Cypselae** golden brown, subcylindric, subterete, 5–6 mm, indistinctly 8–12-ribbed; **pappi** pale yellow, 6–7 mm. $2n = 16$.

Flowering Aug–Sep. Sandy alluvial soils of stream banks, wet meadows, tall-grass prairies, fens, marshy flats, bogs (mainly calcicolous, at least in north); 0–2800 m; Alta., B.C., Man., N.B., Nfld. and Labr. (Nfld.), N.S., Ont., Que., Sask.; Colo., Ill., Ind., Iowa, Ky., Maine, Mich., Minn., Mo., Mont., Nebr., N.J., N.Y., N.Dak., Ohio, Pa., S.Dak., Vt., Wash., Wis., Wyo.

Prenanthes racemosa is recognized by its erect, stout, simple habit, glaucous stems, spatulate proximal leaves with broadly winged petioles and rounded to obtuse apices, heads borne in narrow racemiform arrays, purple and hairy phyllaries, and usually pinkish corollas. It is most similar to *P. aspera*, which differs in its generally smaller stature, more hirsute stems, leaves that are hispid abaxially, proximal leaves usually withered by flowering, and creamy white or yellow corollas. Hybrids between *P. racemosa* and *P. trifoliolata*, known as *P. ×mainensis*, occur in the northeastern United States and southeastern Canada (see discussion under *P. trifoliolata*).

11. **Prenanthes roanensis**   (Chickering) Chickering, Bot. Gaz. 6: 191. 1881 • Roan Mountain rattlesnakeroot   C E

Nabalus roanensis Chickering, Bot. Gaz. 5: 155. 1880; *Prenanthes cylindrica* (Small) E. L. Braun

Plants 18–150 cm; taproots slender or short and thickened, tuberous. **Stems** erect, green or purple, simple, glabrous proximally, crisply tomentulose distally. **Leaves:** proximal present at flowering; petiolate (petioles not winged or distal narrowly winged); blades ovate, 3–22 × 2–15 cm, coriaceous, bases cordate to truncate or rounded, margins often deeply 3–5-palmately lobed or compound, lobes and sinuses angular, ultimate margins irregularly dentate, apices acute, faces usually finely tomentulose on veins; mid cauline short-petiolate or sessile; distal short-petiolate or sessile, reduced, lanceolate. **Heads** in narrowly paniculiform or thyrsiform arrays (nodding). **Involucres** cylindric to narrowly campanulate, 12–14 × 2.5–3 mm. **Calyculi** of 4–6, blackish green, triangular to lanceolate bractlets 1–3 mm, densely setose. **Phyllaries** 5–9, green or dark green, apices blackish, linear to oblanceolate, 9–11 mm, sparsely and coarsely setose. **Florets** 5–7(–13); corollas greenish yellow, 8–10 mm. **Cypselae** tan to yellow, fusiform, subterete or angled, 4–5 mm, indistinctly 8–12-ribbed; **pappi** yellow to whitish, 5–6 mm. **2n** = 16.

Flowering Aug–Sep. Spruce-hardwood forests, wooded slopes, open grassy summits; of conservation concern; 1600–1700 m; N.C., Tenn., Va.

Prenanthes roanensis is recognized by its 5–7 florets and dark setae on the phyllaries. It is found at mid to high elevations in the Blue Ridge Province of the

Appalachian Mountains. It is rather variable and often confused with other species. It is considered rare or threatened throughout its range.

12. **Prenanthes sagittata**   (A. Gray) A. Nelson in J. M. Coulter and A. Nelson, New Man. Bot. Centr. Rocky Mt., 592. 1909 • Arrowleaf snakeroot   E

Prenanthes alata (Hooker) D. Dietrich var. *sagittata* A. Gray in A. Gray et al., Syn. Fl. N. Amer. 1(2): 435. 1884; *Nabalus sagittatus* (A. Gray) Rydberg

Plants 8–75 cm; taproots short, thick and tuberous or long and slender, fascicled. **Stems** erect, greenish and mottled purplish, ± glabrous or proximally glabrate, distally glabrous. **Leaves:** proximal present at flowering; petiolate (petioles narrowly winged, sometimes lobed, 5–15 cm); blades ovate or deltate, 3–10 × 1.5–6 cm, thin, bases hastate to truncate or rounded, sometimes shallowly, palmately lobed, ultimate margins irregularly dentate, faces glabrous; distal reduced. **Heads** in narrow, elongate paniculiform or racemiform arrays. **Calyculi** 5–7, green, subulate to narrowly lanceolate bractlets 3–8 mm (longest 2/3 lengths of phyllaries), glabrous. **Involucres** narrowly campanulate (rounded), 7–13 × 5–8 mm. **Phyllaries** 7–12, green to tan, apices dark, linear-lanceolate, 8–12 mm, margins narrowly scarious, (apices sparsely ciliate), faces glabrous. **Florets** 10–19; corollas white, 9–15 mm. **Cypselae** brown, fusiform to oblanceoloid, subterete or angled, 5 mm, indistinctly 7–10-ribbed; **pappi** pale yellow, 5–6 mm. **2n** = 16.

Flowering Jul–Aug. Stream banks, terraces, moist shady places, talus, rock crevices, mixed conifer woodlands; 1000–1600 m; Alta., B.C.; Idaho, Mont.

Prenanthes sagittata is similar to *P. alata*; it differs in its somewhat smaller stature, heads in narrower and more elongate arrays, and longer calyculi. The ranges of the two species do not overlap.

13. **Prenanthes serpentaria**   Pursh, Fl. Amer. Sept. 2: 499, plate 24. 1813 • Cankerweed, lion's-foot, gall-of-the-earth, butterweed   E F

Nabalus fraseri de Candolle; *N. integrifolius* Cassini; *N. serpentarius* (Pursh) Hooker; *Prenanthes integrifolia* (Cassini) Small

Plants 50–200+ cm; taproots short and thick, with lateral storage roots. **Stems** erect, green to reddish or purple mottled, proximally glabrous, distally sparsely tomentulose. **Leaves:** proximal often withered by flowering; petiolate (petioles 1–10 cm, often with pair of lobes); blades deltate to ovate or elliptic, 5–20 × 4–10 cm, coriaceous, margins usually deeply, pinnately 3–5-lobed, lobes and sinuses large and ± rounded, sometimes deeply cleft to base or palmately divided, apices acute or obtuse, ultimate margins entire or dentate, faces glabrous or finely tomentose on veins; cauline sessile or petiolate; distal reduced in size and lobing, often entire. **Heads** (6–12 in nodding clusters) in broad, paniculiform to corymbiform arrays (often widely branching and subdichotomous, at least some branches elongate). **Calyculi** of 8–10, green to purple, triangular to subulate bractlets 1–4 mm, often tomentulose to setose. **Involucres** cylindric (often attenuate basally to bracteate peduncles), 12–15 × 4–5 mm. **Phyllaries** (7–)8–(10), green or often purple, narrowly lanceolate to elliptic, 10–13 mm, sparsely hispid to appressed, coarsely setose, often reduced to single coarse, appressed seta (setae green or tan). **Florets** (8–)10–14(–19); corollas usually yellow to pale yellow, 9–15 mm. **Cypselae** golden brown to light tan, subcylindric, subterete or angled, 5–8 mm, indistinctly 8–10-ribbed; **pappi** tan, 7–8 mm. *2n* = 16.

Flowering Aug–Oct. Oak-hickory woodlands, borders, oak flats, pine woods, sandy areas; 100–1700 m; Ala., Conn., Del., D.C., Fla., Ga., Ky., Md., Mass., Miss., N.J., N.Y., N.C., Ohio, Pa., R.I., S.C., Tenn., Va., W.Va.

Prenanthes serpentaria is generally recognized by its large, deeply 3–5-lobed proximal leaves with rounded sinuses and lobes, winged petioles, attenuate involucres, sparsely setose phyllaries, and yellow corollas. The leaves are variable in size and lobing, often on the same plant. Some specimens have predominantly ovate to elliptic, unlobed leaves, and these have been variously recognized. Some specimens appear to combine characteristics of *P. crepidinea* or *P. trifoliolata* and may be the result of recent or ancient hybridization. The species boundaries in this group merit further study.

14. **Prenanthes trifoliolata** (Cassini) Fernald, Contr. Bot. Vermont 8: 89. 1900 • Threeleaved rattlesnakeroot, prenanthe trifoliolée [E] [F]

Nabalus trifoliolatus Cassini in F. Cuvier, Dict. Sci. Nat. ed. 2, 34: 95. 1825; *P. trifoliolata* var. *nana* (Bigelow) Fernald

Plants 10–150 cm; taproots thick, with lateral roots. **Stems** erect, green or sometimes mottled purple, usually glabrous, sometimes tomentulose distally. **Leaves:** proximal usually present at flowering; petiolate (petioles winged, 1–25 cm); blades deltate to ovate,

3–12 × 1–15 cm, thin, bases cordate to rounded, margins palmately 3(–5)-lobed to -divided (then leaves compound), lobes and sinuses usually angular (not rounded), lobes short and lanceolate, ultimate margins irregularly serrate, faces glabrous or ciliate along abaxial veins and margins; distal reduced, palmately lobed or entire. **Heads** (2–7, nodding, in irregular clusters) in racemiform or paniculiform arrays. **Involucres** narrowly campanulate (bases attenuate to bracteate peduncles), 10–13 × 4–5 mm. **Calyculi** of 5–7, green to dark green or blackish, triangular bractlets 1–3 mm, glabrous. **Phyllaries** 7–10, green to dark green or blackish proximally, lanceolate to elliptic, 10–11 mm, margins scarious, sometimes ciliate, faces glabrous. **Florets** 8–13; corollas pale yellow, 9–15 mm. **Cypselae** tan to brown, subcylindric, subterete to angled, 4–5 mm, distinctly 8–11-ribbed; **pappi** pale yellow, 7–9 mm. *2n* = 16.

Flowering Aug–Oct. Moist oak-hickory woods, swampy thickets, sandy areas, cliffs, sometimes saline habitats; 0–1400 m; St. Pierre and Miquelon; N.B., Nfld. and Labr., N.S., Ont., P.E.I., Que.; Conn., Ga., Ind., Ky., Maine, Md., Mass., Mich., N.H., N.J., N.Y., N.C., Ohio, Pa., R.I., S.C., Tenn., Vt., Va., W.Va.

Prenanthes trifoliolata is recognized by its relatively large, palmately 3–5-lobed leaves with angular lobes and sinuses, basally attenuate involucres, dark green and glabrous calyculi and phyllaries, and pale yellow corollas. Dwarf plants with deeply parted leaves found in alpine areas of northern New England and Canada have been recognized as *P. nana* or *P. trifoliolata* var. *nana*. This form is probably no more than a phenotypic adaptation to harsh environments. In at least some localities, it intergrades with more typical *P. trifoliolata* at lower elevations.

Hybrids between *Prenanthes trifoliolata* and *P. racemosa*, known as *P. ×mainensis* A. Gray, have been found in Maine, New Brunswick, Nova Scotia, and southern Quebec, usually where the two parents come together in cliff or saline habitats. The leaves of the hybrids are intermediate between 3-lobed and spatulate, the distal are sessile, the heads are nodding, and the phyllaries are glabrous, as in the *P. trifoliolata* parent.

47. LAUNAEA   Cassini in F. Cuvier, Dict. Sci. Nat. ed. 2, 25: 61, 321. 1822

• [For J. Cl. M. Mordant de Launay, 1750–1816, lawyer, later librarian at Musée d'Histoire Naturelle, Paris]   [I]

R. David Whetstone

Kristin R. Brodeur

Brachyramphus de Candolle

Annuals or biennials [perennials, shrubs, sometimes spiny], [5–]30–150 cm; usually taprooted [stoloniferous]. **Stems** erect [prostrate], distally branched, glabrous [± hairy]. **Leaves** basal or basal and proximally cauline; petiolate or sessile; blades ± oblanceolate, often pinnately lobed, ultimate margins usually dentate (teeth usually ± prickly; faces glabrous [± hairy]). **Heads** in spiciform or racemiform to paniculiform arrays [borne singly]. **Peduncles** not inflated distally, bracteolate. **Calyculi** 0 (or bractlets intergrading with phyllaries). **Involucres** cylindric [urceolate, campanulate, or obconic], 3–5[–16] mm diam. **Phyllaries** (persistent, reflexed in fruit) 18–25 in 3–5+ series, unequal, ovate to lanceolate (outer) or linear (inner), margins scarious, apices obtuse to acuminate (faces glabrous [± hairy]). **Receptacles** flat to convex, epaleate. **Florets** 25–30; corollas yellow to ochroleucous [cyanic]. **Cypselae** blackish to grayish, cylindric to fusiform or ± prismatic, sometimes ± compressed, beaks 0 (or lengths 0.05–0.1 times bodies), 4–5-ribbed (or -grooved), ribs usually muricate, faces glabrous; **pappi** persistent or tardily falling [readily falling], double [simple], of 60–100+, outer, white, often ± coiled or crisped (frizzy) hairs or bristles in 2–3 series plus 80–120+, white, coarser, barbellulate to smooth bristles in 2–3+ series, all distinct or some basally connate. *x* = 9.

Species ca. 50 (1 in the flora): introduced; Mexico; West Indies; Central America; South America; introduced also in Europe, n Africa, Atlantic Islands, sw, c Asia.

SELECTED REFERENCE Kilian, N. 1997. Revision of *Launaea* Cass. (Compositae, Lactuceae, Sonchinae). Englera 17.

1. **Launaea intybacea** (Jacquin) Beauverd, Bull. Soc. Bot. Genève, sér. 2, 2: 114. 1910 • Wild-lettuce
[F] [I]

Lactuca intybacea Jacquin in J. A. Murray, Syst. Veg. ed. 14, 713. 1784; *Brachyramphus intybacea* (Jacquin) de Candolle

Leaves (5–)10–25+ × 2–6(–12) cm, usually runcinately to sinuately lobed, bases ± auriculate, ultimate margins spinulodenticulate, apices acute to acuminate. **Involucres** 10–15 mm. **Cypselae** 4–5 mm; **pappi** 6–9 mm. *2n* = 18 (as *Lactuca runcinata*).

Flowering year-round. Disturbed sites, roadsides, coastal strands, agricultural lands; 0–5[–1200+] m; introduced; Fla., Tex.; Mexico; West Indies; Central America; South America; introduced also in Old World.

Launaea intybacea is occasionally used for food. It is a very successful r-strategist and is listed as a noxious plant in *A Global Compendium of Weeds* (R. Randall, http://www.hear.org/gcw/index.html).

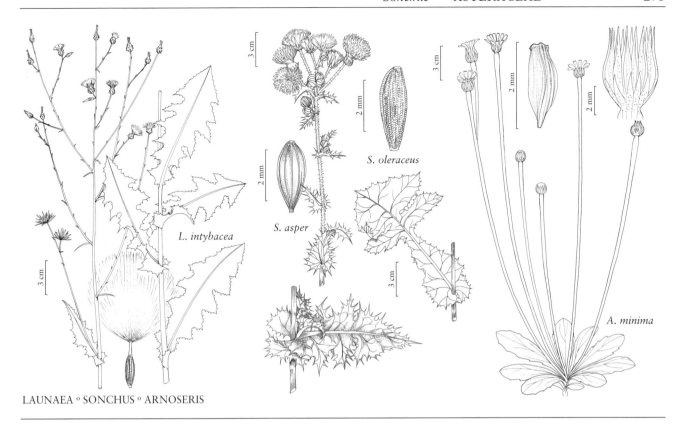

S. oleraceus

S. asper

L. intybacea

A. minima

LAUNAEA ○ SONCHUS ○ ARNOSERIS

48. SONCHUS Linnaeus, Sp. Pl. 2: 793. 1753; Gen. Pl. ed. 5, 347. 1754 • Sow-thistle, laiteron [Greek *sonchos,* ancient name for a kind of thistle] ☐

Philip E. Hyatt

Annuals, biennials, or perennials [shrubs], 3–350+ cm; taprooted, rhizomatous, or stoloniferous. **Stems** erect, branched distally or throughout, usually glabrous, sometimes stipitate-glandular (mostly distally). **Leaves** basal and cauline or mostly cauline; basal usually petiolate (petioles usually winged), cauline often sessile; blades mostly oblong, oblanceolate, or lanceolate, (bases often auriculate) margins usually 1(–2)-pinnately lobed, ultimate margins usually dentate (teeth often ± prickly), sometimes entire. **Heads** borne in corymbiform to subumbelliform arrays. **Peduncles** not notably dilated distally, usually ± bracteolate, glabrous or stipitate-glandular and/or tomentose. **Calyculi** 0. **Involucres** campanulate to urceolate, 5–15+ mm diam. **Phyllaries** 27–50 in 3–5+ series, unequal, green, deltate to lanceolate or linear, margins sometimes narrowly scarious (sometimes ciliate), apices acute (faces glabrous or stipitate- to setose-glandular). **Receptacles** flat to convex, ± pitted, glabrous, epaleate. **Florets** [30–]80–250[–450+]; corollas yellow to orange (not deliquescent; anthers yellowish to brownish apically). **Cypselae** stramineous or reddish to dark brown, ± compressed, mostly oblong or oblanceoloid to ellipsoid, beaks 0, ribs usually 2–4(–5+) on each face, faces sometimes transversely rugulose or tuberculate, glabrous; **pappi** tardily falling or persistent, of 80–100+, white, smooth or barbellulate bristles (some flattened, ± setiform scales), outer usually distinct in ± 1 series, inner basally coherent or connate, in 2–3+ series. *x* = 9.

Species 50–60+ (5 in the flora): introduced; Europe (Mediterranean), w Asia, n Africa, Atlantic Islands; some species introduced nearly worldwide.

Sonchus species are herbs in North America; some are sometimes woody at bases.

SELECTED REFERENCE Boulos, L. 1973. Révision systématique du genre *Sonchus* L. s.l.: IV. Sous-genre: 1. *Sonchus*. Bot. Not. 126: 155–196.

1. Leaf base auricles often recurved or curled, rounded; ligules shorter than tubes; cypselae strongly compressed, ± winged, ribs 3 on each face, faces not rugulose or tuberculate; annuals or biennials . . . . 2. *Sonchus asper*
1. Leaf base auricles usually straight, sometimes curved, obtuse or acute; ligules shorter or longer than tubes; cypselae weakly compressed, ribs 2–5+ on each face, faces transversely rugulose or tuberculate; annuals, biennials, or perennials.
   2. Annuals or biennials; stem bases soft to hard, herbaceous, often hollow.
      3. Leaf blade lobes ± deltate to lanceolate (not constricted at bases), the terminal usually larger than laterals; ligules ± equaling tubes; widespread . . . . 3. *Sonchus oleraceus*
      3. Leaf blade lobes ± rhombic to lanceolate (constricted at bases) or ± linear, the terminal ± equaling laterals; ligules longer than tubes; occasional, mostly historic on ballast . . . . 5. *Sonchus tenerrimus*
   2. Perennials; stem bases hard, sometimes ± woody.
      4. Cauline leaf base auricles rounded; cypselae dark brown, 2.5–3.5 mm; pappi 8–14 mm; widespread . . . . 1. *Sonchus arvensis*
      4. Cauline leaf base auricles acute; cypselae stramineous, 3.5–4 mm; pappi 7–9 mm; Ontario . . . . 4. *Sonchus palustris*

1. Sonchus arvensis Linnaeus, Sp. Pl. 2: 793. 1753

• Field sow-thistle, laiteron des champs [I]

Perennials, 0–150(–200) cm, usually rhizomatous or stoloniferous. **Stem bases** hard, sometimes ± woody. **Leaves:** blades of mid cauline oblong to lanceolate, (3–)6–40 × 2–15 cm, bases auriculate, auricles straight or curved, rounded, margins usually pinnately lobed, lobes ± deltate, not constricted at bases, terminals usually larger than laterals, dentate or entire. **Peduncles** sessile- or stipitate-glandular. **Involucres** 10–17+ mm. **Phyllaries** sessile- or stipitate-glandular. **Corollas:** ligules ± equaling tubes. **Cypselae** dark brown, oblanceoloid to ellipsoid, 2.5–3.5 mm, ribs 4–5(+) on each face, faces transversely rugulose to tuberculate across and between ribs; **pappi** 8–14 mm. $2n = 36, 54$.

Subspecies 2 (2 in the flora): introduced; Europe; introduced also in South America, Asia, Africa, Pacific Islands (New Zealand), Australia.

Sonchus arvensis is introduced in temperate regions of all continents. Plants of the species prefer relatively cooler, moister climates and are more abundant in the northern part of North America.

1. Peduncles stipitate-glandular; phyllaries stipitate-glandular . . . . 1a. *Sonchus arvensis* subsp. *arvensis*
1. Peduncles sessile-glandular; phyllaries usually sessile-glandular, rarely tomentose . . . . 1b. *Sonchus arvensis* subsp. *uliginosus*

1a. Sonchus arvensis Linnaeus subsp. **arvensis** [I] [W]

Peduncles stipitate-glandular. **Phyllaries** stipitate-glandular, longer 14–17 mm. $2n = 54$.

Flowering Jul–Sep. Disturbed sites, mostly wet, particularly in urban areas, roadside ditches, along streams; 0–1600+ m; introduced; St. Pierre and Miquelon; Alta., B.C., Man., N.B., Nfld. and Labr. (Nfld.), N.W.T., N.S., Ont., P.E.I., Que., Sask.; Alaska, Calif., Colo., Conn., Del., Idaho, Ill., Ind., Iowa, Ky., Maine, Md., Mass., Mich., Minn., Mo., Mont., Nev., N.H., N.J., N.Y., N.Dak., Ohio, Oreg., Pa., R.I., S.Dak., Utah, Vt., Wash., Wis., Wyo.; Europe; widely introduced elsewhere.

Subspecies *arvensis* is considered a noxious weed in California, Idaho, Illinois, Iowa, Maryland, Minnesota, Nevada, North Dakota, South Dakota, and Wyoming.

1b. Sonchus arvensis Linnaeus subsp. **uliginosus** (M. Bieberstein) Nyman, Consp. Fl. Eur., 433. 1879 [I]

Sonchus uliginosus M. Bieberstein, Fl. Taur.-Caucas. 2: 238. 1808; *S. arvensis* var. *glabrescens* Günther, Grabowski & Wimmer

Peduncles sessile-glandular. **Phyllaries** usually sessile-glandular, rarely tomentose, the longer 10–15 mm. $2n = 36$.

Flowering Jul–Sep. Mostly wet, disturbed sites, oil-seed or irrigated crops, roadside ditches, along streams, wetlands; 0–2200+ m; introduced; Alta., B.C., Man., N.B., N.S., Ont., P.E.I., Que., Sask., Yukon; Calif., Colo., Conn., Del., Idaho, Ill., Ind., Iowa, Kans., Maine, Md., Mass., Mich., Minn., Miss., Mont., Nebr., Nev., N.J., N.Y., N.C., N.Dak., Ohio, Oreg., Pa., S.Dak., Tenn., Utah, Vt., Va., Wash., W.Va., Wis., Wyo.; Europe; widely introduced elsewhere.

Subspecies *uliginosus* is considered a noxious weed in Alaska, Idaho, Illinois, Iowa, Maryland, Minnesota, North Carolina, North Dakota, South Dakota, and Wyoming.

2. **Sonchus asper** (Linnaeus) Hill, Herb. Brit. 1: 47. 1769 • Spiny-leaf sow-thistle, laiteron rude F I

Sonchus oleraceus Linnaeus var. *asper* Linnaeus, Sp. Pl. 2: 794. 1753

Annuals or biennials, 10–120 (–200+) cm. **Stem bases** soft, herbaceous, hollow. **Leaves:** blades of mid cauline spatulate or oblong to obovate or lanceolate, 6–30 × 1–15 cm, bases auriculate, auricles often recurved or curled, rounded, margins often pinnately lobed, lobes ± deltate (not constricted at bases), terminals usually larger than laterals, usually prickly-dentate. **Peduncles** usually stipitate-glandular, sometimes glabrous. **Involucres** 9–13+ mm. **Phyllaries** usually stipitate-glandular. **Corollas:** ligules mostly shorter than tubes. **Cypselae** stramineous to reddish brown, mostly ellipsoid, strongly compressed, ± winged, 2–3 mm, ribs 3(–5) on each face, faces smooth across and between ribs; **pappi** 6–9 mm. $2n = 18$.

Flowering (Mar–)Jul–Nov (year-round in south). Disturbed sites, roadsides, along streams; 0–2500+ m; introduced; St. Pierre and Miquelon; Alta., B.C., Man., N.B., Nfld. and Labr., N.S., Ont., P.E.I., Que., Sask., Yukon; Ala., Alaska, Ariz., Ark., Calif., Colo., Conn., Del., D.C., Fla., Ga., Idaho, Ill., Ind., Iowa, Kans., Ky., La., Maine, Md., Mass., Mich., Minn., Miss., Mo., Mont., Nebr., Nev., N.H., N.J., N.Mex., N.Y., N.Dak., Ohio, Okla., Oreg., Pa., R.I., S.C., S.Dak., Tenn., Tex., Utah, Vt., Va., Wash., W.Va., Wis., Wyo.; Europe, w Asia, n Africa; introduced also in Mexico, Central America, West Indies, Bermuda, South America, e Asia, s Africa, Pacific Islands (New Zealand), Australia.

L. Boulos (1973) distinguished subsp. *asper* (annuals with leaves mostly cauline, cypselae margins little or not at all curved and/or ciliate, and one pair of chromosomes with small satellites) from subsp. *glaucescens* (biennials with leaves mostly in rosettes, leaves mostly stiffer and more prickly than in subsp. *asper*, cypselae with curved, ciliate margins, and two pairs of chromosomes with large satellites) and noted that the two subspecies are

morphologically rather difficult to distinguish if the specimen in hand lacks the rootstock or stem base.

According to H. N. Barber (1941), crosses between *Sonchus asper* and *S. oleraceus* resulted in sterile hybrids.

3. **Sonchus oleraceus** Linnaeus, Sp. Pl. 2: 794. 1753 • Common sow-thistle, laiteron F I

Annuals or biennials, 10–140 (–200) cm. **Stem bases** soft to hard, herbaceous, often hollow. **Leaves:** blades of mid cauline spatulate or oblong to obovate or lanceolate, 6–35 × 1–15 cm, bases auriculate, auricles deltate to lanceolate, ± straight, acute, margins usually pinnately (often runcinately) lobed, lobes ± deltate to lanceolate, not constricted at bases, terminals usually larger than laterals, entire or dentate. **Peduncles** usually glabrous, sometimes stipi-tate-glandular. **Involucres** 9–13+ mm. **Phyllaries** usually glabrous, sometimes tomentose and/or stipitate-glandular. **Corollas:** ligules ± equaling tubes. **Cypselae** dark brown, mostly oblanceoloid, 2.5–3.5+ mm, ribs 2–4 on each face, faces transversely rugulose or tuberculate across and between ribs; **pappi** 5–8 mm. $2n = 32, 36$.

Flowering (Apr–)Jul–Oct (year-round in south). Disturbed sites, gardens, roadsides, along streams; 0–2000 m; introduced; Greenland; St. Pierre and Miquelon; Alta., B.C., Man., N.B., Nfld. and Labr. (Nfld.), N.W.T., N.S., Ont., P.E.I., Que., Sask.; Ala., Alaska, Ariz., Ark., Calif., Colo., Conn., Del., D.C., Fla., Ga., Idaho, Ill., Ind., Iowa, Kansas, Ky., La., Maine, Md., Mass., Mich., Minn., Miss., Mo., Mont., Nebr., Nev., N.H., N.J., N.Mex., N.Y., N.Dak., Ohio, Okla., Oreg., Pa., R.I., S.C., S.Dak., Tenn., Tex., Utah, Vt., Va., Wash., W.Va., Wis., Wyo.; Europe; introduced also in Mexico, West Indies, Bahamas, Central America, South America, Africa, Asia, Pacific Islands (New Zealand), Australia.

4. **Sonchus palustris** Linnaeus, Sp. Pl. 2: 793. 1753 • Marsh sow-thistle I

Perennials, 100–150(–350+) cm. **Stem bases** hard, sometimes ± woody. **Leaves:** blades of mid cauline oblong to lanceolate or linear, 15–20+ × 2–3(–8+) cm, bases auriculate, auricles lance-olate to linear, ± straight, acute, margins sometimes pinnately lobed, lobes ± deltate to lanceolate or linear (not constricted at bases), terminals usually larger than laterals, entire or dentate (teeth not notably prickly). **Peduncles** usually setose- to stipitate-glandular. **Involucres** 9–13+ mm. **Phyllaries** sparsely to densely setose- to stipitate-glandular. **Corollas:** ligules ± equaling tubes. **Cypselae** stramineous, oblong to ellipsoid, 3.5–4 mm, ribs 4–5 on each face, faces transversely rugulose or tuberculate across and between ribs; **pappi** 7–9 mm. $2n = 18$.

Flowering Jul–Sep. Disturbed, marshy sites, wet roadside ditches on peaty sand, railroad ditches, wet fields; 0–100 m; introduced; Ont.; Europe.

Sonchus palustris was found in Ontario by D. F. Brunton and C. W. Crompton (1993) in the Ottawa area, and by J. K. Morton and J. M. Venn (1995) in the Waterloo region.

The plants in the flora area are subsp. *palustris.* Subspecies *sosnowskyi* (Schchian) Boulos, from a single location in the southern Caucasus of eastern Europe, differs primarily in having glabrous stems, peduncles, and heads.

5. **Sonchus tenerrimus** Linnaeus, Sp. Pl. 2: 794. 1753

• Slender sow-thistle

Annual, biennials, or perennials, 10–80 cm. **Stem bases** soft to hard, herbaceous, often hollow. **Leaves:** blades of mid cauline oblong, 3–20 × 2–6 cm, bases auriculate, auricles ovate to lanceolate or linear, ± straight, obtuse to acute, margins usually pinnately lobed, lobes ± rhombic to lanceolate (constricted at bases) or ± linear, terminals

± equaling laterals, entire or dentate. **Peduncles** usually setose- to stipitate-glandular, often tomentose as well, sometimes glabrous. **Involucres** 10–12+ mm. **Phyllaries** usually setose- to stipitate-glandular, sometimes tomentose as well. **Corollas:** ligules longer than tubes. **Cypselae** reddish brown, oblanceoloid, 2.5–3.5 mm, ribs 1–3 on each face, faces transversely rugulose or tuberculate across and between ribs; **pappi** 5–8 mm. $2n = 14$.

Flowering Mar–Jun. Disturbed sites, often on ballast; 0–200 m; introduced; Calif.; s Europe, w Asia, n Africa.

Sonchus tenerrimus was known historically on ballast from Alabama, Pennsylvania (Philadelphia), New Jersey, New York, and, possibly, Quebec.

49. **ARNOSERIS** Gaertner, Fruct. Sem. Pl. 2: 355, plate 157, fig. 3. 1791 • [Greek *arnos*, sheep, and *seris*, a kind of endive; allusion unclear]

John L. Strother

Annuals, (5–)10–30+ cm; taprooted. **Stems** 1–10+ (scapiform), erect, simple or branched distally, glabrous. **Leaves** mostly basal; obscurely petiolate; blades oblanceolate to spatulate, margins dentate (faces usually scabridulous, mostly near margins). **Heads** borne singly or (2–3) in loose, corymbiform arrays. **Peduncles** inflated (fistulose) distally, not bracteate. **Calyculi** 0 or of 1–10+, lanceolate bractlets. **Involucres** broadly campanulate or urceolate, 4–8 mm diam. **Phyllaries** 10–22+ in 1(–2) series (basally connate), lanceolate to lance-linear (abaxially keeled proximally), equal or subequal, margins little, if at all, scarious, apices acuminate. **Receptacles** flat, pitted, glabrous, epaleate. **Florets** 20–50+; corollas yellow. **Cypselae** dark brown, obovoid, not beaked, ribs 8–10 (3–5 stronger), stramineous, faces smooth or rugose between ribs, glabrous; **pappi** 0. $x = 9$.

Species 1: introduced; Europe.

1. Arnoseris minima (Linnaeus) Schweigger & Körte, Fl. Erlang. 2: 72. 1811 F I

Hyoseris minima Linnaeus, Sp. Pl. 2: 809. 1753

Leaf blades 5–75+ × 3–15(–20+) mm. **Peduncles** 5–3+ cm. **Phyllaries:** keels stramineous and thickened in fruit, abaxial faces scabrellous or glabrate. **Cypselae** 1.5–2+ mm. **2***n* = 18.

Flowering Jun–Sep. Disturbed sites, sandy soils; 10–100 m; introduced; N.S., P.E.I.; Maine, Mich., N.H., N.Y., Ohio, Pa.; Europe.

50. TOLPIS Adanson, Fam. Pl. 2: 112, 612. 1763 • [No etymology in protologue; no readily discernible meaning from Greek or Latin roots] I

John L. Strother

Annuals [perennials, shrubs], 5–100+ cm; taprooted. **Stems** usually 1, erect, branched distally, glabrous or hairy. **Leaves** basal and cauline; basal sessile or petiolate, distal usually sessile; blades ovate-lanceolate or oblanceolate to lanceolate or linear, margins entire or dentate to pinnately lobed or pinnatisect (faces glabrous or sparsely pubescent or arachnose to villous). **Heads** borne singly or in loose, corymbiform arrays (terminal heads often surpassed by others). **Peduncles** not inflated, bracteate. **Calyculi** of 8–13, linear to filiform bractlets (often intergrading with peduncular bracts). **Involucres** campanulate, 5–10[–15+] mm diam. (larger in fruit). **Phyllaries** 20–25+ in 2+ series, lance-linear to linear (± navicular proximally, enfolding ovaries/cypselae of subtended florets, sometimes keeled abaxially), subequal to equal, margins little, if at all, scarious, apices filiform, faces ± arachnose. **Receptacles** flat, pitted, glabrous, epaleate. **Florets** 30–100+; corollas mostly yellow (inner sometimes purplish or brown). **Cypselae** brownish to blackish, ± columnar [obconic], not beaked, ribs or nerves usually 6–8(–10), faces glabrous or ± hirtellous; **pappi** persistent, of 4–15(–30+), whitish, smooth or barbellulate, setiform scales plus bristles (often equal on outer cypselae and unequal on inner ones). *x* = 9.

Species ca. 20 (1 in the flora): introduced; Europe, Africa, Atlantic Islands.

1. Tolpis barbata (Linnaeus) Gaertner, Fruct. Sem. Pl. 2: 372. 1791 F I

Crepis barbata Linnaeus, Sp. Pl. 2: 805. 1753; *Tolpis umbellata* Bertoloni

Leaf blades 30–150 × 5–25 mm, margins dentate to pinnately lobed. **Involucres** 5–15+ × 5–10+ mm. **Cypselae** 1–1.5 mm; **pappi:** scales 0.3–0.5 mm, bristles 3.5–4+ mm. **2***n* = 18.

Flowering Jun. Disturbed sites; 0–200 m; introduced; Calif.; Europe.

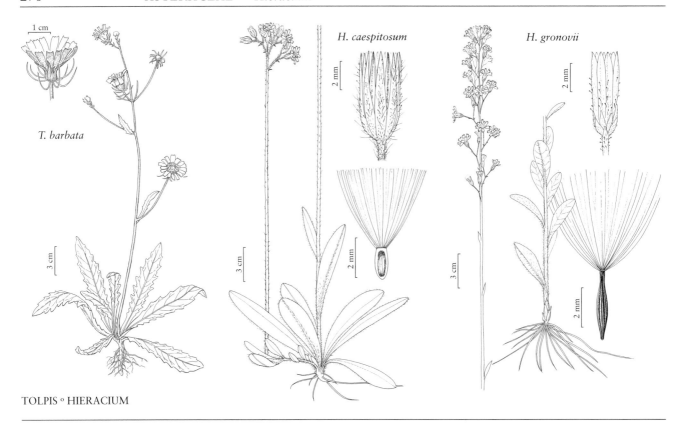

T. barbata

H. caespitosum

H. gronovii

1 cm

2 mm

3 cm

2 mm

2 mm

3 cm

2 mm

3 cm

2 mm

TOLPIS ∘ HIERACIUM

51. HIERACIUM Linnaeus, Sp. Pl. 2: 799. 1753; Gen. Pl. ed. 5, 350. 1754

• Hawkweed, épervière [No etymology in protologue; said to be from Greek *hierax*, hawk]

John L. Strother

Pilosella Hill

Perennials, (5–)20–150+ cm; taprooted (rootstocks sometimes woody, branched; stolons produced in some taxa). **Stems** usually 1, usually erect, usually branched distally, sometimes throughout, sometimes scapiform, glabrous or hairy (induments often complex, see discussion). **Leaves** basal, basal and cauline, or cauline; petiolate or sessile; blades mostly elliptic, lanceolate, oblanceolate, oblong, or spatulate, margins entire, denticulate, or dentate [laciniate to pinnatifid] (faces glabrous or hairy, induments often complex, see discussion). **Heads** borne singly or in corymbiform, paniculiform, thyrsiform, umbelliform, or nearly racemiform arrays. **Peduncles** (terminal and axillary) not inflated, often bracteate. **Calyculi** 0 or of 3–13(–16+), deltate to lanceolate or linear bractlets (in 1–2+ series; sometimes intergrading with phyllaries). **Involucres** hemispheric or campanulate to cylindric, 3–9[–12+] mm diam. **Phyllaries** 5–21(–40+) in 2+ series, lanceolate to linear, subequal to unequal (reflexed in fruit), margins usually little, if at all, scarious, apices obtuse to acute or acuminate. **Receptacles** flat, pitted, glabrous, epaleate. **Florets** 6–150+; corollas usually yellow, sometimes white or ochroleucous, sometimes tinged with cyan or red, rarely orange (then often drying scarlet or purplish). **Cypselae** usually redbrown or black (tan in *H. horridum*), usually ± columnar or prismatic, sometimes ± urceolate (slightly bulbous proximally and narrower distally) or nearly fusiform, not distinctly beaked, ribs (or grooves) usually 10, faces glabrous; **pappi** persistent (fragile), of 20–80+, distinct, white, sordid, stramineous, or rufous, ± equal or unequal, barbellulate bristles in 1–2+ series. *x* = 9.

Species 250–1000+ (36 in the flora): North America, Mexico, Central America, South America, Europe, Asia, Africa.

Most plants referable to *Hieracium* are apomictic (reproducing from asexually produced seeds). Apomictic reproduction often results in perpetuation of morphologic variants at populational and regional levels. Temptation to name such variants as species has proven irresistible to some botanists; upward of 9,000 species names have been published in *Hieracium*. Circumscriptions of "species" in *Hieracium* seem to be more artificial than in most genera of composites. Under these circumstances, I feel it would be irresponsible to recognize any infraspecific taxa here. For the most part, the "species" recognized here are those that have been recognized in local and regional floras of the past 50 or so years. Accepted names and/or changes in synonymies for some species reflect changes suggested by J. H. Beaman (1990), A. Cronquist (1980), and E. G. Voss (1972–1996, vol. 3).

I have maintained a traditional generic circumscription of *Hieracium*. Some other botanists exclude some species (including numbers 1–6 here) from *Hieracium* and treat them in *Pilosella* Hill, which differs from *Hieracium* in having stolons often produced, ribs of cypselae slightly projecting distally, and bristles of the pappi in one series.

For Greenland, T. W. Böcher et al. (1968) reported 18 species of *Hieracium*: 16 said to be endemic to Greenland plus *H. alpinum* Linnaeus (known from Greenland and Eurasia) and *H. groenlandicum* Arvet-Touvet (= *H. vulgatum* Fries; known from Greenland and continental North America). Following H. J. Scoggan (1978–1979, part 4), I have placed 15 names of Greenland "endemics" as synonyms of other names. The type of the 16th "endemic," *H. trigonophorum* Oskarsson, is probably conspecific with that of *H. alpinum*.

In my key and descriptions, "piloso-hirsute" refers to surfaces with scattered to crowded, tapered, whiplike, straight or curly, smooth to ± barbellate hairs mostly (0.5–)2–8(–15+) mm (sometimes called "setae"); "stellate-pubescent" refers to surfaces with scattered to crowded, ± dendritically branched (often called, but seldom truly, "stellate") hairs mostly 0.05–0.2+ mm (such surfaces are sometimes described as "floccose"); and "stipitate-glandular" refers to surfaces with scattered to crowded gland-tipped hairs mostly 0.2–0.8(–1.2+) mm. Surfaces of stems, leaves, peduncles, and phyllaries may be glabrous or may bear one, two, or all three of the types of hairs mentioned here; other induments are seldom encountered in hieraciums of the flora area.

Given the complexity of the reproductive modes among the plants and the likelihood of misidentifications of vouchers, I have not included chromosome numbers for species. Sexual hieraciums are usually diploids ($2n$ = 18) and the apomictic hieraciums are usually triploids ($2n$ = 27).

SELECTED REFERENCES Beaman, J. H. 1990. Revision of *Hieracium* (Asteraceae) in Mexico and Central America. Syst. Bot. Monogr. 29: 1–77. Fernald, M. L. 1943c. Notes on *Hieracium*. Rhodora 45: 317–325.

1. Corollas yellow (often each with abaxial red stripe) or orange (drying scarlet to purple); cypselae 1–2.5 mm; pappi of 25–40+, white or sordid bristles in 1 series (plants sometimes stoloniferous).
    2. Leaf blades elliptic to ± oblanceolate, lengths 2–4(–6+) times widths; heads usually borne singly . . . . 1. *Hieracium pilosella*
    2. Leaf blades obovate or oblanceolate to spatulate or lanceolate, lengths (2–)3–8+ times widths; heads usually (2–)5–30+ in corymbiform to paniculiform or ± umbelliform arrays, rarely borne singly.
        3. Corollas orange (drying scarlet to purple) . . . . 2. *Hieracium aurantiacum*
        3. Corollas yellow (often each with abaxial red stripe).
            4. Heads (1–)2–5+, in ± umbelliform arrays.

5. Stems proximally piloso-hirsute (hairs 2–4+ mm) and stellate-pubescent;
involucres (9–)12–13 mm . . . . 3. *Hieracium flagellare*
5. Stems proximally piloso-hirsute (hairs 1–2+ mm), sometimes stipitate-
glandular as well; involucres 6–8 mm . . . . 4. *Hieracium lactucella*
4. Heads (2–)10–30+, usually in corymbiform to paniculiform, sometimes
± umbelliform, arrays (rarely borne singly in *H. triste*).
6. Leaves usually piloso-hirsute and stellate-pubescent; corollas 8–12+ mm
. . . . 5. *Hieracium caespitosum*
6. Leaves usually glabrous, sometimes piloso-hirsute (on midribs and at
margins) or scabrellous and/or stipitate-glandular (not both piloso-
hirsute and stellate-pubescent); corollas 5–9 mm.
7. Leaves glabrous or piloso-hirsute (on midribs and at margins);
involucres 5–6(–7) mm; florets (40–)60–80+; pappi 3–4 mm . . . . . .
. . . . 6. *Hieracium piloselloides*
7. Leaves glabrous or scabrellous and/or stipitate-glandular; involucres
(6–)7–10 mm; florets 20–60+; pappi 4–5 mm . . . . 29. *Hieracium triste* (in part)
1. Corollas yellow or ochroleucous to white (pinkish in *H. carneum*); cypselae (2–)2.5–7 mm;
pappi of (30–)40–80, white or stramineous to sordid bristles in 1–2+ series (plants not
stoloniferous).
8. Cypselae ± urceolate (slightly bulbous in proximal 1/3–1/2+, narrower in distal 1/3–1/2+,
not distinctly beaked).
9. Corollas yellow.
10. Cauline leaves 0–3; cypselae 5–5.5 mm . . . . 7. *Hieracium bolanderi*
10. Cauline leaves (1–)3–12+; cypselae 3–4.5 mm.
11. Stems proximally piloso-hirsute (hairs 6–15+ mm); blades of cauline leaves
mostly oblanceolate, lengths 4–7+ times widths . . . . 8. *Hieracium longipilum*
11. Stems proximally piloso-hirsute (hairs 1–4+ mm); blades of cauline leaves
elliptic to obovate or oblanceolate, lengths 2–4+ times widths.
12. Heads in ± thyrsiform arrays (lengths of arrays 3–6+ times diams.);
phyllaries usually glabrous or stellate-pubescent, sometimes stipitate-
glandular; florets 12–20+; corollas 8–9 mm . . . . 9. *Hieracium gronovii*
12. Heads in ± corymbiform arrays (lengths of arrays ± 2 times diams.);
phyllaries stellate-pubescent and stipitate-glandular; florets 20–50+;
corollas 10–12+ mm . . . . 10. *Hieracium megacephalum*
9. Corollas usually ochroleucous to whitish, sometimes pale yellow or pinkish.
13. Cauline leaves 0–1(–3+); involucres (10–)12–15+ mm; phyllaries usually piloso-
hirsute (hairs 1–3+ mm) and stellate-pubescent, sometimes stipitate-glandular
as well . . . . 11. *Hieracium fendleri*
13. Cauline leaves (0–)3–8+; involucres 7–11 mm; phyllaries glabrous or stellate-
pubescent and/or stipitate-glandular (not piloso-hirsute).
14. Leaves glabrous or piloso-hirsute (hairs 3–6+ mm); corollas whitish to
pinkish; pappi 4–5 mm . . . . 12. *Hieracium carneum*
14. Leaves glabrous or piloso-hirsute (hairs 0.5–2+ mm); corollas ochroleucous
to whitish or pale yellow; pappi 5–6 mm.
15. Leaf blades elliptic to oblanceolate, 30–85 × 16–35 mm, lengths
2.5–5 times widths; florets 25–40; cypselae 3.5–4.5(–5+) mm
. . . . 13. *Hieracium crepidispermum*
15. Leaf blades oblanceolate to lanceolate, 35–120 × 10–18 mm, lengths
4–10+ times widths; florets 15–25+; cypselae 5–6 mm . . . . . 14. *Hieracium brevipilum*
8. Cypselae columnar or prismatic (little, if at all, bulbous proximally and narrower in
distal 1/3–1/2).
16. Leaves all or mostly cauline (basal leaves usually wanting or withered, cauline
leaves usually 6–45+) at flowering.
17. Corollas white, ochroleucous, or pale yellow.

18. Involucres 6–7 mm; corollas 5–8 mm; cypselae 2–2.5 mm; pappi 4–5 m
. . . . 15. *Hieracium paniculatum*
18. Involucres (7–)8–10(–11) mm; corollas 9–10 mm; cypselae 2.5–4 mm; pappi
(4–)5–7 mm . . . . 30. *Hieracium albiflorum* (in part)
17. Corollas yellow.
19. Phyllaries piloso-hirsute, sometimes stellate-pubescent and/or stipitate-
glandular as well.
20. Stems usually glabrous, sometimes distally stellate-pubescent; leaves
usually glabrous, sometimes scabrous near distal margins (not piloso-
hirsute) . . . . 16. *Hieracium umbellatum* (in part)
20. Stems usually piloso-hirsute, sometimes stellate-pubescent as well,
sometimes glabrous (*H. scouleri*); leaves piloso-hirsute, sometimes
glabrous (*H. scouleri*).
21. Leaf blades: lengths (3–)4–8+ times widths; pappi 6–7 mm . . .
. . . . 20. *Hieracium scouleri* (in part)
21. Leaf blades: lengths 2.5–5+ times widths; pappi 5–6 mm.
22. Involucres 7–10 (× 4–5) mm; phyllaries piloso-hirsute (hairs
2–5 mm) and stellate-pubescent; florets 12–24+ . . . .
. . . . 21. *Hieracium longiberbe* (in part)
22. Involucres 8–12 (× 6–10) mm; phyllaries piloso-hirsute (hairs
2–5 mm) and stellate-pubescent and stipitate-glandular;
florets 25–40+ . . . . 17. *Hieracium schultzii* (in part)
19. Phyllaries usually stellate-pubescent and/or stipitate-glandular (not piloso-
hirsute).
23. Peduncles stipitate-glandular, sometimes piloso-hirsute and/or stellate-
pubescent as well.
24. Leaves: margins of some or all sinuately toothed; florets 15–30+
. . . . 35. *Hieracium argutum* (in part)
24. Leaves: margins usually entire, sometimes denticulate; florets 30–
60+.
25. Peduncles stipitate-glandular (not stellate-pubescent);
involucres 10–12+ mm; phyllaries stipitate-glandular (not
stellate-pubescent) . . . . 24. *Hieracium parryi* (in part)
25. Peduncles stellate-pubescent and stipitate-glandular;
involucres 7–9 mm; phyllaries stellate-pubescent and stipitate-
glandular . . . . 18. *Hieracium scabrum* (in part)
23. Peduncles piloso-hirsute and/or stellate-pubescent (not stipitate-
glandular).
26. Stems distally usually glabrous, sometimes stellate-pubescent (not
piloso-hirsute); leaf lengths (3–)5–10(–15) times widths.
27. Florets 15–30+; pappi white, 3.5–5 mm (California) . . . . .
. . . . 35. *Hieracium argutum* (in part)
27. Florets 30–80+; pappi stramineous to sordid, 6–7 mm
. . . . 16. *Hieracium umbellatum* (in part)
26. Stems distally piloso-hirsute, sometimes stellate-pubescent as well;
leaf lengths 2–4 times widths.
28. Leaves abaxially usually glabrous, sometimes piloso-hirsute
on midribs; involucres 10–15 mm; phyllaries 25–35 (apices
acuminate-caudate) . . . . 27. *Hieracium robinsonii* (in part)
28. Leaves abaxially piloso-hirsute (bases of hairs swollen); involu-
cres 8–9 mm; phyllaries ca. 21 (apices rounded) . . . . . 19. *Hieracium sabaudum*
[16. Shifted to left margin.—Ed.]
16. Leaves all basal or basal and cauline (basal leaves usually 3–12+, cauline leaves usually 0–
8+) at flowering.
29. Phyllaries usually stellate-pubescent and/or stipitate-glandular (rarely, if ever, piloso-
hirsute), rarely glabrous.

30. Florets 4–15+ (15–30+ in 35. *Hieracium argutum* with sinuately toothed leaf margins; California).
   31. Stems proximally piloso-hirsute (hairs at bases of stems 5–8+ mm, often curled and tangled); involucres (7–)8–9 mm; pappi 4–5+ mm . . . . 22. *Hieracium pringlei*
   31. Stems proximally stellate-pubescent, rarely piloso-hirsute (hairs at bases of stems 1–3+ mm, not or rarely curled and tangled); involucres 9–12 mm; pappi 7–9 mm . . . . 23. *Hieracium greenei*
30. Florets 30–80+.
   32. Margins of some or all leaves entire or denticulate.
      33. Stems distally piloso-hirsute and stipitate-glandular . . . . 24. *Hieracium parryi* (in part)
      33. Stems distally usually glabrous or stellate-pubescent, sometimes piloso-hirsute (not stipitate-glandular).
         34. Leaf blades oblanceolate to spatulate, lengths 3–5 times widths; peduncles stellate-pubescent and stipitate-glandular; pappi 6–7 mm . . . . 33. *Hieracium traillii* (in part)
         34. Leaf blades elliptic or obovate to oblanceolate, lengths 2–3(–4) times widths; peduncles usually stipitate-glandular, rarely glabrous or glabrate (not stellate-pubescent); pappi 4–5 mm . . . . 32. *H. venosum* (in part)
   32. Margins of some or all leaves usually dentate.
      35. Leaf blades spatulate or oblanceolate to lance-linear, lengths 5–10(–15+) widths (California) . . . . 35. *Hieracium argutum* (in part)
      35. Leaf blades oblong or elliptic to lance-elliptic or lanceolate, lengths 1.5–4+ widths.
         36. Basal leaves 2–4, cauline (2–)4–10, abaxial faces usually glabrous, sometimes piloso-hirsute on midribs; involucres 10–15 mm; pappi ca. 6 mm . . . . 27. *Hieracium robinsonii* (in part)
         36. Basal leaves 3–8+, cauline (0–)1–5+, abaxial faces usually piloso-hirsute, sometimes stellate-pubescent as well; involucres 8–10 mm; pappi 4–5 mm.
            37. Leaf blades ± elliptic, lengths 1.5–3+ widths, bases usually rounded to truncate, apices ± obtuse (apiculate); corollas 12–13(–16) mm . . . . 25. *Hieracium murorum*
            37. Leaf blades lance-elliptic to lanceolate, lengths 2.5–4 widths, bases usually cuneate, sometimes rounded to truncate, apices acute; corollas 13–18 mm . . . . 26. *Hieracium vulgatum*
[29. Shifted to left margin.—Ed.
29. Phyllaries usually piloso-hirsute, often stellate-pubescent and/or stipitate-glandular as well, rarely glabrous.
38. Heads 1(–2+); involucres 13–18+ mm (Greenland) . . . . 28. *Hieracium alpinum*
38. Heads (1–)3–30(–60+); involucres 6–8(–11) mm.
   39. Leaf blades obovate to spatulate or oblanceolate, (15–)25–40(–60+) mm, lengths 2–3+ times widths, apices rounded to obtuse (often apiculate), abaxial faces usually glabrous, sometimes scabrellous and/or stipitate-glandular (not piloso-hirsute) . . . . 29. *Hieracium triste* (in part)
   39. Leaf blades elliptic, lance-linear, lanceolate, oblanceolate, oblong, or spatulate (15–)50–100(–300) mm, lengths (2–)4–15+ times widths, apices usually rounded, obtuse, or acute (not apiculate), abaxial faces usually piloso-hirsute (hairs 1–7+ mm), sometimes stellate-pubescent as well, rarely glabrous (some *H. scouleri*, *H. albiflorum*, *H. nudicaule*).
      40. Peduncles usually glabrous, sometimes stipitate-glandular (not stellate-pubescent); corollas white . . . . 30. *Hieracium albiflorum* (in part)
      40. Peduncles usually stellate-pubescent and/or stipitate-glandular, sometimes piloso-hirsute as well, rarely glabrous (some *H. scouleri*, *H. venosum*); corollas yellow (sometimes pale in *H. venosum*).

[41. Shifted to left margin.—Ed.

41. Involucres ± cylindric (6–9 × 3–4 mm); florets 6–12(–15+); cypselae 3–3.5 mm (tan to red-brown) .... 31. *Hieracium horridum*
41. Involucres campanulate, hemispheric, obconic, or turbinate (6–10+ × 4–9+ mm); florets (15–)20–60+; cypselae 2–4 mm (usually black, sometimes red-brown).
  42. Cauline leaves 0–1(–2); florets 30–60+ (mostly e of Mississippi River).
    43. Peduncles usually stipitate-glandular, rarely glabrous or glabrate; pappi 4–5 mm
        .... 32. *Hieracium venosum* (in part)
    43. Peduncles stipitate-glandular and stellate-pubescent; pappi 6–7 mm ... 33. *Hieracium traillii* (in part)
  42. Cauline leaves 0–10+; florets 15–45+ (w of Mississippi River).
    44. Faces of leaf blades piloso-hirsute; florets 20–24; corollas 6–9 mm .... 34. *Hieracium abscissum*
    44. Faces of leaf blades piloso-hirsute and stellate-pubescent; florets 15–45+; corollas 8–12 mm.
      45. Margins of some or all leaf blades sinuately toothed .... 35. *Hieracium argutum* (in part)
      45. Margins of leaf blades usually entire, sometimes denticulate.
        46. Cauline leaves (3–)5–10+; pappi 6–7 mm .... 20. *Hieracium scouleri* (in part)
        46. Cauline leaves 0–3(–5); pappi 4–6 mm .... 36. *Hieracium nudicaule*

1. Hieracium pilosella Linnaeus, Sp. Pl. 2: 800. 1753

• Mouse-ear hawkweed, épervière piloselle ☐

Hieracium pilosella var. *niveum* Müller Argoviensis; *Pilosella officinarum* F. W. Schultz & Schultz-Bipontinus

Plants 10–25(–40+) cm. **Stems** proximally piloso-hirsute (hairs 1–3+ mm) and stellate-pubescent, distally piloso-hirsute (hairs 1–2 mm), stellate-pubescent, and stipitate-glandular. **Leaves:** basal (2–)5–10+, cauline 0 (–3); blades elliptic to ± oblanceolate, 10–45(–75+) × 5–12(–18+) mm, lengths 2–4(–6+) times widths, bases cuneate, margins entire, apices acute, faces piloso-hirsute (hairs 2–7+ mm) and stellate-pubescent. **Heads** usually borne singly, rarely 2(–3). **Peduncles** piloso-hirsute and stellate-pubescent. **Calyculi:** bractlets 8–15+. **Involucres** hemispheric to obconic, 7.5–9 mm. **Phyllaries** 20–34+, apices acuminate, abaxial faces stellate-pubescent and stipitate-glandular, sometimes piloso-hirsute as well. **Florets** 60–120+; corollas yellow (often each with abaxial red stripe), 8–13+ mm. **Cypselae** columnar, 1.5–2 mm; **pappi** of 30+, white bristles in 1 series, 4–5 mm.

Flowering May–Aug. Disturbed sites (sandy or gravelly soils, fields, lawns, roadsides); introduced; 10–100+ m; St. Pierre and Miquelon; B.C., N.B., Nfld. and Labr. (Nfld.), N.S., Ont., Que.; Conn., Del., Ga., Ky., Maine, Md., Mass., Mich., Minn., N.H., N.J., N.Y., N.C., Ohio, Oreg., Pa., R.I., Tenn., Vt., Va., Wash., W.Va.; Europe.

2. Hieracium aurantiacum Linnaeus, Sp. Pl. 2: 801. 1753 • Orange hawkweed, épervière orangée ☐

Pilosella aurantiaca (Linnaeus) F. W. Schultz & Schultz-Bipontinus

Plants 15–35(–60+) cm. **Stems** proximally piloso-hirsute (hairs 2–4+ mm), distally piloso-hirsute (hairs 1–4 mm) and stipitate-glandular. **Leaves:** basal 3–8+, cauline 0(–1+); blades spatulate to oblanceolate, 45–70(–160+) × 10–35 mm, lengths 3–5+ times widths, bases cuneate, margins entire, apices acute, faces piloso-hirsute (hairs 1–2+ mm) and stellate-pubescent. **Heads** 3–7(–12+) in ± umbelliform arrays. **Peduncles** stellate-pubescent and stipitate-glandular. **Calyculi:** bractlets 5–8+. **Involucres** campanulate, 6–8 mm. **Phyllaries** 13–30+, apices acuminate, abaxial faces piloso-hirsute, stellate-pubescent, and stipitate-glandular. **Florets** 25–120+; corollas orange (drying scarlet to purplish), 10–14+ mm. **Cypselae** columnar, 1.2–1.5(–2) mm; **pappi** of 25–30+, white bristles in 1 series, 3.5–4 mm.

Flowering (May–)Jun–Aug(–Sep). Disturbed sites (fields, lawns, roadsides), bogs, clays, sands; introduced; 10–300(–1000+) m; St. Pierre and Miquelon; Alta., B.C., Man., N.B., Nfld. and Labr., N.S., Ont., P.E.I., Que., Sask.; Alaska, Ark., Calif., Colo., Conn., Fla., Ga., Idaho, Ill., Ind., Iowa, Ky., Maine, Md., Mass., Mich., Minn., Mont., N.H., N.J., N.Y., N.C., Ohio, Oreg., Pa., R.I., S.Dak., Tenn., Vt., Va., Wash., W.Va., Wis., Wyo.; Europe.

3. Hieracium flagellare Willdenow, Enum. Pl., suppl.: 54. 1814 ⊡

Hieracium flagellare var. *amauracron* (Missback & Zahn) Lepage; *H. flagellare* var. *cernuiforme* (Naegeli & Peter) Lepage; *H. flagellare* var. *pilosius* Lepage; *Pilosella flagellaris* (Willdenow) P. D. Sell & C. West

Plants 5–12(–20+) cm. **Stems** proximally piloso-hirsute (hairs 2–4+ mm) and stellate-pubescent, distally piloso-hirsute (hairs 1–4 mm), stellate-pubescent, and stipitate-glandular. **Leaves:** basal 8–12+, cauline 0(–2+); blades spatulate to oblanceolate, 20–45(–130+) × 8–20(–25+) mm, lengths 2–3+ times widths, bases cuneate, margins entire, apices rounded to acute, abaxial faces piloso-hirsute (hairs 1–4+ mm) and stellate-pubescent, adaxial piloso-hirsute (hairs 1–4+ mm). **Heads** 2–4+ in ± umbelliform to corymbiform arrays. **Peduncles** piloso-hirsute, stellate-pubescent, and stipitate-glandular. **Calyculi:** bractlets 13–15+. **Involucres** hemispheric, (9–)12–13 mm. **Phyllaries** 30–40, apices acuminate, abaxial faces stellate-pubescent, sometimes, piloso-hirsute and stipitate-glandular as well. **Florets** 90–120+; corollas yellow (often each with abaxial red stripe), 6–10+ mm. **Cypselae** columnar, 1–2.5 mm; **pappi** of 25–40+, white bristles in 1 series, 4–5+ mm.

Flowering May. Disturbed sites, roadsides, forest edges; introduced; 10–300(–600+) m; B.C., N.B., N.S., P.E.I., Que.; Conn., Ind., Maine, Mass., Mich., N.H., N.Y., Ohio, Pa., Vt., Va.; Europe.

The type of *Hieracium flagellare* may have resulted from a cross between plants of *H. caespitosum* and *H. pilosella* (A. Cronquist 1980).

4. Hieracium lactucella Wallroth, Sched. Crit., 408. 1822 ⊡

Plants 9–20(–35+) cm. **Stems** proximally piloso-hirsute (hairs 1–2+ mm), sometimes stipitate-glandular as well, distally stellate-pubescent and stipitate-glandular (not piloso-hirsute). **Leaves:** basal 5–8+, cauline 0(–2+); blades spatulate to oblanceolate, 15–40 × 5–12 mm, lengths 3–6+ times widths, bases cuneate, margins entire, apices rounded to acute, abaxial faces glabrous or piloso-hirsute (hairs 1–3+ mm), adaxial glabrous or piloso-hispid (hairs 1–3+ mm), sometimes stellate-pubescent as well. **Heads** usually 2–5+ in ± umbelliform arrays, sometimes borne singly. **Peduncles** stellate-pubescent and stipitate-glandular (not piloso-hirsute). **Calyculi:** bractlets 9–13+.

Involucres hemispheric, 6–8 mm. **Phyllaries** 16–21+, apices acuminate, abaxial faces stellate-pubescent and stipitate-glandular, sometimes piloso-hirsute (hairs 0.8–1.5+) as well. **Florets** 40–60+; corollas yellow, 8+ mm. **Cypselae** columnar, 1–2.5 mm; **pappi** of 25–40, white bristles in 1 series, 4–5 mm.

Flowering May–Jul. Pastures, fields; 10+ m; introduced; N.S.; N.Y.; Europe.

According to A. E. Roland and M. Zinck (1998), *Hieracium auricula* Linnaeus and *Pilosella auricula* (Linnaeus) Schultz-Bipontinus have been misapplied to plants here called *H. lactucella* (see also discussion at 5. *H. caespitosum*).

5. Hieracium caespitosum Dumortier, Fl. Belg., 62. 1827 Ⓕ ⊡

Hieracium pratense Tausch; *Pilosella caespitosa* (Dumortier) P. D. Sell & C. West

Plants 20–75 cm. **Stems** proximally piloso-hirsute (hairs 1–3+ mm) and stipitate-glandular, sometimes stellate-pubescent as well, distally piloso-hirsute (hairs 1–4+ mm), stellate-pubescent, and stipitate-glandular. **Leaves:** basal 3–8+, cauline 0–2(–5+); blades oblanceolate to lanceolate, 35–120(–180+) × 12–20+ mm, lengths 2–6(–10+) times widths, bases cuneate, margins entire or denticulate, apices rounded to acute, faces usually piloso-hirsute (hairs 1–3+ mm) and stellate-pubescent, sometimes glabrate. **Heads** 5–25+ in ± umbelliform or congested, corymbiform arrays. **Peduncles** piloso-hirsute (hairs 1–2.5 mm), stellate-pubescent, and stipitate-glandular. **Calyculi:** bractlets 5–8+. **Involucres** campanulate, 7.5–9 mm. **Phyllaries** 12–18+, apices acute to acuminate, abaxial faces piloso-hirsute (hairs 1–2.5+), stellate-pubescent, and stipitate-glandular. **Florets** 25–50+; corollas yellow, 8–12+ mm. **Cypselae** columnar, 1.5–1.8 mm; **pappi** of 25–30+, white bristles in 1 series, 4–5(–6) mm.

Flowering May–Jul(–Aug). Disturbed sites, stream sides; introduced; 10–300(–1500) m; B.C., Man., N.B., Nfld. and Labr. (Nfld.), N.S., Ont., P.E.I., Que.; Conn., Del., D.C., Ga., Idaho, Ill., Ind., Ky., Maine, Md., Mass., Mich., Minn., Mont., N.H., N.J., N.Y., N.C., Ohio, Pa., R.I., S.C., Tenn., Vt., Va., Wash., W.Va., Wis., Wyo.; Europe.

The type of *Hieracium floribundum* Wimmer & Grabowski probably resulted from a cross between plants of *H. caespitosum* and *H. lactucella* (P. D. Sell and C. West 1976).

6. Hieracium piloselloides Villars, Prosp. Hist. Pl. Dauphiné, 34. 1779 • King devil, épervière des Florentins [I]

Hieracium florentinum Allioni; *Pilosella piloselloides* (Villars) Soják

Plants 15–40(–70+) cm. **Stems** proximally usually piloso-hirsute (hairs 2–4+ mm), rarely glabrous, distally usually glabrous, sometimes piloso-hirsute (hairs 1–3+ mm), stellate-pubescent, and/or stipitate-glandular. **Leaves:** basal 3–8(–20+), cauline 0–2(–4+); blades oblanceolate to lanceolate, 30–100(–150+) × 8–20+ mm, lengths 2.5–8+ times widths, bases cuneate, margins entire or denticulate, apices rounded to acute, faces glabrous or piloso-hirsute (on midribs and at margins, hairs 1–4+ mm). **Heads** (3–)10–30+ in subumbelliform or corymbiform arrays. **Peduncles** piloso-hirsute (hairs 1–2+ mm), stellate-pubescent, and stipitate-glandular. **Calyculi:** bractlets 3–12+. **Involucres** campanulate, 5–7 mm. **Phyllaries** 12–18+, apices acute to acuminate, abaxial faces piloso-hirsute (hairs 0.5–1.5+), stellate-pubescent, and stipitate-glandular. **Florets** (40–)60–80+; corollas yellow, 6–9 mm. **Cypselae** columnar, 1.5–2 mm; **pappi** of 25–40+, white bristles in 1 series, 3–4 mm.

Flowering (May–)Jun–Aug(–Sep). Disturbed sites; introduced; 10–300(–1500) m; B.C., N.B., Nfld. and Labr. (Nfld.), N.S., Ont., P.E.I., Que.; Conn., Del., Ga., Ill., Ind., Iowa, Maine, Md., Mass., Mich., Minn., Mont., N.H., N.J., N.Y., N.C., Ohio, Pa., R.I., S.C., Vt., Va., Wash., W.Va., Wis.; Europe.

Plants called *Hieracium praealtum* Villars ex Gochnat (at least those called *H. praealtum* var. *decipiens* W. D. J. Koch) reputedly differ from members of *H. piloselloides* in having blades of their proximal leaves stellate-pubescent abaxially (M. L. Fernald 1950); such plants may be found in the flora and may merit taxonomic recognition.

7. Hieracium bolanderi A. Gray, Proc. Amer. Acad. Arts 7: 365. 1868

Hieracium siskiyouense M. Peck

Plants 10–60+ cm. **Stems** glabrous or nearly so. **Leaves:** basal 5–9+, cauline 0–3; blades oblanceolate to spatulate, 35–95+ × 12–25+ mm, lengths 2–4+ times widths, bases cuneate, margins entire, apices rounded to acute, abaxial faces glabrous or piloso-hirsute (on midribs, hairs 5–8+ mm), adaxial piloso-hirsute (hairs 5–8+ mm, often tangled). **Heads** (3–)15–15(–40+) in corymbiform arrays. **Peduncles** glabrous

or stipitate-glandular. **Calyculi:** bractlets 3–5+. **Involucres** campanulate to cylindric, (7–)9–10 mm. **Phyllaries** 8–13+, apices obtuse to acute, abaxial faces glabrous or stipitate-glandular. **Florets** 6–12; corollas yellow, 8.5–10 mm. **Cypselae** urceolate, 5–5.5 mm; **pappi** of 50–60+, stramineous to rufous bristles in 2+ series, 5–8 mm.

Flowering Jun–Jul. Pine forests, serpentines; 300–1700 m; Calif., Oreg.; Mexico (Baja California).

8. Hieracium longipilum Torrey ex Hooker, Fl. Bor.-Amer. 1: 298. 1833 [E]

Plants 30–75(–200) cm. **Stems** proximally piloso-hirsute (hairs 6–15+ mm), distally piloso-hirsute (hairs 3–10+ mm), sometimes stellate-pubescent as well. **Leaves:** basal 3–8+, cauline (3–)6–12+; blades oblanceolate, 45–80 (–250+) × 12–30(–40+) mm, lengths 4–7+ times widths, bases cuneate, margins entire, apices rounded to acute, faces piloso-hirsute (hairs 3–8+ mm). **Heads** 10–20+ in paniculiform to nearly racemiform arrays. **Peduncles** stellate-pubescent and stipitate-glandular, sometimes piloso-hirsute as well. **Calyculi:** bractlets 9–13+. **Involucres** campanulate, 6–8(–10) mm. **Phyllaries** 12–21+, apices acuminate, abaxial faces stellate-pubescent and stipitate-glandular. **Florets** 30–40(–60); corollas yellow, ca. 7 mm. **Cypselae** urceolate, 3–4+ mm; **pappi** of 35–40+, stramineous to sordid bristles in 2+ series, 5.5–6.5 mm.

Flowering Jul–Sep. Fields, prairies, roadsides; 100–400 m; Ont.; Ark., Ill., Ind., Iowa, Kans., Ky., La., Mich., Minn., Mo., Nebr., Ohio, Okla., Tenn., Tex., Wis.

Hieracium longipilum may be no longer present in Quebec.

9. Hieracium gronovii Linnaeus, Sp. Pl. 2: 802. 1753 [F]

Plants 30–45(–80) cm. **Stems** proximally piloso-hirsute (hairs 2–4+ mm), sometimes stellate-pubescent as well, distally stellate-pubescent, sometimes piloso-hirsute as well. **Leaves:** basal 0 (–2+), cauline (3–)6–12+; blades elliptic or obovate to oblanceolate, 20–35(–90) × 10–40(–50) mm, lengths 2–4+ times widths, bases cuneate to rounded (sometimes ± clasping), margins entire, apices rounded to acute, abaxial faces piloso-hirsute (hairs 2–4 mm) and stellate-pubescent, adaxial piloso-hirsute. **Heads** (5–)25–50 in usually narrow, thyrsiform arrays (lengths of arrays usually 3–6+ times diams., sometimes shorter). **Peduncles** stellate-pubescent and stipitate-glandular.

Calyculi: bractlets 8–12+. **Involucres** cylindric to campanulate, 7–10 mm. **Phyllaries** 12–15+, apices rounded to acute or acuminate, abaxial faces glabrous or stellate-pubescent, rarely stipitate-glandular as well. **Florets** 12–20+; corollas yellow, 8–9+ mm. **Cypselae** urceolate, 3.5–4.5+ mm; **pappi** of ca. 40+, stramineous bristles in 2+ series, ca. 5 mm.

Flowering (Mar–)Jul–Sep(–Oct). Openings in pine and pine-oak woods, bogs, sands; 30–600 m; Ont.; Ala., Ark., Conn., Del., D.C., Fla., Ga., Ill., Ind., Kans., Ky., La., Maine, Md., Mass., Mich., Minn., Miss., Mo., N.J., N.Y., N.C., Ohio, Okla., Pa., R.I., S.C., Tenn., Tex., Va., W.Va.; Mexico, Central America.

10. **Hieracium megacephalum** Nash, Bull. Torrey Bot. Club 22: 152. 1895 (as megacephalon) [E]

Hieracium argyraeum Small

Plants ca. 40 cm. **Stems** proximally piloso-hirsute (hairs 1–4+ mm) and stellate-pubescent, distally piloso-hirsute and/or stellate-pubescent. **Leaves:** basal 4–6, cauline (1–)3–4+; blades elliptic to obovate, 35–80(–120) × 15–30 (–40+) mm, lengths 2–4+ times widths, bases cuneate to rounded (sometimes ± clasping), margins entire or denticulate, apices rounded to obtuse, abaxial faces piloso-hirsute (hairs 1.5–4 mm), sometimes stellate-pubescent as well, adaxial piloso-hirsute. **Heads** (12–)20–50+ in open, corymbiform arrays (lengths of arrays ± 2 times diams.). **Peduncles** stellate-pubescent and stipitate-glandular. **Calyculi:** bractlets 5–12+. **Involucres** campanulate, 8–11 mm. **Phyllaries** 9–13+, apices acute to acuminate, abaxial faces stellate-pubescent and stipitate-glandular. **Florets** 20–50+; corollas yellow, 10–12+ mm. **Cypselae** urceolate, 3.5–4+ mm; **pappi** of 32–40+, white to stramineous bristles in 1–2+ series, 5–6.5 mm.

Flowering (Jan–)Apr–May(–Oct). Limestone hammocks, pine woods, sands; 0–10+ m; Fla., Ga., S.C.

11. **Hieracium fendleri** Schultz-Bipontinus, Bonplandia (Hanover) 9: 173. 1861

Crepis ambigua A. Gray, Mem. Amer. Acad. Arts, n. s. 4: 114. 1849, not Balbis 1805, not *Hieracium ambiguum* Ehrhart 1790; *Chlorocrepis fendleri* (Schultz-Bipontinus) W. A. Weber; *H. fendleri* var. *discolor* A. Gray

Plants 8–30(–40+) cm (herbage often glaucous). **Stems** proximally piloso-hirsute (hairs 2–5+ mm), distally usually ± piloso-hirsute (hairs 1–3+ mm) and stellate-pubescent, sometimes stipitate-glandular as well. **Leaves:** basal

3–8+, cauline 0–1(–3); blades elliptic or oval to oblanceolate to lanceolate, 20–70 × 12–25(–40) mm, lengths 1.5–6+ times widths, bases cuneate, margins entire or denticulate, apices rounded to acute, faces piloso-hirsute (hairs 1–3+ mm). **Heads** 2–5+ in ± corymbiform arrays. **Peduncles** usually stellate-pubescent, sometimes piloso-hirsute and/or stipitate-glandular as well. **Calyculi:** bractlets 5—8+. **Involucres** cylindric to campanulate, (10–)12–15+ mm. **Phyllaries** 13–16+, apices ± acuminate, abaxial faces usually piloso-hirsute and stellate-pubescent, sometimes stipitate-glandular as well. **Florets** 15–30+; corollas pale yellow to ochroleucous, 7–13+ mm. **Cypselae** urceolate, 5–7 mm; **pappi** of 50–60+, ± stramineous bristles in 2+ series, 5–8+ mm.

Flowering (May–)Jun–Aug(–Sep). Pine woods, near springs; 1400–2900 m; Ariz., Colo., Nev., N.Mex., Tex., Utah, Wyo.; Mexico; Central America (Guatemala).

12. **Hieracium carneum** Greene, Bot. Gaz. 6: 184. 1881

Plants 30–60 cm (herbage sometimes glaucous). **Stems** proximally glabrous or piloso-hirsute (hairs 6–10+ mm), distally glabrous. **Leaves:** basal (0–)3–5, cauline 4–8+; blades oblanceolate to lanceolate or linear, 40–80 (–120) × 5–10(–20) mm, lengths 2.5–8+ times widths, bases cuneate to truncate, margins entire, apices rounded to acute, faces glabrous or piloso-hirsute (hairs 3–6+ mm). **Heads** 6–25+ in ± corymbiform to paniculiform arrays. **Peduncles** usually glabrous, sometimes stellate-pubescent and/or stipitate-glandular. **Calyculi:** bractlets 8–13. **Involucres** campanulate, 7–10 mm. **Phyllaries** 13–16+, apices ± rounded, abaxial faces usually glabrous, sometimes stellate-pubescent and/or stipitate-glandular. **Florets** ca. 20; corollas whitish to pinkish, ca. 8 mm. **Cypselae** weakly urceolate, 3–4.5 mm; **pappi** of 50–60+, white or stramineous bristles in 2+ series, 4–5+ mm.

Flowering Jul. Rocky sites; 2000–2300 m; Ariz., N.Mex., Tex.; Mexico (Chihuahua).

13. **Hieracium crepidispermum** Fries, Nova Acta Regiae Soc. Sci. Upsal. 14: 146. 1848

Hieracium lemmonii A. Gray

Plants 25–45(–100) cm. **Stems** proximally piloso-hirsute (hairs 3–6+ mm), distally piloso-hirsute (hairs 1–2+ mm) and stellate-pubescent. **Leaves:** basal 2–8+, cauline (0–)3–4+; blades elliptic to oblanceolate, 30–85 × 16–35 mm, lengths 2.5–5+ times widths, bases cuneate (basal) or clasping (cauline), margins entire or denticulate, apices obtuse to acute, faces piloso-hirsute (hairs 1–2+ mm). **Heads** 8–12(–25+) in ± paniculiform arrays. **Peduncles** stellate-pubescent. **Calyculi:** bractlets 9–12+. **Involucres** campanulate, 7–9+ mm. **Phyllaries** 13–18+, apices ± rounded to acute, abaxial faces stellate-pubescent. **Florets** 25–40; corollas whitish to pale yellow, 8–9 mm. **Cypselae** urceolate, 3.5–4.5 mm; **pappi** of 45–60+, white or stramineous bristles in 2+ series, 5–5.5+ mm.

Flowering Sep. Springy sites, along streams; 2200–2500 m; Ariz., N.Mex.; Mexico.

14. Hieracium brevipilum Greene, Bull. Torrey Bot. Club 9: 64. 1882 [E]

Hieracium fendleri Schultz-Bipontinus var. *mogollense* A. Gray

Plants 25–65 cm. **Stems** proximally piloso-hirsute (hairs 1–3+ mm), distally stellate-pubescent and stipitate-glandular. **Leaves:** basal 3–6+, cauline 3–6+; blades oblanceolate to lanceolate, 35–120 × 10–18+ mm, lengths (2–)4–10+ times widths, bases cuneate to truncate (± clasping), margins entire, apices obtuse to acute, faces usually piloso-hirsute (hairs 0.5–1.5+ mm), sometimes glabrous. **Heads** 6–10+ in ± paniculiform arrays. **Peduncles** stellate-pubescent and stipitate-glandular. **Calyculi:** bractlets 5–8+. **Involucres** campanulate to cylindric, 10–11 mm. **Phyllaries** 9–13+, apices ± acuminate, abaxial faces stellate-pubescent and stipitate-glandular. **Florets** 15–25+; corollas ochroleucous, ca. 8 mm. **Cypselae** urceolate, 5–6 mm; **pappi** of 50–60+, white to stramineous bristles in 2+ series, 5–6 mm.

Flowering Aug. Pine woods; 2300–3000 m; Ariz., N.Mex.

15. Hieracium paniculatum Linnaeus, Sp. Pl. 2: 802. 1753 [E]

Plants 30–90 cm. **Stems** proximally usually glabrous, sometimes piloso-hirsute (hairs 3–8+ mm) and/or stellate-pubescent, distally usually glabrous. **Leaves:** basal 0(–2), cauline 6–12+; blades elliptic or oblanceolate to lanceolate, 30–80(–150) × 6–25(–35) mm, lengths 3–6+ times widths, bases cuneate, margins usually toothed, sometimes denticulate or entire, apices acute to acuminate, faces glabrous or piloso-hirsute (hairs 1–4+ mm). **Heads** (6–)12–50+ in corymbiform to paniculiform arrays. **Peduncles** usually glabrous, sometimes stipitate-glandular. **Calyculi:** bractlets 8–13+. **Involucres** campanulate to cylindric, 6–7 mm. **Phyllaries** 8–13+, apices acute to acuminate,

abaxial faces usually glabrous, sometimes stipitate-glandular. **Florets** 8–20(–30); corollas ochroleucous to yellow, 5–8 mm. **Cypselae** columnar, 2–2.5 mm; **pappi** of 35–40+, stramineous bristles in ± 2 series, 4–5 mm.

Flowering (Jul–)Aug(–Oct). Openings in forests; (10–)500–800 m; N.B., N.S., Ont., Que.; Ala., Conn., Del., Ga., Ind., Ky., Maine, Md., Mass., Mich., N.H., N.J., N.Y., N.C., Ohio, Pa., R.I., S.C., Tenn., Vt., Va., W.Va.

The type of *Hieracium scribneri* Small may have resulted from a cross between plants of *H. paniculatum* and *H. venosum* (A. Cronquist 1980).

16. Hieracium umbellatum Linnaeus, Sp. Pl. 2: 804. 1753 [F]

Hieracium acranthophorum Omang; *H. canadense* Michaux; *H. canadense* var. *divaricatum* Lepage; *H. canadense* var. *fasciculatum* (Pursh) Fernald; *H. canadense* var. *hirtirameum* Fernald; *H. canadense* var. *subintegrum* Lepage; *H. columbianum* Rydberg; *H. devoldii* Omang; *H.* ×*dutillyanum* Lepage; *H. eugenii* Omang; *H. kalmii* Linnaeus; *H. kalmii* var. *canadense* (Michaux) Reveal; *H. kalmii* var. *fasciculatum* (Pursh) Lepage; *H. musartutense* Omang; *H. nepiocratum* Omang; *H. rigorosum* (Laestadius ex Almquist) Almquist ex Omang; *H. scabriusculum* Schweinitz; *H. scabriusculum* var. *columbianum* (Rydberg) Lepage; *H. scabriusculum* var. *perhirsutum* Lepage; *H. scabriusculum* var. *saximontanum* Lepage; *H. scabriusculum* var. *scabrum* (Schweinitz) Lepage; *H. stiptocaule* Omang; *H. umbellatum* subsp. *canadense* (Michaux) Guppy; *H. umbellatum* var. *scabriusculum* (Schweinitz) Farwell

Plants 15–60+ cm. **Stems** proximally usually glabrous, sometimes piloso-hirsute and/or stellate-pubescent, distally usually glabrous, sometimes stellate-pubescent. **Leaves:** basal 0(–2), cauline (5–)8–15(–45+); blades lance-elliptic to lanceolate, (20–)50–100(–150) × (10–)15–25(–40+) mm, lengths (3–)5–10+ times widths, bases cuneate to rounded or truncate (then sometimes ± clasping), margins usually toothed (to laciniate), denticulate, or entire (often ± revolute, at least distally), apices obtuse to acute, faces glabrous or ± stellate-pubescent, sometimes ± scabrellous (especially at or near distal margins). **Heads** (1–)5–30(–100+) in corymbiform to subumbelliform arrays. **Peduncles** usually stellate-pubescent. **Calyculi:** bractlets 9–15+. **Involucres** campanulate to hemispheric, (8–)9–11+ mm. **Phyllaries** 12–21+, apices rounded to acute, abaxial faces usually glabrous, rarely piloso-hirsute and/or stipitate-glandular. **Florets** 30–80+; corollas yellow, 10–18 mm. **Cypselae** columnar, 2.5–3.5 mm; **pappi** of 50–60+, stramineous to sordid bristles in ± 2 series, 6–7 mm.

Flowering (Jun–)Jul–Sep. Disturbed sites (fields, roadsides), rocky slopes, openings in forests, prairies,

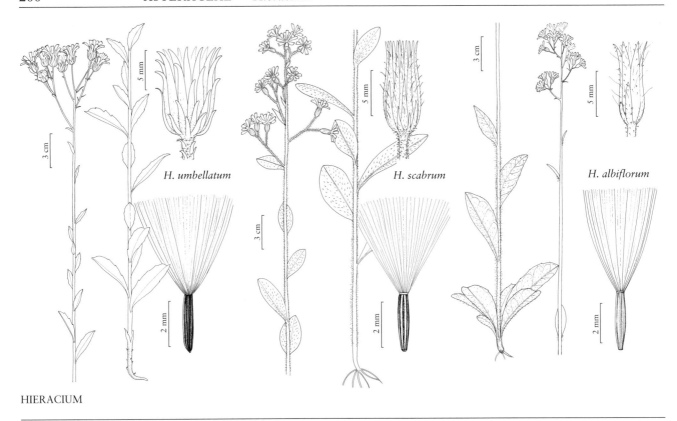

H. umbellatum

H. scabrum

H. albiflorum

HIERACIUM

thickets; 600–3200 m; Greenland; St. Pierre and Miquelon; Alta., B.C., Man., N.B., Nfld. and Labr., N.W.T., N.S., Ont., P.E.I., Que., Sask., Yukon; Alaska, Colo., Conn., Idaho, Ill., Ind., Iowa, Maine, Mass., Mich., Minn., Mo., Mont., Nebr., N.H., N.J., N.Y., N.Dak., Ohio, Oreg., Pa., R.I., S.Dak., Vt., Wash., W.Va., Wis., Wyo.; Europe, Asia.

The circumscription of *Hieracium umbellatum* adopted here is supported by research done by others, especially G. A. Guppy (1978) and E. Lepage (1960). *Hieracium canadense* var. *kalmii* (Linnaeus) Scoggan, referable here, is an illegitimate name.

SELECTED REFERENCE Lepage, E. 1960. *Hieracium canadense* Michx. et ses alliées en Amérique du Nord. Naturaliste Canad. 87: 59–107.

17. **Hieracium schultzii** Fries, Uppsala Univ. Årsskr. 1862: 150. 1862

Hieracium friesii Schultz-Bipontinus, Bonplandia (Hanover) 9: 326. 1861, not Hartman 1838; *H. rusbyi* Greene var. *wrightii* A. Gray; *H. wrightii* (A. Gray) B. L. Robinson & Greenman

Plants 12–70+ cm. **Stems** proximally piloso-hirsute (hairs 4–6+ mm), distally piloso-hirsute (hairs 4–6+ mm). **Leaves:** basal 1–4, cauline 6–9; blades oblanceolate to lanceolate, 25–150 × 10–35+ mm, lengths 2.5–5+ times widths, bases cuneate, margins usually entire,

apices ± acute, faces piloso-hirsute (hairs 3–5+ mm). **Heads** 3–25+ in corymbiform to paniculiform arrays. **Peduncles** piloso-hirsute, stellate-pubescent, and stipitate-glandular. **Calyculi:** bractlets 10–15+. **Involucres** campanulate, 8–9(–12) mm. **Phyllaries** 16–21+, apices acute, abaxial faces piloso-hirsute, stellate-pubescent, and stipitate-glandular. **Florets** 25–40+; corollas yellow, 5–8 mm. **Cypselae** columnar, 2.5–4 mm; **pappi** of 50–60+, white or stramineous bristles in ± 2 series, 5–6 mm.

Flowering Jul–Sep. Grasslands, pine forests; 2000–2700 m; Tex.; Mexico; Central America (Guatemala).

18. **Hieracium scabrum** Michaux, Fl. Bor.-Amer. 2: 86. 1803 E F

Hieracium scabrum var. *intonsum* Fernald & H. St. John; *H. scabrum* var. *leucocaule* Fernald & H. St. John; *H. scabrum* var. *tonsum* Fernald & H. St. John

Plants (15–)30–60+ cm. **Stems** proximally piloso-hirsute (hairs 1.5–5+ mm), sometimes stellate-pubescent as well, distally stellate-pubescent, sometimes stipitate-glandular, rarely piloso-hirsute (hairs 1–2+ mm). **Leaves:** basal 0(–3), cauline 6–24+; blades elliptic or oblanceolate to spatulate or lanceolate, (20–)35–80(–120+) × (10–)20–35(–50+) mm, lengths 2–6+ times widths, bases cuneate to rounded-truncate, margins usually entire, sometimes denticulate, apices rounded to obtuse, faces hirtellous to piloso-hirsute (hairs 0.5–1+ mm). **Heads** (5–)10–25(–50+) in corymbiform to paniculiform arrays. **Peduncles** stellate-pubescent and stipitate-glandular (hairs 0.3–0.6+ mm). **Calyculi:** bractlets 12–15+. **Involucres** campanulate, 7–9 mm. **Phyllaries** 13–21+, apices ± acuminate, abaxial faces stellate-pubescent and stipitate-glandular, rarely, if ever, piloso-hirsute. **Florets** (30–)40–60+; corollas yellow, 9–11 mm. **Cypselae** columnar, 2.5–3 mm; **pappi** of 30–40+, stramineous bristles in ± 2 series, 6–7 mm.

Flowering Jul–Sep. Sandy soils, open, disturbed sites (fields, stream sides), wooded sites; 10–300+ m; N.B., N.S., Ont., P.E.I., Que.; Ark., Conn., Del., D.C., Ga., Ill., Ind., Iowa, Ky., Maine, Md., Mass., Mich., Minn., Mo., N.H., N.J., N.Y., N.C., Ohio, Okla., Pa., R.I., S.C., Tenn., Vt., Va., W.Va., Wis.

19. Hieracium sabaudum Linnaeus, Sp. Pl. 2: 804. 1753

Plants ca. 65 cm. **Stems** proximally piloso-hirsute (hairs 2–3+ mm), distally glabrous or stellate-pubescent, sometimes piloso-hirsute as well (hairs 1–2 mm). **Leaves:** basal 0, cauline 12–20+; blades ± lanceolate, 30–80 × 10–25 mm, lengths 2–4+ times widths, bases cuneate, margins usually toothed, apices acute to acuminate, abaxial faces piloso-hirsute, adaxial glabrous or ± scabrellous (especially distal margins). **Heads** 25–30+ in corymbiform arrays. **Peduncles** stellate-pubescent. **Calyculi:** bractlets 21+. **Involucres** campanulate, 8–9 mm. **Phyllaries** 21+, apices ± rounded, abaxial faces stellate-pubescent and stipitate-glandular. **Florets** 40–60+; corollas yellow, ca. 10 mm. **Cypselae** columnar, 2.5–3 mm; **pappi** of ca. 60+, stramineous bristles in ± 2 series, ca. 6 mm.

Flowering Aug–Sep. Disturbed sites (grasslands, open woods); introduced; 10+ m; B.C., N.S.; Conn., Maine, Mass., N.J., N.Y., Pa., Wis.; Europe.

20. Hieracium scouleri Hooker, Fl. Bor.-Amer. 1: 298. 1833

Hieracium absonum J. F. Macbride & Payson; *H. albertinum* Farr; *H. chapacanum* Zahn; *H. cynoglossoides* Arvet-Touvet; *H. scouleri* var. *albertinum* (Farr) G. W. Douglas & G. A. Allen; *H. scouleri* var. *griseum* A. Nelson

Plants (15–)35–60+ cm. **Stems** proximally usually piloso-hirsute (hairs 1–8+ mm) and stellate-pubescent, sometimes glabrous, distally usually piloso-hirsute (hairs 1–8+ mm)

and stellate-pubescent, rarely stipitate-glandular as well, sometimes glabrous. **Leaves:** basal 0(–5+), cauline (3–)5–10+; blades lanceolate, oblanceolate or narrowly oblong to elliptic, 50–100(–200) × 10–25(–35+) mm, lengths (3–)4–8+ times widths, bases cuneate, margins entire or denticulate, apices obtuse to acute, faces usually piloso-hirsute (hairs 1–5+ mm) and stellate-pubescent, rarely glabrous. **Heads** (3–)9–25+ in corymbiform to thyrsiform arrays. **Peduncles** usually stellate-pubescent, sometimes piloso-hirsute and/or stipitate-glandular as well, rarely glabrous. **Calyculi:** bractlets 5–13+. **Involucres** campanulate, 8–10 mm. **Phyllaries** 12–21+, apices rounded to acute, abaxial faces piloso-hirsute, stellate-pubescent, and stipitate-glandular. **Florets** 20–45+; corollas yellow, 10–12 mm. **Cypselae** (red-brown or black) columnar, ca. 3 mm; **pappi** of 32–40+, white or stramineous bristles in ± 2 series, 6–7 mm.

Flowering Jun–Sep. Disturbed sites, openings in pine forests and sagebrush, borders of meadows; 400–3000 m; Alta., B.C.; Calif., Idaho, Mont., Nev., Oreg., Utah, Wash., Wyo.

21. Hieracium longiberbe Howell, Fl. N.W. Amer., 395. 1901

Plants 25–50+ cm. **Stems** proximally piloso-hirsute (hairs 2–8+ mm), sometimes glabrate, distally glabrous or piloso-hirsute (hairs 2–5+ mm). **Leaves:** basal 0(–3+), cauline 6–12+; blades spatulate to oblanceolate, 25–80(–100+) × 8–15(–25+) mm, lengths 3–5 times widths, bases ± cuneate, margins usually entire, rarely denticulate, apices obtuse to acute, abaxial faces piloso-hirsute (hairs 2–5+ mm), adaxial usually piloso-hirsute at margins (hairs 2–5+ mm), rarely glabrous. **Heads** (3–)6–12+ in corymbiform arrays. **Peduncles** usually stellate-pubescent, sometimes piloso-hirsute and/or stipitate-glandular as well, rarely glabrous. **Calyculi:** bractlets 9–13+. **Involucres** campanulate, 7–10 mm. **Phyllaries** 12–15+, apices ± rounded, abaxial faces piloso-hirsute and stellate-pubescent. **Florets** 12–24+; corollas yellow, 7–12 mm. **Cypselae** columnar, ca. 3.5 mm; **pappi** of 32–40+, white or stramineous bristles in ± 2 series, 5–6 mm.

Flowering Jul. Cliffs; 30–100 m; Oreg., Wash.

Hieracium longiberbe is known only from along the Columbia River.

22. **Hieracium pringlei** A. Gray, Proc. Amer. Acad. Arts 19: 69. 1883

Plants 20–45+ cm. **Stems** proximally usually piloso-hirsute (or lanate, hairs 5–8+ mm, usually curled and tangled), sometimes nearly glabrous, distally piloso-hirsute (hairs 1–2+ mm) and stellate-pubescent. **Leaves:** basal (2–)3–8+, cauline (0–)3+; blades elliptic or spatulate to oblanceolate, lanceolate, or lance-linear, (35–)50–120(–200+) × 10–25(–40+) mm, lengths 2–5(–10+) times widths, bases ± cuneate, margins usually entire, rarely toothed, apices obtuse to acute, faces (proximal leaves) piloso-hirsute (to lanate, hairs 1–3+ mm; distal leaves often ± glabrate). **Heads** 3–12(–20+) in corymbiform arrays. **Peduncles** usually stellate-pubescent and stipitate-glandular. **Calyculi:** bractlets 3–5+. **Involucres** campanulate, (7–)8–9 mm. **Phyllaries** 12–15+, apices rounded to acute, abaxial faces stellate-pubescent and/or stipitate-glandular. **Florets** 12–15+; corollas yellow, 7–9 mm. **Cypselae** columnar, 2.2–4 mm; **pappi** of (40–)60–80, white or stramineous bristles in 2+ series, 4–5+ mm.

Flowering Jul–Aug. Pine, oak, and pine-oak forests; 2000–2300 m; Ariz., N.Mex.; Mexico; Central America (Guatemala).

23. **Hieracium greenei** A. Gray, Proc. Amer. Acad. Arts 19: 69. 1883 [E]

Hieracium oregonicum Zahn

Plants 10–35(–45+) cm. **Stems** proximally stellate-pubescent, rarely piloso-hirsute (hairs 1–3+ mm), distally usually densely, sometimes sparsely, stellate-pubescent. **Leaves:** basal 5–8+, cauline 2–3(–5+); blades spatulate to oblanceolate, 35–80(–150) × 8–15(–25) mm, lengths 2.5–5 times widths, bases ± cuneate, margins usually entire, rarely denticulate, apices rounded to acute, abaxial faces piloso-hirsute (hairs 2–3+ mm) and stellate-pubescent, sometimes stipitate-glandular as well, adaxial piloso-hirsute (hairs 2–3+ mm) and stellate-pubescent. **Heads** (3–)8–25(–50+) in paniculiform arrays (divaricately branched). **Peduncles** densely to sparsely stellate-pubescent. **Calyculi:** bractlets 3–5+. **Involucres** cylindric, 9–12 mm. **Phyllaries** 7–8+, apices rounded to obtuse, abaxial faces sparsely to densely stellate-pubescent. **Florets** 4–10 (–15+); corollas yellow, 8–10 mm. **Cypselae** columnar, 4–5 mm; **pappi** of 60–80, white or stramineous to rufous bristles in 2+ series, 7–9 mm.

Flowering Jul–Sep. Openings in conifer forests, serpentines; 900–2400 m; Calif., Oreg.

24. **Hieracium parryi** Zahn in H. G. A. Engler, Pflanzenr. 79[IV,280]: 1128. 1922 [E]

Plants 15–45+ cm. **Stems** proximally piloso-hirsute (hairs 1–3+ mm) and stipitate-glandular, distally piloso-hirsute (hairs 1–2+ mm) and stipitate-glandular. **Leaves:** basal (0–)3–5, cauline 2–3(–8); blades lance-elliptic to lanceolate, 30–80(–100) × (3–)10–20(–30+) mm, lengths 3–8(–10+) times widths, bases cuneate, margins usually entire, sometimes denticulate, apices obtuse to acute, faces usually piloso-hirsute (hairs 1–3+ mm), sometimes glabrate. **Heads** (1–)3–12+ in corymbiform arrays. **Peduncles** densely stipitate-glandular. **Calyculi:** bractlets 7–9+. **Involucres** campanulate, 10–12 mm. **Phyllaries** 13–21+, apices ± acuminate, abaxial faces stipitate-glandular. **Florets** 30–60+; corollas yellow, 9–11 mm. **Cypselae** columnar, 2.5–3 mm; **pappi** of 50–60+, white bristles in ± 2 series, ca. 5 mm.

Flowering Jun–Jul. Openings in brush, grassy slopes, serpentines; 10–2000 m; Calif., Oreg.

25. **Hieracium murorum** Linnaeus, Sp. Pl. 2: 802. 1753 [I]

Hieracium hyparcticum (Almquist) Elfstraud; *H. lividorubens* (Almquist) Zahn; *H. stelechodes* Omang

Plants 25–60+ cm. **Stems** proximally piloso-hirsute (hairs 1–3+ mm), distally stellate-pubescent and stipitate-glandular. **Leaves:** basal 3–6, cauline (0–)2–3+; blades (often purple-mottled) ± elliptic, 50–110 × 25–45 mm, lengths 1.5–3 times widths, bases rounded to truncate, margins ± dentate, apices ± obtuse (apiculate), abaxial faces piloso-hirsute (hairs 1–3+ mm), adaxial scabrous to piloso-hirsute (hairs 0.5–3 mm). **Heads** 5–8+ in corymbiform arrays. **Peduncles** densely stellate-pubescent and stipitate-glandular. **Calyculi:** bractlets 8–13+. **Involucres** campanulate to obconic, 8–9 mm. **Phyllaries** 18–21+, apices ± acuminate, abaxial faces stellate-pubescent and stipitate-glandular. **Florets** 30–50+; corollas yellow, 12–13(–16) mm. **Cypselae** columnar, 2.5–3 mm; **pappi** of 30–40+, stramineous bristles in ± 2 series, 4–5 mm.

Flowering Jun. Disturbed sites (fields, openings in woods), thickets; introduced; 0–100+ m; B.C., N.B., Nfld. and Labr. (Nfld.), N.S., Ont., Que.; Alaska, Conn., Ill., Maine, Mass., Mich., N.H., N.J., N.Y., Pa., Vt.; Europe.

26. Hieracium vulgatum Fries, Fl. Hall., 128. 1819

Hieracium amitsokense Dahlstedt; *H. groenlandicum* Arvet-Touvet; *H. ivigtutense* Omang; *H. scholanderi* Omang; *H. sylowii* Omang

Plants 30–60(–100+) cm. **Stems** proximally piloso-hirsute (hairs 1–3+ mm), distally stellate-pubescent and stipitate-glandular, sometimes piloso-hirsute as well (hairs 1–2+ mm). **Leaves:** basal 3–8+, cauline 1–5+; blades (often purple-mottled) lance-elliptic to lanceolate, 50–100 × 10–50+ mm, lengths 2–5 times widths, bases usually cuneate, sometimes rounded to truncate, margins ± dentate or entire, apices acute, faces usually ± piloso-hirsute (at least near margins, hairs 1–3+ mm), sometimes glabrate. **Heads** (1–)3–9+ in corymbiform arrays. **Peduncles** stellate-pubescent and stipitate-glandular. **Calyculi:** bractlets 13–16+. **Involucres** ± obconic, 8–10 mm. **Phyllaries** ca. 21+, apices acute to acuminate, abaxial faces stellate-pubescent and stipitate-glandular. **Florets** 40–80+; corollas yellow, 13–18 mm. **Cypselae** columnar, 2.5–3 mm; **pappi** of 30–40+, stramineous bristles in ± 2 series, ca. 5 mm.

Flowering (Jun–)Jul–Aug. Disturbed sites, openings in thickets; 10–100+ m; Greenland; B.C., N.B., Nfld. and Labr., N.S., Ont., P.E.I., Que.; Conn., Del., Maine, Mass., Mich., Minn., N.H., N.J., N.Y., Oreg., Pa., R.I., Vt., Wash., Wis.; Europe.

The correct name for the species here called *Hieracium vulgatum* may be *H. lachenallii* C. C. Gmelin (E. Lepage 1971; E. G. Voss 1972–1996, vol. 3). Plants of *H. groenlandicum* (stems mostly piloso-hirsute and not stipitate-glandular, corolla lobes distally ciliate) are sometimes treated as distinct from *H. vulgatum* in a narrow sense (stems usually stipitate-glandular and not piloso-hirsute, corolla lobes not ciliate distally).

27. Hieracium robinsonii (Zahn) Fernald, Rhodora 45: 317. 1943 [E]

Hieracium smolandicum Almquist ex Dahlstedt subsp. *robinsonii* Zahn in H. G. A. Engler, Pflanzenr. 76[IV,280]: 468. 1921

Plants 10–35+ cm. **Stems** proximally piloso-hirsute (hairs 1–3+ mm) and stellate-pubescent, distally piloso-hirsute (hairs 0.5–1+ mm) and stellate-pubescent. **Leaves:** basal 2–4, cauline (2–)4–10; blades (often purple-mottled) oblong to lanceolate, 20–80 × 7–20+ mm, lengths 3–4 times widths, bases cuneate, margins usually dentate, sometimes entire, apices acute to acuminate, abaxial faces glabrous or piloso-hirsute (especially midribs), adaxial glabrous or stellate-pubescent. **Heads** 1–5(–10) in ± corymbiform arrays. **Peduncles** piloso-hirsute and stellate-pubescent (not, or rarely, stipitate-glandular). **Calyculi:** bractlets 10–16. **Involucres** campanulate to obconic, 10–15 mm. **Phyllaries** 25–35, apices acuminate, abaxial faces stellate-pubescent and stipitate-glandular. **Florets** 30–50+; corollas yellow (lengths unknown). **Cypselae** columnar, 3–5 mm; **pappi** of ca. 30, stramineous bristles in ± 2 series, ca. 6 mm.

Flowering Jul–Aug. Stream banks and lake shores, sands and clays, rock crevices; 10–300+ m; N.B., Nfld. and Labr. (Nfld.), N.S., Que.; Maine, N.H.

28. Hieracium alpinum Linnaeus, Sp. Pl. 2: 800. 1753

Hieracium angmagssalikense Omang

Plants 10–15(–25+) cm. **Stems** proximally piloso-hirsute (hairs 3–5+ mm), stellate-pubescent, and stipitate-glandular, distally piloso-hirsute (hairs 3–5+ mm), stellate-pubescent, and stipitate-glandular. **Leaves:** basal 5–13+, cauline 0 (–2+); blades spatulate to elliptic or narrowly lanceolate, 20–80 × 6–20+ mm, lengths 2–8+ times widths, bases cuneate, margins usually entire, rarely dentate, apices rounded to acute, faces piloso-hirsute (hairs 1–3+ mm). **Heads** borne singly or 2+ in corymbiform arrays. **Peduncles** piloso-hirsute, stellate-pubescent, and stipitate-glandular. **Calyculi:** bractlets 5–8+. **Involucres** ± hemispheric, 13–18 mm. **Phyllaries** 13–21+, apices acute to acuminate, abaxial faces piloso-hirsute and stipitate-glandular. **Florets** 80–120+; corollas yellow, 12–15 mm. **Cypselae** columnar, 3.5–4 mm; **pappi** of 40–60+, stramineous bristles in ± 2 series, 6–7 mm.

Flowering Jul(–Sep). Calcareous stream banks; 0–10+ m; Greenland; Europe.

The type of *Hieracium trigonophorum* Oskarsson is probably conspecific with that of *H. alpinum*.

29. Hieracium triste Willdenow ex Sprengel, Syst. Veg. 3: 640. 1826 [E]

Chlorocrepis tristis (Willdenow ex Sprengel) Á. Löve & D. Löve; *Hieracium gracile* Hooker; *H. gracile* var. *alaskanum* Zahn; *H. gracile* var. *densifloccosum* (Zahn) Cronquist; *H. gracile* var. *detonsum* (A. Gray) A. Gray; *H. gracile* var. *yukonense* A. E. Porsild; *H. triste* var. *fulvum* Hultén; *H. triste* subsp. *gracile* (Hooker) Calder & Roy L. Taylor; *H. triste* var. *gracile* (Hooker) A. Gray; *H. triste* var. *tristiforme* Zahn

Plants (3–)10–20(–40+) cm. **Stems** proximally glabrous or stellate-pubescent, distally usually piloso-hirsute (hairs 1–8+ mm) and/or stellate-pubescent and/or stipitate-glandular, sometimes glabrous. **Leaves:** basal (3–)5–12+, cauline 0–2(–3+); blades obovate to spatulate or oblanceolate, (15–)25–40(–60+) × 5–10(–25+) mm, lengths 2–3+ times widths, bases cuneate, margins usually entire, rarely denticulate, apices rounded to obtuse (often apiculate), faces usually glabrous, sometimes stipitate-glandular and/or scabrellous. **Heads** usually 2–8+ in corymbiform arrays, sometimes borne singly. **Peduncles** stellate-pubescent and stipitate-glandular. **Calyculi:** bractlets 5–8+. **Involucres** ± campanulate, (6–)7–10 mm. **Phyllaries** 13–21+, apices acuminate, abaxial faces piloso-hirsute (hairs 1–3+ mm), stellate-pubescent, and stipitate-glandular. **Florets** 20–60+; corollas yellow, 5–6 mm. **Cypselae** columnar, 1.5–3.5 mm; **pappi** of 30–40+ white or sordid bristles in ± 2 series, 4–5 mm.

Flowering (Jun–)Jul–Aug(–Sep). Rocky slopes, stream sides, conifer forests, drying meadows, subalpine meadows; 100–3500 m; Alta., B.C., N.W.T., Yukon; Alaska, Calif., Colo., Idaho, Mont., N.Mex., Oreg., Utah, Wash., Wyo.

30. Hieracium albiflorum Hooker, Fl. Bor.-Amer. 1: 298. 1833 F

Chlorocrepis albiflora (Hooker) W. A. Weber

Plants 15–40(–90) cm. **Stems** proximally usually piloso-hirsute (hairs 1–6+ mm), rarely glabrous, distally glabrous. **Leaves:** basal (0–)3–8+, cauline 1–5(–12+); blades oblanceolate, 40–100 (–300) × 12–30(–60+) mm, lengths 3–5+ times widths, bases cuneate, margins usually entire, sometimes sinuately toothed, apices obtuse to acute, faces piloso-hirsute (hairs 1–6 mm), rarely glabrous. **Heads** (3–)12–50+ in corymbiform to paniculiform arrays. **Peduncles** usually glabrous, sometimes stipitate-glandular. **Calyculi:** bractlets 5–12+. **Involucres** ± campanulate, (7–)8–10(–11) mm. **Phyllaries** 8–13+, apices acuminate, abaxial faces piloso-hirsute (hairs 1–2+ mm), stellate-pubescent, and stipitate-glandular. **Florets** (6–)12–25+; corollas yellow, 9–10 mm. **Cypselae** columnar, 2.5–4 mm; **pappi** of 30–40+, stramineous bristles in ± 2 series, (4–)5–7 mm.

Flowering (May–)Jun–Sep. Chaparral, conifer forests, meadows, stream beds, serpentines, volcanics, around mineral springs; 10–2900 m; Alta., B.C., N.W.T., Que., Sask.; Alaska, Calif., Colo., Idaho, Mont., Nev., Oreg., S.Dak., Utah, Wash., Wis., Wyo.; Mexico (Chihuahua, Sonora).

31. Hieracium horridum Fries, Uppsala Univ. Årsskr. 1862: 154. 1862 E

Plants 5–25(–40+) cm. **Stems** proximally piloso-hirsute (hairs 2–4+ mm), distally piloso-hirsute (hairs 1–3+ mm) and stellate-pubescent. **Leaves:** basal 3–6+, cauline 3–6+; blades oblong to spatulate or oblanceolate, (15–) 25–75(–150+) × 6–12(–30+) mm, lengths 3–6+ times widths, bases cuneate, margins usually entire, sometimes denticulate, apices obtuse to acute, faces piloso-hirsute (to lanate, hairs 2–4 mm). **Heads** (3–)8–25(–100+) in corymbiform to paniculiform arrays. **Peduncles** usually stellate-pubescent, sometimes piloso-hirsute as well (hairs 1–2 mm). **Calyculi:** bractlets 5–8+. **Involucres** ± cylindric to campanulate, 6–9 (× 3–4) mm. **Phyllaries** 12–15+, apices acute to acuminate, abaxial faces usually densely piloso-hirsute (hairs 1–2+ mm), sometimes stellate-pubescent or glabrous. **Florets** 6–12(–15+); corollas yellow, 8–9 mm. **Cypselae** (tan to red-brown) columnar, 3–3.5 mm; **pappi** of ca. 60+, stramineous bristles in ± 2 series, 4–6 mm.

Flowering (Jun–)Jul–Sep. Boulders, gravels, meadows, pine forests; 1500–3700 m; Calif., Nev., Oreg.

32. Hieracium venosum Linnaeus, Sp. Pl. 2: 800. 1753 E

Hieracium venosum var. *nudicaule* (Michaux) Farwell

Plants ca. 45 cm. **Stems** proximally piloso-hirsute (hairs 1–3+ mm) to glabrate, distally usually glabrous, rarely piloso-hirsute. **Leaves:** basal 3–6+, cauline 0–1+; blades obovate or elliptic to oblanceolate, 40–60(–120) × 15– 35(–50+) mm, lengths 2–3(–4) times widths, bases cuneate, margins usually entire, sometimes denticulate, apices rounded to acute, abaxial faces piloso-hirsute (hairs 2–5 mm) and stellate-pubescent, adaxial glabrous or piloso-hirsute (mostly near margins, hairs 2–3 mm). **Heads** 4–10(–20+) in corymbiform to paniculiform arrays. **Peduncles** usually stipitate-glandular, rarely glabrous or glabrate. **Calyculi:** bractlets 6–10+. **Involucres** obconic to campanulate, 7–9 mm. **Phyllaries** 12–13+, apices acute, abaxial faces usually stellate-pubescent and stipitate-glandular, rarely glabrous. **Florets** 30–45+; corollas yellow (sometimes pale), 7–11 mm. **Cypselae** (usually black, sometimes red-brown) columnar, 3–4 mm; **pappi** of ca. 50+, stramineous bristles in ± 2 series, 4–5 mm.

Flowering (Apr–)May–Jul(–Sep). Openings in forests, sandy hillsides; 10–300 m; Ont.; Ala., Conn., Del., D.C., Ga., Ill., Ind., Ky., Maine, Md., Mass., Mich., Miss., Mo., N.H., N.J., N.Y., N.C., Ohio, Okla., Pa., R.I., S.C., Tenn., Vt., Va., W.Va.

Plants of *Hieracium venosum* with adaxial faces of leaf blades glabrous have been called var. *nudicaule.*

The type of *Hieracium marianum* Willdenow may have resulted from a cross between plants of *H. venosum* and *H. gronovii* or *H. scabrum* (M. L. Fernald 1943c).

33. Hieracium traillii Greene, Pittonia 4: 226. 1900 E

Hieracium greenii Porter & Britton, Bull. Torrey Bot. Club 20: 120. 1893 (not *H. greenei* A. Gray 1883), based on *Pilosella spathulata* F. W. Schultz & Schultz-Bipontinus, Flora 45: 439. 1862, not *H. spathulatum* Scheele 1863

Plants 25–60+ cm. **Stems** proximally glabrous or stellate-pubescent, distally glabrous or stellate-pubescent. **Leaves:** basal 5–6+, cauline 0(–1+); blades oblanceolate to spatulate, 40–80 × 18–45+ mm, lengths 3–5+ times widths, bases cuneate, margins usually entire, apices rounded to obtuse, abaxial faces piloso-hirsute (hairs 1–3+ mm), adaxial glabrous. **Heads** 8–12+ in corymbiform arrays. **Peduncles** stellate-pubescent and stipitate-glandular. **Calyculi:** bractlets 5–8+. **Involucres** hemispheric to campanulate, 8–10 (× 6–9) mm. **Phyllaries** 13–21+, apices acuminate, abaxial faces stellate-pubescent and stipitate-glandular, sometimes piloso-hirsute as well. **Florets** 30–60+; corollas yellow, ca. 12 mm. **Cypselae** (black or red-brown) columnar, 3–4 mm; **pappi** of ca. 40+, stramineous bristles in ± 2 series, 6–7 mm.

Flowering Jun–Aug. Shaley sites; 1000–1300 m; Ky., Md., Ohio, Pa., Va., W.Va.

34. Hieracium abscissum Lessing, Linnaea 5: 132. 1830

Hieracium rusbyi Greene

Plants 40–75+ cm. **Stems** proximally usually glabrous, sometimes piloso-hirsute (hairs 3–6+ mm), distally glabrous or stellate-pubescent, sometimes piloso-hirsute (hairs 1–3+ mm) and stipitate-glandular as well. **Leaves:** basal 3–12+, cauline (1–)3–6; blades oblanceolate to lanceolate, 35–120 (–200) × 10–25(–30+) mm, lengths 3–6+ times widths, bases cuneate to truncate (distal leaves), margins usually entire, sometimes denticulate or toothed, apices

obtuse to acute, faces piloso-hirsute (hairs 2–3+ mm). **Heads** (5–)12–30(–60+) in paniculiform arrays. **Peduncles** piloso-hirsute (hairs 0.5–1+ mm), stellate-pubescent, and stipitate-glandular. **Calyculi:** bractlets 3–8+. **Involucres** ± campanulate, 6–8+ (× 4–5) mm. **Phyllaries** 13–18+, apices ± rounded, abaxial faces piloso-hirsute (hairs 0.5–1+ mm), stellate-pubescent, and stipitate-glandular. **Florets** 20–24+; corollas yellow, 6–9 mm. **Cypselae** (usually black, sometimes red-brown) columnar, 2–3 mm; **pappi** of 36–40+, white or stramineous bristles in ± 2 series, 3.5–5 mm.

Flowering Apr, Jul–Aug. Disturbed sites, openings in pine, oak, and pine-oak forests; 2000–2600 m; Ariz., N.Mex.; Mexico; Central America.

35. Hieracium argutum Nuttall, Trans. Amer. Philos. Soc., n. s. 7: 447. 1841 E

Hieracium argutum var. *parishii* (A. Gray) Jepson

Plants 30–60 cm. **Stems** proximally usually glabrous, sometimes piloso-hirsute (hairs 1–5+ mm) and stellate-pubescent, distally glabrous or stellate-pubescent. **Leaves:** basal 0–7(–12+), cauline (1–)3–8+; blades oblanceolate or spatulate to lance-linear, 60–120(–200) × 8–15(–45+) mm, lengths 5–10(–15+) times widths, bases cuneate, margins of some or all sinuately toothed, apices acute, faces piloso-hirsute (hairs 1–5+ mm) and stellate-pubescent. **Heads** 12–25(–40+) in paniculiform arrays. **Peduncles** usually stellate-pubescent, sometimes stipitate-glandular as well. **Calyculi:** bractlets 8–12+. **Involucres** ± campanulate to obconic, 7–9+ (× 5–6) mm. **Phyllaries** 13–21+, apices acute, abaxial faces usually stellate-pubescent and stipitate-glandular, sometimes piloso-hirsute as well (hairs 0.5–1+ mm), rarely glabrous. **Florets** 15–30+; corollas yellow, 8–10 mm. **Cypselae** (black) columnar, 2.5–3 mm; **pappi** of 40–60+, white bristles in ± 2 series, 3.5–5 mm.

Flowering Jun–Aug(–Oct). Pine and pine-oak forests; 300–1000(–3000) m; Calif.

36. Hieracium nudicaule (A. Gray) A. Heller, Muhlenbergia 2: 149. 1906 E

Hieracium cynoglossoides Arvet-Touvet var. *nudicaule* A. Gray, Proc. Amer. Acad. Arts 19: 68. 1883; *H. scouleri* Hooker var. *nudicaule* (A. Gray) Cronquist

Plants (15–)20–30(–50+) cm. **Stems** proximally usually glabrous, sometimes piloso-hirsute (hairs 1–7+ mm) and/or stellate-pubescent, distally usually glabrous, sometimes piloso-hirsute (hairs 1–7+ mm) and/or stellate-pubescent. **Leaves:** basal (3–)5–8, cauline 0–3(–5); blades lanceolate to lance-elliptic, 50–120 × 6–18(–30) mm, lengths (3–)5–8+ times widths, bases cuneate, margins usually entire, sometimes denticulate, apices acute, faces usually piloso-hirsute (hairs 2–7+ mm) and stellate-pubescent, sometimes glabrous or glabrate. **Heads** (2–)5–12+ in paniculiform to corymbiform or subumbelliform arrays. **Peduncles** ± stellate-pubescent, sometimes stipitate-glandular as well. **Calyculi:** bractlets 8–13+. **Involucres** ± campanulate, 8–10+ (× 4–6) mm. **Phyllaries** 8–13+, apices rounded to acute, abaxial faces piloso-hirsute (hairs 0.5–2+ mm), stellate-pubescent, and stipitate-glandular. **Florets** 20–40+; corollas yellow, 8–10+ mm. **Cypselae** (black) columnar, 2.5–3 mm; **pappi** of 40–60+, white or stramineous bristles in ± 2 series, 4–6 mm.

Flowering Jun–Aug(–Sep). Openings in chaparral, conifer forests; 400–2300 m; Calif., Oreg.

52. LEONTODON Linnaeus, Sp. Pl. 2: 798. 1753; Gen. Pl. ed. 5, 349. 1754, name conserved • Hawkbit [Greek *leon*, lion, and *odons*, tooth, alluding to deeply toothed leaves] [I]

David J. Bogler

Annuals or perennials, 10–80 cm; fibrous-rooted, sometimes tuberous, or with short caudices. **Stems** 1–20+, simple and scapiform or sparingly branched, glabrous, tomentulose, or coarsely hirsute. **Leaves** basal; petiolate (petioles winged); blades oblanceolate, margins entire or dentate or deeply lobed (faces glabrous or hispid, hairs simple or minutely 2–3-fid). **Heads** borne singly or 2–5 in loose, corymbiform arrays. **Peduncles** slightly inflated, naked or minutely bracteate. **Calyculi** of 10–20, subulate to lanceolate bractlets in 1–2 series (unequal), glabrous, tomentulose, or hirsute. **Involucres** campanulate, 4–15 mm diam. **Phyllaries** 16–20 in 2+ series, narrowly lanceolate, subequal, glabrous, tomentulose, or hirsute. **Receptacles** convex, pitted, sometimes slightly villous, epaleate. **Florets** 20–30; corollas yellow to orange (outer sometimes with reddish or greenish stripes). **Cypselae** light to dark brown or reddish brown, fusiform or cylindric, curved, distally narrowed and not beaked, or beaked, ribs 10–14, faces muricate, glabrous; **pappi** of ± distinct, yellowish white, tan, or pale brown bristles in 1–2 series (all uniformly plumose or outer reduced; **pappi** of outer cypselae sometimes reduced to crowns of bristlelike scales). $x = 4, 6, 7$.

Species ca. 50 (3 in the flora): introduced; Europe, n North Africa, Mediterranean, w Asia.

Leontodon is recognized by the basal rosettes of pinnatifid leaves, scapiform stems, loosely imbricate phyllaries, yellow corollas, and plumose pappus bristles. Some species are somewhat doubtfully distinguished by an overlapping mixture of vesture and pappus characters.

1. Heads (1–)2–5 in corymbiform arrays; peduncles minutely bracteate proximal to heads; pappi wholly of plumose bristles .... 1. *Leontodon autumnalis*
1. Head borne singly on scapiform stems; peduncles usually ebracteate proximal to heads; pappi mixed (either outer series different from inner, or pappi of outer cypselae reduced to crowns of bristlelike scales).
   2. Pappi mixed in all cypselae (outer series of bristlelike scales, inner of plumose bristles); phyllaries densely, coarsely hispid or hirsute .... 2. *Leontodon hispidus*
   2. Pappi of 2 types (on outer cypselae, crowns of bristlelike scales; on inner, of plumose bristles); phyllaries glabrate to coarsely hirsute .... 3. *Leontodon saxatilis*

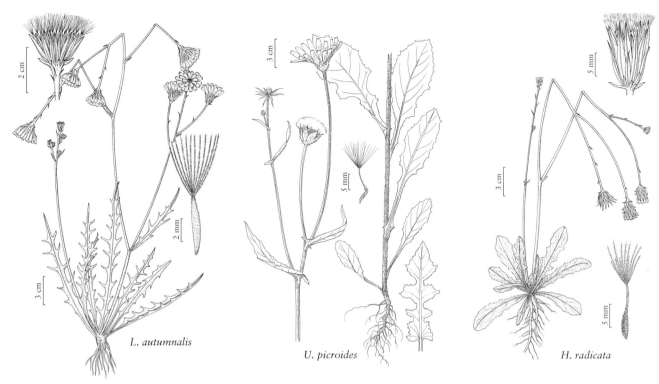

L. autumnalis

U. picroides

H. radicata

LEONTODON ○ UROSPERMUM ○ HYPOCHAERIS

1. **Leontodon autumnalis** Linnaeus, Sp. Pl. 2: 798. 1753 (as autumnale) F I

Leontodon autumnalis var. *pratensis* Koch

Perennials, 10–80 cm. **Stems** 1–20+, decumbent, scapiform, usually branched distally, glabrous proximally, tomentose proximal to heads. **Leaves:** blades narrowly oblanceolate, 4–35 × 0.5–4 cm, entire to deeply dentate or lobed (lobes narrow, straight or slightly recurved), faces glabrous or hirsute, hairs usually simple. **Heads** (1–)2–5 in loose, corymbiform arrays. **Peduncles** bracteate proximal to heads. **Calyculi** of 16–20, narrowly triangular to subulate bractlets 2–4 mm, tomentulose. **Involucres** campanulate, 7–13 × 8–10 mm. **Phyllaries** 18–20, narrowly lanceolate, 10–12 mm, subequal, glabrous, sparsely tomentose, or coarsely hirsute. **Florets** 20–30; corollas deep yellow, 13–16 mm. **Cypselae** cylindric or fusiform, 4–7 mm (not beaked); **pappi** wholly of yellowish white or tan, plumose bristles 5–8 mm. $2n = 12$, 24.

Flowering Jun–Oct. Roadsides, pastures, open fields; 10–1300 m; introduced; Greenland; N.B.; Nfld. and Labr. (Labr.), N.S., P.E.I., Que.; Alaska, Conn., Idaho, Iowa, Maine, Mass., Mich., N.H., N.J., N.Y., Ohio, Oreg., Pa., R.I., Vt., Wash., W.Va.; Eurasia.

Leontodon autumnalis is recognized by the usually branched stems with (1–)2–5 heads, peduncles bracteate proximal to heads, non-beaked cypselae, and pappi wholly of plumose bristles. It is now established in eastern North America and is sporadic in the west. Specimens with coarsely hirsute phyllaries have been recognized as var. *pratensis*; intermediates occur and the characteristic does not seem to correlate with other characters.

2. **Leontodon hispidus** Linnaeus, Sp. Pl. 2: 798. 1753 (as hispidum) • Bristly hawkbit I

Leontodon hastilis Linnaeus

Perennials, 10–60 cm. **Stems** 1–6, usually simple, scapiform, sometimes branched, glabrous or bristly hispid. **Leaves:** blades oblanceolate, 6–30 × 0.5–4 cm, margins coarsely dentate to deeply lobed (lobes straight, often narrowly triangular, terminal lobes usually large), faces usually coarsely hispid or hirsute, hairs often 2–3-fid. **Heads** usually borne singly. **Peduncles** ebracteate. **Calyculi** of 10–12, subulate bractlets 1–3 mm, glabrous or densely hirsute. **Involucres** campanulate, 7–13 × 10–15 mm. **Phyllaries** 12–16, linear-lanceolate, 6–10 mm, subequal, glabrate to coarsely hispid or hirsute. **Florets** 30–50+; corollas bright yellow or outermost orange or reddish, 12–15 mm. **Cypselae** fusiform, 6–12 mm (sometimes narrowed distally and weakly beaked); **pappi** pale brown, mixed: outer series of bristlelike scales, inner of plumose bristles. $2n = 14$.

Flowering Mar–Sep. Fields, lawns, gardens, roadsides; 100–1800 m; introduced; Ont.; Conn., Ga., Kans., N.Y., Ohio, Pa.; Europe.

Leontodon hispidus has been reported in eastern North America. It is recognized by the solitary heads, coarsely hispid leaves and peduncles, and pappi with long plumose and short non-plumose bristles. It is often confused with *L. saxatilis*, in which the pappi of the outermost cypselae are reduced to crowns. *Leontodon hirtus* Linnaeus has been reported from various locations in North America; the specimens appear to be assignable to *L. hispidus* Linnaeus.

3. **Leontodon saxatilis** Lamarck, Fl. Franç. 2: 115. 1779 (as saxatile) • Hairy hawkbit ☐

Annuals or perennials, 10–40 cm. **Stems** 1–15+, ascending, simple, glabrous or coarsely hispid. **Leaves:** blades oblanceolate to oblong, 2–15 × 0.5–2.5 cm, margins entire, dentate, or deeply lobed, faces hispid or hirsute, hairs minutely bifid. **Heads** borne singly. **Peduncles** ebracteate. **Calyculi** of 10–16, subulate bractlets 1–4 mm, glabrous or hirsute. **Involucres** campanulate, 6–13 × 4–9 mm. **Phyllaries** 16–20, narrowly lanceolate, 6–8 mm, subequal, glabrate to coarsely hirsute. **Florets** 20–30; corollas bright yellow to grayish yellow (outer faces), 8–15 mm. **Cypselae** fusiform, 4–5.5 mm, (outer curved, thick, not beaked, often enclosed by phyllaries, inner beaked, beaks 1–3 mm); **pappi** of 2 types: on outer cypselae, crowns of bristlelike scales ca. 0.5 mm; on inner, of dusky white or pale yellow, plumose bristles 5–6 mm. $2n = 8$.

Subspecies 2 (2 in the flora): introduced; Europe.

Leontodon saxatilis is widely introduced in North America. It is identified by heads borne singly, ebracteate

peduncles, glabrous or hairy leaves, and mixed pappi. Some individuals are coarsely hirsute and easily confused with *L. hispidus*.

Leontodon taraxacoides (Villars) Willdenow ex Mérat de Vaumartoise (1831), long used for this species, is a later homonym of *L. taraxacoides* Hoppe & Hornschuch (1821).

1. Perennials, rarely biennials; inner cypselae with beaks ca. 1 mm . . . . . . 3a. *Leontodon saxatilis* subsp. *saxatilis*
1. Annuals, rarely biennials; inner cypselae with beaks 2–3 mm . . . 3b. *Leontodon saxatilis* subsp. *longirostris*

3a. **Leontodon saxatilis** Lamarck subsp. **saxatilis** ☐

Leontodon nudicaulis Mérat; *L. nudicaulis* subsp. *taraxacoides* (Villars) Schinz & Thellung; *L. leysseri* (Wallroth) G. Beck

Perennials, rarely biennials; inner cypselae with beaks ca. 1 mm. $2n = 8$.

Flowering May–Nov. Lawns, waste areas, dunes, grassy areas; 100–1000 m; introduced; B.C.; Ala., Ariz., Calif., Conn., Ill., Ind., Md., Mass., Mich., Mo., Nev., N.J., N.Y., N.C., Ohio, Oreg., Pa., R.I., Tenn., Tex., Vt., Va., Wash., Wis.; Europe.

3b. **Leontodon saxatilis** Lamarck subsp. **longirostris** (Finch & P. D. Sell) P. Silva, Bol. Soc. Brot., ser. 2, 60: 155. 1987 ☐

Leontodon taraxacoides (Villars) Mérat subsp. *longirostris* Finch & P. D. Sell, Bot. J. Linn. Soc. 71: 247. 1976

Annuals, rarely biennials; inner cypselae with beaks 2–3 mm. $2n = 8$

Flowering Jun–Oct. Dry stream beds, riparian areas; 900–1000 m; introduced; Calif.; Europe.

53. **UROSPERMUM** Scopoli, Intr. Hist. Nat., 122. 1777 • [Greek *uro*, tail, and *sperma*, seed, alluding to beaks of cypselae] ☐

John L. Strother

Annuals [perennials], 10–40(–60+) cm; taprooted. **Stems** usually 1, erect, branched distally, setose to hispid or glabrous [pilosulous]. **Leaves** mostly cauline (at flowering) [mostly basal]; proximal ± petiolate, distal sessile; blades of the proximal mostly obovate to oblong-obovate, usually pinnately lobed or dentate, distal ovate to linear (bases often clasping), ultimate margins dentate or entire (veins often setose on abaxial faces). **Heads** borne singly or in loose, corymbiform arrays. **Peduncles** little, if at all, inflated distally, rarely bracteate. **Calyculi** 0. **Involucres** ± urceolate, 10–20+ mm diam. **Phyllaries** 7–8(–12+) in 1(–2) series, ovate-lanceolate to lance-linear (basally connate), subequal, margins scarious, apices acuminate. **Receptacles** flat to convex, pitted, hispid, epaleate. **Florets** 20–50+; corollas yellow, sometimes striped abaxially with red. **Cypselae** brown, bodies flattened-oblong, ± tuberculate, faces glabrous or scabrellous, beaks proximally dilated and tuberculate, distally acuminate and scabrous to scabrellous; **pappi** readily falling, of 18–22+ white [buff to rufous], subequal, plumose bristles in 1(–2) series (basally connate, falling together). $x = 5, 7$?.

Species 2 (1 in the flora): introduced; Europe, adjacent Africa and Asia; adventive in South America, s Africa.

SELECTED REFERENCE Lack, H. W. and B. E. Leuenberger. 1979. Pollen and taxonomy of *Urospermum* (Asteraceae, Lactuceae). Pollen & Spores 21: 415–425.

1. **Urospermum picroides** (Linnaeus) F. W. Schmidt, Samml. Phys.-Oekon. Aufsätze 1: 275. 1795 F I

Tragopogon picroides Linnaeus, Sp. Pl. 2: 790. 1753

Leaves: proximal blades 8–15 × 3–5+ cm, distal 4–8+ × 1–3+ cm. **Peduncles** 5–15+ cm. **Phyllaries:** free portions lance-linear, abaxial faces setose or ± glabrate. **Cypselae** (including beaks) 10–15 mm; **pappi** 9–12 mm. $2n = 8, 10$.

Flowering Apr–Jul. Disturbed sites; 0–400 m; introduced; Calif.; Europe; Asia; also introduced in South America, s Africa.

54. **HYPOCHAERIS** Linnaeus, Sp. Pl. 2: 810. 1753; Gen. Pl. ed. 5, 352. 1754 (as Hypochoeris) • Cat's ear, swine's succory [Greek *hypo*, beneath, and *choiras*, pig, alluding to pigs digging for roots] I

David J. Bogler

Annuals, biennials, or perennials, 10–60 cm; taprooted and with caudices. **Stems** 1–15, erect, branched or unbranched, glabrate or coarsely hirsute. **Leaves** basal or basal and proximally cauline; petiolate or sessile; blades oblanceolate to oblong or elliptic, margins entire or dentate to pinnately lobed (faces glabrous or hirsute). **Heads** borne singly or in loose, cymiform, paniculiform, or corymbiform arrays. **Peduncles** slightly inflated distally, minutely bracteate. **Calyculi** 0 (or indistinguishable from phyllaries). **Involucres** cylindric or campanulate, 5–20 mm diam. (12–25 mm in fruit). **Phyllaries** 20–30 in 3–4 series, unequal, linear-lanceolate, glabrous, glabrate, or coarsely hirsute. **Receptacles** flat, slightly pitted, paleate; paleae linear to subulate, scarious. **Florets** 20–100+; corollas usually yellow or orange, sometimes grayish green or reddish abaxially, rarely white, (not deliquescent). **Cypselae** monomorphic (all beaked) or dimorphic (outer truncate, inner beaked), usually brown to golden, bodies ellipsoid or fusiform, ribs 4–5 or 10, faces ± muricate, otherwise glabrous; **pappi** persistent, of 40–60+, white to tan bristles in 1 (all plumose) or in 2 series (outer barbellate, shorter than plumose inner). $x = 4, 5$.

Species 60+ (4 in the flora): introduced; South America, s Europe, n Africa, Asia.

Hypochaeris is similar to and closely related to *Leontodon*, from which it is distinguished mainly by its paleate receptacles and unequal phyllaries. Plumose pappus bristles are characteristic of both *Hypochaeris* and *Leontodon*. Molecular studies indicate the South American species of *Hypochaeris* are a monophyletic group derived from European stock that has undergone a recent and rapid radiation in the New World (R. Samuel et al. 2003). Two South American species are found as weeds in the flora area.

SELECTED REFERENCE   Samuel, R. et al. 2003. Phylogenetic relationships among species of *Hypochaeris* (Asteraceae, Cichorieae) based on ITS, plastid *trn*L intron, *trn*L-F spacer, and *mat*K sequences. Amer. J. Bot. 90: 496–507.

1. Leaves basal and proximally cauline; pappus bristles in 1 series, all plumose.
  2. Involucres broadly campanulate; phyllaries ± hirsute medially; corollas yellow
    . . . . 1. *Hypochaeris chillensis*
  2. Involucres cylindric or narrowly campanulate; phyllaries glabrous or sparsely
    tomentulose; corollas white . . . . 3. *Hypochaeris microcephala*
1. Leaves all or mostly basal; pappus bristles in 2 series: outer barbellate, shorter than plumose inner.
  3. Annuals; leaves usually glabrous or glabrate, sometimes hirsute on veins; florets
    ± equaling phyllaries at flowering; cypselae dimorphic, outer truncate, inner beaked
    . . . . 2. *Hypochaeris glabra*
  3. Perennials; leaves ± hirsute; florets surpassing phyllaries at flowering; cypselae
    monomorphic, all beaked . . . . 4. *Hypochaeris radicata*

1. Hypochaeris chillensis (Kunth) Britton, Bull. Torrey Bot. Club 19: 371. 1892 (as chilensis) [1]

Apargia chillensis Kunth in A. von Humboldt et al., Nov. Gen. Sp. 4(fol.): 2. 1818; 4(qto.): 3. 1820; *Achyrophorus chillensis* (Kunth) Schultz-Bipontinus; *Hypochaeris brasiliensis* (Lessing) Bentham & Hooker f. ex Grisebach; *Porcellites brasiliensis* Lessing

Biennials or perennials, 30–70 cm; taproots vertical, deep, thick, caudices stout. **Stems** (1–5) erect or ascending, simple or sparingly branched distally, glabrous or pilose proximally. **Leaves** basal and proximally cauline; basal blades elliptic to oblanceolate, 60–200 × 10–50 mm, margins coarsely and sharply dentate or 2-dentate, ciliate, faces glabrous or coarsely hirsute (cauline sessile, blades lanceolate, 50–100 × 10–30 mm, margins sharply dentate or pinnatifid; distal reduced, entire). **Heads** 1–10, in loose, paniculiform to corymbiform arrays. **Involucres** broadly campanulate, 10–20 × 5–20 mm. **Phyllaries** 20–30, linear-lanceolate, 4–15 mm, unequal, ± hirsute (at least medially). **Florets** 50–100+; corollas yellow, 5–7 mm, equaling phyllaries at flowering. **Cypselae** monomorphic, all beaked; bodies golden brown, fusiform, 8–10 mm, muricate, ribs 4–5; **pappi** of white, plumose bristles in 1 series, 6–8 mm. $2n = 8, 10$.

Flowering Apr. Waste areas with sandy soil, roadsides, lawns; 0–100 m; introduced; Ala., Fla., Ga., La., Miss., N.C., S.C., Tex.; South America.

Hypochaeris chillensis is recognized by the sharply dentate or pinnatifid cauline leaves, yellow corollas, and monomorphic, beaked cypselae. Plants in the flora area differ from those found elsewhere mainly in having the outer phyllaries somewhat more hirsute medially.

2. Hypochaeris glabra Linnaeus, Sp. Pl. 2: 811. 1753
  • Smooth cat's ear [1]

Annuals, 10–50 cm; taproots slender, vertical; caudices small, ± herbaceous. **Stems** (1–30), sparingly branched at midstem or distally (lateral branches often short, minutely bracteate or naked), glabrous. **Leaves** usually all basal; blades oblanceolate to oblong, 20–110 × 5–30 mm, margins nearly entire to dentate or pinnatifid, faces usually glabrous or glabrate, sometimes hirsute on veins. **Heads** borne singly or 2–3 in loose, cymiform arrays (terminating branches, not showy). **Involucres** narrowly campanulate, 8–16 × (3–)5–20 mm. **Phyllaries** 18–20, lanceolate, 3–18 mm, unequal, margins scarious, faces glabrous (apices brownish or reddish, sometimes ciliate). **Florets** 20–40; corollas white to yellowish, 5–8 mm, ± equaling phyllaries at flowering. **Cypselae** dimorphic, outer cylindric, stout, truncate, inner fusiform, slender, beaked; bodies dark brown, 10-nerved, 8–10 mm, beaks 3–4 mm; **pappi** of tawny bristles in 2 series, outer barbellate, shorter than plumose inner, longest 9–10 mm. $2n = 8, 10, 12$.

Flowering Feb–Jun(–Dec). Grassy slopes, sage scrub, pine-hardwood forest, disturbed areas, roadsides, commonly in sandy soil; 100–1300 m; introduced; B.C.; Ala., Ark., Calif., Fla., Ga., Ill., La., Maine, Mass., Miss., N.Y., N.C., Oreg., Pa., S.C., Tenn., Tex., Wash., W.Va.; Europe.

Hypochaeris glabra is usually distinguishable by its annual habit and relatively small size, slender and shallow roots, fine stems, often glabrous leaves, and beakless, truncate outer cypselae. Occasional specimens are larger and have induments characteristics of *H. radicata*; they can be distinguished by the dimorphic cypselae.

3. Hypochaeris microcephala (Schultz-Bipontinus) Cabrera, Notas Mus. La Plata, Bot. 2: 200. 1937 (as Hypochoeris) [I]

Achyrophorus microcephalus Schultz-Bipontinus, Jahresber. Pollichia 16–17: 59. 1859

Varieties 2 (1 in the flora): introduced; South America.

Hypochaeris microcephala is distinguished by the perennial habit, presence of cauline leaves, relatively small heads, and white corollas. It is most similar to *H. chillensis*; it usually has smaller heads with fewer florets and glabrous phyllaries.

3a. Hypochaeris microcephala (Schultz-Bipontinus) Cabrera var. albiflora (Kuntze) Cabrera, Notas Mus. La Plata, Bot. 2: 201. 1937 (as Hypochoeris) [I]

Hypochaeris brasiliensis (Lessing) Grisebach var. *albiflora* Kuntze, Revis. Gen. Pl. 3([3]): 159. 1898

Perennials, 10–60 cm; taproots vertical, thick, caudices woody. **Stems** erect, simple or branched distally, glabrous or sparsely hairy proximally. **Leaves** basal and proximally cauline; basal blades narrowly oblanceolate, 40–102(–250) × 10–50 mm, margins entire to sharply dentate or deeply pinnatifid (lobes long and narrow, ciliate), faces glabrous or ± glabrate; (cauline blades not clasping, distally reduced, margins shallowly dentate to coarsely pinnatifid). **Heads** borne singly or 2–13 in paniculiform to corymbiform arrays. **Involucres** cylindric or narrowly campanulate, 8–15(–18) × 5–10(–12) mm. **Phyllaries** 20–24, narrowly lanceolate, 1–14 mm, unequal, margins scarious (apices darkened), faces glabrous or sparsely tomentulose. **Florets** 50–100+; corollas white, 5–7 mm, not surpassing phyllaries at flowering. **Cypselae** monomorphic, all beaked (beaks 4–5 mm); bodies brown, fusiform, 7–8 mm, 10-nerved, muricate-roughened; **pappi** of white, plumose bristles in 1 series, 7–8 mm. $2n = 8, 10$.

Flowering Apr–May. Grazed woodlands, stream bottoms, ditches, sandy or silty soils; 1–50 m; introduced; La., Okla., Tex.; South America.

4. Hypochaeris radicata Linnaeus, Sp. Pl. 2: 811. 1753

• Hairy cat's ear [F] [I]

Perennials, 10–60 cm; taproots vertical, thick, fibrous, caudices woody. **Stems** (1–15) erect, usually branched (2–3 times at midstem and distally, sparsely bracteate or naked), glabrous or coarsely hirsute proximally. **Leaves** all basal; blades oblanceolate, lyrate to slightly runcinate, 50–-350 × 5–30 mm, margins coarsely dentate to pinnatifid, faces ± hirsute (hairs coarse, spreading). **Heads** usually 2–7 in loose arrays, sometimes borne singly. **Involucres** cylindric or campanulate, 10–25 × 10–20 mm. **Phyllaries** 20–30, narrowly lanceolate, 3–20 mm, unequal, margins scarious, green to darkened, faces glabrous or sparsely hirsute medially. **Florets** 10–15 mm, surpassing phyllaries at flowering; corollas bright yellow or grayish green. **Cypselae** monomorphic, all beaked, beaks 3–5 mm; bodies golden brown, fusiform, 6–10 mm, ribs 10–12, muricate; **pappi** of whitish bristles in 2 series, outer barbellate, shorter than plumose inner, longest 10–12 mm. $2n = 8$.

Flowering Apr–Nov. Oak-pine forest, coastal prairie, dunes, waste ground, dry fields, roadside ditches, railroads, lawns; 0–1600 m; introduced; B.C., Nfld. and Labr. (Labr.); Ala., Alaska, Ark., Calif., Colo., Conn., Del., Fla., Ga., Idaho, Ill., Ind., Ky., La., Maine, Md., Mass., Mich., Miss., Mo., Mont., Nev., N.H., N.J., N.Y., N.C., Ohio, Oreg., Pa., R.I., S.C., Tenn., Tex., Utah, Vt., Va., Wash., W.Va., Wis.; Europe; Asia.

Hypochaeris radicata is recognized by the coarse, perennial habit, stout roots, coarsely hirsute leaves and phyllaries, yellow corollas, and monomorphic, beaked cypselae. It is weedy and invasive in some areas.

55. HELMINTHOTHECA Zinn, Cat. Pl. Hort. Gott., 430. 1757 • Oxtongue [Greek *helminthos*, worm, and Latin *theca*, case or container; allusion unclear, perhaps to shapes of cypselae] ⊡

John L. Strother

Annuals or biennials, 10–100+ cm; taprooted. **Stems** usually 1, erect, branched distally or throughout, hirsute to hispid or setose (hair tips 2–4-hooked). **Leaves** basal and cauline; basal ± petiolate, distal sessile; blades elliptic, lanceolate, oblanceolate, oblong, or ovate, margins usually dentate to pinnately lobed (faces hirsute to hispid, hair tips 2–4-hooked). **Heads** borne singly or in loose, corymbiform arrays. **Peduncles** (terminal and axillary) not inflated distally, seldom bracteate. **Calyculi** of (3–)5, cordate or ovate to lance-ovate, foliaceous bracts. **Involucres** ovoid to urceolate, 9–12+ mm diam. (larger in fruit). **Phyllaries** (5–)8–13+ in 1 series, lance-linear to linear, subequal, margins little, if at all, scarious, apices acuminate. **Receptacles** flat, pitted, glabrous, epaleate. **Florets** 30–60+; corollas yellow (often reddish abaxially). **Cypselae** (dimorphic): outer whitish, bodies gibbous, ribs 5–10, beaks about equaling bodies, fragile, faces hirtellous to pilosulous adaxially; inner reddish brown, bodies ± compressed-ellipsoid to fusiform, beaks about equaling or longer than bodies, fragile, faces transversely rugulose to muricate, otherwise glabrous; **pappi** persistent, of 10–15+, whitish, subequal, barbellate or plumose, subulate to setiform (basally connate) scales in 1–2 series. *x* = 5.

Species 4 (1 in the flora): introduced; Europe; widely introduced.

SELECTED REFERENCE Lack, H. W. 1975. A note on *Helminthotheca* Zinn (Compositae). Taxon 24: 111–112.

1. Helminthotheca echioides (Linnaeus) Holub, Folia Geobot. Phytotax. 8: 176. 1973 [F] [I]

Picris echioides Linnaeus, Sp. Pl. 2: 792. 1753

Leaf blades 50–150(–250) × 10–50(–80+) mm. **Peduncles** 1–5+ cm, bristly hispid. **Calyculi:** bracts 9–15+ × 3–5+ mm, margins and tips bristly-ciliate. **Phyllaries** often sigmoid in fruit, abaxial faces and tips bristly. **Cypselae:** bodies 2.5–3 mm, beaks 2.5–5 mm; **pappi** 4–7 mm. *2n* = 10.

Flowering May–Nov. Disturbed sites; 0–50(–500) m; introduced; Alta., N.B., Ont., Sask.; Ariz., Calif., Conn., D.C., Iowa, Maine, Md., Mass., Mo., Mont., N.J., N.Y., N.Dak., Ohio, Oreg., Pa., Vt., Va.; Europe; widely introduced elsewhere.

Reports for British Columbia and Nova Scotia need confirmation.

56. RHAGADIOLUS Jussieu, Gen. Pl., 168. 1789, name conserved • [Greek *rhagado*, crack or split, and *-olus,* diminutive, perhaps alluding to gaps between margins of enfolding phyllaries] ⊡

John L. Strother

Annuals, 5–60+ cm; taprooted. **Stems** usually 1, erect, branched distally, glabrous. **Leaves** basal and cauline (mostly cauline at flowering); basal sessile or ± petiolate, distal ± sessile; blades ovate-lanceolate to lanceolate or linear (often runcinate), margins entire or dentate to pinnately lobed (faces glabrous or ± hispid). **Heads** in ± corymbiform arrays (terminal heads often surpassed by others). **Peduncles** not inflated, rarely bracteolate. **Calyculi** of 5, ovate to deltate

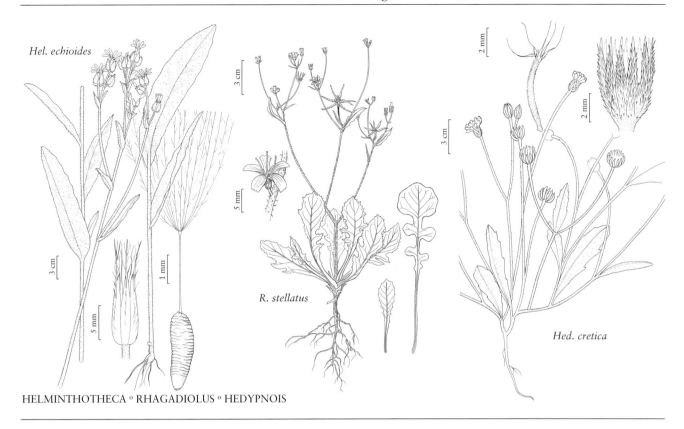

Hel. echioides

R. stellatus

Hed. cretica

HELMINTHOTHECA ∘ RHAGADIOLUS ∘ HEDYPNOIS

bractlets. **Involucres** campanulate to cylindric, 2–4+ mm diam. (larger in fruit). **Phyllaries** (3–) 5–8 in 1 series (closely enfolding ovaries/cypselae of subtended florets; ± patent in fruit), linear, equal, margins often scarious, apices acuminate (abaxially glabrous or ± hispid to scabrous). **Receptacles** ± flat, smooth or ± pitted, glabrous, epaleate. **Florets** 5–6(–10+); corollas yellow. **Cypselae** brownish, heteromorphic; outer (tardily falling with enfolding phyllary) ± terete, narrowed distally, straight to arcuate, not beaked, ribs 0, faces glabrous; inner (readily falling) terete, straight to ± coiled, faces glabrous or closely hirtellous; **pappi** 0. $x = 5$.

Species 1 or 2 (1 in the flora): introduced; Europe, Africa.

1. **Rhagadiolus stellatus** (Linnaeus) Gaertner, Fruct. Sem. Pl. 2: 354. 1791 [F] [I]

Lapsana stellata Linnaeus, Sp. Pl. 2: 811. 1753

Leaf blades 25–100(–150+) × 10–30(–60+) mm. **Phyllaries** linear (ultimately each convolute about a subtended cypsela), glabrous or hispid to setose. **Cypselae:** outer 10–15(–25) mm, inner 5–10 mm. $2n = 10$.

Flowering May. Disturbed sites; 10–200 m; introduced; Calif.; Europe.

57. HEDYPNOIS Miller, Gard. Dict. Abr. ed. 4, vol. 2. 1754 • [Ancient name for an endive-like plant, attributed to Pliny] [I]

John L. Strother

Annuals, (5–)10–60+ cm; taprooted. **Stems** usually 1, erect, branched distally, ± hispid to setose (hair tips often forked). **Leaves** basal and cauline; basal ± petiolate, distal sessile; blades lanceolate, linear, oblanceolate, oblong, or ovate, margins entire or dentate to pinnately lobed (faces ± hispid). **Heads** borne singly or in loose, corymbiform arrays. **Peduncles** ± inflated distally, not bracteate. **Calyculi** of 3–10+, deltate to lanceolate or lance-linear bractlets. **Involucres** campanulate to cylindric, 3–12 mm diam. (larger, ± globose in fruit). **Phyllaries** 5–13+ in 1 series, linear-navicular (± keeled, each ± enfolding subtended ovary or cypsela), subequal, margins little, if at all, scarious, apices acuminate. **Receptacles** flat, ± pitted, glabrous, epaleate. **Florets** 8–30+; corollas yellow (often reddish proximally, greenish abaxially). **Cypselae** dark brown to black, cylindric to fusiform (usually ± arcuate), not beaked, ribs 12–15, faces ± scabrous or barbed; **pappi** persistent, whitish; on outer cypselae often coroniform (distinct or connate, erose to fimbriate scales); on inner cypselae 0–5+, cuneate to lanceolate or subulate outer scales plus 5+, lance-aristate to subulate-aristate, inner scales. $x = 9$.

Species 2 (1 in the flora): introduced; Europe.

1. Hedypnois cretica (Linnaeus) Dumont de Courset, Bot. Cult. 2: 339. 1802 [F] [I]

Hyoseris cretica Linnaeus, Sp. Pl. 2: 810. 1753

Leaf blades 5–150(–250) × 2–25 (–35+) mm. **Peduncles** 2–5(–15+) cm. **Phyllaries** (incurved in fruit), abaxial faces glabrous or scabrous to hispid. **Cypselae** 5–7.5 mm; **pappi** 0.1–0.5(–1) or 2–5 mm. $2n$ = 8, 11, 12, 13, 14, 15, 16, 18 (all Europe).

Flowering (Mar–)Apr–May(–Aug). Disturbed sites; 0–100(–600+) m; introduced; Ariz., Calif., N.Mex., Tex.; Europe.

58. PICRIS Linnaeus, Sp. Pl. 2: 792. 1753; Gen. Pl. ed. 5, 347. 1754 • Oxtongue [Greek *picris*, bitter or sharp; allusion unclear] [I]

John L. Strother

Annuals, biennials, or perennials, 10–100+ cm; tap- or fibrous-rooted, sometimes rhizomatous. **Stems** usually 1, erect, branched distally, hirsute to hispid or setose (hair tips often 2[–4]-hooked). **Leaves** basal and cauline (mostly cauline at flowering); basal ± petiolate, distal sessile; blades oblong, ovate, or lanceolate to oblanceolate or linear, margins entire or sinuate-dentate to pinnately lobed (faces hirsute to hispid or setose, hair tips 2[–4]-hooked). **Heads** usually in ± corymbiform arrays. **Peduncles** not inflated distally, sometimes bracteate. **Calyculi** of 8–13+, lanceolate to lance-linear bractlets (sometimes ± intergrading with phyllaries). **Involucres** campanulate to urceolate, 6–12+ mm diam. (sometimes larger in fruit). **Phyllaries** (8–)13+ in 1–2 series (reflexed in fruit), lanceolate to lance-linear (± flat or navicular proximally, sometimes each ± enfolding its subtended floret), equal, margins often scarious, apices acute. **Receptacles** flat to convex, ± pitted, glabrous, epaleate. **Florets** 30–100+; corollas yellow, often reddish abaxially. **Cypselae** homomorphic [heteromorphic], reddish brown [dark brown], bodies ± fusiform [compressed-ellipsoid], not beaked [beaks ± developed], ribs 5–10, faces transversely rugulose, glabrous; **pappi** falling, of 30–45+, whitish to stramineous, subequal, barbellulate to plumose bristles [scales] in 2–3+ series (basally connate, falling together). $x = 5$.

Species ca. 40 (2 in the flora): introduced; Europe, Asia, n Africa; also introduced in tropical Africa, Australia.

1. Annuals; phyllaries navicular proximally, each enfolding its subtended floret; cypselae 2.5–3 mm . . . . 1. *Picris rhagadioloides*
1. Biennials or perennials; phyllaries ± flat proximally, not each enfolding its subtended floret; cypselae 3–4(–6) mm . . . . 2. *Picris hieracioides*

1. Picris rhagadioloides (Linnaeus) Desfontaines, Tabl. École Bot., 89. 1804 (as rhagadialoides) [I]

Crepis rhagadioloides Linnaeus, Syst. Nat. ed. 12, 2: 527. 1767; Mant. Pl., 108. 1767

Annuals. Leaf blades 20–60(–90) × 5–20+ mm. **Involucres** 8–12+ × 6–9+ mm, larger in fruit. **Phyllaries** proximally navicular, each enfolding its subtended floret, usually bristly and/or tomentulose abaxially. **Cypselae** (outer often arcuate) 2.5–3 mm; **pappi** 1–1.5 (outer cypselae) or 5–6 mm (inner cypselae). $2n = 10$.

Flowering Jun. Disturbed sites; 100–200 m; introduced; Mo.; Europe.

The single report of *Picris rhagadioloides* for the flora area (as *P. sprengerana* from St. Louis; G. Yatskievych, pers. comm.) was confirmed for this study. *Picris sprengerana* (Linnaeus) Poiret has been misapplied to plants here called *P. rhagadioloides* (see W. Greuter 2003).

2. Picris hieracioides Linnaeus, Sp. Pl. 2: 792. 1753 [F] [I]

Picris hieracioides var. *alpina* Koidzumi; *P. hieracioides* subsp. *kamtschatica* (Ledebour) Hultén

Biennials or perennials. Leaf blades 50–150(–300) × 10–25(–50+) mm. **Involucres** 8–15 × 10–16+ mm, larger in fruit. **Phyllaries** proximally ± flat, not each enfolding its subtended floret, usually bristly and/or tomentulose abaxially. **Cypselae** 3–4(–6) mm; **pappi** 5–7 mm. $2n = 10$.

Flowering Jul–Oct. Disturbed sites; 0–50(–100+) m; introduced; Ont.; Alaska, Conn., Ill., Ky., Md., Mass., Mich., Mo., N.J., N.Y., N.C., Ohio, Pa., R.I., Tenn., Vt., Va., Wash.; Europe, Asia; also introduced, Africa, Australia.

In Alaska, *Picris hieracioides* is known only from the Aleutian Islands, where it may be native. This species is evidently no longer present in British Columbia.

59. TRAGOPOGON Linnaeus, Sp. Pl. 2: 789. 1753; Gen. Pl. ed. 5, 346. 1754 • [Greek *tragos*, goat, and *pogon*, beard; probably alluding to pappi] [I]

Pamela S. Soltis

Biennials (sometimes winter annuals) [perennials], (15–)50–150 cm; taprooted. **Stems** 1(–5), erect, usually branched proximally, glabrous or tomentulose to floccose [lanate], often glabrescent. **Leaves** basal and cauline; sessile; blades linear to lance-linear or lance-attenuate (grasslike) [lanceolate to oblong], (bases clasping) margins entire (faces glabrous or tomentulose to floccose [lanate], often glabrescent). **Heads** borne singly (terminal). **Peduncles** often inflated distally (not in *T. pratensis*), ebracteate. **Calyculi** 0. **Involucres** campanulate [cylindric] (at flowering), mostly 10–20+ mm diam. **Phyllaries** usually [5–7] 8–12 [13–16] in 1 series, linear-

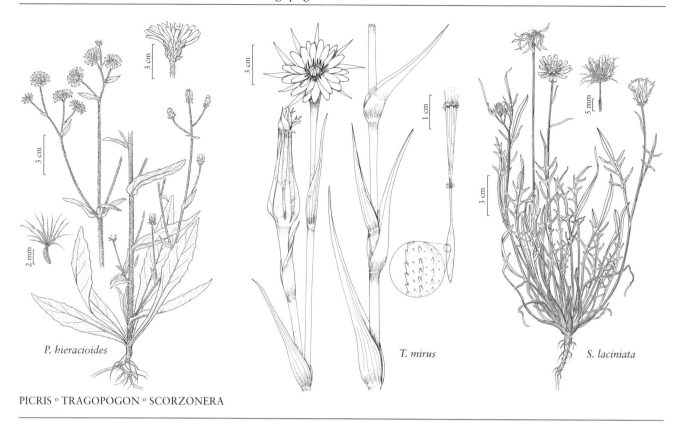

P. *hieracioides*

T. *mirus*

S. *laciniata*

PICRIS ∘ TRAGOPOGON ∘ SCORZONERA

lanceolate, triangular-lanceolate [oblong-lanceolate], linear, ± equal, margins white, narrowly pellucid, apices acute (faces glabrous [with intertwining hairs]). **Receptacles** convex, smooth, glabrous, epaleate. **Florets** (30–)50–180+; corollas yellow or purple (proximally yellow, distally purple in *T. mirus*) (± deliquescent). **Cypselae** dark to pale brown, stramineous, whitish, bodies ± fusiform to cylindric, usually beaked, beaks concolorous with, or paler than bodies, abrupt to gradually tapered, 5–10-ribbed (ribs usually muricate, prickly, or scaly), faces usually glabrous, sometimes scaley or muricate; **pappi** (usually borne on discs at tips of beaks) persistent, of 12–20+, brownish to whitish, basally connate, ± plumose, subequal to unequal awns or subulate scales, in 1 series (lateral barbs or setulae often ± intertwined). $x = 6$.

Species 100–150 (5, including 2 amphidiploids, in the flora): introduced; Eurasia, n Africa; introduced in Australia.

Tragopogon is weedy in North America. Allotetraploids *T. mirus* and *T. miscellus* are native to the United States. The heads of tragopogons usually open early mornings and close by midday.

SELECTED REFERENCES Cook, L. M., P. S. Soltis, S. J. Brunsfeld, and D. E. Soltis. 1998. Multiple independent formations of *Tragopogon* tetraploids (Asteraceae): Evidence from RAPD markers. Molec. Ecol. 7: 1293–1302. Mavrodiev, E. V. et al. 2005. Phylogeny of *Tragopogon* L. (Asteraceae) based on ITS and ETS sequence data. Int. J. Pl. Sci. 166: 117–133. Ownbey, M. 1950c. Natural hybridization and amphiploidy in the genus *Tragopogon*. Amer. J. Bot. 37: 487–499. Novak, S. J., D. E. Soltis, and P. S. Soltis. 1991. Ownbey's tragopogons: Forty years later. Amer. J. Bot. 78: 1586–1600. Soltis, D. E. et al. 2004. Recent and recurrent polyploidy in *Tragopogon* (Asteraceae): Cytogenetic, genomic, and genetic comparisons. Biol. J. Linn. Soc. 82: 485–501. Soltis, P. S., G. M. Plunkett, S. J. Novak, and D. E. Soltis. 1995. Genetic variation in *Tragopogon* species: Additional origins of the allotetraploids *T. mirus* and *T. miscellus* (Compositae). Amer. J. Bot. 82: 1329–1341.

1. Corollas wholly or distally purple (to brownish purple in *T. mirus*); leaf apices straight.
    2. Leaf faces usually glabrous; corollas wholly purple ($2n = 12$) . . . . 4. *Tragopogon porrifolius*
    2. Leaf faces initially tomentulose to floccose, soon glabrescent; corollas each proximally yellowish, distally purple to brownish purple ($2n = 24$) . . . . 5. *Tragopogon mirus*

1. Corollas yellow; leaf apices straight or recurved to coiled.
    3. Leaf apices usually recurved to coiled; peduncles usually little, if at all, inflated (at flowering); involucres urceolate in bud; outer florets equaling or surpassing phyllaries (2*n* = 12) . . . . 1. *Tragopogon pratensis*
    3. Leaf apices straight or recurved to coiled; peduncles distally inflated; involucres conic to urceolate in bud; outer florets usually not surpassing phyllaries (2*n* = 12 or 24).
        4. Leaf apices usually straight (not recurved to coiled); outer florets usually much shorter than phyllaries (2*n* = 12) . . . . 2. *Tragopogon dubius*
        4. Leaf apices usually recurved to coiled; outer florets usually slightly shorter than phyllaries (equaling or surpassing phyllaries in population from Pullman, Washington; 2*n* = 24) . . . . 3. *Tragopogon miscellus*

1. Tragopogon pratensis Linnaeus, Sp. Pl. 2: 789. 1753 (as pratense) • Meadow salsify, Jack-go-to-bed-at-noon, salsifis des prés ⊡

Plants (15–)40–100 cm. **Leaves:** apices usually recurved to coiled, faces initially tomentulose to floccose, soon glabrescent. **Peduncles** usually little, if at all, inflated (at flowering, may be inflated in fruit), initially floccose to tomentulose, soon glabrescent. **Involucres** urceolate in bud. **Outer florets** equaling or surpassing phyllaries; corollas yellow. **2*n* = 12.**

Flowering May–Aug. Disturbed sites; 10–2100 m; introduced; Alta., B.C., Man., N.B., Nfld. and Labr. (Nfld.), N.S., Ont., P.E.I., Que., Sask.; Ariz., Calif., Colo., Conn., Del., D.C., Ga., Idaho, Ill., Ind., Iowa, Kans., Ky., Maine, Md., Mass., Mich., Minn., Mo., Mont., Nebr., Nev., N.H., N.J., N.Mex., N.Y., N.C., Ohio, Okla., Oreg., Pa., R.I., S.Dak., Tenn., Utah, Vt., Va., Wash., W.Va., Wis., Wyo.; Europe.

Tragopogon pratensis is naturalized across much of North America. The circumscription and infraspecific taxonomy of *T. pratensis* in Europe are debated, and the name *T. pratensis* may prove to be inaccurately assigned to the introduced populations in North America.

2. Tragopogon dubius Scopoli, Fl. Carniol. ed. 2, 2: 95. 1772 (as dubium) • Yellow salsify, salsifis majeur ⊡

Plants (30–)40–80(–100) cm. **Leaves:** apices straight (not recurved to coiled), faces initially tomentulose to floccose, soon glabrescent. **Peduncles** distally inflated. **Involucres** conic in bud. **Outer florets** much shorter than phyllaries; corollas yellow. **2*n* = 12.**

Flowering early summer. Disturbed sites and less disturbed sites; 10–2500 m; introduced; Alta., B.C., Man., N.B., Nfld. and Labr. (Nfld.), N.W.T., N.S., Ont., Que., Sask., Yukon; Ariz., Ark., Calif., Colo., Conn., Del., Idaho, Ill., Ind., Iowa, Kans., Ky., La., Maine, Md., Mass., Mich., Minn., Mo., Mont., Nebr., Nev., N.H., N.J., N.Mex., N.Y., N.C., N.Dak., Ohio, Okla., Oreg., Pa., R.I., S.Dak., Tenn., Tex., Utah, Vt., Va., Wash., W.Va., Wis., Wyo.; Europe; Australia; s Africa.

Tragopogon dubius is naturalized across much of North America. It typically grows in sites drier than those where *T. pratensis* is found.

3. Tragopogon miscellus Ownbey, Amer. J. Bot. 37: 498. 1950 • Hybrid or Ownbey's goatsbeard, Moscow salsify ⊡

Plants 60–150+ cm. **Leaves:** apices usually recurved to coiled, faces initially floccose to tomentulose, soon glabrescent. **Peduncles** distally inflated. **Involucres** conic to urceolate in bud. **Outer florets** usually slightly shorter than phyllaries (see discussion for exception); corollas yellow. **2*n* = 24.**

Flowering early summer. Disturbed sites; 700–800 m; Ariz., Idaho, Wash., Wyo.

Tragopogon miscellus has been reported from near Gardiner, Montana; it is no longer present there.

Plants of *Tragopogon miscellus* are larger and more robust than those of *T. pratensis*. They are allotetraploids, formed (probably repeatedly) from hybrids between *T. pratensis* and *T. dubius*. Outer florets are shorter than the phyllaries except at some sites in Pullman, Washington, where outer florets equal or surpass the phyllaries. (The different inflorescence morphs result from reciprocal polyploid origins.) F₁ hybrids between *T. dubius* and *T. pratensis* (= *T. ×crantzii* Dichtl) may resemble *T. miscellus* but are less robust, have low pollen stainability, and set few, if any, seeds. *Tragopogon miscellus* does not occur in Europe, but hybrids between *T. dubius* and *T. pratensis* occur occasionally.

4. Tragopogon porrifolius Linnaeus, Sp. Pl. 2: 789. 1753 (as porrifolium) • Salsify, vegetable oyster, salsifis cultivé [I]

Plants 40–100(–150) cm. **Leaves:** apices straight (not recurved or coiled), faces usually glabrous. **Peduncles** distally inflated. **Involucres** conic in bud. **Outer florets** usually shorter than or equaling phyllaries; corollas purple. $2n = 12$.

Flowering Apr–Aug. Disturbed sites; 200–2000 m; introduced; Alta., B.C., Man., N.B., N.S., Ont., Que.; Ariz., Ark., Calif., Colo., Conn., Del., D.C., Ga., Idaho, Ill., Ind., Iowa, Kans., Ky., Maine, Md., Mass., Mich., Mo., Mont., Nebr., Nev., N.H., N.J., N.Mex., N.Y., N.C., Ohio, Okla., Oreg., Pa., S.Dak., Tenn., Texas, Utah, Vt., Va., Wash., W.Va., Wis., Wyo.; Europe; n Africa; introduced, Pacific Islands (Hawaii); Australia.

Tragopogon porrifolius is occasionally cultivated in Europe and naturalized across much of North America. It grows typically in sites drier than those of *T. pratensis* and in sites shadier and/or moister than those of *T. dubius*. As currently circumscribed, it may not be monophyletic, and nomenclatural changes for the populations here may be required. In North America, *T. porrifolius* hybridizes with both *T. dubius* and *T. pratensis* (= *T.* ×*neohybridus* Farwell, described from North America, and *T.* ×*mirabilis* Rouy, described from Europe).

5. Tragopogon mirus Ownbey, Amer. J. Bot. 37: 497. 1950 • Remarkable goatsbeard [F] [I]

Plants (40–)60–150+ cm. **Leaves:** apices straight (not recurved or coiled), faces initially floccose to tomentulose, soon glabrescent. **Peduncles** distally inflated. **Involucres** conic in bud. **Outer florets** usually shorter than phyllaries; corollas each proximally yellow and distally purple or brownish purple (giving each head a yellow "eye"). $2n = 24$.

Flowering early summer. Disturbed sites; 700–800 m; Ariz., Idaho, Wash.

Tragopogon mirus is allotetraploid, formed from *T. dubius* and *T. porrifolius*. It originated (probably repeatedly) in the United States (eastern Washington, adjacent Idaho, and near Flagstaff, Arizona). F₁ hybrids between *T. dubius* and *T. porrifolius* resemble *T. mirus* but are less robust, have low pollen stainability, and set few, if any, seeds. *Tragopogon mirus* does not occur in Europe, but *T. dubius* and *T. porrifolius* may occasionally hybridize there when sympatric.

---

60. SCORZONERA Linnaeus, Sp. Pl. 2: 790. 1753; Gen. Pl. ed. 5, 346. 1754 • [Perhaps French *scorzonère*, "viper's grass;" allusion unknown] [I]

John L. Strother

Annuals, biennials, or perennials [subshrubs], 5–100+ cm; taprooted. **Stems** 1, erect, branched from bases and/or distally, glabrous or hairy. **Leaves** basal and cauline; basal sessile or petiolate, distal sessile; blades ovate-lanceolate to lanceolate or linear, margins entire or pinnately lobed to pinnatisect (faces glabrous or ± arachnose [tomentose]). **Heads** borne singly or in loose, corymbiform arrays. **Peduncles** not inflated, sometimes bracteate. **Calyculi** 0. **Involucres** ovoid to cylindric, 6–12[–16+] mm diam. (larger in fruit). **Phyllaries** 18–30+ in 3–5+ series, deltate or ovate to lanceolate or lance-linear (± flat proximally, not enfolding subtended florets), unequal, margins scarious, apices obtuse to acute. **Receptacles** flat, pitted, glabrous, epaleate. **Florets** 30–100+; corollas whitish to yellow or purplish. **Cypselae** whitish to brownish, narrowly columnar to obclavate or fusiform (sometimes ± stipitate), not beaked, nerves usually 10, sometimes 0, faces mostly glabrous, sometimes distally villosulous [lanate]; **pappi** persistent, of 28–50+, whitish, subequal, plumose to barbellate, subulate to setiform scales in 2–3 series. $x = 7$.

Species ca. 175 (2 in the flora): introduced; Europe, Asia.

1. Leaf blades 70–200 × (5–)20–40 mm, margins of the basal usually pinnately lobed . . . . . 1. *Scorzonera laciniata*
1. Leaf blades 120–400 × (1–)3–6 mm, margins entire (flat or undulate) . . . . 2. *Scorzonera hispanica*

1. Scorzonera laciniata Linnaeus, Sp. Pl. 2: 791. 1753
[F] [I]

Annuals, biennials, or perennials (monocarpic). **Leaf blades** 70–200 × (5–)20–40 mm, margins of basal usually pinnately lobed, lobes ± linear. **Involucres** 7–20 × 6–12+ mm. **Phyllaries** deltate to lanceolate, ± arachnose. **Cypselae** 8–17 mm (± stipitate); **pappi** 8–17 mm. **2***n* = 14.

Flowering May–Jul. Disturbed sites; 1000–1800 m; introduced; Colo., Kans., Mont., Nebr., N.Mex., Tex., Wyo.; Europe.

2. Scorzonera hispanica Linnaeus, Sp. Pl. 2: 791. 1753
• Black or Spanish salsify   [I]

Perennials. **Leaf blades** 120–400 × (1–)3–6 mm, margins entire (flat or undulate). **Involucres** 20–30 × 8–12+ mm. **Phyllaries** ovate-lanceolate to lanceolate, glabrous. **Cypselae** 10–15(–20) mm; **pappi** 9–15(–20) mm. **2***n* = 14.

Flowering Jun–Jul. Disturbed sites; 10–200 m; introduced; Calif.; Europe.

Scorzonera hispanica sometimes is used culinarily.

---

61. CALYCOSERIS   A. Gray, Smithsonian Contr. Knowl. 5(6): 104, plate 14. 1853

• Tack-stem [Greek *kalyx*, cup, and *seris*, chicory, alluding to shallow cups on apices of cypselae]

L. D. Gottlieb

Annuals, 5–30 cm; taprooted. **Stems** 1–3, erect or ascending, branched from bases; distal stems, branches, and involucres conspicuously dotted with short, stipitate, flat-topped glands that resemble tacks. **Leaves** basal and cauline; alternate; sessile; basal blades pinnately lobed (lobes narrow, linear, often delicate, spreading, margins entire, smooth); distal reduced to linear, entire bracts. **Heads** borne singly or in open, cymiform arrays. **Peduncles** (1–3 cm) not inflated, not bracteate. **Calyculi** of 8–16, reflexed, unequal bractlets (lengths to ½ phyllaries). **Involucres** campanulate, 5–12+ mm diam. **Phyllaries** 12–20 in 1 series, linear-lanceolate, equal, margins scarious, apices acute. **Receptacles** flat, smooth, bristly, epaleate (each floret subtended by 1 fine, capillary bristle). **Florets** ca. 25; corollas white or yellow (showy). **Cypselae** tan to brown, fusiform, beaked, ribs 5, separated by longitudinal grooves, faces glabrous or scabridulous; **pappi** (borne on denticulate cups at beak tips) falling (together), of 50–60+, white, basally connate, smooth bristles 5–8(–9) mm in 1 series. *x* = 7.

Species 2 (2 in the flora): sw United States, nw Mexico.

1. Ligules yellow; tack-shaped glands red- or purple-tipped; cypselae deeply grooved between ribs, faces not rugulose . . . . 1. *Calycoseris parryi*
1. Ligules white; tack-shaped glands straw-colored; cypselae shallowly grooved between ribs, faces weakly cross-rugulose . . . . 2. *Calycoseris wrightii*

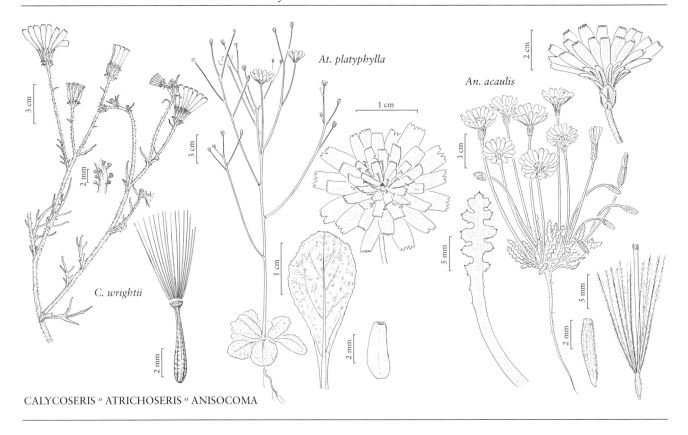

At. platyphylla

An. acaulis

C. wrightii

CALYCOSERIS ° ATRICHOSERIS ° ANISOCOMA

1. Calycoseris parryi A. Gray in W. H. Emory, Rep. U.S. Mex. Bound. 2(1): 106. 1859 • Yellow tack-stem

Stems, distal branches, and phyllaries gland-dotted, glands red- or purple-tipped. **Involucres** 13–17 mm. **Ligules** yellow, 1.5–2.5 cm. **Cypselae** tan or straw-colored, deeply grooved between ribs, faces not rugulose. **2***n* = 14.

Flowering Apr–Jun. Sandy, gravelly soils, sand dunes, slopes, washes, limestone ridges in Lower Sonoran zone, Mohave and Colorado deserts; 200–1800 m; Ariz., Calif., Nev., Utah; Mexico (Baja California).

2. Calycoseris wrightii A. Gray, Smithsonian Contr. Knowl. 5(6): 104, plate 14. 1853 • White tack-stem

F

Stems, distal branches, and phyllaries gland-dotted, glands straw-colored. **Involucres** 10–15 mm. **Ligules** white, often with fine reddish veins abaxially, 2–3 cm. **Cypselae** dark tan to brown, shallowly grooved between ribs, faces weakly cross-rugulose. **2***n* = 14.

Flowering Mar–Jun. Gravels derived from limestone, volcanics, granites, and caliche soils on plains, rocky mesas and slopes, stream bottoms, desert pavement and washes, Upper Sonoran and Lower Sonoran zones, Colorado and Mohave deserts, often in creosote or mesquite associations; 100–1600 m; Ariz., Calif., Nev., N.Mex., Tex., Utah; Mexico (Baja California, Sonora).

62. ATRICHOSERIS   A. Gray in A. Gray et al., Syn. Fl. N. Amer. 1(2): 410. 1884

• Tobacco-weed, gravel-ghost, parachute plant [Greek *a-*, without, *trichos*, hair, and *seris*, chicory, alluding to lack of pappus]

David J. Keil

Malacothrix [unranked] *Anathrix*, A. Gray, Proc. Amer. Acad. Arts 9: 214. 1874

Annuals, 20–180 cm (± scapiform); taprooted. **Stems** 1–3+, erect, branched distally, glabrous. **Leaves** mostly basal or subbasal (usually flat against soil), proximal cauline relatively few; sessile or petiolate (petioles winged); blades obovate, finely dentate (obtuse); distal cauline reduced to triangular scales. **Heads** in corymbiform arrays. **Peduncles** not inflated distally, sometimes bracteate. **Calyculi** of 4–6, triangular to lanceolate bractlets. **Involucres** cylindro-campanulate, 6–13+ mm diam. **Phyllaries** 10–15 in 1–2 series, lanceolate, ± equal, margins broadly hyaline, apices acuminate. **Receptacles** flat or slightly convex, smooth or minutely pitted, epaleate. **Florets** 30–60 (fragrant); corollas white, sometimes tipped with rose or purple (at least outer much longer than phyllaries). **Cypselae** whitish, subcylindric or weakly clavate, not beaked, bluntly 4–5-angled or -ribbed, ribs corky-thickened, faces glabrous or minutely puberulent; **pappi** 0. *x* = 9.

Species 1: sw United States, nw Mexico.

Molecular phylogenetic studies by J. Lee et al. (2003) indicated that *Malacothrix* is polyphyletic and that *Atrichoseris* is basal to the subclade of *Malacothrix* that does not contain the type species of that genus. *Atrichoseris* differs in base chromosome number (*x* = 9) from the species of *Malacothrix* with which it is most closely grouped (*x* = 7). *Atrichoseris* differs from *Malacothrix* in having apically truncate cypselae that are wholly epappose.

1. Atrichoseris platyphylla   (A. Gray) A. Gray in A. Gray et al., Syn. Fl. N. Amer. 1(2): 410.   F

Malacothrix platyphylla A. Gray, Proc. Amer. Acad. Arts 9: 214. 1874

Stems glaucous. **Basal leaves** gray-green, glaucous, often purple-tinged, especially abaxially, often adaxially mottled, 1–12 × 0.5–6 cm. **Heads** 1.5–3.5 cm diam. **Involucres** 6–10 mm. **Florets:** corollas 8–20 mm, inner with bases yellow or purple, much shorter than outer. **Cypselae** 4–4.5 mm. *2n* = 18.

Flowering Feb–May. Dry desert slopes, valleys, and washes; 0–1400 m; Ariz., Calif., Nev., Utah; Mexico (Baja California).

Atrichoseris platyphylla grows in the Mojave and Sonoran deserts.

63. ANISOCOMA   Torrey & A. Gray, Boston J. Nat. Hist. 5: 111, plate 13, figs. 7–11. 1845   • [Greek *anisos*, unequal, and *coma*, hair, alluding to pappus]

David J. Keil

Annuals, 5–25 cm (± scapiform); taprooted. **Stems** 1–25+, erect or ascending, simple, glabrous or densely tomentose, sometimes glabrescent. **Leaves** basal or subbasal; petiolate (petioles winged); blades oblanceolate, margins pinnately lobed, denticulate (teeth callose-tipped). **Heads** borne singly. **Peduncles** not inflated distally, ebracteate. **Calyculi** 0. **Involucres** cylindric to campanulate, 4–20 mm diam. **Phyllaries** 15–25 in 4–5 series, (centers green to reddish purple) unequal, outer ovate to orbiculate, inner linear to oblong, flat, margins papery-transparent, apices rounded or obtuse to subacute. **Receptacles** flat to weakly convex, smooth, paleate (paleae slender, bristlelike). **Florets** ca. 40; corollas cream to bright yellow. **Cypselae** tan, cylindric, (bases tapered) beaks 0 (or ca. 0.5 mm), 10–15-nerved, finely appressed-puberulent; **pappi** readily falling, of 10, distinct, white, plumose bristles in 1 series (adaxial usually shorter than abaxial). $x = 7$.

Species 1: sw United States, nw Mexico.

Molecular phylogenetic studies by J. Lee et al. (2003) indicated that *Malacothrix* is polyphyletic and that *Anisocoma* and *Calycoseris* are nested within the subclade of *Malacothrix* that contains *M. californica,* the type species of that genus. *Anisocoma* shares its base chromosome number ($x = 7$) with *Calycoseris* and the species of *Malacothrix* with which it is most closely grouped. *Anisocoma* differs from *Malacothrix* and *Calycoseris* in having paleate receptacles and pappi of plumose, rather than barbellulate bristles.

1. **Anisocoma acaulis** Torrey & A. Gray, Boston J. Nat. Hist. 5: 111, plate 13, figs. 7–11. 1845 • Scalebud F

Leaves: blades 2–7 cm, lobes oblong to triangular, abaxially ± floccose-tomentose, glabrate in age. **Heads** 3–5+ cm diam., nodding in bud. **Paleae** 7–13 mm, persistent after cypselae have fallen. **Involucres** broadly cylindric to campanulate, 15–35 mm. **Phyllaries** often with reddish tips and dots, outer broadly rounded, 2–6 mm, inner 15–35 mm. **Cypselae** 4–7 mm; **pappi** 20–27 mm. $2n = 14$.

Flowering Mar–May. Sandy washes, dry slopes, chaparral, pine-oak-juniper woodlands, montane coniferous forests; 20–2400 m; Ariz., Calif., Nev.; Mexico (Baja California).

Anisocoma acaulis grows on the southwestern fringe of the Great Basin and in the Mojave and Sonoran deserts.

64. **MALACOTHRIX** de Candolle in A. P. de Candolle and A. L. P. P. de Candolle, Prodr. 7: 192. 1838 • Desertdandelion [Greek *malakos,* soft, and *thrix,* hair]

W. S. Davis

Annuals or perennials, 2–70(–200) cm; taprooted (rhizomatous or taproots becoming caudices in *M. saxatilis*). **Stems** 1–15 (usually from basal rosettes), usually erect, sometimes ± prostrate, usually branched (scapiform in *M. californica*), usually glabrous (sometimes piloso-hirsute, stipitate-glandular, or tomentose to arachnose or puberulent, at least proximally or in leaf axils). **Leaves** usually basal and cauline; sessile; blades mostly oblong or lanceolate to obovate, oblanceolate, or spatulate (often pinnately lobed), ultimate margins entire or ± dentate (faces usually glabrous, sometimes piloso-hirsute or tomentose to arachnose or puberulent). **Heads** usually in corymbiform to paniculiform arrays (borne singly in *M. californica*). **Peduncles** not inflated distally, usually bracteate. **Calyculi** 0 (outer phyllaries intergrading with inner) or of 3–30+, ± deltate or lanceolate to linear or subulate, subequal to unequal bractlets (in 1–2 series distinct from phyllaries, margins usually hyaline, faces usually glabrous, sometimes arachnose, rarely stipitate-glandular). **Involucres** usually broadly to narrowly campanulate, sometimes hemispheric, (5–22+ ×) 2–22+ mm diam. **Phyllaries** either (without calyculi) 25–80+ in 4–6+ series and orbiculate to oblong, lance-oblong, lanceolate, or linear, unequal, or (with calyculi) 12–25+ in 2–3 series and oblong or lanceolate to linear, subequal; margins ± hyaline, 0.05–2.5 mm, apices obtuse to acute or acuminate. **Receptacles** flat to ± convex, pitted or smooth, sometimes bristly, epaleate. **Florets** 15–270; corollas yellow or white (sometimes reddish or lavender abaxially; outer ligules exserted 1–15 mm). **Cypselae** (monomorphic) stramineous to brown or purplish brown, ± prismatic or cylindro-fusiform, not beaked, ribs 15 (often 5 more prominent than others, apices of ribs sometimes projecting and forming coronas subtending pappi), usually glabrous (sometimes minutely hirtellous or muriculate); **pappi** 0, or (single or double) persistent, whitish, crenate crowns or rings of (1–)8–25+ teeth (mostly 0.05–0.1 mm) plus 0–6, coarse, smooth bristles (setiform scales), all in ± 1 series, subtending (i.e., exterior to) the readily falling, inner (or single and only) pappi of 15–35, basally coherent, white, fine, smooth to barbellulate or (proximally) ± plumose bristles in 1 series (falling all together or in groups). $x = 7, 9$.

Species 20 (18 in the flora): w United States, nw Mexico; introduced in South America.

SELECTED REFERENCES Davis, W. S. 1997. The systematics of annual species of *Malacothrix* (Asteraceae: Lactuceae) endemic to the California Islands. Madroño 44: 223–244. Williams, E. W. 1957. The genus *Malacothrix* (Compositae). Amer. Midl. Naturalist 58: 494–512.

1. Perennials (sometimes flowering first year).
    2. Corollas medium yellow; persistent pappi 0 . . . . 8. *Malacothrix incana*
    2. Corollas white (usually each with abaxial purple stripe); persistent pappi of fimbriate
       crowns or 20–25, blunt teeth (0.01–0.05 mm) plus 0 bristles . . . . 12. *Malacothrix saxatilis*
1. Annuals.
    3. Involucres hemispheric; phyllaries orbiculate (outer) to oblong or linear, hyaline
       margins 1–2.5 mm wide . . . . 3. *Malacothrix coulteri*
    3. Involucres usually ± campanulate (to hemispheric in *M. sonchoides*); bractlets of caly-
       culi and/or phyllaries ovate to lanceolate, linear, or subulate, hyaline margins 0.05–
       0.3(–1) mm wide.
        4. Stems seldom branched (heads usually borne singly on scapiform peduncles); cypsela
           ribs: 5 more prominent than others. . . . . 1. *Malacothrix californica*
        4. Stems branched (heads usually in corymbiform to paniculiform arrays); cypsela
           ribs ± equal, or 5 more prominent than others.
            5. Cauline leaves not or seldom notably reduced distally (Channel Islands,
               California).
                6. Persistent pappi usually 0, rarely of 1–2 bristles; pollen 70–100% 3-porate.
                    7. Stems erect or ± prostrate (forming mats); proximal cauline leaves
                       usually fleshy (lobes nearly equal, apices obtuse); San Miguel, Santa
                       Cruz, and Santa Rosa islands . . . . 9. *Malacothrix indecora*
                    7. Stems erect (not forming mats); proximal cauline leaves usually not
                       fleshy (lobes usually unequal, apices acute or obtuse); Anacapa, San
                       Clemente, San Nicolas, and Santa Barbara islands. . . . . 6. *Malacothrix foliosa*
                6. Persistent pappi of 15–20 teeth (0.01–0.1 mm) plus 1–2 bristles; pollen
                   70–100% 4-porate.
                    8. Calyculi 0; phyllaries 31–49 in 5–6+ series, hyaline margins 0.6–1
                       mm wide; Anacapa and Santa Cruz islands, California . . . . . . 16. *Malacothrix squalida*
                    8. Calyculi of 7–11 bractlets; phyllaries 9–13 in 2–3 series, hyaline
                       margins 0.1–0.2 mm wide; Anacapa Island, California . . . . 10. *Malacothrix junakii*
            5. Cauline leaves usually reduced distally (*M. clevelandii* and *M. similis* are known
               from Channel Islands, California).

[9. Shifted to left margin.—Ed.]

9. Corollas 4–10 mm; outer ligules exserted 1–4 mm.
    10. Proximal cauline leaves (pinnately lobed, lobes 3–5 pairs, ± equal, bases white-hairy) ± fleshy; persistent pappi of crenate crowns plus 0 bristles; pollen 70–100% 3-porate . . . . 11. *Malacothrix phaeocarpa*
    10. Proximal cauline leaves (pinnately lobed or not) not fleshy; persistent pappi of 8–24+ teeth plus 1–2 bristles; pollen 3- or 4-porate.
        11. Proximal cauline leaves oblanceolate to lance-linear; corollas pale yellow; cypselae ± cylindro-fusiform or prismatic, 1.2–1.8 mm (ribs extending to apices, 5 more prominent than others); persistent pappi of 14–24+ needlelike teeth plus 1 bristle (mainly California).
            12. Stems branched mostly distally, glabrous; cauline leaves: margins usually dentate; cypselae (stramineous to brown) 1.2–1.8 mm; pollen 3-porate (mean diam. 25 μm) . . . . 2. *Malacothrix clevelandii*
            12. Stems branched mostly proximally; cauline leaves: margins usually entire; cypselae 1.4–1.7 mm (purplish brown); pollen 4-porate (mean pollen diam. 30 μm) . . . . 13. *Malacothrix similis*
        11. Proximal cauline leaves obovate to narrowly oblanceolate; corollas white or pale yellow; cypselae ± cylindro-fusiform, 1.7–2.3 mm (ribs not extending to apices, ± equal); persistent pappi of 8–18 teeth plus 1–2 bristles (Arizona, California, Nevada).
            13. Corollas white or pale yellow; cypselae 1.7–2 mm (bases slightly expanded, distal 0.3 mm smooth); pollen 70–100% 3-porate . . . . 15. *Malacothrix sonorae*
            13. Corollas usually yellow, sometimes white; cypselae 1.7–2.3 mm (bases not expanded, distal 0.1–0.2 mm smooth); pollen 70–100% 4-porate . . . 17. *Malacothrix stebbinsii*
9. Corollas (7–)10–23+ mm; outer ligules exserted 5–15+ mm.
    14. Proximal cauline leaves pinnately lobed (lobes 2–6+ pairs, filiform or triangular to oblong, subequal to unequal, apices acute), not fleshy, ultimate margins dentate or entire; cypsela ribs ± equal.
        15. Proximal cauline leaves usually pinnately lobed (lobes filiform); receptacles bristly; cypselae ± cylindro-fusiform (sometimes weakly 5-angled, ribs extending to apices) . . . . 7. *Malacothrix glabrata*
        15. Proximal cauline leaves sometimes pinnately lobed (lobes relatively broad, triangular to deltate); receptacles not bristly; cypselae ± cylindro-fusiform (ribs not extending to apices, distal 0.3 mm of cypselae smooth) . . . . 4. *Malacothrix fendleri*
    14. Proximal cauline leaves pinnately lobed (lobes 3–8 pairs, ± oblong to triangular, ± equal, apices obtuse or acute), ± fleshy, ultimate margins usually dentate; cypsela ribs ± equal or 5 more prominent than others.
        16. Corollas white or yellow (usually with lavender abaxially); cypselae 1.2–2 mm; persistent pappi 0 . . . . 5. *Malacothrix floccifera*
        16. Corollas lemon or medium yellow; cypselae 1.8–4 mm; persistent pappi of 12–25+ irregular, blunt teeth plus 0–6 bristles.
            17. Distal cauline leaves narrowly triangular to linear (bases usually ± dilated, ± clasping); persistent pappi of crenate crowns or 12–25+ teeth plus 0 bristles . . . . 14. *Malacothrix sonchoides*
            17. Distal cauline leaves ± elliptic to linear (bases narrowly cuneate); persistent pappi of 12–15 teeth plus 0–6 bristles . . . . 18. *Malacothrix torreyi*

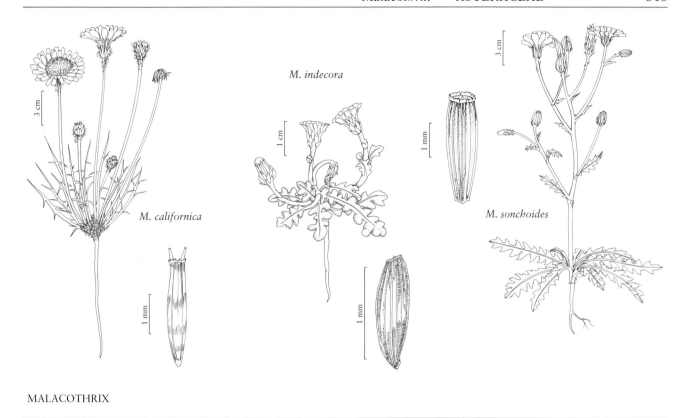

M. indecora

M. californica

M. sonchoides

MALACOTHRIX

1. Malacothrix californica

1. Malacothrix californica de Candolle in A. P. de Candolle and A. L. P. P. de Candolle, Prodr. 7: 192. 1838 • California desertdandelion F

Annuals, 4–45 cm. **Stems** 1 (scapiform), erect or arcuate-ascending, mostly glabrous (shaggily piloso-hirsute proximally and at bases of heads). **Cauline leaves:** proximal oblanceolate to linear, pinnately lobed (lobes usually linear to filiform, sometimes broader), not fleshy, ultimate margins entire or dentate, faces usually shaggily piloso-hirsute (at least proximally), glabrescent; distal usually 0. **Calyculi** of 12–20, lanceolate to linear bractlets, hyaline margins 0.05–0.2 mm, abaxial faces shaggily piloso-hirsute to arachnose. **Involucres** (8–)10–15 × 5–6 mm. **Phyllaries** usually 20–26+ in 2-3+ series, (midstripes often reddish) lanceolate to lance-linear or subulate, unequal, hyaline margins 0.1–0.5 mm wide, abaxial faces (of outermost, at least) shaggily piloso-hirsute to arachnose (at least proximally). **Receptacles** sparsely bristly or glabrous. **Florets** 40–250; corollas usually yellow to pale yellow (often with abaxial reddish stripes), sometimes white, 16–20 mm; outer ligules exserted 11–13 mm. **Cypselae** ± prismatic, 2–3.4 mm, ribs extending to apices, 5 more prominent than others; **persistent pappi** of 12–15+, irregular, lance-deltate teeth plus (1–)2 bristles. **Pollen** 70–100% 3-porate. *2n* = 14.

Flowering Apr–Jun. Open sandy soil in grasslands, oak woodlands, chaparral, or desert margins; 0–1700 m; Calif.; Mexico (Baja California).

Malacothrix californica grows in the San Joaquin Valley, central western California, southwestern California, and the Mojave Desert.

2. Malacothrix clevelandii

2. Malacothrix clevelandii A. Gray in W. H. Brewer et al., Bot. California 1: 433. 1876 (as clevelandi) • Cleveland's desertdandelion

Crepis geisseana Philippi; *Malacothrix senecioides* Reiche

Annuals, 4–36 cm. **Stems** 1–5, erect or ascending, branched mostly distally, glabrous. **Cauline leaves:** proximal oblanceolate to lance-linear, sometimes pinnately lobed, not fleshy, ultimate margins usually dentate, faces glabrous; distal reduced (margins 2–4-dentate near bases or entire). **Calyculi** of 5–12, lance-deltate to lanceolate bractlets, hyaline margins 0.05–0.2 mm. **Involucres** narrowly campanulate, 4–8+ × 2–4+ mm. **Phyllaries** 8–15+ in 2+ series, lance-linear to linear, ± equal, hyaline margins 0.05–0.3 mm wide, abaxial faces glabrous. **Receptacles** not bristly. **Florets** 19–67); corollas pale yellow, 4–7.4 mm; outer ligules exserted 1–3 mm. **Cypselae** fusiform or prismatic, 1.2–1.8 mm, ribs extending to apices, 5 more prominent than others; **persistent pappi** of 15–24+, needlelike teeth plus 1 bristle. Pollen 70–100% 3-porate, mean 25 μm. $2n$ = 14.

Flowering Mar–Jun. Cleared areas (burns, slides), usually chaparral, rarely margins of creosote bush shrub; 20–1500 m; Calif.; Mexico (Baja California); introduced, South America (Argentina, Chile).

Malacothrix clevelandii grows in northwestern California, Sierra Nevada foothills, San Joaquin Valley, central western California, and northern Channel Islands (Santa Rosa Island).

3. **Malacothrix coulteri** Harvey & A. Gray, Mem. Amer. Acad. Arts Sci., n. s. 4: 113. 1849 • Snake's head desertdandelion, snake's head E

Malacolepis coulteri (Harvey & A. Gray) A. Heller; *Malacothrix coulteri* var. *cognata* Jepson; *Zollikoferia elquiensis* Philippi

Annuals, 10–60 cm. **Stems** 1–6, ascending or erect, simple or branched proximally and distally, glaucous or glabrous. **Cauline leaves:** proximal linear to obovate, sometimes pinnately lobed, not fleshy, ultimate margins entire or dentate, faces glabrous; distal reduced (ovate to lanceolate, rarely pinnately lobed, bases clasping). **Calyculi** 0. **Involucres** hemispheric, 10–22+ × 6–22+ mm. **Phyllaries** (25–)40–60+ in 4–6+ series, (midstripes usually reddish or purple) orbiculate to ovate, oblong, lance-oblong, or linear, unequal, hyaline margins 1–2.5 mm wide, faces glabrous. **Receptacles** densely bristly. **Florets** 85–257; corollas usually pale yellow, sometimes white, 8–12 mm; outer ligules exserted 2–5 mm. **Cypselae** ± prismatic, 1.6–3.2 mm, ribs extending beyond apices, 5 more prominent than others; **persistent pappi** of 20–25+, blunt teeth plus 2–6 bristles. Pollen 70–100% 3-porate. $2n$ = 14.

Flowering Mar–May. Sandy, open areas in coastal sage, grasslands, deserts; 100–1800 m; Ariz., Calif., Nev., Utah; introduced, South America (Argentina, Chile).

Variety *cognata*, indistinguishable in floral characters from var. *coulteri* and differing in cauline leaves parted almost to midribs, occurs at some mainland sites in southern California and has been collected on the Channel Islands. In California, *Malacothrix coulteri* grows in the San Joaquin Valley, central western areas, western Transverse Ranges, Peninsular Ranges, and Mojave Desert. In Arizona, it grows in the Sonoran Desert in the Santa Rita and Tucson mountains, and similar places.

4. **Malacothrix fendleri** A. Gray, Smithsonian Contr. Knowl. 5(6): 104. 1853 • Fendler's desertdandelion

Annuals, 3–15(–25+) cm. **Stems** (1–)3–8, ± decumbent or spreading-ascending, branched proximally and distally, glaucous or glabrous. **Cauline leaves:** proximal elliptic to oblong-oblanceolate, sometimes pinnately lobed (lobes 2–4+ pairs, oblong to triangular, unequal, apices acute), not fleshy, ultimate margins usually dentate, faces glabrous; distal reduced (narrowly triangular to linear or filiform, margins dentate or entire). **Calyculi** of 5–12, ovate to lanceolate bractlets, hyaline margins 0.05–0.2 mm wide. **Involucres** ± campanulate, 7–10 × 5–6+ mm. **Phyllaries** 13–25+ in 2–3 series, lance-oblong or lanceolate to lance-linear, subequal, hyaline margins 0.05–0.3 mm wide, faces glabrous. **Receptacles** not bristly. **Florets** 16–88; corollas yellow (usually with red or purplish abaxial stripes), 6–14 mm; outer ligules exserted 5–8 mm. **Cypselae** ± cylindric, 1.8–2.4 mm (distal 0.3 mm slightly expanded, cupped, smooth), ribs not extending to apices, ± equal; **persistent pappi** of 12–15, ± deltate teeth (often hidden within cups at apices of cypselae) plus 1–2 bristles. Pollen 70–100% 3-porate. $2n$ = 14.

Flowering Mar–Jun. Grasslands, pinyon-juniper woodlands, creosote bush associations; 80–2200 m; Ariz., N.Mex., Tex.; Mexico (Baja California, Sonora).

Malacothrix fendleri grows in the Sonoran Desert. "San Bernardino Co." as locality for a specimen from the herbarium of J. G. Lemmon in UC (336493) is evidently an error.

5. **Malacothrix floccifera** (de Candolle) S. F. Blake, Contr. U.S. Natl. Herb. 22: 656. 1924 • Woolly desertdandelion E

Senecio flocciferus de Candolle in A. P. de Candolle and A. L. P. P. de Candolle, Prodr. 6: 426. 1838; *Malacothrix obtusa* Bentham; *M. parviflora* Bentham

Annuals, 10–40 cm. **Stems** 1–8, simple or branched proximally and/or distally, glabrous or proximally puberulent. **Cauline leaves:** proximal oblanceolate to obovate, usually pinnately lobed (lobes 4–6+ pairs, ± equal, apices obtuse), ± fleshy, ultimate margins dentate, abaxial faces usually white-arachnose (usually in patches on lobes); distal reduced (pinnately lobed or dentate proximally, lobes obtuse). **Calyculi** of 3–8+, ovate to lanceolate bractlets, hyaline margins 0.1–0.3 mm wide. **Involucres** campanulate, 5–7(–9) × 2.5–4(–5) mm. **Phyllaries** 13–21+ in 2–3 series, oblong or lanceolate to linear, hyaline margins 0.05–0.2

mm wide, faces glabrous. **Receptacles** bristly. **Florets** 21–60; corollas white or yellow (usually with abaxial lavender stripes), 7–15 mm; outer ligules exserted 5–9 mm. **Cypselae** ± cylindric to prismatic, 1.2–2 mm, ribs extending to apices, 5 more prominent than others; **persistent pappi** 0. **Pollen** 70–100% 3-porate. $2n = 14$.

Flowering Mar–Nov. Burns, slides, road cuts, open areas, usually in loose soil (serpentine, gypsum, or brown-clay) in chaparral, pinyon/juniper woodlands, yellow-pine forests; 60–2000 m; Calif., Nev.

Malacothrix floccifera grows in the Transverse Ranges in Ventura County, in the Coast Ranges to Siskiyou County, on foothills and slopes of Sierra Nevada from Lassen County to Fresno County, and in western Nevada, near Lake Tahoe.

6. **Malacothrix foliosa** A. Gray in A. Gray et al., Syn. Fl. N. Amer. ed. 2, 1: 455. 1886 • Leafy desertdandelion E

Annuals, 4–45 cm. **Stems** 1, erect, simple or branched proximally and/or distally or 2–10, decumbent to ascending, and ± branched distally, glabrous or sparsely arachnose (sometimes only in leaf axils). **Cauline leaves:** proximal oblanceolate to narrowly obovate, usually pinnately lobed (lobes usually unequal, apices acute or obtuse), sometimes ± fleshy, ultimate margins entire or dentate, faces glabrous (proximalmost cauline usually more deeply divided); distal seldom notably reduced (sometimes pinnately lobed). **Calyculi** 0 (i.e., outer phyllaries intergrading with inner), or of 5–12+, oblong or ovate to lanceolate or linear bractlets, hyaline margins 0.1–0.3 mm wide. **Involucres** ± campanulate, 5–12 × 2–7 mm. **Phyllaries** 12–22(–40+, without calyculi) in 2–3(–5+) series, (usually red-tinged) lanceolate to linear, hyaline margins 0.05–0.2 mm wide, faces glabrous. **Receptacles** not bristly. **Florets** 10–123; corollas light to medium yellow, 5–17 mm; outer ligules exserted 1–10 mm. **Cypselae** ± cylindro-fusiform or ± prismatic, 0.9–1.7 mm, ribs extending to apices, ± equal or 5 more prominent than others; **persistent pappi** usually 0, rarely of 1–2 bristles. **Pollen** 70–100% 3-porate. $2n = 14$.

Subspecies 4 (4 in the flora): Channel Islands, California.

1. Outer ligules usually exserted 1–4 mm.
    2. Calyculi 0 or bractlets intergrading with phyllaries; cypselae ± prismatic (5-angled), 1.3–1.6 mm, 5 ribs more prominent than others; Anacapa Island, California .... 6a. *Malacothrix foliosa* subsp. *crispifolia*
    2. Calyculi of 8–12+, oblong to linear bractlets; cypselae ± cylindro-fusiform, 0.9–1.5 mm, ribs ± equal; San Nicolas Island, California .... 6b. *Malacothrix foliosa* subsp. *polycephala*
1. Outer ligules exserted 5–15 mm.
    3. Stems 1, usually erect; distal cauline leaves usually pinnately lobed (near bases, lobes 1–2 pairs, narrow); cypselae ± cylindro-fusiform, 0.9–1.5 mm, ribs ± equal; San Clemente Island, California .... 6c. *Malacothrix foliosa* subsp. *foliosa*
    3. Stems (1–)3–5(–10), erect or decumbent to ascending; distal cauline leaves usually pinnately lobed (from bases to near apices); cypselae ± prismatic (5-angled) 1.3–1.7 mm, 5 ribs more prominent than others; Santa Barbara Island, California .... 6d. *Malacothrix foliosa* subsp. *philbrickii*

6a. **Malacothrix foliosa** A. Gray subsp. **crispifolia** W. S. Davis, Madroño 44: 234. 1997 • Petite desertdandelion C E

Stems 1–6, ascending, branched proximally and distally. **Distal cauline leaves** oblanceolate to obovate, pinnately lobed (lobes 2–5 pairs, narrow, ultimate margins crisped, apices acute). **Calyculi** 0 (or bractlets intergrading with phyllaries). **Involucres** 7–9.2 × 3.4–5.4 mm. **Phyllaries** 22–40 in 4–5 series. **Corollas** medium yellow, 6–10 mm; outer ligules exserted 2.5–4 mm. **Cypselae** ± prismatic (5-angled), 1.3–1.6 mm, 5 ribs more prominent than others. $2n = 14$.

Flowering Mar–Jul. Open sandy areas, dunes, between shrubs in chaparral; of conservation concern; 0–100 m; Calif.

Subspecies *crispifolia* is known only from Anacapa Island.

6b. **Malacothrix foliosa** A. Gray subsp. **polycephala** W. S. Davis, Madroño 44: 237. 1997 • Many-head desertdandelion E

Stems usually 1, erect, branched distally. **Distal cauline leaves** narrowly triangular, usually pinnately lobed (near bases, lobes 1–2 pairs, narrow). **Calyculi** of 8–12+, oblong to linear bractlets. **Involucres** 5–7 × 2–4.5 mm. **Phyllaries** 12–16+ in 2–3 series. **Corollas** medium yellow, 5–9 mm; outer ligules exserted 1.5–3.5 mm. **Cypselae** ± cylindro-fusiform, 0.9–1.5 mm, ribs ± equal. $2n = 14$.

Flowering Mar–Jul. Open sandy areas, dunes, between shrubs in chaparral; 0–100 m; Calif.

Subspecies *polycephala* is known only from San Nicolas Island.

6c. Malacothrix foliosa A. Gray subsp. **foliosa** E

Stems usually 1, usually erect, branched distally. **Distal cauline leaves** ovate to lanceolate, pinnately lobed (at bases, lobes 1–2 pairs, narrow). **Calyculi** of 8–12+, lance-deltate to lanceolate bractlets. **Involucres** 7–11 × 3–8 mm. **Phyllaries** 18–22+ in 3 series. **Corollas** light yellow, 10–17 mm; outer ligules exserted 6–10 mm. **Cypselae** ± cylindro-fusiform, 0.9–1.5 mm, ribs ± equal. $2n = 14$.

Flowering Mar–Jul. Open sandy areas, dunes, between shrubs in chaparral; 10–100 m; Calif.

Subspecies *foliosa* is known only from San Clemente Island.

6d. Malacothrix foliosa A. Gray subsp. **philbrickii** W. S. Davis, Madroño 44: 236. 1997 • Philbrick's desertdandelion C E

Stems (1–)3–5(–10), erect (when single) or decumbent to ascending, branched proximally and/or distally. **Distal cauline leaves** oblanceolate to lanceolate usually pinnately lobed (from bases to near apices). **Calyculi** of 5–12+, oblong to linear bractlets (sometimes ± intergrading with phyllaries). **Involucres** 6–9(–11) × 2.5–5(–7) mm. **Phyllaries** 12–21+ in 2–3 series. **Corollas** medium yellow, 8–16 mm; outer ligules exserted 4–8 mm. **Cypselae** ± prismatic (5-angled), 1.3–1.7 mm, 5 ribs more prominent than others. $2n = 14$.

Flowering Mar–Jul. Open sandy areas, dunes, or between shrubs in island chaparral; of conservation concern; 0–100 m; Calif.

Subspecies *philbrickii* is known only from Santa Barbara Island.

7. Malacothrix glabrata (A. Gray ex D. C. Eaton) A. Gray in A. Gray et al., Syn. Fl. N. Amer. 1(2): 422. 1884 • Smooth desertdandelion

Malacothrix californica de Candolle var. *glabrata* A. Gray ex D. C. Eaton in S. Watson, Botany (Fortieth Parallel), 201. 1871

Annuals, (5–)10–40+ cm. **Stems** (1–)3–5+, ascending to erect, usually branched proximally and distally, glabrous or sparsely arachno-puberulent near bases (sometimes glaucous). **Cauline leaves:** proximal usually pinnately lobed (lobes 3–6+ pairs, usually filiform, subequal to unequal, apices acute), ultimate margins entire, faces glabrous or ± hairy (then usually glabrescent); distal reduced (usually pinnately lobed). **Calyculi** of 12–20+, lanceolate to linear bractlets, hyaline margins 0.05–0.2 mm wide (abaxial faces often ± densely white-hairy). **Involucres** campanulate to hemispheric, 9–17 × 4–7 mm. **Phyllaries** 20–25+ in 2–3 series, lance-linear to linear, hyaline margins 0.05–0.3 mm wide, faces usually glabrous, abaxial sometimes ± white-hairy. **Receptacles** bristly. **Florets** 31–139; corollas usually pale yellow, sometimes white, 15–23+ mm; outer ligules exserted 9–15+ mm. **Cypselae** ± cylindro-fusiform (sometimes weakly 5-angled), 2–3.3 mm, ribs extending to apices, usually ± equal; persistent pappi of 0–12+, blunt to acute teeth plus 1–2(–5) bristles. **Pollen** 70–100% 3-porate. $2n = 14$.

Flowering Mar–Jul. Coarse soils in open areas, or among shrubs, creosote bush scrublands, *Amsinckia*, *Artemisia*, and *Atriplex-Larrea* associations, Joshua tree woodlands; 0–1800 m; Ariz., Calif., Idaho, Nev., N.Mex., Oreg., Utah; Mexico (Baja California, Sonora).

Malacothrix glabrata grows in the Mojave, Great Basin, and Sonoran deserts in California and the Intermountain region in Arizona, Nevada, Oregon, and Utah.

8. Malacothrix incana (Nuttall) Torrey & A. Gray, Fl. N. Amer. 2: 486. 1843 • Dunedelion, dune malacothrix E

Malacomeris incanus Nuttall, Trans. Amer. Philos. Soc., n. s. 7: 435. 1841; *Malacothrix incana* var. *succulenta* (Elmer) E. W. Williams; *M. succulenta* Elmer

Perennials, 12–70 cm (often mounded). **Stems** 1, branched proximally and distally, usually tomentose, sometimes glabrous. **Cauline leaves:** proximal obovate to narrowly spatulate, sometimes pinnately lobed (lobes 1–2+ pairs, subequal, apices obtuse), sometimes ± fleshy, margins usually obtuse-lobed, sometimes entire; distal not notably reduced (similar to others). **Calyculi** of 5–16+, ovate to lanceolate bractlets, hyaline margins 0.05–0.2 mm. **Involucres** 10–14 × 4–8+ mm. **Phyllaries** 16–30 in 2–3 series, (red-tinged) lanceolate or oblong to linear, hyaline margins 0.05–0.1 mm wide, faces glabrous. **Receptacles** not bristly. **Florets** 47–99; corollas medium yellow, 11–20 mm; outer ligules exserted 5–10 mm. **Cypselae** usually cylindro-fusiform, sometimes weakly prismatic, 1.5–2.2 mm, ribs extending to apices, ± equal; persistent pappi 0. **Pollen** 70–100% 3-porate. $2n = 14$.

Flowering Jan–Dec. Coastal dunes; 0–10(–100) m; Calif.

Malacothrix incana, a dune endemic, grows currently on the coastal mainland in Santa Barbara and San Luis Obispo counties, and on San Miguel, San Nicolas, Santa Cruz, and Santa Rosa islands. Its nomenclatural type specimen was collected at San Diego, probably from the Silver Strand dune areas on Coronado Island.

Populations of a glabrous form, var. *succulenta*, occur in Santa Barbara County (e.g., Casmalia Beach) and San Luis Obispo County (e.g., west of Oso Flaco Lake). Glabrous forms and tomentose forms grow together on San Miguel and San Nicolas islands.

Extensive hybridization between *Malacothrix incana* and *M. foliosa* subsp. *polycephala* occurs on San Nicolas Island where dunes have extended into areas of normal soil, particularly along the western and southwestern portions of the island.

Hybridization between *Malacothrix incana* and *M. saxatilis* var. *implicata* occurs on San Miguel Island on east-facing slopes above Cuyler Harbor.

9. **Malacothrix indecora** Greene, Bull. Calif. Acad. Sci. 2: 152. 1886 • Santa Cruz Island desertdandelion C E F

Malacothrix foliosa A. Gray var. *indecora* (Greene) E. W. Williams

Annuals, 2–12(–45) cm. **Stems** usually 1, erect or ± prostrate, usually ± branched proximally (often forming mats, usually leafy throughout), glabrous. **Cauline leaves:** proximal obovate to spatulate, usually pinnately lobed (lobes 1–3+ pairs, nearly equal, apices obtuse), usually fleshy, ultimate margins entire, faces glabrous; distal not reduced (similar to others). **Calyculi** of 8–13+, lance-deltate to lanceolate bractlets (sometimes intergrading with phyllaries), hyaline margins 0.05–0.2 mm wide. **Involucres** ± campanulate, 6–8 × 2–6+ mm. **Phyllaries** 13–22+ in 2–3 series, (often red-tinged) oblong to lanceolate, hyaline margins 0.05–0.2 mm wide, faces glabrous. **Receptacles** glabrous. **Florets** 18–76; corollas medium yellow, 4–8 mm; outer ligules exserted 1–4 mm. **Cypselae** ± prismatic, 1.2–1.5 mm, ribs extending to apices, 5 more prominent than others; **persistent pappi** 0. **Pollen** 70–100% 3-porate. *2n* = 14.

Flowering Apr–Jul. Shallow soils of ocean bluffs, open rocky areas; of conservation concern; 0–30 m; Calif.

Malacothrix indecora is found on San Miguel, Santa Cruz, and Santa Rosa islands.

10. **Malacothrix junakii** W. S. Davis, Madroño 44: 241. 1997 • Junak's desertdandelion, Anacapa Island desertdandelion C E

Annuals, 5–30 cm. **Stems** 1–4, ascending to erect, ± branched proximally and distally (leafy to apices), glabrous. **Cauline leaves:** proximal oblanceolate, pinnately lobed (lobes 2–4 pairs, obtuse), ultimate margins entire or dentate; distal not much reduced (± linear, lobes usually 1–2 near bases, apices acute). **Calyculi** of 7–11, ovate to lanceolate bractlets, hyaline margins 0.05–0.2 mm. **Involucres** 7–8.5 × 3–6 mm. **Phyllaries** 9–13+ in 2–3 series, (red-tinged) oblong to lanceolate or lance-linear, hyaline margins 0.1–0.2 mm wide, faces glabrous. **Receptacles** not bristly. **Florets** 20–85: corollas medium yellow, 7–11 mm; outer ligules exserted 3.5–5.5 mm. **Cypselae** ± prismatic, 1.6–2 mm, ribs extending to apices, ± equal or 5 more prominent than others; **persistent pappi** of 11–14, irregular, ± dentate teeth plus 1–2 bristles. **Pollen** 70–100% 4-porate. *2n* = 28.

Flowering May–Jun. Flats and adjacent coastal bluffs; of conservation concern; 0–30 m; Calif.

Malacothrix junakii is known only from middle Anacapa Island.

11. **Malacothrix phaeocarpa** W. S. Davis, Madroño 40: 101, fig. 1. 1993 • Davis's desertdandelion E

Annuals, 5–44 cm. **Stems** 1–4, erect to ascending, usually branched proximally and distally, glabrous (usually glaucous). **Cauline leaves:** proximal obovate, usually pinnately lobed (lobes 3–8 pairs, ± equal, bases white-hairy), ± fleshy, ultimate margins dentate, faces glabrous; distal reduced (sometimes 2–4-dentate near bases). **Calyculi** of 8–12+, ovate to lanceolate bractlets, hyaline margins 0.05–0.3 mm wide. **Involucres** ± campanulate, 5–8 × 2–6 mm. **Phyllaries** 18–22+ in 2–3 series, oblong or lanceolate to linear, hyaline margins 0.05–0.2 mm wide, faces glabrous. **Receptacles** not bristly. **Florets** 30–65; corollas white, 5–8 mm; outer ligules exserted 1–3 mm. **Cypselae** ± cylindro-funnelform or weakly prismatic, 1.2–2 mm, ribs extending to apices, ± equal or 5 more prominent than others; **pappi** 0 or crenate crowns (0.01–0.1 mm). **Pollen** 70–100% 3-porate. *2n* = 14.

Flowering Apr–Jun. Diatomaceous shale, open chaparral burns, slides, openings in Bishop pine–Douglas fir woodlands; 100–1400 m; Calif.

Malacothrix phaeocarpa grows in Monterey County (near Jolon above Boucher's Gap), San Luis Obispo

County (Santa Lucia Mountains), Santa Barbara County (Purissima Hills and Santa Inez Mountains), and Santa Clara County (Santa Cruz Mountains).

12. Malacothrix saxatilis (Nuttall) Torrey & A. Gray, Fl. N. Amer. 2: 486. 1843 · Cliff desertdandelion E

Leucoseris saxatilis Nuttall, Trans. Amer. Philos. Soc., n. s. 7: 440. 1841

Perennials (sometimes flowering first year), (20–)90–200 cm. **Stems** 1 (from rhizomes) or 2–3+ (from caudices), usually erect (sometimes relatively thick), branched mostly distally (usually relatively leafy proximally, sometimes sparsely leafy distally), glabrous or sparsely arachnose to tomentose. **Cauline leaves:** proximal (somewhat thick, usually withering early) elliptic, oblanceolate, lanceolate, or linear (sometimes 1–2-pinnately lobed, lobes lanceolate or linear to filiform, sometimes antrorse, bases usually tapering), ultimate margins entire or dentate to denticulate, faces glabrous or ± arachnose to tomentose; distal reduced. **Calyculi** of 12–18(–30), lanceolate to subulate bractlets, hyaline margins 0.05–0.2 mm. **Involucres** ± campanulate, 9–12(–16+) × 6–8(–12+) mm. **Phyllaries** 18–30+ in 2–3 series, lanceolate or linear to subulate, hyaline margins 0.05–0.2 mm wide, faces glabrous or ± arachnose and glabrescent. **Receptacles** not bristly. **Florets** 41–100; corollas white (usually each with abaxial purple stripe), 13–20 mm; outer ligules exserted 8–14 mm. **Cypselae** ± prismatic, 1.4–2.5 mm, ribs extending to apices (± muriculate at 30×), 5 more prominent than others; **pappi** persistent, of fimbriate crowns or 20–25, blunt teeth (0.01–0.1 mm). **Pollen** 70–100% 3-porate. $2n = 18$.

Varieties 5 (5 in the flora): California.

The *Malacothrix saxatilis* complex is taxonomically difficult. The varieties intergrade morphologically in leaf shape, indument, blooming time, and growth cycle, depending on relative habitat conditions such as elevation, soil, and temperature regime. Hybridization may occur between varieties where they are sympatric. In addition, arrays of heads in early bloom may be described as congested (peduncles relatively short) and as open (peduncles relatively long) on the same plant at the end of the season.

Varieties *commutata*, *arachnoidea*, and *saxatilis* are similar in leaf morphology; var. *arachnoidea* is distinguished from var. *commutata* mainly by its tomentose herbage; var. *saxatilis* differs from the other two varieties in minor details of leaf morphology and in its area of distribution; var. *implicata* is the most distinctive variety; and var. *tenuifolia* differs from the others primarily in cauline leaf morphology, which is quite variable. Variety *altissima* has been described as similar to vars. *arachnoidea* and *commutata* and different from vars. *saxatilis*, *implicata*, and *tenuifolia* in growth habit. It is similar in leaf morphology to var. *tenuifolia* but differs from vars. *arachnoidea*, *commutata*, *implicata*, and *saxatilis*. Here, var. *altissima* is subsumed in var. *tenuifolia*.

1. Cauline leaves usually 1- or 2-pinnately lobed.
    2. Stems relatively densely leafy distally; cauline leaves mostly (1–)2-pinnately lobed (lobes linear to filiform); California Channel Islands .... 12a. *Malacothrix saxatilis* var. *implicata*
    2. Stems relatively sparsely leafy distally; cauline leaves 1-pinnately lobed or (distalmost) not lobed (linear to filiform, entire) .... 12b. *Malacothrix saxatilis* var. *tenuifolia*
1. Cauline leaves usually not lobed.
    3. Distal cauline leaves usually ovate to linear, apices usually obtuse; coastal bluffs, south slopes of Santa Ynez Mountains, Santa Barbara County .... 12c. *Malacothrix saxatilis* var. *saxatilis*
    3. Distal cauline leaves broadly lance-linear or lanceolate to elliptic, apices usually acute; north slopes of Transverse and Coast ranges to Monterey County.
        4. Herbage tomentose .. 12d. *Malacothrix saxatilis* var. *arachnoidea*
        4. Herbage glabrous or ± arachnose .... 12e. *Malacothrix saxatilis* var. *commutata*

12a. Malacothrix saxatilis (Nuttall) Torrey & A. Gray var. implicata (Eastwood) H. M. Hall, Univ. Calif. Publ. Bot. 3: 269. 1907 E

Malacothrix implicata Eastwood, Proc. Calif. Acad. Sci., ser. 3, 1: 113. 1898

Plants rhizomatous or with woody caudices. **Herbage** glabrous or sparsely hairy (arachnose). **Stems** usually 1, erect, branched distally, proximally woody, relatively densely leafy distally. **Cauline leaves:** proximal (1–)2-pinnately lobed (lobes linear, apices obtuse or acute), ultimate margins entire or denticulate, apices usually obtuse; distal linear to filiform (sometimes pinnately lobed, lobes linear to filiform). **Arrays of heads** relatively crowded. **Cypselae** 1.8–2.2 mm. $2n = 18$.

Flowering Feb–Dec. Clay flats or canyon slopes; 50–400 m; Calif.

Variety *implicata* grows on Anacapa, San Nicolas, San Miguel, Santa Cruz, and Santa Rosa islands. Plants with 1-pinnately parted leaves are present in some populations on Santa Cruz Island.

12b. Malacothrix saxatilis (Nuttall) Torrey & A. Gray var. **tenuifolia** (Nuttall) A. Gray in A. Gray et al., Syn. Fl. N. Amer. 1(2): 423. 1884 [E]

Leucoseris tenuifolia Nuttall, Trans. Amer. Philos. Soc., n. s. 7: 440. 1841; *Malacothrix saxatilis* (Nuttall) Torrey & A. Gray var. *altissima* (Greene) Ferris

Plants with woody caudices. **Herbage** glabrous or sparsely hairy (floccose). **Stems** 1–3+, erect, branched distally, herbaceous, relatively sparsely leafy distally. **Cauline leaves:** proximal usually 1-pinnately lobed (lobes lanceolate to linear or filiform, sometimes antrorse, apices acute); distal linear to filiform (or pinnately lobed, lobes linear to filiform), ultimate margins entire, apices acute. **Arrays of heads** relatively open. **Cypselae** 1.4–2 mm. **2n** = 18.

Flowering Mar–Dec. Canyons, coastal-sage scrub, chaparral; 10–2000 m; Calif.

Variety *tenuifolia* occurs in the Transverse Ranges (Santa Monica, Santa Ynez, San Gabriel, and San Bernardino mountains), south to coastal areas in Orange and San Diego counties, and east to San Bernardino and Riverside counties, and on Santa Catalina Island with var. *saxatilis*. Single populations grow on San Clemente and San Nicolas islands on mounds of sand brought from the mainland.

12c. Malacothrix saxatilis (Nuttall) Torrey & A. Gray var. **saxatilis** [E]

Plants rhizomatous or with woody caudices. **Herbage** glabrous or sparsely hairy (arachnose). **Stems** 1, erect, branched proximally and distally, herbaceous, relatively leafy distally. **Cauline leaves:** proximal not pinnately lobed (lanceolate to linear, margins entire or irregularly dentate, apices obtuse or acute); distal usually ovate to linear, margins denticulate or entire, apices obtuse. **Arrays of heads** relatively crowded to open. **Cypselae** 1.8–2.4 mm. **2n** = 18.

Flowering Mar–Dec. On shelving rocks, road banks in chaparral or coastal sage; 0–200 m; Calif.

Variety *saxatilis* frequently grows in coastal areas of Santa Barbara County on steep bluffs overlooking the ocean, and at lower elevations near beaches, and in canyon mouths along the lower south slopes of the Santa Ynez Mountains.

12d. Malacothrix saxatilis (Nuttall) Torrey & A. Gray var. **arachnoidea** (E. A. McGregor) E. W. Williams, Amer. Midl. Naturalist 58: 509. 1957 • Carmel Valley malacothrix [C][E]

Malacothrix arachnoidea E. A. McGregor, Bull. Torrey Club 36: 605, fig. 3. 1909

Plants rhizomatous or with woody caudices. **Herbage** tomentose. **Stems** 1 (from rhizomes) or 2–3+ (from caudices), erect, freely branched, herbaceous or proximally woody, relatively leafy distally. **Cauline leaves:** proximal not pinnately lobed (lanceolate to elliptic, margins usually entire, apices acute); distal broadly lance-linear to elliptic or linear (firm, bases gradually tapering), margins usually entire, sometimes denticulate, apices usually acute. **Arrays of heads** relatively crowded to open. **Cypselae** 2.2–2.5 mm. **2n** = 18.

Flowering year-round. Rocky open banks, chaparral, coastal sage; of conservation concern; 200–2200 m; Calif.

Variety *arachnoidea* grows in Carmel Valley (Monterey County) and in the San Rafael Mountains (Santa Barbara County).

12e. Malacothrix saxatilis (Nuttall) Torrey & A. Gray var. **commutata** (Torrey & A. Gray) Ferris, Aliso 4: 100. 1958 [E]

Malacothrix commutata Torrey & A. Gray, Fl. N. Amer. 2: 487. 1843

Plants rhizomatous or with woody caudices. **Herbage** glabrous or ± arachnose. **Stems** 1 (plants rhizomatous) or 2–3+ (plants with woody caudices), erect, freely branched, herbaceous, relatively leafy distally. **Cauline leaves:** proximal not pinnately lobed (lanceolate to lance-linear, margins entire or denticulate, apices acute to acuminate); distal lanceolate or elliptic to linear, margins entire or denticulate, apices acute to acuminate. **Arrays of heads** relatively crowded. **Cypselae** 2–2.5 mm. **2n** = 18.

Flowering Mar–Dec. Crumbling shale along roadcuts, in canyons, chaparral, foothill woodlands; 200–1600 m; Calif.

Variety *commutata* grows on the eastern face of the Transverse Range (Ventura County), in canyons in the Coast Ranges to Monterey County, and in the Breckenridge Mountains (Kern County).

13. **Malacothrix similis** W. S. Davis & P. H. Raven, Madroño 16: 262, fig. 2c. 1962 [C]

Annuals, 5–30+ cm. **Stems** 1–11, erect or ascending, branched mostly proximally, glabrous or sparsely arachnose to puberulent and glabrescent. **Cauline leaves:** proximal mostly lance-linear, sometimes pinnately lobed, not fleshy, ultimate margins dentate or entire, faces glabrous or sparsely arachnose to puberulent and glabrescent; distal reduced subentire. **Calyculi** of 8–12+ lance-deltate to lanceolate bractlets, hyaline margins 0.05–0.3 mm wide. **Involucres** narrowly campanulate, 6–10 × 3–6 mm. **Phyllaries** 18–21+ in ± 2 series, lance-linear to linear, ± equal, hyaline margins 0.05–0.2 mm wide, faces glabrous. **Receptacles** not bristly. **Florets** 32–73: corollas yellow, 4–9 mm; outer ligules exserted 2–4 mm. **Cypselae** ± cylindro-fusiform to prismatic, 1.4–1.7 mm, ribs extending to apices, 5 more prominent than others; **pappi** persistent, of ca. 18, irregular needlelike teeth plus 1 bristle. **Pollen** mostly 4-porate (mean diam. 30 µm). $2n = 28$.

Flowering Apr–May. Coastal plains, beaches, dunes; of conservation concern; 0–40 m; Calif.; Mexico (Baja California).

Malacothrix similis has been collected on Santa Cruz Island (in 1888) and at Hueneme Beach (in 1925).

14. **Malacothrix sonchoides** (Nuttall) Torrey & A. Gray, Fl. N. Amer. 2: 486. 1843 · Sow-thistle desertdandelion [E] [F]

Leptoseris sonchoides Nuttall, Trans. Amer. Philos. Soc., n. s. 7: 439. 1841; *Malacothrix runcinata* A. Nelson

Annuals, (5–)10–25(–50) cm. **Stems** 1–5, ascending to erect, branched near bases and distally, usually glabrous (sometimes glaucous), rarely stipitate-glandular. **Cauline leaves:** proximal narrowly oblong to elliptic, pinnately lobed (lobes 3–8+ pairs, oblong to triangular, ± equal, apices obtuse to acute), ± fleshy, ultimate margins dentate to denticulate, faces glabrous; distal reduced (narrowly triangular to linear, bases ± dilated, ± clasping). **Calyculi** of 8–12+, ovate to lanceolate bractlets, hyaline margins 0.05–0.3(–0.7) mm wide, usually glabrous (margins sometimes stipitate-glandular). **Involucres** ± campanulate to hemispheric, 7–13 × 4–6(–12+) mm. **Receptacles** bristly. **Florets** 75–115; corollas lemon yellow, 10–14(–16) mm; outer ligules exserted 6–10(–13) mm. **Cypselae** ± cylindro-fusiform to prismatic, 1.8–3 mm, ribs extending to apices, ± equal

or 5 more prominent than others; **pappi** persistent, crenate crowns of 15–25+, blunt or rounded teeth. **Pollen** 70–100% 3-porate. $2n = 14$.

Flowering Mar–Jun. Usually on dunes or in deep, fine sand in arroyos and on plains in Joshua tree woodlands, grasslands, *Ephedra-Coleogyne* associations; 300–2100 m; Ariz., Calif., Colo., Nev., N.Mex., Utah, Wyo.

Malacothrix sonchoides grows in California in the Mojave Desert (Los Angeles, Kern, San Bernardino, and Riverside counties), the Great Basin Desert (Inyo Mountains), and barely enters the northern margins of the Sonoran Desert. It also grows in the Intermountain Region in Arizona, Colorado, Nevada, New Mexico, Utah, and Wyoming.

15. **Malacothrix sonorae** W. S. Davis & P. H. Raven, Madroño 16: 264, fig. 2d. 1962 · Sonoran desertdandelion

Annuals, 10–35 cm. **Stems** 1(–9), erect, branched from bases and distally, relatively sparsely leafy, glabrous. **Cauline leaves:** proximal narrowly oblanceolate to obovate, usually pinnately lobed (lobes oblong to triangular), not fleshy, ultimate margins ± dentate, faces glabrous; distal greatly reduced (margins entire or basally dentate, apices acute). **Calyculi** of 5–8+, subulate to lanceolate bractlets, hyaline margins 0.05–0.2 mm. **Involucres** ± campanulate, 6–9 × 4–6.6 mm. **Phyllaries** 12–15+ in 2(–3) series, lance-oblong to lance-linear, hyaline margins 0.05–0.2 mm wide, faces glabrous. **Receptacles** not bristly. **Florets** 30–61; corollas white or pale yellow, 6–10+ mm; outer ligules exserted 1–4 mm. **Cypselae** ± cylindro-fusiform, 1.7–2 mm, ribs ending 0.2–0.3 mm short of apices, ± equal (distal 0.2–0.3 mm of cypselae slightly expanded, smooth); **pappi** persistent, of 16–18 needlelike teeth plus 2 bristles. **Pollen** 70–100% 3-porate. $2n = 14$.

Flowering Mar–May. Sandy, open areas among bushes, *Larrea-Lycium-Cercidium-Baccharis* associations, *Quercus, Pinus, Juglans* woodlands; 400–1500 m; Ariz., N.Mex.; Mexico (Sonora).

Malacothrix sonorae is found mainly in the Sonoran Desert (Tucson, Kofa, Pinal, White Tank, Baboquivari, and Waterman mountains).

16. Malacothrix squalida Greene, Bull. Calif. Acad. Sci. 2: 152. 1886 • Island or Santa Cruz desertdandelion C E

Malacothrix foliosa A. Gray var. *squalida* (Greene) E. W. Williams; *M. insularis* Greene var. *squalida* (Greene) Ferris

Annuals, 4–30 cm. **Stems** 1–3+, ascending to erect (stout), branched from bases and distally, ± leafy, glabrous. **Cauline leaves:** proximal obovate to oblanceolate, pinnately lobed (lobes 2–6 pairs), not fleshy, ultimate margins entire or dentate, faces glabrous; distal not notably reduced (narrowly ovate with 5–10 narrow teeth or lobes). **Calyculi** 0. **Involucres** narrowly to broadly campanulate, 9–12 × 4–10 mm. **Phyllaries** 31–49 in 5–6+ series (midstripes green or reddish), broadly ovate (outermost) to lance-oblong or lance-linear, unequal, hyaline margins 0.6–1 mm wide, faces glabrous. **Receptacles** usually not bristly. **Florets** 39–133; corollas light yellow, 12–19 mm; outer ligules exserted 6–11 mm. **Cypselae** ± prismatic or columnar, 1.3–2.1 mm, ribs extending to (and just beyond) apices, 5 more prominent than others; **pappi** persistent, of 15–20, irregular, ± deltate teeth (often hidden by apices of cypselae) plus 0(–1) bristles. **Pollen** 70–100% 4-porate. *2n* = 28.

Flowering Mar–Jun. Open areas between shrubs, on ridges, knife-edges; of conservation concern; 0–30 m; Calif.

Malacothrix squalida is known only from Middle Anacapa and Santa Cruz islands.

17. Malacothrix stebbinsii W. S. Davis & P. H. Raven, Madroño 16: 265, fig. 2e. 1962 • Stebbins's desertdandelion

Malacothrix clevelandii A. Gray var. *stebbinsii* (W. S. Davis & P. H. Raven) Cronquist

Annuals, 5–60 cm. **Stems** 1–5+, ± erect, branched distally, sparsely leafy, glabrous. **Cauline leaves:** proximal obovate, pinnately lobed (lobes 3–5+ pairs, triangular to oblong or linear), not fleshy, ultimate margins entire or dentate, faces glabrous; distal reduced (usually with 2 lobes near bases). **Calyculi** of 5–8+, ovate to lanceolate bractlets, hyaline margins 0.05–0.2 mm. **Involucres** ± campanulate, 7–10 × 3–6 mm. **Phyllaries** 16–20+ in 2–3 series, lanceolate to lance-linear, hyaline margins 0.05–0.1 mm wide, faces glabrous. **Receptacles** not bristly. **Florets** 19–70; corollas usually yellow, sometimes white, 6–7 mm; outer ligules exserted 1–2 mm. **Cypselae** ± cylindro-fusiform, 1.7–

2.3 mm, ribs extending to ca. 0.1–0.2 mm short of apices (minutely hirtellous or muriculate), ± equal (distal 0.1–0.2 mm of cypselae smooth); **pappi** persistent, of 8–15+, needlelike teeth plus 1(–2) bristles. **Pollen** 70–100% 4-porate. *2n* = 28.

Flowering Mar–Jun. Gravelly soils beneath shrubs, along ditches, near streams, in sagebrush steppes, creosote bush scrublands; 300–1300 m; Ariz., Calif., Nev., N.Mex.; Mexico (Sonora).

Malacothrix stebbinsii grows in the Mojave Desert (Borrego area, California) and the Sonoran Desert (Santa Catalina, Mazatzal, Baboquivari, and Santa Rita mountains, and elsewhere in Arizona).

18. Malacothrix torreyi A. Gray, Proc. Amer. Acad. Arts 9: 213. 1874 • Torrey's desertdandelion E

Annuals, (5–)10–25(–40) cm. **Stems** 1–5+, erect to ascending, branched from bases and distally, sparsely leafy, usually ± stipitate-glandular, sometimes sparsely arachnose or glabrous (sometimes glaucous). **Cauline leaves:** proximal obovate to oblong, usually pinnately lobed (lobes 3–8 pairs, oblong or triangular to linear, apices obtuse to acute), ± fleshy, ultimate margins dentate, faces usually glabrous, sometimes sparsely arachnose; distal reduced (± elliptic, then pinnately lobed, to linear, bases narrowly cuneate, margins usually dentate, sometimes entire). **Calyculi** of 8–12+, lance-ovate to lanceolate bractlets, hyaline margins 0.5–0.2 mm wide (faces sometimes stipitate-glandular). **Involucres** ± campanulate, 8–14 × 4–5 mm. **Phyllaries** 12–20+ in 2–3 series, lance-oblong to lance-linear (apices acuminate), hyaline margins 0.05–0.3 mm wide, faces glabrous or stipitate-glandular. **Receptacles** sparsely bristly. **Florets** 58–93; corollas medium yellow, 14–20 mm; outer ligules exserted 7–10 mm. **Cypselae** cylindro-fusiform, 2.5–4 mm, ribs extending to or beyond apices, 5 more prominent than others (often winglike); **pappi** persistent, of 12–15 blunt teeth plus 0–6 bristles. **Pollen** 70–100% 3-porate. *2n* = 14.

Flowering Apr–Jul. Coarse soils, dry sagebrush slopes, *Artemisia-Atriplex-Tetradymia* associations, Juniper grasslands, Pinyon-Juniper woodlands; 700–2000 m; Ariz., Calif., Colo., Idaho, Mont., Nev., Oreg., Utah, Wyo.

Malacothrix torreyi grows primarily in the Great Basin Desert (in the Inyo Mountains and Modoc Plateau, California), and throughout the Intermountain Region in Arizona (Grand Canyon Plateau), Nevada (Central Great Basin, Uintah and Henry mountains, Flaming Gorge area near Pyramid Lake), Utah (northern), Idaho (Snake River Plateau), Wyoming (Ft. Steele, Green River), and Oregon (Steen Mountains, Owyhee Desert).

65. UROPAPPUS Nuttall, Trans. Amer. Philos. Soc., n. s. 7: 424. 1841 • Silver-puffs

[Greek *uro-*, tail, and *pappus*, alluding to slender terminal bristle on each pappus scale]

Kenton L. Chambers

Calaïs de Candolle sect. *Calocalaïs* de Candolle; *Microseris* D. Don sect. *Calocalaïs* (de Candolle) A. Gray

Annuals, 5–70 cm (sometimes acaulescent, glabrous or lightly farinose, usually white-villous proximally); taprooted. **Stems** 1–5+, erect, sometimes well branched proximally (internodes 0.3–6 cm, or plants acaulescent). **Leaves** all or mostly basal; obscurely petiolate; blades (often reddish or purplish), linear or narrowly lanceolate, (bases ± clasping) margins entire or remotely, pinnately lobed or dentate (white-villous-ciliate proximally, apices acuminate, faces glabrous or crisped white-villous throughout). **Heads** borne singly (erect). **Peduncles** (erect from bud to mature fruit) inflated distally, ebracteate. **Calyculi** 0. **Involucres** fusiform to ovoid, 3–15 mm diam. **Phyllaries** 5–26 in 3–4 series, lanceolate, unequal (outer shorter, inner equal), margins scarious, apices long-tapering, acute, faces glabrous. **Receptacles** flat or convex, pitted, glabrous, epaleate. **Florets** 5–150; corollas pale yellow, usually reddish abaxially. **Cypselae** usually black or dark brown, rarely gray (outer sometimes paler), ± columnar to fusiform, often narrowed distally to relatively short beaks, ribs 10, minutely scabrous, hispidulous; **pappi** falling, of 5, distinct, white, lustrous, lanceolate, aristate scales (apices notched, aristae smooth). $x = 9$.

Species 1: w North America, nw Mexico.

Uropappus lindleyi was placed in *Microseris* (K. L. Chambers 1955) because of two allotetraploid species formed by hybridization with annual members of that genus. A number of morphologic features, including narrow, acuminate leaves with villous-ciliate margins, erect heads, relatively long outer phyllaries, cypselae often short-beaked, and pappi of white, lustrous scales suggest a connection with *Nothocalaïs*, especially *N. troximoides*. Phylogenetic studies of chloroplast DNA variation (R. K. Jansen et al. 1991b; J. Whitton et al. 1995) link *Uropappus* with *Nothocalaïs* and *Agoseris* as a sister clade to *Microseris*. Consequently, Jansen et al. separated *Uropappus* from *Microseris* and placed the two allotetraploid species in *Stebbinsoseris*.

SELECTED REFERENCE Chambers, K. L. 1964. Nomenclature of *Microseris lindleyi*. Leafl. W. Bot. 10: 106–108.

1. **Uropappus lindleyi** (de Candolle) Nuttall, Trans. Amer. Philos. Soc., n. s. 7: 425. 1841 • Lindley's silver puff E F

Calaïs lindleyi de Candolle in A. P. de Candolle and A. L. P. P. de Candolle, Prodr. 7: 85. 1838; *Microseris lindleyi* (de Candolle) A. Gray; *M. linearifolia* (Nuttall) Schultz Bipontinus; *Uropappus linearifolius* Nuttall

Leaves 5–30 cm. **Peduncles** 5–40 cm. **Involucres** 10–40 mm after flowering. **Phyllaries** reflexed in fruit, often reddish, outer 2–8, inner 3–18. **Ligules** 2–-10 mm, equaling or barely surpassing phyllaries at flowering. **Cypselae** 7–17 mm; **pappi**: scales 5–15 mm, apices notched 1–2 mm, bristles delicate, 4–6 mm. $2n = 18$.

Flowering Mar–May. Grasslands, shrub steppe, open oak woodlands, chaparral, s coastal scrub, deserts, usually well drained soils on slopes, road banks, serpentine gravels, sandy desert flats; 10–1800 m; B.C.; Ariz., Calif., Idaho, Nev., N.Mex., Oreg., Tex., Utah, Wash.; Mexico (Baja California, Sonora).

Uropappus lindleyi grows in the Columbia-Snake Rivers Plateau Province, Basin and Range Province, Interior Mountains and Plateaus System, and the Pacific Border System.

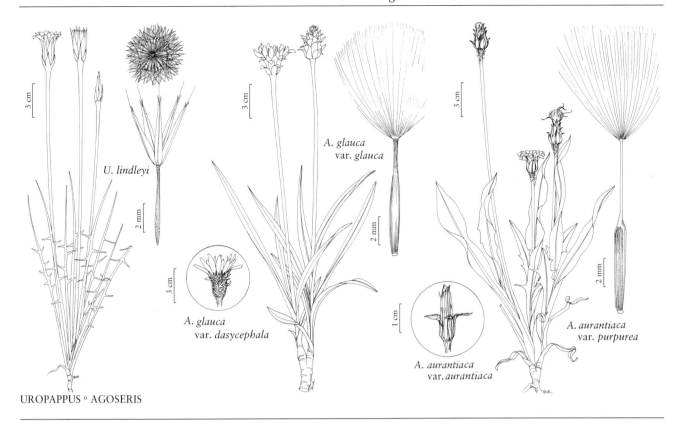

A. glauca
var. glauca

U. lindleyi

A. glauca
var. dasycephala

A. aurantiaca
var. aurantiaca

A. aurantiaca
var. purpurea

UROPAPPUS ∘ AGOSERIS

66. AGOSERIS Rafinesque, Fl. Ludov., 58. 1817 • Mountain- or false dandelion [Greek *agos*, leader, and *seris*, chicory; allusion unclear]

Gary I. Baird

Annuals (*A. heterophylla*) **or perennials**, (2–)5–60(–96) cm (usually acaulescent); taprooted or with caudices. **Stems** usually 0, sometimes 1–5+, erect to decumbent, simple. **Leaves** usually basal (in rosettes), sometimes cauline; petiolate (petioles often purplish, erect to prostrate, ± sheathing); blades linear-lanceolate to spatulate, margins entire or toothed to pinnatifid (sometimes variable on same plant, lobes 2–11 pairs, opposite, subopposite, or irregular, filiform to spatulate, often each with 1, acroscopic, basal lobule, lobules ± triangular, inconspicuous to equaling lobes, faces glabrous and glaucous or pubescent to tomentose). **Heads** borne singly (erect at end of peduncles). **Peduncles** not inflated distally, usually ebracteate. **Calyculi** 0. **Involucres** cylindric to hemispheric, 2–20(–60) mm diam. (bases often broadening in fruit). **Phyllaries** 10–50 in 2–5(–7) series, green or medially rosy purple and often with purple-black midstripes, apices, or spots, rarely nearly all black, narrowly lanceolate to broadly ovate or obovate, unequal to subequal, ± herbaceous, margins usually entire (rarely dentate), sometimes scarious, glabrous or ciliate to lanate, apices obtuse to acuminate, faces glabrous or pubescent to tomentose, often stipitate-glandular, (outer appressed to squarrose, often adaxially hairy (at least on apices), not elongating after flowering; inner erect, adaxially glabrous, often puberulent apically, sometimes elongating in fruit). **Receptacles** flat, pitted, glabrous, rarely paleate (paleae linear-lanceolate, 15–20 mm, ± acuminate, similar to inner phyllaries, ± accrescent). **Florets** 5–500 (± matutinal); corollas yellow (outermost often with purplish abaxial stripe (often drying whitish, purple stripe still evident), or orange, pink, red, or purple (usually drying purple). **Cypselae** monomorphic or dimorphic (outermost differing in color, texture, vestiture, and/or shape from inner), white or tan to dark purple, columnar to fusiform, narrowly conic, or obconic (2–10 mm), ribs usually 10, faces glabrous or pubescent to hirsute, beaked (beak lengths 0.1–4 times bodies); **pappi** persistent, of 50–125, distinct, white, subequal, ± barbellate (sometimes flattened) bristles in 1–6 series. $x = 9$.

Species 11 (10, including 1 hybrid, in the flora): North America, South America.

Agoseris consists of widespread species that individually exhibit great morphologic plasticity. Difficulty in correctly identifying individual specimens is compounded by traits that may vary from region to region, the perpetuation of misleading or inaccurate traits in the literature, and the presence of intermediates. Correct identification of *Agoseris* specimens can be assisted by knowing that species may exhibit variable traits (e.g., pubescence, corolla color, cypsela morphology), some species have leaf lobing variable on single plants (e.g., outermost entire versus inner lobed), and intermediate specimens may occur with any sympatric taxa. Hybridization among members of the genus is common, especially among polyploid taxa, and some hybrid populations appear to be persistent. Autogamy has been demonstrated in some species (K. L. Chambers 1963) and is suspected in others. It appears to be correlated with a reduction in corolla and anther size. Autogamous populations or taxa often exhibit seemingly unique features that appear localized. Attempts at naming these variant populations or regional phases have resulted in a large number of synonyms.

Agoseris has a New World, amphitropical distribution. All of the species are restricted to North America except *A. coronopifolia* (D'Urville) K. L. Chambers, which is found in temperate regions of southern South America. The South American disjunction appears to be the result of long-distance dispersal from North America (K. L. Chambers 1963).

Agoseris appears to be most closely related to *Nothocalaïs*.

Cryptopleura Nuttall, referable here, is a rejected name.

In keys and descriptions, lengths of cypselae include beaks.

SELECTED REFERENCES Baird, G. I. 1996. The Systematics of *Agoseris* (Asteraceae: Lactuceae). Ph.D. dissertation. University of Texas. Chambers, K. L. 1963. Amphitropical species pairs in *Microseris* and *Agoseris* (Compositae: Cichorieae). Quart. Rev. Biol. 38: 124–140. Jones, Q. 1954. Monograph of *Agoseris*, Tribe Cichorieae. Ph.D. dissertation. Harvard University.

1. Annuals . . . . 9. *Agoseris heterophylla*
1. Perennials.
   2. Corollas orange, pink, red, or purplish (often drying purplish).
      3. Peduncles (and phyllaries) eglandular; widespread in North America . . . .
      . . . . 2. *Agoseris aurantiaca* (in part)
      3. Peduncles (and phyllaries) ± stipitate-glandular; California, Oregon, and
      Washington . . . . 10. *Agoseris* ×*elata* (in part)
   2. Corollas yellow, outermost often each with abaxial purplish stripe (often drying whitish, purple stripe still evident).
      4. Cypsela beaks 1–4(–10) mm (lengths to ¹/₂ bodies); inner phyllaries not elongating in fruit.
         5. Leaf margins usually lobed, rarely entire, lobes (3–)5–8 pairs, retrorse to spreading; peduncles (and phyllaries) usually hairy to lanate, sometimes glabrous, eglandular; cypsela beaks (3–)4–10 mm, lengths (¹/₂–)2 times bodies . . . . . .
         . . . . 3. *Agoseris parviflora* (in part)
         5. Leaf margins entire or toothed to lobed, teeth or lobes usually 2–3 pairs, antrorse to spreading, or diverging; peduncles (and phyllaries) glabrous or puberulent to lanate, sometimes stipitate-glandular or eglandular; cypsela beaks 1–4 mm, lengths to ¹/₂ bodies.

6. Peduncles glabrous, or basally glabrate, apically puberulent to lanate, sometimes stipitate-glandular; leaves usually erect, sometimes decumbent, margins usually entire, sometimes dentate, rarely lobed or lacerate; receptacles sometimes paleate; phyllaries in 2–3 series; widespread, various soils and elevations, n Great Plains westward . . . . 1. *Agoseris glauca*

6. Peduncles basally lanate, apically hairy to villous and stipitate-glandular; leaves usually decumbent or prostrate, margins usually dentate or lobed, rarely entire; receptacles rarely paleate; phyllaries in 2–4(–6) series; mostly at high elevations, volcanic or pyroclastic soils, Sierra Nevada and s Cascade Mountains, sporadically eastward to Blue Mountains and Great Basin . . . . 4. *Agoseris monticola*

4. Cypsela beaks 5+ mm, lengths usually equaling or greater than bodies, if less than 5 mm, beaks usually 1/2+ bodies; inner phyllaries elongating in fruit.

7. Peduncles and phyllaries ± stipitate-glandular.

8. Leaves 3–10(–15) cm (plants usually ± caulescent, stems often buried by drifting sand, appearing pseudorhizomatous, sometimes acaulescent); mostly coastal dunes and beach heads, Pacific Coast . . . . . 7. *Agoseris apargioides* (in part)

8. Leaves (7–)10–30 cm (plants acaulescent); mostly grassy hills, meadows, or lowland prairies (not coastal sand dunes).

9. Leaf margins usually pinnately lobed, lobes 5–7(–9) pairs; corolla tubes 3–6 mm; cypselae 9–14 mm; pappus bristles in 3–4 series, 6–10 mm; Coast Ranges of California, especially around San Francisco Bay . . . . 8. *Agoseris hirsuta*

9. Leaf margins entire or laciniately pinnatifid, lobes 2–4 pairs; corolla tubes 8–10 mm; cypselae 14–20 mm; pappus bristles in 2–3 series, 10–14 mm; Washington to California (not Coast Ranges) . . . . 10. *Agoseris ×elata* (in part)

7. Peduncles and phyllaries glabrous or hairy, eglandular.

10. Cypsela beaks (9–)10–25 mm, lengths usually 3–4 times bodies; phyllaries in 3–6 series.

11. Leaf lobes antrorse to spreading; corolla tubes 4–7(–10) mm, ligules 3–7 mm, anthers 1–3 mm; pappus bristles in 2–3 series, 7–15 mm . . . . 5. *Agoseris grandiflora*

11. Leaf lobes retrorse to spreading; corolla tubes (8–)10–20 mm, ligules 6–12(–16) mm, anthers 2–5 mm; pappus bristles in 4–6 series, (11–)15–20 mm . . . . 6. *Agoseris retrorsa*

10. Cypsela beaks (1–)3–10 mm, lengths usually 1/2–2 times bodies; phyllaries in 2–3 series.

12. Corolla tubes 2–5.5 mm, ligules 3–16 mm, anthers 1.5–4.5 mm; cypsela bodies 3–5 mm; leaves 3–10(–15) cm; Pacific shore coastal dunes . . . . 7. *Agoseris apargioides* (in part)

12. Corolla tubes (4–)6–15 mm, ligules (4–)6–20 mm, anthers 2–5 mm; cypsela bodies 5–9 mm; leaves (5–)10–38 cm; widespread North America east of Pacific coast ranges.

13. Corolla ligules 4–12 mm; inner phyllaries elongating in fruit; leaf margins entire or laciniately lobed, lobes 2–4 pairs, spreading to antrorse; montane forests to alpine tundra, often disturbed habitats . . . . 2. *Agoseris aurantiaca* (in part)

13. Corolla ligules 10–20 mm; inner phyllaries not elongating in fruit; leaf margins usually lobed, rarely entire, lobes (3–)5–8 pairs, mostly retrorse; sagebrush steppes, grasslands, pinyon-juniper woodlands, open forests at lower elevations . . . . 3. *Agoseris parviflora* (in part)

1. Agoseris glauca (Pursh) Rafinesque, Herb. Raf., 39. 1833 • Prairie agoseris [E] [F]

Troximon glaucum Pursh, Fl. Amer. Sept. 2: 505. 1813

Stems 0. **Leaves** erect to decumbent; petioles rarely purplish, petiole margins glabrous or pubescent, not usually ciliate; blades lanceolate to oblanceolate, 2–46 cm, margins usually entire, sometimes dentate, rarely lobed or lacerate, lobes 2–3 pairs or irregularly arranged, lanceolate, spreading, lobules 0, faces glabrous and glaucous or sparsely villous to tomentose. **Peduncles** not notably elongating after flowering, 5–60(–90) cm in fruit, glabrous or glabrate, or apically puberulent to lanate, sometimes stipitate-glandular. **Involucres** obconic to hemispheric, 1–3 cm in fruit. **Phyllaries** in 2–3 series, green or medially rosy purple, often with purple-black spots, midstripe, and/or tips, subequal to unequal, margins glabrous or ± hairy, not usually ciliate, faces glabrous or tomentose, sometimes stipitate-glandular or eglandular; outer erect or spreading, apices adaxially glabrous or hirsuto-villous; inner erect, not notably elongating in fruit. **Receptacles** epaleate or paleate. **Florets** 15–150; corollas yellow, tubes 4–18 mm, ligules 6–24 × 2–5 mm; anthers 3–7 mm. **Cypselae** 7–15 mm, bodies fusiform to narrowly conic, 5–9 mm, tapered to stout beaks 1–4 mm, lengths mostly less than ¹/₂ times bodies; ribs flattened to ridged, glabrous, or distally scabrous; **pappi** in 2–3 series, 8–18 mm. **2n** = 18, 36

Varieties 2 (2 in the flora): w North America.

1. Receptacles epaleate; leaf blades usually glabrous and glaucous, rarely sparsely hairy; peduncles (distally, and phyllaries) usually glabrous, sometimes puberulent, eglandular ..... 1a. *Agoseris glauca* var. *glauca*
1. Receptacles usually ± paleate; leaf blades usually puberulent to densely villous, sometimes glabrous; peduncles (distally, and phyllaries) usually villous to lanate or tomentose, sometimes glabrous, often stipitate-glandular .... 1b. *Agoseris glauca* var. *dasycephala*

1a. Agoseris glauca (Pursh) Rafinesque var. **glauca** [E] [F]

Agoseris lacera Greene; *A. lapathifolia* Greene; *A. longissima* Greene; *A. longula* Greene; *A. microdonta* Greene; *A. procera* Greene; *A. vicinalis* Greene

Leaf blades: margins ± straight, usually entire (most) to dentate, sometimes irregularly so, faces glabrous and glaucous, sometimes sparsely hairy. **Peduncles** usually glabrous, sometimes

apically puberulent, eglandular. **Involucres** obconic to campanulate. **Phyllaries** usually green, sometimes medially rose-purple, often purple-black spotted and/or tipped, faces usually glabrous, sometimes proximally puberulent, margins ± straight, glabrous (most) or proximally puberulent; outer erect or spreading, usually abaxially glabrous (most). **Receptacles** epaleate. **Cypselae** usually scabrous distally, sometimes glabrous. **2n** = 18.

Flowering May–Sep. Wet meadows, stream margins, swales, other moist sites, grasslands, steppes, and meadow habitats, slightly alkaline or saline sites, mostly in silts, clays, and other fine-textured soils; 100–2500 (–3300) m; Alta., B.C., Man., Ont., Sask.; Ariz., Calif., Colo., Idaho, Mich., Minn., Mont., Nev., N.Mex., N.Dak., Oreg., S.Dak., Utah, Wash., Wyo.

Variety *glauca* is usually found at lower elevations from the northern prairies westward to valleys and basins of the North American cordillera. Misidentification is often due to falsely assuming this variety is strictly glabrous. Some regional phases have a high percentage of individuals with weakly puberulent peduncles and/or phyllaries. In addition, var. *glauca* intergrades with var. *dasycephala* in some locations.

Hybrids between var. *glauca* and other *Agoseris* species occur. Two have been described (as species). *Agoseris glauca* var. *glauca* × *A. parviflora* (= *Agoseris* ×*agrestis* Osterhout) occurs frequently in the foothills of the Rocky Mountains of Colorado. This is the basis for *A. glauca* var. *agrestis* (Osterhout) Q. Jones, a name that has been widely but mistakenly applied to many specimens of *A. glauca* var. *dasycephala*. *Agoseris glauca* var. *glauca* × *A. monticola* (= *A. ×dasycarpa* Greene) is found primarily on the Modoc Plateau of northeastern California and adjacent Oregon.

1b. Agoseris glauca (Pursh) Rafinesque var. **dasycephala** (Torrey & A. Gray) Jepson, Man. Fl. Pl. Calif., 1005. 1925 • Arctic agoseris [E] [F]

Troximon glaucum Pursh var. *dasycephalum* Torrey & A. Gray, Fl. N. Amer. 2: 490. 1843; *Agoseris altissima* Rydberg; *A. aspera* (Rydberg) Rydberg; *A. eisenhoweri* B. Boivin; *A. glauca* subsp. *aspera* (Rydberg) Piper; *A. glauca* var. *aspera* (Rydberg) Cronquist; *A. glauca* var. *pumila* (Nuttall) Garrett; *A. glauca* subsp. *scorzonerifolia* (Schrader) Piper; *A. glauca* var. *villosa* (Rydberg) G. L. Wittrock; *A. isomeris* Greene; *A. lanulosa* Greene; *A. leontodon* Rydberg var. *aspera* Rydberg; *A. leontodon* var. *pygmaea* Rydberg; *A. maculata* Rydberg; *A. pubescens* Rydberg; *A. pumila* (Nuttall) Rydberg; *A. scorzonerifolia* (Schrader) Greene; *A. turbinata* Rydberg; *A. vestita* Greene; *A. villosa* Rydberg

Leaf blades: margins often undulate, usually entire, sometimes lobed, faces usually puberulent to densely villous, sometimes glabrous, rarely glaucous. **Peduncles** usually apically villous to lanate (glabrous in one phase), sometimes ± stipitate-glandular. **Involucres** obconic to hemispheric. **Phyllaries** often medially rosy purple and usually with some combination of purple-black spots, speckles, midstripes, or tips, rarely all green, margins ± undulate or reflexed, often hairy, faces usually villous to hirsuto-tomentose, sometimes glabrous, mostly stipitate-glandular; outer erect or spreading, usually adaxially villous, sometimes glabrous. **Receptacles** usually ± paleate. **Cypselae** usually glabrous, sometimes scabrous distally. $2n = 36$.

Flowering Jul–Aug. Moist to dry habitats, alpine meadows, montane forests, northern prairies, arctic tundra, in gravelly, rocky, and other coarse-textured soils; 100–3600 m; Alta., B.C., Man., N.W.T., Sask., Yukon; Colo., Idaho, Mont., N.Dak., Oreg., S.Dak., Utah, Wash., Wyo.

Variety *dasycephala* occurs primarily at high elevations in the western cordillera, extending eastward onto the northern prairies, and disjunctively in the Canadian arctic (Caribou Hills). It is more readily distinguished from var. *glauca* southward, where the two varieties are ± elevationally separated. Difficulty in separating them occurs northward, where they are nearer each other and pockets of complete introgression occur, e.g., southeastern British Columbia and southwestern Alberta. Hybrids with *Agoseris aurantiaca* and *A. parviflora* also occur.

Variety *dasycephala* contains regional phases that exhibit a step-clinal distribution. The large number of synonyms reflects the variation. As circumscribed here, var. *dasycephala* encompasses most of what has been called *Agoseris glauca* var. *agrestis* (see discussion under var. *glauca*).

2. Agoseris aurantiaca (Hooker) Greene, Pittonia 2: 177. 1891 • Orange or mountain agoseris Ⓔ Ⓕ

Troximon aurantiacum Hooker, Fl. Bor.-Amer. 1: 300, plate 104. 1833

Stems 0. **Leaves** erect to decumbent; petioles purplish, petiole margins ciliate to hairy; blades 7–38 cm, linear-lanceolate to oblanceolate, margins entire or laciniately pinnatifid, lobes 2–4 pairs, linear to lanceolate, spreading to antrorse, lobules usually inconspicuous to subequaling lobes, rarely lacking, faces glabrous and ± glaucous or sparsely villous. **Peduncles** ± elongating after flowering, 8–40(–80) cm, glabrate, or apically villous to lanate, eglandular. **Involucres** cylindric to obconic or campanulate, 2.5–3 cm at maturity. **Phyllaries** in 2–3 series, green or medially rosy purple, often with purple-black spots, blotches, and/or midstripes, or nearly all black, subequal to unequal, margins ciliate, faces glabrous or villous, eglandular; outer mostly spreading, adaxially glabrous or villous; inner erect, elongating after flowering. **Receptacles** epaleate. **Florets** 15–100; corollas usually orange, sometimes yellow, pink, red, purple, or white, tubes (4–)7–9 mm, ligules 4–12 × 1–3 mm; anthers 2–5 mm. **Cypselae** ± dimorphic, 8–18 mm, bodies cylindric to fusiform or obconic, 6–9(–11) mm, abruptly or gradually tapered to slender beaks (2–)5–10 mm, lengths mostly equaling bodies; ribs strongly ridged, straight, glabrous or scabrous; **pappi** in 2–3 series, 9–15 mm. $2n = 18, 36$.

Varieties 2 (2 in the flora): w North America.

1. Phyllaries ± lanceolate (± herbaceous throughout), margins ± ciliate proximally, usually villous, sometimes glabrous; corollas usually orange, sometimes pinkish or yellow, rarely white; cypselae ± abruptly tapered to beaks, ribs often thicker distally . . . . 2a. *Agoseris aurantiaca* var. *aurantiaca*
1. Phyllaries ± ovate or obovate (somewhat stramineous proximally), margins ± ciliate distally, usually glabrous, rarely villous; corollas usually orange or yellow, sometimes pinkish; cypselae ± gradually tapered to beaks, ribs not thicker distally . . . . 2b. *Agoseris aurantiaca* var. *purpurea*

2a. Agoseris aurantiaca (Hooker) Greene var. **aurantiaca** Ⓔ Ⓕ

Agoseris angustissima Greene; *A. arachnoidea* Rydberg; *A. carnea* Rydberg; *A. gaspensis* Fernald; *A. gracilens* (A. Gray) Greene; *A. gracilens* var. *greenei* (A. Gray) S. F. Blake; *A. greenei* (A. Gray) Rydberg; *A. howellii* Greene; *A. lackschewitzii* Douglas M. Henderson & R. K. Moseley; *A. nana* Rydberg; *A. naskapensis* J. Rousseau & Raymond; *A. prionophylla* Greene; *A. subalpina* G. N. Jones; *A. vulcanica* Greene

Leaf blades: margins entire, dentate, or laciniately pinnatifid, faces glabrous or sparsely villous, mostly not glaucous. **Peduncles:** lengths mostly shorter than leaves at flowering, glabrate, or apically villous. **Phyllaries** green or medially rosy purple, often purple-black spotted or blotched, with or without purple-black midstripes, narrowly to broadly lanceolate, mostly subequal at flowering, or outer longer than inner, herbaceous, faces usually villous, sometimes glabrous, margins ciliate, especially proximally; outer adaxially ± villous. **Corollas** usually orange, sometimes pinkish or yellow, rarely white, mostly subequal to inner phyllaries. **Cypselae:**

bodies ± abruptly contracted to beaks, ribs often more strongly ridged distally than proximally, mostly hirsutellous. $2n = 18$.

Flowering Jun–Sep. Wet meadows, bogs, muskegs, stream and lake margins, montane to alpine habitats, or ± drier sites of open forests and mountain slopes, various soils, often disturbed areas, roadsides; 200–3600 m; Alta., B.C., N.W.T., Que., Yukon; Alaska, Calif., Colo., Idaho, Mont., Nev., N.Mex., Oreg., S.Dak., Utah, Wash., Wyo.

Variety *aurantiaca* is widespread in the western cordillera and is disjunct in Quebec. Two morphologic trends occur within this variety. Plants of wetter habitats represent the typical var. *aurantiaca*; those of drier habitats resemble what past authors have called *Agoseris gracilens* (including *A. gracilens* var. *greenei*). There is a weak geographic trend to this variation, with the *aurantiaca* phase occurring mostly along the Rocky Mountains axis and the *gracilens* phase mostly along the Cascade Mountains-Sierra Nevada axis. In their extremes they appear distinct, but their intergradation is so complete that separation becomes arbitrary. Putative hybrids between var. *aurantiaca* and *A. glauca*, *A. grandiflora*, *A. monticola*, and *A. parviflora* have been collected.

Corolla color in var. *aurantiaca* is variable but most commonly orange. Pink-flowered forms occur sporadically. They have been recognized as *Agoseris lackschewitzii*. Recognition of pink forms is unmerited; if it were, the older name *A. carnea* would have priority.

2b. Agoseris aurantiaca (Hooker) Greene var. **purpurea** (A. Gray) Cronquist, Rhodora 50: 33. 1948 • Colorado Plateau agoseris [E] [F]

Macrorhynchus purpureus A. Gray, Mem. Amer. Acad. Arts, n. s. 4: 114. 1849; *Agoseris arizonica* (Greene) Greene; *A. attenuata* Rydberg; *A. aurantiaca* subsp. *purpurea* (A. Gray) G. W. Douglas; *A. confinis* Greene; *A. frondifera* Osterhout; *A. glauca* (Pursh) Rafinesque var. *cronquistii* S. L. Welsh; *A. graminifolia* Greene; *A. longirostris* Greene; *A. purpurea* (A. Gray) Greene; *A. purpurea* var. *arizonica* (Greene) G. L. Wittrock; *A. roseata* Rydberg; *A. rostrata* Rydberg

Leaf blades: margins usually dentate to laciniately pinnatifid, rarely entire, faces mostly glabrous and often glaucous. **Peduncles** mostly longer than leaves at flowering, glabrate or apically ± villous to lanate. **Phyllaries** ± stramineous proximally, green, often purple-black blotched or spotted, or with a purple-black midstripes, rarely nearly all black, usually ovate or obovate, sometimes lanceolate, subequal to unequal at flowering, margins ± ciliate, especially distally, faces glabrous or slightly pubescent basally or medially; outer mostly glabrous adaxially. **Corollas** usually orange or yellow, sometimes pinkish, subequal to or surpassing inner phyllaries. **Cypselae:** bodies ± gradually tapered to beaks, ribs ± weakly and uniformly ridged (not thickened distally). $2n = 18, 34, 36$.

Flowering Jun–Sep. Moist, subalpine meadows and forests to alpine tundra, often disturbed areas; 1800–3600 m; Ariz., Colo., Nev., N.Mex., Utah, Wyo.

Variety *purpurea* is known mainly from the Colorado Plateau and southern Rocky Mountains. The two varieties are partially sympatric in the mountains of Colorado, Utah, southern Wyoming, and northern New Mexico, var. *aurantiaca* occurring only at very high elevations in that region. Wherever var. *purpurea* and var. *aurantiaca* occur together, they intergrade. Hybrids between var. *purpurea* and *A. glauca* or *A. parviflora* occur. One hybrid has been named (as a species): *Agoseris aurantiaca* var. *purpurea* × *A. glauca* var. *dasycephala* (= *A.* ×*montana* Osterhout) occurs sporadically at high elevations in the Rocky Mountains of Colorado.

Variety *purpurea* tends to exhibit a higher frequency of yellow-flowered populations than var. *aurantiaca*. Plants of var. *purpurea* from the Rocky Mountains usually have orange corollas; those from the plateaus of southern Utah and northern Arizona often have yellow corollas. These more southwestern populations have been called *A. arizonica* (or *A. purpurea* var. *arizonica*); the two regional phases cannot be adequately separated and their segregation is arbitrary.

3. Agoseris parviflora (Nuttall) D. Dietrich, Syn. Pl. 4: 1332. 1847 • Steppe agoseris [E]

Troximon parviflorum Nuttall, Trans. Amer. Philos. Soc., n. s. 7: 434. 1841; *Agoseris caudata* Greene; *A. dens-leonis* Greene; *A. glauca* (Pursh) Rafinesque var. *laciniata* (D. C. Eaton) Kuntze; *A. leptocarpa* Osterhout; *A. rosea* (Nuttall) D. Dietrich; *A. taraxacoides* Greene; *A. tomentosa* Howell

Stems 0. **Leaves** erect to decumbent; petioles sometimes purplish, margins usually ± hairy, sometimes glabrous or ciliate; blades linear-lanceolate to oblanceolate, (5–)10–20(–32) cm, margins usually lobed, sometimes entire (variable within plants, e.g., outer entire, inner lobed), rarely all entire; lobes (3–)5–8 pairs, linear to lanceolate, mostly retrorse, sometimes spreading; lobules often present, faces glabrous and glaucous or densely tomentose. **Peduncles** not notably elongating after flowering, [(6–)10–25(–45) cm in fruit], glabrate, or apically hairy to lanate, eglandular. **Involucres** obconic to hemispheric, 2–3.5 cm in fruit. **Phyllaries** in 2–3 series, usually medially rosy purple, rarely all green or spotted, margins ciliate or lanate, faces glabrous or sparsely villous, eglandular; outer erect or spreading, adaxially ± tomentose (sometimes glabrous); inner erect, not notably elongating after flowering. Receptacles epaleate. **Florets** 30–100; corollas yellow, tubes (4–)6–15 mm, ligules 10–20 × 2–4 mm; anthers 3–5 mm. **Cypselae** 9–18 mm, bodies terete or narrowly conic to obconic, 5–9 mm, beaks 3–10 mm, lengths ($^1/_2$–)2 times bodies; **pappus bristles** in ca. 3 series, 10–20 mm. **2*n*** = 18.

Flowering Apr–Aug. Dry habitats, sandy soils, short-grass prairies, sagebrush steppes, pinyon-juniper woodlands, montane meadows, mixed conifer forests; 1000–3400 m; Ariz., Calif., Colo., Idaho, Mont., Nev., N.Mex., Oreg., S.Dak., Utah, Wyo.

Agoseris parviflora is found in drier habitats from western Great Plains to eastern foothills of the Cascade Mountains and Sierra Nevada. This is *A. glauca* var. *laciniata* of recent authors. The inclusion of *A. parviflora* within an expanded *A. glauca* is based partly on the mistaken perception that those two species readily intergrade and that *A. parviflora* is a xeric variant of the more mesic *A. glauca*. Although hybrids between the two species occur, frequency of intermediates is no greater than that of any other species in the genus. *Agoseris parviflora* is known to form intermediates with *A. aurantiaca*, *A. monticola*, and *A. retrorsa* as well. This species exhibits some regional variations. In their extremes, these phases appear more or less distinct but they so completely intergrade that their separation becomes arbitrary.

4. **Agoseris monticola** Greene, Pittonia 4: 37. 1899
   • Sierra Nevada agoseris [E]

Agoseris covillei Greene; *A. decumbens* Greene; *A. glauca* (Pursh) Rafinesque var. *monticola* (Greene) Q. Jones

Stems 0. **Leaves** mostly decumbent to prostrate; petioles rarely purplish, margins not ciliate; blades oblanceolate to spatulate, 2–10(–14) cm, margins usually dentate to lobed or laciniately pinnatifid, rarely entire, lobes 2–3 pairs, linear to oblanceolate, proximal lobes often retrorse, distal often antrorse, lobules often present, faces mostly puberulent to villous, sometimes glabrous and glaucous. **Peduncles** not elongating after flowering, 2–25 cm in fruit, basally lanate, apically stipitate-glandular. **Involucres** obconic to campanulate, 1–2 cm in fruit. **Phyllaries** in 2–4(–6) series, usually rosy purple, rarely green, sometimes spotted, often with a purple-black midstripes, unequal, faces ± hairy, stipitate-

glandular; outer usually erect, sometimes spreading apically, adaxially glabrous; inner erect, not elongating after flowering. **Receptacles** epaleate, rarely paleate (outer florets only). **Florets** 10–40; corollas yellow, tubes 4–10 mm, ligules 5–11 × 2–4 mm; anthers 3–5 mm. **Cypselae** 6–10 mm; bodies fusiform, 6–9 mm, beaks 1–3 mm, lengths to $^1/_2$ times bodies; ribs ridged to flattened, straight; **pappus bristles** in 2 series, 8–11 mm. **2*n*** = 18, 36.

Flowering Jul–Aug. Mesic subalpine meadows and forests to alpine tundra and rocky slopes, volcanic or pyroclastic soils; 2000–3500 m; Calif., Nev., Oreg., Wash.

Agoseris monticola occurs mainly in the Sierra Nevada and sporadically eastward in the Great Basin (Jarbridge and Ruby Mountains) and northward to the Cascade Range and Blue Mountains of Oregon. It appears to be allied with *A. glauca* and has been treated as a variety of the latter. Ecologically, it approaches *A. glauca* var. *dasycephala*; the two are morphologically and geographically separate from each other. Intermediates between *A. monticola* and *A. aurantiaca*, *A. glauca*, and *A. parviflora* are known.

5. **Agoseris grandiflora** (Nuttall) Greene, Pittonia 2: 178. 1891 • Grassland agoseris [E]

Stylopappus grandiflorus Nuttall, Trans. Amer. Philos. Soc., n. s. 7: 432. 1841

Stems usually 0, rarely weakly developed. **Leaves** erect to ascending; petioles often purplish, margins ciliate to tomentose; blades usually linear to oblanceolate, rarely spatulate, 10–40(–50) cm, margins usually dentate to lobed, or laciniately pinnatifid, rarely subentire, lobes 3–5 pairs, opposite to irregular, linear to lanceolate, usually antrorse, sometimes spreading; lobules often present, faces ± villous to canescent. **Peduncles** elongating after flowering, (15–)25–60(–96) cm in fruit, ± glabrate, or apically villous to tomentose, eglandular, sometimes bracteate. **Involucres** campanulate to hemispheric, 2–5 cm in fruit. **Phyllaries** in 4–5 series, green or medially rosy purple and/or spotted, subequal to unequal, margins occasionally dentate, ciliate or lanate, sometimes glabrous, faces glabrous or hairy, eglandular; outer subequal to or surpassing inner at flowering, spreading, adaxially tomentose; inner erect, greatly elongating after flowering. **Receptacles** epaleate. **Florets** 40–500; corollas yellow, tubes 4–7(–10) mm, ligules 3–7 × ca. 1 mm; anthers 1–3 mm. **Cypselae** 9–28 mm, bodies fusiform, 3–7 mm, beaks 9–21 mm, lengths mostly 3–4 times bodies; ribs ridged to subalate, rarely nerviform, straight, ± glabrous; **pappus bristles** in 2–3 series, 7–15 mm. **2*n*** = 18.

Varieties 2 (2 in the flora): w North America.

1. Leaves: margins dentate to lobed or pinnatifid, mostly 10–35 mm wide (excluding lobes), lobes ± lanceolate; florets (100–)150–300(–500+); outer phyllaries often medially rosy purple, not spotted, often longer than inner at flowering . . . . 5a. *Agoseris grandiflora* var. *grandiflora*
1. Leaves: margins laciniately to finely pinnatifid, mostly 2–4(–8) mm wide (excluding lobes), lobes filiform to linear; florets 40–60+; outer phyllaries often medially rosy purple and spotted, mostly subequal to inner at flowering . . . . 5b. *Agoseris grandiflora* var. *leptophylla*

5a. Agoseris grandiflora (Nuttall) Greene var. **grandiflora** E

Agoseris cinerea Greene; *A. grandiflora* var. *intermedia* (Greene) Jepson; *A. grandiflora* var. *plebeia* (Greene) G. L. Wittrock; *A. intermedia* Greene; *A. marshallii* (Greene) Greene; *A. obtusifolia* (Suksdorf) Rydberg; *A. plebeia* (Greene) Greene

Leaves: blades oblanceolate to spatulate, 10–50 cm × 10–35 mm (excluding lobes), margins dentate to lobed or pinnatifid, lobes linear-lanceolate to lanceolate. **Involucres** campanulate to hemispheric, 3–5.5 cm in fruit. **Phyllaries** medially rosy purple, rarely green, rarely spotted, lanceolate to ovate or obovate; outer often surpassing inner at flowering, margins sometimes dentate. **Florets** (100–)150–300 (–500). $2n = 18$

Flowering Mar–Oct. Mesic to dry meadows, grasslands, sagebrush steppes, open oak or coniferous woodlands, chaparral; 300–2500 m; Calif., Idaho, Mont., Nev., Oreg., Utah, Wash.

Variety *grandiflora* is most commonly found east of the Cascade Mountains and southward into California and occurs primarily in grassland, steppe, or chaparral. It has regional phases, especially southward in its range. These appear more or less distinct but they so completely intergrade that their separation becomes arbitrary. Variety *grandiflora* rarely forms intermediates with other species; putative hybrids with *A. apargioides* have been collected. It is one of the suspected parental taxa of *A. ×elata*, especially the Sierra Nevada populations.

5b. Agoseris grandiflora (Nuttall) Greene var. **leptophylla** G. I. Baird, Sida 21: 267. 2004
• Puget Sound agoseris E

Leaves: blades linear to filiform, 10–36 cm × 2–4(–8) mm (excluding lobes), margins usually laciniately or finely pinnatifid, sometimes toothed, lobes or teeth linear to filiform. **Involucres** campanulate, 2–4 cm in fruit. **Phyllaries** mostly medially rosy purple and spotted, ± lanceolate, margins usually entire, rarely dentate; outer mostly subequal to, sometimes surpassing inner at flowering. **Florets** 40–60+. $2n = 18$.

Flowering Mar–Oct. Moist to dry lowland prairies, coastal forests, open coniferous woodlands; 10–1800 m; B.C.; Calif., Idaho, Oreg., Wash.

Variety *leptophylla* is most commonly found west of the Cascade Mountains from Vancouver Island through the Puget Sound and Willamette Valley to the Siskiyou-Klamath Mountains region of southwestern Oregon and northwestern California. It also occurs sporadically in mesic forest areas on the eastern slopes of the Cascade Mountains, and disjunctively in the Selkirk-Clearwater Mountains region of British Columbia and northern Idaho. In the Selkirk-Clearwater Mountains region, Columbia River Gorge, southern Willamette Valley, and Siskiyou-Klamath Mountains region var. *grandiflora* and var. *leptophylla* are sympatric and appear to be introgressive. In those regions, intermediate specimens are not uncommon. It may be one of the parental taxa of *A. ×elata* (which see), especially the Puget Sound-Willamette Valley populations.

6. Agoseris retrorsa (Bentham) Greene, Pittonia 2: 178. 1891 • Spearleaf agoseris

Macrorhynchus retrorsus Bentham, Pl. Hartw., 320. 1849; *M. angustifolius* Kellogg

Stems 0. **Leaves** erect to ascending; petioles often purplish; blades linear to linear-elliptic, (7–)10–30(–36) cm, margins usually lobed to pinnatifid, rarely toothed or entire, lobes (4–)7–9(–11) pairs, linear to lanceolate, mostly retrorse, sometimes spreading, lobules often present, faces sparsely villous to tomentose. **Peduncles** elongating after flowering, 15–65(–94) cm in fruit, ± glabrate, or apically villous to tomentose, eglandular. **Involucres** cylindric to obconic or campanulate, 4–6 cm in fruit. **Phyllaries** in 3–5 series, medially rosy purple or all green, lacking darker spots or midstripes, margins ciliate to tomentose, faces glabrous or villous, eglandular; outer erect to spreading, adaxially glabrous or tomentose; inner erect, often precociously elongating and much surpassing outer. **Receptacles** epaleate. **Florets** 10–100; corollas yellow, tubes 8–20 mm, ligules 6–15 × 1.5–2.5 mm; anthers 2–5 mm. **Cypselae** (15–)20–31 mm, bodies narrowly obconic, 5–7 mm, beaks (10–)15–25 mm, lengths mostly 3–4 times bodies; ribs strongly ridged, straight, often minutely cinereous-pannose; **pappus bristles** in 4–6 series, (11–)15–20 mm. $2n = 18$.

Flowering Apr–Aug. Mesic to dry habitats in scrublands, chaparral, steppe, and open oak or pine woodlands; 400–2300 m; Ariz., Calif., Nev., Oreg., Utah, Wash.; Mexico (Baja California).

Agoseris retrorsa appears to be most closely related to *A. grandiflora*. It superficially resembles *A. parviflora* and the two are sometimes confused. *Agoseris retrorsa* occurs primarily west of the Sierra Nevada; *A. parviflora* occurs primarily east of the same range. Cypsela characteristics will quickly separate them. Putative hybrids between *A. retrorsa* and *A. grandiflora*, *A. hirsuta*, and *A. parviflora* have been identified.

7. **Agoseris apargioides** (Lessing) Greene, Pittonia 2: 177. 1891 • Seaside agoseris E

Troximon apargioides Lessing, Linnaea 6: 501. 1831;

Stems 0 or 1–5+ (becoming buried by drifting sand and appearing pseudorhizomatous). **Leaves** usually reclining to decumbent, sometimes erect; blades mostly oblanceolate to spatulate, sometimes nearly linear, 3–15 cm, margins usually dentate to lobed or pinnatifid, rarely entire, lobes 3–5(–7) pairs, filiform to spatulate, spreading to antrorse, lobules mostly 0, faces glabrous or densely hairy. **Peduncles** ± elongating after flowering, 7–45 cm in fruit, glabrous or glabrate to hairy, often villous basally, sometimes villous to tomentose apically, sometimes stipitate-glandular. **Involucres** obconic to hemispheric, 1.5–2.5 cm in fruit. **Phyllaries** imbricate (sometimes subequal) in 2–3 series, green or medially rosy purple, often spotted and/or with purple-black midstripes, margins ciliate to tomentose, faces usually ± villous, sometimes glabrous, sometimes stipitate-glandular; outer mostly spreading, adaxially usually ± tomentose, rarely glabrous; inner erect, elongating after flowering. **Receptacles** epaleate. **Florets** 25–200; corollas yellow, tubes 2–5.5 mm, ligules 3–16 × 1–3 mm; anthers 1.5–4.5 mm. **Cypselae** 5–12 mm; bodies fusiform to obconic, 3–5 mm, beaks (1–)3–8 mm, lengths mostly 1–2 times bodies; **pappus bristles** in 2–3 series, 4–9 mm. $2n = 36$.

Varieties 3 (3 in the flora): North America.

A misinterpretation of the type description of *Agoseris apargioides* resulted in its confusion with *A. hirsuta* during the latter half of the twentieth century; the two species are not conspecific. *Agoseris apargioides* (in the strict sense) here includes what most authors of recent floras have called *A. apargioides* subsp. *maritima* and/or var. *eastwoodiae*. It occurs on coastal dunes along the Pacific coast from central California to Washington. A unique feature of *A. apargioides* is that its stems become progressively buried by drifting sand, leaving a terminal rosette of leaves exposed, the plants thus appearing pseudorhizomatous.

Agoseris apargioides is part of a close alliance that includes *A. heterophylla*, *A. hirsuta*, and *A. coronopifolia* from South America. Exact relationships within this group are not clear. Putative hybrids between *A. apargioides* and *A. heterophylla* var. *cryptopleura*, *A. hirsuta*, and *A. grandiflora* var. *grandiflora* have been identified.

1. Ligules 3–6 mm; anthers 1.5–2.5 mm; phyllaries glabrous or villous, eglandular . . . . 7c. *Agoseris apargioides* var. *maritima*
1. Ligules 8–16 mm; anthers 3.5–4.5 mm; phyllaries glabrous or tomentose, ± stipitate-glandular.
  2. Plants often densely villous; leaf blades mostly oblanceolate to spatulate, margins dentate to lobed, lobes oblanceolate to spatulate; phyllaries densely villous to tomentose . . . . . 7b. *Agoseris apargioides* var. *eastwoodiae*
  2. Plants mostly glabrous or sparsely villous; leaf blades usually oblanceolate, sometimes linear, rarely spatulate, margins entire or laciniately pinnatifid, lobes filiform to lanceolate; phyllaries glabrous or villous . . . 7a. *Agoseris apargioides* var. *apargioides*

7a. **Agoseris apargioides** (Lessing) Greene var. **apargioides** E

Agoseris humilis (Bentham) Kuntze

Leaves mostly reclining to prostrate; blades usually linear to oblanceolate, rarely spatulate, 3–10(–15) cm × 1–15(–24) mm, margins toothed to lobed, sometimes laciniately pinnatifid or entire, lobes filiform to oblanceolate, faces glabrous or sparsely villous. **Phyllaries** lanceolate to obovate or oblong, margins ciliate to tomentose, faces glabrous or villous (hairs translucent, whitish or yellowish, sometimes purple-septate), ± stipitate-glandular. **Corollas** surpassing phyllaries at flowering, ligules 8–16 mm; anthers 3.5–4.5 mm. **Cypselae** ± dimorphic, whitish or purplish, ribs ridged to subalate, straight or slightly undulate, often diminishing proximally, glabrous or scabrous. $2n = 36$.

Flowering Apr–May (year-round). Coastal dunes; 0–100 m; Calif.

Variety *apargioides* occurs from San Francisco southward to Point Sur.

7b. Agoseris apargioides (Lessing) Greene var. **eastwoodiae** (Fedde) Munz, Aliso 4: 100. 1958 • Point Reyes agoseris [E]

Agoseris eastwoodiae Fedde, Just's Bot. Jahresber. 31(1): 808. 1904 based on *A. maritima* Eastwood, Bull. Torrey Bot. Club 30: 501. 1903, not E. Sheldon 1903

Leaves mostly reclining to prostrate; blades oblanceolate to spatulate, 4–10(–13) cm × (10–)15–30 mm, margins mostly toothed to lobed, lobes oblanceolate to spatulate, faces often densely villous. **Phyllaries** ovate to obovate, margins mostly densely tomentose, densely villous to tomentose (hairs translucent, mostly yellow), ± stipitate-glandular. **Corollas** surpassing phyllaries at flowering, 8–16 mm; anthers 3.5–4.5 mm. **Cypselae** ± dimorphic, whitish or purplish, ribs ridged to subalate, straight, often diminishing proximally, glabrous or scabrous. $2n = 36$.

Flowering April–May (year-round). Coastal dunes and beach heads; 0–50 m; Calif.

Variety *eastwoodiae* occurs from Point Reyes north to about Point Arena.

7c. Agoseris apargioides (Lessing) Greene var. **maritima** (E. Sheldon) G. I. Baird, Sida 21: 716. 2004 • Oregon agoseris [E]

Agoseris maritima E. Sheldon, Bull. Torrey Bot. Club 30: 310. 1903; *A. apargioides* subsp. *maritima* (E. Sheldon) Q. Jones

Leaves mostly erect; blades oblanceolate, 4–10(–13) cm × 4–20 mm, margins entire, toothed, or lobed, lobes lanceolate to oblanceolate, faces glabrous or sparsely villous. **Phyllaries** lanceolate to oblanceolate, margins glabrous or ciliate, faces glabrous or villous (hairs opaque, mostly whitish), mostly eglandular. **Corollas** ± equal to phyllaries at flowering, ligules 3–6 mm; anthers 1.5–2.5 mm. **Cypselae** ± monomorphic, whitish, not purplish, ribs ± ridged, straight, ± uniform. $2n = 36$.

Flowering Jul–Aug. Coastal dunes, sand hills, and beach heads; 0–50 m; Calif., Oreg., Wash.

Variety *maritima* occurs from Humboldt Bay, California, to Neah Bay, Washington.

8. Agoseris hirsuta (Hooker) Greene, Pittonia 2: 177. 1891 • Coast Range agoseris [E]

Leontodon hirsutum Hooker, Fl. Bor.-Amer. 1: 296. 1833; *Macrorhynchus harfordii* Kellogg; *Taraxacum hirsutum* (Hooker) Torrey & A. Gray

Stems 0. **Leaves** erect to ascending; petioles rarely purplish, petiole margins glabrous or tomentose; blades narrowly to broadly oblanceolate, (7–)10–30 cm, margins usually lobed to pinnatifid, rarely dentate, lobes 5–7(–9) pairs, lanceolate to oblanceolate, spreading to antrorse, lobules sometimes present, faces sparsely to densely hairy. **Peduncles** not notably elongating after flowering, 10–45 cm in fruit, glabrate, or basally and apically pubescent to villous, mostly stipitate-glandular. **Involucres** obconic to campanulate, 1.5–2.5 cm in fruit. **Phyllaries** in 2–3 series, green or medially rosy purple, sometimes purple-black speckled, margins ciliate, faces ± villous, stipitate-glandular; outer erect to spreading, adaxially ± tomentose; inner erect, elongating after flowering. **Receptacles** epaleate. **Florets** 50–250; corollas yellow, tubes 3–6 mm, ligules 6–16 × 2–3 mm; anthers 3–5 mm. **Cypselae** 9–14 mm; bodies fusiform to narrowly conic, 3–5 mm, beaks 6–10 mm, lengths 1–3 times bodies, ribs flattened, ridged, or subalate, straight or slightly undulate; **pappus bristles** in 3–4 series, (4–)6–10 mm. $2n = 18$.

Flowering mostly Apr–Jun, rarely Sep–Dec. Mesic to dry grasslands, oak woodlands, and coastal scrublands; 10–700 m; Calif.

Agoseris hirsuta occurs primarily on grassy hills in the San Francisco Bay area and extends both north and south in the Coast Ranges. It has been treated as *A. apargioides* subsp. or var. *apargioides* in recent floras. *Agoseris hirsuta* is closely related to *A. apargioides* and *A. heterophylla*. Morphologically, it is similar to *A. heterophylla* var. *cryptopleura*. Putative hybrids with *A. apargioides*, *A. grandiflora* var. *grandiflora*, *A. heterophylla* var. *cryptopleura*, and *A. retrorsa* have been collected.

9. Agoseris heterophylla (Nuttall) Greene, Pittonia 2: 178. 1891 • Annual agoseris

Macrorhynchus heterophyllus Nuttall, Trans. Amer. Philos. Soc., n. s. 7: 430. 1841

Annuals. Stems 0 or 1 (erect, 0–5 cm). **Leaves** mostly erect, sometimes prostrate; petioles not purplish, margins glabrous or ciliate; blades usually oblanceolate to spatulate, rarely linear, 1–25 cm, margins entire or lobed; lobes 2–3 pairs, linear to spatulate, spreading to antrorse, lobules mostly 0, glabrous or densely hairy. **Peduncles** elongating after flowering, 3–60 cm in fruit, glabrous or glabrate, or basally puberulent and apically hairy to tomentose, sometimes stipitate-glandular. **Involucres** cylindric to hemispheric, 1–2 cm in fruit. **Phyllaries** in 2–3 series, green or medially rosy purple, sometimes purple-black spotted or tipped, subequal to unequal, margins glabrous or ciliate, faces usually puberulent to villous, mostly stipitate-glandular, sometimes glabrous; outer erect or spreading, adaxially usually villous to lanate, sometimes glabrous; inner erect, ± elongating after flowering. **Receptacles** epaleate. **Florets** 5–100(–300); corollas yellow, tubes 1–5 mm, ligules 2–15 × 1–3 mm; anthers 1–4 mm. **Cypselae** 7–16 mm, bodies mostly fusiform to obconic, sometimes tumid, 2–5(–10) mm, beaks 5–11 mm, lengths 1–4 times bodies, ribs 0 or alate, straight to strongly undulate, uniform or diminishing proximally; **pappus bristles** in 2–3 series, 4–9 mm.

Varieties 3 (3 in the flora): w North America, nw Mexico (including Guadalupe Island); introduced in Europe (Sweden).

1. Ligules 10–15 mm, much surpassing phyllaries; anthers 2–4 mm; leaf blades toothed to lobed, lobes mostly 3–4(–5) pairs . . . . 9b. *Agoseris heterophylla* var. *cryptopleura*
1. Ligules 2–4 mm, subequaling phyllaries; anthers 1–1.5 mm; leaf margins entire, toothed, or lobed, lobes mostly 2–3 pairs.
   2. Peduncle lengths 0.5–3 times leaves at flowering; leaf blades glabrous abaxially, pubescent adaxially; peduncles mostly glabrate, or apically tomentose . . . . 9c. *Agoseris heterophylla* var. *quentinii*
   2. Peduncle lengths mostly 1.5–4.5 times leaves at flowering; leaf blades uniformly glabrous or hairy; peduncles ± glabrate, or apically hairy to villous, sometimes glabrous . . . . 9a. *Agoseris heterophylla* var. *heterophylla*

9a. Agoseris heterophylla (Nuttall) Greene var. heterophylla

Agoseris heterophylla var. *glabra* (Nuttall) Howell; *A. heterophylla* subsp. *glabrata* (Suksdorf) Piper; *A. heterophylla* subsp. *normalis* Piper; *Troximon heterophyllum* (Nuttall) Greene var. *cryptopleuroides* Suksdorf

Stems 0 or 1 (internodes much shorter than subtending leaves). **Leaves** erect to spreading; blades (1–)3–15(–23) cm × (0.5–)2–10(–16) mm, mostly toothed, rarely lobed, lobes or teeth in 2–3 pairs, faces glabrous or densely, ± uniformly hairy. **Peduncles** 6–53 cm, lengths mostly (0.5–)1.5–4.5 times leaves at flowering, (1.5–)2–5 times leaves in fruit,

glabrous or glabrate, or apically puberulent to villous. **Heads** 2–10(–12) mm wide at flowering. **Phyllaries** green or medially rosy purple, sometimes spotted, margins glabrous or ciliate, faces mostly hairy, occasionally glabrous, or villous, stipitate-glandular, translucent, yellowish or purple-septate and often purple-tipped trichomes, or eglandular with whitish opaque trichomes; outer erect to spreading, adaxially glabrous or puberulent and eglandular. **Florets** (5–)20–50(–100); corollas ± equaling phyllaries at flowering, tubes 1–4 mm, ligules 2–6 × 1–2 mm; anthers 1–1.5 mm. **Cypselae**: outermost strongly differing from inner, highly variable in color, shape, ornamentation, and pubescence; ribs flattened to alate, or 0, straight to strongly undulate, not diminishing proximally. $2n = 18, 36$.

Flowering Mar–Sep. Wet to dry, mostly seasonal habitats in deserts, grasslands, chaparral, steppe, oak woodlands, and open pine forests, often disturbed sites; 0–2300 m; B.C.; Ariz., Calif., Idaho, Mont., Nev., N.Mex., Oreg., Utah, Wash.; Mexico (Baja California, Guadalupe Island); introduced, Europe (Sweden).

Variety *heterophylla* is the most widespread of the varieties. It exhibits remarkable variation in cypsela morphology. K. L. Chambers (1963b) demonstrated that this small-flowered variety is strongly autogamous and does not outcross. The same breeding system may be present in other *Agoseris* taxa that exhibit reduced corolla and anther size (e.g., *A. apargioides* var. *maritima*, *A. grandiflora*, *A. aurantiaca*). Hybridization does not appear to be common; apparent intermediates with *A. apargioides* and *A. hirsuta* are known.

9b. Agoseris heterophylla (Nuttall) Greene var. cryptopleura Greene, Pittonia 2: 179. 1891

• California agoseris E

Agoseris californica (Nuttall) Hoover; *A. heterophylla* subsp. *californica* (Nuttall) Piper; *A. heterophylla* var. *californica* (Nuttall) Davidson & Moxley; *A. heterophylla* var. *crenulata* (H. M. Hall) Jepson; *A. heterophylla* var. *turgida* (H. M. Hall) Jepson; *A. major* Jepson ex Greene; *Cryptopleura californica* Nuttall

Stems 0 or 1 (internodes to ¹⁄₂ times lengths of subtending leaves). **Leaves** erect to spreading; blades (2–)5–15 (–24) cm × (1.5–)2–10(–16) mm, mostly toothed to lobed, rarely entire, teeth or lobes (2–)3–5 pairs, faces usually densely, ± uniformly hairy, rarely glabrous. **Peduncles** 6–60 cm, lengths mostly 1.5–4 times leaves at flowering, 2–5 times length of leaves in fruit, glabrous or glabrate, or apically hairy to tomentose. **Heads** (7–)10–18 mm wide (flowering). **Phyllaries** green or medially rosy purple, sometimes spotted, margins glabrous or ciliate, faces mostly hairy, occasionally glabrous or villous, stipitate-glandular, with translucent, yellowish or purple-septate, often purple-tipped hairs or eglandular with whitish-opaque hairs; outer erect to spreading, adaxially glabrous or pubescent and eglandular. **Florets** 20–100(–300); corollas much surpassing phyllaries at flowering, tubes 2–5 mm, ligules 10–15 × 2–3 mm; anthers 2–4 mm. **Cypselae:** outermost strongly differing from inner, highly variable as to color, shape, ornamentation, and pubescence, ribs flattened to alate, or 0, straight to strongly undulate, often strongly diminishing proximally. $2n = 18$.

Flowering Mar–Sep. Mesic to dry habitats in grasslands, chaparral, oak woodlands, and open pine forests; 150–2100 m; Calif.

Variety *cryptopleura* occurs on hills and ranges surrounding the Great Central Valley (it appears to be absent from the valley itself). It is ± sympatric with var. *heterophylla*, except that var. *cryptopleura* almost completely supplants var. *heterophylla* in the south Coast Ranges of California. Older floras treated this taxon as var. *californica*; more recent works have not recognized it at all. K. L. Chambers (1963b) demonstrated that var. *cryptopleura* is strictly allogamous and is not self-fertile. Varieties *heterophylla* and *cryptopleura* are differentiated almost entirely by corolla size, which appears to be correlated with breeding system; otherwise the two exhibit almost complete morphologic overlap. Corolla size in depauperate specimens of var. *cryptopleura* approaches that found in robust specimens of var. *heterophylla*. The two varieties still separate on corollas longer than the phyllaries in var. *cryptopleura* versus subequaling the phyllaries in var. *heterophylla*. The degree of introgression, if any, between var. *cryptopleura* and var. *heterophylla* is not known.

9c. **Agoseris heterophylla** (Nuttall) Greene var. **quentinii** G. I. Baird, Sida 21: 271. 2004 • Arizona agoseris E

Stems 0. **Leaves** spreading to prostrate; blades 2–12 cm × 3–9 (–12) mm, mostly lobed, lobes 2–3 pairs, abaxially glabrous, adaxially pubescent. **Peduncles** 0–26 cm, lengths mostly less than 0.5 times leaves at flowering, 0.5–3 times leaves in fruit, ± glabrate, or apically tomentose. **Florets** 15–30+; corollas ± equaling phyllaries at flowering, tubes 2–3 mm, ligules 2–3 × 0.8–1.5 mm; anthers ca. 1 mm.

Flowering Mar–Jun. Mesic to dry habitats in deserts, grasslands, and oak woodlands; 1200–2000 m; Ariz., N.Mex.

The relatively small corollas and anthers of var. *quentinii* suggest that it, too, may be autogamous, as is var. *heterophylla*.

10. **Agoseris ×elata** (Nuttall) Greene, Pittonia 2: 177. 1891 (as species) • Willamette agoseris E

Stylopappus elatus Nuttall, Trans. Amer. Philos. Soc., n. s. 7: 433. 1841; *Agoseris grandiflora* (Nuttall) Greene var. *laciniata* (Nuttall) Jepson; *A. laciniata* (Nuttall) Greene; *A. tenuifolia* Rydberg; *Stylopappus laciniatus* Nuttall; *S. laciniatus* var. *longifolius* Nuttall; *Troximon grandiflorum* A. Gray var. *laciniatum* (Nuttall) A. Gray; *T. grandiflorum* var. *tenuifolium* A. Gray; *T. nuttallii* A. Gray

Stems 0. **Leaves** erect to ascending; blades usually oblanceolate, sometimes lanceolate or obovate, 15–25 cm, margins usually toothed to lobed or pinnatifid, rarely entire, lobes 2–4 pairs, lanceolate to triangular, antrorse to spreading, lobules mostly 0, faces glabrous and glaucous or sparsely villous. **Peduncles** ± elongating after flowering, 15–65(–90) cm in fruit, glabrate, or apically pubescent or villous, ± stipitate-glandular. **Involucres** obconic to campanulate, 2–4 cm in fruit. **Phyllaries** in 2–4 series, medially rosy purple, sometimes purple-black apically, rarely all green or purple-black spotted, faces pubescent to villous, ± stipitate-glandular; outer mostly spreading, adaxially ± villous and eglandular; inner erect, elongating after flowering. **Receptacles** epaleate. **Florets** (25–)50–150; corollas orange or yellow, tubes 8–10 mm, ligules 6–8(–12) × 1–3 mm; anthers 2–3(–5) mm. **Cypselae** (11–)14–20 mm, bodies ± fusiform, (6–)8–10 mm, beaks 5–10 mm, mostly equaling bodies; ribs broadly ridged, straight; **pappus bristles** in 2–3 series, 10–14 mm.

Flowering Jun–Sep. Lowland prairies; 10–100 m; Oreg., Wash.; and montane meadows or open pine forests; 1400–2800 m; Calif., Oreg., Wash.

Agoseris ×elata has been and continues to be an enigmatic taxon. Perhaps no other name in the genus has been so misunderstood and misapplied. Many herbarium specimens labeled *A. ×elata* are in fact misidentified. Specimens of *A. ×elata* are not abundant in herbaria; the number of actual collections is relatively small compared to those for other *Agoseris*. Specimens that belong to *A. ×elata* represent a complex assemblage that has relatively few defining features and appears to be of hybrid origin. Most specimens appear to be intermediate between *A. grandiflora* and *A. aurantiaca*; most also appear to have characteristics of *A. monticola* or *A. glauca* var. *dasycephala*. The exact parentage remains unclear.

Agoseris ×elata occurs in two geographically separated populations, which cannot be consistently distinguished morphologically: one mainly in scattered lowland prairie locations in the Puget Sound and Willamette Valley areas (the type collection came from this population; no new collections have been taken from this region in over 65 years; it is likely extirpated) and another at high elevations in California, primarily in the Lake Tahoe region and southward in the Sierra Nevada.

67. NOTHOCALAÏS (A. Gray) Greene, Bull. Calif. Acad. Sci. 2: 54. 1886 • False agoseris or dandelion [Greek *notho-*, false, and *Calaïs*, a synonym of *Microseris*]   E

Kenton L. Chambers

Microseris D. Don sect. *Nothocalaïs* A. Gray in A. Gray et al., Syn. Fl. N. Amer. 1(2): 420. 1884

Perennials, 3–45 cm; taprooted, with caudices often multicipital (taproots thick, fleshy, with blackish periderm, lateral rootlets often borne in clusters on knoblike projections). **Stems** 1–5+, erect, scapiform, naked (rarely with 1–3 bracteate nodes near base in *N. troximoides* and *N. nigrescens*), glabrous or white-villous, especially near heads. **Leaves** basal; petiolate (bases attenuate or not); blades linear to oblanceolate, margins usually entire (often pinnately lobed in *N. alpestris*), sometimes undulate (usually white-ciliolate in *N. cuspidata* and *N. troximoides*; faces glabrous or lightly villous). **Heads** borne singly (erect, liguliferous). **Peduncles** not inflated distally, not bracteate. **Calyculi** 0. **Involucres** broadly to narrowly ovoid, campanulate at flowering, 5–20 mm diam. **Phyllaries** 8–50 in 2–5 series, (often longitudinally striped or finely dotted with red or purple) lanceolate to ovate, equal or unequal (outer shorter), herbaceous (thin), midnerves inconspicuous, apices acute to acuminate, faces glabrous or white-villous. **Receptacles** flat, pitted, glabrous, epaleate. **Florets** 13–100; corollas yellow, often reddish abaxially (much surpassing phyllaries in flowering, tubes hairy). **Cypselae** brown or gray, narrowly columnar to fusiform, usually narrowed distally, not beaked, ribs ca. 10, glabrous or distally scabrous; **pappi** fragile, of 10–80, distinct, lustrous, white, ± equal, smooth to barbellulate bristles or aristate scales. $x = 9$.

Species 4 (4 in the flora): c, w North America.

Because pappi of *Nothocalaïs* taxa vary from capillary bristles to aristate scales, the species were earlier assigned to either *Agoseris* or *Microseris*. The totality of morphologic evidence supports the unity of the genus, as well as its segregation from *Agoseris* and *Microseris* (K. L. Chambers 1955, 1957). According to recent phylogenetic studies based on chloroplast DNA (R. K. Jansen et al. 1991b; J. Whitton et al. 1995), *Nothocalaïs* is most closely related to *Uropappus* and *Agoseris*, the three genera together comprising a clade sister to *Microseris*.

SELECTED REFERENCE Chambers, K. L. 1955. A biosystematic study of the annual species of *Microseris*. Contr. Dudley Herb. 4: 207–312.

1. Pappi of 30–50 barbellulate bristles; leaf margins usually coarsely toothed or pinnately lobed, sometimes entire; phyllaries glabrous, evenly and minutely purple-dotted; subalpine, Cascade Range, Klamath Mountains, and (rarely) Sierra Nevada . . . . 1. *Nothocalaïs alpestris*
1. Pappi of 10–30 aristate scales or of 40–80 intergradent, smooth to barbellulate bristles and ± subulate to setiform scales; leaf margins entire, sometimes undulate; phyllaries glabrous or villous, sometimes purple-dotted; Great Plains, n, mid Rocky Mountains, Columbia-Snake Rivers Plateau, n Great Basin.
   2. Pappi of 40–80 intergradent, smooth to barbellulate bristles and ± subulate to setiform scales; Great Plains . . . . 2. *Nothocalaïs cuspidata*
   2. Pappi of 10–30 aristate scales; n, mid Rocky Mountains, Columbia-Snake Rivers Plateau, n Great Basin.

[3. Shifted to left margin.—Ed.]

3. Leaves lanceolate to oblanceolate, margins plane, apices acute; phyllaries broadly lanceolate to ovate, apices acute to acuminate, faces usually glabrous, minutely purple-dotted; n, mid Rocky Mountains . . . . 3. *Nothocalaïs nigrescens*

3. Leaves mostly linear to linear-lanceolate, margins often undulate, apices acuminate; phyllaries lanceolate, apices acuminate, faces glabrous or villous especially on midnerves, margins often ciliolate, green or with purple-lined midnerves, sometimes also minutely purple-dotted; n Great Basin, Columbia-Snake Rivers Plateau . . . . 4. *Nothocalaïs troximoides*

1. **Nothocalaïs alpestris** (A. Gray) K. L. Chambers, Contr. Dudley Herb. 5: 66. 1957 • Alpine lake false dandelion E

Troximon alpestre A. Gray, Proc. Amer. Acad. Arts 19: 70. 1883; *Agoseris alpestris* (A. Gray) Greene; *Microseris alpestris* (A. Gray) Q. Jones ex Cronquist

Plants 3–45 cm. **Stems** (peduncles) ebracteate. **Leaves:** blades linear to oblanceolate, 3–20 cm, bases attenuate to obtuse, margins plane, usually coarsely toothed or pinnately lobed (lobes spreading or retrorse), sometimes entire, sometimes ciliolate, faces glabrous. **Involucres** 10–20 mm. **Phyllaries** 8–35, evenly and minutely purple-dotted, ovate-lanceolate, apices acuminate, faces glabrous. **Florets** 13–75; ligules 10–20 mm. **Cypselae** brown, 5–10 mm, sometimes narrowed distally 1–3 mm, not beaked; **pappi** of 30–50, barbellulate bristles 6–10 mm. $2n = 18$.

Flowering Jul–Sep. Meadows, rocky slopes, pumice flats, mixed subalpine woodland and *Abies* subalpine zone; 1300–2500 m; Calif., Oreg., Wash.

Nothocalaïs alpestris grows in the subalpine Cascade Range, Klamath Mountains, and (rarely) the Sierra Nevada.

2. **Nothocalaïs cuspidata** (Pursh) Greene, Bull. Calif. Acad. Sci. 2: 55. 1886 • Prairie false dandelion E

Troximon cuspidatum Pursh, Fl. Amer. Sept. 2: 742. 1813 (as Troximum); *Agoseris cuspidata* (Pursh) Steudel; *Microseris cuspidata* (Pursh) Schultz-Bipontinus

Plants 7–35 cm. **Stems** (peduncles) ebracteate. **Leaves:** blades linear-lanceolate, 7–30 cm, bases broadly attenuate, margins entire, plane or undulate, ciliolate, (apices acuminate) faces glabrous or villous. **Involucres** 17–27 mm. **Phyllaries** 13–34, often red-striped or dotted, lanceolate, apices acuminate, faces glabrous. **Florets** 13–80; ligules 15–25 mm. **Cypselae** brown, 7–10 mm, narrowed distally, not beaked; **pappi**

of 40–80, intergradent, smooth to barbellulate bristles and ± subulate to setiform scales 8–10 mm. $2n = 18$.

Flowering Apr–Jul. Prairie pastures, slopes, hillsides, and ridges, in sandy, gravelly, or clay soils, various grassland associations; 300–2300 m; Alta., Man., Sask.; Ark., Colo., Ill., Iowa, Kans., Minn., Mo., Mont., Nebr., N.Mex., N.Dak., Okla., S.Dak., Tex., Wis., Wyo.

3. **Nothocalaïs nigrescens** (L. F. Henderson) A. Heller, Muhlenbergia 1: 8. 1900 • Speckled false dandelion E

Microseris nigrescens L. F. Henderson, Bull. Torrey Bot. Club 27: 348. 1900

Plants 5–35 cm. **Stems** (peduncles) often with 1–2, leafy bracts near bases. **Leaves:** blades lanceolate to oblanceolate, 4–20 cm, margins plane, entire, sometimes ciliolate, (apices acute) faces glabrous. **Involucres** 12–22 mm. **Phyllaries** 15–50, minutely purple-dotted except on membranous margins, sometimes purple-lined on midribs, broadly lanceolate to ovate, apices acute to acuminate, faces usually glabrous. **Florets** 13–100; ligules 10–25 mm. **Cypselae** brown, 6–10 mm, tapered distally, not beaked; **pappi** of 10–25, narrow, smooth to barbellulate, aristate scales 8–13 mm.

Flowering May–Jul. Moist meadows, grassy slopes, and forest openings, juniper-ponderosa pine woodland, mixed conifer forest, and spruce-fir subalpine forest; 1500–3000 m; Idaho, Mont., Wyo.

Nothocalaïs nigrescens grows in the northern and middle Rocky Mountains. It is generally well separated in habitat and elevation from *N. cuspidata* and *N. troximoides*; intermediate populations are sometimes found in areas of contact with the two related taxa. In Montana, apparent hybrids with *N. cuspidata* have been found in the upper valleys of the Yellowstone and Missouri rivers. Intermediates between *N. nigrescens* and *N. troximoides* occur rather widely in Idaho, in areas where the sagebrush steppe habitats of the latter species penetrate the coniferous zone. As discussed below, most of the assumed hybrid populations more closely resemble *N. troximoides*.

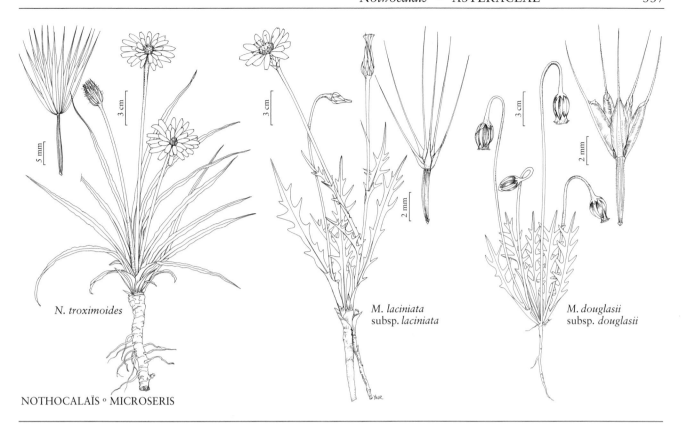

N. troximoides

M. laciniata
subsp. *laciniata*

M. douglasii
subsp. *douglasii*

NOTHOCALAÏS ∘ MICROSERIS

4. Nothocalaïs troximoides (A. Gray) Greene, Bull. Calif. Acad. Sci. 2: 55. 1886 • Sagebrush false dandelion E F

Microseris troximoides A. Gray, Proc. Amer. Acad. Arts 9: 211. 1874; *Scorzonella troximoides* (A. Gray) Jepson

Plants 5–45 cm. **Stems** (peduncles) sometimes with 1–3, leafy bracts proximally. **Leaves:** blades linear to linear-lanceolate, 7–30 cm, margins entire, often undulate, often ciliolate, (apices acuminate) faces glabrous or villous. **Involucres** 14–30 mm. **Phyllaries** 8–25, green to minutely purple-dotted, often with purple-lined midnerves, lanceolate, (apices acuminate) faces glabrous or villous, especially on margins and midribs. **Florets** 13–90; ligules 10–25 mm. **Cypselae** light brown or gray, 7–13 mm, sometimes narrowed distally, not beaked; **pappi** of 10–30, aristate scales 10–17 mm. $2n = 18$.

Flowering Mar–Jun. Vernally moist to dry flats and hillsides, in bunchgrass intermountain grasslands, sagebrush steppe, and open pinyon-juniper and ponderosa pine woodlands, often in stony soils; 50–2000 m; B.C.; Calif., Idaho, Mont., Nev., Oreg., Utah, Wash., Wyo.

Nothocalaïs troximoides grows in the northern Great Basin and in Columbia-Snake Rivers Plateau. It shows increased variability in Idaho and parts of Montana, probably as a result of introgression from *N. nigrescens*. This variability is most evident in the lower Snake River plain and adjacent Oregon, in dryland habitats characteristic of *N. troximoides*, where populations frequently include individuals with roots, leaves, and cypselae typical of *N. troximoides* but with unusually large involucres and phyllaries that are broadly lanceolate, acuminate, and densely speckled with purple.

68. MICROSERIS D. Don, Philos. Mag. Ann. Chem. 11: 388. 1832 • Silverpuffs [Greek *micro-*, small, and *seris*, endive or chicory]

Kenton L. Chambers

Apargidium Torrey & A. Gray; *Calaïs* de Candolle; *Scorzonella* Nuttall

Annuals or perennials, 5–120 cm; taprooted or with caudices (in perennial species; *M. borealis* rhizomatous). **Stems** 1–30+, erect, simple or relatively few- to many-branched (naked or leafy proximally and often distally), glabrous or scurfy-pubescent (especially proximal to heads). **Leaves** mostly basal, cauline 0 or reduced; petiolate (petioles broad to narrow); blades linear to lanceolate or oblanceolate, margins entire, lacerate, dentate, or pinnately lobed (often with narrow rachises and linear lobes; apices acuminate or acute to obtuse, faces glabrous or lightly scurfy-puberulent). **Heads** borne singly (nodding or inclined in bud, erect in flower and fruit). **Peduncles** (erect or curved-ascending) not distally inflated, ebracteate (annuals) or leafy (perennials except *M. borealis*). **Calyculi** 0 (outer phyllaries forming calyculiform series in annuals). **Involucres** fusiform, ovoid, globose, or campanulate, 3–30 mm diam. **Phyllaries** 5–40 in 3–5 series, unequal (outer usually shorter, ± deltate, inner ± lanceolate), herbaceous (midveins often thickened; abaxial faces glabrous or scurfy-puberulent, sometimes black-villous, often adaxially black-villous and minutely white-strigillose). **Receptacles** flat to low-convex, pitted, glabrous, epaleate. **Florets** 5–300; corollas yellow to orange or white, outer often purplish abaxially. **Cypselae** gray to brown or purplish, sometimes purplish-spotted, columnar, obconic, or fusiform (basal callosities knoblike), apices truncate, ribs 10–15, smooth or scabrous (white-villous on marginal cypselae in some species); **pappi** persistent, usually of 5–30, silvery to yellowish, brownish, or blackish aristate scales (often reduced to 0–4 in *M. douglasii*, of 24–48 bristles in *M. borealis*), scale bodies deltate, lanceolate, oblong, ovate, orbiculate, or linear, apices obtuse to acute or lacerate, faces glabrous or villous, aristae barbellulate to barbellate or plumose. $x = 9$.

Species 14 (11 in the flora): w North America, South America, Pacific Islands (New Zealand), Australia.

A broad circumscription of *Microseris*, including *Apargidium* and excluding *Nothocalaïs*, has usually been accepted (e.g., K. L. Chambers 1955, 1960). Recently, molecular data have led to reinstatement of the monotypic genus *Uropappus* and separation of two other species as the allotetraploid genus *Stebbinsoseris* (R. K. Jansen et al. 1991b; Chambers 1993c). A large body of literature has resulted from use of *Microseris* as a model genetic system by K. Bachmann and colleagues (e.g., Bachmann et al. 1979; Bachmann 1992; Bachmann and J. Battjes 1994). Differences in the diploid DNA amount within and between species have been studied by H. J. Price and colleagues (Price and Bachmann 1975; Price et al. 1981, 1983). Additional genetic studies, not referenced here, have involved three species from Australia, New Zealand, and Chile, widely disjunct from the main center of distribution in western North America. Ten of the species are diploid ($2n = 18$); the four tetraploid species ($2n = 36$) are of alloploid origin. The nine North American perennial taxa are closely related and mostly allopatric, occupying different habitats or climatic zones. The five annual species, which sometimes occur in sympatric clusters, are difficult to distinguish without the presence of cypselae.

In keys and descriptions, measurements of pappus scales exclude aristae.

SELECTED REFERENCE Chambers, K. L. 1955. A biosystematic study of the annual species of *Microseris*. Contr. Dudley Herb. 4: 207–312.

1. Perennials (usually caulescent); outer phyllaries shorter than to nearly equaling inner; corollas yellow, surpassing phyllaries by 5+ mm; pappi usually of 5–30 scales (24–48 bristles in *M. borealis*).

    2. Plants rhizomatous; leaves entire or remotely denticulate; pappi of 24–48, brownish bristles; principally in coastal and montane sphagnum bogs . . . . 1. *Microseris borealis*

    2. Plants taprooted; leaves entire or toothed to lacerate or pinnately lobed; pappi of 5–30, aristate scales; marshes, fields, pastures, hillsides, brushlands, and woodlands.

       3. Pappi of 15–30, silvery, aristate scales, aristae plumose; widespread . . . . 2. *Microseris nutans*

       3. Pappi 5–24, silvery to dull yellowish or brownish, aristate scales, aristae barbellulate or barbellate to subplumose; Pacific Coast states.

          4. Pappus scales 4–10 mm, aristae barbellate or subplumose.

             5. Stems branched or simple (usually leafy proximal to midstems); phyllary apices recurved; pappi of 5–10 dull, yellowish brown, aristate scales; c California . . . . 4. *Microseris sylvatica*

             5. Stems simple (leafy proximally); phyllary apices erect; pappi of 9–15, silvery to dull white, aristate scales; s Jackson County, Oregon, adjacent California . . . . 3. *Microseris laciniata* (in part)

          4. Pappus scales 0.5–4 mm (3–6 mm in 6. *M. howellii*), aristae barbellulate to barbellate.

             6. Outer phyllaries often purple-spotted, ovate-lanceolate to broadly ovate, 2.5–9 mm wide, apices acute to cuspidate, faces usually glabrous . . . . 3. *Microseris laciniata* (in part)

             6. Outer phyllaries not or rarely purple-spotted, linear, deltate, or lanceolate, 0.5–2.5 mm wide, apices acute to acuminate, faces often scurfy-pubescent, sometimes black-villous.

                7. Pappi of 8–24, white, aristate scales, aristae barbellate; Klamath Mountains, Oregon, California . . . . 3. *Microseris laciniata* (in part)

                7. Pappi of 5–10, usually white, aristate scales, aristae barbellulate (or pappi brownish and aristae barbellate).

                   8. Pappi dull yellowish brown, aristae barbellate; phyllaries usually black-villous abaxially; coastal c California . . . . 5. *Microseris paludosa*

                   8. Pappi white, aristae usually barbellulate, rarely barbellate; phyllaries sometimes black-villous abaxially.

                      9. Pappus scales 3–6 mm; Klamath Mountains, Oregon . . . . . 6. *Microseris howellii*

                      9. Pappus scales 0.5–2.5 mm; Klamath Mountains, Coast Ranges, Oregon, n California . . . . 3. *Microseris laciniata* (in part)

1. Annuals (acaulous); outer phyllaries notably shorter than inner (forming calyculiform series); corollas yellow, white, or orange, equaling or surpassing phyllaries by 1–3 mm; pappi of 5 aristate scales (0–5 in *M. douglasii*).

    10. Pappus scales linear-lanceolate, 4–11 mm, margins scarcely involute, midveins stout, $^1/_5$–$^1/_3$ widths of bodies, aristae barbellate; principally Sacramento Valley, California, and surrounding foothills . . . . 8. *Microseris acuminata*

    10. Pappus scales lanceolate, ovate, or orbiculate to deltate, 0.2–7 mm (if more than 4 mm, lanceolate to ovate and margins distinctly involute, midveins tapering from thickened base or linear, less than $^1/_5$ widths of bodies, aristae barbellate to barbellulate; Sacramento Valley and elsewhere.

       11. Cypselae usually 1.5–3 mm.

          12. Cypselae columnar to obconic.

             13. Pappus scales 2–7 mm, involute, aristae stout, barbellate; sw California, n Baja California . . . . 7. *Microseris douglasii* (in part)

             13. Pappus scales 0.5–2.5 mm, scarcely involute, aristae fine, barbellulate; n Baja California northward . . . . 11. *Microseris elegans*

          12. Cypselae truncate-fusiform (tapering proximally, widest beyond middles, slightly narrowed distally).

14. Pappus scales 5, 1–4 mm, aristae fine, barbellulate; strictly coastal, c California to British Columbia .... 10. *Microseris bigelovii*

14. Pappus scales 0–5, 0.5–1 mm, aristae ± stout, barbellate; coastal and elsewhere, sw to c California .... 7. *Microseris douglasii* (in part)

[11. Shifted to left margin.—Ed.]

11. Cypselae usually 3–10 mm.

  15. Pappus scales 0.5–1 mm .... 7. *Microseris douglasii* (in part)

  15. Pappus scales 1–7 mm.

    16. Pappus scales plane or slightly curved, arcuate only at bases, 1–4 mm, margins not involute, midveins linear, thicker only at bases; strictly coastal, c California to British Columbia .... 10. *Microseris bigelovii*

    16. Pappus scales ± arcuate throughout, 1–7 mm, margins ± involute, midveins tapering distally from thick bases; absent from immediate coastal area of c California.

      17. Pappus scales 5, glabrous, usually white, rarely brownish, margins plane or slightly involute, aristae barbellulate proximally, barbellate distally; San Joaquin Valley, surrounding foothills, California .... 9. *Microseris campestris*

      17. Pappus scales 0–5, often villous, light to dark, margins usually distinctly involute, aristae wholly barbellate; widespread .... 7. *Microseris douglasii* (in part)

1. **Microseris borealis** (Bongard) Schultz-Bipontinus, Jahresber. Pollichia 22–24: 310. 1866 (as boreale)

• Apargidium, northern microseris [E]

Apargia borealis Bongard, Mém. Acad. Imp. Sci. St.-Pétersbourg, Sér. 6, Sci. Math. 2: 146. 1832; *Apargidium boreale* (Bongard) Torrey & A. Gray; *Scorzonella borealis* (Bongard) Greene

Perennials, 15–70 cm; rhizomatous, with fleshy adventitious roots. **Stems** 0. **Leaves** basal; petiolate; blades mostly oblanceolate, rarely linear, 5–30 cm, margins entire or remotely denticulate, apices acute or acuminate, faces glabrous. **Peduncles** erect (15–70 cm) ebracteate. **Involucres** broadly to narrowly ovoid in fruit, 10–18 mm. **Phyllaries:** (not purple-spotted, apices erect) outer lanceolate to linear-lanceolate, apices acuminate, abaxial faces glabrous or black-villous; inner lanceolate, apices acute, both faces usually lightly black-villous. **Florets** 18–50; corollas yellow-orange, surpassing phyllaries by 5+ mm. **Cypselae** columnar or arcuate near bases, 4–8 mm; **pappi** of 24–48, brownish, barbellate bristles 5–10 mm (bases of bristles sometimes slightly widened). $2n = 18$.

Flowering Jun–-Sep. Mostly coastal and montane sphagnum bogs, other wet sites from lowlands to alpine in n part of range; 0–1800 m; B.C.; Alaska, Calif., Oreg., Wash.

2. **Microseris nutans** (Hooker) Schultz-Bipontinus, Jahresber. Pollichia 22–24: 309. 1866 • Nodding microseris or silverpuffs [E]

Scorzonella nutans Hooker, London J. Bot. 6: 253. 1847

Perennials, 10–70 cm; taprooted. **Stems** branched, leafy distally. **Leaves** basal and cauline; petiolate (proximally, distal often sessile, clasping); blades linear to oblanceolate, 5–30 cm, margins entire or remotely dentate to pinnately lobed (usually with narrow rachises and linear teeth or lobes), apices acuminate, faces glabrous or lightly scurfy-puberulent. **Peduncles** erect or ascending (4–35 cm), ebracteate or leafy. **Involucres** broadly to narrowly ovoid in fruit, 8–22 mm. **Phyllaries:** apices erect, abaxial faces glabrous or scurfy-puberulent; outer lanceolate to triangular or linear, apices acute or acuminate; inner lanceolate, apices acuminate, both faces usually lightly black-villous. **Florets** 10–75; corollas yellow, surpassing phyllaries by 5+ mm. **Cypselae** columnar, 3.5–8 mm; **pappi** of 15–30, silvery, linear to lanceolate or oblong, (flat, glabrous) aristate scales 1–3(–5) mm (margins entire, apices acute or lacerate), aristae (slender) plumose. $2n = 18$.

Flowering Apr–-Jul. Various soils, grasslands, brushlands, woodlands, and coniferous forests; 100–3000 m; Alta., B.C.; Calif., Colo., Idaho, Mont., Nev., Oreg., S.Dak., Utah, Wash., Wyo.

3. **Microseris laciniata** (Hooker) Schultz-Bipontinus, Jahresber. Pollichia 22–24: 309. 1866 • Cutleaf or cut-leaved silverpuffs [E] [F]

Hymenonema laciniatum Hooker, Fl. Bor.-Amer. 1: 301. 1833; *Scorzonella laciniata* (Hooker) Nuttall

Perennials, 15–120 cm; taprooted. **Stems** branched and leafy distally, or simple and leafy only proximally (subsp. *detlingii* and plants of extreme environments). **Leaves** basal and cauline; petiolate (distal often sessile, clasping); blades linear to broadly lanceolate, 10–50 cm, margins entire, dentate, lacerate, or pinnatifid, apices obtuse to acuminate, faces glabrous or scurfy-puberulent. **Peduncles** erect or curved-ascending, ebracteate or leafy (10–70 cm). **Involucres** globose to narrowly ovoid in fruit, 10–30 mm. **Phyllaries:** often purple-spotted (especially in subsp. *laciniata*), apices erect, abaxial faces glabrous or scurfy-puberulent (often black-villous in subsp. *leptosepala*); outer lanceolate to broadly ovate, deltate, or linear, slightly to much shorter than inner, 0.5–9 mm wide, apices cuspidate to acute; inner broadly to narrowly lanceolate, apices acuminate. **Florets** 13–300; corollas yellow, surpassing phyllaries by 5+ mm. **Cypselae** columnar, 3.5–8 mm (tapering to bases); **pappi** of 5–10(–15 in subsp. *detlingii*, or –24 in subsp. *siskiyouensis*), white to dull yellowish, deltate to lanceolate, aristate scales 0.5–8 mm, aristae barbellulate to barbellate. *2n* = 18.

Subspecies 4 (4 in the flora): w North America.

Microseris laciniata comprises four, mostly allopatric subspecies, the diagnostic features of which are found mainly in the phyllaries and pappi. These races intergrade where they come in contact, with the greatest diversity occurring in the Klamath Mountains of northern California and southern Oregon (K. L. Chambers 2004b). S. Mauthe et al. (1981) reported on a detailed morphologic analysis of the heads and pappi of *M. laciniata* and proposed that the observed variation could be explained by the interaction of a limited number of major genes. The species is consistently self-sterile and outcrossing. It also may reproduce clonally by adventitious buds borne on lateral roots.

1. Stems simple; leaves lanceolate or oblanceolate, usually entire, rarely sparingly pinnately lobed; outer phyllaries elliptic-ovate, smallest (2.5–)4 mm wide, apices acute or cuspidate; pappus scales 4–8 mm . . . . . . 3d. *Microseris laciniata* subsp. *detlingii*
1. Stems usually branched; leaves linear to lanceolate or oblanceolate, entire or pinnately lobed; outer phyllaries linear to ovate, smallest 1–9 mm wide, apices acuminate to cuspidate; pappus scales 0.5–3(–4) mm.

[2. Shifted to left margin.—Ed.]
2. Pappus scales 9–24, aristae barbellate; Klamath Mountains, Oregon and California . . . . 3c. *Microseris laciniata* subsp. *siskiyouensis*
2. Pappus scales 5–10, aristae usually barbellulate, rarely barbellate; Klamath Mountains and elsewhere.
   3. Outer phyllaries often purple-spotted, ovate-lanceolate to broadly ovate, smallest 2.5–9 mm wide, apices acute to cuspidate, abaxial faces usually glabrous; widespread . . . . 3a. *Microseris laciniata* subsp. *laciniata*
   3. Outer phyllaries rarely purple-spotted, linear to lanceolate or deltate, smallest 0.5–2.5 mm wide, apices acute to acuminate, abaxial faces often scurfy-puberulent, sometimes black-villous; principally Coast Range and Klamath Mountains, rarely e to Great Basin Region, California, Oregon, Washington . . . . 3b. *Microseris laciniata* subsp. *leptosepala*

3a. **Microseris laciniata** (Hooker) Schultz-Bipontinus subsp. **laciniata** [E] [F]

Calaïs glauca (Hooker) A. Gray var. *procera* A. Gray; *Microseris procera* (A. Gray) A. Gray; *Scorzonella procera* (A. Gray) Greene

Stems usually branched. **Leaves** usually lanceolate or oblanceolate, entire or pinnately lobed. **Outer phyllaries** often purple-spotted, broadly lanceolate to ovate, smallest 2.5–9 mm wide, apices (erect) acute to cuspidate, abaxial faces glabrous. **Pappi** of 5–10, white, deltate to lanceolate, glabrous, aristate scales 0.5–3(–4) mm, aristae barbellulate. *2n* = 18.

Flowering Apr–Aug. Clay, loam, or gravelly soils; open sites, in marshes, meadows, pastures, hillsides, shrublands, and open woods; 10–1900 m; Calif., Oreg., Wash.

Subspecies *laciniata* occurs principally away from the coast, in interior valleys and hills, rarely reaching high elevations. The width of the outer phyllaries is a convenient way to separate it from subsp. *leptosepala*, with which it intergrades in the Klamath Mountains and at various sites east of the Cascade Range. Intergradation with subsp. *detlingii* occurs in Jackson County, Oregon.

3b. Microseris laciniata (Hooker) Schultz-Bipontinus subsp. **leptosepala** (Nuttall) K. L. Chambers, Contr. Dudley Herb. 5: 61. 1957 • Slender-bracted silverpuffs [E]

Scorzonella leptosepala Nuttall, Trans. Amer. Philos. Soc., n. s. 7: 426. 1841; *Microseris leptosepala* (Nuttall) A. Gray

Stems usually branched. **Leaves** linear to narrowly lanceolate, entire or pinnately lobed. **Outer phyllaries** rarely purple-spotted, linear to lanceolate or deltate, smallest 0.5–2.5 mm wide, apices (erect) acute to acuminate, abaxial faces often scurfy-puberulent and black-villous. **Pappi** of 5–10, white, glabrous, aristate scales 0.5–2.5 mm, aristae usually barbellulate, rarely barbellate. $2n = 18$.

Flowering May–Aug. Clay, loam, and gravelly, sometimes serpentine-derived soils, open sites, meadows, hillsides, pine, oak and mixed evergreen woods; 30–2000 m; Calif., Oreg., Wash.

Subspecies *leptosepala* is known from the Klamath Mountains of California and Oregon and rare northward. It also occurs, intergrading with subsp. *laciniata*, in the California North Coast Range and east of the Cascade Range in central Oregon and northeastern California. It intergrades with subsp. *siskiyouensis* in the valleys of the Illinois and Smith rivers, southwestern Oregon and adjacent California (K. L. Chambers 2004b).

3c. Microseris laciniata (Hooker) Schultz-Bipontinus subsp. **siskiyouensis** K. L. Chambers, Sida 21: 195, figs. 1, 2A, C. 2004 • Siskiyou silverpuffs [E]

Stems usually branched. **Leaves** linear to narrowly lanceolate, entire or pinnately lobed. **Outer phyllaries** not spotted, linear to ovate-deltate, smallest 1–1.5 mm wide, apices acute to acuminate, abaxial faces usually scurfy-puberulent. **Pappi** of 9–24, white, glabrous, aristate scales 0.5–2 mm, aristae barbellate. $2n = 18$.

Flowering May–Jul. Loams or gravelly or rocky soils, rarely serpentine-derived, hillsides and valley flats, open grassy sites and woodlands; 100–2100 m; Calif., Oreg.

Subspecies *siskiyouensis* is known only from the Klamath Mountains, principally in Del Norte and Siskiyou counties, California, and Curry, Josephine, and Jackson counties, Oregon. It is found in open, rocky sites as well as woods, mostly at middle elevations. In the Illinois River Valley, Oregon, where it occurs together with *Microseris howellii*, the two are ecologically distinct. *Microseris howellii* is always found on rocky,

open serpentine substrates; subsp. *siskiyouensis* is in mixed evergreen woodlands on better developed loam soils (K. L. Chambers 2004b).

3d. Microseris laciniata (Hooker) Schultz-Bipontinus subsp. **detlingii** K. L. Chambers, Sida 21: 200, figs. 2E, F, 4. 2004 • Detling's silverpuffs [C][E]

Stems simple. **Leaves** lanceolate or oblanceolate, usually entire, rarely sparingly pinnately lobed. **Outer phyllaries** sometimes purple-spotted, elliptic-ovate, smallest (2.5–)4 mm wide, apices (erect), acute or cuspidate, abaxial faces glabrous. **Pappi** of 9–15, silvery to dull white, linear-lanceolate aristate scales 4–8 mm, aristae barbellate. $2n = 18$.

Flowering May–Jun. Heavy clay soils, open hillsides, grasslands, and shrublands; of conservation concern; 600–1500 m; Calif., Oreg.

Subspecies *detlingii* is known only from the Siskiyou Pass region and adjacent foothills of southern Jackson County, Oregon, and Siskiyou County, California. It is found only in clay soils that become very sticky when wet. The plants are unusual in their relatively long pappus scales, simple habit, and deeply penetrating taproots (K. L. Chambers 2004b). Clonal reproduction occurs by buds arising on lateral roots.

4. Microseris sylvatica (Bentham) Schultz-Bipontinus, Jahresber. Pollichia 22–24: 309. 1866 • Sylvan scorzonella, woodland silverpuffs [E]

Scorzonella sylvatica Bentham, Pl. Hartw., 320. 1849; *Calaïs sylvatica* (Bentham) A. Gray

Perennials, 15–75 cm; taprooted. **Stems** branched or simple. **Leaves** basal and cauline; petiolate (distal often sessile, clasping); blades linear to oblong-lanceolate, 8–35 cm, margins entire, dentate, or pinnately lobed, apices acute to acuminate, faces glabrous or scurfy-puberulent. **Peduncles** erect (10–55 cm), ebracteate or leafy. **Involucres** ovoid in fruit, 12–25 mm. **Phyllaries:** abaxial faces glabrous or scurfy-puberulent; outer broadly or narrowly deltate to ovate-lanceolate, apices recurved, acuminate; inner lanceolate, acuminate, faces usually lightly black-villous. **Florets** 25–100; corollas yellow, surpassing phyllaries by 5+ mm. **Cypselae** columnar, 5–12 mm; **pappi** of 5–10 dull, yellowish brown, linear-lanceolate, glabrous aristate scales 4–10 mm, aristae barbellate to subplumose. $2n = 18, 27$.

Flowering Mar–Jun. Clay and loam soils, valley flats and hillsides, grasslands, brushlands, and open oak or conifer woods; 40–1500 m; Calif.

The range of *Microseris sylvatica* includes the Central Valley of California and surrounding foothills. It is becoming rare because of grazing and agriculture (D. P. Tibor 2001). An autotriploid form, $2n = 27$, has been reported from Placer County (A. S. Tomb et al. 1978); it reproduced clonally by adventitious buds on lateral roots.

5. **Microseris paludosa** (Greene) J. T. Howell, Leafl. W. Bot. 5: 108. 1948 • Marsh silverpuffs [C] [E]

Scorzonella paludosa Greene, Bull. Calif. Acad. Sci. 2: 52. 1886

Perennials, 15–70 cm; taprooted. **Stems** branched proximally, leafy proximally. **Leaves** basal and cauline; petiolate (petioles broadly winged, clasping); blades linear to oblanceolate, 6–35 cm, margins entire, dentate, or pinnately lobed, apices acuminate. **Peduncles** erect or arcuate-ascending (15–50 cm), ebracteate. **Involucres** ovoid in fruit, 10–20 mm. **Phyllaries:** not spotted, abaxial faces usually scurfy-puberulent, usually black-villous; outer linear to broadly or narrowly ovate-deltate, apices erect or recurved, acuminate; inner lanceolate, apices erect, acute to acuminate. **Florets** 25–70; corollas yellow-orange, surpassing phyllaries by 5+ mm. **Cypselae** columnar, 4–7 mm; **pappi** of 5–10, dull yellowish brown, lanceolate, glabrous, aristate scales 2–4 mm, aristae barbellate. $2n = 18$.

Flowering Apr–Jun. Sandy, clay, and loam soils, grasslands, brushlands, oak woodlands, and closed-cone pine forests; 10–300 m; Calif.

Microseris paludosa in the central coastal region (D. P. Tibor 2001). It differs from *M. laciniata* subsp. *leptosepala* in its longer, brownish pappus scales and more southern coastal distribution. It is unusual among the perennial taxa of *Microseris* in its self-compatibility and ready self-fertilization in culture.

6. **Microseris howellii** A. Gray, Proc. Amer. Acad. Arts 20: 300. 1885 • Howell's silverpuffs [C] [E]

Scorzonella howellii (A. Gray) Greene

Perennials, 10–50 cm; taprooted. **Stems** branched proximally and often distally. **Leaves** basal and cauline; petiolate (petioles broadly winged, clasping); blades linear to narrowly oblanceolate, 10–30 cm, margins entire, dentate, or pinnately lobed (lobes narrow, often retrorse). **Peduncles** erect (10–50 cm), ebracteate or leafy. **Involucres** narrowly ovoid in fruit, 8–17 mm. **Phyllaries:** sometimes purple-spotted, apices erect, acuminate, abaxial faces glabrous or scurfy-puberulent, often black-villous; outer lanceolate to deltate; inner lanceolate. **Florets** 8–30; corollas yellow, surpassing phyllaries by 5+ mm. **Cypselae** columnar, 4–7 mm; **pappi** of 5–10, white, lanceolate, glabrous, aristate scales 3–6 mm, aristae barbellulate. $2n = 18$.

Flowering May–Jun. Rocky serpentine soils, hillsides and alluvial flats, open shrublands and *Pinus jeffreyi* savannas; of conservation concern; 300–1000 m; Oreg.

Microseris howellii is known only from exposures of peridotite in Josephine County, Oregon. Although related to *M. laciniata*, it is ecologically isolated from the co-occurring members of that complex. Because of its limited range, it is listed as a threatened taxon by the Oregon Natural Heritage Program (2004).

7. **Microseris douglasii** (de Candolle) Schultz-Bipontinus, Jahresber. Pollichia 22–24: 308. 1866 • Douglas's silverpuffs [F]

Calaïs douglasii de Candolle in A. P. de Candolle and A. L. P. P. de Candolle, Prodr. 7: 85. 1838

Annuals, 5–40 cm; taprooted. **Leaves** basal; petiolate; blades linear to oblanceolate, 3–25 cm, margins entire, dentate, or pinnately lobed (lobes slender, tapering), apices acute to acuminate, faces ± scurfy-puberulent. **Peduncles** erect or curved-ascending, ebracteate. **Involucres** globose to fusiform in fruit, 7–16 mm. **Phyllaries:** apices erect, acute to acuminate; outer deltate, glabrous or lightly scurfy-puberulent; inner lanceolate, faces often lightly black-villous on margins (midveins often purple-lined, thickened). **Florets** 5–200; corollas yellow or white, equaling or surpassing phyllaries by 1–3 mm. **Cypselae** columnar or obconic, 3–10 mm; **pappi** of (0–)1–5, white to yellow, brown, or blackish, aristate scales 0.5–7 mm (± arcuate, usually distinctly involute, except subsp. *tenella*, often abaxially villous, midveins usually tapering distally from thick bases, except subsp. *tenella*, widths less than $^1/_5$ bodies), aristae (white or straw-colored, ± stout) barbellate. $2n = 18$.

Subspecies 3 (3 in the flora): w United States; nw Mexico.

The geographic patterns of morphologic variability as well as both chloroplast and nuclear DNA markers in *Microseris douglasii* have been studied by K. Bachmann and J. Battjes (1994) and D. Roelofs and K. Bachmann (1997). Four chloroplast types were identified, two of which were derived by introgression from *M. bigelovii* or its ancestor. Plants in nature are highly inbred and genetically homozygous, as proposed earlier by K. L. Chambers (1955). Subspecies *platycarpha* stands well apart in these studies; subsp. *tenella* is not differentiated molecularly from subsp. *douglasii*.

1. Pappus scales usually 0.5–1 mm, sometimes nearly obsolete; cypselae truncate-fusiform, ribs constricted at apices; s, w-c California, mainly coastal and Coast Ranges . . . .
. . . . 7b. *Microseris douglasii* subsp. *tenella*
1. Pappus scales 1–7 mm; cypselae columnar or obconic, ribs slightly flared at apices.
   2. Pappus scales (glabrous) 0.5 mm shorter to 2 mm longer than cypselae; cypselae obconic or columnar, 3–4.5 mm; sw California, n Baja California. . . . .
       . . . . 7c. *Microseris douglasii* subsp. *platycarpha*
   2. Pappus scales (glabrous or villous) usually 1–6 mm shorter than cypselae; cypselae columnar (slightly tapered distally), 4–10 mm; widespread in cismontane California, mainly n of the Transverse Ranges. . . . .
       . . . . 7a. *Microseris douglasii* subsp. *douglasii*

7a. Microseris douglasii (de Candolle) Schultz-Bipontinus subsp. **douglasii** • Douglas's silverpuffs
E F

Microseris attenuata Greene; *M. parishii* Greene; *M. platycarpha* (A. Gray) Schultz-Bipontinus var. *parishii* (Greene) H. M. Hall

Cypselae columnar, 4–10 mm (tapered to basal callosities, filled by embryos or distal $^1/_4$–$^1/_2$ empty), ribs slightly flared apically, slightly constricted below flaring; **pappi** of orbiculate to lanceolate, arcuate, involute (glabrous or villous) scales 1–6.5 mm, usually 1–6 mm shorter than cypselae. $2n = 18$.

Flowering Mar.–Jun. Clay soils, flats and hillsides, often by vernal pools or near serpentine outcrops, grasslands and open oak and *Pinus sabiniana* woodlands; 10–800 m; Calif., Oreg.

Subspecies *douglasii* is the common subspecies away from the coast in cismontane California, north of the Transverse Ranges; it is relatively rare in southwestern California and southern Oregon. Some populations near the coast appear to be hybrids between it, or subsp. *tenella*, and *Microseris bigelovii*.

7b. Microseris douglasii (de Candolle) Schultz-Bipontinus subsp. **tenella** (A. Gray) K. L. Chambers, Contr. Dudley Herb. 4: 294. 1955 • Tender silverpuffs E

Calaïs tenella A. Gray in War Department [U.S.], Pacif. Railr. Rep. 4(5): 114, plate 17, figs. 6–10. 1857; *C. aphantocarpha* A. Gray; *Microseris aphantocarpha* (A. Gray) Schultz-Bipontinus; *M. tenella* (A. Gray) Schultz-Bipontinus; *M. tenella* var. *aphantocarpha* (A. Gray) S. F. Blake

Cypselae truncate-fusiform, 3–6.5 mm (tapered to basal callosities, widest beyond middles, filled by embryos), ribs constricted at apices; **pappi** usually of deltate or ovate, arcuate, usually plane or barely involute scales 0.5–1 mm, shorter than cypselae, sometimes nearly obsolete. $2n = 18$.

Flowering Mar.–Jun. Sandy or clay soils, flats and hillsides, sometimes on serpentine slopes, grasslands and open oak woodlands; 0–500 m; Calif.

Subspecies *tenella* intergrades completely with subsp. *douglasii* in west-central California. It occurs near the immediate coast; morphologic variation and chloroplast DNA suggest a history of hybridization with *M. bigelovii*.

7c. Microseris douglasii (de Candolle) Schultz-Bipontinus subsp. **platycarpha** (A. Gray) K. L. Chambers, Contr. Dudley Herb. 4: 296. 1955 • San Diego silverpuffs

Calaïs platycarpha A. Gray in War Department [U.S.], Pacif. Railr. Rep. 4(5): 113. 1857; *Microseris platycarpha* (A. Gray) Schultz-Bipontinus

Cypselae columnar to obconic, 3–4.5 mm (tapered to basal callosities, filled by embryos) ribs slightly flared apically; **pappi** of orbiculate to lanceolate, involute, arcuate (glabrous) scales 2.5–7 mm, 0.5–2 mm longer than cypselae. $2n = 18$.

Flowering Mar.–May. Clay soil, mesa flats, hillsides, sometimes near vernal pools, grasslands, southern coastal scrub and Engelmann oak woodlands; 10–1100 m; Calif.; Mexico (Baja California).

Subspecies *platycarpha* is cited as severely declining in abundance because of urban development (D. P. Tibor 2001). It replaces subspp. *tenella* and *douglasii* in southwestern California and northern Baja California.

8. Microseris acuminata Greene, Bull. Torrey Bot. Club 10: 88. 1883 • Sierra foothills silverpuffs [E]

Annuals, 5–35 cm; taprooted. Stems 0. Leaves basal; petiolate; blades linear to narrowly elliptic, 3–20 cm, margins usually pinnately lobed (with narrow rachis and linear lobes), rarely entire, apices acuminate, faces glabrous or lightly scurfy-puberulent. Peduncles erect or curved-ascending, ebracteate. Involucres ovoid to fusiform in fruit, 10–22 mm. Phyllaries: apices erect, acute to acuminate, abaxial faces glabrous; outer deltate; inner lanceolate, (midveins often purple, thickened). Florets 5–50; corollas yellow, equaling or surpassing phyllaries by 1–3 mm. Cypselae columnar, 4.5–7 mm; pappi of 5, white or light brown, linear-lanceolate, aristate scales 4–11 mm (arcuate, scarcely involute, apices acuminate, faces usually glabrous, rarely villous, midveins brownish, stout, widths $^{1}/_{5}$–$^{1}/_{3}$ bodies, tapered distally), aristae (white or brown) barbellate. $2n = 36$.

Flowering Apr–Jun. Clay soils, flats and hillsides, sometimes near vernal pools, grasslands and open oak woodlands; 30–600 m; Calif., Oreg.

Microseris acuminata occurs in the Sacramento and northern San Joaquin valleys and surrounding foothills; it is disjunct in Jackson County, Oregon. K. L. Chambers (1955) proposed that this morphologically distinctive tetraploid species is of alloploid origin and that *M. douglasii* is one of its possible diploid parents. Recent molecular evidence (D. Roelofs et al. 1997) supports that relationship and also favors a relationship, through an extinct common ancestor, with the tetraploid *M. campestris*.

9. Microseris campestris Greene, Pittonia 5: 15. 1902 • San Joaquin silverpuffs [E]

Annuals, 5–50 cm; taprooted. Stems 0. Leaves basal; petiolate; blades linear to narrowly elliptic or oblanceolate, 3–22 cm, margins entire, dentate, or pinnately lobed, apices acute or attenuate, faces glabrous or lightly scurfy-puberulent. Peduncles erect or curved-ascending, ebracteate. Involucres ovoid to fusiform in fruit, 5–20 mm. Phyllaries: apices acute to acuminate, abaxial faces glabrous; outer deltate; inner lanceolate (midveins often purple, thickened). Florets 5–120; corollas yellow or white, equaling or surpassing phyllaries by 1–3 mm. Cypselae columnar, 3–5.5 mm; pappi of 5, white or light brownish, lanceolate or ovate to deltate, aristate scales 1–4.5 mm (straight to slightly arcuate, plane or slightly involute, glabrous, midveins brown, broadened at bases, widths less than $^{1}/_{5}$ bodies, linear distally), aristae (straw-colored or brown) barbellulate proximally, barbellate distally. $2n = 36$.

Flowering Apr–Jun. Clay soils, flats and hillsides, sometimes near vernal pools, grasslands; 30–500 m; Calif.

The tetraploid *Microseris campestris* is morphologically intermediate between *M. elegans* and *M. douglasii*, and the molecular data of D. Roelofs et al. (1997) show a particularly close relationship with the former species. It is known only from the San Joaquin Valley and surrounding foothills. Diploid plants assignable to *M. douglasii* but with fruit morphology similar to *M. campestris*, found much closer to the coast near San Luis Obispo, are thought to be the result of introgression between *M. douglasii* and *M. bigelovii* (K. L. Chambers 1955).

10. Microseris bigelovii (A. Gray) Schultz-Bipontinus, Jahresber. Pollichia 22–24: 308. 1866 (as bigelowii) • Coastal silverpuffs [E]

Calaïs bigelovii A. Gray in War Department [U.S.], Pacif. Railr. Rep. 4(5): 113, plate 17, figs. 1–5. 1857

Annuals, 3–60 cm; taprooted. Stems 0. Leaves basal; petiolate; blades linear to narrowly elliptic or spatulate, 3–25 cm, margins entire, dentate, or pinnately lobed, apices acuminate to obtuse, faces glabrous or lightly scurfy-puberulent. Peduncles erect or curved-ascending, ebracteate. Involucres ovoid to fusiform in fruit, 5–14 mm. Phyllaries: apices acute to acuminate, faces glabrous; outer deltate; inner lanceolate (midveins often purple, thickened). Florets 5–100; corollas yellow or orange, equaling or surpassing phyllaries by 1–3 mm. Cypselae truncate-fusiform, 2.5–5.5 mm; pappi of 5 silvery to blackish, deltate to lanceolate, aristate scales 1–4 mm (slightly arched at bases, flat, glabrous, midveins linear, widths less than $^{1}/_{5}$ bodies, thicker at bases), aristae (brown, fine) barbellulate. $2n = 18$.

Flowering Apr–Jul. Sandy and loam soils, open sites, on coastal terraces, hillsides, rocky headlands, and bird-nesting islands; 0–100 m; B.C.; Calif., Oreg., Wash.

Microseris bigelovii is the most characteristically coastal of the annual taxa and the only one to include plants with obtuse, spatulate leaves (K. Bachmann et al. 1984). A statistical analysis of its morphologic variation was published by Bachmann (1992). It sometimes has been collected at inland sites at 500–600 m, where the cypselae may have been introduced by domestic animals. The northern populations near Victoria, British Columbia, and the San Juan Islands, Washington, are disjunct from the main range, which extends from Oregon to Santa Barbara County, California.

11. Microseris elegans Greene ex A. Gray in A. Gray et al., Syn. Fl. N. Amer. 1(2): 419. 1884 • Elegant silverpuffs

Microseris aphantocarpha (A. Gray) Schultz-Bipontinus var. *elegans* (Greene ex A. Gray) Jepson

Annuals, 5–35 cm; taprooted. **Stems** 0. **Leaves** basal; petiolate; blades linear to narrowly oblanceolate, 2–20 cm, margins entire, dentate, or pinnately lobed, apices acuminate, faces glabrous or lightly scurfy-puberulent. **Peduncles** erect or curved-ascending, ebracteate. **Involucres** globose to ovoid in fruit, 4–8(–10) mm. **Phyllaries:** apices acute to acuminate, faces glabrous; outer deltate; inner lanceolate (midveins often purple, thickened). **Florets** 5–100; corollas yellow or orange, equaling or surpassing phyllaries by 1–2 mm. **Cypselae** columnar to obconic, 1.5–3 mm; **pappi** of (4–)5 white or brownish, ovate to deltate, aristate scales 0.2–2.5 mm (straight or slightly arcuate, scarcely involute, glabrous, midveins linear, widths less than $^1/_5$ bodies, thicker at base), aristae (brown, fine) barbellulate. $2n = 18$.

Flowering Apr–Jun. Mostly clay soils, flats and hillsides, often near vernal pools, grasslands, shrublands; 10–700 m; Calif.; Mexico (Baja California).

Microseris elegans is widespread in interior central California, becoming coastal in the southwestern part of its range. It was hypothesized to be one of the diploid ancestors of *M. campestris* (K. L. Chambers 1955); molecular evidence supporting that relationship was presented by D. Roelofs et al. (1997).

69. STEBBINSOSERIS K. L. Chambers, Amer. J. Bot. 78: 1024. 1991 • Silverpuffs

[For G. Ledyard Stebbins, 1906–2000, California botanist]

Kenton L. Chambers

Microseris D. Don sect. *Brachycarpa* (Nuttall) K. L. Chambers; *Uropappus* Nuttall sect. *Brachycarpa* Nuttall

Annuals, 1–10 cm; taprooted. **Stems** 0, or erect, mostly unbranched, glabrous or lightly scurfy-puberulent. **Leaves** usually all basal; petiolate (petioles narrowly attenuate, usually scurfy-puberulent, especially proximally); blades linear to narrowly oblanceolate, bases slightly clasping, margins entire or irregularly dentate or lobed (teeth and lobes narrow, acute, straight or arcuate, faces glabrous or minutely scurfy-puberulent). **Heads** borne singly (often inclined in bud, erect in flowering and fruit). **Peduncles** not notably inflated, usually ebracteate (glabrous or ± scurfy-puberulent, especially distally). **Calyculi** of (3–)4–14, deltate or ovate to lanceolate bractlets. **Involucres** campanulate, (3–)5–35 mm diam. (fusiform to ovoid in fruit). **Phyllaries** (4–)5–18 in ± 2 series, (green or purple) mostly lanceolate, subequal to equal, herbaceous, apices acute, faces glabrous. **Receptacles** flat, ± pitted, glabrous, epaleate. **Florets** (10–)30–125; corollas yellow or white, outer often purplish abaxially (equaling or surpassing phyllaries by 1–3 mm). **Cypselae** brown, purplish gray, stramineous, or violet, sometimes purple-spotted, columnar or truncate-fusiform, not beaked, ribs 10, ± scabrellous or spiculate, faces glabrous or (on outer) strigose; **pappi** persistent, of 5, usually yellowish or brownish, rarely white, aristate scales (bodies straight or arcuate, lanceolate, usually glabrous, margins plane or involute, apices erose or notched, aristae shorter than to equaling bodies, barbellulate). $x = 18$.

Species 2 (2 in the flora): sw United States, nw Mexico.

Stebbinsoseris comprises two allotetraploid species derived from hybrids between *Microseris* and *Uropappus*. The justification for raising these species to generic rank was given in R. K. Jansen et al. (1991b), where molecular data were presented supporting the separation of *Uropappus* from *Microseris*. *Stebbinsoseris* and *Uropappus* were ranked as sections of *Microseris* in earlier taxonomic treatments (K. L. Chambers 1955, 1960). Because of its hybrid origin, *Stebbinsoseris* is intermediate in critical taxonomic traits of habit, involucre, and fruits (C. Irmler et al. 1982), and it is separated from its parental taxa by rather minor differences. Chloroplast DNA studies show that *Microseris douglasii* and *M. bigelovii* were the maternal parents of the tetraploid hybrids (R. S. Wallace and R. K. Jansen 1990). That *Uropappus lindleyi* was the staminate parent of those crosses is confirmed by nuclear rDNA evidence reported by Wallace and Jansen (1995).

1. Cypselae narrowly truncate-fusiform, brown to purplish, 5–8 mm, apices not widened at bases of pappi; pappus scale bodies 3–5 mm; coastal central California . . . . 1. *Stebbinsoseris decipiens*
1. Cypselae narrowly truncate-fusiform to columnar, gray to pale brown or violet (dark purplish in sw California), 4.5–12 mm, apices slightly widened at bases of pappi; pappus scale bodies 4–11 mm; widespread, rarely coastal except in sw California. . . . . 2. *Stebbinsoseris heterocarpa*

1. Stebbinsoseris decipiens (K. L. Chambers) K. L. Chambers, Amer. J. Bot. 78: 1025. 1991 • Santa Cruz silverpuffs [C][E]

Microseris decipiens K. L. Chambers, Contr. Dudley Herb. 4: 290, fig. 17. 1955

Peduncles 15–60 cm. **Involucres** 6–19 mm. **Florets** 10–80(–100); corollas yellow. **Cypselae** brown to purplish, narrowly truncate-fusiform, 5–8 mm, each filled by embryo or no more than distal 0.5 mm vacant, apices not enlarged at bases of pappi; **pappi** 7–10 mm, scale bodies 3–5 mm, faces glabrous, aristae 4–5 mm. $2n = 36$.

Flowering Apr–May. Sandy, shale, or serpentine soils, grasslands, coastal scrub, chaparral, closed-cone pine woods, roadsides; of conservation concern; 10–500 m; Calif.

Morphologic and molecular evidence (K. L. Chambers 1955; C. Irmler et al. 1982; R. S. Wallace and R. K. Jansen 1990) proves that *Stebbinsoseris decipiens* is an allopolyploid derivative of the hybrid *Microseris bigelovii* × *Uropappus lindleyi*. It occurs in a limited area of central coastal California where the parental taxa are sympatric. Diploid hybrids between the parents, produced experimentally (Chambers), had irregular meiosis and were completely seed-sterile. The species is included in *Inventory of Rare and Endangered Plants of California*, ed. 6 (D. P. Tibor 2001).

SELECTED REFERENCE Irmler, C. et al. 1982. Enzymes and quantitative morphological characters compared between the allotetraploid *Microseris decipiens* and its diploid parental species. Beitr. Biol. Pflanzen 57: 269–289.

2. Stebbinsoseris heterocarpa (Nuttall) K. L. Chambers, Amer. J. Bot. 78: 1024. 1991 • Grassland or derived silverpuffs [F]

Uropappus heterocarpus Nuttall, Trans. Amer. Philos. Soc., n. s. 7: 425. 1841; *Microseris heterocarpa* (Nuttall) K. L. Chambers

Peduncles 5–60 cm. **Involucres** 6–30 mm. **Florets** 10–125; corollas yellow or white. **Cypselae** gray, straw-colored, brown, or violet (dark purplish in southwest California), narrowly truncate-fusiform to columnar, 4.5–12 mm, each filled by embryo or distal 0.5–3 mm vacant and more slender than proximal part, apices slightly widened at bases of pappi; **pappi** 7–19 mm, scale bodies 4–11 mm, faces usually glabrous, rarely villous, aristae 3–8 mm. $2n = 36$.

Flowering Apr–Jun. Clay, gravelly, or rocky soils, open flats and hillsides, grasslands, oak woodlands, chaparral, coastal scrub, desert (rarely), roadsides; 0–1000 m; Ariz., Calif.; Mexico (Baja California).

Stebbinsoseris heterocarpa is widespread in cismontane California and disjunct to Gila and Yavapai counties, Arizona. Variation in enzymatic restriction sites in its chloroplast DNA and nuclear rDNA (R. S. Wallace and R. K. Jansen 1995) indicates that *S. heterocarpa* has arisen independently three or more times from hybrids of *Microseris douglasii* × *Uropappus lindleyi*. Reflecting this multiple origin, the species is more variable in pappi and cypsela size, color, and shape than *S. decipiens*, from which it is largely allopatric. Because *M. douglasii* subsp. *tenella* and *M. bigelovii* intergrade in some coastal areas, forms of *S. heterocarpa* having such hybrids as their maternal parent would be difficult to distinguish from *S. decipiens* (K. L. Chambers 1955; U. Lohwasser and F. R. Blattner 2004).

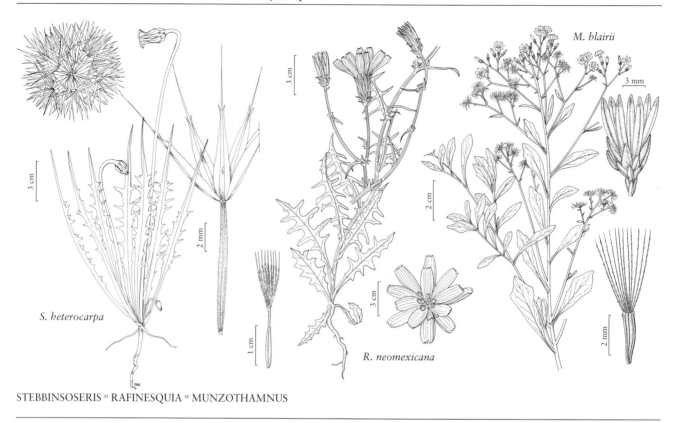

S. heterocarpa

R. neomexicana

M. blairii

STEBBINSOSERIS ∘ RAFINESQUIA ∘ MUNZOTHAMNUS

70. RAFINESQUIA Nuttall, Trans. Amer. Philos. Soc., n. s. 7: 429. 1841, name conserved • Rafinesque's chicory [For Constantin Samuel Rafinesque, 1783–1840, naturalist and polymath who traveled widely in nineteenth-century America]

L. D. Gottlieb

Annuals, 15–150 cm; taprooted. **Stems** 1–3, erect, simple or distally branched (hollow), glabrous. **Leaves** basal and cauline; basal petiolate or sessile, blades oblong to oblanceolate, pinnately lobed (lobes broad or narrow); cauline sessile, sometimes auriculate clasping, distal smaller, becoming entire and bractlike. **Heads** (erect) borne singly (at ends of branches) or in open, paniculiform arrays. **Involucres** cylindro-conic, (4–)6–15+ mm diam. **Calyculi** of 8–14, spreading to reflexed, unequal bractlets (lengths to ¹/₂ phyllaries). **Phyllaries** 7–20 in 1 series, linear-lanceolate, ± equal, margins scarious, apices acuminate. **Receptacles** flat, smooth, glabrous, epaleate. **Florets** 15–30; corollas white, sometimes with rose or purplish veins abaxially (outer surpassing phyllaries). **Cypselae** tan to mottled grayish brown, fusiform, 9–18(–20) mm, bodies tapering to beaks, ribs or ridges 5, outer usually with antrorse hairs or papillate or scaly indument, inner mostly smooth to cross-rugulose, glabrous; **pappi** (borne on small discs) ± persistent, of 5–21, white or sordid, ± plumose (at least proximally, barbs sometimes entangled) bristles in 1 series. $x = 8$.

Species 2 (2 in the flora): sw United States, nw Mexico.

1. Ligules surpassing phyllaries by 5–7 mm; cypselae 9–14 mm, including beaks 5–7 mm; pappus bristles wholly plumose, barbs straight, separate . . . . 1. *Rafinesquia californica*
1. Ligules surpassing phyllaries by 15–20 mm; cypselae 12–18(–20) mm, including beaks 3–6 mm; pappus bristles plumose on proximal 65–80%, barbs crooked, entangled . . . . 2. *Rafinesquia neomexicana*

1. Rafinesquia californica Nuttall, Trans. Amer. Philos. Soc., n. s. 7: 429. 1841 • California chicory

Plants 20–150 cm. Involucres 12–20 mm. Ligules surpassing phyllaries by 5–7 mm. Cypselae light tan or sordid, 9–14 mm, including very slender beaks 5–7 mm; pappi of 6–15, wholly plumose bristles, barbs straight, separate. 2n = 16.

Flowering Apr–Jun. Open places in coastal sage scrub, chaparral, oak woodlands in northern and coastal California and southwestern Oregon (often found after fires), Upper Sonoran Zone; 100–1500 m; Ariz., Calif., Nev., Oreg.; Mexico (Baja California).

2. Rafinesquia neomexicana A. Gray, Smithsonian Contr. Knowl. 5(6): 103. 1853 (as neo-mexicana) • Desert chicory [F]

Plants 15–60 cm. Involucres 18–25 mm. Ligules surpassing phyllaries by 15–20 mm. Cypselae 12–18(–20) mm, including relatively stout beaks 3–6 mm; pappi of 15–21, plumose (proximal 65–80%) bristles, barbs crooked, entangled. 2n = 16.

Flowering Mar–May. Open sites, sandy soils, gravelly clay loams in Lower and Upper Sonoran Zones, often in Creosote Bush Scrub and Joshua Tree Woodland, Mojave and Colorado deserts; 200–1500 m; Ariz., Calif., Nev., N.Mex., Tex., Utah; Mexico (Baja California, Sonora).

71. MUNZOTHAMNUS P. H. Raven, Aliso 5: 345, fig. 6. 1963 • Munz's shrub [For P. A. Munz, 1892–1974, American botanist, and Greek *thamnos*, shrub] [C][E]

L. D. Gottlieb

Shrubs, 100–200 cm; probably taprooted. Stems straggly (± fleshy), branched, tomentose to glabrate. Leaves cauline (tufted at ends of branches); petiolate; blades obovate to oblong-obovate, margins irregularly sinuate or lobulate. Heads in paniculiform arrays (at ends of branches). Peduncles not inflated, bracteolate (usually stipitate-glandular). Calyculi of 7–10, unequal, triangular to ovate bractlets (lengths to ½ phyllaries). Involucres cylindric, 3–5+ mm diam. Phyllaries 8–9 in ± 2 series, lanceolate to linear, margins narrowly scarious, apices obtuse to acute (often sparsely arachnose), faces glabrous or sparsely glandular-puberulent. Receptacles ± flat, pitted, glabrous, epaleate. Florets 9–12; corollas rosy to purplish. Cypselae gray-brown, cylindric, apices truncate (not beaked), faces 5, equal, each with 1–2, narrow, longitudinal lines or shallow grooves, otherwise smooth, glabrous; pappi falling, of 25–35 distinct, white, antrorsely barbellate bristles in 1 series. x = 8?

Species 1: California.

1. Munzothamnus blairii (Munz & I. M. Johnston) P. H. Raven, Aliso 5: 345. 1963 [C][E][F]

Stephanomeria blairii Munz & I. M. Johnston, Bull. Torrey Bot. Club 51: 301. 1924 (as blairi); *Malacothrix blairii* (Munz & I. M. Johnston) Munz & I. M. Johnston

Stems: branches to 1 cm thick, fleshy, studded with persistent petiole bases proximally. Leaves crowded or tufted at ends of branches; petioles ca. 1 cm, narrowly winged, bases thickened; blades bright green, 5–13 × 4–6 cm, relatively thin, bases cuneate, apices rounded or obtuse, faces ± tomentose, soon glabrescent. Heads 20–35. Cypselae 3–3.5 mm. 2n = 16.

Flowering Sep–Nov. Rocky canyon walls; 60–300 m; Calif.

Munzothamnus blairii is known only from San Clemente Island. It has been placed in several taxonomic homes. It was initially placed in *Stephanomeria*; it was transferred to *Malacothrix* (P. A. Munz 1935). P. H. Raven (1963) considered it "clearly a relictual and highly isolated genus," primarily on the basis of its leaf shape and vegetative architecture, which are significantly distinct from those found in any species of *Stephanomeria* or *Malacothrix*, and he erected *Munzothamnus* for it. Others emphasized the similar number and appearance of the chromosomes of *M. blairii* to those of *Stephanomeria* as well as certain similarities between their pappi (number and "coarseness") and concluded that the species belongs in *Stephanomeria* (G. L. Stebbins et al. 1953). In addition, pollen size and sculpturing are similar to that of *Stephanomeria* (A. S. Tomb 1974).

The mode of branch growth was nicely described by S. Carlquist (1974). At the end of the first year, basal lateral buds on the main stem produce side shoots. At the end of the second year, those shoots produce terminal clusters of heads. At the end of the third year, those same side shoots produce new lateral shoots from distal nodes below the heads. Thus, the plants produce shoots from lateral buds both at the base of the stem and from nodes on side shoots below the terminal clusters.

The growth pattern, architecture, and habit of *Munzothamnus blairii* are not at all similar to those of *Stephanomeria guadalupensis* Brandegee, an endemic of Guadalupe Island, Mexico, nor to those of *S. cichoriacea*, which grows in the equable climate of southern coastal California and on off-shore islands. The contrast between *M. blairii* and those undoubted stephanomerias is noteworthy. That *Munzothamnus blairii* is best regarded as a monotypic genus is also supported by recent results of DNA sequencing studies (J. Lee et al. 2002), which showed that it is not in the clade of all stephanomerias.

72. STEPHANOMERIA Nuttall, Trans. Amer. Philos. Soc., n. s., 7: 427. 1841, name conserved • Stickweed, wirelettuce, skeletonweed [Greek *stephanos*, crown, wreath, and *meris*, part, presumably alluding to appearance of plumose bristles of pappus]

L. D. Gottlieb

Annuals, 10–200 cm, taprooted, or **perennials,** 10–100 cm, with deeply seated, woody caudices or stout or slender, creeping rhizomes. **Stems** (1–8) erect, simple or branched, usually glabrous, sometimes hairy (especially when young). **Leaves** basal (withered at flowering in annuals and some perennials) and/or cauline (much reduced, bractlike in annuals and some perennials); usually sessile; blades linear to oblong, oblanceolate, or spatulate, usually runcinate, margins usually pinnately lobed (spinulose-tipped in *S. parryi*), sometimes entire or toothed (*S. lactucina, S. tenuifolia,* and *S. fluminea* (faces glabrous, puberulent, or tomentose); distal bractlike (to 45 mm in *S. fluminea*). **Heads** borne singly or clustered (in paniculiform arrays in some subspecies of *S. exigua*). **Peduncles** not inflated distally, sometimes bracteate. **Calyculi** of 3–5, unequal bractlets (more numerous in some perennials; not distinguishable in *S. cichoriacea*), appressed or reflexed (some annuals). **Involucres** ± cylindric to turbinate, 2–3 (–5+) mm diam. **Phyllaries** usually 5–12 in 1 series, equal (20–25 in 2–3 series, unequal in *S. cichoriacea,* usually glabrous, rarely puberulent, densely stipitate-glandular in *S. exigua* subsp. *deanei*). **Receptacles** flat, usually smooth (pitted in *S. cichoriacea*), glabrous, epaleate. **Florets** (4–)5–16; corollas usually pink or lavender, sometimes white (annuals often purple-tinged abaxially). **Cypselae** light tan to dark brown, columnar, sometimes slightly curved, 5-angled, apices truncate, faces equal, sometimes with ribs between faces, each face with central, narrow, longitudinal groove or furrow (not grooved in *S. virgata*), otherwise smooth or bumpy to tuberculate, usually glabrous (scaberulous in *S. fluminea*); **pappi** persistent (or only widened bases of bristles persistent after distal portions break off) or falling, of 5–40, distinct or basally connate in groups, white to tan, wholly or distally plumose bristles in 1 series. $x = 8$.

Species 16 (14 in the flora): w North America, n Mexico.

Because all the species of *Stephanomeria* have not previously been examined at one time, the present treatment provides the first unified picture of their variability, ecologic specializations, and geographic distributions. The genus includes six annual species (all in the flora) and ten perennial species (eight in the flora, one in the mountains of northern Baja California, and one known only from Guadalupe Island, Mexico).

Taxonomic distinctions among annual species of *Stephanomeria* did not become evident until their morphology and geographic distributions were correlated with their chromosome numbers and reproductive compatibilities (L. D. Gottlieb 1971, 1972). The same studies also provided an hypothesis that satisfactorily accounted for their variability. Studies showed that *S. exigua* and *S. virgata* differed for a relatively large number of characters and that other annual species originated from genetic segregates that were formed by hybridization, at both diploid and tetraploid levels, as well as directly from *S. exigua*.

Stephanomeria exigua has five subspecies; *S. virgata* has two. Within each species, the subspecies share numerous morphologic features as well as chromosomal karyotype. They are recognized as polytypic because reproductive compatibility between any pair of subspecies of *S. exigua* or between subspecies of *S. virgata* is substantially higher than is the compatibility between the two species. The two species appear to represent a fundamental phylogenetic divergence within annuals; nevertheless their different features are combined in different ways in *S. elata* and *S. diegensis*.

Stephanomeria paniculata and *S. malheurensis* probably evolved more or less directly from *S. exigua* subsp. *coronaria*. The speciation process that gave rise to *S. malheurensis* (L. D. Gottlieb 1978) has been examined in a series of studies (Gottlieb 1973b, 1977, 1979; S. Brauner and Gottlieb 1987, 1989). The origin of the highly self-pollinating *S. paniculata* may have been similar but much less evidence is available. *Stephanomeria malheurensis* has served as a model for reintroduction of a species back into its original habitat after local extinction, in its case by competition from invasive cheatgrass (*Bromus tectorum*).

Information about evolution and speciation is not so available for the perennials as for the annuals. Treatment of perennials is based almost entirely on examination of herbarium specimens plus published information describing their chromosome numbers. Although little is known about phylogenetic relationships among perennial species of *Stephanomeria*, a recent DNA sequencing study of nuclear rDNA (J. Lee et al. 2002) showed that the genus does not include either *Munzothamnus blairii* (previously *S. blairii*) or *Pleiacanthus spinosus* (previously *S. spinosa*). Without them, *Stephanomeria* is a well-supported, monophyletic group of species.

The DNA analysis suggested that *Stephanomeria tenuifolia*, *S. runcinata*, *S. fluminea*, and *S. thurberi* comprise a subclade. Those four species are perennial and all have fully plumose, white pappus bristles. They differ markedly in their ecologic specializations, as indicated in their treatments below. The DNA studies also showed a very close relationship between *S. malheurensis* and *S. exigua* subsp. *coronaria* consistent with results of previous studies (cited above). It is to be hoped that taxonomic information presented below will make species of *Stephanomeria* more easily accessible to continuing studies.

SELECTED REFERENCES Gottlieb, L. D. 1971. Evolutionary relationships in the outcrossing diploid annual species of *Stephanomeria* (Compositae). Evolution 25: 312–329. Gottlieb, L. D. 1972. A proposal for classification of the annual species of *Stephanomeria* (Compositae). Madroño 21: 463–481.

1. Perennials.
    2. Florets 8–16.
        3. Calyculi 0 (phyllaries 20–25, unequal); receptacles pitted . . . . 1. *Stephanomeria cichoriacea*
        3. Calyculi of 4–8 bractlets (none longer than the 6–12 phyllaries); receptacles smooth.
            4. Leaf margins thickened, minutely spinose (phyllaries 6–8); pappus bristles 10–15, plumose on distal 80% . . . . 9. *Stephanomeria parryi*
            4. Leaf margins not thickened, not spinose (phyllaries 6–12); pappus bristles 25–40, wholly plumose.

5. Basal leaves runcinate, pinnately lobed; phyllaries 6–8; pappus bristles
   30–40 . . . . 13. *Stephanomeria thurberi*
5. Basal leaves entire or sparsely toothed; phyllaries 8–12; pappus bristles
   25—30 . . . . 6. *Stephanomeria lactucina*
2. Florets 4–6.
   6. Plants with woody caudices; pappus bristles tan or sordid, plumose on distal 80%
      . . . . 10. *Stephanomeria pauciflora*
   6. Plants with rhizomes; pappus bristles white, wholly plumose.
      7. Cauline leaves present (green) at flowering (3–6 cm) . . . . 5. *Stephanomeria fluminea*
      7. Cauline leaves much reduced, bractlike at flowering.
         8. Plants 10–20(–25) cm; basal leaves runcinate, pinnately lobed . . . .
            . . . . 11. *Stephanomeria runcinata*
         8. Plants 20–70 cm; basal leaves entire or sparsely toothed . . . . . 12. *Stephanomeria tenuifolia*
1. Annuals.
   9. Cypselae without longitudinal groove on each face . . . . 14. *Stephanomeria virgata*
   9. Cypselae with longitudinal groove on each face.
      10. Heads in paniculiform arrays; peduncles 10–40 mm . . . . 4. *Stephanomeria exigua* (in part)
      10. Heads borne singly or clustered; peduncles 2–10 mm.
         11. Pappus bristles wholly plumose or at least to widened bases.
            12. Florets 5; calyculus bractlets appressed . . . . 8. *Stephanomeria paniculata*
            12. Florets 9–15; calyculus bractlets usually reflexed, rarely appressed . . . . 3. *Stephanomeria elata*
         11. Pappus bristles plumose on distal 50–85%, bases widened or not.
            13. Calyculus bractlets reflexed.
               14. Florets 6–8; cypselae 5.5–6.8 mm; pappus bristles (persistent)
                  plumose on distal 60–70% . . . . 4. *Stephanomeria exigua* (in part)
               14. Florets 11–13; cypselae 1.9–2.3 mm; pappus bristles (falling) plumose
                  on distal 80–85% . . . . 2. *Stephanomeria diegensis*
            13. Calyculus bractlets appressed.
               15. Florets 5–11; cypselae 2.3–3.1 mm; pappus bristles plumose on distal
                  60–85% . . . . 4. *Stephanomeria exigua* (in part)
               15. Florets 5–6; cypselae 3.3–3.8 mm; pappus bristles plumose on distal
                  50–60% . . . . 7. *Stephanomeria malheurensis*

1. **Stephanomeria cichoriacea** A. Gray, Proc. Amer. Acad. Arts 6: 552. 1865 (as cichoracea) • Chicoryleaf wirelettuce [E]

Perennials, 40–100 cm. **Stems** single, simple or virgately branched, woolly-pubescent when young, glabrescent. **Leaves** green at flowering (spreading at bases); blades oblanceolate to spatulate, 10–20 cm, margins entire or irregularly toothed, teeth remote, faces woolly-pubescent, glabrescent; cauline much reduced distally, margins entire or irregularly toothed. **Heads** borne singly along branches. **Peduncles** ± 0. **Calyculi** 0 (or bractlets intergrading with phyllaries). **Involucres** 12–15 mm (phyllaries 20–25 in 2–3 series, appressed, 2–15 mm, unequal, puberulent; receptacles pitted, each socket 5-sided, surrounded by minute, raised, scaly fringe). **Florets** 10–13. **Cypselae** tan or grayish tan, 5–6 mm, faces smooth, grooved (grooves sometimes absent or only visible as fine lines or striations); **pappi** of 20–25, tan to pale brown bristles (persistent), wholly plumose. $2n = 16$.

Flowering May–Nov. Sandstone, granitic, volcanic, or serpentine soils in coastal scrub and foothill canyons, chaparral, mixed evergreen forests; 50–1500 m; Calif.

Stephanomeria cichoriacea grows primarily in the coastal mountains from southern Monterey County to the San Bernardino Mountains, Santa Ana Mountains, and the Channel Islands.

2. **Stephanomeria diegensis** Gottlieb, Madroño 21: 476, figs. 2,3. 1972 • San Diego wirelettuce [F]

Annuals, 50–200 cm. **Stems** single, branches ascending or spreading, glabrous. **Leaves** withered at flowering (glabrous); basal blades linear to oblanceolate, runcinate, 3–10 cm, margins pinnately lobed; cauline much reduced, bractlike. **Heads** borne singly or clustered along branches. **Peduncles** 3–4 mm. **Calyculi** of reflexed bractlets. **Involucres** 7–9 mm (sparsely glandular-puberulent). **Florets** 11–13. **Cypselae** light tan to brown, 1.9–2.3 mm, faces smooth, slightly bumpy or tuberculate, grooved; **pappi** of 19–21, white bristles (falling), plumose on distal 80–85%. **2n** = 16.

Flowering Aug–Nov. Open, pioneer sites such as old clearings, sand dunes, coastal sage communities, chaparral openings, and sandy roadside embankments; 20–600 m; Calif.; Mexico (Baja California).

The morphologic characteristics of *Stephanomeria diegensis* are a combination of those of *S. exigua* and *S. virgata,* and the species is thought to have evolved from genetic segregates of their hybridization (L. D. Gottlieb 1971; G. P. Gallez and Gottlieb 1982).

3. **Stephanomeria elata** Nuttall, Proc. Acad. Nat. Sci.
   Philadelphia 4: 20. 1848 • Nuttall's wirelettuce [E]

Annuals, 50–150 cm. **Stems** single, branches ascending or spreading, glabrous, puberulent, or glandular-pubescent. **Leaves** withered at flowering (glabrous or puberulent); basal blades linear to oblanceolate, runcinate, 3–10 cm, margins pinnately lobed; cauline much reduced, bractlike. **Heads** borne singly or clustered along branches. **Peduncles** 3–7 mm. **Calyculi** of usually reflexed, rarely appressed bractlets. **Involucres** 5–7 mm (glabrous, puberulent, or stipitate-glandular). **Florets** 9–15. **Cypselae** light tan to dark brown, 2.8–4.5 mm, faces smooth to strongly tuberculate, grooved; **pappi** of 17–22 white or tan bristles (falling or widened bases persistent, bases connate in groups of 2–4, distal portions breaking off), wholly plumose. **2n** = 32.

Flowering Jul–Oct. Chaparral openings, grassy meadows, forest openings, roadsides, often growing as weed; 100–1400 m; Calif., Oreg.

Stephanomeria elata grows in the coastal foothills and mountains, the western slopes of Sierra Nevada, and southwest Oregon.

All the tetraploid populations of annual stephanomerias are placed into *Stephanomeria elata*. The plants are self-compatible and are highly self-pollinating. *Stephanomeria elata* is an allotetraploid species that arose following hybridization between *S. exigua* and *S. virgata* (L. D. Gottlieb 1972). Substantial interpopulation morphologic variability occurs in the length, width, and color of ligules, number of florets, and degree of reflexing of bractlets of the calyculi. Two groups of populations can be distinguished. One group has large cypselae, averaging 3.9–4.5 mm, the bristle bases are widened, and about 30% of the pollen grains have four pores. The second group has smaller cypselae, averaging 2.8–3.3 mm, the bristle bases are not widened, and less than 10% of the pollen grains have four pores. The former group of populations is generally found from southwestern Oregon south to Monterey County in the Coast Ranges of California and on the western slopes of the Sierra Nevada to Fresno County. The latter group is distributed near the coast from Marin County to Santa Barbara County, California. The two groups overlap in Santa Cruz, Santa Clara, and Monterey counties; the distinctions are less evident there.

Stephanomeria elata and its parents *S. exigua* and *S. virgata* form a polyploid complex that perplexed taxonomists for many years. Once the morphologic distinctions between parental species were clarified (L. D. Gottlieb 1972), particularly, the presence versus absence of the longitudinal groove on each face of their cypselae that distinguishes *S. exigua* and *S. elata* from *S. virgata,* and the allotetraploidy of *S. elata* was recognized, it has become much simpler to distinguish the three species in the field.

4. **Stephanomeria exigua** Nuttall, Trans. Amer. Philos.
   Soc., n. s. 7: 428. 1841 • Small wirelettuce,
   skeletonplant [F]

Annuals, 10–200 cm (taproots relatively large). **Stems** single, branches divaricately or freely spreading, glabrous or sparsely pubescent. **Leaves** withered at flowering; basal blades linear to oblanceolate, runcinate, 3–10 cm, margins pinnately lobed; cauline much reduced, bractlike. **Heads** borne singly or clustered along branches or in paniculiform arrays. **Peduncles** 2–10 mm (along branches), or 10–40 mm (in paniculiform arrays). **Calyculi** of appressed or reflexed bractlets (glabrous, pubescent, or stipitate-glandular). **Involucres** 5–7 mm (glabrous, puberulent, or densely stipitate-glandular). **Florets** 5–11. **Cypselae** light tan to dark brown, 2.1–6.8 mm, faces smooth to tuberculate, grooved; **pappi** of 5–24 tan or white bristles (falling or widened bases persistent, bases connate in groups of 2–4 distal portions breaking off), plumose on distal 50–85%. **2n** = 16.

Subspecies 5 (5 in the flora): w United States, nw Mexico.

1. Heads borne singly or clustered along branches; peduncles 2–10 mm.
   2. Calyculus bractlets appressed; peduncles 2–5 mm . . . . 4b. *Stephanomeria exigua* subsp. *coronaria*
   2. Calyculus bractlets reflexed; peduncles 5–10 mm . . . . 4e. *Stephanomeria exigua* subsp. *macrocarpa*
1. Heads in paniculiform arrays; peduncles 10–40 mm.
   3. Calyculus bractlets reflexed . . . . 4a. *Stephanomeria exigua* subsp. *carotifera*
   3. Calyculus bractlets appressed.
      4. Peduncles and involucres densely glandular . . . . . . 4c. *Stephanomeria exigua* subsp. *deanei*
      4. Peduncles and involucres glabrous or sparsely glandular . . . . 4d. *Stephanomeria exigua* subsp. *exigua*

4a. Stephanomeria exigua Nuttall subsp. **carotifera**
(Hoover) Gottlieb, Madroño 21: 472. 1972

· Hoover's wirelettuce   E

Stephanomeria carotifera Hoover,
Leafl. W. Bot. 10: 252. 1966

Heads in paniculiform arrays.
Peduncles 10–25 mm, glabrous or
puberulent. **Calyculi** of reflexed
bractlets. **Involucres** glabrous or
puberulent. **Florets** 7–9. **Cypselae**
3.2–4.3 mm; **pappi** of 18–24
bristles (falling and wholly plu-
mose, or widened bases persistent, bases connate in
groups of 2–4, and bristles plumose to tops of bases).
2*n* = 16.

Flowering Aug–Oct. Open, sandy and shale soils
inland, sand dunes or serpentine near coast; 0–1000 m;
Calif.

Subspecies *carotifera* was originally described as a
species because its large taproot suggested that it is a
perennial and because it differs morphologically from
perennial species of the genus. My field and greenhouse
studies showed it to be an obligate annual and to have
taproots no larger than those of large specimens of other
annual stephanomerias. When it was first described, the
morphologic characteristics of the annuals were not understood and, consequently, its relationships were not
evident. In morphologic features, karyotype, and reproductive compatibilities, subsp. *carotifera* is closely allied
to the other subspecies of *S. exigua.* Coastal and inland
populations of subsp. *carotifera* are morphologically distinguishable, particularly in characters of the pappi. The
bristles of the pappi of inland populations have widened
bases; those of the coastal populations are not widened
or only slightly so. Reproductively, the two types are
fully compatible. Subspecies *carotifera* hybridizes natu-
rally with *S. exigua* subsp. *coronaria* and with both subspecies of *S. virgata* (L. D. Gottlieb 1971, 1972).

4b. Stephanomeria exigua Nuttall subsp. **coronaria**
(Greene) Gottlieb, Madroño 21: 474. 1972 · Small
crown wirelettuce   E   F

Stephanomeria coronaria Greene,
Bull. Calif. Acad. Sci. 1: 194. 1885;
S. exigua var. *coronaria* (Greene)
Jepson

Heads borne singly or clustered
along branches. **Peduncles** 2–5
mm, glabrous or puberulent. **Calyculi** of appressed bractlets. **Involucres** glabrous or puberulent.
Florets 5–11. **Cypselae** 2.3–3.1 mm; **pappi** of 7–20
bristles (widened bases persistent, bases connate in groups
of 2–4, bristles plumose on distal 60–85%). **2***n* = 16.

Flowering Jul–Oct. Equable sites along California
coast, sandy meadows in South Coast Ranges, sandy soils

of Central Valley, openings in yellow pine forest, volcanic soils in eastern Sierra Nevada, sandy, limestone, or
volcanic soils in sagebrush deserts; 0–2800 m; Calif.,
Idaho, Nev., Oreg.

Subspecies *coronaria* occupies the widest range of
habitats of any annual stephanomeria, and it shows striking morphologic variability for some characters including number of florets, lengths and widths of corollas,
numbers and plumosity of pappus bristles, and degree of
"bumpiness" of faces of the cypselae. Populations from
the eastern slopes of the Sierra Nevada in California to
the mountains of Nevada, central Oregon, and Idaho
vary most evidently toward subsp. *exigua*, native to the
deserts, in having heads on longer peduncles, making a
quite different architectural appearance than characteristic of the plants with nearly sessile heads in the Coast
Ranges and elsewhere in California. The variation in subsp.
coronaria presumably results from hybridization with
subsp. *exigua* where the two subspecies make contact at
intermediate elevations. Hybrid individuals are readily
identifiable wherever subsp. *coronaria* makes contact with
other subspecies of *S. exigua* (L. D. Gottlieb 1971, 1972).

4c. Stephanomeria exigua Nuttall subsp. **deanei**
(J. F. Macbride) Gottlieb, Madroño 21: 470. 1972

· Deane's wirelettuce

Stephanomeria exigua var. *deanei*
J. F. Macbride, Contr. Gray Herb.
53: 22. 1918

Heads borne in small paniculiform arrays along branches.
Peduncles 10–40 mm, densely
glandular. **Calyculi** of appressed
bractlets. **Involucres** densely glandular. **Florets** 7–9. **Cypselae** light
to dark tan, 2.1–2.4 mm, faces tuberculate, grooved; **pappi** of 9–14, white to light tan bristles (bases persistent,
connate in groups of 2–4, plumose on distal 55–60%).
2*n* = 16.

Flowering Jun–Oct. Sandy fields and chaparral; 0–
1700 m; Calif.; Mexico (Baja California).

4d. Stephanomeria exigua Nuttall subsp. **exigua**   F

Stephanomeria pentachaeta
D. C. Eaton; *S. exigua* var.
pentachaeta (D. C. Eaton) H. M.
Hall; *S. schottii* (A. Gray) A. Gray

Heads in paniculiform arrays.
Peduncles 10–40 mm, glabrous or
sparsely glandular. **Calyculi** of
appressed bractlets. **Involucres**
glabrous or sparsely glandular.
Florets 5–8. **Cypselae** 2.6–3.2 mm; **pappi** of 5–13 white
to light tan bristles (widened bases persistent, bases
connate in groups of 2–4, if bristles 5, breaking off
completely, bristles plumose on distal 50%). **2***n* = 16.

Flowering May–Jul. Sandy soils, deserts, sagebrush, creosote bush, pinyon-juniper woodlands, Joshua Tree communities; 100–2000 m; Ariz., Calif., Colo., Idaho, Nev., N.Mex., Oreg., Tex., Utah, Wash.; Mexico (Baja California).

Subspecies *exigua* is morphologically variable. Plants with pappi of 5 bristles (often called *Stephanomeria pentachaeta*) are found throughout its distribution; they are fully interfertile with plants having more bristles. *Stephanomeria schottii* A. Gray from southern Yuma County, Arizona, described as having a pappus of 4–6 bristles, "sparsely short-plumose toward the summit," appears to be a synonym of subsp. *exigua*.

4e. Stephanomeria exigua Nuttall subsp. **macrocarpa** Gottlieb, Madroño 21: 473, figs. 2, 3. 1972 • Large seed wirelettuce E

Heads borne singly or clustered along branches. **Peduncles** 5–10 mm, glabrous or puberulent. **Calyculi** of reflexed bractlets. **Involucres** glabrous or puberulent. **Florets** 6–8. **Cypselae** 5.5–6.5 mm; **pappi** of 13–19 tan bristles (widened bases persistent, connate in groups of 2–4, bristles plumose on distal 60–70%. $2n = 16$.

Flowering Aug–Sep. Dry, open sites on western slopes of the Sierra Nevada; 300–1200 m; Calif.

Subspecies *macrocarpa* has rarely been collected; it is common within its limited range on western slopes of the Sierra Nevada. Its cypselae are the largest of any of the annual stephanomerias. Unlike the other subspecies of *Stephanomeria exigua*, which are self-incompatible and obligately outcrossing, subsp. *macrocarpa* is self-compatible and highly self-pollinating. In Kern County, hybrid individuals may be found wherever subsp. *macrocarpa* and subsp. *coronaria* make contact.

5. Stephanomeria fluminea Gottlieb, Madroño 46: 58, fig. 1. 1999 • Creekside wirelettuce C E

Perennials, 15–40 cm (rhizomes slender). **Stems** 1–8, branches ascending, ± tomentose. **Leaves** green (at least cauline) at flowering; blades oblong-oblanceolate, 3–6 cm, margins entire or toothed (teeth remote, faces tomentose). **Heads** borne singly or clustered along stems and branches. **Peduncles** 2–10 mm (glabrous). **Calyculi** of (4–6) appressed bractlets (unequal, lengths to 1/2 phyllaries). **Involucres** 8–10 mm (phyllaries 5, glabrous). **Florets** 5(–6). **Cypselae** tan, 4–4.4 mm, faces smooth, grooved; **pappi** of 30–40, white bristles (persistent), wholly plumose. $2n = 16$.

Flowering Jul–Aug. Spring-flooded flat, gravel stream beds; of conservation concern; 2000–2300 m; Wyo.

Stephanomeria fluminea is known only from northwestern Wyoming. Its habitat is unique among all species of the genus. The plants grow on impermanent, raised cobble benches in flat, gravel beds of creeks that flood and churn after spring snowmelt.

6. Stephanomeria lactucina A. Gray, Proc. Amer. Acad. Arts 6: 552. 1865 • Woodland wirelettuce E

Perennials, 10–60 cm (rhizomes slender). **Stems** single, branches erect or ascending (from near bases), glabrous or sparsely puberulent. **Leaves** green at flowering; blades linear-lanceolate, 3–8 cm, margins entire or toothed (teeth remote, faces glabrous or sparsely puberulent). **Heads** borne singly along branches. **Peduncles** 10–50 mm (bracteate). **Calyculi** of (4–7) appressed bractlets (unequal, lengths to 1/2 phyllaries). **Involucres** 12–14 mm (phyllaries 8–12, glabrous). **Florets** (7–)8–10. **Cypselae** light tan, 5–6 mm, faces smooth, grooved; **pappi** of 25–30, light tan bristles (sometimes connate in groups of 5–6+, persistent), wholly plumose. $2n = 16$.

Flowering Jul–Aug. Sandy soils in dry, open yellow pine and red fir forests; 1100–2300 m; Calif., Oreg.

Stephanomeria lactucina grows in the Cascade Mountains, Sierra Nevada, and North Coast Ranges. It rarely sets seed. The absence of seed is associated with reduced pollen viability, first reported by A. S. Tomb et al. (1978), who noted that "plants [from Siskiyou County, California] were seed-sterile and produced only 45% stainable pollen." The ability of pollen to take up acetocarmine dye is a measure of their viability and, consequently, the fertility of the plant. I examined pollen from five randomly selected specimens collected from Oregon and California and found from 2% to 98% stainable pollen, with a mean of 47%. Pollen stainability and seed set have not been studied systematically in this species. Because these traits directly influence fitness and population persistence, further study would likely be rewarding.

7. Stephanomeria malheurensis Gottlieb, Madroño 25: 44, fig. 1. 1978 • Malheur wirelettuce C E

Annuals, 10–60 cm. **Stems** single, branches ascending, glabrous. **Leaves** withered at flowering; basal blades oblanceolate to spatulate, 5–7 cm, margins entire to pinnately lobed (faces glabrous); cauline much reduced, bractlike. **Heads** borne singly along branches. **Peduncles** 5–10 mm (glabrous). **Calyculi** of appressed bractlets. **Involucres** 8–9.5 mm. **Florets** 5–6 (ligules usually pink, rarely white or orange-yellow). **Cypselae** tan to light brown,

3.3–3.8 mm, faces moderately tuberculate, grooved; **pappi** of 9–12(–15), light tan bristles (connate in groups of 2–4, bristles and/or bases persistent), plumose on distal 50–60%. $2n = 16$.

Flowering Jul–Aug. Soils derived from volcanic tuff, high desert; of conservation concern; 1600 m; Oreg.

Stephanomeria malheurensis has been examined in a series of studies (L. D. Gottlieb 1973b, 1977, 1978b, 1979, 1991; Gottlieb and J. P. Bennett 1983; S. Brauner and Gottlieb 1987, 1989; B. A. Bohm and Gottlieb 1989) because it is one of the very few examples of the recent, natural origin of a diploid, annual plant species. At the type locality, it grows with a population of *S. exigua* subsp. *coronaria* that is thought to be its progenitor.

Stephanomeria malheurensis is known from a single locality in Harney County, Oregon, growing in soil derived from volcanic tuff in the high desert of eastern Oregon. It is a federally listed rare and endangered species, and is in the Center for Plant Conservation's National Collection of Endangered Plants.

8. **Stephanomeria paniculata** Nuttall, Trans. Amer. Philos. Soc., n. s. 7: 428. 1841 • Stiff-branched wirelettuce [E]

Annuals, 0–100 cm. **Stems** single, branched (branches nearly at right angles, stiff), glabrous. **Leaves** withered at flowering; basal blades oblanceolate, 6–10 cm, margins entire or toothed (teeth minute, faces glabrous); cauline much reduced, bractlike. **Heads** borne singly along branches or in paniculiform arrays. **Peduncles** 2–10 mm. **Calyculi** of appressed bractlets. **Involucres** 6–9 mm. **Florets** 5. **Cypselae** light to dark tan, 3.8–4.2 mm, faces slightly bumpy to tuberculate, (grooved); **pappi** of 15–18 tan bristles (connate in groups of 2–4, bases persistent), plumose to tops of bases. $2n = 16$.

Flowering Jun–Sep. Open, sandy or volcanic soils, plains and foothills, often growing as weed along roads; 200–1400 m; Calif., Idaho, Oreg., Wash.

9. **Stephanomeria parryi** A. Gray, Proc. Amer. Acad. Arts 19: 61. 1883 • Parry's wirelettuce [E]

Perennials, 10–30 cm (rhizomes stout). **Stems** 1–3, branches ascending, glabrous. **Leaves** green (at least cauline) at flowering; blades linear to lanceolate, shallowly runcinate, 2–6 cm (relatively thick and firm), margins pinnately lobed (thickened, usually minutely, sharply spinose, faces glabrous). **Heads** borne singly along branches. **Peduncles** 2–10 mm (bracteate). **Calyculi** of (6–8) appressed bractlets

(unequal, lengths to ¹/₂ phyllaries). **Involucres** 10–14 mm. **Florets** (8–)10–13. **Cypselae** tan, 4.5–6 mm, (ribs well developed) faces slightly bumpy, grooved; **pappi** of 10–15, tan bristles (connate in groups of 2–4, bases persistent), plumose on distal 80%. $2n = 32$.

Flowering May–Jun. Open, sandy and gravelly slopes in Upper Sonoran Zone, many plant communities, desert mountains; 700–2000 m; Ariz., Calif. Nev., Utah.

10. **Stephanomeria pauciflora** (Torrey) A. Nelson in J. M. Coulter and A. Nelson, New Man. Bot. Centr. Rocky Mt., 588. 1909 • Prairie skeletonplant, brownplume or few-flowered wirelettuce [F]

Prenanthes pauciflora Torrey, Ann. Lyceum Nat. Hist. New York 2: 210. 1827; *Stephanomeria cinerea* (S. F. Blake) S. F. Blake; *S. lygodesmoides* M. E. Jones ex L. F. Henderson; *S. pauciflora* var. *parishii* (Jepson) Munz

Perennials, 20–50 cm (caudices woody). **Stems** 1–5+, divaricately and intricately branched (often forming dense bushes), usually glabrous, rarely tomentose. **Leaves** withered at flowering; basal blades linear-lanceolate, runcinate, 3–7 cm, margins pinnately lobed (faces glabrous); cauline much reduced and bractlike. **Heads** borne singly along branches. **Peduncles** 3–10 mm. **Calyculi** of appressed bractlets. **Involucres** 8–11 mm (phyllaries 4–6, glabrous). **Florets** 5–6. **Cypselae** tan, 3.5–5 mm, faces tuberculate, grooved; **pappi** of 15–20, usually tan, rarely white, bristles (connate in groups of 2–4, bases persistent), plumose on distal 80%. $2n = 16$.

Flowering May–Sep. Sandy, gravelly washes and slopes in desert shrub communities, juniper woodlands, open, sandy short-grass plains; 200–1500 m; Ariz., Calif., Colo., Kans., Nev., N.Mex., Okla., Tex., Utah, Wyo.; Mexico (Chihuahua, Sonora).

Stephanomeria pauciflora generally grows as an intricately branched, often rounded bush. Occasional plants, usually from Arizona, New Mexico, Texas, and southern Utah, have long, flexuous stems and branches, an architecture that resembles one of the typical forms of *S. tenuifolia*. Some plants of *S. pauciflora* have white pappi, also typical of *S. tenuifolia*. It is not known if these plants represent uncommon and unusual individuals or if they are from populations in which all plants have those traits. It is also not known whether such plants of *S. pauciflora* grow near populations of *S. tenuifolia*; if so, they may result from interspecific hybridization. That is a possibility; experimental hybrid plants produced by crossing individuals from the two species were about 20% fertile. Such fertility suggests the species are sufficiently compatible that fully fertile segregants with variously intermediate morphologies could be expected where they hybridize in nature. The experimental crosses were made

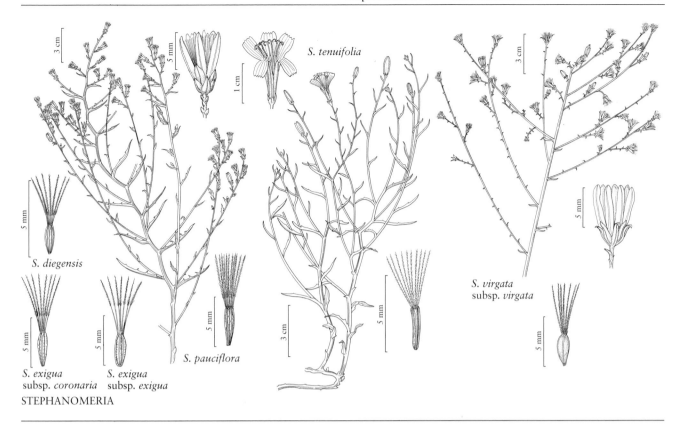

S. tenuifolia

S. diegensis

S. exigua subsp. *coronaria*

S. exigua subsp. *exigua*

S. pauciflora

S. virgata subsp. *virgata*

STEPHANOMERIA

reciprocally between *S. pauciflora* from Riverside County, California (*L. D. Gottlieb 6653*), and *S. tenuifolia* from Wheeler County, Oregon (*L. D. Gottlieb 6692*); specimens of the six F₁ hybrid plants that were produced are deposited at DAV. Plants of *S. pauciflora* that are densely tomentose throughout are occasionally found, particularly in the deserts of California and Nevada, and have been named *S. cinerea* or *S. pauciflora* var. *parishii*.

11. **Stephanomeria runcinata** Nuttall, Trans. Amer. Philos. Soc., n. s. 7: 428. 1841 • Desert or sawtooth wirelettuce E

Perennials, 10–20(–25) cm (rhizomes stout). **Stems** erect or ascending, glabrous or weakly scabrous-puberulent. **Leaves** usually withered at flowering; basal blades narrowly lanceolate, markedly runcinate, 3–7 cm, margins pinnately lobed (faces glabrous or weakly scabrous-puberulent); cauline much reduced, bractlike. **Heads** borne singly on branch tips. **Peduncles** 10–50 mm (usually minutely bracteolate). **Calyculi** of appressed bractlets (unequal, lengths to ¹/₂ phyllaries). **Involucres** 9–12(–13) mm (phyllaries 5–6, glabrous). **Florets** 5–6. **Cypselae** tan, 3–5 mm, faces smooth to slightly bumpy, grooved;

pappi of 15–25, white bristles (persistent), wholly plumose. $2n = 16$.

Flowering June–Sep. Open sandy places, eroded siltstones, clay flats, alkali soils; 600–1800 m; Alta., Sask.; Colo., Mont., Nebr., N.Dak., Utah, Wyo.

Stephanomeria runcinata grows in the upper Great Plains and adjacent intermontane valleys.

12. **Stephanomeria tenuifolia** (Rafinesque) H. M. Hall, Univ. Calif. Publ. Bot. 3: 256. 1907 • Slender wirelettuce, narrow-leaved skeletonplant F

Ptiloria tenuifolia Rafinesque, Atlantic J. 1: 145. 1832, based on *Prenanthes tenuifolia* Torrey, Ann. Lyceum Nat. Hist. New York 2: 210. 1827, not Linnaeus 1753; *Stephanomeria myrioclada* D. C. Eaton; *S. neomexicana* (Greene) Cory; *S. tenuifolia* var. *myrioclada* (D. C. Eaton) Cronquist; *S. tenuifolia* var. *uintaensis* Goodrich & S. L. Welsh; *S. wrightii* A. Gray

Perennials, 20–70 cm (rhizomes stout). **Stems** 1–5+, erect to ascending, relatively densely to sparsely branched, glabrous. **Leaves** withered at flowering; basal blades linear to filiform, 5–8 cm, margins entire or toothed (teeth remote, faces glabrous); cauline much reduced, bractlike.

Heads borne singly on branch tips. **Peduncles** 0, or 1–50+ mm (bracteolate). **Calyculi** of appressed bractlets. **Involucres** 5–14(–15) mm (phyllaries 5–6, glabrous). **Florets** 4–5(–6). **Cypselae** tan, 3–6 mm, faces smooth, grooved; **pappi** of 15–25, white bristles (persistent), wholly plumose. $2n = 16$.

Flowering Jun–Sep. Crevices in volcanic, granitic, and sandstone outcrops, open rocky ridges and slopes, bases of cliffs; 300–3000 m; B.C., Sask.; Ariz., Calif., Colo., Idaho, Mont., Nev., N.Mex., N.Dak., Oreg., Tex., Utah, Wash., Wyo.; nw Mexico.

Stephanomeria tenuifolia is distributed over an immense region and is the most widespread species of the genus. It shows remarkable variability in the form and dimensions of its stems and branches. Plants described as *S. myrioclada*, from the northeasternmost corner of Nevada, present an architecture of relatively numerous, almost threadlike, densely crowded stems (1.5–4 dm) and branches with an irregularly dichotomous pattern. Continuous variation occurs from this form to another in the same region and elsewhere in which the stems are longer (3–7 dm), sparingly branched, and flexuous. The extreme variability in vegetative architecture may be adaptive and deserves further study.

Stephanomeria wrightii was described from western Texas. The type sheet has three specimens, all with a single, slender stem with paniculately disposed branches from a broken-off stub. A. Gray (1884, 1886) stated that those features indicated the plants were "seemingly biennial," and distinguished them from *S. minor* Nuttall (an old synonym of *S. tenuifolia*), which he said has "thick and tortuous roots." The slender stem of the specimens received by Gray may indicate their young age; they do not seem distinctive because other specimens from the same region have the bright white, fully plumose pappi and other features ascribed to *S. wrightii*, but also exhibit large, thick rhizomes. Gray noted that the "achenes are contracted under the summit"; this feature is not evident on the three specimens. Overall, *S. wrightii* does not seem different from *S. tenuifolia*.

The type specimen of *Stephanomeria neomexicana*, originally described as *Ptiloria neomexicana* by E. L. Greene, from New Mexico, exhibits multiple, long, flexuous stems with relatively few branches emerging from a stout rhizome. The pappi consist of plumose bristles, but the proximal 0.5–1 mm is only minutely barbed. Plants with similar pappi are found occasionally in New Mexico and Arizona and differ in no other respect from *S. tenuifolia*.

Stephanomeria tenuifolia var. *uintaensis* grows "in one small isolated stand" (S. Goodrich and S. L. Welsh 1983) and was recognized primarily on the basis of relatively long phyllaries (10–16 mm) and runcinate-pinnatifid basal leaves. I measured phyllary lengths on isotypes; only one, from a terminal head, was 16 mm; others varied from 11–14(–15) mm. Although the lengths are 2–3 mm longer than typical, no other feature of the

plants is unusual. Pinnately lobed leaves may not be typical of the species; because leaves are not often present on *S. tenuifolia* when it is in flower, and most specimens do not include them, the significance of the character is uncertain. The isotype collections are from young individuals and only the paratype from RM has cypselae. The variety does not seem worthy of taxonomic recognition.

13. **Stephanomeria thurberi** A. Gray, Pl. Nov. Thurb., 325. 1854 • Thurber's wirelettuce, skeletonplant

Perennials, 20–50 cm (rhizomes slender). **Stems** single, branches on distal $^{1}/_{3}$–$^{1}/_{2}$, glabrous or sparsely puberulent. **Leaves** green at flowering (at least on plants of spring and early summer, frequently absent in plants of late summer); basal blades oblanceolate to spatulate, runcinate, 4–7 cm, margins pinnately lobed (faces glabrous or sparsely puberulent); cauline reduced, scalelike on plants of spring and early summer, linear and threadlike (to 3 cm) on plants of late summer. **Heads** borne singly on branch tips. **Peduncles** mostly 5–100+ mm (bracteolate). **Calyculi** of (4–6) appressed bractlets (unequal, lengths to $^{1}/_{2}$ phyllaries). **Involucres** 9–11(–12) mm (phyllaries 6–8, glabrous). **Florets** 10–16(–20). **Cypselae** tan, 5–6 mm, faces smooth, grooved; **pappi** of 30–40, white bristles (persistent), wholly plumose. $2n = 16$.

Flowering May–Sep. Open, sandy sites in juniper-mesquite grasslands and in yellow pine forests, sometimes growing as weed along roadsides; 1200–2500 m; Ariz., N.Mex. Tex.; Mexico (Sonora).

Stephanomeria thurberi has been collected most often in May, June, and early July. These specimens have well developed basal rosettes, stems with nodes 3+ cm apart, relatively short branches, usually only on the distal 30–50%, and scalelike, cauline leaves. That is the form described by Gray. Another form has been collected from late July into early September that is morphologically distinct, as first pointed out by A. S. Tomb on the labels of specimens he collected in 1968 and 1970 (see below). That "summer form" lacks basal rosettes, the stems have more numerous nodes, about 1 per cm, beginning at ground level, and threadlike cauline leaves 3–4 cm. Those specimens usually have relatively few heads. Some specimens collected in July are intermediate, having basal rosettes and relatively long, threadlike cauline leaves, or no rosettes and relatively short, scalelike cauline leaves.

It is not known if the different growth forms represent distinct genotypes that initiate growth at different times, or if the same individual produces aboveground parts with differing appearances during the growth season. It is also not known if the two forms commonly grow together as they do in Coconino County, Arizona (*Tomb 280,* August 10, 1968, and *Tomb 631,* June 12, 1970), or if they generally occupy different habitats. The unusual situation calls for study.

14. Stephanomeria virgata Bentham, Bot. Voy. Sulphur, 32. 1844 • Virgate wirelettuce F

Annuals, 50–200 cm. **Stems** single, branches virgate, usually glabrous, rarely tomentose. **Leaves** withered at flowering; basal blades oblanceolate to spatulate, runcinate, 3–10 cm, margins pinnately lobed (faces usually glabrous, rarely tomentose); cauline much reduced, bractlike. **Heads** borne singly or clustered along branches. **Peduncles** 3–10 mm, bracteate. **Calyculi** of appressed or reflexed bractlets. **Involucres** 6–8 mm. **Florets** 5–9. **Cypselae** light tan to dark brown, 2.2–3.6 mm, faces smooth to tuberculate, not grooved; **pappi** (falling) of 23–28, white bristles, wholly plumose. $2n = 16$.

Subspecies 2 (2 in the flora): w United States, nw Mexico.

The two subspecies of *Stephanomeria virgata* are often sympatric in southern California. Hybrids are found frequently.

1. Bractlets of calyculi reflexed; florets 8–9 . . . . .
. . . . 14a. *Stephanomeria virgata* subsp. *virgata*
1. Bractlets of calyculi appressed; florets 5–6 . . . .
. . . . 14b. *Stephanomeria virgata* subsp. *pleurocarpa*

14a. Stephanomeria virgata Bentham subsp. **virgata** E F

Stephanomeria tomentosa Greene; *S. virgata* var. *tomentosa* (Greene) Munz

Calyculi of reflexed bractlets. **Florets** 8–9. $2n = 16$.

Flowering late Jul–Oct. Chaparral openings, dry, sandy hills and grasslands, often growing as weed on roadside embankments; 0–1800 m; Calif.

Subspecies *virgata* grows in the South Coast and Transverse ranges. *Stephanomeria tomentosa* refers to a densely tomentose form occasionally found in some populations of *S. virgata.*

14b. Stephanomeria virgata Bentham subsp. **pleurocarpa** (Greene) Gottlieb, Madroño 21: 480. 1972 • Wand wirelettuce

Ptiloria pleurocarpa Greene, Pittonia 2: 131. 1890

Calyculi of appressed bractlets. **Florets** 5–6. $2n = 16$.

Flowering Jul–Nov. Widespread on soils including those derived from shale, sandstone, serpentine, and volcanics; 50–1800 m; Calif., Nev., Oreg.; Mexico (Baja) California.

73. PRENANTHELLA Rydberg, Bull. Torrey Bot. Club 33: 160. 1906 • [Genus *Prenanthes* and Latin *-ella,* diminutive, alluding to original assignment of type species]

Kenton L. Chambers

Annuals, 5–30 cm; taprooted. **Stems** 1–5+, erect (leafy proximally, often flexuous distally), branched near base bases or distally, glandular-puberulent to glabrescent. **Leaves** basal or proximally cauline; sessile or petiolate; blades spatulate to oblanceolate, often runcinate, margins irregularly dentate to lobed, teeth often spinulose (faces glabrous). **Heads** (10–150+, on flexuous, bracteate, intricately-branched axes) in paniculiform arrays. **Peduncles** not inflated, minutely bracteate. **Calyculi** of 2–3, minute, deltate, unequal bractlets. **Involucres** ovoid, cylindric at flowering, 2–3 mm diam. **Phyllaries** 3–5 in 1 series, linear-lanceolate, equal, herbaceous, margins hyaline, apices acute. **Receptacles** flat, smooth (with scars), glabrous, epaleate. **Florets** 3–4; corollas pink or white (slightly surpassing phyllaries at flowering). **Cypselae** pale brown, columnar, tapering slightly proximally, not beaked, 5-ribbed and grooved, minutely rugulose, glabrous; **pappi** tardily falling, of 80+, basally connate, white, unequal, stiff, smooth bristles. $x = 7$.

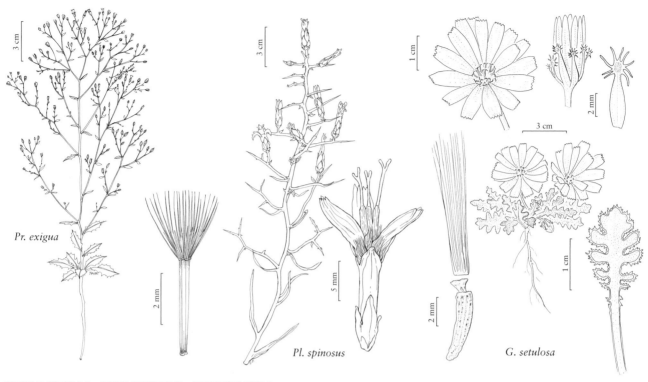

Pr. exigua

Pl. spinosus

G. setulosa

PRENANTHELLA ◦ PLEIACANTHUS ◦ GLYPTOPLEURA

Species 1: sw United States, nw Mexico.

The single species of *Prenanthella* was usually assigned to *Lygodesmia* in treatments prior to A. S. Tomb's (1972b) reassessment. Its base chromosome number of $x = 7$ was taken as evidence of a possible relationship with *Malacothrix*, *Calycoseris*, and *Anisocoma*, also $x = 7$ (Tomb 1974). Recent molecular phylogenetic studies (J. Lee et al. 2003) place *Prenanthella* in a clade with *Pleiacanthus*, *Stephanomeria*, and *Rafinesquia*, all with $x = 8$. This group, together with *Munzothamnus*, constitutes the subtribe Stephanomeriinae of Lee and B. G. Baldwin (2004).

SELECTED REFERENCE   Tomb, A. S. 1972b. Re-establishment of the genus *Prenanthella* Rydb. (Compositae: Cichorieae). Brittonia 24: 223–228.

1. Prenanthella exigua (A. Gray) Rydberg, Bull. Torrey Bot. Club 33: 161. 1906 • Brightwhite, desert prenanthella F

Prenanthes exigua A. Gray, Smithsonian Contr. Knowl. 5(6): 105. 1853; *Lygodesmia exigua* (A. Gray) A. Gray

Leaves 1–3 cm, reduced distally to bracts 0.5–5 mm. **Heads** 10–150+. **Calyculus bractlets** 0.5–1+ mm. **Principal phyllaries** 3–5 mm after flowering. **Corollas** 1.5–2 mm. **Cypselae** 2.5–3.5 mm; **pappi:** longest bristles 2–3 mm. $2n = 14$.

Flowering Mar–Jun. Sandy, gravelly, or clay soils, desert washes and open slopes to sagebrush-juniper steppes; 20–1900 m; Ariz., Calif., Colo., Nev., Oreg., Tex., Utah; Mexico (Baja California, Sonora).

74. PLEIACANTHUS (Nuttall) Rydberg, Fl. Rocky Mts., 1069. 1917 • Thorny skeletonweed [Greek *pleio*, in compounds, more than usual, and *acanthos*, a prickly plant, or *acantha*, thorn] E

L. D. Gottlieb

Lygodesmia D. Don subgen. *Pleiacanthus* Nuttall, Trans. Amer. Philos. Soc., n. s. 7: 444. 1841

Perennials, 10–50 cm; root crowns woody, branched. **Stems** 1–4[–8], branches divaricate, rigid, spine-tipped, with conspicuous tufts of light brown, wool in axils of bud scales at and just below ground level, otherwise glabrous (bud scales lanceolate, 8–30 × 5–8 mm, entire). **Leaves** cauline; sessile; proximal blades linear, margins entire; distal bractlike. **Heads** (± subsessile, erect), borne singly or in paniculiform arrays. **Peduncles** not inflated, not bracteate. **Calyculi** of 4–6 bractlets (unequal, lengths to ½ phyllaries). **Involucres** cylindric, 2–3+ mm diam. **Phyllaries** 3–5 in ±1 series, linear-lanceolate. **Receptacles** flat or convex, smooth, glabrous, epaleate. **Florets** 3–5; corollas pink or lavender. **Cypselae** tan, columnar, apices truncate (beaks 0), with 5 equal faces, separated by weakly rounded ribs, each face with central, narrow, shallow, longitudinal groove, glabrous; **pappi** persistent, of 50–60, distinct, light tan, barbellate bristles (of 2 lengths). *x* = 8.

Species 1: w United States

1. Pleiacanthus spinosus (Nuttall) Rydberg, Fl. Rocky Mts., 1069. 1917 • Thorny skeletonweed E F

Lygodesmia spinosa Nuttall, Trans. Amer. Philos. Soc., n. s. 7: 444. 1841; *Stephanomeria spinosa* (Nuttall) Tomb

Stems: branches 4–8 cm. **Leaves:** proximal 3–7 cm. **Peduncles** 1–4 mm. **Calyculi:** bractlets 4–8 × 1.5–2.5 mm, glabrous. **Involucres** 7–12 mm. **Cypselae** 6–8 mm; **pappus bristles:** the longer 7–11 mm, shorter 5–7 mm. *2n* = 16.

Flowering Jul–Sep. Open, sandy, gravelly washes and slopes, desert shrub, pinyon-juniper communities; 1500–2900 m; Ariz., Calif., Idaho, Mont., Nev., Oreg., Utah.

Pleiacanthus spinosus was first collected by Nuttall, who placed it in *Lygodesmia*. Rydberg elevated it to the genus level; his proposal was not taken up by others. The species remained in *Lygodesmia* until A. S. Tomb (1970) transferred it to *Stephanomeria*. Tomb made the transfer because *P. spinosus* has the same base chromosome number as *Stephanomeria* (*x* = 8) as well as similar echinate pollen grains (Tomb 1974; Tomb et al. 1974) and thereby differs from *Lygodesmia*, which has *x* = 9 and echinolophate pollen. *Pleiacanthus spinosus* also has morphologic traits not shared with any stephanomeria. These include dense, long tufts of wool in the ground-level axils of the bud scales of the stems, sharp-tipped branches and stems, and pappus bristles that are of two lengths and not plumose. Recent results of DNA sequencing studies of *Stephanomeria* and related North American genera showed that *P. spinosus* is not a member of the clade of all stephanomerias (J. Lee et al. 2002).

75. GLYPTOPLEURA D. C. Eaton in S. Watson, Botany (Fortieth Parallel), 207, plate 20, figs. 11–18. 1871 • [Greek *glyptos*, carved, and *pleura*, rib, alluding to cypselae] E

David J. Keil

Annuals, 1–6 cm (low-growing, densely cespitose, herbage glabrous); taprooted. **Stems** 1–25+, ± prostrate, simple or branched, glabrous. **Leaves** basal and cauline, crowded; petiolate or sessile; basal blades ± oblanceolate, margins dentate or pinnately lobed and crustose-denticulate, cauline progressively reduced to oblanceolate, crustose-denticulate bracts. **Heads** (1–3) borne singly or 2–3 in bract axils. **Peduncles** not distally inflated, often bracteate. **Calyculi** of 5–8, linear to oblanceolate bractlets in ± 1 series, apices expanded, crustose-denticulate). **Involucres** cylindric to urceolate, 3–8+ mm diam. **Phyllaries** 5–8+ in 1–2 series, commonly purplish-tinged, linear, equal, margins scarious, apices acute. **Receptacles** ± flat, smooth, glabrous, epaleate. **Florets** 7–18; corollas white to pale yellow, becoming pink-purple (especially when dry). **Cypselae** straw-colored or light brown, subcylindric or slightly flattened, often curved, abruptly beaked, obtusely 4–5-angled, ribs transversely roughened, alternating with 5 rows of pits, glabrous or minutely puberulent; **pappi** falling (outer, individually) or ± persistent (inner, connate at bases in easily fractured rings), of 50–80+, white, barbellulate to smooth bristles in 3–4+ series. $x = 9$.

Species 2 (2 in the flora): w United States.

A molecular phylogenetic investigation by J. Lee et. al (2003) provided evidence that *Glyptopleura* is part of a primarily western North American radiation in Cichorieae. That study did not resolve the relationship of *Glyptopleura* to other genera within the radiation.

1. Ligules 4–10 mm, equaling involucres or exserted 1–5 mm . . . . 1. *Glyptopleura marginata*
1. Ligules 15–25 mm, exserted 10–20 mm. . . . . 2. *Glyptopleura setulosa*

1. **Glyptopleura marginata** D. C. Eaton in S. Watson, Botany (Fortieth Parallel), 207, plate 20, figs. 11–18. 1871 • White-margined wax-plant, carveseed [E]

Plants 1–6 cm. **Leaves** 0.5–5 cm, margins conspicuously white-crustose. **Heads** 1–2 cm diam. **Calyculi:** margins of bractlets crustose-dentate ± throughout. **Involucres** 10–14 mm. **Florets** 9–18; corollas white to cream, aging pink or purple, ligules 4–10 mm, equaling or scarcely exserted beyond involucres. $2n = 18$.

Flowering Apr–Jul. Sandy or rocky deserts, alkali flats, arid grasslands, often with *Atriplex*, sometimes with *Larrea*; 1000–2000 m; Calif., Idaho, Nev., Utah.

Glyptopleura marginata is found in the Great Basin and Mojave deserts.

2. **Glyptopleura setulosa** A. Gray, Proc. Amer. Acad. Arts 9: 211. 1874 • Keyesia, holly-dandelion [E][F]

Glyptopleura marginata D. C. Eaton var. *setulosa* (A. Gray) Jepson

Plants 2–6 cm. **Leaves** 1–6 cm, margins narrowly white-crustose. **Heads** 3–4 cm diam. (often appearing disproportionately large for size of plants). **Calyculi:** margins of outer bractlets crustose-toothed mostly near apices. **Involucres** 10–15 mm. **Florets** 7–14; corollas cream to pale yellow, aging pink or purple, ligules 15–25 mm, exserted 10–20 mm beyond involucres.

Flowering Mar–May. Sandy desert flats, rocky soil, arid grasslands, often with *Larrea*; 1000–1400 m; Ariz., Calif., Nev., Utah.

Glyptopleura setulosa is found in the Mojave desert. Some authors (e.g., P. A. Munz 1974; G. L. Stebbins 1993b) have included *G. setulosa* within *G. marginata*. The taxa appear to be readily distinguishable and worthy of recognition as distinct species.

76. **KRIGIA** Schreber, Gen. Pl. 2: 532. 1791, name conserved • Dwarf dandelion
[For David Krieg, 16??–1713, plant collector in Maryland and Delaware]

Kenton L. Chambers

Robert J. O'Kennon

Apogon Elliott; *Cymbia* (Torrey & A. Gray) Standley; *Cynthia* D. Don; *Serinia* Rafinesque; *Troximon* Gaertner

Annuals or perennials, 3–75 cm; taprooted, fibrous-rooted, or (in *K. dandelion*) with rhizomes bearing globose tubers. **Stems** 1–50+, usually erect, rarely decumbent, scapiform or branched distally, glabrous or sparingly villous (proximally), glandular-villous (especially distally). **Leaves**

mostly basal, sometimes cauline; petiolate (petioles often winged); blades linear to lanceolate, oblanceolate, or spatulate, margins entire, denticulate, or irregularly pinnately lobed, apices acute to obtuse (faces glabrous or glandular-villous, usually glaucous in *K. dandelion* and *K. biflora*); distal cauline usually slightly reduced to bractlike. **Heads** borne singly. **Peduncles** not distally inflated, ebracteate (from rosettes and from axils of cauline leaves or bracts). **Calyculi** 0. **Involucres** turbinate to campanulate, 2–12 mm diam. **Phyllaries** (4–)5–18 in 1–2 series, (sometimes reflexed in fruit) linear-lanceolate to ovate, equal, herbaceous, apices acute (faces glabrous). **Receptacles** flat or low-convex, pitted, glabrous, epaleate. **Florets** 5–60; corollas yellow to orange (equaling or surpassing phyllaries). **Cypselae** brown or reddish brown, columnar, obconic, barrel-shaped, or fusiform, not beaked, nerves or ribs 10–20, glabrous; **pappi** 0, or persistent, often fragile, usually in 2 series, distinct, outer of 5+, yellowish or brownish scales, inner of 5–45, barbellulate bristles (pappi 0 in *K. cespitosa*, 0 or 1 series of tiny scales in *K. wrightii*). $x = (4)\ 5\ (6, 9)$.

Species 7 (7 in the flora): North America, ne Mexico.

Krigia is diverse and limited to North America. On molecular evidence, it stands apart from other clades of Cichorieae and is best placed as a monotypic subtribe (J. Lee et al. 2003; Lee and B. G. Baldwin 2004). Early studies classified the pappose and epappose species as different genera. A unified view of the genus was taken by L. H. Shinners (1947), and this has been supported by recent morphologic and molecular studies (K. J. Kim and T. J. Mabry 1991; Kim and B. L. Turner 1992; Kim et al. 1992b, 1992c; Kim and R. K. Jansen 1994). The most common base number is $x = 5$, with lower and higher numbers having arisen through dysploidy, autoploidy, and both ancient and recent alloploidy (K. L. Chambers 1965, 2004; A. S. Tomb et al. 1978; C. C. Chinnappa 1981; Kim and Turner).

SELECTED REFERENCES Chambers, K. L. 2004. Taxonomic notes on *Krigia* (Asteraceae). Sida 21: 225–236. Chinnappa, C. C. 1981. Cytological studies in *Krigia* (Asteraceae). Canad. J. Genet. Cytol. 23: 671–678. Kim, K. J. et al. 1992b. Evolutionary implications of intraspecific chloroplast DNA variation in dwarf dandelions (*Krigia*–Asteraceae). Amer. J. Bot. 79: 708–715. Kim, K. J. et al. 1992c. Phylogenetic and evolutionary implications of interspecific chloroplast DNA variation in *Krigia* (Asteraceae--Lactuceae). Syst. Bot. 17: 449–469. Kim, K. J. and R. K. Jansen. 1994. Comparisons of phylogenetic hypotheses among different data sets in dwarf dandelions (*Krigia*, Asteraceae): Additional information from internal transcribed spacer sequences of nuclear ribosomal DNA. Pl. Syst. Evol. 190: 157–185. Kim, K. J. and T. J. Mabry. 1991. Phylogenetic and evolutionary implications of nuclear ribosomal DNA variation in dwarf dandelions (*Krigia*–Lactuceae–Asteraceae). Pl. Syst. Evol. 177: 53–69. Kim, K. J. and B. L. Turner. 1992. Systematic overview of *Krigia* (Asteraceae–Lactuceae). Brittonia 44: 173–198. Shinners, L. H. 1947. Revision of the genus *Krigia* Schreb. Wrightia 1: 187–206.

1. Perennials; involucres 7–15 mm; phyllaries reflexed in fruit; pappus bristles 14–45.
    2. Stems scapiform, never leafy or bracteate; peduncles from basal rosettes (1–5+ per plant); tubers (on rhizomes) globose . . . . 1. *Krigia dandelion*
    2. Stems leafy or bracteate; peduncles from axils of cauline leaves and bracts (first-formed peduncle sometimes arising basally in *K. montana*); tubers none.
        3. Peduncles usually in groups of 2–6 from axils of distal cauline, auriculate, clasping bracts (on scapiform stems arising from basal rosettes); corollas orange or yellow-orange (leaves usually glaucous) . . . . 2. *Krigia biflora*
        3. Peduncles usually arising on leafy branches axillary to well-developed cauline leaves; corollas yellow . . . . 3. *Krigia montana*
1. Annuals; involucres 2–9 mm; phyllaries usually erect in fruit (reflexed in *K. virginica*); pappus bristles (1–)4–5, or pappi 0.
    4. Phyllaries reflexed in fruit; pappus bristles surpassing scales by 3+ mm . . . . 4. *Krigia virginica*
    4. Phyllaries erect in fruit; pappus bristles 0 or surpassing scales by no more than 2 mm.
        5. Phyllary midveins evident, not prominent nor forming curved keels; cypselae fusiform, broadest beyond middles (gradually tapering distally), apical areas ± equal to basal areoles; pappi 0 . . . . 7. *Krigia cespitosa*
        5. Phyllary midveins becoming prominent, curving inward at bases to form keels; cypselae obconic to broadly columnar or barrel-shaped, broadest at or just proximal to apices, apical areas broader than basal areoles; pappi 0, or of scales and bristles.

[6. Shifted to left margin.—Ed.]

6. Peduncles from rosettes; cypselae broadly obconic; pappi usually of 5 scales 0.4–0.6 mm plus 5 bristles 1.2–2 mm . . . . 5. *Krigia occidentalis*

6. Peduncles mostly from branching, leafy stems; cypselae broadly columnar or barrel-shaped, slightly constricted at apices; pappi 0, or coroniform (minutes scales, rarely with 1–5 bristles)
. . . . 6. *Krigia wrightii*

1. **Krigia dandelion** (Linnaeus) Nuttall, Gen. N. Amer. Pl. 2: 127. 1818 • Potato dwarfdandelion or dandelion Ⓔ

Leontodon dandelion Linnaeus, Sp. Pl. 2: 798. 1753; *Cynthia dandelion* (Linnaeus) de Candolle

Perennials, 10–50 cm; rhizomes relatively slender, fibrous-rooted, tubers overwintering, globose, 5–15 mm diam., caudices fibrous-rooted. **Stems** scapiform, erect, leafless and ebracteate, glabrous or minutely glandular-villous especially proximal to heads. **Leaves** basal; petioles usually broadly or narrowly winged; blades linear to lanceolate or oblanceolate, 6–24 cm, margins entire or remotely toothed to pinnately lobed, lobes usually entire and acute, apices acute to obtuse, faces usually glabrous, sometimes sparingly villous (glandular or eglandular, often glaucous). **Heads** borne singly. **Peduncles** from basal rosettes. **Involucres** 10–15 mm. **Phyllaries** 12–16, reflexed in fruit, linear-lanceolate, midveins obscure, apices acute. **Florets** 25–34; corollas yellow to yellow-orange, abaxially often purplish-tinged, 15–25 mm. **Cypselae** reddish brown, columnar, 2.5 mm, 10–15-ribbed; **pappi** of ca. 10, outer scales 0.5–1 mm plus 25–45, barbellulate inner bristles 5–8 mm. $2n = 60$.

Flowering Apr–Jun. Sandy or clay-loam soils, open mixed mesophytic and oak-hickory woods, fields, pastures, roadsides; 10–500 m; Ala., Ark., Del., Fla., Ga., Ill., Ind., Kans., Ky., La., Md., Miss., Mo., N.J., N.C., Ohio, Okla., S.C., Tenn., Tex., Va., W.Va.

Krigia dandelion is known from the Eastern deciduous forest biome and tallgrass prairies. It is unique in propagating extensively by means of tubers. These were well described by T. Holm (1891). All chromosome counts to date have shown $2n = 60$, the duodecaploid number based on $x = 5$.

2. **Krigia biflora** (Walter) S. F. Blake, Rhodora 17: 135. 1915 • Orange dwarfdandelion Ⓔ

Hyoseris biflora Walter, Fl. Carol., 194. 1788; *Cynthia virginica* (Linnaeus) D. Don ex de Candolle; *C. viridis* Standley; *H. amplexicaulis* Michaux; *Krigia amplexicaulis* (Michaux) Nuttall; *K. biflora* var. *viridis* (Standley) K. J. Kim; *Tragopogon virginicus* Linnaeus

Perennials, 10–70 cm; caudices stout, fibrous-rooted (sometimes propagating by adventitious buds on roots). **Stems** 1–5+, erect, scapiform, eglandular or glandular-villous distally. **Leaves** mostly basal (rosettes), some cauline (proximal); petioles ± winged; blades oblanceolate to obovate or spatulate, 5–25 cm, margins entire or remotely dentate to pinnately lobed, lobes narrow to bluntly triangular or rounded, apices acute to obtuse or rounded, faces glabrous, eglandular (usually glaucous); cauline 1–4, sessile, lanceolate, bases sheathing or auriculate-clasping, usually entire, distalmost sometimes reduced, bractlike. **Heads** (2–)3–20+. **Peduncles** usually in groups of 2–6 from axils of single or paired distal cauline bracts. **Involucres** 7–11 mm. **Phyllaries** 8–18, reflexed in fruit, lanceolate, midveins obscure, apices acute, faces glabrous. **Florets** 25–60; corollas orange or yellow-orange, 15–25 mm. **Cypselae** reddish brown, columnar, 2–2.5 mm, 12–15-ribbed; **pappi** of ca. 10 outer scales 0.3–0.5 mm plus 20–40, barbellulate inner bristles 4.5–5.5 mm. $2n = 10, 20$.

Flowering Apr–Aug. Sandy, loam, or humus soils, shaded mixed mesophytic, beach-maple, oak-pine, and oak-hickory woods, often near streams, meadows, moist prairies, and Madrean woodlands; 10–2300 m; Man., Ont.; Ala., Ariz., Ark., Colo., Conn., Del., Ga., Ill., Ind., Iowa, Kans., Ky., Md., Mass., Mich., Minn., Miss., Mo., N.J., N.Mex., N.Y., N.C., Ohio, Okla., Pa., R.I., Tenn., Va., W.Va., Wis.

Krigia biflora is known from the Eastern deciduous forest biome, tallgrass prairie, Rocky Mountain forest, and Madrean woodlands. It appears to spread clonally by adventitious buds on the roots. It is related to the more leafy-stemmed, freely branching *K. montana* (K. J. Kim and B. L. Turner 1992); their habital differences are less clear where they are sympatric in the southern Appalachians. An alloploid hybrid between them has become established (see 3. *K. montana*).

3. Krigia montana (Michaux) Nuttall, Gen. N. Amer. Pl. 2: 127. 1818 • Mountain dwarfdandelion [E]

Hyoseris montana Michaux, Fl. Bor.-Amer. 2: 87. 1803; *Cynthia montana* (Michaux) Standley

Perennials, 20–50 cm; caudices fibrous-rooted, stout, often branched. **Stems** 1–5+, usually decumbent proximally, ascending and branched distally, glabrous, eglandular. **Leaves** basal and cauline; petiolate; blades narrowly oblanceolate, apices acute, faces glabrous, eglandular; basal 5–30 cm, margins denticulate to pinnately lobed, lobes broad and blunt or narrow, curving, and acute; cauline becoming linear and reduced distally. **Heads** (2–)3–20+. **Peduncles** terminal on leafy branches or in axils of cauline leaves and bracts (first sometimes scapiform from basal rosettes; usually glabrous, sometimes lightly glandular distally). **Involucres** 7–12 mm. **Phyllaries** 8–16, reflexed in fruit, lanceolate, midveins obscure, apices acute, glabrous. **Florets** 25–60; corollas yellow, 15–25 mm. **Cypselae** brown, columnar, 2–2.8 mm, 12–15-ribbed; **pappi** of 10–16 outer scales 0.5–0.8 mm plus 14–20, scabrous inner bristles 4.5–6 mm. $2n = 20$.

Flowering May–Sep. Gravelly soils and wet mossy crevices on granite cliffs, rock slides, rocky road banks, and heath balds; 700–1900 m; Ga., N.C., S.C., Tenn.

Krigia montana is known from the high-elevation Appalachian ecosystem. In Buncombe County, North Carolina, there is an established population of the hexaploid hybrid between *K. montana* and *K. biflora*, *Krigia* ×*shinnersiana* K. L. Chambers (2004). It has been studied extensively on a molecular basis (K. J. Kim and T. J. Mabry 1991; Kim et al. 1992b, 1992c; Kim and R. K. Jansen 1994). Chloroplast and nuclear DNA data indicate that *K. biflora* was the maternal parent and *K. montana* the paternal parent. *Krigia* ×*shinnersiana* is leafy-stemmed but shows less branching than is typical of *K. montana*, and its flower color is intermediate between the orange of *K. biflora* and the yellow of *K. montana*. The plants propagate extensively by root buds as in the maternal parent.

4. Krigia virginica (Linnaeus) Willdenow, Sp. Pl. 3: 1618. 1803 • Virginia dwarfdandelion [E]

Hyoseris virginica Linnaeus, Sp. Pl. 2: 809. 1753

Annuals, 4–30 cm; taprooted. **Stems** 1–50+, initially scapiform, later elongate and leafy (after rosette leaves wither), erect, eglandular or stipitate-glandular, especially near heads. **Leaves** initially produced basally (rosettes); later leaves cauline; sessile or petiolate; blades oblan-ceolate to spatulate, 1.5–18 cm, usually irregularly pinnately lobed, terminal lobes often denticulate, faces glabrous, eglandular or loosely glandular-villous; cauline (appearing opposite on late-season stems), narrowly oblanceolate to linear, mostly entire. **Heads** borne singly. **Peduncles** from rosettes (early) or axillary or terminal on stems (later). **Involucres** 4.5–8 mm. **Phyllaries** 9–15, reflexed in fruit, lanceolate, midveins obscure, apices acute. **Florets** 8–35; corollas yellow, often purplish abaxially, 5–12 mm. **Cypselae** dark reddish brown, narrowly obconic, 1.5–2.3 mm, 15–20-ribbed; **pappi** of 5, hyaline, rounded outer scales 0.5–1 mm plus 5, scabrous inner bristles 4–6 mm. $2n = 10, 20$.

Flowering Feb–Nov. Sandy, silty, or loam soils, rock outcrops, open mixed mesophytic, northern hardwoods, beach-maple, oak-pine, and oak-hickory woods, pastures, prairies, and roadsides, often in disturbed, weedy sites; 0–1200 m; B.C.; Ala., Ark., Conn., Del., Fla., Ga., Ill., Ind., Iowa, Ky., La., Maine, Md., Mass., Mich., Minn., Miss., Mo., N.H., N.J., N.Y., N.C., Ohio, Okla., Pa., R.I., S.C., Tenn., Tex., Vt., Va., W.Va., Wis.

Krigia virginica is introduced in British Columbia. It grows in the Eastern deciduous forest biome, tallgrass prairie, and southeastern Coastal Plain. Its weedy habit may lead to introductions outside the present range.

The tetraploid populations of *Krigia virginica* are believed to have arisen through autoploidy; they are morphologically indistinguishable from the diploids. A single origin for this autoploid event is suggested by the chloroplast DNA studies by K. J. Kim et al. (1992b, 1992c), with marked genetic divergence of the tetraploids having occurred since their origination. Plants collected late in the season have a branching habit remarkably unlike the scapiform vernal form.

5. Krigia occidentalis Nuttall, J. Acad. Nat. Sci. Philadelphia 7: 104. 1834 • Western dwarfdandelion [E]

Cymbia occidentalis (Nuttall) Standley

Annuals, 4–16 cm; taprooted. **Stems** 1–20+, ± scapiform, erect, eglandular or lightly glandular-villous. **Leaves** basal (rosettes) and proximally cauline (on scarcely elongated branches close to ground); petiolate (petioles sometimes ciliate-glandular); blades linear, oblanceolate, or obovate, 1–7 cm, margins entire or sparingly lobed, lobes linear or triangular to rounded, apices acute or obtuse, faces eglandular. **Heads** borne singly. **Peduncles** from basal rosettes. **Involucres** 2.5–6.5 mm. **Phyllaries** 4–7, erect in fruit, lanceolate in flower, becoming ovate-lanceolate in fruit, midveins and sometimes secondary veins becoming prominent in fruit, curving inward at bases to form keels, apices acute. **Florets** 5–25; corollas yellow, 5–9 mm. **Cypselae** reddish brown, broadly

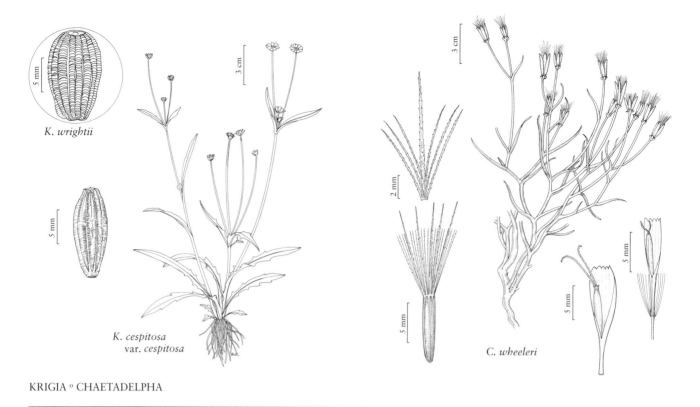

K. wrightii

K. cespitosa
var. *cespitosa*

C. wheeleri

KRIGIA ° CHAETADELPHA

obconic, 1.2–1.8 mm (apical areas broader than basal areoles), 10–15-ribbed; **pappi** of 5, hyaline, rounded outer scales 0.4–0.6 mm plus usually 5, sometimes 0, scabrous inner bristles 1.2–2 mm. $2n = 12$.

Flowering Mar–Jun. Sandy or clay soils, meadows, prairies, edges of open oak-hickory and pine woods; 10–400 m; Ark., Ga., Kans., La., Mo., Okla., Tex.

Krigia occidentalis grows in the Eastern deciduous forest biome, tallgrass prairie, and mixedgrass prairie. It has superficial similarity in pappus form to *K. virginica*; it was consistently placed as the sister species to *K. cespitosa* in chloroplast cpDNA and nuclear rDNA studies by K. J. Kim et al. (1992b, 1992c). In morphology, it is most similar to the polyploid species *K. wrightii*, with $2n = 18$.

6. **Krigia wrightii** (A. Gray) K. L. Chambers ex K. J. Kim, Brittonia 44: 195. 1992 • Wright's dwarfdandelion [E] [F]

Apogon wrightii A. Gray in A. Gray et al., Syn. Fl. N. Amer. 1(2): 411. 1884

Annuals, 4–25 cm; taprooted. **Stems** erect, branching proximally and distally, leafy, eglandular or stipitate-glandular, especially distally. **Leaves** basal (rosettes) and cauline; blades broadly to narrowly oblanceolate, 1–10 cm, margins entire or remotely dentate or lobed, lobes acute or rounded, apices acute to obtuse, faces eglandular or lightly glandular-villous. **Heads** borne singly. **Peduncles** from branching, leafy stems. **Involucres** 3.5–5.5 mm. **Phyllaries** 5–9, erect in fruit, narrowly to broadly lanceolate, midveins becoming prominent in fruit, curving inward at bases to form keels, apices acute. **Florets** 5–25; corollas yellow, 4–7 mm. **Cypselae** reddish brown, broadly columnar or barrel-shaped, 1.3–1.6 mm (apices slightly constricted, apical areas broader than basal areoles), 15-ribbed; **pappi** 0, or coroniform (minute scales, rarely with 1–5 tiny bristles). $2n = 18$.

Flowering Mar–May. Sandy, clay, loam, and rocky soils, fields, pastures, prairies, hillsides, and open oak-hickory and pine woods, sometimes in disturbed areas; 10–300 m; Ark., La., Okla., Tex.

Krigia wrightii grows in the Eastern deciduous forest biome, southeastern Coastal Plain, tallgrass prairie, and mixedgrass prairie. It was confused with *K. cespitosa* by L. H. Shinners (1947); its cypselae, involucres, and chromosome number set it apart. It often grows sympatrically with *K. occidentalis* or *K. cespitosa*, and mixed collections may occur.

7. Krigia cespitosa (Rafinesque) K. L. Chambers, J. Arnold Arbor. 54: 52. 1973 • Common or opposite-leaved dwarfdandelion [F]

Serinia cespitosa Rafinesque, Fl. Ludov., 149. 1817

Annuals, 4–42 cm; taprooted. **Stems** erect or ascending, branched, eglandular or stipitate-glandular, especially distally. **Leaves** basal and cauline; petiolate (at least basal); blades broadly to narrowly oblanceolate, 2–15 cm, margins entire or remotely toothed or lobed, lobes acute or rounded, apices acute to obtuse, faces glabrous or lightly glandular-villous; cauline appearing unequal and opposite proximal to distal peduncles, petiolate or sessile, blades oblanceolate to linear, gradually reduced, bases ± clasping, margins entire or dentate. **Heads** borne singly. **Peduncles** from branching, leafy stems. **Involucres** 2–8 mm. **Phyllaries** 5–10, erect in fruit, narrowly to broadly lanceolate, midveins evident, not prominent or keeled, apices acute. **Florets** 12–35; corollas yellow, 2–11 mm (barely equaling to much surpassing phyllaries). **Cypselae** fusiform, 1.4–1.7 mm, broadest at or ± beyond middles (apical areas ± equal to basal areoles), 15-ribbed; **pappi** 0. $2n = 8$.

Varieties 2 (2 in the flora): se United States, ne Mexico.

The name *Krigia oppositifolia* Rafinesque, long in use for this species, was not accepted by Rafinesque in his original publication and is therefore invalid.

1. Involucres 2–4.5 mm in flower, 3.5–5 mm in fruit; ligules 2–6 mm . . . . . 7a. *Krigia cespitosa* var. *cespitosa*
1. Involucres 4.5–7 mm in flower, 5.5–8.5 mm in fruit; ligules 6.5–11 mm . . . . 7b. *Krigia cespitosa* var. *gracilis*

7a. Krigia cespitosa (Rafinesque) K. L. Chambers var. **cespitosa** [F]

Involucres 2–4.5 mm in flower, 3.5–5 mm in fruit. **Ligules** 2–6 mm. $2n = 8$.

Flowering Feb–Sep. Sandy, clay, gravel, and loam soils, pastures, prairies, hillsides, roadsides, cultivated fields, and open mixed mesophytic, oak-pine, and oak-hickory woods and Coastal Plain pine communities; 0–500 m; Ala., Ark., Fla., Ga., Ill., Kans., Ky., La., Miss., Mo., Nebr., N.C., Okla., S.C., Tenn., Tex., Va.; Mexico (Nuevo León).

Variety *cespitosa* is known from the Eastern deciduous forest biome, southeastern Coastal Plain, tallgrass prairie, and mixedgrass prairie. Throughout its wide range, this variety is recognizable by its relatively small heads and florets barely surpassing the phyllaries. In eastern Texas and adjacent Louisiana and Oklahoma it is sympatric with the larger-headed, more showy-flowered var. *gracilis*, and plants of intermediate size frequently have been observed (K. J. Kim and B. L. Turner 1992). The treatment of the two morphologic races as varieties is an alternative to their being reduced to formae, which was the approach used by Kim and Turner.

7b. Krigia cespitosa (Rafinesque) K. L. Chambers var. **gracilis** (de Candolle) K. L. Chambers, Sida 21: 227. 2004 • Texas dwarfdandelion [E]

Apogon gracilis de Candolle in A. P. de Candolle and A. L. P. P. de Candolle, Prodr. 7: 79. 1838; *Krigia gracilis* (de Candolle) Shinners

Involucres 4.5–7 mm in flower, 5.5–8.5 mm in fruit. **Ligules** 6.5–11 mm. $2n = 8$.

Flowering Apr–Jul. Moist clay and sandy soils, pastures, fields, roadsides, and borders of oak-hickory-pine woods and oak savanna, sometimes in disturbed areas; 50–300 m; La., Okla., Tex.

Variety *gracilis* is known principally from tallgrass prairie and mixedgrass prairie biomes. It was formerly recognized as a species. It was reduced to the rank of forma by K. J. Kim and B. L. Turner (1992), who could find no molecular differences between it and forma *cespitosa* in their cpDNA and rDNA studies (Kim and T. J. Mabry 1991; Kim et al. 1992b, 1992c). The cypsela morphology and chromosome number are identical; the difference in size of the heads and floral parts is well marked, even in areas of sympatry, except for evidently hybrid plants of intermediate morphology.

77. CHAETADELPHA A. Gray ex S. Watson, Amer. Naturalist 7: 301. 1873 • [Greek *chaite*, long hair, bristles, and *adelphe*, sister, alluding to adnation of awns and bristles of pappi] E

L. D. Gottlieb

Perennials, 15–40 cm; rhizomes stout, branching. **Stems** 2–4, ascending to suberect, freely branched, glabrous. **Leaves** basal and cauline (basal withered at flowering); sessile; blades linear to linear-lanceolate, margins entire; distal reduced to scales. **Heads** borne singly (terminal at ends of stems and branches). **Peduncles** not inflated, not bracteate. **Calyculi** of 4–5, minute bractlets. **Involucres** cylindric, 3–4 mm diam. **Phyllaries** 5 in 1 series, linear, equal, margins hyaline, apices acute. **Receptacles** flat, glabrous, epaleate. **Florets** 5; corollas pale lavender to white. **Cypselae** light tan, columnar, 5 ridge-angled, faces smooth, without grooves or striations, glabrous; **pappi** persistent (borne on minute, raised pedestals 0.1 mm), of 5, barbellate awns plus 35–50+, barbellulate bristles in 1–2 series, all ± connate basally (bristles ± in groups alternating with the awns). $x = 9$.

Species 1: sw United States.

SELECTED REFERENCE Tomb, A. S. 1972. Taxonomy of *Chaetadelpha* (Compositae: Cichorieae). Madroño 21: 459–462.

1. Chaetadelpha wheeleri A. Gray ex S. Watson, Amer. Naturalist 7: 301. 1873 E F

Leaves: basal blades 3–50 × 2–4 mm. **Involucres** 11–14 × 3–4 mm. **Calyculi:** bractlets 2–4 mm. **Cypselae** 8–11(–12) mm; **pappi:** awns 8–11 mm, bristles 20–30% shorter. $2n = 18$.

Flowering May–Jul. Dunes, sandy soils and alkali flats in creosote bush scrub, sagebrush scrub; 800–1800 m; Calif., Nev., Oreg.

78. SHINNERSOSERIS Tomb, Sida 5: 186, figs. 1–3. 1974 • Beaked skeleton-weed [For Lloyd H. Shinners, 1918–1971, American botanist] E

David J. Bogler

Annuals, 5–85 cm (herbage glabrous); taprooted (roots deep, slender to thick). **Stems** usually 1, erect, simple proximally, branched distally, glabrous. **Leaves** cauline (opposite proximally); ± sessile; blades linear to filiform, margins entire. **Heads** borne singly. **Peduncles** not inflated distally, usually bracteate. **Calyculi** of 8, ovate to lanceolate bractlets. **Involucres** narrowly cylindric, 4–5 mm diam. **Phyllaries** 8 in 1–2 series, linear, margins scarious, apices acute (keeled). **Receptacles** flat, pitted, glabrous, epaleate. **Florets** 8–11; corollas pale purple or lavender, with white-tips. **Cypselae** pale green or tan, subcylindric to ± fusiform, apices abruptly constricted, not beaked, ribs 8–10, scabrous distally; **pappi** of 30–50, basally connate, white, ± equal, smooth or barbellulate bristles in ± 1 series. $x = 6$.

Species 1: c North America.

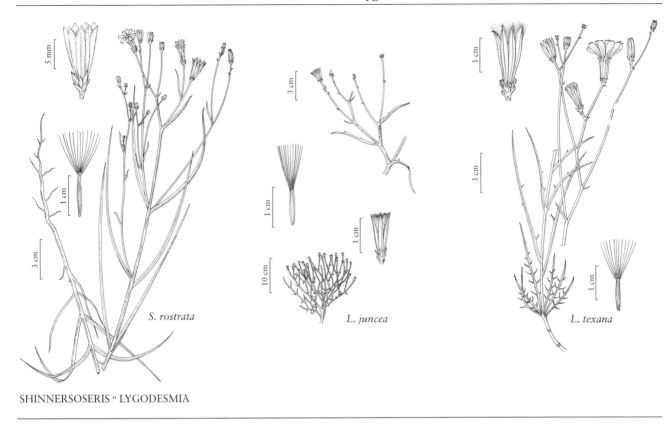

S. rostrata

L. juncea

L. texana

SHINNERSOSERIS ○ LYGODESMIA

Shinnersoseris rostrata was included in *Lygodesmia*; Tomb presented evidence for placing it in a separate genus. It differs from *Lygodesmia* in having opposite proximal leaves, shorter, obscure ligules, shorter style branches, 8–10-ribbed cypselae, echinate pollen, and a different base chromosome number (*x* = 9 in *Lygodesmia*.)

1. **Shinnersoseris rostrata** (A. Gray) Tomb, Sida 5: 186. 1974 • Beaked or annual skeleton-weed E F

Lygodesmia juncea (Pursh) Hooker var. *rostrata* A. Gray, Proc. Acad. Nat. Sci. Philadelphia 15: 69. 1864; *L. rostrata* (A. Gray) A. Gray

Leaves: blades 6–15 cm × 2–4 mm, distalmost reduced to scales. **Calyculus bractlets** 1–5 mm. **Involucres** 12–18 × 4–5 mm. **Phyllaries** 16–20 mm, faces glabrous. **Corollas** 15–16 mm. **Cypselae** 8–10 mm; **pappi** 6–8 mm. **2*n*** = 12.

Flowering Jul–Sep. Sandy soils of stream banks, dunes, and sand hills in prairies; 300–1500 m; Alta., Man., Sask.; Colo., Iowa, Kans., Minn., Nebr., N.Dak., Okla., S.Dak., Tex., Utah, Wyo.

Shinnersoseris rostrata grows in the Prairies and Great Plains region of North America.

79. **LYGODESMIA** D. Don, Edinburgh New Philos. J. 6: 311. 1829 • Skeletonplant, rush pink [Greek *lygos*, twig or stick, and *desme*, bundle, alluding to clumped, sticklike stems with reduced leaves]

David J. Bogler

Perennials, 5–80 cm; rhizomatous or taprooted (roots vertical, rhizomes spreading). **Stems** 1–5+, (green to gray-green, rushlike, ± striate), simple to or much branched proximally and/or distally, usually glabrous, rarely tomentulose. **Leaves** basal (sometimes in rosettes) and cauline; sessile; blades linear to subulate, sometimes reduced to scales, margins entire or sparingly pinnately laciniately lobed (faces usually glabrous, rarely tomentulose). **Heads** borne singly or in loose, corymbiform arrays. **Peduncles** not inflated distally, bracteate. **Calyculi** of 8–16, ovate to subulate or scalelike bracteoles in 1–2 series, unequal, margins scarious to erose-ciliate, faces glabrous or tomentulose, sometimes roughened. **Involucres** cylindric, 5–8 mm diam. (apices truncate, narrow or spreading). **Phyllaries** 5–12 in ± 1 series, grayish green, ± linear, equal, margins scarious, faces glabrous or puberulent, sometimes roughened. **Receptacles** flat, sometimes pitted, sometimes scabrous, epaleate. **Florets** 5–12; corollas usually pink to lavender or purple, rarely white. **Cypselae** pale green to tan, subcylindric, straight or arcuate, subterete or longitudinally angled or sulcate, apices sometimes narrowed, not beaked, faces smooth or rugose-roughened, glabrous; **pappi** persistent, of 60–80, tawny or white, ± connate, smooth bristles in 1–2+ series. *x* = 9.

Species 5 (5 in the flora): North America; n Mexico

Lygodesmia is easily recognized by the green, rushlike stems, narrow and often greatly reduced leaves, and terminal heads of showy, rosy, ligulate florets. It has been considered congeneric with *Stephanomeria*, *Prenanthella*, and *Shinnersoseris*; A. S. Tomb (1980) concluded that those taxa are not closely related. The annual species with plumose pappus bristles that formerly were included in *Lygodesmia* have been removed to those genera.

SELECTED REFERENCE Tomb, A. S. 1980. Taxonomy of *Lygodesmia* (Asteraceae). Syst. Bot. Monogr. 1: 1–51.

1. Involucres 10–16 mm; phyllary apices not appendaged; corollas 18–20 mm, ligules 3–4 mm wide; cypselae 6–10 mm; pappi 6–9 mm .... 3. *Lygodesmia juncea*
1. Involucres 14–25 mm; phyllary apices appendaged (reduced in *L. grandiflora* var. *arizonica*); corollas 12–40 mm, ligules 4–6(–10) mm wide; cypselae 10–19 mm; pappi 10–18 mm.
  2. Basal leaves not forming rosettes; cauline leaves well developed, not reduced to scales; plants 5–25(–60) cm .... 2. *Lygodesmia grandiflora*
  2. Basal leaves forming rosettes (often withered at flowering); cauline leaves mostly reduced to scales; plants 25–65 cm.
    3. Phyllaries 5–7; florets 5(–7) .... 4. *Lygodesmia ramosissima*
    3. Phyllaries 8–10; florets 8–12.
      4. Basal leaves entire; cypselae sulcate on inner faces; stems strongly striate; Florida, Georgia .... 1. *Lygodesmia aphylla*
      4. Basal leaves laciniately lobed; cypselae smooth on inner faces; stems weakly striate; New Mexico, Oklahoma, Texas .... 5. *Lygodesmia texana*

1. Lygodesmia aphylla (Nuttall) de Candolle in A. P. de Candolle and A. L. P. P. de Candolle, Prodr. 7: 198. 1838 • Roserush [E]

Prenanthes aphylla Nuttall, Gen. N. Amer. Pl. 2: 123. 1818; *Erythremia aphylla* (Nuttall) Nuttall

Perennials, 30–80 cm solitary; roots or rhizomes fleshy; caudices woody, branched. **Stems** 1, erect, green, simple or sparsely branched distally, strongly striate. **Leaves** (basal in rosettes, sometimes withered at flowering); proximal blades linear, 100–350 × 2–3 mm, margins entire; cauline similar or reduced to scales. **Heads** (1–5) borne singly or in corymbiform arrays. **Involucres** cylindric, 14–22 × 5–6 mm, apices narrow. **Calyculi** of 8–14, deltate bractlets 1–5 mm, margins scarious, erose-ciliate (faces tomentulose). **Phyllaries** 8, linear, 14–22 mm, margins scarious, apices often purplish, acute, appendaged. **Florets** 8–10; corollas pink to lavender or blue, 30–40 mm, ligules 4–6 mm wide. **Cypselae** (subcylindric) 11–14 mm (faces smooth, abaxial weakly striate, glabrous, adaxial sulcate); **pappi** 12–18 mm. *2n* = 18.

Flowering Feb–Nov. Dry sandy soils, flatwoods, pine barrens, disturbed areas, roadsides; 0–200 m; Fla., Ga.

Lygodesmia aphylla is recognized by its erect, sparingly branched habit, leafless stems, relatively large involucres and florets, apical appendages on phyllaries, and grooved cypselae. The habit is similar to that of *L. texana*. The basal leaves are frequently absent at flowering and are entire, not pinnately lobed as in *L. texana*.

2. Lygodesmia grandiflora (Nuttall) Torrey & A. Gray, Fl. N. Amer. 2: 485. 1843 • Largeflower skeletonplant E

Erythremia grandiflora Nuttall, Trans. Amer. Philos. Soc., n. s. 7: 445. 1841

Perennials 5–25(–60) cm; roots or rhizomes vertical, deep. **Stems** 1–5, erect or ascending, green, simple or branched from bases, obscurely striate (glabrous, puberulent or scabrous). **Leaves** (basal not forming rosettes, cauline present at flowering); basal blades linear to subulate, 5–150 × 1–6 mm, margins entire; cauline similar, sometimes reduced to scales distally. **Heads** (1–30, showy) borne singly or in loose, corymbiform arrays. **Involucres** cylindric, 15–25 × 6–8 mm, apices narrowed or spreading. **Calyculi** of ca. 8, deltate to ovate bractlets 2–5 mm, margins ciliate (faces tomentulose). **Phyllaries** 5–12, linear, 15–24 mm, margins scarious, apices appendaged (faces glabrous or scabrous). **Florets** 5–12; corollas 20–40 mm, lavender, pink, purple, rose, or white, ligules 5–10 mm wide. **Cypselae** (subcylindric, obscurely 4–5-angled) 10–18 mm (faces smooth or rugose, sometimes sulcate); **pappi** 10–13 mm. $2n = 18$.

Varieties 5 (5 in the flora): sw, w United States.

Lygodesmia grandiflora is recognized mainly by its relatively large corollas. Some variants were segregated as distinct species by A. S. Tomb; because of intermediates, putative hybrids, and associated identification problems, it is probably best to recognize these as varieties pending further investigation (A. Cronquist 1994; S. L. Welsh et al. 2003).

1. Phyllaries 8–12; florets 8–12 . . . .
. . . . 2a. *Lygodesmia grandiflora* var. *grandiflora*
1. Phyllaries 5(–6); florets 5(–7).
   2. Corollas white (may turn pinkish when dry); stems (woody) branched from bases; leaves stiff, spreading; involucre apices spreading
   . . . . 2e. *Lygodesmia grandiflora* var. *entrada*
   2. Corollas lavender, pink, purple, rose, or white; stems simple or branched from bases or distally (if branched from bases, either leaves lax or plants from vicinity of Moab, Utah); involucre apices narrow.
      3. Stems much branched from bases; proximal leaves narrow, linear-filiform, 1–3 mm wide, rigid . . . .
      . . . 2d. *Lygodesmia grandiflora* var. *doloresensis*
      3. Stems simple or sparsely branched from bases or distally; proximal leaves lanceolate to linear-subulate, (2–)3–6 mm wide, ± lax (widespread in southwestern states).

[4. Shifted to left margin.—Ed.]
4. Distal leaves not reduced to scales (mostly 10+ mm); cypselae 10–13 mm, abaxial faces rugose, adaxial faces strongly sulcate . . . .
. . . . 2b. *Lygodesmia grandiflora* var. *arizonica*
4. Distal leaves reduced to linear scales (mostly less than 10 mm); cypselae 13–19 mm, abaxial faces smooth, adaxial faces weakly sulcate . . . .
. . . . 2c. *Lygodesmia grandiflora* var. *dianthopsis*

2a. Lygodesmia grandiflora (Nuttall) Torrey & A. Gray var. **grandiflora** E

Lygodesmia grandiflora var. *stricta* Maguire

Plants 5–40 cm. **Stems** erect, simple or sparingly branched from bases, obscurely striate. **Leaves:** proximal blades linear, 50–150 × 1–3 mm, ± lax (margins entire or rarely laciniately-lobed); cauline similar, reduced to scales distally. **Heads** 1–5(–30), in loose, corymbiform arrays. **Involucres** subcylindric, 18–21 × 6–8 mm, apices narrow. **Florets** 6–12; corollas pink to lavender. **Cypselae** 10–13 mm, faces smooth, not sulcate. $2n = 18$.

Flowering May–Jun. Open sites, alluvial, sandy, or gravelly soils, juniper-sagebrush scrub; 1200–2800 m; Colo., N.Mex., Utah, Wyo.

Variety *grandiflora* is recognized by its relatively large, pink corollas and 8–12 phyllaries. The size of the plants varies and stems are often somewhat woody near the bases. This variety appears to intergrade with var. *arizonica* and var. *dianthopsis* (S. L. Welsh et al. 2003).

2b. Lygodesmia grandiflora (Nuttall) Torrey & A. Gray var. **arizonica** (Tomb) S. L. Welsh, Great Basin Naturalist 43: 314. 1983 • Arizona skeletonplant E

Lygodesmia arizonica Tomb, Sida 3: 530, unnumb. fig. p. 531. 1970

Plants 5–15(–25) cm. **Stems** erect to ascending, sparsely branched from bases, smooth. **Leaves:** proximal blades narrowly lanceolate to linear, 40–100 × 2–6 mm, ± lax, glabrous; cauline not reduced to scales (mostly 10+ mm). **Heads** 2–6, borne singly. **Involucres** cylindric, 18–25 mm × 5–7 mm, apices narrow. **Phyllaries** 5–6. **Florets** 5(–7); corollas pale pink to white. **Cypselae** 10–13 mm, abruptly narrowed below apex, faces rugose, adaxial strongly sulcate. $2n = 18$.

Flowering May–June. Arid grasslands on sandy soils; 1100–1600 m; Ariz., Colo., N.Mex., Utah.

Variety *arizonica* is identified by its relatively short size (rarely over 20 cm), branched, leafy stems, and reduced or absent phyllary apical appendages.

2c. Lygodesmia grandiflora (Nuttall) Torrey & A. Gray var. **dianthopsis** (D. C. Eaton) S. L. Welsh, Great Basin Naturalist 43: 314. 1983 • Antelope Island skeletonplant E

Lygodesmia juncea (Pursh) D. Don ex Hooker var. *dianthopsis* D. C. Eaton in S. Watson, Botany (Fortieth Parallel), 200. 1871; *L. dianthopsis* (D. C. Eaton) Tomb

Plants (5–)20–60 cm. **Stems** erect or ascending, purple proximally, slender, simple or sparingly branched from bases or distally, smooth (glabrous or tomentulose). **Leaves:** (proximalmost reduced to scales at ground level) proximal blades linear, 50–110 × 2–6 mm, ± lax; distal linear, less than 10 mm, distally reduced to linear scales. **Heads** 2–13, in loose, corymbiform arrays. **Involucres** subcylindric, 15–22 × 4–5 mm, apices narrow. **Phyllaries** 5–6. **Florets** 5; corollas purple to lavender or white. **Cypselae** 12–19 mm, abaxial faces smooth, adaxial distinctly rugose, weakly sulcate. $2n = 18$.

Flowering Jun–Jul. Sandy and gravelly soils in juniper-pinyon scrub, open fields, sandy roadsides; 1300–2500 m; Colo., Idaho, Nev., Utah.

Variety *dianthopsis* is recognized by its slender, erect, leafy stems, purplish at base, persistent cauline leaves, phyllaries with appendages, 5–6 florets per head, and distinctive rugose-roughened cypselae. It usually is taller than var. *arizonica*. The stems and leaves are occasionally sparsely tomentulose.

2d. Lygodesmia grandiflora (Nuttall) Torrey & A. Gray var. **doloresensis** (Tomb) S. L. Welsh, Rhodora 95: 399. 1993 • Dolores River skeletonplant C E

Lygodesmia doloresensis Tomb, Syst. Bot. Monogr. 1: 48, fig. 49. 1980

Plants (solitary) 15–30 cm. **Stems** erect, much branched proximally, weakly striate proximally. **Leaves:** proximal blades linear-filiform, 30–140 × 1–3 mm; distal similar. **Heads** 2–5, borne singly. **Involucres** subcylindric, 18–20 × 4–5 mm, apices narrow. **Phyllaries** 5. **Florets** 5; corollas lavender, rose, or white. **Cypselae** 18 mm, abaxial faces smooth, adaxial weakly rugose bisulcate with ridge separating sulci. $2n = 18$.

Flowering Jun. Alluvial soil in Juniper grassland; of conservation concern; 1300–1500 m; Colo.

Variety *doloresensis* is distinguished by its much-branched stems, almost filiform leaves, and smooth cypselae. It is known only from the Dolores River valley and is similar to var. *dianthopsis*, which is distinguished by being less branched and by having broader leaves.

Variety *doloresensis* is in the Center for Plant Conservation's National Collection of Endangered Plants.

2e. Lygodesmia grandiflora (Nuttall) Torrey & A. Gray var. **entrada** (S. L. Welsh & Goodrich) S. L. Welsh, Rhodora 95: 399. 1993 C E

Lygodesmia entrada S. L. Welsh & Goodrich, Great Basin Naturalist 40: 83, fig. 4. 1980

Plants 30–45 cm. **Stems** erect or ascending, stiff, woody, branched from bases and distally. **Leaves:** proximal blades linear-acicular, 5–30 × 2–4 mm, stiff, spreading; cauline reduced distally, not scalelike. **Heads** 1–10, in loose, corymbiform arrays. **Involucres** cylindric (V-shaped in fruit), 20–22 × 5–7 mm, apices spreading. **Phyllaries** ca. 6. **Florets** 5, corollas white (or fading to pinkish). **Cypselae** seldom formed (ovaries 4–5 mm, abaxial faces smooth, adaxial weakly rugose, bisulcate with ridge separating sulci). $2n = 27$.

Flowering Jun. Juniper-scrub community, in deep sandy soil, Entrada Sandstone; of conservation concern; 1300–1500 m; Utah.

Variety *entrada* is distinguished by its relatively large size, stiff stems (woody) and leaves, tomentulose phyllaries, and white corollas. It is similar to var. *arizonica*. Unlike other varieties, it has involucre apices somewhat spreading. Chromosome counts indicate that these plants are asexual triploids (A. S. Tomb, pers. comm.). Variety *entrada* is locally abundant in Arches National Park, near Moab.

3. Lygodesmia juncea (Pursh) D. Don ex Hooker, Fl. Bor.-Amer. 1: 295. 1833 • Rush skeletonplant E F

Prenanthes juncea Pursh, Fl. Amer. Sept. 2: 498. 1813

Perennials, 10–35(–70) cm (in bushy clumps); taproots deep, vertical, rhizomes branched, woody. **Stems** erect to ascending or decumbent, green, glaucous, much branched from bases and distally, strongly striate, glabrous (often bearing round galls). **Leaves** (basal not in rosettes, absent at flowering); proximal blades linear, 5–30(–60) × 1–2(–4) mm, margins entire, apices acute, faces glabrous; cauline reduced to subulate scales. **Heads** (1–50+) borne singly or in corymbiform arrays. **Involucres** cylindric, 10–16 × 4–6 mm, apices spreading. **Calyculi** of 8, ovate to linear bractlets 2–4 mm, margins erose-ciliate (faces glabrous). **Phyllaries** 5(–7), linear, 10–15 mm, margins scarious, apices acute or obtuse, not appendaged, faces glabrous. **Florets** usually 5; corollas usually light pink to lavender, rarely white, 18–20 mm, ligules 3–4 mm wide. **Cypselae** 6–10 mm, weakly striate, glabrous; **pappi** 6–9 mm. $2n = 18$.

Flowering Jun–Sep. High Plains, rolling short-grass prairies, blufftop prairies, loess hills, sandy to silty soils, disturbed sites, railroads, roadsides, barren areas; 600–2300 m; Alta., B.C., Man., Sask. ; Ariz., Ark., Colo., Idaho, Ind., Iowa, Kans., Minn., Mo., Mont., Nebr., Nev., N.Mex., N.Dak., Okla., S.Dak., Tex., Utah, Wash., Wis., Wyo.

Lygodesmia juncea is the most widespread species of the genus, occurring throughout the High Plains region of North America. It is easily distinguished by its bushy habit, greatly reduced cauline leaves, relatively small heads and involucres, and phyllaries lacking appendages. Mature cypselae are rarely found on this species, and the plants are presumably sterile and reproduce mainly by vegetative means. Many specimens have round galls to 10 mm diameter on the stems, produced by solitary wasps and apparently unique to this species.

4. **Lygodesmia ramosissima** Greenman, Proc. Amer. Acad. Arts 35: 315. 1900 • Pecos River skeletonplant

Perennials 25–60 cm (dense bushy clumps); taproots deep, vertical, woody. **Stems** ascending to erect, pale green, striate, often much-branched from bases and distally, glabrous. **Leaves:** (basal forming rosettes, sometimes withered and absent at flowering) proximal blades linear, 50–70 × 1–2 mm, entire or sparingly laciniate; cauline 40 mm or less, reduced to scales distally. **Heads** (1–6), borne singly. **Involucres** cylindric, 14–21 × 4–6 mm, apices narrow. **Calyculi** of 8–10, ovate bractlets 2–3 mm, margins ciliate (apices often purple, with small appendages, faces glabrous, sometimes roughened). **Phyllaries** 5–7, linear-oblong, 16–20 mm, margins scarious, faces glabrous, apices appendaged (often dark). **Florets** 5–7; corollas lavender, 12–16 mm, ligules 5–6 mm wide. **Cypselae** 11–14 mm, faces smooth, adaxial weakly sulcate, glabrous; **pappi** 11–13 mm. $2n = 18$.

Flowering Jun–Oct. Rocky grasslands and oak forest, rocky soil, roadsides; 1200–1900 m; Tex.; Mexico (Chihuahua, Durango, Sonora).

Lygodesmia ramosissima is often confused with the more widespread *L. texana*; it can be distinguished by its thinner, more intricately branched stems, 5–7 florets per head, sulcate cypselae, and later daily flowering time (A. S. Tomb 1980).

5. **Lygodesmia texana** (Torrey & A. Gray) Greene ex Small, Fl. S.E. U.S., 1315. 1903 • Texas skeletonplant F

Lygodesmia aphylla (Nuttall) de Candolle var. *texana* Torrey & A. Gray, Fl. N. Amer. 2: 485. 1843

Perennials 25–65 cm (in clumps); taproots thick, fleshy or woody, rhizomes spreading. **Stems** erect, green, stout, branched from bases and distally, weakly striate, glabrous. **Leaves** (basal forming rosettes, sometimes withering before flowering) proximal blades linear, 100–200 × 1–8 mm, margins of usually pinnately laciniately lobed, lobes remote and narrow, 1–15 mm; cauline similar, 5–10 mm, reduced to scales distally. **Heads** borne singly. **Involucres** cylindric, 18–25 mm × 5–8 mm, apices narrow. **Calyculi** of 8–10, ovate bractlets 1–3 mm, margins ciliate-tomentulose. **Phyllaries** 8–10, linear, 18–26 mm, margins scarious, apices appendaged, faces glabrous or tomentulose. **Florets** 8–12; corollas usually pink, purple, or lavender, rarely white, 35–40 mm, ligules 5–6 mm wide. **Cypselae** 11–17 mm, faces smooth, adaxial not sulcate, glabrous; **pappi** 10–15 mm. $2n = 18$.

Flowering Apr–Sep. Rocky, calcareous, alkaline soils in oak-juniper woodlands, mesquite brushlands, open grasslands, red sandy soils, roadsides; 100–1800 m; N.Mex., Okla., Tex.; Mexico (Coahuila).

Lygodesmia texana is easily distinguished by its laciniate-lobed basal leaves that form rosettes in younger stages, relatively large involucres and florets, phyllaries with an apical appendage, and smooth cypselae. It is closely related to *L. aphylla*, which has a more eastern distribution, lacks laciniate leaves in rosettes, and has sulcate cypselae. *Lygodesmia texana* apparently hybridizes with *L. ramossisima* in trans-Pecos Texas, and the two species can be difficult to distinguish (A. S. Tomb 1980).

80. PHALACROSERIS A. Gray, Proc. Amer. Acad. Arts 7: 364. 1868 • [Greek *phalakros*, bald-headed, and *seris*, a kind of endive] E

Kenton L. Chambers

Perennials, 10–45 cm; caudices and taproots fleshy, with blackish periderms. **Stems** 1(–5), erect, simple, scapiform, glabrous. **Leaves** basal (in rosettes); petiolate; blades linear to oblanceolate (± fleshy), margins entire (apices obtuse to acuminate). **Heads** borne singly (erect). **Calyculi** 0. **Involucres** ± campanulate, (3–)5–10 mm diam. **Phyllaries** 8–25 in 2–4 series, basally ± connate, ± lanceolate, equal or some outer shorter, herbaceous, apices acute. **Receptacles** convex, smooth, glabrous, epaleate. **Florets** 13–35; corollas yellow. **Cypselae** brown with darker spots, columnar, truncate, not beaked, 4-nerved, smooth, glabrous; **pappi** 0, or coroniform (less than 0.5 mm). $x = 9$.

Species 1: California.

The simple vegetative and floral morphology, including absence of indument and pappus, have made the subtribal assignment of *Phalacroseris* somewhat problematic. Most recent authors have followed G. L. Stebbins (1953), placing the genus in Microseridinae (H. J. Price and K. Bachmann 1975; R. K. Jansen et al. 1991b). In current molecular phylogenetic studies (J. Lee et al. 2003), *Phalacroseris* forms a sister group to a clade containing both Microseridinae and Stephanomeriinae of Stebbins's classification. It was made a separate subtribe, Phalacroseridinae, by Lee and B. G. Baldwin (2004). The genome of *P. bolanderi* is unusually large, containing 2.5 times the DNA found in perennial taxa of *Microseris* having $2n = 18$ (Price and Bachmann).

1. Phalacroseris bolanderi A. Gray, Proc. Amer. Acad. Arts 7: 364. 1868 • Bolander dandelion E F

Phalacroseris bolanderi var. *coronata* H. M. Hall

Leaves 6–20 cm, faces glabrous. **Involucres** 7–13 mm after flowering. **Phyllaries** not reflexed in fruit, margins often red distally, glabrous or minutely puberulent apically. **Florets** 10–18 mm, much surpassing phyllaries at flowering, glabrous. **Cypselae** straight or slightly curved on one side, 3–4 mm, obtusely 3-angled at 1 adaxial and 2 lateral nerves. $2n = 18$.

Flowering Jun–Aug. Wet meadows and sphagnum bogs, coniferous upper montane forests and mixed subalpine woodlands; 1800–3000 m; Calif.

A vestigial coroniform pappus found in some individuals is too minor a feature to merit varietal designation.

81. PINAROPAPPUS Lessing, Syn. Gen. Compos., 143. 1832 • Rocklettuce [Greek *pinaro*, dirty, squalid, and *pappos*, pappus, alluding to color of pappi]

David J. Bogler

Perennials, 3–40 cm; taprooted (taproots deep, woody) or rhizomatous. **Stems** 1–20+, erect or ascending, simple or branched proximally, ± scapiform, glabrous. **Leaves** basal and cauline; petiolate; basal blades linear to lanceolate, margins entire, toothed, or pinnately lobed (faces glabrous); cauline foliaceous or reduced to minute bracts distally. **Heads** borne singly. **Peduncles** not inflated distally, sometimes bracteate. **Calyculi** 0. **Involucres** cylindric to campanulate, 3–20 mm diam. **Phyllaries** 18–22 in 3–5 series, ovate to lanceolate, unequal,

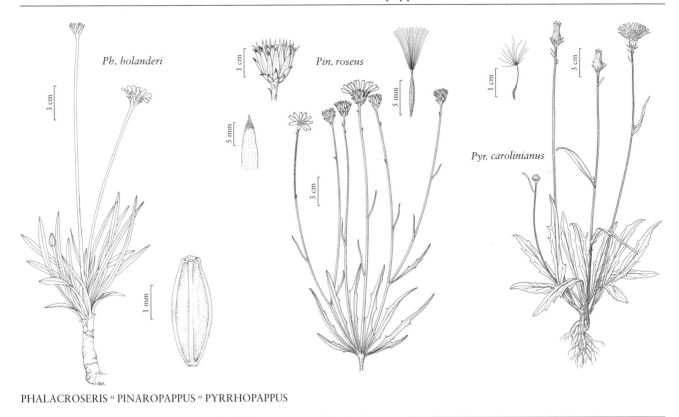

Ph. bolanderi

Pin. roseus

Pyr. carolinianus

PHALACROSERIS ∘ PINAROPAPPUS ∘ PYRRHOPAPPUS

margins scarious, apices acute. **Receptacles** slightly convex, slightly pitted, glabrous, paleate (paleae scarious, acuminate). **Florets** (10–)20–40(–60); corollas pink, purple, lavender, or nearly white. **Cypselae** golden or yellowish brown, cylindric or fusiform, tapered to slender beaks, ribs 5–6, rounded, obscure, scabrous or hispidulous; **pappi** persistent, of 15–60, distinct, tawny or yellowish brown, unequal, barbellulate bristles in 1 series.

Species 7–10 (2 in the flora): North America, Mexico, Central America.

Plants of *Pinaropappus* are recognized by the glabrous leaves in dense rosettes, scapiform stems, graduated phyllaries, and pale lavender and whitish corollas. They are commonly found in dry, rocky, limestone habitats; some species are cliff-dwellers.

1. Plants 3–7 cm (forming dense clumps and mats); involucres cylindric, 8–10 × 3–5 mm; phyllaries purplish, margins white, apices purplish to dull brown (necrotic) . . . . 1. *Pinaropappus parvus*
1. Plants 10–30 cm (forming individual rosettes or clumps); involucres campanulate, 10–15 × 12–20 mm; phyllaries pale green, margins pink, apices dark brown . . . . 2. *Pinaropappus roseus*

1. **Pinaropappus parvus** S. F. Blake, Contr. U.S. Natl. Herb. 22: 655. 1924 • Small rocklettuce E

Perennials, 3–7 cm (forming dense clumps and mats). **Stems** 3–10+, bases relatively think. **Leaf blades** linear-oblanceolate, 2–5 cm × 1–3 mm; cauline progressively reduced to linear bracts. **Involucres** narrowly cylindric, 8–10 × 3–5 mm. **Phyllaries** purplish (margins white), broadly lanceolate, 6–8 mm, apices purple to dull brown (necrotic), acute. **Paleae** 7–8 mm. **Florets** 20–30; corollas pink, 6–8 mm. **Cypselae** 4–5 mm; **pappi** 2–3 mm.

Flowering Jun–Jul. Exposed slopes, rocky ledges, limestone cliffs; 1800–2200 m; N.Mex., Tex.

Pinaropappus parvus is easily recognized by the extremely dense clumping habit, relatively short stems, and relatively small heads. At flowering, the phyllaries are usually purple in the center with scarious margins and purple to dull brown tips.

2. Pinaropappus roseus (Lessing) Lessing, Syn. Gen. Compos., 143. 1832 • White dandelion or rocklettuce F

Achyrophorus roseus Lessing, Linnaea 5: 133. 1830; *Pinaropappus roseus* var. *foliosus* Shinners

Perennials, 10–40 cm. **Stems** 1–20+, bases relatively thin. **Leaf blades** narrowly oblanceolate, 4–12 cm × 2–15 mm; mid cauline often reduced to linear or minute bracts. **Involucres** campanulate, 10–15 × 12–20 mm. **Phyllaries** ovate to narrowly lanceolate, unequal, 2–14 mm, apices dark brown, obtuse to acute. **Paleae** 12–18 mm. **Florets** 20–40; corollas pale pink abaxially, white to yellow adaxially, 15–18 mm. **Cypselae** 5–6 mm; **pappi** 4–7 mm. **2n** = 18.

Flowering Mar–Aug. Open limestone areas, roadsides, cliffs, open grassy flats; 50–2600 m; Ariz., N.Mex., Okla., Tex.; Mexico.

At flowering, phyllaries of *Pinaropappus roseus* are pale green with margins suffused with pink. On fresh specimens, the dark brown tips are distinctive. Specimens from the mountains of southeastern Arizona and southwestern New Mexico with stems leafy beyond the middles have been recognized as var. *foliosus*; that trait appears to be widespread and part of the normal range of variation for the species.

82. PYRRHOPAPPUS de Candolle in A. P. de Candolle and A. L. P. P. de Candolle, Prodr. 7: 144. 1838, name conserved • [Greek *pyrrhos*, yellowish red, and *pappos*, pappus, alluding to color of pappi]

John L. Strother

Annuals or perennials, 5–100+ cm; taprooted or rhizomatous (roots producing tuberiform swellings in *P. grandiflorus*). **Stems** usually 1, sometimes 2–5+, erect, unbranched or branched proximally and/or distally, glabrous or pilosulous. **Leaves** basal or basal and cauline; basal ± petiolate, distal usually sessile; blades oblong, elliptic, or ovate to lanceolate or linear, margins entire or dentate to pinnately lobed (faces usually glabrous, sometimes pilosulous near margins). **Heads** borne singly or in loose, corymbiform arrays. **Peduncles** not inflated distally, sometimes bracteate. **Calyculi** of 3–13+, deltate to subulate or filiform bractlets. **Involucres** cylindric, 4–5[–8+] mm diam. **Phyllaries** 8–21+ in ± 2 series (reflexed in fruit), linear, equal, margins often scarious, apices acute (often thickened or bearing keel-like flaps near tips). **Receptacles** ± convex, pitted, glabrous, epaleate. **Florets** (20–)30–150+; corollas yellow to whitish. **Cypselae** reddish brown to stramineous, bodies ± fusiform, beaks (± concolorous with bodies) ± filiform, fragile, grooves (or broad ribs) 5, faces transversely rugulose, glabrous; **pappi** (borne on discs at tips of beaks) persistent, double: outer coroniform (of whitish, relatively short, spreading, sometimes curly, hairs), inner of 80–120+, rufous to stramineous, subequal, barbellulate bristles in 2–3+ series. *x* = 6.

Species 1, 4, or 5 (4 in the flora): North America, Mexico.

Some plants from Mexico that have been called *Pyrrhopappus multicaulis* de Candolle (type from Mexico) may be distinct at species rank from *P. pauciflorus* (see R. McVaugh 1984).

Almost all botanists who have dealt with the biology or floristics of pyrrhopappuses have remarked similarities among the taxa and/or difficulties with identification of some specimens. Some botanists have noted that "interspecific" hybridizations are common (D. K. Northington 1974; B. L. Turner and K. J. Kim 1990; and works cited therein).

SELECTED REFERENCES Northington, D. K. 1974. Systematic studies of the genus *Pyrrhopappus* (Compositae, Cichorieae). Special Publ. Mus. Texas Tech Univ. 6: 1–38. Turner, B. L. and K. J. Kim. 1990. An overview of the genus *Pyrrhopappus* (Asteraceae: Lactuceae) with emphasis on chloroplast DNA restriction site data. Amer. J. Bot. 77: 845–850.

1. Perennials (roots or rootstocks producing tuberiform swellings 1–15 cm below soil surface); stems usually scapiform; cauline leaves usually 0, sometimes 1–3; anthers 4.5–5 mm . . . . 1. *Pyrrhopappus grandiflorus*
1. Annuals or perennials (not producing tuberiform swellings); stems rarely scapiform; cauline leaves usually 3–9+, seldom 0–2; anthers 2.5–4 mm.
  2. Stems usually sparsely to densely pilosulous proximally, sometimes glabrous; blades of distal cauline leaves usually pinnately (3–)5–7(–9+)-lobed . . . . 2. *Pyrrhopappus pauciflorus*
  2. Stems usually glabrous proximally, sometimes pilosulous; blades of distal cauline leaves usually entire or with 1–2 lobes near bases, sometimes pinnately 3–5(–7+)-lobed.
    3. Perennials; involucres 12–15(–17+) mm; florets 30–60+; cypselae stramineous; pappi 6–7 mm . . . . 3. *Pyrrhopappus rothrockii*
    3. Annuals; involucres 17–24 mm; florets 50–150+; cypselae reddish brown; pappi 9–10 mm . . . . 4. *Pyrrhopappus carolinianus*

1. Pyrrhopappus grandiflorus (Nuttall) Nuttall, Trans. Amer. Philos. Soc., n. s. 7: 430. 1841 E

Barkhausia grandiflora Nuttall, J. Acad. Nat. Sci. Philadelphia 7: 69. 1834 (as Borkhausia)

Perennials (possibly flowering first year), 5–30(–45+) cm (roots or rootstocks producing tuberiform swellings 1–15 cm below soil surface). **Stems** usually scapiform, usually branching from bases, proximally glabrous or sparsely to densely pilosulous. **Cauline leaves** 0(–3), proximal mostly lanceolate, margins usually pinnately lobed, distal linear-filiform, margins entire. **Heads** borne singly or 2–3 in loose, corymbiform arrays. **Calyculi:** bractlets 8 in 1(–2) series, linear to subulate, 3–8 mm. **Involucres** cylindric to turbinate, 17–25 mm. **Phyllaries** 13–22. **Florets** 40–60+; anthers 4.5–5 mm (pollen equatorial diameters 46–52 μm). **Cypselae:** bodies stramineous, 4–5 mm, beaks 6–7 mm; **pappi** 10–12 mm. **2n** = 24.

Flowering Apr–Jun. Disturbed sites, calcareous, loamy, or sandy soils; 100–800 m; Kans., Okla., Tex.

Pyrrhopappus grandiflorus has been reported from Arkansas and New Mexico; I have seen no specimens of it from those states.

2. Pyrrhopappus pauciflorus (D. Don) de Candolle in A. P. de Candolle and A. L. P. P. de Candolle, Prodr. 7: 144. 1838

Chondrilla pauciflora D. Don, Trans. Linn. Soc. London 16: 180. 1830; *Pyrrhopappus geiseri* Shinners; *P. multicaulis* de Candolle; *P. multicaulis* var. *geiseri* (Shinners) Northington

Annuals (sometimes persisting), 5–40(–80+) cm. **Stems** seldom, if ever, scapiform, branching from bases and/or distally, usually sparsely to densely pilosulous proximally, sometimes glabrous. **Cauline leaves** 1–3(–5+), proximal mostly oblanceolate to lanceolate, margins usually pinnately lobed, sometimes dentate or entire, distal ± lanceolate, margins usually pinnately (3–)5–7(–9+)-lobed. **Heads** (1–)3–7+ in loose, corymbiform arrays. **Calyculi:** bractlets 8–13 in 1–2 series, deltate to subulate, 3–5(–6) mm. **Involucres** ± campanulate to cylindric, 16–22 mm. **Phyllaries** 13–21. **Florets** 50–60; anthers 3.5 mm (pollen equatorial diameters 43–46 μm). **Cypselae:** bodies reddish brown, 4–5 mm, beaks 7–9 mm; **pappi** 7–9(–10) mm. **2n** = 12.

Flowering (Feb–)Apr–May. Disturbed sites, prairies, clay soils; 10–500 m; Tex.; Mexico (Coahuila, Nuevo León, Tamaulipas).

Some specimens of *Pyrrhopappus* from the Panhandle of Texas and from New Mexico are intermediate for traits used here to distinguish *P. pauciflorus* and *P. rothrockii*. Some authors (e.g., B. L. Turner and K. J. Kim 1990) have included *P. rothrockii* within *P. pauciflorus*.

3. Pyrrhopappus rothrockii A. Gray, Proc. Amer. Acad. Arts 11: 80. 1876 E

Perennials (possibly flowering first year), 15–40 cm. **Stems** seldom, if ever, scapiform, branching from bases and/or distally, glabrous or pilosulous proximally. **Cauline leaves** (1–)3–9+, proximal mostly spatulate or oblanceolate to linear, margins entire, dentate, or pinnately lobed, distal usually narrowly lanceolate to lance-attenuate, margins usually entire or with 1–2 lobes near bases, sometimes pinnately 3–5(–7+)-lobed. **Heads** (1–)3–5+ in loose, corymbiform arrays. **Calyculi:** bractlets 3–5+ in 1–2 series, deltate to subulate, 2–5 mm. **Involucres** ± cylindric, 12–15(–20+) mm. **Phyllaries** 13–16+. **Florets** (20–)30–60+; anthers 3.5–4 mm (pollen equatorial diameters unknown). **Cypselae:** bodies stramineous, 3–4 mm, beaks 6–7 mm; **pappi** 6–7 mm. **2n** = 12.

Flowering May–Sep. Meadows, stream banks, flood plains; 1100–2700 m; Ariz., N.Mex., Tex.

For the present, I concur with D. K. Northington (1974) and I treat *Pyrrhopappus rothrockii* as distinct from *P. pauciflorus*.

4. Pyrrhopappus carolinianus (Walter) de Candolle in A. P. de Candolle and A. L. P. P. de Candolle, Prodr. 7: 144. 1838 [E] [F]

Leontodon carolinianum Walter, Fl. Carol., 192. 1788; *Pyrrhopappus carolinianus* var. *georgianus* (Shinners) H. E. Ahles; *P. georgianus* Shinners

Annuals (sometimes persisting), (5–)20–50(–100+) cm. **Stems** usually branching from bases and/or distally, rarely scapiform, usually glabrous proximally, sometimes pilosulous. **Cauline leaves** (1–)3–9+, proximal mostly lanceolate, margins usually dentate, sometimes pinnately lobed, distal narrowly lanceolate to lance-attenuate, margins entire or with 1–2 lobes near bases. **Heads** (1–)3–5+ in loose, corymbiform arrays. **Calyculi:** bractlets 13–16+ in 2–3 series, subulate to filiform, 8–12+ mm. **Involucres** ± cylindric to campanulate, 17–24+ mm. **Phyllaries** 16–21+. **Florets** 50–150+; anthers 2.5–3.5 mm (pollen equatorial diameters mostly 43–47 µm). **Cypselae:** bodies reddish brown, 4–6 mm, beaks 8–10 mm; **pappi** 7–10+ mm. $2n = 12$.

Flowering (Feb–)May–Jun(–Sep). Disturbed sites, edges of woods, prairies, sandy soils; 10–600 m; Ala., Ark., Fla., Ga., Kans., Mo., Nebr., N.C., Pa., S.C., Tenn., Tex., Va., W.Va.

187f. ASTERACEAE Martinov tribe CALENDULEAE Cassini, J. Phys. Chim. Hist. Nat. Arts 88: 161. 1819 ☐

Annuals, perennials, shrubs, or trees. **Leaves** cauline [rosulate]; usually alternate, sometimes opposite; petiolate or sessile; margins usually entire or dentate, sometimes lobed to dissected. **Heads** heterogamous (radiate), borne singly or in ± corymbiform arrays. **Calyculi** 0 (peduncular bractlets sometimes intergrade with phyllaries). **Phyllaries** persistent, in (1–)2(–3) series, distinct [connate], usually ± equal, usually herbaceous (sometimes fleshy), margins and/or apices ± scarious. **Receptacles** flat to conic, epaleate (sometimes bristly-setose). **Ray florets** in 1–2+ series (more in horticultural "doubles"), pistillate, fertile [styliferous and sterile or neuter]; corollas usually yellow to orange and/or cyanic to white (abaxial and adaxial faces often different colors). **Disc florets** usually functionally staminate, sometimes bisexual and fertile; corollas yellow, orange, or cyanic (sometimes combinations within corollas), lobes 5, ± deltate (tips sometimes terete or dilated); anther bases ± tailed, apical appendages ovate to deltate; styles abaxially papillate (at least near tips), branches sometimes barely discernible (0.5–1 mm), adaxially stigmatic in 2 lines from bases to apices, apices rounded to truncate, appendages penicillate or essentially none. **Cypselae** usually polymorphic within heads (straight, arcuate, contorted, or ± coiled), ± columnar to prismatic, sometimes obcompressed, compressed, or flattened, sometimes ± beaked, bodies usually tuberculate, ridged, and/or winged (usually glabrous; blue-black and drupelike in *Chrysanthemoides*); **pappi** 0 [bristles].

Genera ca. 8, species 100+ (4 genera, 7 species in the flora): introduced; most in Africa, also Atlantic Islands, Europe, and sw Asia; some are widely cultivated and become ± established in local floras.

Plants of *Dimorphotheca cuneata* (Thunberg) Lessing, a native of southern Africa, have been collected outside cultivation in Arizona (Gila County, ca. 1000 m): subshrubs or shrubs to 100 cm, leaves cuneate to obovate, 1–2(–3)cm, margins dentate to denticulate, faces ± glutinous, phyllaries usually 13, 6–9 mm, ray corollas abaxially bluish to violet, adaxially white, disc cypselae obovate to nearly orbiculate, 10 mm.

SELECTED REFERENCE Norlindh, T. 1943. Studies in the Calenduleae 1. Monograph of the Genera *Dimorphotheca, Castalis, Osteospermum, Gibbaria,* and *Chrysanthemoides.* Lund.

1. Shrubs or trees; cypselae fleshy (drupelike) . . . . 83. *Chrysanthemoides*, p. 379
1. Annuals, perennials, or shrubs; cypselae not fleshy (often tuberculate or ridged and/or winged).
   2. Disc florets bisexual (some or all fertile) . . . . 84. *Dimorphotheca*, p. 380
   2. Disc florets all functionally staminate.
      3. Cypselae arcuate to ± coiled, abaxially tuberculate, sometimes winged . . . . . . 85. *Calendula*, p. 381
      3. Cypselae triquetrous-prismatic to clavate, ± tuberculate and/or winged . . . 86. *Osteospermum*, p. 382

83. CHRYSANTHEMOIDES Fabricius, Enum., 79. 1759 • [Generic name *Chrysanthemum* and Latin -*oides*, resembling] ☐

John L. Strother

Shrubs or trees (evergreen), 50–300+ cm. **Stems** erect, glabrous or ± tomentose [spiny]. **Leaves** ± petiolate; blades ovate or elliptic to orbiculate, obovate, or oblanceolate, margins usually denticulate, sometimes entire, faces usually ± arachnose and glabrate, sometimes woolly or glabrous. **Heads** borne singly or in loose, corymbiform arrays. **Involucres** ± hemispheric or

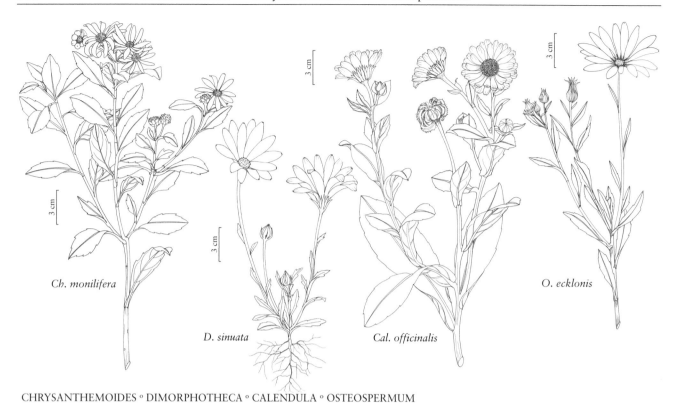

Ch. monilifera

D. sinuata

Cal. officinalis

O. ecklonis

CHRYSANTHEMOIDES ◦ DIMORPHOTHECA ◦ CALENDULA ◦ OSTEOSPERMUM

broader, 9–12+ mm diam. **Phyllaries** 12–24+ in 2–3 series, deltate or lanceolate to ovate, oblong, or linear. **Receptacles** flat to convex. **Ray florets** 5–13 in 1 series; corollas yellow, laminae ± ovate to elliptic or linear. **Disc florets** 30–80+, functionally staminate; corollas yellow, tubes (± pilosulous) shorter than ± campanulate throats. **Cypselae** ± globose, fleshy (blue-black, becoming brown, drupelike). *x* = 10.

Species 2 (1 in the flora): introduced; e tropical Africa, South Africa.

1. Chrysanthemoides monilifera (Linnaeus) Norlindh, Stud. Calenduleae, 374. 1943 [F] [I]

Osteospermum moniliferum Linnaeus, Sp. Pl. 2: 923. 1753

Leaf blades 20–70+ × 8–25(–45+) mm, ± coriaceous. **Peduncles** 1–3 cm. **Involucres** 3–5 × 9–12 mm. **Ray corolla laminae** 5–15+ mm. **Disc corollas** 3–4 mm. **Cypselae** 4–7 mm. *2n* = 20.

Flowering Apr. Disturbed places; 0–200 m; introduced, Calif.; Africa; introduced elsewhere.

Chrysanthemoides monilifera is cultivated and rarely escapes and/or persists in the flora. It is reputed to be a pest in Australia and New Zealand.

84. DIMORPHOTHECA Moench, Methodus, 585. 1794, name conserved • African daisy, Cape marigold [Greek *di-*, two, *morphe*, shape, and *theca*, case or container, alluding to two forms of cypselae within each head] [I]

John L. Strother

Annuals [perennials, subshrubs, shrubs], 5–40[150+] cm. **Stems** procumbent to erect [prostrate], glabrous or arachnose to piloso-hirtellous and/or stipitate-glandular. **Leaves** ± sessile or petiolate;

blades oblong or oblanceolate to linear, margins entire or dentate [pinnately lobed], faces sparsely arachnose and/or stipitate-glandular. **Heads** borne singly. **Involucres** campanulate to hemispheric or broader, 5–20+ mm diam. **Phyllaries** 15–21 in 2(–3) series, lanceolate to lance-linear. **Receptacles** flat to conic. **Ray florets** 10–21+ in ± 1 series; corollas usually yellow to orange or white, sometimes purplish abaxially and/or at bases or apices, laminae oblong-elliptic to oblanceolate. **Disc florets** 15–50+, bisexual, all or mostly fertile (inner sometimes functionally staminate); corollas whitish or yellow, red, or purplish, tubes much shorter than ± campanulate throats (lobes sometimes with terete or dilated appendages). **Cypselae** (ray) triquetrous-prismatic to clavate, ± tuberculate or ridged; (disc) compressed, often winged, ± smooth. *x* = 9.

Species 7–18+ (2 in the flora): introduced; s Africa.

1. Ray corollas mostly yellow to orange adaxially . . . . 1. *Dimorphotheca sinuata*
1. Ray corollas mostly whitish adaxially . . . . 2. *Dimorphotheca pluvialis*

1. Dimorphotheca sinuata de Candolle in A. P. de Candolle and A. L. P. P. de Candolle, Prodr. 6: 72. 1838 F I

Annuals, 5–30+ cm. **Leaf blades** narrowly oblong or oblanceolate to linear, 10–50(–100+) × 2–20 (–30+) mm, margins usually sinuately denticulate, sometimes serrate or entire, rarely pinnatifid. **Peduncles** (2–)5–15+ cm. **Phyllaries** 6–12+ mm. **Ray corolla** laminae abaxially yellow to orange (often marked with purple), adaxially mostly yellow to orange (sometimes purplish at bases and/or apices), 15–20(–30+) mm. **Disc florets** 15–50+; corollas yellow to orange, usually purplish distally, 4–5 mm (lobe apices acute, terete, or dilated). **Ray cypselae** 4–5 mm; disc cypselae 6–8 mm. *2n* = 18.

Flowering (Dec–)Mar–May. Disturbed places; 0–1000 m; introduced; Ariz., Calif.; South Africa.

Plants treated here (and in horticultural trade) as *Dimorphotheca sinuata* are sometimes called "*D. aurantiaca* Hort., non de Candolle" and/or "*D. calendulacea* Harvey." But for corolla colors, plants of *D. sinuata* are not easily distinguished from plants called *D. pluvialis*. They may prove to be better considered a color form of *D. pluvialis*.

2. Dimorphotheca pluvialis (Linnaeus) Moench, Methodus, 585. 1794 I

Calendula pluvialis Linnaeus, Sp. Pl. 2: 921. 1753

Annuals, 5–20(–40+) cm. **Leaf blades** narrowly oblong or oblanceolate to linear, 15–50 (–100+) × 3–15(–25+) mm, margins usually dentate to pinnatifid, sometimes entire. **Peduncles** (3–) 5–12+ cm. **Phyllaries** 6–12+ mm. **Ray corolla** laminae abaxially white to ochroleucous or yellowish (often marked with blue to purple), adaxially mostly whitish (often blue to purplish at bases and/or apices), 20–30+ mm. **Disc florets** 30–50+; corollas whitish to yellowish, usually bluish to purplish distally, 4–6 mm (lobe apices acute, terete, or dilated). **Ray cypselae** 4–6 mm; disc cypselae 6–8 mm. *2n* = 18.

Flowering Mar–Apr. Disturbed places; 0–100 m; introduced; Calif.; South Africa.

Plants with "intermediate" ray-corolla colors have been treated as hybrids between *Dimorphotheca pluvialis* and *D. sinuata* (e.g., N. T. Norlindh 1943).

85. CALENDULA Linnaeus, Sp. Pl. 2: 921. 1753; Gen. Pl. ed. 5, 393. 1754 • Pot marigold [Latin *calends*, first day of the month, and *-ula*, tendency; perhaps meaning "through the months" and alluding to ± year-round flowering] I

John L. Strother

Annuals, perennials [subshrubs], 5–50(–70+) cm. **Stems** procumbent to erect, glabrous or arachnose and/or stipitate-glandular. **Leaves** sessile or petiolate; blades elliptic, obovate, or oblong to oblanceolate, spatulate, or linear (bases sometimes clasping), margins entire or denticulate, faces sparsely arachnose and/or stipitate-glandular. **Heads** borne singly. **Involucres** campanulate to hemispheric or broader, 10–30+ mm diam. **Phyllaries** 13–40+ in 2+ series, lanceolate to lance-linear. **Receptacles** flat. **Ray florets** 13–50(–100+) in 1–3+ series; corollas yellow to orange, laminae linear to oblanceolate. **Disc florets** 20–60(–150+), functionally staminate; corollas yellow, reddish, or purplish, tubes shorter than ± campanulate throats. **Cypselae** arcuate to ± coiled (usually some beaked, some 2–3-winged, some both beaked and winged), abaxial faces usually tuberculate and/or transversely ridged. $x = 11$.

Species 15 or fewer (2 in the flora): introduced; Macaronesia, Mediterranean Europe, Near East, North Africa; widely cultivated, sometimes escaping and/or persisting.

SELECTED REFERENCE Heyn, C. C., O. Dagan, and B. Nachman. 1974. The annual *Calendula* species: Taxonomy and relationships. Israel J. Bot. 23: 169–201.

1. Phyllaries 13–21+, 6–10+ mm; ray florets 13–18+, corolla laminae 5–8(–12+) mm . . . . 1. *Calendula arvensis*
1. Phyllaries 12–40+, (8–)10–12+ mm; ray florets 30–50(–100+), corolla laminae 12–20+ mm
. . . . 2. *Calendula officinalis*

1. Calendula arvensis Linnaeus, Sp. Pl. ed. 2, 2: 1303. 1763 [I]

Leaf blades (1–)3–8(–10+) cm × 4–15(–20+) mm. **Peduncles** 3–5 cm. **Phyllaries** 13–21+, 6–10+ mm. **Ray florets** 13–18+; corolla laminae 5–8(–12+) mm. **Disc florets** 20–30+; corollas 2.5–4 mm. **Cypselae** 3–12(–25+) mm. $2n = 44$ (18 and 36 have been reported).

Flowering Feb–Apr. Disturbed places; 0–200 m; introduced; Calif.; Eurasia; Africa; Atlantic Islands.

2. Calendula officinalis Linnaeus, Sp. Pl. 2: 921. 1753 [F] [I]

Leaf blades 3–12(–18+) cm × 10–30(–60+) mm. **Peduncles** 5–8 (–12+) cm. **Phyllaries** 12–40+, (8–)10–12+ mm. **Ray florets** 30–50(–100+); corolla laminae 12–20+ mm. **Disc florets** (30–)60–150+; corollas (4–)5–6+ mm. **Cypselae** 9–15(–25+) mm. $2n = 14, 32$.

Flowering year-round. Disturbed places; 0–500 m; introduced; Calif., Conn., Maine, Mass., Mich., N.H., Ohio, Pa.; Eurasia; Africa; Atlantic Islands.

Cultivars of *Calendula officinalis* are widely used horticulturally, medicinally, and as pot herbs.

86. **OSTEOSPERMUM** Linnaeus, Sp. Pl. 2: 923. 1753; Gen. Pl. ed. 5, 395. 1754, name conserved · [Greek *osteon*, bone, and *sperma*, seed, alluding to hard fruits of original species] [I]

John L. Strother

Perennials, subshrubs, or shrubs [annuals], 5–150+ cm. **Stems** procumbent to erect [prostrate], glabrous or arachnose to piloso-hirtellous and/or stipitate-glandular. **Leaves** sessile or petiolate; blades orbiculate, elliptic, or oblong to oblanceolate, lanceolate, or linear, margins entire or denticulate [pinnately lobed], faces glabrous or sparsely arachnose and/or stipitate-glandular, often glabrate. **Heads** borne singly [in corymbiform to umbelliform arrays]. **Involucres** campanulate to hemispheric or broader, 5–20+ mm diam. **Phyllaries** 5–21+ in 1–2(–3+) series, lanceolate to lance-linear (apices ± attenuate). **Receptacles** flat to conic. **Ray florets** 10–21+ in ± 1 series; corollas whitish to purplish or yellow to orange, laminae ± oblong-elliptic to oblanceolate. **Disc florets** 12–50+, functionally staminate; corollas yellow or purplish, tubes shorter than the ± campanulate throats. **Cypselae** triquetrous-prismatic to clavate, often ± tuberculate or ridged and/or winged. $x = 10$.

Species ca. 75 (2 in the flora): introduced; Africa; widely cultivated, sometimes escaping and/or persisting.

1. Leaf blades obovate or oblong to oblanceolate, 30–50(–100+) × 0–20(–40+) mm; ray corollas purplish to whitish .... 1. *Osteospermum ecklonis*
1. Leaf blades cuneate to clavate, 3–12(–20+) × 1–4+ mm; ray corollas yellow to orange .... 2. *Osteospermum spinescens*

1. **Osteospermum ecklonis** (de Candolle) Norlindh, Stud. Calenduleae, 244. 1943 [F][I]

Dimorphotheca ecklonis de Candolle in A. P. de Candolle and A. L. P. P. de Candolle, Prodr. 6: 71. 1838

Perennials, subshrubs, or shrubs, 10–100+ cm. **Leaf blades** obovate or oblong to elliptic or oblanceolate, 30–50(–100+) × 10–20(–40+) mm, margins entire or denticulate, faces stipitate-glandular (at least distal leaves), glabrescent. **Peduncles** (25–)50–150+ mm. **Phyllaries** 12–16+, 10–15+ mm. **Ray florets** 12–21+; corollas abaxially usually violet to blue or purplish, sometimes cream to pink or salmon, adaxially whitish to blue or purplish, laminae (15–)20–45+ mm. **Disc florets** 40–60+; corollas ± purplish, 3.5–4.5 mm. **Cypselae** 6–7 mm. **2*n*** = 20.

Flowering Dec–Apr. Disturbed places; 0–200 m; introduced; Calif.; Africa.

2. **Osteospermum spinescens** Thunberg, Prodr. Pl. Cap., 166. 1800

Shrubs, 10–30+ cm (stems becoming thorns). **Leaf blades** cuneate to clavate, 3–12(–20+) × 1–4+ mm, margins entire or distally toothed [lobed], faces ± arachnose, glabrescent, often ± stipitate-glandular distally. **Peduncles** 5–15+ mm. **Phyllaries** 5–8(–12+), 4–8 mm (margins and tips often stipitate-glandular). **Ray florets** 5–8(–13+); corollas yellow to orange, laminae 4–8(–15+) mm. **Disc florets** 12–20+; corollas yellow to orange, 3–4 mm. **Cypselae** 8–10 mm (usually 3-winged). **2*n*** = 20.

Flowering Mar. Disturbed places; 1800 m; introduced; N.Mex.; Africa.

As of 2003, *Osteospermum spinescens* is known in the flora from a single plant that has persisted near Silver City for at least five years.

187g. ASTERACEAE Martinov tribe GNAPHALIEAE Cassini ex Lecoq & Juillet, Dict. Rais. Term. Bot., 296. 1831

Annuals, perennials, subshrubs, or shrubs (often ± woolly annuals 1–10 cm). **Leaves** basal and/or cauline; usually alternate, rarely opposite; petiolate or sessile (bases often decurrent onto stems); blade margins usually entire, rarely denticulate (often revolute or involute, faces often tomentose or woolly and/or glandular-pubescent). **Heads** usually heterogamous (usually disciform, rarely "quasi-radiate"), sometimes homogamous (discoid, sometimes pistillate or functionally staminate), usually borne in corymbiform, paniculiform, or racemiform arrays, sometimes in glomerules, sometimes aggregated into second-order heads, rarely borne singly. **Calyculi** 0. **Phyllaries** sometimes 0 (apparent phyllaries interpreted as outer receptacular paleae), usually persistent (often wholly or partially white or brightly colored, sometimes woolly, at least proximally and/or medially), usually (12–30+) in 3–10+ series, distinct, and unequal, sometimes in 1–2 series, distinct and subequal to equal, medially herbaceous to membranous or scarious, margins and/or apices usually notably scarious. **Receptacles** usually flat to convex, sometimes conic to ± columnar, usually epaleate, sometimes paleate (paleae sometimes enfolding pistillate florets). **Ray florets** usually 0 (sometimes peripheral florets pistillate and corollas ± zygomorphic and bearing a ± flat limb and interpreted by some as ray florets). **Peripheral (pistillate) florets** (often 100+ in disciform heads) in 1–3+ series; corollas (usually present) yellow or purplish to whitish, sometimes red-tipped (actinomorphic and filiform). **Disc (inner) florets** (sometimes 1–10, usually more) bisexual or functionally staminate; corollas yellow or purplish, sometimes red-tipped, usually actinomorphic, lobes usually (4–)5, usually ± deltate, rarely lance-ovate to lanceolate; anther bases usually ± tailed [not tailed], apical appendages mostly ovate to lance-ovate or linear (usually flat); styles abaxially glabrous or (mostly distally) papillate, branches ± linear, adaxially stigmatic in 2 lines from bases to apices, apices truncate or truncate-penicillate, appendages essentially none. **Cypselae** usually monomorphic within heads, mostly ovoid to obovoid and compressed or obcompressed, not beaked or apically attenuate, bodies smooth or ± papillate or muriculate, often 2-, 3-, or 5-ribbed (glabrous or hairy, seldom glandular, sometimes with myxogenic hairs or papillae); **pappi** (rarely 0) persistent or readily falling, usually of barbellulate, sometimes plumose bristles, sometimes of scales (scales often aristate), sometimes combinations of bristles and scales.

Genera ca. 187, species ca. 1240 (19 genera, 111 species in the flora): nearly worldwide, with centers of concentration in southern Africa and Australia; in both Old World and New World, the greater numbers of genera and species in the southern hemisphere.

Traditionally, taxa included here in Gnaphalieae have been treated in Inuleae within Gnaphaliinae (cf. A. A. Anderberg 1991; K. Bremer 1994). Gnaphalieae include everlastings and helichrysums, which have brightly colored, persistent phyllaries and are much used in dried floral arangements, and other ornamentals, e.g., species or cultivars from *Anaphalis*, *Antennaria*, and *Leontopodium* (edelweiss). Some species of *Facelis*, *Gamochaeta*, *Gnaphalium*, *Helichrysum*, and *Pseudognaphalium* are widespread as weeds.

Phyllaries in most Gnaphalieae are usually more or less herbaceous to more or less cartilaginous medially and/or proximally and membranous to scarious laterally and distally. The herbaceous to cartilagionous area of a phyllary is usually somewhat thicker than the rest and such areas are called stereomes. Stereomes may be more or less divided or not and may be more or less glandular or not. The membranous to scarious portion of a phyllary distal to the stereome is sometimes called a lamina (not to be confused with corolla laminae of other groups). The phyllaries, or laminae are often colored and may be more or less opaque or more or less hyaline.

Surfaces of cypselae of Gnaphalieae may be smooth, longitudinally ridged, or papillate (with minute bumps or projections from one or both ends of each epidermal cell; see A. A. Anderberg 1991). In addition, the cypselae may be glabrous or may bear myxogenic (producing mucilage when wetted) or non-myxogenic "twin-hairs." The twin-hairs may be relatively long and form sericeous to strigillose induments. Very short, globose to clavate twin-hairs (lengths equaling or not much greater than diameters) are characteristic of some taxa and have sometimes been called "papillae" in descriptions of members of Gnaphalieae. Cypselae with such very short twin-hairs are described here as minutely hairy and the hairs are referred to as papilliform.

Stuartina hamata Philipson, a member of Gnaphalieae, was collected near a wool mill in South Carolina in 1957 (G. L. Nesom 2004c). It is native to Australia and may be characterized as annuals, prostrate, mostly 6–12 cm across, ± woolly, leaves cauline (crowded near heads), petioles basally dilated, blades suborbiculate, heads ca. 3 mm (borne in glomerules), phyllaries ovate to lanceolate, inner uncinate, florets 5 (outer 4 pistillate, inner bisexual), cypselae 0.8–1 mm, epappose.

Genera 97–105 below (genera following second lead 3 in key to genera, members of Filagininae in a narrow sense) have exceptionally small heads and florets (even for composites) and are closely similar in expressions of some characters. Together, as found in the flora, they may be characterized as:

Annuals, taprooted, usually arachnoid-sericeous to lanuginose throughout, sometimes glabrescent proximally on stems and/or on adaxial faces of leaves. **Leaves** usually sessile, sometimes obscurely petiolate (usually gradually larger and more crowded distally, sometimes again smaller among heads, where referred to as capitular leaves); blades 1-nerved, bases usually ± cuneate, sometimes rounded, margins entire. **Heads** disciform. **Involucres** absent, vestigial, or inconspicuous, often simulated by leaves or paleae. **Phyllaries** 0, vestigial, or 1–6. **Receptacles** paleate (at least peripherally), usually glabrous among paleae (bristly in *Hesperevax*). **Florets** pistillate, functionally staminate (usually referred to as staminate), or bisexual; corollas whitish, usually distally yellowish, reddish, brownish, or purplish.

Leaves of Filagininae that immediately subtend heads and/or glomerules are here called **capitular leaves.** Flowering branches may also immediately subtend heads or glomerules; if so, capitular leaves collectively subtend such branches and their heads/glomerules, and heads/glomerules appear to be sessile in forks of pseudo-dichotomies or -polytomies. Sometimes capitular leaves subtend only glomerules and not individual heads and individual heads may be difficult to distinguish within glomerules.

Paleae subtend all or at least some florets in members of Filagininae. They are referred to as bisexual, pistillate, or staminate paleae depending on sorts of florets subtended. **Pistillate paleae** persistent or shed with cypselae, usually incurved over inner (bisexual or functionally staminate) florets at flowering; margins usually thinner, ± scarious, forming wings (sometimes gradually and obscurely so); wings recurved to erect to incurved or inflexed; abaxial faces usually ± lanuginose to sericeous, sometimes glabrate or glabrous; apices rounded or obtuse to acuminate or aristate. **Bisexual and/or staminate paleae** usually persistent, sometimes falling in fruit, sometimes 0 (most *Filago*, *Logfia*, and *Micropus*, all *Psilocarphus*; then simulated by pistillate paleae), usually erect at flowering, incurved, erect, or spreading in fruit, sometimes enlarging as cypselae mature (then adaxially lanuginose to sericeous), shorter than or surpassing pistillate paleae; bodies ± ovate or lanceolate to spatulate (saccate in *Micropsis*); margins rarely forming wings; abaxial faces lanuginose to sericeous or nearly glabrous; apices usually entire, sometimes 2–3-fid, sometimes aristate to spinose (uncinate in *Ancistrocarphus filagineus*).

Corolla scars on cypselae of Filagininae may be offset adaxially to subapical or ± median positions and may be diagnostic for certain taxa (corolla attachments usually appear to be apical before ovaries mature).

All or outer pistillate florets of Filagininae lack **pappi.** Inner pistillate, bisexual, and staminate florets have pappi 0 or of 1–28+, whitish, fragile (easily broken or detached) or readily falling, ± barbellate to barbellulate or smooth (sometimes smooth only distally) bristles in 1 series. At bases of bristles, barbs are sometimes notably longer and finer, sometimes wavy or curled, and more patent than antrorse and may interweave, resulting in proximal coherence of adjacent bristles (e.g., in *Logfia*); cohering bristles may be shed in groups or complete rings and separate subsequently.

In descriptions of Filagininae, **median** refers to areas or positions about midway between a base and corresponding apex, and **medial** refers to areas or positions ± centered between opposing lateral edges.

As noted by A. Cronquist (1950) for *Psilocarphus*, the wings or apices of pistillate paleae of Filagininae are incurved during flowering, guide styles over bisexual or functionally staminate florets, and, likely, enforce nearly obligate within-head geitonogamy. Reproductive isolation created by this self-pollinating syndrome may allow interspecific or intergeneric hybrids, when fertile, to persist and become independently reproducing species among their parental taxa (J. D. Morefield 1992, 1992b).

Birds harvest shoots of *Logfia, Micropus, Psilocarphus,* and *Stylocline* species, presumably for nesting materials (neststraw is common name for *Stylocline*) and may be significant in shorter-distance dispersal of some taxa. The light, fluffy paleae enclosing epappose cypselae of the same genera aid in wind dispersal, as suggested by A. Cronquist (1950).

Different species and genera of Filagininae often grow together and are frequently mixed and/or misidentified on herbarium sheets. Young or stunted plants often will not fit keys and descriptions; whenever practicable in attempting identifications of members of Filagininae, use well-developed plants with at least some heads in fruit. Some diagnostic characters require careful evaluation of structures within heads, usually with magnification; for example, anther tips and corolla lobes may be similar in color and shape and may be difficult to distinguish.

Good illustrations of most North American Filagininae may be found in L. Abrams and R. S. Ferris (1923–1960, vol. 4), G. Beauverd (1913), or J. C. Hickman (1993).

SELECTED REFERENCES Anderberg, A. A. 1991. Taxonomy and phylogeny of the tribe Gnaphalieae (Asteraceae). Opera Bot. 104: 1–195. Drury, D. G. 1970. A fresh approach to the classification of the genus *Gnaphalium* with particular reference to the species present in New Zealand (Inuleae–Compositae). New Zealand J. Bot. 9: 157–185. Hilliard, O. M. and B. L. Burtt. 1981. Some generic concepts in Compositae–Gnaphaliinae. Bot. J. Linn. Soc. 82: 181–232.

1. Heads usually discoid (unisexual or nearly so, staminate or pistillate; plants unisexual or nearly so; predominantly pistillate heads rarely with 1–9 central, functionally staminate florets; predominantly staminate heads rarely with 1–4+ peripheral, pistillate florets; involucres mostly 6–10 mm).
    2. Plants (0.2–)4–25(–70) cm; basal leaves usually present at flowering (withering before in *A. geyeri*); pappus bristles (at least pistillate) usually basally connate or coherent .... 87. *Antennaria*, p. 388
    2. Plants mostly 20–80(–120+) cm; basal leaves usually withering before flowering; pappus bristles distinct or basally connate .... 90. *Anaphalis*, p. 426

1. Heads usually disciform (plants not unisexual; heads mostly alike, each with 4–200+ pistillate and 1–200+ bisexual or functionally staminate florets; heads rarely discoid in *Xerochrysum*, which has involucres 10–30 mm and brightly colored phyllaries in 3–8+ series).

   3. Annuals, biennials, perennials, or subshrubs; receptacles epaleate; cypselae all pappose.

      4. Pistillate florets fewer than bisexual.

         5. Subshrubs; heads in glomerules in corymbiform arrays; involucres campanulate, 4–8 mm . . . . 89. *Helichrysum*, p. 425

         5. Annuals, biennials, or perennials; heads borne singly or 2–3 in loose, corymbiform arrays; involucres ± hemispheric, 10–30 mm . . . . 91. *Xerochrysum*, p. 427

      4. Pistillate florets more numerous than bisexual.

         6. Annuals; pappi persistent; pappus bristles ± plumose . . . . 96. *Facelis*, p. 442

         6. Annuals, biennials, or perennials; pappi readily falling; pappus bristles barbellate to barbellulate.

            7. Heads in spiciform or subcapitate arrays or in glomerules in continuous or interrupted, usually spiciform, sometimes paniculiform, arrays (terminal glomerules in depauperate plants); cypselae ± papillate (papillae or papilliform hairs myxogenic) or strigillose (hairs not myxogenic); pappus bristles basally connate, falling readily (in groups or rings; distinct in *Omalotheca supina*).

               8. Annuals or short-lived perennials; phyllaries in 3–7 series; pistillate florets 50–130+; cypselae ± papillate (papillae or papilliform hairs myxogenic) . . . . 93. *Gamochaeta*, p. 431

               8. Perennials; phyllaries in 2–3 series; pistillate florets 35–70+; cypselae strigillose (hairs not myxogenic) . . . . 94. *Omalotheca*, p. 438

            7. Heads usually in ± capitate clusters (subtended by leafy bracts) or corymbiform or paniculiform (often bracteate) arrays; cypselae usually glabrous or minutely hairy or papillate (papillae or papilliform hairs not myxogenic), sometimes minutely roughened and/or with 4–6 longitudinal ridges; pappus bristles distinct (falling separately) or basally coherent (falling in groups or rings).

               9. Involucres narrowly campanulate to cylindric; phyllaries mostly stramineous to brownish, sometimes purplish to pinkish (hyaline, stereomes not glandular) . . . . 95. *Euchiton*, p. 440

               9. Involucres narrowly to broadly campanulate to cylindric; phyllaries white, rosy, tawny, or brown (opaque or hyaline, stereomes usually glandular).

                  10. Annuals, (1–)3–30 cm; heads usually in ± capitate clusters (in axils of leaves or bracts), sometimes in spiciform glomerules; cypselae usually glabrous, sometimes minutely hairy or papillate (hairs or papillae not myxogenic) . . . . 92. *Gnaphalium*, p. 428

                  10. Annuals, biennials, or perennials, (4–)15–150(–200) cm; heads usually in glomerules in corymbiform or paniculiform arrays, sometimes in terminal clusters; cypselae usually smooth, sometimes papillate-roughened and/or with 4–6 longitudinal ridges, usually glabrous (papilliform hairs in *P. luteoalbum*) . . . . . . 88. *Pseudognaphalium*, p. 415

   3. Annuals; receptacles ± paleate (all or at least the outermost florets each subtended by a palea); cypselae (all or at least the outermost) epappose.

      11. Bisexual florets (1–)2–10(–11), pappi of (11–)13–28+ bristles visible in heads; functionally staminate florets 0.

         12. Receptacles fungiform to obovoid (heights 0.4–1.6 times diams.); most pistillate paleae ± saccate, each ± enclosing a floret, apices blunt; innermost paleae spreading in fruit; cypselae dimorphic (outer longer than inner) . . . . 97. *Logfia*, p. 443

         12. Receptacles cylindric to clavate (heights 5–15 times diams.); most pistillate paleae open to ± folded (at most each enfolding, not enclosing a floret; apices acuminate to aristate); innermost paleae erect to ascending in fruit; cypselae monomorphic (outer ± equaling inner) . . . . 98. *Filago*, p. 447

[11. Shifted to left margin.—Ed.]

11. Bisexual florets 0 or 2–7, pappi 0; functionally staminate florets 0 or 2–12, pappi 0 or of 1–10(–13) bristles hidden in heads.
  13. Pistillate paleae open most of lengths, flat or concave to loosely folded (not enclosing florets); pappi 0.
    14. Bisexual paleae saccate, each enclosing a floret, apices 2-fid or 3-fid; cypselae (at least outer) strigose; coastal Texas . . . . 103.   *Micropsis*, p. 463
    14. Bisexual or staminate paleae flat to concave, not enclosing florets, apices entire; cypselae glabrous; central and western North America.
      15. Receptacles glabrous; pistillate paleae falling (all or the inner together); staminate (or bisexual) paleae: bodies ± spatulate (apices scarcely enlarged); central North America . . . . 102.   *Diaperia*, p. 460
      15. Receptacles bristly; pistillate paleae persistent; staminate paleae: bodies obovate (apices enlarged); California and Oregon . . . . 105.   *Hesperevax*, p. 467
  13. Pistillate paleae saccate most of lengths (each enclosing a floret, outermost rarely open); pappi 0 or of 1–10(–13) bristles.
    16. Staminate paleae 5(–7), ± spreading proximally, enlarged in fruit (apices incurved to uncinate); pistillate paleae with 3, ± parallel (prominent) nerves; cypselae: corolla scars apical . . . . 104.   *Ancistrocarphus*, p. 465
    16. Staminate paleae 0 or 1–4, erect, not enlarged in fruit (apices erect); pistillate paleae with 5+, reticulate (and prominent) or ± parallel (and obscure) nerves; cypselae: corolla scars subapical to ± median.
      17. Cauline leaves mostly opposite; phyllaries 0; pistillate paleae (nerves reticulate, prominent): wings inflexed (and ± lateral); staminate paleae 0; pappi 0 . . . . 101.   *Psilocarphus*, p. 456
      17. Cauline leaves mostly alternate; phyllaries 0 or 1–6; pistillate paleae (nerves parallel, obscure): wings ± erect (or subapical); staminate paleae 1–4 and/or staminate pappi of (0–)1–10(–13) bristles.
        18. Pistillate paleae (obcompressed to terete, not galeate): wings ± erect (and apical); receptacles cylindric to clavate (heights 2.8–8 times diams.); phyllaries 0 or 1–4 (similar to paleae); cypselae: corolla scars subapical . . . 99.   *Stylocline*, p. 450
        18. Pistillate paleae (compressed, galeate): wings ± erect (and lateral) or inflexed (and subapical); receptacles depressed-spheric to obovoid (heights 0.5–1.8 times diams.); phyllaries 4–6 (unlike paleae); cypselae: corolla scars ± lateral . . . . 100.   *Micropus*, p. 454

87. ANTENNARIA   Gaertner, Fruct. Sem. Pl. 2: 410, plate 167, fig. 3. 1791, name conserved • Pussytoes, everlasting, ladies' tobacco, antennaire [Latin *antenna*, and *-aria*, connection to or possession of, alluding to similarity of clavate pappus bristles in staminate florets to antennae of some insects]

Randall J. Bayer

Perennials or subshrubs (dioecious, gynoecious, or polygamodioecious), (0.2–)4–25(–70) cm (sometimes cespitose, sometimes stoloniferous, sometimes rhizomatous). **Stems** erect. **Leaves** basal and cauline; alternate; petiolate or sessile; blades (1–7-nerved) mostly cuneate, elliptic, lanceolate, linear, oblanceolate, or spatulate, margins entire, abaxial faces usually tomentose, adaxial glabrous or ± tomentose to sericeous or glabrescent. **Heads** discoid (unisexual), borne singly or in corymbiform, paniculiform, racemiform, or subcapitate arrays. **Involucres:** staminate campanulate to hemispheric, 2–6+ mm diam.; pistillate turbinate or campanulate to cylindric, 3–7(–9+) mm diam. **Phyllaries** in 3–6+ series, usually relatively narrow, unequal (proximally papery or membranous; distally ± scarious, often black, brown, castaneous, cream, gray, green, olivaceous, pink, red, white, or yellow), apices usually acute, sometimes obtuse to ± truncate. **Receptacles** flat to convex or ovoid, foveolate, epaleate. **Ray florets** 0. **Disc florets** mostly 20–100+, (functionally) staminate or pistillate; staminate corollas white, yellow, or red, narrowly funnelform or tubular (lobes usually 5, erect to recurved); pistillate corollas white, yellow, or red, narrowly tubular to filiform. **Cypselae** mostly ellipsoid to ovoid, faces usually glabrous, often papillate (stout, myxogenic twin-hairs); **pappi:** falling (bristles basally connate or coherent, shed together in rings or in groups); staminate usually of 10–20+ (usually ± clavate, sometimes capillary, barbellate to barbellulate) bristles; pistillate usually of 12–20+ (capillary, barbellulate to smooth) bristles. $x = 14$.

Species 45 (34 in the flora): temperate and arctic/alpine regions, North America, Mexico, South America, Eurasia.

Some species of *Antennaria*, especially the stoloniferous, mat-forming species, are cultivated as rock-garden ornamentals. Among the more suitable species widely used for that purpose are *A. dioica*, *A. microphylla*, *A. parvifolia*, *A. rosea*, and *A. suffrutescens*. Clones with red or pink phyllaries have been selected as prized for cultivation. Some species are used in the dried-flower trade.

Phylogenetic relationships within *Antennaria*. *Antennaria* is composed of two major lineages: the Leontipes group, mostly restricted to western North America, and the Catipes group, occurring throughout the Northern Hemisphere and South America (R. J. Bayer et al. 1996). The Leontipes group consists of five smaller groups (the Geyerae, Arcuatae, Argenteae, Dimorphae, and Pulcherrimae) and comprises species that are primarily diploid (tetraploids are known only in *A. dimorpha* and *A. pulcherrima*, Bayer and G. L. Stebbins 1987, and, as far as is known, always amphimictic, sexually reproducing). Most of the species of the Leontipes group lack horizontal stoloniferous growth (except *A. flagellaris* and *A. arcuata*). Morphologically, the Leontipes group is considered primitive in the genus, based on unspecialized morphologic features such as non-stoloniferous growth, lack of extensive polyploidy, and general lack of well-developed sexual dimorphism; the Catipes group has amphimictic diploids and tetraploids. Derived from them are all of the polyploid agamic complexes (fig. 1). Most species of the Catipes group have horizontal stolons, an effective means of asexual reproduction; it is considered more specialized than the Leontipes group.

For the most part, the smaller monophyletic groups composing the Leontipes group correspond to traditionally recognized groups (R. J. Bayer 1990; Bayer et al. 1996). The Geyerae group is monotypic, consisting of *Antennaria geyeri*, and the tendency toward polygamodioecy in that species, along with its lack of basal leaves, makes it more similar morphologically to *Anaphalis* than to the remainder of *Antennaria*. *Antennaria arcuata* is the only member of the newly recognized Arcuatae group, and it was previously considered to be a portion of the Argenteae along with *A. luzuloides* and *A. argentea* (Bayer), a relationship that was always considered weak. The Argenteae group comprises *A. argentea*, *A. luzuloides*, and *A. stenophylla* and is sister to the *A. arcuata*–*A. geyeri* clade (Bayer et al.). The Dimorphae group, *A. dimorpha* and *A. flagellaris*, is sister to the Geyerae-Arcuatae-Argenteae clade (Bayer et al.), and the Pulcherrimae group comprises *A. pulcherrima*, *A. anaphaloides*, and *A. lanata* (Bayer; Bayer et al.).

The Catipes group is well supported in both morphologic and molecular phylogenetic trees (R. J. Bayer et al. 1996); support for subclades within Catipes is weak. Traditionally, members of Catipes were split into the Alpinae, distributed in tundra, with black or olivaceous phyllaries, and the Dioicae with lighter phyllaries. Based on DNA sequence data and morphology, the two groups are artificial and should be abandoned (Bayer et al.). Amphimixis, apomixis (agamospermy), and high levels of polyploidy (Bayer and T. M. Minish 1993) are prevalent among polyploid derivatives of the Catipes group, which consists of diploids and some tetraploids in which sexual dimorphism is highly evolved (Bayer 1990). Some species of the Catipes group are specialized as edaphic endemics, e.g., *Antennaria virginica* on Devonian-age shale barrens (Bayer and G. L. Stebbins 1987, 1993), *A. suffrutescens* on serpentine (Bayer and Stebbins 1993), and *A. aromatica* and *A. densifolia* on limestone talus (Bayer 1989). Five polyploid agamic complexes, *A. alpina* (together with the smaller *A. media, A. monocephala*, and *A. friesiana* complexes), *A. howellii, A. parlinii, A. parvifolia*, and *A. rosea*, have evolved via multiple hybridization among members of the Catipes group (Bayer 1987, 1997). The great success of the Catipes group seems to be correlated with their ability to grow in diverse habitats throughout their range across Eurasia and North America to Tierra del Fuego in South America, and to their acquisition of characteristics such as strong sexual dimorphism, polyploidy, agamospermy, and vegetative reproduction (stolons). The amphimictic taxa of the Catipes group include *A. aromatica, A. corymbosa, A. densifolia, A. dioica, A. friesiana* subsp. *alaskana, A. friesiana* subsp. *neoalaskana, A. marginata, A. microphylla, A. monocephala* subsp. *monocephala, A. neglecta, A. plantaginifolia, A. pulchella, A. racemosa, A. rosulata, A. solitaria, A. suffrutescens, A. umbrinella*, and *A. virginica*. Some of those have contributed to the genetic makeup of the polyploid complexes, whose morphologic variation is correlated to the number of diploid genomes contributed to the origin of the complex. Morphologic overlap between the complexes is a direct consequence of pivotal genomes recurring in some complexes. For example, the *A. parlinii* and *A. howellii* complexes share two pivotal genomes from *A. plantaginifolia* (PLA) and *A. racemosa* (RAC). Some apomictic clones (identified under the name *A. howellii* subsp. *howellii* in part) appear to bridge the morphologic gap between the two complexes. The *A. parlinii* complex has three diploid progenitors: *A. solitaria* (SOL), *A. plantaginifolia* (PLA), and *A. racemosa* (RAC); the *A. howellii* complex has five: *A. marginata* (MAR), *A. neglecta* (NEG), *A. plantaginifolia* (PLA), *A. racemosa* (RAC), and *A. virginica* (VIR). *Antennaria parvifolia* has three major progenitors: *A. dioica* (DIO), *A. neglecta* (NEG), and *A. marginata* (MAR); it is likely that high elevation segregates of the complex also contain genomic contributions from *A. pulchella* (PUL) and/or *A. media. Antennaria rosea* is morphologically the most diverse of the polyploid complexes and has as its primary progenitors: *A. aromatica* (ARO), *A. corymbosa* (COR), *A. microphylla* (MIC), *A. racemosa* (RAC), and *A. umbrinella* (UMB). It is likely that *A. marginata* (MAR), *A. rosulata* (ROS), and possibly *A. suffrutescens* (SUF) have also contributed to the origins of some *A. rosea* clones. The circumpolar allopolyploid *A. alpina* complex appears to have its origins from the amphimictic, dark-phyllaried, arctic-alpine taxa including *A. aromatica* (ARO), *A. densifolia* (DEN), *A. friesiana* subsp. *alaskana* (ALA), *A. friesiana* subsp. *neoalaskana* (NEO), *A. monocephala* subsp. *monocephala* (MON), and *A. pulchella* (PUL). Three polyploid complexes, *A. friesiana* subsp. *friesiana, A. media*, and *A. monocephala* subsp. *angustata* (ANG) appear to be of non-hybrid, autopolyploid origin and are direct polyploid derivatives of *A. friesiana* (subspp. *alaskana* and *neoalaskana*), *A. monocephala* subsp. *monocephala*, and *A. pulchella*, respectively; most polyploids are of multiple hybrid origin from among multiple amphimicts. *Antennaria marginata* has also given rise to apparent autopolyploid apomictic derivatives.

Key pivotal genomes involved in the origins of the polyploid complexes include *Antennaria aromatica, A. marginata, A. neglecta, A plantaginifolia, A. pulchella*, and *A. racemosa*; significant contributions have also been made by *A. corymbosa, A. densifolia, A. dioica, A. friesiana, A. microphylla, A. monocephala, A. solitaria, A. umbrinella*, and *A. virginica*.

Classification of *Antennaria*. Past practice has been to attempt to recognize each agamospecies as a distinct taxonomic entity, usually at species rank. That has led to unwieldy classifications that can be used only by experts on the group. Clearly, that method is unsatisfactory and a more

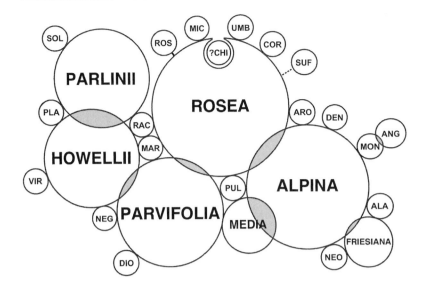

FIGURE. 1. Relationship of the polyploid agamic complexes in *Antennaria* to each other and to sexual progenitors. Relative size of the taxonomic group (bubble) is indicative of the relative amounts of morphologic variation within each taxon. Shading indicates areas of morphologic overlap between the polyploid complexes. The polyploid complexes are *Antennaria alpina, A. friesiana* subsp. *friesiana* (FRIESIANA), *A. howellii, A. media, A. monocephala* subsp. *angustata* (ANG), *A. parlinii, A. parvifolia,* and *A. rosea* (perhaps including *A. chilensis* of Patagonia (?CHI); see text under *A. rosea* for explanation). Sexual progenitors are labeled with the first three letters of their specific or subspecific epithets: *A. aromatica* (ARO), *A. corymbosa* (COR), *A. densifolia* (DEN), *A. dioica* (DIO), *A. friesiana* subsp. *alaskana* (ALA), *A. marginata* (MAR), *A. microphylla* (MIC), *A. monocephala* subsp. *monocephala* (MON), *A. neglecta* (NEG), *A. friesiana* subsp. *neoalaskana* (NEO), *A. plantaginifolia* (PLA), *A. pulchella* (PUL), *A. racemosa* (RAC), *A. rosulata* (ROS), *A. solitaria* (SOL), *A. suffrutescens* (SUF), *A. umbrinella* (UMB), and *A. virginica* (VIR). Contact between sexual progenitors and the polyploid complexes indicates possible contributions of genes from that sexual to the polyploid. Some sexuals (e.g. NEG, RAC, MAR, and PUL) are said to have pivotal genomes because they have probably contributed genes to more than one complex.

reasonable scheme for classifying polyploid agamic complexes, such as the one advocated by E. Babcock and G. L. Stebbins (1938), should be adopted. R. J. Bayer and Stebbins (1982) were the first to use the Babcock and Stebbins method in *Antennaria*.

Because the sexual diploids are morphologically discrete, they are each recognized as species. Polyploids that are morphologically identical with sexual diploid (nonhybrid- or auto-polyploid) taxa, whether they are agamospermous or amphimictic, are treated as conspecific with their sexual diploids, e.g., tetraploid cytotypes of *Antennaria virginica* and some other taxa are treated as conspecific with their corresponding sexual diploids because they are morphologically (R. J. Bayer and G. L. Stebbins 1982) and, in the case of *A. virginica*, genetically (Bayer and D. J. Crawford 1986) inseparable from the sexual diploids. Sexual and asexual polyploids that are of hybrid origin (segmental and genomic allopolyploids) are recognized as species because their genetic composition is not attributable to any single diploid origin. For example, Bayer and Stebbins classified *A. parlinii* as distinct from its sexual diploid progenitors, *A. plantaginifolia, A. racemosa,* and *A. solitaria*. A. Cronquist (1945) recognized *A. parlinii* (sensu Bayer and Stebbins) as two varieties of *A. plantaginifolia,* a view Stebbins and I opposed because the polyploids, while containing genes from *A. plantaginifolia* in their genetic background, also have genes from *A. racemosa* and *A. solitaria* (Bayer 1985b; Bayer and Crawford). The polyploid complexes are each defined primarily by assessing their genomic composition through the use of genetic markers, as well as through morphologic studies. This philosophy and method of classification has been extended to the other polyploid agamic complexes.

Identifying *Antennaria* specimens. Users of this treatment should be aware that multiple details must be kept in mind when collecting and trying to identify species of *Antennaria*. For example, assigning specimens to species in the "mat-forming," stoloniferous Catipes group is particularly difficult because of widespread polyploidy and apomixis. One determinative taxonomic character (whether populations are gynoecious or dioecious) may not be readily observed on herbarium specimens but is readily determined in the field by gender ratios. On herbarium specimens, assuming pistillates are always present in populations, absence of staminates could mean either that they were not collected or that they were actually absent from the population. This character comes into use in separating the infraspecific taxa within both *A. monocephala* and *A. friesiana*. If this character cannot be readily determined on herbarium material, i.e., when staminates are absent, then such specimens are best keyed to the specific level only.

Another feature of importance in identifying specimens of *Antennaria* is the presence or absence of well-developed stolons that root at their tips. Some *Antennaria* species produce stiff, semi-erect stolons that do not root at the tips, and those stolons should not be confused with the typical stolons that are more elongate and horizontal and root at their tips.

The final feature of importance in identifying specimens of *Antennaria* is the presence or absence of flags on tips of mid and distal cauline leaves. Flags are flat, linear, scarious appendages of the leaf tips that are similar to the tips of the phyllaries; they are not to be confused with ordinary subulate or blunt leaf tips that are essentially green and herbaceous. In keys and descriptions, leaves are referred to as flagged or not flagged.

SELECTED REFERENCES Bayer, R. J. 1984. Chromosome numbers and taxonomic notes for North American species of *Antennaria* (Asteraceae: Inuleae). Syst. Bot. 9: 74–83. Bayer, R. J. 1987. Evolution and phylogenetic relationships of the *Antennaria* (Asteraceae: Inuleae) polyploid agamic complexes. Biol. Zentralbl. 106: 683–698. Bayer, R. J. 1990. A phylogenetic reconstruction of *Antennaria* Gaertner (Asteraceae: Inuleae). Canad. J. Bot. 68: 1389–1397. Bayer, R. J. 1993. A taxonomic revision of the genus *Antennaria* (Asteraceae: Inuleae: Gnaphaliinae) of Alaska and Yukon Territory, northwestern North America. Arctic Alpine Res. 25: 150–159. Bayer, R. J., D. E. Soltis, and P. S. Soltis. 1996. Phylogenetic inferences in *Antennaria* (Asteraceae: Inuleae: Gnaphaliinae) based on sequences from the nuclear ribosomal DNA internal transcribed spacers (ITS). Amer. J. Bot. 83: 516–527. Bayer, R. J. and G. L. Stebbins. 1982. A revised classification of *Antennaria* (Asteraceae: Inuleae) of the eastern United States. Syst. Bot. 7: 300–313. Bayer, R. J. and G. L. Stebbins. 1987. Chromosome numbers, patterns of distribution, and apomixis in *Antennaria* (Asteraceae: Inuleae). Syst. Bot. 12: 305–319. Bayer, R. J. and G. L. Stebbins. 1993. A synopsis with keys for the genus *Antennaria* (Asteraceae: Inuleae: Gnaphaliinae) for North America. Canad. J. Bot. 71: 1589–1604. Fernald, M. L. 1945c. Key to *Antennaria* of the "Manual range." Rhodora. 47: 221–239. Malte, M. O. 1934. *Antennaria* of arctic America. Rhodora 36: 101–117. Nelson, E. E. 1901. A revision of certain species of plants of the genus *Antennaria*. Proc. U.S. Natl. Mus. 23: 697–713. Porsild, A. E. 1950. The genus *Antennaria* in northwestern Canada. Canad. Field-Naturalist 64: 1–25. Porsild, A. E. 1965. The genus *Antennaria* in eastern arctic and subarctic America. Bot. Tidsskr. 6l: 22–55. Stebbins, G. L. 1932. Cytology of *Antennaria*. I. Normal species. Bot. Gaz. 94: 134–151. Stebbins, G. L. 1932b. Cytology of *Antennaria*. II. Parthenogenetic species. Bot. Gaz. 94: 322–345.

Key to Groups of *Antennaria* Species

1. Heads usually borne singly, rarely in 2s or 3s . . . . Group 1, p. 393
1. Heads usually (2–)3–15(–110+), rarely borne singly.
    2. Stolons none (or erect and not rooting at tips; plants not forming mats; in fruit, heights
        of pistillate plants ± equal to staminates) . . . . Group 2, p. 393
    2. Stolons mostly 1–5(–18) cm (usually prostrate, sometimes ascending, usually rooting at
        tips; plants forming mats; in fruit, heights of pistillate plants usually greater than or
        equal to heights of staminates).
        3. Basal leaves 3–5(–7)-nerved . . . . Group 3, p. 384
        3. Basal leaves mostly 1-nerved (1–3-nerved in *A. arcuata* and *A. marginata*).

[4. Shifted to left margin.—Ed.]
4. Phyllaries (proximally dark) distally mostly dark to light brown, black, or olivaceous (some
   times inner whitish); arctic or alpine tundra to just below treeline . . . . Group 4, p. 394
4. Phyllaries (proximally light) distally mostly light brown, cream, gray, green, ivory, pink,
   red, rose, or white; seldom arctic or alpine tundra (*A. corymbosa* sometimes alpine) . . . . Group 5, p. 395

Group 1

1. Basal leaves 3–5-nerved; se United States . . . . 12. *Antennaria solitaria*
1. Basal leaves 1-nerved; w North America, Arctic.
   2. Plants 0.2–1.5(–2) cm (heads subsessile among basal leaves); basal leaves silvery gray-
      pubescent; Arizona, Colorado, New Mexico, Utah . . . . 23. *Antennaria rosulata*
   2. Plants 0.5–13 cm; basal leaves abaxially gray-tomentose (sometimes none at flowering,
      A. suffrutescens); usually not Arizona, Colorado, New Mexico, Utah (*A. dimorpha*
      sometimes Colorado, New Mexico, Utah).
      3. Cauline leaves spatulate (apices emarginate or obtuse), adaxial faces green, gla-
         brous (bases of plants ± woody); nw California and sw Oregon . . . . . . 25. *Antennaria suffrutescens*
      3. Cauline leaves linear or oblanceolate (apices acute to obtuse), adaxial faces green
         or gray, glabrous or pubescent, sericeous, tomentose, or villous; not nw California
         or sw Oregon.
         4. Leaves: adaxial faces green, usually glabrous, rarely villous or pubescent (flags
            of mid and distal cauline leaves brown); arctic and alpine tundra . . . . . . 32. *Antennaria monocephala*
         4. Leaves: adaxial faces gray, sericeous or tomentose (flags of mid and distal cauline
            leaves none); sagebrush steppe or talus near treeline.
            5. Stolons none (plants cespitose, not forming mats); semidesert . . . . . 6. *Antennaria dimorpha*
            5. Stolons 0.5–2 or 3–10 cm (plants forming mats); semidesert or limestone
               talus.
               6. Gynoecious; stolons 0.5–2 cm (leafy); basal leaves spatulate, rhombic-
                  spatulate, or cuneate; s Nevada . . . . 28. *Antennaria soliceps*
               6. Dioecious; stolons 3–10 cm (not leafy); basal leaves linear-oblanceolate;
                  not s Nevada . . . . 7. *Antennaria flagellaris*

Group 2

1. Plants 7–15 cm (mid and distal cauline leaves flagged); low and high arctic . . . .
   . . . . 33. *Antennaria friesiana* (in part)
1. Plants either 15–65 cm (low arctic or subalpine) OR 0–70 cm (desert steppe or alpine,
   except *A. pulcherrima* arctic).
   2. Plants (3–)10–15 cm; basal leaves linear to narrowly oblanceolate (1–3 mm wide); phyl-
      laries distally light brown, dingy brown, or olivaceous . . . . 5. *Antennaria stenophylla*
   2. Plants 3–70 cm; basal leaves elliptic, lanceolate, linear, oblanceolate, or spatulate (3–25
      mm wide; sometimes none at flowering); phyllaries distally usually brown, cream, pink,
      red, or white, rarely black, dark brown, castaneous, or olivaceous.
      3. Phyllaries (scarious, glabrous).
         4. Basal leaves oblanceolate to elliptic; cauline leaves oblanceolate; pistillate
            involucres 4–5 mm; phyllaries (usually pale green proximally) distally silvery
            white . . . . 3. *Antennaria argentea*
         4. Basal leaves linear to narrowly spatulate; cauline leaves narrowly oblanceolate,
            narrowly spatulate, or linear; pistillate involucres 3.5–6.5 mm; phyllaries (usu-
            ally light brown to golden proximally) distally white (often red- or pink-flecked)
            . . . . 4. *Antennaria luzuloides*
      3. Phyllaries (scarious distally, hairy proximally).

[5. Shifted to left margin.—Ed.]

5. Basal leaves (absent at flowering; plants woody at bases); phyllaries (densely pubescent to well distal of middle), distally usually pink to red, sometimes light brown or white . . . . . 1. *Antennaria geyeri*
5. Basal leaves elliptic, lanceolate, oblanceolate, or spatulate; phyllaries (moderately pubescent proximal to middle) distally usually black, brown, castaneous, cream, olivaceous, or white (rarely red- or pink-flecked).
   6. Plants 3–20 cm; phyllaries (light brown, dark brown, or olivaceous proximally) distally whitish or light brown; alpine slopes . . . . 9. *Antennaria lanata*
   6. Plants 8–65 cm; phyllaries (outer usually each with a dark spot at base) distally black, brown, castaneous, cream, olivaceous, or white; mostly subalpine, montane, or subarctic.
      7. Pistillate involucres 4.5–7 mm; phyllaries (bases each with dark spot 0.1–1 mm) distally cream or white (apices obtuse); dry montane or steppe . . . . 8. *Antennaria anaphaloides*
      7. Pistillate involucres 7–12 mm; phyllaries (bases each with dark spot 1–3 mm) distally black, brown, castaneous, or olivaceous (apices acute); wet sites, subalpine or subarctic (*A. pulcherrima* subsp. *pulcherrima*) or limestone near sea level (*A. pulcherrima* subsp. *eucosma*) . . . . 10. *Antennaria pulcherrima*

Group 3

1. Basal leaves adaxially glabrous; heads in loose racemiform or paniculiform arrays (stems proximal to heads with purple glandular hairs); w North America . . . . 13. *Antennaria racemosa*
1. Basal leaves adaxially gray-pubescent, floccose-glabrescent, or green-glabrescent; heads in tight, corymbiform arrays (stems proximal to heads with or without purple glandular hairs); e North America to e Manitoba, Minnesota, Missouri, Texas.
   2. Basal leaves abaxially tomentose; pistillate involucres 5–7 mm; staminate corollas 2–3.5 mm; pistillate corollas 3–4 mm (young stolons mostly ascending); Appalachians, Piedmont, Atlantic seaboard, and driftless area of Wisconsin, Minnesota . . . . 11. *Antennaria plantaginifolia*
   2. Basal leaves adaxially tomentose or glabrous; pistillate involucres (7–)8–13 mm; staminate corollas 3.5–5 mm; pistillate corollas 4–7 mm (young stolons mostly decumbent); e North America from Atlantic seaboard to e margin of Great Plains . . . . 14. *Antennaria parlinii*

Group 4

1. Stolons 0.5–2.5 cm (prostrate); basal leaves usually cuneate, cuneate-spatulate, or spatulate, sometimes oblanceolate (lengths 1–2 times widths); limestone talus, n Wyoming to Yukon and Northwest territories.
   2. Basal leaves mostly 5–16 × 3–10 mm; distal cauline leaves not flagged; pistillate involucres 5–7(–9) mm (living plants with odor of citronella when crushed; flowering stems, leaves, and bases of phyllaries stipitate-glandular) . . . . 26. *Antennaria aromatica*
   2. Basal leaves mostly 3–7 × 2–5 mm; mid and distal cauline leaves flagged; pistillate involucres 4.5–7.5 mm (living plants odorless; flowering stems, leaves, and bases of phyllaries not stipitate-glandular) . . . . 31. *Antennaria densifolia*
1. Stolons 0.1–16 cm (erect or decumbent); basal leaves cuneate, linear-cuneate, oblanceolate, or spatulate (lengths 2–6+ times widths); not limestone talus, w North America, Arizona, New Mexico to circumpolar Arctic.
   3. Stolons (usually erect, slightly woody); phyllaries distally pale brown, white, or yellowish (sometimes streaked with pink or rose, usually blunt); montane, rarely above treeline . . . . 24. *Antennaria umbrinella* (in part)
   3. Stolons (usually decumbent, herbaceous); phyllaries distally black, brown, or olivaceous (sometimes whitish at tips, usually acute); arctic, alpine, rarely subalpine.

[4. Shifted to left margin.—Ed.]

4. Cauline leaves 3–11(–13) mm; staminate corollas 1.9–2.8 mm; pistillate corollas 2–3 mm (stems, leaves, and phyllaries often stipitate-glandular); Sierra Nevada (Lake Tahoe to Mt. Whitney), California, adjacent Nevada . . . . 29. *Antennaria pulchella*
4. Cauline leaves 4–20 mm; staminate corollas 2.5–4.5 mm; pistillate corollas 3–4.5 mm (stems, leaves, and phyllaries stipitate-glandular or not); w North America (Arizona, California, New Mexico n to Yukon, Northwest Territories, e Arctic).
   5. Basal leaves: faces gray-pubescent; mid and distal cauline leaves mostly not flagged (sometimes flagged near heads) . . . . 30. *Antennaria media*
   5. Basal leaves: abaxial faces tomentose, abaxial green-glabrescent to gray-pubescent; mid and distal cauline leaves flagged.
      6. Stolons 0.1–4 cm; involucres: pistillate 5.5–8 mm; corollas: staminate 2.5–3 mm, pistillate 3–4.5 mm (stems, leaves, and phyllaries stipitate-glandular, hairs purple) . . . . 33. *Antennaria friesiana* (in part)
      6. Stolons 1–7 cm; involucres: pistillate 4–7(–10) mm; corollas: staminate 3–3.5 mm, pistillate 3.5–5 mm (stems, leaves, and phyllaries not stipitate-glandular) . . . . 34. *Antennaria alpina*

Group 5

1. Basal leaves adaxially usually green-glabrous, sometimes gray-pubescent; phyllaries distally brown, pink, or white.
   2. Basal leaves 1–3-nerved, 9–12 mm wide; distal cauline leaves flagged; phyllaries distally usually light brown, sometimes white . . . . 17. *Antennaria howellii* (in part)
   2. Basal leaves 1-nerved, 3–9 mm wide; distal cauline leaves flagged or not; phyllaries distally whitish or pink.
      3. Distal cauline leaves flagged; phyllaries distally white or cream; e, c North America . . . . 7. *Antennaria howellii* (in part)
      3. Distal cauline leaves not flagged; phyllaries distally white or light to dark pink; sw United States, Alaska (Aleutian Islands).
         4. Stolons (pubescent) 2–5 cm (stems not stipitate-glandular); basal leaves adaxially green-glabrous (margins not white-woolly); Alaska (Aleutian Islands) . . . 19. *Antennaria dioica*
         4. Stolons (densely woolly, hairs obscuring surfaces) 2–7 cm (stems sometimes stipitate-glandular, hairs white or purplish); basal leaves adaxially green-glabrous (margins white-woolly); Arizona, se California, sw Colorado, New Mexico . . . . 18. *Antennaria marginata*
1. Basal leaves usually adaxially pubescent, sometimes glabrous or glabrate with age (*A. neglecta*); phyllaries distally black, brown, cream, green, ivory, pink, red, stramineous, white, or yellow (sometimes streaked with pink or rose).
   5. Basal leaves (largest 20–65 × 6–20 mm); phyllaries distally light brown, ivory, or white (never black, dark brown, dark green, pink, or red).
      6. Basal leaves cuneate-oblanceolate, narrowly to broadly ovate, spatulate, or spatulate-obovate, 20–48(–65) × 2.5–20 mm, abaxially tomentose, adaxially gray-pubescent or green-glabrous; mid and distal cauline leaves mostly not flagged (sometimes flagged near heads) . . . . 17. *Antennaria howellii* (in part)
      6. Basal leaves cuneate-oblanceolate to spatulate, 15–65 × 6–18 mm, abaxially tomentose, adaxially gray-pubescent (green-glabrescent in age); mid and distal cauline leaves flagged . . . . 15. *Antennaria neglecta*
   5. Basal leaves (largest mostly 16–45 × 4–15 mm; if 20+ mm long, less than 6.5 mm wide; if 6.5+ mm wide, less than 20 mm long); phyllaries distally light to dark brown, ivory, pink, red, rose, or white.
      7. Stolons either 4–10 cm (herbaceous, arched) or 4–16 cm (slightly woody, erect); phyllaries distally usually pale brown, sometimes whitish or yellowish (rarely white- or pink-flecked).

8. Stolons 4–16 cm (somewhat woody, usually erect); basal leaves narrowly spatulate to cuneate, 10–17 × 2–5.4 mm (plants woody at bases); Alberta and British Columbia to California and Colorado . . . . 24. *Antennaria umbrinella* (in part)
8. Stolons 4–10 cm (herbaceous, arched); basal leaves narrowly to broadly spatulate or rhombic-ovate, 20–45 × 3–15 mm; Blaine County, Idaho, Elko County, Nevada, and Fremont County, Wyoming . . . . 2. *Antennaria arcuata*
[7. Shifted to left margin.—Ed.]
7. Stolons 1–10 cm (herbaceous, usually decumbent); phyllaries distally usually cream, gray, green, pink, red, stramineous, white (ivory to pure), or yellow (if light brown, staminate plants absent from populations).
   9. Plants 2–8(–15) cm; pistillate involucres 8–10(–15) mm . . . . 20. *Antennaria parvifolia*
   9. Plants 4–30 cm; pistillate involucres 4–10 mm.
      10. Basal leaves spatulate; phyllaries (each with chestnut brown spot near base) distally white or light brown (willow thickets, similar moist habitats, subalpine to alpine zones, Rocky Mountains and c Sierra Nevada) . . . . 21. *Antennaria corymbosa*
      10. Basal leaves cuneate-oblanceolate, spatulate, or linear; phyllaries (uniformly or combinations of) light brown, cream, gray, green, pink, red, white, or light yellow.
         11. Gynoecious (staminate plants very rare); basal leaves linear; phyllaries distally usually (combinations of) light brown, cream, gray, green, pink, red, white, or yellow (if solid white, not Appalachian) . . . . 27. *Antennaria rosea*
         11. Dioecious (stems sometimes stipitate-glandular distally); basal leaves cuneate-oblanceolate or spatulate; phyllaries distally stramineous, white, or light yellow.
            12. Basal leaves 10–25 × 3–9 mm, faces gray-pubescent; shale barrens, e Ohio, w Pennsylvania, Maryland, Virginia, West Virginia . . . . 16. *Antennaria virginica*
            12. Basal leaves 6–16 × 2–6 mm, faces silvery white-pubescent (stems stipitate-glandular distally, hairs purple or white, moniliform); w North America e to Ontario . . . . 22. *Antennaria microphylla*

1. **Antennaria geyeri** A. Gray, Mem. Amer. Acad. Arts, n. s. 4: 107. 1849 • Pinewoods pussytoes [E] [F]

Dioecious. Plants 3–14 cm (bases woody). **Stolons** none. **Basal leaves** absent at flowering. **Cauline leaves** linear-lanceolate to cuneate-oblanceolate, 11–35 × 2–6 imm, acute, not flagged (apices acute, faces gray-pubescent. **Heads** 3–25 in corymbiform to paniculiform arrays. **Involucres:** staminate 6–8 mm; pistillate 6–8 mm. **Phyllaries** distally red to pink, light brown, or white. **Corollas:** staminate 3–4.5 mm; pistillate 5–6 mm. **Cypselae** 2–2.5 mm, pubescent and papillate; **pappi:** staminate 6–7 mm (capillary); pistillate 6–7 mm. $2n = 28$.

Flowering summer. Dry lower montane to montane coniferous forests, usually in ± thick duff under *Pinus ponderosa*; 600–2400 m; Calif., Nev., Oreg., Wash.

Antennaria geyeri is distinctive because it has woody upright branches and is not stoloniferous. It lacks basal leaves at flowering and has heads that are often described as subdioecious (central flowers are often bisexual). As the only member of the Geyerae group, *A. geyeri* is not closely related to any other species of *Antennaria*; it bears strong similarities to some species of *Anaphalis* (R. J. Bayer 1990; Bayer et al. 1996).

2. **Antennaria arcuata** Cronquist, Leafl. W. Bot. 6: 41. 1950 • Box or meadow pussytoes [C] [E]

Dioecious. Plants 5–15(–20) cm (stems woolly). **Stolons** 4–10 cm (arched). **Basal leaves** 1–3-nerved, narrowly to broadly spatulate, or narrowly rhombic-obovate, 20–45 × 3–15 mm, tips mucronate, faces densely white-woolly. **Cauline leaves** linear, (2–)5–40 mm, not flagged. **Heads** (4–)7–25, in racemiform to paniculiform or corymbiform arrays. **Involucres:** staminate 3–5 mm; pistillate 4.5–6(–7) mm. **Phyllaries** distally whitish (mostly staminate) or grayish stramineous to light brown. **Corollas:** staminate 2.5–4 mm; pistillate 3.5–5 mm. **Cypselae** 1–1.8 mm, glabrous; **pappi:** staminate 3–4.5 mm; pistillate 4–6 mm. $2n = 28$.

Flowering summer. Moist alkaline basins in sagebrush steppe; 1500–2300 m; Idaho, Nev., Wyo.

Antennaria arcuata is known from three widely disjunct areas in Blaine County, Idaho; Elko County, Nevada; and Fremont County, Wyoming (R. J. Bayer 1992). It is characterized by arching stolons and white-

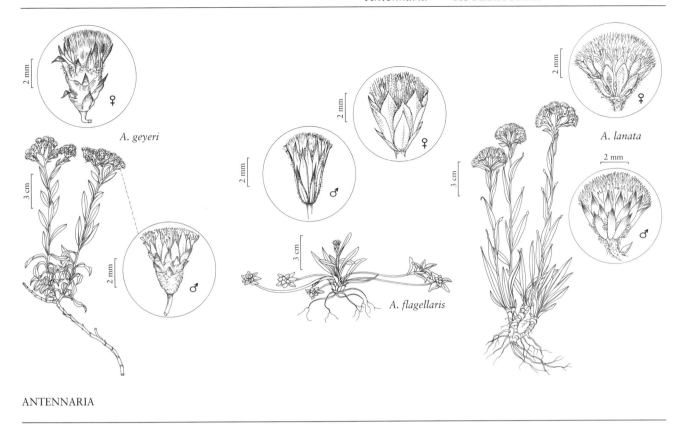

A. geyeri

A. flagellaris

A. lanata

ANTENNARIA

woolly indument (Bayer) and is not easily confused with other species of *Antennaria*.

3. Antennaria argentea Bentham, Pl. Hartw., 319. 1849 • Silver pussytoes E

Dioecious. Plants 18–40 cm. **Stolons** none. **Basal leaves:** 1–3-nerved, oblanceolate to elliptic, 20–50 × 4–15 mm, tips acute, faces ± gray-tomentose. **Cauline leaves** lanceolate, 15–45 mm, not flagged. **Heads** 10–75 in paniculiform arrays. **Involucres:** staminate 4–5 mm; pistillate 4–5 mm. **Phyllaries** (relatively broad) distally silvery white. **Corollas:** staminate 2.5–3.5 mm; pistillate 3–4 mm. **Cypselae** 1–1.5 mm, glandular; **pappi:** staminate 4–5 mm; pistillate 3–4 mm. **2n** = 28.

Flowering summer. Openings in dry coniferous forests; 600–2000 m; Calif., Nev., Oreg., Wash.

Antennaria argentea is distinguished by its robustly stoloniferous habit and silvery white phyllaries.

4. Antennaria luzuloides Torrey & A. Gray, Fl. N. Amer. 2: 430. 1843 • Rush or silvery brown pussytoes E

Dioecious. Plants 7–35(–70) cm (often viviparous in late season, bearing propagules in distal and, sometimes, proximal leaf axils, sometimes woody at bases). **Stolons** none. **Basal leaves** 1–3-nerved, linear to narrowly spatulate, 18–55 × 1–10 mm, tips acuminate, faces gray-tomentose. **Cauline leaves** narrowly oblanceolate to linear, 5–60 mm, flagged. **Heads** 10–110+ in racemiform to paniculiform or corymbiform arrays. **Involucres:** staminate 3.5–5.5 mm; pistillate 3.5–6.5 mm. **Phyllaries** (relatively narrow, proximally green or golden brown, glabrous) distally white, acute. **Corollas:** staminate 2.5–4 mm; pistillate 2–4 mm. **Cypselae** 1–2 mm, sparingly papillate or papillate-strigose (hairs clavate); **pappi:** staminate 3–4.5 mm; pistillate 2.5–4 mm. **2n** = 28.

Subspecies 2 (2 in the flora); w North America.

Some authors have recognized *Antennaria microcephala* (= *A. luzuloides* subsp. *aberrans*) as a distinct species. Given the intergradation between *A. luzuloides* in the strict sense and *A. microcephala* in the strict sense, one species with two subspecies seems justified. Perhaps the most significant difference between the subspecies is ecologic. *Antennaria luzuloides* is a member of the Argenteae group.

1. Heads 10–110+ in corymbiform arrays; basal leaves (1–)3-nerved; dry sagebrush-ponderosa pine communities . . . 4b. *Antennaria luzuloides* subsp. *luzuloides*
1. Heads 10–30 in racemiform to paniculiform arrays; basal leaves 1(–3)-nerved; moist meadows or along moist drainages in ponderosa pine communities . . . . 4a. *Antennaria luzuloides* subsp. *aberrans*

4a. Antennaria luzuloides Torrey & A. Gray subsp. **aberrans** (E. E. Nelson) R. J. Bayer & Stebbins, Canad. J. Bot. 71: 1597. 1993 • Small-headed rush pussytoes E

Antennaria argentea Bentham subsp. *aberrans* E. E. Nelson, Bot. Gaz. 34: 124. 1902; *A. luzuloides* Torrey & A. Gray var. *microcephala* (A. Gray) Cronquist; *A. microcephala* A. Gray; *A. pyramidata* Greene

Basal leaves 1(–3)-nerved, linear, 1–5 mm. **Heads** 10–30 in racemiform to paniculiform arrays. **Involucres:** staminate 3.5–4 mm; pistillate 3.5–4.5 mm. **Phyllaries** proximally green, distally white. **Corollas:** staminate 2.5–3.2 mm; pistillate 2–2.5 mm. **Cypselae** 1–1.5 mm, sparingly papillate-strigose; **pappi:** staminate 3–3.5 mm; pistillate 2.5–3 mm.

Flowering summer. Moist, open, meadows and drainages, lower montane, ponderosa pine zone; 1000–2200 m; Calif., Nev., Oreg.

Subspecies *aberrans* is characterized by its nonstoloniferous habit and relatively small heads in racemiform to paniculiform arrays.

4b. Antennaria luzuloides Torrey & A. Gray subsp. **luzuloides** E

Antennaria luzuloides var. *oblanceolata* (Rydberg) M. Peck; *A. oblanceolata* Rydberg

Basal leaves (1–)3-nerved, linear to narrowly spatulate, 2–10 mm. **Heads** 10–110+ in corymbiform arrays. **Involucres:** staminate 4–5.5 mm; pistillate 5–6.5 mm. **Phyllaries** proximally golden brown, distally white. **Corollas:** staminate 3–4 mm; pistillate 2.5–4 mm. **Cypselae** 1–2 mm, sparingly papillate; **pappi:** staminate 3–4.5 mm; pistillate 3–4 mm.

Flowering summer. Semidry meadows and drainages, lower montane, ponderosa pine zone; 900–3200 m; Alta., B.C.; Calif., Colo., Idaho, Mont., Nev., Oreg., S.Dak., Utah, Wash., Wyo.

Subspecies *luzuloides* is characterized by nonstoloniferous habit and relatively small heads in corymbiform arrays.

5. Antennaria stenophylla (A. Gray) A. Gray, Proc. Amer. Acad. Arts 17: 213. 1882 • Narrowleaf pussytoes E

Antennaria alpina (Linnaeus) Gaertner var. *stenophylla* A. Gray in C. Wilkes et al., U.S. Expl. Exped. 17: 366. 1874; *A. leucophaea* Piper

Dioecious. Plants (3–)10–15 cm. **Stolons** none. **Basal leaves** 1-nerved, linear to narrowly oblanceolate, 15–50 × 1–2(–4) mm, tips acute, not flagged, faces ± gray tomentose. **Cauline leaves** (gradually reduced distally) narrowly linear, 5–60 mm, distalmost flagged. **Heads** 2–8(–10) in subcapitate arrays. **Involucres:** staminate 4–5 mm; pistillate 4–6.5 mm. **Phyllaries** distally light brown, dingy brown, or olivaceous (apices acute-acuminate). **Corollas:** staminate 2.5–3.5 mm; pistillate 2.5–4 mm. **Cypselae** 1–1.8 mm, glandular-puberulent; **pappi:** staminate 3–4.5 mm (bristles barbellate at tips); pistillate 3–4.5 mm. **2n** = 56.

Flowering in late spring–early summer. Dry, often sagebrush (*Artemisia*) covered hillsides and dry margins around seasonally moist depressions in sagebrush steppe of the Great Basin and Columbia Plateau; 1500–2300 m; Idaho, Nev., Oreg., Wash.

Antennaria stenophylla is a xerophyte in the Argenteae group. It is distinguished by relatively narrow leaves, heads in subcapitate clusters, and light brown, dingy brown, or olivaceous phyllary tips.

6. Antennaria dimorpha (Nuttall) Torrey & A. Gray, Fl. N. Amer. 2: 431. 1843 • Low or two-form or cushion pussytoes E

Gnaphalium dimorphum Nuttall, Trans. Amer. Philos. Soc., n. s. 7: 405. 1841; *Antennaria dimorpha* var. *integra* L. F. Henderson; *A. dimorpha* var. *macrocephala* D. C. Eaton; *A. dimorpha* var. *nuttallii* D. C. Eaton; *A. latisquama* Piper; *A. macrocephala* (D. C. Eaton) Rydberg

Dioecious. Plants 0.5–4 cm. **Stolons** none. **Basal leaves:** 1-nerved, linear to narrowly spatulate, 8–11 × 1–1.2 mm, tips acute, faces ± gray-tomentose. **Cauline leaves** linear or oblanceolate, 7–12 mm, not flagged (apices acute). **Heads** borne singly. **Involucres:** staminate 6–8 mm; pistillate 10–11 mm. **Phyllaries** distally dingy brown (apices acute-acuminate). **Corollas:** staminate 3–5 mm; pistillate 8–10 mm. **Cypselae** 2–3.5 mm, pubescent; **pappi:** staminate 4.5–6 mm; pistillate 10–12 mm. **2n** = 28, 56.

Flowering early–mid spring. Sagebrush steppe, plains, foothills of mountains; 600–3400 m; Alta., B.C., Sask.;

Calif., Colo., Idaho, Mont., Nebr., Nev., N.Mex., Oreg., Utah, Wash., Wyo.

Antennaria dimorpha is characterized by narrowly oblanceolate leaves and relatively large heads (borne singly). It is, perhaps, the most xerophytic of spring-blooming *Antennaria* species. It belongs to the Dimorphae group.

7. **Antennaria flagellaris** (A. Gray) A. Gray, Proc. Amer. Acad. Arts 17: 212. 1882 • Whip or stoloniferous pussytoes E F

Antennaria dimorpha (Nuttall) Torrey & A. Gray var. *flagellaris* A. Gray in C. Wilkes et al., U.S. Expl. Exped. 17: 366. 1874

Dioecious. Plants 0.5–1.5 cm. **Stolons** 3–10 cm (leafless except tips, relatively slender). **Basal leaves** 1-nerved, linear-oblanceolate, 16–18 × 1.5–2 mm, tips acute, faces ± gray-tomentose. **Cauline leaves** linear or oblanceolate, 7–15 mm, not flagged. **Heads** borne singly. **Involucres:** staminate 6–7 mm; pistillate 7–9 mm. **Phyllaries** (relatively wide) distally brown to blackish or whitish. **Corollas:** staminate 3–4.5 mm; pistillate 5–7 mm. **Cypselae** 2–3 mm, papillate; **pappi:** staminate 3.5–4.5 mm; pistillate 6–8 mm. $2n = 28$.

Flowering mid–late spring. Seasonally dry basins in foothills of mountains, often associated with sagebrush flats; 900–2700 m; B.C.; Calif., Idaho, Nev., Oreg., S.Dak., Wash., Wyo.

Antennaria flagellaris is among the more distinctive species of *Antennaria*, with its flagelliform stolons (whip-like with leaves only at the very end) and heads borne singly. It belongs to the Dimorphae group (R. J. Bayer 1990; Bayer et al. 1996).

8. **Antennaria anaphaloides** Rydberg, Mem. New York Bot. Gard. 1: 409. 1900 • Pearly or handsome or tall pussytoes E

Antennaria anaphaloides var. *straminea* B. Boivin; *A. pulcherrima* (Hooker) Greene subsp. *anaphaloides* (Rydberg) W. A. Weber; *A. pulcherrima* var. *anaphaloides* (Rydberg) G. W. Douglas

Dioecious. Plants 15–35(–50) cm. **Stolons** none. **Basal leaves** (ephemeral) 3–5-nerved, narrowly oblanceolate or narrowly elliptic, 25–150(–200) × 4–20 (–25) mm, tips mucronate, faces gray-pubescent. **Cauline**

leaves oblanceolate or linear, 10–80 mm, usually flagged. **Heads** 8–30(–50+) in corymbiform arrays. **Involucres:** staminate (4–)5–6.5 mm; pistillate 4.5–7 mm. **Phyllaries** (each with dark brown or blackish spot in middle) distally white or cream (sometimes suffused pink to rose). **Corollas:** staminate 2.5–4 mm; pistillate 3–4.5 mm. **Cypselae** 1–1.8 mm, glabrous; **pappi:** staminate 3–4.5 mm; pistillate 3.5–4.5(–5.5) mm. $2n = 28$.

Flowering summer. Dry meadows and aspen forest openings; 1000–3400 m; Alta., B.C., Sask.; Colo., Idaho, Mont., Nev., Oreg., Utah, Wash., Wyo.

Antennaria anaphaloides is native to the northern Rocky Mountains and is characterized by whitish phyllaries, each with a black spot at the base. Some morphologic overlap occurs between *A. anaphaloides* and *A. pulcherrima*; the two occur in different habitats: *A. anaphaloides* grows in dry meadows and aspen forest openings; *A. pulcherrima* is usually found in moist willow thickets along streams (K. M. Urbanska 1983). *Antennaria anaphaloides* is closely related to the other members of the Pulcherrimae group (R. J. Bayer 1990; Bayer et al. 1996).

9. **Antennaria lanata** (Hooker) Greene, Pittonia 3: 288. 1898 • Woolly pussytoes E F

Antennaria carpatica (Wahlenberg) Hooker var. *lanata* Hooker, Fl. Bor.-Amer. 1: 329. 1834

Dioecious. Plants 3–20 cm (caudices branching or rhizomes stout). **Stolons** none. **Basal leaves** 3-nerved, narrowly oblanceolate, 10–60(–100) × 3–12 mm, tips acute, faces gray-woolly or tomentose. **Cauline leaves** linear, 5–40 mm, mid and distal flagged. **Heads** 3–9 in corymbiform arrays. **Involucres:** staminate 4.5–6 mm; pistillate 5–8 mm. **Phyllaries** (proximally light brown, dark brown, or olivaceous) distally whitish or light brown. **Corollas:** staminate 3–4.5 mm; pistillate 2.5–4 mm. **Cypselae** 1–1.6 mm, glabrous; **pappi:** staminate 4–5 mm; pistillate 3.5–5 mm. $2n = 28$ (under *A. neodioica*).

Flowering summer. Protected alpine and subalpine sites, gravelly or sandy soils near conifers at timberline; 1400–3400 m; Alta., B.C.; Calif., Idaho, Mont., Oreg., Utah, Wash., Wyo.

10. **Antennaria pulcherrima** (Hooker) Greene, Pittonia 3: 176. 1897 • Showy or handsome pussytoes, antennaire magnifique [E]

Antennaria carpatica (Hooker) R. Brown var. *pulcherrima* Hooker, Fl. Bor.-Amer. 1: 329. 1834; *A. pulcherrima* var. *angustisquama* A. E. Porsild; *A. pulcherrima* var. *sordida* B. Boivin

Dioecious. **Plants** (8–)30–65 cm (caudices branching or rhizomes stout). **Stolons** none. **Basal leaves** 3–5-nerved, spatulate to oblanceolate or lanceolate, 50–200 × 4–25 mm, tips acute, mucronate, faces gray-pubescent or silvery sericeous. **Cauline leaves** linear, 8–140 mm, distal flagged or not. **Heads** 3–30 in corymbiform to paniculiform arrays. **Involucres:** staminate 5–8 mm; pistillate 7–12 mm. **Phyllaries** distally black, dark brown, light brown, castaneous, or olivaceous. **Corollas:** staminate 3–5 mm; pistillate 3–6 mm. **Cypselae** 1–1.5 mm, glabrous; **pappi:** staminate 4–6 mm; pistillate (5–)8–10 mm. $2n = 28, 56$.

Subspecies 2 (2 in the flora): North America.

Antennaria pulcherrima is characterized by relatively large basal leaves, rhizomatous growth, and dark-colored phyllaries, each with a relatively large black spot at the base (R. J. Bayer and G. L. Stebbins 1987).

1. Cauline leaves: distal usually flagged; corollas: staminate 3.5–5 mm; pistillate 4–6 mm; wet sites, willow thickets, subalpine or subarctic, Colorado to Alaska, e to Ontario, w Quebec . . . .
. . . . . . 10a. *Antennaria pulcherrima* subsp. *pulcherrima*

1. Cauline leaves: mostly not flagged (sometimes flagged near heads); corollas: staminate 3–4 mm; pistillate 3–4.3 mm; limestone near sea level, w Newfoundland, Anticosti Island . . . .
. . . . 10b. *Antennaria pulcherrima* subsp. *eucosma*

10a. **Antennaria pulcherrima** (Hooker) Greene subsp. **pulcherrima** [E]

Plants (15–)30–65 cm. **Basal leaves** spatulate to oblanceolate, 50–200 × 4–25 mm, faces gray-pubescent. **Cauline leaves** 8–140 mm, distal usually flagged. **Heads** 3–30. **Involucres:** staminate 5–8 mm; pistillate 7–12 mm. **Phyllaries** distally black, dark brown, light brown, or olivaceous. **Corollas:** staminate 3.5–5 mm; pistillate 4–6 mm. **Cypselae** 1–1.5 mm; **pappi:** pistillate (7–)8–10 mm. $2n = 28, 56$.

Flowering summer. Alluvial soils in moist willow thickets subject to intermittent flooding along streams, montane to subalpine zone; 0–3000 m; Alta., B.C., Man.,

N.W.T., Ont., Que., Sask., Yukon; Alaska, Colo., Idaho, Mont., Utah, Wash., Wyo.

10b. **Antennaria pulcherrima** (Hooker) Greene subsp. **eucosma** (Fernald & Wiegand) R. J. Bayer, Sida 21: 768. 2004 • Elegant pussytoes [E]

Antennaria eucosma Fernald & Wiegand, Rhodora 13: 23. 1911; *A. carpatica* (Wahlenberg) Hooker var. *humilis* Hooker

Plants 8–25 cm. **Basal leaves** lanceolate to oblanceolate, 50–170 × 5–18 mm, faces silvery sericeous. **Cauline leaves** 7–30 mm, mostly not flagged (sometimes flagged near heads). **Heads** 3–15. **Involucres:** staminate 5.5–7 mm; pistillate 7–10 mm. **Phyllaries** distally light brown or castaneous. **Corollas:** staminate 3–4 mm; pistillate 3–4.3 mm. **Cypselae** 1.2–1.5 mm; **pappi:** staminate 4–6 mm; pistillate 5–7 mm. $2n = 56$.

Flowering summer. Limestone barrens, serpentine; 0–300 m; Nfld. and Labr. (Nfld.), Que. (Anticosti Island).

Subspecies *eucosma* is known from limestone barrens in western Newfoundland and Anticosti Island (K. M. Urbanska 1983). It is morphologically similar to subsp. *pulcherrima*; the two are separated mainly by the presence of prominent flags on cauline leaves in subsp. *pulcherrima* and their absence in subsp. *eucosma*.

11. **Antennaria plantaginifolia** (Linnaeus) Hooker, Fl. Bor.-Amer. 1: 330. 1834 • Plantain-leaved pussytoes, antennaire à feuilles de plantain [E] [F]

Gnaphalium plantaginifolium Linnaeus, Sp. Pl. 2: 850. 1753; *Antennaria caroliniana* Rydberg; *A. decipiens* Greene; *A. denikeana* B. Boivin; *A. nemoralis* Greene; *A. pinetorum* Greene; *A. plantaginifolia* var. *petiolata* (Fernald) A. Heller

Dioecious. **Plants** 6.5–20(–25) cm. **Stolons** 2.5–7.5 cm (mostly ascending when young). **Basal leaves** (petiolate) 3–5(–7)-nerved, obovate to suborbiculate, 35–75 × 15–35 mm, tips minutely mucronate, abaxially tomentose, adaxially green-glabrescent to gray-pubescent. **Cauline leaves** linear, 6.5–35 mm, distal flagged. **Heads** 4–17(–30) in tight corymbiform arrays. **Involucres:** staminate 5–7(–8) mm; pistillate 5–7 mm. **Phyllaries** distally white. **Corollas:** staminate 2–3.5 mm; pistillate 3–4 mm. **Cypselae** 0.5–1.6 mm, slightly papillate; **pappi:** staminate 2.5–4 mm; pistillate 3.5–5.5 mm. $2n = 28$.

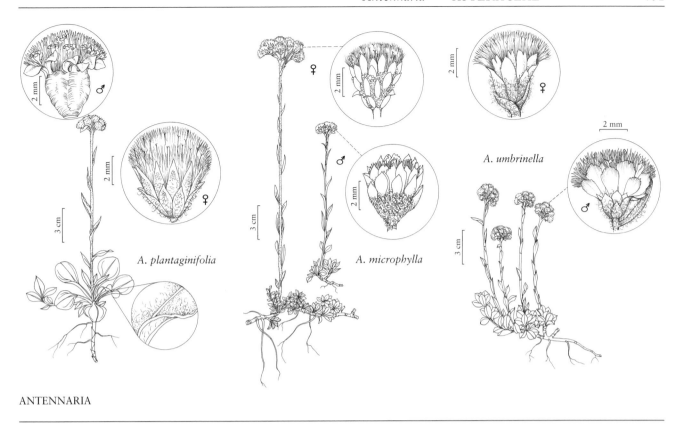

A. plantaginifolia

A. microphylla

A. umbrinella

ANTENNARIA

Flowering mid–late spring. Dry, open, deciduous woodlands, tops of banks, ridges, and bluffs, sandstone formations, slopes in openings in woodlands; 0–1500 m; Man., N.B., N.S., Que.; Ala., Ark., Conn., Del., Fla., Ga., Ill., Ind., Iowa, Ky., Maine, Md., Mass., Minn., Miss., Mo., N.H., N.J., N.Y., N.C., Okla., Pa., R.I., S.C., Tenn., Vt., Va., W.Va., Wis.

Antennaria plantaginifolia is a diploid progenitor of the *A. parlinii* complex and is similar to that species except for smaller heads and adaxially gray-pubescent basal leaves (R. J. Bayer and G. L. Stebbins 1982; Bayer 1985b; Bayer and D. J. Crawford 1986). It is a diploid ancestor of the *A. howellii* complex. It is found in the Appalachian region; disjunct populations occur in the driftless area of Wisconsin and Minnesota (Bayer and Stebbins).

pubescent to floccose-glabrate. **Cauline leaves** linear, 1–17 mm, distal flagged. **Heads** borne singly. **Involucres:** staminate 8–11 mm; pistillate 8–14 mm. **Phyllaries** (bases green or brown) distally white. **Corollas:** staminate 3.8–5.5 mm; pistillate 4.5–7 mm. **Cypselae** 1–2 mm, papillate; **pappi:** staminate 4.5–7 mm; pistillate 6–9 mm. **2***n* = 28.

Flowering early–mid spring. Slopes or stream banks in moist, rich, deciduous woodlands, forests, sometimes forest openings; 0–1500 m; Ala., Ark., Ga., Ind., Ky., La., Md., Miss., N.C., Ohio, Pa., S.C., Tenn., Va., W.Va.

With its relatively large, 3–5-nerved, basal leaves and relatively large heads borne singly, *Antennaria solitaria* is an easily recognized amphimictic member of the Catipes group (R. J. Bayer and G. L. Stebbins 1982). It is a sexual diploid progenitor of the *A. parlinii* polyploid complex.

12. **Antennaria solitaria** Rydberg, Bull. Torrey Bot. Club 24: 304. 1897 • Singlehead pussytoes E

Antennaria plantaginifolia (Linnaeus) Richardson var. *monocephala* Torrey & A. Gray, Fl. N. Amer. 2: 431. 1843, not *A. monocephala* de Candolle 1838; *A. monocephala* (Torrey & A. Gray) Greene

Dioecious. Plants 2–25(–35) cm. **Stolons** 5.5–20 cm (filiform). **Basal leaves:** 3–5-nerved, obovate to broadly oblong-spatulate, 20–75 × 15–45 mm, tips mucronate, abaxially tomentose, adaxially gray-

13. **Antennaria racemosa** Hooker, Fl. Bor.-Amer. 1: 330. 1834 • Racemose pussytoes E

Antennaria petasites Greene; *A. piperi* Rydberg

Dioecious. Plants 12–50 cm (stems stipitate-glandular distally). **Stolons** 3–8 cm. **Basal leaves:** 3-nerved, elliptic to oblong, 30–100 × 10–40 mm, tips mucronate, abaxially tomentose, adaxially glabrous. **Cauline leaves** linear, 10–30 mm, not flagged (apices obtuse to acute). **Heads** 3–12 in loose, racemiform to paniculiform arrays.

Involucres: staminate 4–8 mm; pistillate 7–9 mm. **Phyllaries** (relatively wide) distally white or light brown (apices blunt). **Corollas:** staminate 3–4 mm; pistillate 3–4 mm. **Cypselae** 1–1.5 mm, glabrous or slightly papillate; **pappi:** staminate 3–4.5 mm; pistillate 4.5–7 mm. $2n = 28$.

Flowering summer. Moist, cool, montane and subalpine coniferous forests and roadcuts in forests; 1200–3000 m; Alta., B.C.; Calif., Idaho, Mont., Oreg., Wash., Wyo.

Antennaria racemosa is characterized by adaxially glabrous basal leaves and open, racemiform to paniculiform arrays of heads (R. J. Bayer 1985b). The young leaves have a slight odor of citronella when crushed. *Antennaria racemosa* has a pivotal sexual genome of the Catipes group and has contributed to the origin of clones in the *A. howellii, A. parlinii,* and *A. rosea* polyploid agamic complexes (Bayer 1985, 1985b, 1990b).

14. Antennaria parlinii Fernald, Gard. & Forest 10: 284. 1897 • Parlin's pussytoes [E]

Dioecious or gynoecious (staminate plants in equal frequencies as pistillates or none in populations, respectively). **Plants** 12–35(–45) cm. **Stolons** 3.5–11(–14) cm (mostly decumbent when young). **Basal leaves** 3–5-nerved, obovate-spatulate, obovate, rhombic-obovate, or suborbiculate, 30–95 × 12–45 mm, tips mucronate, faces gray-pubescent to floccose-glabrescent. **Cauline leaves** oblong-lanceolate, 3.5–45 mm, distalmost flagged. **Heads** 4–12(–15) in tight corymbiform arrays. **Involucres:** staminate 6–9 mm; pistillate (7–)8–13 mm. **Phyllaries** distally white. **Corollas:** staminate 3.5–5 mm; pistillate 4–7 mm. **Cypselae** 1–2 mm, minutely papillate; **pappi:** staminate 4–5 mm; pistillate 5–8 mm. $2n = 56, 84, 70, 112$.

Subspecies 2 (2 in the flora): e North America.

The *Antennaria parlinii* complex consists of two fairly distinct subspecies that differ in induments of basal leaves (tomentose in subsp. *fallax*; glabrous in subsp. *parlinii*) and other characters (R. J. Bayer and G. L. Stebbins 1982). *Antennaria parlinii* is the most common eastern North American species (Bayer and Stebbins 1982, 1983). This complex of polyploid sexual and apomictic populations is the result of multiple hybridizations among sexual diploid species including *A. plantaginifolia, A. racemosa,* and *A. solitaria* (Bayer 1985b; Bayer and D. J. Crawford 1986). A. Cronquist (1945; H. A. Gleason and Cronquist 1991) included *A. parlinii* within his circumscription of *A. plantaginifolia.* By not including the hybrid polyploid within the circumscription of a single one of its sexual progenitors, the circumscription here better portrays the evolutionary relationships between *A. parlinii* and its sexual progenitors.

1. Stems usually glandless; basal leaves adaxially tomentose . . . . . 14a. *Antennaria parlinii* subsp. *fallax*
1. Stems usually with purple glandular hairs (at least near summits of young flowering stems); basal leaves adaxially green-glabrous . . . . 14b. *Antennaria parlinii* subsp. *parlinii*

14a. Antennaria parlinii Fernald subsp. fallax (Greene) R. J. Bayer & Stebbins, Syst. Bot. 7: 310. 1982 • Deceitful pussytoes, antennaire litigieuse [E]

Antennaria fallax Greene, Pittonia 3: 321. 1898; *A. ambigens* (Greene) Fernald; *A. ampla* Bush; *A. arkansana* Greene; *A. arnoglossa* Greene var. *ambigens* Greene; *A. bifrons* Greene; *A. brainerdii* Fernald; *A. calophylla* Greene; *A. elliptica* Greene; *A. fallax* var. *calophylla* (Greene) Fernald; *A. farwellii* Greene; *A. greenei* Bush; *A. mesochora* Greene; *A. munda* Fernald; *A. occidentalis* Greene; *A. parlinii* var. *ambigens* (Greene) Fernald; *A. parlinii* var. *farwellii* (Greene) B. Boivin; *A. plantaginifolia* (Linnaeus) Richardson var. *ambigens* (Greene) Cronquist; *A. umbellata* Greene

Stems usually glandless. **Basal leaves** abaxially and adaxially gray-tomentose to floccose-glabrescent.

Flowering early–mid spring. Clearings, fields, roadsides, and open deciduous woods; 0–1500 m; N.B., Ont., Que.; Ala., Ark., Conn., Del., Ga., Ill., Ind., Iowa, Kans., Ky., La., Maine, Md., Mass., Mich., Minn., Miss., Mo., Nebr., N.H., N.J., N.Y., N.C., Ohio, Okla., Pa., R.I., S.C., S.Dak., Tenn., Tex., Vt., Va., W.Va., Wis.

The indument of subsp. *fallax* is undoubtedly inherited from two of its sexual progenitors, *Antennaria plantaginifolia* and *A. solitaria* (R. J. Bayer 1985b; Bayer and D. J. Crawford 1986).

14b. Antennaria parlinii Fernald subsp. parlinii • Antennaire de Parlin [E]

Antennaria arnoglossa Greene; *A. parlinii* var. *arnoglossa* (Greene) Fernald; *A. plantaginifolia* (Linnaeus) Richardson var. *arnoglossa* (Greene) Cronquist; *A. propinqua* Greene

Stems usually with purple glandular hairs (at least near summits of young flowering stems). **Basal leaves** abaxially gray-tomentose, adaxially green-glabrous.

Flowering early–mid spring. Clearings, fields, roadsides, and open woods; 0–1500 m; Man., N.B., N.S., Ont., Que.; Ala., Ark., Conn., Del., Ga., Ill., Ind., Iowa,

Ky., Maine, Md., Mass., Mich., Miss., Mo., N.H., N.J., N.Y., N.C., Ohio, Pa., R.I., S.C., Tenn., Tex., Vt., Va., W.Va., Wis.

Glabrosity of subsp. *parlinii* is undoubtedly inherited from *Antennaria racemosa* (R. J. Bayer 1985b; Bayer and D. J. Crawford 1986).

15. Antennaria neglecta Greene, Pittonia 3: 173. 1897

- Field pussytoes, antennaire négligée [E]

Antennaria athabascensis Greene; *A. campestris* Rydberg; *A. campestris* var. *athabascensis* (Greene) B. Boivin; *A. chelonica* Lunell; *A. erosa* Greene; *A. howellii* Greene var. *athabascensis* (Greene) B. Boivin; *A. howellii* var. *campestris* (Rydberg) B. Boivin; *A. longifolia* Greene; *A. lunellii* Greene; *A. nebrascensis* Greene; *A. neglecta* Greene var. *athabascensis* (Greene) Roy L. Taylor & MacBryde; *A. neglecta* var. *campestris* (Rydberg) Steyermark; *A. neglecta* var. *simplex* Peck; *A. parvula* Greene; *A. wilsonii* Greene

Dioecious. Plants 4–25 cm. **Stolons** 2.5–18 cm. **Basal leaves:** 1-nerved, narrowly spatulate to cuneate-oblanceolate, 15–65 × 6–18 mm, tips mucronate, faces abaxially tomentose, adaxially gray-pubescent (green-glabrescent with age). **Cauline leaves** linear, 1.5–25 mm, distal flagged. **Heads** (1–)2–8 in corymbiform to spiciform or racemiform arrays. **Involucres:** staminate 4–7 mm; pistillate 6–10 mm. **Phyllaries** distally white. **Corollas:** staminate 2.7–5 mm; pistillate 4.5–6.5(–7) mm. **Cypselae** 0.9–1.4 mm, minutely papillate; **pappi:** staminate 3.5–6.5 mm; pistillate 6–8.5(–9.5) mm. **2n = 28.**

Flowering early–mid spring. Plains, grasslands, pastures, and open woodlands; 0–2500 m; Alta., B.C., Man., N.W.T., N.S., Ont., Que., Sask.; Ark., Colo., Conn., Del., Ill., Ind., Iowa, Kans., Ky., Maine, Md., Mass., Mich., Minn., Mo., Mont., Nebr., N.H., N.J., N.Y., N.Dak., Ohio, Okla., Pa., R.I., S.Dak., Vt., Va., W.Va., Wis., Wyo.

Antennaria neglecta is a sexual progenitor of both the *A. howellii* and *A. parvifolia* polyploid complexes and has one of the more widespread ranges among the amphimictic species in the genus in North America. Amphimicts generally have small ranges compared to those of the polyploid agamic complexes derived from them. Characteristic features of *A. neglecta* are its lashlike stolons that bear reduced leaves (except at the ends), flags on the distal cauline leaves, and basal leaves that are green-glabrescent with age (R. J. Bayer and G. L. Stebbins 1982).

16. Antennaria virginica Stebbins, Rhodora 37: 230, figs. 1, 2. 1935 • Shalebarren pussytoes [E]

Antennaria neglecta Greene var. *argillicola* (Stebbins) Cronquist; *A. neodioica* Greene var. *argillicola* (Stebbins) Fernald; *A. virginica* var. *argillicola* Stebbins

Dioecious. Plants 4–25 cm. **Stolons** 2–8 cm. **Basal leaves** 1-nerved, spatulate to cuneate-oblanceolate, 10–25 × 3–9 mm, tips mucronate, faces greenish gray, moderately pubescent. **Cauline leaves** linear, 4–20 mm, not flagged (apices acute). **Heads** 3–6(–9) in corymbiform arrays. **Involucres:** staminate 3.8–6 mm; pistillate 5–7 mm. **Phyllaries** distally white or stramineous. **Corollas:** staminate 2.2–3.5 mm; pistillate 2.8–4.5 mm. **Cypselae** 0.8–1.3 mm, slightly papillate; **pappi:** staminate 2.8–4(–5) mm; pistillate 3.5–5.2 mm. **2n = 28, 56.**

Flowering early–mid spring. Devonian shale barrens and argillaceous soils derived from them, open deciduous woods and fields; 300–600 m; Md., Ohio, Pa., Va., W.Va.

G. L. Stebbins (1936) and R. J. Bayer and Stebbins (1982) maintained that *Antennaria virginica* is a distinct species. After previously recognizing the taxon as a variety of *A. neglecta*, A. Cronquist (1945; H. A. Gleason and Cronquist 1991) agreed. It is a sexual progenitor of the *A. howellii* complex and is most closely related to *A. howellii* subsp. *neodioica* (Bayer 1985). *Antennaria virginica* is dioecious and is characterized by its relatively small, spatulate, basal leaves and subulate-tipped cauline leaves, which separate it from *A. neglecta* and the gynoecious *A. howellii* complex (Stebbins 1935; Bayer and Stebbins).

17. Antennaria howellii Greene, Pittonia 3: 174. 1897

- Howell's pussytoes, antennaire de Howell [E]

A. neglecta Greene subsp. *howellii* (Greene) Hultén; *A. neglecta* var. *howellii* (Greene) Cronquist; *A. neodioica* Greene subsp. *howellii* (Greene) R. J. Bayer

Gynoecious (staminate plants very uncommon). **Plants** (6–)8–35 cm (stems sometimes stipitate-glandular). **Stolons** 1–9(–12) cm. **Basal leaves** 1-nerved, spatulate to oblanceolate, spatulate-obovate, narrowly to broadly ovate, or cuneate-oblanceolate, 20–48(–65) × 2.5–20 mm, tips mucronate, faces abaxially tomentose, adaxially green-glabrous or gray-pubescent. **Cauline leaves** linear, 8–40 mm, distal sometimes flagged (apices acute). **Heads** 3–15 in corymbiform arrays. **Involucres:** staminate 6–6.5 mm; pistillate 6–11 mm. **Phyllaries** (bases sometimes rose) distally white, cream, or light brown. **Corollas:** staminate 3–4 mm; pistillate 3.5–6.5(–8) mm. **Cypselae** 0.8–2 mm, ± papillate; **pappi:** staminate 4–4.5 mm; pistillate 5.5–9 mm. **2***n* = 56, 84, 140 (under *A. neodioica*).

Subspecies 4 (4 in the flora): North America.

The *Antennaria howellii* (previously *A. neodioica*) polyploid complex is highly variable morphologically; four more or less distinct subspecies can be recognized within it. The sexual progenitors of the complex are *A. neglecta*, *A. plantaginifolia*, *A. racemosa*, and *A. virginica* (see R. J. Bayer 1985). *Antennaria marginata* may also be a minor contributor to the origins of the complex. A. Cronquist (H. A. Gleason and Cronquist 1991) included members of this complex in *A. neglecta*; I maintain, because these apomicts are of hybrid polyploid origin from among multiple sexual progenitors, they best not be included within the circumscription of any one sexual progenitor (Bayer 1989d).

1. Basal leaves green-glabrous adaxially.
  2. Basal leaves 1-nerved, 20–40 × 6–9 mm; distal cauline leaves flagged; phyllaries distally white or cream . . . .
    . . . . 17a. *Antennaria howellii* subsp. *canadensis*
  2. Basal leaves 1–3-nerved, 25–40 × 9–12 mm; distal cauline leaves not flagged; phyllaries distally light brown or white . . . .
    . . . . 17b. *Antennaria howellii* subsp. *howellii*
1. Basal leaves pubescent adaxially (sometimes glabrescent with age).
  3. Basal leaves spatulate to narrowly or broadly obovate (petiolate); stolons 3–8(–12) cm (leaves along stolons almost equal to those in rosettes at ends) . . 17c. *Antennaria howellii* subsp. *neodioica*
  3. Basal leaves cuneate-oblanceolate, spatulate, or spatulate-obovate (without distinct petioles); stolons 4–9 cm (leaves along stolons smaller than in rosettes at ends) . . . .
    . . . . 17d. *Antennaria howellii* subsp. *petaloidea*

17a. Antennaria howellii Greene subsp. **canadensis** (Greene) R. J. Bayer, Brittonia 41: 397. 1989
• Canadian pussytoes, antennaire du Canada   E

Antennaria canadensis Greene, Pittonia 3: 275. 1898; *A. canadensis* var. *randii* Fernald; *A. canadensis* var. *spathulata* Fernald; *A. neglecta* Greene var. *randii* (Fernald) Cronquist; *A. neodioica* Greene subsp. *canadensis* (Greene) R. J. Bayer & Stebbins; *A. neodioica* Greene var. *randii* (Fernald) B. Boivin; *A. spathulata* (Fernald) Fernald

Plants 15–35 cm (stems stipitate-glandular). **Stolons** 3–8 cm. **Basal leaves** 1-nerved, spatulate to oblanceolate, 20–40 × 6–9 mm, tips mucronate, faces abaxially tomentose, adaxially green-glabrous. **Cauline leaves** linear, 12–30 mm, distal flagged. **Heads** 3–7 in corymbiform arrays. **Involucres:** staminate unknown; pistillate 7–10 mm. **Phyllaries** (sometimes rose at bases) distally white or cream. **Corollas:** pistillate 4–6.5 mm. **Cypselae** 1–1.5 mm, papillate; **pappi:** pistillate 7–9 mm. **2***n* = 56, 84 (under *A. neodioica*).

Flowering mid spring–early summer. Pastures, dry fields, openings in woodlands and forests, and rock barrens; 0–1500 m; Man., N.B., Nfld. and Labr. (Nfld.), N.S., Ont., P.E.I., Que., Sask., Yukon; Conn., Del., Ind., Maine, Md., Mass., Mich., Minn., N.H., N.J., N.Y., Ohio, Pa., R.I., Vt., Va., W.Va., Wis.

Subspecies *canadensis* is almost restricted to the eastern half of North America. It is probably most closely related to *Antennaria racemosa* of the northern Rockies and *A. neglecta* of the Great Plains; see R. J. Bayer 1985).

17b. Antennaria howellii Greene subsp. **howellii**   E

Antennaria callilepis Greene; *A. exima* Greene

Plants 15–30 cm. **Stolons** 1–4 cm. **Basal leaves** 1–3-nerved, spatulate to oblanceolate, 25–40 × 9–12 mm, tips mucronate, faces abaxially tomentose, adaxially green-glabrous. **Cauline leaves** linear, 20–40 mm, not flagged. **Heads** 5–12 in corymbiform arrays. **Involucres:** staminate unknown; pistillate 6–7.5 mm. **Phyllaries** distally light brown or white. **Corollas:** pistillate 5–6 mm. **Cypselae** 1.5–2 mm, notably papillate; **pappi:** pistillate 6–8 mm. **2***n* = 56, 84, 140 (under *A. neodioica*).

Flowering mid spring–early summer. Pastures, dry fields, openings in deciduous woodlands and coniferous forests, and rock barrens; 0–2200 m; Alta., B.C., Ont.,

Sask., Yukon; Calif., Colo., Idaho, Minn., Mont., Oreg., S.Dak., Utah, Wash., Wis., Wyo.

Subspecies *howellii* is most common in the western half of the range of *Antennaria howellii*. Based on morphology, this group of apomicts is closely related to *A. racemosa* of the northern Rockies (R. J. Bayer 1985) and, perhaps, to *A. marginata* of the southern Rockies.

17c. Antennaria howellii Greene subsp. **neodioica** (Greene) R. J. Bayer, Brittonia 41: 397. 1989

• Antennaire néodioïque   E

Antennaria neodioica Greene, Pittonia 3: 184. 1897; *A. alsinoides* Greene; *A. grandis* (Fernald) House; *A. neglecta* Greene var. *attenuata* (Fernald) Cronquist; *A. neglecta* var. *neodioica* (Greene) Cronquist; *A. neodioica* var. *attenuata* Fernald; *A. neodioica* var. *chlorophylla* Fernald; *A. neodioica* var. *grandis* Fernald; *A. neodioica* var. *interjecta* Fernald; *A. neodioica* var. *rupicola* (Fernald) Fernald; *A. obovata* E. E. Nelson; *A. rupicola* Fernald

Plants (6–)15–35 cm. **Stolons** 3–8(–12) cm (leaves along stolons almost equal to those of rosettes at ends). **Basal leaves** 1-nerved (sometimes obscurely 3-nerved), spatulate to narrowly or broadly obovate (petiolate), 14–48 × 2.5–20 mm, tips mucronate, faces abaxially tomentose, adaxially gray-pubescent to floccose-glabrescent. **Cauline leaves** linear, 8–35 mm, distal flagged. **Heads** 4–8(–13) in corymbiform arrays. **Involucres:** staminate (very uncommon) 6 mm; pistillate 6–9 mm. **Phyllaries** distally white or cream **Corollas:** staminate 3.5 mm; pistillate 3.5–6 mm. **Cypselae** 0.9–1.5 mm, minutely papillate; **pappi:** staminate 4 mm; pistillate 5.5–7 mm. **2n** = 56, 84 (under *A. neodioica*).

Flowering mid spring–early summer. Pastures, dry fields, openings in woodlands and forests, and rock barrens and dry lake shores; 0–2200 m; St. Pierre and Miquelon; Alta., B.C., Man., N.B., Nfld. and Labr. (Nfld.), N.W.T., N.S., Ont., P.E.I., Que., Sask.; Colo., Conn., Del., Idaho, Ill., Ind., Iowa, Kans., Ky., Maine, Md., Mass., Mich., Minn., Mont., Nebr., N.H., N.J., N.Y., N.C., N.Dak., Ohio, Oreg., Pa., R.I., S.Dak., Tenn., Vt., Va., Wash., W.Va., Wis., Wyo.

Subspecies *neodioica* is most common in the eastern half of the range of *Antennaria howellii*; it is also found sporadically as far west as Washington and British Columbia. *Antennaria virginica* is likely the primary sexual progenitor of apomicts in subsp. *neodioica* (R. J. Bayer 1985).

17d. Antennaria howellii Greene subsp. **petaloidea** (Fernald) R. J. Bayer, Brittonia 41: 397. 1989

• Petaloid pussytoes, antennaire pétaloïde   E

Antennaria neodioica Greene var. *petaloidea* Fernald, Proc. Boston Soc. Nat. Hist. 28: 245. 1898; *A. appendiculata* Fernald; *A. concolor* Piper; *A. neglecta* Greene var. *petaloidea* (Fernald) Cronquist; *A. neglecta* var. *subcorymbosa* Fernald; *A. neodioica* Greene subsp. *petaloidea* (Fernald) R. J. Bayer & Stebbins; *A. neodioica* var. *petaloidea* Fernald; *A. pedicellata* Greene; *A. petaloidea* (Fernald) Fernald; *A. petaloidea* var. *novaboracensis* Fernald; *A. petaloidea* var. *scariosa* Fernald; *A. petaloidea* var. *subcorymbosa* (Fernald) Fernald; *A. stenolepis* Greene

Plants 8–35 cm. **Stolons** 4–9 cm (leaves along stolons smaller than in rosettes at ends). **Basal leaves** 1(–3)-nerved, cuneate-oblanceolate, spatulate, or spatulate-obovate (without distinct petioles), 15–45(–65) × 5–20 mm, tips mucronate, faces abaxially tomentose, adaxially gray-pubescent to floccose-glabrescent. **Cauline leaves** linear, 10–35 mm, mid and distal flagged. **Heads** 4–10 (–15) in corymbiform arrays. **Involucres:** staminate (uncommon) 6–6.5 mm; pistillate 7–11 mm. **Phyllaries** distally white or cream. **Corollas:** staminate 3–4 mm; pistillate 4–6.5(–8) mm. **Cypselae** 0.8–1.7 mm, minutely papillate; **pappi:** staminate 4–4.5 mm; pistillate 5.5–8 mm. **2n** = 56, 84 (under *A. neodioica*).

Flowering mid spring–early summer. Pastures, dry fields, openings in woodlands and forests, and rock barrens; 0–2200 m; St. Pierre and Miquelon; Alta., B.C., N.B., Nfld. and Labr. (Nfld.), N.S., Ont., P.E.I., Que.; Colo., Conn., Del., Ill., Ind., Maine, Md., Mass., Mich., Minn., Mont., N.H., N.J., N.Y., N.C., N.Dak., Ohio, Oreg., Pa., R.I., S.Dak., Vt., Va., Wash., W.Va., Wis., Wyo.

Subspecies *petaloidea* is most common in the eastern half of the range of *Antennaria howellii* and is frequent as far west as British Columbia and Washington. Its primary sexual progenitors include *A. plantaginifolia* and *A. neglecta* (R. J. Bayer 1985).

18. Antennaria marginata Greene, Pittonia 3: 290. 1898 • Whitemargin pussytoes

Antennaria dioica (Linnaeus) Gaertner var. *marginata* (Greene) Jepson; *A. fendleri* Greene; *A. marginata* var. *glandulifera* A. Nelson; *A. peramoena* Greene

Dioecious or gynoecious (staminate plants in equal frequency as pistillates or none in populations, respectively). **Plants** 5–20 cm (stems sometimes stipitate-glandular, especially in dioecious diploids). **Stolons** 2–7 cm (woolly). **Basal leaves** 1–3-nerved, spatulate, 15–20 × 4–6 mm, tips mucronate, abaxial faces gray-tomentose, adaxial green-glabrous (margins white woolly). **Cauline leaves** linear, 7–16 mm, (apices acute) not flagged. **Heads** 5–8 in corymbiform arrays. **Involucres:** staminate 4.5–7 mm; pistillate 5–7(–9) mm. **Phyllaries** (relatively wide), distally white (apices acuminate). **Corollas:** staminate 3–5 mm; pistillate 4.5–6.5 mm. **Cypselae** 0.8–2 mm, glabrous or slightly papillate; **pappi:** staminate 3.5–5.5 mm; pistillate 5.5–8.5 mm. $2n$ = 28, 56, 84, 112, 140.

Flowering summer. Moist forests, slopes and tops of ridges under Douglas fir, ponderosa pine, Engelmann spruce or Gambel oaks, openings in the forests; 1500–2900 m; Ariz., Calif., Colo., N.Mex., Tex.; Mexico (Chihuahua, Coahuila).

Antennaria marginata has rims of white hairs (from the abaxial faces) around its adaxially glabrous leaves. It has both dioecious and gynoecious populations and cytotypes ranging from diploid to decaploid (R. J. Bayer and G. L. Stebbins 1987). It is probably a primary sexual progenitor of the *A. parvifolia* polyploid complex; the two taxa sometimes overlap morphologically; they differ in induments of basal leaves. *Antennaria marginata* may also be a contributor to the parentage of the *A. howellii* and *A. rosea* agamic complexes.

19. **Antennaria dioica** (Linnaeus) Gaertner, Fruct. Sem. Pl. 2: 410. 1791 • Stoloniferous pussytoes

Gnaphalium dioicum Linnaeus, Sp. Pl. 2: 850. 1753; *Antennaria hyperborea* D. Don; *A. insularis* Greene

Dioecious. Plants 3–10 cm. **Stolons** 2–5 cm. **Basal leaves** 1-nerved, spatulate or rhombic-spatulate, 3–18 × 3–6 mm, tips mucronate, abaxial faces gray-tomentose, adaxial green-glabrous. **Cauline leaves** linear, 7–13 mm, not flagged (apices acute). **Heads** 3–7 in corymbiform arrays. **Involucres:** staminate 5–6.5 mm; pistillate 5–7 mm. **Phyllaries** distally dark pink to light pink or white. **Corollas:** staminate 3–4 mm; pistillate 4–5 mm. **Cypselae** 0.5–1 mm, papillate; **pappi:** staminate 3.5–4.5 mm; pistillate 5–6 mm. $2n$ = 28.

Flowering summer. Dry slopes on tundra; 0–600 m; Alaska (Aleutian Islands); Eurasia.

Antennaria dioica ranges from the British Isles to Japan and into the Aleutian Islands (R. J. Bayer 2000). It is characterized by glabrous adaxial leaf faces and distally pink or white phyllaries. The circumscription of *A. dioica* in North America has long been debated; *A. marginata* of southwestern states bears a remarkable similarity to *A. dioica*. DNA sequence data (Bayer et al. 1996) indicate that the two taxa are not sisters; they are only

distantly related. They are allopatric. *Antennaria dioica* may be a sexual progenitor of the *A. parvifolia* complex.

20. **Antennaria parvifolia** Nuttall, Trans. Amer. Philos. Soc., n. s. 7: 406. 1841 • Small-leaf or Nuttall's pussytoes

Antennaria aprica Greene; *A. aprica* var. *aureola* (Lunell) J. W. Moore; *A. aprica* var. *minuscula* (B. Boivin) B. Boivin; *A. aureola* Lunell; *A. dioica* (Linnaeus) Gaertner var. *parvifolia* (Nuttall) Torrey & A. Gray; *A. holmii* Greene; *A. latisquamea* Greene; *A. minuscula* B. Boivin; *A. recurva* Greene; *A. rhodantha* Suksdorf

Dioecious or gynoecious (staminate plants uncommon or in equal frequency as pistillates, respectively). **Plants** 2–8(–15) cm. **Stolons** 1–6 cm. **Basal leaves** 1-nerved, narrowly spatulate to spatulate or oblanceolate, 8–35 × 2–15 mm, tips mucronate, faces gray-tomentose. **Cauline leaves** linear to narrowly oblanceolate, 8–20 mm, not flagged (apices acute). **Heads** 2–7 in corymbiform arrays. **Involucres:** staminate 5.5–7.5 mm; pistillate 8–10(–15) mm (gynoecious), 7–7.2 mm (dioecious). **Phyllaries** distally white, pink, green, red, or brown. **Corollas:** staminate 3.5–4.5 mm; pistillate 5–8 mm. **Cypselae** 1–1.8 mm, glabrous or minutely papillate; **pappi:** staminate 4–5.5 mm; pistillate 6.5–9 mm. $2n$ = 56, 84, 112, 140.

Flowering late spring–summer. Prairies, pastures, roadsides, mountain parks, open deciduous woods, and drier coniferous forests, usually ponderosa or lodgepole pine; 100–3400 m; Alta., B.C., Man., Ont., Sask.; Ariz., Colo., Idaho, Iowa, Mich., Minn., Mont., Nebr., Nev., N.Mex., N.Dak., Okla. (expected in panhandle), Oreg., S.Dak., Tex., Utah, Wash., Wis. (expected), Wyo.; Mexico (Chihuahua, Nuevo León).

Antennaria parvifolia is a widespread, polyploid complex of sexual (dioecious) and asexual (gynoecious) populations (G. L. Stebbins 1932b; R. J. Bayer and Stebbins 1987). Although variable morphologically, no infraspecific taxa seem warranted at this time. Sexual (dioecious) populations are known primarily from New Mexico and Colorado; apomictic plants occur throughout the range of the species. Probable sexual diploid/tetraploid progenitors of the *A. parvifolia* complex include *A. dioica*, *A. marginata*, *A. neglecta*, and *A. pulchella*/*A. media*. *Antennaria parvifolia* is characterized by relatively short stature and relatively small numbers of relatively large heads. The epithet *parvifolia* has been rendered as "*parviflora*" in floras, e.g., key in Great Plains Flora Association (1986); E. H. Moss (1959); H. J. Scoggan (1978–1979, part 4). In some floras, *A. parvifolia* has been confused with *A. microphylla*; the two are probably not closely related.

21. **Antennaria corymbosa** E. E. Nelson, Bot. Gaz. 27: 212. 1899 • Flat-top or meadow pussytoes [E]

Antennaria acuta Rydberg; *A. dioica* (Linnaeus) Gaertner var. *corymbosa* (E. E. Nelson) Jepson; *A. hygrophila* Greene; *A. nardina* Greene

Dioecious. Plants 6–15 cm. **Stolons** 1–10 cm. **Basal leaves** 1-nerved, spatulate, 18–45 × 2–4 mm, tips mucronate, faces ± gray-tomentose. **Cauline leaves** linear, 8–13 mm, not flagged (apices acuminate). **Heads** 3–7 in corymbiform arrays. **Involucres:** staminate 4–5.3 mm; pistillate 4–5 mm. **Phyllaries** (bases each with distinct dark brown or blackish spot) distally white or light brown. **Corollas:** staminate 2–3.2 mm; pistillate 2.5–3.5 mm. **Cypselae** 0.5–1 mm, slightly papillate; **pappi:** staminate 2.5–3.5 mm; pistillate 3.5–4.5 mm. $2n = 28$.

Flowering early–mid summer. Moist subalpine-alpine willow thickets in the Rocky and Cascade mountains, the Sierra Nevada and mountains of the Great Basin; 1900–3500 m; Calif., Colo., Idaho, Mont., Nev., N.Mex., Oreg., Utah, Wash., Wyo.

Antennaria corymbosa is characterized by linear-oblanceolate basal leaves and white-tipped phyllaries, each with a distinct black spot near the base of the scarious portion. A form with black phyllaries (*A. acuta*) occurs sporadically throughout the range of the species (R. J. Bayer 1988). *Antennaria corymbosa* is a sexual progenitor of the *A. rosea* complex.

22. **Antennaria microphylla** Rydberg, Bull. Torrey Bot. Club 24: 303. 1897 • Littleleaf pussytoes [E] [F]

Antennaria bracteosa Rydberg; *A. microphylla* Lunell var. *solstitialis* Lunell; *A. nitida* Greene; *A. rosea* Greene var. *nitida* (Greene) Breitung; *A. solstitialis* Lunell

Dioecious. Plants 9–30 cm (stems stipitate-glandular distally). **Stolons** 1–5 cm. **Basal leaves** 1-nerved, spatulate, 6–16 × 2–6 mm, tips mucronate, faces silvery gray-pubescent. **Cauline leaves** linear, 5–25 mm, not flagged (apices acute). **Heads** 6–13 in corymbiform arrays. **Involucres:** staminate 5–6.5 mm; pistillate 5.5–7 mm. **Phyllaries** distally bright white to light yellow. **Corollas:** staminate 2.5–3 mm; pistillate 3–4.3 mm. **Cypselae** 0.7–1.2 mm, glabrous or sparingly papillate; **pappi:** staminate 3–4 mm; pistillate 3–5 mm. $2n = 28$.

Flowering early–mid summer. Moist open areas, flood plains of streams, margins of alkaline depressions, lower

montane to subalpine (subarctic); 0–3200 m; Alta., B.C., Man., N.W.T., Nunavut, Ont., Que., Sask., Yukon; Alaska, Ariz., Calif., Colo., Idaho, Minn., Mont., Nebr., Nev., N.Mex., N.Dak., Oreg., S.Dak., Utah, Wash., Wyo.

Antennaria microphylla is a primary sexual progenitor of the *A. rosea* polyploid agamic complex (R. J. Bayer 1990b). A. Cronquist (1955) included *A. rosea* within his circumscription of *A. microphylla*. It is preferable to recognize sexual diploids as distinct from their morphologically discrete hybrid apomictic derivatives. *Antennaria microphylla* is always dioecious and has stems distally stipitate-glandular and white phyllaries; *A. rosea* is always gynoecious and has stems without glandular hairs and phyllaries only occasionally white.

Some authors (A. E. Porsild 1950; E. H. Moss 1959; Porsild and W. J. Cody 1980) have recognized *A. nitida* as distinct; comparisons of the nomenclatural types of the two show that they are conspecific. *Antennaria microphylla* has allelopathic properties (G. D. Manners and D. S. Galitz 1985).

23. **Antennaria rosulata** Rydberg, Bull. Torrey Bot. Club 24: 300. 1897 • Kaibab or woolly pussytoes [E]

Antennaria sierrae-blancae Rydberg

Dioecious. Plants 0.2–1.5(–2) cm. **Stolons** 1–2(–3.5) cm. **Basal leaves** 1-nerved, spatulate, spatulate-obovate, or oblanceolate, 6.5–13 × 2–5 mm, tips mucronate, faces silvery gray-pubescent (often obscurely stipitate-glandular). **Cauline leaves** linear, 2–9 mm, not flagged (apices acute). **Heads** usually borne singly (rarely 2–3; subsessile among basal leaves). **Involucres:** staminate 5–7.5 mm; pistillate 6–10 mm. **Phyllaries** distally white. **Corollas:** staminate 2.5–4.5 mm; pistillate 3.5–5.5 mm. **Cypselae** 0.8–1.5 mm, papillate (bases puberulent); **pappi:** staminate 3.5–5 mm; pistillate 5.5–6.5 mm. $2n = 28$.

Flowering summer. Open slopes and dry meadows, lower montane to montane, or subalpine zone, usually with big sagebrush, *Artemisia tridentata*; 2200–3300 m; Ariz., Colo., N.Mex., Utah.

Antennaria rosulata is easily recognizable by its silvery gray leaves, dense, humifuse growth form, and heads borne singly (R. J. Bayer 1987b). Its distribution is centered on the four corners area (Bayer and G. L. Stebbins 1987). It has probably contributed to the origins of some of the clones of *A. rosea* with low stature and low numbers of flowering heads that are found in Arizona, Colorado, New Mexico, and Utah.

24. Antennaria umbrinella Rydberg, Bull. Torrey Bot. Club 24: 302. 1897 • Umber or brown or brown-bracted pussytoes [E] [F]

Antennaria aizoides Greene; *A. flavescens* Rydberg; *A. reflexa* E. E. Nelson

Dioecious. Plants 7–16 cm (bases somewhat woody). **Stolons** 7–16 cm (usually erect, slightly woody). **Basal leaves** 1-nerved, narrowly spatulate to cuneate, 10–17 × 2–5.4 mm, tips mucronate, faces gray-tomentose. **Cauline leaves** linear, 8–18 mm, not flagged (apices acute). **Heads** 3–8 in corymbiform arrays. **Involucres:** staminate 3–6 mm; pistillate 4–6.5 mm. **Phyllaries** distally whitish, yellowish, or pale brownish (often streaked with pink or rose). **Corollas:** staminate 2.5–3.5 mm; pistillate 2.5–3.5 mm. **Cypselae** 0.5–1.2 mm, glabrous; **pappi:** staminate 3–4.5 mm; pistillate 3–5 mm. 2*n* = 28, 56.

Flowering summer. Sagebrush steppe to open, dry, coniferous montane forests to subalpine meadows; 1100–3400 m; Alta., B.C., Sask.; Ariz., Calif., Colo., Idaho, Mont., Nev., Oreg., Utah, Wash., Wyo.

Antennaria umbrinella is a primary sexual progenitor of the *A. rosea* complex (R. J. Bayer 1990b). It is characterized by somewhat erect, slightly woody stolons and phyllaries that are usually various shades of brown, sometimes white, or streaked with pink or rose (Bayer 1987b).

25. Antennaria suffrutescens Greene, Pittonia 3: 277. 1898 • Evergreen or everlasting pussytoes [E]

Dioecious. Plants 5–12 cm (densely tufted, bases woody; root crowns relatively slender). **Stolons** none. **Basal leaves** absent at flowering. **Cauline leaves** spatulate, 5–12 × 2–4 mm, not flagged (apices emarginate or obtuse, abaxial faces tomentose, adaxial green). **Heads** borne singly. **Involucres:** staminate 5–9 mm; pistillate 10–15 mm. **Phyllaries** (relatively wide) distally white. **Corollas:** staminate 4–5 mm; pistillate 5–8 mm. **Cypselae** 1–2 mm, papillate; **pappi:** staminate 4.5–5.5 mm; pistillate 7–9 mm. 2*n* = 28.

Flowering early summer. Dry, open coniferous woods or barren slopes on serpentine; 500–1600 m; Calif., Oreg.

Antennaria suffrutescens is characterized by suffrutescent growth form, relatively small, emarginate, adaxially glabrous, coriaceous leaves, and relatively large heads borne singly. It is known only from serpentine soils in open montane pine forests in Curry and Josephine counties, Oregon, and neighboring Del Norte and Humboldt counties, California (R. J. Bayer and

G. L. Stebbins 1987). *Antennaria suffrutescens* may have contributed to the origin of some of the clones of the *A. rosea* complex (e.g., *J. T. Howell 27718*, NY).

26. Antennaria aromatica Evert, Madroño 31: 109, fig. 1. 1984 • Scented or aromatic pussytoes [E]

Dioecious. Plants 2–7 cm (stems stipitate-glandular). **Stolons** 0.5–2.5 cm. **Basal leaves** 1-nerved, usually cuneate-spatulate, sometimes oblanceolate, 5–16 × 3–10 mm, tips mucronate, faces gray-pubescent (and stipitate-glandular; fresh leaves citronella scented). **Cauline leaves** linear, 3–14 mm, not flagged (apices acute). **Heads** borne singly or 2–5 in corymbiform arrays. **Involucres:** staminate 4.5–6.5 mm; pistillate 5–7(–9) mm. **Phyllaries** distally light brown, dark brown, or olivaceous. **Corollas:** staminate 2.5–3 mm; pistillate 3.5–4.5 mm. **Cypselae** 0.9–2 mm, sparingly papillate; **pappi:** staminate 3–4 mm; pistillate 4.5–5.5 mm. 2*n* = 28, 56, 84.

Flowering mid summer. Subalpine limestone talus; 1600–3000 m; Alta.; Idaho, Mont., Wyo.

Known only from the northern Rockies, *Antennaria aromatica* is characterized by glandulosity, cuneate leaves, and odor of citronella in crushed leaves of living material. It is most closely related to *A. densifolia* of the Northwest Territories and Yukon (R. J. Bayer 1989c). Some collections of pistillate plants from Colorado and other areas of the Rockies superficially resemble *A. aromatica* and undoubtedly have *A. aromatica* in their parentage. They are non-glandular and odorless and are closer to the type of *A. pulvinata*, which is included in the circumscription of *A. rosea*, as *A. rosea* subsp. *pulvinata* (Bayer). *Antennaria aromatica* is a sexual progenitor of the *A. rosea* and *A. alpina* polyploid complexes.

27. Antennaria rosea Greene, Pittonia 3: 281. 1898 • Rosy pussytoes, antennaire rosée [E]

Gynoecious (staminate plants uncommon). **Plants** 4–30 cm. **Stolons** 1–7 cm. **Basal leaves** 1-nerved, 8–40 × 2–10 mm, spatulate, oblanceolate, or cuneate, tips mucronate, faces usually gray-pubescent, adaxial sometimes green-glabrous. **Cauline leaves** linear, 6–36 mm, usually not flagged (apices acute to subulate or with lanceolate flags). **Heads** 3–20 in corymbiform arrays. **Involucres:** staminate unknown; pistillate 4–10 mm. **Phyllaries** distally brown, cream, gray, green, pink, red, white, or yellow (apices acute or erose-obtuse). **Corollas:** staminate unknown; pistillate 2.5–6 mm. **Cypselae**

0.7–1.8 mm, glabrous or papillate; **pappi:** staminate unknown; pistillate 3.5–6.5 mm. $2n$ = 42, 56, (70).

Subspecies 4 (4 in the flora): North America.

Antennaria rosea is the most widespread *Antennaria* of North America, occurring in dry to moist habitats from near sea level to the alpine zone. The *A. rosea* polyploid agamic complex is one of the more morphologically diverse complexes of North American *Antennaria*. It occurs from the western cordillera of North America from southern California, Arizona, and New Mexico north to subarctic Alaska and east to Greenland and, disjunctly, in the Canadian maritime provinces, eastern Quebec, and immediately north of and adjacent to Lake Superior (R. J. Bayer et al. 1991). *Antennaria chilensis* (including *A. chilensis* var. *magellanica*) is a Patagonian endemic that morphologically fits within the circumscription of *A. rosea* and may well be an amphitropical disjunct member of the complex.

Antennaria rosea is taxonomically confusing; it includes agamospermous microspecies that have been recognized as distinct taxonomic species. Morphometric and isozyme analyses have demonstrated that the primary source of morphologic variability in the complex derives from six sexually reproducing progenitors, *A. aromatica*, *A. corymbosa*, *A. pulchella*, *A. microphylla*, *A. racemosa*, and *A. umbrinella* (R. J. Bayer 1989b, 1990b, 1990c). Additionally, three other sexually reproducing species, *A. marginata*, *A. suffrutescens*, and *A. rosulata*, may have contributed to the genetic complexity of the *A. rosea* complex (Bayer 1990b). Here, four reasonably distinct subspecies are recognized within the complex.

1. Basal leaves 20–40 mm; phyllaries distally usually green, pink, red or white, seldom brown . . . . 27d. *Antennaria rosea* subsp. *rosea*
1. Basal leaves 8–20 mm; phyllaries distally brown, cream, gray, green, pink, red, white, or yellow.
  2. Pistillate: involucres 4–6.5 mm, corollas 2.5–4, pappi 3.5–5; cauline leaves 6–20 mm (tips subulate); phyllaries usually distally brown, sometimes cream, gray, or yellow . . . . 27b. *Antennaria rosea* subsp. *confinis*
  2. Pistillate: involucres 6.5–10 mm, corollas 3.5–6 mm, pappi 5–6.5 mm; cauline leaves 6–19 or 9–26 mm (tips sometimes with flat, lanceolate scarious appendages); phyllaries distally brown, green, pink, red, or white.
    3. Plants 19–30 cm; cauline leaves 9–26 mm (proximalmost usually 19+ mm); heads usually 6–12 . . . 27a. *Antennaria rosea* subsp. *arida*
    3. Plants 4–17 cm; cauline leaves 6–19 mm (proximalmost usually less than 19 mm); heads usually 3–5 . . . . 27c. *Antennaria rosea* subsp. *pulvinata*

27a. Antennaria rosea Greene subsp. **arida** (E. E. Nelson) R. J. Bayer, Brittonia 41: 57. 1989 • Desert pussytoes E

Antennaria arida E. E. Nelson, Bot. Gaz. 27: 210. 1899; *A. arida* subsp. *viscidula* E. E. Nelson; *A. scariosa* E. E. Nelson; *A. viscidula* (E. E. Nelson) A. Nelson ex Rydberg

Plants 19–30 cm. **Stolons** 1.5–4.5 cm. **Basal leaves** spatulate to narrowly cuneate, 10–20 mm, faces gray-pubescent. **Cauline leaves** 9–26 mm. **Heads** usually 6–12. **Involucres:** pistillate 6.5–8 mm. **Phyllaries** distally white, pink, green, red, or brown. **Corollas:** pistillate 3.5–6 mm. **Pappi:** pistillate 5–6 mm. $2n$ = 42, 56, (70).

Flowering summer. Dry to moist habitats, tundra, rock outcrops, fields, meadows, forests, savannas, and roadcuts, other similarly disturbed places; 0–3800 m; Alta., B.C., Nfld. and Labr. (Nfld.), Que., Sask., Yukon; Alaska, Ariz., Calif. (unconfirmed), Colo., Idaho, Maine, Mont., Nev., N.Mex., Oreg., S.Dak., Utah, Wash., Wyo.

Subspecies *arida* is most closely related to *Antennaria microphylla* (R. J. Bayer 1989e), as shown by their similar morphologies.

27b. Antennaria rosea Greene subsp. **confinis** (Greene) R. J. Bayer, Brittonia 41: 57. 1989 E

Antennaria confinis Greene, Pittonia 4: 40. 1899; *A. affinis;* Fernald, *A. albicans* Fernald; *A. angustifolia* Rydberg; *A. arida* E. E. Nelson var. *humilis* (Rydberg) E. E. Nelson; *A. breitungii* A. E. Porsild; *A. brevistyla* Fernald; *A. concinna* E. E. Nelson; *A. dioica* (Linnaeus) Gaertner var. *kernensis* Jepson; *A. elegans* A. E. Porsild; *A. foliacea* Greene var. *humilis* Rydberg; *A. incarnata* A. E. Porsild; *A. laingii* A. E. Porsild; *A. leontopodioides* Cody; *A. leuchippii* Porsild; *A. polyphylla* Greene ex C. F. Baker (name published without description); *A. rosea* var. *angustifolia* (Rydberg) E. E. Nelson; *A. sedoides* Greene; *A. sordida* Greene; *A. subviscosa* Fernald; *A. tomentella* E. E. Nelson

Plants 9–25 cm. **Stolons** 1.5–4.5 cm. **Basal leaves** spatulate to cuneate, 10–20 mm, faces gray-pubescent. **Cauline leaves** 6–20 mm. **Heads** 4–11. **Involucres:** pistillate 4–6.5 mm. **Phyllaries** distally usually brown, sometimes cream, gray, or pale yellow. **Corollas:** pistillate 2.5–4 mm. **Pappi:** pistillate 3.5–5 mm. $2n$ = 42, 56, (70).

Flowering summer. Dry to moist habitats, tundra, rock outcrops, fields, meadows, forests, savannas, and roadcuts, other similarly disturbed places; 0–3800 m; Greenland; Alta., B.C., Nfld. and Labr. (Labr.), N.W.T., Nunavut, Ont., Que., Yukon; Alaska, Ariz., Calif., Colo.,

Idaho, Mont., Nev., N.Mex. (expected), Oreg., Utah, Wash., Wyo.

Subspecies *confinis* is closely related to *Antennaria pulchella* and *A. umbrinella* (R. J. Bayer 1989e), as shown by its relatively small basal leaves and smallish heads that usually have dark phyllaries.

27c. Antennaria rosea Greene subsp. **pulvinata** (Greene) R. J. Bayer, Brittonia 41: 59. 1989   • Pulvinate pussytoes E

Antennaria pulvinata Greene, Pittonia 3: 287. 1898; *A. albescens* (E. E. Nelson) Rydberg; *A. fusca* E. E. Nelson; *A. gaspensis* (Fernald) Fernald; *A. howellii* Greene subsp. *gaspensis* (Fernald) Chmielewski; *A. isolepis* Greene; *A. maculata* Greene; *A. manicouagana* P. Landry; *A. media* Greene subsp. *fusca* (E. E. Nelson) Chmielewski; *A. neglecta* Greene var. *gaspensis* (Fernald) Cronquist; *A. neodioica* Greene var. *gaspensis* Fernald; *A. peasei* Fernald; *A. pulvinata* Greene subsp. *albescens* E. E. Nelson; *A. sansonii* Greene; *A. straminea* Fernald

Plants 4–17 cm. **Stolons** 1–6 cm. **Basal leaves** spatulate to cuneate, 8–18 mm, faces gray-pubescent. **Cauline leaves** 6–19 mm. **Heads** 3–5. **Involucres:** pistillate 6.5–10 mm. **Phyllaries** distally brown, green, pink, red, or white. **Corollas:** pistillate 3.5–5 mm. **Pappi:** pistillate 5–6.5 mm. 2*n* = 42, 56.

Flowering summer. Dry to moist open habitats, usually on rock outcrops or barrens; 0–3800 m; Alta., B.C., Man., Nfld. and Labr., N.W.T., Nunavut, Que., Sask., Yukon; Alaska, Ariz., Calif., Colo., Idaho, Maine, Mont., Nev., Oreg., Utah, Wash., Wyo.

Antennaria aromatica is undoubtedly a sexual progenitor of *A. rosea* subsp. *pulvinata* (R. J. Bayer 1989e), as evidenced by its comparatively short stature and relatively low number of relatively large heads.

27d. Antennaria rosea Greene subsp. **rosea** E

Antennaria acuminata Greene; *A. alborosea* A. E. Porsild; *A. chlorantha* Greene; *A. formosa* Greene; *A. hendersonii* Piper; *A. imbricata* E. E. Nelson; *A. lanulosa* Greene; *A. neodioica* Greene var. *chlorantha* (Greene) B. Boivin; *A. oxyphylla* Greene; *A. rosea* subsp. *divaricata* E. E. Nelson; *A. speciosa* E. E. Nelson

Plants 10–40 cm. **Stolons** 2–7 cm. **Basal leaves** spatulate, oblanceolate, or cuneate, 20–40 mm, faces usually gray-pubescent, adaxial sometimes green-glabrous. **Cauline leaves** 8–36 mm. **Heads** 6–20. **Involucres:** pistillate 5–8 mm. **Phyllaries** distally usually green, pink,

red, or white, seldom brown. **Corollas:** pistillate 3–4.5 mm. **Pappi:** pistillate 4–6 mm. 2*n* = 42, 56.

Flowering summer. Dry to moist habitats, tundra, rock outcrops, fields, meadows, forests, savannas, and roadcuts, other similarly disturbed places; 400–3800 m; Alta., B.C., Man., N.W.T., Ont., Sask., Yukon; Alaska, Calif., Colo., Idaho, Mich., Minn., Mont., N.Dak., Oreg. Utah, Wash., Wyo.

Subspecies *rosea* is most closely related to *Antennaria corymbosa* and *A. racemosa* (R. J. Bayer 1989e), as shown by its relatively long basal leaves ranging from gray-pubescent to adaxially green-glabrous.

Subspecies *rosea* is expected to occur in Nevada.

28. Antennaria soliceps S. F. Blake, Proc. Biol. Soc. Wash. 51: 7. 1938   • Charleston Mountain or Charleston pussytoes C E

Gynoecious (staminate plants unknown). **Plants** 1–4 cm. **Stolons** 0.5–2 cm. **Basal leaves** 1-nerved, spatulate, rhombic-spatulate, or cuneate, 4–13 × 2–8 mm, tips mucronate, faces densely gray-tomentose. **Cauline leaves** linear, 4–10 mm, distalmost flagged. **Heads** usually borne singly, rarely 2–3 in corymbiform arrays. **Involucres** staminate unknown; pistillate 8–11 mm. **Phyllaries** distally white, light brown, dark brown, or olivaceous. **Corollas:** staminate unknown; pistillate 4–5.5 mm. **Cypselae** 1.5–1.8 mm, glabrous; **pappi:** staminate unknown; pistillate 5–6 mm. 2*n* = ca. 168.

Flowering summer. Talus areas on limestone ridge at treeline in the subalpine zone; 3000–3400 m; Nev.

Antennaria soliceps is a high-polyploid apomict known only from limestone talus at treeline in the Spring (Charleston) Mountains, Nevada (R. J. Bayer and T. M. Minish 1993). It is probably most closely related to *A. aromatica*, an amphimictic species occurring in the northern Rockies, and is characterized by a cushion-plant growth form and heads borne singly (Bayer and Minish).

29. Antennaria pulchella Greene, Leafl. Bot. Observ. Crit. 2: 149. 1911   • Sierra pussytoes E

Antennaria alpina (Linnaeus) Gaertner var. *scabra* (Greene) Jepson; *A. media* Greene subsp. *ciliata* E. E. Nelson; *A. media* subsp. *pulchella* (Greene) Chmielewski; *A. scabra* Greene

Dioecious. Plants (1–)3–12 cm (stems usually stipitate-glandular). **Stolons** 1–4(–9) cm. **Basal leaves** 1-nerved, spatulate to linear-cuneate, 6–12 × 1.5–4.5 mm, tips mucronate, faces glabrescent-scabrous to gray-pubescent (often with purple glandular hairs). **Cauline**

leaves linear, 3–11(–13) mm, usually not flagged (apices acute to acuminate), rarely distal flagged. **Heads** 4–6 in corymbiform arrays. **Involucres:** staminate 4–5 mm; pistillate 3.5–4.5 mm. **Phyllaries** (relatively wide) distally dark brown-black (sometimes light brown or whitish at very tips; apices blunt). **Corollas:** staminate 1.9–2.8 mm; pistillate 2–3 mm. **Cypselae** 0.7–1.3 mm, glabrous or slightly papillate; **pappi:** staminate 2.5–3.5 mm; pistillate 2.5–3.5 mm. **2***n* = 28 (as *A. media*).

Flowering summer. Moist subalpine-alpine meadows, snow basins, margins of tarns, streams, or run-off from snow masses; 2800–3700 m; Calif., Nev.

Antennaria pulchella is the diploid progenitor of *A. media* and, consequently, a progenitor of the *A. alpina* complex (R. J. Bayer 1990d). The *A. rosea* and *A. parvifolia* complexes also have the genome of *A. pulchella*, shown in the high elevation clones with dark phyllaries in these two polyploid complexes. *Antennaria pulchella* is differentiated from *A. media* by shorter pistillate or staminate corollas and shorter cauline leaves (Bayer). This sexually reproducing diploid ranges from the area around Lake Tahoe to the Mt. Whitney region (Bayer).

30. Antennaria media Greene, Pittonia 3: 286. 1898 • Rocky Mountain pussytoes [E]

Antennaria alpina (Linnaeus) Gaertner var. *media* (Greene) Jepson; *A. austromontana* E. E. Nelson; *A. candida* Greene; *A. densa* Greene; *A. modesta* Greene; *A. mucronata* E. E. Nelson

Dioecious or gynoecious (staminate plants rare or in equal frequency to pistillates, respectively). **Plants** 5–13 cm. **Stolons** 1–4 cm. **Basal leaves** 1-nerved, spatulate to oblanceolate, 6–19 × 2.5–6 mm, tips mucronate, faces gray-pubescent. **Cauline leaves** linear, 5–20 mm, not flagged (apices acute). **Heads** 2–5(–9) in corymbiform arrays. **Involucres:** staminate (3.5–)4.5–6.5 mm; pistillate 4–8 mm. **Phyllaries** distally dark brown, black, or olivaceous. **Corollas:** staminate 2.5–4.5 mm; pistillate 3–4.5 mm. **Cypselae** 0.6–1.6 mm, glabrous or papillate; **pappi:** staminate 2.5–4.5 mm; pistillate 4–5.5 mm. **2***n* = 56, 98, 112.

Flowering summer. Dry, rocky to moist alpine tundra; 1500–3800 m; Alta., B.C., N.W.T., Yukon; Alaska, Ariz., Calif., Colo., Idaho, Mont., Nev., N.Mex., Oreg., Utah, Wash., Wyo.

Antennaria media ranges from Arizona to Alaska; dioecious and gynoecious populations are encountered (R. J. Bayer and G. L. Stebbins 1987). The dioecious (sexual) populations are restricted primarily to California and Oregon (Bayer et al. 1990). The main distinction between *A. media* and *A. alpina* is flags on distal

cauline leaves present in *A. alpina* and mostly absent in *A. media* (Bayer 1990d). Phyllaries of the pistillate plants in *A. alpina* tend to be acute; they are blunter in *A. media*. At some point, it may be preferable to follow W. L. Jepson ([1923–1925]) and some later authors and treat *A. media* as a subspecies of *A. alpina*. *Antennaria media* appears to be an autopolyploid derivative of *A. pulchella*; genes from *A. pulchella* may have introgressed into the *A. alpina* and *A. parvifolia* complexes indirectly through *A. media*.

31. Antennaria densifolia A. E. Porsild, Bull. Natl. Mus. Canada 101: 26. 1945 • Denseleaf pussytoes [E]

Antennaria ellyae A. E. Porsild

Dioecious. Plants 3.5–16 cm. **Stolons** 1–2 cm. **Basal leaves** 1-nerved, spatulate to cuneate, 3–7 × 2–5 mm, tips mucronate, faces gray-tomentose. **Cauline leaves** linear, 2–13 mm, distal flagged. **Heads** 2–5 in corymbiform arrays. **Involucres:** staminate 3–6.5 mm; pistillate 4.5–7.5 mm. **Pistillate involucres** 4.5–7.5 mm. **Phyllaries** distally light brown, dark brown, or black. **Corollas:** staminate 2–3.5 mm; pistillate 2.5–4.5 mm. **Cypselae** 0.8–1.5 mm, glabrous; **pappi:** staminate 2.5–3.5 mm; pistillate 2.5–3.5 mm. **2***n* = 28.

Flowering summer. Subalpine-alpine limestone talus; 700–2800 m; B.C., N.W.T., Yukon; Alaska, Mont.

Antennaria densifolia is found on limestone talus below treeline in the MacKenzie, Richardson, and Ogilvie mountains of the District of MacKenzie and Yukon Territory and in Granite County, Montana (R. J. Bayer 1989c). It differs from *A. aromatica* in being non-glandular and in other characters. Herbarium specimens (in DAO) from British Columbia that morphologically appear to be a strictly gynoecious form of *A. densifolia* may be apomicts related to *A. alpina* that are derived from *A. densifolia*, a sexual progenitor of the complex.

32. Antennaria monocephala de Candolle in A. P. de Candolle and A. L. P. P. de Candolle, Prodr. 6: 269. 1838 • Pygmy pussytoes [F]

Antennaria alpina (Linnaeus) Gaertner var. *monocephala* (de Candolle) Torrey & A. Gray

Dioecious or gynoecious (staminates uncommon or in equal frequencies as pistillates, respectively). **Plants** 5–13 cm (stems usually stipitate-glandular). **Stolons** 2–4 cm. **Basal leaves** 1-nerved, spatulate to narrowly spatulate or oblanceolate, 9–18 × 2–4 mm, tips mucronate, abaxial faces tomentose, adaxial glabrous or green-glabrescent, or both gray-pubescent.

Cauline leaves linear, 4–11 mm, flagged. **Heads** usually borne singly (rarely 2–3). **Involucres:** staminate 5–7 mm; pistillate 5–8 mm. **Phyllaries** distally brown, dark brown, black, or olivaceous. **Corollas:** staminate 2.5–3.5 mm; pistillate 3.5–4 mm. **Cypselae** 1–1.3 mm, usually glabrous; **pappi:** staminate 3–4 mm (none in gynoecious populations); pistillate 4–5 mm. $2n = 28, 56, 60?, 70.$

Subspecies 2 (2 in the flora): n North America, Russian Far East (Chukotka Peninsula).

It seems reasonable to follow in part E. Hultén's (1968) broad concept of *Antennaria monocephala* (R. J. Bayer 1991). Hultén circumscribed it as containing three subspecies. The sexual phase of *A. monocephala* (i.e., subsp. *monocephala* and subsp. *philonipha*) is known from southern Alaska, south of the Brooks Range, and to Yukon Territory and adjacent areas of the Northwest Territories and across the Bering Strait on the Chukotka Peninsula. Within his concept of *A. monocephala*, Hultén also circumscribed the presumably autopolyploid apomictic form of the species as *A. monocephala* subsp. *angustata*, thereby extending the range of the species across the Canadian arctic into Greenland and down the western Cordillera into Montana and Wyoming.

Antennaria monocephala subsp. *monocephala* is an amphimictic progenitor of the *A. alpina* agamic complex, as well as the sexual progenitor of the apomicts of subsp. *angustata*.

1. Plants gynoecious (staminate plants unknown) . . . . . . 32a. *Antennaria monocephala* subsp. *angustata*
1. Plants dioecious (staminates and pistillates in equal frequencies in populations) . . . . 32b. *Antennaria monocephala* subsp. *monocephala*

32a. **Antennaria monocephala** de Candolle subsp. **angustata** (Greene) Hultén, Ark. Bot., n. s. 7: 135. 1968 • Narrow-leaved pygmy pussytoes; antennaire étroite E

Antennaria angustata Greene, Pittonia 3: 284. 1898; *A. alpina* (Linnaeus) Gaertner var. *megacephala* (Fernald ex Raup) S. L. Welsh; *A. burwellensis* Malte; *A. congesta* Malte; *A. fernaldiana* Polunin; *A. hudsonica* Malte; *A. megacephala* Fernald ex Raup; *A. pygmaea* Fernald; *A. tansleyi* Polunin; *A. tweedsmuirii* Polunin

Gynoecious (staminate plants unknown). **Stems** usually stipitate-glandular. **Basal leaves** spatulate to oblanceolate, faces usually gray-pubescent, adaxial sometimes green-glabrescent. $2n = 56, 60?, 70.$

Flowering summer. Moist tundra, disturbed margins of solifluction lobes, unstable, gravelly slopes; 0–2900 m; Greenland; Alta., B.C., Nfld. and Labr. (Labr.), N.W.T., Nunavut, Que., Yukon; Alaska, Mont., Wyo.

Subspecies *angustata* is apomictic and has a wider geographic range than the amphimictic subsp. *monocephala*. It is easily recognized by heads borne singly, dark phyllaries, and gynoecious population structure.

32b. **Antennaria monocephala** de Candolle subsp. **monocephala** F

A. exilis Greene; *A. monocephala* var. *exilis* (Greene) Hultén; *A. monocephala* var. *latisquamea* Hultén; *A. monocephala* de Candolle subsp. *philonipha* (A. E. Porsild) Hultén; *A. nitens* Greene; *A. philonipha* A. E. Porsild; *A. shumaginensis* A. E. Porsild

Dioecious (staminates and pistillates in equal frequencies in populations). **Stems** stipitate-glandular. **Basal leaves** narrowly spatulate to oblanceolate, mostly abaxially tomentose, adaxially green-glabrous to glabrescent, sometimes both gray-pubescent. $2n = 28.$

Flowering summer. Moist tundra, often on disturbed margins of solifluction lobes or on unstable, gravelly slopes; 0–1800 m; B.C., N.W.T., Yukon; Alaska; Russian Far East (Chukotka Peninsula).

E. Hultén's (1968) key distinctions between subsp. *monocephala* and subsp. *philonipha* are obscure and seemingly arbitrary (R. J. Bayer 1991). *Antennaria exilis* is a pubescent form of *A. monocephala*; complete intergradation occurs between the forms. Taxonomic recognition for the two forms does not seem warranted. *Antennaria monocephala* subsp. *monocephala* is the diploid progenitor of the apomictic clones that make up subsp. *angustata*, and a progenitor of the *A. alpina* agamic complex.

33. **Antennaria friesiana** (Trautvetter) E. Ekman, Svensk Bot. Tidskr. 22: 416. 1928 (as frieseana) • Fries's pussytoes, antennaire de Fries

Antennaria alpina (Linnaeus) Gaertner var. *friesiana* Trautvetter, Trudy Imp. S.-Petersburgsk. Bot. Sada 6: 24. 1879

Dioecious or gynoecious (staminate plants uncommon or in equal frequencies to pistillates, respectively). **Plants** 7–15 cm (stems stipitate-glandular, hairs purple). **Stolons** 0.1–4 cm. **Basal leaves** 1-nerved, narrowly spatulate to oblanceolate, 11–30 × 2–4 mm, tips mucronate, abaxial faces tomentose, adaxial green-glabrescent to gray-pubescent. **Cauline leaves** linear, 4–20 mm, flagged. **Heads** 2–6 in corymbiform arrays. **Involucres:** stami-

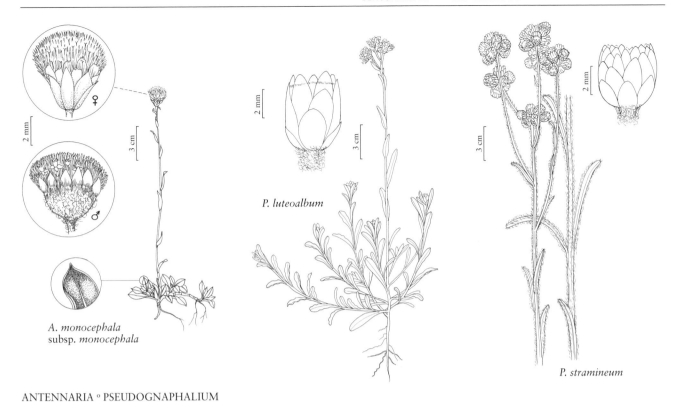

P. luteoalbum

A. monocephala
subsp. *monocephala*

P. stramineum

ANTENNARIA ∘ PSEUDOGNAPHALIUM

nate 4–6.5 mm; pistillate 5.5–8 mm. **Phyllaries** distally usually black, light brown, dark brown, or olivaceous, sometimes. **Corollas:** staminate 2.5–3 mm; pistillate 3–4.5 mm. **Cypselae** 1.2–2 mm, glabrous or slightly papillate; **pappi:** staminate 3–4 mm; pistillate 3.5–5 mm. **2n** = 28, 56, 63, 100+.

Subspecies 3 (3 in the flora): arctic North America, arctic Siberia.

The *Antennaria friesiana* complex consists of subsp. *alaskana*, subsp. *neoalaskana*, and subsp. *friesiana*, the former two are dioecious (sexual) phases of the latter gynoecious (asexual) form. The sexual populations are known from Alaska and cordilleran areas of northern Yukon and adjacent Northwest Territories (R. J. Bayer 1991). The apomictic phase is almost circumpolar, occurring from the central and eastern Siberian plateau eastward across the North American arctic to Greenland (Bayer). E. Hultén (1968) circumscribed a fourth subspecies, *A. friesiana* subsp. *compacta*. After studying its morphology, in the field and herbarium, it is apparent that Hultén's taxon contains at least three incongruous entities that are probably not at all related to the other two subspecies of *A. friesiana*. Hultén's subsp. *compacta* included *A. densifolia*, which is recognized as a distinct species, and *A. crymophila* and *A. neoalaskana* as taxonomic synonyms. *Antennaria compacta* in the strict sense and *A. crymophila* are perhaps hybrid apomicts and are treated here in *A. alpina* (see Bayer for details).

Antennaria neoalaskana is treated here as a subspecies of *A. friesiana*.

1. Stolons 1–4 cm (usually prostrate, sometimes ascending, usually rooting at tips, plants forming mats); involucres: pistillate 7–8 mm . . . .
. . . . 33c. *Antennaria friesiana* subsp. *neoalaskana*
1. Stolons 0.5–1 cm (erect, usually not rooting at tips, plants not forming mats); involucres: pistillate 5.5–7 mm.
  2. Plants gynoecious (staminates unknown)
  . . . . 33b. *Antennaria friesiana* subsp. *friesiana*
  2. Plants dioecious (staminates and pistillates in equal frequencies in populations) . . . .
  . . . . 33a. *Antennaria friesiana* subsp. *alaskana*

33a. Antennaria friesiana (Trautvetter) E. Ekman subsp. **alaskana** (Malte) Hultén, Ark. Bot., n. s. 7: 134. 1968 • Alaskan pussytoes [E]

Antennaria alaskana Malte, Rhodora 36: 107. 1934; *A. friesiana* var. *beringensis* Hultén

Dioecious (staminates and pistillates in equal frequencies in populations). **Plants** 7.5–14 cm. **Stolons** 0.5–1 cm. **Cauline leaves** 4–20 mm. **Heads** 2–6. **Involucres:** staminate 4–6.5 mm; pistillate 5.5–7 mm. **Corollas:** staminate 2.5–3 mm; pistillate 3–

4.5 mm. **Cypselae** 1.2–1.8 mm; **pappi:** staminate 3–4 mm; pistillate 3.5–5 mm. $2n = 28, 56$.

Flowering summer. Arctic and alpine tundra, on or near dry rocky outcrops or sand ridges; 500–1500 m; N.W.T., Yukon; Alaska.

Subspecies *alaskana* is dioecious (sexual) and known from Alaska and cordilleran areas of northern Yukon and adjacent Northwest Territories (R. J. Bayer 1991, 1993). It is a probable progenitor of the *Antennaria alpina* complex.

33b. Antennaria friesiana (Trautvetter) E. Ekman subsp. friesiana

Antennaria angustifolia E. Ekman; *A. ekmaniana* A. E. Porsild

Gynoecious (staminates unknown). **Plants** 3.5–12.5 cm. **Stolons** 0.5–1 cm. **Cauline leaves** 7–14 mm. **Heads** 2–4. **Involucres:** staminate unknown; pistillate 6–7 mm. **Corollas** staminate unknown; pistillate 3.5–4 mm. **Cypselae** 1–2 mm; **pappi:** staminate unknown; pistillate 4–4.5 mm. $2n = 56, 63, 100+$.

Flowering summer. Arctic and alpine tundra, on or near dry rocky outcrops or sand ridges; 100–1500 m; Greenland; Nfld. and Labr. (Labr.), N.W.T., Nunavut, Que., Yukon; Alaska; Siberia (arctic).

The sexual progenitors of subsp. *friesiana* are subspp. *alaskana* and *neoalaskana*.

33c. Antennaria friesiana (Trautvetter) E. Ekman subsp. neoalaskana (A. E. Porsild) R. J. Bayer & Stebbins, Canad. J. Bot. 71: 1596. 1993 • Frost boil or outcrop pussytoes E

Antennaria neoalaskana A. E. Porsild, Sargentia 4: 71. 1943

Dioecious (staminates and pistillates in equal frequencies in populations). **Plants** 7.5–14 cm. **Stolons** 1–4 cm. **Cauline leaves** 4–20 mm. **Heads** 2–6. **Involucres:** staminate 4–6.5 mm; pistillate 7–8 mm. **Corollas:** staminate 2.5–3 mm; pistillate 3–4.5 mm. **Cypselae** 1.2–1.8 mm; **pappi:** staminate 3–4 mm; pistillate 3.5–5 mm. $2n = 56$.

Flowering summer. Arctic and alpine tundra, on dry rocky outcrops, fell fields, or gravelly frost boils; 600–1500 m; N.W.T., Yukon; Alaska.

Subspecies *neoalaskana* is dioecious (sexual) and occurs from the eastern Brooks Range, Alaska, to the Richardson Mountains and into the central MacKenzie Mountains, on the Yukon-Northwest Territories boundary (R. J. Bayer 1993). It intergrades somewhat with the other two subspecies of *Antennaria friesiana*; it can be separated from other arctic members of *Antennaria* because it is dioecious and has 2–6 heads and well-developed stolons. It is a likely progenitor of the *A. alpina* complex.

34. Antennaria alpina (Linnaeus) Gaertner, Fruct. Sem. Pl. 2: 410. 1791 • Alpine pussytoes

Gnaphalium alpinum Linnaeus, Sp. Pl. 2: 856. 1753; *Antennaria alpina* subsp. *canescens* (Lange) Chmielewski; *A. alpina* subsp. *porsildii* (E. Ekman) Chmielewski; *A. alpina* var. *cana* Fernald & Wiegand; *A. alpina* var. *canescens* Lange; *A. alpina* var. *compacta* (Malte) S. L. Welsh; *A. alpina* var. *glabrata* J. Vahl; *A. alpina* var. *intermedia* Rosenvinge; *A. alpina* var. *porsildii* (E. Ekman) T. J. Sørensen; *A. alpina* var. *stolonifera* (A. E. Porsild) S. L. Welsh; *A. alpina* var. *ungavensis* Fernald; *A. arenicola* Malte; *A. atriceps* Fernald ex Raup; *A. bayardi* Fernald; *A. boecheriana* A. E. Porsild; *A. brevistyla* Fernald; *A. brunnescens* Fernald; *A. cana* Fernald & Wiegand (Fernald); *A. canescens* (Lange) Malte; *A. canescens* subsp. *porsildii* (E. Ekman) Á. Löve & D. Löve; *A. canescens* var. *pseudoporsildii* Böcher; *A. columnaris* Fernald; *A. compacta* Malte; *A. confusa* Fernald; *A. crymophila* A. E. Porsild; *A. foggii* Fernald; *A. friesiana* (Trautvetter) E. Ekman subsp. *compacta* (Malte) Hultén; *A. glabrata* (J. Vahl) Greene; *A. intermedia* (Rosenvinge) Porsild; *A. labradorica* Nuttall; *A. longii* Fernald; *A. media* Greene subsp. *compacta* (Malte) Chmielewski; *A. pallida* E. E. Nelson; *A. pedunculata* A. E. Porsild; *A. porsildii* E. Ekman; *A. sornborgeri* Fernald; *A. stolonifera* A. E. Porsild; *A. subcanescens* Ostenfeld ex Malte; *A. ungavensis* (Fernald) Malte; *A. vexillifera* Fernald; *A. wiegandii* Fernald

Gynoecious (staminate plants uncommon). **Plants** 3–18 cm. **Stolons** 1–7 cm. **Basal leaves:** 1-nerved, spatulate to oblanceolate, 6–25 × 2–7 mm, tips mucronate, abaxial faces tomentose, adaxial green-glabrescent to gray-pubescent. **Cauline leaves** linear, 5–20 mm, at least mid and distal flagged. **Heads** 2–5 in corymbiform arrays. **Involucres:** staminate 5–6.5 mm; pistillate 4–7(–10) mm. **Phyllaries** distally dark brown, black, or olivaceous. **Corollas:** staminate 3–3.5 mm; pistillate 3.5–5 mm. **Cypselae** 1–1.8 mm, sparingly papillate; **pappi:** staminate 3.5–4 mm; pistillate 4.5–6 mm. $2n = 56, 84, 98, 112$.

Flowering mid–late summer. Dry to moist alpine tundra; 100–2400 m; Greenland; Alta., B.C., Nfld. and Labr., N.W.T., Nunavut, Ont., Que., Yukon; Alaska, Mont., Wyo.; Eurasia.

Antennaria alpina is one of the more morphologically variable agamic complexes in the genus. Some taxonomists have argued that true *Antennaria alpina* does not occur in North America, because none of the North American material exactly matches the type of *A. alpina*, which is from Lapland (M. O. Malte 1934; A. E. Porsild 1965). If one uses a strict typological species concept, then this is true; I recognize that this species complex is composed of innumerable apomictic clones and am circumscribing a broad species concept for *A. alpina*. The potential morphologic overlap between the *A. media* and *A. alpina* complexes is a major taxonomic problem. The chief difference between members of the two complexes is the presence of prominent flags on cauline leaves in *A. alpina* and their absence in *A. media*. *Antennaria alpina* of North America is gynoecious and characterized by its dark green to black phyllaries and conspicuous flags on the distal cauline leaves. The basal leaves vary from glabrous, as in the type material, to pubescent. The

primary progenitors of the *A. alpina* complex include *A. aromatica*, *A. densifolia*, *A. friesiana* subsp. *alaskana*, *A. friesiana* subsp. *neoalaskana*, *A. monocephala* subsp. *monocephala*, and *A. pulchella*.

Excluded names:

Some *Antennaria* names are based on early-generation interspecific hybrids, including:

Antennaria ×*erigeroides* Greene = *A. corymbosa* × *A. racemosa*

A. ×*foliacea* Greene = *A. microphylla* × *A. racemosa*

A. ×*macounii* Greene = *A. media* × *A. umbrinella*

A. ×*oblancifolia* E. E. Nelson = *A. racemosa* × *A. umbrinella*

A. ×*rousseaui* A. E. Porsild = ? *A. alpina* × *A. rosea*

88. PSEUDOGNAPHALIUM Kirpicznikov, Trudy Bot. Inst. Akad. Nauk S.S.S.R., Ser. 1, Fl. Sist. Vyssh. Rast. 9: 33. 1950 • [Greek *pseudo-*, deceptively similar, and genus name *Gnaphalium*, alluding to resemblance]

Guy L. Nesom

Annuals, biennials, or perennials (sometimes aromatic), (4–)15–150(–200) cm (usually taprooted, sometimes fibrous-rooted). **Stems** 1+, usually erect, sometimes decumbent to procumbent (± woolly-tomentose, sometimes stipitate- or sessile-glandular). **Leaves** basal and cauline or mostly cauline; alternate; usually sessile; blades mostly narrowly lanceolate to oblanceolate, bases often clasping and/or decurrent, margins entire, faces bicolor or concolor, abaxial white to gray and tomentose to velutinous, adaxial usually greenish and glabrous or glabrescent, sometimes grayish and loosely arachnose (sometimes stipitate- or sessile-glandular). **Heads** disciform, usually in glomerules in corymbiform or paniculiform arrays, sometimes in terminal clusters. **Involucres** mostly campanulate to cylindric, (3–)4–7 mm. **Phyllaries** in (2–)3–7(–10) series, whitish, rosy, tawny, or brownish (opaque or hyaline, dull or shiny; stereomes usually green, usually sessile-glandular distally), unequal, usually chartaceous toward tips. **Receptacles** flat, smooth, epaleate. **Peripheral (pistillate) florets** (15–)25–250+ (more numerous than bisexual); corollas yellowish. **Inner (bisexual) florets** (1–)5–20(–40+); corollas yellowish (red-tipped in *P. luteoalbum*). **Cypselae** oblong-compressed or cylindric, faces usually smooth, sometimes papillate-roughened and/or with 4–6 longitudinal ridges, usually glabrous (papilliform hairs in *P. luteoalbum*); **pappi** readily falling, of 10–12 distinct (coherent basally in *Pseudognaphalium luteoalbum* and *P. stramineum*), barbellate bristles in 1 series. *x* = 7.

Species ca. 100 (21 in the flora): worldwide, mostly South America to North America, mostly in temperate regions.

Fifteen of the species treated here occur also in Mexico; those that do not are *Pseudognaphalium obtusifolium*, *P. saxicola*, *P. micradenium*, and *P. helleri* (eastern United States and adjacent Canada), and *P. ramosissimum* and *P. thermale* (western United States and adjacent Canada).

Basal and proximal leaves of *Pseudognaphalium* species often wither before plants reach flowering. In the key and descriptions here, references to leaves are to cauline leaves of plants at flowering unless otherwise indicated.

SELECTED REFERENCE Nesom, G. L. 2004d. *Pseudognaphalium canescens* (Asteraceae: Gnaphalieae) and putative relatives in western North America. Sida 21: 781–790.

1. Leaf faces strongly to weakly bicolor (abaxial gray to white, tomentose, adaxial green, not tomentose, sometimes glandular).
   2. Bases of leaf blades not clasping, not decurrent.
      3. Stems white (tomentose, rarely glandular near bases); pistillate florets 38–96; bisexual florets 4–8(–11) . . . . 7. *Pseudognaphalium obtusifolium*
      3. Stems greenish (lacking or soon losing most tomentum, usually densely stipitate-glandular); pistillate florets 47–107; bisexual florets 7–20.
         4. Stems glandular-villous (stipitate glands mostly 0.3–1 mm, variable in height on any portion of stem, stalks broadened toward bases, about equaling gland widths); leaf blades mostly oblong-lanceolate, 2.5–7 cm × 4–20 mm; pistillate florets 83–107; bisexual florets 9–15 . . . . 9. *Pseudognaphalium helleri*
         4. Stems glandular-puberulent (stipitate glands 0.1–0.2 mm, stalks narrower than gland widths); leaf blades linear to linear-lanceolate or linear-oblanceolate, 1.5–5.5 cm × 1.5–10 mm; pistillate florets 47–78; bisexual florets (7–)11–20 . . . . 10. *Pseudognaphalium micradenium*
   2. Bases of leaf blades clasping and/or decurrent.
      5. Leaves crowded (internodes usually 1–3, sometimes to 10 mm), blades linear to linear-lanceolate, margins strongly revolute.
         6. Phyllaries bright white, opaque, dull; cypselar faces smooth . . . . 15. *Pseudognaphalium leucocephalum*
         6. Phyllaries tawny to silvery white, ± hyaline, shiny; cypselar faces papillate-roughened.
            7. Stems stipitate-glandular; leaf bases not subclasping (proximal usually decurrent 3–10 mm); phyllaries ovate-lanceolate (apices of inner not thickened along midribs, not apiculate); pistillate florets ca. 200–250; bisexual florets (13–)16–29 . . . . 13. *Pseudognaphalium viscosum*
            7. Stems not glandular; leaf bases subclasping (not decurrent); phyllaries narrowly ovate to oblong or elliptic (apices of inner thickened and slightly raised along midribs, apiculate); pistillate florets [46–]76–102; bisexual florets (6–)8–11 . . . . 14. *Pseudognaphalium austrotexanum*
      5. Leaves not crowded (internodes mostly more than 5 mm), blades elliptic, elliptic-ovate, lanceolate, oblanceolate, or oblong, margins flat or slightly revolute.
         8. Stems not glandular; leaf bases auriculate-clasping, faces strongly bicolor . . . . 16. *Pseudognaphalium biolettii*
         8. Stems glandular; leaf bases weakly, if at all, clasping, faces strongly to weakly bicolor.
            9. Involucres 4.5–5.5 mm; phyllaries in 4–5 series; bisexual florets 7–12 . . . . 11. *Pseudognaphalium macounii*
            9. Involucres 3.5–4 mm; phyllaries in 2–3 series; bisexual florets (1–)2–6 . . . . 12. *Pseudognaphalium pringlei*
1. Leaf faces concolor or weakly bicolor (both usually gray to gray-green or greenish, tomentose, adaxial sometimes glandular beneath tomentum).
   10. Leaf bases clasping to subclasping, seldom decurrent (decurrent 1–2 mm in *P. luteoalbum* and *P. stramineum*, 2–10 mm in *P. californicum*).
      11. Heads in terminal glomerules; involucres 3–6 mm; phyllaries silver-gray to yellowish (hyaline); pistillate florets 135–160 (pappus bristles loosely coherent basally, released in clusters or easily fragmented rings).
         12. Involucres 3–4 mm; bisexual florets 5–10 (corollas red-tipped; cypselae with papilliform hairs) . . . . 1. *Pseudognaphalium luteoalbum*
         12. Involucres 4–6 mm; bisexual florets mostly 18–28 (corollas evenly yellowish, not red-tipped; cypselae glabrous) . . . . 2. *Pseudognaphalium stramineum*
      11. Heads in corymbiform arrays; involucres 4–7 mm; phyllaries usually silvery white to white, sometimes pink (mostly opaque); pistillate florets 35–140 (pappus bristles not coherent basally, released singly).

13. Stems stipitate-glandular; abaxial faces of leaf blades green, stipitate-glandular; phyllaries in 7–10 series . . . . 17. *Pseudognaphalium californicum*
13. Stems not glandular; abaxial faces of leaf blades white-tomentose or woolly-tomentose; phyllaries in 5–6 series . . . . 18. *Pseudognaphalium roseum*

[10. shifted to left margin.—Ed.]

10. Leaf bases not clasping or subclasping, decurrent or not.
  14. Leaf bases not decurrent.
    15. Annuals 4–15(–30) cm; faces of leaf blades concolor, green, thinly arachnoid-tomentose to glabrate, not glandular (veiny reticulum evident); heads in terminal capitate clusters; cypselae smooth . . . . 8. *Pseudognaphalium saxicola*
    15. Annuals, biennials, or perennials, 20–100+ cm; faces of leaf blades concolor or weakly bicolor, white to gray, ± tomentose, sometimes glandular beneath tomentum; heads in corymbiform arrays; cypselae smooth or papillate-roughened.
      16. Adaxial faces of leaf blades sometimes sessile-glandular beneath adaxial tomentum; involucres 4–5 mm; phyllaries in 3–4 series; bisexual florets (1–)2–5(–6) . . . . 3. *Pseudognaphalium canescens*
      16. Adaxial faces of leaf blades not glandular; involucres 5–6 mm; phyllaries in 4–6 series; bisexual florets 5–9 . . . . 4. *Pseudognaphalium microcephalum*
  14. Leaf bases decurrent.
    17. Stems and leaves stipitate-glandular; heads in corymbiform or paniculiform arrays; phyllaries usually white or pinkish, sometimes greenish.
      18. Leaf blades mostly narrowly oblong-lanceolate, 5–10(–20) mm wide; heads in corymbiform arrays; involucres campanulo-globose; phyllaries in 7–10 series, white . . . . 17. *Pseudognaphalium californicum*
      18. Leaf blades linear to lanceolate, oblong, or narrowly spatulate, 3–5(–7) mm wide; heads in paniculiform (usually elongate to broadly columnar) arrays; involucres turbinate to short-cylindric; phyllaries in 4–5 series, usually pinkish, sometimes white to greenish . . . . 21. *Pseudognaphalium ramosissimum*
    17. Stems and leaves not glandular; heads usually in corymbiform or paniculiform arrays (sometimes borne singly or in glomerules in *P. arizonicum*); phyllaries usually white, whitish, or brownish to tawny, rarely rosy.
      19. Phyllaries usually brownish to tawny, rarely rosy, ovate-lanceolate to lanceolate (hairs of stems and leaves commonly with reddish or purplish cross walls) . . . . 19. *Pseudognaphalium arizonicum*
      19. Phyllaries white or whitish, ovate to ovate-oblong (hairs of stems and leaves without colored cross walls).
        20. Pistillate florets 80–115[–180]; bisexual florets (6–)8–12[–30] (bases of hairs on leaves persistent, enlarged) . . . . 20. *Pseudognaphalium jaliscense*
        20. Pistillate florets 35–69; bisexual florets 3–8(–11) (bases of hairs on leaves not persistent and enlarged).
          21. Blades of basal and proximal cauline leaves linear (proximal and distal similar in size and shape); heads usually in paniculiform arrays; phyllaries in (4–)5–6(–7) series (usually opaque, dull to shiny) . . . . 5. *Pseudognaphalium beneolens*
          21. Blades of basal and proximal cauline leaves linear-oblanceolate (the distal shorter and narrower); heads in corymbiform to paniculiform arrays; phyllaries in 3–4(–5) series, hyaline or opaque, usually shiny, sometimes dull . . . . 6. *Pseudognaphalium thermale*

1. **Pseudognaphalium luteoalbum** (Linnaeus) Hilliard & B. L. Burtt, Bot. J. Linn. Soc. 82: 206. 1981 (as luteo-album) • Red-tip rabbit-tobacco [F] [I]

Gnaphalium luteoalbum Linnaeus, Sp. Pl. 2: 851. 1753

Annuals, 15–40 cm; taprooted or fibrous-rooted. **Stems** loosely white-tomentose, not glandular. **Leaf blades** (crowded, internodes 1–5, sometimes to 10 mm) narrowly obovate to subspatulate, 1–3(–6) cm × 2–8 mm (distal smaller, oblanceolate to narrowly oblong or linear), bases subclasping, usually decurrent 1–2 mm, margins weakly revolute, faces mostly concolor to weakly bicolor, abaxial gray-tomentose, adaxial usually gray-tomentose, sometimes glabrescent, neither glandular. **Heads** in terminal glomerules (1–2 cm diam.). **Involucres** broadly campanulate, 3–4 mm. **Phyllaries** in 3–4 series, silvery gray to yellowish (hyaline), ovate to ovate-oblong, glabrous. **Pistillate florets** 135–160. **Bisexual florets** 5–10 (corollas red-tipped). **Cypselae** not evidently ridged (conspicuously dotted with whitish, papilliform hairs; pappus bristles loosely coherent basally, released in clusters or easily fragmented rings). $2n$ = 14, 16, 28.

Flowering Apr–Oct. Roadsides, fields and pastures, ditches, streambanks, seasonal ponds, gardens, and other disturbed sites; 5–2000 m; introduced; Ariz., Ark., Calif., Fla., La., Nev., N.Mex., N.Y., Oreg., Tex., Utah, Wash.; Mexico; Europe; Asia; Africa; Pacific Islands (New Zealand); Australia.

Pseudognaphalium luteoalbum is native to Eurasia. It is similar in overall habit to *P. stramineum* but distinctive in its larger heads and red-tipped corollas (visible through the translucent phyllaries). Cypselae of *P. luteoalbum* have papilliform hairs; cypselae of other North American species of *Pseudognaphalium* are glabrous.

2. **Pseudognaphalium stramineum** (Kunth) Anderberg, Opera Bot. 104: 148. 1991 • Cotton-batting-plant [F]

Gnaphalium stramineum Kunth in A. von Humboldt et al., Nov. Gen. Sp. 4(fol.): 66. 1818; 4(qto.): 85. 1820; *G. chilense* Sprengel; *G. chilense* var. *confertifolium* Greene; *G. gossypinum* Nuttall; *G. lagopodioides* Rydberg; *G. proximum* Greene; *G. sulphurescens* Rydberg

Annuals or biennials, 30–60(–80) cm; taprooted. **Stems** (1+ from base, erect to ascending) loosely tomentose, not glandular. **Leaf blades** (crowded, internodes usually 1–5, sometimes to 10 mm) oblong to narrowly oblanceolate or subspatulate, 2–8(–9.5) cm × 2–5(–10) mm (smaller distally, narrowly lanceolate to linear), bases subclasping, usually not decurrent, sometimes decurrent 1–2 mm, margins flat or slightly revolute, faces concolor, loosely and persistently gray-tomentose, not glandular. **Heads** in terminal glomerules (1–2 cm diam.). **Involucres** subglobose, 4–6 mm. **Phyllaries** in 4–5 series, whitish (often yellowish with age, hyaline, shiny), ovate to oblong-obovate, glabrous. **Pistillate florets** 160–200. **Bisexual florets** [8–]18–28. **Cypselae** weakly, if at all, ridged (otherwise smooth or papillate-roughened, glabrous, without papilliform hairs; pappus bristles loosely coherent basally, released in clusters or easily fragmented rings). $2n$ = 28.

Flowering Mar–Oct. Sandy fields, streamsides, washes, swales, dunes, chaparral slopes, roadsides, fields, disturbed places, moist disturbed places; 10–1600 m; B.C.; Ariz., Calif., Colo., Idaho, Mont., Nebr., Nev., N.Mex., N.Y., N.C., Okla., Oreg., S.C., Tex., Utah, Va., Wash., Wyo.; Mexico; South America

Pseudognaphalium stramineum is probably native from South America to western North America; it is adventive in sandy fields on the Atlantic coastal plain, where it flowers May–Aug.

3. **Pseudognaphalium canescens** (de Candolle) Anderberg, Opera Bot. 104: 147. 1991 • Wright's rabbit-tobacco

Gnaphalium canescens de Candolle in A. P. de Candolle and A. L. P. P. de Candolle, Prodr. 6: 228. 1838; *G. sonorae* I. M. Johnston; *G. texanum* I. M Johnston; *G. viridulum* I. M. Johnston; *G. wrightii* A. Gray

Annuals or perennials, 20–70 (–100+) cm; taprooted. **Stems** persistently tomentose, not glandular (2–3 mm diam. near bases). **Leaf blades** narrowly to broadly oblanceolate, mostly 2–4(–5) cm × 2–8(–15) mm, bases not clasping, not decurrent, margins flat, faces weakly bicolor, tomentose (adaxial less densely tomentose, sometimes sessile-glandular beneath tomentum). **Heads** usually in loose, corymbiform arrays. **Involucres** turbinate-campanulate, 4–5 mm. **Phyllaries** in 3–4 series, white (opaque to hyaline, dull to shiny), narrowly ovate-lanceolate, glabrous. **Pistillate florets** (16–)24–44. **Bisexual florets** (1–)2–5 (–6), 5–6 more common in northern part of range. **Cypselae** ridged, weakly papillate-roughened. $2n$ = 28.

Flowering Aug–Nov(–Jan). Lava beds, rocky sites, grasslands, oak, pine-oak, and pine woodlands; 1100–2500(–2700) m; Ariz., Calif., Colo., N.Mex., Okla., Tex., Utah; Mexico.

Most plants of *Pseudognaphalium canescens* produce white, opaque, keeled, apiculate phyllaries; in the southern portion of its range (Jalisco southeastward) and scattered localities elsewhere, the phyllaries may be more hyaline and lack a pronounced keel and apiculum.

4. **Pseudognaphalium microcephalum** (Nuttall) Anderberg, Opera Bot. 104: 147. 1991 • San Diego rabbit-tobacco

Gnaphalium microcephalum Nuttall, Trans. Amer. Philos. Soc., n. s. 7: 404. 1841; *G. albidum* I. M. Johnston; *G. canescens* de Candolle subsp. *microcephalum* (Nuttall) Stebbins & D. J. Keil; *Pseudognaphalium canescens* (de Candolle) Anderberg subsp. *microcephalum* (Nuttall) Kartesz

Perennials, (30–)50–100 cm; taprooted. **Stems** persistently grayish tomentose, not glandular, (3–5 mm diam. near bases). **Leaf blades** narrowly oblanceolate, 2–5(–8) cm × 5–10(–18) mm (gradually smaller distally, becoming lanceolate), bases not clasping, not decurrent, margins flat, faces weakly bicolor, tomentose (adaxial less densely), not glandular. **Heads** in loose, corymbiform arrays. **Involucres** turbinate-campanulate, 5–6 mm. **Phyllaries** in 4–5 series, white (opaque, dull), ovate to oblong-ovate (inner narrower, all usually with filiform but definitely thickened keel and slight apiculum), tomentose (at least bases). **Pistillate florets** 29–49. **Bisexual florets** 5–9. **Cypselae** ridged, smooth to weakly papillate-roughened. $2n = 28$.

Flowering (Apr–)Jun–Aug(–Nov). Grassy hillsides, gravelly canyon bottoms, chaparral, coastal sage scrub; 50–900(–1800) m; Calif.; Mexico (Baja California).

Pseudognaphalium microcephalum is characterized by stems commonly stiffly erect and slightly zigzag distally, relatively thick (3–5 mm diam. near bases), and closely grayish tomentose, leaves oblanceolate, sessile, sometimes clasping, not decurrent, and weakly bicolor, and heads usually in open, corymbiform arrays.

5. **Pseudognaphalium beneolens** (Davidson) Anderberg, Opera Bot. 104: 147. 1991 • Fragrant rabbit-tobacco

Gnaphalium beneolens Davidson, Bull. S. Calif. Acad. Sci. 17: 17, unnumb. fig. p. 16. 1918; *G. canescens* de Candolle subsp. *beneolens* (Davidson) Stebbins & D. J. Keil; *Pseudognaphalium canescens* (de Candolle) Anderberg subsp. *beneolens* (Davidson) Kartesz

Annuals or short-lived perennials, 30–80(–110) cm; taprooted. **Stems** persistently tomentose, not glandular. **Leaf blades** mostly linear, 3–6 cm × 1.5–3.5 mm (sometimes smaller distally), bases not clasping, decurrent 5–15 mm, margins flat, faces concolor,

loosely tomentose, not glandular. **Heads** usually in loose, paniculiform arrays. **Involucres** turbinate-campanulate, 5–6 mm. **Phyllaries** in (4–)5–6(–7) series, white (opaque, dull to shiny), ovate to ovate-oblong (inner usually with filiform keel and slight apiculum), glabrous. **Pistillate florets** (39–)44–69. **Bisexual florets** 5–8(–11). **Cypselae** ridged, smooth or weakly papillate-roughened. $2n = 14$.

Flowering (Apr–)Jun–Oct. Dry, open slopes and ridges, streambeds, road banks and other disturbed sites, sandy flats, dunes, coastal sage scrub, chaparral, yellow pine, foothill pine, blue oak woodland; (1–)50–800 (–2000) m; Calif.; Mexico (Baja California).

Pseudognaphalium beneolens differs from *P. thermale* in its leaves linear throughout, heads usually in elongate, paniculiform arrays, larger heads (greater numbers of phyllaries in greater numbers of series) with phyllaries more opaque and duller, and greater numbers of bisexual florets. The cauline leaves of *P. beneolens* tend to become curving-coiling. In areas of sympatry, habitats of *P. beneolens* are characteristically at lower elevations than those of *P. thermale*.

6. **Pseudognaphalium thermale** (E. E. Nelson) G. L. Nesom, Sida 21: 781. 2004 • Northwestern rabbit-tobacco E F

Gnaphalium thermale E. E. Nelson, Bot. Gaz. 30: 121. 1900; *G. canescens* de Candolle subsp. *thermale* (E. E. Nelson) Stebbins & D. J. Keil; *G. johnstonii* G. N. Jones; *G. microcephalum* Nuttall var. *thermale* (E. E. Nelson) Cronquist; *G. microcephalum* subsp. *thermale* (E. E. Nelson) G. W. Douglas; *Pseudognaphalium canescens* (de Candolle) Anderberg subsp. *thermale* (E. E. Nelson) Kartesz; *P. microcephalum* (Nuttall) Anderberg var. *thermale* (E. E. Nelson) Dorn

Perennials, (20–)30–70 cm; taprooted. **Stems** loosely tomentose, not glandular. **Leaf blades** narrowly oblanceolate, 3–8 cm × 3–6 mm (gradually smaller distally, becoming linear), bases not clasping, decurrent 5–14 mm, margins flat, faces concolor, loosely tomentose, sessile-glandular beneath tomentum. **Heads** in loose to dense, corymbiform to paniculiform arrays. **Involucres** turbinate-campanulate, (4–)5–6 mm. **Phyllaries** in 3–4 (–5) series, whitish (hyaline or opaque, usually shiny, sometimes dull), ovate to ovate-oblong (outer broadly acute, inner rounded-apiculate), glabrous. **Pistillate florets** 35–55. **Bisexual florets** (2–)4–7. **Cypselae** ridged, densely papillate-roughened.

Flowering Jun–Sep(–Oct). Dry, sandy road banks, roadside ditches, streambeds and banks, lakeshores, granitic sand, open woods of yellow pine, Jeffrey pine, red fir, Douglas fir, mixed conifer, and mixed evergreen; (50–)300–2300(–2500) m; Alta., B.C.; Calif., Idaho, Mont., Nev., Oreg., Utah, Wash., Wyo.

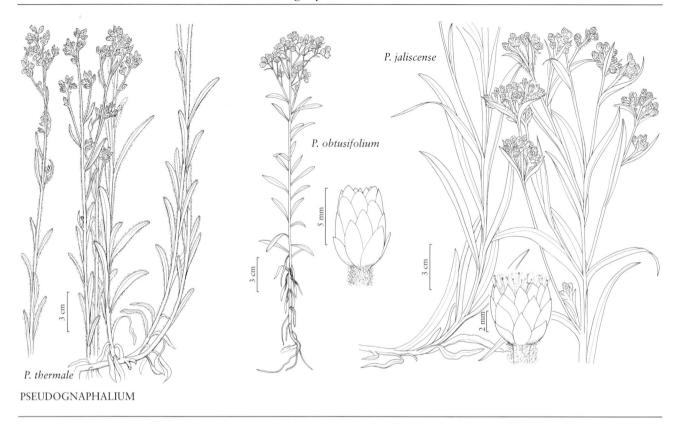

P. jaliscense

P. obtusifolium

P. thermale

PSEUDOGNAPHALIUM

7. **Pseudognaphalium obtusifolium**   (Linnaeus) Hilliard & B. L. Burtt, Bot. J. Linn. Soc. 82: 205. 1981   •   Eastern rabbit-tobacco, gnaphale à feuilles obtuses   E F

Gnaphalium obtusifolium Linnaeus, Sp. Pl. 2: 851. 1753; *G. obtusifolium* var. *praecox* Fernald

Annuals or winter annuals (sometimes faintly fragrant), (10–)30–100 cm; taprooted. **Stems** white-tomentose, sometimes lightly so, usually not glandular, rarely glandular near bases. **Leaf blades** linear-lanceolate to elliptic or oblanceolate, 2.5–10 cm × 2–10 mm (relatively even-sized), bases not clasping, not decurrent, margins flat, faces bicolor, abaxial white-tomentose, adaxial green, usually glabrous or slightly glandular, sometimes with persistent light tomentum. **Heads** in corymbiform (sometimes rounded to elongate) arrays. **Involucres** broadly campanulate, 5–7 mm. **Phyllaries** in 4–6 series, white (opaque, usually shiny, sometimes dull), ovate to ovate-oblong, glabrous or tomentose (bases). **Pistillate florets** 38–96. **Bisexual florets** 4–8(–11). **Cypselae** ridged, smooth.

Flowering Aug–Oct. Open sites, often disturbed, roadsides, fields, pastures, open woods, in various soils, most abundantly in sand; 5–200 m; N.B., N.S., Ont., P.E.I., Que.; Ala., Ark., Conn., Del., Fla., Ga., Ill., Ind., Iowa, Kans., Ky., La., Maine, Mass., Mich., Minn., Miss., Mo.,

Nebr., N.H., N.Y., N.C., Ohio, Okla., Pa., R.I., S.C., Tenn., Tex., Vt., Va., W.Va., Wis.

SELECTED REFERENCE   Nesom, G. L. 2001c. Notes on variation in *Pseudognaphalium obtusifolium* (Asteraceae: Gnaphalieae). Sida 19: 615–619.

8. **Pseudognaphalium saxicola**   (Fassett) H. E. Ballard & Feller, Sida 21: 777. 2004   •   Cliff cudweed   C E

Gnaphalium saxicola Fassett, Rhodora 33: 75. 1931; *G. obtusifolium* Linnaeus var. *saxicola* (Fassett) Cronquist; *Pseudognaphalium obtusifolium* (Linnaeus) Hilliard & B. L. Burtt var. *saxicola* (Fassett) Kartesz

Annuals, 4–15(–30) cm; taprooted. **Stems** (filiform) persistently tomentose (indument a loose, envelope-like, transparent haze of extremely thin hairs, doubling apparent stem width), not glandular. **Leaf blades** elliptic-oblanceolate to oblanceolate, 0.5–3 cm × 2–6 mm (largest at midstem), bases not clasping, not decurrent, margins flat, faces concolor, green, thinly arachnoid-tomentose to glabrate, not glandular (veiny reticulum evident). **Heads** (2–4) in terminal, capitate clusters (usually immediately subtended by distalmost cauline leaf, clusters sometimes in subcorymbiform arrays). **Involucres** turbinate, 4–5 mm. **Phyllaries** in 3(–4) series, whitish to slightly tawny (hyaline, shiny), narrowly triangular to narrowly oblong-triangular, glabrous. **Pistillate florets** 25–28. **Bisexual florets** 6–7. **Cypselae** not ridged, smooth.

Flowering (Jul–)Aug–Sep. Mostly bare sandstone cliff faces, ledges, and cracks, s- to e-facing, commonly shaded; 200–300 m; Wis.

Pseudognaphalium saxicola probably is an evolutionary derivative of *P. obtusifolium*. Plants of *P. saxicola* are relatively small and have relatively few, relatively small heads and occur in a specialized habitat; they constitute the only narrowly endemic species of *Pseudognaphalium* in the United States. Depauperate individuals of *P. obtusifolium* from localities over its whole geographic range may sometimes be as short as 5–10 cm and similar in habit to *P. saxicola*; such plants differ from *P. saxicola* in their close and denser stem vestiture, bicolor and relatively narrow leaves, larger heads with greater numbers of pistillate florets, and broader phyllaries with rounded apices.

9. Pseudognaphalium helleri (Britton) Anderberg,
Opera Bot. 104: 147. 1991 • Heller's rabbit-tobacco
E

Gnaphalium helleri Britton, Bull. Torrey Bot. Club 20: 280. 1893; *G. obtusifolium* Linnaeus var. *helleri* (Britton) S. F. Blake

Annuals (fragrant), 30–100 cm; fibrous-rooted (roots relatively thick and lignescent). **Stems** greenish, glandular-villous (without persistent tomentum, stipitate glands mostly 0.3–1 mm, often variable, stalks broadened toward bases, about equaling gland widths). **Leaf blades** mostly oblong-lanceolate, 2.5–7 cm × 4–20 mm, bases not clasping, not decurrent, margins flat, faces bicolor, abaxial white to gray with lightly persistent tomentum, adaxial green, both minutely stipitate-glandular. **Heads** in corymbiform arrays. **Involucres** campanulate, 6–7 mm. **Phyllaries** in 4–6 series, white (opaque, shiny), ovate to ovate-oblong or oblong, ± tomentose. **Pistillate florets** 83–107. **Bisexual florets** 9–15. **Cypselae** ridged, smooth.

Flowering Sep–Oct(–Nov). Dry woods and openings, clay and sandy clay, sand hills; 10–300 m; Ala., Ark., Fla., Ga., La., Miss., N.C., Okla., S.C., Tenn., Tex., Va.

Pseudognaphalium helleri and *P. micradenium* are similar to *P. obtusifolium* in most features; both differ in their glandular stems without the persistent whitish tomentum of *P. obtusifolium*.

10. Pseudognaphalium micradenium (Weatherby)
G. L. Nesom, Sida 19: 618. 2001 • Delicate rabbit-tobacco E

Gnaphalium obtusifolium Linnaeus var. *micradenium* Weatherby, Rhodora 25: 22. 1923; *G. helleri* Britton var. *micradenium* (Weatherby) Mahler; *Pseudognaphalium helleri* (Britton) Anderberg subsp. *micradenium* (Weatherby) Kartesz

Annuals (fragrant), 15–60 cm; taprooted or fibrous-rooted. **Stems** glandular-puberulent (without persistent tomentum, stipitate glands 0.1–0.2 mm, stalks narrower than gland widths. **Leaf blades** linear to linear-lanceolate or linear-oblanceolate, 1.5–5.5 cm × 1.5–10 mm, bases not clasping, not decurrent, margins flat, faces bicolor, abaxial white to gray, tomentose, adaxial green, both minutely stipitate-glandular. **Heads** in corymbiform arrays. **Involucres** turbinate-campanulate, 5–6 mm. **Phyllaries** in 4–6 series, white to tawny white (hyaline, shiny), narrowly ovate to oblong, glabrous. **Pistillate florets** 47–78. **Bisexual florets** (7–)11–20. **Cypselae** ridged, smooth.

Flowering Sep–Oct. Dry woods and openings, roadsides; 10–600 m; Ga., Ind., Ky., Maine, Md., Mass., Mich., Minn., Mo., N.H., N.J., N.Y., N.C., Pa., S.C., Tenn., Va., Wis.

Pseudognaphalium micradenium has a more northern and Appalachian distribution than *P. helleri*. A report of *P. micradenium* for Louisiana probably was based on specimens of *P. helleri*. The two species differ in vestiture and other features; stems of *P. micradenium* are more slender than those of its close relatives.

11. Pseudognaphalium macounii (Greene) Kartesz in
J. T. Kartesz and C. A. Meacham, Synth. N. Amer. Fl., nomencl. innov. 30. 1999 • Macoun's rabbit-tobacco, gnaphale de Macoun

Gnaphalium macounii Greene, Ottawa Naturalist 15: 278. 1902; *G. decurrens* Ives 1819, not Linnaeus 1759

Annuals or biennials (often sweetly fragrant), 40–90 cm; taprooted. **Stems** stipitate-glandular throughout (usually persistently lightly white-tomentose distally). **Leaf blades** (not crowded, internodes mostly 5+ mm) lanceolate to oblanceolate,

3–10 cm × 3–13 mm (distal linear), bases not clasping, decurrent 5–10 mm, margins flat to slightly revolute, faces weakly bicolor, abaxial tomentose, adaxial stipitate-glandular, otherwise glabrescent or glabrous. **Heads** in corymbiform arrays. **Involucres** campanulo-subglobose, 4.5–5.5 mm. **Phyllaries** in 4–5 series, stramineous to creamy (hyaline, shiny), ovate to ovate-oblong, glabrous. **Pistillate florets** 47–101(–156). **Bisexual florets** 5–12 [–21]. **Cypselae** not ridged, ± papillate-roughened.

Flowering July–Oct. Dry, open habitats, pastures, open woods or edges, roadsides; 50–2600(–3000) m; Alta., B.C., Man., N.B., N.S., Ont., P.E.I., Que., Sask.; Ariz., Calif., Colo., Conn., Idaho, Ill., Ind., Maine, Mass., Mich., Minn., Mont., N.H., N.Mex., N.Y., Ohio, Oreg., Pa., S.Dak., Tenn., Utah, Vt., Va., Wash., W.Va., Wis., Wyo.; Mexico.

Pseudognaphalium macounii is recognized by its stipitate-glandular, proximally glabrescent stems, bicolor and decurrent leaves, relatively large and many-flowered heads, and hyaline, shiny phyllaries. Reports of *P. macounii* from Texas are based on specimens of *P. viscosum.*

12. Pseudognaphalium pringlei (A. Gray) Anderberg, Opera Bot. 104: 147. 1991 • Pringle's rabbit-tobacco

Gnaphalium pringlei A. Gray, Proc. Amer. Acad. Arts 21: 387. 1886

Annuals or perennials, 30–80 cm; taprooted. **Stems** lightly white-tomentose and/or glabrescent and green, minutely stipitate- or sessile-glandular beneath other induments. **Leaf blades** (not crowded, internodes mostly 5+ mm) oblanceolate-spatulate to obovate- or petiolate-spatulate, 5–10 cm × 10–20 mm (distal oblong to lanceolate or oblanceolate, 2–8 cm, slightly smaller), bases not clasping and decurrent 3–20 mm or clasping and decurrent 1–3 mm or not decurrent at all, margins flat to slightly revolute, faces bicolor, abaxial thinly white-tomentose, adaxial minutely stipitate- or sessile-glandular, otherwise glabrous or glabrate (bases of hairs persistent, enlarged). **Heads** in loose, corymbiform arrays. **Involucres** campanulate to turbinate, 3.5–4 mm. **Phyllaries** in 2–3 series, silvery white to tawny, oblong to oblong-ovate, (hyaline, shiny), glabrous. **Pistillate florets** 15–40(–64). **Bisexual florets** (1–)2–6. **Cypselae** ridged, papillate-roughened.

Flowering (Aug–)Sep–Nov. Rock outcrops and slopes, crevices and thin soil on cliffs, oak or oak-pine woodlands; 1500–2300 m; Ariz., N.Mex., Tex.; Mexico (Chihuahua, Durango, Sonora).

13. Pseudognaphalium viscosum (Kunth) Anderberg, Opera Bot. 104: 148. 1991 • Sticky rabbit-tobacco

Gnaphalium viscosum Kunth in A. von Humboldt et al., Nov. Gen. Sp. 4(fol.): 64. 1818; 4(qto.): 82. 1820

Annuals (viscid and unpleasantly aromatic), 30–100 cm; taprooted. **Stems** persistently white-tomentose and stipitate-glandular. **Leaf blades** (crowded, internodes mostly 1–3, sometimes to 10, mm) linear-lanceolate, (2–)4–8 cm × 3–10 mm, bases not clasping, usually (at least the proximal) decurrent 3–10 mm, margins strongly revolute to revolute-undulate, faces bicolor, abaxial densely white-tomentose, adaxial densely stipitate-glandular. **Heads** in corymbiform arrays. **Involucres** campanulate, 5–6 mm. **Phyllaries** in 5–6 series, tawny-silvery to silvery white (hyaline, shiny), ovate-lanceolate (not keeled or thickened along midribs, not apiculate), glabrous. **Pistillate florets** 200–250. **Bisexual florets** (13–)16–29. **Cypselae** not ridged, papillate-roughened. $2n = 28$.

Flowering Jul–Sep. Rocky open sites, roadsides; 1400–1800 m; Tex.; Mexico; Central America.

Reports of *Pseudognaphalium viscosum* from the flora for states other than Texas are based on plants of *P. macounii*. *Pseudognaphalium viscosum* is similar to *P. leucocephalum*, which has broader and white-opaque phyllaries, longer bisexual corollas, and smooth cypselae.

14. Pseudognaphalium austrotexanum G. L. Nesom, Sida 19: 507, fig. 1. 2001 • South Texas rabbit-tobacco

Annuals, 30–70 cm; taprooted. **Stems** densely and closely white-tomentose-floccose, glabrescent, not glandular. **Leaf blades** (crowded, internodes mostly 1–3, sometimes to 10 mm) linear to linear-lanceolate, 2–5 cm × 1–3 mm, bases subclasping, not decurrent, margins strongly revolute, sometimes closely sinuate, faces bicolor, abaxial densely and closely white-tomentose, adaxial green, densely stipitate-glandular, otherwise glabrate. **Heads** in corymbiform arrays. **Involucres** broadly campanulate, 4.5–5 mm. **Phyllaries** in 5–7 series, silvery white (hyaline, shiny), narrowly ovate to oblong or elliptic (apices of inner thickened and slightly raised along midribs, apiculate), glabrous. **Pistillate florets** [46–]76–102. **Bisexual florets** (6–)8–11. **Cypselae** ridged, papillate-roughened.

Flowering Oct–Dec(–Jan). Sandy soil in pastures, grasslands, open disturbed sites; 0–10[–600] m; Tex.; Mexico (Nuevo León).

Pseudognaphalium austrotexanum is similar to *P. viscosum* in general appearance: taprooted annuals with stems white-tomentose, strictly erect, and mostly unbranched proximal to heads, leaves linear to linear-lanceolate, strongly bicolor (green and glandular adaxially, white-tomentose abaxially), loosely to strictly ascending, crowded on relatively short internodes and continuing to immediately proximal to the heads, and basally subclasping but not strongly auriculate, phyllaries silvery white, thin, and hyaline, and cypselae minutely papillate-roughened.

15. **Pseudognaphalium leucocephalum** (A. Gray) Anderberg, Opera Bot. 104: 147. 1991 • White rabbit-tobacco

Gnaphalium leucocephalum A. Gray, Smithsonian Contr. Knowl. 5(6): 99. 1853

Biennials or short-lived perennials, 30–60 cm; taprooted. **Stems** densely and persistently white-tomentose, usually with stipitate-glandular hairs protruding through tomentum. **Leaf blades** (crowded, internodes mostly 1–3, sometimes to 10 mm) linear-lanceolate, 3–7 cm × 1–5(–6) mm, bases subclasping, not decurrent, margins strongly revolute, faces bicolor, abaxial densely white-tomentose, adaxial green, densely stipitate-glandular. **Heads** in corymbiform arrays. **Involucres** broadly campanulate, 5–6 mm. **Phyllaries** in 5–7 series, bright white (opaque, dull), oblong to oblong-ovate, glabrous. **Pistillate florets** 66–85. **Bisexual florets** (6–14, California)29–44. **Cypselae** ridged, smooth. $2n = 28$.

Flowering (Jul–)Aug–Nov(–Dec). Sandy or gravelly slopes, stream bottoms, arroyos, areas of oak-sycamore, oak-pine, to pine woodlands, commonly in riparian vegetation; 50–2100 m; Ariz., Calif., N.Mex.; Mexico (Baja California, Baja California Sur, Chihuahua, Durango, Sinaloa, Sonora).

Pseudognaphalium leucocephalum is similar to *P. viscosum*, which has shiny, hyaline, ovate-lanceolate phyllaries, 200–250 pistillate florets, (13–)16–29 bisexual florets, and papillate-roughened cypselae. Some plants of *P. leucocephalum* also appear to approach *P. biolettii* in general appearance, and it is possible that some of them may represent hybrids. Plants of *P. biolettii* differ from *P. leucocephalum* in their typically eglandular stems, broader, basally ampliate, clasping, more widely spaced, and less densely glandular leaves, and thinner, shiny phyllaries.

16. **Pseudognaphalium biolettii** Anderberg, Opera Bot. 104: 147. 1991 • Bioletti's rabbit-tobacco

Gnaphalium bicolor Bioletti, Erythea 1: 16. 1893, not (Lindley) Schultz-Bipontinus 1845

Perennials (fragrant), 20–70(–120) cm; taprooted. **Stems** (sometimes lignescent near bases) proximally glabrescent, distally persistently tomentose, at least in the region of the heads, not glandular. **Leaf blades** (not crowded, internodes mostly 5+ mm), oblong-oblanceolate to oblanceolate (the distal lanceolate), 1.5–5(–8) cm × 4–10(–15) mm, bases auriculate-clasping, not decurrent, margins flat or slightly revolute, often undulate, faces bicolor (at least basal and proximal cauline), abaxial white-tomentose, adaxial bright green and ± densely glandular. **Heads** in corymbiform arrays. **Involucres** turbinate-campanulate, 5–5.5(–6) mm. **Phyllaries** in 4–5 series, white or sometimes slightly pinkish (opaque, shiny), ovate to oblong-ovate or oblong (often longitudinally wrinkled or grooved), glabrous. **Pistillate florets** 41–73. **Bisexual florets** 5–13. **Cypselae** ridged, smooth.

Flowering Apr–Jun(–Oct). Rocky slopes, roadsides, sandy plains with *Larrea*, coastal strand, matorral, and chaparral; 5–600(–1200) m; Calif.; Mexico (Baja California, Baja California Sur).

17. **Pseudognaphalium californicum** (de Candolle) Anderberg, Opera Bot. 104: 147. 1991 • California rabbit-tobacco

Gnaphalium californicum de Candolle in A. P. de Candolle and A. L. P. P. de Candolle, Prodr. 6: 224. 1838; *G. decurrens* Ives var. *californicum* (de Candolle) A. Gray

Annuals, biennials, or perennials, 20–40 cm; taprooted. **Stems** stipitate-glandular, sometimes lightly villous as well. **Leaf blades** mostly narrowly oblong-lanceolate, 4–10 cm × 5–10 (–20) mm (relatively even-sized distally), bases auriculate-clasping to subclasping or not, decurrent (2–10 mm) or not, margins flat or slightly revolute, faces concolor, mostly green, stipitate-glandular, viscid, sometimes lightly villous as well. **Heads** in corymbiform arrays. **Involucres** campanulo-globose, 5.5–7 mm. **Phyllaries** in 7–10 series, white (opaque, shiny or dull), broadly ovate to oblong-obovate, glabrous. **Pistillate florets** 105–140. **Bisexual florets** 7–12. **Cypselae** ridged, smooth. $2n = 28$.

Flowering Apr–Jul. Sandy canyons, dry hills, coastal chaparral; 60–800 m; Calif., Oreg.; Mexico (Baja California).

Leaf insertion in *Pseudognaphalium californicum* is unusually variable: bases may be clasping to subclasping without decurrent margins or decurrent and not show any clasping tendency.

18. Pseudognaphalium roseum   (Kunth) Anderberg, Opera Bot. 104: 148. 1991 • Rosy rabbit-tobacco

Gnaphalium roseum Kunth in A. von Humboldt et al., Nov. Gen. Sp. 4(fol.): 63. 1818; 4(qto.): 81. 1820

Annuals or perennials, 50–200 cm; taprooted. **Stems** persistently woolly-tomentose, not glandular. **Leaf blades** oblong-lanceolate to oblanceolate, mid-cauline 3–7 cm × (3–)6–15(–20) mm, bases clasping to subclasping, not decurrent, margins usually undulate, faces concolor or weakly bicolor, usually woolly-tomentose, sometimes tardily glabrescent adaxially, stipitate- or sessile-glandular beneath tomentum. **Heads** in corymbiform arrays. **Involucres** campanulate, 4–4.5 mm. **Phyllaries** in 5–6 series, usually white, sometimes pink (opaque or hyaline, dull to shiny), ovate to ovate-oblong, glabrous. **Pistillate florets** 45–90(–110). **Bisexual florets** (5–)6–12 (–18). **Cypselae** weakly ridged, smooth.

Flowering Mar–Jun. Open, disturbed sites; 10–50 [–1000+] m; introduced; Calif.; Mexico; Central America.

Pseudognaphalium roseum usually grows above 1000 m in Mexico; it grows below 50 m in California, where it is probably adventive. The closest collections of the species southward from California are from Sinaloa and southern Chihuahua. It is abundant in Mexico only in the eastern and southern states.

Pseudognaphalium roseum is recognized by its persistently tomentose stems and leaves, the leaves clasping to subclasping and non-decurrent, weakly bicolor and sessile-glandular beneath the tomentum, often relatively thick stems, relatively large heads with relatively numerous, white or pink, opaque phyllaries, relatively numerous florets, and smooth-faced cypselae. It has been confused with *P. canescens*; plants of *P. roseum* with relatively few bisexual florets can be distinguished from *P. canescens* by their subclasping leaves commonly with closely wavy margins, broader and more numerous phyllaries, and smooth-faced cypselae. Plants from southern California are atypical in their slightly smaller heads.

19. Pseudognaphalium arizonicum   (A. Gray) Anderberg, Opera Bot. 104: 147. 1991 • Arizona rabbit-tobacco

Gnaphalium arizonicum A. Gray, Proc. Amer. Acad. Arts 19: 3. 1883

Annuals or perennials, 20–50 cm; taprooted. **Stems** loosely and densely woolly-tomentose (hairs usually with reddish or purplish cross walls), not glandular. **Leaf blades** linear-oblanceolate to linear-lanceolate, 2–6 cm × 2–7 mm, bases not clasping, decurrent 3–15(–20) mm, margins weakly and narrowly revolute, faces concolor to weakly bicolor, tomentose (hairs commonly with reddish or purplish cross walls), not glandular. **Heads** borne singly or in terminal glomerules or corymbiform arrays. **Involucres** turbinate-campanulate, 5–6 mm. **Phyllaries** in 4–5 series, usually brownish to tawny, rarely slightly rosy (opaque, shiny), ovate-lanceolate to lanceolate, glabrous. **Pistillate florets** (25–)30–49. **Bisexual florets** (1–)3–6. **Cypselae** ridged, papillate-roughened.

Flowering Aug–Sep. Open woodlands and chaparral [wide ranging habitats in Mexico, agricultural land to oak and pine woodlands]; 1600–2300 m; Ariz., Tex.; w Mexico.

Pseudognaphalium arizonicum is superficially similar to *P. stramineum* in its narrow, concolor leaves; *P. stramineum* has non-decurrent leaves, light yellowish phyllaries, and more pistillate and bisexual florets.

20. Pseudognaphalium jaliscense   (Greenman) Anderberg, Opera Bot. 104: 147. 1991 • Jalisco rabbit-tobacco   F

Gnaphalium jaliscense Greenman, Proc. Amer. Acad. Arts 39: 96. 1903

Annuals or biennials, 30–70 cm; taprooted. **Stems** (branched among heads) densely and persistently loosely woolly-tomentose-sericeous, not glandular. **Leaf blades** narrowly lanceolate to nearly linear, 3–10 cm × 3–6 mm, bases not clasping, decurrent 4–8 mm, margins flat or slightly revolute, faces concolor, tomentose-sericeous (bases of hairs enlarged), sessile-glandular beneath tomentum. **Heads** in corymbiform arrays. **Involucres** campanulate, 5–6 mm. **Phyllaries** in 5–6(–7) series, white (opaque, dull), ovate or elliptic (keeled, apiculate), glabrous. **Pistillate florets** (80–)115[–180]. **Bisexual florets** (6–)8–12[–30]. **Cypselae** weakly ridged, papillate-roughened or smooth.

Flowering Jul–Oct. Grasslands, chaparral, openings in oak-pine-juniper, oak, and ponderosa pine woodlands, roadsides, disturbed sites; 1500–2300 m; Ariz., Colo., Nebr., N.Mex., Tex.; Mexico.

Pseudognaphalium jaliscense is recognized by its relatively long, narrow, concolor to weakly bicolor leaves with non-clasping, short-decurrent bases, relatively large heads with white, opaque, dull phyllaries, and relatively large numbers of pistillate and bisexual florets. Counts of pistillate and bisexual florets from the United States collections are mostly 90–115 and (6–)8–12 (fewer than in Mexico).

21. Pseudognaphalium ramosissimum   (Nuttall) Anderberg, Opera Bot. 104: 147. 1991   • Pink rabbit-tobacco   E

Gnaphalium ramosissimum Nuttall, Proc. Acad. Nat. Sci. Philadelphia 4: 20. 1848

Biennials (sweetly fragrant), 50–120(–150) cm; taprooted. **Stems** (erect) gray-tomentose, glabrescent, stipitate-glandular beneath tomentum. **Leaf blades** linear to lanceolate, oblong, or narrowly spatulate, (1–)3–7 cm × 3–5(–7) mm, bases not clasping, decurrent 2–10 mm, margins revolute and closely undulate, faces concolor, greenish, loosely tomentose, stipitate-glandular. **Heads** usually in paniculiform (broadly columnar or at least as long as broad, sometimes pyramidal) arrays. **Involucres** turbinate to short-cylindric, 5–6 mm. **Phyllaries** in 4–5 series, usually pinkish, sometimes white or greenish (hyaline, dull), ovate to ovate-oblong, loosely tomentose (bases). **Pistillate florets** 38–62. **Bisexual florets** 2–7. **Cypselae** ridged, smooth. **2*n*** = 28.

Flowering Jul–Sep. Dry, open slopes, sparsely wooded, sandy fields, dunes; 20–600 m; Calif.

89. HELICHRYSUM   Miller, Gard. Dict. abr. ed. 4, vol. 2. 1754 (as Elichrysum), name and orthography conserved   •   [Greek *helios*, sun, and *chrysos*, gold, and *helichrysos*, Greek name for a local species of Asteraceae]   I

Guy L. Nesom

[**Annuals, biennials, perennials,**] **subshrubs, or shrubs** (often aromatic), mostly 20–80 cm; taprooted. **Stems** usually 1, usually erect, sometimes decumbent to procumbent (± woolly-tomentose, usually stipitate- or sessile-glandular as well). **Leaves** cauline; alternate; petiolate [sessile]; blades ovate [spatulate to lanceolate or linear], bases cuneate to truncate [usually clasping and/or decurrent], margins entire (sometimes revolute), faces concolor [bicolor], usually gray to white and tomentose or sericeous [adaxial sometimes greenish and glabrescent], sometimes stipitate- or sessile-glandular as well. **Heads** disciform or discoid, in glomerules in corymbiform arrays. **Involucres** campanulate, 4–8 mm. **Phyllaries** in 3–5[–7] series, whitish [stramineous, orange, reddish, or pinkish] (opaque or hyaline, usually shiny; stereomes green, usually sessile-glandular distally). **Receptacles** flat, glabrous, epaleate. **Peripheral (pistillate) florets** 0 or 1–2 (fewer than bisexual): corollas yellowish. **Inner (bisexual) florets** 3–30[–50+]; corollas usually yellowish. **Cypselae** ± columnar, faces usually smooth, sometimes papillate (roughened by raised, imbricate tips of epidermal cells), sometimes with 4–6 longitudinal ridges, glabrous [± strigose or myxogenic, papilliform hairs]; **pappi** readily falling, of 12–20 distinct or loosely coherent basally, barbellate [subplumose] bristles in 1 series. ***x*** = 7.

Species about 600 (1 in the flora): introduced; mostly Old World, especially s Africa and Madagascar.

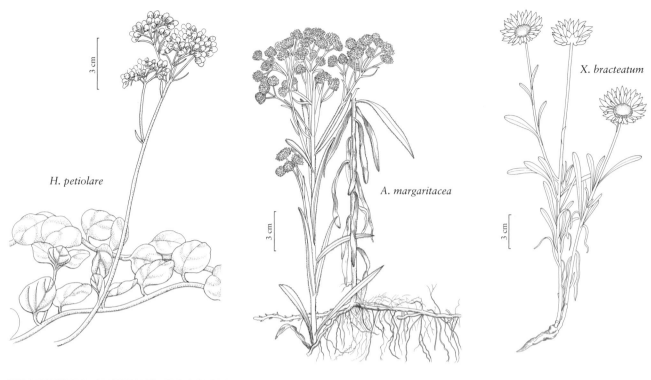

HELICHRYSUM ∘ ANAPHALIS ∘ XEROCHRYSUM

1. Helichrysum petiolare Hilliard & B. L. Burtt, Notes Roy. Bot. Gard. Edinburgh 32: 357. 1973 • Licorice plant, trailing dusty miller F I

Shrubs or subshrubs, aromatic, to 60 cm. **Stems** loosely branched, straggling. **Leaves** petiolate; blades ovate, 1–2 cm, apices obtuse to subacute, faces concolor, silvery green, woolly-tomentose. **Phyllaries** creamy white.

Flowering summer. "Shrubby plants forming a dense thicket on the slope of Bolinas Ridge above Stinson Beach" (J. T. Howell 1970); 300–600 m; introduced; Calif; s Africa (Cape region); introduced in Europe.

The species is included among the invasive species of California (C. C. Bossard et al. 2000); it is native to the Cape region of South Africa, long cultivated in Europe, and now widely offered for sale on the internet. In contrast to the assessment of invasiveness in California, one commercial website selling the species notes this "forms tangled mounds of trailing grey-felted leaves. Given something to scramble up, it will climb. And although this description makes it sound fairly rampant, it is definitely not invasive." It is often recommended for planters and hanging baskets, where the silvery leaves provide backdrop for bright colors.

The name *Helichrysum petiolatum* D. Don has been misapplied to plants of *H. petiolare*.

90. ANAPHALIS de Candolle in A. P. de Candolle and A. L. P. P. de Candolle, Prodr. 6: 271. 1838 • [An ancient name or, perhaps, derived from generic name *Gnaphalium*]

Guy L. Nesom

Perennials [subshrubs] (dioecious or subdioecious), 20–80(–120+) cm; fibrous-rooted (rhizomatous, not stoloniferous). **Stems** usually 1, usually erect. **Leaves** basal and cauline; alternate; petiolate or sessile; blades oblanceolate or lanceolate to linear, bases ± cuneate, margins entire, faces usually bicolor [concolor], abaxial usually white to gray and tomentose (sometimes glandular as well, proximal leaves sometimes ± glabrate), adaxial usually greenish and glabrate or glabrous, sometimes grayish and sparsely arachnose. **Heads** usually discoid (unisexual or nearly so) or disciform, in glomerules in corymbiform or paniculiform arrays. **Involucres** subglobose, 6–8(–10) mm. **Phyllaries** in 8–12 series, bright white (opaque, at least toward tips, often proximally woolly; stereomes not glandular), unequal, ± papery (at least toward tips). **Peripheral (pistillate) florets** 50–150 (more numerous than staminate; sometimes a few pistillate florets peripheral in predominantly staminate heads or 1–9 staminate florets central in predominantly pistillate heads); corollas yellowish. **Inner (functionally staminate) florets** 30–55; corollas yellowish. **Cypselae** oblong [obclavate, ovoid, or cylindric] (2-nerved), faces ± scabrous (hairs clavate, not myxogenic); **pappi** usually readily falling, of 10–20 distinct or basally connate, barbellate bristles (tips of bristles ± clavate in bisexual or functionally staminate florets). $x = 14$.

Species ca. 110 or fewer (1 in the flora): North America, mostly central Asia and India.

1. **Anaphalis margaritacea** (Linnaeus) Bentham & Hooker f., Gen. Pl. 2: 303. 1873 • Pearly everlasting, anaphale marguerite, immortelle blanche [F]

Gnaphalium margaritaceum Linnaeus, Sp. Pl. 2: 850. 1753; *Anaphalis margaritacea* var. *occidentalis* Greene; *A. margaritacea* var. *subalpina* (A. Gray) A. Gray

Perennials; rhizomes relatively slender. **Stems** white, densely and closely tomentose, not glandular. **Leaf blades** 1–3-nerved, 3–10(–15) cm, bases subclasping, decurrent, margins revolute, abaxial faces tomentose or glabrescent (proximal leaves), not glandular or very sparsely and inconspicuously glandular, adaxial faces green, glabrate. **Involucres** 5–7 × 6–8(–10) mm. **Phyllaries** ovate to nearly linear (innermost), subequal to unequal, apices white, opaque. **Cypselae** 0.5–1 mm, bases constricted into stipiform carpopodia. $2n = 28$.

Flowering Jul–Oct (sporadically longer). Dry woods, often with aspen or mixed conifer-hardwood, borders and trails, dunes, fields, roadsides, other open, often disturbed sites; 0–3200 m; St. Pierre and Miquelon; Alta., B.C. , Man., N.B., Nfld. and Labr., N.W.T., N.S., Ont., P.E.I. Que., Sask., Yukon; Alaska, Ariz., Ark., Calif., Colo., Conn., Del., Idaho, Ill., Ind., Iowa, Ky., Maine, Md., Mass., Mich., Minn., Mont., Nebr., Nev., N.H., N.J., N.Mex., N.Y., N.C., Ohio, Oreg., Pa., R.I., S.Dak., Tenn., Tex., Utah, Vt., Va., Wash., W.Va., Wis., Wyo.; Mexico (Baja California); Asia; introduced in Europe.

Anaphalis margaretacea was widely planted as an ornamental and escaped. It apparently naturalized from its native range in both Asia and North America; it is cultivated and naturalized in Europe.

Anaphalis margaritacea has the aspect of *Pseudognaphalium*; it differs in being subdioecious (polygamo-dioecious; the heads either staminate or primarily pistillate) and in its distinctive cypselar vestiture. It is further recognized by its combination of rhizomatous habit, subclasping-decurrent, bicolor, revolute leaves, and distally white phyllaries. Segregate species and varieties have been described among the North American plants (in addition to the two cited above), based on variation in habit, vesture, and leaf morphology and density, but the variants appear to be more like a complex series of ecotypes rather than broader evolutionary entities.

91. **XEROCHRYSUM** Tzvelev, Novosti Sist. Vyssh. Rast. 27: 151. 1990 • [Greek *xeros*, dry, and *chrysos*, gold, perhaps alluding to phyllaries] [I]

Guy L. Nesom

Bracteantha Anderberg & Haegi

Annuals, biennials, or perennials, 20–90+ cm; taprooted (rhizomatous, not stoloniferous). **Stems** usually 1, erect (rarely 1–2 or more times branched; usually arachnose and ± stipitate-glandular). **Leaves** cauline; alternate; sessile (or nearly so); blades elliptic or spatulate to oblanceolate, lanceolate, or linear, bases cuneate, margins entire, faces concolor, usually arachnose and ± stipitate-glandular. **Heads** disciform, borne singly or (2–3) in loose, corymbiform arrays. **Involucres**

± hemispheric, 10–30 mm. **Phyllaries** in 3–8+ series, usually yellow or brown to purple, sometimes white or pinkish (opaque, stereomes not glandular), unequal, usually chartaceous toward tips (spreading at flowering, deflexed in age). **Receptacles** flat, glabrous, epaleate. **Peripheral (pistillate) florets** (0–)25–50 (fewer than bisexual); corollas yellow. **Inner (bisexual) florets** 200–400; corollas yellow (lobes erect). **Cypselae** columnar to ± prismatic (4-angled), faces smooth, glabrous; pappi readily falling, of 25–35+, distinct or loosely basally ± coherent, subplumose to barbellate bristles in 1 series (falling separately or in groups or rings). *x* = 12, 13, 14, 15.

Species 6 (1 in the flora): introduced; Australia; cultivated and escaping in many areas elsewhere.

1. **Xerochrysum bracteatum** (Ventenat) Tzvelev, Novosti Sist. Vyssh. Rast. 27: 151. 1990 • Strawflower, golden everlasting, paper daisy F I

Xeranthemum bracteatum Ventenat, Jard. Malmaison 1: plate 2. 1803; *Bracteantha bracteata* (Ventenat) Anderberg & Haegi; *Helichrysum bracteatum* (Ventenat) Andrews

Annuals or perennials, 20–90 cm (dwarf varieties mostly less than 30 cm); taprooted. **Stems** scabrellous and minutely glandular, sometimes lightly arachnoid as well. **Leaf blades** narrowly elliptic or oblong to oblong-lanceolate, 2–5

(–10) cm, sometimes slightly fleshy, vestiture like stems. **Heads** 3–5 cm diam. (across open involucres). **Phyllaries** yellow (wild form) or white to pink, red, and purple (cultivars), 10–20 mm. **2n** = 24, 26, 28.

Flowering summer. Disturbed sites; 20–50 m; introduced; Conn., Mass.; Australia.

Known in horticulture as *Helichrysum bracteatum*, *Xerochrysum bracteatum* is widely cultivated for its relatively large, spreading, and spectacularly colored involucral bracts. It apparently rarely escapes from cultivation.

92. **GNAPHALIUM** Linnaeus, Sp. Pl. 2: 850. 1753; Gen. Pl. ed. 5, 368. 1754 • [Greek *gnaphalion*, a downy plant, the name anciently applied to these or similar plants]

Guy L. Nesom

Filaginella Opiz

Annuals [biennials or perennials], (1–)3–30 cm; usually taprooted, sometimes fibrous-rooted. **Stems** usually 1, erect (often with decumbent-ascending branches from bases; ± woolly-tomentose, not glandular). **Leaves** mostly cauline; alternate; ± sessile; blades oblanceolate to spatulate or linear, bases ± cuneate, margins entire, faces concolor, gray and tomentose. **Heads** disciform, usually in ± capitate clusters (in axils of leaves or bracts), sometimes in spiciform glomerules. **Involucres** narrowly to broadly campanulate, 2.5–4 mm. **Phyllaries** in 3–5 series, usually white or tawny to brown (opaque or hyaline, often shiny; stereomes usually glandular distally), ± equal to unequal, chartaceous toward tips (inner phyllaries narrowly oblong, usually white-tipped and protruding distal to outer). **Receptacles** flat, smooth, epaleate. **Peripheral (pistillate) florets** 40–80 (more numerous than bisexual); corollas purplish or whitish. **Inner (bisexual) florets** 4–7; corollas purplish or whitish. **Cypselae** oblong, faces usually glabrous, sometimes minutely papillate (hairs ± papilliform, not myxogenic); pappi readily falling, of 8–12 distinct, barbellate bristles in 1 series. *x* = 7.

Species ca. 38 (3 in the flora): North America, Mexico, Central America, South America, Asia, Africa, Australia.

Generic segregations have reduced *Gnaphalium* from hundreds of species to ca. 38. North American species (north of Mexico) not included here have been segregated to *Euchiton*, *Gamochaeta*, *Omalotheca*, and *Pseudognaphalium*. Species of *Gnaphalium* in the strict sense (adopted here) are usually ca. 3–30 cm, loosely tomentose and not glandular, and have loosely glomerulate heads, involucres 2–3(–4) mm diam., white-tipped inner phyllaries, papillate cypselae, and readily falling pappi of distinct bristles, features especially contrasting with *Pseudognaphalium*, to which most North American species have been transferred. Because of their relatively small stature and tendency to produce loosely spiciform arrays of heads, gnaphaliums sometimes are identified as gamochaetas, which have different cypselar vestitures and different pappi. The lectotype species of *Gnaphalium* is *G. uliginosum* Linnaeus; discussion of this choice rather than *Pseudognaphalium* (*Gnaphalium*) *luteoalbum* (Linnaeus) Hilliard & Burtt is given in C. Jeffrey (1979), O. M. Hilliard and B. L. Burtt (1981), and J. McNeill et al. (1987).

Gnaphalium polycaulon Persoon is included in the key because it probably will be found in warmer coastal localities in the United States (perhaps Florida or California). It is a cosmopolitan weed (Old World native) and occurs in Mexico. It has sometimes been identified by the misapplied name *Gnaphalium indicum* Linnaeus.

1. Heads in relatively elongate, interrupted, spiciform glomerules, not subtended by foliaceous bracts; leaf blades oblanceolate, mostly 2–4 cm; tips of inner phyllaries brownish . . . . [*Gnaphalium polycaulon*]
1. Heads in terminal glomerules, subtended by foliaceous bracts; leaf blades spatulate to oblanceolate-oblong or linear to narrowly oblanceolate, mostly 0.5–2.5 cm; tips of inner phyllaries white.
  2. Leaf blades spatulate to oblanceolate-oblong, 3–8(–10) mm wide; bracts subtending heads oblanceolate to obovate, longest 4–12 × 1.5–4 mm, shorter than or equaling to slightly surpassing glomerules; inner phyllaries narrowly oblong, apices blunt . . . . 1. *Gnaphalium palustre*
  2. Leaf blades linear to narrowly oblanceolate, 0.5–3 mm wide; bracts subtending heads linear, oblanceolate, or obovate, 5–25 × 0.5–2 mm, surpassing glomerules; inner phyllaries narrowly triangular, apices acute.
    3. Leaf blades linear, the largest 0.4–5 cm; bracts subtending heads linear, 10–25 × 0.5–1 mm; heads in spiciform arrays of spikelike, axillary glomerules . . . . 2. *Gnaphalium exilifolium*
    3. Leaf blades oblanceolate, the largest 1–5 cm; bracts subtending heads linear, oblanceolate, or obovate, 5–15 × 1–2 mm; heads in terminal, capitate glomerules, sometimes in axillary glomerules . . . . 3. *Gnaphalium uliginosum*

1. Gnaphalium palustre Nuttall, Trans. Amer. Philos. Soc., n. s. 7: 403. 1841 • Western marsh cudweed

Filaginella palustris (Nuttall) Holub; *Gnaphalium palustre* var. *nanum* Jepson; *G. heteroides* Klatt

Annuals, (1–)3–15(–30) cm; taprooted or fibrous-rooted. **Stems** commonly with decumbent branches produced from bases, densely or loosely and persistently woolly-tomentose. **Leaf blades** spatulate to oblanceolate-oblong, 1–3.5 cm × 3–8(–10) mm. **Bracts** subtending heads oblanceolate to obovate, 4–12 × 1.5–4 mm, shorter than or surpassing glomerules. **Heads** in capitate glomerules (at stem tips and in distalmost axils). **Involucres** 2.5–4 mm. **Phyllaries** brownish, bases woolly, the inner narrowly oblong with white (opaque), blunt apices. $2n = 14$.

Flowering May–Oct. Arroyos, sandy streambeds, pond edges, potholes, other moist, open sites; 100–2900 m; Alta., B.C., Sask.; Ariz., Calif., Colo., Idaho, Mont., Nebr., Nev., N.Mex., N.Dak., Oreg., S.Dak., Utah, Wash., Wyo.; Mexico.

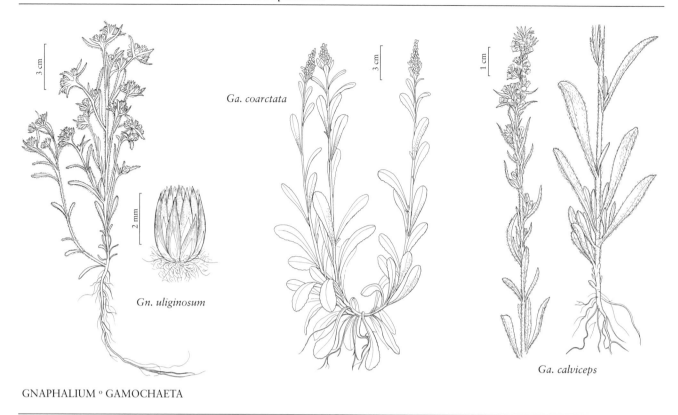

Ga. coarctata

Gn. uliginosum

Ga. calviceps

GNAPHALIUM ○ GAMOCHAETA

2. Gnaphalium exilifolium   A. Nelson, Bull. Torrey Bot. Club 29: 406. 1902 • Slender cudweed

Gnaphalium angustifolium A. Nelson, Bull. Torrey Bot. Club 26: 357. 1899, not Lamarck 1788; *G. grayi* A. Nelson & J. F. Macbride; *G. strictum* A. Gray 1857, not Lamarck 1788

Annuals, 3–15(–25) cm; taprooted or fibrous-rooted. **Stems** commonly branched from bases, erect to ascending, tomentose. **Leaf blades** linear, 0.4–5 cm × 0.5–3 mm. **Bracts** subtending heads linear, 10–25 × 0.5–1 mm, surpassing glomerules. **Heads** in spiciform glomerules (along distal $^1/_3$–$^2/_3$ of main stems, sometimes appearing loosely spiciform). **Involucres** 2.5–3.5 mm. **Phyllaries** brownish, bases woolly, inner narrowly triangular with whitish, acute apices. **2***n* = 14.

Flowering Jul–Oct. Lake and pond margins, streamsides, seeps, moist meadows; 1400–3000 m; Ariz., Colo., N.Mex., S.Dak., Utah, Wyo.; Mexico (Chihuahua).

Gnaphalium exilifolium, a New World native, differs from the Old World *G. uliginosum* chiefly in its linear, slightly longer cauline leaves and bracts and in its spiciform arrangement of heads. Pistillate corollas of *G. exilifolium* are sometimes red-tipped. Cypselae of both taxa are variably smooth to papillate; local populations apparently are consistent in this feature; the variation is otherwise geographically inconsistent.

3. Gnaphalium uliginosum   Linnaeus, Sp. Pl. 2: 856. 1753 • Marsh cudweed   F I

Filaginella uliginosa (Linnaeus) Opiz

Annuals, 3–15(–25) cm; taprooted or fibrous-rooted. **Stems** erect, usually branched from bases, sometimes simple, closely to loosely tomentose. **Leaf blades** oblanceolate, 1–5 cm × 1–3 mm. **Bracts** subtending heads linear, oblanceolate, or obovate, 5–15 × 1–2 mm, usually surpassing glomerules. **Heads** borne singly or in terminal, capitate glomerules, sometimes in axillary glomerules. **Involucres** 2–4 mm. **Phyllaries** brownish, bases woolly, inner narrowly triangular with whitish, acute apices. **2***n* = 14.

Flowering Jul–Oct. Lake and pond margins, stream banks, wet meadows, other permanently or sporadically moist sites, disturbed sites; 1400–3000 m; Greenland; St. Pierre and Miquelon; Alta., B.C., Man., N.B., Nfld. and Labr. (Nfld.), N.S., Ont., P.E.I., Que., Sask., Yukon; Alaska, Colo., Conn., Del., Idaho, Ill., Iowa, Kans., Ky., Maine, Md., Mass., Minn., Mont., N.H., N.J., N.Y., Ohio, Oreg., Pa., R.I., S.Dak., Utah, Vt., Va., Wash., W.Va., Wis., Wyo.; Europe.

Gnaphalium uliginosum is native to Europe; it is not clear whether some or all of the North American plants may have been introduced into the flora.

93. GAMOCHAETA Weddell, Chlor. Andina 1: 151. 1856 • [Greek *gamos*, union, and *chaete*, loose and flowing hair, alluding to basally connate pappus bristles]

Guy L. Nesom

Gnaphalium Linnaeus sect. *Gamochaeta* (Weddell) O. Hoffmann; *G.* subg. *Gamochaeta* (Weddell) Grenier

Annuals, biennials, or perennials, (1–)5–65 cm; taprooted or fibrous-rooted [subrhizomatous]. **Stems** 1+, usually erect, sometimes decumbent-ascending. **Leaves** basal and cauline; alternate; sessile; blades mostly linear to oblanceolate or spatulate, bases cuneate to ± cordate, margins entire, sometimes sinuate, abaxial faces mostly white or gray and tomentose or pannose-tomentose, adaxial green and glabrescent or glabrous, or grayish and arachnose, loosely tomentose, or subpannose. **Heads** disciform, usually in glomerules borne in continuous or interrupted, usually spiciform, sometimes paniculiform, arrays (reduced to terminal glomerules in depauperate individuals). **Involucres** narrowly to broadly campanulate, 2.5–5 mm. **Phyllaries** in 3–7 series, unequal, mostly brownish to stramineous, sometimes purplish, hyaline, often shiny, distally chartaceous to scarious, eglandular. **Receptacles** flat (concave in fruit), glabrous, epaleate. **Pistillate florets** 50–130, more numerous than bisexual florets; corollas all yellow or purplish-tipped. **Bisexual florets** 2–7; corollas all yellow or distally purplish. **Cypselae** oblong, slightly flattened, faces with papilliform hairs (myxogenic, their lengths about equaling diams.); **pappi** readily falling, of 12–28 barbellulate bristles in 1 series (basally connate in smooth rings, falling as units). $x = 7$.

Species ca. 50 (12 in the flora): North America, Mexico, West Indies, Central America, South America; some species adventive and naturalized in Europe, Asia, Australia, and elsewhere.

Gamochaeta comprises about 50 species (A. L. Cabrera 1961; S. E. Freire and L. Iharlegui 1997) or about 80 species (Cabrera 1977+, part 10), all native to the Americas. Most are known only from South America; some apparently are native to Mexico and the flora area (G. L. Nesom 2004b). Some species are strongly weedy and have extended non-native ranges. Because of inconsistencies in the identification of those species, it has been difficult to evaluate the overall distributions of the widespread species.

The distinctiveness of *Gamochaeta* was emphasized by A. L. Cabrera (1961, and in later floristic treatments of South American species) and by other botanists who have treated it as a separate genus in the last decade . The genus is distinguished by its combination of relatively small heads in spiciform arrays, concave post-fruiting receptacles, truncate collecting appendages of style branches in bisexual florets, relatively small cypselae with minute, mucilage-producing papilliform hairs on the faces, and pappus bristles basally connate in smooth rings and released as single units. It seems likely that most species are primarily autogamous, in view of the tiny, non-showy heads that barely open through flowering. The consistency of vegetative and floral features in some of the species supports this hypothesis.

In the flora area, some species have commonly been treated as variants within *Gamochaeta purpurea*; distinctions are evident in the field, where it is common to find as many as five species growing in proximity without intergradation. Species of *Gamochaeta* are distinguished by differences primarily in root form, leaf shape, nature and distribution of indument, and phyllary morphology. Chromosome counts have been reported for some species; because of the unreliability of identifications, vouchers for those counts should be restudied.

SELECTED REFERENCES Freire, S. E. and L. Iharlegui. 1997. Sinopsis preliminar del género *Gamochaeta* (Asteraceae, Gnaphalieae). Bol. Soc. Argent. Bot. 33: 23–35. Nesom, G. L. 1990f. The taxonomic status of *Gamochaeta* (Asteraceae: Inuleae) and the species of the United States. Phytologia 68: 186–198.

1. Leaves bicolor (abaxial faces closely white-pannose to pannose-tomentose, indument obscuring epidermis, adaxial faces glabrous or glabrate to sparsely arachnose).
    2. Basal and proximal cauline leaves usually withering before flowering (clusters of smaller leaves usually present in cauline axils); stems erect or ascending; plants (30–)50–85 cm; apices of inner phyllaries acute-acuminate; flowering mostly Jul–Aug . . . . 3. *Gamochaeta simplicicaulis*
    2. Basal and proximal cauline leaves present or not at flowering; stems erect to decumbent-ascending; plants mostly 10–50 cm; apices of inner phyllaries acute to obtuse, rounded, or blunt; flowering mostly Apr–Jun(–Jul in *G. calviceps*).
        3. Adaxial leaf faces glabrous or glabrate; involucres (± purplish) 2.5–3 mm, bases glabrous; outer phyllaries elliptic-obovate to broadly ovate-elliptic, apices rounded to obtuse; bisexual florets 2–3 . . . . 7. *Gamochaeta coarctata*
        3. Adaxial leaf faces sparsely arachnose (hairs persistent, evident at 10×); involucres (sometimes purplish) 3–4.5(–5) mm, bases (imbedded in tomentum) often sparsely arachnose on proximal $^{1}/_{5}$–$^{1}/_{2}$; outer phyllaries ovate, ovate-triangular, or ovate-lanceolate, apices acute to acuminate; bisexual florets 2–6.
            4. Stems not pannose (induments whitish, like closely appressed, polished cloth, hairs usually not individually evident); involucres 3–3.5(–4) mm; apices of inner phyllaries acute to acute-acuminate; bisexual florets 2–4 (cypselae purple) . . . . 4. *Gamochaeta chionesthes*
            4. Stems usually ± pannose or pannose-tomentose (hairs individually evident, longitudinally arranged); involucres 3–4.5 mm; apices of inner phyllaries acute, obtuse, or truncate-rounded, sometimes apiculate; bisexual florets 3–6 (cypselae tan to brownish).
                5. Blades of cauline leaves oblanceolate to spatulate (basal cells of hairs on adaxial faces persistent, expanded, glassy); involucres 4–4.5 mm; laminae of inner phyllaries triangular, apices acute (not apiculate); bisexual florets 3–4; plants fibrous-rooted or taprooted . . . . 1. *Gamochaeta purpurea*
                5. Blades of cauline leaves oblanceolate to oblanceolate-oblong or oblanceolate-obovate; involucres 4.5–5 or 3–3.5 mm; laminae of inner phyllaries elliptic-oblong to oblong, apices truncate-rounded or obtuse and apiculate; bisexual florets (3–)4–6; plants usually fibrous-rooted, rarely taprooted.
                    6. Arrays of heads uninterrupted (1–5 cm in early flowering) to strongly interrupted, mostly 5–18 cm × 10–12 mm (pressed); involucres 3–3.5 mm; outer phyllaries (tawny-transparent, never dark brown) ovate to ovate-lanceolate; cypselae 0.5–0.6 mm . . . . 5. *Gamochaeta argyrinea*
                    6. Arrays of heads usually continuous, rarely interrupted (then proximally), mostly 1–6(–8) cm × 12–18 mm (pressed); involucres 4.5–5 mm; outer phyllaries (and, often, laminae of inner, dark or greenish brown) broadly ovate-triangular (mid phyllaries ± keeled near apices); cypselae 0.7–0.8 mm . . . . 6. *Gamochaeta ustulata*
1. Leaves concolor or weakly bicolor (abaxial and adaxial faces ± equally greenish to gray-greenish, indument usually loosely tomentose or arachnose, sometimes subpannose, adaxial sometimes glabrescent in *G. stagnalis*).
    7. Leaf blades linear to narrowly oblanceolate (often folded along midveins, distally becoming arcuate, ± patent bracts surpassing the heads; basal cells of hairs on adaxial faces expanded, glassy); apices of outer phyllaries (brown) acute-acuminate (usually involute and spreading or recurving); Texas, near Mexican border . . . . 2. *Gamochaeta sphacelata*
    7. Leaf blades linear or linear-oblanceolate to spatulate, narrowly lanceolate, oblong-oblanceolate, oblanceolate-obovate, or oblanceolate (folded or not; if distal leaves or bracts surpassing heads, not arcuate and ± patent); apices of outer phyllaries (sometimes brownish) sometimes acute to acute-acuminate or attenuate-apiculate (not involute, except in *G. calviceps*); mostly (not always) se United States.

[8. shifted to left margin.—Ed.]

8. Blades of basal and proximal cauline leaves 4–16 mm wide (bracts among heads spatulate to oblanceolate, at least the proximal surpassing the glomerules of heads) . . . . 11. *Gamochaeta pensylvanica*
8. Blades of basal and proximal cauline leaves 2–6(–10) mm wide (mid and distal cauline becoming oblanceolate to linear, bracts among heads linear, oblanceolate, or oblong-oblanceolate, surpassing glomerules or not).
   9. Blades of mid and distal cauline leaves oblanceolate (bases subclasping; bracts among heads mostly shorter than glomerules); involucres (not purplish) 3.5–4 mm; phyllaries in 4–5 series . . . . 12. *Gamochaeta stachydifolia*
   9. Blades of mid and distal cauline leaves oblong-oblanceolate or oblanceolate to spatulate, narrowly lanceolate, linear-oblanceolate, or linear (bases not subclasping; bracts among heads shorter or longer than glomerules); involucres (sometimes purplish) 2.5–3.5 mm; phyllaries in 5–7 or 3–4(–5) series.
      10. Involucres (not purplish) 3–3.5 mm, bases sparsely arachnose or glabrous; arrays of heads interrupted (at least distally, main axes visible between heads); phyllaries in 5–7 series, outer ovate-triangular, lengths $^1/_3$–$^1/_2$ inner, apices acute-acuminate; flowering May–Jul . . . . 8. *Gamochaeta calviceps*
      10. Involucres (usually purplish, at stereome-lamina junctions of phyllaries) 2.5–3 mm, bases sparsely arachnose; arrays of heads initially capitate clusters or cylindric and uninterrupted (at least distally, main axes obscured by heads); phyllaries in 3–4(–5) series, outer ovate-lanceolate, lengths $^1/_2$–$^2/_3$ inner, apices narrowly to broadly acute; flowering (Feb–)Mar–May (later with moisture).
         11. Cauline leaves and bracts among heads narrowly lanceolate, linear-oblanceolate, or linear; arrays of heads initially uninterrupted cylindro-spiciform (becoming glomerulate-interrupted in late flowering) . . . 9. *Gamochaeta antillana*
         11. Cauline leaves and bracts among heads mostly oblanceolate to oblong-oblanceolate (± uniform in size and shape); arrays of heads capitate clusters or interrupted spiciform arrays (from early through late flowering) . . . . 10. *Gamochaeta stagnalis*

1. **Gamochaeta purpurea** (Linnaeus) Cabrera, Bol. Soc. Argent. Bot. 9: 377. 1961 • Spoon-leaf cudweed

Gnaphalium purpureum Linnaeus, Sp. Pl. 2: 854. 1753; *Gamochaeta rosacea* (I. M. Johnston) Anderberg; *Gnaphalium rosaceum* I. M. Johnston

Annuals (sometimes winter annuals), 10–40(–50) cm; fibrous-rooted or taprooted. **Stems** erect to decumbent-ascending, densely but loosely pannose or pannose-tomentose. **Leaves** basal and cauline, basal and proximal cauline usually withering before flowering; blades oblanceolate to spatulate, 1–6 cm × 5–14 mm (distal similar, at least among proximal heads, margins sometimes sinuate), faces usually bicolor, abaxial closely white-pannose, adaxial usually sparsely arachnose (basal cells of hairs persistent, expanded, glassy), sometimes glabrescent. **Heads** initially in continuous spiciform arrays 1–4(–5) cm × (5–)10–15 mm, later interrupted (glomerules widely separated, bracteate, the proximal often on relatively long peduncles). **Involucres** turbinate-cylindric, 4–4.5 mm, bases sparsely arachnose. **Phyllaries** in 4–5 series, outer ovate-triangular, lengths $^1/_3$–$^2/_3$ inner, apices acute-acuminate, inner triangular-lanceolate (usually striate), laminae purplish (in bud) to whitish or silvery (in fruit), apices acute (not apiculate). **Florets:** bisexual 3–4; all corollas usually purplish distally. **Cypselae** (tan) 0.6–0.7 mm. $2n = 14, 28$.

Flowering Apr–May(–Jun). Open, usually disturbed, commonly sandy habitats, roadsides, fields, woodland clearings and edges; 5–300 m; Ont.; Ala., Ariz., Ark. Conn., Del., D.C., Fla., Ga., Ill., Ind., Iowa, Kans., Ky., La., Maine, Md., Mass., Mich., Miss., Mo., N.J., N.Y., N.C., Ohio, Okla., Pa., R.I., S.C., Tenn., Tex., Va., W.Va.; Mexico; West Indies; Central America (Nicaragua); South America; Pacific Islands (Hawaii).

Gamochaeta purpurea apparently is native to North America and adventive elsewhere.

Basal cells of hairs on adaxial faces of leaves are expanded and glassy (versus hairs filiform to bases in most other species) and are diagnostic for *Gamochaeta purpurea*. From Maryland northward, plants of *G. purpurea* produce relatively small basal rosettes and relatively shallow fibrous roots or a filiform taproot; southward and southwestward, the basal rosettes often are larger and the fibrous roots are denser.

Gamochaeta purpurea apparently occurs widely through the world as a weed; it is fairly clearly native to eastern North America, where it is the least weedy of the gamochaetas. Plants of *G. purpurea* in southern Arizona along perennial streams at the base of the Santa Catalina Mountains were first collected in 1903 (G. L. Nesom 2004) and were, perhaps, accidentally established through visitation; the same sites are heavily infested by other, more aggressive, nonnative species. Collections of *G. purpurea* also have been made at higher elevations in the Santa Catalina, Rincon, and Chiricahua mountains, where the species is less likely to have been introduced by human activity. It also seems unlikely that plants in scattered Mexican localities were introduced there by human activity.

2. **Gamochaeta sphacelata** (Kunth) Cabrera, Bol. Soc. Argent. Bot. 9: 380. 1961 (as sphacilata) • Owl's crown

Gnaphalium sphacelatum Kunth in A. von Humboldt et al., Nov. Gen. Sp. 4(fol.): 67. 1818; 4(qto.): 86. 1820 (as sphacilatum)

Annuals, 10–35(–50) cm; usually taprooted, rarely fibrous-rooted. **Stems** erect to ascending, densely gray-white pannose. **Leaves** basal and cauline, basal usually withering before flowering; blades linear to narrowly oblanceolate (often folded along midveins), 1–4 cm × 1–3(–5) mm (distally becoming ± patent, arcuate bracts surpassing the heads), faces concolor or weakly bicolor, pannose-tomentose (± equally grayish to whitish, basal cells of hairs on adaxial faces persistent, expanded, glassy). **Heads** in terminal glomerules 1 cm or in interrupted, spiciform arrays 2–14 cm × 10–12(–14) mm (pressed; glomerules sometimes axillary). **Involucres** cylindro-campanulate, 3.5–4(–5) mm, sparsely arachnose. **Phyllaries** in (3–)4–5 series, outer triangular, lengths ¹⁄₃–¹⁄₂ inner, apices (brown) acute-acuminate (usually inrolled and spreading or recurved), inner triangular-lanceolate, laminae purplish (in bud) to whitish or silvery (in fruit), apices (usually striate) acute (not apiculate). **Florets:** bisexual 3–5; all corollas usually purplish distally. **Cypselae** (tan) 0.5–0.6 mm.

Flowering Jul–Sep. Grasslands, pine-oak woodlands, dry and wet sites; 2000–2800 m; Tex.; Mexico; South America (scattered, fide A. L. Cabrera 1977+, part 10).

Gamochaeta sphacelata is recognized by its linear to narrowly oblanceolate, concolor leaves, interrupted, spiciform arrays of heads in compact axillary glomerules or on lateral branches, glomerules subtended by ± patent, arcuate bracts, dark brown involucres, and acute-acuminate, recurving tips of outer and mid phyllaries. Roots are typically lignescent taproots. As in *G. purpurea*, the basal cells of each hair on adaxial leaf faces are persistent, expanded, and glassy.

3. **Gamochaeta simplicicaulis** (Willldenow ex Sprengel) Cabrera, Bol. Soc. Argent. Bot. 9: 379. 1961 • Simple-stem cudweed [I]

Gnaphalium simplicicaule Willldenow ex Sprengel, Syst. Veg. 3: 481. 1826; *G. purpureum* Linnaeus var. *simplicicaule* (Willldenow ex Sprengel) Klatt

Annuals or biennials, (30–)50–85 cm; fibrous-rooted. **Stems** erect or ascending (usually 1, sometimes 2–5), densely and closely white-pannose. **Leaves** basal and cauline, basal usually withering before flowering, blades oblanceolate to oblanceolate-spatulate, 5–9 cm × 6–18 mm (gradually smaller distally, margins closely undulate, nearly crenulate; distal cauline linear-lanceolate to linear-oblanceolate, apices long-acute; sessile clusters of smaller leaves produced in axils of mid and distal cauline leaves), faces bicolor, abaxial closely white-pannose, adaxial glabrous (shiny). **Heads** in interrupted, spiciform arrays (8–)16–30 cm × 10–14 mm (pressed; sometimes with ascending, lateral branches, glomerules usually subtended by ± patent linear bracts longer than the glomerules). **Involucres** cylindro-campanulate, 3–3.5 mm, bases glabrous. **Phyllaries** in 4–6 series, outer ovate to oblong, lengths ¹⁄₃–¹⁄₂ inner, apices acute-acuminate, inner narrowly oblong, laminae brownish to tan (not purplish), apices acuminate-apiculate. **Florets:** bisexual (2–)3; all corollas yellowish distally. **Cypselae** (tan) 0.5–0.6 mm.

Flowering (Jun–)Jul–Aug(–Oct). Open sites, sandy soil, roadsides, fields, open woods, dunes; 0–10 m; introduced; Ala., Fla., Ga., N.C., S.C.; South America; naturalized in New Zealand, Australia, Java.

Gamochaeta simplicicaulis was reported from North America by G. L. Nesom (1999b, 2000b) as an apparently recent adventive.

4. **Gamochaeta chionesthes** G. L. Nesom, Sida 21: 725, figs. 2-4. 2004 • White-cloaked cudweed [E]

Annuals (winter annuals), 1–45 cm; fibrous-rooted. **Stems** erect to decumbent-ascending, not pannose (indument whitish, like closely appressed, polished cloth, hairs usually not individually evident). **Leaves** basal and cauline, basal present through flowering, blades oblanceolate to oblanceolate-spatulate, 2–6(–7) cm × 5–13 mm (gradually smaller, becoming linear bracts distally), faces bicolor, abaxial closely white-pannose, adaxial sparsely arachnose (light green, hairs persistent, closely appressed, nearly microscopic). **Heads** initially in ± continuous, cylindric arrays 3–5(–7) cm × 10–12 mm (pressed), later sometimes interrupted and 7–20 cm (producing axillary glomerules from proximal nodes). **Involucres** cylindro-campanulate, 3.5–4 mm, bases sparsely arachnose. **Phyllaries** in 4–5 series, outer ovate, lengths ⅓ inner, apices acute to acute-acuminate, inner oblong-lanceolate, laminae (± striate) purplish (at stereome and on distal margins or not at all), apices acute to acute-acuminate (not apiculate, slightly flaring outward in fruit). **Florets:** bisexual 2–4; all corollas brownish yellow to purple distally (sometimes purple only on adaxial faces of lobes in bisexual corollas). **Cypselae** (purple) 0.5–0.6 mm.

Flowering (Mar–)Apr–May(–Jun). Disturbed, open sites, roadsides, banks, woods edges and clearings, fields, flood plains, pastures, sandy, loamy, and clay soils; 0–200 m; Ala., Ark., Fla., Ga., La., Miss., N.C.

5. Gamochaeta argyrinea G. L. Nesom, Sida 21: 718, figs. 1-4. 2004 • Silvery cudweed

Annuals (winter annuals), 12–40 cm; usually fibrous-rooted, rarely taprooted. **Stems** decumbent-ascending, closely white-pannose (hairs usually individually evident, seldom forming clothlike induments). **Leaves** basal and cauline, basal present through flowering, blades oblanceolate to oblanceolate-oblong or oblanceolate-obovate, 1.5–5(–8) cm × 5–12(–18) mm (gradually smaller distally), faces bicolor, abaxial closely white-pannose, adaxial sparsely arachnose (evident at 10×). **Heads** initially in continuous, cylindric arrays 1.5–5 cm × 10–12 mm (pressed), later sometimes interrupted, 5–18 cm × 10–12 mm (pressed; producing axillary glomerules from proximal nodes). **Involucres** campanulate, 3–3.5 mm, bases sparsely arachnose. **Phyllaries** in 4–6 series, outer (tawny-transparent, never dark brown) ovate to ovate-lanceolate, lengths ⅓–⅘ inner, apices acute to acuminate, inner elliptic-oblong to oblong, laminae often purplish tinged (around stereome/lamina junction, otherwise hyaline and slightly brownish), apices truncate-rounded, apiculate (flexing slightly outward in fruit). **Florets:** bisexual 4–5(–6); all corollas purple- to yellow-brown distally. **Cypselae** (tan) 0.5–0.6 mm.

Flowering Mar–Jun(–Oct). Roadsides, fields, lawns, open woods, sand or clayey soils, open, disturbed areas; 0–300 m; Ala., Ark., Fla., Ga., Kans., Ky., La., Md., Miss., N.C., Okla., Pa., S.C., Tenn., Tex., Va., W.Va.; West Indies.

Gamochaeta argyrinea has been confused with *G. purpurea*, which also occurs across the coastal states of eastern United States (G. L. Nesom 2004).

6. Gamochaeta ustulata (Nuttall) Holub, Folia Geobot. Phytotax. 11: 83. 1976 • Featherweed, Pacific cudweed E

Gnaphalium ustulatum Nuttall, Trans. Amer. Philos. Soc., n. s. 7: 404. 1841; *G. pannosum* Gandoger; *G. purpureum* Linnaeus var. *ustulatum* (Nuttall) B. Boivin

Annuals, biennials, or perennials, 10–40 cm; fibrous-rooted. **Stems** erect to ascending (commonly decumbent-ascending and rhizome like), densely white-pannose. **Leaves** basal and cauline, basal usually withering before flowering, blades spatulate to oblanceolate, 2–5 cm × 6–12(–35) mm (little smaller distally), faces bicolor, abaxial white-pannose, adaxial sparsely to densely arachnose-tomentose. **Heads** in usually continuous, rarely interrupted (proximally), cylindric arrays 1–6(–8+) cm × 12–18 mm (pressed). **Involucres** campanulo-urceolate, 4.5–5 mm, bases sparsely arachnose. **Phyllaries** in 4–6 series, outer (brown or greenish brown) broadly ovate-triangular, lengths ca. ½ inner, apices acute to acute-acuminate (mid phyllaries ± keeled near apices), inner oblong, laminae usually dark brown, sometimes purplish (at stereome-lamina junction), apices rounded to obtuse, apiculate. **Florets:** bisexual (3–)4–6; all corollas usually yellowish, sometimes purplish distally. **Cypselae** (tan to brownish) 0.7–0.8 mm.

Flowering Apr–Jul(–Oct). Mostly coastal and near-coastal sites, dunes, ocean bluffs, sandy fields, and roadsides, clay-loam, roadcuts, ditches, cliffs, pine woods, chaparral slopes, tidal marsh edges; 0–700(–1100) m; B.C.; Calif., Oreg., Wash.

Gamochaeta ustulata usually has been included in *G. purpurea*; it differs mostly in its longer duration, thicker and shorter stems, larger, more compact arrays of larger, brown heads, and aspects of phyllary morphology.

7. Gamochaeta coarctata (Willdenow) Kerguélen, Lejeunia 120: 104. 1987 • Elegant cudweed F I

Gnaphalium coarctatum Willdenow, Sp. Pl. 3: 1886. 1803, based on *G. spicatum* Lamarck in J. Lamarck et al., Encycl. 2: 757. 1788, not Miller 1768; *Gamochaeta spicata* Cabrera

Winter annuals or biennials, 15–35(–50) cm; fibrous-rooted. **Stems** decumbent-ascending, white-pannose (tomentum usually sheath-like). **Leaves** basal and cauline, basal present (in rosettes) at flowering, blades spatulate to oblanceolate-obovate, (1.5–)3–8(–12) cm × 6–15(–22) mm (gradually or little smaller distally, slightly succulent, margins often crenulate on drying), faces bicolor, abaxial closely white-pannose, adaxial glabrous or glabrate. **Heads** initially usually in dense, continuous spiciform arrays 2–20 cm × 10–14 mm (pressed), later branched, interrupted. **Involucres** cylindro-campanulate, 2.5–3 mm, bases glabrous. **Phyllaries** in 4–5 series, outer (purplish or rosy) elliptic-obovate to broadly ovate-elliptic, lengths 1/3–1/4 inner, apices rounded to obtuse, inner oblong, laminae brown-hyaline, apices rounded to obtuse or blunt, apiculate. **Florets:** bisexual 2–3; all corollas usually purplish distally. **Cypselae** (tan) 0.5–0.6 mm. *2n* = 28, 40.

Flowering Apr–Jun. Ditches, low roadsides, sidewalk cracks, shaded spots around buildings, other shaded, moist habitats; 0–150 m; introduced; Ala., Ark., Calif., Fla., Ga., La., Miss., N.C., S.C., Tex., Va.; South America; also introduced in Mexico, West Indies, Europe, Asia, Pacific Islands (New Zealand), Australia.

Gamochaeta coarctata is native to South America and is also introduced in Mexico. R. K. Godfrey (1958) identified specimens of *G. coarctata* as *Gnaphalium spicatum* Lamarck. Some specimens from the flora area were misidentified by G. L. Nesom (1990f) as *Gamochaeta americana* (Miller) Weddell, which was described from Jamaica; it is widespread in Central America and Mexico, and has not been observed in the flora area.

8. **Gamochaeta calviceps** (Fernald) Cabrera, Bol. Soc. Argent. Bot. 9: 368. 1961 [F]

Gnaphalium calviceps Fernald, Rhodora 37: 449, plate 405, figs. 1–4. 1935 (as calvescens)

Annuals, 8–45(–55) cm; taprooted or fibrous-rooted. **Stems** erect, ascending (usually branched ± throughout), subpannose (hairs silver-gray, longitudinally arranged). **Leaves** mostly cauline, basal usually withering before flowering, blades spatulate to oblanceolate, mostly 2–6 cm × 2–9 mm (becoming linear-oblanceolate to linear distally, commonly folded along midveins), faces concolor or weakly bicolor, subpannose (hairs closely appressed). **Heads** initially in continuous or interrupted, spiciform arrays, 2–4 cm × 8–12 mm (pressed), later in paniculiform arrays 4–18 cm (main axes usually visible between heads, peduncles usually evident). **Involucres** campanulate, 3–3.5 mm, bases usually glabrous or glabrate. **Phyllaries** in 5–7 series, outer ovate-triangular, lengths 1/3–1/2 inner, apices acute-acuminate (involute and spreading to recurved), inner oblong, laminae slightly brown (not purple), apices obtuse-apiculate. **Florets:** bisexual 2–4; all corollas purple distally. **Cypselae** (tan) 0.4–0.5 mm.

Flowering (Apr–)May–Jul. Disturbed sites, sandy or clay soils, roadsides, fields, clearing and edges of woods, flower beds; 0–500 m; Ala., Ark., Calif., La., Miss., N.C.,

Okla., S.C., Tex., Va.; South America; Europe; Pacific Islands (New Zealand).

Gamochaeta calviceps is recognized by its subpannose cauline and foliar indument (perhaps intermediate between the looser tomentum of *G. antillana* and the tight, pannose covering of *G. argyrinea*) and the contrast of its spatulate proximal leaves with the much narrower cauline ones, glabrous or glabrescent involucres, and phyllaries in 5–7 series, lacking purple color, the outer and mid with acute-acuminate apices commonly becoming subulate (by inrolled margins). The distal cauline leaves usually are folded along the midveins (at least when pressed). The relatively late flowering also is distinctive. Plants on the Atlantic coastal plain usually produce 2–3 bisexual florets per head, those on the Gulf coast 3–4.

Heads of older plants are borne in paniculiform arrays resulting from development of lateral branches, the heads usually on evident peduncles and with very little tomentum at the base of the outer phyllaries, thus appearing discrete. In early-season plants, lateral branches may not have formed or lengthened and the arrays of heads may appear continuous-cylindric at stem apices; in such plants, the species can usually still be recognized by the relatively numerous, relatively shorter, axillary shoots along the main stems.

9. **Gamochaeta antillana** (Urban) Anderberg, Opera Bot. 104: 157. 1991 (as antillarum) • Delicate everlasting

Gnaphalium antillanum Urban, Repert. Spec. Nov. Regni Veg. 13: 482. 1915; *Gamochaeta subfalcata* (Cabrera) Cabrera; *Gnaphalium subfalcatum* Cabrera

Annuals, 6–40 cm; taprooted. **Stems** erect to decumbent-ascending, loosely arachnose-tomentose. **Leaves** basal and cauline, basal usually withering before flowering, blades spatulate to oblanceolate, narrowly lanceolate, linear-oblanceolate, or linear, 2–3(–4) cm × 2–3.5(–5) mm (distal rarely folded along midveins), faces concolor, loosely tomentose. **Heads** initially in uninterrupted, cylindro-spiciform arrays (1–)3–4(–10) cm × 8–12 mm (pressed), usually becoming glomerulate-interrupted in late flowering (equally leafy-bracted throughout, bracts linear to narrowly lanceolate, smaller distally). **Involucres** campanulate, 2.5–3 mm, bases sparsely arachnose. **Phyllaries** in 3–4(–5) series, outer ovate-lanceolate, lengths 1/2–2/3 inner, apices (sometimes purplish-tinged) narrowly to broadly acute, inner usually purple (immediately beyond stereome and along proximal margins), oblong, laminae usually purple (at stereome and along proximal margins), apices (whitish, tinged with brown) rounded-obtuse. **Florets:** bisexual 3–5; all corollas usually purple distally. **Cypselae** (tan) 0.4–0.5 mm.

Flowering (Feb–)Mar–May, sometimes later with moisture. Open sites in sandy soils, commonly in roadsides and other disturbed sites, stream and pond banks; 10–100 m; Ala., Ark., Fla., Ga., La., Miss., N.C., Okla., S.C., Tenn., Tex., Va.; South America; Europe; New Zealand.

Gamochaeta antillana and *G. calviceps* have been combined in concept and often misidentified as *Gamochaeta falcata* (Lamarck) Cabrera; the latter name applies to a South American species that has not been recorded from the flora area. *Gamochaeta subfalcata*, which has been attributed to the United States (e.g., S. E. Freire and L. Iharlegui 1997), almost certainly applies to the same species as *G. antillana*.

10. Gamochaeta stagnalis (I. M. Johnston) Anderberg, Opera Bot. 104: 157. 1991 (as stagnale) • Desert cudweed

Gnaphalium stagnale I. M. Johnston, Contr. Gray Herb., 68: 100. 1923

Annuals, 2.5–20(–35) cm; usually taprooted, sometimes fibrousrooted. **Stems** erect to decumbentascending, densely and loosely arachnose-tomentose. **Leaves** mostly cauline, basal usually withering before flowering, blades mostly oblanceolate to oblong-oblanceolate (± uniform in size and shape), 1–2.5(–3) cm × 2–6 mm, faces concolor or weakly bicolor, both loosely tomentose or adaxial glabrescent and greener. **Heads** in capitate clusters (in smallest plants) ca. 1 cm or interrupted, spiciform arrays 1–3(–12) cm × 8–12 mm (pressed, sometimes branching at proximal nodes, glomerules subtended by divergent-ascending bracts similar to distal cauline leaves). **Involucres** campanulate, 2.5–3 mm, bases sparsely arachnose. **Phyllaries** in 3–4(–5) series, outer ovate-triangular, lengths $^{1}/_{2}$–$^{2}/_{3}$ inner, apices broadly acute, inner oblong, laminae usually purple (immediately beyond stereome and along proximal margins), apices (whitish) rounded-obtuse. **Florets:** bisexual (2–)3(–4); all corollas purplish distally. **Cypselae** (tan) 0.3–0.5 mm.

Flowering (Mar–)Apr(–May). Sandy, often moist soils, washes, permanent streams, canyon bottoms, flower beds, riparian, desert grasslands, juniper-grasslands, creosote bush-mesquite-cholla, oak woodlands; 900–1800 m; Ariz., N.Mex.; Mexico.

Morphologic differences between *Gamochaeta antillana* and *G. stagnalis* are subtle but consistent; the two are distinct in geography and ecology. The previous attribution of *G. falcata* (Lamarck) Cabrera to Arizona (G. L. Nesom 1990f) was based on specimens of *G. stagnalis*. Those plants have been misidentified as *G. purpurea* also.

11. Gamochaeta pensylvanica (Willdenow) Cabrera, Bol. Soc. Argent. Bot. 9: 375. 1961 • Pennsylvania cudweed F

Gnaphalium pensylvanicum Willdenow, Enum. Pl., 867. 1809, based on *G. spathulatum* Lamarck in J. Lamarck et al., Encycl. 2: 758. 1788, not Burman f. 1768; *G. peregrinum* Fernald

Annuals, 10–50 cm; taprooted. **Stems** erect to decumbent or procumbent, loosely arachnose-tomentose. **Leaves** basal and cauline, proximal usually present at flowering, blades spatulate to oblanceolate-obovate, 2–7 cm × 4–16 mm (becoming spatulate to oblanceolate bracts among proximal heads, surpassing glomerules, bases narrowed to petiolar regions, margins sinuate, apices often apiculate), faces concolor or weakly bicolor, loosely tomentose. **Heads** in glomerules in continuous or interrupted, spiciform arrays 1–12 cm × 10–15 mm (pressed). **Involucres** cupulate-campanulate, 3–3.5 mm, bases sparsely arachnose. **Phyllaries** in 3–4 series, outer ovate-triangular, lengths $^{1}/_{2}$–$^{2}/_{3}$ inner, apices attenuate-apiculate, inner oblong, laminae often purple-tinged (at stereome), apices (transparent, sometimes golden) acute to obtuse. **Florets:** bisexual 3–4; all (or at least bisexual) corollas usually purplish distally. **Cypselae** (tan) 0.4–0.5 mm. $2n = 28$.

Flowering Mar–Jun(–Aug). Disturbed sites, exposed, moist soils, commonly partially shaded; 0–500 m; Ala., Calif., Fla., Ga., La., Md., Mass., Miss., N.C., Okla., Pa., S.C., Tex., Va.; Mexico; Central America; South America; Europe; Asia; Africa; Australia.

Gamochaeta pensylvanica is recognized by its obovate-spatulate, loosely tomentose and concolor or weakly bicolor basal and proximal cauline leaves, and similarly shaped spreading bracts among the heads. Occasional plants appear intermediate between *G. pensylvanica* and *G. antillana*. The latter differs in its more erect stems, linear to oblanceolate basal and proximal cauline leaves, and more nearly continuous arrays of heads with linear to narrowly oblanceolate bracts.

12. Gamochaeta stachydifolia (Lamarck) Cabrera, Bol. Soc. Argent. Bot. 9: 382. 1961 I

Gnaphalium stachydifolium Lamarck in J. Lamarck et al., Encycl. 2: 757. 1788; *G. purpureum* Linnaeus var. *stachydifolium* (Lamarck) Baker

Annuals, 4–15 cm; taprooted. **Stems** erect, densely and loosely gray-white tomentose-arachnose. **Leaves** basal and cauline, basal mostly withering before flowering, blades oblanceolate,

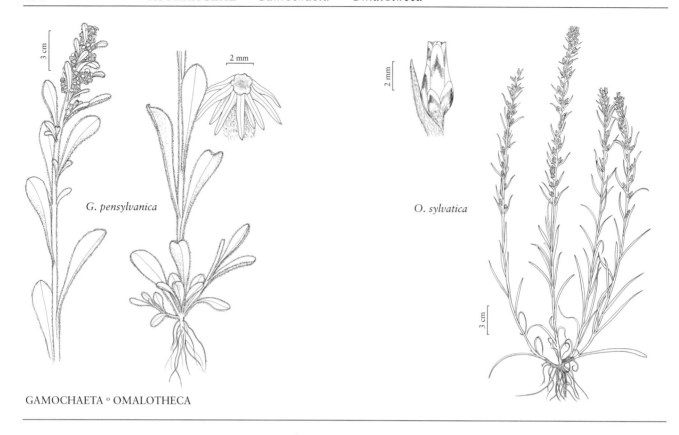

GAMOCHAETA ° OMALOTHECA

1–3 cm × 2–6 mm (usually folded along midveins, bases subclasping, not auriculate; nearly unreduced among proximal heads, none longer than glomerules), faces concolor, tomentose-arachnose. **Heads** in continuous cylindric arrays 2–3(–4) cm × 10–12 mm (pressed). **Involucres** campanulate, 3.5–4 mm, bases sparsely tomentose. **Phyllaries** in 4–5 series, outer narrowly ovate-triangular, lengths ca. $1/3$ inner, apices acute to acute-acuminate, inner oblong, laminae tan to brownish (not purple), apices (brownish) rounded-apiculate. **Florets:** bisexual 2–4; all corollas yellowish. **Cypselae:** not seen.

Flowering Mar–May. Hillsides, riparian woodlands, disturbed sites, sandy soils; 30–500 m; introduced; Calif.; South America.

94. OMALOTHECA   Cassini in F. Cuvier, Dict. Sci. Nat. ed. 2, 56: 218. 1828 [Greek *omalo*, even or equal, and *theke*, container, envelope, or sheath, perhaps alluding to involucres]

Guy L. Nesom

Perennials, 2–70 cm (fibrous-rooted, rhizomatous, not stoloniferous). **Stems** usually 1, erect (branched from bases or distally, woolly-tomentose to sericeous). **Leaves** basal (persistent in rosettes) and cauline; alternate; sessile; blades mostly narrowly lanceolate to oblanceolate, bases cuneate, margins entire, faces bicolor or concolor, abaxial white to gray, thinly tomentose, adaxial white to grayish and sericeous to thinly woolly or greenish and glabrate. **Heads** disciform, in spiciform or subcapitate arrays. **Involucres** campanulate to turbinate, 5–6 mm. **Phyllaries** in 2–3 series, stramineous to brownish (sometimes mottled; hyaline, stereomes not glandular), unequal, chartaceous toward apices. **Receptacles** flat to concave, smooth, epaleate. **Peripheral (pistillate) florets** 35–70+ (more numerous than bisexual); corollas purplish or whitish. **Inner (bisexual) florets** 3–4; corollas purplish or whitish, distally purplish or reddish. **Cypselae**

obovoid to cylindric or fusiform, sometimes slightly compressed, faces strigillose (hairs not myxogenic, lengths 6–12 times diams.) and papillate (carpopodia forming minute stipes); **pappi** falling readily, of 15–25 distinct (falling separately) or basally connate (falling together), barbellate bristles in 1 series. x = 14.

Species 8–10 (3 in the flora): mostly Eurasian; three species reaching North America in native distribution (*Omalotheca sylvatica* perhaps not native, see below).

The species of *Omalotheca* have been placed in subg. *Omalotheca* (capitulescences of 1–10 heads, cypselae compressed-obovoid, and pappus bristles distinct and falling separately) and subg. *Gamochaetiopsis* Schultz-Bipontinus & F. W. Schultz (capitulescences of 10–100 heads, cypselae cylindric, and pappus bristles basally connate and falling together). In the flora, *O. norvegica* and *O. sylvatica* belong in subg. *Gamochaetiopsis*; *O. supina* is in subg. *Omalotheca*.

SELECTED REFERENCE Nesom, G. L. 1990b. Taxonomic summary of *Omalotheca* (Asteraceae: Inuleae). Phytologia 68: 241–246.

1. Plants 2–8(–12) cm; heads in subcapitate to loose, spiciform arrays; pappus bristles distinct, falling separately . . . . 1. *Omalotheca supina*
1. Plants 10–70 cm; heads in compact or loose, spiciform arrays; pappus bristles basally connate, falling together.
  2. Leaf blades: basal and proximal cauline 3-nerved, 6–30 mm wide, distal cauline oblanceolate, faces concolor or weakly bicolor; arrays of heads 1.5–5 cm, rarely interrupted; alpine sites . . . . 2. *Omalotheca norvegica*
  2. Leaf blades: basal and proximal cauline 1-nerved, 2–10 mm wide, distal cauline linear, faces bicolor; arrays of heads 4–35 cm, usually interrupted; lower elevations . . . . . 3. *Omalotheca sylvatica*

1. **Omalotheca supina** (Linnaeus) de Candolle in A. P. de Candolle and A. L. P. P. de Candolle, Prodr. 6: 245. 1838
   • Alpine Arctic-cudweed, gnaphale couché

Gnaphalium supinum Linnaeus, Syst. Nat. ed. 12, 3: 234. 1768

Plants 2–8(–12) cm. **Leaves** mostly basal (in persistent rosettes); blades 1-nerved, linear to linear-oblanceolate, 5–25 × 3 mm, cauline similar, faces concolor, gray-green, thinly woolly. **Heads** (usually 1–7) in subcapitate to loose, spiciform arrays. **Involucres** campanulate, 5–6 mm. **Phyllaries** light green to tan, oblong to lanceolate, outer obtuse, inner mostly acute, margins and tips dark brown. **Cypselae** obovoid, strigose; **pappus** bristles distinct, falling separately. $2n$ = 28.

Flowering Jul–Sep. Granite outcrops, gravelly slopes, other alpine sites; 200–1300 m; Greenland; Nfld. and Labr., Que.; Maine, N.H., Vt.; Europe; Asia (Caucasus, Iran).

2. **Omalotheca norvegica** (Gunnerus) Schultz-Bipontinus & F. W. Schultz in F. W. Schultz, Arch. Fl., 311. 1861
   • Norwegian Arctic-cudweed, gnaphale de Norvège

Gnaphalium norvegicum Gunnerus, Fl. Norveg. 2: 105. 1772

Plants 10–40 cm. **Leaves** basal and cauline, basal petiolate, blades 3-nerved, lanceolate to oblanceolate, 5–12 cm × 6–30 mm, distal cauline slightly smaller, oblanceolate, faces concolor or weakly bicolor, grayish, thinly woolly. **Heads** (10–60+) in compact, spiciform (leafy-bracteate, sometimes interrupted) arrays (1.5–14 cm, occupying ⅛–¼ of plant heights, primary axes usually not visible). **Involucres** cylindro-campanulate, 5.5–6 mm. **Phyllaries** brown to reddish brown with narrow pale center and base. **Cypselae** cylindric, minutely strigose; **pappus** bristles basally connate, falling together. $2n$ = 56.

Flowering Jul–Sep. Wet or peaty slopes, alpine and subalpine meadows, cliff ledges, rocky slopes; 400–1300 m; Greenland; Nfld. and Labr., Que.; Europe.

3. Omalotheca sylvatica (Linnaeus) Schultz-Bipontinus & F. W. Schultz in F. W. Schultz, Arch. Fl., 311. 1861 • Woodland Arctic-cudweed, gnaphale des bois F

Gnaphalium sylvaticum Linnaeus, Sp. Pl. 2: 856. 1753

Plants 10–70 cm. **Leaves** basal and cauline; blades 1-nerved, linear to narrowly oblanceolate or lanceolate, 2–8 cm × 2–10 mm, distal cauline smaller, linear, faces bicolor, abaxial gray, silvery sericeous, adaxial green, glabrescent. **Heads** (20–90) in loose, spiciform (leafy-bracteate, interrupted) arrays (4–35 cm, occupying ⅓–⅚ plant heights, simple or branched at bases, primary axes mostly visible). **Involucres** campanulo-turbinate, 5–6.5 mm. **Phyllaries** some or all with conspicuous dark brown spot distal to middle. **Cypselae** cylindric to fusiform, minutely strigose; **pappus** bristles basally connate, falling together. $2n = 56$.

Flowering Jul–Sep(–Oct). Open woods, boggy woods, rocky slopes, clearings, fields, borders of woods, roadsides, muddy banks, disturbed sites; 10–500 m; St. Pierre and Miquelon; B.C., N.B., Nfld. and Labr. (Nfld.), N.S., Ont., P.E.I., Que.; Maine, Mich., N.H., N.Y., Pa., Vt., Wis.; Europe; Asia (Caucasus, Iran, Siberia).

The circumboreal *Omalotheca sylvatica* may have been introduced from Eurasia (Frère Marie-Victorin 1995). *Omalotheca alpigena* (K. Koch) Holub and *O. caucasica* (Sommier & Levier) S. K. Cherepanov were treated as synonyms of *O. sylvatica* by A. J. C. Grierson (1975); they have been recognized as distinct species in other treatments.

95. EUCHITON Cassini in F. Cuvier, Dict. Sci. Nat. ed. 2, 56: 214. 1828 [Greek *eu-*, good or true, and *chiton*, tunic, alluding to 'close-fitting' clusters of bracts subtending clusters of heads] I

Guy L. Nesom

Annuals or perennials, 5–80 cm (usually fibrous-rooted, sometimes rhizomatous, usually stoloniferous). **Stems** usually 1, erect. **Leaves** basal and cauline (sometimes in rosettes); alternate; petiolate or sessile; blades oblanceolate, spatulate, lanceolate, or linear, bases cuneate or ampliate, margins entire (sometimes undulate and/or revolute), faces bicolor, abaxial usually silvery, tomentose, adaxial usually green, glabrate or glabrous. **Heads** disciform, usually in terminal clusters (subtended by leafy bracts, sometimes with axillary clusters), rarely borne singly. **Involucres** narrowly campanulate to cylindric, 3–5 mm. **Phyllaries** in 3–4+ series, mostly stramineous to brownish, sometimes purplish to pinkish (hyaline, stereomes not glandular), unequal, chartaceous toward tips. **Receptacles** flat, smooth, epaleate. **Peripheral (pistillate) florets** 16–150 (more numerous than bisexual); corollas purple or distally purplish. **Inner (bisexual) florets** 1–7; corollas purple or distally purplish. **Cypselae** obovoid-ellipsoid, slightly flattened, faces minutely hairy or papillate (papilliform hairs or papillae ± clavate, not myxogenic); **pappi** readily falling (singly or in groups), of 12–20, distinct or basally coherent, barbellate bristles in 1 series. $x = 14$.

Species 17 (3 in the flora): introduced; Australia, New Zealand, New Guinea, e Asia; some species widely naturalized.

SELECTED REFERENCE Nesom, G. L. 2002. *Euchiton* (Asteraceae: Gnaphalieae) in North America and Hawaii. Sida 20: 515–521.

1. Annuals; taprooted; leaf bases not clasping; bracts subtending heads 4–8; heads in globose clusters; bisexual florets 1 . . . . 3. *Euchiton sphaericus*
1. Perennials or biennials; fibrous-rooted; leaf bases subclasping; bracts subtending heads 2–5; heads in hemispheric clusters; bisexual florets 3–7.
  2. Stolons usually present; basal leaves in rosettes at flowering; cauline leaves 2–4(–6), blades linear to oblanceolate, 1–2 cm × 1–2 mm; bracts subtending heads 2–3, not surpassing heads; pistillate florets 40–60 . . . . 1. *Euchiton gymnocephalus*
  2. Stolons usually absent; basal leaves withering before flowering; cauline leaves 6–10, blades mostly linear, 3–8 cm × 2–3 mm; bracts subtending heads 3–5, surpassing heads; pistillate florets 80–150 . . . . 2. *Euchiton involucratus*

1. **Euchiton gymnocephalus** (de Candolle) Holub, Folia Geobot. Phytotax. 9: 271. 1974 • Creeping-cudweed

Gnaphalium gymnocephalum de Candolle in A. P. de Candolle and A. L. P. P. de Candolle, Prodr. 6: 235. 1838

Perennials or biennials, 5–40 cm; fibrous-rooted (rhizomatous); stolons usually present (leafy, rooting at nodes). **Aerial stems** simple or branched, thinly and persistently white-tomentose. **Leaves:** basal persistent in rosettes (blades oblanceolate to spatulate, 2–10 cm, bases narrowly cuneate, petioliform); cauline 2–4(–6), blades linear to oblanceolate, 1–2 cm × 1–2 mm (even-sized), bases subclasping, margins revolute, abaxial faces silvery, tomentose, adaxial faces green, glabrous (shiny). **Bracts** subtending heads 2–3, 6–10 mm, not surpassing heads, plus some shorter. **Heads** in hemispheric clusters 10–20 mm diam. (sometimes with axillary clusters). **Involucres** 4–4.5 mm. **Phyllaries** tawny or rosy-tinged (shiny), oblong, apices rounded to obtuse. **Pistillate florets** 40–60. **Bisexual florets** 3–5. **Pappus bristles** distinct.

Flowering May–Oct. Grassy open places in wooded areas, roadsides; 100–800 m; introduced; Calif., Oreg.; Pacific Islands (New Zealand); Australia.

2. **Euchiton involucratus** (G. Forster) Anderberg, Opera Bot. 104: 167. 1991 • Common-cudweed

Gnaphalium involucratum G. Forster, Fl. Ins. Austr., 55. 1786

Biennials or perennials, 30–40 cm; fibrous-rooted; stolons usually absent [reportedly present in Australia and New Zealand]. **Aerial stems** erect, simple, thinly and persistently white-tomentose. **Leaves:** basal withering before flowering; cauline 6–10, blades linear to linear-oblanceolate or linear-lanceolate, 3–8 cm × 2–3 mm (largest at midstem), bases subclasping (not ampliate), margins revolute, abaxial faces silvery, tomentose, adaxial faces green, glabrate (shiny). **Bracts** subtending heads 3–5, 10–15 mm, surpassing heads, plus some shorter. **Heads** in hemispheric clusters 10–15 mm diam. (sometimes with axillary clusters). **Involucres** 4–4.5 mm. **Phyllaries** tawny or rosy-tinged (shiny), oblong, apices rounded to obtuse. **Pistillate florets** 80–150. **Bisexual florets** 3–5(–7). **Pappus bristles** distinct or basally coherent (falling in groups).

Flowering Jul–Oct. Grassy open places, often moist or wet; 50–700 m; introduced; Calif., Mass.; Pacific Islands (New Zealand); Australia.

3. **Euchiton sphaericus** (Willdenow) Anderberg, Opera Bot. 104: 167. 1991 • Star-cudweed

Gnaphalium sphaericum Willdenow, Enum. Pl., 867. 1809

Annuals, 5–80 cm; taprooted; stolons absent. **Aerial stems** simple or branched from bases (sometimes branched from leaf axils), thinly and persistently white-tomentose. **Leaves:** basal and proximal cauline withering before flowering (blades 1-nerved, oblanceolate to spatulate); cauline 8–12, blades linear, 2–4 cm × 1–2 mm (largest at midstem), bases not clasping, margins revolute, sometimes undulate, abaxial faces white, tomentose, adaxial faces green, glabrous. **Bracts** subtending heads 4–8, 10–30 mm, surpassing heads. **Heads** in globose clusters 10–20 mm diam. **Involucres** 3.5–4 mm. **Phyllaries** brownish to tawny, sometimes purple-tinged (shiny), elliptic-lanceolate, apices acute. **Pistillate florets** 16–26. **Bisexual florets** 1. **Pappus bristles** distinct or basally coherent (falling in groups). $2n = 28$.

Flowering Jul–Oct. Grassy open places in wooded areas, disturbed soils, recent clearings; 30–700 m; introduced; Calif., Oreg.; se Asia; Pacific Islands (Hawaii, New Guinea, New Zealand); Australia.

In California and Hawaii, plants of *Euchiton sphaericus* have been identified as *Gnaphalium japonicum*; annual duration, slender taproots, nonclasping leaf bases, and single bisexual florets establish the correct identity of *E. sphaericus*. It "varies enormously in length, position and degree of branching, and the branches may be either vegetative or flower-bearing" (D. G. Drury 1972).

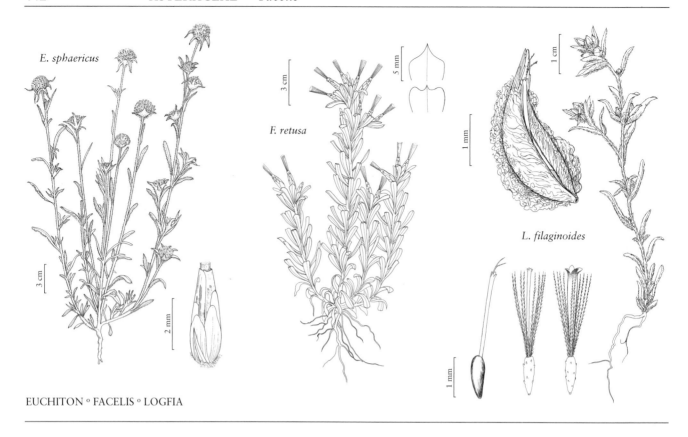

E. sphaericus

F. retusa

L. filaginoides

EUCHITON ° FACELIS ° LOGFIA

96. FACELIS Cassini, Bull. Sci. Soc. Philom. Paris 1819: 94. 1819 [Etymology unknown]

Guy L. Nesom

Annuals, 3–30 cm (taprooted or fibrous-rooted, not stoloniferous). **Stems** usually 1, erect to decumbent. **Leaves** cauline; alternate; sessile; blades spatulate or oblanceolate to lance-linear, bases ± cuneate, margins entire (usually revolute), faces bicolor, abaxial gray, tomentose, adaxial green, glabrous. **Heads** disciform, usually in ± capitate to loose, spiciform arrays, rarely borne singly (in leaf axils). **Involucres** narrowly ovoid or nearly cylindric [campanulate or broadly turbinate], 8–11 mm. **Phyllaries** in 3–5+ series, usually greenish to stramineous, sometimes purplish (hyaline, stereomes not glandular), unequal, chartaceous toward tips. **Receptacles** flat, smooth, epaleate. **Peripheral (pistillate) florets** 10–25 (more numerous than bisexual); corollas whitish to purplish. **Inner (bisexual) florets** 3–5; corollas white to purplish. **Cypselae** obovoid, ± compressed (2–3-nerved), faces silvery sericeous (hairs not myxogenic); **pappi** persistent, of 20–30+, distinct or basally connate, ± plumose bristles in 1 series. $x = 14$.

Species 3 or 4 (1 in the flora): North America, South America, Africa, Australia.

SELECTED REFERENCE Beauverd, G. 1913. Le genre *Facelis* Cassini (emend. Beauverd). Bull. Soc. Bot. Genève, sér. 2, 5: 212–220.

1. Facelis retusa (Lamarck) Schultz-Bipontinus, Linnaea 34: 532. 1866 • Annual trampweed F

Gnaphalium retusum Lamarck in J. Lamarck et al., Encycl. 2: 758. 1788

Stems usually with decumbent to nearly procumbent branches arising from bases, 3–30 cm, loosely tomentose. **Leaves** crowded; blades 7–20(–30) × 1.5–4 mm, apices truncate-apiculate to retuse. **Heads:** clusters subtended by bractlike leaves. **Cypselae** 1.6 mm; **pappi** 10–11 mm. $2n = 14$.

Flowering Mar–Jun. Lawns, roadsides, pastures, other disturbed sites, usually on sandy soils; 0–100 m; Ala., Ark., Fla., Ga., Kans., La., Miss., N.C., Okla., S.C., Tex., Va.; South America; Australia.

97. LOGFIA Cassini, Bull. Sci. Soc. Philom. Paris 1819: 143. 1819 (as "genre ou sous-genre"); in F. Cuvier, Dict. Sci. Nat. ed. 2, 23: 564. 1822 • Cottonrose, cotonnière, filzkraut, fluffweed, cottonweed [anagram of generic name *Filago*]

James D. Morefield

Oglifa (Cassini) Cassini

Annuals, 1–50(–70) cm. **Stems** 1, erect, or 2–10+, ascending to prostrate. **Leaves** cauline; alternate; blades subulate to obovate. **Heads** usually in glomerules of 2–10(–14) in racemiform to paniculiform or dichasiiform arrays, or some [all] borne singly. **Involucres** 0 or inconspicuous. **Phyllaries** 0, vestigial, 1–4 (unequal), or 4–6 (equal). **Receptacles** fungiform to obovoid (heights 0.4–1.6 times diams.), glabrous. **Pistillate paleae** (except usually innermost) readily or tardily falling, erect to ascending; bodies with 5+ nerves (nerves ± parallel, obscure), lanceolate to ovate or boat-shaped, ± saccate most of lengths (obcompressed to terete, sometimes ± galeate, each ± enclosing a floret); wings erect to incurved (apices blunt). **Innermost paleae** usually all pistillate, in some species bisexual and pistillate, persistent, usually 5 or 8, spreading (and enlarged) in fruit, surpassing other pistillate paleae; bodies lanceolate to ovate. **Pistillate florets** 14–45+. **Functionally staminate florets** 0. **Bisexual florets** 2–10; corolla lobes 4–5, ± equal. **Cypselae** brown, dimorphic: outer compressed to obcompressed, obovoid to ± cylindric, straight or curved, longer than inner, abaxially gibbous, faces glabrous, smooth, shiny; inner ± terete, faces glabrous, usually papillate to muriculate, sometimes smooth, dull; corolla scars apical to subapical; **pappi:** outer pistillate 0, inner pistillate and bisexual of (11–)13–28+ bristles (visible in heads). $x = 14$.

Species 12 (6 in the flora): North America, n Mexico, Europe, Asia, n Africa, introduced in South America, Pacific Islands, probably elsewhere.

See discussion of Filagininae following the tribal description (p. 385).

Logfia occurs in dry open habitats of Mediterranean, semiarid, arid, and, sometimes, humid temperate climates. The three introduced species (*L. arvensis, L. gallica, L. minima*) do not appear to be aggressively invasive. *Logfia filaginoides* (formerly known as *Filago californica*) is somewhat weedy in the more mesic, coastal portions of its range, where it often mixes about equally with *L. gallica.* Some specimens of *L. filaginoides, L. depressa,* and *L. minima* may be difficult to identify without full comparison to descriptions.

Usually included in *Filago* in North America, *Logfia* (including *Oglifa*) has been separated for some decades in Old World treatments. Contrary to G. Wagenitz (1969), I agree with J. Holub (1976, 1998) that the two genera warrant separation. Priority of *Logfia* over *Oglifa* at generic rank is ambiguous, however, depending on whether *Logfia* was validated in 1819 or 1822 (J. D. Morefield 2004). Because uncertainty remains, I here preserve current usage of *Logfia*, pending a proposal to conserve the name.

Morphologic evidence suggests *Logfia* is basal in Filagininae, with ancestral, sister, and/or reticulate relationships to *Filago* and *Stylocline* (J. D. Morefield 1992). *Logfia depressa* appears to be transitional toward *Stylocline*. Ancestors of *Logfia* probably resembled *Gnaphalium palustre* Nuttall, which is frequently misidentified as a member of Filagininae. *Logfia* is most easily recognized by outer epappose florets subtended by saccate paleae, prominent pappi on inner pistillate and bisexual florets, and innermost paleae open, persistent, spreading.

1. Pistillate florets: inner (0–)1–2 pappose; branches usually leafless between proximal forks, becoming purplish to black, glabrescent (longest capitular leaves ± linear, 2–5 times head heights) . . . . 4. *Logfia arizonica*
1. Pistillate florets: inner (4–)8–35 pappose; branches ± leafy between proximal forks, remaining greenish to grayish or whitish, arachnoid-sericeous to lanuginose.
  2. Pistillate paleae ± vertically ranked, outer inflexed 70–90° proximally, gibbous, ± galeate, bodies ± bony; leaves subulate or broader; phyllaries usually 5, equal.
    3. Leaves narrowly elliptic to narrowly ovate; longest capitular leaves 0.8–1.5 times head heights; innermost paleae ± 8, spreading in 2 series; bisexual corollas 1.6–2.1 mm; pappi mostly of 13–16 bristles falling in 1s or 2s . . . . 5. *Logfia minima*
    3. Leaves mostly subulate; longest capitular leaves 2–5 times head heights; innermost paleae ± 5, spreading in 1 series; bisexual corollas 2.2–3 mm; pappi of 18–28+ bristles falling in complete or partial rings . . . . 6. *Logfia gallica*
  2. Pistillate paleae spirally ranked, outer incurved 20–60°, somewhat gibbous, not galeate, bodies ± cartilaginous or chartaceous; leaves not subulate; phyllaries 0, vestigial, or 1–4, unequal.
    4. Outer pistillate paleae: distal 5–10% of lengths glabrous abaxially, wings obscured by indument; outer 2–4(–6) pistillate florets epappose; innermost paleae ± 8, spreading in 2 series . . . . 1. *Logfia arvensis*
    4. Outer pistillate paleae: distal 15–50% of lengths glabrous abaxially; wings prominent; outer 7–13 pistillate florets epappose; innermost paleae ± 5, spreading in 1 series.
      5. Outer pistillate paleae: bodies ± cartilaginous; bisexual corolla lobes mostly 4, bright reddish to purplish; inner cypselae mostly papillate, pappi of 17–23+ bristles falling in complete or partial rings; stems typically ± erect; capitular leaves mostly acute . . . . 2. *Logfia filaginoides*
      5. Outer pistillate paleae: bodies (except midnerves) chartaceous; bisexual corolla lobes mostly 5, yellowish to brownish; inner cypselae mostly smooth, pappi of 11–15 bristles falling in 1s or 2s; stems typically ± prostrate; capitular leaves obtuse . . . . 3. *Logfia depressa*

1. **Logfia arvensis** (Linnaeus) Holub, Notes Roy. Bot. Gard. Edinburgh 33: 432. 1975 • Field cottonrose, cotonnière des champs [I]

Filago arvensis Linnaeus, Sp. Pl. 2: add. 1753; *Oglifa arvensis* (Linnaeus) Cassini

Plants 3–50[–70] cm. **Stems** usually 1, erect; branches leafy between proximal forks, remaining grayish to whitish, lanuginose to sericeous. **Leaves** oblanceolate to lanceolate, largest 14–20(–40) × 3–4(–5) mm, pliant; longest capitular leaves 0.8–1.5(–2) times head heights, acute. **Heads** mostly in glomerules of 2–10(–13) in racemiform to paniculiform arrays, broadly pyriform to ± cylindric, largest 4–6 × 3.5–5 mm. **Phyllaries** 0, vestigial, or 1–4, unequal, ± like paleae. **Receptacles** ± fungiform, 0.4–0.7 mm, heights 0.4–0.5 times diams. **Pistillate paleae** (except innermost) 2–4 (–6) in 1(–2) series, spirally ranked, loosely saccate, incurved 20–60°, scarcely gibbous, not galeate, longest 3.3–4.5 mm, distal 5–10% of lengths glabrous abaxially; bodies ± cartilaginous, ± terete; wings obscured by indument. **Innermost paleae** ± 8, spreading in 2 series, pistillate. **Pistillate florets:** outer 2–4(–6) epappose, inner 15–20 pappose. **Bisexual florets** 3–4; corollas 2.3–3 mm, lobes mostly 4, reddish to purplish. **Cypselae:** outer nearly straight, ± erect, compressed, 0.9–1.1 mm; inner papillate; **pappi** of 17–23 bristles falling in complete or partial rings, 2.5–3.5 mm. $2n$ = 28 (Caucasus, Finland, Germany, Slovakia).

Flowering and fruiting mid Jun–mid Sep. Open, sandy to gravelly soils, disturbed sites (road and ditch banks, lakeshores, clear cuts, old fields, dwellings, grazed lands); 100–1700 m; introduced; B.C., Sask.; Alaska, Idaho, Mich., Minn., Mont., Nebr., Oreg., S.Dak., Wash., Wyo.; Eurasia; nw Africa.

Logfia arvensis appears to be basal or nearly so in *Logfia* and Filagininae (J. D. Morefield 1992); only 2–4 epappose florets are present in most heads. Reports of *L. arvensis* from Ontario and New York have not been confirmed by me. A report from the Desert National Wildlife Range in southern Nevada (T. L. Ackerman et al. 2003) was likely based on specimens of *L. filaginoides*. The earliest specimen confirmed from the flora area was from Bonner County, Idaho, in 1934. The label on one nineteenth-century specimen (mixed with *Diaperia verna*) identifying it as coming from Dallas, Texas, is probably in error; no other collections of *L. arvensis* are known from in or near Texas.

2. **Logfia filaginoides** (Hooker & Arnott) Morefield, Novon 14: 473. 2004 • California cottonrose, fluffweed [F]

Gnaphalium filaginoides Hooker & Arnott, Bot. Beechey Voy., 359. 1839; *Filago californica* Nuttall; *Logfia californica* (Nuttall) Holub; *Oglifa californica* (Nuttall) Rydberg

Plants 1–30(–55) cm. **Stems** 1(–7), typically ± erect; branches leafy between proximal forks, remaining grayish to greenish, arachnoid-sericeous. **Leaves** mostly oblanceolate, largest 10–15 (–20) × 2–3(–4) mm, pliant; longest capitular leaves 1–2(–3) times head heights, mostly acute. **Heads** mostly in glomerules of 2–4 in racemiform, paniculiform, or distally dichasiiform arrays, ± pyriform, largest 3.5–4.5 × 2.5–3 mm. **Phyllaries** 0, vestigial, or 1–4, unequal, ± like paleae. **Receptacles** ± fungiform, mostly 0.6–0.7 mm, heights 0.7–0.9 times diams. **Pistillate paleae** (except innermost) 7–13 in 2(–3) series, spirally ranked, loosely saccate, incurved 20–60°, somewhat gibbous, not galeate, longest 2.7–3.3 mm, distal 15–30% of lengths glabrous abaxially; bodies ± cartilaginous, ± terete; wings prominent. **Innermost paleae** ± 5, spreading in 1 series, pistillate. **Pistillate florets:** outer 7–13 epappose, inner 14–35 pappose. **Bisexual florets** 4–7; corollas 1.9–2.8 mm, lobes mostly 4, bright reddish to purplish. **Cypselae:** outer nearly straight, ± erect, compressed, mostly 0.9–1 mm; inner mostly papillate; **pappi** of 17–23+ bristles falling in complete or partial rings, 1.9–3 mm. $2n$ = 28.

Flowering and fruiting mid Feb–early Jul. Mediterranean climates: open slopes, flats, diverse substrates (including serpentine), old disturbances (chaparral burns) or seasonally moist sites, or warm deserts: protected slopes or higher elevations, among rocks, boulders (often granitic), less disturbed; 0–1800(–2000) m; Ariz., Calif., Nev., N.Mex., Tex., Utah; Mexico (Baja California, Baja California Sur, Sonora).

Long known as *Filago californica*, *Logfia filaginoides* is relatively common in the Californian Floristic Province south of Humboldt County, California, to northern Baja California Sur (including Channel Islands, and Angel de la Guarda, Cedros, and Guadalupe islands in Mexico). Eastward, it is scattered to southwestern Utah and western Texas. An 1893 gathering labeled "Blue Lakes, Snake Plains" is of uncertain origin.

3. **Logfia depressa** (A. Gray) Holub, Preslia 70: 107. 1998 • Dwarf cottonrose, spreading cottonrose, hierba limpia

Filago depressa A. Gray, Proc. Amer. Acad. Arts 19: 3. 1883; *Oglifa depressa* (A. Gray) Chrtek & Holub

Plants 1–5(–10) cm. **Stems** (1–) 3–10+, typically ± prostrate; branches ± leafy between proximal forks, remaining grayish to whitish, lanuginose. **Leaves** elliptic to obovate, largest 6–8(–10) × 1–2 mm, pliant; longest capitular leaves mostly 0.8–1.5 times head heights, obtuse. **Heads** in glomerules of 2–5 in ± dichasiiform arrays, ± pyriform, largest 3–4 × 2–2.5 mm. **Phyllaries** 0, vestigial, or 1–4, unequal, ± like paleae. **Receptacles** obovoid, 0.9–1.2 mm, heights 1.4–1.6 times diams. **Pistillate paleae** (except innermost) 7–13 in 2(–3) series, spirally ranked, loosely saccate, incurved 20–60°, somewhat gibbous, not galeate, longest 2.1–3.1 mm, distal 20–50% of lengths glabrous abaxially; bodies (except midnerves) chartaceous, ± terete; wings prominent. **Innermost paleae** ± 5, spreading in 1 series, pistillate. **Pistillate florets:** outer 7–13 epappose, inner (4–)10–21 pappose. **Bisexual florets** 2–5; corollas 1.3–2 mm, lobes mostly 5, yellowish to brownish. **Cypselae:** outer nearly straight, ± erect, compressed, 0.7–0.9 mm; inner mostly smooth; **pappi** of (11–)13–15 bristles falling in 1s or 2s, 1.3–2.4 mm.

Flowering and fruiting mostly Feb–May. Desert flats, alluvial slopes, loose sandy to gravelly soils, openings among shrubs, often with extra moisture (dry drainages, roadsides), rarely outside deserts; 0–1500 m; Ariz., Calif., Nev.; Mexico (Baja California, Sonora).

Reported to be eaten by desert tortoises (*Gopherus agassizii*), *Logfia depressa* is known from the Mojave Desert and most of the Sonoran Desert (except islands and lower Colorado River valley) including arid Baja California; isolated collections are known from southwestern California and Carson City, Nevada (where recently extirpated).

4. **Logfia arizonica** (A. Gray) Holub, Preslia 70: 107. 1998 • Arizona cottonrose

Filago arizonica A. Gray, Proc. Amer. Acad. Arts 8: 652. 1873; *Oglifa arizonica* (A. Gray) Chrtek & Holub

Plants 2–10(–20) cm. **Stems** (1–)3–10, spreading to sometimes ascending; branches usually leafless between proximal forks, becoming purplish to black, glabrescent. **Leaves** linear to narrowly oblanceolate, largest 15–20(–25) × 1–1.5 mm, pliant; longest capitular leaves 2–5 times head heights, acute. **Heads** in glomerules of 4–10 in strictly dichasiiform arrays, ± pyramidal, largest ± 4 × 3 mm. **Phyllaries** 0 or vestigial. **Receptacles** obovoid, 0.4–0.9 mm, heights 1–1.6 times diams. **Pistillate paleae** (except innermost) 9–13 (–17) in 2–3 series, vertically ranked, loosely saccate, incurved 20–60°, somewhat gibbous, not galeate, longest 2.2–2.7 mm, distal 15–25% of lengths glabrous abaxially; bodies cartilaginous, obcompressed; wings prominent. **Innermost paleae** ± 5, spreading in 1 series, bisexual and (usually) pistillate. **Pistillate florets:** outer 9–13(–17) epappose, inner (0–)1–2 pappose. **Bisexual florets** 4–10; corollas 1.2–1.7 mm, lobes mostly 5, brownish to yellowish. **Cypselae:** outer incurved, proximally ascending, distally erect, obcompressed, mostly 0.9–1 mm; inner densely muriculate; **pappi** mostly of 17–23 bristles falling in complete or partial rings, 1.3–2 mm. **2n** = 28.

Flowering and fruiting mid Feb–mid May(–mid Jun). Seasonally moist clay flats, sandy drainages, coastal slopes, flood plains, rocky places, roadsides, Mediterranean to arid climates; 0–1000(–1400) m; Ariz., Calif.; Mexico (Baja California, Baja California Sur, Sonora).

Logfia arizonica is known from southwestern California to northern Baja California Sur to southern Arizona and adjacent Sonora (except Colorado River valley) and from the Channel Islands (Santa Catalina, San Clemente) and Angel de la Guarda, Cedros, and Guadalupe islands in Mexico.

5. **Logfia minima** (Smith) Dumortier, Fl. Belg., 68. 1827 • Little cottonrose, cotonnière naine 1

Gnaphalium minimum Smith, Fl. Brit., 873. 1804; *Filago minima* (Smith) Persoon; *Oglifa minima* (Smith) Reichenbach f.

Plants 2–20[–30] cm. **Stems** 1, ± erect, or 2–5, ascending to spreading; branches leafy between proximal forks, remaining grayish, arachnoid-sericeous. **Leaves** narrowly elliptic to narrowly ovate, largest 7–10(–12) × 1–1.5(–2) mm, pliant; longest capitular leaves 0.8–1.5 times head heights, acute. **Heads** in glomerules of 3–7 in mostly dichasiiform, sometimes racemiform or paniculiform arrays, ± pyriform, largest 3–3.5 × 2 mm. **Phyllaries** usually 5, equal, ± like paleae. **Receptacles** ± fungiform, mostly 0.3–0.4 mm, heights 0.4–0.6 times diams. **Pistillate paleae** (except innermost) 8–12 in 1–2 series, ± vertically ranked, saccate, inflexed 70–90° proximally, gibbous, ± galeate, longest 2.6–3.1 mm, distal 15–30% of lengths glabrous abaxially; bodies ± bony, ± terete; wings prominent. **Innermost paleae** ± 8, spreading in 2 series, pistillate. **Pistillate florets:** outer 8–12 epappose, inner 18–30+ pappose. **Bisexual florets** 3–5; corollas 1.6–2.1 mm, lobes mostly 4, brownish to yellowish. **Cypselae:** outer incurved, proximally ± horizontal, distally erect, compressed, 0.8–0.9 mm; inner papillate; **pappi** mostly of 13–16 bristles falling in 1s or 2s, 1.8–2.1 mm. $2n = 28$ (Byelorussia, former Czechoslovakia, Germany)

Flowering and fruiting Jun–Aug. Ballast dumps (probably temporary), sometimes open disturbed sites away from coast; 0–400 m; introduced; Mass., N.Y., Oreg., Pa.; Eurasia.

Reports of *Logfia minima* from British Columbia and Washington have not been confirmed by me. A report from California was erroneous. The earliest known specimen from the flora area came from Girard Point, Pennsylvania, in 1878.

6. **Logfia gallica** (Linnaeus) Cosson & Germain, Ann. Sci. Nat., Bot., sér. 2, 20: 291. 1843 • Daggerleaf or narrowleaf cottonrose, cotonnière de France ⊡

Filago gallica Linnaeus, Sp. Pl. 2: add. 1753; *Oglifa gallica* (Linnaeus) Chrtek & Holub

Plants 2–50[–30] cm. **Stems** 1–5, ± erect; branches ± leafy between proximal forks, remaining grayish to greenish, arachnoid-sericeous. **Leaves** mostly subulate, largest 20–30(–40) × 1–1.5(–2) mm, ± stiff; longest capitular leaves 2–5 times head heights, acute or subspinose. **Heads** in glomerules of (2–)3–10 (–14) in strictly dichasiiform arrays, narrowly ampulliform, largest (3–)3.5–4.5 × 2–3 mm. **Phyllaries** usually 5, equal, unlike paleae (hyaline, obovate). **Receptacles** fungiform to obovoid, 0.7–0.9 mm, heights 0.8–1.1 times diams. **Pistillate paleae** (except innermost) 9–12 in 2 series, ± vertically ranked, tightly saccate, inflexed 70–90° proximally, gibbous, ± galeate, longest 3.3–4.1 mm, distal 15–30% of lengths glabrous abaxially; bodies ± bony, ± terete; wings prominent. **Innermost paleae** ± 5, spreading in 1 series, pistillate. **Pistillate florets:** outer 9–12 epappose, inner 8–14(–30) pappose. **Bisexual florets** [2–]3–5; corollas 2.2–3 mm, lobes mostly 4, brownish to yellowish. **Cypselae:** outer incurved, proximally ± horizontal, distally erect, compressed, [0.8–]0.9–1 mm; inner ± sparsely papillate; **pappi** of 18–28+ bristles falling in complete or partial rings, 2.2–3 mm. $2n = 28$ (former USSR, Portugal).

Flowering and fruiting mid Mar–early Jul(–Aug). Mediterranean climates, open slopes, flats, diverse substrates (including serpentine), often ruderal or disturbed sites (especially chaparral burns); 0–1100(–1400) m; introduced; Calif., Oreg.; Eurasia; n Africa; also introduced in Mexico (Baja California); South America; Atlantic Islands; Pacific Islands; Australia.

Logfia gallica is introduced in South America, Atlantic Islands, Pacific Islands (Hawaii), Australia, and probably elsewhere.

Logfia gallica is readily recognized by its relatively long and stiff awl-shaped leaves. In the flora, *L. gallica* is relatively common in the Californian Floristic Province from southwestern Oregon to northwestern Baja California (including the Channel Islands). It is often so well integrated with indigenous vegetation as to appear native. The first known collection in the flora area was from Newcastle, California, around 1883. It had been collected throughout central California by 1935 and had occupied most of its present North American range by 1970.

In the flora area, *Logfia gallica* tends to grow larger than in its native range.

98. **FILAGO** Linnaeus, Sp. Pl. 2: 927, 1199. 1753; Gen. Pl. ed. 5, 397. 1754, name conserved • Herba impia, cotonnière, cottonrose [Latin *filum*, thread, and *-ago*, possessing or resembling, alluding to abundant cottony indument] ⊡

James D. Morefield

Annuals, (1–)5–40 cm. **Stems** [0] 1, ± erect, or 2–7[–10+], ± ascending [prostrate]. **Leaves** cauline [basal]; alternate; blades lanceolate to oblanceolate [spatulate or ± round]. **Heads** in (dense, spheric [hemispheric]) glomerules of [2–]8–35+ in ± dichasiiform arrays [borne singly]. **Involucres** 0 or inconspicuous. **Phyllaries** usually 0, rarely 1–4, unequal (similar to paleae). **Receptacles** cylindric to clavate (heights [2–]5–15 times diams.), glabrous. **Pistillate paleae** (except usually innermost) ± persistent [falling], ± erect to ascending; bodies with 5+ nerves (nerves ± parallel, obscure), lanceolate to ovate, open to ± folded (each at most enfolding, not enclosing a floret); wings erect to recurved (apices acuminate to aristate). **Innermost paleae** usually all pistillate, in some species bisexual and pistillate, persistent or tardily falling, usually 5, erect to ascending [spreading] (scarcely enlarged) in fruit, shorter than other pistillate paleae; bodies lanceolate to ovate. **Pistillate florets** [12–]27–40+. **Functionally staminate florets** 0. **Bisexual florets** (1–)2–9(–11); corolla lobes 4, ± equal. **Cypselae** brown, ± monomorphic: terete to ± compressed, cylindric to ± obovoid, usually straight, not gibbous, faces papillate to muricate [glabrous, smooth], dull; corolla scars apical [subapical]; **pappi:** outer pistillate 0, inner pistillate and bisexual of [3–]13–21 bristles (visible in heads). $x = 14$.

Species 12(–23) (2 in the flora): introduced; Europe, w Asia, n Africa, Atlantic Islands, introduced in North America, Australia.

See discussion of Filagininae following the tribal description (p. 385).

The name *Filago* has been used also for the genus now usually recognized as *Evax* Gaertner. Here *Filago,* in the narrow sense, contains twelve Old World species. Six species long included in *Filago* in North America are here separated as *Logfia*.

Filago species grow in open, dry or somewhat moist habitats of arid, semiarid, Mediterranean, and humid-temperate to subtropical climates. The species in the flora grow in disturbed habitats; neither appears to be aggressively weedy.

Filago appears to be sister to or derived from *Logfia* and is probably ancestral to *Evacopsis* and *Evax* (J. D. Morefield 1992). *Filago* is most easily recognized by outer epappose florets subtended by open or ± folded, persistent, acuminate to aristate paleae, and prominent pappi on inner pistillate and bisexual florets.

SELECTED REFERENCES   Wagenitz, G. 1969. Abgrenzung und Gliederung der Gattung *Filago* L. s.l. (Compositae–Inuleae). Willdenowia 5: 395–444. Wagenitz, G. 1976. Two species of the "*Filago germanica*" group (Compositae–Inuleae) in the United States. Sida 6: 221–223.

1. Largest leaves oblong to lanceolate, widest in proximal ²/₃; heads in glomerules of (15–)20–35+, narrowly ± ampulliform, largest 1.5–2 mm diam.; pistillate paleae spirally ranked, longest 3.5–4.2 mm, innermost surrounding 14–25+ florets (10–20+ pistillate); longest distal capitular leaves 0.8–1.1 times head heights, acute. . . . . 1. *Filago vulgaris*
1. Largest leaves oblanceolate to narrowly spatulate, widest in distal ¹/₃; heads in glomerules of mostly (8–)12–16(–20), ± bipyramidal, largest 2.5–4 mm diam.; pistillate paleae vertically ranked, longest 4.5–6 mm, innermost surrounding 8–13 florets (2–7 pistillate); longest distal capitular leaves 1.3–2 times head heights, obtuse . . . . 2. *Filago pyramidata*

1. Filago vulgaris Lamarck, Fl. Franç. 2: 61. 1779
  • Common cottonrose, cotonnière commune   [F] [I]

Filago germanica Linnaeus, Sp. Pl. ed. 2, 2: 1311. 1763, not Hudson 1762; *Gifola germanica* Dumortier

Leaves: largest oblong to lanceolate, widest in proximal ²/₃, 15–25(–30) × 2–3(–4) mm, ± undulate; longest distal capitular leaves 0.8–1.1 times head heights, acute. **Heads** in glomerules of (15–)20–35+, narrowly ± ampulliform, largest (4–)5–6 × 1.5–2 mm; largest glomerules 9–13 mm diam. **Receptacles** clavate. **Pistillate paleae** (except innermost) ± (10–)15 in (2–)3 series, spirally ranked, rounded in cross section, longest 3.5–4.2 mm; wings yellowish tinged reddish, apices erect. **Innermost paleae** pistillate, surrounding 14–25+ florets. **Pistillate florets:** outer ± (10–)15 epappose, inner 10–

20+ pappose. **Bisexual florets** (1–)2–3(–4). $2n = 28$ (British Isles, Bulgaria, Czechoslovakia, Greece).

Flowering and fruiting (Apr–)Jun–Sep. Relatively dry, usually sandy soils, old fields, pastures, usually disturbed; 10–1000 m; introduced; B.C.; Del., D.C., Fla., La., Md., Mass., N.J., N.Y., N.C., Ohio, Oreg., Pa., S.C., Va., W.Va.; Eurasia; n Africa.

Filago vulgaris arrived in North America before 1739 (G. Wagenitz 1976). Paucity of modern collections suggests that it may not spread much or persist for long in the flora area. Reports from southern Ontario, Alabama, Georgia, and Kentucky have not been confirmed by me. The illustration of "*Filago germanica*" in L. Abrams and R. S. Ferris (1923–1960, vol. 4) depicts *F. pyramidata* var. *pyramidata*.

I agree with G. Wagenitz's (1965) nomenclatural analysis. *Filago vulgaris* replaced *F. germanica* Linnaeus, which is a later homonym of *F. germanica* Hudson, which in turn is a superfluous name for *F. pyramidata* Linnaeus.

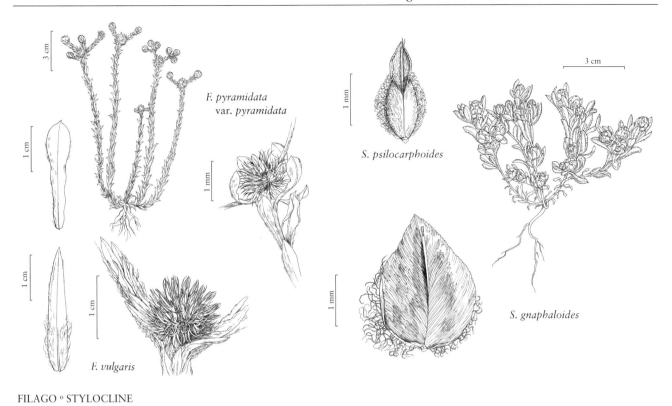

F. pyramidata
var. *pyramidata*

F. vulgaris

S. psilocarphoides

S. gnaphaloides

FILAGO ° STYLOCLINE

2. Filago pyramidata Linnaeus, Sp. Pl. 2: 1199. 1753

- Broadleaf cottonrose, cotonnière spatulée F I

Filago germanica Hudson

Varieties 4 (1 in the flora): s Europe, sw Asia, n Africa, Atlantic Islands, introduced.

Filago pyramidata was first collected in North America in 1935 (G. Wagenitz 1976). It is known in the flora only from northwestern California and from Saturna Island, British Columbia. It is variable in its native range; additional named varieties may eventually turn up in the flora. The name *F. vulgaris* was misapplied to this species by P. A. Munz (1968).

2a. Filago pyramidata Linnaeus var. pyramidata F I

Leaves: largest oblanceolate to narrowly spatulate, widest in distal ¹/₃, (10–)15–20 × 3–5 mm, ± plane; longest distal capitular leaves 1.3–2 times head heights, obtuse. **Heads** in glomerules of (8–)12–16(–20), ± bipyramidal, largest 5–7 × 2.5–4 mm; largest glomerules 11–15 mm diam. **Receptacles** filiform-cylindric. **Pistillate paleae** (except innermost) (15–)20–25 in (3–)4–5 series, vertically ranked, carinate in cross section, longest 4.5–6 mm; wings uniformly yellowish, apices recurved. **Innermost paleae** pistillate and often bisexual, surrounding 8–13 florets. **Pistillate florets:** outer (15–)20–25 epappose, inner 2–7 pappose. **Bisexual florets** 5–9(–11). $2n = 28$ (Balearic Islands, Canary Islands, Yugoslavia).

Flowering and fruiting late Apr–mid Aug. Mediterranean and cold-temperate climates, dry, open, sometimes rocky slopes, disturbed sites (chaparral burns, roadsides); 10–800 m; introduced; B.C.; Calif.; Eurasia; n Africa; Atlantic Islands; Australia.

Variety *pyramidata* is also introduced in Australia.

99. STYLOCLINE Nuttall, Trans. Amer. Philos. Soc., n. s. 7: 338. 1840 • Neststraw

[Greek *stylos*, column, pillar, or pole, and *cline*, couch or bed (or *gyne*, female, specified by Nuttall in protologue), alluding to narrowly cylindric receptacles of the type species]

James D. Morefield

Annuals, 1–10(–20) cm. **Stems** 1, ± erect, or 2–10+, ascending to ± prostrate. **Leaves** cauline; mostly alternate; blades oblanceolate to lanceolate or broader. **Heads** in glomerules of 2–10 (rarely borne singly) in ± dichasiform (sometimes ± paniculiform) arrays. **Involucres** 0 or inconspicuous. **Phyllaries** 0, vestigial, or 1–4, unequal or subequal (similar to paleae). **Receptacles** cylindric to clavate (heights 2.8–8 times diams.), glabrous. **Pistillate paleae** falling, erect to ascending; bodies with 5+ nerves (nerves ± parallel, obscure), ± ovate or boat-shaped, saccate most of lengths (obcompressed to terete, not galeate, each usually enclosing a floret, outermost open in 2 spp.); wings ± erect (apical). **Staminate paleae** readily falling, mostly 2–4, erect in fruit (not enlarged), shorter than pistillate paleae; bodies linear-lanceolate to oblanceolate. **Pistillate florets** 12–25+. **Functionally staminate florets** 2–6; corolla lobes (4–)5, ± equal. **Bisexual florets** 0. **Cypselae** brown, monomorphic: compressed to obcompressed, ± obovoid, slightly incurved, abaxially gibbous, faces glabrous, smooth, shiny; corolla scars subapical; **pappi:** pistillate 0, staminate rarely 0, usually of 1–10(–13) bristles (hidden in heads). $x = 14$.

Species 7 (7 in the flora): sw United States, nw Mexico.

See discussion of Filagininae following the tribal description (p. 385).

Stylocline occurs in Mediterranean, desert, and semi-desert climates; most species appear only after exceptionally wet winters, or in moisture-accumulating microsites (e.g., rock bases, washes, shrub drip-lines). Plants usually grow in undisturbed soils (often with soil crusts) and sometimes colonize stabilized disturbances.

In some species of *Stylocline*, the outermost bracts of heads are merely concave, not saccate; these are paleae (if they subtend and fall with florets) or phyllaries (if they persist and subtend only adjacent saccate paleae). Texture of the palea bodies is diagnostic for each species. In dried specimens, chartaceous bodies tear easily and irregularly when the abaxial indument is gently scraped. Cartilaginous bodies can be scraped clean without tearing and split lengthwise only if forced.

Stylocline appears to be ancestral to *Micropus* and *Psilocarphus*, and derived from, sister to, and/or reticulate with *Logfia* (J. D. Morefield 1992). *Stylocline citroleum*, *S. sonorensis*, and *L. depressa* show some transitional traits between the genera.

SELECTED REFERENCE Morefield, J. D. 1992. Evolution and Systematics of *Stylocline* (Asteraceae: Inuleae). Ph.D. dissertation. Claremont Graduate School. California.

1. Longest pistillate paleae winged proximally and distally, wings widest in proximal $^2/_3$ of palea lengths; phyllaries ± persistent, elliptic or broader, 1–3.5 mm.
   2. Longest pistillate paleae: wings broadly ovate (bases rounded or cordate); staminate ovaries vestigial, 0–0.2 mm, pappi usually of 1–5 bristles; heads arachnoid to thinly lanuginose (often shiny, indument obscured by palea wings) . . . . 1. *Stylocline gnaphaloides*
   2. Longest pistillate paleae: wings elliptic to slightly obovate (bases acute); staminate ovaries partially developed, (0.2–)0.3–0.6 mm, pappi of (5–)6–12(–13) bristles; heads thickly lanuginose (dull, indument evident) . . . . 2. *Stylocline citroleum*

1. Longest pistillate paleae winged distally, wings widest in distal ¹/₃ of palea lengths; phyllaries 0, vestigial, or falling, ± subulate, mostly 0.1–0.5 mm.
    3. Receptacles clavate, heights 2.8–3.5 times diams.; staminate ovaries partially developed, 0.3–0.6 mm (cypselae 0.6–0.8 mm; heads ± spheric, diams. 3–4 mm; longest pistillate paleae 1.9–3.1 mm; proximal leaves blunt) . . . . 3. *Stylocline sonorensis*
    3. Receptacles ± cylindric, heights 4–8 times diams.; staminate ovaries ± vestigial, 0–0.3 (–0.4) mm (cypselae 0.8–1.6 mm, heads ovoid to ellipsoid or diams. 5–9 mm, longest pistillate paleae 3.4–4.5 mm and/or proximal leaves acute).
        4. Heads ± spheric, thickly lanuginose, largest diams. 5–9 mm; pistillate paleae: longest 3.4–4.5 mm, outermost saccate.
            5. Bodies of longest pistillate paleae (except midnerves) chartaceous; cypselae compressed; largest capitular leaves (some or all) subulate to lanceolate (widest in proximal ¹/₃), (7–)11–17 mm (distalmost mainly 1.5–2 times head heights) . . . . 4. *Stylocline micropoides*
            5. Bodies of longest pistillate paleae cartilaginous; cypselae obcompressed; largest capitular leaves (all) ± elliptic to ± oblanceolate (widest in distal ²/₃), 4–11 mm (distalmost mainly 0.8–1.2 times head heights) . . . . 5. *Stylocline intertexta*
        4. Heads ovoid to ellipsoid, thinly lanuginose, largest diams. 1.5–4 mm; pistillate paleae: longest 2–3.3 mm, outermost open, concave.
            6. Heads 2.5–4 mm diam.; longest pistillate paleae 2.8–3.3 mm; cypselae 1.1–1.6 mm; staminate corollas 1.1–1.7 mm (lobes usually 5); leaves ± acute . . . . 6. *Stylocline psilocarphoides*
            6. Heads 1.5–2.5 mm diam.; longest pistillate paleae 2–2.7 mm; cypselae 0.7–1 mm; staminate corollas 0.8–1.1 mm (lobes usually 4); leaves blunt . . . . . . 7. *Stylocline masonii*

1. **Stylocline gnaphaloides** Nuttall, Trans. Amer. Philos. Soc., n. s. 7: 338. 1840 • Mountain or everlasting neststraw F

Plants 1–15(–20) cm. **Leaves** mostly blunt, not or scarcely mucronate, longest 6–14 mm; largest capitular leaves spatulate to ± elliptic or oblong, 4–13 × 1.5–3 mm. **Heads** in dichasiform or proximally ± racemiform arrays, ± spheric, largest 3–6 mm, arachnoid to thinly lanuginose (often shiny, appearing nearly glabrous, indument obscured by palea wings). **Phyllaries** ± persistent, ± ovate, 1–3.5 mm, subequal. **Receptacles** narrowly cylindric, 1.3–2.2 mm, heights 5–8 times diams.; scars ± evenly distributed, slightly sunken. **Pistillate paleae:** longest 1.8–4.5 mm, winged proximally and distally; wings broadly ovate (bases rounded or cordate), widest in proximal ¹/₃ of palea lengths; bodies (except midnerves) chartaceous; outermost paleae ± saccate. **Functionally staminate florets** 2–5; ovaries vestigial, 0–0.2 mm; corollas 1–1.8 mm. **Cypselae** 0.8–1 mm, compressed; **pappi:** staminate of 1–5(–6) smooth to barbellulate bristles 1.3–1.9 mm. $2n = 28$.

Flowering and fruiting (Jan–)Mar–May(–Jul). Dry, open, sandy slopes, flats, dry drainages (relatively mesic sites in deserts), often on old disturbances; 0–1200 (–1700) m; Ariz., Calif.; Mexico (Baja California, Sonora).

Stylocline gnaphaloides (often misspelled "gnaphalioides") is relatively common in southwestern California, extending to the nearer Channel Islands, the San Francisco Bay area, and most of Baja California; it is not known from the Colorado River valley; it occurs in (mainly) disturbed areas of south-central Arizona and adjacent Sonora, Mexico. A specimen from Zion National Park, Utah, in 1937 is likely either mislabeled or from an introduction that did not persist.

Forms with larger, often hairier heads (*Stylocline gnaphaloides* var. *bigelovii* A. Gray, *S. arizonica* Coville) are more frequent inland but also occur (with intermediates) on the California Channel Islands and elsewhere; no taxonomic segregation seems warranted. Sterile hybrids with *Logfia filaginoides* do not much resemble *S. citroleum*, which is suspected to have originated from such parentage (J. D. Morefield 1992, 1992b). Sterile hybrids between *S. gnaphaloides* and *L. arizonica* superficially resemble *S. psilocarphoides*.

2. Stylocline citroleum Morefield, Madroño 39: 125, fig. 4. 1992 • Oil neststraw C E

Plants 2–9(–13) cm. **Leaves** broadly acute, mucronate, longest 6–13 mm; largest capitular leaves broadly oblanceolate to ± elliptic, 4–12 × 2–3.5 mm. **Heads** in dichasiform or proximally ± racemiform arrays, ± spheric, largest 4–5.5 × 3.5–5 mm, thickly lanuginose (dull, indument evident). **Phyllaries** ± persistent, elliptic to obovate, 1.5–2.5 mm, subequal. **Receptacles** clavate, 1.5–2.5 mm, heights 4–6 times diams.; scars ± evenly distributed, slightly sunken. **Pistillate paleae:** longest 2.5–3.5 mm, winged proximally and distally; wings elliptic to slightly obovate (bases acute), widest in median ¹/₃ of palea lengths; bodies (except midnerves) chartaceous; outermost paleae ± saccate. **Functionally staminate florets** 3–6; ovaries partially developed, (0.2–)0.3–0.6 mm; corollas 1–1.6 mm. **Cypselae** 0.8–1 mm, compressed; **pappi:** staminate of (5–)6–12(–13) barbellate bristles 1.4–1.8 mm (proximal barbs longer, spreading).

Flowering and fruiting Mar–Apr. Dry, open, relatively undisturbed, often crusted sandy or clay soils, gentle slopes, dry drainage banks, clearings in saltbush (*Atriplex*) vegetation; of conservation concern; 60–300 m; Calif.

Stylocline citroleum is known from western Kern County; basis for an 1883 San Diego County report is likely extirpated or plants were mislabeled. It is fairly widespread on and near the former Elk Hills Naval Petroleum Reserve (D. F. Williams et al. 1998).

Stylocline citroleum almost always occurs with *S. gnaphaloides* and/or *Logfia filaginoides*. It is thought to be descended from hybrid ancestors (J. D. Morefield 1992, 1992b) and now appears to be fertile, morphologically uniform, and reproductively isolated from sympatric taxa. Character states place it sister to *S. gnaphaloides*.

3. Stylocline sonorensis Wiggins, Contr. Dudley Herb. 4: 26. 1950 • Sonoran or mesquite neststraw

Plants 2–10(–15) cm. **Leaves** blunt (proximal) or acute (median and distal), mucronate, longest 6–13 mm; largest capitular leaves ± elliptic to narrowly ovate, 3–10 × 2–3 mm. **Heads** in cymiform to ± paniculiform, sometimes dichasiform arrays, ± spheric, largest 3.5–4.5 × 3–4 mm, thickly lanuginose. **Phyllaries** 0, vestigial, or falling, ± subulate, mostly 0.1–0.5 mm, unequal. **Receptacles** clavate, 1.2–2.2 mm, heights 2.8–3.5 times diams.; scars ± evenly distributed,

mamillate. **Pistillate paleae:** longest 1.9–3.1 mm, winged distally; wings ± elliptic, widest in distal ¹/₃ of palea lengths; bodies (except midnerves) chartaceous; outermost paleae ± saccate. **Functionally staminate florets** 2–5; ovaries partially developed, 0.3–0.6 mm; corollas 0.9–1.4 mm. **Cypselae** 0.6–0.8 mm, slightly compressed; **pappi:** staminate of (1–)3–8 barbellate bristles 0.9–1.3 mm (proximal barbs longer, spreading).

Flowering and fruiting Mar–May. Grassy hillsides, sandy drainages, with mesquite (*Prosopis*); 400–1400 m; Ariz., Calif.; Mexico (Sonora).

Stylocline sonorensis is known from southeastern Arizona and northeastern Sonora. A disjunct, 1930 California occurrence was from apparently suitable habitat; recent searches have not relocated it.

Stylocline sonorensis is illustrated in J. D. Morefield (1992). It is superficially similar to *S. citroleum*; its closest relative is *S. micropoides*. Its ancestors may have been hybrid products involving *Logfia depressa* or its progenitors (Morefield).

4. Stylocline micropoides A. Gray, Smithsonian Contr. Knowl. 5(6): 84. 1853 • Woollyhead or desert or woollyhead fanbract

Plants 2–14(–20) cm. **Leaves** acute, mucronate, longest 8–20 mm; largest capitular leaves (some or all) subulate to lanceolate (widest in proximal ¹/₃), (7–)11–17 × 1.5–2.5 mm (distalmost mainly 1.5–2 times head heights). **Heads** in ± paniculiform to cymiform, rarely dichasiform arrays, ± spheric, largest 5–9 mm, thickly lanuginose. **Phyllaries** 0, vestigial, or falling, ± subulate, mostly 0.1–0.5 mm, unequal. **Receptacles** narrowly cylindric, 2–3 mm, heights 5–8 times diams.; scars ± evenly distributed, ± flat. **Pistillate paleae:** longest 3.4–4.5 mm, winged distally; wings elliptic to ovate, widest in distal ¹/₃ of palea lengths; bodies (except midnerves) chartaceous; outermost paleae ± saccate. **Functionally staminate florets** 3–6; ovaries vestigial, 0–0.3 mm; corollas 1.2–1.9 mm. **Cypselae** 1–1.4 mm, compressed; **pappi:** staminate of 2–5(–10) smooth to barbellulate bristles 1.1–2 mm. $2n = 28$.

Flowering and fruiting Feb–May(–Aug). Relatively stable sandy or gravelly desert soils, often under shrubs, at rock bases; 70–1600 m; Ariz., Calif., Nev., N.Mex., Tex., Utah; Mexico (Baja California, Chihuahua, Sonora).

Stylocline micropoides occurs throughout warm deserts of the flora area, through the Grand Canyon to southern Utah and up Owens Valley to the White Mountains of California. Sterile and malformed natural hybrids between *S. micropoides* and *Logfia arizonica* have been seen. The description of *S. micropoides* by D. S. Correll and M. C. Johnston (1970) was clearly based on specimens of *L. filaginoides*; both species occur near El Paso, Texas.

5. **Stylocline intertexta** Morefield, Madroño 39: 121, fig. 3. 1992 • Mojave or Morefield neststraw [E]

Plants 2–8(–11) cm. **Leaves** acute, mucronate, longest 6–15 mm; largest capitular leaves (all) ± elliptic to ± oblanceolate (widest in distal ²/₃), 4–11 × 1–2.5 mm (distalmost mainly 0.8–1.2 times head heights). **Heads** in ± paniculiform to cymiform, rarely dichasiform, arrays, ± spheric, largest 5–6 mm, thickly lanuginose. **Phyllaries** 0, vestigial, or falling, ± subulate, mostly 0.1–0.5 mm, unequal. **Receptacles** cylindric, 1.4–2.7 mm, heights 4–7 times diams.; scars ± evenly distributed, mamillate. **Pistillate paleae:** longest 3.4–4.2 mm, winged distally; wings elliptic to ovate, widest in distal ¹/₃ of palea lengths; bodies cartilaginous; outermost paleae ± saccate. **Functionally staminate florets** 3–6; ovaries vestigial, 0–0.3 mm; corollas 1.1–2.3 mm. **Cypselae** 1–1.4 mm, obcompressed; **pappi:** staminate rarely 0, usually of 1–4(–8) smooth to barbellulate bristles 1.1–2 mm.

Flowering and fruiting Feb–May. Open, stable, often calcareous desert gravels, sands, often with extra moisture (rock bases, shrub drip lines, dry drainages, depressions); 40–1400 m; Ariz., Calif., Nev., Utah.

Stylocline intertexta is known from the Mojave and northwestern Sonoran deserts. It combines character states of *S. micropoides* and *S. psilocarphoides*, is often sympatric with both, and appears to be stable, uniform, and reproducing independently. *Stylocline intertexta* shares most character states with *S. micropoides*. Presence of some subulate to lanceolate capitular leaves in *S. micropoides* helps distinguish the species in the field.

6. **Stylocline psilocarphoides** M. Peck, Leafl. W. Bot. 4: 185. 1945 • Peck or baretwig neststraw, Malheur stylocline [E] [F]

Plants 1–8(–18) cm (branches ± leafless between proximal forks). **Leaves** ± acute, mucronate, longest 8–18 mm; largest capitular leaves elliptic to spatulate, 3–10 × 1–2 mm. **Heads** in strictly dichasiform arrays, ovoid, largest 3.5–5 × 2.5–4 mm, thinly lanuginose. **Phyllaries** 0, vestigial, or falling, ± subulate, mostly 0.1–0.5 mm, unequal. **Receptacles** ± cylindric, 2–3 mm, heights 5–8 times diams.; scars concentrated proximally and distally, raised, peglike. **Pistillate paleae:** longest 2.8–3.3 mm, winged distally; wings oblanceolate to oblacrimate, widest in distal ¹/₃ of palea lengths; bodies cartilaginous; outermost paleae open, concave. **Functionally staminate florets** 2–5; ovaries ± vestigial, 0.1–0.3(–0.4) mm; corollas 1.1–1.7 mm (lobes usually 5). **Cypselae** 1.1–1.6 mm, obcompressed; **pappi:** staminate rarely 0, usually of 1–3 smooth bristles 1.1–1.5 mm.

Flowering and fruiting Feb–Jun. Open, relatively stable, often granitic, sandy to gravelly desert soils, often at rock bases, under shrubs; 100–2000 m; Calif., Idaho, Nev., Oreg., Utah.

Stylocline psilocarphoides is centered in the Mojave and western Great Basin deserts and extends to southeastern Oregon, southwestern Idaho, and through the western Sonoran Desert nearly to Mexico. Disjunct collections from southern Arizona in 1876 and 1941 are likely to be either labeling errors or from introductions that did not persist.

Circumscription of *Stylocline psilocarphoides* became clearer after recognition of *S. intertexta* and *S. masonii* (J. D. Morefield 1992b).

7. **Stylocline masonii** Morefield, Madroño 39: 117, fig. 1. 1992 • Mason neststraw [C] [E]

Plants 1–7(–10) cm. **Leaves** blunt, not mucronate, longest 5–9 mm; largest capitular leaves linear-oblong to narrowly elliptic, 2–5 × ± 1 mm. **Heads** in dichasiform or proximally racemiform arrays, ± ellipsoid, largest 2–5 × 1.5–2.5 mm, thinly lanuginose. **Phyllaries** 0, vestigial, or falling, ± subulate, mostly 0.1–0.5 mm, unequal. **Receptacles** ± cylindric, 2–3 mm, heights 5–8 times diams.; scars concentrated proximally and distally, raised, peglike. **Pistillate paleae:** longest 2–2.7 mm, winged distally; wings oblanceolate to oblacrimate, widest in distal ¹/₃ of palea lengths; bodies cartilaginous; outermost paleae open, concave. **Functionally staminate florets** 2–4; ovaries vestigial, 0–0.1 mm; corollas 0.8–1.1 mm (lobes usually 4). **Cypselae** 0.7–1 mm, obcompressed; **pappi:** staminate 0 or of 1 smooth bristle 0.7–1 mm.

Flowering and fruiting late Mar–early Jun. Open, loose, sandy soils, usually in dry washes; of conservation concern; 100–1200 m; Calif.

Stylocline masonii is known from the southern San Joaquin Valley and adjacent mountains of Kern, Los Angeles, Monterey, and San Luis Obispo counties. Reportedly, it is in identifiable condition for only 2–4 weeks in wet years. On average, it is the least conspicuous *Stylocline*.

Stylocline masonii is sister to, and largely allopatric with, *S. psilocarphoides* (J. D. Morefield 1992), from which it remains distinct when sympatric (T. S. Ross and S. Boyd 1996).

100. MICROPUS Linnaeus, Sp. Pl. 2: 927. 1753; Gen. Pl. ed. 5, 398. 1754 • Cottonseed, micrope, falzblume [Greek *micros*, small, and *pous*, foot, perhaps alluding to tiny receptacles]

James D. Morefield

Bombycilaena (de Candolle) Smoljaninova

Annuals, 1–50 cm. **Stems** 1, ± erect, or 2–5[–10], ascending to erect [prostrate]. **Leaves** cauline; mostly alternate [opposite]; blades narrowly oblanceolate to elliptic [spatulate]. **Heads** usually in glomerules of 2–5 in racemiform to paniculiform or distally ± dichasiform [axillary] arrays, sometimes borne singly. **Involucres** inconspicuous. **Phyllaries** 4–6, ± equal (unlike paleae, scarious, hyaline). **Receptacles** depressed-spheric or obovoid (heights 0.5–1.8 times diams.), glabrous. **Pistillate paleae** falling, erect to incurved; bodies with 5+ nerves (nerves ± parallel, obscure), obovoid, saccate most of lengths (trigonously [evenly] compressed, galeate, abaxially rounded [corniculate-crested], each enclosing a floret); wings ± erect (and lateral) or inflexed (and subapical). **Staminate paleae** 0 or 1–3, falling, erect in fruit (not enlarged), shorter than pistillate paleae; bodies linear-lanceolate to oblanceolate. **Pistillate florets** 4–12. **Functionally staminate florets** 2–5; corolla lobes 4–5, ± equal. **Bisexual florets** 0. **Cypselae** brown, monomorphic: ± trigonously [evenly] compressed, ± obovoid, curved, gibbous abaxially, faces glabrous, smooth, shiny, corolla scars ± lateral; **pappi:** pistillate 0, staminate 0 or of 1–5 bristles (hidden in heads). *x* = 14.

Species 5 (2 in the flora): w United States, nw Mexico, s Europe, sw Asia, n Africa.

See discussion of Filagininae following the tribal description (p. 385).

Micropus species are found mostly in dry, open habitats of Mediterranean climates. In the flora, they are known only from west-draining portions of the Californian Floristic Province and the Willamette Valley in Oregon.

The two North American species constitute *Micropus* sect. *Rhyncholepis* Nuttall. Recent European workers (e.g., J. Holub 1998) have included sect. *Rhyncholepis* in *Bombycilaena*, leaving *M. supinus* Linnaeus in a monotypic genus. Based on phylogenetic data (J. D. Morefield 1992), that approach would include in *Bombycilaena* species ancestral to, and derived from ancestors of, *Micropus*. I maintain *Micropus* in its traditional sense here. *Micropus* and *Psilocarphus* appear to be monophyletic sister genera derived from near or within *Stylocline*. A malformed specimen from Monterey County, California, appears to be a sterile hybrid between *M. californicus* and a species of *Psilocarphus*.

1. Pistillate paleae 8–12 in 2 series, bodies mainly chartaceous, cartilaginous medially, wings prominent, subapical, inflexed, plane to concave; cypselae: corolla scars in distal $^{1}/_{4}$; receptacle heights 1.2–1.8 times diams.; staminate paleae mostly 1–3; staminate corolla lobes 4 (–5); staminate pappi of 1–5 bristles . . . . 1. *Micropus amphibolus*
1. Pistillate paleae 4–7(–8) in 1 series, bodies cartilaginous to bony throughout, wings obscure, lateral, ± erect, involute; cypselae: corolla scars ± median; receptacle heights 0.5–0.8 times diams.; staminate paleae 0; staminate corolla lobes usually 5; staminate pappi 0 or of 1 bristle . . . . 2. *Micropus californicus*

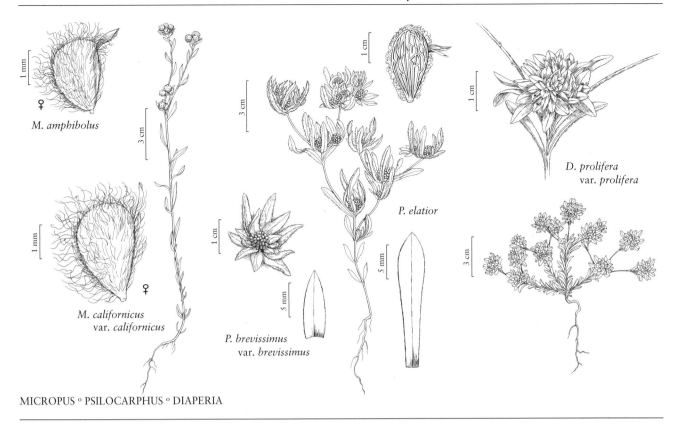

M. *amphibolus*

M. *californicus*
var. *californicus*

P. *brevissimus*
var. *brevissimus*

P. *elatior*

D. *prolifera*
var. *prolifera*

MICROPUS ° PSILOCARPHUS ° DIAPERIA

1. **Micropus amphibolus** A. Gray, Proc. Amer. Acad. Arts 17: 214. 1882 • Mount Diablo cottonseed E F

Stylocline amphibola (A. Gray) J. T. Howell

Plants 2–20 cm. **Heads** ± spheric, 3.5–5 mm diam. **Receptacles** obovoid, mostly 0.7–1 mm, heights 1.2–1.8 times diams. **Pistillate paleae** 8–12 in 2 series, longest 2–3 mm; wings prominent, not withering, subapical, inflexed, ± obovate, plane to concave; bodies somewhat galeate, mainly chartaceous, cartilaginous medially, ± lanuginose. **Staminate paleae** mostly 1–3. **Staminate corollas** 1.2–1.9 mm, lobes 4(–5). **Cypselae** 1–1.5 mm, corolla scars in distal ¹/₄; **pappi:** staminate of 1–5, ± barbellulate bristles 1.7–2 mm.

Flowering and fruiting late Mar–Jun. Slopes, ridges, on and near rock outcrops, shallow gravelly soils derived from shale, mudstone, sandstone, volcanic rocks (not recorded on serpentine), edges of or in clearings, grasslands, chaparral, woodlands; 40–900 m; Calif.

Micropus amphibolus is relatively uncommon; it is known from mountains and hills of the San Francisco Bay area and northern Coast Ranges, plus populations in outer southern Coast Ranges and in foothills of the northern Sierra Nevada. It is superficially similar to and often confused with *M. californicus* (see discussion there).

Relatively few recent collections of *M. amphibolus* are known, raising significant conservation concerns.

Though consistently distinct, *Micropus amphibolus* combines some character states of *Stylocline* and *Micropus* (J. T. Howell 1948), suggesting that it may have descended from a hybrid between, or common ancestor of, the two genera. The best available morphologic evidence consistently places it basal within *Micropus*, with which it shares a majority of character states (J. D. Morefield 1992).

2. **Micropus californicus** Fischer & C. A. Meyer, Index Seminum (St. Petersburg) 2: 42. 1836 • Slender cottonseed, Q-tips F

Bombycilaena californica (Fischer & C. A. Meyer) Holub

Plants 1–50 cm. **Heads** depressed-spheric, 2–4 × 3–6 mm. **Receptacles** depressed-spheric, mostly 0.3–0.6 mm, heights 0.5–0.8 times diams. **Pistillate paleae** 4–7(–8) in 1 series, longest 2–4 mm; wings obscure, withering, lateral, ± erect, narrowly oblanceolate, involute; bodies galeate, cartilaginous to bony throughout, sericeous to lanuginose. **Staminate paleae** 0. **Staminate corollas** 1–2 mm, lobes usually 5. **Cypselae** 1.4–2.6 mm, corolla scars ± median; **pappi:** staminate 0 or of 1, ± smooth bristle 0.9–1.5 mm.

Varieties 2 (2 in the flora): w United States, nw Mexico.

Though common and often found on recent disturbances within its range, *Micropus californicus* does not appear to be particularly invasive or weedy. The mature pistillate paleae, with roughly the profile of a harp or of an inverted lower-case letter "q," are distinctive. That shape and the accompanying dense, cottony indument of the common variety explain the vernacular names. By contrast, pistillate paleae of *M. amphibolus* have roughly the profile of a human head wearing a billed cap.

Within the more limited range of var. *subvestitus*, the two varieties are broadly sympatric, usually in separate populations, occasionally in mixed populations that sometimes include intermediate plants.

1. Pistillate paleae densely and loosely lanuginose, longest usually 3–4 mm . . . .
    . . . . 2a. *Micropus californicus* var. *californicus*
1. Pistillate paleae thinly sericeous-lanuginose, longest usually 2–3 mm . . . .
    . . . . 2b. *Micropus californicus* var. *subvestitus*

2a. **Micropus californicus** Fischer & C. A. Meyer var. **californicus** • Cottontop F

Plants 1–50 cm. **Heads** mostly 3–4 × 4.5–6 mm. **Pistillate paleae** densely and loosely lanuginose, longest usually 3–4 mm. **Staminate corollas** 1.3–2 mm. **Cypselae** 1.8–2.6 mm; **pappi:** staminate 1.3–1.5 mm.

Flowering and fruiting mid Mar–early Jul. Slopes, flats, dry or seasonally moist soils of various substrates (including serpentine), often recently disturbed, in grasslands, chaparral, wooded areas, clearings, mostly inland from coastal fog belt; 10–1600 m; Calif., Oreg.; Mexico (Baja California).

Variety *californicus* is known nearly throughout the west-draining portion of the Californian Floristic Province and the Willamette Valley, from western Oregon to northwestern Baja California, including the California Channel Islands and Isla Guadalupe.

2b. **Micropus californicus** Fischer & C. A. Meyer var. **subvestitus** A. Gray in W. H. Brewer et al., Bot. California 1: 335. 1876 • Silky cottonseed

Plants 1–40 cm. **Heads** mostly 2–3 × 3–4.5 mm. **Pistillate paleae** thinly sericeous-lanuginose, longest usually 2–3 mm. **Staminate corollas** 1–1.5 mm. **Cypselae** 1.4–2.2 mm; **pappi:** staminate 0.9–1.3 mm.

Flowering and fruiting late Apr–early Jun. Ridges, hilltops, slopes, relatively dry, shallow, exposed soils of various substrates (including serpentine), often recently disturbed, usually within coastal fog belt; 50–1100 m; Calif.

Variety *subvestitus* is known from the southern San Francisco Bay area through outer southern Coast Ranges to San Luis Obispo County, plus populations in foothills of the central Sierra Nevada.

101. PSILOCARPHUS Nuttall, Trans. Amer. Philos. Soc., n. s. 7: 340. 1840 • Woolly marbles, woollyheads [Greek *psilos*, slender, and *karphos*, chaff, alluding to papery paleae of heads]

James D. Morefield

Annuals, 1–15(–20) cm. **Stems** 1, erect, or 2–10, ascending to ± prostrate. **Leaves** cauline; mostly opposite; blades linear to ovate or obovate. **Heads** borne singly or in glomerules of 2–4 in ± dichasiform (sometimes ± paniculiform) arrays. **Involucres** 0. **Phyllaries** 0. **Receptacles** ± obovoid (sometimes lobed, heights 1–2 times diams.), glabrous. **Pistillate paleae** falling, radiating in all directions; bodies with 5+ nerves (nerves reticulate, prominent), obovoid to ± cylindric, saccate most of lengths (terete, galeate or cucullate, each loosely enclosing a floret); wings inflexed (± lateral). **Staminate paleae** 0. **Pistillate florets** (8–)20–100+. **Functionally staminate florets** 2–10; corolla lobes 4–5, ± equal. **Bisexual florets** 0. **Cypselae** brown, monomorphic: terete to ± compressed, narrowly obovoid (then slightly incurved, abaxially gibbous) to ± cylindric, faces glabrous, smooth, shiny; corolla scars usually ± subapical; **pappi** 0. $x = 14$.

Species 5 (5 in the flora): amphitropical, w North America, nw Mexico, s South America.

See discussion of Filagininae following the tribal description (p. 385).

In the flora area, *Psilocarphus* inhabits sites with Mediterranean, semiarid, and cool-temperate climates. In the south, it remains within the Californian Floristic Province, not entering the Mojave and Sonoran deserts; to the north, it extends from the Pacific Northwest eastward across the Great Basin, Columbia Plateau, and northern Rocky Mountains to the western edge of the northern Great Plains. Ongoing degradation of vernal pool habitats in California may soon justify conservation concern for *P. chilensis* and *P. brevissimus* var. *multiflorus*.

The amphitropical species of *Psilocarphus* and *Micropsis* (*P. brevissimus* var. *brevissimus*, *P. chilensis*, *M. dasycarpa*) occupy littoral habitats; migratory shorebirds probably facilitate occasional long-distance dispersal of their light cypsela-palea complexes (A. Cronquist 1950). Populations of these self-pollinating species can establish from one cypsela.

Psilocarphus is monophyletic and probably sister to *Micropus*, with ancestors in or near *Stylocline* (J. D. Morefield 1992). *Psilocarphus* is easily recognized by leaves opposite and paleae cucullate or galeate, reticulately nerved; the clusters of heads resemble compact bunches of woolly grapes or marbles. Differences between species are slight but consistent in most specimens.

SELECTED REFERENCE Cronquist, A. 1950. A review of the genus *Psilocarphus*. Res. Stud. State Coll. Wash. 18: 71–89.

1. Largest heads 6–14 mm; pistillate paleae collectively hidden by indument and/or longest 2.8–4 mm.
   2. Heads ± ovoid, largest 9–14 mm; receptacles deeply lobed; pistillate paleae ± cylindric, lengths mostly 3.5–6 times longest diams. (wings ± median) . . . . . . 1. *Psilocarphus brevissimus* (in part)
   2. Heads spheric, largest 6–9 mm; receptacles unlobed or shallowly lobed; pistillate paleae obovoid, lengths 1.5–3 times longest diams. (wings supramedian to subapical).
      3. Capitular leaves mostly lanceolate to ovate, widest in proximal $^2/_3$, longest mostly 8–15 mm, lengths mostly 1.5–4 times widths; plants usually densely lanuginose; cypselae narrowly obovoid . . . . 1. *Psilocarphus brevissimus* (in part)
      3. Capitular leaves mostly oblanceolate to nearly linear, widest in distal $^1/_3$, longest mostly 17–35 mm, lengths mostly 4.5–9 times widths; plants ± sericeous; cypselae ± cylindric. . . . . 2. *Psilocarphus elatior*
1. Largest heads mostly 3–6 mm; pistillate paleae usually individually visible through indument, longest mostly 1.5–2.7 mm.
   4. Capitular leaves linear to narrowly oblanceolate, lengths mostly 6–12 times widths, (3–)3.5–5 times head heights; cypselae ± cylindric . . . . 3. *Psilocarphus oregonus*
   4. Capitular leaves mostly spatulate to obovate or ovate, lengths mostly 1.2–5 times widths, 1–2.5(–3) times head heights; cypselae ± obovoid.
      5. Capitular leaves mostly not appressed to heads, spatulate to obovate, lengths mostly 2–5 widths; proximal internode lengths mostly 1–2(–3) times leaf lengths; staminate corolla lobes mostly 5 . . . . 4. *Psilocarphus tenellus*
      5. Capitular leaves appressed to heads, ovate to broadly elliptic, lengths mostly 1.2–1.8(–2) times widths; proximal internode lengths (2–)3–6 times leaf lengths; staminate corolla lobes mostly 4 . . . . 5. *Psilocarphus chilensis*

1. Psilocarphus brevissimus Nuttall, Trans. Amer. Philos. Soc., n. s. 7: 340. 1840 • Short woollyheads F

Plants greenish to grayish, sericeous to densely lanuginose. **Stems** (1–)2–10, erect to prostrate; proximal internode lengths mostly 0.5–1.5(–2) times leaf lengths. **Capitular leaves** ± erect, appressed to heads (sometimes spreading), linear-lanceolate to ovate, widest in proximal $^2/_3$, longest 8–25 mm, lengths mostly 1.5–6 times widths, 1–2.5(–3) times head heights. **Heads** ± spheric, rarely ovoid, largest 6–14 mm. **Receptacles** unlobed or ± lobed. **Pistillate paleae** hidden by or visible through indument, longest 2.8–4 mm (lengths 1.5–6 times longest diams.; wings subapical to ± median). **Staminate corollas** 0.8–1.6 mm, lobes mostly 5. **Cypselae** narrowly obovoid, ± compressed, 0.8–1.9 mm. $2n = 28$.

Varieties 2 (2 in the flora): w North America (including nw Mexico); s South America.

1. Heads ± spheric, largest 6–9 mm, receptacles unlobed or shallowly lobed; pistillate paleae obovoid, lengths 1.5–3 times longest diams., wings supramedian to subapical . . . .
. . . . 1a. *Psilocarphus brevissimus* var. *brevissimus*
1. Heads ovoid, largest 9–14 mm, receptacles deeply lobed; pistillate paleae ± cylindric, lengths mostly 3.5–6 times longest diams., wings ± median . . .
. . . . 1b. *Psilocarphus brevissimus* var. *multiflorus*

1a. Psilocarphus brevissimus Nuttall var. **brevissimus** • Dwarf woollyheads F

Psilocarphus globiferus Nuttall

Plants lanuginose to sericeous. **Stems** (1–)2–10, ascending to ± prostrate (sometimes erect). **Capitular leaves:** longest mostly 8–15 mm, lengths 1.5–4 times widths. **Heads** ± spheric, largest 6–9 mm, usually densely lanuginose. **Receptacles** unlobed or shallowly lobed. **Pistillate paleae** obovoid, lengths 1.5–3 times longest diams.; wings supramedian to subapical. $2n = 28$.

Flowering and fruiting mid Mar–mid Aug. Drying margins of seasonally inundated sites (vernal pools, ditches), sometimes alkaline; 10–2500 m; Alta., B.C., Sask.; Calif., Idaho, Mont., Nev., Oreg., Utah, Wash., Wyo.; Mexico (Baja California); South America (Argentina, Chile).

Variety *brevissimus* occupies nearly the full range of the genus (uncommon west of the Cascade Range); some occurrences toward the northeast appear to be recent introductions.

Plants named *Psilocarphus globiferus* are superficially somewhat intermediate between *P. brevissimus* and *P. tenellus*, as are populations from southern California mountains with leaves somewhat shorter, broader, and less hairy than typical *P. brevissimus*.

1b. Psilocarphus brevissimus Nuttall var. **multiflorus** Cronquist, Res. Stud. State Coll. Wash. 18: 80. 1950 • Delta woolly marbles E

Plants ± thinly arachnoid-sericeous. **Stems** usually 1, erect. **Capitular leaves:** longest mostly 14–25 mm, lengths 3–6 times widths. **Heads** ovoid, largest 9–14 mm, ± sericeous. **Receptacles** deeply lobed. **Pistillate paleae** ± cylindric, lengths mostly 3.5–6 times longest diams.; wings ± median.

Flowering and fruiting May–mid Jun. Drying margins of vernal pools, seasonally moist flats; 10–100(–500) m; Calif.

Variety *multiflorus* is known from the San Francisco Bay area and adjacent Central Valley delta and may soon warrant conservation concern. Its deeply lobed receptacles, each lobe bearing a set of staminate florets, appear to represent fusion of multiple heads into second-order heads (A. Cronquist 1950). Less fusion occurs in plants intermediate with var. *brevissimus*.

Variety *multiflorus* shares traits with *Psilocarphus elatior*. An Idaho specimen from within the range of *P. elatior* was cited (but not mapped) as *P. brevissimus* var. *multiflorus* by A. Cronquist (1950); it was probably either mislabeled or from an introduction that did not persist.

2. Psilocarphus elatior (A. Gray) A. Gray in A. Gray et al., Syn. Fl. N. Amer. ed. 2, 1: 448. 1886 • Tall or meadow woollyheads E F

Psilocarphus oregonus Nuttall var. *elatior* A. Gray, Proc. Amer. Acad. Arts 8: 652. 1873

Plants greenish gray to silvery, ± sericeous. **Stems** 1(–3), ± erect; proximal internode lengths mostly 0.5–1.5(–2) times leaf lengths. **Capitular leaves** ± erect, appressed to heads, mostly oblanceolate to nearly linear, widest in distal $^1/_3$, longest mostly 17–35 mm, lengths mostly 4.5–9 times widths, 2.5–5 times head heights. **Heads** ± spheric, largest 6–8 mm. **Receptacles**

unlobed. **Pistillate paleae** collectively ± hidden by indument, longest 2.8–3.8 mm (lengths 1.5–3 times longest diams.; wings supramedian). **Staminate corollas** 1.3–1.9 mm, lobes 5. **Cypselae** ± cylindric, terete, 0.9–1.7 mm.

Flowering and fruiting mid May–mid Aug. Mainly coastal or montane, relatively dry or seasonally flooded, wooded, grassy, or barren slopes, flats, often disturbed sites (roadsides, trails, drainages), rarely near vernal pools; 0–1700 m; B.C.; Calif., Idaho, Mont., Oreg., Wash.

Psilocarphus elatior occurs west of the Cascade Range from California to Vancouver Island, British Columbia, and in scattered areas eastward (northwestern Montana, mountains surrounding the border area common to Oregon, Washington, and Idaho). Reports of *P. elatior* from Alberta and Saskatchewan were based on relatively erect forms of *P. brevissimus* var. *brevissimus*. *Psilocarphus elatior* has been of conservation concern in Canada (J. M. Illingworth and G. W. Douglas 1994).

Where sympatric, *Psilocarphus elatior* tends to inhabit relatively dry or seasonally flooded sites in more mesic coastal or montane climates and *P. brevissimus* var. *brevissimus* occurs mainly in wetter, seasonally inundated sites in semiarid climates. Some specimens appear to be intermediate; further study may show the two taxa to be better treated as varietally distinct. See also under *P. brevissimus* var. *multiflorus*.

3. **Psilocarphus oregonus** Nuttall, Trans. Amer. Philos. Soc., n. s. 7: 341. 1840 • Oregon woollyheads or woolly marbles E

Plants silvery to whitish, densely sericeous to somewhat lanuginose. **Stems** (1–)2–10, ascending to ± prostrate; proximal internode lengths mostly 0.5–1.5(–2) times leaf lengths. **Capitular leaves** ± erect, appressed to heads, linear to narrowly oblanceolate, widest in distal ²/₃, longest 12–20 mm, lengths mostly 6–12 times widths, (3–)3.5–5 times head heights. **Heads** ± spheric, largest 4–6 mm. **Receptacles** unlobed. **Pistillate paleae** individually visible through indument, longest mostly 1.5–2.7 mm. **Staminate corollas** 0.7–1.4 mm, lobes mostly 4. **Cypselae** narrowly ± cylindric, terete, 0.6–1.2 mm.

Flowering and fruiting late Mar–mid Aug. Seasonally inundated or flooded clay soils (vernal pool margins, drainages, moist rocky slopes); 10–1800(–2400) m; Calif., Idaho, Nev., Oreg., Wash.

Psilocarphus oregonus occurs from west-central California through most of Oregon to southeastern Washington, western Idaho, and northern Nevada. Relatively narrow-leaved, montane forms of *P. tenellus* account for reports of *P. oregonus* from the southern Sierra Nevada

to Baja California; further study may show these to be intermediates between the two taxa.

A malformed plant collected in Merced County, California, appears to have been a sterile hybrid between *P. oregonus* and *Hesperevax caulescens* (J. D. Morefield 1992c).

4. **Psilocarphus tenellus** Nuttall, Trans. Amer. Philos. Soc., n. s. 7: 341. 1840 • Slender woolly marbles

Plants greenish to grayish, arachnoid to ± sericeous. **Stems** (1–)2–10, ascending to ± prostrate; proximal internode lengths mostly 1–2(–3) times leaf lengths. **Capitular leaves** ± spreading, mostly not appressed to heads, spatulate to obovate, widest in distal ¹/₃, longest 6–15 mm, lengths mostly 2–5 times widths, 1.5–2.5(–3) times head heights. **Heads** ± spheric, largest 3–5.5 mm. **Receptacles** unlobed. **Pistillate paleae** individually visible through indument, longest mostly 1.5–2.7 mm. **Staminate corollas** 0.8–1.5 mm, lobes mostly 5. **Cypselae** narrowly obovoid, somewhat compressed, 0.7–1.2 mm.

Flowering and fruiting late Mar–early Aug. Dry or seasonally moist, barren to wooded slopes, flats, often disturbed sites (foot paths, road beds, burns), sometimes near vernal pools toward s; 0–2100 m; B.C.; Calif., Idaho, Oreg., Wash.; Mexico (Baja California).

Psilocarphus tenellus is relatively common in the Californian Floristic Province from northwestern Baja California to southwestern Oregon; northward it is scattered to northern Idaho and Vancouver Island, British Columbia. It has been of conservation concern in Canada (J. M. Illingworth and G. W. Douglas 1994b).

5. **Psilocarphus chilensis** A. Gray in A. Gray et al., Syn. Fl. N. Amer. ed. 2, 1: 448. 1886 • Round woolly marbles

Micropus globiferus Bertero ex de Candolle in A. P. de Candolle and A. L. P. P. de Candolle, Prodr. 5: 460. 1836, not *Psilocarphus globiferus* Nuttall 1840; *P. tenellus* Nuttall var. *globiferus* (Bertero ex de Candolle) Morefield; *P. tenellus* var. *tenuis* (Eastwood) Cronquist

Plants mostly greenish, thinly arachnoid-sericeous (in coastal forms grayish to whitish, ± lanuginose). **Stems** mostly (1–)2–7, ascending to ± prostrate; proximal internode lengths (2–)3–6 times leaf lengths. **Capitular leaves** erect to incurved, appressed to heads, ovate to broadly elliptic, widest in proximal ²/₃, longest 5–12 mm, lengths mostly 1.2–1.8(–2) times widths, 1–2(–2.5) times head heights. **Heads** ± spheric, largest 3–5.5 mm. **Receptacles** unlobed. **Pistillate paleae** usually individually visible through indument, longest mostly 1.5–2.7 mm. **Staminate corollas** 0.8–1.3 mm, lobes mostly 4. **Cypselae** narrowly obovoid, somewhat compressed, 0.6–1.2 mm.

Flowering and fruiting mid Mar–early Jul. Saturated to drying vernal pool margins, seasonally inundated sites, coastal interdune areas; 0–600 m; Calif.; South America (Chile).

Psilocarphus chilensis occurs mainly in west-central California and central Chile; one recent collection is from southern California (western Riverside County). Ecotypes from coastal interdune areas are more lanuginose with shorter stems and internodes than intergrading populations farther inland; they are indistinguishable from the type of *Micropus globiferus* from Chile (J. D. Morefield 1992d). *Psilocarphus chilensis* and *P. tenellus* are at least as distinct as the other species of *Psilocarphus*; contrary to suggestions by A. Cronquist (1950), intermediates between the two are at most very uncommon.

Psilocarphus berteri I. M. Johnston is a superfluous name for *P. chilensis*. I. M. Johnston (1938) erroneously applied *P. chilensis* to a species not including the type of *Micropus globiferus*; such plants are here included in *P. brevissimus* var. *brevissimus*.

102. **DIAPERIA** Nuttall, Trans. Amer. Philos. Soc., n. s. 7: 337. 1840 • Rabbit-tobacco, dwarf cudweed [Greek *diapero*, to pass through, alluding to pseudo-polytomous branching pattern ("proliferous inflorescence") of type species]

James D. Morefield

Evax Gaertner sect. *Diaperia* (Nuttall) A. Gray

Annuals, 3–25 cm. **Stems** 1, erect, or 2–10, ascending to ± prostrate. **Leaves** basal and cauline; alternate; blades oblanceolate to obovate. **Heads** borne singly or in glomerules of 2–40+ in ± dichasiform, pseudo-polytomous, spiciform, or racemiform arrays. **Involucres** inconspicuous. **Phyllaries** (2–)4–6, ± equal (similar to paleae). **Receptacles** pulvinate to conic (heights 0.2–2.4 times diams.), glabrous. **Pistillate paleae** readily falling (all or inner together, ± coherent distally by tangled indument) or outermost sometimes persistent, erect to ascending; bodies with 5+ nerves (nerves ± parallel, obscure), oblanceolate to oblong, flat to concave most of lengths (not enclosing florets); wings 0. **Staminate or bisexual paleae** readily falling (coherent with pistillate), (1–)3–5, erect to apically somewhat spreading or incurved (scarcely enlarged) in fruit, slightly surpassing pistillate paleae; bodies ± spatulate (apices entire, sometimes involute and ± gibbous). **Pistillate florets** 13–35+. **Functionally staminate or bisexual florets** 2–5; corolla lobes mostly 4, equal or unequal. **Cypselae** light to dark brown, monomorphic: terete to obcompressed, ± obovoid, ± straight, not gibbous, faces glabrous, minutely papillate, dull or ± shiny; corolla scars apical; **pappi** 0. $x = 7$.

Species 3 (3 in the flora): c United States, n Mexico.

See discussion of Filagininae following the tribal description (p. 385).

Diaperia occurs in open, moist or dry habitats of humid to semiarid, temperate to subtropical climates. Though apparently not aggressively invasive in their native range, the species are competitive in disturbed habitats (vacant lots, fallow fields, lawns, cemeteries, and roadsides). *Diaperia verna* var. *verna*, in particular, is widely regarded as a weed; the species are potentially invasive outside the flora.

Diaperia appears to be monophyletic, with ancestors near *Evax* sect. *Filaginoides* Smoljaninova of the Mediterranean basin and central Asia (particularly *E. eriosphaera* Boissier & Heldreich; J. D. Morefield 1992). It is separated from *Evax* by stems well-developed, leafy, usually branched, paleae falling together (coherent distally by tangled indument), and staminate paleae somewhat enlarged, apices obtuse, ± herbaceous, uniformly hairy (Morefield 2004). Species of *Diaperia* are sharply distinct by size, shape, and arrangement of branches, glomerules, heads, and capitular leaves.

Diaperia candida is aberrant by its inner florets bisexual, bisexual paleae distally gibbous, and reported chromosome complement of $2n = 14$ (D. J. Keil and D. J. Pinkava 1976). These traits might eventually justify resurrection of the monotypic *Calymmandra* Torrey & A. Gray, after further study and confirmation of the chromosome number. While $2n = 14$ is common elsewhere in Gnaphalieae, all other 25 counted species of Filagininae have $2n = 28$ (species of *Evax*, *Filago*, *Logfia*, *Micropus*, *Psilocarphus*, and *Stylocline*) or $2n = 26$ (*Diaperia* and *Evax*). The implication that *D. candida* retains an ancestral diploid condition has no phylogenetic support (J. D. Morefield 1992).

SELECTED REFERENCE Shinners, L. H. 1951. The Texas species of *Evax* (Compositae). Field & Lab. 19: 125–126.

1. Heads in racemiform or spiciform arrays, 1.5–2 mm; branches proximal or none; longest pistillate paleae 0.9–1.3 mm; bisexual florets 3–5 (corollas 0.5–0.9 mm, protruding from heads); functionally staminate florets usually 0 . . . . 1. *Diaperia candida*
1. Heads in ± dichasiform or pseudo-polytomous arrays, 2–4.5 mm; branches proximal and distal, rarely none; longest pistillate paleae 1.9–4 mm; bisexual florets 0; functionally staminate florets 2–5 (corollas 1.4–2.5 mm, hidden in heads).
    2. Heads in subdichasiform arrays, ± campanulate to spheric, 2–3.3 mm, heights ± equal to diams.; capitular leaves ± hidden between and surpassed by heads; pistillate paleae scarcely imbricate; cypselae mostly 0.7–0.9 mm . . . . 2. *Diaperia verna*
    2. Heads in strictly dichasiform or pseudo-polytomous arrays (sometimes appearing monochasiform), ellipsoid to ± cylindric, 3.5–4.5 mm, heights 2–3 times diams.; capitular leaves visible between and surpassing heads; pistillate paleae imbricate; cypselae mostly 0.9–1.2 mm . . . . 3. *Diaperia prolifera*

1. **Diaperia candida** (Torrey & A. Gray) Bentham & Hooker f., Gen. Pl. 2: 298. 1873 • Silver rabbit-tobacco [E]

Calymmandra candida Torrey & A. Gray, Fl. N. Amer. 2: 262. 1842; *Evax candida* (Torrey & A. Gray) A. Gray

Plants grayish silvery, 3–25 cm, densely sericeous. **Stems** mostly 1; branches proximal or none. **Leaves:** largest 10–18 × 2–3 mm; capitular leaves subtending glomerules only, or sometimes also hidden between and surpassed by heads. **Heads** proximal and distal, in spiciform or racemiform arrays, ± spheric, 1.5–2 mm, heights ± equal to diams. **Receptacles** ± spheric, 0.3–0.5 mm, heights ± equal to diams. **Pistillate paleae** scarcely imbricate, longest 0.9–1.3 mm. **Bisexual paleae** mostly 1–3, apices incurved, ± involute, gibbous. **Functionally staminate florets** usually 0. **Bisexual florets** 3–5; corollas protruding from heads, ± zygomorphic, 0.5–0.9 mm, glabrous, lobes unequal (1–2 enlarged). **Cypselae** rounded, ± terete, mostly 0.5–0.6 mm (bisexual slightly longer). $2n = 14$.

Flowering and fruiting late Mar–early Jun. Open, dry, deep sandy soils, oak and pine woodlands, prairies, coastal areas, sometimes disturbed sites (fields, lawns, road beds); 10–400 m; Ark., La., Okla., Tex.

Diaperia candida is the most restricted of the three species, occupying most of eastern Texas (including the coast) and extending to adjacent corners of southeastern Oklahoma, southwestern Arkansas, and northwestern Louisiana.

2. **Diaperia verna** (Rafinesque) Morefield, Novon 14: 468. 2004 • Spring or many-stem rabbit-tobacco [W]

Evax verna Rafinesque, Atlantic J. 1: 178. 1833; *E. multicaulis* de Candolle

Plants greenish to grayish, 2–15 (–25) cm, ± lanuginose. **Stems** mostly 2–10; branches proximal and distal (distal subopposite), rarely none. **Leaves:** largest 7–13 × 2–4 mm; capitular leaves subtending glomerules, also ± hidden between and surpassed by heads. **Heads** mostly distal, in subdichasiform arrays, campanulate to ± spheric, 2–3.3 mm, heights ± equal to diams. **Receptacles** pulvinate, 0.3–0.6 mm, heights ± 0.2–0.5 times diams. **Pistillate paleae** scarcely imbricate, longest 1.9–2.7 mm. **Staminate paleae** mostly 3–5, apices somewhat spreading, ± plane. **Functionally staminate florets** 3–5; ovaries vestigial, 0–0.1 mm; corollas hidden in heads, actinomorphic, 1.8–2.5 mm, often ± spreading-arachnoid, lobes equal. **Bisexual florets** 0. **Cypselae** rounded, ± terete, mostly 0.7–0.9 mm. $2n = 26$.

Varieties 2 (2 in the flora): s United States, n Mexico.

The two varieties of *Diaperia verna* intergrade within a broad band inland from the Gulf of Mexico in southeastern Texas. Though some specimens are difficult to assign with confidence, the varieties show enough correlated geographic and ecologic segregation to warrant taxonomic recognition.

As neotypified by J. D. Morefield (2004), the name *Evax verna* now applies to the taxon that de Candolle named *E. multicaulis*.

1. Pistillate paleae collectively hidden by thick lanuginose indument; heads ± campanulate, largest mostly 2–2.5 mm .... 2a. *Diaperia verna* var. *verna*
1. Pistillate paleae individually visible through thin sericeous indument; heads ± spheric, largest mostly 2.5–3.3 mm ..... 2b. *Diaperia verna* var. *drummondii*

2a. Diaperia verna (Rafinesque) Morefield var. **verna**

Heads campanulate, largest mostly 2–2.5 mm. **Pistillate paleae** collectively hidden by thick lanuginose indument. $2n = 26$.

Flowering and fruiting early Mar–late Jun(–Aug). Open, barren to grassy, brushy, or wooded slopes, plains, often disturbed substrates, toward sw usually with extra moisture (playas, drainages, roadsides, urban areas); 10–1600 m; Ariz., Ark., Ga., La., N.Mex., Okla., S.C., Tex.; Mexico (Chihuahua, Coahuila, Nuevo León, Sonora, Tamaulipas).

Variety *verna* occurs nearly throughout Texas inland from the coast, extending to northern Louisiana, central Oklahoma, and southern New Mexico. Some disjunct populations in southern Arizona, Georgia, Louisiana, and South Carolina are known from disturbed habitats and may be introduced. Variety *verna* is known throughout the northern tier of states in mainland Mexico; it probably also occurs in the next tier south. Two collections of this variety collected in 1875 and 1903 are purportedly from southern California; one is mixed with other Californian Filagininae. These might represent introductions that did not persist, or accidental admixtures from other collections.

2b. Diaperia verna (Rafinesque) Morefield var. **drummondii** (Torrey & A. Gray) Morefield, Novon 14: 469. 2004 • Gulf rabbit-tobacco [E]

Filaginopsis drummondii Torrey & A. Gray, Fl. N. Amer. 2: 263. 1842; *Evax multicaulis* de Candolle var. *drummondii* (Torrey & A. Gray) A. Gray; *E. verna* Rafinesque var. *drummondii* (Torrey & A. Gray) Kartesz & Gandhi

Heads ± spheric, largest mostly 2.5–3.3 mm. **Pistillate paleae** individually visible through thin sericeous indument.

Flowering and fruiting mid Feb–mid May. Dunes, beaches, sandy soils, often where disturbed; 0–100 m; Ala., Tex.

Variety *drummondii* is known from coastal Alabama and Texas (nearly to Mexico), including islands; it may be found with var. *verna* up to about 300 km inland in Texas (most plants starting about 50–150 km inland are intermediate to some degree).

3. Diaperia prolifera (Nuttall ex de Candolle) Nuttall, Trans. Amer. Philos. Soc., n. s. 7: 338. 1840 • Bighead rabbit-tobacco [E] [F]

Evax prolifera Nuttall ex de Candolle in A. P. de Candolle and A. L. P. P. de Candolle, Prodr. 5: 459. 1836

Plants grayish green to silvery, 3–15 cm, sericeous to lanuginose. **Stems** mostly 2–10; branches proximal and distal (distal opposite or, sometimes, appearing alternate when unequal), rarely none. **Leaves:** largest 7–15 × 2–4 mm; capitular leaves subtending glomerules, also visible between and surpassing heads. **Heads** in strictly dichasiform or pseudo-polytomous arrays (sometimes appearing monochasiiform), cylindric to ± ellipsoid, 3.5–4.5 mm, heights 2–3 times diams. **Receptacles** broadly or narrowly conic, 0.4–0.6 mm or ± 0.9–1.1 mm, heights 0.5–0.7 or 2–2.4 times diams. **Pistillate paleae** imbricate, longest 2.5–4 mm. **Staminate paleae** ± 3, apices erect to somewhat spreading, ± plane. **Functionally staminate florets** 2–4; ovaries partly developed, 0.4–0.6 mm; corollas hidden in heads, actinomorphic, 1.4–2 mm, glabrous, lobes equal. **Bisexual florets** 0. **Cypselae** ± angular, obcompressed, mostly 0.9–1.2 mm.

Varieties 2 (2 in the flora): sw United States.

Intermediates between the two varieties of *Diaperia prolifera* occur where their ranges meet in central Texas and central Oklahoma. The strictly dichasiform or pseudo-polytomous branching pattern of *D. prolifera* is distinctive and diagnostic within the genus. Specimens of *D. prolifera* from introductions around a wool mill in South Carolina (G. L. Nesom 2004c, as *Evax prolifera*) are as yet undetermined to variety and are not included in the distributions below.

1. Plants grayish to greenish, loosely lanuginose; heads 4–40+ in largest glomerules; receptacle heights mostly 0.5–0.7 times diams.; capitular leaves usually ± spreading, scarcely involucral, not or scarcely carinate, pliant to somewhat rigid; distal branches mostly spreading to ascending; longest pistillate paleae 3.3–4 mm . . . .
. . . . 3a. *Diaperia prolifera* var. *prolifera*
1. Plants silvery white, tightly sericeous; heads borne singly, or 2–3 in largest glomerules; receptacle heights mostly 2–2.4 times diams.; capitular leaves erect, involucral, proximally carinate, becoming indurate; distal branches strictly ascending to erect; longest pistillate paleae 2.5–3.2 mm . . . .
. . . . 3b. *Diaperia prolifera* var. *barnebyi*

3a. Diaperia prolifera (Nuttall ex de Candolle) Nuttall var. **prolifera** [E] [F]

Plants mostly grayish green, 3–15 cm, loosely lanuginose. **Stems** mostly 2–10; branches ± equal, distal mostly spreading to ascending. **Leaves:** largest 9–15 × 2–4 mm; capitular leaves usually ± spreading, scarcely involucral, not or scarcely carinate, pliant to somewhat rigid. **Heads** 4–40+ in largest glomerules. **Receptacles** broadly conic, 0.4–0.6 mm, heights mostly 0.5–0.7 times diams. **Pistillate paleae:** longest 3.3–4 mm.

Flowering and fruiting (Apr–)May–Jun(–Sep). Dry, open, often disturbed silty to clay soils, barren to grassy, brushy, or wooded slopes, plains, prairies, toward s and e usually over carbonate (limestone, chalk); 90–1500 (–2200) m; Ala., Ark., Colo., Kans., La., Miss., Mo., Mont., Nebr., N.Mex., Okla., S.Dak., Tex., Wyo.

Variety *prolifera* occupies a broad crescent from western South Dakota and southeastern Montana to northeastern New Mexico, central and eastern Texas (nearly to Mexico), southern Missouri, and southwestern Arkansas, with outliers in chalk prairies of southern Mississippi and southern Alabama. A specimen of it (mixed with other Californian Filagininae) collected in 1903 is purportedly from southern California; it might represent an introduction that did not persist or an accidental admixture from another collection.

3b. Diaperia prolifera (Nuttall ex de Candolle) Nuttall var. **barnebyi** Morefield, Novon 14: 470. 2004
• Barneby rabbit-tobacco [E]

Plants silvery white, 3–9 cm, tightly sericeous. **Stems** mostly 1–5; branches equal to unequal, distal strictly ascending to erect. **Leaves:** largest 7–11 × 2–3 mm; capitular leaves erect, involucral, proximally carinate, becoming indurate. **Heads** borne singly, or 2–3 in largest glomerules. **Receptacles** narrowly conic, 0.9–1.1 mm, heights mostly 2–2.4 times diams. **Pistillate paleae:** longest 2.5–3.2 mm.

Flowering and fruiting mid Apr–mid Jun. Open, dry, shallow rocky or gravelly soils, usually over limestone or gypsum, sometimes with extra moisture (dry drainages, disturbed places); 500–1500 m; N.Mex., Okla., Tex.

Variety *barnebyi* is not particularly weedy. It occurs in southwestern Oklahoma, southeastern New Mexico, and western Texas (nearly to Mexico), largely allopatric with var. *prolifera*.

103. MICROPSIS   de Candolle in A. P. de Candolle and A. L. P. P. de Candolle, Prodr. 5: 459. 1836   • Straitjackets [Generic name *Micropus* and Greek *-opsis*, resembling]   [I]

James D. Morefield

Annuals, 1–10 cm. **Stems** 1, ± erect, or 2–7, ascending to erect [decumbent]. **Leaves** mostly cauline; alternate; blades oblanceolate to spatulate [± linear]. **Heads** borne singly or in pairs in dense [loose] spiciform [axillary] arrays or second-order glomerules. **Involucres** 0 or inconspicuous. **Phyllaries** 0, vestigial, or 1–5, unequal (similar to paleae). **Receptacles** flat to pulvinate (heights 0–0.3 times diams.), glabrous. **Pistillate paleae** persistent, ± erect or incurved, yellowish to brownish; bodies with 5+ nerves (nerves ± parallel, ± prominent), obovate to oblanceolate, open most of lengths (not enclosing florets); wings incurved. **Bisexual paleae** persistent or tardily falling, 2–7, erect (somewhat enlarged) in fruit, shorter than or equal to pistillate paleae; bodies broadly lanceoloid to oblanceoloid (saccate [involute], each enclosing a floret, apices 2–3-fid [entire or erose]). **Pistillate florets** [2–8]15–30. **Functionally staminate florets** 0. **Bisexual**

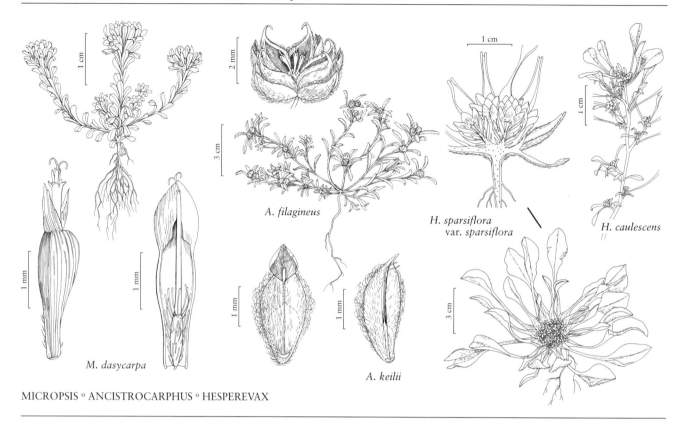

A. filagineus

H. sparsiflora
var. sparsiflora

H. caulescens

M. dasycarpa

A. keilii

MICROPSIS ° ANCISTROCARPHUS ° HESPEREVAX

florets 2–7; corolla lobes 4[–5], equal. **Cypselae** ± brownish, dimorphic: pistillate obcompressed, ellipsoid to obovoid (somewhat angular), incurved, not gibbous, faces densely strigose; bisexual terete, straight, faces sparsely strigose [glabrescent]; corolla scars apical; **pappi** 0 (simulated by hairs of cypselae) [coroniform].

Species 5 (1 in the flora: introduced (possibly native) in s-c United States; s South America.

See discussion of Filagininae following the tribal description (p. 385).

Micropsis species occur in moist or dry, often sandy or alkaline habitats of Mediterranean to humid temperate climates. Other than a recently discovered population in the flora area, *Micropsis* is known only from temperate South America. It appears to be monophyletic, with ancestors possibly near *Evax* or *Stuartina* Sonder (J. D. Morefield 1992). *Micropsis* is most easily recognized by bisexual paleae enlarged, ± saccate, usually apically lacerate, often distally gibbous, resembling straitjackets tightly enveloping their florets and cypselae, and pistillate cypselae densely strigose (known in other Filagininae only in Old World *Evax* and relatives).

SELECTED REFERENCE Beauverd, G. 1913b. Contribution à l'étude des Composées, suite VIII. IV. Le genre *Micropsis* DC. (emend. Beauverd). Bull. Soc. Bot. Genève, sér. 2, 5: 221–228.

1. Micropsis dasycarpa (Grisebach) Beauverd, Bull. Soc. Bot. Genève, sér. 2, 5: 224. 1913 • Bighead straitjackets F I

Filago dasycarpa Grisebach, Abh. Königl. Ges. Wiss. Göttingen 24: 185. 1879

Plants mostly 5–8 cm. **Leaves** 6–9 × 2–4 mm. **Heads** congested distally, mostly 4–5 mm. **Bisexual corollas:** lobes usually 4.

Flowering and fruiting Mar–May. Drying mud at edge of pool; ± 20 m; introduced; Tex.; se South America.

Micropsis dasycarpa is known in the flora area from one gathering (Victoria County, Gulf Coast prairie with *Mimosa strigillosa* Torrey & A. Gray, in 1987). For this treatment, I have assumed the Texas population to be a recent introduction; R. Barneby provided evidence that *M. dasycarpa* could be an overlooked amphitropical native (J. D. Morefield 2004; see also discussion under 101. *Psilocarphus*).

104. ANCISTROCARPHUS A. Gray, Proc. Amer. Acad. Arts 7: 355. 1868 • Groundstar [Greek *ankistros*, fishhook, and *karphos*, chaff, alluding to staminate paleae of type species]

James D. Morefield

Stylocline Nuttall sect. *Ancistrocarphus* (A. Gray) A. Gray

Annuals, 0.5–14 cm. **Stems** 0 (*A. keilii*) or, usually, 2–10 and ascending to prostrate, sometimes 1 and ± erect. **Leaves** basal or basal and cauline; alternate; blades oblanceolate to obovate. **Heads** borne singly or in glomerules of 2–5 in ± cymiform arrays. **Involucres** 0 or inconspicuous. **Phyllaries** 0, vestigial, or 3–6, ± equal (unlike paleae, hyaline). **Receptacles** distally expanded, ± fungiform or hourglass-shaped (heights 1–2 times diams.), glabrous. **Pistillate paleae** tardily falling or ± persistent, erect to ascending; bodies with 3 nerves (nerves ± parallel, prominent, or midnerves sometimes thinning with age, obscure), ± boat-shaped to ovoid, saccate most of lengths (obcompressed, not galeate, each enclosing a floret); wings erect or incurved (apical). **Staminate paleae** persistent, 5(–7), ± spreading proximally (and enlarged) in fruit, shorter than or surpassing pistillate paleae; bodies lanceolate or spatulate (apices incurved to uncinate). **Pistillate florets** 5–11. **Functionally staminate florets** 3–6; corolla (zygomorphic, ± gibbous in *A. keilii*) lobes 4–5(–6), equal or unequal. **Bisexual florets** 0. **Cypselae** uniformly brown or proximally black-banded, monomorphic: obcompressed, obovoid, slightly incurved, not gibbous, faces glabrous, smooth, dull; corolla scars apical; **pappi** 0.

Species 2 (2 in the flora): sw United States, nw Mexico.

See discussion of Filagininae following the tribal description (p. 385).

Ancistrocarphus is found in temperate Mediterranean to semiarid climates. It consistently resolved as the sister of *Hesperevax* in phylogenies based on morphology (J. D. Morefield 1992). *Ancistrocarphus* is not closely related to *Stylocline*, where it has been placed based on sharing saccate, winged pistillate paleae. The pistillate paleae of *Ancistrocarphus* differ from those of *Stylocline* and all other Filagininae in having two prominent lateral nerves in addition to the (sometimes obscure) medial one.

Addition of *Ancistrocarphus keilii* made the genus more heterogeneous and somewhat bridged the gap with *Hesperevax*. That species may be derived from a common ancestor of, or a hybrid between, the two genera. It shares most of its character states with *A. filagineus*; the two species consistently resolved as sister taxa in all phylogenetic analyses (J. D. Morefield 1992).

SELECTED REFERENCE   Morefield, J. D. 1992. Evolution and Systematics of *Stylocline* (Asteraceae: Inuleae). Ph.D. dissertation. Claremont Graduate School.

1. Stems (1–)2–10; leaves basal and cauline; staminate paleae 2.7–4.1 mm, surpassing pistil-late, broadly lanceolate, apices uncinate, acuminate, spinose; cypselae 1.4–2 × 0.6–0.9 mm, transverse bands 1; pistillate paleae: wings prominent, ovate . . . .   1. *Ancistrocarphus filagineus*
1. Stems 0; leaves basal; staminate paleae 1.8–2.8 mm, slightly surpassed by pistillate, broadly spatulate, apices incurved, obtuse to rounded, apiculate, not spinose; cypselae 1–1.4 × 0.5– 0.6 mm, transverse bands 0; pistillate paleae: wings obscure, lanceolate . . . .   2. *Ancistrocarphus keilii*

1. **Ancistrocarphus filagineus**   A. Gray, Proc. Amer. Acad. Arts 7: 356. 1868 • Hooked groundstar, woolly fishhooks   F

Stylocline filaginea (A. Gray) A. Gray

Plants 2–10(–14) cm. **Stems** usu-ally 2–10, ascending to prostrate, sometimes 1, ± erect; branches mostly distal (sometimes absent). **Leaves** basal and cauline, sessile or obscurely petiolate, largest 8–16 (–28) × 2–3(–4) mm, very unlike pistillate paleae; bases (of leaves or petioles) ± attenuate, 1-nerved, greenish, herbaceous, scarcely involucral. **Heads** in glomerules of 2–5 in ± cymiform arrays, 3.5–5 mm (excluding staminate paleae). **Phyllaries** usually 3–6, minute, ± equal, hyaline, unlike pistillate paleae, sometimes vestigial or apparently 0. **Receptacles** fungiform, 0.9–1.5 mm. **Pistillate paleae:** wings prominent, erect, ovate, ± plane. **Staminate paleae** 5, broadly lanceolate, 2.7–4.1 mm, surpassing pistillate; apices uncinate, terete, acuminate, cartilaginous, spinose. **Staminate corollas** actinomorphic, 1–1.5 mm, lobes (4–) 5(–6), ± equal. **Cypselae** 1.4–2 × 0.6–0.9 mm, trans-verse bands 1, proximal, black.

Flowering and fruiting mid Mar–late Jun. Open, dry or vernally moist slopes, flats, drainages, shallow rocky to deep sandy or clay soils (often over serpentine or basalt), sometimes disturbed sites (burns, road beds); 60–1900(–2100) m; Calif., Idaho, Nev., Oreg.; Mexico (Baja California).

Ancistrocarphus filagineus occurs nearly throughout the Californian Floristic Province inland from the imme-diate coast; in the south, it stops at the western fringes of the Mojave and Sonoran deserts; farther north, it con-tinues into the northern Great Basin and Columbia Plateau of Oregon, northern Nevada, and southwestern Idaho. The rigid whorl of inwardly-hooked staminate paleae holds tenaciously to clothing, skin, animal coats, even vehicle tires, providing an effective dispersal mecha-nism. It is surprising that it has not spread farther.

Plants of populations farther inland, on drier sites and/ or in deeper sandy soils, tend to be more compact with leaves shorter, more rounded, and indument denser and whitish. *Stylocline filaginea* var. *depressa* has sometimes been misapplied to these unnamed forms. The two forms intermingle enough so as not to warrant formal taxonomic status.

2. **Ancistrocarphus keilii**   Morefield, Novon 14: 464, fig. 1. 2004 • Santa Ynez groundstar   C   E   F

Plants 0.5–1(–2) cm. **Stems** 0. **Leaves** basal, petiolate, largest 8–10(–18) × 1(–2) mm, some ± grading into pistillate paleae (by reduction or loss of blades and expansion of petiole bases, resem-bling aristate phyllaries); bases (of petioles) expanded, 3-nerved, yellowish tan, parchment-like, closely involucral. **Heads** borne singly amidst surround-ing leaves (see discussion below), 2.5–3.5 mm. **Phyllar-ies** 0. **Receptacles** broadly hourglass-shaped, 0.9–1.3 mm. **Pistillate paleae:** wings obscure, incurved, lanceolate, ± involute. **Staminate paleae** 5(–7), broadly spatulate, 1.8–2.8 mm, slightly surpassed by pistillate; apices incurved, ± involute, obtuse to rounded, apiculate, shortly scarious, not spinose. **Staminate corollas** ± zygomorphic, 0.9–1.3 mm, lobes 4, unequal (1–2 enlarged). **Cypselae** 1–1.4 × 0.5–0.6 mm, transverse bands 0.

Flowering and fruiting Mar–Apr. Sandy soils, chaparral bordering oak woodlands, under shrubs; of conservation concern; 40–130 m; Calif.

Unlike *Ancistrocarphus filagineus*, *A. keilii* has no obvious dispersal mechanism. Its cypselae appear to germinate more or less in place from previous years' plants, often producing dense, turflike growths of doz-ens to hundreds of individuals (these sometimes resemble leafy glomerules of heads; each head is on a separate plant with its own root). Lack of dispersal may explain its very limited geographic range in the Santa Ynez River drainage of Santa Barbara County. Its range does not appear to overlap with that of *A. filagineus*, which is found farther inland to the east and north. No other Filagininae have been found mixed with *A. keilii*, sug-gesting that its habitat is fairly distinctive.

Ancistrocarphus keilii is known from only three collections and appears to occur in relatively vulnerable habitats (J. D. Morefield 2004).

105. HESPEREVAX (A. Gray) A. Gray, Proc. Amer. Acad. Arts 7: 356. 1868 • [Greek *hesperos*, western, and genus name *Evax,* alluding to first discoveries from western limits of *Evax* distribution] E

James D. Morefield

Evax Gaertner sect. *Hesperevax* A. Gray in War Department [U.S.], Pacif. Railr. Rep. 4(5): 101. 1857

Annuals, 0.5–10(–20) cm. **Stems** 0, or 1, ± erect, or 2–10, ascending-erect to prostrate. **Leaves** basal or cauline; alternate (leaf or petiole bases yellowish, enlarged); blades oblanceolate to ± round. **Heads** usually in loose to dense glomerules of 2–40+ in cymiform or spiciform arrays, sometimes borne singly. **Involucres** 0. **Phyllaries** 0 or vestigial. **Receptacles** proximally conic, distally cylindric or slightly expanded (heights 0.8–1.3 or 4–6 times diams.), bristly (bristle lengths ± ¹/₂ palea lengths). **Pistillate paleae** persistent, erect to ascending or spreading; bodies with 5+ nerves (nerves ± parallel, obscure), obovate to oblanceolate, not gibbous, flat to concave or loosely folded most of lengths (almost closed near bases, not enclosing florets); wings 0. **Staminate paleae** persistent, usually 5, erect or distally spreading (and enlarged, thickened) in fruit, shorter than or surpassing pistillate paleae; bodies obovate (apices entire, blunt, adaxially green and lanuginose). **Pistillate florets** 3–25. **Functionally staminate florets** 2–6(–12); corolla (zygomorphic, ± gibbous) lobes (3–)4(–5), unequal. **Bisexual florets** 0. **Cypselae** brown, proximally black-banded, monomorphic: obcompressed, obovoid (somewhat angular), slightly incurved, not gibbous, faces glabrous, smooth, dull to ± shiny; corolla scars apical; **pappi** 0.

Species 3 (3 in the flora): sw United States.

See discussion of Filagininae following tribal description (p. 385).

Hesperevax is known from sites in Mediterranean climates of western California and southwestern Oregon. Most of its taxa grow in distinctive habitats, where they tolerate or exploit moderate levels of disturbance.

Hesperevax is monophyletic and probably sister to *Ancistrocarphus*; *Gifolaria* (Cosson & Kralik) Pomel may also be related (J. D. Morefield 1992). *Hesperevax* is sharply distinct from *Evax* and relatives by receptacles bristly (unique in Filagininae), staminate paleae enlarged, thickened, apices erect or spreading, blunt, adaxially green and lanuginose, and staminate corollas zygomorphic (Morefield 1992c). The species are relatively easy to distinguish by the sizes, shapes, and arrangements of leaves, glomerules, and heads.

SELECTED REFERENCE Morefield, J. D. 1992c. Resurrection and revision of *Hesperevax* (Asteraceae: Inuleae). Syst. Bot. 17: 293–310.

1. Terminal heads in loose glomerules of 3–5, cylindric, heights 1.8–2.5 times diams.; capitular leaves 1–4 per glomerule or head; staminate paleae: apices ± erect . . . . 1. *Hesperevax sparsiflora*
1. Terminal heads borne singly or in dense glomerules of (2–)10–40+, campanulate or ± obpyramidal, heights 1–1.5 times diams.; capitular leaves 6–20 per glomerule or head; staminate paleae: apices spreading.
   2. Largest leaves 4–22(–32) × 0.5–4(–5) mm; petiole lengths 0–1.5 times blade lengths, bases scarcely thickened, pliant to ± cartilaginous; heads borne singly or, rarely, in glomerules (3–7 mm diam.) of 2–8 mixed with leaves . . . . 2. *Hesperevax acaulis*
   2. Largest leaves (25–)33–90 × 7–20 mm; petiole lengths 2–3 times blade lengths, bases thickened, indurate; heads in glomerules (10–25 mm diam.) of 10–40+ not mixed with leaves . . . . 3. *Hesperevax caulescens*

1. **Hesperevax sparsiflora** (A. Gray) Greene, Fl.
Francisc., 402. 1897 • Erect evax [E] [F]

Evax caulescens (Bentham) A. Gray
var. *sparsiflora* A. Gray in A. Gray
et al., Syn. Fl. N. Amer. 1(2): 229.
1884; *E. sparsiflora* (A. Gray)
Jepson

Plants 2–17 cm. **Stems** 1–10,
ascending to erect (rarely 0);
branches 0 or proximal. **Leaves**
mostly cauline, petiolate, distal
scarcely congested, ± equal to proximal, largest 6–32
× 3–8(–10) mm; petioles: lengths 0.9–1.5 times blade
lengths, bases thickened, cartilaginous to ± indurate;
capitular leaves 1–4 per glomerule or head, approximate,
not whorled, ascending to erect, some sometimes grading
into pistillate paleae (by reduction of blades, expansion of
petiole bases). **Heads** terminal in loose glomerules (3–4
mm diam.) of 3–5 mixed with leaves and some borne
singly in axils, cylindric, 3–4.5 × 1.5–2 mm, heights 1.8–
2.5 times diams. **Receptacles** distinct, 1.4–1.8 × 0.2–0.4
mm. **Pistillate paleae** in 1–3 series, spirally ranked, oblanceolate, 2.5–4.5 mm. **Staminate paleae** not or scarcely
surpassing pistillate, 1.1–1.8 mm, lengths 0.3–0.4 times
head heights; apices ± erect. **Functionally staminate florets** 2–5; corollas 0.8–1.1 mm. **Cypselae** mostly 1–1.7
mm.

Varieties 2 (2 in the flora): sw United States.

The varieties of *Hesperevax sparsiflora* occur in distinct habitats and might be treated as separate species
but for some morphologically intermediate specimens
from areas of sympatry in the San Francisco Bay area.

1. Largest leaves (10–)13–32 × 4–8(–10) mm, blades
broadly oblanceolate to obovate, arachnoid-sericeous . . . . . . 1a. *Hesperevax sparsiflora* var. *sparsiflora*
1. Largest leaves 6–12(–14) × 3–5(–6) mm, blades
± round, ± densely lanuginose . . . .
. . . . 1b. *Hesperevax sparsiflora* var. *brevifolia*

1a. **Hesperevax sparsiflora** (A. Gray) Greene var.
sparsiflora [E] [F]

Plants greenish to grayish, (2–)8–
17 cm. **Leaves** mostly cauline,
largest (10–)13–32 × 4–8(–10)
mm; blades broadly oblanceolate
to obovate, arachnoid-sericeous.
Heads: longest 3.6–4.5 mm.
Pistillate paleae 3–4.5 mm.
Cypselae mostly 1.3–1.7 mm.

Flowering and fruiting mostly
late Mar–early Jun. Open soils, often shallow and rocky,
usually over serpentine; 10–900 m; Calif.

Variety *sparsiflora* is known from the Coast Ranges
and has outliers in western San Diego County and on
the nearer Channel Islands. In the southern half of its
range, it grows mostly in coastal marine environments,
where it remains distinct from the north-coastal var.
brevifolia. Consistently stemless plants from Santa Rosa
Island may warrant taxonomic separation.

1b. **Hesperevax sparsiflora** (A. Gray) Greene var.
brevifolia (A. Gray) Morefield, Syst. Bot. 17: 302.
1992 • Seaside or short-leaved evax [E]

Evax caulescens (Bentham) A. Gray
var. *brevifolia* A. Gray in A. Gray
et al., Syn. Fl. N. Amer. 1(2): 229.
1884; *E. sparsiflora* (A. Gray)
Jepson var. *brevifolia* (A. Gray)
Jepson

Plants ± greenish, mostly 3–9 cm.
Leaves cauline, largest 6–12(–14)
× 3–5(–6) mm; blades ± round,
± densely lanuginose. **Heads:** longest 3–3.7 mm. **Pistillate paleae** 2.5–3.7 mm. **Cypselae** mostly 1–1.7 mm.

Flowering and fruiting late Mar–early Jul. Sandy,
grassy or wooded coastal bluffs, terraces, dunes; 0–300
m; Calif., Oreg.

Variety *brevifolia* occurs in a coastal band from the
San Francisco Bay area to southwestern Oregon. It is of
potential conservation concern in both states. See discussion under *Hesperevax acaulis* var. *robustior*.

2. **Hesperevax acaulis** (Kellogg) Greene, Fl. Francisc.,
402. 1897 • Dwarf evax [E]

Stylocline acaulis Kellogg, Proc.
Calif. Acad. Sci. 7: 112. 1877 (as
acaule); *Evax acaulis* (Kellogg)
Greene

Plants 0.5–7 cm. **Stems** 0 or 1–10,
erect to prostrate; branches usually 0, sometimes proximal and/
or distal. **Leaves** basal or cauline,
sessile or petiolate, distal con-
gested, larger than proximal (if present), largest 4–22
(–32) × 0.5–4(–5) mm; petioles: lengths 0–1.5 times blade
lengths, bases scarcely thickened, pliant to somewhat
cartilaginous; capitular leaves 6–12 (rarely more) per
glomerule or head, whorled, ± erect or distally spreading, unlike pistillate paleae. **Heads** terminal, borne singly or, rarely, in dense glomerules (3–7 mm diam.) of 2–8
mixed with leaves and, rarely, some borne singly in axils
(then smaller), campanulate, 2–4 × 1.5–3.5 mm, heights
1–1.5 times diams. **Receptacles** distinct, 0.9–1.9 × 0.8–
1.7 mm. **Pistillate paleae** in 2–5 series, spirally ranked,
broadly spatulate, 1–3 mm. **Staminate paleae** surpassing pistillate, 1.6–3.2 mm, lengths 0.6–0.8 times head
heights; apices spreading. **Functionally staminate florets**
2–5(–12); corollas 0.6–1 mm. **Cypselae** 0.6–1.6 mm.

Varieties 3 (3 in the flora): sw United States.

The varieties of *Hesperevax acaulis* show enough geographic and ecologic segregation correlated with morphologic differences to warrant taxonomic recognition. Across west-central California, where all three are broadly sympatric, the varieties tend to occur in different habitats and/or elevation zones. Intermediate specimens are difficult to assign with confidence.

1. Stems ± erect; largest leaves (9–)12–22(–32) × 2–4 mm; longest staminate paleae 2.5–3.2 mm . . . . 2a. *Hesperevax acaulis* var. *robustior*
1. Stems usually ± prostrate or 0; largest leaves mostly 4–12 × 0.5–2 mm; longest staminate paleae 1.6–2.4 mm.
   2. Leaves petiolate, largest mostly 4–7 × 1–2 mm, blades ± round, obtuse . . . . 2b. *Hesperevax acaulis* var. *ambusticola*
   2. Leaves sessile or obscurely petiolate, largest mostly 7–12 × 0.5–2 mm, blades oblanceolate, acute . . . . 2c. *Hesperevax acaulis* var. *acaulis*

2a. Hesperevax acaulis (Kellogg) Greene var. **robustior** Morefield, Syst. Bot. 17: 308, fig. 5N. 1992

• Robust evax [E]

Plants mostly 2–7 cm. **Stems** 1(–7), ± erect; branches usually 0. **Leaves** cauline, sessile or obscurely petiolate, largest (9–)12–22(–32) × 2–4(–5) mm; petiole lengths mostly 0–0.8 times blade lengths; blades oblanceolate to obovate, acute to obtuse; capitular leaves erect or distally ± spreading. **Heads** borne singly or, sometimes, in glomerules of 2–8, largest 3–4 × 2.5–3.5 mm. **Receptacles** 1.4–1.9 × 1.2–1.7 mm. **Pistillate paleae** in 3–5 series, 2.5–4 mm. **Staminate paleae:** longest 2.5–3.2 mm. **Functionally staminate florets** 2–5(–10); corollas 0.7–1 mm. **Cypselae** mostly 1–1.6 mm.

Flowering and fruiting mid Apr–mid Jun. Dry slopes, flats, woodlands, chaparral, in clearings or under shrubs, often with extra moisture (swales, canyons, roadsides, path edges); 60–1100 m; Calif., Oreg.

Variety *robustior* is known from the mountains of west-central California to interior southwestern Oregon (most known Oregon collections occurred before 1925). Like var. *ambusticola*, it tends to occur higher than var. *acaulis* where sympatric. Toward the south, it intergrades about equally with the other varieties (J. D. Morefield 1992c). The largest sizes described above are from a garden-grown specimen; field-collected plants from the same gathering were depauperate but otherwise typical. Variety *robustior* and *Hesperevax sparsiflora* var. *brevifolia* are superficially similar and have been confused; they are not known to intergrade or hybridize.

2b. Hesperevax acaulis (Kellogg) Greene var. **ambusticola** Morefield, Syst. Bot. 17: 306, fig. 5A–L. 1992 • Fire evax [E]

Plants 0.5–2(–4) cm. **Stems** (0–)4–10, ± prostrate (in depauperate plants sometimes 0, or 1, erect, unbranched); branches proximal and/or distal. **Leaves** cauline, petiolate, largest mostly 4–7 × 1–2 mm; petiole lengths mostly 0.5–1.5 times blade lengths; blades obovate to ± round, obtuse; capitular leaves mostly distally spreading. **Heads** usually borne singly, largest 2–2.5 × 1.5–2 mm. **Receptacles** 1–1.4 × 0.8–1.1 mm. **Pistillate paleae** in 3–4 series, 1.7–2.5 mm. **Staminate paleae:** longest 1.6–2 mm. **Functionally staminate florets** 2–5(–9); corollas 0.6–0.8 mm. **Cypselae** mostly 0.8–1 mm.

Flowering and fruiting mostly Apr–early Jun. Dry or vernally moist slopes, relatively barren or in clearings, recent chaparral burns; 200–1300 m; Calif.

Variety *ambusticola* is known from mountains of western Riverside County to the San Francisco Bay area and northern Sierra Nevada foothills. It occurs at about the same elevations as, and intergrades extensively with, var. *robustior* in northern California (J. D. Morefield 1992c).

Depauperate individuals of var. *ambusticola* are among the smaller and more inconspicuous Compositae, sometimes consisting of just one grayish head embedded in a leaf rosette on a root 1–2 cm long.

2c. Hesperevax acaulis (Kellogg) Greene var. **acaulis**

• Stemless evax [E]

Plants 1–2(–4) cm. **Stems** usually 0, sometimes 1(–7), ± prostrate; branches 0. **Leaves** basal or cauline, sessile or obscurely petiolate, largest mostly 7–12 × 0.5–2 mm; petiole lengths 0–0.8 times blade lengths; blades oblanceolate, acute; capitular leaves ± erect. **Heads** borne singly or, sometimes, in glomerules of 2–8, largest 2.5–3 × 2–2.5 mm. **Receptacles** 0.8–1.4 × 0.8–1.4 mm. **Pistillate paleae** in 3–5 series, 2–3 mm. **Staminate paleae:** longest 1.9–2.4 mm. **Functionally staminate florets** 2–12; corollas 0.7–1 mm. **Cypselae** mostly 0.6–0.8 mm.

Flowering and fruiting mid Mar–mid May. Open, dry or vernally moist, sandy to gravelly soils in swales, drainages, grasslands, chaparral, woodland clearings, sometimes in moss or other turf; 30–900 m; Calif., Oreg.

Geographically, var. *acaulis* is mainly intermediate with the other two varieties, usually occurring at lower elevations where sympatric. It is also morphologically intermediate in some characters. Typical forms, most frequent along the Sierra Nevada foothills, are distinguished by stems 0 and leaves narrowest relative to lengths. A recent collection from southwestern Oregon is intermediate toward var. *robustior*; it may be introduced, or it may represent previously unsampled variation in the region.

Kellogg's protologue described pistillate paleae like those of *Stylocline*, suggesting that he may have been looking also at plants of *Ancistrocarphus keilii*. The two taxa are not known to be sympatric. The type material clearly belongs to *Hesperevax*. Besides its saccate pistillate paleae, *A. keilii* differs by petioles longer than blades with bases expanded, 3-nerved, and parchment-like.

3. **Hesperevax caulescens** (Bentham) A. Gray, Proc. Amer. Acad. Arts 7: 356. 1868 • Hogwallow starfish, involucrate evax [E] [F]

Psilocarphus caulescens Bentham, Pl. Hartw., 319. 1849; *Evax caulescens* (Bentham) A. Gray; *E. caulescens* var. *humilis* (Greene) Jepson; *E. involucrata* Greene

Plants (2–)3–8(–17) cm. **Stems** usually 0, sometimes 1–4, erect to decumbent; branches 0 or proximal. **Leaves** basal or cauline, petiolate, distal congested, larger than proximal (if any), largest (25–)33–90 × 7–20 mm; petioles: lengths 2–3 times blade lengths, bases thickened, indurate; capitular leaves (6–)10–20 per glomerule, whorled, mostly spreading, unlike pistillate paleae. **Heads** terminal in dense glomerules (10–25 mm diam.) of 10–40+ not mixed with leaves, never in axils, ± obpyramidal, 3–5 × 2.5–4 mm, heights 1–1.5 times diams. **Receptacles** of adjacent heads proximally connate, 0.4–0.7 × 0.5–0.7 mm. **Pistillate paleae** in 2–4 series, ± vertically ranked, broadly spatulate, 2.5–4.5 mm. **Staminate paleae** surpassing pistillate, 3–3.9 mm, lengths 0.7–0.9 times head heights, apices spreading. **Functionally staminate florets** 3–6; corollas 1.1–1.6 mm. **Cypselae** mostly 1.5–2 mm.

Flowering and fruiting mid Mar–early Jun. Drying shrink-swell clay soils, vernal pools, other vernally moist places; 0–200(–500) m; Calif.

Hesperevax caulescens occurs mainly in the Sacramento and northern San Joaquin valleys, with southern outliers in the southern San Joaquin and upper Salinas River valleys and the Otay Mesa area of San Diego County. The Otay Mesa populations are thought to have been naturalized (R. M. Beauchamp 1986) and are now considered extirpated. An 1869 specimen from northern Oregon is likely either mislabeled or from an introduction that did not persist.

Caulescent forms are scattered among predominantly stemless populations throughout the northern range of *Hesperevax caulescens*. The tallest forms (*Evax involucrata*) are often grayer and more densely lanuginose. The various growth forms appear to be environmentally induced and taxonomically insignificant. Plants from the southern outliers are all stemless, tend to be smaller in sizes and numbers of structures, and may warrant varietal status (J. D. Morefield 1992c). A malformed plant from Merced County appears to have been a hybrid between *H. caulescens* and *Psilocarphus oregonus*.

Hesperevax caulescens may soon be of conservation concern in California as its vernal pool habitats continue to decline. The vernacular "hogwallow starfish" is widely used for *H. caulescens* by students and enthusiasts of California's vernal pools.

187h. ASTERACEAE Martinov tribe INULEAE Cassini, J. Phys. Chim. Hist. Nat. Arts 88: 193. 1819 ▢

Annuals or perennials [subshrubs, shrubs, or trees]. **Leaves** basal and/or cauline (basal often withering before flowering); alternate [opposite]; petiolate or sessile; margins entire or dentate to serrate [pinnately divided]. **Heads** heterogamous (radiate [disciform]) [homogamous (discoid)], usually in corymbiform, paniculiform, or racemiform arrays, sometimes borne singly (on ± leafy stems). **Calyculi** 0. **Phyllaries** persistent [falling], in (2–)3–7+ series, distinct, unequal to subequal, herbaceous to chartaceous or membranous, margins and/or apices usually scarious. **Receptacles** flat to convex [concave], epaleate [paleate]. **Ray florets** [0] in 1(–2+) series, pistillate and fertile [neuter]; corollas usually yellow, sometimes reddish [ochroleucous or purplish] (laminae often linear). **Disc florets** bisexual, fertile; corollas usually yellow, sometimes reddish [ochroleucous or purplish], not 2-lipped, lobes (4–)5, usually ± deltate; anther bases ± tailed, apical appendages ovate to lance-ovate or linear; styles abaxially glabrous or papillate (distally), branches ± linear, adaxially stigmatic in 2 lines from bases to apices (lines often confluent distally), apices rounded to truncate, appendages essentially none. **Cypselae** usually monomorphic within heads, usually ellipsoid or columnar to prismatic [compressed or obcompressed], not beaked (sometimes abruptly constricted at each end), bodies often ribbed (glabrous or hairy, often glandular, hairs not myxogenic); **pappi** persistent (fragile), of ± barbellate [plumose] scales (sometimes setiform or aristate) and/or bristles.

Genera ca. 40, species ca. 500 (3 genera, 5 species in the flora): introduced; Old World, especially Eurasia and n Africa; some species widely introduced and established in local floras.

Following A. A. Anderberg (1994) and others, some 180 genera and 2000 species have been segregated from traditionally circumscribed Inuleae as Gnaphalieae, and another 25+ genera and 200+ species as Plucheeae.

SELECTED REFERENCES Anderberg, A. A. 1994. Tribe Inuleae. In: K. Bremer. 1994. Asteraceae: Cladistics & Classification. Portland. Pp. 273–291. Arriagada, J. E. 1998. The genera of Inuleae (Compositae: Asteraceae) in the southeastern United States. Harvard Pap. Bot. 3: 1–48.

1. Pappi of outer, ± connate, ± erose scales in 1 series and inner, distinct bristles in 1 series . . . . 106. *Pulicaria*, p. 471
1. Pappi of distinct or basally connate bristles and/or setiform-scales in 1 series.
  2. Annuals (pilosulous to hispid and stipitate-glandular, viscid); involucres 3–8 mm diam.; laminae of ray corollas 2–5(–7) mm . . . . 107. *Dittrichia*, p. 473
  2. Perennials; involucres 10–40 mm diam.; laminae of ray corollas 10–30+ mm . . . . 108. *Inula*, p. 473

106. PULICARIA Gaertner, Fruct. Sem. Pl. 2: 461, plate 173, fig. 7. 1791 • False fleabane [Latin *pulex*, flea, and *-aria*, pertaining to; alluding to use of the plants as flea repellent] ▢

Robert E. Preston

Annuals (biennials, or perennials) [shrubs, subshrubs], (5–)20–120 cm (sometimes rhizomatous). **Leaves** basal and/or cauline (mostly cauline at flowering), alternate; usually sessile; blade margins entire or ± dentate to serrate. **Heads** radiate [disciform or discoid], in corymbiform, racemiform, or paniculiform arrays. **Involucres** hemispheric to campanulate, [3–]5–10[–20+] mm diam. **Phyllaries** persistent (reflexed in fruit), in (2–)3–4+ series, unequal to subequal. **Receptacles**

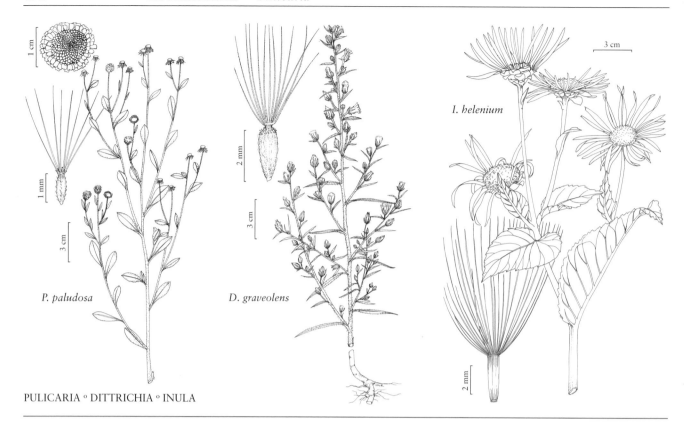

I. *helenium*

P. *paludosa*

D. *graveolens*

PULICARIA ∘ DITTRICHIA ∘ INULA

flat, smooth or minutely alveolate, epaleate. **Ray florets** (10–)20–30[–60+], pistillate, fertile; corollas yellow, laminae 1.5–2+ mm. **Disc florets** (9–)40–100[–150+]; corollas yellow, lobes 5. **Cypselae** ellipsoid (abruptly constricted distally; often glandular distally); **pappi** persistent, outer of basally connate, ± erose scales (usually forming cups), inner of distinct (fragile), barbellate or flattened bristles. *x* = 7, 9, 10.

Species 100+ (1 in the flora): introduced; Europe, Asia, Africa.

Pulicaria arabica (Linnaeus) Cassini (*Vicoa auriculata* Cassini) was collected in Alabama, California, and Florida in the late 1800s. It does not appear to have become naturalized at any of those locations (A. Cronquist 1980; J. E. Arriagada 1998).

Pulicaria dysenterica (Linnaeus) Bernhardi was collected in the late 1800s as a ballast weed in New Jersey and in Pennsylvania. It was collected in the 1920s growing on the margins of a marsh in Maryland. Although it is widely cultivated for its insecticidal properties, there is no evidence that it has ever become established in the flora (J. E. Arriagada 1998).

Pulicaria vulgaris Gaertner was collected as a ballast weed in New Jersey in 1879. No other collections of the species from North America are known to me.

1. Pulicaria paludosa Link, Neues J. Bot. 1(3): 142. 1806 • Spanish false fleabane [F] [I]

Pulicaria hispanica (Boissier) Boissier

Annuals (biennials, or short-lived perennials), (5–)20–120 cm. **Leaf blades** oblong to narrowly oblanceolate, 1–3 cm × 2–7 mm, bases clasping, margins entire, ± revolute. **Phyllaries** lance-linear to linear, 3–5 mm, pilosulous. **Ray florets** (10–)20–30+; corolla laminae 1.5–2+ mm. **Disc florets** (9–)40–100+; corollas 2–3 mm. **Cypselae** 0.8–1 mm, ± hirsutulous; **pappi:** outer cups 0.1–0.2(–0.4) mm, inner 12–20+ bristles 2–3 mm. *2n* = 18 (Spain).

Flowering Jul–Oct. Roadways, dry streambeds, seasonal wetlands; 30–600 m; introduced; Calif.; Europe (Portugal, Spain).

Pulicaria paludosa was first recognized as a naturalized weed in southern California in the early 1960s (P. H. Raven 1963b); it had been collected earlier by P. A. Munz (in 1946).

SELECTED REFERENCE Raven, P. H. 1963b. *Pulicaria hispanica* (Compositae: Inuleae), a weed new to California. Aliso 5: 251–253.

107. DITTRICHIA Greuter, Exsicc. Genav. Conserv. Bot. Distrib. Fasc. 4: 71. 1973

• [For Manfred Dittrich, b. 1934, German botanist] [I]

Robert E. Preston

Annuals [perennials], mostly 20–130 cm (glandular, viscid). **Leaves** mostly cauline (at flowering), alternate; sessile; blade margins entire or dentate [serrate]. **Heads** radiate, in racemiform or paniculiform arrays. **Involucres** ± campanulate, 3–8[–10+] mm diam. **Phyllaries** persistent (spreading to reflexed in fruit), in 3–4 series, unequal. **Receptacles** flat, smooth or alveolate, epaleate. **Ray florets** (6–)10–12(–16), pistillate, fertile; corollas yellow, aging reddish, laminae 2–7 mm. **Disc florets** 8–20+; corollas yellow, aging reddish, lobes 4–5. **Cypselae** ellipsoid to terete (abruptly constricted distally; usually glandular distally); **pappi** persistent (fragile), of basally connate, barbellate bristles in 1 series. x = 9, 10.

Species 2 (1 in the flora): introduced; Mediterranean; introduced also in Asia, Africa, Australia.

Dittrichia viscosa (Linnaeus) Greuter [*Inula viscosa* Linnaeus, *Cupularia viscosa* (Linnaeus) Godron & Grenier] was collected in the late 1800s as a ballast weed in Florida, New Jersey, and Pennsylvania. It does not appear to have become naturalized at any of those sites (A. Cronquist 1980; J. E. Arriagada 1998).

1. Dittrichia graveolens (Linnaeus) Greuter, Exsicc. Genav. Conserv. Bot. Distrib. Fasc. 4: 71. 1973

• Stinkwort [F] [I]

Erigeron graveolens Linnaeus, Cent. Pl. I, 28. 1755; *Inula graveolens* (Linnaeus) Desfontaines

Plants viscid, rank smelling, 20–130 cm; stems ± pilose and stipitate-glandular. **Leaf blades** linear to lance-linear, 1–3(–7) cm × 1–3(–10) mm, margins entire or denticulate, apices acute, faces pilosulous to hirtellous and minutely stipitate-glandular. **Phyllaries** 1–8 mm. **Ray florets** (6–)10–12(–16); corolla laminae 2–5(–7) mm. **Disc florets** 9–14+; corollas 3–4 mm. **Cypselae** 1.5–2 mm; **pappi** 3–4(–5) mm. $2n$ = 18 (Morocco).

Flowering Sep–Nov. Disturbed fields, roadways, estuarine borders; 0–200 m; introduced; Calif., N.J., N.Y.; Europe (Balkan Peninsula, sw Italy); Asia (India); Africa (South Africa); Australia.

Dittrichia graveolens was collected as a ruderal in Connecticut in the 1930s, and collections from New York in the late 1940s recorded it as abundant in areas where road construction was underway. It appears to be recently introduced in California, where it has spread rapidly and has the potential to become a noxious weed (R. E. Preston 1997). *Dittrichia graveolens* has been shown to cause allergic contact dermatitis (J. N. Burry and P. M. Kloot 1982). The plants produce sesquiterpene lactones (G. S. d'Alcontres et al. 1973; A. Rustaiyan et al. 1987; R. Lanzetta et al. 1991), which have been shown for many composites to be linked to allergic contact dermatitis in humans (J. C. Mitchell and G. Dupuis 1971). Little evidence exists that the plants are toxic, although oxalate poisoning has been reported to be associated with grazing (kind of animals not reported) of *D. graveolens* (C. Lamp and F. Collet 1979), and fishermen in southern Italy reportedly use the macerated leaves to stun fish (Lanzetta et al.). Livestock deaths due to ingestion of *D. graveolens* have been linked to enteritis caused by the barbed pappus bristles puncturing the small intestine (C. A. Gardner and H. W. Bennetts 1956; D. J. Schneider and J. L. Du Plessis 1980).

108. INULA Linnaeus, Sp. Pl. 2: 881. 1753; Gen. Pl. ed. 5, 375. 1754 • [Greek *inaein*, to clean, alluding to medicinal effects; or Latin *inula*, an ancient name for elecampane] [I]

Neil A. Harriman

Perennials [annuals], 20–200 cm. **Leaves** basal (usually withering before flowering) and cauline; petiolate (proximal) or sessile (distal); blade margins usually serrate to dentate, sometimes entire. **Heads** radiate [disciform, discoid], borne singly or in open, corymbiform arrays. **Involucres** hemispheric or campanulate, [5–]10–40 mm diam. **Phyllaries** persistent, in 4–7+ series. **Receptacles** flat or convex, smooth or alveolate, epaleate. **Ray florets** (15–)50–150+, pistillate, fertile; corollas yellow [orange], laminae 10–30+ mm. **Disc florets** mostly (50–)100–250+; corollas yellow, lobes 5. **Cypselae** ± columnar (subterete) or prismatic (± 4–5-ribbed or -angled); **pappi** persistent, of basally connate, barbellate bristles or setiform scales in 1 series. *x* = 8, 9, 10.

Species ca. 100 (3 in the flora): introduced; Old World.

The three species in the flora are probably escapes from cultivation. Formerly, *Inula* was circumscribed more broadly.

1. Blades of basal (and proximal cauline) leaves 100–200 mm wide; involucres (20–)30–40 mm diam.; outer phyllaries 6–8(–20+) mm wide .... 1. *Inula helenium*
1. Blades of basal (and proximal cauline) leaves mostly 10–30 mm wide; involucres 7–15 (–20) mm diam.; outer phyllaries mostly 0.5–2.5 mm wide.
   2. Blades of cauline leaves lance-elliptic to lance-linear (venation not raised adaxially, reticulation not evident); outer phyllaries 4–6 × 0.5–0.8 mm .... 2. *Inula brittanica*
   2. Blades of cauline leaves broadly elliptic to lanceolate (venation raised adaxially, reticulation prominent); outer phyllaries 5–7 × 1.5–2.5 mm .... 3. *Inula salicina*

1. Inula helenium Linnaeus, Sp. Pl. 2: 881. 1753

F I

Plants 50–100(–200) cm. **Leaves:** basal blades ± elliptic, mostly 15–40 cm × 100–200+ mm (bases decurrent onto strongly ribbed petioles, margins callose-denticulate, otherwise entire, abaxial faces velvety-woolly, adaxial thinly hairy); cauline blades ovate or elliptic to lanceolate, 10–30 cm × 45–120 mm, bases cordate, clasping, margins serrate. **Involucres** (20–)30–40 mm diam. **Outer phyllaries** ovate, oblong, or ± deltate to lanceolate, 12–20(–25+) × 6–8(–20+) mm (abaxially velvety-hairy); inner phyllaries progressively narrower, less hairy, more scarious. **Ray florets** (15–)50–100+; corolla laminae (10–)20–30+ mm. **Disc corollas** 9–11 mm. **Cypselae** 3–4 mm, glabrous; **pappi** of (40–)50–60 basally connate, barbellate bristles or setiform scales 6–10 mm. **2*n* = 20.**

Flowering mid–late summer. Roadsides, waste places, streamsides; 0–300(–600+) m; introduced; B.C., Man., N.B., N.S., Ont., P.E.I., Que.; Calif., Conn., Del., Ill., Ind., Iowa, Ky., Maine, Md., Mass., Mich., Minn., Mo., N.H., N.J., N.Y., N.C., Ohio, Oreg., Pa., R.I., Tenn., Utah, Vt., Va., Wash., Wis.; Europe; introduced, Asia and beyond.

Inula helenium is widespread in the Old World.

2. Inula brittanica Linnaeus, Sp. Pl. 2: 882. 1753   I

Plants 10–40(–75) cm. **Leaves:** basal blades lanceolate, (3–)6–7 cm × 8–20(–30+) mm; cauline blades lance-elliptic to lance-linear, 2–5+ cm × 5–12(–20+) mm, bases ± cordate, clasping, margins entire or serrulate (abaxial faces usually villous, adaxial sparsely strigillose to

glabrate). **Involucres** 7–9(–15) mm diam. **Outer phyllaries** lance-linear, 4–6 × 0.5–0.8 mm (bases sericeous); inner phyllaries similar, more scarious. **Ray florets** 40–70+; corolla laminae 10–15+ mm. **Disc corollas** 4–6 mm. **Cypselae** 1–1.5 mm, puberulent or glabrate; **pappi** of 15–25 distinct or basally connate bristles 4–6 mm. **2*n* = 32.**

Flowering mid–late summer. Roadsides, waste places; introduced; Ont., Que.; N.Y.; Europe.

Inula brittanica is probably an occasional escape from garden trash and not truly established in the flora.

3. Inula salicina Linnaeus, Sp. Pl. 2: 882. 1753   I

Plants 20–80+ cm. **Leaves:** basal blades lanceolate, 2–6 cm × 5–15 (–30) mm; cauline blades broadly elliptic to lanceolate, (3–)5–7+ cm × 12–20 mm, bases cordate, clasping (faces usually glabrous, adaxial venation raised, reticulation prominent). **Involucres** 8–12(–20) mm diam. **Outer phyllaries** lance-linear, 5–7 × 1.5–2.5 mm; inner similar, more scarious. **Ray florets** 35–70; corolla laminae 10–15+ mm. **Disc corollas** 5–7+ mm. **Cypselae** 1.5–2 mm, glabrous; **pappi** of 30–40+, distinct or basally connate, barbellate bristles 7–8 mm. **2*n* = 16.**

Flowering mid–late summer. Roadsides, waste places; introduced; Mass., N.Y.; Europe.

Inula salicina is probably an occasional escape from garden trash and not truly established in the flora.

187i. ASTERACEAE Martinov tribe PLUCHEEAE (Cassini ex Dumortier) Anderberg, Canad. J. Bot. 67: 2293. 1989 (as Plucheae)

Plucheinae Cassini ex Dumortier, Anal. Fam. Pl., 31. 1829 (as Plucheae)

Herbs, subshrubs, shrubs, or trees. Leaves usually cauline, sometimes mostly basal; alternate; petiolate or sessile; margins entire or denticulate to serrate or dentate [1–2-pinnately divided]. **Heads** heterogamous (usually disciform, rarely obscurely radiate), usually in corymbiform, paniculiform, or spiciform arrays, sometimes borne singly (on scapiform stems in *Sachsia*). **Calyculi** 0. **Phyllaries** persistent or falling, (12–30+) in 3–6+ series, distinct, unequal, usually ± herbaceous to chartaceous, sometimes indurate, margins and/or apices seldom notably scarious. **Receptacles** flat to convex, epaleate [paleate]. **Ray florets** 0 (whitish corollas of peripheral pistillate florets sometimes with minute, 3-toothed laminae in *Sachsia*) [in 1(–2+) series, pistillate and fertile or neuter]. **Peripheral (pistillate) florets** [0] (in disciform heads) in 1–10+ series; corollas (usually present) usually pink to purplish, sometimes whitish or ochroleucous, rarely yellowish. **Disc (inner) florets** bisexual or functionally staminate; corollas usually pink to purplish, sometimes whitish or ochroleucous, rarely yellowish, not 2-lipped, lobes (4–)5, deltate; anther bases ± tailed, apical appendages ovate to lance-ovate or linear; styles abaxially papillate or hairy (sweeping hairs usually obtuse, usually present from proximal to the separation of the branches to near the tips), branches linear, adaxially stigmatic in 2 lines from bases to apices (lines ± confluent distally), apices obtuse to rounded, appendages essentially none. **Cypselae** mostly monomorphic within heads, usually columnar, cylindric, ellipsoid, or fusiform, sometimes ± prismatic, sometimes compressed or flattened, not beaked, bodies smooth or ribbed (glabrous or ± hirsutulous, hairs straight-tipped, uncinate, or glochidiform, faces sometimes glandular as well); **pappi** [sometimes 0] persistent or tardily falling, usually of smooth to barbellate [plumose] bristles or setiform scales (in 1–2 series).

Genera ca. 27, species ca. 219 (3 genera, 12 species in the flora): mostly tropical and subtropical areas of Central America, South America, Africa, Asia, and Australia; some species are widely introduced and established in local floras.

Plucheeae were segregated from traditionally circumscribed Inuleae by Anderberg in 1989 (see discussion in A. A. Anderberg 1994b).

SELECTED REFERENCES Anderberg, A. A. 1991b. Taxonomy and phylogeny of the tribe Plucheae (Asteraceae). Pl. Syst. Evol. 176: 145–177. Anderberg, A. A. 1994b. Tribe Plucheeae. In: K. Bremer. 1994. Asteraceae: Cladistics & Classification. Portland. Pp. 292–303

1. Stems winged; heads in spiciform arrays . . . . 109. *Pterocaulon*, p. 476
1. Stems seldom winged (see *Pluchea sagittalis*); heads usually in corymbiform or paniculiform arrays, rarely borne singly.
   2. Leaves all or mostly basal . . . . 110. *Sachsia*, p. 477
   2. Leaves all or mostly cauline . . . . 111. *Pluchea*, p. 478

109. PTEROCAULON Elliott, Sketch Bot. S. Carolina 2: 323. 1823 • [Greek *pteron*, wing, and *kaulos*, stem, alluding to stems winged by decurrent leaf bases]

Guy L. Nesom

Perennials, 20–150 cm; usually rhizomatous and/or lignescent-tuberous-rooted. **Stems** erect, simple, internodes winged (by decurrent leaf bases), lanate-tomentose and/or glandular. **Leaves** cauline, alternate; sessile; blades linear to elliptic or obovate, bases decurrent, margins usually serrate to serrulate or denticulate, rarely entire, abaxial faces densely whitish-tomentose [puberulent or glabrescent], adaxial (green) glabrous or glabrescent, both faces usually stipitate- or sessile-glandular. **Heads** disciform, (sessile) in spiciform arrays (at ends of branches). **Involucres** cylindro-campanulate to campanulate, 2–3[–5] mm diam. **Phyllaries** persistent, in 4–6 series, distinct, narrowly lanceolate, unequal (subindurate to scarious). **Receptacles** flat, epaleate. **Ray florets** 0. **Peripheral (pistillate) florets** in 1–3+ series, fertile; corollas yellowish. **Inner (functionally staminate [bisexual]) florets** [1–]2–15[+]; corollas yellowish, lobes 5. **Cypselae** cylindric to fusiform, angled or slightly compressed, ribs 6–9 (white, narrow), faces usually sparsely strigose to hispidulous, minutely sessile-glandular between ribs; **pappi** persistent, of distinct, barbellulate bristles in 1–2 series. $x = 10$.

Species 17 (2 in the flora): North America, South America, Australasia.

The 11 primarily South American species of *Pterocaulon* sect. *Pterocaulon* have 1–17 functionally staminate florets per head, 1–2-seriate pappi, and hairs of tomentum with the relatively long, aseptate portion arising from clusters of basal cells. The 6 Australasian species of sect. *Monenteles* (Labillardière) Cabrera have a single functionally staminate floret per head, uniseriate pappi, and hairs of the tomentum equally septate throughout.

SELECTED REFERENCE Cabrera, A. L. and A. M. Ragonese. 1978. Revisión del género *Pterocaulon* (Compositae). Darwiniana 21: 185–257.

1. Heads in dense, usually continuous, rarely interrupted, narrow, ± ovoid arrays (2–)3–8 (–10) cm (axes completely hidden by crowded heads); lengths of leaf blades mostly 2–7 times widths; functionally staminate florets 6–10(–15); flowering May–Jun; Alabama, Florida, Georgia, Mississippi, North Carolina, South Carolina . . . . 1. *Pterocaulon pycnostachyum*
1. Heads in open, ± interrupted, ± cylindric arrays (5–)8–20 cm (main axes visible between glomerules of heads); lengths of leaf blades mostly 6–8 times widths; functionally staminate florets 2–4(–5); flowering Aug–Oct; Louisiana, Texas . . . . 2. *Pterocaulon virgatum*

1. **Pterocaulon pycnostachyum** (Michaux) Elliott, Sketch Bot. S. Carolina 2: 324. 1823 • Fox-tail or coastal blackroot E

Conyza pycnostachya Michaux, Fl. Bor.-Amer. 2: 126. 1803; *Pterocaulon undulatum* C. Mohr

Plants 2–8 dm. **Leaf blades** lanceolate to obovate-lanceolate, oblong, or elliptic, 3–11 × 1–3 (–3.5) cm, lengths mostly 2–7 times widths, margins usually dentate or denticulate, slightly repand, sometimes nearly entire. **Heads** in dense, usually continuous, rarely interrupted (then near bases), narrow, ± ovoid arrays (2–)3–8(–10) cm (usually single, sometimes with 1–2 basal branches). **Involucres** campanulate, 4–5 mm. **Pistillate florets** 23–44.

Functionally staminate florets 6–10(–15). **Cypselae** 1–1.3 mm. $2n = 20$.

Flowering May–Jun. Sandy pinelands, sandy fields, depressions, ditches; 0–20 m; Ala., Fla., Ga., Miss., N.C., S.C.

Differences between *Pterocaulon pycnostachyum* and *P. alopecuroides* (Lamarck) de Candolle, which is widespread in the West Indies and South America, are these: plants 50–70 cm high in *P. pycnostachyum* (versus 70–150 cm in *P. alopecuroides*), arrays of heads 4–8 cm (versus 3–17 cm) long, involucres 3.5–4 mm (versus 4.5–5 mm) high, and 6–15 (versus 1–3) functionally staminate florets (A. L. Cabrera and A. M. Ragonese 1978). In *P. alopecuroides*, the arrays of heads are almost always interrupted proximally, commonly producing sessile to subsessile branches. A count of functionally staminate florets provides a clear determinant for plants that might appear ambiguous in other features.

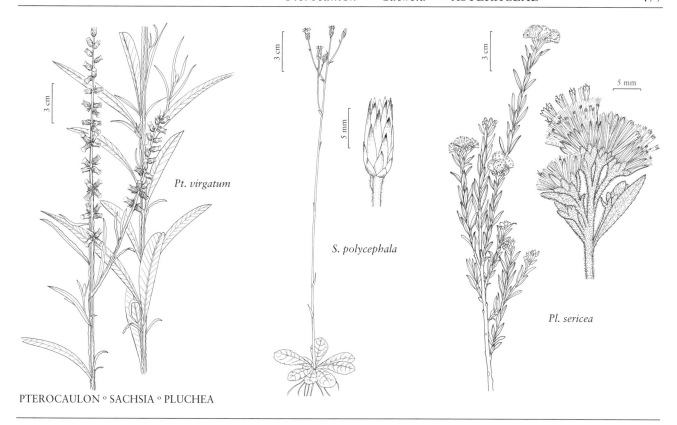

PTEROCAULON ○ SACHSIA ○ PLUCHEA

2. **Pterocaulon virgatum** (Linnaeus) de Candolle in A. P. de Candolle and A. L. P. P. de Candolle, Prodr. 5: 454. 1836 • Wand blackroot [F] [I]

Gnaphalium virgatum Linnaeus, Syst. Nat. ed. 10, 2: 1211. 1759

Plants 4–15 dm. **Leaf blades** linear to narrowly elliptic or linear-lanceolate, 5–10(–15) cm × (2–)5–10(–14) mm, lengths mostly 6–8 times widths, margins entire or minutely denticulate (revolute). **Heads** in open, interrupted, ± cylindric arrays (5–)8–20 cm (main axes visible between glomerules of heads). **Involucres** campanulate to cylindric, 4–5 mm. **Pistillate florets** 25–50. **Functionally staminate florets** 2–4(–5). **Cypselae** 1–1.4 mm.

Flowering Aug–Oct. Marshy areas, ditches, moist places in woods, in sand, sandy loam, and sandy clay; 0–20 m; La., Tex.; Mexico (Tamaulipas); West Indies; Central America; South America.

110. **SACHSIA** Grisebach, Cat. Pl. Cub., 150. 1866 • [For F. G. J. von Sachs, 1832–1897, German plant physiologist, noted by Grisebach to be "ingeniosi"]

Guy L. Nesom

Perennials, 10–60+ cm; fibrous-rooted, sometimes rhizomatous. **Stems** erect (scapiform), simple or distally branched, finely glandular, bases sericeous-woolly. **Leaves** all or mostly basal, alternate; sessile; blades obovate to oblanceolate or spatulate, bases not clasping, margins denticulate to dentate [pinnatifid], abaxial faces ± sericeous and glandular, adaxial glabrous (shiny, venation raised, reticulate). **Heads** radiate or disciform, in loose, corymbiform to paniculiform arrays. **Involucres** cylindric-ovoid to urceolate, 2–3 mm diam. **Phyllaries** falling in fruit, in 5–6 series, lance-ovate to lance-linear, unequal. **Receptacles** flat, epaleate. **Ray or**

peripheral (pistillate) florets in 1(–3) series, fertile; corollas whitish (tubes filiform, laminae absent or ± developed and apically 3-toothed). **Inner (functionally staminate or bisexual) florets** 6–18; corollas whitish, lobes (4–)5. **Cypselae** cylindric to narrowly ellipsoid, ribs 8–12 (white, raised), faces strigillose; **pappi** persistent (fragile), of distinct or basally connate, smooth or barbellulate bristles in 1 series. $x = 10$.

Species 3 (1 in the flora): Florida, West Indies (Bahamas, Cuba, Hispaniola, Jamaica).

The monotypic *Rhodogeron* Grisebach was transferred to *Sachsia* by A. A. Anderberg, who noted (1991b) that "...it is a derived relative of *Sachsia*. The two differ only in the shape of the leaves and the female florets" (leaves pinnatifid, ray florets with laminae well developed and relatively broad in *Rhodogeron*). Anderberg placed *Sachsia* as "Plucheeae insertae sedis," and noted that "...the genus is anomalous, but could prove to belong in the Plucheeae, in spite of its acute sweeping hairs." The two species of *Sachsia* other than *S. polycephala* are endemic to Cuba.

1. **Sachsia polycephala** Grisebach, Cat. Pl. Cub., 151. 1866 • Bahama sachsia [F]

Sachsia bahamensis Urban; *S. divaricata* Grisebach

Plants 10–60 cm. **Leaf blades** 2–12 × 1–4 cm, margins denticulate to coarsely and regularly dentate or repand-dentate with callous-tipped teeth. **Involucres** 5–8 mm. **Phyllaries:** apices acute, abaxial faces ± glandular. **Ray or pistillate corollas:** laminae ca. 0.5 mm or none. **Cypselae** 1.5–2 mm; **pappi** 3–4 mm. $2n = 20$.

Flowering Dec–Apr(–May), sporadically year-round. Rocky pinelands, pine-palm woods, commonly moist habitats; 0–10 m; Fla.; West Indies (Bahamas, Cuba, Hispaniola, Jamaica).

111. **PLUCHEA** Cassini, Bull. Sci. Soc. Philom. Paris 1817: 31. 1817 • [For Abbé N. A. Pluche, 1688–1761, French naturalist]

Guy L. Nesom

Annuals, perennials, subshrubs, shrubs, or trees (usually fetid-aromatic), (20–)50–200(–500) cm; taprooted or fibrous-rooted. **Stems** erect, simple or branched, seldom winged (see *P. sagittalis*), usually puberulent to tomentose and stipitate- or sessile-glandular, sometimes glabrous. **Leaves** cauline, alternate; petiolate or sessile; blades mostly elliptic, lanceolate, oblanceolate, oblong, obovate, or ovate, bases clasping or not, margins entire or dentate, abaxial faces mostly arachnose, puberulent, sericeous, strigose, or villous and/or stipitate- or sessile-glandular, adaxial similar or glabrate or glabrous. **Heads** disciform, in corymbiform or paniculiform arrays (flat-topped or ± elongate). **Involucres** mostly campanulate, cupulate, cylindric, hemispheric, or turbinate, 3–10(–12) mm diam. **Phyllaries** persistent or falling, in 3–6+ series, mostly ovate to lanceolate or linear, unequal. **Receptacles** flat, epaleate. **Peripheral (pistillate) florets** in 3–10+ series, fertile; corollas creamy white, whitish, yellowish, pinkish, lavender, purplish, or rosy. **Inner (functionally staminate) florets** 2–40+; corollas creamy white, whitish, yellowish, pinkish, lavender, purplish, or rosy, lobes (4–)5. **Cypselae** oblong-cylindric, ribs 4–8, faces strigillose and/or minutely sessile-glandular or glabrous (in the flora, only *P. sericea*); **pappi** persistent or tardily falling, of distinct or basally connate, barbellate bristles in 1 series. $x = 10$.

Species 40–60 (9 in the flora): tropical and warm-temperate regions, North America, West Indies, South America, se Asia, Africa, Australia, Pacific Islands.

As currently treated, *Pluchea* is a heterogeneous group of species, variable in habit (trees and shrubs to herbs) and foliar, floral, and fruit morphology. The American, primarily herbaceous, species are divided into groups (G. L. Nesom 1989): sect. *Pluchea*, sect. *Amplectifolium* G. L. Nesom, and sect. *Pterocaulis* G. L. Nesom. Among the woody species, segregate genera have been recognized (*Tessaria* Ruiz & Pavón, *Berthelotia* de Candolle, *Eremohylema* A. Nelson); boundaries among segregates have not been clearly drawn.

SELECTED REFERENCES Ariza E., L. 1979. Contribución al conocimiento del género *Tessaria* (Compositae), I. Consideraciónes sobre los géneros *Tessaria* y *Pluchea*. Kurtziana 12–13: 47–62. Cabrera, A. L. 1939. Las especies Argentinas del género "*Tessaria*." Lilloa 4: 181–189. Godfrey, R. K. 1952. *Pluchea*, section *Stylimnus*, in North America. J. Elisha Mitchell Sci. Soc. 68: 238–271. Keeley, S. C. and R. K. Jansen. 1991. Evidence from chloroplast DNA for the recognition of a new tribe, the Tarchonantheae, and the tribal placement of *Pluchea* (Asteraceae). Syst. Bot. 16: 173–181. King-Jones, S. 2001. Revision of *Pluchea* Cass. (Compositae, Plucheeae) in the Old World. Englera 23. Nesom, G. L. 1989. New species, new sections, and a taxonomic overview of American *Pluchea* (Compositae: Inuleae). Phytologia 67: 158–167. Robinson, H. and J. Cuatrecasas. 1973. The generic limits of *Pluchea* and *Tessaria*. Phytologia 27: 277–285.

1. Shrubs or trees; leaves and stems sericeous, not glandular . . . . 1. *Pluchea sericea*
1. Annuals, perennials, or subshrubs; leaves and stems not sericeous, usually glandular.
   2. Stems (winged by decurrent leaf bases) . . . . 2. *Pluchea sagittalis*
   2. Stems (not winged, leaf bases sometimes clasping, not decurrent).
      3. Leaves petiolate, blades mostly elliptic, lanceolate, oblanceolate, oblong-elliptic, oblong-ovate, or ovate (bases not clasping).
         4. Subshrubs, 100–400 cm; leaf margins entire or denticulate (teeth callous-tipped) . . . . 3. *Pluchea carolinensis*
         4. Annuals or perennials, 50–200+ cm; leaf margins serrate.
            5. Involucres 4–6 × 3–4 mm; phyllaries usually cream, sometimes purplish, usually minutely sessile-glandular, sometimes glabrate (the outermost puberulent); stems usually closely arachose (hairs appressed); arrays of heads paniculiform (of rounded-convex corymbiform clusters terminating branches from distal nodes, arrays usually resulting from strongly ascending, bracteate branches, the central axis longest, first to flower, and, rarely, the only component of an array); leaves petiolate; inland, non-saline habitats . . . . 4. *Pluchea camphorata*
            5. Involucres 5–6 × 4–8(–10) mm; phyllaries usually cream, sometimes purplish, minutely sessile-glandular (outer also puberulent, hairs multicellular, viscid), sometimes glabrate; stems not arachose; arrays of heads corymbiform (flat-topped to rounded, often layered, sometimes incorporating relatively long, leafy, lateral branches, clusters of heads terminal on leafy branches, some lateral branches nearly equaling or surpassing central portion); leaves sessile or petiolate; primarily coastal salt marshes, also inland habitats west of Mississippi River . . . . 5. *Pluchea odorata*
      3. Leaves sessile, blades mostly elliptic, lanceolate, oblong, or ovate (bases clasping to subclasping).
         6. Leaves mostly 8–20 × 3–7 cm; involucres cylindro-campanulate, 9–12 mm (mid phyllaries 2.5–3 mm wide); phyllaries and corollas creamy white . . . . 9. *Pluchea longifolia*
         6. Leaves mostly 3–10 × 1–3 cm; involucres turbinate-campanulate to cylindro-campanulate, 5–8 mm (mid phyllaries 1–1.5 mm wide); phyllaries and corollas yellowish or creamy white to lavender, pale pink, pinkish, purplish, or rosy.
            7. Stems and leaves (slightly succulent, shiny) glandular, otherwise mostly glabrous; involucres 5–6 × 4–5 mm; phyllaries and corollas pink to lavender or cream or pinkish to rosy . . . . 8. *Pluchea yucatanensis*
            7. Stems and leaves at least puberulent or arachose as well as glandular; involucres 4–10 × 5–12 mm; phyllaries and corollas rose-pink to purplish, rose-purple, greenish, cream, or creamy white to yellowish, or pale pink.

[8. Shifted to left margin.—Ed.]

8. Phyllaries and corollas usually creamy white, sometimes cream, greenish, rose-purple, purplish, pale pink, or yellowish; involucres 5–10 × 6–9(–12) mm (bases rounded to impressed); phyllaries thinly arachnose and sessile-glandular . . . . 6. *Pluchea foetida*

8. Phyllaries and corollas rose-pink to purplish; involucres 4–6 × 5–9 mm (bases obtuse to barely acute); phyllaries usually arachnose (sometimes also with viscid hairs) . . . . 7. *Pluchea baccharis*

1. **Pluchea sericea** (Nuttall) Coville, Contr. U.S. Natl. Herb. 4: 128. 1893 • Arrowweed F

Polypappus sericeus Nuttall, Proc. Acad. Nat. Sci. Philadelphia 4: 22. 1848; *Tessaria sericea* (Nuttall) Shinners

Shrubs or trees (not aromatic), 150–300(–500) cm; roots not seen. **Stems** (densely leafy) sericeous, not glandular. **Leaves** sessile; blades lanceolate to narrowly lanceolate or narrowly oblanceolate, 1–5 × 0.2–1 cm, margins entire, faces sparsely to densely silvery sericeous, not glandular (minutely punctate). **Heads** in cymiform clusters. **Involucres** ± campanulate, 4–6 × 3–5 mm. **Phyllaries** pink to purplish, tomentose to villosulous (outer) to arachnose-ciliate or glabrate (inner). **Corollas** pink to purplish. **Pappi** persistent, bristles distinct (distally dilated in functionally staminate florets; cypselae glabrous). $2n = 20$.

Flowering mostly Mar–Jul (sometimes year-round). Floodplains, streambanks, dry lake beds, dunes, sand flats; 40–1000 m; Ariz., Calif., Nev., N.Mex., Tex., Utah; Mexico (Baja California, Chihuahua, Sonora).

Pluchea sericea (with its woody habit and eglandular, densely arranged, sericeous leaves) is isolated among North American *Pluchea*. Torrey and Gray recognized its close similarity to the Asian *Pluchea lanceolata* (de Candolle) Oliver & Hiern, the type species of the Asian genus *Berthelotia*. It has been treated within *Berthelotia* and the South American and Central American segregate *Tessaria*, and a distinct genus (*Eremohylema* A. Nelson) has been created for it.

Pluchea sericea is more similar to the segregates than to herbaceous American groups; the delimitations of the genera are not clear. *Tessaria* was restricted to *T. integrifolia* Ruiz & Pavón, based on its single functionally staminate flower per head and corolla lobes cut nearly to the base of the limb, by H. Robinson and J. Cuatrecasas (1973). The sister species of *T. integrifolia* appears to be *T. absinthoides* (Hooker & Arnott) Cabrera [= *Pluchea absinthoides* (Hooker & Arnott) H. Robinson & Cuatrecasas]. Both species have inner phyllaries with reflexing tips, alveolate and paleate receptacles, pappus bristles basally united into a thick cup, and other distinctive floral features (G. L. Nesom 1989). It seems likely that the accepted definition of *Tessaria* may be expanded to two species or more broadly

to include *P. sericea* and the species placed in *Berthelotia*. Alternately, *Berthelotia* (including *P. sericea*) might be accepted at generic rank, coordinate with *Tessaria*. In any case, it seems likely that *P. sericea* ultimately will be treated outside a more strictly circumscribed *Pluchea*.

2. **Pluchea sagittalis** (Lamarck) Cabrera, Bol. Soc. Argent. Bot. 3: 36. 1949 • Wing-stem camphorweed I

Conyza sagittalis Lamarck in J. Lamarck et al., Encycl. 2: 94. 1786; *Pluchea quitoc* de Candolle; *Pluchea suaveolens* (Vellozo) Kuntze

Perennials, 50–200 cm; fibrous-rooted. **Stems** minutely hirtellous to strigillose and sessile-glandular (winged by decurrent leaf bases). **Leaves** sessile; blades usually lanceolate to lance-elliptic (proximal sometimes spatulate or oblanceolate), mostly 5–15 × 1–3(–4) cm, margins shallowly and closely toothed, faces minutely hirtellous to strigillose and sessile-glandular. **Heads** in corymbiform arrays. **Involucres** hemispheric to cupulate, 4–7 × 8–10 mm. **Phyllaries** greenish to cream, ± stipitate-glandular (outer oval-oblong to linear-attenuate). **Corollas** white or rose-purple. **Pappi** persistent, bristles distinct. $2n = 20$.

Flowering Jul–Aug. Moist or wet, open habitats, ballast deposit areas; 0–10 m; introduced; Ala., Fla.; West Indies; South America.

Pluchea sagittalis is adventive, probably a waif; it was collected as a ballast weed by C. Mohr near Mobile (1891, 1894, 1896) and by A. H. Curtiss near Pensacola (1886, 1901).

3. **Pluchea carolinensis** (Jacquin) G. Don in R. Sweet, Hort Brit. ed. 3, 350. 1839 • Cure-for-all, cough bush, wild tobacco, sourbush I

Conyza carolinensis Jacquin, Collectanea 2: 271. 1789

Subshrubs, 100–400 cm; taprooted. **Stems** matted-villous with viscid, vitreous hairs, proximally glabrescent, not evidently glandular. **Leaves** petiolate (petioles 10–40 mm); blades (thickish, strongly bicolor) elliptic to oblong-obovate or ovate, 5–16(–20) × 2–6(–8) cm, margins entire or denticulate (teeth callous-tipped), abaxial faces moderately or sparsely matted-villous to crinkly-puberulent, adaxial (green) glabrate. **Heads** in dense, corymbiform arrays (held beyond the leaves, axes minutely bracteate, bracts abruptly differentiated from cauline leaves). **Involucres** broadly campanulate to cupulate, 4.5–6 × 5–10 mm. **Phyllaries** greenish to creamy or tan, sometimes slightly purple, glandular-tomentose. **Corollas** whitish to pink-lavender. **Pappi** tardily falling, bristles distinct. $2n = 20$.

Late Feb–Jun. Roadsides, borders of hammocks; 0 m; introduced, Fla.; Mexico; West Indies; Bermuda; Central America; South America; introduced in Pacific Islands.

Pluchea carolinensis is naturalized in the Hawaiian Islands and other Pacific Islands.

The names *Pluchea odorata* of authors, not (Linnaeus) Cassini, and *P. symphytifolia* of authors, not *Conyza symphytifolia* Miller in the sense of W. T. Gillis (1977), have been used for plants here called *Pluchea carolinensis*. The taxon was long identified as *P. odorata* (R. K. Godfrey 1952) and was known as *P.* [*Conyza*] *symphytifolia* (Miller) Gillis for a while. *Conyza symphytifolia* Miller is a synonym of *Neurolaena lobata* (Linnaeus) Cassini (R. Khan and C. E. Jarvis 1989).

4. **Pluchea camphorata** (Linnaeus) de Candolle in A. P. de Candolle and A. L. P. P. de Candolle, Prodr. 5: 452. 1836 • Plowman's-wort E

Erigeron camphoratus Linnaeus, Sp. Pl. 2: 864. 1753

Annuals or perennials, 50–200+ cm; fibrous-rooted. **Stems** minutely puberulent and sessile-glandular, usually also closely arachnose (hairs ap-pressed). **Leaves** petiolate (peti-oles 10–20 mm); blades elliptic to oblong-elliptic, 6–15 × 3–7 cm, margins dentate-serrate or entire, faces glandular-puberulent or puberulent and sessile-glandular. **Heads** in paniculiform arrays (of rounded-convex, corymbiform clusters terminating branches from distal nodes, arrays usually resulting from axillary, strongly ascending, bracteate branches, the central axis longest and first to flower and, rarely, the only component of an array). **Involucres** campanulate, 4–6 × 3–4 mm. **Phyllaries** usually cream, sometimes purplish, minutely sessile-glandular (the outer also sparsely puberulent), sometimes glabrate. **Corollas** rose purplish. **Pappi** persistent, bristles distinct.

Flowering Aug–Oct (year-round in south). Flatwoods, bottomland channels, other wet or moist freshwater habitats; 0–30 m; Ala., Ark., Del., Fla., Ga., Ill., Ind., Kans., Ky., La., Md., Miss., Mo., N.J., N.C., Ohio, Okla., Pa., S.C., Tenn., Tex., Va., W.Va.

Pluchea camphorata is similar to *P. odorata* and rarely may hybridize with it. In *P. camphorata*, the phyllaries of the inner 2–3 series are thin and nearly translucent, lanceolate, and more than twice as long as deltate-ovate phyllaries of the outer series. The inner may be glandular but they are otherwise glabrous, prominently different in vestiture from the outer. The phyllaries of *P. odorata* are more strongly graduated and the inner are glandular and also clearly puberulent as well.

5. **Pluchea odorata** (Linnaeus) Cassini in F. Cuvier, Dict. Sci. Nat. ed. 2, 42: 3. 1826 • Sweetscent, shrubby camphor-weed F

Conyza odorata Linnaeus, Syst. Nat. ed. 10, 2: 1213. 1759

Annuals or perennials, 20–200 cm; fibrous-rooted. **Stems** stipitate- to sessile-glan-dular (commonly with eglandular but viscid hairs as well), not arachnose. **Leaves** petiolate or sessile; blades (succulent, drying thin) lance-ovate to ovate or ovate-elliptic, mostly 4–15 × 1–7 cm, margins shallowly serrate, faces glabrate to moderately or densely pubescent (hairs crinkly). **Heads** in corymbiform arrays (flat-topped to rounded, often layered, sometimes incorporating relatively long, leafy, lateral branches, clusters of heads terminal on branches, some lateral branches nearly equaling or surpassing central portions). **Involucres** cylindro-campanulate, 5–6 × 4–8(–10) mm. **Phyllaries** usually cream, sometimes purplish, minutely sessile-glandular (outer usually also puberulent), sometimes glabrate. **Corollas** pink to rosy or purple. **Pappi** persistent, bristles distinct. $2n = 20$.

Varieties 2 (2 in the flora): North America, South America, w Africa, Pacific Islands.

1. Involucres 5–6 × 4–6(–7) mm; functionally staminate florets 6–13(–19); plants 20–80(–200) cm . . . . 5a. *Pluchea odorata* var. *odorata*
1. Involucres 5–6 × 7–8(–10) mm; functionally staminate florets (14–)21–34; plants 20–60 cm . . . . 5b. *Pluchea odorata* var. *succulenta*

5a. **Pluchea odorata** (Linnaeus) Cassini var. **odorata** F

Pluchea petiolata Cassini; *P. purpurascens* (Swartz) de Candolle

Plants 20–80(–200) cm. **Involucres** 5–6 × 4–6(–7) mm. **Functionally staminate florets** 6–13(–19).

Flowering Aug–Oct (year-round in south). Salt or brackish marshes and estuaries, near coast, less commonly in inland saline habitats, inland freshwater springs and ephemerally

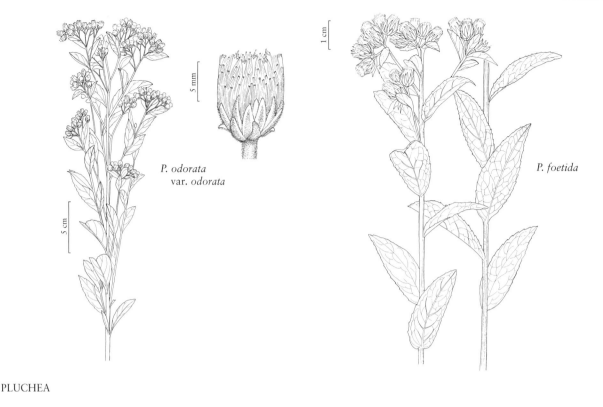

P. odorata
var. odorata

P. foetida

PLUCHEA

moist drainages; 0–50(–1400) m; Ala., Ariz., Ark., Calif., Fla., Ga., Kans., La., Md., Miss., Nev., N.Mex., N.C., Okla., S.C., Tex., Utah, Va.; Mexico; West Indies; Central America; South America (Colombia, Venezuela); w Africa; Pacific Islands.

Pluchea odorata var. *odorata* is naturalized in Hawaii. According to S. King-Jones (2001) *Pluchea senegalensis* Klatt is a synonym of *Pluchea odorata* var. *odorata*.

5b. Pluchea odorata (Linnaeus) Cassini var. **succulenta** (Fernald) Cronquist in A. E. Radford et al., Vasc. Fl. S.E. U.S. 1: 175. 1980 • Saltmarsh fleabane [E]

Pluchea purpurascens (Swartz) de Candolle var. *succulenta* Fernald, Rhodora 44: 227. 1942

Plants 20–60 cm. **Involucres** 5–6 × 7–8(–10) mm. **Functionally staminate florets** (14–)21–34.

Flowering Aug–Oct. Salt or brackish marshes, rarely inland; 0–10 m; Ont.; Conn., Del., Ill., Ind., Maine, Md., Mass., Mich., N.J., N.Y., N.C., Pa., R.I., Va.

The larger size of the heads of *Pluchea odorata* var. *succulenta* (versus var. *odorata*) is correlated with the greater number of functionally staminate florets. The two taxa apparently intergrade across a wide area, from southeastern Virginia through South Carolina, and identifications are arbitrary in that region. In the Carolinas, heads appear to be smaller (even heads with relatively numerous florets), and in northern localities (Virginia northward) the smaller size of plants may simply reflect a shorter growing season. It may be more realistic to identify plants of all localities simply as *P. odorata*, noting that a trend toward increasing head size occurs northward along the Atlantic Coast.

6. Pluchea foetida (Linnaeus) de Candolle in A. P. de Candolle and A. L. P. P. de Candolle, Prodr. 5: 452. 1836 • Stinking camphorweed [F]

Baccharis foetida Linnaeus, Sp. Pl. 2: 861. 1753; *Pluchea eggersii* Urban; *P. foetida* var. *imbricata* Kearney; *P. imbricata* (Kearney) Nash; *P. tenuifolia* Small

Annuals or perennials, 40–100 cm; fibrous-rooted, sometimes rhizomatous. **Stems** (often dark purplish) arachnose, glandular. **Leaves** sessile; blades (thick, reticulate-veined) oblong to elliptic, lance-ovate, or ovate, mostly 3–10(–13) × 1–4 cm (bases clasping), margins denticulate (apices rounded to acute), faces minutely sessile-glandular. **Heads** in loose to dense, corymbiform arrays. **Involucres** usually cupulate to campanulate, sometimes turbinate-campanulate, 5–10 × 6–9(–12) mm (bases mostly rounded to impressed, sometimes obtuse). **Phyllaries** usually creamy white, sometimes cream, greenish, pinkish, rose-purplish, purplish, yellowish, or pale pink, thinly arachnoid-pubescent and sessile-glandular (the outer ovate to ovate-lanceolate, lengths mostly 0.2–0.6 times inner). **Corollas** creamy white to yellowish or pale pink. **Pappi** persistent, bristles distinct.

Late Jul–Oct (year-round in south). Seasonally wet soil, pond and lake edges, ditches, borrow pits, swampy woods, bogs, other freshwater wetlands; 0–20 m; Ala., Ark., Del., Fla., Ga., La., Md., Miss., Mo., N.J., N.C., Okla., S.C., Tex., Va.; Mexico; West Indies (Hispaniola).

Pluchea foetida var. *imbricata* has not been treated as distinct from typical *P. foetida* by recent authors (e.g., A. Cronquist 1980; R. K. Godfrey and J. W. Wooten 1981; R. P. Wunderlin et al. 1996). Although plants similar to the type can be found scattered in Florida and Georgia, a populational integrity does not appear to occur, and intermediate forms exist. Nevertheless, field biologists should be aware of the putative distinctions of var. *imbricata* to make more critical observations regarding its status.

1. Distalmost leaves triangular-ovate; heads in tight, rounded clusters at ends of branches; involucres turbinate-campanulate, 9–10 mm; phyllaries pinkish to rose-purplish . . . . . *Pluchea foetida* var. *imbricata*
1. Distalmost leaves mostly oblong-elliptic; heads in paniculiform arrays of usually flat-topped cymiform clusters; involucres broadly campanulate, 5–8 mm; phyllaries cream to greenish . . . . *Pluchea foetida* var. *foetida*

7. **Pluchea baccharis** (Miller) Pruski, Sida 21: 2035. 2005 • Rosy camphorweed

Conyza baccharis Miller, Gard. Dict. ed. 8, Conyza no. 16. 1768; *Pluchea rosea* R. K. Godfrey

Perennials, 40–60 cm; fibrous-rooted, sometimes rhizomatous. **Stems** puberulent to sparsely villous and stipitate- to sessile-glandular (sometimes viscid). **Leaves** sessile; blades ovate to ovate-oblong or elliptic-oblong, 2–7 × 0.5–3 cm (bases cuneate to truncate or subcordate, clasping to subclasping), margins shallowly apiculate-toothed, faces puberulent to sparsely villous and stipitate- to sessile-glandular (sometimes viscid). **Heads** in corymbiform arrays. **Involucres** campanulate to turbinate-campanulate or turbinate, 4–6 × 5–9 mm (bases obtuse to barely acute). **Phyllaries** rose-pink to purplish, moderately appressed-villous to puberulous or arachnose, usually viscid-hairy as well (outer phyllaries ovate-acuminate to ovate-lanceolate, lengths 0.5–1 times inner). **Corollas** rose-pink to purplish. **Pappi** persistent, bristles distinct. $2n = 20$.

Flowering Jun–Jul. Wet savannas, flatwoods, pond edges, borrow pits, ditches; 0–20 m; Ala., Fla., Ga., La., Miss., N.C., S.C., Tex.; Mexico; West Indies (Bahamas); Central America (Belize, Honduras, Nicaragua).

Pluchea baccharis has been reported from Arkansas; I have not seen a specimen.

Pluchea rosea var. *mexicana* R. K. Godfrey, endemic to inland gypseous-saline habitats in east-central Mexico, has been treated at specific rank (G. L. Nesom 1989).

The geographic ranges of *Pluchea baccharis* and *P. foetida* are nearly congruent and the taxa intergrade in morphology. The distinction between them is based primarily on corolla and phyllary color. Features of involucral vestiture also appear to be relatively constant. Head size and shape are not reliable diagnostic features.

8. **Pluchea yucatanensis** G. L. Nesom, Phytologia 67: 160. 1989 • Yucatan camphorweed [I]

Perennials, 20–60 cm; probably fibrous-rooted. **Stems** ± stipitate- or sessile-glandular, otherwise glabrous. **Leaves** sessile; blades (leathery, slightly succulent, shiny) oblong-obovate to oblong-oblanceolate or broadly lanceolate, 3–5 × (0.6–)1.5–2 cm (bases subclasping and sub-auriculate), margins serrulate, faces ± stipitate- or sessile-glandular, otherwise glabrous or distalmost minutely puberulent. **Heads** in corymbiform arrays. **Involucres** turbinate to campanulate, 5–6 × 4–5 mm. **Phyllaries** pink to lavender or cream, proximally stipitate- or sessile-glandular, distally densely stipitate-glandular (outermost ovate-lanceolate, lengths usually 1 times inner, rarely only 0.5 times inner). **Corollas** pink to lavender or cream or pinkish to rosy. **Pappi** persistent, bristles distinct.

Flowering late May–Aug. Low woods; 0–10 m; introduced; Ala., Miss.; Mexico; Central America (Belize).

Pluchea yucatanensis apparently is native along the Gulf and Caribbean coasts of Mexico and Central America, most commonly on the Yucatan Peninsula and in Belize. In the United States, it is known from collections made from 1896 to 1969 in coastal Alabama and Mississippi; it appears to be naturalized in the flora.

Pluchea yucatanensis is similar in habit and general appearance to *P. foetida* and *P. baccharis* and has been identified as both; the rosy tinted phyllaries and florets are more similar to those of *P. baccharis*. The glabrous, slightly thickened, shiny leaves and glabrous phyllaries are recognition traits for the species.

9. Pluchea longifolia Nash, Bull. Torrey Bot. Club 23: 108. 1896 • Long-leaf camphorweed [E]

Perennials, 60–150(–250) cm; fibrous-rooted. **Stems** sparsely arachnose. **Leaves** (crowded) sessile; blades oblong to elliptic, lance-ovate, or ovate, mostly 8–20 × 3–7 cm (bases clasping to subclasping), margins coarsely and irregularly toothed, abaxial faces villous and sessile-glandular, adaxial hirtellous and sessile-glandular. **Heads** in corymbiform arrays. **Involucres** cylindro-campanulate, 9–12 × 6–9 mm (lengths mostly 2 times diams.). **Phyllaries** creamy white, powdery puberulent, sometimes sparsely glandular (mid phyllaries 2.5–3 mm wide). **Corollas** creamy white. **Pappi** persistent, bristles basally connate.

Flowering Jun–Oct(–Nov). Brackish to fresh swamps, marshes, hammocks, lake shores, ditches, and canals; 0–10 m; Fla.

187j. ASTERACEAE Martinov tribe ANTHEMIDEAE Cassini, J. Phys. Chim. Hist. Nat. Arts 88: 192. 1819

Annuals, biennials, perennials, subshrubs, or shrubs (herbage often aromatic). **Leaves** usually cauline, sometimes mostly basal; alternate [opposite]; usually petiolate, sometimes sessile; margins usually 1–3-palmately or -pinnately lobed (ultimate segments usually linear to filiform), sometimes not lobed, margins dentate or entire. **Heads** homogamous (discoid) or heterogamous (radiate or disciform), usually in corymbiform, paniculiform, racemiform, or spiciform arrays, sometimes borne singly or in subcapitate clusters. **Calyculi** 0. **Phyllaries** persistent or falling, usually in 3–5+ series, distinct, unequal, and wholly scarious or with margins and/or apices notably scarious, sometimes in 1–2 series, distinct, subequal, and herbaceous with margins and/or apices barely scarious. **Receptacles** flat to columnar or conic, epaleate or paleate (sometimes partially paleate; paleae usually falling, oblong to linear or filiform, flat or ± conduplicate). **Ray florets** 0 or in 1(–2+) series, usually pistillate and fertile, rarely neuter or styliferous and sterile; corollas usually yellow or white, sometimes reddish to cyanic (sometimes combinations within corollas). **Peripheral (pistillate) florets** 0 or (in disciform heads) in 1–3+ series; corollas usually present, usually yellow, sometimes ochroleucous or reddish to cyanic (sometimes combinations within corollas), sometimes lacking. **Disc (inner) florets** usually bisexual and fertile, rarely functionally staminate; corollas usually yellow, sometimes ochroleucous or reddish to cyanic (sometimes combinations within corollas); anther bases obtuse, not tailed, apical appendages ± ovate; styles abaxially glabrous or papillate (distally), branches linear, adaxially stigmatic in 2 lines from bases to apices, apices ± truncate, appendages rings of papillae or essentially 0. **Cypselae** monomorphic or dimorphic within heads, usually obovoid or columnar to prismatic, sometimes 3-angled or -winged, or compressed or obcompressed (then often winged); **pappi** usually 0 (cypselar wall tissues sometimes produced as pappus-like, entire or lobed to toothed wings, coronas, or tubes), if present, pappi persistent or falling, usually of scales, rarely of bristles.

Genera 100+, species 1700+ (26 genera, 99 species in the flora): almost worldwide, mostly Old World.

Centers of concentration for anthemids are in the Mediterranean regions of Europe and Africa, in central Asia, and in southern Africa.

Studies by L. E. Watson et al. (2000) and syntheses by K. Bremer and C. J. Humphries (1993) and by C. Oberprieler et al. (unpubl.) have resulted in changes in circumscriptions for some genera and in better understandings of relationships among genera of anthemids.

From 1753 until near the end of the 20th century, the mums and florists' chrysanthemums of horticulture, along with marguerites, Paris daisies, Shasta daisies, and others, were included by most botanists in *Chrysanthemum*. Then, traditional *Chrysanthemum* came to be dismembered (see, e.g., K. Bremer and C. J. Humphries 1993) and the mums, florists' chrysanthemums, and their closest relatives came to be treated botanically in the genus *Dendranthema*. A great hue and cry ensued from horticulturalists, especially in Europe. The upshot was that a proposal to typify *Chrysanthemum* on *C. indicum* (florists' chrysanthemum) succeeded, and that generic name is once again applied to the mums and florists' chrysanthemums of horticulture, none of which is known to be well established in the flora area. Here, the marguerites and Paris daisies are treated in *Argyranthemum*, Shasta daisies in *Leucanthemum*, and corn chrysanthemums and crown daisies in *Glebionis*.

Some cultivated members of Anthemideae have been reported as ephemeral or ± persistent escapes in the flora area. Some of those are documented in generic treatments below; others include florists' chrysanthemum (*Chrysanthemum indicum*, which may be encountered as abandoned plantings almost anywhere in the flora area) and Shasta daisy (reported for Michigan as *Chrysanthemum* ×*superbum* J. K. Ingram by E. G. Voss 1972–1996, vol. 3; see treatment of *Leucanthemum*).

Centipeda minima (Linnaeus) A. Braun & Ascherson has been recorded from Massachusetts and Pennsylvania as non-persistent introductions. The plants are somewhat succulent, gland-dotted annuals with stems decumbent to procumbent, leaves oblanceolate to cuneate, mostly 5–15 mm, usually lobed or toothed distally, heads disciform, involucres hemispheric or broader (2–3+ mm diam.), phyllaries in 2–4 series, pistillate florets in 2–8 series, bisexual florets 10 (–30), corollas yellow, 4-lobed, and cypselae obovoid, 1–1.5 mm, epappose. Traditionally, *Centipeda* has been included in Anthemideae; in more recent studies, it has been placed in Astereae (e.g., E. V. Boyko 2003; G. L. Nesom 2000) or in Centipedinae within Athroismeae (J. L. Panero 2005).

SELECTED REFERENCES Arriagada, J. E. and N. G. Miller. 1997. The genera of Anthemideae (Compositae: Asteraceae) in the southeastern United States. Harvard Pap. Bot. 2: 1–46. Bremer, K. and C. J. Humphries. 1993. Generic monograph of the Asteraceae–Anthemideae. Bull. Nat. Hist. Mus. London, Bot. 23: 71–177. Oberprieler, C. 2002. A phylogenetic analysis of *Chamaemelum* Mill. (Compositae: Anthemideae) and related genera based upon nrDNA ITS and cpDNA *trn*L/*trn*F IGS sequence variation. Bot. J. Linn. Soc. 138: 255–273. Oberprieler, C. and R. Vogt. 2000. The position of *Castrilanthemum* Vogt & Oberprieler and the phylogeny of Mediterranean Anthemideae (Compositae) as inferred from nrDNA ITS and cpDNA *trn*L/*trn*F IGS sequence vaiation. Pl. Syst. Evol. 225: 145–170. Watson, L. E. et al. 2002. Molecular phylogeny of subtribe Artemisiinae (Asteraceae), including Artemisia and its allied and segregate genera. B. M. C. Evol. Biol. 2: 17 (12 pp.). Watson, L. E., T. M. Evans, and T. Boluarte. 2000. Molecular phylogeny of the tribe Anthemideae (Asteraceae), based on chloroplast gene *ndh*F. Molec. Phylogen. Evol. 15: 59–69.

1. Heads discoid or disciform (ray florets 0; if peripheral florets pistillate, lacking corollas or corollas lacking laminae).
  2. Heads disciform (peripheral florets pistillate, disc florets bisexual or functionally staminate).
    3. Pistillate corollas lacking.
      4. Pistillate corollas lacking; disc florets 12–200+, bisexual and fertile (heads pedunculate) . . . . 126. *Cotula*, p. 543
      4. Pistillate corollas lacking (styles sheathed by pericarp tissues, soon indurate, becoming spinelike); disc florets 2–8+, functionally staminate (heads sessile) . . . . 127. *Soliva*, p. 545
    3. Pistillate corollas ± tubular (distally 3–5-lobed or truncate).
      5. Heads usually in lax to dense, corymbiform arrays, rarely borne singly; involucres mostly hemispheric or broader, (3–)5–22+ mm diam.; phyllaries (20–)30–60+ in (2–)3–5+ series . . . . 112. *Tanacetum* (in part), p. 489
      5. Heads usually in paniculiform, racemiform, or spiciform arrays, sometimes in subcapitate clusters, sometimes borne singly; involucres hemispheric to campanulate, obconic, ovoid, or turbinate, 1.5–5(–12) mm diam.; phyllaries 5–20+ in 1–4(–7) series.
        6. Heads borne singly or (2–20+) in usually corymbiform, rarely paniculiform, arrays or in subcapitate clusters; disc floret corollas bright yellow or ochroleucous (leaves usually gland-dotted) . . . . 118. *Sphaeromeria*, p. 499
        6. Heads (usually 20–200+, 2–12+ in *Picrothamnus*) in paniculiform, racemiform, or spiciform arrays, rarely borne singly; disc floret corollas usually pale yellow, rarely red.
          7. Subshrubs or shrubs (thorny); disc florets functionally staminate (corollas ± villous) . . . . 117. *Picrothamnus*, p. 498
          7. Annuals, perennials, subshrubs, or shrubs (not thorny); disc florets bisexual and fertile or functionally staminate (corollas usually glabrous, sometimes hairy, not villous) . . . . 119. *Artemisia* (in part), p. 503

2. Heads discoid (all florets bisexual, fertile).
    8. Subshrubs or shrubs.
        9. Leaf blades usually 1-pinnately lobed (lobes usually crowded, not in one plane, overall effect ± vermiform); receptacles paleate; disc florets 60–250+ . . . . 116. *Santolina*, p. 497
        9. Leaf blades not lobed or 1–2+ pinnately and/or palmately lobed, sometimes apically 3-lobed or -toothed (not vermiform); receptacles epaleate (except *Artemisia palmeri*); disc florets 3–20(–30+) or 40–100+.
            10. Heads borne singly or in loose, corymbiform arrays; phyllaries 22–40 in 3–4+ series; disc florets 40–100+; cypselae narrowly obovoid to oblong or elliptic, ± terete or flattened, ribs 5; pappi of 1 (often cupped or earlike) or 3–5 (erose to lacerate) scales . . . . 125. *Pentzia*, p. 543
            10. Heads in paniculiform, racemiform, or spiciform arrays; phyllaries 12–20+ in 4–7 series; disc florets 2–20(–30+); cypselae columnar, obovoid, or fusiform, ribs 0 (and faces finely striate) or 2–5; pappi usually 0 (coroniform in *A. californica* and *A. papposa*) . . . . 119. *Artemisia* (in part), p. 503
    8. Annuals or perennials.
        11. Perennials.
            12. Receptacles convex, epaleate; disc corolla tubes ± cylindric (proximally swollen, becoming spongy in fruit, not clasping apices of cypselae) . . . . 137. *Leucanthemum* (in part), p. 557
            12. Receptacles hemispheric to conic, paleate; disc corolla tubes ± cylindric (bases saccate, weakly clasping apices of cypselae) . . . . 115. *Chamaemelum* (in part), p. 496
        11. Annuals (scarious tips of inner phyllaries not notably dilated; receptacles conic or oblong-ovoid to ± hemispheric or globose).
            13. Phyllaries lance-triangular to ± ovate or elliptic (carinate); cypselae columnar to prismatic, ribs 4 (pericarps without myxogenic cells); pappi coroniform (shorter adaxially) . . . . 123. *Oncosiphon*, p. 539
            13. Phyllaries oblong or ovate to spatulate or linear-spatulate (not carinate); cypselae obconic, slightly compressed, ribs 5 (pericarps sometimes with myxogenic cells); pappi 0, coroniform, or adaxial auricles . . . . 124. *Matricaria* (in part), p. 540
1. Heads radiate (ray florets 3–35+, corollas with laminae).
    14. Receptacles wholly or partially paleate.
        15. Heads in compact to open (± flat-topped), corymbiform or compound-corymbiform arrays; ray florets 3–5(–12); disc florets (5–)15–30+ (cypselae obcompressed) . . . 113. *Achillea*, p. 492
        15. Heads borne singly or in lax, corymbiform arrays; ray florets 5–30+; disc florets 40–300+ (cypselae mostly not, sometimes weakly, obcompressed).
            16. Disc corolla tubes ± cylindric (bases ± saccate or spurred, ± clasping apices of ovaries and/or cypselae); cypselae: ribs or nerves (weak): 2 lateral, 1 adaxial.
                17. Phyllaries 16–24 in 2–3+ series; ray corollas orange, yellow, or white with yellow bases . . . . 114. *Cladanthus*, p. 495
                17. Phyllaries 22–45+ in 3–4+ series; ray corollas white . . . . 115. *Chamaemelum* (in part), p. 496
            16. Disc corolla tubes ± compressed or cylindric (bases sometimes proximally dilated, not saccate or spurred); cypselae: ribs 9–10, or 0, or 2 lateral (sometimes ± winged) plus 3–10 finer ribs on each face.
                18. Perennials; ray corollas yellow . . . . 128. *Cota*, p. 547
                18. Annuals (biennials); ray corollas usually white, rarely yellow or pink . . . . 122. *Anthemis*, p. 537
    14. Receptacles epaleate.
        19. Cypselae dimorphic: outer (ray) usually 3-angled and ± winged (except *Mauranthemum*); inner (disc) ± compressed-prismatic or ± flattened (angles winged), or ± quadrate (1 or 2 angles sometimes winged), or columnar (not winged).
            20. Subshrubs or shrubs . . . . 131. *Argyranthemum*, p. 552
            20. Annuals.
                21. Stems and leaves hirtellous to pilosulous (some hairs gland-tipped, plants sticky, viscid) . . . . 130. *Heteranthemis*, p. 551

21. Stems and leaves glabrous or hairy (hairs not gland-tipped, plants not viscid).

    22. Ray corollas proximally white or red to purple, distally yellow or white; disc corollas proximally ochroleucous, distally red to purple (phyllaries ± carinate) . . . . 132. *Ismelia*, p. 552

    22. Ray corollas mostly white (usually yellowish at bases) or mostly yellow (sometimes paler distally); disc corollas ± yellow (phyllaries not carinate).

        23. Involucres 8–12(–15+) mm diam.; ray floret corollas white (usually yellowish at bases, drying pinkish), laminae oblong to ovate (6–12+ mm); disc corolla lobes (2–)5 (without resin sacs) . . . . 133. *Mauranthemum* (in part), p. 554

        23. Involucres 15–25+ mm diam.; ray floret corollas mostly yellow, sometimes paler distally, laminae linear, oblong, or ovate (8–25 mm); disc corolla lobes 5 (each with a resin sac) . . . . 134. *Glebionis*, p. 554

[19. Shifted to left margin.—Ed.]

19. Cypselae ± monomorphic, outer and inner similar (0 notably winged).

    24. Shrubs (leaves mostly clustered distally on stems; pappi crowns of scales) . . . . . . 135. *Nipponanthemum*, p. 555

    24. Annuals, biennials, perennials.

        25. Annuals.

            26. Leaves usually irregularly 1-pinnately lobed or toothed; involucres ± hemispheric, 8–12(–15+) mm diam; disc florets 60–100+; cypselae: ribs 7–10 . . . . 133. *Mauranthemum* (in part), p. 554

            26. Leaves (1–)2–3-pinnately lobed; involucres 4–14 mm diam.; disc florets 120–750+; cypselae: ribs 3–5.

                27. Cypselae obconic, slightly compressed (usually asymmetric, apices oblique), ribs 5, faces smooth between ribs (pericarps without resin sacs, or resin sacs within lateral ribs) . . . . 124. *Matricaria* (in part), p. 540

                27. Cypselae trigonous, ± compressed, ribs 3–5 (0–2 abaxial, 2 lateral, 1 adaxial, usually whitish, relatively thick, smooth), faces smooth or rugose to tuberculate between ribs (pericarps usually with 2–3, sometimes 1–5, mostly abaxial-apical, resin sacs) . . . . 129. *Tripleurospermum* (in part), p. 548

        25. Biennials or perennials.

            28. Heads usually in lax to dense, corymbiform arrays, rarely borne singly; pappi usually coroniform, rarely 0 . . . . 112. *Tanacetum* (in part), p. 489

            28. Heads usually borne singly or in 2s or 3s, sometimes in corymbiform arrays; pappi usually 0, sometimes crowns of 6–12 irregular teeth.

                29. Ray florets styliferous and sterile . . . . 136. *Leucanthemella*, p. 557

                29. Ray florets pistillate and fertile.

                      30. Cypselae trigonous, ± compressed, ribs 3–5 (0–2 abaxial, 2 lateral, 1 adaxial, usually whitish, relatively thick, smooth), faces smooth or rugose to tuberculate between ribs (pericarps usually with 2–3, sometimes 1–5, mostly abaxial-apical, resin sacs) . . . . 129. *Tripleurospermum* (in part), p. 548

                      30. Cypselae ± columnar, cylindro-obconic, obconic, or obovoid, ribs 5–10 (pericarps without apical resin sacs).

                          31. Plants (10–)50–130(–200+) cm; disc corolla tubes ± cylindric (proximally swollen, becoming spongy in fruit); cypselae ± columnar to obovoid, ribs ± 10 (pericarps with myxogenic cells)137. *Leucanthemum* (in part), p. 557

                          31. Plants mostly (0.6–)5–40+ cm; disc corolla tubes ± cylindric (not swollen, not becoming spongy in fruit); cypselae ± obconic to cylindro-obconic, ribs 5 or 5–8(–10) (pericarps without myxogenic cells).

[32. Shifted to left margin.—Ed.]

32. Leaves not lobed, margins entire; involucres (4–)4.5–6(–6.5) mm diam.; phyllaries 20–26(+) in 2(–3+) series (receptacles ± villous) . . . . 120. *Hulteniella*, p. 534
32. Leaves usually pinnati-palmately lobed (lobes 3–7), ultimate margins coarsely crenate, dentate, or entire; involucres 13–29 mm diam. (flattened); phyllaries (22–)25–34(–44) in 3(–4) series (receptacles glabrous) . . . . 121. *Arctanthemum*, p. 535

112. TANACETUM Linnaeus, Sp. Pl. 2: 843. 1753; Gen. Pl. ed. 5, 366. 1754 • Tansy, tanaisie [Derivation unknown; possibly Greek *athanasia*, immortality, through Medieval Latin *tanazita*]

Linda E. Watson

Perennials [annuals, subshrubs], 5–150 cm (usually rhizomatous; usually aromatic). **Stems** 1 or 2–5+, erect or prostrate to ascending, branched proximally and/or distally, glabrous or hairy (hairs basifixed and/or medifixed, sometimes stellate). **Leaves** basal and/or cauline; alternate; petiolate or sessile; blades mostly obovate to spatulate, usually 1–3-pinnately lobed, ultimate margins entire, crenate, or dentate, faces glabrous or hairy. **Heads** usually radiate, sometimes disciform (or quasi-radiate or -radiant), usually in lax to dense, corymbiform arrays, rarely borne singly. **Involucres** mostly hemispheric or broader, (3–)5–22+ mm diam. **Phyllaries** persistent, (20–)30–60+ in (2–)3–5+ series, distinct, ± ovate to oblong or oblong to lanceolate or lance-linear (sometimes carinate), unequal, margins and apices (pale to dark brown or blackish) scarious (tips sometimes dilated). **Receptacles** flat to conic or hemispheric (sometimes hairy), epaleate. **Ray florets** usually 10–21+ (pistillate and fertile or neuter; corollas pale yellow to yellow or white, usually with yellowish bases [pink], laminae oblong to flabellate), sometimes 0 (in disciform or quasi-radiate or -radiant heads, peripheral pistillate florets 8–30+; corollas pale yellow, ± zygomorphic, lobes 3–4, sometimes ± raylike). **Disc florets** 60–300+, bisexual, fertile; corollas yellow, tubes ± cylindric, throats narrowly funnelform to campanulate, lobes (4–)5, ± deltate. **Cypselae** obconic or ± columnar (circular in cross section), ribs (4–)5–10 (–12+), faces usually gland-dotted, sometimes glabrous (pericarps without myxogenic cells or resin sacs, embryo sac development tetrasporic); **pappi** usually coroniform, rarely 0 [distinct scales or each pappus an adaxial auricle]. *x* = 9 (polyploidy).

Species 160 (4 in the flora): North America, Europe, Asia, n Africa; some species widely cultivated.

1. Leaves usually not pinnately lobed (sometimes with 1–4+ lateral lobes near bases of blades), ultimate margins ± crenate. . . . . 1. *Tanacetum balsamita*
1. Leaves usually 1–3-pinnately lobed, ultimate margins entire or dentate.
    2. Leaf blades 1–2-pinnately lobed (primary lobes 3–5 pairs, ± ovate), faces (at least abaxial) usually puberulent; ray florets 10–21+ (more in "doubles"), corollas white, laminae 2–8(–12+) mm; pappi 0 or coroniform (0.1–0.2+ mm) . . . . 2. *Tanacetum parthenium*
    2. Leaf blades 2–3-pinnately lobed (primary lobes 4–24+ pairs, ± oblong to elliptic or linear), faces usually arachno-villous to villous, sometimes glabrescent or glabrous; ray florets 0 or 8–30+, corollas pale yellow to yellow, laminae 1–8+ mm; pappi coroniform (0.1–0.5+ mm).
        3. Leaves: faces glabrous or sparsely hairy; heads (disciform) 20–200 in corymbiform arrays; involucres 5–10 mm diam . . . . 3. *Tanacetum vulgare*
        3. Leaves: faces usually ± villous or arachno-villous to lanate, sometimes glabrescent or glabrate; heads (radiate, quasi-radiant or -radiate, or disciform) (2–)5–12(–20+) in corymbiform arrays or borne singly; involucres 8–22+ mm diam . . . . 4. *Tanacetum bipinnatum*

1. Tanacetum balsamita Linnaeus, Sp. Pl. 2: 845. 1753

• Costmary, chrysanthème balsamique [I]

Balsamita major Desfontaines; *Chrysanthemum balsamita* (Linnaeus) Baillon; *C. balsamita* var. *tanacetoides* Boissier; *Pyrethrum majus* (Desfontaines) Tzvelev

Perennials, 30–80(–120) cm. **Stems** 1, erect, simple or branched (strigose, glabrate). **Leaves** basal and cauline; petiolate (proximal) or sessile (distal); blades (basal and proximal cauline) mostly elliptic to oblong, 10–20 × 2–8 cm, usually not pinnately lobed (sometimes with 1–4+ lateral lobes near bases), margins ± crenate, faces usually silvery strigose or sericeous (at least when young), glabrescent, ± gland-dotted. **Heads** (3–)10–60+ in corymbiform arrays. **Involucres** (3–)5–8(–10) mm diam. (phyllaries 40–60+ in 3–4+ series, tips usually ± dilated). **Receptacles** flat to convex. **Ray florets** usually 0 [sometimes 12–15, pistillate, fertile; corollas white, laminae 4–6+ mm]. **Disc corollas** ca. 2 mm. **Cypselae** ± columnar, 1.5–2 mm, 5–8-ribbed (with non-mucilaginous glands); **pappi** coroniform, 0.1–0.4 mm (entire or ± toothed). $2n = 18, 54$.

Flowering Aug–Sep. Disturbed sites, abandoned plantings; 0–1800 m; introduced; N.S., Ont., Que., Sask.; Calif., Colo., Conn., Del., Idaho, Ill., Ind., Kans., Maine, Md., Mass., Mich., Mo., Mont., Nev., N.H., N.Y., Ohio, Oreg., Pa., R.I., S.Dak., Utah, Vt., Wash., Wis., Wyo.; Asia.

2. Tanacetum parthenium (Linnaeus) Schultz-Bipontinus, Tanaceteen, 55. 1844 • Feverfew [I]

Matricaria parthenium Linnaeus, Sp. Pl. 2: 890. 1753; *Chrysanthemum parthenium* (Linnaeus) Bernhardi

Perennials, (20–)30–60(–80) cm. **Stems** 1–3+ (ridged), erect, branched (usually glabrous proximally, puberulent distally). **Leaves** mainly cauline; petiolate; blades ovate to rounded-deltate, 4–10+ × 1.5–4 cm, usually 1–2-pinnately lobed (primary lobes 3–5+ pairs, ± ovate), ultimate margins pinnatifid to dentate, faces (at least abaxial) usually puberulent, gland-dotted. **Heads** 5–20 (–30) in corymbiform arrays. **Involucres** 5–7 mm diam. **Ray florets** 10–21+ (more in "doubles"), pistillate, fertile; corollas white, laminae 2–8(–12) mm. **Disc corollas** ca. 2 mm. **Cypselae** ± columnar, 1–2 mm, 5–10-ribbed; **pappi** 0 or coroniform, 0.1–0.2+ mm. $2n = 18$.

Flowering Jun–Nov. Disturbed sites, urban areas, roadsides, fields, abandoned plantings; 10–1900 m; introduced; B.C., Ont.; Ala., Colo., Conn., Del., Idaho, Ill., Ind., Ky., Maine, Md., Mass., Mich., Miss., Mo.,

Mont., Nev., N.H., N.J., N.Y., N.C., Ohio, Oreg., Pa., R.I., S.C., Utah, Vt., Wash., W.Va., Wis., Wyo.; Eurasia; n Africa; widely naturalized in New World and Old World.

Tanacetum parthenium is widely cultivated throughout North America.

3. Tanacetum vulgare Linnaeus, Sp. Pl. 2: 844. 1753

• Common tansy, tanaisie vulgaire [F] [I]

Perennials, mostly 40–150 cm. **Stems** 1–2+ (ridged), erect, branched distally (glabrous or sparsely hairy). **Leaves** basal (soon withering) and cauline; petiolate or sessile; blades broadly oblong or oval to elliptic, 4–20 × 2–10 cm, pinnately lobed (rachises ± winged, primary lobes 4–10 pairs, lance-linear to lanceolate or narrowly elliptic, often pinnately lobed or toothed), ultimate margins dentate, faces glabrous or sparsely hairy, gland-dotted. **Heads** 20–200 in compact, corymbiform arrays. **Involucres** 5–10 mm diam. **Receptacles** convex to conic, epaleate. **Ray florets** 0 (heads disciform, peripheral pistillate florets ca. 20; corollas yellow, lobes 3–4). **Disc corollas** 2–3 mm. **Cypselae** 1–2 mm, 4–5-angled or -ribbed, gland-dotted; **pappi** coroniform, 0.2–0.4 mm. $2n = 18$.

Flowering Jul–Sep. Disturbed sites (often moist), abandoned plantings; 10–1600 m; introduced; Alta., B.C., Man., N.B., Nfld. and Labr (Nfld.), N.S., Ont., N.W.T., P.E.I., Que., Sask.; Alaska, Ariz., Ark., Calif., Colo., Conn., Ga., Idaho, Ill., Ind., Iowa, Kans., Ky., La., Maine, Md., Mass., Mich., Minn., Mo., Mont., Nebr., Nev., N.H., N.J., N.Mex., N.Y., N.C., N.Dak., Ohio, Okla., Oreg., Pa., R.I., S.Dak., Tenn., Utah, Vt., Va., Wash., W.Va., Wis., Wyo.; Eurasia; widely introduced in New World and Old World.

Tanacetum vulgare escapes from and/or persists after cultivation. In the flora area, it is naturalized mostly in the northeastern and Pacific Coast states and provinces and sporadically elsewhere.

4. Tanacetum bipinnatum (Linnaeus) Schultz-Bipontinus, Tanaceteen, 48. 1844 • Tanaisie bipennée

Chrysanthemum bipinnatum Linnaeus, Sp. Pl. 2: 890. 1753; *C. bipinnatum* subsp. *huronense* (Nuttall) Hultén; *Tanacetum bipinnatum* subsp. *huronense* (Nuttall) Breitung; *T. camphoratum* Lessing; *T. douglasii* de Candolle; *T. huronense* Nuttall; *T. huronense* var. *bifarium* Fernald; *T. huronense* var. *floccosum* Raup; *T. huronense* var. *johannense* Fernald; *T. huronense* var. *terrae-novae* Fernald

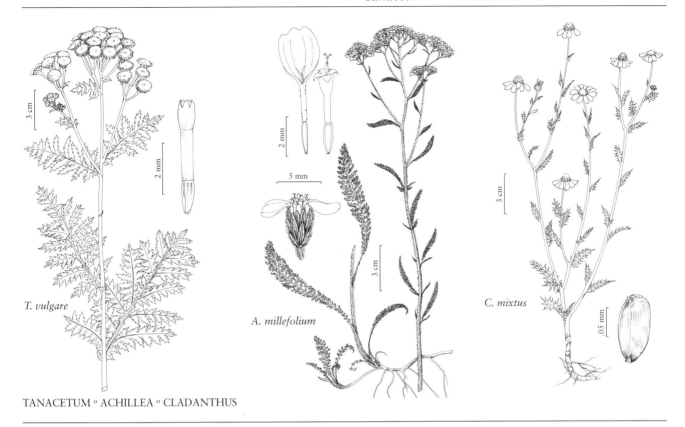

3 cm

2 mm

T. vulgare

2 mm

5 mm

3 cm

A. millefolium

3 cm

C. mixtus

.05 mm

TANACETUM ° ACHILLEA ° CLADANTHUS

Perennials, 5–30(–80) cm. **Stems** (sometimes purple-tinged) 1–2+, ± decumbent to ascending or erect, branched. **Leaves** basal (soon withering) and cauline; petiolate (bases often clasping) or sessile; blades ± ovate or elliptic to obovate or spatulate, mostly 7–25+ × 3–5(–10+) cm, 2–3-pinnately lobed (primary lobes mostly 6–24+ pairs, narrowly oblong to linear-elliptic or linear, lobules oblong or ovate to ± lanceolate, sometimes curled), ultimate margins entire or ± dentate, faces usually ± villous or arachno-villous to lanate, sometimes glabrescent or glabrate, usually gland-dotted (in pits). **Heads** (2–)5–12(–20+) in corymbiform arrays or borne singly. **Involucres** 8–22+ mm diam. **Receptacles** flat to hemispheric. **Ray florets** 8–21+ (pistillate, fertile; corollas pale yellow to yellow, laminae mostly 1–7+ mm, usually 3-lobed) or 0 (heads quasi-radiant or -radiate or ± disciform, peripheral pistillate florets 15–30+; corollas pale yellow, ± zygomorphic, lobes 3–5, abaxial more pronounced). **Disc corollas** (2–)3(–4) mm. **Cypselae** 2–3(–4) mm, weakly 5-ribbed or -angled, gland-dotted; **pappi** coroniform, 0.1–0.5+ mm (entire or erose to lacerate). $2n = 54$.

Flowering May–Sep. Dunes, other sandy sites, calcareous soils, coastal scrub; 0–200+ m; Alta., B.C., Man., N.B., Nfld. and Labr. (Nfld.), Ont., Que., Sask., Yukon; Alaska, Calif., Maine, Mich., Oreg, Wash., Wis.; Eurasia.

The circumscription of *Tanacetum bipinnatum* adopted here includes not only *T. huronense* (see E. Hultén 1941–1950, vol. 10, 1968) but *T. camphoratum* and *T. douglasii* as well (see D. W. Kyhos and P. H. Raven 1982; C. J. Mickelson and H. H. Iltis 1966). Subspecies *bipinnatum* has been distinguished from subsp. *huronense* by having heads borne singly or 2–4 together versus (1–)3–12(–20+) in corymbiform arrays, phyllary margins dark brown versus pale brown, and laminae of ray corollas mostly 3–7 mm versus 1–3 mm. Relatively low plants, 10–20(–40 cm) from dune habitats along the southern shore of Lake Athabasca, Saskatchewan, with mostly 1–4, lanate cauline leaves and 1(–2) heads per flowering stem have been called *T. huronense* var. *floccosum*.

113. ACHILLEA Linnaeus, Sp. Pl. 2: 896. 1753; Gen. Pl. ed. 5, 382. 1754 • Milfoil, achillée [for Greek god Achilles, who is supposed to have used the plants to treat his wounds]

Debra K. Trock

Perennials [subshrubs], 6–80 cm (usually rhizomatous, sometimes fibrous rooted or taprooted; usually aromatic). Stems 1(–4+, clustered), usually erect, branched mostly distally, glabrous or sparsely to densely lanate (hairs usually basifixed). Leaves basal (often withering before flowering) and cauline; alternate; petiolate or sessile (bases ± clasping); blades (cauline equaling basal or slightly smaller distally) linear to oblong-lanceolate, usually 1–2[–4]-pinnately lobed, ultimate margins entire, abaxial faces sparsely to densely lanate, adaxial faces glabrate to sparsely tomentose. Heads radiate [discoid], in compact to open (± flat-topped), simple or compound, corymbiform arrays [borne singly]. Involucres campanulate to hemispheric, mostly 2–3(–5+) mm diam. Phyllaries persistent, 10–30 in (1–)2–3(–4) series, oblong, ovate, or oblanceolate to lanceolate (midribs conspicuous), unequal, margins and apices (pale to black) scarious. Receptacles usually flat to slightly convex, rarely conic, paleate; paleae membranous, ± folded (sometimes each with central resin duct). Ray florets [0] 3–5(–12+), usually pistillate and fertile; corollas usually white (laminae yellow at bases), sometimes pale yellow to pink or purple (tubes ± flattened), laminae orbiculate to suborbiculate (becoming reflexed). Disc florets usually (5–)15–75+, rarely 0, bisexual, fertile; corollas white to grayish or yellowish [yellow, pink], tubes ± flattened (bases ± saccate, clasping apices of cypselae), throats ± campanulate, lobes 5, ± deltate. Cypselae obcompressed, oblong to obovate (margins sometimes winged, apices rounded); ribs usually 2, lateral (sometimes plus 1 adaxial), faces glabrous (pericarps with myxogenic cells, sometimes with resin sacs; embryo sac development monosporic). $x = 9$.

Species ca. 115 (4 in the flora): subtropic to temperate and arctic regions of North America and Eurasia.

Centers of diversity for *Achillea* are in Europe and Asia. *Achillea ageratum*, *A. distans*, and *A. ligustica* have been reported as occurring in North America. Labels on herbarium specimens examined indicated that those reports were based on cultivated plants; there is no evidence that any of the three has become established in our flora. *Achillea filipendulina* may be persistent or established in California (F. Hrusa et al. 2002) and in Michigan (E. Voss 1972–1996, vol. 3).

Achillea includes aromatic herbs with diverse vegetative morphologies. Floral characters show much less variation. Some species are widely cultivated both in Eurasia and North America. Interspecific hybridization has made identifications difficult and has evidently contributed to long lists of synonyms for some species.

Plants of *Achillea* contain secondary metabolites with purported therapeutic and pharmacologic uses. Native Americans used the plants to treat earaches, diarrhea, and hemorrhages.

SELECTED REFERENCES Clausen, J., D. D. Keck, and W. M. Hiesey. 1948. Experimental studies on the nature of species. III. Environmental responses of climatic races of *Achillea*. Publ. Carnegie Inst. Wash. 581. Pollard, C. L. 1899. The genus *Achillea* in North America. Bull. Torrey Bot. Club, 26: 365–372. Tyrl, R. J. 1975. Origin and distribution of polyploid *Achillea* (Compositae) in western North America. Brittonia 27: 187–196.

1. Leaf blades 1–2-pinnately lobed (lobes of single leaves often arrayed in multiple planes).
   2. Phyllaries 20–30 in 3 series; ray florets (3–)5–8, laminae 1.5–3 × 1.5–3 mm; cypselae 1–2 mm. . . . . 1. *Achillea millefolium*
   2. Phyllaries 10–13 in (1–)2 series; ray florets 8–10(–13), laminae 1–1.5 × 2–2.5 mm; cypselae 0.75–1 mm . . . . 2. *Achillea nobilis*

1. Leaf blades not lobed (margins usually serrulate, rarely subentire or serrate to doubly serrate).
    3. Ray laminae 4–5 mm; disc florets 45–75+; leaf margins usually serrulate, rarely subentire .... 3. *Achillea ptarmica*
    3. Ray laminae 1–3 mm; disc florets 25–30+; leaf margins serrate or doubly serrate (teeth antrorse) .... 4. *Achillea alpina*

1. Achillea millefolium Linnaeus, Sp. Pl. 2: 899. 1753

• Common yarrow, achillée millefeuille, herbe-à-dinde F I

Achillea alpicola (Rydberg) Rydberg; *A. arenicola* A. Heller; *A. borealis* Bongard subsp. *arenicola* (A. Heller) D. D. Keck; *A. borealis* subsp. *californica* (Pollard) D. D. Keck; *A. californica* Pollard; *A. gigantea* Pollard; *A. lanulosa* Nuttall; *A. lanulosa* subsp. *alpicola* (Rydberg) D. D. Keck; *A. laxiflora* Pollard & Cockerell; *A. megacephala* Raup; *A. millefolium* var. *alpicola* (Rydberg) Garrett; *A. millefolium* var. *arenicola* (A. Heller) Nobs; *A. millefolium* var. *asplenifolia* (Ventenat) Farwell; *A. millefolium* subsp. *borealis* (Bongard) Breitung; *A. millefolium* var. *borealis* (Bongard) Farwell; *A. millefolium* var. *californica* (Pollard) Jepson; *A. millefolium* var. *gigantea* (Pollard) Nobs; *A. millefolium* subsp. *lanulosa* (Nuttall) Piper; *A. millefolium* var. *lanulosa* (Nuttall) Piper; *A. millefolium* var. *litoralis* Ehrendorfer ex Nobs; *A. millefolium* var. *maritima* Jepson; *A. millefolium* var. *megacephala* (Raup) B. Boivin; *A. millefolium* var. *nigrescens* E. Meyer; *A. millefolium* var. *occidentalis* de Candolle; *A. millefolium* var. *pacifica* (Rydberg) G. N. Jones; *A. millefolium* var. *puberula* (Rydberg) Nobs; *A. nigrescens* (E. Meyer) Rydberg; *A. occidentalis* (de Candolle) Rafinesque ex Rydberg; *A. pacifica* Rydberg; *A. puberula* Rydberg; *A. rosea* Desfontaines; *A. subalpina* Greene

Perennials, 6–65+ cm (usually rhizomatous, sometimes stoloniferous). **Stems** 1(–4), erect, simple or branched, densely lanate-tomentose to glabrate. **Leaves** petiolate (proximally) or sessile (distally, weakly clasping and gradually reduced); blades oblong or lanceolate, 3.5–35+ cm × 5–35 mm, 1–2-pinnately lobed (ultimate lobes ± lanceolate, often arrayed in multiple planes), faces glabrate to sparsely tomentose or densely lanate. **Heads** 10–100+, in simple or compound, corymbiform arrays. **Phyllaries** 20–30 in ± 3 series, (light green, midribs dark green to yellowish, margins green to light or dark brown) ovate to lanceolate, abaxial faces tomentose. **Receptacles** convex; paleae lanceolate, 1.5–4 mm. **Ray florets** (3–)5–8, pistillate, fertile; corollas white or light pink to deep purple, laminae 1.5–3 × 1.5–3 mm. **Disc florets** 10–20; corollas white to grayish white, 2–4.5 mm. **Cypselae** 1–2 mm (margins broadly winged). $2n$ = 18, 27, 36, 45, 54, 63, 72 (including counts from Europe).

Flowering late Apr–early Jul (south), mid Jul–mid Sep (north). Pastures, meadows, roadsides, stream sides, woodlands, waste grounds, dry or sandy soils, also in damp, clayey, and salty soils; 0–3600 m; Greenland; St. Pierre and Miquelon; Alta., B.C., Man., N.B., Nfld. and Labr., N.W.T., N.S., Nunavut, Ont., P.E.I., Que., Sask., Yukon; Ala., Alaska, Ariz., Ark., Calif., Colo., Conn., Del., D.C., Fla., Ga., Idaho, Ill., Ind., Iowa, Kans., Ky., La., Maine, Md., Mass., Mich., Minn., Miss., Mo., Mont., Nebr., Nev., N.H., N.J., N.Mex., N.Y., N.C., N.Dak., Ohio, Okla., Oreg., Pa., R.I., S.C., S.Dak., Tenn., Tex., Utah, Vt., Va., Wash., W.Va., Wis., Wyo.; Mexico; Eurasia.

Achillea millefolium is morphologically variable and has been treated as either a single species with varieties or as multiple distinct species. At least 58 names have been used for North American specimens. Some early workers (e.g., J. Clausen et al. 1948) thought the native North American plants were taxonomically distinguishable from introduced, Old World plants. Other workers (e.g., R. J. Tyrl 1975) have treated *A. millefolium* as a cosmopolitan, Northern Hemisphere polyploid complex of native and introduced plants that have hybridized, forming diploid, tetraploid, pentaploid, hexaploid, septaploid, and octoploid plants and/or populations constituting a single, variable species.

Morphologic characters that have been used to segregate these populations into species and/or varieties include: (1) degree and persistence of tomentum; (2) phyllaries with greenish, light brown, or dark brown margins; (3) shapes of capitulescences (rounded or flat-topped); and (4) degrees of leaf dissection and shapes of lobes.

While examining specimens for this treatment, two general trends were noted: (1) Plants growing either at high latitudes or high elevations tend to have darker colored margins on the phyllaries. (2) Plants at high latitudes or elevations or from extreme desert locations tend to be more densely lanate than plants from less extreme habitats. These are only trends; variations in local populations due to local environmental conditions are to be expected.

An eco-morphotype adapted to the Athabasca sand dunes of northern Saskatchewan has been known as *A. megacephala* or *A. millefolium* var. *megacephala* and has been treated as a taxon of special concern in Canada (V. L. Harms 1999).

2. Achillea nobilis Linnaeus, Sp. Pl. 2: 899. 1753

• Noble yarrow   I

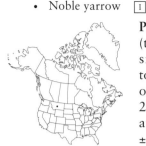

Perennials, (10–)15–60 cm (taprooted). **Stems** 1(–4), erect, simple, glabrous or sparsely tomentose. **Leaves** sessile; blades ovate, 1.5–3 cm × 10–15 mm, 1–2-pinnately lobed, (bases auriculate, slightly clasping) faces ± hairy. **Heads** 30–100+, in simple or compound, corymbiform arrays. **Phyllaries** 10–13+ in (1–)2 series, (midribs light green) oblong (apices papillose, translucent), abaxial faces glabrous. **Receptacles** flat to slightly convex; paleae narrowly oblong, 2–2.5 mm. **Ray florets** 8–10(–13), pistillate, fertile; corollas white, laminae 1–1.5 × 2–2.5 mm. **Disc florets** 10–25+; corollas grayish white, 1.5–2.5 mm. **Cypselae** 0.75–1 mm. $2n$ = 18, 27.

Flowering mid Jul–early Sep. Pastures, disturbed roadsides; 900–1500 m; introduced; Mont.; Europe.

Achillea nobilis was introduced in the flora area through cultivation. It has been reported in Minnesota, Massachusetts, and New York; herbarium sheets indicate that those reports were based on cultivated specimens. It was first collected in Montana in 1902; it appears to have become established in that state, where it is known from Lake and Flathead counties.

3. Achillea ptarmica Linnaeus, Sp. Pl. 2: 898. 1753

• Sneezeweed, sneezewort, achillée ptarmique, herbe-à-éternuer   I

Perennials, 30–60+ cm (rhizomatous). **Stems** 1, erect, branched distally, proximally glabrate, distally villous or tomentose. **Leaves** sessile; blades linear to narrowly lanceolate, 3–10 × 3–5 mm, (bases slightly clasping, margins usually serrulate, rarely subentire), faces glabrous or sparsely hairy on midveins adaxially. **Heads** 3–15+, in simple or compound, corymbiform arrays. **Phyllaries** 25–30+ in ± 3 series, (light green, midribs yellowish, margins light brown) lanceolate to oblanceolate, faces tomentose. **Receptacles** flat to slightly convex; paleae oblanceolate, 3–4 mm. **Ray florets** 8–10(–13), styliferous and sterile (45–70, pistillate, fertile, in horticultural doubles);

corollas white, laminae suborbiculate, 4–5 × 4–5 mm (5–7 × 3.5–4.5 mm in doubles). **Disc florets** 45–75+ (sometimes 0 in doubles); corollas grayish white, ca. 3 mm. **Cypselae** 1.5–2 mm. $2n$ = 18.

Flowering late Jun–mid Sep. Roadsides, disturbed sites, open fields and pastures, in sandy or gravelly soils or in moist to drying silty soils; 0–2400 m; introduced; Greenland; Alta., Man., N.B., Nfld. and Labrador (Nfld.), N.S., Ont., P.E.I., Que.; Alaska, Colo., Conn., Idaho, Ind., Maine, Mass., Mich., Minn., Mo., Mont., N.H., N.J., N.Y., N.Dak., Ohio, Pa., R.I., Vt., Wash., W.Va., Wis.; Eurasia.

Achillea ptarmica is naturalized from Eurasia. "Double-flowered" plants originated as cultivars; apparently, they persist outside of cultivation.

4. Achillea alpina Linnaeus, Sp. Pl. 2: 899. 1753

• Siberian yarrow, achillée de Sibérie

Achillea sibirica Ledebour

Perennials, 50–80 cm (fibrous-rooted and rhizomatous). **Stems** 1, erect, branched or unbranched distally, sparsely villous to glabrate. **Leaves** sessile; blades linear-lanceolate to oblong-lanceolate, 5–10 cm × 4–8 mm, (margins serrate to doubly serrate, teeth antrorse) faces sparingly villous or glabrate. **Heads** 10–25+, in crowded, simple or compound, corymbiform arrays. **Phyllaries** 20–30 in ± 3 series, (light green, margins light to dark brown, midribs dark green or yellow-green) lanceolate to oblanceolate, faces (abaxial) sparingly tomentose. **Receptacles** convex; paleae oblong, 3.5–4.5 mm (apices dark, rounded). **Ray florets** 6–8 (–12), pistillate, fertile; corollas white, laminae 1–3 × 2–3 mm. **Disc florets** 25–30+; corollas grayish or yellowish white, 2–3 mm. **Cypselae** 2.5 mm. $2n$ = 36.

Flowering early Jul–early Sep. Meadows, forest edges, roadsides, lakeshores, along streams, moist soils; 100–600 m; Alta., B.C., Man., N.W.T., Ont., Que., Sask., Yukon; Alaska, Minn., N.Dak.; Asia.

Achillea alpina has been reported (as *A. sibirica*) as occurring in New Jersey and Missouri. Specimens examined from those states were from plants cultivated in botanical gardens; there is no evidence that *Achillea alpina* has escaped in those states.

114. CLADANTHUS Cassini, Bull. Sci. Soc. Philom. Paris 1816: 199. 1816 • [Derivation not given; possibly Greek *klados*, branch, and *anthos*, flower, alluding to branching of stems at bases of sessile heads in original species] ☐

Linda E. Watson

Ormenis (Cassini) Cassini

Annuals [perennials, subshrubs], 10–60+ cm (usually aromatic). **Stems** usually 1, usually erect [prostrate], branched [immediately proximal to sessile, terminal heads], puberulent or villous to arachnose (hairs basifixed), glabrescent. **Leaves** mostly cauline; alternate; petiolate (proximal) or sessile (distal); blades obovate or spatulate to oblong or linear, 1–2(–3)-pinnately lobed (ultimate lobes ± linear to filiform), ultimate margins entire or dentate, faces villous to arachnose, glabrescent. **Heads** radiate, borne singly or in lax, corymbiform arrays. **Involucres** hemispheric or broader, 5–8[–12+] mm diam. **Phyllaries** persistent, 16–24+ in 2–3+ series, lance-linear or lanceolate to oblong or obovate, subequal, margins and apices (hyaline) scarious (apices ± dilated, rounded, abaxial faces ± villous or arachnose, glabrescent). **Receptacles** hemispheric to narrowly columnar or conic, paleate; paleae ± folded (carinate, each with central, red-brown resin duct). **Ray florets** 12–18+, neuter or styliferous and sterile; corollas orange, yellow, or white with yellow bases, laminae ± oblong (spreading to reflexed, ± marcescent). **Disc florets** 40–150[–200+], bisexual, fertile; corollas orange or yellow, tubes ± cylindric (bases saccate, each obliquely spurred, adaxially clasping distal 0.5+ of cypsela), throats campanulate to funnelform, lobes 5, deltate (apices minutely crested or dilated). **Cypselae** ± obovoid (apices oblique), weakly flattened (stylopodia sublateral), ribs or nerves (weak): 2 lateral, 1 adaxial, faces finely striate, glabrous [hairy] (pericarps with myxogenic cells in longitudinal rows, without resin sacs); **pappi** 0. $x = 9$.

Species ca. 5 (1 in the flora): introduced; s Europe, sw Asia, n Africa; introduced also in South America, elsewhere in Old World.

SELECTED REFERENCE Oberprieler, C. and R. Vogt. 2002. *Cladanthus*. In: Med Checklist notulae, 21. Willdenowia 32: 195. 208.

1. **Cladanthus mixtus** (Linnaeus) Chevallier, Fl. Gén. Env. Paris ed. 2, 2: 576. 1836 F ☐

Anthemis mixta Linnaeus, Sp. Pl. 2: 894. 1753; *Chamaemelum mixtum* (Linnaeus) Allioni; *Ormenis mixta* (Linnaeus) Dumortier

Stems often puberulent. **Leaves:** blades 20–60(–80+) × 3–25(–35+) mm, lobes lanceolate to linear or filiform, apices apiculate. **Phyllaries** greenish, faces arachnose to villous. **Ray laminae** 5–6(–10) mm. **Disc corollas** 2–2.5 mm (including spurs). **Cypselae** 0.8–1.3 mm. $2n = 18$ (Europe, Africa).

Flowering Jun. Disturbed sites; 0–200 m; introduced; B.C.; Fla., N.J., N.Y., N.C., Ohio, Oreg., Pa.; Europe.

Cladanthus mixtus is sporadically encountered in the flora area.

115. CHAMAEMELUM   Miller, Gard. Dict. Abr. ed. 4, vol. 1. 1754 • [Greek *chamae*-, on the ground, lowly, creeping, and *melon*, orchard, alluding to common habitat] ☐

Linda E. Watson

Annuals or perennials, 5–20(–35+) cm, (aromatic). **Stems** usually 1, erect, ascending, or prostrate, usually branched, glabrous or glabrate, puberulent, or villous to strigoso-sericeous (hairs basifixed). **Leaves** mostly cauline (at flowering); alternate; petiolate or sessile; blades oblong, ovate, elliptic, or spatulate, 1–3-pinnately lobed (ultimate lobes narrowly spatulate to linear or filiform, apices apiculate), ultimate margins entire, faces glabrous or glabrate, puberulent, or villous to strigoso-sericeous. **Heads** radiate or discoid, borne singly or in lax corymbiform arrays. **Involucres** hemispheric or broader, 6–10 mm diam. **Phyllaries** persistent, 22–45+ in 3–4+ series (sometimes reflexed in fruit), mostly ovate to oblong, unequal, margins and apices (colorless, brownish, or greenish) scarious. **Receptacles** hemispheric to conic, paleate; paleae weakly navicular to ± flat (medially chartaceous, margins scarious, apices rounded). **Ray florets** 0 or 12–21+, pistillate and fertile or styliferous and sterile; corollas white, laminae oblong (often marcescent, reflexed in fruit). **Disc florets** 100–200+, bisexual, fertile; corollas yellow, tubes ± cylindric (somewhat dilated, bases saccate, weakly clasping apices of cypselae), throats funnelform, lobes 5, deltate. **Cypselae** ± obovoid, weakly obcompressed, ribs or nerves (weak): 2 lateral, 1 adaxial, faces finely striate, glabrous (pericarps with myxogenic cells in longitudinal rows, without resin sacs); **pappi** 0. $x = 9$.

Species 2 (2 in the flora): introduced; s, w Europe, n Africa.

1. Annuals; stems erect or ascending, glabrous or puberulent; margins and apices of phyllaries brownish . . . . 1. *Chamaemelum fuscatum*
1. Perennials; stems mostly prostrate (much branched, often forming mats), ± strigoso-sericeous to villous; margins and apices of phyllaries greenish or lacking pigment . . . . 2. *Chamaemelum nobile*

1. Chamaemelum fuscatum (Brotero) Vasconcellos, Anais Inst. Vinho Porto 20: 276. 1966

• Chamomile ☐

Anthemis fuscata Brotero, Fl. Lusit. 1: 394. 1804

Annuals, 5–20(–35+) cm. **Stems** erect to ascending (simple or branched, not forming mats), glabrous or puberulent. **Leaves:** proximal petiolate, ± ovate to elliptic, 1–4 cm, usually 2-pinnately lobed; distal sessile, ± elliptic, 1–2 cm, 0–1-pinnately lobed. **Involucres** 3–4 × 6–10 mm. **Phyllaries:** margins and apices brownish, abaxial faces glabrous or glabrate. **Paleae** 2–3 mm, margins brownish. **Ray florets** 12–15+; laminae 8–15 mm. **Disc corollas** 2.5–3 mm. **Cypselae** 1–1.3 mm. $2n = 18$.

Flowering Mar–Apr. Disturbed sites; 10–100 m; introduced; Calif.; Europe.

Chamaemelum fuscatum is found on the Outer North Coast ranges, especially in vineyards. It is native to the Mediterranean.

2. Chamaemelum nobile (Linnaeus) Allioni, Fl. Pedem. 1: 185. 1785 ☐ ☐

Anthemis nobilis Linnaeus, Sp. Pl. 2: 894. 1753

Perennials, 10–20(–30) cm across. **Stems** mostly prostrate (much branched, often forming mats), ± strigoso-sericeous to villous. **Leaves** sessile; blades oblong, 1–3(–5) cm, 2–3-pinnately lobed. **Involucres** 4–6 × 7–10+ mm. **Phyllaries:** margins and apices greenish or lacking pigment, abaxial faces ± villous. **Paleae** 3–4+ mm, margins greenish or lacking pigment. **Ray florets** usually 13–21+, rarely 0; laminae 7–10+ mm. **Disc corollas** 2–3 mm. **Cypselae** 1–1.5 mm. $2n = 18$.

Flowering Jul–Aug. Disturbed sites; 10–300 m; introduced; Calif., Conn., Del., Ill., Ind., Md., N.J., N.Y., N.C., Ohio, Wis.; Europe.

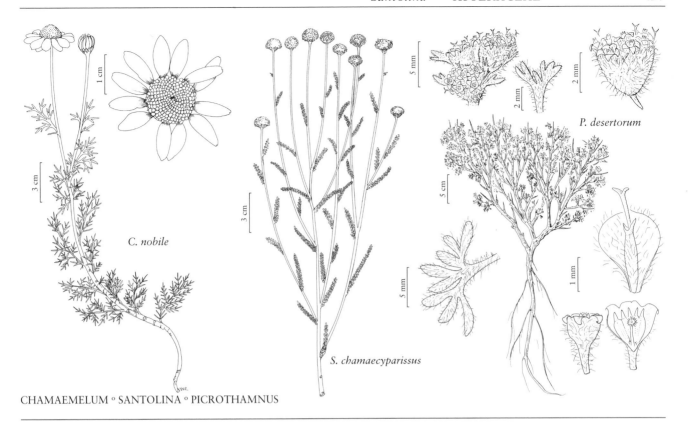

C. nobile

S. chamaecyparissus

P. desertorum

CHAMAEMELUM ∘ SANTOLINA ∘ PICROTHAMNUS

116. SANTOLINA Linnaeus, Sp. Pl. 2: 842. 1753; Gen. Pl. ed. 5, 365. 1754 • [Latin *sanctus*, holy, and *linum*, flax; ancient name for a species of the genus] ☐

Linda E. Watson

Subshrubs, [5–]10–60 cm (sometimes ± rhizomatous; aromatic). Stems usually 1, ± erect [decumbent to ascending], branched mostly from bases [± throughout], often tomentose to lanate, sometimes glabrate or glabrous [sericeous] (hairs mostly medifixed) and gland-dotted. Leaves cauline [mostly basal]; alternate; petiolate or sessile; blades narrowly oblong or spatulate to linear, usually 1-pinnately lobed (lobes usually crowded and overall effect ± vermiform), ultimate margins entire or ± crenate, faces usually arachnose or tomentose to lanate, sometimes glabrate or glabrous [sericeous]. Heads discoid, borne singly (pedunculate). Involucres campanulate to hemispheric or broader, [3–]6–10[–12+] mm diam. Phyllaries persistent (soon indurate), 18–45+ in 3(–5+) series, distinct, lanceolate to elliptic (usually carinate), unequal, margins and apices (usually light to dark brown, sometimes purple) scarious (abaxial faces glabrous or ± arachnose, glabrescent). Receptacles convex to hemispheric, paleate; paleae ± lanceolate, ± navicular (each with central resin duct). Ray florets 0. Disc florets mostly 60–250+, bisexual, fertile; corollas pale to bright or deep yellow [whitish], tubes often compressed (± winged and/or saccate, ± clasping apices of cypselae), throats ± funnelform, lobes 5, lance-ovate. Cypselae obconic to obovoid, sometimes slightly obcompressed, 3–5-angled, faces glabrous (pericarps sometimes with myxogenic cells, without resin sacs); pappi 0. $x = 9$.

Species 8–12+ (1 in the flora): introduced; s Europe, n Africa.

1. Santolina chamaecyparissus Linnaeus, Sp. Pl. 2: 842. 1753 • Lavender-cotton F I

Subshrubs, silvery-gray to white. **Leaf blades** 1-pinnately lobed, 10–20(–40) × 1–3(–5) mm. **Peduncles** 3–6 cm. **Phyllaries** carinate, apices of inner rounded, ± lacerate to fimbrillate. **Corollas** 3–4 mm, gland-dotted. **Cypselae** 2.5–3 mm, angles sometimes narrowly winged. $2n = 18$.

Flowering Mar–Oct. Disturbed sites, abandoned plantings; 0–400 m; introduced; Ark., Calif., Ga., La., Mass., N.C., S.C.; s Europe; n Africa.

Santolina chamaecyparissus is widely cultivated and probably persists in the flora area in states other than those listed here.

117. PICROTHAMNUS Nuttall, Trans. Amer. Philos Soc., n. s. 7: 417. 1841 • Budsage [Greek *picro-*, bitter, and *thamnos*, bush, alluding to bitterness of the plants] E

Leila M. Shultz

Subshrubs or shrubs, 5–30(–50) cm (strongly aromatic). **Stems** 1–10+, usually erect, diffusely branched from bases and throughout (some laterals persistent, forming thorns), villous to arachnose (hairs medifixed). **Leaves** mostly cauline; alternate; petiolate or sessile; blades ± orbiculate to flabellate, simple or 1–2-pedately lobed (lobes orbiculate to spatulate or linear), ultimate margins entire, faces ± villous and gland-dotted. **Heads** disciform, usually (2–12+) in ± leafy, racemiform to spiciform arrays, rarely borne singly. **Involucres** ± obconic, 2–3(–5) mm diam. **Phyllaries** persistent, 5–8 in ± 2 series, distinct, ± obovate, subequal, margins and apices (hyaline) narrowly scarious. **Receptacles** convex (glabrous), epaleate. **Ray florets** 0 (peripheral pistillate florets 2–8; corollas pale yellow, ± filiform, ± villous). **Disc florets** 5–13(–15), functionally staminate; corollas pale yellow (± villous), tubes ± cylindric, throats campanulate, lobes 5, ± deltate. **Cypselae** (brown) obovoid to ellipsoid, ribs 0, faces ± villous and obscurely nerved (pericarps without myxogenic cells or resin sacs); **pappi** 0. $x = 9$.

Species 1: w North America.

Separation of *Picrothamnus* from *Artemisia* calls attention to differing views of generic circumscription within Anthemideae (see discussion under *Artemisia*). Distinguished by its spinescent branches and relatively large heads held among the leaves, *Picrothamnus* is among the more distinct of proposed segregates. The diffuse-porous woods (S. Carlquist 1966) correspond to the early spring-blooming phenology of the plants and provide an anatomic feature that helps to distinguish *Picrothamnus* from *Artemisia*.

1. **Picrothamnus desertorum** Nuttall, Trans. Amer. Philos. Soc., n. s. 7: 417. 1841 E F

Artemisia spinescens D. C. Eaton

Stems diffusely branched from bases, some laterals ending in spinelike tips. **Leaf blades or lobes** orbiculate to linear, 1–5(–20) × 1–5(–20) mm. **Phyllaries** whitish green. **Cypselae** 1–1.5 mm. $2n$ = 18, 36.

Flowering Apr–Jun. Arid slopes and valleys, sands or clays, sometimes saline soils; 1800–2200 m; Ariz., Calif., Colo., Idaho, Mont., Nev., N.Mex., Oreg., Utah, Wyo.

Budsage provides nutritious forage for wildlife and domestic sheep in winter; it can be poisonous or fatal to calves and lambs, if consumed in great quantity during spring months.

118. **SPHAEROMERIA** Nuttall, Trans. Amer. Philos. Soc., n. s. 7: 401. 1841 • False sagebrush, chickensage [Greek *sphaira*, sphere, and *meros*, a part, alluding to the capitate arrays of heads in *S. capitata*]

Timothy K. Lowrey

Leila M. Shultz

Perennials or subshrubs, (1–)5–60(–70) cm (usually aromatic). **Stems** 1–5+ (with interxylary cork), lax to erect, branched from bases or throughout, glabrous or sericeous to tomentose (hairs medifixed), usually gland-dotted as well. **Leaves** mostly basal (sometimes in rosettes) or basal and cauline; alternate; petiolate or sessile; blades obovate to cuneate or linear, usually 1–2-pinnately or -palmately lobed (ultimate lobes oblanceolate to linear) or apically 3-toothed or -lobed, ultimate margins entire, faces glabrous or ± sericeous to tomentose and usually gland-dotted. **Heads** disciform, borne singly or (2–60+) in usually corymbiform, rarely paniculiform, arrays or in subcapitate clusters. **Involucres** hemispheric to campanulate, 3–5(–12) mm diam. **Phyllaries** persistent, (8–)12–20+ in 2–3+ series, mostly obovate, unequal to subequal, margins and apices (colorless, pinkish, or brownish) scarious. **Receptacles** conic to nearly flat (villous in *S. potentilloides*), epaleate. **Ray florets** 0 (peripheral pistillate florets 4–15+; corollas ± filiform, lobes usually 3). **Disc florets** 30–50+, bisexual, fertile; corollas usually bright yellow (ochroleucous in *S. cana*), tubes ± cylindric, throats ± campanulate, lobes 5, ± deltate (tips glabrous or hairy). **Cypselae** columnar or obovoid to obconic, ribs 2–3 or 5–10, faces glabrous or gland-dotted (pericarps with myxogenic cells only in *S. potentilloides*, without resin sacs); **pappi** usually 0 (of 3–5 subulate scales in *S. compacta*). x = 9.

Species 9 (8 in the flora): w United States, nw Mexico.

SELECTED REFERENCE Holmgren, A. H., L. M. Shultz, and T. K. Lowrey. 1976. *Sphaeromeria*, a genus closer to *Artemisia* than to *Tanacetum* (Asteraceae: Anthemideae). Brittonia 28: 255–262.

1. Receptacles villous; cypselae becoming mucilaginous when wet . . . . 1. *Sphaeromeria potentilloides*
1. Receptacles glabrous; cypselae not mucilaginous when wet.
    2. Subshrubs, 20–70 cm; leaves mostly cauline (not forming basal clusters).
        3. Corollas ochroleucous, lobes villous . . . . 2. *Sphaeromeria cana*
        3. Corollas yellow, lobes glabrous or gland-dotted.
            4. Leaves glabrous; heads 5–20 in corymbiform or subumbelliform arrays (as wide as or wider than long) . . . . 3. *Sphaeromeria diversifolia*
            4. Leaves tomentose; heads (8–)10–30(–60) in paniculiform arrays (longer than wide) . . . . 4. *Sphaeromeria ruthiae*

[2. Shifted to left margin.—Ed.]
2. Perennials, 1–20 cm (often cespitose); leaves mostly basal (commonly forming clusters).
   5. Heads 1–7 (*S. argentea*) or 8–20.
      6. Leaf blades (± cuneate, 7–15 mm) entire or apices 3(–5)-toothed or -lobed, faces silvery-canescent; heads usually in subcapitate to corymbiform arrays (1–1.5 cm across), sometimes borne singly . . . . 5. *Sphaeromeria argentea*
      6. Leaf blades (± cuneate, 8–20 mm) usually 1–2-pinnati-palmately lobed (ultimate lobes ± linear), faces ± tomentose; heads in tight, capitate arrays (1–1.3 cm across) . . . . 6. *Sphaeromeria capitata*
   5. Heads 1–3(–5).
      7. Leaves: blades (10–25 mm) pinnati-palmately lobed (lobes 3–6+, ± oblanceolate to linear) or linear, faces sericeous (pappi of 3–5, subulate scales) . . . . 7. *Sphaeromeria compacta*
      7. Leaves: blades (15–30 × 1.5–8 mm) pinnati-palmately lobed (lobes 2–3, linear, 1–2 mm wide) or linear, faces silvery-canescent . . . . 8. *Sphaeromeria simplex*

1. **Sphaeromeria potentilloides** (A. Gray) A. Heller, Muhlenbergia 1: 7. 1900 • Fivefinger chickensage, cinquefoil false sagebrush E F

Artemisia potentilloides A. Gray, Proc. Amer. Acad. Arts 6: 551. 1866; *Tanacetum potentilloides* (A. Gray) A. Gray; *Vesicarpa potentilloides* (A. Gray) Rydberg

Perennials, 3–30 cm. **Stems** lax. **Leaves** basal and cauline; basal (20–70 mm) 2-pinnately lobed (ultimate lobes linear to filiform, ca. 1 mm wide); cauline (5–18 mm) usually pinnately lobed, faces silky-tomentose. **Heads** 2–20 in corymbiform to paniculiform arrays or borne singly. **Involucres** 2.3–5.2 mm. **Phyllaries** 8–10, hairy. (**Receptacles** villous, hairs white, curly, 0.5–1 mm.) **Cypselae** 0.8–1.5 mm (faces obscurely nerved, becoming mucilaginous when wet).

Varieties 2 (2 in the flora): w United States.

1. Heads (3–)4–20; involucres 3–4(–4.5) mm; basal leaves 2-pinnately lobed . . . .
   . . . . 1a. *Sphaeromeria potentilloides* var. *potentilloides*
1. Heads 1–3(–4); involucres 4–5.1 mm; basal leaves 1-pinnately lobed . . . .
   . . . . 1b. *Sphaeromeria potentilloides* var. *nitrophila*

1a. **Sphaeromeria potentilloides** (A. Gray) A. Heller var. **potentilloides** E

Plants 10–30 cm. **Leaves** basal and cauline; basal mostly 2-pinnately lobed; cauline usually pinnately lobed, rarely 3-fid. **Heads** (3–)4–20. **Involucres** 3–4(–4.5) mm. $2n = 18$.

Flowering May–Jul. Wet, ± alkaline meadows, flats, seepage areas; 1300–2100 m; Calif., Nev., Oreg.

1b. **Sphaeromeria potentilloides** (A. Gray) A. Heller var. **nitrophila** (Cronquist) A. H. Holmgren, L. M. Shultz & Lowrey, Brittonia 28: 261. 1976 E F

Tanacetum potentilloides (A. Gray) A. Gray var. *nitrophilum* Cronquist, Leafl. W. Bot. 6: 49. 1950; *Vesicarpa potentilloides* (A. Gray) Rydberg var. *nitrophilum* (Cronquist) Kartesz

Plants 3–15 cm. **Leaves** mostly basal; basal mostly 1-pinnately lobed; cauline (2–)3-fid or entire. **Heads** 1–3(–4). **Involucres** 4–5.1 mm. $2n = 18$.

Flowering May–Jul. Highly alkaline sites; 1600–2200 m; Idaho, Nev.

Variety *nitrophila* grows in northern and central Nevada and in the hills north of Snake River Plains in Idaho.

2. **Sphaeromeria cana** (D. C. Eaton) A. Heller, Muhlenbergia 1: 7. 1900 • Gray chickensage E F

Tanacetum canum D. C. Eaton in S. Watson, Botany (Fortieth Parallel), 179, plate 19, figs. 8–14. 1871

Subshrubs (flowering twigs dying back each year), (5–)15–60 cm (aromatic). **Stems** erect, sericeous-canescent. **Leaves** mostly cauline; blades (linear or linear-oblanceolate, 10–40 mm) entire or pinnately or subpalmately lobed, faces ± sericeous. **Heads** usually 3–12 in compact, corymbiform cymes (1–3 cm across), sometimes 1–2. **Involucres** 2.5–4(–6) mm. **Phyllaries** 10–12, canescent. (**Disc corollas** ochroleucous, lobes villous.) **Cypselae** 1.8–1.9 mm (apices with irregular, thickened rims, faces glabrous or slightly gland-dotted). $2n = 18$.

Flowering Jul–Sep. Crevices in cliffs, dry rocky slopes; 1800–4000 m; Calif., Nev., Oreg.

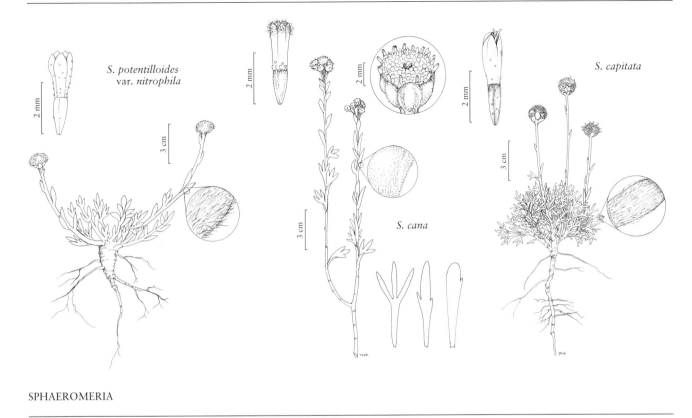

S. potentilloides var. nitrophila

2 mm

3 cm

2 mm

2 mm

S. cana

3 cm

2 mm

S. capitata

3 cm

SPHAEROMERIA

Sphaeromeria cana grows in western United States, with northern extension in the Steens Mountains of Oregon.

3. **Sphaeromeria diversifolia** (D. C. Eaton) Rydberg in N. L. Britton et al., N. Amer. Fl. 34: 242. 1916 · Separateleaf chickensage or false sagebrush [E]

Tanacetum diversifolium D. C. Eaton in S. Watson, Botany (Fortieth Parallel), 180, plate 19, figs. 1–7. 1871

Subshrubs, 15–40 cm (aromatic). **Stems** erect, glabrous. **Leaves** mostly cauline; blades (green, mostly linear, 15–80 mm) entire or irregularly toothed or lobed, faces glabrous. **Heads** 5–20(–30) in corymbiform or subumbelliform arrays (1.3–5.6 cm across, as wide as or wider than long). **Involucres** 2–4 mm. **Phyllaries** 8–12(–14), glabrate or sparsely hairy. **Cypselae** 0.5–1 mm (apices with irregular, thickened rims, faces glabrous or gland-dotted).

Flowering Jul–Sep. Crevices of cliffs, steep rocky slopes, usually on limestone, sometimes on quartzite; 1700–2700 m; Nev., Utah.

4. **Sphaeromeria ruthiae** A. H. Holmgren, L. M. Shultz & Lowrey, Brittonia 28: 257, fig. 1. 1976 · Zion chickensage or tansy [E]

Subshrubs, 30–60(–70) cm (aromatic). **Stems** erect to ± decumbent at bases, tomentose. **Leaves** mostly cauline; blades (20–90 mm) 1-pinnately lobed to 3-fid (proximal) or entire (distal), faces tomentose-canescent and gland-dotted. **Heads** (8–)10–30 (–60) in paniculiform arrays (longer than wide). **Involucres** 4–4.5(–5) mm. **Phyllaries** 12–16, densely tomentose. **Cypselae** 1.5–2 mm (obscurely 10-ribbed, apices with thickened rims, faces gland-dotted).

Flowering Sep. Crevices in sandstone cliffs; 1900–2000 m; Utah.

5. Sphaeromeria argentea Nuttall, Trans. Amer. Philos. Soc., n. s. 7: 402. 1841 • Silver chickensage, Nuttall's false sagebrush E

Tanacetum nuttallii Torrey & A. Gray

Perennials, 5–20 cm (cespitose). **Stems** erect, canescent. **Leaves** mostly basal; blades (± cuneate, 7–15 mm) entire or apices 3(–5)-toothed or -lobed, faces silvery-canescent. **Heads** usually 2–7 in subcapitate to corymbiform arrays (1–1.5 cm across), sometimes borne singly. **Phyllaries** 12–16, sparsely tomentose. **Involucres** 3–5 (–7) mm. **Cypselae** 2–2.5 mm (3–5-ribbed, apices with thickened rims, faces glabrous or sparsely gland-dotted). $2n = 18$.

Flowering Jun–Jul. Dry hills; 2000–2600 m; Colo., Idaho, Nev., Wyo.

Sphaeromeria argentea is moderately tolerant of alkali.

6. Sphaeromeria capitata Nuttall, Trans. Amer. Philos. Soc., n. s. 7: 402. 1841 • Rock tansy E F

Tanacetum capitatum (Nuttall) Torrey & A. Gray

Perennials, 5–20 cm (cespitose, bases sometimes woody). **Stems** erect, ± tomentose. **Leaves** mostly basal; blades (± cuneate, 8–20 mm) usually 1–2-pinnati-palmately lobed (ultimate lobes ± linear), faces ± tomentose. **Heads** usually 8–20 in tight, capitate arrays (1–1.3 cm across), sometimes with solitary heads proximal to main arrays. **Involucres** (2–)3–4 mm. **Phyllaries** 5–8, tomentose. **Receptacles** glabrous. **Cypselae** 1–2 mm (3–5-ribbed, apices with thickened rims, faces glabrous or sparsely gland-dotted).

Flowering May–Jul. Dry, rocky hills; 1500–2400 m; Colo., Mont., Utah, Wyo.

7. Sphaeromeria compacta (H. M. Hall) A. H. Holmgren, L. M. Shultz & Lowrey, Brittonia 28: 261. 1976 • Compact chickensage, Charleston tansy C E

Tanacetum compactum H. M. Hall, Muhlenbergia 2: 343. 1916

Perennials, 1–3+ cm (cespitose; caudices woody). **Stems** erect, sericeous. **Leaves** mostly basal; blades (10–25 mm) pinnati-palmately lobed (lobes 3–6+, ± oblanceolate to linear) or linear, faces sericeous. **Heads** borne singly or 2–3(–5) in corymbiform arrays. **Involucres** 4–6.5 mm. **Phyllaries** (14–)18–20, sericeous. (**Disc corollas:** tubes glandular, lobes villous). **Cypselae** 1.5–2 mm (2–3-ribbed, glabrous; pappi of 3–5 subulate scales ca. 0.4 mm).

Flowering Jul–Sep. Rocky ledges and slopes; of conservation concern; 2700–3500 m; Nev.

Sphaeromeria compacta grows in the Charleston Mountains, Clark County.

8. Sphaeromeria simplex (A. Nelson) A. Heller, Muhlenbergia 1: 7. 1900 • Laramie chickensage C E

Tanacetum simplex A. Nelson, Bull. Torrey Bot. Club 26: 484. 1899

Perennials, 1.5–2.5(–8) cm (cespitose, bases sometimes woody). **Stems** erect to decumbent, sericeous. **Leaves** mostly basal; blades (15–30 × 1.5–8 mm) pinnati-palmately lobed (lobes 2–3, linear, 1–2 mm wide) or linear, faces silvery-canescent. **Heads** borne singly. **Involucres** 4.5–7 mm. **Phyllaries** (12–)14–20+, midribs canescent or glabrous. **Cypselae** (light brown) 1.8–2.5 mm (faces finely striate).

Flowering May–Aug. Dry hills and slopes; of conservation concern; 2200–2700 m; Wyo.

Sphaeromeria simplex is in the Center for Plant Conservation's National Collection of Endangered Plants.

119. ARTEMISIA Linnaeus, Sp. Pl. 2: 845. 1753; Gen. Pl. ed. 5, 367. 1754 • Felon-herb, mugwort, sagebrush, sailor's-tobacco, wormwood, armoise, herbe Saint-Jean [Greek *Artemis,* goddess of the hunt and namesake of Artemisia, Queen of Anatolia]

Leila M. Shultz

Annuals, biennials, perennials, subshrubs, or shrubs, 3–350 cm (usually, rarely not, aromatic). **Stems** 1–10+, usually erect, usually branched, glabrous or hairy (hairs basi- or medifixed). **Leaves** basal or basal and cauline; alternate; petiolate or sessile; blades filiform, linear, lanceolate, ovate, elliptic, oblong, oblanceolate, obovate, cuneate, flabellate, or spatulate, usually pinnately and/or palmately lobed, sometimes apically ± 3-lobed or -toothed, or entire, faces glabrous or hairy (hairs multicelled and filled with aromatic terpenoids and/or 1-celled and hollow, dolabriform, T-shaped). **Heads** usually discoid, sometimes disciform (subradiate in *A. bigelovii*), in relatively broad, paniculiform arrays, or in relatively narrow, racemiform or spiciform arrays. **Involucres** campanulate, globose, ovoid, or turbinate, 1.5–8 mm diam. **Phyllaries** persistent, 2–20+ in 4–7 series, distinct, (usually green to whitish green, rarely stramineous) ovate to lanceolate, unequal, margins and apices (usually green or white, rarely dark brown or black) ± scarious (abaxial faces glabrous or hairy). **Receptacles** flat, convex, or conic (glabrous or hairy), epaleate (except paleate in *A. palmeri*). **Ray florets** 0 (peripheral pistillate florets in disciform heads usually 1–20, their corollas filiform; corollas of 1–3 pistillate florets in heads of *A. bigelovii* sometimes ± 2-lobed, weakly raylike). **Disc florets** 2–20(–30+), bisexual and fertile, or functionally staminate; corollas (glabrous or ± hirtellous) usually pale yellow, rarely red, tubes ± cylindric, throats subglobose or funnelform, lobes 5, ± deltate. **Cypselae** (brown) fusiform, ribs 0 (and faces finely striate) or 2–5, faces glabrous or hairy (not villous), often gland-dotted (pericarps sometimes with myxogenic cells, without resin sacs; embryo sac development monosporic); **pappi** usually 0 (coroniform in *A. californica* and *A. papposa,* sometimes on outer in *A. rothrockii*). *x* = 9.

Species ca. 350–500 (50 in the flora), mostly Northern Hemisphere (North America, Eurasia), some in South America and Africa.

As circumscribed here, there are five subgenera in *Artemisia;* four are represented in the flora area.

Etymologies of the common names used for *Artemisia* species provide glimpses of their uses and demonstrate the rich diversity within the genus. The common name 'mugwort' is from the Old English *mucgwyrt, mucg* meaning 'midge,' and refers to the use of Old World herbaceous species in repelling flies and midges. *Artemisia* was called Motherwort in nineteenth century Maine (as an indication of the high esteem for this otherwise rather pedestrian plant), and in the herbal by R. Banckes (1525): "This herb helpeth a woman to conceyve a chylde, and clenseth the mother, and maketh a woman to have her flowers." Early settlers in North America brought European plants of *A. dracunculus, A. vulgaris, A. absinthium,* and *A. abrotanum* into their herb gardens for seasoning and medicinal uses; they would also have learned about aboriginal uses of *Artemisia* species native to North America, uses that included fertility rites (sagebrush in western North America) and antihelminthics (wormwoods of grasslands and mountain habitats). Immigrants used *A. annua* (sweet Annie) in *potpourris* and later recognized its utility as an anti-malarial drug, a use that was well known in oriental medicine. 'Bulwand' is the local name used for herbaceous wormwoods in Scotland, and 'green-ginger' and 'Sailor's tobacco' are local names in England (T. Coffey 1993). Use of the names 'sagewort' and 'sagebrush' in North America arise from the familiar aroma of culinary sage, *Salvia officinalis* (Lamiaceae). Because true sages (*Salvia*) and sagewort/sagebrushes (*Artemisia*) are in separate families, the chemical similarities are an example of convergent evolution. The intense aroma and bitter taste of the plants from terpenoids and sesquiterpene lactones discourages herbivory and undoubtedly has contributed to the remarkable evolutionary success (measured by abundance as well as diversity) of species in this genus.

Members of *Artemisia* are wind-pollinated and their heads and florets are exceptionally small (even for composites) and, consequently, difficult to examine and assess. Nevertheless, the sexual constitution of floral heads is important in recognition of subgenera. Plant habits and ornamentations of receptacles have also figured in arriving at subgeneric circumscriptions; additional characteristics are enumerated in the descriptions.

Artemisia has a well-deserved reputation for being taxonomically difficult. The number of subgenera varies from four to five in modern treatments, and the number of taxa recognized at the species or subspecific levels varies between 250 and 500 (K. Bremer and C. J. Humphries 1993; H. M. Hall and F. E. Clements 1923; Y. R. Ling 1982, 1995; P. P. Poljakov 1961; M. Torrell et al. 1999). In this treatment, I recognize four native subgenera; subg. *Seriphidium* is endemic to Asia. In the flora area, the greatest diversity occurs in subg. *Artemisia*. Subgenus *Absinthium* can be segregated on the basis of hairs on the receptacle; it may be not phylogenetically distinct (L. E. Watson et al. 2002; J. Valles and E. D. McArthur 2001). Subgenus *Dracunculus* is clearly distinguished by molecular differences, and subg. *Tridentatae* is well defined with the exception of *A. pygmaea*.

This treatment is based on extensive fieldwork, review of recent research, and examination of thousands of specimens; taxonomic circumscriptions remain controversial. Molecular analyses have helped define subgenera but have not clarified relationships between closely related species. The morphologic characters useful in distinguishing species tend to be variable and are often hard to assess (i.e., the sexuality of microscopic florets). Users of the keys will meet with frustrations; descriptions of subgenera and illustrations will help in defining the major groupings of species.

The subgenera are arranged in approximate phylogenetic order; species are arranged alphabetically within the subgenera. Molecular studies define subg. *Dracunculus* as a major clade that is ancestral to the majority of *Artemisia*. The subgenera *Absinthium*, *Tridentatae*, and *Artemisia* can be classified as clades; they are weakly supported by molecular evidence.

SELECTED REFERENCES   Hall, H. M. and F. E. Clements. 1923. The phylogenetic method in taxonomy: The North America species of *Artemisia*, *Chrysothamnus*, and *Atriplex*. Publ. Carnegie Inst. Wash. 326. Ling, Y. R. 1982. On the system of the genus *Artemisia* L. and the relationship with its allies. Bull. Bot. Lab. N. E. Forest. Inst., Harbin 2: 1–60. Ling, Y. R. 1995. The New World *Artemisia* L. In: D. J. N. Hind et al., eds. 1995. Advances in Compositae Systematics. Kew. Pp. 225–281. Torrell, M., N. Garcia-Jacas, A. Susanna, and J. Valles. 1999. Phylogeny in *Artemisia* (Asteraceae, Anthemideae) inferred from nuclear ribosomal DNA (ITS) sequences. Taxon 48: 721–736. Valles, J. and E. D. McArthur. 2001. *Artemisia* systematics and phylogeny: Cytogenetic and molecular insights. In: E. D. McArthur and D. J. Fairbanks, comps. 2001. Shrubland Ecosystem Genetics and Biodiversity: Proceedings: Provo, UT, June 13–15, 2000. Ogden. Pp. 67–74.

1. Shrubs; leaves in lateral fascicles (on vegetative shoots); heads discoid (except in *A. bigelovii* with, rarely, 1–2 raylike florets): florets bisexual (corollas 5-lobed); receptacles glabrous .... 119b. *Artemisia* subg. *Tridentatae*, p. 509
1. Annuals, biennials, perennials, or subshrubs (shrubs in *A. filifolia*, *A. californica*, and *A. nesiotica*); leaves not in fascicles; heads usually disciform, rarely discoid; receptacles glabrous or villous.
    2. Disc florets functionally staminate (not setting fruits), corollas subglobose .... 119a. *Artemisia* subg. *Dracunculus*, p. 505
    2. Disc florets usually bisexual and fertile (sometimes functionally staminate in *A. packardiae* in subg. *Artemisia*), corollas funnelform.
        3. Receptacles villous .... 119c. *Artemisia* subg. *Absinthium*, p. 518
        3. Receptacles glabrous (paleate in *A. palmeri*) .... 119d. *Artemisia* subg. *Artemisia*, p. 520

119a. ARTEMISIA Linnaeus subg. DRANCUNCULUS Besser, Bull. Soc. Imp. Naturalistes Moscou 1: 223. 1829 (as Dracunculi)

Biennials, perennials, or subshrubs (shrubs in *A. filifolia*); fibrous rooted or taprooted, caudices woody, rhizomes absent. **Stems** wandlike (new stems may sprout from caudices). **Leaves** deciduous (persistent in *A. aleutica* and *A. borealis*), usually cauline, sometimes basal, not in fascicles. **Heads** disciform. **Receptacles** epaleate, glabrous. **Florets:** peripheral 1–25 pistillate and fertile; central 3–32 functionally staminate (not setting fruits); corollas subglobose.

Species ca. 80 (8 in the flora): North America, Eurasia.

1. Plants 5–30(–80+) cm (often cespitose and/or mounded).
    2. Perennials; leaves 2–3-palmately or -pinnately lobed.
        3. Leaves 2-palmately lobed; corollas purplish red; Aleutian Islands . . . . 1. *Artemisia aleutica*
        3. Leaves 2–3-pinnately or -ternately lobed; corollas (at least lobes) usually yellow-orange or deep red; n latitudes and w mountains . . . . 2. *Artemisia borealis*
    2. Perennials or subshrubs; leaves 1–2-pinnately or -ternately lobed.
        4. Leaves gray-green, lobes 1–2 mm wide; involucres 3–4 × 3–4 mm; corollas yellow, usually red-tinged, glabrous . . . . 6. *Artemisia pedatifida*
        4. Leaves silver-green, lobes mostly 2–3 mm wide; involucres 4–5(–7) × 2–3 mm wide; corollas pale yellow, glandular . . . . 7. *Artemisia porteri*
1. Plants (10–)50–180 cm (not cespitose).
    5. Plants tarragon-scented or not aromatic; leaves mostly entire, sometimes (basal) irregularly lobed, faces usually glabrous, sometimes glabrescent (deserts) . . . . . 4. *Artemisia dracunculus*
    5. Plants faintly to strongly aromatic (not tarragon-scented); leaves lobed, faces hairy.
        6. Shrubs, 60–180 cm (rounded, stems wandlike); involucres 1.5–2 mm diam . . . . 5. *Artemisia filifolia*
        6. Biennials or perennials, (10–)30–80(–150) cm; involucres 2–4.5(–7) mm diam.
            7. Stems usually 1–5; heads in (mostly leafless) paniculiform arrays . . . . 3. *Artemisia campestris*
            7. Stems usually 10+; heads (clustered in glomerules) in (densely leafy) paniculiform to spiciform arrays . . . . 8. *Artemisia pycnocephala*

1. Artemisia aleutica Hultén, Bot. Not. 1939: 829, fig. 2. 1939 • Aleutian wormwood [C][E]

Perennials, 5–10 cm (cespitose), mildly aromatic; caudices branched. **Stems** usually 1, reddish brown to gray, tomentose to glabrate. **Leaves** persistent, mostly basal, gray-green; (petioles often expanded) blades (at least proximal) obovate, 1.5–5 × 0.5–1 cm, 2-palmately lobed, lobes relatively narrow, apices acute, faces densely white-villous (brownish in age); cauline smaller, distally 1-ternate. **Heads** (sessile or peduncles 2–15 mm) in racemiform or spiciform arrays, 1.5–3 × 0.5–1 cm. **Involucres** hemispheric or globose, (2–)5–7 × (2–)6–8 mm. **Phyllaries** villous. **Florets:** pistillate 4–6; functionally staminate 15–30; corollas purplish red, 1.5–2 mm, hairy. **Cypselae** oblong, ca. 1 mm, faintly nerved, glabrous.

Flowering mid–late summer. Open areas, fellfield tundra; of conservation concern; 0–100 m; Alaska.

Artemisia aleutica is known only from the western Aleutian Islands. It is morphologically similar to *A. borealis,* and the relationships of these species complexes warrant further study.

2. Artemisia borealis Pallas, Reise Russ. Reich. 3: 755. 1776

Artemisia campestris Linnaeus subsp. *borealis* (Pallas) H. M. Hall & Clements

Perennials, (6–)8–20(–40) cm (cespitose), mildly aromatic; taprooted, caudices branched. **Stems** (1–)2–5, gray-green, tomentose. **Leaves** persistent, basal rosettes persistent, gray-green to white; blades ovate, 2–4 × 0.5–1 cm, 2–3-pinnately or -ternately lobed, lobes linear to narrowly oblong, apices acute, faces moderately to densely sericeous. **Heads** (proximal sessile, distal pedunculate) in (leafy) spiciform arrays 4–9(–12) × (0.5–)1–5 cm. **Involucres** hemispheric, 3–4 × 3.5–4 mm. **Phyllaries** (obscurely scarious) densely tomentose-villous. **Florets:**

pistillate 8–10; functionally staminate 15–30; corollas (or lobes) yellow-orange or deep red, 2.2–3.5. **Cypselae** oblong-lanceoloid, somewhat compressed, 0.4–1 mm, faintly nerved, glabrous.

Subspecies 2 (2 in the flora): w North America, especially at high elevations and northern latitudes; Eurasia.

1. Herbage villoso-tomentose, glabrate, or glabrous; corollas (at least lobes) usually yellow-orange, 2.2–3 mm; mountains, w North America . . . . 2a. *Artemisia borealis* subsp. *borealis*
1. Herbage white-hoary; corollas (at least lobes) deep red, 3–3.5 mm; w arctic North America . . . . 2b. *Artemisia borealis* subsp. *richardsoniana*

2a. Artemisia borealis Pallas subsp. **borealis**

• Boreal sage

Artemisia campestris Linnaeus var. *purshii* (Besser) Cronquist; *A. campestris* var. *spithamaea* (Pursh) M. Peck; *A. campestris* var. *strutziae* S. L. Welsh; *A. purshii* Besser; *A. spithamaea* Pursh; *Oligosporus borealis* (Pallas) Polyakov; *Oligosporus groenlandicus* (Hornemann) Á. Löve & D. Löve

Plants 10–20(–40) cm, herbage villoso-tomentose, glabrate, or glabrous. **Corollas** (at least lobes) usually yellow-orange (sometimes red-tinged), 2.2–3 mm. $2n = 18, 36$.

Flowering mid–late summer. Open meadows, usually on well-drained soils; 0–3500 m; Alta., B.C., N.W.T., Nunavut, Sask., Yukon; Alaska, Calif., Colo., Idaho, Mont., Utah, Wash., Wyo.; Eurasia.

Subspecies *borealis* is widespread in the mountains of western North America. Some high-elevation populations have corollas tinged with red.

2b. Artemisia borealis Pallas subsp. **richardsoniana**

(Besser) Korobkov in A. I. Tolmatchew, Fl. Arct. URSS 10: 178. 1987 • Richardson's sagewort E

Artemisia richardsoniana Besser, Bull. Soc. Imp. Naturalistes Moscou 9: 64. 1836, based on *A. arctica* Besser in W. J. Hooker, Fl. Bor.-Amer. 1: 323. 1833, not Lessing 1831; *A. caudata* Michaux var. *richardsoniana* (Besser) B. Boivin; *A. desertorum* Sprengel var. *richardsoniana* (Besser) Besser

Plants (6–)8–15(–20) cm, herbage white-hoary. **Corollas** (at least lobes) deep red, 3–3.5 mm.

Flowering mid–late summer. River terraces, tundra; 0–600 m; N.W.T., Nunavut, Yukon; Alaska.

Long overlooked, subsp. *richardsoniana* is easily distinguished by its relatively short stature, dense, white indument, and deep red corollas. It is known only from western arctic North America.

3. Artemisia campestris Linnaeus, Sp. Pl. 2: 846. 1753

• Field sagewort, sand wormwood F

Biennials or perennials, (10–)30–80(–150) cm, faintly aromatic; taprooted, caudices branched. **Stems** usually 1–5, turning reddish brown, (often ribbed) tomentose or glabrous. **Leaves** persistent or deciduous, mostly basal; basal blades 4–12 cm, cauline gradually reduced, 2–4 × 0.5–1.5 cm, 2–3-pinnately lobed, lobes linear to narrowly oblong, apices acute, faces densely to sparsely white-pubescent. **Heads** (pedunculate) in (mostly leafless) paniculiform arrays. **Involucres** broadly turbinate, 2.5–3(–5) × 2–3.5(–7) mm. **Phyllaries** (margins scarious) glabrous or villous-tomentose. **Florets:** pistillate 5–20; functionally staminate 12–30; corollas pale yellow, sparsely hairy or glabrous. **Cypselae** oblong-lanceoloid, somewhat compressed, 0.8–1 mm, faintly nerved, glabrous.

Subspecies ca. 7 (3 in the flora): North America, especially mountains and high latitudes; Eurasia.

Artemisia campestris varies; each morphologic form grades into another. The present circumscription is conservative in that only three subspecies are recognized; the subspecies usually can be separated geographically as well as morphologically. Populations in western North America consist primarily of subsp. *pacifica*; east of the continental divide, plants are assigned to subsp. *canadensis* in northern latitudes and to subsp. *caudata* in southern latitudes.

1. Perennials; stems 2–5; basal rosettes persistent. . . . . 3c. *Artemisia campestris* subsp. *pacifica*
1. Biennials; stems 1(–3); basal rosettes not persistent (withering before flowering).
    2. Involucres globose, 3–4 × 3.5–5 mm; n of 50°, primarily Canada . . . . 3a. *Artemisia campestris* subsp. *canadensis*
    2. Involucres turbinate, 2–3 × 2–3 mm; s of 50°, e from Rocky Mountains to coastal North America . . . . 3b. *Artemisia campestris* subsp. *caudata*

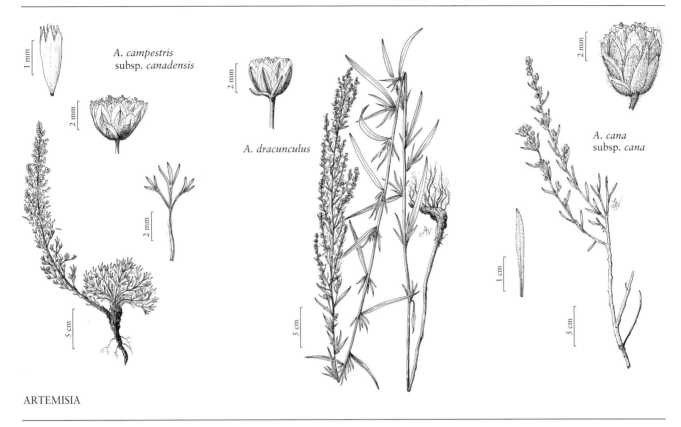

A. campestris subsp. canadensis

A. dracunculus

A. cana subsp. cana

ARTEMISIA

3a. Artemisia campestris Linnaeus subsp. **canadensis** (Michaux) Scoggan, Fl. Canada 1: 52. 1978

• Armoise du Canada E F

Artemisia canadensis Michaux, Fl. Bor.-Amer. 2: 128. 1803

Biennials, 20–40 cm. **Stems** usually 1. **Leaves:** basal rosettes not persistent. **Heads** in arrays 8–10(–12) × 1–2(–3) cm. **Involucres** globose, 3–4 × 3.5–5(–6) mm.

Flowering early–late summer. Open meadows; 0–1000 m; Greenland; Man., N.B., Nfld. and Labr., Nunavut, Ont., Que., Sask.; Maine.

A more broadly circumscribed interpretation of the subsp. *caudata* complex would encompass subsp. *canadensis*.

3b. Artemisia campestris Linnaeus subsp. **caudata** (Michaux) H. M. Hall & Clements, Publ. Carnegie Inst. Wash. 326: 122. 1923 • Armoise caudée E

Artemisia caudata Michaux, Fl. Bor.-Amer. 2: 129. 1803; *A. forwoodii* A. Gray

Biennials, 20–80(–150) cm. **Stems** usually 1. **Leaves:** basal rosettes not persistent (faces green and glabrous or sparsely white-pubescent). **Heads** in arrays 12–30(–35) × 1–8(–12) cm. **Involucres** turbinate, 2–3 × 2–3 mm.

Flowering early–late summer. Open meadows, usually moist soils, sometimes sandy or rocky habitats; 10–1000; Ont., Que., Sask.; Ark., Colo., Conn., Fla., Ill., Ind., Iowa, Kans., Maine, Mass., Mich., Minn., Miss., Mo., Mont., Nebr., N.H., N.J., N.Mex., N.Y., N.C., N.Dak., Ohio, Okla., Pa., R.I., S.C., S.Dak., Tex., Vt., Wis.

A population of *Artemisia campestris* found in Massachusetts differs from populations of subsp. *caudata* by its smaller heads and multiple branched stems. That population is typical of subsp. *campestris*, formerly believed to be restricted to Europe.

3c. Artemisia campestris Linnaeus subsp. **pacifica** (Nuttall) H. M. Hall & Clements, Publ. Carnegie Inst. Wash. 326: 122. 1923 • Western sagewort [E]

Artemisia pacifica Nuttall, Trans. Amer. Philos. Soc., n. s. 7: 401. 1841; *A. campestris* var. *petiolata* S. L. Welsh; *A. campestris* var. *scouleriana* (Besser) Cronquist; *A. desertorum* Sprengel var. *scouleriana* Besser; *Oligosporus campestris* (Linnaeus) Cassini subsp. *pacificus* (Nuttall) W. A. Weber; *O. pacificus* (Nuttall) Poljakov

Perennials, 30–100 cm. **Stems** 2–5. **Leaves:** basal rosettes persistent (faces green and glabrous or gray-green and sparsely hairy). **Heads** in arrays 10–22 × 1–3(–7) cm. **Involucres** turbinate, 2–3 × 2(–3) mm.

Flowering mid–late summer. Sandy soils, arid regions; 100–2500 m; Alta., B.C., Man., Sask.; Alaska, Ariz., Calif., Colo., Idaho, Mont., Nebr., Nev., N.Mex., Oreg., Tex., Utah, Wash., Wyo.

Throughout western North America, members of the *Artemisia campestris* complex can be assigned to subsp. *pacifica*. Although morphologically distinct through much of its range, subsp. *pacifica* may intergrade with subsp. *canadensis* and *A. borealis* in areas where their ranges overlap.

4. Artemisia dracunculus Linnaeus, Sp. Pl. 2: 849. 1753 • Wild tarragon [F]

Artemisia aromatica A. Nelson; *A. dracunculina* S. Watson; *A. dracunculoides* Pursh; *A. dracunculoides* subsp. *dracunculina* (S. Watson) H. M. Hall & Clements; *A. glauca* Pallas ex Willdenow; *A. glauca* var. *megacephala* B. Boivin

Perennials or subshrubs, 50–120 (–150) cm, strongly tarragon-scented or not aromatic; rhizomatous, caudices coarse. **Stems** relatively numerous, erect, green to brown or reddish brown, somewhat woody, glabrous. **Leaves:** proximal blades bright green and glabrous or gray-green and sparsely hairy, 5–8 cm; cauline blades bright green (gray-green in desert forms), linear, lanceolate, or oblong, 1–7 × 0.1–0.5(–0.9) cm, mostly entire, sometimes irregularly lobed, acute, usually glabrous, sometimes glabrescent (deserts). **Heads** in terminal or lateral, leafy, paniculiform arrays 15–45 × 6–30 cm; appearing ball-like on slender, sometimes nodding peduncles. **Involucres** globose, 2–3 × 2–3.5(–6) mm. **Phyllaries** (light brown, broadly lanceolate, membranous): margins broadly hyaline, glabrous. **Florets:** pistillate 6–25; functionally staminate

8–20; corollas pale yellow, 1.8–2 mm, eglandular or sparsely glandular. **Cypselae** oblong, 0.5–0.8 mm, faintly nerved, glabrous. **2*n*** = 18.

Flowering mid summer–late fall. Open meadows and fields, desert scrub, moist drainages, roadsides; 500–3000 m; Alta., B.C., Man., Ont., Sask., Yukon; Alaska, Ariz., Calif., Colo., Idaho, Ill., Iowa, Kans., Minn., Mo., Mont., Nebr., Nev., N.Mex., N.Dak., Okla., Oreg., S.Dak., Tex., Utah, Wash., Wis., Wyo.; Eurasia.

Artemisia dracunculus is widely cultivated as a culinary herb and may be introduced in parts of its range. It is easily cultivated from rootstocks, and while establishment from seeds is rare, seedlings can be found with amenable environmental conditions. Because of its popularity as an herb, it may suffer from overcollecting. Its scarcity in Missouri, Iowa, and Illinois (J. T. Kartesz and C. A. Meacham 1999) may have been caused by overly enthusiastic collecting as well as habitat loss.

5. Artemisia filifolia Torrey, Ann. Lyceum Nat. Hist. New York 2: 211. 1827 • Sand sage [E]

Artemisia plattensis Nuttall; *Oligosporus filifolius* (Torrey) Poljakov

Shrubs, 60–180 cm (rounded), faintly aromatic. **Stems** green or gray-green, wandlike (usually slender, curved, sometimes stout and stunted in harsh habitats), glabrous or sparsely hairy. **Leaves** gray-green; blades linear if entire, obovate if lobed, (1.5–)2–5(–6) × 0.1–2.5 cm, entire to 3-lobed, lobes filiform (less than 1 mm wide), apices acute, glabrous or sparsely hairy. **Heads** (mostly sessile) in paniculiform arrays 8–15(–17) × 2–4(–5) cm (branches erect to somewhat recurved). **Involucres** globose, 1.5–2 × 1.5–2 mm. **Phyllaries** (ovate, inconspicuous, margins scarious) densely hairy. **Florets:** pistillate 1–4; functionally staminate 3–6; corollas pale yellow, 1–1.5 mm, glabrous. **Cypselae** oblong (distally incurved-falcate and oblique), 0.2–0.5 mm, obscurely nerved, glabrous. **2*n*** = 18.

Flowering late summer–early winter. Open prairies, dunes, sandy soils; 500–2000 m; Ariz., Colo., Kans., Nebr., Nev., N.Mex., Okla., S.Dak., Tex., Utah, Wyo.

One of the more easily distinguished of the shrubby *Artemisia* species, *A. filifolia* occurs in sandy soils and cohabits with species of *Yucca,* Cactaceae, and *Salvia dorrii,* the purple sage of western literary fame. Its filiform leaves and faintly aromatic foliage distinguish it from members of subg. *Tridentatae.*

6. Artemisia pedatifida Nuttall, Trans. Amer. Philos. Soc., n. s. 7: 399. 1841 • Matted sagewort E

Perennials or subshrubs, 5–15 cm (cespitose), aromatic. Stems 5–20), gray-green, glabrescent. Leaves persistent, gray-green, mostly basal; proximal blades reduced, mostly less than 1 cm, lobed or entire; distal blades 1–2 × 0.5–0.8 cm, 1–2-ternately lobed, lobes 1–2 mm wide, apices acute, faces densely tomentose. Heads (mostly 6–15, 1 or 3–4 on lateral branches; mostly erect, sessile or pedunculate) in racemiform-paniculiform arrays, 5–8 × 0.5–0.8 cm. Involucres globose, 3–4 × 3–4 mm. Phyllaries (margins scarious, obscured) white-tomentose. Florets: pistillate 4–7; functionally staminate 5–9; corollas yellow, usually red-tinged, 2–3 mm, glabrous. Cypselae (brown) ellipsoid (angled), 0.8–1 mm, (sometimes with white ribs) glabrous.

Flowering early spring–mid summer. High plains, grasslands; 1600–1800 m; Colo., Idaho, Mont., Wyo.

7. Artemisia porteri Cronquist, Madroño 11: 145. 1951 • Porter mugwort E

Perennials or subshrubs, (7–)8–14 cm (cespitose), faintly aromatic. Stems 5–8, silver-gray, densely tomentose. Leaves persistent, silver-green, mostly basal; proximalmost blades 3–4 × 1–1.5 cm, 1-pinnately lobed, lobes mostly 2–3 mm wide; blades of flowering stems somewhat reduced, (1–)2–3(–5) × 0.15 cm, mostly entire; apices rounded, faces densely hairy. Heads borne singly or (clustered in 2s and 3s on lateral branches; peduncles 0 or to 5 mm) in paniculiform arrays, (2–)4–9 × 1–1.5(–2) cm. Involucres broadly campanulate, 4–5(–7) × 2–3 mm. Phyllaries (ovate, margins broadly scarious) densely tomentose. Florets: pistillate 8–10 (2–2.8 mm); functionally staminate 22–32; corollas pale yellow, 2.2–4.5 mm, glandular. Cypselae (light brown) ellipsoid, flattened (faintly nerved), 1.5–2 mm, sparsely hairy, glabrous or resinous.

Flowering mid–late summer. Barren clay and gravelly soils; 1800–2000 m; Mont., Wyo.

Although Cronquist observed that *Artemisia porteri* may be an autopolyploid derivative of *A. pedatifida*, morphologic similarities to northerly cespitose taxa suggest a more complex origin.

Artemisia porteri is in the Center for Plant Conservation's National Collection of Endangered Plants.

8. Artemisia pycnocephala (Lessing) de Candolle in A. P. de Candolle and A. L. P. P. de Candolle, Prodr. 6: 99. 1838 • Coastal sagewort E

Oligosporus pycnocephalus Lessing, Linnaea 6: 524. 1831; *Artemisia campestris* Linnaeus subsp. *pycnocephala* (Lessing) H. M. Hall & Clements

Perennials, 30–70(–100) cm, faintly aromatic. Stems usually 10+ (rising beyond basal leaves, decumbent), whitish gray, (ca. 5 mm diam., densely leafy) densely hairy. Leaves persistent, gray-green; blades broadly lanceolate, faces woolly-hairy; proximalmost blades 3–8 × 2–6 cm, 2–3-pinnatifid, lobes linear (to 2 mm wide); cauline somewhat reduced, 2–3 × 0.8–1.2 cm; apices acute, faces hairy. Heads (sessile, clustered in glomerules) in (densely leafy) paniculiform to spiciform arrays 10–20(–30) × 1–4 cm. Involucres globose, 3–4.5 × 3–4.5 mm. Phyllaries (lanceolate, margins obscured by indument, hairs straight. Florets: pistillate 5–20; functionally staminate 12–25; corollas pale yellow (broadly tubular), ca. 2 mm, glabrous. Cypselae ellipsoid (faintly nerved), 1–1.5 mm, glabrous.

Flowering late spring–mid summer. Rocky or sandy soils of coastal beaches; 0–200 m; Calif.

119b. ARTEMISIA Linnaeus subg. **TRIDENTATAE** (Rydberg) McArthur, Amer. J. Bot. 68: 590. 1981

Artemisia [unranked] *Tridentatae* Rydberg in N. L. Britton et al., N. Amer. Fl. 34: 282. 1916

Shrubs; fibrous rooted, caudices woody, rhizomes absent. Stems not wandlike (relatively numerous; new stems may sprout from caudices). Leaves (pungently aromatic) deciduous or persistent, cauline (in lateral fascicles on vegetative shoots). Heads discoid (except *A. bigelovii* with, rarely, 1–2 raylike florets). Receptacles epaleate, glabrous. Pappi 0. Florets: 3–20, bisexual, fertile; corollas (pale yellow) funnelform.

Species 10 (10 in the flora): North America, nw Mexico.

Difficulty in classification of *Artemisia* subg. *Tridentatae* has been complicated by transfer of North American species to *Seriphidium* (Y. R. Ling 1995b; W. A. Weber 1984b), a disposition not followed here. Species circumscription varies among authors, but most modern treatments recognize the species as defined here. The most useful field characteristics in sagebrush taxonomy are size of the plant, shape and lobing of the vegetative leaves, and size and shape of the flowering heads (A. A. Beetle 1960; A. H. Winward 1970). Differences in chromosome number are more useful in defining subspecies than species (E. D. McArthur et al. 1981; G. H. Ward 1953), and introgression among subspecies is common (McArthur et al. 1988; McArthur and S. C. Sanderson 1999). The following key relies on vegetative characteristics, and unless noted, descriptions of leaf size and lobing refer to the leaves found in the vegetative shoots proximal to arrays of heads. These 'vegetative leaves' occur in bundles, or fascicles that are part of the lateral shoots. They are subtended by an elongate leaf (termed 'ephemeral'), which is attached to the primary stem and falls off early in the season. With the exception of *Artemisia spiciformis*, which retains its ephemeral leaves through most of the growing season, ephemeral leaves normally drop from the plant before the onset of flowering.

SELECTED REFERENCES Beetle, A. A. 1960. A study of sagebrush. The section *Tridentatae* of *Artemisia*. Wyoming Agric. Exp. Sta. Bull. 368. Ling, Y. R. 1995b. The New World *Seriphidium* (Besser) Fourr. In: D. J. N. Hind et al., eds. 1995. Advances in Compositae Systematics. Royal Botanic Gardens, Kew. Pp. 283–291. McArthur, E. D., C. L. Pope, and D. C. Freeman. 1981. Chromosomal studies of subgenus *Tridentatae* of *Artemisia*: Evidence for autopolyploidy. Amer. J. Bot. 68: 589–605. McArthur, E. D. et al. 1998. Randomly amplified polymorphic DNA analysis (RAPD) of *Artemisia* subgenus *Tridentatae* species and hybrids. Great Basin Naturalist 58: 12–27. McArthur, E. D. and S. C. Sanderson. 1999. Cytogeography and chromosome evolution of subgenus *Tridentatae* of *Artemisia*. Amer. J. Bot. 86: 1754–1775. Shultz, L. M. 1983. Systematics and Anatomical Studies of *Artemisia* subgenus *Tridentatae*. Ph.D. dissertation. Claremont Graduate School. Shultz, L. M. 1986. Taxonomic and geographic limits of *Artemisia* subgenus *Tridentatae* (Beetle) McArthur. In: E. D. McArthur and B. L. Welch, eds. 1986. Proceedings, Symposium on the Biology of *Artemisia* and *Chrysothamnus*, Provo, Utah, July 9–13, 1984. Ogden. Pp. 20–28. Shultz, L. M. 1986b. Comparative leaf anatomy of sagebrush. In: E. D. McArthur and B. L. Welch, eds. 1986. Proceedings, Symposium on the Biology of *Artemisia* and *Chrysothamnus*, Provo, Utah, July 9–13, 1984. Ogden. Pp. 253–264. Ward, G. H. 1953. *Artemisia* section *Seriphidium* in North America, a cytotaxonomical study. Contr. Dudley Herb. 4: 155–206. Winward, A. H. 1970. Taxonomic and Ecological Relationships of the Big Sagebrush Complex in Idaho. Ph.D. dissertation. University of Idaho.

1. Leaves deciduous, blades usually entire, sometimes irregularly lobed; moist habitats . . . . . 11. *Artemisia cana*
1. Leaves deciduous or persistent, blades usually lobed, sometimes entire; dry habitats.
    2. Leaves bright green, pinnately lobed, lobes 3–7 (gypsum or shale) . . . . 13. *Artemisia pygmaea*
    2. Leaves gray-green, usually palmately lobed, lobes 0 or 3 or 3–6.
        3. Leaf lobe lengths $^1/_3+$ blade lengths, widths 1–1.5 mm.
            4. Leaves rigid (lava scablands, Oregon and Washington) . . . . 14. *Artemisia rigida*
            4. Leaves not rigid. . . . . 18. *Artemisia tripartita*
        3. Leaf lobe lengths to $^1/_3$ blade lengths, widths (1–)1.5–5 mm.
            5. Shrubs, 50–200(–250) cm.
                6. Leaves mostly deciduous (variable in size and shape, entire or irregularly 3–6-lobed, lobes rounded or acute); involucres broadly campanulate . . . . 16. *Artemisia spiciformis*
                6. Leaves persistent (lobes 3, uniform, lengths to $^1/_3$ blade lengths); involucres lanceoloid or ovoid.
                    7. Leaves light or dark gray-green, sticky-resinous; involucres ovoid, 3–5 × 4–6 mm; florets 12–20 (California) . . . . 15. *Artemisia rothrockii* (in part)
                    7. Leaves gray-green, not sticky-resinous (widespread, including California); involucres lanceoloid, (1–)1.5–4 × 1–3 mm; florets 3–8 . . . . 17. *Artemisia tridentata*
            5. Shrubs, 10–50 cm.
                8. Leaves silver-green, blades narrowly cuneate, lobes acute; heads mostly nodding; involucres globose. . . . . 10. *Artemisia bigelovii*
                8. Leaves dark green to gray-green, blades broadly cuneate, lobes obtuse or rounded; heads mostly erect; involucres campanulate, globose-ovoid, or turbinate.

[9. Shifted to left margin.—Ed.]

9. Leaves on flowering stems entire (heads mostly pedunculate); involucres narrowly turbinate; phyllaries sparsely hairy or glabrous . . . . 12. *Artemisia nova*
9. Leaves on flowering stems entire or lobed (heads mostly sessile); involucres campanulate or globose-ovoid; phyllaries densely pubescent or tomentose.
   10. Leaves on flowering stems 3-lobed, not sticky-resinous . . . . 9. *Artemisia arbuscula*
   10. Leaves on flowering stems entire, sticky-resinous or densely hairy and not sticky . . .
    . . . . 15. *Artemisia rothrockii* (in part)

9. Artemisia arbuscula Nuttall, Trans. Amer. Philos. Soc., n. s. 7: 398. 1841 • Low sagebrush [E]

Artemisia tridentata Nuttall subsp. *arbuscula* (Nuttall) H. M. Hall & Clements; *A. tridentata* var. *arbuscula* (Nuttall) McMinn; *Seriphidium arbusculum* (Nuttall) W. A. Weber

Shrubs, 10–30(–50) cm, aromatic; root-sprouting. **Stems** gray-green to brown, glabrate (diffusely branched from bases, brittle). **Leaves** (vegetative stems) persistent, gray-green; blades broadly to narrowly cuneate, 3–10 × 2–5 mm, lobed (lobes 3, oblong-linear, to $^1/_3$ blade lengths, mostly 1–3 mm wide, flat, obtuse, laterals sometimes 2–3-fid; leaves on flowering stems deciduous, blades narrowly cuneate, deeply 3-lobed), faces densely hairy (not sticky resinous). **Heads** usually borne singly, rarely (1–4, erect, mostly sessile, in pedunculate clusters) in spiciform or paniculiform arrays 2–9 × 0.5–2 cm (branches slender). **Involucres** campanulate or globose-ovoid, (1.5–)2–4(–5) × 1.5–4.5 mm. **Phyllaries** (margins green) ovate (outer) to oblong, pubescent or tomentose. **Florets** 4–6(–10); corollas 1.5–2 mm, glabrous. **Cypselae** (light brown) 0.7–0.8 mm, resinous.

Subspecies 3 (3 in the flora): w North America.

Artemisia arbuscula is one of the more perplexing species in the *Tridentatae* complex. Anatomic and morphologic characteristics suggest multiple hybrid origins for the subspecies. Deciduous leaves of flowering stems in plants that otherwise have persistent leaves suggest a hybrid origin involving plants of the *A. tridentata* and *A. cana* lineages. In most instances, populations of *A. arbuscula* appear to be reproductively stable. The disposition of *Artemisia arbuscula* subsp. *longicaulis* Winward & McArthur (with $2n = 54$) has not been determined.

1. Involucres 2–4.5 mm diam.; usually in rocky soils; flowering mid–late summer . . . .
    . . . . 9a. *Artemisia arbuscula* subsp. *arbuscula*
1. Involucres 1.5–2.5 mm diam.; clays or stony soils; flowering early spring–late summer.

[2. Shifted to left margin.—Ed.]

2. Leaves broadly cuneate (4–10 × 2–5 mm, often irregularly lobed, lobes rounded, middle lobes overlapping lateral lobes); usually in clay soils; flowering early spring. . . . . 9b. *Artemisia arbuscula* subsp. *longiloba*
2. Leaves narrowly cuneate (5–10 × 3–6 mm, lobed, lobes $^1/_2$+ blade lengths, laterals to 1 mm wide, often acute); usually in stony soils; flowering mid–late summer . . . . 9c. *Artemisia arbuscula* subsp. *thermopola*

9a. Artemisia arbuscula Nuttall subsp. **arbuscula** [E]

Leaves broadly cuneate (lobed, lobes less than $^1/_2$ blade lengths, 1–3 mm wide, rounded). **Involucres** 3.5–4(–5) × 2–4.5 mm. $2n = 18$, 36.

Flowering mid–late summer. Rocky sedimentary soils, high valleys, mountain slopes; 1500–3800 m; Calif., Colo., Idaho, Mont., Nev., Utah, Wash., Wyo.

The relatively large heads of *Artemisia arbuscula* subsp. *arbuscula* suggest a relationship with *A. cana*; the extreme morphologic variability of this subspecies from east to west may be the result of hybridization with various subspecies within the *A. cana* complex.

9b. Artemisia arbuscula Nuttall subsp. **longiloba** (Osterhout) L. M. Shultz, Sida 21: 1637. 2005 [E]

Artemisia spiciformis Osterhout var. *longiloba* Osterhout, Muhlenbergia 4: 69. 1908; *A. longiloba* (Osterhout) Beetle; *Seriphidium arbusculum* (Nuttall) W. A. Weber subsp. *longilobum* (Osterhout) W. A. Weber

Leaves broadly cuneate (4–10 × 2–5 mm, often irregularly lobed, lobes rounded, middle lobes overlapping lateral lobes). **Involucres** 2–3 × 1.5–2.5 mm. $2n = 18$, 36.

Flowering early–late spring. Clay soils of alkaline basins and valleys, occasionally on outwash plains of mountains; 1500–2500 m; Calif., Colo., Idaho, Mont., Oreg., Utah, Wyo.

Subspecies *longiloba* is distinguished from other members of the *Artemisia arbuscula* complex by its early blooming time. It is the only member of subg. *Tridentatae* to begin flowering as snow melts in early spring, and it is ecologically distinguished from other subspecies by its occurrence at low elevations, in fine-grained clay soils.

9c. Artemisia arbuscula Nuttall subsp. **thermopola** Beetle, Rhodora 61: 83. 1959 • Hot Springs sagebrush E

Seriphidium arbusculum (Nuttall) W. A. Weber var. *thermopolum* (Beetle) Y. R. Ling

Leaves narrowly cuneate (5–10 × 3–6 mm, lobed, lobes ¹/₂+ blade lengths, laterals to 1 mm wide, often acute). **Involucres** (1.5–)2–2.5 × 1.5–2 mm.

Flowering mid–late summer. Rocky soils of igneous origin; 2200–2500 m; Idaho, Utah, Wyo.

Because of its deeply lobed leaves, subsp. *thermopola* can be confused with *Artemisia tripartita*. The habit, leaf morphology, and geographic distribution of subsp. *thermopola* suggest introgression between typical *A. arbuscula* and *A. tripartita*.

10. Artemisia bigelovii A. Gray in War Department [U.S.], Pacif. Railr. Rep. 4(5): 110. 1857 • Bigelow sagebrush E

Artemisia petrophila Wooton & Standley; *Seriphidium bigelovii* (A. Gray) K. Bremer & Humphries

Shrubs, 20–40(–60) cm (branched from bases, rounded), mildly aromatic; not root-sprouting. **Stems** silvery, canescent (bark gray-brown). **Leaves** persistent, light gray-green; blades narrowly cuneate, 0.5–3 × 0.2–0.5 cm, entire or 3(–5)-lobed (lobes 1.5–2 mm, less than ¹/₃ blade lengths, acute), faces silvery canescent. **Heads** (usually nodding) in arrays 6–25 × 1–4 cm (branches erect, somewhat curved). **Involucres** globose, 2–3 × 1.5–2.5 mm. **Phyllaries** (8–15) ovate, canescent or tomentose. **Florets:** pistillate 0–2 (raylike, laminae to 1 mm); bisexual 1–3; corollas 1–1.5 mm (style branches of ray florets elongate, exsert, epapillate, tips acute; of disc florets, short, truncate, papillate). **Cypselae** (ellipsoid, 5-ribbed) 0.8–1 mm, glabrous. $2n = 18, 36, 72$.

Flowering early summer–late fall. Deserts, sandy or alkaline soils, rock outcrops; 1000–2500 m; Ariz., Calif., Colo., Nev., N.Mex., Tex., Utah.

Artemisia bigelovii of the southwestern deserts is easily confused in the field with *A. tridentata*, even though it is well distinguished ecologically and morphologically. Systematic placement within subg. *Tridentatae* remains problematic. Presence of "ray" florets (though rare) and vestigial spines on the pollen (R. P. Wodehouse 1935) suggest a relationship with groups ancestral to *Tridentatae*. The species also has the unusual characteristic of lignified trichomes (L. M. Shultz 1986b). Further research may help to determine proper placement; its affinities may be with members of subg. *Artemisia*.

11. Artemisia cana Pursh, Fl. Amer. Sept. 2: 521. 1813 E F

Seriphidium canum (Pursh) W. A. Weber

Shrubs, 50–150 cm (trunks definite, freely branched from bases, branches erect), pleasantly aromatic; root-sprouting. **Stems** light brown to gray-green (woody, somewhat pliable, leafy), persistently canescent to glabrescent. **Leaves** deciduous, whitish gray or green to dark gray-green; blades narrowly elliptic to lanceolate, 1.5–8 × 0.2–1 cm, usually entire, sometimes irregularly lobed, sparsely to densely hairy. **Heads** in (congested, leafy) paniculiform arrays 10–20 × 0.2–7 cm. **Involucres** (subtended by green, leaflike bracts) narrowly to broadly campanulate, 3–4 × 2–5 mm. **Phyllaries** ovate or lanceolate (scarious margins nearly invisible), densely canescent. **Florets** 4–20; corollas 2–3 mm, resinous (style branches ellipsoid, to 2.3 mm, exsert, gland-dotted). **Cypselae** (light brown) 1–2.3 mm, resinous.

Subspecies 3 (3 in the flora): w North America.

1. Shrubs 100–150 cm; leaves 2–8 cm (entire); primarily e of continental divide . . . . 11a. *Artemisia cana* subsp. *cana*
1. Shrubs 50–90 cm; leaves 1.5–4 cm (usually some with irregular lobes); w of continental divide.
   2. Stems felty-tomentose; leaves green to gray-green; involucres 4–5 mm diam.; California, Nevada, Oregon . . . . 11b. *Artemisia cana* subsp. *bolanderi*
   2. Stems hairy (not felty-tomentose); leaves green to dark green; involucres 2–3(–4) mm diam.; Arizona, Colorado, Idaho, Montana, Nevada, New Mexico, Utah, Wyoming . . . . 11c. *Artemisia cana* subsp. *viscidula*

11a. Artemisia cana Pursh subsp. **cana** • Silver wormwood [E] [F]

Artemisia columbiensis Nuttall

Shrubs, 100–150 cm. **Stems** white to light gray or brown. **Leaves** whitish gray, blades narrowly elliptic to lanceolate, 2–8 × 0.3–1 cm, usually entire, sometimes irregularly lobed, densely silvery-canescent. **Heads** in (leafy) arrays 10–20 × 5–7 cm. **Involucres** broadly campanulate, 3–4 × 3–5 mm. **Phyllaries** broadly ovate (mostly obtuse), densely hairy. **Florets** 10–20. **Cypselae** 1–1.2 mm. $2n = 54$.

Flowering mid–late summer. Sandy loam soils, often along streams; 1000–1500 m; Alta., B.C., Man., Sask.; Colo., Mont., Nebr., N.Dak., S.Dak., Wyo.

Subspecies *cana* is found primarily in the grasslands of Canada and the west-central United States. It is unusual within the species in that there is no morphologic evidence of hybridization with other species in subg. *Tridentatae.*

11b. Artemisia cana Pursh subsp. **bolanderi** (A. Gray) G. H. Ward, Contr. Dudley Herb. 4: 192. 1953 • Bolander sagebrush [E]

Artemisia bolanderi A. Gray, Proc. Amer. Acad. Arts 19: 50. 1883; *A. tridentata* Nuttall subsp. *bolanderi* (A. Gray) H. M. Hall & Clements; *Seriphidium bolanderi* (A. Gray) Y. R. Ling; *S. canum* (Pursh) W. A. Weber subsp. *bolanderi* (A. Gray) W. A. Weber

Shrubs, 50–60(–80) cm. **Stems** white (felty-tomentose when young). **Leaves** sometimes bright green, blades linear to narrowly lanceolate, (1.5–)3–4 × 0.2–0.6 cm, usually entire, sometimes with irregular lobes. **Heads** (2–3 per branch) in (sparsely leafy) arrays 12–18 × 1–2 cm. **Involucres** broadly campanulate, 3–4 × 4–5 mm. **Phyllaries** narrowly ovate-lanceolate, acute (outer) or obtuse, densely hairy. **Florets** 8–16. **Cypselae** 1–1.5 mm. $2n = 18, 36$.

Flowering mid–late summer. Gravel soils, mountain meadows, stream banks; 1600–3300 m; Calif., Nev., Oreg.

Subspecies *bolanderi* is known only from the western United States.

11c. Artemisia cana Pursh subsp. **viscidula** (Osterhout) Beetle, Rhodora 61: 84. 1959 • Sticky sagebrush [E]

Artemisia cana var. *viscidula* Osterhout, Bull. Torrey Bot. Club 27: 507. 1900; *A. argillosa* Beetle; *A. viscidula* (Osterhout) Rydberg

Shrubs, 50–70(–90) cm. **Stems** white (sparsely tomentose) or brown (glabrous). **Leaves** bright to dull green, blades linear to narrowly lanceolate, (1.5–)2–3 × 0.2–0.4 cm, often with irregular lobes, sparsely hairy or glabrescent, viscid. **Heads** (2–3 per branch, erect, sessile) in (sparsely leafy) arrays 12–20 × 1–2 cm. **Involucres** narrowly campanulate, 3–4 × 2–3(–4) mm. **Phyllaries** narrowly lanceolate, acute (outer) or obtuse, sparsely hairy. **Florets** 4–8. **Cypselae** 1–2.3 mm. $2n = 18, 36, 72$.

Flowering mid–late summer. Wet mountain meadows, stream banks, rocky areas with late-lying snows; 2000–3300 m; Ariz., Colo., Idaho, Mont., Nev., N.Mex., Utah, Wyo.

Subspecies *viscidula* is the common silver sagebrush of the intermountain region of western North America. In New Mexico, it is known only from Rio Arriba County. It is distinguished from subsp. *bolanderi* by geography as well as its darker green foliage and sparsely (rather than densely) tomentose or glabrous stems. Usually restricted to wet meadows and stream banks, it is distinctive in the late summer and fall by its yellowing ephemeral leaves.

12. Artemisia nova A. Nelson, Bull. Torrey Bot. Club 27: 274. 1900 • Black sagebrush, black sage [E]

Artemisia arbuscula Nuttall subsp. *nova* (A. Nelson) G. H. Ward; *A. arbuscula* var. *nova* (A. Nelson) Cronquist; *A. tridentata* Nuttall subsp. *nova* (A. Nelson) H. M. Hall & Clements; *Seriphidium novum* (A. Nelson) W. A. Weber

Shrubs, 10–30(–50) cm (trunks relatively short, widely and loosely branched), pungently aromatic; not root-sprouting. **Stems** brown, glabrescent (vegetative of approximately equal heights, giving plants a 'hedged' appearance; bark dark gray, exfoliating with age). **Leaves** persistent, usually bright green to dark green, sometimes gray-green; blades cuneate, 3-lobed (lobes to ¹/₃ blade lengths, 0.5–2 × 0.2–1 cm, rounded), faces sparsely hairy, gland-dotted. **Heads** in paniculiform arrays 4–10 × 0.5–3 cm (branches ± erect; peduncles slender). **Involucres** narrowly turbinate, 2–3 × 2 mm. **Phyllaries** (straw-colored

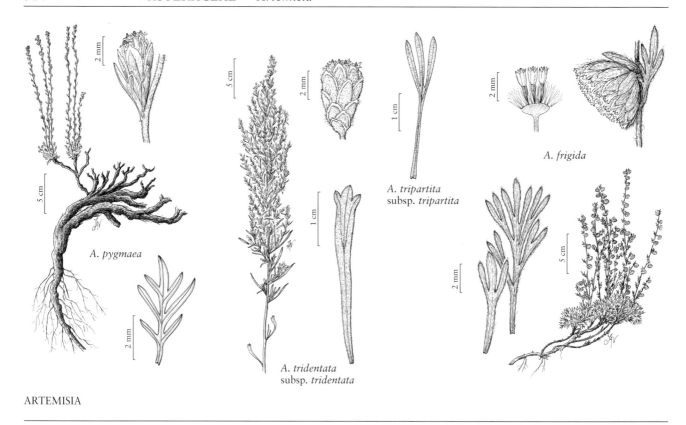

A. pygmaea

A. tridentata
subsp. *tridentata*

A. tripartita
subsp. *tripartita*

A. frigida

ARTEMISIA

or light green) ovate to elliptic (margins hyaline, shiny-resinous), sparsely hairy or glabrous. **Florets** 2–6; corollas 2–3 mm, glabrous (style branches scarcely exsert). **Cypselae** (ribbed) 0.8–1.5 mm, glabrous or resinous. $2n = 18, 36$.

Flowering mid summer–late fall. Shallow soils, desert valleys, exposed mountain slopes; 1500–2300 m; Ariz., Calif., Colo., Idaho, Mont., Nev., N.Mex., Utah, Wyo.

Artemisia nova is the common low-growing dark-green ("black") sagebrush of desert valleys or south-southwest-facing slopes. It is prized by sheep ranchers as forage in areas where little else is available for grazing. It is conspicuous by its low growth habit, dark green foliage, and, in late season, by its pale orange to light brown flowering branches that rise beyond the vegetative growth. Often confused in herbarium collections with *A. arbuscula*, *A. nova* is easily distinguished by the entire leaves of the flowering stems, pedunculate heads, narrowly turbinate involucres, and often straw-colored, glabrous or sparsely hairy phyllaries.

13. Artemisia pygmaea A. Gray, Proc. Amer. Acad. Arts 21: 413. 1886 • Pygmy sage [E] [F]

Seriphidium pygmaeum (A. Gray) W. A. Weber

Shrubs, 5–10 cm, slightly aromatic; not root-sprouting (caudices coarsely woody, branched). **Stems** pale to light brown (stiffly erect, densely clothed with appressed foliage), sparsely tomentose. **Leaves** persistent (sessile, rigid), bright green; blades oblong to ovate, 0.3–0.5 × 0.2–0.3 cm, pinnately lobed (nearly to midribs, $^1/_3$+ widths of blades, lobes 3–7, divergent), faces glabrous or sparsely tomentose, resinous. **Heads** (sessile, erect) in paniculiform to racemiform arrays (1–)2–3 × 0.5–1 cm. **Involucres** narrowly turbinate, 2–3 × 3–4 mm. **Phyllaries** (green) narrowly lanceolate (midribs prominent), glabrous or sparsely tomentose. **Florets** 2–6; corollas 2.5–3 mm, glandular (style branches flat, erose, exsert). **Cypselae** (prismatic) 0.4–0.5 mm, glabrous, resinous. $2n = 18$.

Flowering mid summer–fall. Fine-textured soils of gypsum or shale; 1500–1800 m; Ariz., Colo., Nev., N.Mex., Utah.

Artemisia pygmaea is a distinctive, faintly aromatic shrublet, often mistaken for something other than a sagebrush. In early spring its stiff, bright green, deeply pinnatifid leaves are reminiscent of some prickly member of Polemoniaceae. After flowering, its heads and narrow panicles easily identify it as a member of *Artemisia*; it is unlike other members of the subgenus (which typically have 3-lobed leaves in fascicled lateral shoots). The molecular analysis by L. E. Watson et al. (2002) supported its phylogenetic alignment within subg. *Tridentatae*.

14. Artemisia rigida (Nuttall) A. Gray, Proc. Amer. Acad. Arts 19: 49. 1883 • Scabland sagebrush E

Artemisia trifida Nuttall var. *rigida* Nuttall, Trans. Amer. Philos. Soc., n. s. 7: 398. 1841; *Seriphidium rigidum* (Nuttall) W. A. Weber

Shrubs, 20–40 cm (branches widely spreading), mildly aromatic; root-sprouting (caudices stout). **Stems** gray (coarse, brittle), hairy (bark gray, exfoliating). **Leaves** deciduous, silver-gray (rigid); blades broadly spatulate, 1.5–4 × 0.5–0.7 cm (bases narrow), 3-lobed (lobes $^{1}/_{2}$+ blade lengths, ca. 1 mm wide), faces densely hairy. **Heads** borne singly or (in glomerules) in (densely leafy) spiciform or paniculiform arrays 2–20 × 2 cm. **Involucres** narrowly campanulate, 4–5 × 2.5–3.5 mm. **Phyllaries** elliptic (acute to obtuse), densely canescent. **Florets** 4–8; corollas yellowish red to red, 2–2.8 mm (style branches oblong, truncate, exsert). **Cypselae** (4–5-ribbed) 1–1.5 mm, glabrous. $2n = 18, 36.$

Flowering mid summer–early fall. Dry rocky scablands, volcanic plains; 1500–1800 m; Idaho, Mont., Oreg., Wash.

Artemisia rigida is an important successional species following fires because the plants form new shoots from the underground caudices. This characteristic aligns the species with other 'sprouters' in the subgenus, namely *A. cana, A. tripartita,* and *A. arbuscula.*

15. Artemisia rothrockii A. Gray in W. H. Brewer et al., Bot. California 1: 618. 1876 • Rothrock or sticky sagebrush E

Artemisia tridentata Nuttall subsp. *rothrockii* (A. Gray) H. M. Hall & Clements; *Seriphidium rothrockii* (A. Gray) W. A. Weber

Shrubs, 20–50 cm (sticky-resinous and dark green throughout), pungently aromatic; not root-sprouting (trunks relatively narrow). **Stems** white (becoming dark gray with age), canescent (bark exfoliating). **Leaves** persistent, light or dark gray-green; blades long-cuneate to lanceolate, (0.4–)1–1.5(–2) × 0.2–0.4 cm, 3-lobed (lobes to $^{1}/_{3}$ blade lengths, rounded, margins sometimes entire, somewhat wavy), faces densely to sparsely canescent, gland-dotted, sticky-resinous. **Heads** (erect, sessile or pedunculate) in paniculiform arrays, 5–15 × 1–2(–3) cm. **Involucres** broadly ovoid, 3–5 × 4–6 mm. **Phyllaries** (usually gray-green) ovate, densely or sparsely canescent. **Florets** 12–20; corollas 2.5–3.5 mm. **Cypselae** 0.8–2 mm, (smooth), resinous. $2n = 36, 54, 72.$

Flowering mid summer–all. Clay soils of mountain meadows; 2500–3100 m; Calif.

Artemisia rothrockii is known only from the central and southern Sierra Nevada and the White Mountains of California. In the Rocky Mountains, *A. spiciformis* has been confused with *A. rothrockii*. Distinctive chemistry and anatomical structure of the leaves support the distinctness of *A. rothrockii* (L. M. Shultz 1986b). Intermediate characteristics suggest a hybrid origin from races of *A. cana* and *A. tridentata.*

16. Artemisia spiciformis Osterhout, Bull. Torrey Bot. Club 27: 507. 1900 • Snowfield sagebrush E

Artemisia tridentata Nuttall subsp. *spiciformis* (Osterhout) Kartesz & Gandhi; *Seriphidium spiciforme* (Osterhout) Y. R. Ling

Shrubs, 30–80 cm (widely branched, gray-tomentose), aromatic; root-sprouting. **Stems** relatively numerous, brown or grayish green. **Leaves** ± deciduous (by late summer, turning yellow); blades lanceolate, oblanceolate, or cuneate, 2.5–5.5 × 0.8–1.2+ cm, entire or irregularly 3–6-lobed (lobes to $^{1}/_{3}$ blade lengths, 1.5+ mm wide, rounded or acute; leaves of flowering stems usually smaller, entire), faces ± sericeous or tomentose. **Heads** (erect) in (leafy) paniculiform arrays 8–15(–25) × 0.5–3(–4) cm. **Involucres** ovoid or lanceoloid, (2.5–)4–6 (–7) mm. **Phyllaries** lanceolate, sparsely to densely hairy. **Florets** 8–18(–27); corollas 2.5–3.5, glabrous. **Cypselae** 1–1.5 mm, glabrous or resinous. $2n = 18, 36, 54, 72.$

Flowering mid summer–fall. Moist open slopes, rocky meadows, streamsides, woodlands, late-lying snowfields; 2100–3700 m; Calif., Colo., Idaho, Nev., Oreg., Utah, Wash., Wyo.

Often confused with *Artemisia rothrockii, A. spiciformis* has been recognized only recently as a widespread, high-elevation sagebrush of late-lying snowfields. Molecular analysis has not yet determined the degree to which this species intergrades with *A. cana* subsp. *viscidula* and *A. tridentata* subsp. *vaseyana,* the presumed parents of this putative hybrid. Because snowfield sagebrush produces fertile seeds and forms a stable community type, it is treated here as a distinct species.

17. Artemisia tridentata Nuttall, Trans. Amer. Philos.
Soc., n. s. 7: 398. 1841 [F]

Seriphidium tridentatum (Nuttall)
W. A. Weber

Shrubs, 40–200(–300) cm (herbage
gray-haired), aromatic; not root-
sprouting (trunks relatively thick).
Stems gray-brown, glabrate (bark
gray, exfoliating in strips). **Leaves**
persistent, gray-green; blades
usually cuneate, (0.4–)0.5–3.5 ×
0.1–0.7 cm, 3-lobed (lobes to ¹/₃ blade lengths, 1.5+ mm
wide, rounded), faces densely hairy. **Heads** (usually erect,
on slender peduncles) in paniculiform arrays 5–30 × 1–6
cm. **Involucres** lanceolate, (1–)1.5–4 × 1–3 mm. **Phyllaries**
oblanceolate to widely obovate, densely tomentose. **Florets**
3–8; corollas 1.5–2.5 mm, glabrous. **Cypselae** 1–2 mm,
hairy or glabrous, glandular.

Subspecies 4 (4 in the flora): w North America, nw
Mexico.

Artemisia tridentata has undergone considerable
taxonomic revision in the past century and
circumscription of subspecies remains a topic of
considerable controversy. Workers in the field should be
aware of the morphologic variation within the subspecies
across the range of the species (i.e., approximately from
the Sierra Nevada in the west to the plains of the Rocky
Mountains in the east). Because rangeland managers and
conservationists can often identify local morphologic and
chemical races based on grazing or habitat preferences
of wildlife and domestic animals, some impetus exists
to further subdivide the subspecies within *A. tridentata*
at the varietal level. This treatment of the species
complex remains conservative in light of the need for
further study. As to chemical differences among the
subspecies, aroma is often used to distinguish subspecies
in the field. Volatile resins in the plants are strongly
aromatic and, when crushed, leaves have very distinctive
(although not easily described) aromas.

SELECTED REFERENCES McArthur, E. D. 1984. Natural and artificial
hybridization among *Artemisia tridentata* populations. [Abstract.]
Amer. J. Bot. 71(suppl.): 105.   McArthur, E. D., B. L. Welch, and
S. C. Sanderson. 1988. Natural and artificial hybridization between
big sagebrush (*Artemisia tridentata*) subspecies. J. Heredity 79: 268–
276.

1. Shrubs 100–200(–300) cm (leaf blades: lengths
   usually 3+ times widths); heads in relatively broad,
   paniculiform arrays.
   2. Involucres 1.5–2.5 × 1–2 mm; deep, well drained
      (usually sandy) soils in valley bottoms, lower
      montane slopes along drainages . . . .
      . . . . 17a. *Artemisia tridentata* subsp. *tridentata*
   2. Involucres 2–4 × 1–2 mm; loose, sandy soils
      of valleys and foothills . . . .
      . . . . 17b. *Artemisia tridentata* subsp. *parishii*

1. Shrubs 30–150 cm (leaf blades: lengths usually
   less than 3 times widths); heads in relatively
   narrow, paniculiform arrays.
   3. Shrubs, 60–80(–150) cm (crowns flat-topped);
      heads in arrays 10–15 cm; involucres 2–3 ×
      1.5–3 mm; mountains . . . .
      . . . . 17c. *Artemisia tridentata* subsp. *vaseyana*
   3. Shrubs, 30–50(–150) cm (crowns rounded);
      heads in arrays 2–6(–8) cm; involucres (1–)1.5–
      2 × 1.5–2 mm; usually cold-desert basins and
      high plateaus, sometimes foothills . . . .
      . . . . 17d. *Artemisia tridentata* subsp. *wyomingensis*

17a. Artemisia tridentata Nuttall subsp. **tridentata**
• Great Basin sagebrush, big sage   [E] [F]

Artemisia angustifolia (A. Gray)
Rydberg; *A. tridentata* subsp.
xericensis Winward ex
R. Rosentreter & R. G. Kelsey

Shrubs, 100–200(–300) cm.
Vegetative branches nearly equaling
flowering branches. **Leaves** cuneate
or lanceolate, 0.5–1.2(–2.5) × 0.2–
0.3(–0.6) cm, 3-lobed (lobes to ¹/₃
lengths of blades, rounded). **Heads** in paniculiform arrays
5–15(–20) × (1.5–)5–6 cm. **Involucres** 1.5–2.5 × 1–2 mm.
Florets 4–6. **Cypselae** glabrous. $2n = 18, 36$.

Flowering mid summer–late fall. Deep, well-drained
(usually sandy) soils in valley bottoms, lower montane
slopes, along drainages; 1300–2200 m; Alta., B.C.; Ariz.,
Calif., Colo., Idaho, Mont., Nev., N.Mex., Oreg., Utah,
Wash., Wyo.

Subspecies *tridentata* is the common sagebrush of
deep, well-drained soils in the Great Basin of western
North America, where it is often the dominant shrub of
valleys and open grasslands. On drier sites and on high
plateaus, it is replaced by subsp. *wyomingensis*, a taxon
that appears to be increasing with prolonged droughts
and disturbance from grazing. In dry valley bottoms of
the Great Basin, subsp. *tridentata* is conspicuous by its
great height and wide arrays of heads along roadways,
fencerows, and other areas where moisture is more
readily available through runoff or reduced competition.

17b. **Artemisia tridentata** Nuttall subsp. **parishii** (A. Gray) H. M. Hall & Clements, Publ. Carnegie Inst. Wash. 326: 137. 1923 (as parishi) • Mojave sagebrush

Artemisia parishii A. Gray, Proc. Amer. Acad. Arts 17: 220. 1882; *A. tridentata* var. *parishii* (A. Gray) Jepson; *Seriphidium tridentatum* (Nuttall) W. A. Weber *subsp. parishii* (A. Gray) W. A. Weber

Shrubs, 100–200(–300) cm (crowns rounded). **Vegetative branches** interspersed among flowering stems. **Leaves** cuneate or lanceolate (1–)1.5–2(–2.5) × 0.1–0.3 cm, usually 3-lobed, sometimes entire. **Heads** in paniculiform arrays 15–30 × 2–6 cm (branches widely spreading or drooping). **Involucres** 2–4 × 1–2 mm. **Florets** 3–7. **Cypselae** hairy or glabrous. $2n = 36$.

Flowering mid summer–late fall. Loose sandy soils of valleys and foothills; 300–1800 m; Ariz., Calif., Nev., Utah; Mexico (Baja California).

Subspecies *parishii* is found in coastal ranges in southern California and Baja California, and inland to areas south of the Great Basin. It has been distinguished traditionally by the presence of drooping flowering branches and hairy cypselae, characteristics found on the type specimen. These characteristics occur sporadically in populations of other subspecies throughout the warm desert regions of southern California, Nevada, and Utah; the characteristically longer leaves and distinctive aroma support recognition of this subspecies. This treatment is the first to include Mojave Desert, Owens Valley, and Colorado Plateau populations within subsp. *parishii*.

17c. **Artemisia tridentata** Nuttall subsp. **vaseyana** (Rydberg) Beetle, Rhodora 61: 83. 1959 • Mountain sagebrush [E]

Artemisia vaseyana Rydberg in N. L. Britton et al., N. Amer. Fl. 34: 283. 1916; *A. tridentata* var. *pauciflora* Winward & Goodrich; *A. tridentata* var. *vaseyana* (Rydberg) B. Boivin; *Seriphidium vaseyanum* (Rydberg) W. A. Weber

Shrubs, 60–80(–150) cm (plants highly aromatic, crowns flat-topped). **Vegetative branches** of nearly equal lengths. **Leaves** (vegetative branches) broadly cuneate, 1.2–3.5 × 0.3–0.7 cm, regularly 3-lobed to irregularly toothed. **Heads** in paniculiform arrays 10–15 × 2–4 cm. **Involucres** 2–3 × 1.5–3 mm. **Florets** 3–9. **Cypselae** glabrous. $2n = 18, 36$.

Flowering mid summer–late fall. Montane meadows, usually in rocky soils, sometimes in forested areas; 2000–

2800 m; B.C.; Calif., Colo., Idaho, Mont., Nev., N.Dak., Oreg., S.Dak., Utah, Wash., Wyo.

Subspecies *vaseyana* is the common sagebrush of mountain slopes and is the most abundant of all the subspecies of *Artemisia tridentata*. A. A. Beetle (1960) estimated that it dominates an area of approximately 260,000 square kilometers. That estimate remains reasonably accurate today even though sagebrush is often cleared (by burning, herbicide spray, or the practice of 'chaining') and replaced by grasses (especially crested wheatgrass) suitable for livestock grazing. The acreage in which sagebrush has been removed appears to be more than compensated by acreage where it has increased in abundance because of overgrazing. While there may be evidence of introgression with other subspecies of *A. tridentata*, the subsp. *vaseyana* is usually well-separated geographically and ecologically from the other three subspecies. Variation within subsp. *vaseyana* may warrant the recognition of two varieties. A few-flowered (6 or fewer florets) form occurs at lower elevations (usually less than 2300 m) than the more robust form (with more than 6 florets per head), occurring at higher elevations (generally more than 2300 m). The type specimen of *A. vaseyana* is the large-headed variant. Pending further study, I am including var. *pauciflora* Winward & McArthur as part of subsp. *vaseyana*. In areas where populations of subsp. *vaseyana* co-occur with subspecies of *A. cana*, introgression is common.

17d. **Artemisia tridentata** Nuttall subsp. **wyomingensis** Beetle & A. M. Young, Rhodora 67: 405. 1965 • Wyoming sagebrush [E]

Artemisia tridentata var. *wyomingensis* (Beetle & A. M. Young) S. L. Welsh; *Seriphidium tridentatum* (Nuttall) W. A. Weber subsp. *wyomingense* (Beetle & A. M. Young) W. A. Weber

Shrubs, 30–50(–150) cm (crowns rounded). **Vegetative branches** (stiffly spreading, often persisting, giving mature plants a twiggy appearance) interspersed among flowering stems. **Leaves** narrowly to broadly cuneate, (0.4–)0.7–1.1(–2) × (0.1–)0.2–0.3 cm, lobed (lobes rounded). **Heads** in paniculiform arrays 2–6(–8) × 1–3 cm (often immersed in vegetative branches). **Involucres** (1–)1.5–2 × 1.5–2 mm. **Florets** 4–8. **Cypselae** glabrous. $2n = 36, 54$.

Flowering mid summer–late fall. Rocky or fine-grained soils, cold-desert basins to high plateaus, foothills; 800–2200 m; Ariz., Calif., Colo., Idaho, Mont., Nebr., Nev., N.Mex., N.Dak., Oreg., S.Dak., Utah, Wash., Wyo.

Subspecies *wyomingensis* is the common sagebrush of rocky or fine-grained soils from valleys to high plateaus in the Great Basin. It is an allopolyploid that may be derived from the populations of subsp. *tridentata* with which it occurs. Identification is based primarily on the shorter leaves of subsp. *wyomingensis*, its usually shorter stature, and its shorter flowering branches that are retained from year to year. Wyoming sagebrush may be increasing in abundance in response to increased grazing pressure and drought in the high valleys of the Great Basin.

18. **Artemisia tripartita** Rydberg, Mem. New York Bot. Gard. 1: 432. 1900 • Three-tipped sagebrush [E] [F]

Artemisia trifida Nuttall, Trans. Amer. Philos. Soc., n. s. 7: 398. 1841, not Turczaninow 1832; *A. tridentata* Nuttall subsp. *trifida* H. M. Hall & Clements; *Seriphidium tripartitum* (Rydberg) W. A. Weber

Shrubs, 5–15 or 20–150(–200) cm, aromatic; root-sprouting (caudices with adventitious buds, fibrous rooted). **Stems** pale gray, glabrous. **Leaves** deciduous, gray-green; blades broadly cuneate, 1.5–4 × 0.5–2 cm, deeply 3-lobed (lobes 1–1.4 mm wide, acute; cauline leaves smaller, mostly 3-lobed). **Heads** in paniculiform or spiciform arrays (5–)8–15(–35) × (0.5–)1–5 cm. **Involucres** globose or turbinate, 2–4 × 1.5–3 mm. **Phyllaries** broadly lanceolate (margins scarious, obscured by indument), canescent. **Florets** 3–11; corollas 2–2.5 mm, glandular (style branches included). **Cypselae** (columnar, unequally ribbed) 1.8–2.3 mm, glabrous or resinous.

Subspecies 2 (2 in the flora): w North America.

1. Shrubs 20–150(–200) cm; lobes of leaves linear, to 0.5 mm wide; loamy soils, w of continental divide . . . . 18a. *Artemisia tripartita* subsp. *tripartita*
1. Shrubs 5–15 cm; lobes of leaves lanceolate, 1–1.5 mm wide; stony grasslands, e Wyoming . . . . 18b. *Artemisia tripartita* subsp. *rupicola*

18a. **Artemisia tripartita** Rydberg subsp. **tripartita** [E] [F]

Shrubs, 20–150(–200) cm. **Leaves** 1.5–4 × 0.5–1.5 cm, lobes linear, to 0.5 mm wide. **Heads** in spiciform arrays (6–)8–15(–35) × (1–)4–5 cm. **Involucres** 2–3 × 2 mm. **Florets** 4–8. **Cypselae** 1.8–2.3 mm. $2n = 18, 36$.

Flowering mid summer–late fall. Deep loam soils, usually igneous in origin; 900–1900 m; B.C.; Idaho, Nev., Oreg., Wash., Wyo.

Subspecies *tripartita* ranges throughout the Snake River and Columbia River basins, extending north through central British Columbia, where average annual precipitation is 375–800 mm. Because much of the range includes fertile agricultural land, much of the habitat has been lost to farming, and populations of subsp. *tripartita* occur as isolated islands along drainages and at the bases of mountain slopes. It may be one of the parents involved in the presumed hybrid origin of *Artemisia arbuscula* subsp. *thermopola*.

18b. **Artemisia tripartita** Rydberg subsp. **rupicola** Beetle, Rhodora 61: 82. 1959 [E]

Shrubs, 5–15 cm. **Leaves** 1.5–3.5 × 1–2 cm; lobes lanceolate, 1–1.5 mm wide. **Heads** in paniculiform arrays (5–)8–12(–15) × (0.5–)1–3 cm. **Involucres** 2–4 × 1.5–3 mm. **Florets** 3–11. **Cypselae** 1.8–2 mm.

Flowering early–late summer. Shallow rocky soils, grasslands; 2500–2900 m; Wyo.

Subspecies *rupicola* is known only from the high, windy plains of south-central Wyoming. The continental divide separates subsp. *tripartita* to the west and subsp. *rupicola* to the east. Because of its limited distribution and scant representation in herbaria, this distinct morphologic form escaped notice until it was described in 1959. Its dwarf habit and ecology distinguish it from subsp. *tripartita*.

119c. ARTEMISIA Linnaeus subg. ABSINTHIUM (Miller) Lessing, Syn. Gen. Compos., 264. 1832

Absinthium Miller, Gard. Dict. Abr. ed. 4, vol. 1. 1754; *Artemisia* subsect. *Absinthium* (Miller) Darijma

Perennials; fibrous rooted, caudices woody, rhizomes absent. **Stems** not wandlike. **Leaves** deciduous or persistent, basal and/or cauline (petiolate or sessile, not in fascicles). **Heads** disciform. **Receptacles** epaleate, villous (hairs relatively long). **Florets:** peripheral 6–27 pistillate and fertile; central 15–100 bisexual and fertile; corollas (pale yellow) funnelform.

Species ca. 40 (5 in the flora): temperate regions, North America, South America, Eurasia.

SELECTED REFERENCE Besser, W. S. J. G. von. 1829. Lettre [on *Artemisia* including *Absinthium*]...au Directeur. Bull. Soc. Imp. Naturalistes Moscou 1: 219–265.

1. Plants 10–40 or 40–60(–100) cm; leaves pinnately lobed (basal 2–3-pinnatifid) or 1–2-ternately lobed.
    2. Leaves pinnately lobed (basal 2–3-pinnatifid, lobes obovate) .... 19. *Artemisia absinthium*
    2. Leaves 1–2-ternately lobed (lobes filiform, to 0.5 mm wide) .... 20. *Artemisia frigida* (in part)
1. Plants 5–50 cm; leaves entire or 1–3-pinnately lobed.
    3. Involucres 3–5 mm diam. .... 20. *Artemisia frigida* (in part)
    3. Involucres 4–8 mm diam.
        4. Leaves bright green, faces glabrous or sparsely hairy; phyllary margins light green
        .... 22. *Artemisia rupestris*
        4. Leaves gray-green, faces canescent to villous; phyllary margins black to brown.
            5. Heads borne singly or (2–5) in paniculiform to racemiform arrays; corolla lobes
            glabrous .... 21. *Artemisia pattersonii*
            5. Heads (5–22) in spiciform arrays; corolla lobes hairy .... 23. *Artemisia scopulorum*

19. Artemisia absinthium Linnaeus, Sp. Pl. 2: 848.
1753 • Common wormwood, armoise absinthe ⬚I

Perennials, 40–60(–100) cm (mat-forming), aromatic. **Stems** gray-green (sometimes woody proximally), densely canescent to glabrescent (hairs appressed). **Leaves** deciduous, gray-green; blades broadly ovate, 3–8 × 1–4 cm, mostly pinnately lobed (basal 2–3-pinnatifid, lobes obovate), faces densely canescent. **Heads** (nodding) in open (diffusely branched), paniculiform arrays 10–20(–35) × (2–)10–13 (–15) cm. **Involucres** broadly ovoid, 2–3 × 3–5 mm. **Phyllaries** gray-green, densely sericeous. **Florets:** pistillate 9–20; bisexual 30–50; corollas 1–2 mm, glandular. **Cypselae** (± cylindric, slightly curved, obscurely nerved), ± 0.5 mm, glabrous (shiny). $2n = 18$.

Flowering mid summer–fall. Widely cultivated, persisting from plantings, disturbed areas; 0–1000 m; introduced; Alta., B.C., Man., N.B., Nfld. and Labr. (Nfld.), N.S., Ont., P.E.I., Que., Sask.; Calif., Colo., Conn., Idaho, Ill., Ind., Iowa, Kans., Maine, Md., Mass., Mich., Minn., Mo., Mont., Nebr., N.H., N.J., N.Y., N.C., N.Dak., Ohio, Oreg., Pa., R.I., S.C., S.Dak., Tenn., Utah, Vt., Wash., Wis., Wyo.; Europe.

Artemisia absinthium provides the flavoring as well as the psychoactive ingredient for *absinthe* liquor, a beverage that is illegal in some markets. Known as a powerful neurotoxin, absinthe in large quantities is addictive as well as deadly. The species is popular in the horticultural trade. Prized by gardeners for its gracefully scalloped leaves and gray-green foliage, it creates an attractive and winter-hardy flower border.

20. Artemisia frigida Willdenow, Sp. Pl. 3: 1838. 1803
• Fringed sage, prairie sagewort, armoise douce ⬚F

Artemisia frigida var. *gmeliniana* (Besser) Besser; *A. frigida* var. *williamsiae* S. L. Welsh

Perennials, 10–40 cm (forming silvery mats or mounds), strongly aromatic. **Stems** gray-green or brown, glabrescent. **Leaves** persistent, silver-gray; blades ovate, 0.5–1.5(–2.5) cm, 1–2-ternately lobed (lobes 0.2–0.5 mm wide), faces densely whitish-pubescent. **Heads** in (leafy) paniculiform arrays 0.5–2(–4) × 4–15(–20) cm. **Involucres** globose, (3–)5 × (2–)5–6 mm. **Phyllaries** gray-green (margins sometimes brownish), densely tomentose. **Florets:** pistillate 10–17; bisexual 20–50; corollas 1.5–2 mm, glabrous. **Cypselae** 1–1.5 mm, glabrous. $2n = 18$.

Flowering summer–fall. Fields, meadows, dry grasslands, steppes, usually stony, well-drained soils; 500–3300 m; Alta., B.C., Man., N.B., N.W.T., Nunavut, Ont., Que., Sask., Yukon; Alaska, Ariz., Colo., Idaho, Ill., Iowa, Kans., Minn., Mont., Nebr., N.Mex., N.Dak., S.Dak., Tex., Utah, Wash., Wis., Wyo.; Eurasia.

Reports of *Artemisia frigida* from eastern Canada (Ontario eastward), the eastern United States (e.g., Connecticut, Massachusetts, Michigan, Vermont), and Arkansas and Missouri appear to be from old garden sites where the plants may persist. The similarity of this native species to cultivars from eastern Asia (especially Siberia) has led to a number of reports that are apparently based on other cultivars. As a plant with attractive silver foliage, this species has good potential as a drought-hardy plant for flower gardens in cold climates.

21. Artemisia pattersonii A. Gray in A. Gray et al., Syn. Fl. N. Amer. ed. 2, 1(2): 453. 1886 (as pattersoni)

• Patterson sagewort [E]

Artemisia monocephala (A. Gray) A. Heller; *A. scopulorum* A. Gray var. *monocephala* A. Gray

Perennials, 8–20 cm, mildly aromatic. **Stems** gray-brown, glabrate or finely pubescent. **Leaves** deciduous, gray-green; petiolate; blades (basal) broadly spatulate, 2–4 × 0.5 cm, pinnately lobed (lobes ca. 1.5 mm wide; cauline smaller, 1-pinnately lobed or entire), faces silky-hairy. **Heads** borne singly or (2–5, spreading to nodding, pedunculate) in paniculiform or racemiform arrays 1–5 × 0.5–1 cm. **Involucres** broadly hemispheric, 5–8 × 5–8(–10) mm. **Phyllaries** gray (margins dark brown to black), villous. **Florets:** pistillate 7–27; bisexual 32–100; corollas (yellow tinged with red), 2–3 mm (including exsert anthers), mostly glabrous (embedded in tangled receptacular hairs). **Cypselae** 1.5–2 mm, glabrous. $2n = 14$.

Flowering mid–late summer. Alpine meadows; 3500–4000 m; Colo., N.Mex., Wyo.

Artemisia pattersonii can be distinguished from the closely related *A. scopulorum* by its heads being borne singly and narrower phyllary margins.

22. Artemisia rupestris Linnaeus, Sp. Pl. 2: 847. 1753

Absinthium viridifolium (Ledebour) Besser var. *rupestre* (Linnaeus) Besser; *Artemisia rupestris* subsp. *woodii* Neilson

Perennials, 5–15(–25) cm (cespitose), faintly aromatic. **Stems** brownish purple, glabrous. **Leaves** deciduous, bright green; blades (proximalmost petiolate) ovate, 1.5–5 × 1–2.5 cm, 2–3-pinnately lobed (cauline sessile, ternately or pinnately lobed, terminal lobes lance-linear, 1–6 × 0.5–1 mm), faces glabrous or sparsely hairy, glandular. **Heads** (5–9, pedunculate or sessile, spreading or drooping) in spiciform arrays 3–9 × 0.5–1 cm. **Involucres** globose, 4–5(–7) × 4–5(–7) mm. **Phyllaries** green (margins light green), ± hairy. **Florets:** pistillate 14–16 (glandular, style branches exsert, linear, spreading); bisexual 40–70; corollas 1.5–2 mm, glabrous or glandular (styles shorter than corollas). **Cypselae** ca. 1 mm (apices flat), glabrous.

Flowering late summer–fall. Steppes, alkaline meadows, stony slopes; 0–1400 m; Yukon; Asia.

The sole North American occurrence of *Artemisia rupestris* in southwestern Yukon is a remarkable disjunction from the Asiatic range of this species.

23. Artemisia scopulorum A. Gray, Proc. Acad. Nat. Sci. Philadelphia 15: 66. 1863 [E]

Perennials, 10–25 cm (cespitose), mildly aromatic (caudices relatively slender). **Stems** gray-green, glabrate. **Leaves** persistent, gray-green; blades (basal) oblanceolate, 2–7 × 0.1 cm, 2-pinnately lobed (lobes linear or oblanceolate; cauline blades smaller, 1–2-pinnate or entire), faces silky-canescent. **Heads** (5–22) in spiciform arrays 5–9 × 1–1.5 cm. **Involucres** broadly globose or subglobose, 4 × 4–7 mm. **Phyllaries** green (margins black or dark brown), densely villous. **Florets:** pistillate 6–13; bisexual 15–30; corollas 1.5–2.5 mm, hairy (at least on lobes). **Cypselae** 0.8–1 mm, glabrous. $2n = 18$.

Flowering mid–late summer. Alpine meadows, protected areas, bases of rocks; 3100–4200 m; Colo., Mont., Nev., N.Mex., Utah, Wyo.

119d. ARTEMISIA Linnaeus subg. ARTEMISIA

Artemisia sect. *Abrotanum* Besser

Annuals, biennials, or perennials (shrubs in *A. californica*, subshrubs in *A. nesiotica*); usually fibrous-rooted, sometimes taprooted, caudices sometimes woody, rhizomes sometimes present. **Stems** usually not wandlike (wandlike in *A. californica*, *A. nesiotica*, *A. palmeri*). **Leaves** usually deciduous, rarely persistent, basal (rosettes) and/or cauline (not in fascicles). **Heads** usually disciform (discoid in *A. nesiotica* and *A. palmeri*). **Receptacles** glabrous (paleate in *A. palmeri*). **Florets:** usually peripheral 3–20 pistillate and fertile (0 pistillate in *A. nesiotica*, *A. palmeri*); central (or all) 14–70 bisexual and fertile; corollas funnelform.

Species ca. 220 (27 in the flora): widespread in Northern Hemisphere, especially North America, Europe, and central and northern Asia, sporadic in South America and northern Africa.

SELECTED REFERENCES Estes, J. R. 1969. Evidence for autoploid evolution in the *Artemisia ludoviciana* complex of the Pacific Northwest. Brittonia 21: 29–43. Keck, D. D. 1946. A revision of the *Artemisia vulgaris* complex in North America. Proc. Calif. Acad. Sci., ser. 4, 25: 421–468.

1. Subshrubs or shrubs (stems wandlike).
    2. Plants 100–350 cm; leaves relatively deeply and coarsely pinnately lobed (lobes 3–7+; coastal California and Baja California) . . . . 42. *Artemisia palmeri*
    2. Plants 10–250 cm; leaves pinnately lobed or 3-lobed.
        3. Shrubs (20–250 cm); leaves pinnately lobed (lobes 0.5–1 mm wide); California (chaparral) . . . . 28. *Artemisia californica*
        3. Subshrubs (10–60 cm, stems mostly prostrate); leaves 3-lobed (lobes 1–2 mm wide); Channel Islands, California . . . . 39. *Artemisia nesiotica*
1. Annuals, biennials, perennials, subshrubs, or shrubs (stems sometimes brittle, not wandlike).
    4. Annuals or biennials; leaves among heads (relatively deeply) lobed.
        5. Annuals, 30–200(–300) cm, sweetly aromatic; leaves 2–3-pinnatifid; arrays of heads 10–20 cm diam . . . . 26. *Artemisia annua*
        5. Annuals or biennials, (10–)30–80(–150) cm, not aromatic; leaves 1–2-pinnately lobed; arrays of heads 2–4 cm diam . . . . 27. *Artemisia biennis*
    4. Perennials, subshrubs, or shrubs; leaves among heads mostly entire.
        6. Perennials, subshrubs, or shrubs (not rhizomatous).
            7. Heads in capitate or dense, corymbiform arrays (plants cespitose).
                8. Leaves pinnately lobed; involucres 3–5 mm diam.; phyllaries lanceolate to ovate (margins white) . . . . 45. *Artemisia senjavinensis*
                8. Leaves 1–2-palmatifid; involucres 3.5–11 mm diam.; phyllaries lanceolate (margins brown or white).
                    9. Involucres 3.5–6 × 6–11 mm; phyllaries lanceolate (margins brown); corollas yellow or reddish black, glabrous or glandular (not pilose) . . . . 33. *Artemisia globularia*
                    9. Involucres 3–4 × 3.5–5 mm; phyllaries lanceolate (margins white); corollas yellow, glabrous or pilose . . . . 34. *Artemisia glomerata*
            7. Heads in paniculiform or racemiform arrays (plants not cespitose).
                10. Leaves entire, irregularly palmatifid, or palmately 3-lobed to 2-ternately lobed.
                    11. Pappi coroniform; Idaho, Nevada, Oregon . . . . 43. *Artemisia papposa*
                    11. Pappi 0; Alberta, British Columbia, Northwest Territories, Nunavut; Alaska, Washington.
                        12. Plants 15–70 cm; leaves palmately 3-lobed to 2-ternately lobed . . . . 25. *Artemisia alaskana*
                        12. Plants (5–)10–40 cm; leaves 1–3-palmately lobed . . . . 32. *Artemisia furcata*
                10. Leaves 2–3-pinnatifid.
                    13. Perennials or subshrubs, 50–170 cm (widely branched, stems brittle); leaf lobes less than 1 mm wide; heads erect; involucres 1.5–3 mm diam. (gardens, waste places, much of North America) . . . . 24. *Artemisia abrotanum*
                    13. Perennials, 10–50 cm (erect, stems not brittle); leaf lobes 1+ mm wide (margins coarsely toothed); heads nodding; involucres 4–10 mm diam.
                        14. Peduncles 0 or to 10 mm . . . . 35. *Artemisia laciniata*
                        14. Peduncles to 50 mm . . . . 40. *Artemisia norvegica*
        6. Perennials (usually rhizomatous, stems sometimes woody at bases).
            15. Leaves entire, serrate, toothed, or lobed (sinuses to 1/2 blade widths).
                16. Plants 20–50(–80, rarely more) cm.
                    17. Leaves usually entire, sometimes toothed or lobed; involucres 4–5 mm diam. . . . . 36. *Artemisia longifolia*
                    17. Leaves usually lobed, sometimes entire; involucres (1–)2–5 mm diam.

18. Involucres 2–4(–5.5) mm; phyllaries (gray-green) densely tomentose
. . . . 37. *Artemisia ludoviciana*
18. Involucres 3.5–4 mm; phyllaries (violet-brown) sparsely tomentose
(w North America) . . . . 49. *Artemisia tilesii*
16. Plants 50–300 cm.
19. Phyllaries glabrous or sparsely hairy (coast, n California to British
Columbia) . . . . 48. *Artemisia suksdorfii*
19. Phyllaries usually densely hairy.
20. Leaves densely hairy (both faces, broadly lanceolate, mostly entire,
the proximal lobed; w North America, mostly inland grasslands)
. . . . 30. *Artemisia douglasiana*
20. Leaves (bicolor) hairy abaxially, glabrate or glabrous adaxially.
21. Leaves serrate (teeth ca. 2 mm; inland grasslands and barren
areas, high plains) . . . . 46. *Artemisia serrata*
21. Leaves mostly deeply lobed (lobes 4–20 mm; mostly e North
America, introduced w coast) . . . . 50. *Artemisia vulgaris*
[15. Shifted to left margin.—Ed.]
15. Leaves (relatively deeply) lobed (sinuses ¹/₂+ blade widths).
22. Leaves not bicolor (both faces bright green or silvery), lobes acute or rounded.
23. Leaves silver-gray, lobes rounded (coastal dunes) . . . . 47. *Artemisia stelleriana*
23. Leaves bright green, lobes acute (not coastal dunes).
24. Involucres (4–)5–8 × 4–10 mm; 0–3800 m . . . . 40. *Artemisia norvegica*
24. Involucres 2.5–3.5 × 2–4.5 mm; 100–2400 m . . . . 41. *Artemisia packardiae*
22. Leaves bicolor (abaxial faces silvery, adaxial green), lobes acute.
25. Plants 30–100 cm, lemon-scented; heads usually erect (subalpine and alpine)
. . . . 38. *Artemisia michauxiana*
25. Plants 15–70 cm, not lemon-scented; heads usually nodding.
26. Perennials (widely spreading, stems brittle); garden escapes, c, e North America
. . . . 44. *Artemisia pontica*
26. Biennials or perennials (erect).
27. Leaves pinnatifid (lobes 3–5, 0.5–1 mm wide); grasslands or deserts, 600–
2900 m . . . . 29. *Artemisia carruthii*
27. Leaves 2–3-pinnately lobed (lobes elliptic, 2–6 mm wide); w mountains,
2200–3100 m . . . . 31. *Artemisia franserioides*

24. **Artemisia abrotanum** Linnaeus, Sp. Pl. 2: 845.
1753 • Southernwood, lad's love, old man, armoise
aurone [I]

Perennials or subshrubs, 50–130(–170) cm (not cespitose), aromatic (roots thick, woody). **Stems** relatively numerous, erect, brown, branched, (woody, brittle), glabrous or sparsely hairy. **Leaves** cauline, dark green; blades broadly ovate, (2–)3–6 × 0.02–0.15 cm, 2–3-pinnatifid (lobes linear or filiform), faces sparsely hairy (abaxial) or glabrous (adaxial). **Heads** (nodding at maturity) in open, widely branched arrays 10–30 × 2–10 cm. **Involucres** ovoid, (1–)2–3.5 × (1–)2–2.5 mm. **Phyllaries** oblong-elliptic, sparsely hairy. **Florets:** pistillate 4–8(–15); bisexual 14–16(–20); corollas yellow, 0.5–1 mm, glandular. **Cypselae** (light brown) ellipsoid (2–5-angled, flattened, furrowed), 0.5–1 mm, glabrous. $2n$ = 18.

Flowering late summer–fall. Waste places; 0–3000 m; introduced; Alta., Man., N.B., Ont., Que., Sask.; Colo., Conn., Del., D.C., Ill., Iowa, Kans., Maine, Md., Mass., Mich., Minn., Nebr., N.H., N.J., N.Y., N.C., Oreg., Pa., S.C., Utah, Vt., Wis., Wyo.; Eurasia; Africa.

Artemisia abrotanum has been widely cultivated in gardens for old-time uses such as a fly and parasite repellent. It has had a renewed popularity in xeriscape gardening; it is drought tolerant and can fill difficult garden spaces (e.g., dry rocky slopes). Reports of naturalization may be exaggerated; it is not known to become weedy in any of its known locations in North America.

25. Artemisia alaskana Rydberg in N. L. Britton et al., N. Amer. Fl. 34: 281. 1916 • Siberian wormwood E

Artemisia tyrrellii Rydberg

Perennials or subshrubs, 15–30 (–60) cm (not cespitose), aromatic (caudices woody). **Stems** 1–10, erect, gray-green, simple (suffrutescent from woody offsets), densely hairy to glabrescent. **Leaves** basal and cauline, mostly gray-green; blades obovate, 1.5–5 × 0.5–1.5 cm, 3-lobed to 2-ternately lobed (lobes 0.5–3 mm wide, margins flat; cauline leaves smaller, sometimes entire), faces tomentose. **Heads** (peduncles 0 or to 30 mm) in (leafy) paniculiform to racemiform arrays 12–25 × 1–4.5 cm. **Involucres** broadly campanulate, 3.5–5 × 6–9 mm. **Phyllaries** ovate (margins brownish or hyaline), tomentose. **Florets:** pistillate 8–10; bisexual 20–45; corollas yellow, 2–2.5 mm, glabrous or glandular. **Cypselae** ellipsoid (flattened), 1–1.5 mm, glabrous. $2n = 18$.

Flowering early–late summer. Well-drained soils, flood plains, gravel stream banks, roadsides, dry, rocky slopes, forest openings, alpine and arctic tundras; 100–2500 m; B.C., N.W.T., Yukon; Alaska.

As circumscribed here, *Artemisia alaskana* is known from northwestern North America. The type specimen of *A. alaskana* is atypical, with longer peduncles and narrower leaf lobes than are found in most populations.

26. Artemisia annua Linnaeus, Sp. Pl. 2: 847. 1753 • Sweet Annie, sweet sagewort, armoise annuelle I

Artemisia chamomilla C. Winkler

Annuals, 30–200(–300) cm, sweetly aromatic. **Stems** mostly 1, erect, green, turning to reddish brown with age, simple (smooth or ribbed), glabrous or sparsely hairy. **Leaves** cauline, bright green; blades triangular to broadly ovate, 2–5(–10) × 2–4 cm, 2–3-pinnatifid (lobes relatively narrow, ± toothed), faces glabrous, gland-dotted. **Heads** (nodding, peduncles 2–5 mm) in open, (diffusely branched, leafy) arrays 15–30 (–40) × 10–20 cm. **Involucres** globose, 1.5–2.5 × 1.5–2.5 mm. **Phyllaries** (green) lanceolate, glabrous. **Florets:** pistillate (0–)10–20; bisexual 18–24; corollas pale yellow (broadly campanulate), 0.5–1 mm, glabrous. **Cypselae** oblong (flattened), 0.3–0.8 mm, glabrous. $2n = 18$.

Flowering late summer–fall. Moist waste areas, sandy soils; 0–2000 m; introduced; N.B., Ont., Que.; Ala., Ariz., Ark., Calif., Colo., Conn., Del., D.C., Idaho, Ill., Ind.,

Iowa, Kans., Ky., La., Maine, Md., Mass., Mich., Miss., Mont., Nebr., N.H., N.J., N.Y., N.C., Ohio, Okla., Oreg., Pa., Tenn., Tex., Vt., Va., Wash., W.Va., Wis., Wyo.; Eurasia.

Widely cultivated for aromatic oils, *Artemisia annua* often persists in gardens, becoming naturalized in moist-temperate areas (especially in eastern United States). Reports of naturalization may be exaggerated (reported for Prince Edward Island, but not established).

The systematic placement of this species appears to align most closely with species of the Eurasian subg. *Seriphidium* (L. E. Watson et al. 2002). Molecular evidence suggests that the *Artemisia annua* lineage may be ancestral to woody species in the Old World.

27. Artemisia biennis Willdenow, Phytographia, 11. 1794 • Biennial wormwood, armoise bisannuelle E

Artemisia biennis var. *diffusa* Dorn

Annuals or biennials, (10–)30–80(–150) cm, not aromatic. **Stems** 1, erect, often reddish, simple (finely striate), glabrous. **Leaves** cauline, green or yellow-green (sessile); blades broadly lanceolate to ovate, 4–10(–13) × 1.5–4 cm, 1–2-pinnately lobed (ultimate lobes coarsely toothed), faces glabrous. **Heads** (erect, subsessile) in (leafy) paniculiform to spiciform arrays 12–35(–40) × 2–4 cm (lateral branches relatively short). **Involucres** globose, 2–4 × (1.5–)2–4 mm. **Phyllaries** (green) broadly elliptic to obovate, glabrous. **Florets:** pistillate 6–25; bisexual 15–40; corollas pale yellow, ca. 2 mm, glabrous. **Cypselae** ellipsoid (4–5-nerved), 0.2–0.9 mm, glabrous. $2n = 18$.

Flowering mid summer–late fall. Disturbed habitats, margins of vernal pools, desert flats, usually clay or silty soils; 600–2000 m; Alta., B.C., Man., N.B., N.W.T., N.S., Ont., P.E.I., Que., Sask., Yukon; Alaska, Ariz., Calif., Colo., Conn., Del., Idaho, Ill., Ind., Iowa, Kans., Ky., Maine, Md., Mass., Mich., Minn., Mo., Mont., Nebr., Nev., N.H., N.J., N.Mex., N.Y., N.Dak., Ohio, Oreg., Pa., R.I., S.Dak., Utah, Vt., Wash., W.Va., Wis., Wyo.; introduced in Europe, Pacific Islands (New Zealand).

Artemisia biennis is naturalized and weedy in the eastern portion of its range. It is morphologically similar to *A. annua*, differing primarily in the coarser leaf lobes and larger heads that are sessile in axils of leaflike bracts. *Artemisia biennis* is considered native to the northwest United States; it may be introduced in other parts of its range. The type specimen is a horticultural specimen from New Zealand.

28. Artemisia californica Lessing, Linnaea 6: 523. 1831 • California sagebrush

Artemisia abrotanoides Nuttall; *A. fischeriana* Besser; *A. foliosa* Nuttall; *Crossostephium californicum* (Lessing) Rydberg

Shrubs, (20–)150–250 cm (rounded), pungently aromatic. **Stems** relatively numerous, arched, green or brown, branched (slender, wandlike, bases brittle), densely canescent to glabrate. **Leaves** cauline, light green to gray; blades filiform or spatulate to obovate, 3–5(–9) × 0.5–2 cm, sometimes pinnately lobed (lobes filiform, 0.5–1 mm wide), faces sparsely to densely hairy. **Heads** (nodding at maturity, pedunculate) in paniculiform arrays 6–20 × 1–3 cm (branches erect to broadly spreading). **Involucres** globose, 2–3(–4) × 2–4(–5) mm. **Phyllaries** broadly ovate, sparsely canescent. **Florets:** pistillate 6–10; bisexual 18–25; corollas pale yellow, 0.8–1.2 mm, glabrous. **Cypselae** ellipsoid, 0.5–1.5 mm, resinous (**pappi** coroniform). $2n = 18$.

Flowering early–late summer. Coastal scrub, dry foothills; 0–800 m; Calif.; Mexico (Baja California).

Artemisia californica is the common sagebrush of chaparral in southern California. Its threadlike leaves and green flowering heads distinguish it from any other shrub in California. *Artemisia nesiotica*, an endemic of the Channel Islands that was initially considered a morphologic variant of *A. californica*, is distinct in size and form. Systematic placement of the complex may be problematic. The molecular phylogeny of L. E. Watson et al. (2002) suggests an alignment of *A. californica* within subg. *Tridentatae*. Based on this finding, a subgeneric realignment of this species may be in order. The odor of *A. californica* is markedly like that of the culinary mints known as common sage (*Salvia* species).

29. Artemisia carruthii Alph. Wood ex Carruth, Trans. Kansas Acad. Sci. 5: 51. 1877 • Carruth wormwood

Artemisia bakeri Greene; *A. coloradensis* Osterhout; *A. kansana* Britton; *A. vulgaris* Linnaeus subsp. *wrightii* (A. Gray) H. M. Hall & Clements; *A. wrightii* A. Gray

Perennials, 15–40(–70) cm, faintly aromatic (rhizomatous). **Stems** mostly 3–8, ascending, brown to gray-green, simple (bases curved, somewhat woody), sparsely to densely tomentose. **Leaves** cauline, bicolor (± gray-green); blades narrowly elliptic, 0.1–2.5(–3) × 0.5–1 cm (gradually smaller distally), relatively deeply pinnatifid (lobes 3–5), faces densely tomentose (abaxial)

to sparsely hairy (adaxial). **Heads** (usually nodding) in (leafy) paniculiform arrays 10–30 × 3–9 cm (branches erect). **Involucres** campanulate, 2–2.5(–3) × 1.5–3 mm. **Phyllaries** lanceolate, gray-tomentose. **Florets:** pistillate 1–5; bisexual 7–25; corollas pale yellow, 1–2 mm, glandular-pubescent. **Cypselae** (light brown) cylindro-elliptic, ca. 0.5 mm, (curved at summits, scarcely nerved), glabrous (shining). $2n = 18$.

Flowering mid summer–early fall. Open sites, usually sandy soils, wooded areas, grasslands, railroads; 600–2900 m; Ariz., Colo., Kans., Mich., Mo., N.Mex., Okla., Tex., Utah; Mexico (Chihuahua, Sonora).

Artemisia carruthii is closely related to members of the *A. ludoviciana* complex, with which it may intergrade.

30. Artemisia douglasiana Besser in W. J. Hooker, Fl. Bor.-Amer. 1: 323. 1833 • Northwest mugwort, Douglas sagewort E

Artemisia campestris Linnaeus var. *douglasiana* (Besser) B. Boivin; *A. caudata* Michaux var. *douglasiana* (Besser) B. Boivin; *A. commutata* Besser var. *douglasiana* (Besser) Besser; *A. desertorum* Sprengel var. *douglasiana* Besser; *A. heterophylla* Nuttall; *A. ludoviciana* Nuttall var. *douglasiana* (Besser) D. C. Eaton; *A. vulgaris* Linnaeus var. *douglasiana* (Besser) H. St. John

Perennials, 50–180(–250) cm, aromatic (rhizomatous). **Stems** 1–20, erect, brown to gray-green, simple, hairy or glabrescent. **Leaves** cauline, bicolor (white and green to light gray-green); blades narrowly elliptic to widely oblanceolate, (1–)3–11(–15) × 0.5–2(–6) cm (proximal with 3–5 lateral lobes, distal mostly entire), faces sparsely tomentose (abaxial) to sparsely hairy (adaxial). **Heads** (usually nodding) in (leafy) paniculiform arrays 10–30 × 3–9 cm (branches widely spreading, ascending, stout). **Involucres** narrowly turbinate to campanulate, 2–3 × 2–4 mm. **Phyllaries** (green to gray) ovate, tomentose to pubescent. **Florets:** pistillate 6–10; bisexual 6–25; corollas pale yellow, 1–1.5 mm, glabrous, sometimes glandular. **Cypselae** ellipsoid, 0.5–1 mm, glabrous. $2n = 54$.

Flowering mid spring–late fall. Meadows, shaded sites, along drainages; 100–2200 m; Calif., Nev., Oreg., Wash.

Artemisia douglasiana is sometimes weedy. Reports from areas outside the northwestern portion of the United States are based on misidentifications of plants in the *A. ludoviciana* complex.

31. **Artemisia franserioides** Greene, Bull. Torrey Bot. Club 10: 42. 1883 • Bursage mugwort

Biennials or perennials, 30–100 cm, faintly aromatic (rhizomatous). **Stems** 1–3, erect, reddish brown, simple (leafy), glabrous or glabrate. **Leaves** basal (in rosettes, petiolate) and cauline, bicolor (white and green); blades ovate, 3–7(–20) × 2–4(–6) cm, 2–3-pinnately-lobed (lobes elliptic, 2–6 mm wide; cauline sessile, smaller), faces tomentose (abaxial) or glabrous or glabrescent (adaxial), glandular. **Heads** (nodding, peduncles 0 or 2) in paniculiform to racemiform arrays 10–35 × 2–4 cm (often 1-sided). **Involucres** broadly ovate, 3–5 × 4–5(–6) mm. **Phyllaries** broadly ovate, sparsely hairy. **Florets:** pistillate 4–5 (–13), (1–1.5 mm); bisexual 25–35; corollas yellow, 1.5–2 mm, glabrous. **Cypselae** elliptic, 0.5–0.8 mm, glabrous.

Flowering late summer–early fall. Open coniferous forests, mid to upper montane; 2200–3100 m; Ariz., Colo., N.Mex.; Mexico (Chihuahua).

32. **Artemisia furcata** M. Bieberstein, Fl. Taur.-Caucas. 3: 567. 1819

Artemisia furcata var. *heterophylla* (Besser) Hultén; *A. hyperborea* Rydberg; *A. tacomensis* Rydberg; *A. trifurcata* Stephani ex Sprengel

Perennials, 7–35 cm (not cespitose), faintly aromatic (not rhizomatous, taproots stout, caudices simple or branched, branches clothed with persistent leaf bases). **Stems** (flowering) 1–5, erect, light brown, simple, strigillose or glabrate. **Leaves** basal (in rosettes) and cauline, gray-green; blades oval, 2–10(–12) cm (basal) or 1–1.5 × 0.4–0.6 cm (cauline), 1–3-palmately lobed, faces sparsely to densely strigillose. **Heads** (erect or spreading, some nodding, peduncles 0 or to 30 mm) in racemiform or spiciform arrays 1–6 × 1–2 cm. **Involucres** broadly campanulate, 3–6 × 4.5–8 mm. **Phyllaries** (greenish, color often obscured by indument) ovate or lanceolate (margins dark brown), sparsely to densely tomentose. **Florets:** pistillate 6–7; bisexual 15–26; corollas mostly yellow, sometimes red-tinged, 1–2 mm, glabrous or glabrate. **Cypselae** oblong (ribbed), 1–1.5 mm, glabrous. $2n$ = 18, 36, 72, 90.

Flowering late summer. Talus slopes or tundra; 500–2700 m; Alta., B.C., N.W.T., Nunavut, Yukon; Alaska, Wash.; Asia.

Artemisia furcata extends from the islands of the Bering Sea into southern and interior Alaska, parts of Canada (disjunct in British Columbia and the northernmost Rocky Mountains of Alberta), and on Mt. Rainier in Washington. The array of names applied to *A. furcata* shows the taxonomic confusion arising from a myriad of morphologic variants that may indicate introgression with other species.

33. **Artemisia globularia** Chamisso ex Besser, Nouv. Mém. Soc. Imp. Naturalistes Moscou 3: 64. 1833

Ajania globularia (Besser) Poljakov; *Artemisia norvegica* Fries subsp. *globularia* (Besser) H. M. Hall & Clements

Perennials, (3–)5–16(–30) cm (cespitose), faintly aromatic (not rhizomatous, taproots stout, caudices simple or branched, proximal branches clothed with persistent leaf bases). **Stems** 1–5, erect, whitish gray, densely tomentose. **Leaves** mostly basal (cauline 1–4), greenish to whitish green; blades (basal) 1–4.5 × 0.6–1.5 cm, 1–2-ternately to palmately lobed (flowering-stem blades 3-lobed), faces sparsely hairy. **Heads** (2–20, peduncles 0 or to 25 mm) in subcapitate to capitate arrays 2–3 × 2–3 cm. **Involucres** campanulate or hemispheric, 3.5–6 × 6–11 mm. **Phyllaries** lanceolate (margins brown), pilose. **Florets:** pistillate 9–10; bisexual 20–30; corollas yellow or reddish black, 2–3 mm, sometimes glandular. **Cypselae** oblong, 1.5–2.5 mm, (apices flattened) glabrous.

Subspecies 2 (2 in the flora): nw North America, Asia.

1. Corollas reddish black; cypselae ca. 2.5 mm, margins with relatively narrow ribs. 33a. *Artemisia globularia* subsp. *globularia*
1. Corollas yellow; cypselae 1.5–2 mm, margins with relatively broad ribs. . . . . 33b. *Artemisia globularia* subsp. *lutea*

33a. **Artemisia globularia** Chamisso ex Besser subsp. **globularia**

Flowering stems 2–12(–16) cm. **Corollas** reddish black, eglandular. **Cypselae** ca. 2.5 mm, margins with relatively narrow ribs. $2n$ = 18, 36.

Flowering late summer. Arctic and alpine tundra, interior mountains; 200–1300 m; Yukon; Alaska; e Asia (Russian Far East).

The red color of the glabrous corollas and the larger heads of subsp. *globularia* distinguish it from the morphologically similar *Artemisia glomerata*.

33b. Artemisia globularia Chamisso ex Besser subsp. **lutea** (Hultén) L. M. Shultz, Sida 21: 1638. 2005 E

Artemisia globularia var. *lutea* Hultén, Fl. Alaska Yukon 10: 1567. 1950; *A. flava* Jurtzev

Flowering stems 5–9 cm. **Corollas** yellow, orange-glandular. **Cypselae** 1.5–2 mm, margins with relatively broad ribs.

Flowering late spring–mid summer. Rocky ridges and slopes; 0–100 m; Alaska.

Subspecies *lutea* is common on St. Mathew Island and infrequent on surrounding islands and the mainland (Seward Peninsula). It might be confused with *Artemisia furcata*.

34. Artemisia glomerata Ledebour, Mém. Acad. Imp. Sci. St. Pétersbourg Hist. Acad. 5: 564. 1815

• Congested sagewort E

Ajania glomerata (Ledebour) Poljakov; *Artemisia glomerata* var. *subglabra* Hultén; *A. norvegica* Fries var. *glomerata* (Ledebour) H. M. Hall & Clements

Perennials, 30–50(–100) cm (densely cespitose), mildly aromatic (not rhizomatous, caudices subterranean, branches clothed with persistent leaf bases). **Stems** relatively numerous, ascending, brown, simple, hairy. **Leaves** mostly basal (cauline mostly 2–4, smaller), whitish; blades (basal) flabellate, 0.5–1(–2) × 0.5–0.8(–1.5) cm, relatively deeply lobed (lobes 5–9, linear), faces strigillose. **Heads** (3–10, erect or nodding, peduncles 0 or to 15 mm) in subcapitate to corymbiform arrays 1–5 × 2–4 cm. **Involucres** broadly campanulate, 3–4 × 3.5–5 mm. **Phyllaries** lanceolate (margins brownish), densely pilose. **Florets:** pistillate 4–5; bisexual 10–15; corollas yellow (3–5-toothed), 2–2.5 mm, glabrous or sparsely pilose. **Cypselae** ellipsoid (flattened, margins ribbed), 1–1.5 mm, glabrous. **2n** = 18, 27, 36, 54.

Flowering mid summer. Arctic and alpine tundra and sandy slopes; 0–1000 m; Yukon; Alaska.

Artemisia glomerata is similar to *A. senjavinensis*; it can be distinguished by its more deeply lobed leaves and sparser indument.

35. Artemisia laciniata Willdenow, Sp. Pl. 3: 1843. 1803

Perennials, 5–15 cm (not cespitose), sometimes mildly aromatic. **Stems** 1–3, erect, reddish brown, simple, strigillose to spreading-hairy, or glabrous. **Leaves** basal (in rosettes, petioles to 12 cm) and cauline, greenish; blades (basal) 2–3-pinnate, relatively deeply lobed (cauline sessile, 1–2-pinnately lobed to entire), faces sparsely hairy to pilose. **Heads** (10–70, spreading to nodding, peduncles 0 or to 10 mm) in spiciform arrays 2–5 × 0.5–1 or 8–18 × 1–4 cm. **Involucres** globose, 3–5 × 4–8 mm. **Phyllaries** (greenish or yellowish) elliptic (margins hyaline, brownish), glabrous or sparsely hairy. **Florets:** pistillate 6–8; bisexual 20–50; corollas yellowish or yellow to reddish-tinged, 1–2 mm, hairy (hairs tangled). **Cypselae** oblong, 0.5–1 mm, glabrous.

Subspecies 2 (2 in the flora): nw, w North America; Eurasia.

1. Stems hairy; leaves mostly cauline, blades 5–20 × 1–2 cm, sparsely hairy; arrays of heads 8–18 × 1–4 cm; corollas yellowish, 1–1.5 mm . . . . 35a. *Artemisia laciniata* subsp. *laciniata*
1. Stems glabrous; leaves mostly basal, blades of proximalmost 4–8 × 0.5–1 cm, of cauline 1.5–0.8 × 0.2 cm, sericeous; arrays of heads 2–5 × 0.5–1 cm; corollas yellow or reddish-tinged, 1.5–2 mm . . . . 35b. *Artemisia laciniata* subsp. *parryi*

35a. Artemisia laciniata Willdenow subsp. **laciniata**

Artemisia laciniatiformis Komarov; *A. macrobotrys* Ledebour

Stems 20–50 cm, hairy. **Leaves** mostly cauline, blades 5–20 × 1–2 cm, sparsely hairy. **Heads** in arrays 8–18 × 1–4 cm. **Involucres** 3–5 × 4–8 mm. **Corollas** yellowish, 1–1.5 mm. **Cypselae** 0.5–1 mm. **2n** = 18.

Flowering mid–late summer. Dry, gravelly stream banks, grassy flats, forested areas, dry hillsides; 100–1500 m; Yukon; Alaska; Eurasia.

Widely distributed in Eurasia and often confused with *Artemisia norvegica*, *A. laciniata* subsp. *laciniata* can be distinguished by its well-developed basal leaves and smaller heads.

35b. Artemisia laciniata subsp. **parryi** (A. Gray) W. A. Weber, Phytologia 58: 382. 1985 • Parry sagewort [E]

Artemisia parryi A. Gray, Proc. Amer. Acad. Arts 7: 361. 1868; *A. saxicola* Rydberg var. *parryi* A. Nelson

Stems 10–40 cm, glabrous. **Leaves** mostly basal, faces sericeous; proximalmost 4–8 × 0.5–1 cm, cauline 1.5–0.8 × 0.2 cm. **Heads** in 2–5 × 0.5–1 cm arrays. **Involucres** 3–4 × 4–5 mm. **Corollas** yellow or reddish-tinged, 1.5–2 mm.

Flowering late summer–fall. Subalpine and alpine meadows, rocky soils; 2700–3900 m; Colo., Idaho, N.Mex., Utah.

36. Artemisia longifolia Nuttall, Gen. N. Amer. Pl. 2: 142. 1818 • Long-leaved sage [E]

Artemisia falcata Rydberg; *A. ludoviciana* Nuttall var. *integrifolia* A. Nelson; *A. natronensis* A. Nelson; *A. vulgaris* Linnaeus subsp. *longifolia* (Nuttall) H. M. Hall & Clements; *A. vulgaris* var. *longifolia* (Nuttall) M. Peck

Perennials, 20–80 cm (not cespitose), pleasantly aromatic (fibrous-rooted, rootstocks relatively short, horizontal, layered stems sometimes sprouting). **Stems** 3–20+, erect, gray-green, usually simple, sometimes branched (bases woody), densely tomentose. **Leaves** cauline, bicolor (white and green); blades linear to lanceolate, 3–12 × 1 cm, margins usually entire, sometimes toothed or lobed, faces densely tomentose (abaxial) or glabrate (adaxial). **Heads** (peduncles 0 or to 2 mm) in mostly racemiform arrays 8–13 × 1–2 cm. **Involucres** campanulate, 4–5 × 4–5 mm. **Phyllaries** ovate-lanceolate (margins hyaline), densely to sparsely tomentose. **Florets:** pistillate 3–10; bisexual 8–26; corollas pale yellow, 1–2 mm, sparsely glandular. **Cypselae** ellipsoid, 0.5–0.8 mm, glabrous. $2n = 36$.

Flowering mid summer–early fall. Alkaline flats, grasslands, barren areas, high plains; 500–1800 m; Alta., B.C., Man., Sask.; Idaho, Minn., Mont., Nebr., N.Dak., S.Dak., Wyo.

Artemisia longifolia appears to be more salt-tolerant than most species of the genus. It is closely related to *A. ludoviciana*.

37. Artemisia ludoviciana Nuttall, Gen. N. Amer. Pl. 2: 143. 1818 • Silver wormwood, white or silver sage [F]

Artemisia vulgaris Linnaeus var. *ludoviciana* (Nuttall) Kuntze

Perennials, 20–80 (rarely to 120 in desert washes) cm, aromatic (rhizomatous). **Stems** relatively few to relatively numerous, erect, gray-green, simple or widely branched, hairy. **Leaves** cauline, uniformly gray-green, green, or white, or bicolor (white and green); blades linear to broadly elliptic, 1.5–11 × 0.5–4 cm, entire or lobed to relatively deeply pinnatifid, faces hairy. **Heads** (erect to nodding, peduncles 0 or 2–5 mm) in congested to open (widely branched) arrays. **Involucres** campanulate or turbinate, (1–)2–4(–5) × 2–5(–8) mm. **Phyllaries** (gray-green), lanceolate to ovate or obovate (margins narrowly hyaline), densely tomentose. **Florets:** pistillate 5–12; bisexual 6–45; corollas yellow, sometimes red-tinged, 1.5–2.8 mm, glabrous. **Cypselae** ellipsoid ca. 0.5 mm, (obscurely nerved) glabrous. $2n = 18, 36, 54$.

Subspecies ca. 7 (6 in the flora): North America, Mexico.

1. Leaves usually relatively deeply lobed ($^1/_3$+ widths, nearly to midrib, proximal leaves sometimes entire); involucres 3–8 mm diam.; mountain meadows and slopes.
　2. Involucres 4–5 × 4–8 mm . . . .
　　. . . . . . . 37c. *Artemisia ludoviciana* subsp. *candicans*
　2. Involucres 3–4 × 3–5 mm . . . .
　　. . . . . . . 37d. *Artemisia ludoviciana* subsp. *incompta*
1. Leaves entire or relatively shallowly lobed (lobes to $^1/_3$ widths); involucres 2–3(–4) mm diam.; desert valleys and mountains.
　3. Heads in paniculiform arrays (4–)8–30 cm diam.; leaves mostly 1.5–2 cm . . . .
　　. . . . . 37b. *Artemisia ludoviciana* subsp. *albula*
　3. Heads in paniculiform or racemiform arrays 1–6 cm diam.; leaves 1.5–11 cm.
　　4. Leaf margins plane . . . .
　　　. . 37a. *Artemisia ludoviciana* subsp. *ludoviciana*
　　4. Leaf margins revolute.
　　　5. Leaves bicolor (gray-green and bright green), margins mostly entire, abaxial faces glabrous . . . .
　　　　. . . 37f. *Artemisia ludoviciana* subsp. *redolens*
　　　5. Leaves gray-green, margins usually lobed, abaxial faces hairy . . . .
　　　　. . . . . 37e. *Artemisia ludoviciana* subsp. *mexicana*

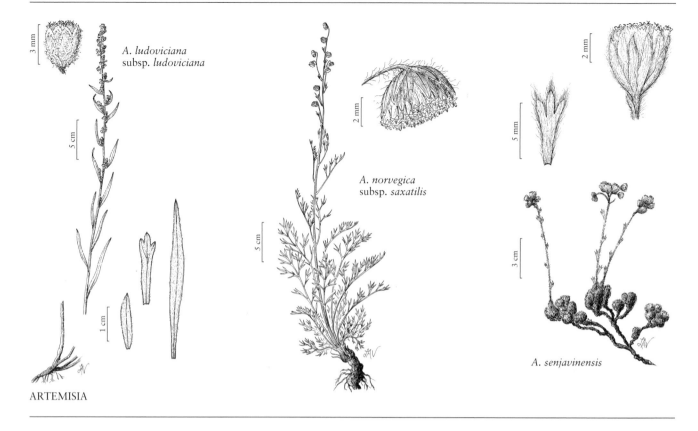

A. *ludoviciana*
subsp. *ludoviciana*

A. *norvegica*
subsp. *saxatilis*

A. *senjavinensis*

ARTEMISIA

37a. Artemisia ludoviciana Nuttall subsp. **ludoviciana**

• Armoise de l'ouest E F

Artemisia brittonii Rydberg;
A. cuneata Rydberg; *A. diversifolia*
Rydberg, *A. falcata* Rydberg;
A. gnaphaloides Nuttall;
A. herriotii Rydberg;
A. lindheimeriana Scheele;
A. pabularis (A. Nelson) Rydberg;
A. paucicephala A. Nelson;
A. pudica Rydberg; *A. purshiana*
Besser; *A. rhizomata* A. Nelson

Stems (gray to white) mostly simple, 30–80 cm, tomentose. **Leaves** gray; blades linear to narrowly elliptic, 3–11 cm, margins plane (proximalmost entire or apically lobed, lobes to ¹/₃ blade lengths; cauline 1.5–11 × 1–1.5 cm, entire or lobed to pinnatifid), faces densely tomentose. **Heads** in congested, paniculiform or racemiform arrays 5–30 × 1–4 cm. **Involucres** 3–4 × 2–4 mm. **Florets:** pistillate 5–12; bisexual 6–30; corollas 1.9–2.8 mm. **2*n*** = 18, 36.

Flowering mid summer–late fall. Disturbed roadsides, open meadows, rocky slopes; 100–3000 m; Alta., B.C., Man., N.B., Ont., P.E.I., Que., Sask.; Ala., Ariz., Ark., Calif., Colo., Conn., Del., Ga., Idaho, Ill., Ind., Iowa, Kans., Ky., La., Maine, Md., Mass., Mich., Minn., Miss., Mo., Mont., Nebr., N.H., N.J., N.Mex., N.Y., N.C., N.Dak., Ohio, Okla., Penn., R.I., S.C., S.Dak., Tenn., Tex., Utah, Vt., Va., Wash., Wis., Wyo.

Subspecies *ludoviciana* is widespread in North America in diverse habitats. It is the most common subspecies and the most variable morphologically.

37b. Artemisia ludoviciana Nuttall subsp. **albula** (Wooton) D. D. Keck, Proc. Calif. Acad. Sci., ser. 4, 25: 446. 1946 • White wormwood

Artemisia albula Wooton, Contr. U.S. Natl. Herb. 16: 193. 1913, based on *A. microcephala* Wooton, Bull. Torrey Bot. Club 25: 455. 1898, not Hillebrand 1888

Stems 30–80 cm, widely branched, tomentose or glabrous. **Leaves** uniformly whitish green; blades lance-linear (and entire), or obovate to elliptic (with antrorse teeth or lobes to ¹/₃ blade lengths, usually 1–2 cm, margins revolute), faces ± tomentose. **Heads** in open, paniculiform arrays (9–)15–40 × (4–)8–30 cm. **Involucres** 1–2 × 2–3 mm. **Florets:** pistillate 8–11; bisexual 8–13; corollas 1–1.5 mm.

Flowering early summer–fall. Desert drainages, sandy soils; 1500–2000 m; Ariz., Calif., Colo., Nev., N.Mex., Tex., Utah; Mexico.

37c. Artemisia ludoviciana Nuttall subsp. **candicans** (Rydberg) D. D. Keck, Proc. Calif. Acad. Sci., ser. 4, 25: 447. 1946

Artemisia candicans Rydberg, Bull. Torrey Bot. Club 24: 296. 1897; *A. gracilenta* A. Nelson; *A. latiloba* (Nuttall) Rydberg; *A. ludoviciana* var. *candicans* (Rydberg) H. St. John; *A. ludoviciana* var. *latiloba* Nuttall; *A. vulgaris* Linnaeus subsp. *candicans* (Rydberg) H. M. Hall & Clements; *A. vulgaris* var. *candicans* (Rydberg) M. Peck

Stems 30–50(–80) cm, mostly simple, sparsely tomentose. **Leaves** ± white; blades broadly obovate to oblong, 4–10 × 1.5–4 cm, relatively deeply lobed (lobes, lanceolate, 1/3+ blade lengths, acute), faces tomentose (usually less densely adaxially). **Heads** in racemiform arrays 4–7 × 2–4 cm. **Involucres** (broadly campanulate), 4–5 × 4–8 mm. **Florets:** pistillate 8–10; bisexual 20–30; corollas 1.5–2 mm.

Flowering mid–late summer. Mountains, usually in loamy soils; 1800–3000 m; Alta., B.C.; Calif., Idaho, Mont., Nev., Oreg., Utah, Wash., Wyo.

Subspecies *candicans* is broadly circumscribed in this treatment. Variants within the complex, especially var. *latiloba*, may merit segregation.

37d. Artemisia ludoviciana Nuttall subsp. **incompta** (Nuttall) D. D. Keck, Publ. Carnegie Inst. Wash. 520: 327. 1940 • Mountain wormwood

Artemisia incompta Nuttall, Trans. Amer. Philos. Soc., n. s. 7: 400. 1841; *A. arachnoidea* E. Sheldon; *A. atomifera* Piper; *A. lindleyana* Besser; *A. ludoviciana* var. *incompta* (Nuttall) Cronquist; *A. prescottiana* Besser; *A. vulgaris* Linnaeus var. *incompta* (Nuttall) H. St. John

Stems 20–50(–80) cm, mostly simple, hairy. **Leaves** bicolor (white and green, or gray-green and green); blades narrowly to broadly lanceolate, 1.5–11 × 1–1.5 cm, irregularly lobed (lobes usually 1/3+ blade widths), faces hairy (abaxial) or glabrous or glabrescent. **Heads** in paniculiform arrays (10–)15–25(–35) × (2–)3–8(–10) cm. **Involucres** (broadly campanulate) 3(–4) × 3–5) mm. **Florets:** pistillate 5–12; bisexual 6–45; corollas yellow, sometimes red-tinged, 1.5–2 mm. $2n$ = 36, 54.

Flowering mid summer–late fall. Open meadows, mountain slopes; 1900–3500 m; Alta., B.C.; Calif., Colo., Idaho, Mont., Nev., Oreg., Utah, Wash., Wyo.; Mexico.

Subspecies *incompta* varies. It could be (and has been) subdivided. The variation found within one population,

as well as from one part of its range to another, makes the recognition of discrete taxa difficult as well as questionable. The morphologic variant treated as *Artemisia lindleyana* by some authors may warrant infraspecific status.

37e. Artemisia ludoviciana Nuttall subsp. **mexicana** (Willdenow ex Sprengel) D. D. Keck, Proc. Calif. Acad. Sci., ser. 4, 25: 452. 1946 • Mexican wormwood

Artemisia mexicana Willdenow ex Sprengel, Syst. Veg. 3: 490. 1826; *A. neomexicana* Greene ex Rydberg; *A. revoluta* Rydberg

Stems 30–80(–100) cm, widely branched, sparsely hairy or glabrous. **Leaves** strongly bicolor (white and green, or gray-green and green), narrowly lanceolate, 4–10 × 0.5–1 cm, lobed; faces ± tomentose (abaxial) or glabrescent. **Heads** in paniculiform or racemiform arrays 1–8 × 2–5 cm. **Involucres** (campanulate) 3–4 × 2–3 mm. **Florets:** pistillate (5–)8–10; bisexual (6–)10–20; corollas ca. 1.5 mm. $2n$ = 18, 36.

Flowering late summer–late fall. Mountain slopes, rocky soils; 500–1000 m; Ariz., Colo., Nev., N.Mex., Okla., Tex.; Mexico.

37f. Artemisia ludoviciana Nuttall subsp. **redolens** (A. Gray) D. D. Keck, Proc. Calif. Acad. Sci., ser. 4, 25: 454. 1946

Artemisia redolens A. Gray, Proc. Amer. Acad. Arts 21: 393. 1886; *A. vulgaris* Linnaeus subsp. *redolens* (A. Gray) H. M. Hall & Clements

Stems 20–80 cm, mostly unbranched, glabrous. **Leaves** bicolor (white and green); blades lanceolate to ovate, 5–7 × 0.5–1 cm, irregularly pinnatifid; abaxial faces tomentose (abaxial) or glabrate (adaxial). **Involucres** campanulate, 2–3 × 2–3 mm. **Heads** in paniculiform or racemiform arrays 10–15 × 2–6 cm. **Florets:** pistillate ca. 10; bisexual 3–8; corollas 2 mm.

Flowering mid–late summer. Rocky loam soils; 100–300 m; Ariz., N.Mex., Tex.; Mexico.

38. Artemisia michauxiana Besser in W. J. Hooker, Fl. Bor.-Amer. 1: 324. 1833 • Lemon sagewort [E]

Artemisia discolor Douglas ex Besser; *A. vulgaris* Linnaeus subsp. *michauxiana* (Besser) H. St. John

Perennials, 30–100 cm, strongly aromatic (lemon-scented; rhizomatous). **Stems** relatively many, erect, green, simple, glabrous. **Leaves** cauline, green; blades broadly lanceolate to narrowly elliptic, 1.5–11 × 1–1.5 cm, 2-pinnately lobed, (ultimate lobes toothed), faces white-tomentose (abaxial) or glabrous (adaxial), yellow-gland-dotted. **Heads** (erect to nodding, peduncles 0 or to 10 mm) in paniculiform to spiciform arrays 8–15 × 1–1.5 cm. **Involucres** campanulate, 3(–4) × 2–5.5 mm. **Phyllaries** (yellow-green, rarely purplish) broadly ovate, glabrous or sparsely hairy, yellow-gland-dotted. **Florets:** pistillate 9–12; bisexual 15–35; corollas yellow, 1–1.5 mm, glandular. **Cypselae** (yellow to light brown) ellipsoid, ca. 0.5 mm, glabrous or glandular. *2n* = 18, 36.

Flowering mid summer–early fall. Talus slopes, alpine and subalpine drainages; 1900–3700 m; Alta., B.C., Yukon; Calif., Colo., Idaho, Mont., Nev., Oreg., Utah, Wash., Wyo.

Members of the *Artemisia ludoviciana* complex with deeply lobed leaves are sometimes confused with *A. michauxiana,* and there is evidence that plants hybridize in some locations. *Artemisia michauxiana* is distinguished by its glabrous, bright green to yellow-green foliage and lemony-sweet fragrance.

39. Artemisia nesiotica P. H. Raven, Aliso 5: 341. 1963 • Island sagebrush [E]

Crossostephium insulare Rydberg in N. L. Britton et al., N. Amer. Fl. 34: 244. 1916, not *Artemisia insularis* Kitamura 1936; *A. californica* Lessing var. *insularis* (Rydberg) Munz

Subshrubs, 10–60 cm (rounded), aromatic. **Stems** relatively numerous, ascending or prostrate, gray, simple or branched (slender, wandlike, soft, bases woody and brittle), densely canescent. **Leaves** cauline, gray-green; blades linear-oblong, 3–5 × 1–2 cm, mostly 3-lobed (lobes 1–2 mm wide), faces gray-hairy. **Heads** (usually erect, sometimes nodding) in (leafy) paniculiform arrays 10–25 × 3–5(–7) cm. **Involucres** broadly campanulate, 2.5 × 4–4.5 mm. **Phyllaries** broadly ovate, densely hairy. **Florets:** pistillate 0; bisexual 20–50; corollas pale yellow, 1.2–1.5 mm, glandular. **Cypselae** (light brown) ellipsoid (ribbed), 0.5 mm, resinous.

Flowering mid–late summer. Rocky slopes, often fog-shrouded hillsides; 0–100 m; Calif.

Artemisia nesiotica is known only from the Channel Islands of California. It differs from the closely related *A. californica* by its shorter stature, wider leaf lobes, and larger heads.

40. Artemisia norvegica Fries, Novit. Fl. Svec., 56. 1817 • Alpine sagewort [F]

Subspecies 2 (1 in the flora): w North America, Eurasia.

40a. Artemisia norvegica Fries subsp. **saxatilis** (Besser) H. M. Hall & Clements, Publ. Carnegie Inst. Wash. 326: 58. 1923 [F]

Artemisia chamissoniana Besser var. *saxatilis* Besser in W. J. Hooker, Fl. Bor.-Amer. 1: 324. 1833; *A. arctica* Lessing; *A. arctica* subsp. *beringensis* (Hultén) Hultén; *A. arctica* subsp. *comata* (Rydberg) Hultén; *A. arctica* subsp. *ehrendorferi* Korobkov; *A. arctica* var. *saxatilis* (Besser) Y. R. Ling; *A. comata* Rydberg; *A. norvegica* var. *piceetorum* S. L. Welsh & Goodrich

Perennials, 25–40(–60) cm (not cespitose), mildly aromatic (roots often horizontal, woody). **Stems** 1–3, erect to ascending, green or reddish, simple, glabrous or sparsely tomentose. **Leaves** mostly basal (in rosettes, petiolate), bright green; blades (basal) broadly lanceolate, 5–8(–10) × 2–3(–4) cm, 1–3-pinnately lobed (apical lobes 1–7 × 1.5–3 mm; mid cauline sessile, pinnately lobed; on flowering stems, sessile, linear, entire), faces glabrous or hairy. **Heads** (nodding, proximalmost on peduncles to 50 mm) in racemiform arrays 10–17 × 1–2 cm. **Involucres** globose, (4–)5–8 × 4–10 mm. **Phyllaries** ovate-lanceolate to elliptic (margins dark brown to black), sparsely hairy to villous. **Florets:** pistillate 6–20; bisexual (30–)50–70; corollas yellow or red-tinged, 1.5–2.5(–3.5) mm, long-hairy. **Cypselae** ovoid-oblong (angular), ca. 2.5 mm, glabrous or villous. *2n* = 18, 36.

Flowering mid–late summer. Coastal, arctic, subalpine to alpine habitats, boreal forests, moist soils; 0–3800 m; Alta., B.C., Yukon; Alaska, Calif., Colo., Mont., Utah, Wash., Wyo.; e Asia (Russian Far East).

Variation within *Artemisia norvegica* in North America is not well understood and, for that reason, this treatment represents a conservative taxonomy with only one subspecies for the flora area. Subspecies *saxatilis* differs from typical *A. norvegica* primarily by its larger heads. European plants have involucres less than 10 mm in diameter. Chromosome number may be used to justify separation of taxa either at the level of subspecies or species. If separated as distinct species, then *A. arctica* is the name for North American plants.

The diploid *A. arctica* (2*n* = 18) and tetraploid *A. comata* (2*n* = 36) are treated as separate species by R. Elven et al. (pers. comm.).

41. Artemisia packardiae J. W. Grimes & Ertter, Brittonia 31: 454, fig. 1. 1979 • Succor Creek mugwort E

Perennials, 20–50(–60) cm, strongly aromatic (rhizomatous, fibrous-rooted). **Stems** 3–20, erect, light brown, simple or branched, glabrous. **Leaves** cauline, dark green; blades lanceolate, 1.5–5 × 1–2.5 cm, 2-pinnatifid (primary lobes 5–9, 0.4–1.5 cm; cauline smaller, pinnatifid to entire), faces tomentose (abaxial) or glabrous (adaxial). **Heads** (peduncles 0 or to 3 mm) in usually paniculiform, sometimes racemiform, arrays 5–20 × 1.5–4 cm. **Involucres** campanulate to hemispheric, 2.5–3.5 × 2–4.5 mm. **Phyllaries** broadly ovate, glandular (at least at bases). **Florets:** pistillate 3–8; bisexual, sometimes functionally staminate, (15–)20–35; corollas bright yellow, 1.3–2.2 mm, glandular. **Cypselae** (light brown) ellipsoid (± arcuate, ribs 4, prominent), ca. 1 mm, glandular. **2*n* = 18.**

Flowering late summer. Coarse taluses, alkaline soils, erosion gullies; 1000–2400 m; Idaho, Nev., Oreg.

Artemisia packardiae is known only from southeastern Oregon, western Idaho, and northeastern Nevada. It is closely related to *A. michauxiana* and could be considered an ecologic variant.

42. Artemisia palmeri A. Gray, Proc. Amer. Acad. Arts 11: 79. 1876 • Palmer sagewort C

Artemisiastrum palmeri (A. Gray) Rydberg

Subshrubs, 100–350 cm, mildly aromatic. **Stems** usually 1–15, erect, brown, simple (wandlike, brittle, bases woody), glabrous. **Leaves** cauline (petiolate), bicolor (gray-green and dark green); blades broadly lanceolate, 3.5–12(–15) × 0.2–10 cm, relatively deeply and coarsely pinnately lobed (lobes 3–7+), faces canescent (abaxial) or glabrous or sparsely hairy (adaxial). **Heads** (erect or nodding, peduncles relatively slender) in open, paniculiform arrays, 15–40 × 3–10 cm (widely branched). **Involucres** globose, 2.5–3.5 × 2–5 mm. **Phyllaries** (pale green to stramineous) broadly ovate, glabrous or sparsely hairy (receptacles paleate). **Florets:** pistillate 0; bisexual 8–30; corollas pale yellow, 1.5–2.2 mm, resinous-glandular (style branches exsert, truncate, erose). **Cypselae** (light brown, shiny) ellipsoid, 1–1.2 mm, (4-angled), glabrous or glandular. **2*n* = 18.**

Flowering early–mid summer. Ravines, coastal areas, sandy soils; of conservation concern; 100–300 m; Calif.; Mexico (Baja California).

Artemisia palmeri is known only from drainages near the coast, from northeast of San Diego to just south of Ensenada. Most of its habitat has been destroyed by urban development. It is of particular interest because of its paleate receptacles, an anomalous trait that confounds our understanding of its evolutionary relationship to other species of *Artemisia*.

43. Artemisia papposa S. F. Blake & Cronquist, Leafl. W. Bot. 6: 43, plate 1. 1950 • Owyhee sage E

Shrubs, 5–15(–20) cm (not cespitose), aromatic. **Stems** relatively numerous, erect, gray, simple (annual flowering branches leafy), loosely sericeous. **Leaves** (semideciduous) cauline (sessile), gray-green; blades oblanceolate, 0.5–3 × 0.2–1.5 cm (bases attenuate), 3-lobed or irregularly palmatifid (lobes narrow, apices acute), sparsely sericeous-lanate. **Heads** (mostly erect, peduncles 0 or to 25 mm) in racemiform arrays (4–)8–12(–14) × (0.5–)1–2 (–4) cm. **Involucres** globose, 3.5–5 × 4–5 mm. **Phyllaries** ovate, sparsely sericeous. **Florets:** pistillate 8; bisexual 20–35; corollas yellow (tubular with broad throats), ca. 2 mm, glandular. **Cypselae** (light brown) oblanceoloid (4–5-angled, broadest at truncate apices), 0.3–0.5 mm, glandular-pubescent (pappi coroniform, 0.3–0.6 mm, irregularly lacerate).

Flowering early spring–mid summer. Rocky swales, dry meadows, alkaline mud flats; 1400–2100 m; Idaho, Nev., Oreg.

The pappose cypselae make *Artemisia papposa* anomalous within *Artemisia*. *Artemisia papposa* has capitulescence characteristics that suggest a relationship to *Sphaeromeria*.

44. Artemisia pontica Linnaeus, Sp. Pl. 2: 847. 1753 • Roman wormwood, green-ginger, armoise de la mer Noire I

Perennials, 40–100 cm, somewhat aromatic; rhizomes creeping, woody. **Stems** relatively numerous, erect, brown, mostly simple (brittle, bases woody) canescent or glabrate. **Leaves** cauline, grayish green; sessile (proximalmost short-petiolate); blades triangular to ovate, 1–5 × 1–3 cm, 2–3-pinnatifid (lobes 0.5–1 mm wide, acute), faces pubescent (abaxial) or hairy to glabrate (adaxial). **Heads** (nodding) in paniculiform arrays 10–22 × 2–4 cm.

Involucres spheric, 1.5–2(–3) mm. **Phyllaries** (subequal) linear, hairy. **Florets:** pistillate 10–12; bisexual 40–45; corollas pale yellow, 0.2–0.3 mm, sometimes gland-dotted (stigma lobes relatively short, not emerging from tubes, short-ciliate). **Cypselae** ellipsoid (angled), 0.1–0.2 mm, glabrous. $2n = 18$.

Flowering late summer–fall. Disturbed areas, valleys, shaded thickets; 100–500 m; introduced; Man., N.S., Ont., Que.; Conn., Del., Ill., Ky., Maine, Md., Mass., Mich., Minn., N.H., N.J., N.Y., Ohio, Pa., R.I., Vt., Wis.; Eurasia.

Artemisia pontica has finely dissected gray foliage and is widely planted as an ornamental. It escapes locally; it has not been reported as problematic. The only species with which it has been confused in North America is *A. abrotanum*, which has dark green (not gray) foliage.

45. Artemisia senjavinensis   Besser, Nouv. Mém. Soc. Imp. Naturalistes Moscou 3: 35. 1834   [F]

Ajania senjavinensis (Besser) Poljakov; *Artemisia androsacea* Seemann

Perennials, 30–90 cm (densely cespitose), mildly aromatic (caudices branched, woody, taprooted). **Stems** 1–9, erect, gray-green, lanate. **Leaves** mostly basal (in rosettes, cauline 2–5, scattered on flowering stems); blades (basal) broadly oblanceolate, 0.5–0.8 × 0.5–0.7 cm, relatively deeply lobed (lobes 3–5, acute; cauline blades 0.5–1 cm, entire or pinnately lobed, lobes 3–5), faces densely tomentose to sericeous (hairs 1–2 mm). **Heads** in corymbiform arrays 0.5–2.5 × 0.5–2.5 cm (subtended by white-sericeous bracts). **Involucres** turbinate, 3–4 × 3–5 mm. **Phyllaries** lanceolate or ovate, hairy. **Florets:** pistillate 4–5; bisexual 3–4; corollas yellow or tan, 1.5–2, glandular (style branches blunt, not fringed). **Cypselae** (brown) linear-oblong, ca. 2 mm, (apices flat), glabrous. $2n = 36, 54$.

Flowering mid–late summer. Open calcareous gravelly slopes in tundra or heath, sandy slopes above high tide; 0–600 m; Alaska; e Asia (Russian Far East, Chukotka).

Artemisia senjavinensis is known only from western Alaska (Seward Peninsula) and the Chukchi Peninsula of the Russian Far East.

46. Artemisia serrata   Nuttall, Gen. N. Amer. Pl. 2: 142. 1818   •   Serrate-leaved sage   [E]

Artemisia vulgaris Linnaeus subsp. *serrata* (Nuttall) H. M. Hall & Clements

Perennials, 50–100(–300) cm (not cespitose), pleasantly aromatic (fibrous-rooted, rhizomes horizontal, relatively short). **Stems** 2–5, erect, brown, mostly simple (bases woody), sparsely tomentose. **Leaves** cauline, bicolor (white and green); blades lanceolate, 7–15 × 1–2.5 cm, serrate (teeth ca. 2 mm), faces densely tomentose (abaxial) or glabrate (adaxial). **Heads** (peduncles 0 or to 2 mm) in racemiform arrays 10–15 × 5–15 cm. **Involucres** campanulate, 2.5–3 × 2–2.5 mm. **Phyllaries** lanceolate (margins hyaline), densely tomentose. **Florets:** pistillate 3–5; bisexual 9–10; corollas pale yellow, 1.5–2 mm, sparsely glandular. **Cypselae** ellipsoid, ca. 1 mm, glabrous. $2n = 36$.

Flowering mid summer–early fall. Grasslands and barren areas on high plains; 500–1800 m; Ill., Iowa, Minn., N.Y., N.Dak., Wis.

Artemisia serrata is closely related to *A. ludoviciana* and *A. longifolia*; it is distinguished by its prominent, serrated leaf margins. It is apparently native to the upper Mississippi Valley and naturalized in New York, presumably following introduction as a garden plant. Reports from Kansas and Missouri may be based on collections of *A. ludoviciana*.

47. Artemisia stelleriana   Besser, Nouv. Mém. Soc. Imp. Naturalistes Moscou 3: 79, plate 5. 1834   •   Beach wormwood, armoise de Steller

Perennials, (15–)20–60(–70) cm (mat-forming), sometimes faintly aromatic (rhizomes creeping, relatively thin). **Stems** 1–3, erect or ascending, white, simple (stout), densely tomentose to floccose. **Leaves** basal and cauline (petiolate), silver-gray; blades oblanceolate, (proximalmost) 3–10 × 1–5 cm, pinnatifid (lobes relatively broad, rounded; distal leaves, on flowering stems, smaller), faces densely tomentose. **Heads** (erect or spreading, peduncles 0 or to 3 mm) in dense, paniculiform, racemiform, or spiciform arrays 8–20 × 2–4 cm. **Involucres** broadly campanulate, 5–8 × 6–7 mm. **Phyllaries** broadly lanceolate, tomentose. **Florets:** pistillate 12–16; bisexual 25–30; corollas yellow (narrow or tubular), 3.2–4 mm (unusually large), glabrous or sparsely hairy (style branches prominent, erect, blunt). **Cypselae** (dark brown) narrowly oblong-linear (slightly flattened, smooth), 3–4 mm, glabrous. $2n = 18$.

Flowering early spring–fall. Sandy soils, coastal strand; 0–200 m; St. Pierre and Miquelon; N.B., Nfld. and Labr. (Nfld.), N.S., Ont., P.E.I., Que.; Alaska, Conn., Del., Fla., La., Maine, Md., Mass., Mich., Minn., N.H., N.J., N.Y., N.C., Ohio, Pa., R.I., Vt., Va., Wash., W.Va., Wis.; n Europe; e Asia (Japan, Kamchatka).

Artemisia stelleriana is apparently native along the western tip of the Aleutian islands (D. F. Murray, pers. comm.). It is an attractive ornamental and, in parts of its range in the flora area, it appears to have escaped from cultivation and is naturalized in beach dunes and other sandy habitats.

48. Artemisia suksdorfii Piper, Bull. Torrey Bot. Club 28: 42. 1901 • Suksdorf sagewort E

Artemisia heterophylla Nuttall, Trans. Amer. Philos. Soc., n. s. 7: 400. 1841, not Besser 1834; *A. vulgaris* Linnaeus var. *littoralis* Suksdorf

Perennials, 50–170(–200) cm, aromatic (rhizomes woody, coarse). **Stems** usually 10+, erect, light brown, simple, usually glabrous. **Leaves** cauline (sessile), bicolor (white and dark green); blades lanceolate, 5–10(–15) × 1–5 cm (bases strongly tapered, attenuate, coarsely and irregularly lobed, faces tomentose (abaxial) or glabrous (adaxial). **Heads** (erect) in crowded (proximally leafy), paniculiform or racemiform arrays 17–30 × (2–)4–5 cm (lateral branches stiff, erect). **Involucres** narrowly turbinate or globose, 1.5–2.5 × 1–1.5 mm. **Phyllaries** (straw-colored to yellow-green, shiny) lanceolate, glabrous or sparsely hairy. **Florets:** pistillate 2–5; bisexual 2–7; corollas yellow, 1.5–3 mm, glabrous. **Cypselae** ellipsoid, 0.8–1.5 mm, glabrous. **2*n* = 18.**

Flowering mid summer–fall. Coastal habitats, often along roads or drainages; 0–200 m; B.C.; Calif., Oreg., Wash.

Artemisia suksdorfii is similar morphologically to *A. douglasiana*; it has more and smaller heads, and glabrous phyllaries. The two species hybridize where their ranges overlap.

49. Artemisia tilesii Ledebour, Mém. Acad. Imp. Sci. St. Pétersbourg Hist. Acad. 5: 568. 1814

Artemisia hookeriana Besser; *A. hultenii* M. M. Maximova; *A. tilesii* var. *aleutica* (Hultén) S. L. Welsh; *A. tilesii* var. *elatior* Torrey & A. Gray; *A. tilesii* subsp. *gormanii* (Rydberg) Hultén; *A. tilesii* subsp. *hultenii* (M. M. Maximova) V. G. Sergienko; *A. tilesii* var. *unalaschcensis* Besser; *A. unalaskensis* Rydberg var. *aleutica* Hultén; *A. vulgaris* Linnaeus subsp. *tilesii* (Ledebour) H. M. Hall & Clements

Perennials, 20–60(–80) cm, mildly aromatic (rhizomes coarse). **Stems** 1–3, erect, white, tomentose (on distal branches, hairs appressed) or glabrate. **Leaves** basal and cauline, bicolor (white and green); blades (basal) linear to broadly lanceolate, 3–7(–10) × 2–5(–6) cm, coarsely pinnately lobed (cauline becoming linear distally), faces tomentose (abaxial) or glabrous (adaxial). **Heads** in compact to broadly branched, paniculiform arrays 1–20 × 2–6 cm. **Involucres** broadly campanulate, 4–5 × 3.5–4 mm. **Phyllaries** (violet-brown) oval (outer) to elliptic or lanceolate, sparsely tomentose. **Florets:** pistillate 9; bisexual 25–60; corollas yellow, 1.5–3 mm, glabrous (style branches included, erect, linear, relatively short, short-ciliate). **Cypselae** oblong-linear (angular), 1.2–1.5 mm, glabrous. **2*n* = 18, 36.**

Flowering mid summer–early fall. Arctic and alpine tundra, sandy, rocky slopes near shorelines; 0–2000 m; Alta., B.C., Man., N.W.T., Nunavut, Ont., Que., Sask., Yukon; Alaska, Idaho, Mont., Oreg., Wash.; Asia (Russia).

Artemisia tilesii has a bewildering array of variation in leaf and inflorescence morphology that has been separated into four infraspecific taxa recognized in some floras. I am unable to separate these taxa consistently and am including them within a broad circumscription of the species.

50. Artemisia vulgaris Linnaeus, Sp. Pl. 2: 848. 1753 • Common mugwort, felon-herb, green-ginger, armoise vulgaire I

Artemisia opulenta Pampanini; *A. vulgaris* var. *glabra* Ledebour; *A. vulgaris* var. *kamtschatica* Besser

Perennials, (40–)60–190 cm, sometimes faintly aromatic (rhizomes coarse). **Stems** relatively numerous, erect, brownish to reddish brown, simple proximally, branched distally (angularly ribbed), sparsely hairy or glabrous. **Leaves** basal (petiolate) and cauline (sessile), uniformly green or bicolor;

blades broadly lanceolate, ovate, or linear, (2–)3–10 (–12) × 1.8–8 cm (proximal reduced and entire, distal pinnately dissected, lobes to 20 mm wide), faces pubescent or glabrescent (abaxial) or glabrous (adaxial). **Heads** in compact, paniculiform or racemiform arrays (10–)20–30(–40) × (5–)7–15(–20) cm. **Involucres** ovoid to campanulate, 2–3(–4) mm. **Phyllaries** lanceolate, hairy or glabrescent. **Florets:** pistillate 7–10; bisexual (5–) 8–20; corollas yellowish to reddish brown, 1.5–3 mm, glabrous (style branches arched-curved, truncate, ciliate). **Cypselae** ellipsoid, 0.5–1(–1.2) mm, glabrous, sometimes resinous. $2n$ = 18, 36, 40, 54.

Flowering mid summer–late fall. Sandy or loamy soils, forested areas, coastal strands, roadsides; 0–500 m; introduced; Greenland; Alta., B.C., Man., N.B., Nfld. and Labr. (Nfld.), N.S., Ont., P.E.I., Que., Sask.; Ala., Alaska, Conn., Del., D.C., Fla., Ga., Idaho, Ill., Ind., Iowa, Kans., Ky., La., Maine, Md., Mass., Mich., Minn., Mo., Mont., N.H., N.J., N.Y., N.C., Ohio, Oreg., Pa., R.I., S.C., Tenn., Vt., Va., Wash., W.Va., Wis.; Eurasia.

Grown as a medicinal plant, most commonly as a vermifuge, *Artemisia vulgaris* is widely established in eastern North America and is often weedy in disturbed sites. Populational differences in morphologic forms are reflected in size of flowering heads, degree of dissection of leaves, and overall color of plants (from pale to dark green), suggesting multiple introductions that may date back to the first visits by Europeans. It is tempting to recognize the different forms as subspecies and varieties; the array of variation in the field is bewildering. If genetically distinct forms exist in native populations, the differences appear to have been blurred by introgression among the various introductions in North America. A case could be made for recognizing var. *kamtschatica* in Alaska based on its larger heads and shorter growth form; apparent introgression with populations that extend across Canada confounds that taxonomic segregation.

120. HULTENIELLA   Tzvelev in A. I. Tomatchew, Fl. Arct. URSS 10: 117. 1987

• [For Eric Hultén, 1894–1981, Swedish botanist, specialist of the circumpolar flora]

Luc Brouillet

Dendranthema Des Moulins sect. *Haplophylla* Tzvelev in V. L. Komarov et al., Fl. URSS 26: 388, 880. 1961

Perennials, (0.6–)1–12 cm (sterile basal rosettes 1–10+; rhizomes at or below ground, ± woody, giving rise to branched, woody caudices). **Stems** (flowering) 1–10+ (1 per rosette), erect, not branched (scapiform), ± villous to woolly, particularly near heads (hairs basifixed). **Leaves** (basal marcescent, erect) all or mostly basal; alternate; sessile; blades (appearing 1-nerved or nerves inconspicuous) linear, not lobed, margins entire (villous-ciliate), faces glabrous. **Heads** radiate, borne singly. **Involucres** hemispheric or broader, (4–)4.5–6(–6.5) mm diam. **Phyllaries** persistent (green), 20–26(+) in 2(–3+) series, distinct, oblong to lance-oblong (not carinate, 1-nerved), margins and apices (dark brown, fimbriate) scarious (apices obtuse to acute, abaxial faces glabrous or villous). **Receptacles** convex (± villous), epaleate. **Ray florets** 11–19, pistillate, fertile; corollas white, laminae obovate. **Disc florets** 60–80, bisexual, fertile; corollas yellow, tubes ± cylindric, throats campanulate, lobes 5, deltate (without resin sacs). **Cypselae** ± obconic, ribs 5, faces glabrous (pericarps without myxogenic cells or resin sacs); **pappi** crowns of 6–12 irregular teeth. x = 9.

Species 1: n Canada, Alaska, n Eurasia.

K. Bremer and C. J. Humphries (1993) included *Hulteniella* in *Arctanthemum*. *Hulteniella* has not been included in molecular phylogenetic studies.

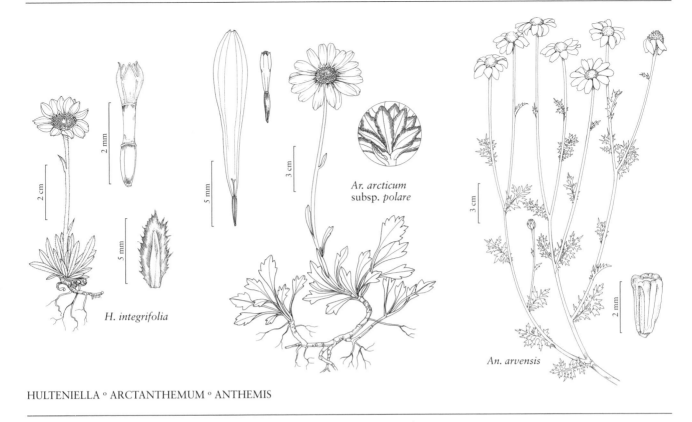

HULTENIELLA ° ARCTANTHEMUM ° ANTHEMIS

1. Hulteniella integrifolia (Richardson) Tzvelev in A. I. Tolmatchew, Fl. Arct. URSS 10: 118. 1987 • Entire-leaved daisy [F]

Chrysanthemum integrifolium Richardson in J. Franklin et al., Narr. Journey Polar Sea, 749. 1823; *Arctanthemum integrifolium* (Richardson) Tzvelev; *Dendranthema integrifolium* (Richardson) Tzvelev; *Leucanthemum integrifolium* (Richardson) de Candolle

Rhizomes 1–3 mm diam. **Leaves:** blades (2–)4–18(–30) × 0.8–1.2(–2) mm. **Peduncular bracts** 0–3, remote, distal (mostly beyond midstem), 4.5–7 mm, margins and apices brown, hyaline, erose, scarious. **Phyllaries** 5–8 × 2–3 mm. **Ray laminae** (4–)6–7 × 2–3.2 mm, with 2–5 hairs inside opening of tube. **Disc corollas** 2.5–3 mm, tubes greenish, lobes 0.35–0.5 mm, reflexed, tips brown and callous. **Cypselae** brown, 1.5–2 mm, ribs pale. $2n = 18$.

Flowering summer. Alpine zones, moist to mesic gravelly sites, arctic tundra, moist to well-drained, exposed areas with low organic content, on rock, gravel, sand, silt, clay, till, or moss, calcareous or not, seldom on sea beaches or areas exposed to salt spray; 0–500+ m; B.C., N.W.T., Nunavut, Que., Yukon; Alaska; e Asia (Russian Far East).

121. ARCTANTHEMUM (Tzvelev) Tzvelev, Novosti Sist. Vyssh. Rast. 22: 274. 1985

• [Greek *arktos*, northern, and *anthemon*, flower, alluding to arctic range]

Luc Brouillet

Dendranthema Des Moulins sect. *Arctanthemum* Tzvelev in V. L. Komarov et al., Fl. URSS 26: 384, 879. 1961

Perennials, (2.5–)5–40 cm (taller in fruit, rhizomatous). **Stems** 1 (per rosette), ascending, usually simple, sometimes branched distally, glabrous or ± woolly (hairs basifixed). **Leaves** (not marcescent) basal and cauline; alternate; petiolate or sessile (distal); blades (± fleshy) fan-shaped, cuneate, or spatulate to narrowly lanceolate (distally), usually pinnati-palmately lobed (lobes 3–7), ultimate margins coarsely crenate, dentate (at apices of lobes), or entire (distally), faces ± woolly, glabrescent. **Heads** radiate, borne singly or in 2s or 3s. **Involucres** patelliform to hemispheric, 13–29 mm diam. **Phyllaries** persistent, (22–)25–34(–44) in 3(–4) series, distinct, lanceolate to oblong, unequal, margins and apices (hyaline and colorless or brownish) scarious. **Receptacles** dome-shaped (glabrous), epaleate. **Ray florets** (9–)14–25(–30), pistillate, fertile; corollas white, laminae lance-elliptic or elliptic to oblong or oblong-lanceolate. **Disc florets** 140–360+, bisexual, fertile; corollas yellow, tubes broadly cylindric, throats campanulate, lobes 5, deltate (without resin sacs). **Cypselae** cylindro-obconic, ribs 5–8(–10), faces glabrous, usually gland-dotted (pericarps without myxogenic cells or resin sacs; embryo sac development monosporic); **pappi** 0. *x* = 9.

Species 1: n North America; n Eurasia (Russia), ne Asia (Pacific coast).

C. Oberprieler et al. (pers. comm.) have placed *Arctanthemum* in their Asian grade (Group II) based on the study by L. E. Watson et al. (2002), distinct from *Chrysanthemum* (conserved name, syn. *Dendranthema*); *Arctanthemum* was also recognized as distinct by K. Bremer and C. J. Humphries (1993), who included *Hulteniella* in it. *Hulteniella* is kept separate here (as done by Oberprieler et al.). In both papers, three species were recognized within *Arctanthemum*, as done by N. N. Tzvelev when he established the genus. Later, Tzvelev (1987) placed *A. hultenii* in synonymy with *A. arcticum* subsp. *polare* (as here). The status of the remaining species, *A. kurilense* (Tzvelev) Tzvelev, which has not been reported for North America, appears dubious and the taxon is not recognized here; it may belong with *A. arcticum* subsp. *arcticum*.

SELECTED REFERENCE Tzvelev, N. N. 1987. *Arctanthemum*. In: A. I. Tolmatchew, ed. 1960–1987. Flora Arctica URSS. 10 vols. Moscow and Leningrad. Vol. 10, pp. 114–117.

1. **Arctanthemum arcticum** (Linnaeus) Tzvelev in A. I. Tolmatchew, Fl. Arct. URSS 10: 115. 1987 F

Chrysanthemum arcticum Linnaeus, Sp. Pl. 2: 889. 1753; *Dendranthema arctica* (Linnaeus) Tzvelev; *Leucanthemum arcticum* (Linnaeus) de Candolle

Perennials, forming tufts of 1–10+ rosettes; rhizomes at or below ground, fleshy, tough, branched; taproots (young plants) 3–6(–10) mm diam. **Stems** glabrous or sparsely woolly proximally to ± densely white-woolly near heads. **Leaves** sometimes glaucous; petioles 0–95 mm; bases cuneate to truncate, sheathing to clasping (distal), margins revolute, teeth obtuse to rounded, sometimes mucronulate, faces glabrous or sparsely woolly, glabrescent; basal and proximal cauline blades ± fan-shaped to cuneate or spatulate, 6–50 × 4–35 mm, lobes usually 3–7, apices blunt; distal blades linear, 4–35 × 1–28(–40) mm, bases attenuate, margins crenate, dentate, or entire, apices acute or obtuse; faces ± woolly, glabrescent. **Peduncles** usually bracteate, bracts 0–4, linear-lanceolate to linear, 6–19 mm. **Ray laminae** 8–12 mm. **Disc corollas:** tubes greenish yellow, (1.1–)1.4–2 mm, throats and lobes yellow turning brownish, throats 0.7–1 mm, sometimes

very sparsely glandular, lobes (0.25–)0.4–0.6 mm. **Cypselae** 1.5–2.5 mm. *2n* = 18 (72?).

Subspecies 2 (2 in the flora): coastal Alaska and low-arctic Canada, arctic Eurasia, ne Asia (s to Japan).

When Hultén described *Chrysanthemum arcticum* subsp. *polare*, he mapped the two subspecies. Along the Bering Strait in Alaska, all the material appears to belong to subsp. *polare*, except at Norton Bay, where a specimen (ALA) could be attributed to subsp. *arcticum*. That population would be very disjunct from the range of the subspecies along the south coast of Alaska. The two subspecies do not overlap widely in North America.

1. Plants 10–40 cm (more in fruit); stems sometimes branched; blades of basal leaves fan-shaped to cuneate or spatulate, 3–5(–7)-lobed; ray laminae (15–)17–25(–31) mm, veins (5–)8–10(–12). . . . . 1a. *Arctanthemum arcticum* subsp. *arcticum*
1. Plants (2.5–)5–20(–26) cm (more in fruit); stems not branched; blades of basal leaves cuneate to spatulate, 0–3-lobed; ray laminae (7–)9–18(–21) mm, veins 4–5(–7) . . . . 1b. *Arctanthemum arcticum* subsp. *polare*

1a. Arctanthemum arcticum (Linnaeus) Tzvelev subsp. **arcticum** • Arctic daisy

Plants 10–40 cm (more in fruit). Stems sometimes branched. Leaves basal and cauline (cauline to beyond midstem, regularly distributed); basal and proximal: petioles 15–95 mm, blades fan-shaped to cuneate or spatulate, 15–50 × 4–35 mm, 3–5(–7)-lobed; cauline: petioles 0–55 mm, blades 6–35 × 4–28 mm. Heads 1–2(–3), 19–29 mm diam. (flattened, excluding rays). Peduncular bracts 9–19 × 1–2 mm. Ray laminae (15–)17–25(–31) × (3.2–)3.4–5.5(–8.2) mm, veins (5–)8–10(–12).

Flowering summer. Open, coastal, wet, brackish habitats on clay, sand, gravel or rocks, upper tidal marshes, brackish coastal meadows, coastal herbaceous or heath tundras, sloughs, coastal rocks, flood plains and bars at mouths of streams; 0–10 m; B.C.; Alaska; ne Asia (Japan, Russian East Coast).

Subspecies *arcticum* is coastal from northern Japan and Russia to North America as far as northwestern British Columbia. It is cultivated as an ornamental. The chromosome number 2*n* = 72 has been reported from Russia for the species; it is uncertain to what subspecies this number may belong. No report apparently exists for this subspecies in North America.

1b. Arctanthemum arcticum (Linnaeus) Tzvelev subsp. **polare** (Hultén) Tzvelev in A. I. Tolmatchew, Fl. Arct. URSS 10: 116. 1987 • Polar daisy, chrysanthème polaire [F]

Chrysanthemum arcticum Linnaeus subsp. *polare* Hultén, Svensk Bot. Tidskr. 43: 776. 1949; *Arctanthemum hultenii* (Á. Löve & D. Löve) Tzvelev; *Dendranthema arcticum* (Linnaeus) Tzvelev subsp. *polare* (Hultén) Heywood; *D. hultenii* (Á. Löve & D. Löve) Tzvelev

Plants (2.5–)5–20(–26) cm (more in fruit). Stems not branched. Leaves mostly basal (cauline to midstem at most); basal and proximal: petioles 4–50 mm, blades cuneate to spatulate, 6–30 × 4–16 mm, mostly 0–3-lobed; cauline: petioles 0 or 10–25 mm, blades 4–23 × 1–8 mm. Heads 1(–2), 13–20 mm diam. (flattened, excluding rays). Peduncular bracts 6–12 × 1 mm. Ray laminae (7–)9–18 (–21) × (1.5–)1.8–4.3 mm, veins 4–5(–12). 2*n* = 18.

Flowering summer. Open, coastal, wet, brackish habitats on clay, sand, gravel, or rocks, beaches near tideline, coastal dunes, beach meadows, carex salt marshes, estuarine marshes and flood plains, coastal tundra; 0–10 m; Man., N.W.T., Nunavut, Ont., Que., Yukon; Alaska; arctic Russia.

In North America, subsp. *polare* ranges more or less continuously from St. Lawrence Island and the Seward Peninsula (Alaska) in the west, to the Ungava Peninsula (Quebec) in the east, along the low arctic continental coast, the coast of Hudson Bay, and the southern coast of the lower islands of the Arctic Archipelago.

122. ANTHEMIS Linnaeus, Sp. Pl. 2: 893. 1753; Gen. Pl. ed. 5, 381. 1754 • Chamomile [Greek *anthemon*, flower] [I]

Linda E. Watson

Annuals (biennials) [perennials, subshrubs], mostly 5–90 cm (often aromatic). Stems 1–5+, erect to decumbent, usually branched, strigillose or strigoso-sericeous to villous (hairs medifixed), glabrescent [glabrous or sericeous to lanate]. Leaves mostly cauline; alternate; petiolate or sessile; blades ± obovate to spatulate, 1–3-pinnately lobed, ultimate margins dentate to lobed, faces glabrous or strigillose to villous [glabrous or sericeous to lanate]. Heads radiate [discoid], borne singly or in lax, corymbiform arrays (peduncles sometimes clavate and/or curved in fruit). Involucres obconic to hemispheric or broader, 5–13[–20] mm diam. Phyllaries persistent, mostly 21–35+ in 3–5 series, distinct, deltate to lanceolate, oblong, or elliptic, unequal, margins and apices (hyaline and colorless or brownish [black]) scarious. Receptacles hemispheric to narrowly conic, paleate (wholly or only distally); paleae ± flat, scarious to indurate (subulate or elliptic to obovate with mucronate to acuminate-spinose tips). Ray florets [0 or 2–]5–20 [–30+], pistillate and fertile or styliferous and sterile; corollas usually white, rarely yellow or pink, laminae mostly oblong (tubes sometimes hairy). **Disc florets** (60–)100–300+, bisexual, fertile; corollas usually yellow, rarely pink, tubes ± cylindric (usually proximally dilated, ± spongy in fruit, sometimes hairy, not saccate), throats funnelform, lobes 5, ± triangular (abaxially minutely crested). **Cypselae** obovoid to obconic or turbinate (circular or 4-angled in cross section), ribs usually 9–10 (0) and smooth or tuberculate, faces glabrous (pericarps with myxogenic cells); **pappi** 0 or coroniform. $x = 9$.

Species 175 (2 in the flora): introduced; Europe, sw Asia, n, e Africa; introduced in s Africa, Pacific Islands (New Zealand), Australia.

Anthemis secundiramea Bivona-Bernardi, a European species, was collected once in Virginia; it differs from *A. arvensis* in having peduncles to 8 cm and ribs on cypselae ± tuberculate.

SELECTED REFERENCE   Oberprieler, C. 2001. Phylogenetic relationships in *Anthemis* L. (Compositae, Anthemideae) based on nrDNA ITS sequence variation. Taxon 50: 745–762.

1. Stems branched mostly proximally; receptacles paleate throughout, paleae lanceolate to oblanceolate, weakly navicular (± carinate, tips acuminate-spinose); cypselae: ribs smooth or weakly tuberculate .... 1. *Anthemis arvensis*
1. Stems branched mostly distally or ± throughout; receptacles paleate distally, paleae subulate to acerose; cypselae: ribs ± tuberculate .... 2. *Anthemis cotula*

1. Anthemis arvensis Linnaeus, Sp. Pl. 2: 894. 1753

• Corn chamomile, fausse camomille, anthémis des champs F I

Anthemis arvensis var. *agrestis* (Wallroth) de Candolle; *Chamaemelum arvense* (Linnaeus) Hoffmannsegg & Link

Annuals (sometimes persisting), (5–)10–30+[–80] cm, not notably scented. **Stems** green or reddish, decumbent (sometimes rooting at nodes) or ascending to erect, branched mostly proximally, ± strigoso-sericeous or villous, glabrescent. **Leaf blades** 15–35 × 8–16 mm, 1–2-pinnately lobed (ultimate lobes triangular to narrowly elliptic or linear). **Peduncles** mostly 4–15 cm (sometimes clavate in fruit). **Involucres** 6–13 mm diam., ± villous. **Receptacles** paleate throughout; paleae lanceolate to oblanceolate (± carinate), 3–4+ mm (including acuminate-spinose tips). **Ray florets** 5–20, pistillate, fertile; corollas white, rarely tinged with pink, laminae 5–15 mm. **Disc corollas** (sometimes tinged with purple) 2–3(–4) mm. **Cypselae** 1.7–2+ mm, ribs smooth or weakly tuberculate (sometimes separated by relatively deep furrows); **pappi** 0 or coroniform (0.01+ mm). $2n = 18$.

Flowering May–Jul(–Sep). Disturbed sites, abandoned plantings; 10–400+ m; introduced; B.C., N.B., Nfld. and Labr. (Nfld.), N.S., Ont., P.E.I., Que.; Ala., Calif., Colo., Conn., Del., D.C., Fla., Ga., Idaho, Ill., Ind., Iowa, Ky., La., Maine, Md., Mass., Miss., Mo., Nebr., N.H., N.J., N.Y., N.C., N.Dak., Ohio, Oreg., Pa., R.I., S.C., Tenn., Vt., Va., Wash., W.Va., Wyo.; Europe.

Anthemis arvensis is morphologically variable; it is found throughout much of North America.

2. Anthemis cotula Linnaeus, Sp. Pl. 2: 894. 1753

• Mayweed, stinking chamomile, camomille maroute I

Anthemis foetida Lamarck; *Chamaemelum cotula* (Linnaeus) Allioni; *Maruta cotula* (Linnaeus) de Candolle

Annuals, (5–)15–35(–90) cm, usually ill-scented. **Stems** green (sometimes red-tinged), usually erect, branched mostly distally or ± throughout, glabrous, glabrate, puberulent, or sparsely strigillose to strigoso-sericeous (glabrescent, hairs mostly medifixed) and gland-dotted. **Leaf blades** 25–55 × 15–30 mm, 1–2-pinnately lobed. **Peduncles** mostly (2–)4–6(–15) cm. **Involucres** 5–9 mm diam., ± villosulous to arachnose. **Receptacles** paleate mostly distally; paleae subulate to acerose 2–3+ mm (often gland-dotted). **Ray florets** 10–15, styliferous and sterile; corollas white, laminae 5–15+ mm. **Disc corollas** 2–2.5 mm (sparsely gland-dotted). **Cypselae** 1.3–2 mm, ribs ± tuberculate (furrows often gland-dotted); **pappi** 0. $2n = 18$.

Flowering (Apr–)May–Aug(–Oct). Disturbed sites, clearings, fields, roadsides; 10–600(–1500+) m; introduced; Alta., B.C., Man., N.B., Nfld. and Labr. (Nfld.), N.S., Ont., P.E.I., Que., Sask., Yukon; Ala., Alaska, Ariz., Ark., Calif., Colo., Conn., Del., D.C., Fla., Ga., Idaho, Ill., Ind., Iowa, Kans., Ky., La., Maine, Md., Mass., Minn., Miss., Mo., Mont., Nebr., Nev., N.H., N.J., N.Mex., N.Y., N.C., N.Dak., Ohio, Okla., Oreg., Pa., R.I., S.C., S.Dak., Tenn., Tex., Utah, Vt., Va., Wash., W.Va., Wis., Wyo.; Eurasia.

Anthemis cotula is a weed throughout North America.

123. ONCOSIPHON Källersjö, Bot. J. Linn. Soc. 96: 310, figs. 2F, 6, 16E1, 16E2. 1988

• [Greek *onkos*, swelling, and *siphon*, tube; allusion unclear] I

David J. Keil

Annuals, 5–15(–50+) cm (usually strongly scented). **Stems** 1–5+ from bases, erect to ascending, branched ± throughout, ± puberulent or sparsely strigillose (hairs basifixed) and gland-dotted [glabrous]. **Leaves** mostly cauline; alternate; petiolate or sessile; blades ovate to obovate [linear], usually [1–]2(–3)-pinnately lobed, ultimate margins entire, faces ± puberulent or sparsely strigillose and gland-dotted [glabrous]. **Heads** discoid [radiate], borne singly or in corymbiform arrays (pedunculate). **Involucres** ± hemispheric or broader [cylindric], [3–]4–6 [–8+] mm diam. **Phyllaries** persistent or tardily falling (± reflexed in fruit), [15–]22–35[–60+] in 3–4+ series, distinct, lance-triangular to ± ovate or elliptic (± navicular, carinate, sometimes each with central resin canal), unequal, margins and apices scarious. **Receptacles** [± flat] hemispheric, ± globose, or conic, epaleate (often muricate). **Ray florets** 0 [to 10, pistillate, fertile; corollas white, laminae ± obovate]. **Disc florets** [20–]100–250+, bisexual, fertile; corollas yellow, tubes ± cylindric (often basally invaginated and ± decurrent onto ovaries; tubes and proximal portions of throats externally continuous and ± dilated with relatively large, quadrate cells, often oblique, fragile when dry, distally constricted and texturally distinct from ± campanulate distal portions of throats), lobes 4, triangular. **Cypselae** columnar to prismatic (proximally tapered), ribs 4, faces gland-dotted between ribs, especially adaxially (pericarps without myxogenic cells or resin sacs; embryo sac development monosporic); **pappi** coroniform (shorter adaxially), dentate or entire. $x = 6, 8$.

Species 8 (1 in the flora): introduced; s Africa; also introduced in Australia.

Oncosiphon suffruticosum (Linnaeus) Källersjö [*Matricaria suffruticosa* (Linnaeus) Druce] was collected on ballast at Portland, Oregon, in 1902. It differs from *O. piluliferum* by smaller (3–5 mm diam.) heads clustered in dense, ± flat-topped, corymbiform arrays. Both *O. piluliferum* and *O. suffruticosum* are common weeds in their native southern Africa and in Australia. Both have the potential to become serious pests in western North America. They are unpalatable to livestock and tend to dominate infested pasturelands. When consumed by animals, they taint the flavor of milk and meat.

SELECTED REFERENCE Källersjö, M. 1988. A generic re-classification of *Pentzia* Thunb. (Compositae: Anthemideae) from southern Africa. Bot. J. Linn. Soc. 96: 299–322.

1. Oncosiphon piluliferum (Linnaeus f.) Källersjö, Bot. J. Linn. Soc. 96: 314. 1988 • Stinknet, globe chamomile F I

Cotula pilulifera Linnaeus f., Suppl. Pl., 378. 1782; *Matricaria globifera* (Thunberg) Fenzl ex Harvey; *Pentzia globifera* (Thunberg) Hutchinson

Herbage with vile odor. **Leaf blades** mostly obovate, 20–35 mm, ultimate lobes linear to spatulate or oblanceolate, 0.5–1 mm wide, sparsely puberulent or strigillose, minutely gland-dotted (in pits). **Heads** borne singly or 2–4 in corymbiform arrays. **Peduncles** 1–2(–8) cm, ebracteate. **Involucres** 3–3.5 × 4–6 mm. **Receptacles** ± conic to subspheric, muricate. **Disc corollas** 1.5–2 mm. **Cypselae** slightly curved, 0.6–0.8 mm, sparingly dotted with minute, glistening oil glands; **pappi** coroniform, 0.05–0.1 mm, margins subentire or minutely dentate. $2n = 12$.

Flowering Mar–Jun. Disturbed sites, coastal scrub; 500–900 m; introduced; Ariz., Calif.; South Africa.

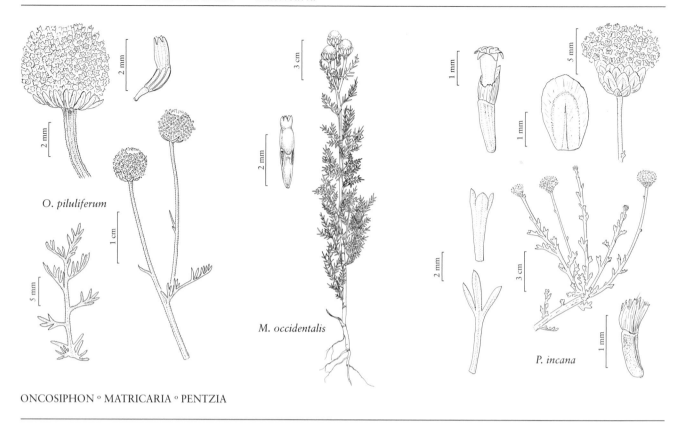

O. piluliferum

M. occidentalis

P. incana

ONCOSIPHON ∘ MATRICARIA ∘ PENTZIA

124. MATRICARIA Linnaeus, Sp. Pl. 2: 890. 1753; Gen. Pl. ed. 5, 380. 1754, name conserved • Mayweed, chamomile, matricary, matricaire, chamomille [Greek *matrix*, womb, and *-aria*, pertaining to; alluding to reputed medicinal properties]

Luc Brouillet

Chamomilla Gray

Annuals, (1–)10–25(–80) cm (taprooted; often aromatic). **Stems** 1–10+, usually erect or ascending, sometimes decumbent, branched or not, glabrous or glabrate to sparsely hairy (hairs basifixed). **Leaves** (not marcescent) basal (soon withering) and cauline; alternate; subpetiolate or sessile; blades spatulate to oblong or ovate (bases sheathing or clasping, often pinnately auriculate), (1–)2–3-pinnately lobed (lobes linear, often curved distally), ultimate margins entire (± recurved), mucronate, faces glabrous or glabrate to sparsely hairy (hairs basifixed). **Heads** radiate or discoid, borne singly or in open, corymbiform arrays. **Involucres** patelliform, 4–14 mm diam. **Phyllaries** persistent, 25–50 in [2–]3–4[–5] series, distinct, oblong or ovate to spatulate or linear-spatulate (membranous, not carinate, bases not indurate), subequal, margins and apices (hyaline) scarious (apices rounded to obtuse). **Receptacles** conic to oblong-ovoid [subulate] (hollow), epaleate. **Ray florets** 0 or 10–22, pistillate, fertile; corollas white, laminae elliptic-ovate. **Disc florets** 120–750+, bisexual, fertile; corollas (persisting in fruit, often slightly asymmetric, sometimes with scattered, sessile, golden glands) yellow to greenish yellow or yellowish green, tubes ± dilated), throats urceolate to campanulate, lobes 4–5 (spreading), deltate [with resin sacs]. **Cypselae** obconic, slightly compressed (usually asymmetric, apices oblique), ribs [3–]5, faces glabrous, smooth between ribs (pericarps sometimes with myxogenic cells abaxially and/or in ribs; embryo sac development monosporic); **pappi** 0, coroniform, or (sometimes on ray cypselae) adaxial auricles. *x* = 9.

Species 7 (3 in the flora): North America, Eurasia, North Africa, some species widespread weeds in the southern hemisphere.

Matricaria has been confused with *Tripleurospermum* (see discussion under the latter). Typification of *Matricaria* was discussed by K. Bremer and C. J. Humphries (1993), who rejected the arguments of S. Rauschert (1974) in favor of the use of *Chamomilla* over *Matricaria*.

1. Heads radiate (ray corollas white); disc corollas yellow to greenish yellow, lobes 5; cypselae 0.75–0.9 mm; pappi usually 0, sometimes coroniform (entire or lobed) or (on ray cypselae) each a toothed auricle . . . . 3. *Matricaria chamomilla*
1. Heads discoid; disc corollas greenish yellow, lobes 4(–5); cypselae 1–1.5 mm (ribs, at least 2 lateral, each with longitudinal mucilage gland); pappi coroniform, entire or lobed.
  2. Heads (1–)4–50(–300); discs 4–7(–11) × 4–7.5(–10) mm; cypselae: 2 lateral ribs each with mucilage gland along ± entire length; pappi coroniform, entire; plants aromatic (pineapple odor when bruised); stems 1–10+, usually erect or ascending, sometimes decumbent, branched from bases . . . . 1. *Matricaria discoidea*
  2. Heads 1–15+; discs 5–12.5 × 6–14 mm; cypselae: 2 lateral ribs each with mucilage gland in distal ¹/₂ (glands expanding into lobes); pappi coroniform, lobed (lobes 2, abaxio-lateral); plants not notably aromatic; stems usually erect, sometimes ascending, simple or branched mostly distally, sometimes proximally (then ascending) . . . . . 2. *Matricaria occidentalis*

1. **Matricaria discoidea** de Candolle in A. P. de Candolle and A. L. P. P. de Candolle, Prodr. 6: 50. 1838
  • Pineappleweed, disc mayweed, rayless chamomile, matricaire odorante E

Chamomilla suaveolens (Pursh) Rydberg; *C. discoidea* (de Candolle) J. Gay ex A. Braun; *Santolina suaveolens* Pursh 1813, not *Matricaria suaveolens* Linnaeus 1755

Annuals, (1–)4–40(–50) cm; aromatic (pineapple odor when bruised). **Stems** 1–10+, usually erect or ascending, sometimes decumbent, branched from bases. **Leaf blades** (5–)10–65(–85) × 2–20 mm. **Heads** discoid, (1–)4–50(–300), usually borne singly, sometimes in open, corymbiform arrays. **Peduncles** 2–25(–30) mm (sometimes villous near heads). **Involucres** 2.5–3.8 mm. **Phyllaries** 29–47+ in 3 series, margins mostly entire. **Receptacles** 2.5–7.5 mm, ± acute or obtuse. **Ray florets** 0. **Discs** hemispheric to broadly ovoid, 4–7(–11) × 4–7.5(–10) mm. **Disc florets** 125–535+; corollas greenish yellow, 1.1–1.3 mm (± glandular), lobes 4(–5). **Cypselae** pale brown to tan, ± cylindric-obconic (asymmetric, abaxially ± gibbous distally), 1.15–1.5 mm, ribs white (lateral 2 each with reddish brown mucilage gland along ± entire length, glands sometimes distally expanded, abaxial 1–2 weak, sometimes each with elongate mucilaginous gland), faces not glandular; **pappi** coroniform, entire. **2n** = 18. [as *M. matricarioides*]

Flowering early summer–fall. Open areas, bare disturbed areas and rural or urban waste grounds, sometimes alkaline, roadsides, railroads, footpaths, cultivated and abandoned fields and gardens, irrigation ditches, stream banks, sandbars; 0–2700 m; Greenland; St. Pierre and Miquelon; Alta., B.C., Man., N.B., Nfld. and Labr., N.W.T., N.S., Ont., P.E.I., Que., Sask., Yukon; Alaska, Ariz., Ark., Calif., Colo., Conn., Del., Idaho, Ill., Ind., Iowa, Kansas, Ky., La., Maine, Md., Mass., Mich., Minn., Miss., Mo., Mont., Nev., N.H., N.J., N.Mex., N.Y., N.C., N.Dak., Ohio, Okla., Oreg., Pa., R.I., S.C., S.Dak., Tenn., Tex., Utah, Vt., Va., Wash., W.Va., Wis., Wyo.; introduced in Eurasia, Australia.

Matricaria discoidea has been used as a medicinal and aromatic plant by Native American tribes (D. E. Moerman 1998). It also is considered a weed, and it is resistant to a photosystem II inhibitor herbicide in the United Kingdom (www.weedscience.org). It is a northwestern North American native that has spread to eastern and northern North America and elsewhere (E. McClintock 1993b; E. G. Voss 1972–1996, vol. 3; A. Cronquist 1994). NatureServe (www.natureserve.org) and Natural Resources Conservation Service (plants.usda.gov) erroneously present *M. discoidea* as introduced on the continent. Its natural habitat is ill-defined because the species has become ruderal even in its native range. For discussion of the nomenclature of this taxon, see S. Rauschert (1974); K. N. Gandhi and R. D. Thomas (1991); Cronquist; and Voss.

"*Matricaria matricarioides* (Lessing) Porter" cannot be applied to the American taxon; *M. matricarioides* was originally published as *Artemisia matricarioides* Lessing, a new name for *Tanacetum pauciflorum* Richardson (see S. Rauschert 1974), itself a synonym of *T. huronense* Nuttall. W. Greuter (pers. comm.), who accepts

M. discodea, considers Rauschert's treating *Artemisia matricarioides* as homotypic with *T. pauciflorum* as equivalent to a lectotype designation.

2. Matricaria occidentalis Greene, Bull. Calif. Acad. Sci. 2: 150. 1886 • Valley mayweed E F

Chamomilla occidentalis (Greene) Rydberg

Annuals, 8–45(–70) cm; not notably aromatic. **Stems** 1–5+, usually erect, sometimes ascending, simple or branched mostly distally, sometimes proximally (then ascending). **Leaf blades** 1560(–80) × 5–28 mm. **Heads** discoid, 1–15+, borne singly or in open, corymbiform arrays. **Peduncles** 1–56(–90) mm (glabrous or villosulous near heads, bracts 0–3, simple or 1-pinnate). **Involucres** 3.5–4.5 mm. **Phyllaries** 25–55+ in 3–4 series, margins erose. **Receptacles** 4.5–6.6 mm, rounded to obtuse or ± acute. **Ray florets** 0. **Discs** ovoid to hemispheric, 5.5–12.5 × 6–14 mm. **Disc florets** 400–750+; corollasgreenish yellow, 1.2–1.7 mm (sparsely glandular), lobes 4. **Cypselae** tan to brownish, obtriangular-obconic (angular, abaxial faces rounded, adaxial faces slightly convex), 1–1.5 mm, ribs white (2 abaxial, 2 lateral, each with reddish brown mucilage gland in distal $\frac{1}{2}$, glands expanding into lobes, 1 adaxial, usually obscure, sometimes with a gland), faces not glandular; **pappi** coroniform, lobed (lobes 2, spreading, abaxio-lateral, ± obtuse to rounded). **2n** = 18.

Flowering spring–summer, fruiting summer–fall. Undisturbed alkali flats, vernal pools, edges of salt marshes; 0–2400 m; Calif., Oreg.

E. McClintock (1993b) reported *Matricaria occidentalis* as being used as a substitute for chamomile. In California, the species is found in the Coast Ranges to the South Coast, in parts of the San Joaquin Valley, and in the High Cascade, Sierra Nevada, and Desert Ranges and surrounding areas. Although listed as secure, the species appears to have been eradicated from some counties in California and its range may be shrinking due to disappearance of its habitat.

3. Matricaria chamomilla Linnaeus, Sp. Pl. 2: 891. 1753 • German chamomile, matricaire camomille I

Chamomilla recutita (Linnaeus) Rauschert; *Matricaria courrantiana* de Candolle; *M. recutita* Linnaeus

Annuals, (2–)8–60(–80) cm; aromatic (when bruised). **Stems** 1–8+, erect or ascending, branched distally. **Leaf blades** 5–78 × 3–18 mm. **Heads** radiate, (1–)8–120 (–900), borne singly. **Peduncles** (5–)20–50(–75+) mm. **Involucres** 2–3.2 mm. **Phyllaries** 34–42+ in 3 series, margins entire or distally erose. **Receptacles** (with reddish brown, longitudinal mucilaginous glands), 4–6 mm, acute to obtuse. **Ray florets** [1] (10–)14–26; corollas white, tubes narrowly winged, laminae soon deflexed, 7–8.5 × 2.4–3.3 mm. **Discs** obovoid or spheroid to ovoid, 5–7 × 5–9.5 mm. **Disc florets** 250–570+; corollas yellow to greenish yellow, 1.6–1.8 mm, lobes 5. **Cypselae** tan, obconic, 0.75–0.9 mm, ribs white, 3 abaxial, 2 nearly marginal, faces glandular; **pappi** usually 0, sometimes coroniform (entire or lobed) or (ray florets) toothed auricles as long as or longer than cypselae [minute crowns]. **2n** = 18.

Flowering spring. Dry roadsides, railroads, other waste places; 0–2700 m; introduced; Greenland; Alta., B.C., Man., Nfld. and Labr. (Nfld.), N.S., Ont., Que., Sask.; Ala., Ariz., Ark., Calif., Conn., D.C., Ill., Ind., Iowa, Kans., Ky., Maine, Md., Mass., Mich., Minn., Miss., Mo., N.J., N.Y., N.Dak., Ohio, Oreg., Pa., R.I., Tenn., Tex., Utah, Vt., Va., Wash., Wis.; Eurasia.

Matricaria chamomilla has numerous, and ages-old, usages, particularly as herb tea, as a natural medicine, and for pharmaceutical extracts. It has anti-inflammatory, antibacterial, anti-allergic, and sedative properties. It is grown commercially on all continents. Reports for New Brunswick have not been confirmed, all specimens having been redetermined to *Anthemis cotula* (H. R. Hinds 2000). Although the name *Matricaria chamomilla* has been considered to be misapplied (e.g., S. Rauschert 1974; A. Cronquist 1994; E. G. Voss 1972–1996, vol. 3), W. L. Applequist (2002) argued convincingly that the name is indeed correctly applied to the taxon described here. Among the North American material, specimens with coronate ray cypselae (var. *chamomilla*), or wholly without coronas [var. *recutita* (Linnaeus) Grierson] have been encountered but none with fully coronate cypselae (var. *coronata* J. Gay ex Boissier), even though synonymy under this name includes *M. courrantiana*, reported for Texas and New Mexico (specimens not seen). The varieties may not be worth recognizing (Applequist; Q. O. N. Kay 1976) and are not treated formally here.

125. PENTZIA Thunberg, Prodr. Pl. Cap., 145. 1800 • [For Hendrik Christian Pentz, 1738–1803, Swedish plant collector] ☐

David J. Keil

Shrubs [subshrubs], [5–]15–35[–50] cm (usually aromatic). **Stems** 1–10+, usually much branched, ± tomentose (hairs medifixed) [glabrate or glabrous]. **Leaves** cauline (often fascicled, ± ericoid); alternate [opposite]; petiolate or sessile; blades linear to obovate or cuneate, often 1[–2]-pinnately lobed, ultimate margins entire or toothed, faces ± tomentose [glabrate or glabrous]. **Heads** discoid, borne singly or in loose, corymbiform arrays (pedunculate). **Involucres** spheric to hemispheric or broader, 3–6[–12+] mm diam. **Phyllaries** persistent or tardily falling, 22–40[–60+] in 3–4+ series, distinct, deltate-ovate to elliptic or obovate (sometimes each with central resin canal and/or keel), unequal, margins and apices (often hyaline, sometimes brownish) ± scarious (tips of inner ± dilated). **Receptacles** conic [convex to hemispheric], epaleate. **Ray florets** 0. **Disc florets** 40–100[–200+], bisexual, fertile; corollas yellow, tubes ± cylindric (usually dilated, with relatively thick vascular strands), throats abruptly dilated, lobes 5, triangular. **Cypselae** narrowly obovoid to oblong or elliptic, ± terete or flattened, ribs 5, faces gland-dotted between ribs (pericarps with myxogenic cells abaxially and along ribs); **pappi** of 1 (often cupped or earlike) or 3–5 (erose to lacerate) scales (usually more developed adaxially) [asymmetrically coroniform or 0]. *x* = 9.

Species ± 23 (1 in the flora): introduced; Africa (mostly South Africa, also Namibia and Mediterranean Africa and adjacent areas).

1. Pentzia incana (Thunberg) Kuntze, Revis. Gen. Pl. 3(2): 166. 1898 • African sheepbush ☐ ☐

Chrysanthemum incanum Thunberg, Prodr. Pl. Cap., 161. 1800

Stem branches ascending, finely appressed-tomentulose. **Leaves** often fascicled; blades ± obovate to cuneate, 3–9 mm, 1(–2) times divided, ultimate lobes linear to narrowly spatulate, revolute, inconspicuously appressed-puberulent to tomentulose or glabrate, minutely gland-dotted. **Peduncles** slender, 2–3 cm, proximally bracteate, bracts remote, appressed, leaflike, distally usually ebracteate. **Involucres** 3–4 × 4–5 mm. **Phyllaries:** outer with relatively narrow scarious margins, inner with relatively broad, hyaline tips. **Corollas** ± goblet-shaped, 1.8–2+ mm. **Cypselae** gray-brown, 1.2–1.8 mm; **pappi** of 1–3, adaxial, ± truncate, erose to lacerate scales 0.6–1 mm.

Flowering Apr–Sep. Grasslands, pinyon-juniper-oak scrub; 900–1600 m; introduced; Ariz., N.Mex.; s Africa; also introduced in Australia.

Pentzia incana was introduced in Arizona from southern Africa by the U.S. Soil Conservation Service. It is valued in its native region as a browse plant.

126. COTULA Linnaeus, Sp. Pl. 2: 891. 1753; Gen. Pl. ed. 5, 380. 1754 • [Greek *kotule*, small cup] ☐

Linda E. Watson

Annuals or perennials, 2–25[–50+] cm (sometimes aromatic). **Stems** usually 1, erect or prostrate to decumbent or ascending (sometimes rooting at nodes), usually branched, glabrous or ± strigillose to villous (hairs mostly basifixed). **Leaves** usually mostly cauline [basal]; alternate [opposite]; petiolate or sessile; blades obovate or spatulate to lanceolate or linear, sometimes 1–3-pinnately [palmati-pinnately] lobed, ultimate margins entire or irregularly toothed, faces glabrous or ± strigillose to villous [lanate] (hairs mostly basifixed). **Heads** disciform [discoid or radiate], borne singly (peduncles sometimes dilated). **Involucres** broadly hemispheric to saucer-shaped, 3–12+[–15+] mm diam. **Phyllaries** persistent, 13–30+ in 2–3+ series, margins and apices (colorless, light to dark brown, or purplish) scarious. **Receptacles** flat to convex [conic], epaleate (sometimes ± covered with persistent stalks of florets). **Ray florets** 0 [5–8+, pistillate, fertile; corollas white] (peripheral pistillate florets 8–80+ in 1–3+ series; corollas usually none). **Disc florets** 12–200+[–600+], bisexual, fertile [functionally staminate]; corollas ochroleucous or yellow, tubes ± cylindric (bases sometimes adaxially saccate), throats abruptly ampliate, lobes (3–)4, ± deltate (sometimes one larger than others, usually each with central resin canal). **Cypselae** obovoid to oblong, ob-compressed or -flattened, ribs 2, lateral, sometimes becoming wings, faces ± papillate (pericarps relatively thin, sometimes with myxogenic cells and/or 2 lateral resin sacs); **pappi** 0. $x = 10$.

Species 55 (2 in the flora): introduced; s Old World; introduced also (perhaps some native) in Mexico, South America, s Oceanic Islands.

Some species of *Cotula* are widely naturalized. F. Hrusa et al. (2002) reported *Cotula mexicana* (de Candolle) Cabrera as established on golf courses in California; it is similar to *C. australis* and differs in leaf blades mostly 1-pinnate, receptacles pilose, and disc florets functionally staminate.

1. Annuals (± villous); leaf blades 2–3-pinnately lobed; involucres 3–4(–6) mm diam.; pistillate florets 8–40(–80+) in ± 1–3+ series . . . . 1. *Cotula australis*
1. Perennials (glabrous); leaf blades entire or irregularly toothed or lobed; involucres 6–9 (–12+) mm diam.; pistillate florets 12–40+ in 1 series . . . . 2. *Cotula coronopifolia*

1. Cotula australis (Sieber ex Sprengel) Hooker f., Fl. Nov.-Zel. 1: 128. 1852 ⬚

Anacyclus australis Sieber ex Sprengel, Syst. Veg. 3: 497. 1826

Annuals, 2–10(–25+) cm. **Stems** branched ± throughout, ± strigillose to villous, glabrescent. **Leaves** petiolate or sessile; blades obovate to spatulate, 2–3-pinnately lobed (ultimate lobes narrowly spatulate to linear), (1–)2–3(–6) cm. **Involucres** 3–4(–6) mm diam. **Phyllaries** 13–22+ in 2–3 series, subequal (apices brown). **Ray florets** 0 (peripheral pistillate florets 8–80+ in 1–3+ series; corollas 0). **Disc corollas** ochroleucous to pale yellow, 0.5–0.8 mm. **Cypselae:** outer stalked, 1–1.2 mm, ± winged, faces ± papillate; inner sessile, 0.8–1 mm, not winged, not papillate. $2n = 36, 40$.

Flowering Jan–Sep. Disturbed sites; 0–800 m; introduced; Ariz., Calif., Fla., Maine, Oreg., Tex.; Australia; also introduced in Northern Hemisphere: Mexico (Baja California); South America; Africa.

2. Cotula coronopifolia Linnaeus, Sp. Pl. 2: 892. 1753

• Brass-buttons, cotule ⬚ ⬚

Perennials, (3–)5–15(–25+) cm tall or across. **Stems** prostrate (rooting at nodes, ± fleshy) to decumbent or erect, glabrous. **Leaves** sessile (bases sheathing stems); blades linear to lanceolate or oblong (sometimes lobed, lobes lanceolate to linear), (1–)2–3(–7) cm, ultimate margins entire or irregularly toothed or lobed, faces gland-dotted. **Involucres** 6–9(–12+) mm diam. **Phyllaries** 21–30+ in 2–3+ series. **Ray florets** 0 (peripheral pistillate florets 12–40+ in 1 series; corollas 0). **Disc corollas** yellow, 1–1.5 mm. **Cypselae:** outer 1.2–1.8 mm, winged, adaxial faces papillate; inner 0.7–1 mm, scarcely, if at all, winged, not papillate. $2n = 20$.

Flowering Mar–Dec. Saline and freshwater marshes, along streams; 0–900 m; presumed introduced; B.C., N.B., N.S., P.E.I., Que.; Alaska, Ariz., Calif., Md., Mass., Nev., Oreg., Wash.; S Africa, Australia; also introduced in Mexico (Baja California); South America; Europe.

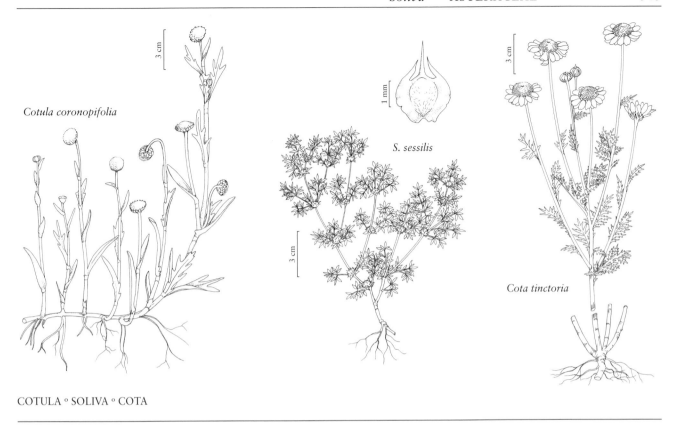

Cotula coronopifolia

S. sessilis

Cota tinctoria

COTULA ∘ SOLIVA ∘ COTA

127. **SOLIVA** Ruiz & Pavón, Fl. Peruv. Prodr., 113, plate 24. 1794 • Burrweed [For Salvador Soliva, an 18th-Century physician to the Spanish court] ⊡

Linda E. Watson

Gymnostyles Jussieu

Annuals, (1–)2–10(–30+) cm (taprooted, often stoloniferous and/or mat-forming). **Stems** 1–5+, ± procumbent to erect, simple or branched, glabrous or strigillose to villous [piloso-hispid] (hairs mostly basifixed). **Leaves** basal or basal and cauline; alternate; sessile or petiolate (bases ± clasping); blades ± obovate to spatulate, usually (1–)2–3-pinnately or -pinnati-palmately lobed (ultimate lobes oblong or lanceolate to oblanceolate or spatulate), ultimate margins entire or dentate, faces glabrous or strigillose to villous. **Heads** disciform, borne singly (sessile in leaf axils). **Involucres** ± hemispheric or broader, 2–8+ mm diam. **Phyllaries** persistent, 5–8+ in 1–2+ series, distinct, lanceolate to ovate, subequal, margins and apices (usually colorless, sometimes brownish) scarious. **Receptacles** flat to hemispheric or conic, epaleate. **Ray florets** 0 (peripheral, pistillate florets 5–100+ in 1–8+ series; corollas 0, styles sheathed by pericarp tissues, soon indurate, becoming spinelike). **Disc florets** 2–8+, functionally staminate; corollas yellowish or whitish, tubes ± cylindric, throats campanulate to funnelform, lobes (2–)4, ± deltate (without resin sacs). **Cypselae:** (bodies) ± obovate to oblanceolate or oblong-cuneate, ob-compressed or -flattened, winged (wings entire or sculpted), faces glabrous or ± scabrellous to hirtellous or (distally) villous to pilose or arachnose, glabrescent (pericarps without myxogenic cells or resin sacs); **pappi** 0 (persistent stylar sheaths indurate, spinelike). *x* = unknown.

Species 4–8 (3 in the flora): introduced; South America; introduced also in Europe, Asia, Pacific Islands (New Zealand), Australia.

SELECTED REFERENCES Cabrera, A. L. 1949. Sinopsis del género *Soliva* (Compositae). Notas Mus. La Plata, Bot. 14: 123–139. Webb, C. J. 1986. Variation in achene morphology and its implications for taxonomy in *Soliva* subgenus *Soliva* (Anthemideae, Asteraceae). New Zealand J. Bot. 24: 665–669.

1. Cypselar bodies obovate to oblanceolate, margins usually ± winged (wings not transversely rugulose or ribbed), faces glabrous or ± scabrellous to hirtellous . . . . 1. *Soliva sessilis*
1. Cypselar bodies oblong-cuneate, margins winged (wings transversely rugulose or ribbed), faces glabrous or distally villous to pilose or arachnose, glabrescent.
   2. Leaves mostly basal, blades 3–8(–15) cm, 2–3-pinnati-palmately lobed; cypselar wings rugulose or ribbed in proximal ²/₃, shoulders not spinose . . . . 2. *Soliva anthemifolia*
   2. Leaves basal and cauline, blades 1–2(–3) cm, 1(–2)-pinnati-palmately lobed; cypselar wings rugulose or ribbed in proximal ⁹/₁₀, shoulders ± spinose laterally . . . . 3. *Soliva stolonifera*

1. Soliva sessilis Ruiz & Pavón, Syst. Veg. Fl. Peruv. Chil., 113, plate 24. 1798 • Field burrweed, lawn burweed F I

Soliva daucifolia Nuttall; *S. pterosperma* (Jussieu) Lessing

Plants (1–)2–5(–25+) cm (high or across), ± villous, glabrescent (not stoloniferous, stems purplish, prostrate to ascending, often rooting at nodes). **Leaves** basal and cauline; blades ± oblanceolate, 1–2(–3)+ cm, 2(–3)-pinnati-palmately lobed. **Heads** mostly scattered along stems. **Involucres** 2–4(–5) mm diam. **Pistillate florets** 5–8(–17+) in 1–2+ series. **Disc florets** 4–8+; corollas 1.5–2.5 mm. **Cypselae:** bodies ± obovate to lanceolate, (1.5–)2.5–3+ mm, usually winged (wings entire or ± sinuate to incised, each shoulder usually distally projecting as spinelike tooth), faces glabrous or ± scabrellous to hirtellous; **pappi** 0 (persistent stylar sheaths indurate, spinelike, 1–2+ mm, erect or slightly inflexed). **2n** = ca. 92 (as *S. pterosperma*), 110+ (from Portugal), 118–120.

Flowering Mar–Jun(–Dec). Disturbed sites, lawns, roadsides; 0–600 m; introduced; B.C.; Ala., Ark., Calif., Fla., Ga., La., Miss., N.C., Okla., Oreg., S.C., Tex., Va., Wash.; South America; introduced also in Europe.

2. Soliva anthemifolia (Jussieu) Sweet, Hort. Brit., 243. 1826 • Button burrweed I

Gymnostyles anthemifolia Jussieu, Ann. Mus. Natl. Hist. Nat. 4: 262, plate 61, fig. 1. 1804; *Soliva mutisii* Kunth

Plants mostly 3–15(–30+) cm (high or across), ± villous, glabrescent (sometimes stoloniferous, ± mat-forming). **Leaves** mostly basal; blades ± obovate to spatulate, 3–8(–15) cm, 2–3-pinnati-palmately lobed. **Heads** mostly clustered in leaf axils (at ground level), rarely scattered along stems. **Involucres** 4–8+ mm diam. **Pistillate florets** (20–)50–100+ in 1–8+ series. **Disc florets** 2–4+; corollas 1.5–2 mm. **Cypselae:** bodies oblanceolate to cuneate-oblong, 1.5–2+ mm, wings transversely rugulose or ribbed on proximal ²/₃, shoulders not spinose laterally, faces distally villous to pilose, sometimes glabrescent; **pappi** 0 (persistent stylar sheaths indurate, spinelike, 1.5–3 mm, usually inflexed). **2n** = 118 (Punjab).

Flowering Mar–May(–Aug). Disturbed sites; 10–100+ m; introduced; Ark., Fla., La., Tex.; South America; introduced also in Mexico, Asia, Australia.

3. Soliva stolonifera (Brotero) Sweet, Hort. Brit., 243. 1826 • Carpet burrweed I

Hippia stolonifera Brotero, Phytogr. Lusitan. Select., no. 14. 1801; *Gymnostyles nasturtifolia* Jussieu; *G. stolonifera* (Brotero) Tutin; *Soliva nasturtifolia* (Jussieu) de Candolle

Plants 2–5(–15+) cm (high or across), ± strigillose to villous (stems procumbent, rooting at nodes). **Leaves** basal and cauline, blades ± spatulate, 1–2(–3) cm, 1(–2)-pinnati-palmately lobed. **Heads** mostly scattered along stems. **Involucres** 3–6(–8) mm diam. **Pistillate florets** (20–)40–60+ in 2–4+ series. **Disc florets** 4–6+; corollas 1.5–2 mm. **Cypselae:** bodies oblanceolate to oblong-cuneate, 1.8–2.2 mm, wings transversely rugulose or ribbed on proximal ⁹/₁₀, shoulders ± spinose laterally, faces distally villous to pilose or arachnose; **pappi** 0 (persistent stylar sheaths indurate, spinelike, 0.8–1.5 mm, usually inflexed). **2n** = ca. 114 (as *S. nasturtifolia*).

Flowering Mar–Apr(–Aug). Disturbed sites; 10–100+ m; introduced; Ala., Ark., Fla., Ga., La., S.C., Tex.; South America; introduced also in Europe.

128. COTA   J. Gay ex Gussone, Fl. Sicul. Syn. 2: 866. 1845 • [Possibly from pre-Linnaean generic name used as epithet in *Anthemis cota* Linnaeus]   ⊡

Linda E. Watson

Perennials [annuals, biennials], [5–]30–90 cm (rhizomatous and/or stoloniferous; not notably aromatic). **Stems** 1–5+, ascending to erect, not much branched, ± villous to sericeous (hairs usually medifixed, rarely basifixed) [glabrescent, glabrate]. **Leaves** cauline; alternate; petiolate or sessile; blades obovate to oblanceolate or spatulate, (1–)2(–3)-pinnately[-subpalmately] lobed (primary lobes often pectinately divided), ultimate margins serrate or entire, faces ± villous to sericeous [glabrescent, glabrate]. **Heads** radiate [discoid], borne singly or in lax, corymbiform arrays (peduncles sometimes distally dilated). **Involucres** hemispheric or broader [obconic], [5–]10–12[–20+] mm diam. **Phyllaries** persistent, 40–75+ in 3–5 series, distinct, deltate to lanceolate (often cartilaginous medially), unequal, margins and apices scarious (apices often mucronate to spinose, abaxial faces usually villous to arachnose). **Receptacles** ± hemispheric, paleate (throughout); paleae (persistent) ± flat, often cartilaginous (each proximally elliptic to oblong, distally ± subulate to spinose). **Ray florets** [0] 12–21[–35+], pistillate and fertile, or styliferous and sterile; corollas yellow [white or pinkish], laminae oblong (often reflexed, marcescent in fruit). **Disc florets** 80–200+, bisexual, fertile; corollas yellow [pinkish], tubes ± compressed (bases not dilated, not clasping cypselae), throats funnelform, lobes 5, deltate (minutely crested abaxially near tips). **Cypselae** obconic to prismatic (± 4-angled), ± ob-compressed or -flattened, ribs 0 or 2 lateral (sometimes winged) plus 3–10 finer ribs or nerves on each face, faces glabrous (pericarps with myxogenic cells; embryo sac development tetrasporic); **pappi** [0] coroniform. *x* = 9.

Species 40 (1 in the flora): introduced; Europe, sw Asia, n Africa.

Cota, sometimes a subgenus of *Anthemis*, is distinguished by its obcompressed and indistinctly ribbed cypselae; plants of *Anthemis* in narrow sense usually have ± terete cypselae with 10 distinct ribs. Additional differences include cytological, phytochemical, and molecular traits.

Apparently, *Cota altissima* (Linnaeus) Gay (*Anthemis altissima* Linnaeus) was reported from Oregon in error (L. Abrams and R. S. Ferris 1923–1960, vol. 4). No specimens of *C. altissima* from Oregon have been located, and the taxon has been specifically excluded from the Oregon Flora checklist (K. L. Chambers and S. Sundberg, www.oregonflora.org).

Cota austriaca (Jacquin) Schultz-Bipontinus (*Anthemis austriaca* Jacquin), a European species, has been reported as established near Pullman, Washington (specimen in UC); it differs from *C. tinctoria* in having an annual or biennial habit, subhirsute phyllaries, and white ray laminae (R. Fernandes 1976).

1. **Cota tinctoria**   (Linnaeus) J. Gay ex Gussone, Fl. Sicul. Syn. 2: 867. 1845 • Golden marguerite, yellow chamomile   F I

Anthemis tinctoria Linnaeus, Sp. Pl. 2: 896. 1753

Leaves 1–3(–5) cm, ultimate lobes narrowly oblong to spatulate or linear, ultimate margins entire or serrate (teeth apiculate). **Phyllaries:** abaxial faces sericeous to arachnose, margins often ciliolate. **Paleae** 4–5 mm (including spinose tips). **Ray laminae** yellow, 6–12+ mm. **Disc corollas** 3.5–4 mm. **Cypselae** 1.8–2.2 mm; **pappi** usually 2–2.5 mm. *2n* = 18.

Flowering Jul–Sep. Disturbed sites, fields, roadsides; 10–600+ m; introduced; Alta., B.C., Man., N.B., Nfld. and Labr. (Nfld.), N.S., Ont., Que., Sask.; Alaska, Ark., Calif., Conn., Idaho, Ill., Iowa, Maine, Md., Mass., Mich., Minn., Mo., N.H., N.J., N.Y., N.Dak., Ohio, Oreg., Pa., R.I., S.Dak., Utah, Vt., Va., Wash., W.Va., Wis., Wyo.; Europe, Asia.

129. TRIPLEUROSPERMUM Schultz-Bipontinus, Tanaceteen, 31. 1844 • Mayweed

[Greek *tri-*, three-, *pleuro-*, ribbed, and *sperma*, seed, alluding to strongly 3-ribbed cypselae]

Luc Brouillet

Annuals, biennials, or perennials, 5–80 cm (taprooted; scentless or nearly so). **Stems** 1–5+, usually erect or ascending, sometimes procumbent, branched or simple, glabrous or sparsely hairy (hairs basifixed). **Leaves** (not marcescent) basal (usually withering by flowering) and cauline; petiolate or sessile; blades oblong, 1–3-pinnately lobed, ultimate margins crenate or serrate, faces glabrous or sparsely hairy. **Heads** radiate [discoid or disciform], borne singly or in corymbiform arrays. **Involucres** hemispheric to patelliform, 8–12+ mm diam. **Phyllaries** persistent, 28–60+ in 2–5 series, distinct, broadly ovate, unequal to subequal, margins and apices (pale to dark brown or black) narrowly to widely scarious (abaxial faces glabrous). **Receptacles** convex to conic (± solid), epaleate. **Ray florets** [0] 10–34+, pistillate, fertile; corollas white [seldom pinkish], laminae mostly oblong. **Disc florets** 300–500, bisexual, fertile; corollas yellow [greenish], tubes cylindric (sessile-glandular), throats narrowly campanulate (cylindric or compressed), lobes 5, deltate (with resin sacs). **Cypselae** trigonous, ± compressed (apices truncate), ribs 3–5 (0–2 abaxial, 2 lateral, 1 adaxial, usually whitish, relatively thick, smooth), faces often rugose or tuberculate abaxially and between ribs and glabrous (pericarps sometimes with myxogenic cells and usually 2–3, sometimes 1–5, abaxial-apical resin sacs; embryo sac development tetrasporic); **pappi** 0 [coroniform or each an adaxial auricle]. *x* = 9.

Species ca. 40 (2 in the flora): North America, Eurasia, North Africa; introduced in New Zealand; one species widespread as a weed.

Tripleurospermum often has been included within *Matricaria*. It is distinguished from the latter by 3-ribbed cypselae with 2 abaxial-apical resin sacs versus 5-ribbed cypselae without apical sacs, and by production of 7-glycoside flavonols versus 3-glycosides (K. Bremer and C. J. Humphries 1993). C. Oberprieler (2001) showed that *Tripleurospermum* is closely related to *Anthemis* in the strict sense and not to *Matricaria*. This is supported by the tetrasporic embryo sac shared by *Tripleurospermum* and *Anthemis* (versus monosporic in *Matricaria* and other anthemid genera), similarity of karyotype, and the presence in both genera of matricaria ester in the polyacetylene pathway, otherwise unknown in the tribe.

Although W. L. Applequist (2002) favored a single species, *Tripleurospermum maritimum*, for the three taxa (as subspecies) present in North America, based in large parts on the findings of A. Vaarama (1953) on crossing ability among these taxa, on the existence of hybrid populations, and on the (mistaken) supposition that they form a polyploid series, some recent authors (e.g., Q. O. N. Kay 1976b, 1994; A. R. Clapham et al. 1987; E. G. Voss 1972–1996, vol. 3) have maintained two species based on morphologic characters, notably perennial versus annual habit, spacing of ribs (closer versus wider), elongate versus ± circular resin sacs on cypselae, relative infrequency of hybrid populations, and relative sterility of some hybrid individuals, as well as both species harboring diploid and tetraploid populations (Clapham et al.; Kay 1994). Success in interbreeding among plants does not always indicate conspecific status. In North America, authors such as M. L. Fernald (1950) and H. A. Gleason and A. Cronquist (1991) treated (in *Matricaria*) *T. inodorum* and *T. maritimum* subsp. *maritimum* as a single species with varieties or as a single entity, resulting in confusion about the distinctness of the two taxa in their ranges of introduction, and also difficulty in recognizing the presence of *T. inodorum* in areas where *T. maritimum* subsp. *maritimum* is present in coastal northeastern

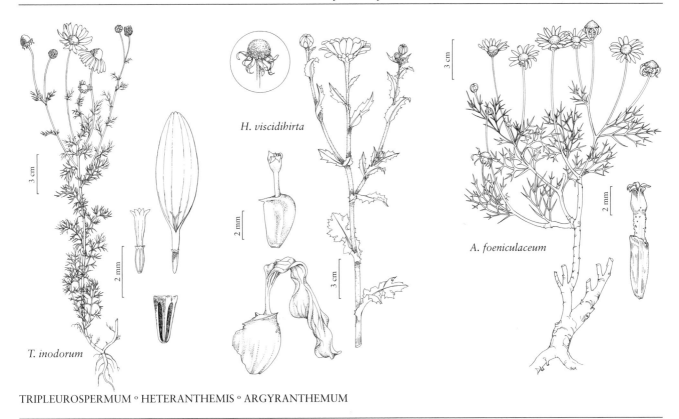

H. viscidihirta

A. foeniculaceum

T. inodorum

TRIPLEUROSPERMUM ∘ HETERANTHEMIS ∘ ARGYRANTHEMUM

United States (e.g., D. W. Magee and H. E. Ahles 1999). N. N. Tzvelev (2002b) segregated the subspecies of *T. maritimum* as species; under his scheme, three species are present in North America: *T. inodorum*, *T. maritimum* in the strict sense, and *T. hookeri* (= *T. maritimum* subsp. *phaeocephalum*). Pending further taxonomic work on this complex, I have here retained the taxonomy (except generic assignment) adopted by Kay (1976b), with two species, one including two subspecies.

1. Annuals (sometimes biennials or perennials); stems ascending to erect; ultimate leaf lobes not fleshy; cypselae: ribs separated by at least ¹/₃ their widths, resin glands ± circular (lengths mostly 1–1.5 times widths); scarious margins of phyllaries light brown, relatively narrow; cultivated and waste grounds . . . . 1. *Tripleurospermum inodorum*
1. Perennials (or biennials); stems usually procumbent, sometimes ascending; ultimate leaf lobes fleshy; cypselae: ribs contiguous or separated by less than ¹/₄ their widths, resin glands ± elongate (lengths 2+ times widths); scarious margins of phyllaries pale to dark brown, relatively wide; open coastal habitats. . . . . 2. *Tripleurospermum maritimum*

1. **Tripleurospermum inodorum** (Linnaeus) Schultz-Bipontinus, Tanaceteen, 32. 1844 • Scentless or false mayweed, false chamomile, matricaire inodore F I

Matricaria inodora Linnaeus, Fl. Suec. ed. 2, 297. 1755; *Chamomilla inodora* (Linnaeus) K. Koch; *M. maritima* Linnaeus var. *agrestis* (Knaf) Wilmott; *M. maritima* subsp. *inodora* (Linnaeus) Soó; *M. maritima* var. *inodora* (Linnaeus) Soó; *M. perforata* Mérat; *Tripleurospermum maritimum* (Linnaeus) W. D. J. Koch subsp. *inodorum* (Linnaeus) Applequist; *T. perforatum* (Mérat) M. Lainz

Annuals (sometimes biennials or perennials), (5–)30–60(–80) cm. **Stems** 1, ascending to erect, usually branched distally, sometimes proximally, glabrous or glabrate (sparsely hairy when young). **Leaf blades** 2–8 cm, ultimate lobes filiform, 4–20 mm, not fleshy, apices apiculate. **Heads** (1–)10–200+, 3–4.5 cm diam., in corymbiform arrays of solitary heads at ends of branches. **Phyllaries** centrally dark greenish or brownish, oblong, subequal, scarious margins colorless to light brown, 0.1–0.2 mm wide. **Ray florets** 10–25; corollas (4–)10–13(–20) mm. **Disc corollas** 1–2.5 mm. **Cypselae** pale brown, ribs separated by ¹/₃+ their widths, abaxial-apical resin glands ± circular, faces minutely roughened between ribs. **2*n*** = 18, 36.

Flowering May–Sep. Fields, dry shorelines, waste places; 0–1500+ m; introduced; Greenland; Alta., B.C., Man., N.B., Nfld. and Labr., N.W.T., N.S., Ont., P.E.I., Que., Sask., Yukon; Ala., Alaska, Calif., Colo., Conn., Fla., Idaho, Ill., Iowa, Kans., Ky., Maine, Md., Mass., Mich., Minn., Mo., Mont., Nebr., Nev., N.H., N.J., N.Y., N.Dak., Ohio, Oreg., Pa., S.Dak., Utah, Wash., Wis., Wyo.; also introduced in Europe; Pacific Islands (New Zealand).

Tripleurospermum inodorum has been classified as a noxious weed (class C) in the state of Washington and is considered invasive in other states (it is resistant to some herbicides); it is a weed of cereals in western Canada. W. L. Applequist (2002) has shown that the name *Matricaria inodora* is not a superfluous new name for *M. chamomilla* as earlier stated by S. Rauschert (1974). Therefore, the appropriate name under *Tripleurospermum* is *T. inodorum*. She also considered its type to belong in *T. maritimum* and formally recognized it there as subsp. *inodorum*, on the basis of hybridization with other *T. maritimum* subspecies (A. Vaarama 1953); on the same basis, however, Hämet-Ahti maintained the species distinction between *T. inodorum* and *T. maritimum*, while making *T. phaeocephalum* a subspecies of the latter. Q. O. N. Kay (1994), in a more extensive review of the literature and of hybridization data, also maintained *T. inodorum* and *T. maritimum* as distinct species, a conclusion followed here. From the standpoint of weed science, taxonomic merging of *T. inodorum* and *T. maritimum* has the inconvenience of grouping under a single specific name taxa that have different physiologies, ecologies, weed potentials, and, possibly, reactions to weed control measures.

The name *Matricaria inodora* var. *agrestis* Weiss was not validly published.

SELECTED REFERENCE   Kay, Q. O. N. 1994. Biological flora of the British Isles: *Tripleurospermum inodorum* (L.) Schultz Bip. (*Matricaria inodora* L., *Matricaria maritima* auct. p.p. non L., *Matricaria perforata* Merat, *Tripleurospermum perforatum* (Merat) Wagenitz, *Tripleurospermum maritimum* (L.) Schultz Bip. p.p. non L.). J. Ecol. 82: 681–698.

2. **Tripleurospermum maritimum** (Linnaeus) W. D. J. Koch, Syn. Fl. Germ. Helv. ed. 2, 1026. 1845 • False mayweed

Matricaria maritima Linnaeus, Sp. Pl. 2: 891. 1753; *Chamomilla maritima* (Linnaeus) Rydberg

Perennials (sometimes biennials), 10–50(–80) cm. **Stems** 1–5+, usually procumbent, sometimes ascending (± erect), branched distally, glabrous or sparsely hairy. **Leaf blades** 2–8 cm, ultimate lobes ± cylindric, 4–20 mm, ± fleshy, apices obtuse or mucronate. **Heads** (3–)10–50 in corymbiform arrays or

borne singly. **Phyllaries** centrally dark greenish or brownish, oblong to narrowly triangular or broadly triangular, unequal, scarious margins pale to dark brown or blackish brown, 0.2–1 mm wide. **Ray florets** 10–34+; corollas 10–16[–20] mm. **Disc corollas** ca. 2.5 mm. **Cypselae** pale to dark blackish brown, ribs ± contiguous or separated by less than $^1/_4$ their widths (1–2 supernumerary ribs sometimes present), abaxial-apical resin glands elongate (lengths 2+ times widths), abaxial faces transversely rugose, minutely roughened on faces and between ribs. **2n** = 18, 36.

Subspecies 3 (2 in the flora): North America, Eurasia.

1. Phyllaries oblong or narrowly triangular, scarious margins pale to dark brown, 0.2–0.4 mm wide; cypselae: resin gland lengths often much more than twice widths; coastal ne North America . . . . . .
. . . . 2a. *Tripleurospermum maritimum* subsp. *maritimum*
1. Phyllaries broadly triangular, scarious margins blackish brown, 0.4–1 mm wide; cypselae: resin gland lengths usually little more than twice widths; coastal arctic . . . . . . 2b. *Tripleurospermum maritimum* subsp. *phaeocephalum*

2a. **Tripleurospermum maritimum** (Linnaeus) W. D. J. Koch subsp. **maritimum** • False chamomile, matricaire maritime [1]

Phyllaries oblong or narrowly triangular, scarious margins pale to dark brown, 0.2–0.4 mm wide. **Cypselae:** resin gland lengths often much more than twice widths. **2n** = 18, 36.

Flowering Jun–Sep. Roadsides, fields, waste places; 0–10 m; introduced; St. Pierre and Miquelon; N.B., Nfld. and Labr. (Nfld.), N.S., Que.; Calif., Maine, Mass., Pa.; Europe.

Specimens examined from inland North America that had been attributed to *Tripleurospermum maritimum* subsp. *maritimum* were all *T. inodorum*. The identity of the plant called *T. maritimum* in St. Pierre and Miquelon is uncertain; I did not have access to the voucher specimens and some plants so labeled in maritime eastern Canada were in fact *T. inodorum*. I am uncertain as to whether or not this taxon persists in northeastern North America. Some specimens identified to this taxon, even on the coast, may be *T. inodorum* individuals that have become multi-stemmed through damage or via other mechanisms, particularly on sand dunes. Such specimens are difficult to classify as annual or perennial if the taproot is not dug out, and the lack of cypselae may prevent positive identification.

2b. Tripleurospermum maritimum (Linnaeus) W. D. J. Koch subsp. **phaeocephalum** (Ruprecht) Hämet-Ahti, Acta Bot. Fenn. 75: 9. 1967 • Seashore or wild chamomile, matricaire à capitules brunâtres

Matricaria inodora Linnaeus var. *phaeocephala* Ruprecht, Fl. Samojed. Cisural., 42. 1845; *M. ambigua* (Ledebour) Krylov; *M. maritima* Linnaeus subsp. *phaeocephala* (Ruprecht) Rauschert; *M. maritima* var. *phaeocephala* (Ruprecht) Rauschert; *Tripleurospermum hookeri* Schultz-Bipontinus; *T. phaeocephalum* (Ruprecht) Pobedimova

Phyllaries broadly triangular, scarious margins blackish brown, 0.4–1 mm wide. **Cypselae:** resin gland lengths usually little more than twice widths. $2n = 18$.

Flowering Jul–Oct. Low Arctic seashores and beaches (near high tide strand line), moist to moderately well-drained areas, gravel, sand, moss; 0–50 m; Greenland; Man., N.W.T., Nunavut, Ont., Que., Yukon; Alaska; arctic Eurasia.

Subspecies *phaeocephalum* occurs along the low Arctic shore (both continental and insular) from Alaska to Quebec; it does not penetrate north into the Canadian Arctic Archipelago; it does not reach Labrador; it is present in Greenland. If treated as a species, the name *T. hookeri* Schultz-Bipontinus has priority (N. N. Tzvelev 2002b).

130. HETERANTHEMIS Schott, Isis (Oken) 1818(5): 822. 1818 • [Greek *heteros-*, different, and *anthemis*, a genus name] ⊡

John L. Strother

Annuals, 10–30(–100+) cm (plants sticky, viscid). **Stems** 1, erect, usually branched distally, hirtellous to pilosulous (hairs basifixed, some gland-tipped). **Leaves** mostly cauline; alternate; petiolate or sessile; blades obovate to oblong (bases sometimes ± clasping), often pinnately lobed, ultimate margins usually dentate, rarely entire, faces hirtellous to pilosulous (some hairs gland-tipped). **Heads** radiate, borne singly or in 2s or 3s. **Involucres** hemispheric or broader, 12–25+ mm diam. **Phyllaries** persistent, 20–30+ in 2–3+ series, distinct, ovate or obovate to lance-deltate or lanceolate (not keeled abaxially, usually each with central resin canal), unequal, margins and apices scarious (tips of inner usually dilated). **Receptacles** convex to conic, epaleate. **Ray florets** 13–21+, pistillate, fertile; corollas yellow, laminae ± linear. **Disc florets** 40–80(–100+), bisexual, fertile; corollas proximally ochroleucous, distally yellow, tubes cylindric (usually gland-dotted), throats funnelform, lobes 5, deltate. **Cypselae** dimorphic: outer 3-angled (each angle ± winged, wings ± spine-tipped); inner ± compressed (adaxial angles ± winged, wings ± spine-tipped); ribs 0, faces glabrous (peric)arps without myxogenic cells or resin sacs; embryo sac development tetrasporic); **pappi** 0. $x = 9$.

Species 1: introduced; Mediterranean Europe, n Africa, widely adventive elsewhere.

1. Heteranthemis viscidihirta Schott, Isis (Oken) 1818(5): 822, fig. 5. 1818 (as viscide-hirta) ⊞ ⊡

Leaf blades 25–55+ × 12–30+ mm. **Ray laminae** 12–20+ mm. **Ray cypselae** 3.5–4.5 mm; disc cypselae 2.5–3.5 mm. $2n = 18$.

Flowering Apr. Disturbed, usually damp, sites with sandy soils; 0–50 m; introduced; Calif; Mediterranean Europe, n Africa.

131. ARGYRANTHEMUM Webb ex Schultz-Bipontinus in P. B. Webb and S. Berthelot, Hist. Nat. Îles Canaries 3(2,75): 245, 258. 1844 • Marguerite [Greek *argyros*, silver, and *anthemon*, flower; allusion unclear] □

John L. Strother

Subshrubs or shrubs, 10–80+[–150] cm. **Stems** usually 1, procumbent to erect, usually branched, glabrous [hairy]. **Leaves** mostly cauline; alternate; petiolate or sessile; blades ± obovate [oblong to lanceolate or linear] (bases sometimes ± clasping), [0](1–)2–3-pinnately lobed (lobes cuneate to linear), ultimate margins dentate [entire], faces glabrous [hairy]. **Heads** radiate [discoid], borne singly or in open, corymbiform arrays. **Involucres** hemispheric or broader, [6–]10–18[–22+] mm diam. **Phyllaries** persistent, 28–45+ in 3–4 series, distinct, oblanceolate or ovate to lance-deltate or lanceolate (not keeled abaxially), unequal, margins and apices (stramineous to brown) scarious (tips of inner often ± dilated). **Receptacles** convex to conic, epaleate. **Ray florets** 12–35+, pistillate, fertile; corollas usually white, sometimes yellow or pink, laminae ± ovate to linear. **Disc florets** [50–]80–150+, bisexual, fertile; corollas yellow [red, purple], tubes ± cylindric (not basally dilated, ± gland-dotted), throats campanulate, lobes 5, deltate (without resin sacs). **Cypselae** dimorphic: outer (ray) 3-angled, each angle usually ± winged (wings not spine-tipped); inner (disc) compressed-prismatic (± quadrate, sometimes 2 angles winged, wings not spine-tipped); all ± ribbed or nerved, faces usually glabrous, sometimes gland-dotted between ribs (pericarps without myxogenic cells or resin sacs; embryo sac development bisporic); **pappi** 0 (cypselar wall tissue sometimes produced as teeth, crowns, or oblique tubes similar in texture to cypselar wings). $x = 9$.

Species 24 (1 in the flora): introduced; Atlantic Ocean Islands (Macaronesia).

SELECTED REFERENCE Humphries, C. J. 1976. A revision of the Macaronesian genus *Argyranthemum* Webb ex Schultz Bip. (Compositae–Anthemideae). Bull. Brit. Mus. (Nat. Hist.), Bot. 5: 147–240.

1. **Argyranthemum foeniculaceum** (Willdenow) Webb ex Schultz-Bipontinus in P. B. Webb and S. Berthelot, Hist. Nat. Îles Canaries 3(2,75): 262. 1844 F □

Pyrethrum foeniculaceum Willdenow, Enum. Pl., 903. 1809; *Argyranthemum anethifolium* Webb; *Chrysanthemum anethifolium* Willdenow; *C. foeniculaceum* (Willdenow) Steudel

Leaves scattered or crowded at bases of peduncles, ± succulent; blades ± obovate, 20–45(–100) × 10–30(–65) mm, (1–)2–3-pinnately lobed, ultimate margins dentate or entire, faces glabrous, ± glaucous. **Peduncles** 5–10 cm. **Cypselae** 2–4(–6) mm; **pappi** 0 (cypselar wall tissue sometimes produced as teeth or crowns at apices of some cypselae). $2n = 18$.

Flowering May. Disturbed sites; 0–10 m; introduced; Calif.; Macaronesia.

Cultivars of *Argyranthemum foeniculaceum* and/or *A. frutescens* (Linnaeus) Schultz-Bipontinus (similar, not glaucous) are widely used horticulturally in landscaping and for cut flowers and are often sold as florists' daisies, marguerites, or Paris daisies.

132. ISMELIA Cassini in F. Cuvier, Dict. Sci. Nat. ed. 2, 41: 40. 1826 • [Etymology unknown] □

John L. Strother

Annuals, 10–30(–50+) cm. **Stems** 1, erect, usually branched distally, glabrous or sparsely hairy (hairs basifixed). **Leaves** mostly cauline; alternate; petiolate or sessile; blades obovate to oblong (bases sometimes clasping), usually 2-pinnately lobed, ultimate margins usually dentate, rarely

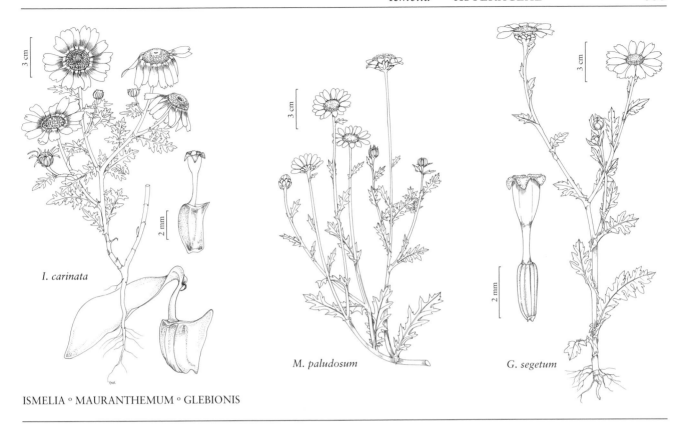

I. *carinata*

M. *paludosum*

G. *segetum*

ISMELIA ○ MAURANTHEMUM ○ GLEBIONIS

entire, faces glabrous. **Heads** radiate, borne singly or in 2s or 3s. **Involucres** hemispheric or broader, 12–25+ mm diam. **Phyllaries** persistent, 20–30+ in 2–3 series, distinct, ovate or obovate to lance-deltate or lanceolate (± carinate), unequal, margins and apices scarious (tips of inner often ± dilated). **Receptacles** convex to conic, epaleate. **Ray florets** 13–21+, pistillate, fertile; corollas proximally white or red to purple, distally yellow or white, laminae ± linear. **Disc florets** 80–150+, bisexual, fertile; corollas proximally ochroleucous, distally red to purple, tubes cylindric (stipitate-glandular or gland-dotted), throats funnelform, lobes 5, deltate. **Cypselae** dimorphic: outer 3-angled (each angle ± winged, wings not spine-tipped); inner compressed, adaxial angles ± winged (wings not spine-tipped); ribs 0, faces glabrous (pericarps without myxogenic cells or resin sacs); **pappi** 0. $x = 9$.

Species 1: introduced; n Africa; widely adventive elsewhere.

1. **Ismelia carinata** (Schousboe) Schultz-Bipontinus in P. B. Webb and S. Berthelot, Hist. Nat. Îles Canaries 3(2,2): 271. 1844 F I

Chrysanthemum carinatum Schousboe, Iagttag. Vextrig. Marokko, 198, plate 6. 1800

Leaf blades 25–50+ × 15–30+ mm. **Ray laminae** 20–30+ mm. **Cypselae:** ray 3.5–4 mm; disc 2.5–3.5 mm. $2n = 18$.

Flowering Mar. Disturbed sites; 0–200 m; introduced; Calif.; n Africa.

133. MAURANTHEMUM Vogt & Oberprieler, Taxon 44: 377. 1995 • [Latin *Mauros*, a native of North Africa, and Greek *anthemon*, flower] □

John L. Strother

Leucoglossum B. H. Wilcox, K. Bremer & Humphries, Bull. Nat. Hist. Mus. London, Bot. 23: 142. 1993, not S. Imai 1942

Annuals [perennials], 2–25(–40) cm. **Stems** 1–5+, erect, usually branched near bases, glabrous. **Leaves** cauline; alternate; petiolate or sessile; blades obovate or oblong to lanceolate or linear (bases sometimes clasping), usually irregularly 1-pinnately lobed or toothed, ultimate margins toothed or entire, faces glabrous (minutely gland-dotted abaxially). **Heads** radiate, borne singly or in open, corymbiform arrays. **Involucres** ± hemispheric, 8–12(–15+) mm diam. **Phyllaries** persistent, 25–40+ in 3–4 series, distinct, ovate or oblong to lance-deltate or lanceolate (not carinate), unequal, margins and apices (pale brown to black) scarious. **Receptacles** ± conic, epaleate. **Ray florets** 10–21+, pistillate and fertile, or styliferous and sterile; corollas mostly white (usually yellowish at bases, drying pinkish), laminae oblong to ovate. **Disc florets** 60–100+, bisexual, fertile; corollas ± yellow, tubes ± cylindric (basally dilated, not gland-dotted), throats ± campanulate, lobes (2–)5, deltate (without resin sacs). **Cypselae** mostly columnar to obovoid or fusiform, outer sometimes 3-angled, ribs 7–10, faces glabrous (pericarps with myxogenic cells on ribs and resin sacs between ribs); **pappi** 0 (sterile ovaries of rays usually with apical coronas). $x = 9$.

Species ca. 4 (1 in the flora): introduced; sw Europe, n Africa.

1. **Mauranthemum paludosum** (Poiret) Vogt & Oberprieler, Taxon 44: 377. 1995 [F] [I]

Chrysanthemum paludosum Poiret, Voy. Barbarie 2: 241. 1789; *Leucanthemum paludosum* (Poiret) Pomel; *Leucoglossum paludosum* (Poiret) B. H. Wilcox, K. Bremer & Humphries

Leaf blades 15–35(–55) × 3–12 (–20) mm. **Ray laminae** 6–12+ mm. **Cypselae** 2–3 mm; **pappi** 0 (cypselar wall tissues sometimes produced as tubular or coroniform collars on ray ovaries). $2n = 18$.

Flowering spring–summer. Disturbed places; 0–200 m; introduced; Calif.; sw Europe; n Africa (cultivated, sparingly adventive).

134. GLEBIONIS Cassini in F. Cuvier, Dict. Sci. Nat. ed. 2, 41: 41. 1826 • [Latin *gleba*, soil, and *-ionis*, characteristic of; allusion unclear, perhaps to agricultural association] □

John L. Strother

Xantophtalmum Schultz-Bipontinus

Annuals, 10–30(–80+) cm. **Stems** usually 1, erect to ascending, usually branched distally, glabrous. **Leaves** mostly cauline; alternate; petiolate or sessile; blades obovate to oblong (bases sometimes ± clasping), usually 1–3-pinnately lobed, ultimate margins usually dentate, rarely entire, faces glabrous. **Heads** radiate, borne singly or in 2s or 3s. **Involucres** ± hemispheric or broader, 15–25+ mm diam. **Phyllaries** persistent, 25–60+ in 3–4 series, distinct, ovate or obovate to lance-deltate or lanceolate (not carinate, usually each with a resin canal), margins and apices (colorless or stramineous to pale brown) scarious (tips of inner often ± dilated). **Receptacles** convex to hemispheric, epaleate. **Ray florets** 10–21+, pistillate, fertile; corollas mostly yellow, sometimes paler distally, laminae linear, oblong, or ovate. **Disc florets** 60–150+, bisexual, fertile; corollas ± yellow, tubes ± cylindric (basally ± dilated, ± gland-dotted), throats funnelform, lobes 5, narrowly deltate (each with a resin sac). **Cypselae** dimorphic: outer (ray) 3-angled (each angle ± winged, wings not spine-tipped); inner (disc) compressed-prismatic to columnar (adaxial, rarely abaxial, angles sometimes ± winged, wings not spine-tipped); ribs usually 10, faces glabrous, sometimes gland-dotted between ribs (pericarps without myxogenic cells or resin sacs; embryo sac development monosporic); **pappi** 0. $x = 9$.

Species 2 (2 in the flora): introduced; Eurasia (especially Mediterranean and Macaronesia) and northern Africa (widely cultivated and adventive).

1. Leaf blades coarsely 1-pinnate or not lobed; disc cypselae ± compressed-columnar, 10-ribbed, not winged . . . . 1. *Glebionis segetum*
1. Leaf blades mostly 2–3-pinnately lobed; disc cypselae ± prismatic, ± compressed, obscurely ribbed, sometimes with adaxial (rarely the abaxial) rib ± winged . . . . 2. *Glebionis coronaria*

1. **Glebionis segetum** (Linnaeus) Fourreau, Ann. Soc. Linn. Lyon, n. s. 17: 90. 1869 • Corn marigold F I

Chrysanthemum segetum Linnaeus, Sp. Pl. 2: 889. 1753; *Xantophtalmum segetum* (Linnaeus) Schultz-Bipontinus

Leaf blades oblong to obovate or spatulate, 25–65+ × 8–25+ mm, coarsely 1-pinnate or not lobed, ultimate margins dentate or entire. **Ray corollas** golden yellow, laminae oblong-ovate, 8–20 mm. **Ray cypselae** weakly triquetrous, 2.5–3 mm, obscurely, if at all, winged, lateral faces 2-ribbed, abaxial faces 3-ribbed; disc cypselae ± compressed-columnar, 2–3 mm, 10-ribbed, not winged. $2n = 18$.

Flowering Apr–Aug(–Oct). Disturbed sites; 0–100 m; introduced; Ala., Alaska, Calif., Conn., La., Maine, Mass., N.Y., N.C., Ohio, Oreg., Pa., S.C.; Eurasia; n Africa.

2. **Glebionis coronaria** (Linnaeus) Cassini ex Spach, Hist. Nat. Vég. 10: 181. 1841 • Crown daisy, garland chrysanthemum I

Chrysanthemum coronarium Linnaeus, Sp. Pl. 2: 890. 1753

Leaf blades oblong to obovate, mostly 30–55+ × 15–30+ mm, mostly 2–3-pinnately lobed, ultimate margins dentate. **Ray corollas** pale yellow, sometimes white-tipped, laminae oblong to linear, 15–25 mm. **Ray cypselae** triquetrous, 2.5–3 mm, angles ± winged, faces obscurely nerved or ribbed; disc cypselae ± prismatic, ± compressed, 2.5–3 mm, obscurely ribbed, sometimes with adaxial (rarely the abaxial) rib ± winged. $2n = 18$.

Flowering Apr–Jul. Disturbed sites; 0–200 m; introduced; Ariz., Calif., Fla.; Eurasia; n Africa.

135. **NIPPONANTHEMUM** (Kitamura) Kitamura, Acta Phytotax. Geobot. 29: 168. 1978

• [Japanese *Nippon*, name of Japan, and Greek *anthemon*, flower] I

John L. Strother

Chrysanthemum Linnaeus sect. *Nipponanthemum* Kitamura, Kiku, 115. 1948

Shrubs, 20–100 cm. **Stems** 1+, erect, branched distally, puberulent, glabrescent. **Leaves** cauline (mostly clustered distally on stems); alternate; sessile; blades oblong or spatulate to lanceolate

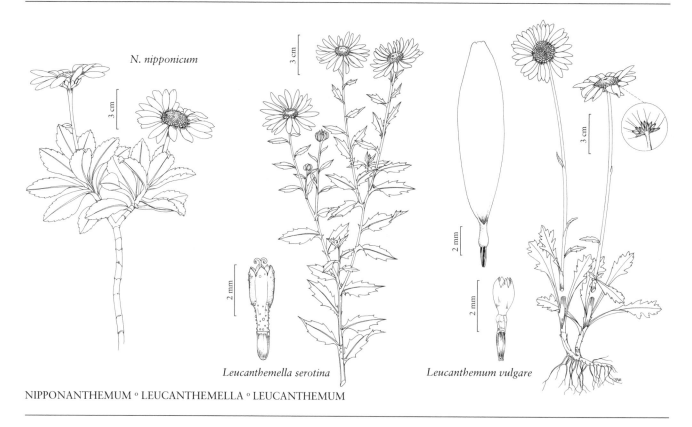

N. *nipponicum*

Leucanthemella serotina   *Leucanthemum vulgare*

NIPPONANTHEMUM ∘ LEUCANTHEMELLA ∘ LEUCANTHEMUM

(bases cuneate), margins distally toothed or entire, faces glabrous or hirtellous. **Heads** radiate, borne singly. **Involucres** hemispheric or broader, 15–20 mm diam. **Phyllaries** persistent, 30–50+ in ± 4 series, distinct, lanceolate, oblong, obovate, or ovate (not carinate), unequal, margins and apices scarious. **Receptacles** convex, epaleate. **Ray florets** 21–34+, pistillate, fertile; corollas white, laminae ± ovate to linear. **Disc florets** [50–]80–250+, bisexual, fertile; corollas yellow [red or purple], tubes cylindric (not basally dilated, usually gland-dotted), throats ± campanulate, lobes 5, deltate (without resin sacs). **Cypselae** ± columnar, ribs 8–10 (not winged), faces probably glabrous (pericarps without myxogenic cells or resin sacs); **pappi** crowns of scales. $x = 9$.

Species 1: introduced; Asia (Japan).

1. **Nipponanthemum nipponicum**   (Franchet ex Maximowicz) Kitamura, Acta Phytotax. Geobot. 29: 169. 1978 • Nippon daisy [F] [I]

Leucanthemum nipponicum Franchet ex Maximowicz, Bull. Acad. Imp. Sci. Saint-Pétersbourg 17: 420. 1872; *Chrysanthemum nipponicum* (Franchet ex Maximowicz) Sprenger

Leaf blades 20–75(–90) × 12–15 (–20) mm, ultimate margins distally dentate or entire. **Ray laminae** 20–30 mm. **Cypselae** 3–4 mm. $2n = 18$.

Flowering Oct–Nov. Disturbed sites; 0–10+ m; introduced; N.J., N.Y.; Asia (Japan).

136. LEUCANTHEMELLA Tzvelev in V. L. Komarov et al., Fl. URSS 26: 137. 1961

• [*Leucanthemum*, a genus name, plus Latin *-ella*, diminutive] ⊡

John L. Strother

Decaneurum Schultz-Bipontius, Tanaceteen, 44. 1844, not de Candolle 1833

Perennials, (20–)50–150 cm (not aromatic; not rhizomatous). **Stems** usually 1, erect, usually branched distally, ± hairy (hairs basi- or medi-fixed). **Leaves** cauline; alternate; sessile; blades oblong to lanceolate [pinnate] (bases sometimes ± clasping), margins often with 2–4 lobes proximally, serrate distally, faces hairy or glabrous, gland-dotted. **Heads** radiate, borne singly or in 2s or 3s. **Involucres** hemispheric or broader, 12–25+ mm diam. **Phyllaries** persistent, 30–50+ in 2–3+ series, distinct, ovate or obovate to oblanceolate (not keeled abaxially), margins and apices scarious (tips not notably dilated). **Receptacles** convex, epaleate. **Ray florets** 21–34+, styliferous and sterile; corollas usually white, sometimes reddish, laminae ± ovate to linear. **Disc florets** 100–150+, bisexual, fertile; corollas usually yellow, sometimes reddish, tubes ± cylindric (± gland-dotted), throats campanulate, lobes 5, deltate (without resin sacs). **Cypselae** ± columnar, ribs 7–12, faces probably glabrous (pericarps without myxogenic cells or resin sacs); **pappi** 0 (cypselar wall tissues sometimes distally produced as coronas). $x = 9$.

Species 2 (1 in the flora): introduced; e Europe, Asia.

1. Leucanthemella serotina (Linnaeus) Tzvelev in V. L. Komarov et al., Fl. URSS 26: 139. 1961

• Leucanthémelle tardive Ⓕ ⊡

Chrysanthemum serotinum Linnaeus, Sp. Pl. 2: 888. 1753; *C. uliginosum* Persoon

Leaf blades 50–120+ × 15–30+ mm. **Ray laminae** 10–25 mm. **Cypselae** 2–3 mm. $2n = 18$.

Flowering Jun–Oct. Disturbed sites, fields, roadsides; 0–200 m; introduced; Ont., Que.; Conn., Mass., Mich., Minn.; e Europe; Asia.

137. LEUCANTHEMUM Miller, Gard. Dict. Abr. ed. 4, vol. 2. 1754 • [Greek *leuco-*, white, and *anthemon*, flower] ⊡

John L. Strother

Perennials, (10–)40–130(–200+) cm (rhizomatous, roots usually red-tipped). **Stems** usually 1, erect, simple or branched, glabrous or hairy (hairs basifixed). **Leaves** mostly basal or basal and cauline; petiolate or sessile; blades obovate to lanceolate or linear, often 1[–2+]-pinnately lobed or toothed, ultimate margins dentate or entire, faces glabrous or sparsely hairy. **Heads** usually radiate, rarely discoid, borne singly or in 2s or 3s. **Involucres** hemispheric or broader, 12–35+ mm diam. **Phyllaries** persistent, 35–60+ in 3–4+ series, distinct, ovate or lance-ovate to oblanceolate, unequal, margins and apices (colorless or pale to dark brown) scarious (tips not notably dilated; abaxial faces glabrous or sparsely hairy). **Receptacles** convex, epaleate. **Ray florets** usually 13–34+, rarely 0, pistillate, fertile; corollas white (drying pinkish), laminae ovate to linear. **Disc florets** 120–200+, bisexual, fertile; corollas yellow, tubes ± cylindric (proximally swollen, becoming spongy in fruit), throats campanulate, lobes 5, deltate (without resin sacs). **Cypselae** ± columnar to obovoid, ribs ± 10, faces glabrous (pericarps with myxogenic cells on ribs and resin sacs between ribs; embryo sac development monosporic); **pappi** 0 (wall tissue of ray cypselae sometimes produced as coronas or auricles on some cypselae). $x = 9$.

Species 20–40+ (3 in the flora): introduced; mostly temperate Europe (some widely cultivated and sparingly adventive).

The three leucanthemums recognized here are weakly distinct and are sometimes included (with a dozen or more others) in a single, polymorphic *Leucanthemum vulgare*.

SELECTED REFERENCE   Vogt, R. 1991. Die Gattung *Leucanthemum* (Compositae–Anthemideae) auf der Iberischen Halbinsel. Ruizia 10: 1–261.

1. Blades of basal leaves usually pinnately lobed (lobes 3–7+) and/or irregularly toothed; margins of mid-stem leaves usually irregularly toothed proximally and distally . . . . 1. *Leucanthemum vulgare*
1. Blades of basal leaves not lobed, usually toothed, rarely entire; margins of mid-stem leaves usually entire proximally, regularly serrate distally.
  2. Blades of cauline leaves oblanceolate to lanceolate or linear, 50–120+ × 8–22+ mm; larger phyllaries 2–3 mm wide; ray cypselae 2–3(–4 mm), apices usually bare, rarely adaxially auriculate. . . . . 2. *Leucanthemum maximum*
  2. Blades of cauline leaves elliptic to oblanceolate, 30–120+ × 12–25(–35+) mm; larger phyllaries 4–5 mm wide; ray cypselae 3–4 mm, apices usually adaxially auriculate
  . . . . 3. *Leucanthemum lacustre*

1. **Leucanthemum vulgare** Lamarck, Fl. Franç. 2: 137. 1779 • Ox-eye daisy, marguerite blanche [F] [I]

Chrysanthemum leucanthemum Linnaeus, Sp. Pl. 2: 888. 1753; *C. leucanthemum* var. *pinnatifidum* Lecoq & Lamotte

Perennials, 10–30(–100+) cm. **Stems** simple or distally branched. **Basal leaves:** petioles 10–30(–120) mm, expanding into obovate to spatulate blades 12–35(–50+) × 8–20(–30) mm, margins usually pinnately lobed (lobes 3–7+) and/or irregularly toothed. **Cauline leaves** petiolate or sessile; blades oblanceolate or spatulate to lanceolate or linear, 30–80+ × 2–15+ mm, margins of mid-stem leaves usually irregularly toothed proximally and distally. **Involucres** 12–20+ mm diam. **Phyllaries** (the larger) 2–3 mm wide. **Ray florets** usually 13–34+, rarely 0; laminae 12–20(–35+) mm. **Ray cypselae** 1.5–2.5 mm, apices usually coronate or auriculate. $2n = 18$, 36, 54, 72, 90.

Flowering spring–fall. Disturbed places, meadows, seeps, clearings; 0–2000 m; introduced; Alta., B.C., Ont., Que., Sask.; Alaska, Ariz., Ark., Calif., Colo., Conn., Fla., Idaho, Ill., Ind., Iowa, Kans., Mass., Mich., Mo., Mont., Nev., N.Mex., N.Y., N.Dak., Ohio, Okla., Oreg., Pa., S.C., S.Dak., Tenn., Utah, Va., Wash., W.Va., Wis., Wyo.; Europe, widely adventive.

Some botanists (e.g., W. J. Cody 1996) have treated *Leucanthemum ircutianum* de Candolle, with blades of mid and distal cauline leaves oblong to oblong-lanceolate and not ± pinnate at bases, as distinct from *L. vulgare*.

2. **Leucanthemum maximum** (Ramond) de Candolle in A. P. de Candolle and A. L. P. P. de Candolle, Prodr. 6: 46. 1838 • Shasta daisy [I]

Chrysanthemum maximum Ramond, Bull. Sci. Soc. Philom. Paris 2: 140. 1800

Perennials, 20–60(–80+) cm. **Stems** simple or distally branched. **Basal leaves:** petioles 50–80 (–200+) mm, expanding into obovate to spatulate blades 50–80 (–120+) × 15–25(–35+) mm, margins not lobed, usually toothed, rarely entire. **Cauline leaves** petiolate or sessile; blades oblanceolate to lanceolate or linear, 50–120+ × 8–22+ mm, margins of mid-stem leaves usually entire proximally, regularly serrate distally. **Involucres** 18–28+ mm diam. **Phyllaries** (the larger) 2–3 mm wide. **Ray florets** 21–34+; laminae 20–30(–40+) mm. **Ray cypselae** 2–3(–4) mm, apices usually bare, rarely obscurely auriculate. $2n = 90$, 108.

Flowering spring–summer. Disturbed sites, meadows, seeps, clearings; 0–1500+ m; introduced; Ala., Calif., Wyo.; w Europe (widely cultivated, sparingly adventive).

The name Shasta daisy of horticulture is associated also with *Leucanthemum* ×*superbum* (Bergmans ex J. Ingram) Bergmans ex D. H. Kent, which is generally thought to have been derived from hybrids between *L. maximum* and *L. lacustre*. Cultivars of "Shasta daisy" number in the dozens, including "single," "double," "quill," and "shaggy" forms; they may be encountered as waifs or persisting from abandoned plantings.

3. **Leucanthemum lacustre** (Brotero) Sampaio, Lista Espécies Herb. Português, 132. 1913 • Portugese daisy [1]

Chrysanthemum lacustre Brotero, Fl. Lusit. 1: 379. 1804

Perennials, 30–120(–200+) cm. **Stems** simple or distally branched. **Basal leaves:** petioles 30–60+ mm, expanding into lanceolate blades 50–100+ × 15–30+ mm, margins not lobed, usually toothed, sometimes entire. **Cauline leaves** ± petiolate or sessile; blades elliptic to oblanceolate, 30–120+ × 12–25(–35+) mm, margins of mid-stem leaves usually entire proximally, regularly serrate distally. **Involucres** (18–)25–35 mm diam. **Phyllaries** (the larger) 3–5 mm wide. **Ray florets** 21–34+; laminae 20–25 (–35) mm. **Ray cypselae** 3–4 mm, apices usually auriculate adaxially. $2n = 198$.

Flowering spring–summer. Disturbed places, meadows, seeps, clearings; 0–50 m; introduced; Calif.; Europe (Portugal).

Literature Cited

Robert W. Kiger, Editor

A consolidated list of all works cited in volumes 19, 20, and 21 appears in Volume 21 beginning on page 554.

Index

Names in *italics* are synonyms, casually mentioned hybrids, or plants not established in the flora. Page numbers in **boldface** indicate the primary entry for a taxon. Page numbers in *italics* indicate an illustration. Roman type is used for all other entries, including author names, vernacular names, and accepted scientific names for plants treated as established members of the flora.

Abrams, L., 386, 448, 547
Absinthium, 518
 viridifolium
  rupestre, 520
Acamptopappus, 50, 68
Acanthospermum, 17, 18
Achillea, 14, 487, **492**
 ageratum, 492
 alpicola, 493
 alpina, 493, **494**
 arenicola, 493
 borealis
  arenicola, 493
  californica, 493
 californica, 493
 distans, 492
 filipendulina, 492
 gigantea, 493
 lanulosa, 493
  alpicola, 493
 laxifolia, 493
 ligustica, 492
 megacephala, 493
 millefolium, *491*, 492, **493**
  alpicola, 493
  arenicola, 493
  asplenifolia, 493
  borealis, 493
  californica, 493
  gigantea, 493
  lanulosa, 493
  litoralis, 493
  maritima, 493
  megacephala, 493
  nigrescens, 493
  occidentalis, 493
  pacifica, 493
  puberula, 493
 nigrescens, 493

nobilis, 492, **494**
 occidentalis, 493
 pacifica, 493
 ptarmica, 493, **494**
 puberula, 493
 rosea, 493
 sibirica, 494
 subalpina, 493
Achillée, 492
Achillée de Sibérie, 494
Achillée millefeuille, 493
Achillée ptarmique, 494
Achyrachaena, 15, 19
Achyrophorus
 chillensis, 298
 microcephalus, 299
 roseus, 376
Ackerman, T. L., 445
Acmella, 13, 18, 25, 27
Acosta, 181, 189
 diffusa, 190
 maculosa, 189
Acourtia, 12, 14, 57, 70, 71, **72**, 73
 microcephala, 73, **74**
 nana, 73, **74**
 runcinata, 72, **73**, 74
 thurberi, 73, **74**
 wrightii, 72, 73, **74**
Acroptilon, 7, 58, 67, 83, **171**
 picris, 172
 repens, *170*, **172**
Adenocaulon, 12, 52, 70, 71, **77**, 78
 bicolor, **77**, 78
Adenophyllum, 45
Adobe Hills thistle, 128, 129
African daisy, 380
African sheepbush, 543

African-daisy, 198
Ageratina, 54
Ageratum, 51, 66
Agnorhiza, 19, 23, 27, 28
Agoseris, 215, 217, 322, **323**, 324, 326, 333, 334, 335
 ×agrestis, 326
 alpestris, 336
 altissima, 326
 angustissima, 327
 apargioides, 325, 330, **331**, 332, 333
  apargioides, **331**, 332
  eastwoodiae, 331, **332**
  maritima, 331, **332**, 333
 arachnoidea, 327
 arizonica, 328
 aspera, 326
 attenuata, 328
 aurantiaca, *323*, 324, 325, **327**, 328, 329, 333, 334
  aurantiaca, *323*, **327**, 328
  dasycephala, 328
  purpurea, *323*, 327, **328**
 californica, 333
 carnea, 327, 328
 caudata, 328
 cinerea, 330
 confinis, 328
 coronopifolia, 324, 331
 covillei, 329
 cuspidata, 336
 ×*dasycarpa*, 326
 decumbens, 329
 dens-leonis, 328
 eastwoodiae, 332
 eisenhoweri, 326
 ×elata, 324, 325, 330, **334**
 frondifera, 328

gaspensis, 327
glauca, *323*, 325, **326**, 328, 329
 agrestis, 326, 327
 aspera, 326
 cronquistii, 328
 dasycephala, *323*, **326**, 327, 328, 329, 334
 glauca, *323*, **326**, 327
 laciniata, 328, 329
 monticola, 329
 pumila, 326
 scorzoneraefolia, 326
 villosa, 326
gracilens, 327, 328
 greenei, 327, 328
graminifolia, 328
grandiflora, 325, 328, **329**, 331, 333, 334
 grandiflora, **330**, 331, 332
 intermedia, 330
 laciniata, 334
 leptophylla, **330**
 plebeia, 330
greenei, 327
heterophylla, 323, 324, 331, **332**
 californica, 333, 334
 crenulata, 333
 cryptopleura, 331, 332, **333**, 334
 glabra, 333
 glabrata, 333
 heterophylla, **333**, 334
 normalis, 333
 quentinii, 333, **334**
 turgida, 333
hirsuta, 325, 331, **332**, 333
howellii, 327

Agoseris (*continued*)
  humilis, 331
  intermedia, 330
  isomeris, 326
  lacera, 326
  laciniata, 334
  lackschewitzii, 327, 328
  lanulosa, 326
  lapathifolia, 326
  leontodon
    aspera, 326
    pygmaea, 326
  leptocarpa, 328
  longirostris, 328
  longissima, 326
  longula, 326
  maculata, 326
  major, 333
  maritima, 332
  marshallii, 330
  microdonta, 326
  ×*montana*, 328
  monticola, 325, 326, 328, **329**, 334
  nana, 327
  naskapensis, 327
  obtusifolia, 330
  parviflora, 324, 325, 326, 327, **328**, 329, 331
  plebeia, 330
  prionophylla, 327
  procera, 326
  pubescens, 326
  pumila, 326
  purpurea, 328
    arizonica, 328
  retrorsa, 325, 329, **330**, 331, 332
  rosea, 328
  roseata, 328
  rostrata, 328
  scorzonerifolia, 326
  subalpina, 327
  taraxacoides, 328
  tenuifolia, 334
  tomentosa, 328
  turbinata, 326
  vestita, 326
  vicinalis, 326
  villosa, 326
  vulcanica, 327
Ahles, H. E., 549
Aiken, S. G., 240, 241
Ajania
  globularia, 525
  glomerata, 526
  senjavinensis, 532
Alameda County thistle, 160
Alaska dandelion, 251
Alaskan pussytoes, 413
Alcove thistle, 162
Aleutian wormwood, 505
Almutaster, 38
Alpinae, 389
Alpine Arctic-cudweed, 439
Alpine dandelion, 250
Alpine lake false dandelion, 336
Alpine pussytoes, 414
Alpine sagewort, 530

Alpine thistle, 152
Alseuosmiaceae, 4
Amauriopsis, 34
Amberboa, 67, 84, **172**
  amberboi, 173
  moschata, *170*, **173**
  muricata, 175
Amblyolepis, 21, 50, 51
Amblyopappus, 65
Ambrosia, 25, 27
  artemisiifolia, 7
  trifida, 7
Ambrosiinae, 4
American basketflower, 176
American saw-wort, 166
American star-thistle, 176
American threefold, 75
Ampelaster, 35
Amphiachyris, 34, 49
Amphipappus, 42, 49, 61, 68
Anacapa Island desertdandelion, 317
Anacyclus
  australis, 544
Anaphale marguerite, 427
Anaphalis, 59, 384, 386, 389, 396, **426**
  margaritacea, 426, **427**
    occidentalis, 427
    subalpina, 427
Ancistrocarphus, 26, 388, **465**, 467
  filagineus, 385, *464*, 465, **466**
  keilii, *464*, 465, **466**, 470
Anderberg, A. A., 384, 385, 471, 475, 478
Anderson's hawksbeard, 235
Anderson's thistle, 145
Anisocarpus, 20, 28
Anisocoma, 217, **309**, 310, 360
  acaulis *308*, **310**
Annual agoseris, 332
Annual skeleton-weed, 369
Annual trampweed, 443
Antelope Island skeletonplant, 372
Antennaire, 388
Antennaire à feuilles de plantain, 400
Antennaire de Fries, 412
Antennaire de Howell, 403
Antennaire de Parlin, 402
Antennaire du Canada, 404
Antennaire étroite, 414
Antennaire litigieuse, 402
Antennaire magnifique, 400
Antennaire négligée, 403
Antennaire néodioïque, 405
Antennaire pétaloïde, 405
Antennaire rosée, 408
Antennaria, 59, 384, 386, **388**, 389, 390, 391, 392, 396, 397, 399, 409, 414
  [group] Arcuatae, 389
  [group] *Argenteae*, 389, 397, 398
  [group] *Catipes*, 389, 390, 392, 401, 402
  [group] *Dimorpae*, 389, 399

[group] *Geyerae*, 389, 396
[group] *Leontipes*, 389
[group] *Pulcherrimae*, 389, 399
  acuminata, 410
  acuta, 407
  affinis, 409
  aizoides, 408
  alaskana, 413
  albescens, 410
  albicans, 409
  alborosea, 410
  alpina, 390, 395, 408, 411, 412, 413, **414**, 415
    cana, 414
    canescens, 414
    compacta, 414
    friesiana, 412
    glabrata, 414
    intermedia, 414
    media, 411
    megacephala, 412
    monocephala, 411
    porsildii, 414
    scabra, 410
    stenophylla, 398
    stolonifera, 414
    typica, 414
    ungavensis, 414
  alsinoides, 405
  ambigens, 402
  ampla, 402
  anaphaloides, 389, 394, **399**
    straminea, 399
  angustata, 412
  angustifolia, 409, 414
  appendiculata, 405
  aprica, 406
    aureola, 406
    minuscula, 406
  arcuata, 389, 392, **396**
  arenicola, 414
  argentea, 389, 393, **397**
    aberrans, 398
  arida, 409
    humilis, 409
    viscidula, 409
  arkansana, 402
  arnoglossa, 402
    ambigens, 402
  aromatica, 390, 394, **408**, 409, 410, 411, 415
  athabascensis, 403
  atriceps, 414
  aureola, 406
  austromontana, 411
  bayardi, 414
  bifrons, 402
  boecheriana, 414
  bracteosa, 407
  brainerdii, 402
  breitungii, 409
  brevistyla, 409, 414
  brunnescens, 414
  burwellensis, 412
  callilepis, 404
  calophylla, 402
  campestris, 403
    athabascensis, 403

cana, 414
canadensis, 404
  randii, 404
  spathulata, 404
candida, 411
canescens, 414
  fastigiata, 414
  porsildii, 414
  pseudoporsildii, 414
caroliniana, 400
carpatica
  humilis, 400
  lanata, 399
  pulcherrima, 400
chelonica, 403
chilensis, 409
  magellanica, 409
chlorantha, 410
columnaris, 414
compacta, 414
concinna, 409
concolor, 405
confinis, 409
confusa, 414
congesta, 412
corymbosa, 390, 393, 396, **407**, 409, 410, 415
crymophila, 413, 414
decipiens, 400
denikeana, 400
densa, 411
densifolia, 390, 394, 408, **411**, 413, 415
dimorpha, 389, 393, **398**, 399
  flagellaris, 399
  integra, 398
  macrocephala, 398
  nuttallii, 398
dioica, 389, 390, 395, **406**
  corymbosa, 407
  kernensis, 409
  marginata, 405
  parvifolia, 406
ekmaniana, 414
elegans, 409
elliptica, 402
ellyae, 411
×*erigeroides*, 415
erosa, 403
eucosma, 400
exilis, 412
exima, 404
fallax, 402
  calophylla, 402
farwellii, 402
fendleri, 405
fernaldiana, 412
flagellaris, 389, 393, *397*, **399**
flavescens, 408
foggii, 414
foliacea
  humilis, 409
  ×*foliacea*, 415
formosa, 410
friesiana, 390, 392, 393, 395, **412**, 413, 414
  alaskana, 390, **413**, 414, 415

beringensis, 413
compacta, 413, 414
friesiana, 390, 413, **414**
neoalaskana, 390, 413, **414**, 415
fusca, 410
gaspensis, 410
geyeri, 59, 386, 389, 394, **396**, 397
glabrata, 414
grandis, 405
greenei, 402
hendersonii, 410
holmii, 406
howellii, 390, 395, 401, 402, **403**, 404, 405, 406
    athabascensis, 403
    campestris, 403
    canadensis, **404**
    gaspensis, 410
    howellii, 390, **404**, 405
    neodioica, 403, 404, **405**
    petaloidea, 404, **405**
hudsonica, 412
hygrophila, 407
hyperborea, 406
imbricata, 410
incarnata, 409
insularis, 406
intermedia, 414
isolepis, 410
labradorica, 414
laingii, 409
lanata, 389, 394, 397, **399**
lanulosa, 410
latisquama, 398
latisquamea, 406
leontopodioides, 409
leuchippii, 409
leucophaea, 398
longifolia, 403
longii, 414
lunellii, 403
luzuloides, 389, 393, **397**
    aberrans, 397, **398**
    luzuloides, **398**
    microcephala, 398
    oblanceolata, 398
×*macounii*, 415
macrocephala, 398
maculata, 410
manicouagana, 410
marginata, 390, 392, 395, 404, **405**, 406, 409
    glandulifera, 405
media, 390, 395, 406, **411**, 415
    ciliata, 410
    compacta, 414
    fusca, 410
    pulchella, 410
megacephala, 412
mesochora, 402
microcephala, 397, 398
microphylla, 389, 390, 396, **401**, 406, **407**, 409, 415
solstitialis, 407
minuscula, 406
modesta, 411

monocephala, 390, 392, 393, 401, **411**, 412, *413*
    angustata, 390, 412
    exilis, 412
    latisquamea, 412
    monocephala, 390, 412, *413*, 415
    philonipha, 412
mucronata, 411
munda, 402
nardina, 407
nebrascensis, 403
neglecta, 390, 395, **403**, 404, 405, 406
    argillicola, 403
    athabascensis, 403
    attenuata, 405
    campestris, 403
    gaspensis, 410
    howellii, 403
    neodioica, 405
    petaloidea, 405
    randii, 404
    simplex, 403
    subcorymbosa, 405
nemoralis, 400
neoalaskana, 413, 414
neodioica, 399, 404, 405
    argillicola, 403
    attenuata, 405
    canadensis, 404
    chlorantha, 410
    chlorophylla, 405
    gaspensis, 410
    grandis, 405
    howellii, 403
    interjecta, 405
    neodioica, 405
    petaloidea, 405
    randii, 404
    rupicola, 405
nitens, 412
nitida, 407
oblanceolata, 398
×*oblancifolia*, 415
obovata, 405
occidentalis, 402
oxyphylla, 410
pallida, 414
parlinii, 390, 391, 394, 401, **402**
    ambigens, 402
    arnoglossa, 402
    fallax, **402**
    farwellii, 402
    parlinii, **402**, 403
parvifolia, 389, 390, 396, 403, **406**, 411
parvula, 403
peasei, 410
pedicellata, 405
pedunculata, 414
peramoena, 405
petaloidea, 405
    novaboracensis, 405
    scariosa, 405
    subcorymbosa, 405
petasites, 401
philonipha, 412

pinetorum, 400
piperi, 401
plantaginifolia, 390, 391, 394, **400**, *401*, 402, 404, 405
    ambigens, 402
    arnoglossa, 402
    monocephala, 401
    petiolata, 400
polyphylla, 409
porsildii, 414
propinqua, 402
pulchella, 390, 395, 406, 409, **410**, 411, 415
pulcherrima, 389, 393, 394, 399, **400**
    anaphaloides, 399
    angustisquama, 400
    eucosma, 394, **400**
    pulcherrima, 394, **400**
    sordida, 400
pulvinata, 408, 410
    albescens, 410
pygmaea, 412
pyramidata, 398
racemosa, 390, 391, 394, **401**, 402, 403, 404, 405, 409, 410, 415
recurva, 406
reflexa, 408
rhodantha, 406
rosea, 389, 390, 396, 402, 406, 407, **408**, 409, 411, 415
    angustifolia, 409
    arida, **409**
    confinis, **409**, 410
    divaricata, 410
    nitida, 407
    pulvinata, 408, 409, **410**
    rosea, 409, **410**
rosulata, 390, 393, **407**, 409
×*rousseaui*, 415
rupicola, 405
sansonii, 410
scabra, 410
scariosa, 409
sedoides, 409
shumaginensis, 412
sierrae-blancae, 407
soliceps, 393, **410**
solitaria, 390, 391, 393, **401**, 402
solstitialis, 407
sordida, 409
sornborgeri, 414
spathulata, 404
speciosa, 410
stenolepis, 405
stenophylla, 389, 393, **398**
stolonifera, 414
straminea, 410
subcanescens, 414
subviscosa, 409
suffrutescens, 389, 390, 393, **408**, 409
tansleyi, 412
tomentella, 409
tweedsmuirii, 412
umbellata, 402

umbrinella, 390, 394, 396, *401*, **408**, 409, 410, 415
ungavensis, 414
vexillifera, 414
virginica, 390, 391, 396, **403**, 404, 405
    argillicola, 403
viscidula, 409
wiegandii, 414
wilsonii, 403
Anthemideae, 5, 6, 8, 11, **485**, 486, 498
Anthemis, 14, 487, **537**, 547, 548
    altissima, 547
    arvensis, *535*, **538**
        agrestis, 538
    austriaca, 547
    cotula, **538**, 542
    foetida, 538
    fuscata, 496
    mixta, 495
    nobilis, 496
    secundiramea, 538
    tinctoria, 547
Anthémis des champs, 538
Antonio, T. M., 7
Apargia
    borealis, 340
    chillensis, 298
Apargidium, 338, 340
    boreale, 340
Aphanostephus, 33, 49
Apogon, 362
    gracilis, 367
    wrightii, 366
Applequist, W. L., 542, 548, 550
Aquifoliales, 4
Arctanthemum, 31, 489, 534, **535**, 536
    arcticum, *535*, **536**
        arcticum, 536, **537**
        polare, *535*, 536, **537**
    hultenii, 536, 537
    integrifolium, 535
    kurilense, 536
Arctic agoseris, 326
Arctic daisy, 537
Arctium, 28, 58, 83, **168**, 169
    lappa, **169**
    minus, 169, *170*, 171
    nemorosum, 170, 171
    tomentosum, **169**, 170
    vulgare, 171
Arctoteae, 6
Arctotheca, 48, 195, **197**
    calendula, *196*, **197**
Arctotideae, 10, **195**, 200
Arctotis, 34, 48, 195, **198**
    calendula, 197
    fastuosa, **198**
    grandis, 198
    stoechadifolia, *196*, **198**, 199
        grandis, 198
    venusta, 198, 199
Arcuatae, 389
Arènes, J., 170
Argenteae, 389, 397, 398

Argophyllaceae, 4
Argyranthemum, 30, 485, 487, **552**
  anethifolium, 552
  foeniculaceum, *549*, **552**
  frutescens, 552
Arida, 36, 37, 38, 61
Arizona agoseris, 334
Arizona cottonrose, 446
Arizona rabbit-tobacco, 424
Arizona skeletonplant, 371
Arizona thistle, 141
Armoise, 503
Armoise absinthe, 519
Armoise annuelle, 523
Armoise aurone, 522
Armoise bisannuelle, 523
Armoise caudée, 507
Armoise de la mer Noire, 531
Armoise de l'ouest, 528
Armoise de Steller, 532
Armoise douce, 519
Armoise du Canada, 507
Armoise vulgaire, 533
Arnica, 32, 40, 54
Arnoglossum, 56
Arnoseris, 216, **276**
  minima, *273*, **277**
Aromatic pussytoes, 408
Arriagada, J. E., 472, 473
Arrowleaf snakeroot, 270
Arrowweed, 480
Artemisia, 6, 26, 53, 398, 486, 487, 498, **503**, 504, 508, 515, 531
  [unranked] *Tridentatae*, 509
  sect. *Abrotanum*, 520
  subg. Absinthium, 504, **518**
  subg. Artemisia, 504, 512, **520**
  subg. Dracunculus, 504, **505**
  subg. *Seriphidium*, 504
  subg. Tridentatae, 504, 508, **509**, 510, 511, 512, 513, 515, 524
  subsect. *Absinthium*, 518
  abrotanoides, 524
  abrotanum, 503, 521, **522**, 532
  absinthium, 503, **519**
  alaskana, 521, **523**
  albula, 528
  aleutica, **505**
  androsacea, 532
  angustifolia, 516
  annua, 503, 521, **523**
  arachnoidea, 529
  arbuscula, **511**, 512, 514, 515
    arbuscula, **511**
    longicaulis, 511
    longiloba, **511**, 512
    nova, 513
    thermopola, 511, **512**, 518
  arctica, 506, 530, 531
    beringensis, 530
    comata, 530
    ehrendorferi, 530
    saxatilis, 530
  argillosa, 513
  aromatica, 508

atomifera, 529
bakeri, 524
biennis, 521, **523**
  diffusa, 523
bigelovii, 503, 504, 509, 510, **512**
bolanderi, 513
borealis, **505**, 508
  borealis, **506**
  richardsoniana, **506**
brittonii, 528
californica, 487, 503, 504, 520, 521, **524**, 530
  insularis, 530
campestris, 505, **506**, *507*, 508
  borealis, 505
  campestris, 507
  canadensis, 506, **507**, 508
  caudata, 506, **507**
  douglasiana, 524
  pacifica, 506, **508**
  petiolata, 508
  purshii, 506
  pycnocephala, **509**
  scouleriana, 508
  spithamaea, 506
  strutzae, 506
cana, *507*, 510, 511, **512**, 515, 517
  bolanderi, 512, **513**
  cana, *507*, 512, **513**
  viscidula, 512, **513**, 515
canadensis, 507
candicans, 529
carruthii, 522, **524**
caudata, 507
  douglasiana, 524
  richardsoniana, 506
chamissoniana
  saxatilis, 530
chamomilla, 523
coloradensis, 524
columbiensis, 513
comata, 530, 531
commutata
  douglasiana, 524
  cuneata, 528
desertorum
  douglasiana, 524
  richardsoniana, 506
  scouleriana, 508
discolor, 530
diversifolia, 528
douglasiana, 522, **524**, 533
dracunculina, 508
dracunculoides, 508
  dracunculina, 508
dracunculus, 503, 505, *507*, **508**
falcata, 527, 528
filifolia, 504, 505, **508**
fischeriana, 524
flava, 526
foliosa, 524
forwoodii, 507
franserioides, 522, **525**
frigida, *514*, **519**
  gmeliniana, 519
  williamsae, 519

furcata, 521, **525**, 526
  heterophylla, 525
glauca, 508
  megacephala, 508
globularia, 521, **525**
  globularia, **525**
  lutea, 525, **526**
glomerata, 521, 525, **526**
  subglabra, 526
gnaphaloides, 528
gracilenta, 529
herriotii, 528
heterophylla, 524
hookeriana, 533
hultenii, 533
hyperborea, 525
incompta, 529
insularis, 530
kansana, 524
laciniata, 521, **526**
  laciniata, **526**
  parryi, 526, **527**
laciniatiformis, 526
latiloba, 529
lindheimeriana, 528
lindleyana, 529
longifolia, 521, **527**, 532
longiloba, 511
ludoviciana, 522, 524, **527**, *528*, 530, 532
  albula, 527, **528**
  candicans, 527, **529**
  douglasiana, 524
  incompta, 527, **529**
  integrifolia, 527
  latiloba, 529
  ludoviciana, 527, **528**
  mexicana, 527, **529**
  redolens, 527, **529**
macrobotrys, 526
matricarioides, 541, 542
mexicana, 529
michauxiana, 522, **530**, 531
microcephala, 528
monocephala, 520
natronensis, 527
neomexicana, 529
nesiotica, 504, 520, 521, 524, **530**
norvegica, 521, 522, 526, 528, **530**
  globularia, 525
  glomerata, 526
  piceetorum, 530
  saxatilis, 528, **530**
nova, 511, **513**, 514
opulenta, 533
pabularis, 528
pacifica, 508
packardiae, 504, 522, **531**
palmeri, 487, 503, 504, 520, 521, **531**
papposa, 487, 503, 521, **531**
parishii, 517
parryi, 527
pattersonii, 519, **520**
paucicephala, 528
pedatifida, 505, **509**
petrophila, 512

plattensis, 508
pontica, 522, **531**, 532
porteri, 505, **509**
potentilloides, 500
prescottiana, 529
pudica, 528
purshiana, 528
purshii, 506
pycnocephala, 505, **509**
pygmaea, 504, 510, *514*
redolens, 529
revoluta, 529
rhizomata, 528
richardsoniana, 506
rigida, 510, **515**
rothrockii, 503, 510, 511, **515**
rupestris, 519, **520**
  woodii, 520
saxicola
  parryi, 527
scopulorum, 519, **520**
  monocephala, 520
senjavinensis, 521, 526, *528*, **532**
serrata, 522, **532**
spiciformis, 510, **515**
  longiloba, 511
spinescens, 499
spithamaea, 506
stelleriana, 522, **532**, 533
suksdorfii, 522, **533**
tacomensis, 525
tilesii, 522, **533**
  aleutica, 533
  elatior, 533
  gormanii, 533
  hultenii, 533
  unalaschcensis, 533
tridentata, 407, 510, 511, 512, *514*, 515, **516**, 517
  arbuscula, 511
  bolanderi, 513
  nova, 513
  parishii, 516, **517**
  pauciflora, 517
  rothrockii, 515
  spiciformis, 515
  tridentata, *514*, **516**, 518
  trifida, 518
  vaseyana, 515, 516, **517**
  wyomingensis, 516, **517**, 518
  xericensis, 516
trifida, 518
  rigida, 515
trifurcata, 525
tripartita, 510, 512, *514*, 515, **518**
  rupicola, **518**
  tripartita, *514*, **518**
tyrellii, 523
unalaskensis
  aleutica, 533
vaseyana, 517
viscidula, 513
vulgaris, 503, 522, **533**, 534
  candicans, 529
  douglasiana, 524
  glabra, 533

incompta, 529
kamtschatica, 533, 534
littoralis, 533
longifolia, 527
ludoviciana, 527
michauxiana, 530
redolens, 529
serrata, 532
tilesii, 533
wrightii, 524
wrightii, 524
Artemisiastrum
palmeri, 531
Artichoke, 89, 90
Artichoke thistle, 89, 90
Asanthus, 54
Ashland thistle, 121
Aster, 7, 37
Asteraceae, **3,** 4, 5, 6, 7, 8, 9, 10,
     11, 12, 13, 16, 70, 76, 82,
     95, 172, 176, 177, 195,
     200, 214, 216, 257, 360,
     362, 363, 374, 379, 384,
     385, 386, 444, 445, 448,
     450, 454, 457, 460, 461,
     463, 464, 465, 467, 471,
     475, 478, 485, 486, 498
  subfam. *Asteroideae,* 5, 6
  subfam. *Barnadesioideae,* 5,
     70
  subfam. *Carduoideae,* 6
  subfam. *Cichorioideae,* 6, 195,
     200
  subfam. *Gochnatioideae,* 6
  subfam. *Hecastocleioideae,* 6
  subfam. *Mutisioideae,* 6
  subtribe Ambrosiinae, 4
  subtribe *Centaureinae,* 172,
     176, 177
  subtribe *Centipedinae,* 486
  subtribe *Filagininae,* 385, 386,
     444, 445, 448, 450, 454,
     457, 460, 461, 463, 464,
     465, 467
  subtribe *Gnaphaliinae,* 384
  subtribe *Gochnatiinae,* 76
  subtribe Madiinae, 6
  subtribe *Microseridinae,* 374
  subtribe Milleriinae, 16
  subtribe *Phalacroseridinae,*
     374
  subtribe *Stephanomeriinae,*
     360, 374
  tribe Anthemideae, 5, 6, 8, 11,
     485, 486, 498
  tribe *Arctoteae,* 6
  tribe Arctotideae, 10, **195,** 200
  tribe *Argenteae,* 389, 397
  tribe Astereae, 5, 6, 11, 486
  tribe *Athroismeae,* 486
  tribe *Bahieae,* 6
  tribe Calenduleae, 6, 12, 379
  tribe *Cardueae,* 6
  tribe *Carlineae,* 82
  tribe *Chaenactideae,* 6
  tribe Cichorieae, 5, 6, 8, 9, 10,
     13, 195, 200, **214,** 257,
     362, 363
  tribe *Coreopsideae,* 6

tribe Cynareae, 5, 6, 8, 12, **82,**
     95, 172, 176, 195
  tribe *Echinopeae,* 82
  tribe Eupatorieae, 5, 6, 11
  tribe Gnaphalieae, 5, 6, 12,
     384, 385, 461, 471
  tribe *Gochnatieae,* 6
  tribe *Gundelieae,* 195, 200
  tribe *Hecastocleideae,* 6
  tribe *Helenieae,* 6
  tribe Heliantheae, 3, 5, 6, 11
  tribe Inuleae, 6, 12, 384, **471,**
     475
  tribe *Lactuceae,* 214
  tribe *Liabeae,* 195, 200
  tribe *Madieae,* 6
  tribe *Millereae,* 6
  tribe Mutisieae, 5, 6, 8, 9, 12,
     70, 76
  tribe *Perityleae,* 6
  tribe Plucheeae, 5, 6, 12, 471,
     475, 478
  tribe *Polymnieae,* 6
  tribe Senecioneae, 6, 11
  tribe *Tageteae,* 6
  tribe Vernonieae, 5, 6, 11, 195,
     200
Asterales, 4, 5
Astereae, 5, 6, 11, 486
Asteridae, 4
Asteroideae, 5, 6
Astranthium, 33
Athroismeae, 486
Atrichoseris, 216, **309**
  platyphylla, *308,* **309**

Babcock, E. B., 222, 225, 226,
     227, 228, 229, 230, 232,
     233, 235, 238, 391
Baccharis, 8, 11, 58, 60
  foetida, 482
Bachelor's-button, 184
Bachmann, K., 338, 343, 345,
     374
Bahama Sachsia, 478
Bahia, 46, 49, 50
Bahieae, 6
Bahiopsis, 22
Baileya, 34, 35, 49
Baird, G. I., 323
Baker's hawksbeard, 226
Balduina, 50
Baldwin, B. G., 5, 95, 360, 363,
     374
Balsamita
  major, 490
Balsamorhiza, 19
Banckes, R., 503
Barbeau, 184
Barbed rattlesnakeroot, 267
Barber, H. N., 275
Barber's hawksbeard, 236
Barbour, M. G., 6
Bardane, 168
Bardane tomenteuse, 169
Baretwig neststraw, 453
Barkhausia
  grandiflora, 377
Barkley, T. M., 3

Barkleyanthus, 40
Barlow-Irick, P. L., 131, 140,
     142, 146
Barnaby star-thistle, 193
Barnadesia, 70
Barnadesioideae, 5, 70
Barneby rabbit tobacco, 463
Barneby's thistle, 124
Basketflower, 175
Batra, S. W. T., 94
Battaglia, E., 253
Battjes, J., 338, 343
Bayer, R. J., 5, 388, 389, 390,
     391, 396, 399, 400, 401,
     402, 403, 404, 405, 406,
     407, 408, 409, 410, 411,
     412, 413, 414
Beach wormwood, 532
Beaked hawksbeard, 238
Beaked skeleton-weed, 368, 369
Beaman, J. H., 279
Bearded creeper, 178
Beauchamp, R. M., 470
Beauverd, G., 386
Bebbia, 11, 29
Beetle, A. A., 510, 517
Bellis, 33
Benitoa, 44, 49
Bennett, J. P., 356
Bennetts, H. W., 473
Bentham, G., 5
Berlandiera, 17, 22
Berthelotia, 479, 480
Bidens, 16, 20, 21, 25, 29
  aristosa, 21
  polylepis, 21
Biennial wormwood, 523
Big sage, 516
Bigelow sagebrush, 512
Bigelowia, 60
Big-head knapweed, 185
Big-head rabbit tobacco, 462
Bighead straitjackets, 465
Bigspine thistle, 115
Bioletti's rabbit-tobacco, 423
Black knapweed, 187
Black sage, 513
Black sagebrush, 513
Black salsify, 307
Black thistle, 114
Black-eyed Susan, 7
Blackland thistle, 117
Blake, S. F., 6
Blattner, F. R., 347
Blaver, 184
Blazingstar, 7
Blennosperma, 32
Blepharipappus, 15, 19
Blepharizonia, 20
Blessed milkthistle, 164
Blessed thistle, 192
Bleuet, 184
Bloom, W. L., 112, 113
Bluebonnets, 184
Bluebottle, 184
Blue-eyed African-daisy, 198
Blue-poppy, 184
Böcher, T. W., 240, 279

Bogler, D. J., 222, 254, 257, 264,
     294, 297, 368, 369, 374
Bohm, B., 356
Bolander dandelion, 374
Bolander sagebrush, 513
Boltonia, 47, 48
Bombycilaena, 454
  californica, 455
Boot's rattlesnakeroot, 268
Boreal sage, 506
Borrichia, 24
Bossard, C. C., 426
Boulette commune, 86
Boulette de Hongrie, 86
Boulos, L., 275
Bourrier, 170
Box pussytoes, 396
Boyd, S., 453
Boyko, E. V., 486
Brachyramphus, 272
  intybacea, 272
Bracteantha, 427
  bracteata, 428
Bradburia, 43, 48
Brass-buttons, 544
Brauner, S., 351, 356
Breea
  arvensis, 109
Bremer, K., 4, 5, 70, 76, 82, 183,
     195, 255, 257, 258, 384,
     485, 504, 534, 536, 541,
     548
Brewer's thistle, 133
Brickellia, 54
Brickelliastrum, 55
Brightwhite, 360
Brintonia, 59
Briquet, J., 177
Bristly hawkbit, 295
Bristly hawksbeard, 237
Bristly thistle, 114
Britton, N. L., 6
Broadleaf cottonrose, 449
Brock, M. T., 241
Brodeur, K. R., 272
Bromus
  tectorum, 351
Brooks, R. E., 112, 126
Brouillet, L., 3, 239, 534, 535,
     540, 548
Brown knapweed, 186
Brown pussytoes, 408
Brown-bracted pussytoes, 408
Brownfoot, 74
Brownie thistle, 160
Brownplume wirelettuce, 356
Brown-ray knapweed, 186
Brunton, D. F., 242, 246, 276
Brushes, 184
Budsage, 498, 499
Bull cottonthistle, 88
Bull thistle, 96, 109, 114
Bulwand, 503
Burdock, 168
Burks, K. A., 163
Burrweed, 545
Burry, J. N., 473
Bursage mugwort, 525
Burtt, B. L., 429

Butterweed, 270
Button burrweed, 546

Cabrera, A. L., 431, 434, 476
Cacaliopsis, 56
Calaïs, 338
    sect. *Calocalaïs*, 322
    aphantocarpha, 344
    bigelovii, 345
    douglasii, 343
    glauca
        procera, 341
    lindleyi, 322
    platycarpha, 344
    sylvatica, 342
    tenella, 344
Calendula, 30, 379, **381**
    arvensis, **382**
    officinalis, *380*, **382**
    pluvialis, 381
Calenduleae, 6, 12, **379**
California agoseris, 333
California chicory, 349
California cottonrose, 445
California dandelion, 248
California desertdandelion, 313
California rabbit-tobacco, 423
California sagebrush, 524
California swamp thistle, 133
California thistle, 139
California threefold, 75
Caltrops, 191
Calycadenia, 15, 20
Calyceraceae, 4, 5
Calycerales, 4
Calycoseris, 216, **307**, 310, 360
    parryi, 307, **308**
    wrightii, 307, *308*
Calymmandra, 461
    candida, 461
Calyptocarpus, 24
Camomille maroute, 538
Campanulaceae, 4
Canada thistle, 96, 109, 110
Canadanthus, 38
Canadian pussytoes, 404
Cankerweed, 264, 270
Cape marigold, 380
Capeweed, 197
Cardo del valle, 176
Cardon, 146
Cardoon, 89
Cardueae, 6, 82
Carduoideae, 6
Carduus, 57, 66, 83, **91**, 96, 97, 122
    acanthoides, *90*, 91, **92**, 93
        acanthoides, *90*, **92**
    altissimus, 111
    arvensis, 109
    candidissimus, 140
    canovirens, 136
    carolinianus, 118
    coloradensis, 156
    crassicaulis, 132
    crispus, 91, **92**
    cymosus, 136
    discolor, 112
    flodmanii, 120

foliosus, 159
glaber, 119
helenioides, 110
hookerianus
    eriocephalus, 152
hydrophilus, 132
inamoenus, 134
lanceolatus, 109
lecontei, 114
macounii, 147
macrocephalus, 93
macrolepis, 93
marianus, 164
maritima, 161
mohavensis, 134
muticus, 113
nevadensis, 134, 153
nutans, 91, **93**
    leiocephalus, 93
    leiophyllus, 93
    macrocephalus, 93
    macrolepis, 93
    nutans, 93
    vestitus, 93
nuttallii, 119
occidentalis, 137
×*orthocephalus*, 93
osterhoutii, 128
palustris, 110
perplexans, 128
pulcherrimus, 125
pumilus, 116
pycnocephalus, 91, **93**, 94
    pycnocephalus, **93**, 94
    tenuiflorus, 94
remotifolius, 129
repandus, 113
revolutus, 118
smallii, 116
spinosissimus, 115
tenuiflorus, 91, **94**
×*theriotii*, 94
thoermeri, 93
tracyi, 121
undulatus, 120
validus, 158
venustus, 140
vinaceus, 163
virginianus, 118
vittatus, 116
vulgaris, 109
Carlina, 28, 83, **84**
    vulgaris, *85*
        longifolia, 85
        vulgaris, *85*
Carlineae, 82
Carline-thistle, 84
Carlquist, S., 5, 350, 498
Carlquistia, 65
Carmel Valley malacothrix, 319
Carminatia, 54, 66
Carolina thistle, 118
Carpet burrweed, 546
Carphephorus, 28, 59
Carpochaete, 66
Carr's rattlesnakeroot, 268
Carruth wormwood, 524
Carthamus, 52, 67, 83, **178**
    baeticus, 180

creticus, 179, **180**
    creticus, 180
    laevis, 201
lanatus, *178*, 179, **180**
    baeticus, 180
    creticus, 180
leucocaulos, **179**, 180
    oxyacantha, 181
tinctorius, 179, **180**, 181
Carveseed, 362
Casse lunette, 184
Cassini, H. , 5
Catipes, 389, 390, 392, 401, 402
Catling, P. M., 242, 245
Cedar Rim thistle, 126
Centaurè du solstice, 193
Centaurea, 52, 57, 58, 67, 83, 84, 96, 171, 172, 176, 177, **181**, 182, 183, 186, 192, 194
    sect. *Acrolophus*, 189
    sect. *Jacea*, 186, 187
    sect. *Plectocephalus*, 175
    subg. *Amberboa*, 172
    subg. *Crupina*, 177
    subg. *Jacea*, 186
    americana, 176, 182
    aspera, 182, 191
    austriaca, 188, 189
    babylonica, 182
    benedicta, 181, 183, **190**, **192**
    biebersteinii, 189
    bovina, 194
    calcitrapa, 183, **191**, 192
    calcitrapoides, 191
    centaurium, 182
    cineraria, 182
    crupina, 178
    cyanus, *178*, 183, **184**
    debeauxii
        thuillieri, 186, 187
    depressa, 183, **184**
    diffusa, 183, **190**, 194
    diluta, 182, 183, 184, **192**, 193
    dubia, 186, 187, 188
        nigrescens, 188
        vochinensis, 188
    eriophora, 182
    iberica, 183, **192**
    jacea, 182, 184, **186**, 187, 188, 189
        jacea, 187, 188
        nigra, 187, 188
        nigrescens, 187, 188
        pratensis, 187
        ×*pratensis*, 187
    macrocephala, 183, **185**
    maculosa, 189
        micranthos, 189
    melitensis, 183, **193**
    ×*moncktonii*, 184, 186, **187**, 188
    montana, 183, **185**
    moschata, 173
    muricata, 175
    nemoralis, 186, 187
    nervosa, 189

nigra, 184, 186, **187**, 188, 189
    radiata, 187
nigrescens, 184, 186, 187, **188**, 189
    paniculata, 194
    phrygia, 182, 183, **188**, 189
        phrygia, 189
    picris, 172
    ×*pouzinii*, 191
    pratensis, 186, 188
    ×*pratensis*, 186
    ×*psammogena*, 190
    repens, 172
    rhenana, 189
    rothrockii, 176, 182
    salmantica, 174
    scabiosa, 183, **185**
    solstitialis, 183, **190**, **193**
    stoebe, 182, 183, **189**, *190*, 194
        micranthos, **189**, *190*, 194
        stoebe, 189
    sulphurea, 182, 183, **194**
    thuillieri, 186, 187
    transalpina, 188
    trichocephala, 182
    triumfettii, 191
    uniflora
        nervosa, 189
    variegata, 191
    virgata, 183, **191**
        squarrosa, **191**
    vochinensis, 188
Centaurée, 181
Centaurée à gros capitules, 185
Centaurée chausse-trappe, 191
Centaurée de Russie, 171
Centaurée des montagnes, 185
Centaurée diffuse, 190
Centaurée jacée, 186
Centaurée maculée, 189
Centaurée noirâtre, 188
Centaurée noire, 187
Centaurée scabieuse, 185
Centaurée tachetée, 189
Centaureinae, 172, 176, 177
Centipeda, 486
    minima, 486
Centipedinae, 486
Centratherum, 67, 201, **206**
    punctatum, *205*, **206**
Centromadia, 15, 16, 20
Chaenactideae, 6
Chaenactis, 8, 29, 52, 68, 69
Chaetadelpha, 218, **368**
    wheeleri, *366*, **368**
Chaetopappa, 33, 36, 38, 47
Chaffless saw-wort, 167
Chamaechaenactis, 68
Chamaemelum, 14, 26, 487, **496**
    arvense, 538
    cotula, 538
    fuscatum, **496**
    mixtum, 495
    nobile, **496**, *497*
Chambers, K. L., 322, 324, 333, 334, 335, 338, 341, 342, 344, 345, 346, 347, 359, 362, 363, 365, 374, 547

Chamomile, 537, 540, 542
Chamomilla, 540, 541
   discoidea, 541
   inodora, 549
   maritima, 550
   occidentalis, 542
   recutita, 542
   suaveolens, 541
Chamomille, 540
Chaptalia, 12, 14, 57, 70, 71, **78**, 80
   albicans, **79**
   alsophila, 81
   dentata, 79
   leiocarpa, 79
   leucocephala, 81
   lyrata, 81
   nutans
      texana, 79
   texana, **79**
   tomentosa, 77, 79, **80**
Chardon, 91, 95
Chardon bénit, 192
Chardon crépu, 92
Chardon de Flodman, 120
Chardon des champs, 109
Chardon des marais, 110
Chardon discolore, 112
Chardon du Canada, 109
Chardon écailleux, 154
Chardon épineux, 92
Chardon lancéolé, 109
Chardon Marie, 164
Chardon mutique, 113
Chardon penché, 93
Chardon vulgaire, 109
Charleston Mountain pussytoes, 410
Charleston pussytoes, 410
Charleston tansy, 502
Chausse-trappe, 191
Cheatgrass, 351
Chickensage, 499
Chicorée, 222
Chicory, 222
Chicoryleaf wirelettuce, 352
Chinnappa, C. C., 363
Chloracantha, 36, 37
Chlorocrepis
   albiflora, 292
   fendleri, 286
   tristis, 291
Chomonque, 77
Chondrilla, 216, **252**
   juncea, *253*
   pauciflora, 377
Chorisiva, 27
Chorro Creek bog thistle, 162
Chou bourache, 170
Christensen, N. L., 6
Chromolaena, 27, 55
Chrysactinia, 40
Chrysanthème balsamique, 490
Chrysanthème polaire, 537
Chrysanthemoides, 12, 30, **379**
   monilifera, *380*
Chrysanthemum, 485, 486, 536
   sect. *Nipponanthemum*, 555
   anethifolium, 552

arcticum, 536
   polare, 536, 537
balsamita, 490
   tanacetoides, *490*
bipinnatum, 490
   huronense, 490
carinatum, 553
coronarium, 555
foeniculaceum, 552
incanum, 543
indicum, 485, 486
integrifolium, 535
lacustre, 559
leucanthemum, 558
   pinnatifidum, 558
maximum, 558
nipponicum, 556
paludosum, 554
parthenium, 490
segetum, 555
serotinum, 557
×superbum, 486
uliginosum, 557
Chrysocoma
   acaulis, 208
   gigantea, 212
Chrysogonum, 18, 23
Chrysoma, 42, 60
Chrysopsis, 42, 43, 47
Chrysothamnus, 60, 68
Cibourroche, 170
Cicerbita, 261
   floridana, 261
Cichorieae, 5, 6, 8, 9, 10, 13, 195, 200, **214**, 362, 363
Cichorioideae, 6, 195, 200
Cichorium, 218, **221**
   endivia, 221
   intybus, 221, **222**
Cinquefoil false sagebrush, 500
Circium
   edule, 146
Cirse des champs, 109
Cirse des marais, 110
Cirsium, 57, 66, 82, 83, 93, **95**, 96, 97, 100, 102, 117, 130, 155
   subsect. *Americana*, 130
   acanthodontum, 130, 131
   acaule
      americanum, 156
   acaulescens, 154, 156
   acuatum, 121
   altissimum, 100, **111**, 112
      biltmoreanum, 111
   amblylepis, 130, 131
   americanum, 156
      callilepis, 130
   andersonii, 101, 131, **145**, 146
   andrewsii, 104, **141**, 160
   araneans, 128
   arcuum, 140
   aridum, 126
   arizonicum, 100, 101, 104, 131, **141**, 142, 143, 144, *145*
      arizonicum, 142, **143**, 144, 145
      bipinnatum, 142, **143**

chellyense, 142, **144**
   nidulum, 143
   rothrockii, 142, **143**
   tenuisectum, 142, **144**, *145*, 151
   arvense, *95*, *96*, *97*, 102, **109**, 110
   argenteum, 109
   horridum, 109
   integrifolium, 109
   mite, 109
   vestitum, 109
   austrinum, 119
   barnebyi, 106, 107, 108, **124**
   bernardinum, 139
   bipinnatum, 143
   blumeri, 125
   botrys, 136
   brevifolium, 101, 121, **124**
   brevistylum, 102, 104, 107, 146, 147, **148**
   breweri, 133
   butleri, 155
   calcareum, 143
      bipinnatum, 143
      pulchellum, 143
   californicum, 139
      bernardinum, 139
      pseudoreglense, 139
   callilepis, 130, 131
      oregonense, 130
      pseudocarlinoides, 130
   campylon, 162
   canescens, 97, 100, 101, 121, **122**, 150, 157
   canovirens, 136
   canum, 163
   carolinianum, 98, **118**
   centaureae, 127, 128, 130
   chellyense, 144
   chuskaense, 144
   ciliolatum, 105, 108, **121**
   clavatum, 103, 105, 108, **126**, 127, 129, 130
      americanum, 126, 127, **128**
      clavatum, **127**, 128
      markaguntense, 127
      osterhoutii, 127, **128**
   clokeyi, 150, 151
   coccinatum, 153
   coloradense, 154, 156
      acaulescens, 156
      longissimum, 156
   congdonii, 154, 158
   coulteri, 138
   crassicaule, 103, 104, **132**
   cymosum, 105, 108, **136**
      canovirens, **136**
      cymosum, **136**
   davisii, 135
   diffusum, 143
   discolor, *90*, 98, 99, **112**, 113
   douglasii, 104, 108, *115*, **133**, 160
      breweri, *115*, **133**
      canescens, 133
      douglasii, **133**
   drummondii, 99, 100, **153**, 154

acaulescens, 156
latisquamum, 156, 158
oregonense, 156
vexans, 156
eatonii, 100, 102, 103, 104, 107, 127, 128, 135, *145*, **150**
   clokeyi, 144, **151**
   eatonii, 127, **151**
   eriocephalum, 128, 151, **152**
   harrisonii, 151
   hesperium, 151, **152**
   murdockii, 126, *145*, 151, **152**, 156
   peckii, 135, 151, **153**
   viperinum, 135, **151**
edule, 102, 103, 104, 131, **146**, 147, 148
   edule, **147**
   macounii, **147**
   wenatchense, **147**
engelmannii, 99, **117**
eriocephalum, 152
erosum, 156
filipendulum, 117
flaccidum, 118
floccosum, 121
flodmanii, 98, 100, 101, 106, 107, 113, **120**, 121
foliosum, *95*, *97*, 107, 154, 155, 156, 157, **159**, 160
fontinale, 102, **161**
   campylon, 161, **162**
   fontinale, 160, 161, **162**
   obispoense, 161, **162**
gilense, 149, 150
glabrum, 119
grahamii, 103, 114, **124**, 125, 150, 156
griseum, 128
hallii, 146, 147
helenioides, 98, **110**, 111
helleri, 119
hesperium, 152
heterophyllum, 110
hillii, 117
hookerianum, 100, 101, 102, 121, 126, **148**, 149
   scariosum, 154
horridulum, 99, 113, **114**, *115*
   elliottii, 115
   horridulum, 114, **115**, 116
   megacanthum, 114, *115*
   vittatum, 114, **116**
howellii, 121
humboldtense, 135, 140, 153
hydrophilum, 103, 108, **132**
   hydrophilum, **132**
   vaseyi, 132, **133**
inamoenum, 105, 106, 108, **134**, 135, 136, 141, 151, 153
   davisii, 108, **135**
   inamoenum, **135**
incanum, 109
inornatum, 149, 150
iowense, 111, 112
×iowense, 112

Cirsium (*continued*)
joannae, 102, 103, 105, **163**
kamtschaticum, 107, **111**
kelseyi, 148, 149
lacerum, 155
lactucinum, 162
lanceolatum, 109
laterifolium, 128
lecontei, 99, **114**
loncholepis, 158, 159
longistylum, 104, **149**
macounii, 146, 147
magnificum, 155
maritimum, 161
megacanthum, 115
megacephalum, 120
minganense, 155, 156
modestum, 128
mohavense, 108, **134**
montigenum, 133
murdockii, 152
muticum, 98, 99, 112, **113**, 120
monticola, 113
navajoense, 144
nebraskense, 122
nelsonii, 122
neomexicanum, 105, 135, **140**, 141
utahense, 140
nevadense, 134
nidulum, 143, 144
nuttallii, 98, 114, **119**
oblanceolatum, 120
occidentale, 101, 102, 104, 105, **137**, *138*, 140, 141, 159, 160
californicum, 137, *138*, **139**, 140
candidissimum, 131, 137, *138*, **140**
compactum, 137, **139**
coulteri, 137, *138*, 139
lucianum, 137, **139**
occidentale, **137**, 138, 139, 140, 159, 161
venustum, 137, 138, 139, **140**
ochrocentrum, 100, 101, 106, **123**
martinii, **123**
ochrocentrum, **123**
odoratum, 116
olivescens, 125, 156
oreganum, 131
osterhoutii, 128
ownbeyi, 105, **153**
pallidum, 149, 150
palousense, 124
palustre, 96, 97, 102, **110**
parryi, 104, 122, 125, **149**, 150
mogollonicum, 149, 150
pastoris, 140
peckii, 150, 153
perennans, 125
perplexans, 103, **128**, 129
pitcheri, 98, 100, **122**, 123
plattense, 122

polyphyllum, 152
praeteriens, 101, **160**, 161
proteanum, 140
pulchellum, 143
bipinnatum, 143
diffusum, 143
glabrescens, 143
pulcherrimum, 98, 105, 108, **125**, 126
aridum, **126**
pulcherrimum, **126**, 128, 152
pumilum, 99, **116**, 117
hillii, 116, **117**
pumilum, 113, 115, **116**, 117
quercetorum, 101, 106, 107, 131, 141, **160**
citrinum, 158
walkerianum, 160
xerolepis, 160
remotifolium, 103, 105, 107, *115*, 128, **129**, 130, 131, 147
odontolepis, *115*, **130**, 131, 147, 160
oregonense, 130
pseudocarlinoides, 130, 131
remotifolium, **130**, 131, 147
rivulare, 130, **131**, 147
repandum, 98, 99, **113**, 115
revolutum, 118
rhothophilum, 102, 103, 159, **161**
rothrockii, 143
rushyi, 134
rydbergii, 103, 105, **162**, 163
scabrum, 164
scapanolepis, 128
scariosum, 99, 101, 103, 104, 106, 107, 122, *145*, **154**, 155, 156, *157*, 158, 159, 160
americanum, 155, **156**, *157*
citrinum, 155, **158**, 159, 161
coloradense, 121, 125, 155, **156**, 157
congdonii, 155, **158**
robustum, 155, **159**
scariosum, *145*, 149, 152, 154, **155**, 156, 157, 158, 159, 160
thorneae, 154, 155, 156, **157**, 158
toiyabense, 154, 155, **158**
scopulorum, 152
setosum, 109
smallii, 116
spathulifolium, 128
spinosissimum, 115
subniveum, 135, 136
terrae-nigrae, 117
texanum, 98, **119**
stenolepis, 119
tioganum, 154, 156
coloradense, 156
tracyi, 106, 107, **121**, 122
triacanthum, 136

turneri, 99, 101, 142, **144**, 145
tweedyi, 152
undulatum, 100, 101, 106, 107, 119, **120**, 121, 122, 149, 156, 157
albescens, 140
ciliolatum, 121
megacephalum, 120
undulatus
tracyi, 121
utahense, 135, 140, 141
×*vancouveriense*, 147, 148
vaseyi, 133
vernale, 128
vinaceum, 102, 131, *157*, **163**
virginense, 134
virginianum, 98, **118**
filipendulum, 117
vittatum, 116
vulgare, 96, 99, 100, **109**
walkerianum, 160
wallowense, 135
wheeleri, 108, **125**, 143
salinense, 125
wrightii, 97, 102, **131**, 163
Cladanthus, 14, 487, **495**
mixtus, *491*, **495**
Clapham, A. R., 548
Clappia, 46
Clarionia
runcinata, 73
Clausen, J., 493
Clements, F. E., 504
Cleveland's desertdandelion, 313
Cliff cudweed, 420
Cliff desertdandelion, 318
Cliff thistle, 144
Clokey thistle, 151
Clotbur, 168
Clustered thistle, 148
Cnicus, 181, 183
andersonii, 145
andrewsii, 141
arizonicus, 141
benedictus, 182, 192
benedictus, 192
kotschyi, 192
breweri, 133
vaseyi, 133
carlinoides
americanus, 128, 130
clavatus, 126
discolor, 112
drummondii
bipinnatus, 143
eatonii, 150
eriocephalus, 152
fontinalis, 161
giganteus, 164
hallii, 146
hesperius, 152
hillii, 117
parryi, 149
pitcheri, 122
quercetorum, 160
rothrockii, 143
wheeleri, 125
Coast Range agoseris, 332

Coastal blackroot, 476
Coastal plain thistle, 113
Coastal sagewort, 509
Coastal silverpuffs, 345
Cobwebby thistle, 137
Cody, W. J., 240, 407, *558*
Coffey, T., 503
Collet, F., 473
Colorado Plateau agoseris, 328
Colorado thistle, 156
Columbiadoria, 42
Common burdock, 170
Common cottonrose, 448
Common crupina, 178
Common dandelion, 244
Common dwarfdandelion, 367
Common mugwort, 533
Common saussurea, 166
Common sow-thistle, 275
Common tansy, 490
Common thistle, 109
Common wormwood, 519
Common yarrow, 493
Common-cudweed, 441
Compact chickensage, 502
Compact cobwebby thistle, 139
Compositae, 3, 4, 469
Composite, 3, 7, 8
Coneflower, 7
Congdon, J. W., 161
Congested sagewort, 526
Conoclinium, 54
Constancea, 49
Conyza, 36, 62
carolinensis, 480
cinerea, 205
odorata, 481
pycnostachya, 476
sagittalis, 480
symphytifolia, 481
Coreocarpus, 16, 18, 21, 22, 25, 29
Coreopsideae, 6
Coreopsis, 7, 8, 9, 16, 21
Corethrogyne, 36, 38
Corn pinks, 184
Corn chamomile, 538
Corn chrysanthemum, 485
Corn marigold, *555*
Cornflower, 181, 184
Correll, D. S., 112, 119, 452
Cosmos, 16, 20
Costmary, 490
Cota, 14, 487, **547**
altissima, 547
austriaca, 547
tinctoria, *545*, **547**
Cotonnière, 443, 447
Cotonnière commune, 448
Cotonnière de France, 447
Cotonnière des champs, 445
Cotonnière naine, 446
Cotonnière spatulée, 449
Cotton thistle, 87
Cotton-batting-plant, 418
Cottonrose, 443, 447
Cottonseed, 454
Cottontop, 456
Cottonweed, 443

Cotula, 52, 486, **543**, 544
 australis, **544**
 coronopifolia, **544**, *545*
 mexicana, 544
 pilulifera, 539
Cotule, 544
Cough bush, 480
Coulter's thistle, 138
Crawford, D. J., 391, 401, 402, 403
Creekside wirelettuce, 355
Creeping knapweed, 171
Creeping thistle, 109
Creeping-cudweed, 441
Crepidium
 glaucum, 236
Crepis, 214, 216, 217, 219, **222**, 223, 228
 acuminata, *221*, 224, **225**, 227, 229, 234
 intermedia, 229
 pleurocarpa, 234
 pluriflora, 225
 ambigua, 286
 andersonii, 235
 angustata, 225
 atribarba, 224, **225**, 227, 229
 originalis, 225
 bakeri, 225, **226**, 233
 bakeri, **226**
 cusickii, **226**
 idahoensis, **226**
 barbata, 277
 barberi, 236
 barbigera, 224, **227**
 biennis, 223, **227**, 232
 bursifolia, 224, **227**
 capillaris, 223, **228**, 232
 cooperi, 228
 cusickii, 226
 elegans, 224, **228**, 231
 exilis, 225
 originalis, 225
 foetida, 223, **228**, 229
 geisseana, 313
 glareosa, 230
 intermedia, 224, 225, **229**, 233, 234
 pleurocarpa, 234
 japonica, 256
 modocensis, 224, 227, **229**
 glareosa, 229, **230**
 modocensis, 229, **230**
 rostrata, 229, **230**
 subacaulis, 229, **230**
 monticola, 224, 226, **230**, 231
 nana, 224, 228, *231*
 clivicola, 231
 lyratifolia, 231, 256
 ramosa, 231
 nicaeënsis, 223, **232**
 occidentalis, 225, 226, *231*, **232**
 conjuncta, 232, **233**
 costata, 232, **233**
 crinita, 230
 gracilis, 225
 occidentalis, *231*, **232**, 233
 pumila, 232, **233**

 subacaulis, 230
 pannonica, 224, **233**
 pleurocarpa, 224, 225, 229, **234**
 pulchra, 223, *231*, **234**
 pumila, 233
 rhagdioloides, 303
 rostrata, 230
 rubra, 223, **234**
 runcinata, 224, **235**, *236*
 andersonii, **235**, 236
 barberi, 235, **236**
 glauca, 235, **236**, 237
 hallii, 235, **237**
 hispidulosa, 235, **237**
 imbricata, 235, **237**
 runcinata, **235**, *236*, 237
 scopulorum, 230
 seselifolia, 225
 setosa, 223, **237**
 tectorum, 223, **238**
 vesicaria, 223, **238**
 haenseleri, 238
 taraxaciflora, 238
 virens, 228
 zacintha, 223, **238**, 239
Crépis, 222
Crépis capillaire, 228
Crépis des troits, 238
Crested wheatgrass, 517
Crocidium, 32, 40
Croix de Malte, 193
Crompton, C. W., 276
Cronquist, A., 4, 5, 6, 79, 92, 112, 130, 134, 135, 146, 148, 154, 155, 187, 188, 240, 242, 279, 284, 287, 371, 386, 391, 402, 403, 404, 407, 457, 458, 460, 472, 473, 483, 541, 542, 548
Croptilon, 43
Crossostephium
 californicum, 524
 insulare, 530
Crown daisy, 485, 555
Crupina, 67, 83, *177*, *178*
 vulgaris, *178*
Cryptopleura, 324
 californica, 333
Cryptostemma
 calendula, 197
Cuatrecasas, J., 480
Cuesta Ridge thistle, 139
Cuniculotinus, 61
Cupularia
 viscosa, 473
Cure-for-all, 480
Curled thistle, 92
Curtiss, A. H., 480
Cushion pussytoes, 398
Cusick's hawksbeard, 226
Cutleaf silverpuffs, 341
Cut-leaved silverpuffs, 341
Cuvier, F., 5
Cyanopsis, 174
 muricata, 175
Cyanthillium, 67, 201, **204**
 cinereum, *205*

Cyclachaena, 25, 51
Cymbia, 362
 occidentalis, 365
Cynara, 57, 58, 66, 67, 83, **89**
 cardunculus, **89**, *90*
 cardunculus, 89, **90**
 flavescens, 89, *90*
 scolymus, 90
 scolymus, 89, 90
Cynareae, 5, 6, 8, 12, **82**, 95, 172, 176, 195
Cynthia, 362
 dandelion, 364
 montana, 365
 virginica, 364
 viridis, 364

d'Alcontres, G. S., 473
Dabydeen, S., 112, 120
Dagger-flower, 174
Dahlstedt, H., 240, 242
Daisy, 552
Dandelion, 239
Dandelion hawksbeard, 235
Dark hawksbeard, 225
Davidson, R. A., 112
Davis, W. S., 310
Davis's desertdandelion, 317
Davis's thistle, 135
Dawson, H. W., 5
de Candolle, A. P., 5
de Mirbel, C., 9
de Villiers, S. E., 5
Deane, W., 194
Deane's wirelettuce, 354
Decaneurum, 557
Deceitful pussytoes, 402
Deinandra, 16, 20
Delairea, 55
Delicate everlasting, 436
Delicate rabbit-tobacco, 421
Delta woolly marbles, 458
den Nijs, J. C. M., 241
Dendranthema, 485, 536
 sect. *Arctanthemum*, 535
 sect. *Haplophylla*, 534
 arctica, 536
 arcticum
 polare, 537
 hultenii, 537
 integrifolium, 535
Denseleaf pussytoes, 411
Derived silverpuffs, 347
Desert chicory, 349
Desert cudweed, 437
Desert fanbract, 452
Desert holly, 74
Desert mountains thistle, 144
Desert neststraw, 452
Desert paeonia, 73
Desert prenanthella, 360
Desert pussytoes, 409
Desert thistle, 140, 141
Desert wirelettuce, 357
Desertdandelion, 310
Desertpeony, 72
Desrochers, A. M., 92, 93
Detling's silverpuffs, 342
DeVore, M. L., 5

Diaperia, 26, 388, **460**, 461
 candida, **461**
 prolifera, *455*, 461, **462**, 463
 barnebyi, **463**
 prolifera, *455*, **463**
 verna, 445, **461**, 462
 drummondii, **462**
 verna, 460, **462**
Dichaetophora, 49
Dicoria, 25, 51
Dicranocarpus, 20
Dieteria, 37, 39, 62, 64
Diffuse knapweed, 190
Dimeresia, 64
Dimorphae, 389, 399
Dimorphotheca, 30, 379, **380**
 aurantiaca, 381
 calendulacea, 381
 ecklonis, 383
 pluvialis, 381
 sinuata, *380*, **381**
Dinnerplate thistle, 156
Dipsacales, 4
Disc mayweed, 541
Distaff thistle, 178
Dittrich, M., 82
Dittrichia, 40, 471, **473**
 graveolens, *472*, **473**
 viscosa, 473
Dodd, J., 253
Doellingeria, 39
Doll, R., 240
Dolores River skeletonplant, 372
Doronicum, 31, 41
Dostál, J., 186, 187, 189, 194
Douglas sagewort, 524
Douglas, G. W., 459
Douglas's silverpuffs, 343, 344
Douglas's thistle, 133
Drummond's thistle, 153
Drury, D. G., 441
Du Plessis, J. L., 473
Dudman, A. A., 240, 241
Duistermaat, H., 170
Dunce-nettle, 113
Dune malacothrix, 316
Dune thistle, 122
Dunedelion, 316
Dupuis, G., 473
Dusky chamomile, 496
Dusty miller, 182
Dwarf alpine hawksbeard, 231
Dwarf cottonrose, 446
Dwarf cudweed, 460
Dwarf dandelion, 362
Dwarf desertpeony, 74
Dwarf evax, 468
Dwarf saw-wort, 167
Dwarf thistle, 153
Dwarf woollyheads, 458
Dysodiopsis, 45
Dyssodia, 45

Eastern rabbit-tobacco, 420
Eastwoodia, 11, 29
Eatonella, 46
Eaton's thistle, 150
Echinacea, 7, 17, 22
Echinopeae, 82

Echinops, 12, 28, 64, 82, 83, **85**
  commutatus, 86
  exaltatus, *85*, **86**
  ritro, 86, **87**
    ruthenicus, **87**
  ruthenicus, 87
  sphaerocephalus, **86**
Eclipta, 18, 24
Edible thistle, 146
Egletes, 33
Eldenäs, P. K., 5
Elegant cudweed, 435
Elegant hawksbeard, 228
Elegant pussytoes, 400
Elegant silverpuffs, 346
Elephant's foot, 202
Elephantopus, 64, 201, **202**
  carolinianus, *202*, **203**
  elatus, **203**
  nudatus, **203**
  spicatus, 204
  tomentosus, **203**
Elk thistle, 154, 159
Ellstrand, N. C., 241, 245
Emilia, 56
Encelia, 19, 23, 27, 30
  nutans, 23, 30
Enceliopsis, 19, 23
Engelmannia, 23
Engelmann's thistle, 117
Entire-leaved daisy, 535
Épervière, 278
Épervière des Florentins, 285
Épervière orangée, 283
Épervière piloselle, 283
Erechtites, 55
Erect evax, 468
Eremohylema, 479, 480
Ericameria, 42, 61
Erigeron, 13, 36, 37, 43, 47, 62,
  63, 69
  camphoratus, 481
  graveolens, 473
Eriophyllum, 18, 22, 34, 46, 65
  mohavense, 65
Erythremia
  aphylla, 370
  grandiflora, 371
Eucephalus, 37, 63
Euchiton, 58, 387, 429, **440**
  gymnocephalus, **441**
  involucratus, **441**
  sphaericus, **441**, *442*
Eupatorieae, 5, 6, 11
Eupatorium, 54
  capillifolium, 54
European swamp thistle, 96, 110
Eurybia, 39
Euthamia, 42, 44
Eutrochium, 54
Evacopsis, 448
Evax, 448, 460, 461, 464, 467
  sect. *Diaperia*, 460
  sect. *Filaginoides*, 460
  sect. *Hesperevax*, 467
  acaulis, 468
  candida, 461
  caulescens, 470
    brevifolia, 468

humilis, 470
  sparsiflora, 468
  eriosphaera, 460
  involucrata, 470
  multicaulis, 461, 462
    drummondii, 462
  prolifera, 462, 463
  sparsiflora, 468
    brevifolia, 468
  verna, 461, 462
    drummondii, 462
Evergreen pussytoes, 408
Everlasting, 388
Everlasting neststraw, 451
Everlasting pussytoes, 408

Facelis, 58, 384, 387, **442**
  retusa, *442*, **443**
False agoseris, 335
False chamomile, 549, 550
False dandelion, 323, 335
False elephant's foot, 204
False fleabane, 471
False mayweed, 550
False sagebrush, 499, 501
Falzblume, 454
Fausse camomille, 538
Featherhead knapweed, 182
Featherleaf desertpeony, 73
Featherweed, 435
Felon-herb, 503, 533
Fendler's desertdandelion, 314
Fernald, M. L., 6, 240, 285, 293,
  548
Fernandes, R., 547
Ferris, R. S., 6, 386, 448, 547
Feverfew, 490
Few-flowered wirelettuce, 356
Fewleaf thistle, 129
Fiddleleaf hawksbeard, 235
Field burrweed, 546
Field cottonrose, 445
Field pussytoes, 403
Field sagewort, 506
Field sow-thistle, 274
Field thistle, 109, 112
Filaginella, 428
  palustris, 429
  uliginosa, 430
Filagininae, 385, 386, 444, 445,
  448, 450, 454, 457, 460,
  461, 463, 464, 465, 467
Filaginopsis
  drummondii, 462
Filago, 26, 28, 385, 387, 444,
  447, 448, 461
  arizonica, 446
  arvensis, 445
  californica, 443, 445
  dasycarpa, 465
  depressa, 446
  gallica, 447
  germanica, 448, 449
  minima, 446
  pyramidata, 448, *449*
    pyramidata, 448, *449*
  vulgaris, 448, *449*
Filzkraut, 443
Fire evax, 469

Fish Lake thistle, 127
Fivefinger chickensage, 500
Flat-top pussytoes, 407
Flaveria, 32, 46, 51
Fleischmannia, 55
Flodman's thistle, 120
Florestina, 66
Florida thistle, 116
Flourensia, 23, 30
Fluffweed, 443, 445
Flyriella, 54, 55
Foliose thistle, 159
Ford, H., 241
Fountain thistle, 161
Four Corners thistle, 143
Fox-tail blackroot, 476
Fragrant rabbit-tobacco, 419
Franciscan thistle, 141
Frankton, C., 92, 112, 113, 114,
  115, 116, 117, 127, 142,
  144, 146, 149, 150, 154,
  156, 168, 169, 170, 187
Freire, S. E., 431, 437
Fries's pussytoes, 412
Fringed sage, 519
Fringe-scaled thistle, 130
Frost boil pussytoes, 414
Funk, V. A., 5, 6, 70, 82, 195,
  200

Gabrielian, E., 173
Gaillardia, 6, 8, 22, 29, 34, 50,
  69
Galinsoga, 16, 21
Galitz, D. S., 407
Gall-of-the-earth, 264, 270
Gallez, G. J., 353
Gamochaeta, 60, 387, 429, **431**
  americana, 436
    antillana, 433, **436**, 437
    argyrinea, 432, **435**, 436
    calviceps, *430*, 432, 433, **436**,
      437
    chionesthes, 432, **434**
    coarctata, *430*, 432, **435**, 436
    falcata, 437
    pensylvanica, 433, **437**, *438*
    purpurea, 431, 432, **433**, 434,
      435, 437
    rosacea, 433
    simplicicaulis, 432, **434**
    sphacelata, 432, **434**
    spicata, 435
    stachydifolia, 433, **437**
    stagnalis, 432, 433, **437**
    subfalcata, 436, 437
    ustulata, 432, **435**
Gandhi, K. N., 541
Garberia, 59
Garcia-Jacas, N., 177, 181, 182,
  183
Garden cornflower, 184
Gardner, C. A., 473
Gardou, C., 186, 187
Garland chrysanthemum, 555
Gates, F. C., 241
Gayfeather, 7
Gazania, 48, 195, **196**
  linearis, *196*, **197**

longiscapa, 197
Gentianales, 4
Geraea, 23, 30
Gerbera
  jamesonii, 70
  lyrata, 81
German chamomile, 542
Geyerae, 389, 396
Ghost thistle, 153
Gifola
  germanica, 448
Gifolaria, 467
Gillis, W. T., 481
Glaucous rattlesnakeroot, 269
Glaucous white lettuce, 269
Gleason, H. A., 6, 92, 112, 187,
  188, 240, 242, 402, 403,
  404, 548
Glebionis, 30, 485, 488, **554**
  coronaria, **555**
  segetum, *553*, **555**
Globe centaurea, 185
Globe chamomile, 539
Globe thistle, 85
Glyptopleura, 216, **361**, 362
  marginata, **362**
  setulosa, 362
  setulosa, *360*, **362**
Gmelin, C. C., 291
Gnaphale à feuilles obtuses, 420
Gnaphale couché, 439
Gnaphale de macoun, 421
Gnaphale de Norvège, 439
Gnaphale des bois, 440
Gnaphalieae, 5, 6, 12, **384**, 385,
  461, 471
Gnaphaliinae, 384
Gnaphalium, 58, 384, 387, **428**,
  429
  sect. *Gamochaeta*, 431
  subg. *Gamochaeta*, 431
  albidum, 419
  alpinum, 414
  angustifolium, 430
  antillanum, 436
  arizonicum, 424
  beneolens, 419
  bicolor, 423
  californicum, 423
  calviceps, 436
  canescens, 418
    beneolens, 419
    microcephalum, 419
    thermale, 419
  chilense, 418
    confertifolium, 418
  coarctatum, 435
  decurrens, 421
    californicum, 423
  dimorphum, 398
  dioicum, 406
  exilifolium, 429, **430**
  filaginoides, 445
  gossypinum, 418
  grayi, 430
  gymnocephalus, 441
  helleri, 421
    micradenium, 421
  heteroides, 429

indicum, 429
involucratum, 441
jaliscense, 424
japonicum, 441
johnstonii, 419
lagopodioides, 418
leucocephalum, 423
luteoalbum, 418, 429
macounii, 421
margaritaceum, 427
microcephalum, 419
  thermale, 419
minimum, 446
norvegicum, 439
obtusifolium, 420
  helleri, 421
  micradenium, 421
  praecox, 420
  saxicola, 420
palustre, **429**, 444
  nanum, 429
pannosum, 435
pensylvanicum, 437
peregrinum, 437
plantaginifolium, 400
polycaulon, 429
pringlei, 422
proximum, 418
purpureum, 433
  simplicicaule, 434
  stachydifolium, 437
  ustulatum, 435
ramosissimum, 425
retusum, 443
rosaceum, 433
roseum, 424
saxicola, 420
simplicicaule, 434
sonorae, 418
spathulatum, 437
sphacelatum, 434
sphaericum, 441
spicatum, 436
stachydifolium, 437
stagnale, 437
stramineum, 418
strictum, 430
subfalcatum, 436
sulphurescens, 418
supinum, 439
sylvaticum, 440
texanum, 418
thermale, 419
uliginosum, 429, *430*
ustulatum, 435
virgatum, 477
viridulum, 418
viscosum, 422
wrightii, 418
Gochnatia, 12, 58, 70, 71, **76**
  hypoleuca, 77
    obtusata, 77
  obtusata, 77
Gochnatieae, 6
Gochnatiinae, 76
Gochnatioideae, 6
Godfrey, R. K., 436, 481, 483
Golden everlasting, 428
Golden marguerite, 547

Golden thistle, 220
Goldenrod, 7
Goodeniaceae, 4, 5
Goodrich, S., 358
Gopherus
  agassizii, 446
Gorteria
  linearis, 197
Gottlieb, L. D., 252, 307, 348, 349, 350, 351, 353, 354, 356, 361, 368
Graham, A., 5
Graham's thistle, 124
Grande bardane, 169
Grassland agoseris, 329
Grassland silverpuffs, 347
Gravel-ghost, 309
Gray chickensage, 500
Gray hawksbeard, 232
Gray thistle, 120
Graygreen thistle, 136
Gray, A., 358, 465
Great Basin sagebrush, 516
Great burdock, 169
Great globe-thistle, 86
Greater knapweed, 185
Greene, E. L., 6, 240, 358
Greene's thistle, 135
Green-ginger, 503, 531, 533
Greuter, W., 182, 303
Grierson, A. J. C., 440
Grindelia, 33, 34, 47, 53, 69
Groh, H., 168
Gros chardon, 109
Grossheimia, 181
  macrocephala, 185
Groundstar, 465
Guardiola, 18
Guayule, 7
Guizotia, 16
Gulf of St. Lawrence dandelion, 248
Gulf rabbit tobacco, 462
Gum succory, 252
Gundelieae, 195, 200
Gundlachia, 61
Guppy, G. A., 288
Gustafsson, M. H. G., 4, 5
Gutierrezia, 48, 49, 50
Gymnosperma, 33
Gymnostyles, 545
  anthemifolia, 546
  nasturtifolia, 546
  stolonifera, 546
Gynura, 55

Haglund, G., 240
Hairy cat's ear, 299
Hairy hawkbit, 296
Hairy-head knapweed, 182
Hall, H. M., 6, 504
Hall's hawksbeard, 237
Hall's thistle, 147
Hämet-Ahti, 550
Handel-Mazzetti, H., 240, 246, 251
Handsome pussytoes, 399, 400
Hanelt, P., 180

Haplocarpa
  lyrata, 195
Haploësthes, 40
Hardheads, 185
Harmonia, 20
Harms, V. L., 493
Harriman, N. A., 473
Hartwrightia, 28, 29, 52, 59, 67
Hasteola, 56
Hawkbit, 294
Hawksbeard, 222
Hawkweed, 278
Hayley, D. E., 241
Hazardia, 36, 42, 44, 61, 64
Hecastocleideae, 6
Hecastocleioideae, 6
Hecastocleis, 12, 56, 64, 70, 71
  shockleyi, **72**
Hedosyne, 27
Hedypnois, 218, **302**
  cretica, *301*, **302**
Heidel, B. L., 149
Helenieae, 6
Helenium, 50, 52, 69
Heliantheae, 3, 5, 6, 11
Helianthella, 18, 19, 23, 24
Helianthus, 7, 13, 17, 22
  annuus, 7
  tuberosus, 7
Helichrysum, 59, 384, 387, 425
  bracteatum, 428
  petiolare, **426**
  petiolatum, 426
Heliomeris, 17
Heliopsis, 18, 22
Heller's rabbit-tobacco, 421
Hellwig, F. H., 183
Helminthotheca, 216, **300**
  echioides, **300**, *301*
Hemizonella, 15
Hemizonia, 16
Herba impia, 447
Herbe Saint-Jean, 503
Herbe-à-dinde, 493
Herbe-à-éternuer, 494
Herrickia, 36, 39
Hesperevax, 26, 385, 388, 465, 467, 470
  acaulis, 467, **468**, 469
    acaulis, **469**
    ambusticola, **469**
    robustior, 468, **469**, 470
  caulescens, 459, 464, 467, **470**
  sparsiflora, *464*, 467, **468**
    brevifolia, **468**, 469
    sparsiflora, *464*, **468**
Heteranthemis, 30, 487, **551**
  viscidihirta, *549*, **551**
Heterosperma, 16, 21
Heterotheca, 43, 48, 63, 69
Heywood, V. H., 6
Hickman, J. C., 386
Hieracium, 219, **278**, 279
  abscissum, 283, **293**
  absonum, 289
  acranthophorum, 287
  albertinum, 289
  albiflorum, 281, 282, **288**, **292**
  alpinum, 279, 282, **291**

ambiguum, 286
amitsokense, 291
angmagssalikense, 291
argutum, 281, 282, 283, **293**
  parishii, 293
argyraeum, 286
aurantiacum, 279, **283**
auricula, 284
bolanderi, 280, **285**
brevipilum, 280, **287**
caespitosum, 278, 280, **284**
canadense, 287
  divaricatum, 287
  fasciculatum, 287
  hirtirameum, 287
  kalmii, 288
  subintegrum, 287
carneum, 280, **286**
chapacanum, 289
columbianum, 287
crepidispermum, 280, **286**
cynoglossoides, 289
  nudicaule, 293
devoldii, 287
×*dutillyanum*, 287
eugenii, 287
fendleri, 280, **286**
  discolor, 286
  mogollense, 287
flagellare, 280, **284**
  amauracron, 284
  cernuiforme, 284
  pilosius, 284
florentinum, 285
floribundum, 284
friesii, 288
gracile, 291
  alaskanum, 291
  densifloccosum, 291
  detonsum, 291
  yukonense, 291
greenei, 282, **290**, 293
greenii, 293
groenlandicum, 279, **291**
gronovii, 278, 280, **285**, 293
horridum, 278, 283, **292**
hyparcticum, 290
ivigtutense, 291
kalmii, 287
  canadense, 287
  fasciculatum, 287
lachenallii, 291
lactucella, 280, **284**
lemmonii, 286
lividorubens, 290
longiberbe, 281, **289**
longipilum, 280, **285**
marianum, 293
megacephalum, 280, **286**
murorum, 282, **290**
musartutense, 287
nepiocratum, 287
nudicaule, 282, 283, **293**
oregonicum, 290
paniculatum, 281, **287**
pannonicum, 233
parryi, 281, 282, **290**
pilosella, 279, **283**, 284
  niveum, 283

Hieracium (*continued*)
  piloselloides, 280, **285**
  praealtum, 285
    decipiens, 285
  pratense, 284
  pringlei, 282, **290**
  rigorosum, 287
  robinsonii, 281, 282, **291**
  runcinatum, 235
  rusbyi, 293
    wrightii, 288
  sabaudum, 281, **289**
  scabriusculum, 287
    columbianum, 287
    perhirsutum, 287
    saximontanum, 287
    scabrum, 287
  scabrum, 281, **288**, 293
    intonsum, 288
    leucocaule, 288
    tonsum, 288
  scholanderi, 291
  schultzii, 281, **288**
  scouleri, 281, 282, 283, **289**
    albertinum, 289
    griseum, 289
    nudicaule, 293
  scribneri, 287
  siskiyouense, 285
  smolandicum
    robinsonii, 291
  spathulatum, 293
  stelechodes, 290
  stiptocaule, 287
  sylowii, 291
  traillii, 282, 283, **293**
  trigonophorum, 279, 291
  triste, 280, 282, **291**
    fulvum, 291
    gracile, 291
    tristiforme, 291
  umbellatum, 281, **287**, 288
    canadense, 287
    scabriusculum, 287
  venosum, 282, 283, 287, **292**, 293
    nudicaule, 292, 293
  vulgatum, 279, 282, **291**
  wrightii, 288
Hierba del aire, 75
Hierba limpia, 446
High-arctic dandelion, 249
Hilliard, O. M., 429
Hill's thistle, 117
Hinds, D. J. N., 6, 76
Hinds, H. R., 542
Hippia
  stolonifera, 546
Hogwallow starfish, 470
Holly-dandelion, 362
Holm, T., 364
Holmen's dandelion, 251
Holocarpha, 16
Holozonia, 15, 19
Holub, J., 189, 444, 454
Hooked groundstar, 466
Hooker's thistle, 148
Hoover's wirelettuce, 354
Horned dandelion, 247

Horrid thistle, 114, 115
Horsetops, 113
Hot Springs sagebrush, 512
Howell, J. T., 121, 130, 131, 135, 138, 141, 144, 146, 160, 164, 426, 455
Howell's pussytoes, 403
Howell's silverpuffs, 343
Hrusa, F., 182, 544
Hsi, Y.-T., 112
Hughes, J., 241
Hulsea, 50
Hultén, E., 167, 240, 256, 412, 413, 491
Hulteniella, 31, 48, 489, **534**, 536
  integrifolia, *535*
Humphries, C. J., 485, 504, 534, 536, 541, 548
Hurtsickle, 184
Hyatt, P. E., 273
Hybrid goatsbeard, 305
Hymenonema
  laciniatum, 341
Hymenopappus, 18, 22, 33, 35, 48, 68
Hymenothrix, 46, 65
Hymenoxys, 35, 49, 51, 69
Hyoseris
  amplexicaulis, 364
  biflora, 364
  cretica, 302
  minima, 277
  montana, 365
  virginica, 365
Hypochaeris, 214, 216, **297**, 298
  brasiliensis, 298
    albiflora, 299
  chillensis, **298**, 299
  glabra, **298**, 299
  microcephala, 298, **299**
    albiflora, **299**
  radicata, *295*, 298, **299**

Ianthopappus, 76
Iberian knapweed, 192
Iberian star thistle, 192
Idaho hawksbeard, 226
Iharlegui, L., 431, 437
Illingworth, J. M., 459
Illyrian thistle, 88
Iltis, H. H., 491
Immortelle blanche, 427
Indian thistle, 148
Ingram, J. K., 486
Inula, 40, 471, **473**, 474
  brittanica, **474**
  graveolens, 473
  helenium, 472, **474**
  salicina, **474**
  trixis, 76
  viscosa, 473
Inuleae, 6, 12, 384, **471**, 475
Involucrate evax, 470
Ionactis, 36, 37, 47
Irmler, C., 347
Ironweed, 206
Island desertdandelion, 321
Island sagebrush, 530

Ismelia, 30, 488, **552**
  carinata, *553*
Isocarpha, 25
Isocoma, 61, 64
Italian hawksbit, 227
Italian thistle, 93, 94
Iva, 25, 51
Ixeris, 216, **254**
  stolonifera, *253*, **254**

Jacea, 181
  pratensis, 182, 186
Jacée des prés, 186
Jack-go-to-bed-at-noon, 305
Jalisco rabbit-tobacco, 424
Jamesianthus, 32, 40
Jansen, R. K., 5, 70, 322, 335, 338, 346, 347, 363, 365, 374
Japanese nipplewort, 255
Jarvis, C. E., 173, 481
Jaumea, 32
Jefea, 24
Jeffrey, C., 5, 429
Jensia, 20
Jepson, W. L., 138, 411
Jersey knapweed, 194
Jerusalem artichoke, 7
Joanna's thistle, 163
Johnson, A. W., 241
Johnston, I. M., 460
Johnston, M. C., 112, 119, 452
Jones, S. B., 207
Judd, W. S., 4
Junak's desertdandelion, 317

Kaibab pussytoes, 407
Kamchatka thistle, 111
Kartesz, J. T., 112, 182, 188, 191, 194, 508
Kay, Q. O. N., 542, 548, 549, 550
Kazmi, S. M. A., 92
Keck, D. D, 6
Keil, D. J., 75, 77, 84, 85, 87, 89, 91, 95, 136, 138, 139, 144, 155, 164, 165, 168, 171, 172, 173, 174, 175, 177, 178, 181, 309, 361, 461, 539, 543
Kelch, D. G., 95
Kentrophyllum
  baeticum, 180
Keyesia, 362
Khan, R., 481
Khidir, M. O., 180
Kim, H. G., 73, 76
Kim, K. J., 5, 363, 364, 365, 366, 367, 376, 377
Kim, S. C., 264
King devil, 285
King, L. M., 241, 245
King, R. M., 5, 73
King-Jones, S., 482
Kirschner, J., 239, 240, 241, 245
Klamath thistle, 131
Kloot, P. M., 473
Knapweed, 181
Knowles, P. F., 180

Knowlton, C. H., 194
Koanophyllon, 54, 55
Krigia, 217, 219, **362**, 363
  amplexicaulis, 364
  biflora, 363, **364**, 365
    viridis, 364
  cespitosa, 363, *366*, **367**
    cespitosa, *366*, **367**
    gracilis, **367**
  dandelion, 362, 363, **364**
  gracilis, 367
  montana, 363, 364, **365**
  occidentalis, 364, **365**, 366
  oppositifolia, 367
  ×shinnersiana, 365
  virginica, 363, **365**, 366
  wrightii, 363, 364, **366**
Kubitzki, K., 6
Kuntze, K., 76
Kupicha, F. K., 192
Kyhos, D. W., 491
Kyhosia, 15, 20

La Graciosa thistle, 158
Labillardière, 476
Lactuca, 216, 258, **259**
  biennis, *259*, 260, **261**
  campestris, 262
  canadensis, 259, 260, **261**
    latifolia, 261
    longifolia, 261
    obovata, 261
  floridana, 260, **261**
    villosa, 261
  graminifolia, 260, **261**, 262
    arizonica, 262
  hirsuta, 260, **262**
    albiflora, 262
    sanguinea, 262
  intybacea, 272
  ludoviciana, 260, **262**
  muralis, 258
  oblongifolia, 259
  pulchella, 259
  runcinata, 272
  sagittifolia, 261
  saligna, 260, **262**
  sativa, 260, **263**
  scariola, 263
    integrata, 263
  serriola, 260, **262**, 263
    integrata, 263
    integrifolia, 263
  stolonifera, 254
  tatarica
    pulchella, 259
  terrae-novae, 261
  virosa, 260, **262**
Lactuceae, 214
Ladies' tobacco, 388
Lad's love, 522
Laënnecia, 36, 43, 62, 63
Lagascea, 25, 28, 51, 64
Lagophylla, 15
Laiteron, 273, 275
Laiteron des champs, 274
Laiteron rude, 275
Laitue, 259
Lamiaceau, 503

Lamp, C., 473
Lanzetta, R., 473
Lapland dandelion, 246
Lappa
  minor, 170
Lapsana, 218, 255, **257**
  apogonoides, 255
  capillaris, 228
  communis, 255, *256*, **257**
  stellata, 301
  zacintha, 238
Lapsanastrum, 216, **254**, 255, 257
  apogonoides, *253*, **255**
Laramie chickensage, 502
Large seed wirelettuce, 355
Largeflower hawksbeard, 232
Largeflower skeletonplant, 371
Large-lobed dandelion, 245
Larkdaisy, 206
Lasianthaea, 24
Lasthenia, 6, 32, 46, 51
  minor, 46
  platycarpha, 46
Launaea, 214, 215, 216, 217, 219, **272**
  intybacea, **272**, *273*
Lavender-cotton, 498
Lawn burweed, 546
Layia, 15, 19, 28
Le Conte' thistle, 114
Leafy desertdandelion, 315
Leafy thistle, 159
Lee, J., 309, 310, 350, 351, 360, 361, 362, 363, 374
Leibnitzia, 12, 14, 57, 70, 71, **80**
  lyrata, 77, 80, **81**
  seemannii, 81
Lemon sagewort, 530
Leontipes, 389
Leontodon, 215, 217, 240, **294**, 298
  autumnalis, 294, *295*
    pratensis, 295
  carolinianum, 378
  ceratophorus, 247
  dandelion, 364
  hastilis, 295
  hirsutum, 332
  hirtus, 296
  hispidus, 294, **295**, 296
  leysseri, 296
  nudicaulis, 296
    taraxacoides, 296
  palustris, 246
  saxatilis, 294, **296**
    longirostris, **296**
    saxatilis, **296**
  scopulorum, 250
  taraxacoides, 296
    longirostris, 296
  taraxacum, 239, 244
Leontopodium, 384
Lepage, E., 288, 291
Lepidospartum, 11, 53, 55
Leptoseris
  sonchoides, 320
Lesser burdock, 170
Lesser knapweed, 187

Lessing, C. F., 5
Lessingia, 8, 37, 44, 59, 62
Lettuce, 259
Leucacantha, 181
  cyanus, 184
Leucanthemella, 31, 488, **557**
  serotina, *556*, **557**
Leucanthémelle tardive, 557
Leucanthemum, 31, 53, 485, 486, 487, 488, **557**, 558
  ×superbum, 559
  arcticum, 536
  integrifolium, 535
  ircutianum, 558
  lacustre, 558, **559**
  maximum, **558**, 559
  nipponicum, 556
  paludosum, 554
  vulgare, *556*, **558**
Leuciva, 52
Leucoglossum, 554
  paludosum, 554
Leucoseris
  saxatilis, 318
  tenuifolia, 319
Liabeae, 195, 200
Liatris, 7, 59
Licorice plant, 426
Ligularia, 41
Limestone hawksbeard, 229
Lindheimera, 17, 22
Lindley' silver puff, 322
Linnaeus, C., 76
Lion's-foot, 270
Lipschitz, S. J., 168
Little cottonrose, 446
Littleleaf pussytoes, 407
Lobeliaceae, 4
Logfia, 26, 28, 385, 386, 387, **443**, 444, 445, 448, 450, 461
  arizonica, 444, **446**, 451, 452
  arvensis, 443, 444, **445**
  californica, 445
  depressa, 443, 444, **446**, 450, 452
  filaginoides, *442*, 443, 444, **445**, 451, 452, 453
  gallica, 443, 444, **447**
  minima, 443, 444, **446**, 447
Lohwasser, U., 347
Long-leaf camphorweed, 484
Longleaf hawksbeard, 225
Long-leaved sage, 527
Long-style thistle, 149
Lorandersonia, 28, 42, 60
Lost thistle, 160
Louda, S. M., 97, 122
Low cornflower, 184
Low pussytoes, 398
Low sagebrush, 511
Lowrey, T. K., 499
Luina, 56
Lygodesmia, 219, 360, 361, **369**, 370
  subgen. *Pleiacanthus*, 361
  aphylla, **370**, 373
    texana, **373**

arizonica, 371
dianthopsis, 372
doloresensis, 372
entrada, 372
exigua, 360
grandiflora, 370, **371**
  arizonica, 370, **371**, 372
  dianthopsis, 371, **372**
  doloresensis, 371, **372**
  entrada, 371, **372**
  grandiflora, **371**
  stricta, 371
juncea, 369, 370, **372**, 373
  dianthopsis, 372
  rostrata, 369
ramosissima, 370, **373**
  rostrata, 369
  spinosa, 361
texana, *369*, 370, **373**
Lyman, J. C., 241, 245

Mabry, T. J., 6, 363, 365, 367
Machaeranthera, 37
Macoun's rabbit-tobacco, 421
Macoun' thistle, 147
Macrorhynchus
  angustifolius, 330
  harfordii, 332
  heterophyllus, 332
  purpureus, 328
  retrorsus, 330
Madia, 15, 24, 25, 51
Madieae, 6
Madiinae, 6
Magee, D. W., 549
Mahoney, A. M., 196, 197, 198
Malacolepis
  coulteri, 314
Malacomeris
  incanus, 316
Malacothrix, 219, 309, **310**, 349, 360
  [unranked] *Anathrix*, 309
  arachnoidea, 319
  blairii, 349
  californica, 310, 311, *313*
    glabrata, 316
  clevelandii, 311, 312, **313**, 314
    stebbinsii, 321
  commutata, 319
  coulteri, 311, **314**
    cognata, 314
    coulteri, 314
  fendleri, 312, **314**
  floccifera, 312, **314**, 315
  foliosa, 311, **315**
    crispifolia, **315**
    foliosa, 315, **316**
    indecora, 317
    philbrickii, 315, **316**
    polycephala, **315**, 316, 317
    squalida, 321
  glabrata, 312, **316**
  implicata, 318
  incana, 311, **316**, 317
    succulenta, 316, 317
  indecora, 311, *313*, **317**
  insularis
    squalida, 321

junakii, 311, **317**
  obtusa, 314
  parviflora, 314
  phaeocarpa, 312, **317**
  platyphylla, 309
  runcinata, 320
  saxatilis, 310, 311, **318**
    altissima, 318, 319
    arachnoidea, 318, **319**
    commutata, 318, **319**
    implicata, **318**
    saxatilis, 318, **319**
    tenuifolia, 318, **319**
  senecioides, 313
  similis, 311, 312, **320**
  sonchoides, 311, 312, *313*, **320**
  sonorae, 312, **320**
  squalida, 311, **321**
  stebbinsii, 312, **321**
  succulenta, 316
  torreyi, 312, **321**
Malheur stylocline, 453
Malheur wirelettuce, 355
Malperia, 66
Malte, M. O., 415
Maltese centaury, 193
Maltese star thistle, 193
Manners, G. D., 407
Mantisalca, 57, 67, 84, **173**
  salmantica, *174*
Many-head desertdandelion, 315
Many-stem rabbit tobacco, 461
Marguerite, 485, 552
Marguerite blanche, 558
Marie-Victorin, F., 156, 440
Marsden-Jones, E. M., 186, 187
Marsh cudweed, 430
Marsh dandelion, 246
Marsh silverpuffs, 343
Marsh sow-thistle, 275
Marsh thistle, 110
Marshallia, 29
Martin's thistle, 123
Maruta
  cotula, 538
Masi, S., 7
Mason neststraw, 453
Matricaire, 540
Matricaire à capitules brunâtres, 551
Matricaire camomille, 542
Matricaire inodore, 549
Matricaire maritime, 550
Matricaire odorante, 541
Matricaria, 31, 48, 53, 68, 487, 488, **540**, 541, 548
  ambigua, 551
  chamomilla, 541, **542**, 550
    chamomilla, 542
    coronata, 542
    recutita, 542
  courrantiana, 542
  discoidea, **541**
  globifera, 539
  inodora, 549, 550
    agrestis, 550
    phaeocephala, 551

Matricaria (*continued*)
  maritima, 550
    agrestis, 549
    inodora, 549
    phaeocephala, 551
  matricarioides, 541
  occidentalis, 540, 541, **542**
  parthenium, 490
  perforata, 549
  recutita, 542
  suaveolens, 541
  suffruticosa, 539
Matricary, 540
Matted sagewort, 509
Mauranthemum, 30, 31, 487, 488, **554**
  paludosum, 553, **554**
Mauthe, S., 341
Mayweed, 538, 540, 548
McArthur, E. D., 504, 510
McClintock, E., 541, 542
McGregor, R. L., 192
McNeill, J., 429
McPherson, G. D., 241
McVaugh, R., 376
Meacham, C. A., 112, 182, 188, 191, 194, 508
Meadow hawksbeard, 237
Meadow knapweed, 187, 188
Meadow pussytoes, 396, 407
Meadow salsify, 305
Meadow thistle, 154
Meadow woollyheads, 458
Melampodium, 17, 18
Melancholy thistle, 110
Melanthera, 27, 29
Menken, S. B. J., 241
Menyanthaceae, 4
Mesquite neststraw, 452
Mexican basketflower, 176, 177
Mexican trixis, 75
Mexican wormwood, 529
Mickelson, C. J., 491
Micrope, 454
Micropsis, 26, 385, 388, 457, **463**, 464
  dasycarpa, 457, *464*, **465**
Micropus, 26, 28, 385, 386, 388, 450, 454, 455, 457, 461
  sect. *Rhyncholepis*, 454
  amphibolus, 454, **455**, 456
  californicus, 454, **455**, 456
    californicus, *455*, **456**
    subvestitus, **456**
  globiferus, 459, 460
  supinus, 454
Microseridinae, 374
Microseris, 217, 218, 322, 335, **338**, 343, 346, 374
  sect. *Brachycarpa*, 346
  sect. *Calocalaïs*, 322
  sect. *Nothocalaïs*, 335
  acuminata, 339, **345**
  alpestris, 336
  aphantocarpha, 344
    elegans, 346
  attenuata, 344

bigelovii, 340, 343, 344, **345**, 347
borealis, 338, 339, **340**
campestris, 340, **345**, 346
cuspidata, 336
decipiens, 347
douglasii, *337*, 338, 339, 340, **343**, 345, 347
  douglasii, *337*, **344**
  platycarpha, **344**
  tenella, 343, **344**, 347
elegans, 339, 345, **346**
heterocarpa, 347
howellii, 339, 342, **343**
laciniata, *337*, 339, **341**, 343
  detlingii, 341, **342**
  laciniata, *337*, **341**, 342
  leptosepala, 341, **342**, 343
  siskiyouensis, 341, **342**
leptosepala, 342
lindleyi, 322
linearifolia, 322
nigrescens, 336
nutans, 339, **340**
paludosa, 339, **343**
parishii, 344
platycarpha, 344
  parishii, 344
procera, 341
sylvatica, 339, **342**, 343
tenella, 344
  aphantocarpha, 344
troximoides, 337
Mikania, 54
Milfoil, 492
Milk thistle, 164
Millereae, 6
Milleriinae, 16
Milstead, W. L., 266
Mimosa
  strigillosa, 465
Minish, T. M., 390, 410
Mitchell, J. C., 473
Modoc hawksbeard, 229, 230
Moerman, D. E., 541
Mogie, M., 241
Mohave sagebrush, 517
Mohr, C., 480
Mojave neststraw, 453
Mojave thistle, 134
Monarch-of-the-veld, 198
Monolopia, 34, 46
Monoptilon, 35, 47
Moore, R. J., 92, 110, 112, 113, 114, 115, 116, 117, 127, 142, 144, 146, 149, 150, 154, 156, 168, 169, 170, 187, 188, 189
Moquinia
  hypoleuca, 77
Morefield neststraw, 453
Morefield, J. D., 386, 443, 444, 445, 447, 448, 450, 451, 452, 453, 454, 455, 456, 457, 459, 460, 461, 462, 463, 464, 465, 466, 467, 469, 470
Morley, T., 112
Morocco knapweed, 175

Morton, J. K., 276
Moscow salsify, 305
Mosquin, T., 241
Moss, E. H., 406, 407
Motherwort, 503
Mount Diablo cottonseed, 455
Mount Tamalpais thistle, 133
Mountain agoseris, 327
Mountain bluet, 185
Mountain cornflower, 185
Mountain dwarfdandelion, 365
Mountain hawksbeard, 230
Mountain neststraw, 451
Mountain sagebrush, 517
Mountain thistle, 152
Mountain wormwood, 529
Mountain-bluet, 171
Mountain-dandelion, 323
Mountaintop thistle, 150
Mouse-ear hawkweed, 283
Mt. Hamilton thistle, 162
Mugwort, 503
Mulgedium, 215, 219, **258**, 259
  floridanum, 261
  pulchellum, **259**
  tataricum, 259
Mulligan, G. A., 241
Munz, P. A., 138, 144, 349, 362, 449, 472
Munzothamnus, 218, **349**, 360
  blairii, *348*, **349**, 350, 351
Munz' shrub, 349
Musk thistle, 93
Mutisieae, 5, 6, 8, 9, 12, **70**, 76
Mutisioideae, 6
Mycelis, 215, 216, **257**
  muralis, 256, **258**

Nabalus, 264
  alatus, 266
  albus, 266
  altissimus, 266
  asper, 267
  bootii, 268
  crepidineus, 268
  fraseri, 270
    barbatus, 267
  integrifolius, 270
  racemosus, 269
  roanensis, 270
  sagittatus, 270
  serpentarius, 270
  trifoliolatus, 271
  virgatus, 267
Naked hawksbeard, 234
Naked-stem hawksbeard, 235
Napa thistle, 193
Narrowleaf cottonrose, 447
Narrowleaf hawksbeard, 238
Narrowleaf pussytoes, 398
Narrow-leaved pygmy pussytoes, 412
Narrow-leaved saw-wort, 166
Narrow-leaved skeletonplant, 357
Navajo thistle, 144
Neonesomia, 36, 42

Nesom, G. L., 78, 79, 80, 81, 385, 415, 425, 426, 427, 428, 431, 434, 435, 436, 437, 438, 440, 442, 463, 476, 477, 478, 479, 480, 483, 486
Nestotus, 42
Neststraw, 386, 450
Neurolaena
  lobata, 481
New Mexico thistle, 140
Nicolletia, 45
Nipplewort, 257
Nippon daisy, 556
Nipponanthemum, 488, **555**
  nipponicum, *556*
Noble yarrow, 494
Nodding microseris, 340
Nodding rattlesnakeroot, 268
Nodding silverpuffs, 340
Nodding thistle, 93
Norlindh, N. T., 195, 199, 381
North African knapweed, 192
Northern dandelion, 250
Northern microseris, 340
Northern mountain thistle, 152
Northington, D. K., 376, 378
Northwest mugwort, 524
Northwestern rabbit-tobacco, 419
Norwegian Arctic-cudweed, 439
Nothocalaïs, 217, 322, 324, **335**, 338
  alpestris, 335, **336**
  cuspidata, 335, **336**
  nigrescens, 335, **336**, 337
  troximoides, 322, 335, 336, *337*
Nuttall's false-sagebrush, 502
Nuttall's pussytoes, 406
Nuttall's thistle, 119
Nuttall's wirelettuce, 353

O'Kennon, R. J., 362
Oberprieler, C., 536, 548
Ochsmann, J., 181, 189
Oclemena, 39
Oglifa, 443, 444
  arizonica, 446
  arvensis, 445
  californica, 445
  depressa, 446
  gallica, 447
  minima, 446
Oil neststraw, 452
Old man, 522
Oligosporus
  borealis, 506
  campestris
    pacificus, 508
  filifolius, 508
  groenlandicus, 506
  pacificus, 508
  pycnocephalus, 509
Olmstead, R. G., 4
Omalotheca, 58, 387, 429, **438**, 439
  subg. *Gamochaetiopsis*, 439
  subg. *Omalotheca*, 439

alpigena, 440
caucasica, 440
norvegica, **439**
supina, 58, 387, **439**
sylvatica, *438*, 439, **440**
Oncosiphon, 53, 487, **539**
piluliferum, **539**, *540*
suffruticosum, 539
Onoporde, 87
Onoporde acanthe, 88
Onopordum, 57, 66, 83, **87**, 96
acanthium, **88**
acanthium, 87, **88**
illyricum, *85*, 87, **88**
tauricum, 87, **88**
Oönopsis, 44, 45, 64
Opposite-leaved dwarfdandelion, 367
Orange agoseris, 327
Orange dwarfdandelion, 364
Orange hawkweed, 283
Orchidaceae, 7
Oregon agoseris, 332
Oregon woolly marbles, 459
Oregon woollyheads, 459
Oreochrysum, 44
Oreostemma, 38
Oriental false hawksbeard, 256
Ormenis, 495
mixta, 495
Orochaenactis, 65
Osmadenia, 15, 20
Osteospermum, 30, 379, **382**
ecklonis, *380*, **383**
moniliferum, 380
spinescens, **383**
Osterhout's thistle, 128
Outcrop pussytoes, 414
Owl's crown, 434
Ownbey, G. B., 112, 113
Ownbey's goatsbeard, 305
Ownbey's thistle, 153
Owyhee sage, 531
Ox-eye daisy, 558
Oxtongue, 300, 302
Oxytenia, 27

Pacific cudweed, 435
Pacific fringed thistle, 130
Packer, J. G., 241
Packera, 41, 56
Pak, J. H., 255, 257
Palafoxia, 46, 66
Palmer sagewort, 531
Palmer, J. D., 5
Palo Alto thistle, 160
Palouse thistle, 124
Panero, J. L., 5, 6, 70, 82, 195, 200, 486
Panetta, F. D., 253
Paper daisy, 428
Parachute plant, 309
Parasenecio, 56
Paris daisies, 485, 552
Parlin's pussytoes, 402
Parry sagewort, 527
Parry thistle, 149
Parry's wirelettuce, 356
Parthenice, 15

Parthenium, 21, 27, 29
argentatum, 7
Pascalia, 17, 18, 22, 24
Pasture thistle, 116, 120
Patterson sagewort, 520
Pearly everlasting, 427
Pearly pussytoes, 399
Peck neststraw, 453
Pecos River skeletonplant, 373
Pectis, 32, 35, 40, 45
Pemberton, R. W., 97
Pennsylvania cudweed, 437
Pentachaeta, 36, 44, 49, 50, 59, 62, 69
Pentaphragmaceae, 4
Pentzia, 53, 68, 487, **543**
globifera, 539
incana, *540*, **543**
Perdicium
radiale, 76
Peregrine thistle, 136
Perezia, 73
sect. *Acourtia*, 73
microcephala, 74
nana, 74
runcinata, 74
thurberi, 74
wrightii, 74
Pericallis, 31, 35, 47, 48
Pericome, 51, 65
Perityle, 32, 33, 46, 51, 65
Perityleae, 6
Petaloid pussytoes, 405
Petasites, 35, 55, 56
Petite bardane, 170
Petite desertdandelion, 315
Petradoria, 41
Petrak, F., 130, 141, 142, 154, 160
Peucephyllum, 60, 68
Phalacroseridinae, 374
Phalacroseris, 216, **374**
bolanderi, **374**, *375*
coronata, 374
Phellinaceae, 4
Philbrick's desertdandelion, 316
Phoebanthus, 19, 23
Picradeniopsis, 46
Picris, 219, **302**
echioides, 300
hieracioides, **303**, *304*
alpina, 303
kamtschatica, 303
rhagadioloides, **303**
sprengerana, 303
Picrothamnus, 52, 53, 486, **498**
desertorum, **497**, **499**
Pilosella, 278, 279
aurantiaca, 283
auricula, 284
caespitosa, 284
flagellaris, 284
officinarum, 283
piloselloides, 285
spathulata, 293
Pinaropappus, 215, **374**, 375
parvus, **375**
roseus, *375*, **376**
foliosus, 376

Pineappleweed, 541
Pinewoods pussytoes, 396
Pink dandelion, 252
Pink rabbit-Tobacco, 425
Pinkava, D. J., 461
Piqueux, 109
Pissenlit, 239
Pissenlit à graines rouges, 245
Pissenlit à lobes larges, 245
Pissenlit de Laponie, 246
Pissenlit du golfe du Saint-Laurent, 248
Pissenlit officinal, 244
Pissenlit palustre, 246
Pissenlit tuberculé, 247
Pityopsis, 43, 47
Plantain-leaved pussytoes, 400
Plateilema, 51
Platte thistle, 122
Platyschkuhria, 50
Plectocephalus, 57, 84, **175**, 176, 182
americanus, **176**, 177
rothrockii, *174*, **176**, 177
Pleiacanthus, 219, 360, **361**
spinosus, 351, *360*, **361**
Pleurocoronis, 66
Plowman's-wort, 481
Pluchea, 58, 475, **478**, 479, 480
sect. *Amplectifolium*, 479
sect. *Pluchea*, 479
sect. *Pterocaulis*, 479
absinthoides, 480
baccharis, 480, **483**
camphorata, 479, **481**
carolinensis, 479, **480**, 481
eggersii, 482
foetida, 480, **482**, 483
foetida, 483
imbricata, 482, 483
imbricata, 482
lanceolata, 480
longifolia, 479, **484**
odorata, *frontispiece*, 479, **481**, 482
odorata, *frontispiece*, **481**, 482
succulenta, 481, **482**
petiolata, 481
purpurascens, 481
succulenta, 482
quitoc, 480
rosea, 480, **483**
mexicana, 483
sagittalis, 475, 478, 479, **480**
senegalensis, 482
sericea, 477, 478, 479, **480**
suaveolens, 480
symphytifolia, 481
tenuifolia, 482
yucatanensis, 479, **483**
Plucheeae, 5, 6, 12, 471, **475**, 478
Plucheinae, 475
Plumeless thistle, 91, 92
Point Reyes agoseris, 332
Polar daisy, 537
Polemoniaceae, 515
Poljakov, P. P., 504

Polymnia, 18, 25
Polymnieae, 6
Polypappus
sericeus, 480
Poole, J. M., 149
Porcellites
brasiliensis, 298
Porophyllum, 54, 60
Porsild, A. E., 240, 251, 407, 415
Porsild, M. P., 240
Porter mugwort, 509
Portugese daisy, 559
Pot marigold, 381
Potato dandelion, 364
Potato dwarfdandelion, 364
Powderpuff thistle, 176
Powell, A. M., 94, 191
Prairie agoseris, 326
Prairie false dandelion, 336
Prairie sagewort, 519
Prairie skeletonplant wirelettuce, 356
Prairie thistle, 120, 122
Prenanthe, 264
Prenanthe à grappe, 269
Prenanthe blanche, 266
Prenanthe élevée, 266
Prenanthe trifoliolée, 271
Prenanthella, 219, **359**, 360, 370
exigua, *360*
Prenanthes, 219, **264**
alata, *263*, 265, **266**, 268, 270
sagittata, 270
alba, 265, **266**
pallida, 266
altissima, 264, **266**, 267
cinnamomea, 266, 267
hispidula, 266, 267
aphylla, 370
aspera, 265, **267**, 269, 270
autumnalis, 264, **267**
barbata, 265, **267**, 268
bootii, 264, 265, **268**
carrii, 265, **268**
crepidinea, *263*, 265, **268**, 269, 271
cylindrica, 270
exigua, 360
hastata, 266
integrifolia, 270
japonica, 256
juncea, 372
lessingii, 266
×mainensis, 270, 271
muralis, 258
nana, 271
pauciflora, 356
racemosa, 265, **269**, 270, 271
multiflora, 269
pinnatifida, 269
roanensis, 265, **270**
sagittata, 265, 268, **270**
serpentaria, 265, 269, **270**, 271
barbata, 267

Prenanthes (*continued*)
tenuifolia, 357
trifoliolata, 265, 269, 270,
271
nana, 271
virgata, 267
Preston, R. E., 471, 473
Price, H. J., 338, 374
Prickle-leaf, 72
Pringle's rabbit-tobacco, 422
Protean knapweed, 187
Proustia, 73
Psacalium, 56
Psathyrotes, 62, 63
Psathyrotopsis, 63
Pseudelephantopus, 64, 201, **204**
spicatus, *202*, **204**
Pseudobahia, 34
Pseudoclappia, 22, 46
Pseudognaphalium, 60, 384,
387, **415**, 418, 421, 427,
429
arizonicum, 417, **424**
austrotexanum, 416, **422**, 423
beneolens, 417, **419**
biolettii, 416, **423**
californicum, 416, 417, **423**,
424
canescens, 417, **418**, 424
beneolens, 419
microcephalum, 419
thermale, 419
helleri, 415, 416, **421**
micradenium, 421
jaliscense, 417, **420**, **424**, 425
leucocephalum, 416, **422**, **423**
luteoalbum, 387, *413*, 415,
416, **418**, 429
macounii, 416, **421**, 422
micradenium, 415, 416, **421**
microcephalum, 417, **419**
thermale, 419
obtusifolium, 415, 416, *420*,
421
saxicola, 420
pringlei, 416, **422**
ramosissimum, 415, 417, **425**
roseum, 417, **424**
saxicola, 415, 417, **420**, 421
stramineum, *413*, 415, 416,
418, 424
thermale, 415, 417, **419**, *420*
viscosum, 416, **422**, 423
Pseudogynoxys, 40
Psilactis, 33, 36
Psilocarphus, 8, 24, 385, 386,
388, 450, 454, **456**, 457,
460, 461, 465
berteri, 460
brevissimus, *455*, 457, **458**
brevissimus, *455*, 457, **458**,
459, 460
multiflorus, 457, **458**, 459
caulescens, 470
chilensis, 457, **459**, 460
elatior, *455*, 457, **458**, 459
globiferus, 458, 459
oregonus, 457, **459**, 470
elatior, 458

tenellus, 457, 458, **459**, 460
globiferus, 459
tenuis, 459
Psilochenia
occidentalis, 232
Psilostrophe, 49
Pterocaulon, 59, 475, **476**
sect. *Monenteles*, 476
sect. *Pterocaulon*, 476
alopecuroides, 476
pycnostachyum, **476**
undulatum, 476
virgatum, 476, **477**
Ptiloria
neomexicana, 358
pleurocarpa, 359
Puget Sound agoseris, 330
Pulcherrimae, 389, 399
Pulicaria, 47, **471**
arabica, 472
dysenterica, 472
hispanica, 472
paludosa, **472**
vulgaris, 472
Pulvinate pussytoes, 410
Purple rattlesnakeroot, 269
Purple star-thistle, 191
Purple thistle, 118
Pussytoes, 388
Pygmy pussytoes, 411
Pygmy sage, 514
Pyrethrum
foeniculaceum, 552
majus, 490
Pyrrhopappus, 215, **376**, 377
carolinianus, *375*, 377, **378**
georgianus, 378
geiseri, 377
georgianus, 378
grandiflorus, 376, **377**
multicaulis, 376, 377
geiseri, 377
pauciflorus, 376, **377**, 378
rothrockii, **377**, 378
Pyrrocoma, 45, 62

Q-tips, 455

Rabbit tobacco, 460
Racemose pussytoes, 401
Rafinesque's chicory, 348
Rafinesquia, 215, **348**, 360
californica, 348, **349**
neomexicana, *348*, **349**
Ragonese, A. M., 476
Ragweed, 7
Raillardella, 15, 19, 28
Rainiera, 56
Ratibida, 17, 21
Rattlesnakeroot, 264
Rauschert, S., 541, 542, 550
Raven, P. H., 349, 472, 491
Rayjacksonia, 43
Rayless chamomile, 541
Red hawksbeard, 234
Red-seeded dandelion, 245
Red-tip rabbit-tobacco, 418
Remarkable goatsbeard, 306
Remote-leaved thistle, 129, 130

Reveal, J. L., 73
Rhagadiolus, 214, 218, **300**
stellatus, *301*
zacintha, 238
Rhinocyllus
conicus, 93, 97, 122
Rhodogeron, 478
Richards, A. J., 239, 240, 241,
245, 246
Richardson's sagewort, 506
Rigiopappus, 11, 20
Roadside hawksbeard, 228
Roadside thistle, 111
Roan Mountain rattlesnakeroot,
270
Robinson, H., 5, 200, 207, 480
Robust evax, 469
Roché, B. F., 182
Roché, C. T., 182
Rock tansy, 502
Rocklettuce, 374
Rocky Mountain fringed thistle,
128
Rocky Mountain pussytoes, 411
Rocky Mountain thistle, 129
Roelofs, D., 343, 345, 346
Roland, A. E., 284
Roldana, 41
Roman wormwood, 531
Roque, N., 76
Rose thistle, 145
Roserush, 370
Rosette thistle, 158
Ross, T. S., 453
Rosy camphorweed, 483
Rosy pussytoes, 408
Rosy rabbit-tobacco, 424
Rothrock sagebrush, 515
Rothrock's basketflower, 176
Rothrock's knapweed, 176
Rothrock's thistle, 143
Rubialels, 4
Rough hawksbeard, 227
Rough rattlesnakeroot, 267
Rough star thistle, 182
Round woolly marbles, 459
Rousseauaceae, 4
Rudbeckia, 7, 17, 21, 29
Rugelia, 56
Rush pink, 369
Rush pussytoes, 397
Rush skeletonplant, 372
Rushlike skeleton weed, 253
Russian centaurea, 171
Russian knapweed, 171, 172
Rustaiyan, A., 473
Rydberg, P. A., 6, 240
Rydberg's thistle, 162

Sachsia, 12, 38, 58, 475, **477**,
478
bahamensis, 478
divaricata, 478
polycephala, *477*, **478**
Sacramento Mountains thistle,
163
Safflower, 180, 181
Sage, 503
Sagebrush, 398, 503, 514, 517

Sagebrush false dandelion, 337
Sagewort, 503
Sailor's tobacco, 503
Salsifis cultivé, 306
Salsifis des prés, 305
Salsifis majeur, 305
Salsify, 306
Saltmarsh fleabane, 482
Salvia, 503, 524
officinalis, 503
Samuel, R., 298
San Diego rabbit-tobacco, 419
San Diego silverpuffs, 344
San Diego wirelettuce, 352
San Joaquin silverpuffs, 345
Sand sage, 508
Sand wormwood, 506
Sand-dune thistle, 122
Sanderson, S. C., 510
Sand-hill thistle, 113
Santa Cruz desertdandelion, 321
Santa Cruz Island
desertdandelion, 317
Santa Cruz silverpuffs, 347
Santa Ynez groundstar, 466
Santolina, 26, 487, **497**
chamaecyparissus, *497*, **498**
suaveolens, 541
Sanvitalia, 22
Sartwellia, 46
Saussurea, 58, 83, **165**
alpina, 167
densa, 167
ledebourii, 167
amara, 165, **168**
glomerata, **168**
americana, *157*, 165, **166**
angustifolia, 165, **166**, 168
angustifolia, 166, **167**
viscida, 166, **167**
yukonensis, 166, **167**
densa, 166, 167
foliosa, 166
glomerata, **168**
nuda, 165, **167**, 168
densa, 167
nuda, 167
triangulata, 165, **166**
×tschuktschorum, 168
viscida, 167
yukonensis, 167
weberi, 165, **168**
Sawtooth wirelettuce, 357
Saw-wort, 165
Scabland sagebrush, 515
Scabrethia, 19, 23
Scalebud, 310
Scapose hawksbeard, 235
Scented pussytoes, 408
Scentless mayweed, 549
Schaal, B. A., 241, 245
Schkuhria, 46, 65
Schneider, D. J., 473
Sclerocarpus, 17, 22
Sclerolepis, 66
Scoggan, H. J., 240, 279, 406
Scolymus, 214, 217, 218, **220**
hispanicus, **221**, **220**
maculatus, **220**

Scorzonella, 338
  borealis, 340
  howellii, 343
  laciniata, 341
  leptosepala, 342
  nutans, 340
  paludosa, 343
  procera, 341
  sylvatica, 342
  troximoides, 337
Scorzonera, 218, **306**
  hispanica, 306, **307**
  laciniata, *304*, 306, **307**
Scotch thistle, 88
Seashore chamomile, 551
Seaside agoseris, 331
Seaside evax, 468
Seedhead weevil, 93, 97, 122
Seemann's sunbonnet, 81
Sell, P. D., 240, 246, 284
Senecio, 35, 40, 41, 55, 56
  flocciferus, 314
  mohavensis, 55
Senecioneae, 6, 11
Separateleaf chickensage, 501
Sericocarpus, 39
Serinia, 362
  cespitosa, 367
Seriphidium, 510, 523
  arbusculum, 511
    longilobum, 511
    thermopolum, 512
  bigelovii, 512
  bolanderi, 513
  canum, 512
    bolanderi, 513
  novum, 513
  pygmaeum, 514
  rigidum, 515
  rothrockii, 515
  spiciforme, 515
  tridentatum, 516
    parishii, 517
    wyomingensis, 517
  tripartitum, 518
  vaseyanum, 517
Serrate-leaved sage, 532
Serratula
  alpina
    angustifolia, 166
  amara, 168
  angustifolia, 166
  arvensis, 109
  glauca, 209
  noveboracensis, 209
Sessile thistle, 156
Shalebarren pussytoes, 403
Shasta daisy, 485, 558, 559
Shasta Valley thistle, 159
Shinners, L. H., 207, 213, 246, 363, 366
Shinnersia, 51
Shinnersoseris, 214, 219, **368**, 370
  rostrata, *369*
Short woollyheads, 458
Short-fringed knapweed, 188
Short-leaved evax, 468
Short-stemmed thistle, 153

Short-style thistle, 148
Showy pussytoes, 400
Shrubby bullseye, 77
Shrubby camphor-weed, 481
Shultz, L. M., 498, 499, 503, 512, 515
Siberian wormwood, 523
Siberian yarrow, 494
Sicilian star-thistle, 194
Sierra foothills silverpuffs, 345
Sierra Nevada agoseris, 329
Sierra pussytoes, 410
Silky cottonseed, 456
Silphium, 18, 23
Silver chickensage, 502
Silver pussytoes, 397
Silver rabbit tobacco, 461
Silver sage, 527
Silver wormwood, 513, 527
Silverpuff, 79
Silverpuffs, 338, 346
Silver-puffs, 322
Silvery cudweed, 435
Silvery-brown pussytoes, 397
Silybum, 66, 82, 83, **164**
  marianum, *157*, **164**
Simple-stem cudweed, 434
Simpson, B. B., 71, 72, 76
Simsia, 17, 22
Singhurst, J. R., 268
Singlehead pussytoes, 401
Sinosenecio, 41
Siskiyou hawksbeard, 229, 230
Siskiyou silverpuffs, 342
Skeletonplant, 353, 358, 369
Skeletonweed, 350
Sledge, W. A., 110
Slender cottonseed, 455
Slender cudweed, 430
Slender hawksbeard, 225
Slender rattlesnakeroot, 267
Slender sow-thistle, 276
Slender wirelettuce, 357
Slender woolly marbles, 459
Slender-bracted silverpuffs, 342
Slender-flowered thistle, 94
Slough thistle, 132
Small crown wirelettuce, 354
Small rocklettuce, 375
Small wirelettuce, 353
Small, E., 242, 245
Small, J. K., 5, 6
Smallanthus, 18
Smallflower hawksbeard, 229, 234
Smallhead thistle, 118
Small-headed rush pussytoes, 398
Small-leaf pussytoes, 406
Smooth cat's ear, 298
Smooth desertdandelion, 316
Smooth distaff thistle, 180
Smooth hawksbeard, 228, 236
Snake Range thistle, 151
Snake's head, 314
Snake's head desertdandelion, 314
Sneezeweed, 494
Sneezewort, 494

Snowfield sagebrush, 515
Snowy thistle, 140
Soft thistle, 118
Solbrig, O. T., 241, 245
Solidago, 6, 7, 38, 39, 44, 63
Soliva, 52, 486, **545**
  anthemifolia, **546**
  daucifolia, 546
  mutisii, 546
  nasturtifolia, 546
  pterosperma, 546
  sessilis, **545**, **546**
  stolonifera, **546**
Soltis, P. S., 303
Sonchus, 219, **273**, 274
  arvensis, **274**
    arvensis, **274**
      glabrescens, 274
      uliginosus, **274**, 275
  asper, *273*, 274, **275**
    asper, 275
    glaucescens, 275
  biennis, 261
  floridanus, 261
  hastatus, 266
  ludovicianus, 262
  oleraceus, *273*, 274, **275**
    asper, 275
  palustris, 274, **275**, 276
    palustris, 276
    sosnowskyi, 276
  pulchellus, 259
  tenerrimus, 274, **276**
  uliginosus, 274
Sonoran desert dandelion, 320
Sonoran neststraw, 452
Sourbush, 480
South Texas rabbit-tobacco, 422
Southern globe-thistle, 87
Southern thistle, 119
Southernwood, 522
Sow-thistle, 273
Sow-thistle desertdandelion, 320
Spanish false fleabane, 472
Spanish oyster, 220
Spanish salsify, 220, 307
Spanish succory, 252
Spear thistle, 109
Spearleaf agoseris, 330
Speckled false dandelion, 336
Sphaeromeria, 52, 68, 486, **499**, 531
  argentea, 500, **502**
  cana, 499, **500**, *501*
  capitata, 500, *501*, **502**
  compacta, 499, 500, **502**
  diversifolia, 499, **501**
  potentilloides, 499, **500**, *501*
    nitrophila, 500, *501*
    potentilloides, **500**
  ruthiae, 499, **501**
  simplex, 500, **502**
Sphagneticola, 18, 24
Spiny-leaf sow-thistle, 275
Spoon-leaf cudweed, 433
Spotted knapweed, 189
Spreading cottonrose, 446
Spring Mountains thistle, 151
Spring rabbit-tobacco, 461

Spurr, P. L., 255
Squarrose knapweed, 191
St. Barnaby's thistle, 193
Stace, C. A., 186
Star thistle, 181
Star-cudweed, 441
Starr, J. R., 5
Stebbins, G. L., 350, 362, 374, 389, 390, 391, 400, 401, 402, 403, 406, 407, 408, 411
Stebbins's desertdandelion, 321
Stebbinsoseris, 217, 218, 322, 338, **346**, 347
  decipiens, **347**
  heterocarpa, **347**, *348*
Steens Mountain thistle, 153
Stemless evax, 469
Stemless thistle, 156
Stenotus, 45
'tpánek, J., 239, 240, 245
Stephanomeria, 219, 349, **350**, 351, 354, 360, 361, 370
  blairii, 349, 351
  carotifera, 354
  cichoriacea, 350, 351, **352**
  cinerea, 356, 357
  coronaria, 354
  diegensis, 351, **352**, 353, *357*
  elata, 351, 352, **353**
  exigua, 350, 351, 352, **353**, 354, 355, *357*
    carotifera, 353, **354**
    coronaria, 351, 353, **354**, 355, 356, *357*
    deanei, 350, 353, **354**
    exigua, 353, **354**, 355, *357*
    macrocarpa, 353, **355**
    pentachaeta, 354
  fluminea, 350, 351, 352, **355**
  guadalupensis, 350
  lactucina, 350, 352, **355**
  lygodesmoides, 356
  malheurensis, 351, 352, **355**, 356
  minor, 358
  myrioclada, 357, 358
  neomexicana, 357, 358
  paniculata, 351, 352, **356**
  parryi, 350, 351, **356**
  pauciflora, 352, **356**, *357*
    parishii, 356, 357
  pentachaeta, 354, 355
  runcinata, 351, 352, **357**
  schottii, 354, 355
  spinosa, 351, 361
  tenuifolia, 350, 351, 352, 356, *357*, 358
    myrioclada, 357
    uintaensis, 357, 358
  thurberi, 351, 352, **358**
  tomentosa, 359
  virgata, 350, 351, 352, 353, 354, *357*, **359**
    pleurocarpa, **359**
    tomentosa, 359
    virgata, *357*, **359**
  wrightii, 357, 358
Stephanomeriinae, 360, 374

Steppe agoseris, 328
Stevia, 51, 52, 54, 66, 67
Steyermark, J. A., 269
Stickweed, 350
Sticky rabbit-tobacco, 422
Sticky sagebrush, 513, 515
Sticky saw-wort, 167
Stiff-branched wirelettuce, 356
Stinking camphorweed, 482
Stinking chamomile, 538
Stinking hawksbeard, 228
Stinknet, 539
Stinkwort, 473
Stokesia, 11, 67, 200, **201**
    laevis, 200, **201**, *202*
Stoloniferous pussytoes, 399,
    406
Straitjackets, 463
Strawflower, 428
Striped hawksbeard, 238
Strother, J. L., 3, 201, 202, 204,
    206, 220, 221, 254, 257,
    258, 259, 276, 277, 278,
    296, 300, 302, 306, 376,
    379, 380, 381, 382, 551,
    552, 554, 555, 557
Stuartina, 464
    hamata, 385
Stuessy, T. F., 5
Stylidiaceae, 4
Stylocline, 26, 27, 386, 388, 444,
    450, 453, 454, 455, 457,
    461, 465, 470
    sect. *Ancistrocarphus*, 465
    acaulis, 468
    amphibola, 455
    arizonica, 451
    citroleum, 450, 451, **452**
    filaginea, 466
        depressa, 466
    gnaphaloides, *449*, 450, **451**,
        452
        bigelovii, 451
    intertexta, 451, **453**
    masonii, 451, **453**
    micropoides, 451, **452**, 453
    psilocarphoides, *449*, 451, **453**
    sonorensis, 450, 451, **452**
Stylopappus
    elatus, 334
    grandiflorus, 329
    laciniatus, 334
    longifolius, 334
Succor Creek mugwort, 531
Suisun thistle, 132
Suksdorf sagewort, 533
Sulphur knapweed, 194
Sulphur-colored Sicilian thistle,
    194
Sundberg, S., 547
Sunflower, 7
Surf thistle, 161
Susanna, A., 171, 172, 176, 177,
    182
Swamp thistle, 113
Sweet Annie, 503, 523
Sweet sagewort, 523
Sweetscent, 481
Sweet-sultan, 173

Swine's succory, 297
Sylvan scorzonella, 342
Symphyotrichum, 6, 7, 36, 38,
    62
    pilosum, 7
Synedrella, 24
Syntrichopappus, 32, 33, 34,
    35, 40, 48, 50

Tack-stem, 307
Tageteae, 6
Tagetes, 45, 65
Tall globe-thistle, 86
Tall mountain thistle, 152
Tall pussytoes, 399
Tall rattlesnakeroot, 266
Tall saw-wort, 168
Tall thistle, 111
Tall woollyheads, 458
Tamaulipa, 55, 59
Tanacetum, 31, 48, 49, 52, 53,
    68, 486, 488, **489**
    balsamita, 489, **490**
    bipinnatum, 489, **490**, 491
        bipinnatum, 491
        huronense, 490, 491
    camphoratum, 490, 491
    canum, 500
    capitatum, 502
    compactum, 502
    diversifolium, 501
    douglasii, 490, 491
    huronense, 490, 491, 541
        bifarium, 490
        floccosum, 490, 491
        johannense, 490
        terrae-novae, 490
    nuttallii, 502
    parthenium, 489, **490**
    pauciflorum, 541, 542
    potentilloides, 500
        nitrophilum, 500
    simplex, 502
    vulgare, 489, **490**, *491*
Tanaisie, 489
Tanaisie bipennée, 490
Tanaisie vulgaire, 490
Tansy, 489
Tansy chickensage, 501
Tapertip hawksbeard, 225
Taraxacum, 8, 215, **239**, 240,
    241, 242
    sect. *Arctica*, 242, 243
    sect. *Borealia*, 242
    sect. *Crocea*, 239
    sect. *Erythrosperma*, 242
    sect. *Palustria*, 242, 246
    sect. *Ruderalia*, 239, 241,
        242
    sect. *Spectabilia*, 242, 246
    sect. *Taraxacum*, 239
    subg. *Ixeris*, 254
    alaskanum, 242, 244, **251**,
        252
    alukense, 246
    ambigens, 246
        fultius, 247
    angulatum, 247
    arctogenum, 247

    atroglaucum, 246
    brachyceras, 247
    californicum, 242, 243, **248**,
        249
    campylodes, 239, 246
    carneocoloratum, 242, 244,
        252
    carthamopsis, 247
    ceratophorum, 236, 241, 242,
        247, 248, 249
        bernardinum, 248
    cognatum, 246
    coverum, 247
    croceum, 246
    curvidens, 246
    cyclocentrum, 246
    davidssonii, 246
    dilutisquameum, 246
    dumetorum, 247
    eriophorum, 247
    erythrospermum, 239, 240,
        241, 242, 243, **245**, 246
    eurylepium, 247
    firmum, 246
    groenlandicum, 247
    hirsutum, 332
    holmenianum, 242, 244, **251**
    hyparcticum, 242, 244, **249**,
        250
    hyperboreum, 247, 248
    integratiforme, 247
    integratum, 247
    islandiciforme, 247
    kamtschaticum, 251, 252
    lacerum, 247, 248
    laevigatum, 241, 246
        erythrospermum, 245
        scapifolium, 241
    lapponicum, 242, 243, 245,
        246, 247
    lateritium, 247
    latilobum, 242, 243, **245**
    latispinulosum, 246
    laurentianum, 242, **248**
    longii, 247, 248
    lyratum, 251
    mackenziense, 247
    malteanum, 247
    maurolepium, 247
    microcerum, 247
    naevosum, 246
    obtusatum, 246
    officinale, 7, 239, 240, 241,
        242, 243, **244**, 245, 246,
        248
        erythrospermum, 245
        palustre, 244
        scopulorum, 250
    ovinum, 247
    palustre, 242, 243, **246**
    pellianum, 247
    phymatocarpum, 242, 243,
        244, **250**, 251
    pleniflorum, 246
    pseudokamtschaticum, 251,
        252
    pseudonorvegicum, 247
    pumilum, 251
    scanicum, 245, 246

    scopulorum, *236*, 242, 244,
        250, 251
    sibiricum, 252
    spectabile, 242, 246
    sylvanicum, 244
    torngatense, 246
    trigonolobum, 242, 248, **249**
    turfosum, 246
    umbriniforme, 247
    umbrinum, 247
Taurian thistle, 88
Taylor, R. J., 241, 245
Tender silverpuffs, 344
Tephroseris, 35, 41, 56
Tessaria, 479, 480
    absinthoides, 480
    integrifolia, 480
    sericea, 480
Tetradymia, 11, 53, 55, 68
Tetragonotheca, 16, 21
Tetraneuris, 49, 69
Texas dwarfdandelion, 367
Texas purple thistle, 119
Texas skeletonplant, 373
Texas thistle, 119
Thelesperma, 16, 20, 25, 28
Thimbles, 184
Thistle, 85, 95, 96
Thomas, R. D., 541
Thorne's thistle, 157
Thornless thistle, 176
Thorny skeletonweed, 361
Threefold, 75
Threeleaved rattlesnakeroot, 271
Three-tipped sagebrush, 518
Thurber's desertpeony, 74
Thurber's wirelettuce, 358
Thurovia, 68
Thymophylla, 8, 45
Tibor, D. P., 343, 344, 347
Tickseed, 7
Tithonia, 17, 22
Tobacco-weed, 309
Tocalote, 193
Toiyabe thistle, 158
Toiyabea, 63
Tolpis, 218, **277**
    barbata, **277**, 278
    umbellata, 277
Tomb, A. S., 343, 350, 355, 358,
    360, 361, 363, 370, 373
Tonestus, 45, 63
Torrell, M., 504
Torrey, J., 465
Torrey's desertdandelion, 321
Townsendia, 37, 45, 49, 50
Tracyina, 44
Tracy's thistle, 121
Tragopogon, 215, **303**, 304
    ×crantzii, 305
    dubius, **305**, 306
    ×mirabilis, 306
    mirus, *304*, **306**
    miscellus, 304, **305**
    ×neohybridus, 306
    picroides, 297
    porrifolius, 304, **306**
    pratensis, 303, **305**, 306
    virginicus, 364

Trail plant, 78
Trailing dusty miller, 426
Transvaal daisy, 70
Treasure-flower, 196
Trichocoronis, 54, 66
Trichoptilium, 69
Tridax, 21
Triniteurybia, 64
Tripleurospermum, 31, 488,
    541, **548**, 550
    hookeri, 549, 551
    inodorum, 548, *549*, 550
    maritimum, 548, 549, **550**
        inodorum, 549, 550
        maritimum, 548, **550**
        phaeocephalum, 549, 550,
        551
    perforatum, 549
    phaeocephalum, 550, **551**
Trixis, 12, 14, 59, 70, 71, 73,
    75, 76
    californica, 72, **75**
        californica, 72, **75**
    inula, **75**, 76
    radialis, 76
Trock, D. K., 492
Tropical threefold, 75
Troximon, 362
    alpestre, 336
    apargioides, 331
    auranticacum, 327
    cuspidatum, 336
    glaucum, 326
        dasycephalum, 326
    grandiflorum
        laciniatum, 334
        tenuifolium, 334
    heterophyllum
        cryptopleuroides, 333
    nuttallii, 334
    parviflorum, 328
Tumble knapweed, 190
Turkestan thistle, 171
Turkish hawksbeard, 232
Turner, B. L., 73, 363, 364,
    367, 376, 377
Turner, C. E., 136, 138, 139,
    144, 155
Turrill, W. B., 186, 187
Tussilago, 41
    albicans, 79
Tutin, T. G., 238
Two-form pussytoes, 398
Tyrl, R. J., 493
Tyrol knapweed, 188
Tzvelev, N. N., 536, 549,
    551

Umber pussytoes, 408
Urbanska, K. M., 399, 400
Uropappus, 215, 216, **322**,
    335, 338, 346
    sect. *Brachycarpa*, 346
    heterocarpus, 347
    lindleyi, **322**, *323*, 347
    linearifolius, 322
Urospermum, 215, **296**
    picroides, *295*, **297**

Vaarama, A., 548, 550
Valles, J., 504
Valley mayweed, 542
Varilla, 26
Vasey's thistle, 133
Vegetable oyster, 306
Venegasia, 32, 34
Venidium
    fastuosum, 198
Venn, J. M., 276
Venus thistle, 140
Verbesina, 9, 19, 23, 24, 25, 27,
    29
Vernonia, 58, 67, 200, 201, **206**,
    207
    acaulis, 207, **208**
    altissima, 207, 212
        marginata, 211
        taeniotricha, 212
    angustifolia, 207, **209**
        mohrii, 209
        scaberrima, 209
        texana, 211
    arborea, 200
    arkansana, *205*, 207, **208**
    baldwinii, 200, 207, 208, *210*,
        213
        interior, 213
    blodgettii, 208, **212**
    cinerea, 205
    ×concinna, 207
    crinita, 208
    ×*dissimilis*, 207
    fasciculata, 208, **211**
        corymbosa, 211
    flaccidifolia, 208, **212**
    ×*georgiana*, 207
    gigantea, 208, **212**
        ovalifolia, 212
    glauca, 207, **209**
    guadalupensis, 207
    harperi, 209
    interior, 213
    larseniae, **210**
    larsenii, 207
    lettermannii, 208, **211**
    lindheimeri, 207, *210*
        leucophylla, 210
    marginata, 207, 208, **211**
    missurica, 208, **212**
    noveboracensis, 207, **209**, *210*
        tomentosa, 209
    ovalifolia, 207, 212
    pulchella, 206, 207, **209**
    ×*recurva*, 207
    scaberrima, 209
    texana, 208, **211**
    vulturina, 207
Vernonieae, 5, 6, 11, 195, **200**
Vesicarpa
    potentilloides, 500
    nitrophilum, 500
Vicoa
    auriculata, 472
Viguiera, 17, 22
Virgate wirelettuce, 359
Virginia dwarfdandelion, 365
Virginia thistle, 118

Vochin knapweed, 188
Volutarella, 174
    muricata, 175
Volutaria, 67, 84, **174**
    muricata, *174*, **175**
Voss, E. G., 110, 188, 279, 291,
    486, 492, 541, 542, 548

Wagenitz, G., 5, 6, 177, 181,
    183, 186, 444, 448, 449
Wallace, R. S., 347
Wand blackroot, 477
Wand wirelettuce, 359
Ward, G. H., 510
Watson, L. E., 485, 489, 495,
    496, 497, 504, 515, 523,
    524, 536, 537, 543, 545,
    547
Watson, S., 61
Wavyleaf thistle, 120, 121
Weber, W. A., 182, 189, 510
Wedelia, 24, 29
Weedy hawksbeard, 238
Wells, H., 139
Welsh, S. L., 127, 134, 141, 143,
    154, 166, 358, 371
Welted thistle, 92
Wenatchee thistle, 147
West, C., 284
Western dwarfdandelion, 365
Western hawksbeard, 232
Western marsh cudweed, 429
Western rattlesnakeroot, 266
Western sagewort, 508
Western thistle, 137
Western white lettuce, 266
Wheeler's thistle, 125
Whetstone, R. D., 272
Whip pussytoes, 399
White dandelion, 376
White knapweed, 190
White lettuce, 266
White rabbit-tobacco, 423
White rattlesnakeroot, 266
White rocklettuce, 376
White sage, 527
White sunbonnet, 79
White tack-stem, 308
White thistle, 148
White wormwood, 528
White-cloaked cudweed, 434
Whitemargin pussytoes, 405
White-margined wax-plant, 362
White-spine thistle, 151
White-stem distaff thistle, 179
Whitton, J., 322, 335
Wig knapweed, 188
Wiklund, A., 89
Wild chamomile, 551
Wild tarragon, 508
Wild tobacco, 480
Wild-lettuce, 272
Willamette agoseris, 334
Williams, D. F., 452
Wing-leaved rattlesnakeroot, 266
Wing-stem camphorweed, 480
Winward, A., 510
Wirelettuce, 350

Witch's bells, 184
Wittmann, R. C., 182
Wodehouse, R. P., 512
Woodland Arctic-cudweed, 440
Woodland silverpuffs, 342
Woodland wirelettuce, 355
Woolly burdock, 169
Woolly desertdandelion, 314
Woolly distaff thistle, 180
Woolly fishhooks, 466
Woolly marbles, 456
Woolly pussytoes, 399
Woolly sunbonnet, 80
Woollyhead fanbract, 452
Woollyhead neststraw, 452
Woollyheads, 456
Wooly pussytoes, 407
Wooten, J. W., 483
Wormwood, 503
Wright's dwarfdandelion, 366
Wright's marsh thistle, 131
Wright's rabbit-tobacco, 418
Wright's thistle, 131
Wunderlin, R. P., 483
Wyethia, 19, 23
Wyoming sagebrush, 517
Wyoming thistle, 125, 126

Xanthisma, 36, 38, 42, 43, 61
Xanthium, 25, 27
Xanthocephalum, 34, 50
Xantophtalmum, 554
    segetum, 555
Xeranthemum
    bracteatum, 428
Xerochrysum, 59, 387, **427**
    bracteatum, *426*, **428**
Xylorhiza, 36, 39

Yellow bachelor's button, 185
Yellow chamomile, 547
Yellow cornflower, 185
Yellow salsify, 305
Yellow star-thistle, 193
Yellow tack-stem, 308
Yellow thistle, 114
Yellowspine thistle, 123
Yermo, 55
Youngia, 217, **255**
    americana, 256
    japonica, *256*
        elstonii, 256
        japonica, 256
Yucatan camphorweed, 483

Zacintha, 239
    verrucosa, 238
Zaluzania, 17
Zavada, M. S., 5
Zinck, F. M., 284
Zinnia, 17, 21
Zion chickensage, 501
Zion tansy, 501
Zoegea, 5
Zollikoferia
    elquiensis, 314
Zyégée, 5